M000238648

Phase-Locking in
High-Performance Systems

IEEE Press
445 Hoes Lane
Piscataway, NJ 08855

IEEE Press Editorial Board
Stamatios V. Kartalopoulos, *Editor in Chief*

M. Akay	M. E. El-Hawary	M. Padgett
J. B. Anderson	R. J. Herrick	W. D. Reeve
R. J. Baker	D. Kirk	S. Tewksbury
J. E. Brewer	R. Leonardi	G. Zobrist
	M. S. Newman	

Kenneth Moore, *Director of IEEE Press*
Catherine Faduska, *Senior Acquisitions Editor*
John Griffin, *Acquisitions Editor*
Tony VenGraitis, *Project Editor*

Technical Reviewers

R. Jacob Baker, Boise State University
Stewart K. Tewksbury, Stevens Institute of Technology

Books of Related Interest from the IEEE Press

Monolithic Phase-Locked Loops and Clock Recovery Circuits
Edited by Behzad Razavi
1996   Hardcover   532 pp   ISBN 0-7803-1149-2

CMOS: Mixed-Signal Circuit Design
R. Jacob Baker
2002   Hardcover   508 pp   ISBN 0-471-22754-4

Advanced Semiconductor Memories
Ashok K. Sharma
2003   Hardcover   682 pp   ISBN 0-471-20813-2

Phase-Locking in High-Performance Systems

From Devices to Architectures

Edited by

Behzad Razavi

Professor of Electrical Engineering
University of California, Los Angeles

A Selected Reprint Volume

IEEE Press

A JOHN WILEY & SONS PUBLICATION

Copyright © 2003 by the Institute of Electrical and Electronics Engineers, Inc. All rights reserved.

Published by John Wiley & Sons, Inc., Hoboken, New Jersey.

No part of this publication may be reproduced, stored in a retrieval system or transmitted in any form or by any means, electronic, mechanical, photocopying, recording, scanning or otherwise, except as permitted under Section 107 or 108 of the 1976 United States Copyright Act, without either the prior written permission of the Publisher, or authorization through payment of the appropriate per-copy fee to the Copyright Clearance Center, Inc., 222 Rosewood Drive, Danvers, MA 01923, (978) 750-8400, fax (978) 750-4744, or on the web at www.copyright.com. Requests to the Publisher for permission should be addressed to the Permissions Department, John Wiley & Sons, Inc., 111 River Street, Hoboken, NJ 07030, (201) 748-6011, fax (201) 748-6008, e-mail: permreq@wiley.com.

Limit of Liability/Disclaimer of Warranty: While the publisher and author have used their best efforts in preparing this book, they make no representation or warranties with respect to the accuracy or completeness of the contents of this book and specifically disclaim any implied warranties of merchantability or fitness for a particular purpose. No warranty may be created or extended by sales representatives or written sales materials. The advice and strategies contained herein may not be suitable for your situation. You should consult with a professional where appropriate. Neither the publisher nor author shall be liable for any loss of profit or any other commercial damages, including but not limited to special, incidental, consequential, or other damages.

For general information on our other products and services please contact our Customer Care Department within the U.S. at 877-762-2974, outside the U.S. at 317-572-3993 or fax 317-572-4002.

Wiley also publishes its books in a variety of electronic formats. Some content that appears in print, however, may not be available in electronic format.

Library of Congress Cataloging-in-Publication Data is available.

ISBN 0-471-44727-7

Printed in the United States of America.

10 9 8 7 6 5 4 3 2 1

Contents

Preface

Since the publication of *Monolithic Phase-Locked Loops and Clock and Data Recovery Circuits* in 1996, the field of integrated phase-locked systems has made tremendous advances. Owing to the rapid growth of high-speed semiconductor and communication technologies, phase-locked loops (PLLs) have become an essential part of memories, microprocessors, radio-frequency (RF) transceivers, and broadband data communication systems.

The recent developments in PLL design have encompassed extensive research at three levels of abstraction: devices, circuits, and architectures. For example, to achieve a low phase noise, a great deal of effort has been expended on the optimization of inductors, oscillator topologies, and synthesizer architectures that suppress the oscillator phase noise. Another important trend has been the emergence of CMOS PLLs in the frequency range of tens of gigahertz.

This book addresses the need for a new compilation of recent work on PLLs. Intended as a complement to the 1996 volume, the book consists of five original tutorials and 83 papers selected to represent the advances in the past six years.

The tutorials in Part I deal with devices, delay-locked loops (DLLs), fractional-N synthesizers, bang-bang PLLs, and simulation of phase noise and jitter, aiming to provide a solid foundation for the reader.

Part II is concerned with passive devices such as inductors, transformers, and varactors. Part III presents papers on the analysis of phase noise and jitter in various types of oscillators.

Part IV concentrates on building blocks, describing the design of oscillators, frequency dividers, and phase/frequency detectors. Part V addresses the problem of clock generation by phase-locking for timing and digital applications.

Part VI deals with RF synthesis, presenting both integer-N and fractional-N realizations. Part VII is concerned with the application of phase-locking to clock and data recovery circuits in the range of 2.5 Gb/s to 40 Gb/s.

I wish to express my gratitude to Ian Galton, Ken Kundert, Rick Walker, and Ken Yang for writing and revising the tutorials in a short time. I would also like to thank Anthony VenGraitis of IEEE Press and Andrew Prince of Wiley for their kind support.

BEHZAD RAZAVI
November 2002

About the Author

Behzad Razavi received the B.Sc. degree in electrical engineering from Sharif University of Technology in 1985 and the M.Sc. and Ph.D. degrees in electrical engineering from Stanford University in 1988 and 1992, respectively. He was with AT&T Bell Laboratories and Hewlett-Packard Laboratories until 1996. Since Sept. 1996, he has been Associate Professor and subsequently Professor of electrical engineering at the University of California, Los Angeles. His current research includes wireless transceivers, frequency synthesizers, phase-locking and clock recovery for high-speed data communications, and data converters.

Professor Razavi was an Adjunct Professor at Princeton University, Princeton, NJ, from 1992 to 1994, and at Stanford University in 1995. He served on the Technical Program Committee of the International Solid-State Circuits Conference (ISSCC) from 1993 to 2002 and is presently a member of the Technical Program Committee of Symposium on VLSI Circuits. He has also served as Guest Editor and Associate Editor of the *IEEE Journal of Solid-State Circuits, IEEE Transactions on Circuits and Systems,* and *International Journal of High Speed Electronics.*

Professor Razavi received the Beatrice Winner Award for Editorial Excellence at the 1994 ISSCC, the best paper award at the 1994 European Solid-State Circuits Conference, the best panel award at the 1995 and 1997 ISSCC, the TRW Innovative Teaching Award in 1997, and the best paper award at the IEEE Custom Integrated Circuits Conference in 1998. He was the co-recipient of both the Jack Kilby Outstanding Student Paper Award and the Beatrice Winner Award for Editorial Excellence at the 2001 ISSCC. He is an IEEE Distinguished Lecturer and the author of *Principles of Data Conversion System Design* (IEEE Press, 1995), *RF Microelectronics* (Prentice Hall, 1998) (also translated to Japanese), *Design of Analog CMOS Integrated Circuits* (Mc-Graw-Hill, 2001) (also translated to Chinese), and *Design of Integrated Circuits for Optical Communications* (McGraw-Hill, 2002), and the editor of *Monolithic Phase-Locked Loops and Clock Recovery Circuits* (IEEE Press, 1996).

PART I

Original Contributions

Devices and Circuits for Phase-Locked Systems

Behzad Razavi

Abstract—This turtorial deals with the design of devices such as varactors and inductors and circuits such as ring and LC oscillators. First, MOS varactors are introduced as a means of frequency control for low-voltage circuits and their modeling issues are discussed. Next, spiral inductors are studied and various geometries targeting improved Q or higher self-resonance frequencies are presented. Noise-tolerant ring oscillator topologies are then described. Finally, a procedure for the design of LC oscillators is outlined.

The design of phase-locked systems requires a thorough understanding of devices, circuits, and architectures. Intended as a continuation of [1], this tutorial provides an overview of concepts in device and circuit design for phase-locking in digital, broadband, and RF systems.

I. PASSIVE DEVICES

The demand for low-noise PLLs has encouraged extensive research on active and passive devices. In this section, we study varactors and inductors as essential components of LC oscillators.

A. Varactors

As supply voltages scale down, pn junctions become a less attractive choice for varactors. Specifically, two factors limit the dynamic range of pn-junction capacitances: (1) the weak dependence of the capacitance upon the reverse bias voltage, e.g., $C_j = C_{j0}/(1 + V_R/\phi_B)^m$, where $m \approx 0.3$.; and (2) the narrow control voltage range if forward-biasing the varactor must be avoided.

As an example, consider the LC oscillator shown in Fig. 1. It is desirable to maximize the voltage swings at nodes X and

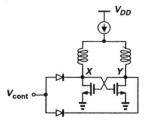

Fig. 1. LC oscillator using pn-junction varactors.

Y so as to both minimize the relative phase noise and ease the

design of the stage(s) driven by the VCO. On the other hand, to avoid forward-biasing the varactors significantly, V_X and V_Y must remain above approximately $V_{cont} - 0.4$ V. Thus, the peak-to-peak swing at each node is limited to about 0.8 V. Note that the cathode terminals of the varactors also introduce substantial n-well capacitance at X and Y, further constraining the tuning range.

In contrast to pn junctions, MOS varactors are immune to forward biasing while exhibiting a sharper C-V characteristic and a wider dynamic range. If configured as a capacitor [Fig. 2(a)], a MOSFET suffers from both a nonmonotonic C-V be-

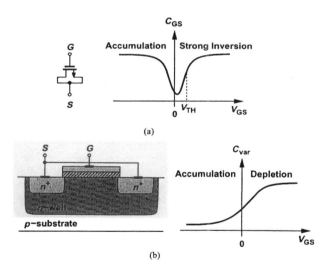

Fig. 2. (a) Simple MOSFET operating as capacitor, (b) MOS varactor.

havior and a high channel resistance in the region between accumulation and strong inversion. To avoid these issues, an "accumulation-mode" MOS varactor is formed by placing an NMOS device inside an n-well [Fig. 2(b)]. Providing an ohmic connection between the source and drain for all gate voltages, the n-well experiences depletion of mobile charges under the oxide as the gate voltage becomes more negative. Thus, the varactor capacitance, C_{var}, (equal to the series combination of the oxide capacitance and the depletion region capacitance) varies as shown in Fig. 2(b). Note that for a sufficiently positive gate voltage, C_{var} approaches the oxide capacitance.

The design of MOS varactors must deal with two important issues: (1) the trade-off between the dynamic range and the channel resistance, and (2) proper modeling for circuit simulations. We now study each issue.

Dynamic Range Deep-submicron MOSFETs exhibit susbtantial overlap capacitance between the gate and source/drain terminals. For example, in a typical 0.13-μm technology, a transistor having minimum channel length, L_{min}, displays an overlap capacitance of 0.4 fF/μm and a gate-channel capacitance of 12 fF/μm^2. In other words, for an effective channel length of 0.12 μm and a given width, the overlap capacitance between the gate and source/drain terminals of a varactor constitutes 2×0.4 fF /$(0.12 \times 12$ fF$+2 \times 0.4$ fF$) \approx 36\%$ of the total capacitance. Thus, even if the gate-channel component varies by a factor of two across the allowable voltage range, the overall dynamic range of the capacitance is given by $(0.12 \times 12$ fF$+2 \times 0.4$ fF$)/(0.12 \times 6$ fF $+2 \times 0.4$ fF$) = 1.47$.

In order to widen the varactor dynamic range, the transistor length can be increased, thereby raising the voltage-dependent component while maintaining the overlap capacitance relatively constant. This remedy, however, leads to a greater resistance between the source and drain, lowering the Q. The resistance reaches a maximum for the most negative gate-source voltage, at which the depletion region's width is maximum and the path through the n-well the longest (Fig. 3).[1] Note that

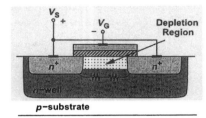

Fig. 3. Effect of n-well resistance in MOS varactor.

the total equivalent resistance that appears in series with the varactor is equal to $1/12$ of the drain-source resistance. This is because shorting the drain and source lowers the resistance by a factor of 4 and the distributed nature of the capacitance and resistance reduces it by another factor of 3 [2]. Depending on both the phase noise requirements and the Q limitations imposed by inductors, the varactor length is typically chosen between L_{min} and $3L_{min}$.

Modeling The C-V characteristics of MOS varactors can be approximated by a hyperbolic tangent function with reasonable accuracy. Using the characteristic shown in Fig. 4 and noting that $\tanh(\pm\infty) = \pm1$, we can write

$$C_{var}(V_{GS}) = \frac{C_{max} - C_{min}}{2} \tanh(a + \frac{V_{GS}}{V_0}) + \frac{C_{max} + C_{min}}{2}. \quad (1)$$

Here, a and V_0 allow fitting for the intercept and the slope, respectively, and C_{min} includes the overlap capacitance.

The above model yields different characteristics in different circuit simulation programs! Simulation tools that analyze

[1]Fortunately, the capacitance reaches a minimum at this point, and the Q degrades only gradually.

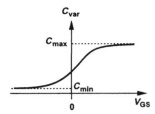

Fig. 4. Typical MOS varactor characteristic.

circuits in terms of voltages and currents (e.g., SPICE) interpret the nonlinear capacitance equation correctly. On the other hand, programs that represent the behavior of capacitors by charge equations (e.g., Cadence's Spectre) require that the model be transformed to a Q-V relationship [3]:

$$Q_{var} = \int C_{var} dV_{GS} \quad (2)$$

$$= \frac{C_{max} - C_{min}}{2} V_0 \ln \left[\cosh(a + \frac{V_{GS}}{V_0}) \right]$$

$$+ \frac{C_{max} + C_{min}}{2} V_{GS}, \quad (3)$$

which is then used to compute

$$I_{var} = \frac{dQ_{var}}{dt}. \quad (4)$$

If used in charge-based analyses, Eq. (1) typically overestimates the tuning range of oscillators.

B. Inductors

The design of monolithic inductors has been studied extensively. The parameters of interest include the inductance, the Q, the parasitic capacitance (i.e., the self-resonance frequency, f_{SR}), and the area, all of which trade with each other to some extent. For a spiral structure such as that in Fig. 5, the line width, the line spacing, the number of turns, and the outer

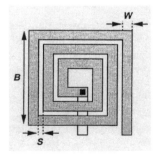

Fig. 5. Spiral inductor.

dimension are under the designer's control, chosen so as to obtain the required performance.

Quality Factor The quality factor of monolithic inductors has been the subject of many studies. Before considering the phenomena that limit the Q, it is important to select a useful and clear definition for this quantity. For a simple inductor operating at low frequencies, the Q is defined as

$$Q = \frac{L_S \omega}{R_S}, \tag{5}$$

where R_S denotes the metal series resistance. In analogy with this expression, a more general definition is sometimes given as

$$Q = \frac{\text{Im}(Z_L)}{\text{Re}(Z_L)}, \tag{6}$$

where Z_L represents the overall impedance of the inductor at the frequency of interest. While reducing to Eq. (5) at low frequencies, this definition yields $Q = 0$ if the inductor resonates with its own capacitance and/or any other capacitance. This is because at resonance, the impedance is purely resistive. Since nearly all circuits employ inductors in a resonance mode,[2] this expression fails to provide a meaningful measure of inductor performance in circuit design. A more versatile definition assumes that a resonant tank can be represented by a *parallel* combination [Fig. 6(a)], yielding

$$Q = \frac{R_P}{L_P \omega_R}, \tag{7}$$

where ω_R is the resonance frequency. Note that the tank reduces to R_P at $\omega = \omega_R$, exhibiting a finite (rather than zero) Q. Hereafter, we consider the behavior of inductors at or near resonance.

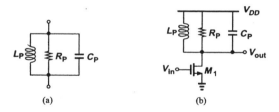

(a)   (b)

Fig. 6. (a) Parallel tank for definition of Q, (b) common-source stage using a tank.

The utility of Eq. (7) can be seen in the example illustrated in Fig. 6(b). Here, the knowledge of L_P and $Q = R_P/(L_P \omega_R)$ directly provides the voltage gain and the output swing, whereas the Q given by Eq. (6) serves no purpose.

The Q of inductors is limited by resistive losses: parasitic resistances dissipate a fraction of the energy that is reciprocated between the inductor and the capacitor in a tank. Note that the finite Q is also accompanied by generation of *noise*. For example, in the circuit of Fig. 6(b), R_P produces an output noise voltage of $\overline{V_n^2} = 4kTR_P = 4kTQL_P\omega_R$ per unit bandwidth if L_P resonates with C_P.

The losses in inductors arise from three mechanisms (Fig. 7): (1) the series resistance of the spiral, including both low-frequency resistance and current crowding due to skin effect;

<hr/>

[2]One exception is inductive degeneration in low-noise amplifiers.

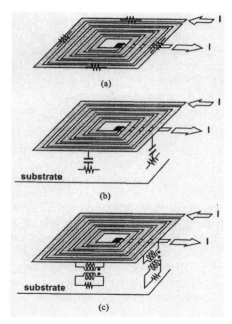

Fig. 7. Inductor loss mechanisms: (a) metal resistance, (b) substrate loss due to electric coupling, (c) substrate loss due to magnetic coupling.

(2) the flow of displacement current through the series combination of the inductor's parasitic capacitance and the substrate resistance; (3) the flow of magnetically-induced ("eddy") currents in the substrate resistance. At low frequencies, the dc resistance is dominant, and as the frequency rises, the other components begin to manifest themselves.

With the above observations in mind, let us construct a circuit model for inductors. Depicted in Fig. 8(a) is a simple model where R_S denotes the series resistance at the frequency

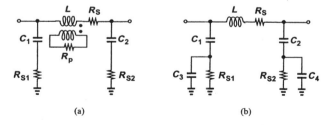

(a)   (b)

Fig. 8. (a) Inductor model including magnetic coupling to substrate, (b) simplified model.

of interest, R_{S1} and R_{S2} represent the substrate resistance through which the diplacement current flows, the transformer models magnetic coupling to the substrate, and R_P is the substrate resistance through which the eddy currents flow. This model reveals how the Q drops at high frequencies. As the impedance of C_1 and C_2 falls, R_{S1} and R_{S2} appear as a constant resistance in *parallel* with the inductor, lowering the Q as ω rises. Similarly, at high frequencies, the effect of R_P becomes relatively constant, shunting L_P and further reducing the Q.

In practice, the model of Fig. 8(a) is modified as shown in Fig. 8(b) to both allow an easier fit to measured data and account for the substrate capacitance. The model is usually assumed to be symmetric, i.e., $C_1 = C_2, C_3 = C_4$, and $R_{S1} = R_{S2}$, implying that the equivalent parasitic capacitance, C_{eq}, is one-half of the total capacitance, C_{tot}, if one end of the inductor is grounded. This result, however, is not correct because the distributed nature of the structure yields $C_{eq} = C_{tot}/3$ in this case [5]. To avoid this inaccuracy, the inductor must be modeled as a distributed network [5].

Characterization Most inductor modeling programs provide limited capabilities in terms of the type of structure that they can analyze or the maximum frequency at which their results are valid. For this reason, it is often necessary to fabricate and characterize monolithic inductors and use the results to revise the simulated models, thereby obtaining a better fit.

Owing to the need for precise measurements at high frequencies, inductors are typically characterized by direct on-wafer probing. High-speed coaxial probes having a tightly-controlled 50-Ω characteristic impedance and a low loss are positioned on pads connected to the inductor. Figure 9(a) shows an example where one end of the spiral is tied to the

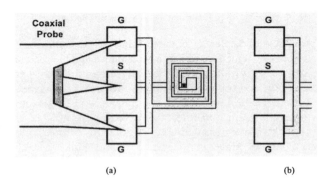

(a)   (b)

Fig. 9. (a) On-wafer measurement of inductor using coaxial probe, (b) calibration structure.

"signal" (S) pad and the other to the "ground" (G) pads. The signal pad is sensed by the center conductor and the ground pads by the outer shield of the coaxial probe.

Since the capacitance of the pads and the wires connecting to the spiral is typically significant, the test device is accompanied by a calibration structure [Fig. 9(b)], where the spiral itself is omitted. The scattering (S) parameters of both structures are measured by means of a network analyzer across the band of interest and subsequently converted to Y parameters. Subtraction of the Y parameters of the calibration geometry from those of the device under test yields the actual characteristics of the spiral.

An alternative method of measuring the Q of inductors is illustrated in Fig. 10. Here, inductors are incorporated in an oscillator and the tail current can be controlled externally. In the laboratory measurement, the output is monitored on a spectrum analyzer while I_{SS} is reduced so as to place the circuit at the edge of oscillation. Next, the value of I_{SS} thus obtained is used in the simulation of the oscillator and the equivalent parallel resistance of each tank, R_P, is lowered

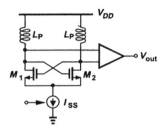

Fig. 10. Setup for "in-situ" measurement of Q.

until the circuit fails to oscillate. For such value of R_P, we have $Q = R_P/(L\omega)$. Of course, this technique assumes that the value of the inductor and the oscillation frequency are known.

The above method proves useful if (a) the frequency of interest is so high and/or the inductance so low that direct measurements are difficult, or (b) an oscillator has been fabricated but the inductors are not available individually, requiring "in-situ" measurement of the Q. Note that other oscillator parameters such as phase noise and ouput swing are also functions of Q, but it is much more straightforward to place the circuit at the edge of oscillation than to calculate the Q from phase noise or output swing measurements.

Choice of Geometry The design of inductors begins with the choice of the geometry. Shown in Fig. 11 are two commonly-used structures. The asymmetric spiral of Fig. 11(a) exhibits

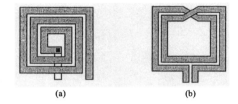

(a)   (b)

Fig. 11. (a) Asymmetric and (b) symmetric inductors.

a moderate Q, about 5 to 6 at 5 GHz, and its interwinding capacitance does not limit the self-resonance frequency because adjacent turns sustain a small potential difference. The line spacing is therefore set to the minimum allowed by the technology.

The symmetric geometry of Fig. 11(b) provides a greater Q if stimulated differentially [4], about 7 to 10 at 5 GHz, but its interwinding capacitance is typically quite significant because of the large voltage difference between adjacent turns. For this reason, the line spacing is chosen to be twice or three times the minimum allowable value, lowering the fringe capacitance considerably but degrading the Q slightly.

In differential circuits, the use of symmetric inductors appears to save area as well. For example, two asymmetric 1-nH inductors can be replaced by a symmetric 2-nH structure, which occupies less area. However, a cascade of differential stages employing multiple symmetric inductors [Fig. 12(a)] faces routing difficulties. As illustrated in Fig. 12(b), the signal lines must travel across the spirals, impacting the

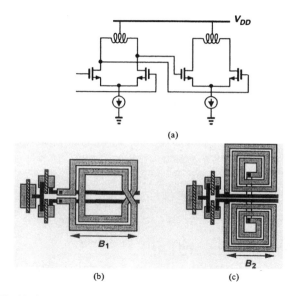

(a)

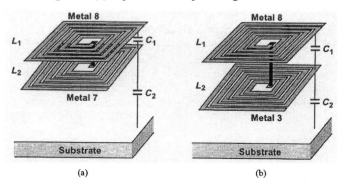

(b)   (c)

Fig. 12. (a) Cascade of inductively-loaded differential pairs, (b) layout of first stage using a symmetric inductor, (c) layout of first stage using asymmetric inductors.

performance of the inductors. Furthermore, the power and ground lines must either cross the spirals or go around with adequate spacing. With asymmetric inductors, on the other hand, the lines can be routed as shown in Fig. 12(c), leaving the inductors undisturbed. Note that B_1 is quite larger than B_2 because the symmetric structure must provide an inductance twice that of each asymmetric spiral. Thus, the signal lines in Fig. 12(b) are longer.

The two geometries of Fig. 11 can also be converted to stacked structures, wherein spirals in different metal layers are placed in series so as to achieve a greater inductance per unit area. Figure 13(a) depicts an example using metal 8 and metal

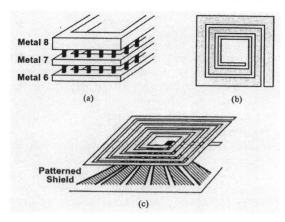

(a)   (b)

Fig. 13. Stack of (a) metal 8 and metal 7 spirals, (b) metal 8 and metal 3 spirals.

7 spirals. The total inductance is equal to $L_1 + L_2 + 2M$, where M denotes the mutual coupling between L_1 and L_2. Owing to the strong magnetic coupling, the value of M is close to L_1 and L_2, suggesting a fourfold increase in the overall inductance as a result of stacking. In the general case, n stacked identical spirals raise the inductance by a factor of approximately n^2.

Stacking reduces the area occupied by inductors signifi-

cantly. However, the capacitance between the spirals may limit the self-resonance frequency. For the two-layer structure of Fig. 13(a), the overall equivalent capacitance is given by [5]

$$C_{eq} = \frac{4C_1 + C_2}{12}. \qquad (8)$$

Thus, if the bottom layer is moved down [Fig. 13(b)], then C_{eq} falls considerably. For example, in a typical 0.13-μm CMOS technology having eight metal layers, the geometry of Fig. 13(b) exhibits one-fifth as much as capacitance as the structure in Fig. 13(a) does.

Stacked structures use lower metal layers, which typically suffer from a greater sheet resistance than the topmost layer. As explained below, the resistance can be reduced by placing spirals in parallel.

Figure 14 illustrates three other configurations aiming to improve the quality factor. In Fig. 14(a), multiple spirals are

Fig. 14. (a) Parallel combination of spirals to reduce metal resistance, (b) tapered metal width, (c) patterned shield.

placed in parallel so as to reduce the series resistance, but at the cost of larger capacitance to the substrate. Nonetheless, in a typical process having eight metal layers, metal 6 capacitance is about 30% greater than that of metal 8. Since metal 8 is typically twice as thick as metal 6 or metal 7, this topology lowers the series resistance by twofold while raising the parasitic capacitance by 30%. By the same token, in the stacked structure of Fig. 13(b), addition of a metal 2 spiral in parallel with the metal 3 layer decreases the overall resistance by 30% while increasing the equivalent capacitance by about 15%.

At frequencies above 5 GHz, the skin depth of aluminum falls below 2 μm, making the parallel combination of spirals less effective. Electromagnetic field simulations may therefore be necessary to determine the optimum configuration.

The structure in Fig. 14(b) employs tapering of the line width to reduce the resistance of the outer turns. The idea is to maintain a relatively constant inductance-resistance product per turn, achieving a slightly higher Q for a given inductance and capacitance. Unfortunately, most inductor simulation programs cannot analyze such a geometry.

Shown in Fig. 14(c) is a method of lowering the loss due to the electric coupling to the substrate. A heavily-conductive shield is placed under the spiral and connected to ground so that the displacement current flowing through the inductor's bottom-plate capacitance does not experience resistive loss. To stop the flow of magnetically-induced currents, the shield is broken regularly. Note that eddy currents still flow through the substrate, dissipating energy.

The conductive shield in Fig. 14(c) may be realized in n-well, n^+, or p^+ diffusion, polysilicon, or metal, thus bearing a trade-off between the parasitic capacitance and the Q enhancement. The resulting increase in the Q depends on the frequency of operation and the type of shield material, falling in the range of 5 to 10%.

Thus far, we have studied square spirals. However, for a given inductance value, a circular structure exhibits less series resistance. Since mask generation for circles is more difficult, some inductors are designed as octagonal geometries to benefit from a slightly higher Q.

C. MOS Transistors

The modeling of MOSFETs for analog and high-frequency design continues to pose challenging problems as sub-0.1-μm generations emerge. BSIM models provide reasonable accuracy for phase-locked system design, with the exception that their representation of thermal and flicker noise may err considerably. This issue becomes critical in the prediction of oscillator phase noise.

The thermal noise arising from the channel resistance is usually represented by a current source tied between the source and drain and having a spectral density $\overline{I_n^2} = 4kT\gamma g_m$, where γ is the excess noise coefficient. For long-channel devices, $\gamma = 2/3$, but for submicron transistors, γ may reach 2.5 to 3. Since some MOS models lack an explicit γ parameter that the user can set, it is often necessary to artificially raise the effective value of γ in circuit simulations. For linear, time-variant circuits, this can be accomplished using a noise copying technique [6]. However, the time variance of currents and voltages in oscillators make it difficult to apply this method. As a first-order approximation, the contribution of the transistors to the overall phase noise can be increased by a factor equal to $2.5/(2/3)$ before all of the noise components are summed.[3]

The flicker noise parameters are usually obtained by measurements. It is therefore important to check the validity of the device models by comparing measured and simulated results. Owing to their buried channel, PMOS transistors exhibit substantially less flicker noise than NMOS devices even in deep submicron technologies.

II. RING OSCILLATORS

Despite their relative high noise and poor drive capability, ring oscillators are used in many high-speed applications. Several reasons justify this popularity: (1) in some cases, the oscillator must be tuned over a wide frequency range (e.g., one decade) because the system must support different data

---

[3]In reality, the effective value of γ also depends on the drain-source voltage to some extent, further complicating the matter.

rates or retain a low-frequency clock in the "sleep" mode; (2) ring oscillators occupy substantially less area than LC topologies do, an important issue if many oscillators are used; (3) the behavior of ring oscillators across process, supply, and temperature corners is predicted with reasonable accuracy by standard MOS models, whereas the design of LC oscillators heavily relies on inductor and varactor models.

In mostly-digital systems such as microprocessors, ring oscillators experience considerable supply and substrate noise, making differential topologies desirable. Figure 15(a) shows an example of a differential gain stage that allows several

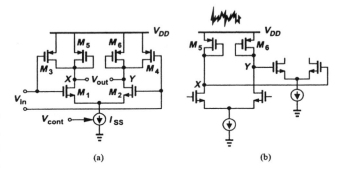

Fig. 15. (a) Differential stage for use in a ring oscillator, (b) effect of supply noise.

decades of frequency tuning with relatively constant voltage swings. Here, M_5 and M_6 define the output common-mode (CM) level while M_3 and M_4 pull nodes X and Y to V_{DD}, maintaining a constant voltage swing even at low current levels.

Unlike a simple differential pair, the stage of Fig. 15(a) does respond to input CM noise even with an ideal I_{SS}. This is because the gate voltages of M_3 and M_4 are referenced to V_{DD}, introducing a change in the drain currents if the input CM level varies. In the presence of asymmetries, such a change results in a differential component at the output. Nevertheless, since the input CM level of each stage in the ring is referenced to V_{DD} by the diode-connected PMOS devices in the preceding stage [Fig. 15(b)], the oscillator exhibits low sensitivity to supply voltage.

Figure 16(a) depicts another ring oscillator topology that has become popular in low-voltage digital systems. Here, the

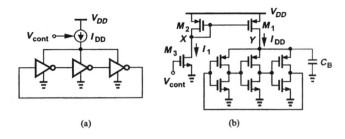

Fig. 16. (a) Constant-current ring oscillator, (b) transistor-level implementation of (a).

inverters in the ring are supplied by a current source, I_{DD}, rather than a voltage source, and frequency tuning is also accomplished through I_{DD}. If I_{DD} is designed for low sensitivity to V_{DD}, then the oscillator remains relatively immune to supply noise—the principal advantage of this configuration over standard inverter-based rings that are directly connected to the supply voltage.

In practice, the nonidealities associated with I_{DD} limit the supply rejection. Shown in Fig. 16(b) is a transistor implementation where M_1 operates as a contolled current source. If I_1 is constant, V_X tracks V_{DD} variations whereas V_Y does not, yielding a change in I_{DD} through channel-length modulation in M_1. Choosing long channels for M_1 and M_2 alleviates this issue while necessitating wide channels as well to allow a relatively small drain-source voltage for M_1. However, the resulting high drain junction capacitance of M_1 at Y creates a low-impedance path from V_{DD} to this node at high frequencies. To suppress both resistive and capacitive feedthrough of V_{DD} noise, a bypass capacitor, C_B, is tied from Y to ground. However, the pole associated with this node now enters the VCO transfer function, complicating the design of the PLL.

Let us now study the response of the circuit of Figs. 15(a) and 16 to substrate noise, V_{sub}. In the former, V_{sub} manifests itself through two mechanisms (Fig. 17): (1) by modulating the drain junction capacitance of M_1 and M_2 and hence

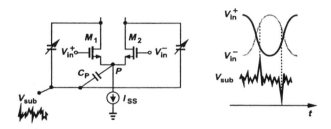

Fig. 17. Effect of substrate noise on a differential stage.

the delay of the stage (a static effect); and (2) by injecting a common-mode displacement current through C_P (a dynamic effect). If injected slightly before or after the zero crossings of the oscillation waveform, such a current gives rise to a differential component at the drains of M_1 and M_2 because these transistors display unequal transconductances as they depart from equilibrium.

In the circuit of Fig. 16(b), V_{sub} modulates both the drain junction capacitance of the NMOS devices and their threshold voltage (and hence the transition points of the waveform). Both effects are static, making the circuit susceptible even to low-frequency noise.

It is instructive to determine the minimum supply voltage for the above two circuits. At the midpoint of switching, where the input and output differential voltages are around zero, the stage of Fig. 15(a) requires that $V_{DD} \geq |V_{GSP}| + V_{GSN} + V_{ISS}$, where V_{GSP} abd V_{GSN} denote the gate-source voltages of M_3-M_4 and M_1-M_2, respectively, and V_{ISS} is the minimum voltage necessary for I_{SS}. Interestingly, the circuit of Fig. 16(b) imposes the same minimum supply voltage.

Another critical issue in the circuits of Figs. 15(a) and

16 relates to frequency tuning by means of current sources. The voltage-to-current (V/I) conversion required here presents difficulties at low supply voltages. In the example of Fig. 16(b), as V_{cont} rises and V_X falls, transistor M_3 eventually enters the triode region, thus making I_1 supply-dependent. The useful range of V_{cont} is therefore given by $V_{THN} < V_{cont} < V_{DD} - |V_{GSP}| - V_{THN}$, suggesting the use of a wide device for M_2 to minimize $|V_{GSP}|$.

III. LC OSCILLATORS

LC oscillators have found wide usage in high-speed and/or low-noise systems. Extensive research on inductors, varactors, and oscillator topologies has provided the grounds for systematic design, helping to demystify the "black magic."

LC oscillators offer a number of advantages over ring structures: (a) lower phase noise for a given frequency and power dissipation; (b) greater output voltage swings, with peak levels that can exceed the supply voltage; and (c) ability to operate at higher frequencies.

However, LC VCO design requires precise device and circuit modeling because (a) the narrow tuning range calls for accurate prediction of the center frequency; (b) the phase noise is greatly affected by the quality of inductors and varactors and the noise of transistors. Also, occupying a large area, spiral inductors pick up noise from the substrate and make it difficult to incorporate many such oscillators on one chip.

The design of LC VCOs targets the following parameters: center frequency, phase noise, tuning range, power dissipation, voltage headroom, startup condition, output voltage swing, and drive capability. The last two have often received less attention, but they directly determine the design difficulty and power consumption of the stages following the oscillator. That is, a buffer placed after the VCO may consume more power than the VCO itself!

A. Design Example

As an example of VCO design, let us consider the topology shown in Fig. 18. Here, M_1 and M_2 present a small-signal negative resistance of $-2/g_{m1,2}$ between nodes X and Y,

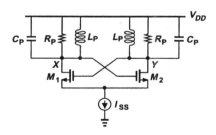

Fig. 18. LC oscillator.

compensating for the resistive loss in the tanks and sustaining oscillation. Each tank is modeled by a parallel RLC network, with all loss mechanisms lumped in R_P.[4]

---

[4]For a narrow frequency range, series resistances in the tank elements can be transformed to parallel components.

The design process begins with a power budget and hence a maximum value for I_{SS}. This is justified by the following observation. Once completed and optimized for a given power budget, the design can readily be scaled for different power levels, bearing a linear trade-off with phase noise while maintaining all other parameters constant. For example, if I_{SS}, the width of M_1 and M_2, and the total tank capacitance are doubled and the inductance value is halved, the phase noise power falls by a factor of two but the frequency of oscillation and the output voltage swings remain unchanged.[5]

Since subsequent stages typically require the VCO core to provide a minimum voltage swing, V_{min}, we assume M_1 and M_2 steer nearly all of I_{SS} to their correponding tanks and write $I_{SS}R_P = V_{min}$. Thus, the minimum inductance value is given by

$$L_P = \frac{R_P}{Q\omega} \qquad (9)$$

$$= \frac{V_{min}}{I_{SS}Q\omega}, \qquad (10)$$

where it is assumed the tank Q is limited by that of the inductor. Note that this calculation demands knowledge of the Q *before* the inductance is computed, a minor issue because for a given geometry and frequency of operation, the Q is relatively independent of the inductance.

We now determine the dimensions of M_1 and M_2. Increasing the channel length beyond the minimum value allowed by the technology does not significantly lower γ unless the length exceeds approximately 0.5 μm. For this reason, the minimum length is usually chosen to minimize the capacitance contributed by the transistors. The transistors must be wide enough to steer most of I_{SS} while experiencing a voltage swing of V_{min} at nodes X and Y. Viewing M_1 and M_2 as a differential pair, we note that M_1 must turn off as $V_X - V_Y$ reaches V_{min}. For square-law devices,

$$V_{min} = \sqrt{\frac{2I_{SS}}{\mu_n C_{ox} W/L}}, \qquad (11)$$

and hence

$$W = \frac{2I_{SS}}{\mu_n C_{ox} V_{min}^2 / L}, \qquad (12)$$

but for short-channel devices, W must be obtained by simulations using proper device models. This choice of W typically guarantees a small-signal loop gain greater than unity, enabling the circuit to start at power-up.

With L_P computed from Eq. (10), the total capacitance at nodes X and Y is calculated as $C_{tot} = (L_P\omega^2)^{-1}$. This capacitance includes the following *fixed* components: (1) the parasitic capacitance of L_P, C_{LP}; (2) the drain junction, gate-source, and gate-drain capacitances of M_1 and M_2, $C_{DB} + C_{GS} + 4C_{GD}$;[6] and (3) the input capacitance of the next state

[5]We assume that, at a given frequency, the Q is relatively independent of the inductance value.

[6]Since C_{GD} experiences a total voltage swing of $2V_{min}$, its Miller effect translates to a factor of two for each transistor.

(typically a buffer), C_L. Thus, the allowable varactor capacitance is given by the difference between C_{tot} and the sum of these components:

$$C_{var} = (L_P\omega^2)^{-1} - C_{LP} - C_{DB} - C_{GS} - 4C_{GD} - C_L. \quad (13)$$

This expression gives the center value of the tolerable varactor capacitance. Of course, a negative C_{var} means the inductance is excessively large, calling for a lower L_P, a smaller R_P, and hence a larger I_{SS}. However, to steer a greater tail current, the circuit must employ wider MOS transistors, thus incurring a larger capacitance at nodes X and Y and approaching diminishing returns. This ultimately limits the frequency of oscillation in a given technology.

For a given supply voltage and oscillator topology, the varactor capacitance exhibits a known dynamic range $C_{var,min} \leq C_{var} \leq C_{var,max}$, yielding a tuning range of $\omega_{min} \leq \omega_{osc} \leq \omega_{max}$, where

$$\omega_{min} = \frac{1}{\sqrt{L_P(C_{var,max} + C_{fixed})}} \qquad (14)$$

$$\omega_{max} = \frac{1}{\sqrt{L_P(C_{var,min} + C_{fixed})}}, \qquad (15)$$

and $C_{fixed} = C_{LP} + C_{DB} + C_{GS} + 4C_{GD} + C_L$.

Figure 19(a) depicts the oscillator with MOS varactors directly tied to X and Y. Since the output common-mode level

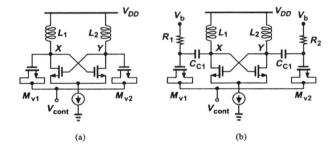

(a)   (b)

Fig. 19. LC oscillator with (a) direct coupling and (b) capacitive coupling of varactors to tanks.

is near V_{DD}, M_3 and M_4 sustain only a positive gate-source voltage (if $0 < V_{cont} < V_{DD}$). As seen from the C-V characteristic of Fig. 2(b), this limitation reduces the dynamic range of the capacitance by about a factor of two. As a remedy, the varactors can be capacitively coupled to X and Y, allowing independent choice of dc levels. Illustrated in Fig. 19(b), such an arrangement defines the gate voltage of M_{v1} and M_{v2} by $V_b \approx V_{DD}/2$ through large resistors R_1 and R_2.

The coupling capacitors, C_{C1} and C_{C2}, must be chosen much greater than the maximum value of C_{var} so as not to limit the tuning range. For example, if $C_{C1} = C_{C2} = 5C_{var,max}$, then the equivalent series capacitance reaches only $5C_{var,max}^2/(6C_{var,max}) = 0.83C_{var,max}$, suffering from a 17% reduction in dynamic range. On the other hand, large coupling capacitors display significant bottom-plate capacitance, thereby loading the oscillator and limiting the tuning range.[7] It is possible to realize C_{C1} and C_{C2} as "fringe" capacitors (Fig. 20) [7] to exploit the lateral field between adjacent metal

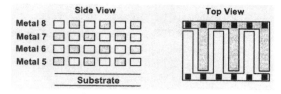

Fig. 20. Fringe capacitor.

lines. This structure exhibits a bottom-plate parasitic of a few percent, but its value must usually be calculated by means of field simulators.

The tuning range of LC VCOs must be wide enough to encompass (a) process and temperature variations, (b) uncertainties due to model inaccuracies; and (c) the frequency band of interest. In wireless communications, the last component makes the design particularly difficult, especially if a single VCO must cover more than one band. For example, in the Global System for Mobile Communication (GSM) standard, the transmit and receive bands span 890-915 MHz and 935-960 MHz, respectively. For one VCO to operate from 890 MHz to 960 MHz, the tuning range must exceed 7.8%. With another 7 to 10% required for variations and model inaccuracies, the overall tuning rang reaches 15 to 18%, a value difficult to achieve. In such cases, two or more oscillators may prove necessary, but at the cost of area and signal routing issues.

The phase noise of each oscillator topology must be quantified carefully. The reader is referred to the extensive literature on the subject.

B. Digital Tuning

Our study thus far implies that it is desirable to maximize the tuning range. However, for a given supply voltage, a wider tuning range inevitably translates to a greater VCO gain, K_{VCO}, thereby making the circuit more sensitive to disturbance ("ripple") on the control line. This effect leads to larger reference sidebands in RF synthesizers and higher jitter in timing applications. With the scaling of supply voltages, the problem of high K_{VCO} has become more serious, calling for alternative solutions.

A number of circuit and architecture techniques have been devised to lower the sensitivity of the VCO to ripple on the control line. For example, a *digital* tuning mechanism can be added to perform coarse adjustment of the frequency, allowing the analog (fine) control to cover a much narrower range. Illustrated in Fig. 21(a), the idea is to switch constant capacitors into or out of the tanks, thereby introducing discrete frequency steps. The varactors then tune the frequency within each step, leading to the characteristic shown in Fig. 21(b). Note that the switches are placed between the capacitors and ground - rather than between the tank and the capacitors. This permits

[7] This is relatively independent of whether the bottom plates are connected to nodes X and Y or to R_1 and R_2.

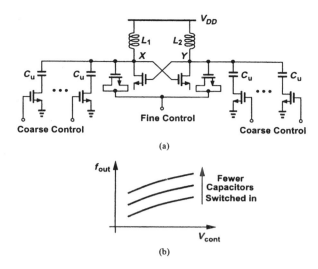

(a)

(b)

Fig. 21. (a) VCO with fine and coarse digital control, (b) resulting characteristics.

the use of NMOS devices with a gate-source voltage equal to V_{DD}, minimizing their on-resistance.

The above technique entails three critical issues. First, the trade-off between the on-resistance and junction capacitance of the MOS switches translates to another between the Q and the tuning range. When on, each switch limits the Q of its corresponding capacitor to $(R_{on}C_u\omega)^{-1}$. When off, each switch presents its drain junction and gate-drain capacitances, $C_{DB} + C_{GD}$, in series with C_u, constraining the lower bound of the capacitance to $C_u(C_{DB} + C_{GD})/(C_u + C_{DB}C_{GD})$ rather than zero. In other words, wider switches degrade the overall Q to a lesser extent but at the cost of narrowing the discrete frequency steps.

The second issue relates to potential "blind" zones in the characteristic of Fig. 21(b). As exemplified by Fig. 22, if the

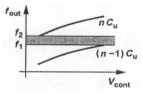

Fig. 22. Blind zone resulting from insufficient fine tuning range.

discrete step resulting from switching out one unit capacitor is greater than the range spanned continuously by the varactors, then the oscillator fails to assume the frequency values between f_1 and f_2 for any combination of the digital and analog controls. For this reason, the discrete steps must be sufficiently small to ensure overlap between consecutive bands.[8]

The third issue stems from the loop settling speed. As described below, the PLL takes a long time to determine how

[8] With a finite overlap, however, more than one combination of digital and analog controls may yield a given frequency. To avoid this ambiguity, the loop must begin with a minimum (or maximum) value of the digital control and adjust it monotonically.

many capacitors must be switched into the tanks. Thus, if a change in temperature or channel frequency requires a discrete frequency step, then the system using the PLL must remain idle while the loop settles.

When employed in a phase-locked loop, the oscillator of Fig. 21(a) requires additional mechanisms for setting the digital control. Figure 23 depicts an example for frequency synthesis.

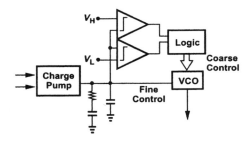

Fig. 23. Synthesizer using fine and coarse frequency control.

Here, the oscillator control voltage is monitored and compared with two low and high voltages, V_L and V_H, respectively. If V_{cont} falls below V_L, the oscillation frequency is excessively low[9], and one unit capacitor is switched out. Conversely, if V_{cont} exceeds V_H, one unit capacitor is switched in. After each switching, the loop settles and, if still unlocked, continues to undergo discrete frequency steps.

REFERENCES

[1] B. Razavi, "Design of Monolithic Phase-Locked Loops and Clock Recovery Circuits - A Tutorial," in *Monolithic Phase-Locked Loops and Clock Recovery Circuits*, B. Razavi, Ed., Piscataway, NJ: IEEE Press, 1996.

[2] P. Larsson, "Parasitic Resistance in an MOS Transistor Used as On-Chip Decoupling Capacitor," *IEEE J. Solid-State Circuits*, vol. 32, pp. 574-576, April 1997.

[3] K. Kundert, Private Communication.

[4] M. Danesh et al., "A Q-Factor Enhancement Technique for MMIC Inductors," *Proc. IEEE Radio Frequency Integrated Circuits Symp.*, pp. 217-220, April 1998.

[5] A. Zolfaghari, A. Y. Chan, and B. Razavi, "Stacked Inductors and Transformers in CMOS Technology," *IEEE Journal of Solid-State Circuits*, vol. 36, pp. 620-628, April 2001.

[6] F. Behbahani, et al., "A 2.4-GHz Low-IF Receiver for Wideband WLAN in 0.6-μm CMOS," *IEEE Journal of Solid-State Circuits*, vol. 35, pp. 1908-1916, December 2000.

[7] O. E. Akcasu, "High-Capacity Structures in a Semiconductor Device," US Patent 5,208,725, May 1993.

---

[9] We assume the frequency increases with V_{cont}.

Delay-Locked Loops - An Overview

Chih-Kong Ken Yang

Abstract — **Phase-locked loops have been used for a wide range of applications from synthesizing a desired phase or frequency to recovering the phase and frequency of an input signal. Delay-locked loops (DLLs) have emerged as a viable alternative to the traditional oscillator-based phase-locked loops. With its first-order loop characteristic, a DLL both is easier to stabilize and has no jitter accumulation. The paper describes design considerations and techniques to achieve high performance in a wide range of applications. Issues such as avoiding false lock, maintaining 50% clock duty cycle, building unlimited phase range for frequency synthesis, and multiplying the reference frequency are discussed.**

I. INTRODUCTION

Many applications require accurate placement of the phase of a clock or data signal. Although simply delaying the signal could shift the phase, the phase shift is not robust to variations in processing, voltage, or temperature. For more precise control, designers incorporate the phase shift into a feedback loop that locks the output phase with an input reference signal that indicates the desired phase shift. In essence, the loop is identical to a phase-locked loop (PLL) except that phase is the only state variable and that a variable-delay line replaces the oscillator. Such a loop is commonly referred to as a delay-line phase-locked loop or delay-locked loop (DLL). As with a PLL, the goals are (1) accurate phase position or low static-phase offset, and (2) low phase noise or jitter.

Because a DLL does not contain an element of variable frequency, it historically has fewer applications than PLLs. Bazes in [1] demonstrated an example of precisely delaying a signal in generating the timing of the row and column access strobe signals for a DRAM. Another common application uses a DLL to generate a buffered clock that has the same phase as a weakly-driven input clock. Johnson in [2] synchronizes the timing of the buffered clock of a floating-point unit with the clock of a microprocessor. A similar application recovers the data of a parallel bus by generating a properly positioned sampling clock. Typically, these systems provide a sampling clock with the same sampling rate but with an arbitrary phase as compared to the data (i.e. a "mesochronous" system [4]). A clocked DRAM data bus is an example of such a system. A clock propagates with the data as one of the signals in the bus and therefore has a nominally known phase relationship with the data. However, in order to receive and buffer the clock to sample

C.K. Ken Yang is with University of California at Los Angeles, yang@ee.ucla.edu.

the data bus, the actual sampling clock is no longer properly aligned with the data. A DLL is commonly used to lock the phase of the buffered clock to that of the input data. The phase locking significantly reduces timing uncertainty in sampling the data, which then enables higher data rates as in [3].

Although aperiodic signals can also be delayed by the delay line in a DLL, the inputs to delay lines are typically clock signals. By using a periodic signal, the delay lines do not need arbitrarily long delays and typically only need to span the period of the clock to generate all possible phases. A data signal can be delayed by sampling the data with the appropriately delayed clock.

The motivation for using DLLs is that the design of the control loop is simplified by having only phase as the state variable. Section II reviews how such a loop is unconditionally stable and has better jitter characteristics. However, a DLL is not without its own limitations. The variable delay line has a finite delay range and finite bandwidth. Section II also discusses these design considerations. Section III describes different implementations of the variable delay line. Within the past ten years, modifications to the basic DLL architecture have enabled clock and data recovery applications in "plesiochronous" systems [4] where the sampling rates for clock and data differ by a few hundred parts-per-million in frequency. Delay lines with effectively infinite delay are also addressed in Section III.

More recently, several researchers such as [5] and [6] have introduced architectures that permit frequency multiplication based on delay lines which further extends their use in clock generation and frequency synthesis. Section IV describes these architectures.

II. DLL CHARACTERISTICS

The basic loop building blocks are similar to that of a PLL: a phase detector, a filter, and a variable-delay line. Figure 1 illustrates the three main functional blocks. Since phase is the only state variable, a control loop higher than first-order is not needed to compensate a fixed phase error. The resulting transient impulse response is a simple exponential. Although the simple loop characteristics are an advantage that DLLs have over PLLs, the design is complicated by the additional circuitry that is needed to overcome having a limited delay range and not producing its own frequency.

A. First-order Loop

A phase detector compares the phase of the reference input and the delay-line output. The comparison yields a signal proportional to the phase error. The error is low-pass

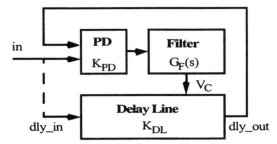

Figure 1: DLL architecture.

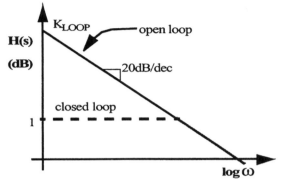

Figure 2: Open- and closed-loop transfer characteristics.

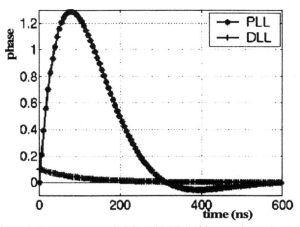

Figure 3: Step response of PLL and DLL (with same loop characteristics).

filtered to produce a control voltage or current that adjusts the delay of the delay line. The delay-line input can be either the reference input or a clean clock signal.

The s-domain representation of each loop element is depicted within each block in Fig. 1. The open-loop transfer function can be written as $T(s) = K_{PD}K_{DL}G_F(s)$ where K_{PD} is the phase-detector gain, $G_F(s)$ is the filter transfer function, and K_{DL} is the delay-line gain. If the loop has finite gain at dc, the resulting output signal will exhibit a static phase error as shown in the following equation.

$$H(s)\big|_{s=0} = \frac{1}{1 + 1/(K_{PD}K_{DL}G_F(s))}\bigg|_{s=0} \quad (1)$$

To eliminate the static phase error, the filter is often an integrator to store the phase variable. This results in a first-order closed-loop transfer function.

$$H(s) = \frac{1}{1 + (s/K_{PD}K_{DL}G_F)} \quad (2)$$

The equation assumes that the delay-line input is a clean reference as opposed to the reference input. Higher-order loop filters have not commonly been used but can enable better tracking of a phase ramp (i.e. a frequency difference).

Figure 2 shows the open-loop and closed-loop transfer functions. With only a single integrator, the open-loop phase margin is 90°. The loop is unconditionally stable as long as the delay in the loop does not degrade the phase margin excessively. The closed-loop transfer function illustrates that

the tracking of the phase of the input clock changes at different frequencies. Based on the transfer function, the loop bandwidth is $\omega_{bw} = K_{PD}K_{DL}G_F$. For frequencies within the loop bandwidth the phase of the output clock will track that of the reference input and reject noise within the loop. The phase characteristics of the output clock above the bandwidth of the loop depend on the phase behavior of the delay-line input and the noise from the delay line. The noise transfer function from a noise source lumped at the delay-line output is a high-pass response.

$$H(s) = \frac{(s/K_{PD}K_{DL}G_F)}{1 + (s/K_{PD}K_{DL}G_F)} \quad (3)$$

In some degenerate cases, the delay-line input is also the reference input. The feedback loop would guarantee a fixed phase relationship between the delay-line output and the reference so any phase variations in the reference would directly appear at the delay-line output in an all-pass response. However, noise due to the delay line is still high-pass filtered.

B. Advantages over a PLL

The loop characteristics are considerably simpler than those of a PLL. A PLL would contain at least two states to store both the frequency and phase information. In order to maintain loop stability, an additional zero is needed. A DLL is less constrained with only a single pole. The loop gain directly determines the desired bandwidth. The only stability consideration is when the loop bandwidth is very near the reference frequency. The periodic sampling nature of the phase detection and the delay in the feedback loop degrade the phase margin. For instance, if the feedback delay is one reference cycle, the loop bandwidth should not exceed 1/4 of the reference frequency.

Figure 3 illustrates the response to a noise step applied to the control voltage for both a PLL and a DLL. A PLL accumulates phase error due to its higher-order loop characteristic. In response to a phase error, the control

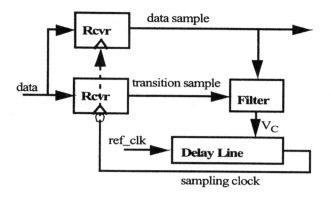

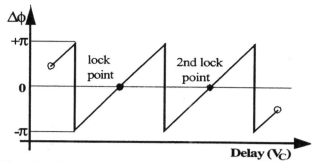

Figure 5: Delay line phase/delay characteristic.

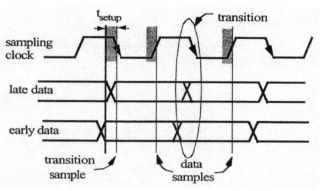

Figure 4: Early-late receiver architecture using the receiver as the phase detector. Timing diagram showing early and late data.

voltage alters the frequency of an oscillator. The output phase is an integration of the frequency change. In response to a noise perturbation, the loop accumulates a phase error before correcting. In contrast, a DLL attenuates the phase error by the time constant of the loop. In the figure, both loops are designed with the same 3-dB bandwidth, the same delay elements, and the PLL is a 2nd-order loop with a damping factor of unity. Clearly, the PLL suffers from larger phase errors due to the phase accumulation.

A second advantage relates to clock and data recovery applications. An effective way to recover the timing for sampling a data input is to use the data receiver as a phase detector. The architecture, depicted in Fig. 4, uses the 180°-shifted clock to sample the data transitions in addition to sampling the data values [7]. Whenever data changes values, the sampled transition and the data values can be combined to indicate whether the sampling clock edge is earlier or later than the data transition. Phase information is only present with data transitions. The feedback loop locks when the transition sampling clock samples a metastable value. This commonly used design is known as an early-late or bang-bang architecture. The timing diagram in Fig. 4 illustrates examples of the data being early and late. Due to the inherent setup time of the data receiver, the transition sampling clock may not occur at the same time as the data transition. The phase shift compensates for the receiver setup time and maximizes the margin of error for the data sampling.

However, the data receiver is ultimately a binary comparator and the phase detector does not indicate an error that is proportional to the phase difference. Hence, the timing-recovery loop is nonlinear. Although a higher-order PLL using early-late control can be made conditionally stable [8], the resulting phase dithers with a limit cycle translating into jitter. The oscillation depends on the loop parameters and can be considerable for high bandwidth loops. With an early-late DLL, the phase of the clock output also dithers. But because the stability only depends on the delay within the loop, the dithering would only be a few cycles and can be significantly less than the dithering of a PLL.

C. Design Considerations in a DLL

A typical DLL involves several design considerations. First, the delay line usually has a finite delay range. If the desired phase of the output signal is beyond the delay range, the loop will not lock properly. Second, the output of the DLL also depends greatly on the input to the delay line. Since the delay-line input propagates to the DLL output, tracking jitter and the output's duty cycle depend not only on the delay-line design but also on the delay-line input. Third, the basic DLL cannot generate new frequencies different from that of the delay-line input.

A variable-delay line adjusts the delay by varying the RC time constant of a buffer and often has limited adjustment range. Section III will describe several techniques in greater detail. Even though the delay range is limited, DLLs for a periodic clock signal only need the range to exceed 2π in phase across process and systematic variations to cover all possible phases. For systems with a range of operating frequencies, the delay line must span 2π for the lowest input frequency.

An issue known as false-locking occurs when the delay range exceeds 2π. There can be several secondary lock points repeating every 2π. Figure 5 depicts an example of the characteristic of a delay line with two lock points. Since phase detectors must be periodic, if the delay line initializes within π of the second lock point, the phase detector will push the delay line toward lock with a longer than necessary delay. Long delays require large RC time constants for a given variable-delay buffer element. The bandlimiting by the filter would significantly attenuate a high-frequency input clock. The attenuation increases the jitter and may even prohibit the input from reaching the output.

Even if the delay line is constrained to span only one lock point but greater than 2π, a second similar issue exists. It is difficult to design a delay line such that the adjustable range is exactly $-\pi$ to $+\pi$ across different operating and processing conditions. If initialized at the minimum or maximum delay, the phase detector may push the loop toward either the maximum or minimum delay limit and "false-lock" to an incorrect phase.

To address false-locking, designers employ several techniques depending on the application. For systems that require a delay line with a known fixed delay, operating condition variations may be small enough such that the delay line only needs a small variable range that is less than $+\pi$ and $-\pi$. For systems that lock to a fixed phase over a wide range of frequencies, one design [9] uses an auxiliary frequency-sensing loop that generates a voltage to coarsely set the delay for the given input frequency. Then DLL only fine tunes the delay for the desired phase. For data recovery applications where the clock phase can be arbitrary with respect to the data, a common design uses a startup circuit for the DLL that initializes the delay line at its minimum delay to avoid any secondary lock points. However, as mentioned earlier, the phase detector may keep the delay line at the minimum delay. A sensing circuit or a state machine detects when the delay line is at its limit and optionally inverts the feedback clock. The phase would flip by 180° and the loop would lock properly. As will be discussed in Section III, a more robust alternative reconfigures the delay line such that the delay only spans 2π and wraps back to 0° when the delay exceeds 360°.

The jitter and duty cycle of the delay-line output clock depend on the input, the coupling of the input to the delay line, and the delay line itself. Often the input is from off-chip and, therefore, it must be carefully received to prevent supply and substrate noise from coupling onto the signal as jitter. In contrast, the high-frequency phase noise of the clock output of an oscillator-based PLL depends primarily on the oscillator design. An improperly received input clock can often result in worse jitter performance in a DLL as compared to a PLL. Similarly, while the duty cycle from an oscillator is only modestly distorted (by the difference between the rising edge and falling edge delays), the duty cycle of the DLL's input clock can be significantly distorted as it propagates to the output. Since duty cycle is a systematic error, a good design corrects duty cycle using an explicit block instead of compounding the difficulty of the delay-line design.

A duty-cycle corrector (DCC) is commonly added to either the DLL input or output. Figure 6 illustrates the basic components of the feedback loop: an input with finite slew rate, a buffer element with adjustable threshold, a comparator, and an integrator. The comparator determines the threshold crossing of the clock waveform. The result is integrated and used to skew the threshold of the buffer stage.

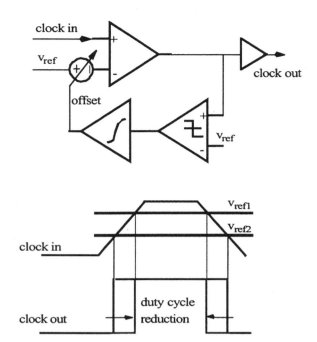

Figure 6: Duty-cycle corrector block diagram. Timing diagram shows change in duty cycle with changing offset.

Because the buffer input has finite slew rate, changing the threshold effectively adjusts the output high and low half-periods. The loop settles when the high and low half-periods are equal. Figure 6 illustrates the reduction in duty cycle as the threshold shifts from V_{ref1} to V_{ref2}. Since random variation of the duty cycle effectively appears as jitter, single-ended implementations such as that shown in the figure can be very sensitive to common-mode noise. For this reason, differential architectures are preferred [3].

For low jitter on the output clock, the loop components must be carefully designed. Many of the loop components are very similar to that of a PLL and are well described in [10]. For a charge-pump based loop filter, since the filter is only first-order, a simple capacitor replaces the RC filter. As in a PLL, noise on the control voltage directly translates into jitter. Designers may use additional filtering to suppress the noise. The loop element that has deviated the most from PLL design and is critical for functionality and performance is the design of the delay line.

III. DELAY-LINE ARCHITECTURES

The primary characteristics of a delay line are (1) gain (i.e. change in delay for a given change in voltage), and (2) delay range. For most applications using periodic inputs, the absolute delay is not critical as long as the range spans 2π. Because delay lines are relatively short, they do not contribute significant thermal or $1/f$ phase noise. However, for large digital systems, low supply/substrate sensitivity is needed to reject the on-chip switching noise.

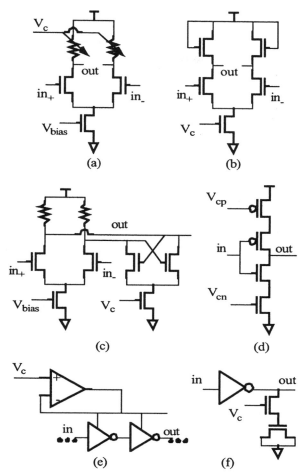

(a)　　　　　　(b)

(c)　　　　　　(d)

(e)　　　　　　(f)

Figure 7: Six different delay elements.

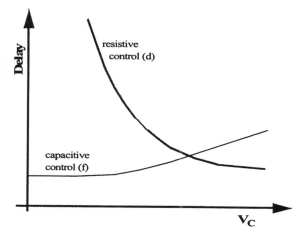

Figure 8: Delay versus voltage for two different delay buffer elements: types (d) and (f) of Fig. 7.

A. Basic Delay Line

A delay line comprises of a chain of variable-delay elements. Each element is controllable by either a voltage or a current. The delay of each element is proportional to its RC time constant and changing the effective resistance or capacitance adjusts the delay.

Figure 7 depicts several examples of buffer elements. For a differential buffer, the load resistance can be an MOS transistor in the triode region [Fig. 7-(a)] where the resistance is proportional to V_{GS}-V_{th}. Varying the gate voltage adjusts the delay of the element. A non-linear device such as a diode can also serve as a load resistance [Fig. 7-(b)]. Since the resistance varies with the current, varying the bias current of the buffer would adjust the delay. Similarly, a negative transconductance that changes with the bias current can be placed in parallel with a fixed load resistance [Fig. 7-(c)]. The varying negative transconductance changes the effective load resistance and hence varies the delay. Because nonlinear elements have resistances that depend on both voltage and current, they can be more sensitive to supply noise.

For push-pull type elements such as inverters, the delay can be changed by changing the rate at which the output capacitance is charged [Fig. 7-(d)]. An adjustable current source limits the peak current of an inverter and varies the delay. An alternative method regulates the supply voltage of the inverters and uses the control voltage to set the supply voltage [Fig. 7-(e)]. The effective switching resistance varies with the supply voltage. Instead of changing the resistance, the effective capacitance can also be made adjustable [Fig. 7-(f)]. A transistor that behaves as an adjustable resistance can be used to decouple an explicit output capacitance. The larger the resistance the less capacitance is seen at the output.

Figure 8 illustrates the delay versus control voltage for a resistively-controlled delay element. For the element of Fig. 7-(d), either V_{GS}-V_{th} or the bias current can be zero and, therefore, a single element's delay can span from the minimum buffer delay to infinite. However, since the time constant is proportional to the delay, a long delay setting would significantly attenuate a high-frequency clock. Delay lines with a wide range for high clock frequencies require a large number of broadband delay elements.

Unlike resistive control, the maximum delay in a capacitively-controlled element [Fig. 7-(f)] is proportional to $R(C_{int}+C_{exp})$ and the minimum delay is proportional to RC_{int} where C_{int} is the intrinsic capacitance of the buffer and the load of the subsequent stage, and C_{exp} is the explicit capacitance added to the circuit. Because of the limited range per buffer, obtaining a wide delay range involves a large number of buffers. The maximum delay of each buffer is chosen to avoid attenuating the signal. In designs where the clock has a large voltage swing, the transistor in series with the explicit capacitance no longer appears as a variable resistor because the device enters saturation and cut-off. For these buffers, the control voltage determines the fraction of current and period of time in which the buffer's current charges the explicit capacitance.

An example of the delay versus control voltage for a capacitively-controlled element is overlaid in Fig. 8. Most

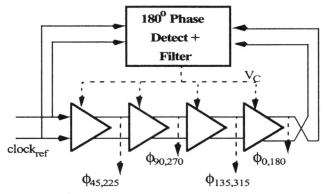

Figure 9: 180°-locked DLL to generate intermediate phases that are a fraction of a cycle.

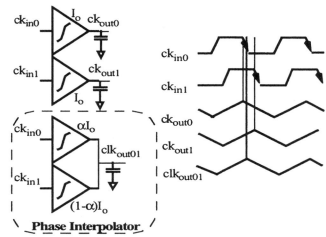

Figure 10: Phase interpolator design by shorting of the output of two integrators/buffers. .

delay elements exhibit some nonlinearity. As a result, the delay-line gain, K_{DL}, is a function of the delay. Because a DLL is unconditionally stable, the loop still functions with the varying loop parameter. However, more linear elements are better for designs that require a constant loop bandwidth. To compensate for the variable K_{DL}, designers add programmability to the loop-filter capacitor.

The control signal for either type of delay elements can be digital. In a digital implementation [11], the current source is binary weighted and switched by a digital word. For capacitively-controlled elements, the capacitance can be binary weighted and switched. A nearly all-digital DLL is then possible by using a simple counter to replace the analog integrating filter.

B. Phase Interpolation

Instead of only using the clock phase at the end of a delay line, an earlier clock phase can be tapped from the middle of a delay line. Some applications require the delay line to produce a delay that is a fixed *fraction* of the input-clock period. Figure 9 shows one implementation that uses a DLL to lock the input clock to the output. An 180° phase detector would guarantee the absolute delay of a delay line to be a half-cycle. Tapping from different points on the delay line provides different phases. As shown in Fig. 9, for a 45° phase shift, the clock can be tapped from the first delay stage of a 4-stage differential delay line. If an arbitrary phase is needed, each delay stage can be tapped and multiplexers can select the nearest desired phase. The number of delay elements quantizes the phase step and limits the resolution [12]. Fine phase resolution requires longer delay lines. Yet, the resolution is limited at high clock frequencies because the maximum number of delay elements needed to span 180° is limited.

An arbitrary intermediate phase can be obtained by "interpolating" between two clock phases that are tapped from a delay line. Depending on the weighting, an interpolator produces a clock that has a programmable output phase in between the input clock phases. As long as discrete clock phases that span the entire cycle are available as inputs, any phase for the interpolator's output is possible.

Multiplexers are needed to select the phases to interpolate between. For example, with phases tapped from a 4-stage delay line, if the desired output clock phase is 120°, the interpolator inputs would be from the second and third delay elements.

Interpolators essentially perform a weighted average of the input phases. As shown in Fig. 10, ideally, the two input phases drive two integrators which charge a single output. The weighting of the average is by the relative currents of the two integrators. When $\alpha=1$, the output clock phase depends only on ck_{in0}. When $\alpha=0.5$, i.e. the current is split equally between the two integrators, the output phase is additionally delayed by half the phase difference. As illustrated in Fig. 10, the phase of the interpolated output (ck_{out01}) falls between the phases of the non-interpolated outputs (ck_{out0} and ck_{out1}).

With ideal integrators, the interpolation is linear, resulting in a constant K_{DL}. Alternatively, an interpolator can effectively be formed with buffer elements instead of integrators. By weighting the drive strength or current of two buffer elements whose outputs are shorted together, one can adjust the output phase. Because the output is not integrated, the resulting interpolation is slightly nonlinear and depends on (1) the phase difference between the inputs and (2) the slew rate (or time constant) of the input and output signals [13]. Figure 11 depicts the linearity of the interpolation for two different input phase separations, $s=\tau$ and $s=2\tau$ where τ is the buffer's time constant. The larger phase spacing results in greater nonlinearity. Similar to RC delay elements, the interpolation can be digitally controlled. Since the weighting of the interpolation depends on the proportional current, the current sources of the integrators or buffers can be digitally weighted and programmed.

In a design for clock and data recovery by [3], quadrature clocks are interpolated to generate an intermediate clock phase within a quadrant. Figure 12 illustrates the mostly analog architecture. An analog control

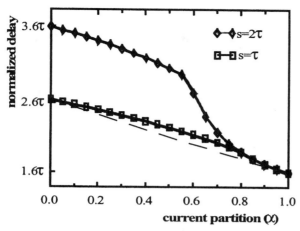

Figure 11: Buffer based phase interpolator linearity.

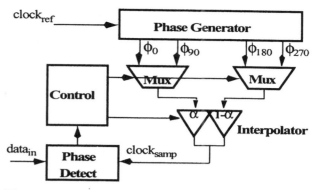

Figure 12: Infinite-range delay line based on phase rotation.

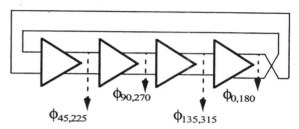

Figure 13: Oscillator with tapped outputs for multiple phases.

voltage produced by the phase detector and filter determines the interpolator currents. Comparators indicate when the current is fully steered to one integrator. A finite state machine driven by the comparators selects the appropriate quadrant by switching the interpolator inputs such that all 360° phases are possible. The quadrature input clocks is generated from an external reference clock through the use of a divide-by-two circuit.

Interestingly, because the phase rotates from one quadrant to the next, the architecture effectively has an unlimited delay range. If the input data rate and the reference clock frequency are slightly different, a DLL would continually increase or decrease the delay in order to track the accumulating input phase. A typical DLL with a finite delay range would run out of delay or lose lock. On the other hand, plesiochronous operation is possible with an interpolator-based delay line since the phase smoothly rotates between quadrants.

Interpolating between clocks with large phase spacings such as quadrature clocks results in an output clock with slow slew rate. Such waveforms are more susceptible to noise and result in higher jitter. An enhancement uses more closely-spaced phases that span the cycle. The finer phases spacing is possible using a multi-stage ring oscillator. As shown in Fig. 13, a 4-stage differential oscillator would generate 8 phases 45° apart. To guarantee a correct period for each clock phase, the ring oscillator is locked to the external reference clock using a PLL. The role of the PLL is solely for generating the phases. A purely DLL-based architecture is also possible by replacing by using the DLL in Fig. 9 that locks the delay-line output with a 180° phase shift [13].

The architecture is commonly known as a dual-loop design because the first loop, a PLL or DLL, generates the phases and the second loop, the interpolation-based DLL, recovers the data and phase. Since the first loop is not in the feedback of the second loop (or vice versa), the overall system is stable as long as each loop is individually stable. A dual-loop design is possible with the second loop within the feedback of the first loop [15] as long as the stability of the loop is carefully considered.

The data recovery portion of a dual-loop design is conducive to a digital implementation. The binary output of the receiver-replica phase detector can be accumulated using a digital counter. The counter output selects the appropriate phase from the oscillator and controls the digitally programmable interpolators [13],[14]. As long as the quantized phase step is small, the small error only minimally impacts the data recovery.

C. Oversampled Implementation

An alternative purely digital approach to clock and data recovery can be implemented by oversampling the data. Figure 14 illustrates an example of a digital architecture. Multiple finely-spaced clock phases oversample the data input. The sampled results are digitally processed to determine both the correct data value and the optimal phase of the data sample. The digital processing can vary in complexity. Simple implementations use the optimal data sample as the received data [18] or take a majority vote from the samples of a single bit [17]. The bit boundaries determine the samples associated with a bit. Transitions that are detected in the samples from the prior or current bits indicate the bit boundaries.

The sampling rate limits the timing error margin. Greater amount of oversampling reduces the data-recovery timing error, but increases the number of clock phases. Low data rate UARTs [16] typically use 8 to 16 times oversampling. For high data rates, generating accurate clock phases separated by sub-100ps is very challenging. More

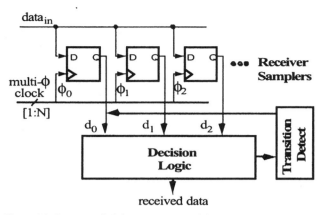

Figure 14: Oversampled data recovery architecture.

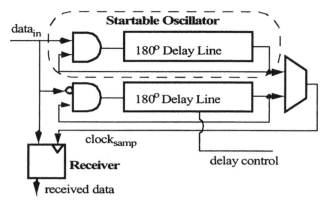

Figure 15: Clock/data recovery using startable oscillator.

aggressive designs with the least amount of clocking overhead and high data rates use a minimum of 3x oversampling [17], [18].

Even though phase spacing scales with the gate delay of a technology, so does the bit time in each generation of applications. For oversampling of the data bits, finely-spaced clock phases are needed. Tapping from a delay line produces phases separated by a buffer delay. For even finer phases, several techniques are commonly used. For example, several interpolators can be used where each interpolator has slightly different weighting to generate intermediate phases with spacing less than a buffer delay [19]. An alternative method uses a chain or array of coupled oscillators [20]. By taking a chain of oscillators and coupling them such that the output and input of the chain are separated by only one gate delay, sub-gate-delay phase spacings result from the outputs of each oscillator. Lastly, if the data can be delayed with a chain of delay buffers along with the clock, the clock at each delay stage can be used to sample the data of the corresponding stage. As long as the data and the clock delay lines have slightly different delays, the sub-sampled outputs are effectively an oversampling of the data. The effective phase spacing depends only on the difference between the data delay and clock delay [21]. The architecture has a drawback in that it requires delaying the data and clock by long delays of several cycles, which can significantly increase jitter.

IV. CLOCK MULTIPLICATION

With a dual-loop architecture, a DLL can produce a frequency plesiochronous to the delay-line input. However, the rate at which the interpolator weight changes limits the frequency difference. Generating a significantly different or multiplied frequency from a low-frequency input reference is not possible with the architecture.

Recently designers have explored several methods of using DLLs for frequency multiplication. One method uses a delay line that is locked to 180°. With the phases that span an entire cycle, the tapped clock edges are combined to form a clock with multiplied frequency. The most direct method

uses logical AND-ORs to combine the multiple phased clocks into a single high-frequency clock [23]. Alternatively, the method in [24] converts each phase into a small pulse and ORs the pulses together to form the output clock. In cases where the output capacitance of the logic gates limits the output frequency, one design [6] uses phases to excite a tuned LC tank to combine the clock phases.

Instead of edge combining, the multiplied clock can be the direct output of a delay line. The architecture is similar to a technique for clock and data recovery that uses a startable oscillator [22]. As shown in Fig. 15, the architecture uses data transitions to trigger startable oscillators: high-value data triggers one oscillator and low-value data triggers another. Each startable oscillator comprises of a delay line and an AND gate. The data value enables the AND gate and the triggered oscillator propagates an edge through the delay elements and produces a clock edge delayed by a half-cycle. The edge is used to sample the data. In the absence of input transitions, the delay line is configured an oscillator and generates a sampling edge every cycle. Whenever a new data transition occurs, the oscillator resynchronizes its phase to that of the input. In the implementation by [22], the natural oscillation frequency of the oscillator is determined by an external plesiochronous clock reference. The architecture has not been widely applied to higher data rate designs because the sampling phase is directly derived from the input data without any filtering. The deterministic and random jitter inherent in the data are effectively doubled and can be considerable.

If the input is a low-jitter reference clock, a similar architecture can be used for clock multiplication [5]. As illustrated in Fig. 16, a lower frequency but clean reference clock is one input to a multiplexer that feeds into a delay line. The output of the delay line is fed back to the multiplexer as the second input. When a reference clock edge is available, the multiplexer selects the reference input. Otherwise, the multiplexer configures the delay line as an oscillator with the output frequency controlled by the delay. The multiplexer inputs are selected by a counter circuit that determines the number of cycles to oscillate before accepting the next reference clock edge. A phase detector compares the

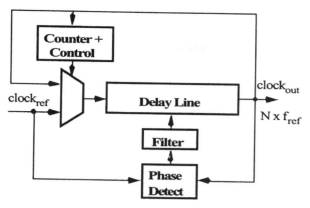

Figure 16: DLL-based clock multiplication.

reference input with the oscillator output and tunes the delay of the delay elements. Once locked, the resulting output clock frequency is a multiple of the input reference frequency. Recent designs [26] extend the frequency range and use an interpolator instead of a multiplexer to blend the delay-line feedback and the low-frequency reference clocks.

Both edge-combining multiplication and delay-line multiplication reduce the phase noise of the output clock because the core DLL does not have an oscillator that accumulates phase error. After N cycles, where N is the divide ratio, a new clean reference clock edge arrives and resets any accumulated phase error to zero. The architecture potentially lowers jitter by eliminating the peaking in the transfer function and allows a high tracking bandwidth. However, matching is critical in these designs. Mismatches in the phase detector or charge pump result in a static phase error that modulates the output frequency at the input reference frequency. Similarly, in the edge combining implementations, if the delay line is mismatched, the output clock would contain significant reference tones. Designers either choose the reference frequency carefully so that the tones do not impact the system performance or employ additional circuitry to compensate for the mismatches.

V. CONCLUSION

DLLs have been commonly used for generating precise phase delays of a signal and have been increasingly popular in clock generation and data recovery applications. Most importantly, because of the first-order loop characteristics that controls the phase directly, DLLs can be designed with high tracking bandwidths and do not exhibit the phase accumulation of an oscillator-based PLL.

The more simple loop characteristics belie many subtleties in DLL design. The delay-line input clock must have low-jitter and good duty-cycle. Furthermore, it must be carefully received and coupled to the input of the delay line to maintain good jitter performance. This source of jitter counter-balances the jitter accumulation of PLLs and results in less jitter improvement. Additional circuitry is often needed to prevent false-locking. Since a delay line does not restore a clock's duty cycle, the output clock requires correction circuitry. To use DLLs in plesiochronous systems, the delay line must have even more circuitry to achieve an unlimited delay range. In clock multiplication applications, very careful matching in the DLL components is critical to eliminate reference tones. In the many designs that have addressed these subtleties, DLLs have demonstrated low-jitter clock outputs for a variety of clock generation and data recovery applications.

REFERENCES

[1] Bazes, M., "A Novel Precision MOS Synchronous Delay Line," *IEEE Journal of Solid-State Circuits*, vol sc-20, no 6, Dec. 1985, pp. 1265-71

[2] Johnson, M.G., E.L. Hudson, "A Variable Delay Line PLL for CPU-Coprocessor Synchronization," *IEEE Journal of Solid-State Circuits*, vol 23, no 5, Oct. 1988, pp. 1218-23

[3] Lee, T.H., et. al., "A 2.5V CMOS Delay-Locked Loop for an 18 Mbit, 500Megabytes/s DRAM," *IEEE Journal of Solid-State Circuits*, vol 29, no 12, Dec. 1994, pp. 1491-6

[4] Messerschmitt, D.G., "Synchronization in Digital System Design," *IEEE Journal on Selected Areas in Communications*, Oct. 1990, pp. 1404-1420

[5] Waizman, A., "A Delay Line Loop for Frequency Synthesis of De-Skewed Clock," *IEEE ISSCC Dig. of Tech. Papers*, Feb. 1994, San Francisco, Session 18.5

[6] Chien, G., P.R. Gray, "A 900-MHz Local Oscillator Using a DLL-Based Frequency Multiplier Technique for PCS Applications," *IEEE Journal of Solid-State Circuits*, vol 35, no 12, Dec. 2000, pp. 1996-9

[7] Alexander, J.D., "Clock Recovery from Random Binary Data," *Electronic Letters*, vol 11, Oct. 1975, pp 541-2

[8] D'Andrea, N.A., F. Russo, "A binary quantized digital phase locked loop: a graphical analysis," *IEEE Transactions on Communications*, vol.COM-26, (no.9), Sept. 1978. p.1355-64

[9] Moon, Y., "An All-Analog Multiphase Delay-Locked Loop Using A Replica Delay Line for Wide-Range Operation and Low-Jitter Performance," *IEEE Journal of Solid-State Circuits*, vol 35, no 3, Mar. 2000, pp. 377-84

[10] Razavi, B., "Design of Monolithic Phase-Locked Loops and Clock Recovery Circuits - A Tutorial," *Monolithic Phase-locked Loops and Clock Recovery Circuits*, IEEE Press 1996 New Jersey, pp. 1-28

[11] Dunning, J., et. al. "An All-Digital Phase-Locked Loop with 50-Cycle Lock Time Suitable for High-Performance microprocessors," *IEEE Journal of Solid-State Circuits*, vol 30, no 4, Apr. 1995, pp. 412-22

[12] Efendovich, A., et. al., "Multifrequency Zero-Jitter Delay-Locked Loop," *IEEE Journal of Solid-State Circuits*, vol 29, no 1, Jan. 1994, pp. 67-70

[13] Sidiropoulos, S., M.A. Horowitz, "A Semidigital Dual Delay-Locked Loop," *IEEE Journal of Solid-State Circuits*, vol 32, no 11, Nov. 1997, pp. 1683-92

[14] Garlepp, B., et. al., "A Portable Digital DLL for High-Speed CMOS Interface Circuits," *IEEE Journal of Solid-State Circuits*, vol 34, no 5, May 1996, pp. 632-44

[15] Larsson, P., "A 2-1600-MHz CMOS Clock Recovery PLL with Low-Vdd Capability," *IEEE Journal of*

Solid-State Circuits, vol 34, no 12, Dec. 1999, pp. 1951-60

[16] Cordell, R., "A 45-Mbit/s CMOS VLSI Digital Phase Aligner," *IEEE Journal of Solid-State Circuits*, vol 23, no 2, Apr. 1988, pp. 323-28

[17] Lee, K., et. al., "A CMOS Serial Link For Fully Duplexed Data Communication," *IEEE Journal of Solid-State Circuits*, vol 30, no 4, Apr. 1995, pp. 353-64

[18] Yang, C.K., et al., "A 0.5-μm CMOS 4.0-Gb/s Serial Link Transceiver with Data Recovery Using Oversampling," *IEEE Journal of Solid-State Circuits*, vol 33, no 5, May 1998, pp. 713-22

[19] Weinlader, D., et al., "An Eight Channel 36-GS/s CMOS Timing Analyzer," *IEEE ISSCC Dig. of Tech. Papers*, Feb. 2000, San Francisco, pp. 170-1

[20] Maneatis, J., M. Horowitz, "Precise Delay Generation Using Coupled Oscillators," *IEEE Journal of Solid-State Circuits*, vol 28, no 12, Dec. 1993, pp. 1273-82

[21] Gray, C., et. al., "A Sampling Technique and Its CMOS Implementation with 1Gb/s Bandwidth and 25ps Resolution", *IEEE Journal of Solid-State Circuits*, vol 29, no 3, Mar. 1994, pp. 340

[22] Ota, Y. et. al., "High-Speed, Burst-Mode, Packet Capable Optical Receiver and Instantaneous Clock Recovery for Optical Bus Operation," *IEEE Journal of Lightwave Technology*, vol 12, no 2, Feb. 1994, pp. 325-330

[23] Foley, D., M.P. Flynn, "CMOS DLL-Based 2-V 3.2ps Jitter 1-GHz Clock Synthesizer and Temperature-Compensated Tunable Oscillaor," *IEEE Journal of Solid-State Circuits*, vol 36, no 3, Mar. 2001, pp. 417-23

[24] Kim, C., I. Hwang, S.M. Kang, "Low-Power Small-Area +/-7.28ps Jitter 1GHz DLL-Based Clock Generator," *IEEE ISSCC Dig. of Tech. Papers*, Feb. 2002, San Francisco, Session 8.3

[25] Farjad-rad, R., et. al., "A 0.2-2GHz 12mW Multiplying DLL for Low-Jitter Clock Synthesis in Highly-Integrated Data Communication Chips," *IEEE ISSCC Dig. of Tech. Papers*, Feb. 2002, San Francisco, Session 4.5

[26] Ye, S., L. Jansson, I. Galton, "A Multiple-Crystal Interface PLL with VCO Realignment to Reduce Phase Noise," *IEEE ISSCC Dig. of Tech. Papers*, Feb. 2002, San Francisco, Session 4.6

[27] Kim, J., et. al., "A Low-Jitter Mixed-Mode DLL for High-Speed DRAM Applications," *IEEE Journal of Solid-State Circuits*, vol 35, no 10, Oct. 2000, pp. 1430-3

Delta-Sigma Fractional-N Phase-Locked Loops

Ian Galton

Abstract—**This paper presents a tutorial on delta-sigma fractional-N PLLs for frequency synthesis. The presentation assumes the reader has a working knowledge of integer-N PLLs. It builds on this knowledge by introducing the additional concepts required to understand $\Delta\Sigma$ fractional-N PLLs. After explaining the limitations of integer-N PLLs with respect to tuning resolution, the paper introduces the delta-sigma fractional-N PLL as a means of avoiding these limitations. It then presents a self-contained explanation of the relevant aspects of delta-sigma modulation, an extension of the well known integer-N PLL linearized model to delta-sigma fractional-N PLLs, a design example, and techniques for wideband digital modulation of the VCO within a delta-sigma fractional-N PLL.**

I. INTRODUCTION

Over the last decade, delta-sigma ($\Delta\Sigma$) fractional-N phase locked loops (PLLs) have become widely used for frequency synthesis in consumer-oriented electronic communications products such as cellular phones and wireless LANs. Unlike an integer-N PLL, the output frequency of a $\Delta\Sigma$ fractional-N PLL is not limited to integer multiples of a reference frequency. The core of a $\Delta\Sigma$ fractional-N PLL is similar to an integer-N PLL, but it incorporates additional digital circuitry that allows it to accurately interpolate between integer multiples of the reference frequency. The tuning resolution depends only on the complexity of the digital circuitry, so considerable flexibility and programmability is achieved. A single $\Delta\Sigma$ fractional-N PLL often can be used for local oscillator generation in applications that would otherwise require a cascade of two or more integer-N PLLs. Moreover, the fine tuning resolution makes it possible to perform digitally-controlled frequency modulation for generation of continuous-phase (e.g., FSK and MSK) transmit signals, thereby simplifying wireless transmitters. These benefits come at the expense of increased digital complexity and somewhat increased phase noise relative to integer-N PLLs. However, with the relentless progress in silicon VLSI technology optimized for digital circuitry, this tradeoff is increasingly attractive, especially in consumer products which tend to favor cost reduction over performance.

This paper presents a tutorial on $\Delta\Sigma$ fractional-N PLLs. It is assumed that the reader has a working knowledge of integer-N PLLs. The paper builds on this knowledge by presenting the additional concepts required to understand $\Delta\Sigma$ fractional-N PLLs. The limitations of integer-N PLLs with respect to tuning resolution are described in Section II. The key ideas

The author is with the Department of Electrical and Computer Engineering, University of California at San Diego, La Jolla, CA, USA.

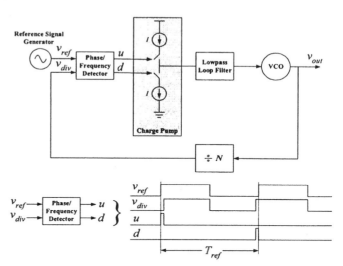

Figure 1: A typical integer-N PLL.

underlying fractional-N PLLs in general and $\Delta\Sigma$ fractional-N PLLs in particular are presented in Section III. The primary innovation in $\Delta\Sigma$ fractional-N PLLs relative to other types of fractional-N PLLs is the use of $\Delta\Sigma$ modulation. Therefore, a self-contained introduction to $\Delta\Sigma$ modulation as it relates to $\Delta\Sigma$ fractional-N PLLs is presented in Section IV. A $\Delta\Sigma$ fractional-N PLL linearized model is derived in Section V and compared to the corresponding model for integer-N PLLs. A design example is presented to demonstrate how the model is used in practice. Design issues that arise in $\Delta\Sigma$ fractional-N PLLs but not integer-N PLLs are presented in Section VI, and recently developed enhancements to $\Delta\Sigma$ fractional-N PLLs that allow wideband digital modulation of the VCO are presented in Section VII.

II. INTEGER-N PLL LIMITATIONS

An example of a typical integer-N PLL for frequency synthesis is shown in Figure 1 [1], [2]. Its purpose is to generate a spectrally pure periodic output signal with a frequency of N f_{ref}, where N is an integer, and f_{ref} is the frequency of the reference signal. The example PLL consists of a phase-frequency detector (PFD), a charge pump, a lowpass loop filter, a voltage controlled oscillator (VCO), and an N-fold digital divider. The PFD compares the positive-going edges of the reference signal to those from the divider and causes the charge pump to drive the loop filter with current pulses whose widths are proportional to the phase difference between the two signals. The pulses are lowpass filtered by the loop filter and the resulting waveform drives the VCO. Within the loop bandwidth phase noise from the VCO is suppressed and outside the loop bandwidth most of the other noise sources are suppressed, so the

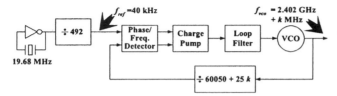

Figure 2: An example integer-*N* PLL for generation of the Bluetooth wireless LAN RF channel frequencies.

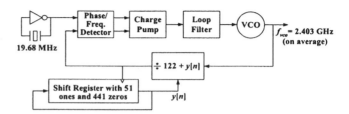

Figure 3: A fractional-*N* PLL that generates non-integer multiples of the reference frequency, but has phase noise consisting of large spurious tones.

PLL can be designed to generate a spectrally pure output signal at any integer multiple of the reference frequency, f_{ref}.

As indicated by the timing diagram in Figure 1, the loop filter is updated by the charge pump once every reference period. This discrete-time behavior places an upper limit on the loop bandwidth of approximately $f_{ref}/10$ above which the PLL tends to be unstable [1]. In integrated circuit PLLs, it is common to further limit the bandwidth to approximately $f_{ref}/20$ to allow for process and temperature variations.

The output frequency can be changed by changing *N*, but *N* must be an integer, so the output frequency can be changed only by integer multiples of the reference frequency. If finer tuning resolution is required the only option is to reduce the reference frequency. Unfortunately, this tends to reduce the maximum practical loop bandwidth, thereby increasing the settling time of the PLL, the noise contributed by the VCO, and the in-band portions of the noise contributed by the reference source, the PFD, the charge pump, and the divider.

This fundamental tradeoff between bandwidth and tuning resolution in integer-*N* PLLs creates problems in many applications. For example, a PLL that can be tuned from 2.402 GHz to 2.480 GHz in steps of 1 MHz is required to generate the local oscillator signal in a direct conversion Bluetooth transceiver [3]. An integer-*N* PLL capable of generating the local oscillator signal from a commonly used crystal oscillator frequency, 19.68 MHz, is shown in Figure 2. A reference frequency of f_{ref} = 40 kHz—the greatest common divisor of the crystal frequency and the set of desired output frequencies—is obtained by dividing the crystal oscillator signal by 492. The resulting PLL output frequency is 60050 + 25*k* times the reference frequency, where *k* is an integer used to select the desired frequency step.

The PLL achieves the desired output frequencies, but its bandwidth is limited to approximately 2 kHz, i.e., $f_{ref}/20$. Unfortunately, with such a low bandwidth the settling time exceeds the 200 μS limit specified in the Bluetooth standard, and the phase noise contributed by the VCO would be unacceptably high if it were implemented in present-day CMOS technology. One solution is to use a 1 MHz reference signal, but this requires the crystal frequency to be an integer multiple of 1 MHz, or another PLL to generate a 1 MHz reference frequency. Unfortunately, in low cost consumer electronics applications such as Bluetooth, it is often desirable to be compatible with all of the popular crystal frequencies, so restricting the crystal frequencies to multiples of 1 MHz is not always an option. In such cases, an additional PLL capable of generating the 1 MHz reference signal with very little phase noise from any of the crystal frequencies is required, or, as described in the next section, a single fractional-*N* PLL can be used.

III. THE IDEA BEHIND ΔΣ FRACTIONAL-*N* PLLs

In this section, the example problem of generating the second Bluetooth channel frequency, 2.403 GHz, with a reference frequency of 19.68 MHz is used as a vehicle with which to explain the idea behind ΔΣ fractional-*N* PLLs. First, a pair of "bad" fractional-*N* PLLs are presented that achieve the desired frequency but have poor phase noise performance. Then the ΔΣ fractional-*N* PLL technique is presented as a means of improving the phase noise performance.

The output frequency of an integer-*N* PLL with a reference frequency of 19.68 MHz is 2.40096 GHz when the divider modulus, *N*, is set to 122 and 2.42064 GHz when *N* is set to 123. The problem is that to achieve the desired frequency of 2.403 GHz, *N* would have to be set to the non-integer value of 122 + 51/492. This cannot be implemented directly because the divider modulus must be an integer value. However the divider modulus can be updated each reference period, so one option is to switch between *N* = 122 and *N* = 123 such that the average modulus over many reference periods converges to 122 + 51/492. In this case, the resulting average PLL output frequency is 2.403 GHz as desired. This is the fundamental idea behind most fractional-*N* PLLs [4].

While dynamically switching the divider modulus solves the problem of achieving non-integer multiples of the reference frequency, a price is paid in the form of increased phase noise. During each reference period the difference between the actual divider modulus and the average, i.e., ideal, divider modulus represents error that gets injected into the PLL and results in increased phase noise. As described below, the amount by which the phase noise is increased depends upon the characteristics of the sequence of divider moduli.

For example, in the fractional-*N* PLL shown in Figure 3, the divider modulus is set each reference period to 122 or 123 such that over each set of 492 consecutive reference periods it is set to 122 a total of 441 times and 123 a total of 51 times. Thus, the average modulus is 122 + 51/492 as required. The sequence of moduli is periodic with a period of 492, so it repeats at a rate of 40 kHz. Consequently, the difference between the actual divider moduli and their average is a periodic sequence with a repeat rate of 40 kHz, so the resulting phase noise is periodic and is comprised of spurious tones at integer multiples of 40 kHz. Many of the spurious tones occur at low frequencies, and they can be very large. Unfortunately, the

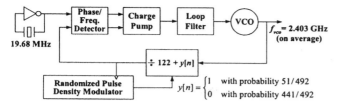

Figure 4: A fractional-N PLL that generates non-integer multiples of the reference frequency, but has a large amount of in-band phase noise.

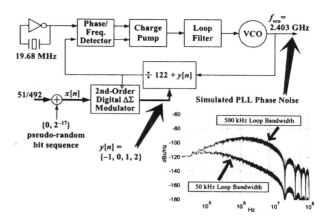

Figure 5: A $\Delta\Sigma$ fractional-N PLL example.

only way to suppress the tones is have a very small PLL bandwidth, which negates the potential benefit of the fractional-N technique.

One way to eliminate spurious tones is to introduce randomness to break up the periodicity in the sequence of moduli while still achieving the desired average modulus. For example, as shown in Figure 4, a digital block can be used to generate a sequence, $y[n]$, that approximates a sampled sequence of independent random variables that take on values of 0 and 1 with probabilities 441/492 and 51/492, respectively. During the nth reference period the divider modulus is set to $122 + y[n]$, so the sequence of moduli has the desired average yet its power spectral density (PSD) is that of white noise. Thus, instead of contributing spurious tones, the modified technique introduces white noise. Unfortunately, the portion of the white noise within the PLL's bandwidth is integrated by the PLL transfer function, so the overall phase noise contribution again can be significant unless the PLL bandwidth is small.

In each fractional-N PLL example presented above, the sequence, $y[n]$, can be written as $y[n] = x + e_m[n]$, where x is the desired fractional part of the modulus, i.e., $x = 51/492$, and $e_m[n]$ is undesired zero-mean *quantization noise* caused by using integer moduli in place of the ideal fractional value. In the first example, $e_m[n]$ is periodic and therefore consists of spurious tones at multiples of 40 kHz. In the second example, $e_m[n]$ is white noise. Each PLL attenuates the portion of $e_m[n]$ outside its bandwidth, but the portion within its bandwidth is not significantly attenuated. Unfortunately, in each example $e_m[n]$ contains significant power at low frequencies, so it contributes substantial phase noise unless the PLL bandwidth is very low.

A $\Delta\Sigma$ fractional-N PLL avoids this problem by generating the sequence of moduli such that the quantization noise has most of its power in a frequency band well above the desired bandwidth of the PLL [5], [6], [7]. An example $\Delta\Sigma$ fractional-N PLL is shown in Figure 5. The PLL core is similar to those of the previous fractional-N PLL examples, but in this case $y[n]$ is generated by a digital $\Delta\Sigma$ modulator. The details of how the $\Delta\Sigma$ modulator works are presented in the next section, but its purpose is to coarsely quantize its input sequence, $x[n]$, such that $y[n]$ is integer-valued and has the form: $y[n] = x[n - 2] + e_m[n]$, where $e_m[n]$ is dc-free quantization noise with most of its power outside the PLL bandwidth. In this example, $x[n]$ consists of the desired fractional modulus value, 51/492, plus a small, pseudo-random, 1-bit sequence. As described in the next section, the pseudo-random sequence is necessary to avoid spurious tones in the $\Delta\Sigma$ modulator's quantization noise, but its amplitude is very small so it does not appreciably in-

crease the phase noise of the PLL.

Also shown in Figure 5 are PSD plots of the output phase noise arising from $\Delta\Sigma$ modulator quantization noise, $e_m[n]$, in two computer simulated versions of the example $\Delta\Sigma$ fractional-N PLL, one with a 50 kHz loop bandwidth and the other with a 500 kHz loop bandwidth. As shown in the next section, the PSD of $e_m[n]$ increases with frequency, so the phase noise PSD corresponding to the 50 kHz bandwidth PLL is significantly smaller than that corresponding to the 500 kHz bandwidth PLL. For example, the former easily meets the requirements for a local oscillator in a direct conversion Bluetooth transceiver, but the latter falls short of the requirements by at least 23 dB.

IV. DELTA-SIGMA MODULATION OVERVIEW

As mentioned above, a digital $\Delta\Sigma$ modulator performs coarse quantization in such a way that the inevitable error introduced by the quantization process, i.e., the quantization noise, is attenuated in a specific frequency band of interest. There are many different $\Delta\Sigma$ modulator architectures. Most use coarse uniform quantizers to perform the quantization with feedback around the quantizers to suppress the quantization noise in particular frequency bands. Therefore, to illustrate the $\Delta\Sigma$ modulator concept, first a specific uniform quantizer example is considered in isolation, and then a specific $\Delta\Sigma$ modulator architecture that incorporates the uniform quantizer is presented.

A. An Example Uniform Quantizer

The input-output characteristic of the example uniform quantizer is shown in Figure 6. It is a 9-level quantizer with integer valued output levels. For each input value with a magnitude less than 4.5, the quantizer generates the corresponding output sample by rounding the input value to the nearest integer. For each input value greater than 4.5 or less than –4.5, the quantizer sets its output to 4 or –4, respectively; such values are said to *overload* the quantizer. By defining the quantization noise as $e_q[n] = y[n] - r[n]$, the quantizer can be viewed without approximation as an additive noise source as illustrated in the figure.

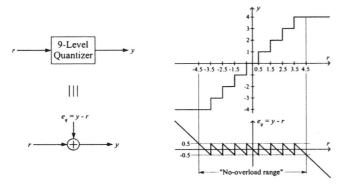

Figure 6: A 9-level quantizer example.

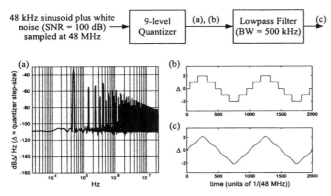

Figure 7: (a) A power spectral density plot of the quantizer output in dB, relative to the quantization step-size of $\Delta = 1$, per Hz, (b) a time domain plot of the quantizer output, and (c) a time domain plot of the quantizer output filtered by a sharp lowpass filter with a cutoff frequency of 500 kHz.

To illustrate some properties of the example quantizer, consider a 48 Msample/s input sequence, $x[n]$, consisting of a 48 kHz sinusoid with an amplitude of 1.7 plus a small amount of white noise such that the input signal-to-noise ratio (SNR) is 100 dB. Figure 7(a) shows the PSD plot of the resulting quantizer output sequence, and Figure 7(b) shows a time domain plot of the quantizer output sequence over two periods of the sinusoid. Given the coarseness of the quantization, it is not surprising that the quantizer output sequence is not a precise representation of the quantizer input sequence. As evident in Figure 7(a), the quantization noise for this input sequence consists primarily of harmonic distortion as represented by the numerous spurious tones distributed over the entire discrete-time frequency band. Even in the relatively narrow frequency band below 500 kHz, significant harmonic distortion corrupts the desired signal. To illustrate this in the time domain, Figure 7(c) shows the sequence obtained by passing the quantizer output sequence through a sharp lowpass discrete-time filter with a cutoff frequency of 500 kHz. The significant quantization noise power in the zero to 500 kHz frequency band causes the sequence shown in Figure 7(c) to deviate significantly from the sinusoidal quantizer input sequence.

B. An Example ΔΣ Modulator

The example ΔΣ modulator architecture shown in Figure 8

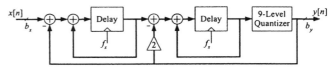

Figure 8: A ΔΣ modulator example.

can be used to circumvent this problem. The structure incorporates the same 9-level quantizer presented above, but in this case the quantizer is preceded by two delaying discrete-time integrators (i.e., accumulators), and surrounded by two feedback loops [8], [9]. Each discrete-time integrator has a transfer function of $z^{-1}/(1-z^{-1})$ which implies that its nth output sample is the sum of all its input samples for times $k < n$. With the quantizer represented as an additive noise source as depicted in Figure 6, the ΔΣ modulator can be viewed as a two-input, single-output, linear time-invariant, discrete-time system. It is straightforward to verify that

$$y[n] = x[n-2] + e_m[n], \qquad (1)$$

where $e_m[n]$ is the overall quantization noise of the ΔΣ modulator and is given by

$$e_m[n] = e_q[n] - 2e_q[n-1] + e_q[n-2]. \qquad (2)$$

To illustrate the behavior of the ΔΣ modulator, suppose that the same 48 Msample/s input sequence considered above is applied to the input of the ΔΣ modulator, and that the discrete-time integrators in the ΔΣ modulator are clocked at 48 MHz. Figure 9(a) shows the PSD plot of the resulting ΔΣ modulator output sequence, $y[n]$, and Figure 9(b) shows a time domain plot of $y[n]$ over two periods of the sinusoid. Two important differences with respect to the uniform quantization example shown in Figure 7 are apparent: the quantization noise PSD is significantly attenuated at low frequencies, and no spurious tones are visible anywhere in the discrete-time spectrum. For instance, the SNR in the zero to 500 kHz frequency band is approximately 84 dB for this example as opposed to 14 dB for the uniform quantization example of Figure 7. Consequently, subjecting the ΔΣ modulator output sequence to a lowpass filter with a cutoff frequency of 500 kHz results in a sequence that is very nearly equal to the ΔΣ modulator input sequence as demonstrated in Figure 9(c).

Below about 120 kHz, the PSD shown in Figure 9(a) is dominated by the two components of the ΔΣ modulator input sequence: the 48 kHz sinusoid component, and the input noise component. Above 120 kHz, the PSD is dominated by the ΔΣ modulator quantization noise, $e_m[n]$, and rises with a slope of 40 dB per decade. It follows from (2) that $e_m[n]$ can be viewed as the result of passing the additive noise from the quantizer, $e_q[n]$, through a discrete-time filter with transfer function $(1-z^{-1})^2$. Since this filter has two zeros at dc, the smooth 40 dB per decade increase of the PSD of $e_m[n]$ indicates that $e_q[n]$ is very nearly white noise, at least for the example shown in Figure 9.

It can be proven that $e_q[n]$ is indeed white noise; it has a variance of 1/12 and is uncorrelated with the ΔΣ modulator input sequence [10]. Moreover, this situation holds in general for the example ΔΣ modulator architecture provided that the

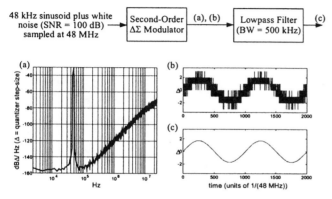

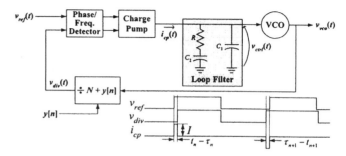

Figure 10: The $\Delta\Sigma$ fractional-N PLL with the details of a commonly used loop filter and a timing diagram relating to the charge pump output.

Figure 9: (a) A power spectral density plot of the $\Delta\Sigma$ modulator output in dB, relative to the quantization step-size of $\Delta = 1$, per Hz, (b) a time domain plot of the $\Delta\Sigma$ modulator output, and (c) a time domain plot of the $\Delta\Sigma$ modulator output filtered by a sharp lowpass filter with a cutoff frequency of 500 kHz.

input sequence satisfies two conditions: 1) its magnitude is sufficiently small that the quantizer within the $\Delta\Sigma$ modulator never overloads, and 2) it consists of a signal component plus a small amount of independent white noise. It can be shown that the first condition is satisfied if the input signal is bounded in magnitude by 3Δ where Δ is the step-size of the quantizer (for this example, $\Delta = 1$) [11]. Input sequences with values even slightly exceeding 3Δ in magnitude generally cause the quantizer to overload with the result that $e_q[n]$ contains spurious tones and the SNR in the frequency band of interest is degraded. For this reason, the range between -3Δ and 3Δ is said to be the *input no-overload range* of the $\Delta\Sigma$ modulator. For the second condition to be satisfied, the power of the $\Delta\Sigma$ modulator input sequence's white noise component may be arbitrarily small, but if it is absent altogether, $e_q[n]$, is not guaranteed to be white. For instance, in the example shown in Figure 9 the input sequence contains a white noise component with 100dB less power than the signal component. If this tiny noise component were not present, the resulting $\Delta\Sigma$ modulator output PSD would contain numerous spurious tones. Since the $\Delta\Sigma$ modulators used in $\Delta\Sigma$ fractional-N PLLs are all-digital devices, the noise must be added digitally. As shown in [12], it is sufficient to add a 1-bit, sub-LSB, independent, white noise *dither sequence* with zero mean at the input node. In practice, a 1-bit pseudo-random dither sequence is typically used in place of a truly random dither sequence. Such a sequence can be generated easily using a linear feedback shift register, and has the desired result with respect to the quantization noise despite not being truly random [13], [14].

C. Other $\Delta\Sigma$ Modulator Options

To this point, the $\Delta\Sigma$ modulation concept has been illustrated via the particular example $\Delta\Sigma$ modulator architecture shown in Figure 8, namely a second-order multi-bit $\Delta\Sigma$ modulator. While this type of $\Delta\Sigma$ modulator is widely used in $\Delta\Sigma$ fractional-N PLLs, there exist other types of $\Delta\Sigma$ modulators that can be applied to $\Delta\Sigma$ fractional-N PLLs. Most of the other architectures are higher-order $\Delta\Sigma$ modulators that perform higher than second-order quantization noise shaping,

thereby more aggressively suppressing quantization noise in particular frequency bands relative to the example second-order $\Delta\Sigma$ modulator. Some of these higher-order $\Delta\Sigma$ modulators incorporate a higher than second-order loop filter (e.g., more than two discrete-time integrators) and a single quantizer surrounded by one or more feedback loops [15], [16]. In many cases, these $\Delta\Sigma$ modulators are designed specifically to allow one-bit quantization [7], [17], [18]. This simplifies the design of the divider in that only two moduli are required, but such $\Delta\Sigma$ modulators tend to have spurious tones in their quantization noise that cannot be completely suppressed even with elaborate dithering techniques. Others of these higher-order $\Delta\Sigma$ modulators, often referred to as MASH, cascaded, or multistage $\Delta\Sigma$ modulators, are comprised of multiple lower-order $\Delta\Sigma$ modulators, such as the second-order $\Delta\Sigma$ modulator presented above, cascaded to obtain the equivalent of a single higher-order $\Delta\Sigma$ modulator [5], [19], [20].

V. $\Delta\Sigma$ FRACTIONAL-N PLL DYNAMICS

A $\Delta\Sigma$ fractional-N PLL linearized model is derived in this section in the form of a block diagram that describes the output phase noise in terms of the component parameters and noise sources in the PLL. As in the case of an integer-N PLL the model provides an accurate tool with which to predict the total phase noise, bandwidth, and stability of the PLL.

A. Derivation of a $\Delta\Sigma$ fractional-N PLL Linearized Model

In PLL analyses it is common to assume that each periodic signal within the PLL has the form $v(t) = A(t)\sin(\omega t + \theta(t))$, where $A(t)$ is a positive *amplitude* function, ω is a constant *center frequency* in radians/sec, and $\theta(t)$ is zero-mean *phase noise* in radians. In most cases of interest for PLL analysis, the amplitude is well modeled as a constant value, and the phase noise is very small relative to π with a bandwidth that is much lower than the center frequency. Solving for the time of the nth positive-going zero crossing, γ_n, of $v(t)$ gives $\gamma_n = [n - \theta(\gamma_n)/(2\pi)]\cdot T$, where $T = 2\pi/\omega$ is the period of the signal. Therefore, the sequence, γ_n, is a sampled version of the phase noise with very little aliasing, so knowing the sequence and T is approximately equivalent to knowing the phase noise. This approximation is made throughout the following analysis.

The relationship between the charge pump output current and the PFD input signals is shown in Figure 10. Ideally, during the nth reference period the charge pump output is a current pulse of amplitude I or $-I$ and duration $|t_n - \tau_n|$, where t_n and τ_n are the times of the charge pump output transitions triggered by the positive-going edges of the divider output and reference signal, respectively. Therefore, the average current sourced or sunk by the charge pump during the nth reference period is $I \cdot (t_n - \tau_n)/T_{ref}$. In practice, the PFD is usually designed such that, except for a possible constant offset, this result holds even though the current sources have finite rise and fall times [2].

The first step in deriving the model is to develop an expression for $t_n - \tau_n$. Ideally, $\tau_n = nT_{ref}$, but phase noise introduced by the reference source and PFD cause it to have the form

$$\tau_n = nT_{ref} - \frac{T_{ref}}{2\pi}\Big[\theta_{ref}(\tau_n) + \theta_{PFD}(\tau_n)\Big], \quad (3)$$

where $\theta_{ref}(t)$ and $\theta_{PFD}(t)$ are the reference source and PFD phase noise functions, respectively. If the VCO output were ideal its positive-going edges would be spaced at uniform intervals of $T_{ref} / (N + \alpha)$, where α is the fractional part of the modulus (e.g., $\alpha = 51/492$ in Figure 5). Therefore, ideally,

$$t_n = \frac{T_{ref}}{N+\alpha}\sum_{k=0}^{n-1}(N + y[k]),$$

but in practice it deviates because of VCO phase noise, $\theta_{VCO}(t)$, divider phase noise, $\theta_{div}(t)$, and instantaneous deviations of the VCO control voltage from its ideal average value of $\bar{v}_{ctrl} = (N+\alpha)/(T_{ref}K_{VCO})$, where K_{VCO} is the VCO gain in units of Hz/Volt. As a result,

$$t_n = \frac{T_{ref}}{N+\alpha}\left[\sum_{k=0}^{n-1}(N + y[k]) - k_{VCO}\int_0^{t_n}\left(v_{ctrl}(t) - \bar{v}_{ctrl}\right)dt - \frac{\theta_{VCO}(t_n)}{2\pi}\right]$$
$$- \frac{T_{ref}}{2\pi}\theta_{div}(t_n),$$

which reduces to

$$t_n = nT_{ref} + \frac{T_{ref}}{N+\alpha}\left[\sum_{k=0}^{n-1}(y[k] - \alpha)\right.$$
$$\left. - k_{VCO}\int_0^{t_n}\left(v_{ctrl}(t) - \bar{v}_{ctrl}\right)dt - \frac{\theta_{VCO}(t_n)}{2\pi}\right] \quad (4)$$
$$- \frac{T_{ref}}{2\pi}\theta_{div}(t_n).$$

Subtracting (3) from (4) yields an expression for the average current sourced or sunk by the charge pump during the nth reference period:

$$I(t_n - \tau_n)/T_{ref} =$$
$$I\left[\frac{\sum_{k=0}^{n}(y[k] - \alpha) - k_{VCO}\int_0^{t_n}\left(v_{ctrl}(t) - \bar{v}_{ctrl}\right)dt - \frac{\theta_{VCO}(t_n)}{2\pi}}{N+\alpha}\right.$$
$$\left. - \frac{\theta_{div}(t_n)}{2\pi} + \frac{\theta_{ref}(\tau_n)}{2\pi} + \frac{\theta_{PFD}(\tau_n)}{2\pi}\right]. \quad (5)$$

As mentioned above, the phase noise terms are assumed to have bandwidths that are much smaller than the reference frequency. Consequently, the sampling of the phase noise func-

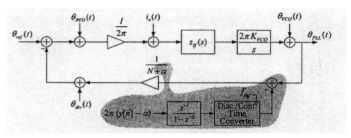

Figure 11: The $\Delta\Sigma$ fractional-N PLL linearized model. Except for the shaded region the model is identical to the corresponding integer-N PLL model.

tions in (5) can be neglected, and the charge pump output can be modeled as a smoothly varying function of time with an average value over each reference period equal to that of (5). With these approximations, (5) implies that

$$i_{cp}(t) = I\left[\frac{u_m(t) - k_{VCO}\int_0^t\left(v_{ctrl}(t) - \bar{v}_{ctrl}\right)dt - \frac{\theta_{VCO}(t)}{2\pi}}{N+\alpha}\right.$$
$$\left. - \frac{\theta_{div}(t)}{2\pi} + \frac{\theta_{ref}(t)}{2\pi} + \frac{\theta_{PFD}(t)}{2\pi}\right], \quad (6)$$

where $u_m(t)$ is the result of discrete-time integrating and converting to continuous-time the quantity, $y[n] - \alpha$.

The $\Delta\Sigma$ fractional-N PLL linearized model follows directly from (6) and Figure 10. It is shown in Figure 11, where $i_n(t)$ represents the noise contributed by the charge pump current sources and the loop filter, and $z_{lf}(s)$ is the transfer function of the loop filter. The model specifies the phase noise transfer functions and loop dynamics of the PLL. For example, the model implies that

$$\frac{\theta_{PLL}(s)}{\theta_{ref}(s)} = (N+\alpha)\frac{T(s)}{1+T(s)}, \quad \text{and} \quad \frac{\theta_{PLL}(s)}{\theta_{VCO}(s)} = \frac{1}{1+T(s)} \quad (7)$$

where

$$T(s) = \frac{IK_{VCO}z_{lf}(s)}{s(N+\alpha)} \quad (8)$$

is the *loop gain* of the PLL. For the loop filter shown in Figure 10, the transfer function is

$$z_{lf}(s) = \frac{1}{C_1 + C_2}\frac{1 + sRC_2}{s\big[1 + sRC_1C_2/(C_1 + C_2)\big]}. \quad (9)$$

B. Differences Between the $\Delta\Sigma$ Fractional-N and Integer-N PLL Models

The shaded region in Figure 11 indicates the part of the model that is specific to $\Delta\Sigma$ fractional-N PLLs; except for the shaded region the model is identical to the corresponding model for integer-N PLLs. Therefore, each phase noise transfer function in an integer-N PLL is identical to the corresponding phase noise transfer function in a $\Delta\Sigma$ fractional-N PLL, except every occurrence of N in the former is replaced by $N+\alpha$ in the latter. In most cases, $N \gg 1$ and $\alpha < 1$, so $N + \alpha \approx N$ and the corresponding transfer functions in integer-N and $\Delta\Sigma$ fractional-N PLLs are nearly identical in practice. Similarly, the loop dynamics and stability issues are nearly the same in $\Delta\Sigma$ fractional-N PLLs and integer-N PLLs.

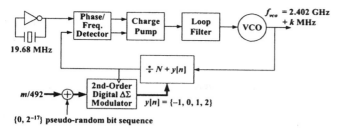

Frequency Plan:
- To get $k = 0, 1, \ldots,$ or 18:   set $N = 122$, $m = k \cdot 25 + 26$
- To get $k = 19, 21, \ldots,$ or 38:   set $N = 123$, $m = (k - 19) \cdot 25 + 9$
- To get $k = 39, 41, \ldots,$ or 57:   set $N = 124$, $m = (k - 39) \cdot 25 + 17$
- To get $k = 58, 60, \ldots,$ or 79:   set $N = 125$, $m = (k - 58) \cdot 25$

Figure12: The example ΔΣ fractional-N PLL and frequency plan for generation of the Bluetooth wireless LAN RF channel frequencies.

The primary difference between the ΔΣ fractional-N and integer-N PLL models is the signal path corresponding to the ΔΣ modulator shown in the shaded region of Figure 11. The sequence, $y[n] - \alpha$, consists of ΔΣ modulator quantization noise, $e_m[n]$, which, as described previously, gives rise to phase error in the PLL output. For the example second-order ΔΣ modulator it follows from the results presented in Section IV and the ΔΣ fractional-N PLL model equations presented above that the PLL phase noise component resulting from $e_m[n]$ has a PSD given by

$$S_{\theta_{PLL}}(f)\Big|_{\Delta\Sigma \; only} =$$

$$10 \cdot \log\left[\frac{\pi}{12 f_{ref}} \left[2 \cdot \sin\left(\frac{\pi f}{f_{ref}} \right) \right]^2 \left| \frac{1}{N+\alpha} \cdot \frac{\theta_{PLL}(j2\pi f)}{\theta_{ref}(j2\pi f)} \right|^2 \right] \; \text{dBc/Hz}. \quad (10)$$

The argument of the log function has the form of a highpass function times a lowpass function, which is consistent with the claim in Section III that the PLL lowpass filters the primarily high frequency quantization noise from the ΔΣ modulator. It follows from (10) that the phase noise resulting from $e_m[n]$ can be decreased by reducing the PLL bandwidth or increasing the reference frequency. If a higher-order ΔΣ modulator is used, an equation similar to (10) results except that the exponent of the sinusoid is greater than two. This reduces the in-band portion of the quantization noise, but increases the out-of-band portion, which, depending upon the loop parameters of the PLL, can result in a somewhat lower overall phase noise. However, the PLL loop filter is highly constrained to maintain PLL stability, so the phase noise reduction that can be achieved by increasing the order of the ΔΣ modulator is limited in most applications [16].

C. A System Design Example

The PLL bandwidth and the phase margin both depend upon the loop gain, $T(s)$, which, for the loop filter shown in Figure 10, depends upon the parameters f_{ref}, N, I, K_{VCO}, R, C_1, and C_2. Usually, f_{ref} and N are dictated by the application, and I and K_{VCO} are, at least partially, dictated by circuit design choices. This leaves the loop filter components as the main variables with which to set the desired PLL bandwidth, phase

margin, and ΔΣ modulator quantization noise suppression.

The process is demonstrated below for the ΔΣ fractional-N PLL presented in Section III to generate the local oscillator frequencies in a direct conversion Bluetooth wireless LAN transceiver. The PLL is shown in Figure 12 with additional detail regarding the frequency plan. As described previously, the desired output frequencies are $f_{VCO} = 2.402$ GHz + k MHz for $k = 0, \ldots, 78$, and the crystal reference frequency is 19.68 MHz. Each of the 79 possible output frequencies is chosen by selecting m and N as indicated in the figure. In each case, the divider modulus is restricted to the set of four integers $\{N - 1, N, N + 1, N + 2\}$. The combinations of m and N were chosen to achieve the desired output frequencies yet keep the signals at the input of the ΔΣ modulator sufficiently small so as not to overload the ΔΣ modulator [11].

Typical requirements for such a PLL are that the loop bandwidth must be greater than 40 kHz, the phase margin must be greater than 60°, and the PLL phase noise be less than −120 dBc/Hz at offsets from the carrier of 3 MHz and above. Assume that the VCO, divider, PFD, and charge pump circuits have been designed such that the overall PLL phase noise specification can be met provided the phase noise contributed by the ΔΣ modulator and loop filter are each less than −130 dBc/Hz at offsets from the carrier of 3 MHz and above. Furthermore, assume that the VCO and charge pump circuits are such that K_{VCO} and I are 200 MHz/V and 200 μA, respectively, and that the loop filter has the form shown in Figure 10. Thus, the remaining design task is to choose the loop filter components such that the bandwidth, phase margin, and phase noise specifications are met.

The PLL phase margin, bandwidth, and phase noise arising from ΔΣ modulator quantization noise can be derived from the linearized model equations, (7) through (10). While this can be done directly, it involves the solution of third order equations which can be messy. Alternatively, approximate solutions of the equations can be derived that provide better intuition [21]. A particularly convenient set of approximate solutions are

$$PM = \tan^{-1}\left(\frac{b-1}{2\sqrt{b}} \right), \quad (11)$$

$$f_{BW} = \frac{I K_{VCO} R}{2\pi N} \cdot \frac{b-1}{b}, \quad (12)$$

$$RC_2 = \frac{\sqrt{b}}{2\pi f_{BW}}, \quad (13)$$

and

$$S_{\theta_{PLL}}(f)\Big|_{\Delta\Sigma \; only} \approx 10 \cdot \log\left[\frac{2\pi b}{3 f_{ref}} \sin^2\left(\frac{\pi f}{f_{ref}} \right)\left(\frac{f_{BW}}{f} \right)^4 \right] \; \text{dBc/Hz}, \quad (14)$$

where PM is the phase margin of the PLL, f_{BW} is the 3 dB bandwidth of the PLL, and $b = 1 + C_2/C_1$ is a measure of the separation between the two loop filter capacitors [22]. The derivations assume that b is greater than about 10, and (14) is valid for frequencies greater than $(C_2+C_1)/(2\pi RC_2C_1)$.

These equations are sufficient to determine appropriate loop filter component values. For example, suppose b is set to

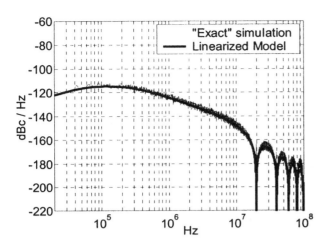

Figure13: Simulated and calculated PSD plots of the phase noise arising from ΔΣ modulator quantization noise for the example ΔΣ fractional-N PLL.

49, so, as indicated by (11), the phase margin is approximately 70°. Solving (14) with the phase noise set to −130 dBc/Hz at f = 3 MHz indicates that $f_{BW} \approx$ 50 kHz. Therefore, the phase noise resulting from ΔΣ modulator quantization noise is sufficiently suppressed with a 50 kHz bandwidth and a phase margin of 70°. With this information (12) can be solved to find R = 960 Ω with which (13) and the definition of b can be used to calculate C_2 = 23 nF and C_1 = 480 pF. It is straightforward to verify that the phase noise introduced by the loop filter resistor (the only noise source in the loop filter) is well below −130 dBc/Hz at offsets from the carrier of 3 MHz and above as required.

Figure 13 shows PSD plots of the phase noise arising from ΔΣ modulator quantization noise for the example PLL with the loop filter component values derived above. The heavy curve was calculated directly from the linearized model equations (7) through (10). The light curve was obtained through a behavioral computer simulation of the PLL. As is evident from the figure, the two curves agree very well which suggests that the approximations made in obtaining the linearized model are reasonable.

An effect that does not have a counterpart in integer-N PLLs is the presence of zeros in the PSD of the phase noise arising from ΔΣ modulator quantization noise at multiples of the reference frequency. These zeros are a result of the discrete-to-continuous-time conversion of the ΔΣ modulator quantization noise; each zero is a sampling image of the dc zero imposed on the quantization noise by the ΔΣ modulator.

VI. ΔΣ FRACTIONAL-N PLL SPECIFIC PROBLEMS

One of the most significant problems specific to ΔΣ fractional-N PLLs is that they can be sensitive to modulus-dependent divider delays. In practice, each positive-going divider edge is separated from the VCO edge that triggered it by a propagation delay. Ideally, this propagation delay is independent of the corresponding divider modulus, in which case it introduces a constant phase offset but does not otherwise contribute to the phase noise. However, if the propaga-

tion delay depends upon the divider modulus and the number of ΔΣ modulator output levels is greater than two, the effect is that of a hard non-linearity applied to the ΔΣ modulator quantization noise. This tends to fold out-of-band ΔΣ modulator quantization noise to low frequencies and introduce spurious tones, which can significantly increase the PLL phase noise. The problem is analogous to that of multi-bit digital-to-analog converter step-size mismatches in analog ΔΣ data converters [23]. Unfortunately, circuit simulations are required to evaluate the severity of the problem on a case by case basis as both the extent of any modulus-dependent delays and their affect on the PLL phase noise are difficult to predict using hand analysis.

There are two well-known solutions to this problem. One solution is to resynchronize the divider output to the nearest VCO edge or at least a higher-frequency edge obtained from within the divider circuitry [22], [24]. The resynchronization erases memory of modulus-dependent delays and noise introduced within the divider circuitry, but care must be taken to ensure that the signal used for resynchronization is itself free of modulus dependent delays. The primary drawback of the approach is that it increases power consumption.

The other solution is to use a ΔΣ modulator with single-bit (i.e., two level) quantization. In this case, modulus-dependent delays give rise to phase error at the output of the divider that consists of a constant offset plus a scaled version of the ΔΣ modulator quantization noise. Since, by design, the ΔΣ modulator quantization noise has most of its power outside the PLL bandwidth, the modulus-dependent delays increase the phase noise only slightly. Unfortunately, ΔΣ modulators with single-bit quantization tend not to perform as well as ΔΣ modulators with multi-bit (i.e., more than two-level) quantization. For example, if the 9-level quantizer in the 48 Msample/s ΔΣ modulator example presented in Section IV were replaced by a one-bit quantizer, the dynamic range of the ΔΣ modulator in the zero to 500 kHz band would be reduced from 88.5 dB to approximately 65 dB. Moreover, unlike the 9-level quantizer case, the additive noise from the single-bit quantizer would not be white and would be correlated with the input sequence. Its variance would be input dependent and it would contain spurious tones.

These problems can be mitigated by using a higher-order ΔΣ modulator architecture to more aggressively suppress the in-band portion of the additive noise from the two-level quantizer. However, to maintain stability in a higher-order ΔΣ modulator with single-bit quantization, the useful input range of the ΔΣ modulator input signal must be reduced and more poles and zeros must be introduced within the feedback loop as compared to a multi-bit design with a comparable dynamic range. Even then, the problem of spurious tones persists, and it is difficult to predict where they will appear except through extensive simulation. Furthermore, to compensate for the restricted input range of the ΔΣ modulator the reference frequency must be large enough that all of the desired PLL output frequencies can be achieved. This can severely limit design flexibility. For example, if the magnitude of the ΔΣ modulator input signal were limited to less than 0.5 in the case of the Bluetooth local oscillator application considered above, the reference frequency would have to be greater than 79 MHz. Otherwise, it would not be possible to generate all the Bluetooth channel frequencies.

Another issue specific to $\Delta\Sigma$ fractional-N PLLs is that modulus switching increases the average duration over which the charge pump current sources are turned on each period relative to integer-N PLLs. For comparison, consider a $\Delta\Sigma$ fractional-N PLL and an integer-N PLL with the same N (where $N >> \alpha$), the same f_{ref}, and identical loop components. It follows from (5) that

$$\left(t_n - \tau_n\right)\big|_{Fractional-N} = \left(t_n - \tau_n\right)\big|_{Integer-N} + \frac{T_{ref}}{N+\alpha}\sum_{k=0}^{n-1}\left(y[k] - \alpha\right).$$

(15)

The last term in (15), which is caused by having the $\Delta\Sigma$ modulator switch the divider modulus, represents a significant increase in the time during which the charge pump current sources are turned on each reference period. Consequently, the phase noise arising just from charge pump current source noise is larger in the $\Delta\Sigma$ fractional-N PLL by

$$A \cdot \log\left[\frac{\text{Average fractional-}N \text{ PLL charge pump "on time"}}{\text{Average integer-}N \text{ PLL charge pump "on time"}}\right]$$

where A is a constant between 10 and 20. The value of A depends upon the autocorrelation of the charge pump current source noise. For example, if the current source noise in successive charge pump pulses is completely uncorrelated, then A is 10. Near the other extreme, A is close to 20.

VII. TECHNIQUES TO WIDEN $\Delta\Sigma$ FRACTIONAL-N PLL LOOP BANDWIDTHS

A transmitter with virtually any modulation format can be implemented using D/A conversion to generate analog baseband or IF signals and upconversion to generate the final RF signal. However, many of the commonly used modulation formats in wireless communication systems such as MSK and FSK involve only frequency or phase modulation of a single carrier [25]. In such cases, the transmitted signal can be generated by modulating a radio frequency (RF) VCO, thereby eliminating the need for conventional upconversion stages and much of the attendant analog filtering. At least two approaches have been successfully implemented in commercial wireless transmitters to date. One is based on open-loop VCO modulation, and the other is based on $\Delta\Sigma$ fractional-N synthesis.

An example of a commercial transmitter that uses the open-loop VCO modulation technique is presented in [26] and [27], in this case for a DECT cordless telephone. Between transmit bursts, the desired center frequency is set relative to a reference frequency by enclosing the VCO within a conventional PLL. During each transmit burst the VCO is switched out of the PLL and the desired frequency modulation is applied directly to its input. The primary limitation of the approach is that it tends to be highly sensitive to noise and interference from other circuits. For example, in [27], the required level of isolation precluded the implementation of a single-chip transmitter. Furthermore, the modulation index of the transmitted signal depends upon the absolute tolerances of the VCO components which are often difficult to control in low-cost VLSI technologies and can also drift rapidly over time.

In principle, $\Delta\Sigma$ fractional-N PLLs can avoid these problems by modulating the VCO within the PLL. This can be done by driving the input of the digital $\Delta\Sigma$ modulator with the desired frequency modulation of the transmitted signal. The primary limitation is that bandwidth of the PLL must be narrow enough that the quantization noise from the $\Delta\Sigma$ modulator is sufficiently attenuated, but sufficiently high to allow for the modulation. For instance, the phase noise PSD of the example $\Delta\Sigma$ fractional-N PLL shown in Figure 5 with a 50 kHz loop bandwidth meets the necessary phase noise specifications when used as a local oscillator in a conventional upconversion stage within a Bluetooth wireless LAN transmitter. However, if the Bluetooth transmitter is to be implemented by modulating the VCO through the digital $\Delta\Sigma$ modulator, then the loop bandwidth of the PLL must be approximately 500 kHz. Unfortunately, when the loop bandwidth of the fractional-N PLL shown in Figure 5 is widened to 500 kHz, the resulting phase noise becomes too large to meet the Bluetooth transmit requirements.

Nevertheless, commercial transmitters with VCO modulation through $\Delta\Sigma$ fractional-N synthesizers are beginning to be deployed, especially in low-performance, low-cost wireless systems such as Bluetooth wireless LANs [28]. Facilitating this trend are various solutions that have been devised in recent years to allow for wideband VCO modulation in $\Delta\Sigma$ fractional-N PLLs without incurring the phase noise penalty mentioned above. One of the solutions is to keep the loop bandwidth relatively low, but pre-emphasize (i.e., highpass filter) the digital phase modulation signal prior to the digital $\Delta\Sigma$ modulator [29]. Unfortunately, this approach requires the highpass response of the digital pre-emphasis filter to be a reasonably close match to the inverse of the closed-loop filtering imposed by the largely analog PLL. Another of the solutions is to use a high-order loop filter in the PLL with a sharp lowpass response [30]. Increasing the order of the loop filter increases the attenuation of out-of-band quantization noise which allows for higher-order $\Delta\Sigma$ modulation to reduce in-band quantization noise thereby allowing the loop bandwidth to be increased without increasing the total phase noise. However, as described in [30], this necessitates the use of a Type 1 PLL which significantly complicates the design of the phase detector. Yet another solution is to use a narrow loop bandwidth but modulate the VCO both through the digital $\Delta\Sigma$ modulator and through an auxiliary modulation port at the VCO input [28]. The idea is to apply the low-frequency modulation components at the $\Delta\Sigma$ modulator input and the high frequency modulation components directly to the VCO. Again, matching is an issue, but it has proven to be manageable at least for low-end applications such as Bluetooth transceivers.

VIII. CONCLUSION

The additional concepts and issues associated with $\Delta\Sigma$

fractional-N PLLs for frequency synthesis relative to integer-N PLLs have been presented. It has been shown that $\Delta\Sigma$ fractional-N PLLs provide tuning resolution limited only by digital logic complexity, and, in contrast to integer-N PLLs, increased tuning resolution does not come at the expense of reduced bandwidth. Since one of the main innovations in a $\Delta\Sigma$ fractional-N PLL is the use of a $\Delta\Sigma$ modulator to control the divider modulus, the relevant concepts underlying $\Delta\Sigma$ modulation have been described in detail. A linearized model has been derived from first principles and a design example has been presented to illustrate how the model is used in practice. Techniques for wideband digital modulation of the VCO within a delta-sigma fractional-N PLL have also been presented.

ACKNOWLEDGEMENTS

The author is grateful to Sudhakar Pamarti, Eric Siragusa, and Ashok Swaminathan for their helpful discussions and advice regarding this paper.

REFERENCES

1. P. M. Gardner, "Charge-pump phase-lock loops," *IEEE Transactions on Communications*, vol. COM-28, pp. 1849-1858, November 1980.

2. B. Razavi, *Design of Analog CMOS Integrated Circuits*, McGraw Hill, 2001.

3. Bluetooth Wireless LAN Specification, Version 1.0, 2000.

4. U. L. Rohde, *Microwave and Wireless Synthesizers Theory and Design*, John Wiley & Sons, 1997.

5. B. Miller, B. Conley, "A multiple modulator fractional divider," *Annual IEEE Symposium on Frequency Control*, vol. 44, pp. 559-568, March 1990.

6. B. Miller, B. Conley, "A multiple modulator fractional divider," *IEEE Transactions on Instrumentation and Measurement*, vol. 40, no. 3, pp. 578-583, June 1991.

7. T. A. Riley, M. A. Copeland, T. A. Kwasniewski, "Delta-sigma modulation in fractional-N frequency synthesis," *IEEE Journal of Solid-State Circuits*, vol. 28, no. 5, pp. 553-559, May, 1993.

8. S. K. Tewksbury, R. W. Hallock, "Oversampled, linear predictive and noise-shaping coders of order $N>1$," *IEEE Transactions on Circuits and Systems*, vol. CAS-25, pp. 436-447, July 1978.

9. G. Lainey, R. Saintlaurens, P. Senn, "Switched-capacitor second-order noise-shaping coder," *IEE Electronics Letters*, vol. 19, pp. 149-150, February 1983.

10. I. Galton, "Granular quantization noise in a class of delta-sigma modulators," *IEEE Transactions on Information Theory*, vol. 40, no. 3, pp. 848-859, May 1994.

11. N. He, F. Kuhlmann, A. Buzo, "Multiloop sigma-delta quantization," *IEEE Transactions on Information Theory*, vol. 38, no.3, pp.1015-1028, May 1992.

12. I. Galton, "One-bit dithering in delta-sigma modulator-based D/A conversion," Proc. of the IEEE International Symposium on Circuits and Systems, 1993.

13. S. W. Golomb, *Shift Register Sequences*. Laguna Hills, CA: Aegean Park Press, 1982

14. E. J. McCluskey, *Logic Design Principles*. Englewood Cliffs, NJ: Prentice-Hall, 1986.

15. S. K. Tewksbury, R. W. Hallock, "Oversampled, linear predictive and noise-shaping coders of order $N>1$," *IEEE Transactions on Circuits and Systems*, vol. CAS-25, pp. 436-447, July 1978.

16. W. Rhee, B. S. Song, A. Ali, "A 1.1-GHz CMOS fractional-N frequency synthesizer with a 3-b third-order $\Delta\Sigma$ modulator," *IEEE Journal of Solid-State Circuits*, vol. 35, no. 10 , pp. 1453-1460, October 2000.

17. W. L. Lee, C. G. Sodini, "A topology for higher order interpolative coders," *Proceedings of the 1987 IEEE International Symposium on Circuits and Systems*, vol. 2, pp.459-462, May 1987.

18. K. C.-H. Chao, S. Nadeem, W. L. Lee, C. G. Sodini, "A higher order topology for interpolative modulators for oversampling A/D converters," *IEEE Transactions on Circuits and Systems*, vol. 37, no.3, p.309-318, March 1990.

19. Y. Matsuya, K. Uchimura, A. Iwata, T. Kobayashi, M. Ishikawa, T. Yoshitome, "A 16-bit oversampling A-to-D conversion technology using triple integration noise shaping," *IEEE Journal of Solid-State Circuits*, vol. SC-22, pp. 921-929, December 1987.

20. K. Uchimura, T. Hayashi, T. Kimura, A. Iwata, "Oversampling A-to-D and D-to-A converters with multistage noise shaping modulators," *IEEE Transactions on Acoustics, Speech, and Signal Processing*, vol. AASP-36, pp. 1899-1905, December 1988.

21. J. Craninckx, M. S. J. Steyaert, "A fully integrated CMOS DCS-1800 frequency synthesizer," *IEEE Journal of Solid-State Circuits*, vol. 33, pp. 2054=2065, December 1998.

22. S. Pamarti, "Techniques for Wideband Fractional-N Phase-Locked Loops," PhD Dissertation, University of California, San Diego, 2003.

23. S. R. Norsworthy, R. Schreier, G. C. Temes, Eds. *Delta-Sigma Data Converters, Theory, Design, and Simulation*, New York: IEEE Press, 1997.

24. L. Lin, L. Tee, P. R. Gray, "A 1.4 GHz differential low-noise CMOS frequency synthesizer using a wideband PLL architecture", *IEEE ISSCC Digest of Technical Papers*, pp. 204-205, Feb. 2000.

25. J. G. Proakis, *Digital Communications*, fourth ed., McGraw Hill, 2000.

26. S. Heinen, S. Beyer, J. Fenk, "A 3.0 V 2 GHz transmitter IC for digital radio communication with integrated VCO's," Digest of Technical Papers, *IEEE International Solid-State Circuits Conference*, vol. 38, pp. 150-151,

Feb. 1995.

27. S. Heinen, K. Hadjizada, U. Matter, W. Geppert, V. Thomas, S. Weber, S. Beyer, J. Fenk, E. Matshke, "A 2.7 V 2.5 GHz bipolar chipset for digital wireless communication," Digest of Technical Papers, *IEEE International Solid-State Circuits Conference*, vol. 40, pp. 306-307, Feb. 1997.

28. N. Filiol, et. al., "A 22 mW Bluetooth RF transceiver with direct RF modulation and on-chip IF filtering," Digest of Technical Papers, *IEEE International Solid-State Circuits Conference*, vol. 43, pp. 202-203, Feb. 2001.

29. M. H. Perrott, T. L. Tewksbury III, C. G. Sodini, "A 27-mW CMOS fractional-N synthesizer using digital compensation for 2.5-Mb/s GFSK modulation," *IEEE Journal of Solid-State Circuits*, vol. 32, no. 12, pp. 2048-2059, Dec. 1997.

30. S. Willingham, M. Perrott, B. Setterberg, A. Grzegorek, B. McFarland, "An integrated 2.5GHz $\Sigma\Delta$ frequency synthesizer with 5μs settling and 2Mb/s closed loop modulation," Digest of Technical Papers, *IEEE International Solid-State Circuits Conference*, vol. 43, pp. 200-201, Feb. 2000.

Ian Galton received the Sc.B. degree from Brown University in 1984, and the M.S. and Ph.D. degrees from the California Institute of Technology in 1989 and 1992, respectively, all in electrical engineering.

Since 1996 he has been a professor of electrical engineering at the University of California, San Diego where he teaches and conducts research in the field of mixed-signal integrated circuits and systems for communications. Prior to 1996 he was with UC Irvine, the NASA Jet Propulsion Laboratory, Acuson, and Mead Data Central. His research involves the invention, analysis, and integrated circuit implementation of key communication system blocks such as data converters, frequency synthesizers, and clock recovery systems. The emphasis of his research is on the development of digital signal processing techniques to mitigate the effects of non-ideal analog circuit behavior with the objective of generating enabling technology for highly integrated, low-cost, communication systems. In addition to his academic research, he regularly consults at several communications and semiconductor companies and teaches portions of various industry-oriented short courses on the design of data converters, PLLs, and wireless transceivers. He has served on a corporate Board of Directors and several corporate Technical Advisory Boards, and his is the Editor-in-Chief of the IEEE Transactions on Circuits and Systems II: Analog and Digital Signal Processing.

Designing Bang-Bang PLLs for Clock and Data Recovery in Serial Data Transmission Systems

Richard C. Walker

Abstract - **Clock recovery using phase-locked loops (PLL) with binary (bang-bang) or ternary-quantized phase detectors has become increasingly common starting with the advent of fully monolithic clock and data recovery (CDR) Circuits in the late 1980's. Bang-bang CDR circuits have the unique advantages of inherent sampling phase alignment, adaptability to multi-phase sampling structures, and operation at the highest speed at which a process can make a working flip-flop. This paper gives insight into the behavior of the nonlinear bang-bang PLL loop dynamics, giving approximate equations for loop jitter, recovered clock spectrum, and jitter tracking performance as a function of various design parameters. A novel analysis shows that the bang-bang loop output jitter grows as the square-root of the input jitter as contrasted with the linear dependence of the linear PLL.**

I. INTRODUCTION

Prior to the advent of fully monolithic designs, clock recovery was traditionally performed with some variant of the circuit in Fig. 1. The clock frequency component was typically extracted from

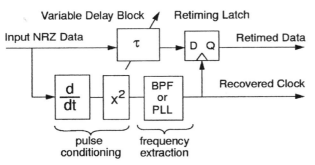

Fig. 1. Traditional non-monolithic clock and data recovery architecture.

the data stream using some combination of differentiation, rectification and filtering. The bandpass frequency filtering was provided by LC tank, surface acoustic wave (SAW) filter, dielectric resonator or PLL. Because the clock recovery path was separate from the data retiming path, it was difficult to maintain optimum sampling phase alignment over process, temperature, data-rate, and voltage variations. Even the PLL techniques had the drawback of using phase detectors with different set-up times than the retiming flip-flop so that the recovered clock was not intrinsically aligned to the optimum sampling point in the data eye. Circuits utilizing SAW resonator filtering typically required hand matching of SAW and circuit temperature coefficients along with custom cut

R. Walker is with Agilent Laboratories, 3500 Deer Creek Road, MS 26-U4, Palo Alto CA 94304. (e-mail: rick_walker@labs.agilent.com).

coaxial delay lines for setting the timing of the recovered sampling clock with respect to the data eye [1].

Early monolithic CDR designs imitated these discrete block diagrams. The propagation delay differences between data and clock paths could be ignored as long as the gate delay skew was a negligible fraction of the total bit time, or unit interval. The need for higher link speeds grew faster than Moore's law, and as clock frequencies approached the effective f_T of the active devices, it became increasingly difficult to maintain an optimum sampling phase alignment between the recovered clock and the data over process, temperature, data-rate, and voltage variations.

A second problem was that most linear phase detectors produced narrow pulses with widths proportional to the phase error between the timing of the data and the clock [2], [3]. These narrow pulses required a process speed in excess of that required to simply sample data at a given rate. The timing skew and speed of linear phase detector circuits then became the limiting factor for aggressive designs.

Both these difficulties are eliminated by a family of circuits which simultaneously retime data and measure phase error by using matched flip-flops to sample both the middle of each data bit and the transitions between the data bits. Fig. 2 shows such an

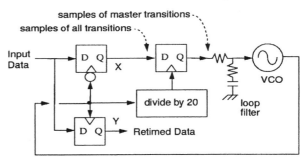

Fig. 2. A simple bang-bang loop using a flip-flop for a phase detector to lock onto a data stream with a guaranteed "0" to "1" transition every 20 bits.

early gigabit-rate monolithic example of such a circuit [4] which samples data with two matched flip-flops. Flip-flop "Y" samples the middle of each data bit on the rising edge of the VCO clock to produce retimed data, while flip-flop "X" samples the transition of each bit using the falling edge of the VCO clock.

The loop is designed to use the 16B/20B line code of Fig. 3 which guarantees a "01" "master transition" every 20 bits. The divide by 20 circuit and associated flip-flop in Fig. 2 discard every

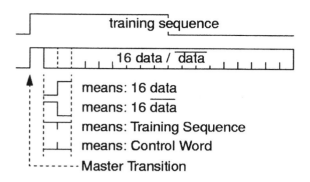

Fig. 3. Format of 16B/20B line code used with bang-bang CDR of. Fig. 2.

transition sample except for this master transition sample. During link start-up a training sequence is sent that has only one rising transition at the location of the master transition. Once the loop is locked, arbitrary data is allowed to be sent at the other 18 bits of the frame, while the transition sampler pays attention only to the data stream in the vicinity of the master transition. If the VCO frequency is too high, the transition flip-flop starts sampling prior to the master transition and outputs a "0" to the loop filter. A slightly lower VCO frequency, on the other hand, will cause the loop to be driven by 1's.

The loop drives the falling edge of the VCO into alignment with the data transitions based on the binary-quantized phase error. Because the clock-to-Q delay of the retiming flip-flop is monolithically matched with the phase detector flip-flop, the PLL aligns the recovered clock precisely in the middle of the data eye with no first-order timing skew over process and temperature variations. Because the narrowest pulse is the output of a flip-flop, such detectors operate at the full speed at which a process is capable of building a functioning flip-flop. This ensures that the phase detector will not be the limiting factor in building the fastest possible retiming circuit.

An additional advantage of flip-flop-based phase detectors is that since they only require simple processing of digital values, they easily generalize to multi-phase sampling structures allowing CDR operation at frequencies in which it would be impossible to build a working full-speed flip-flop. In contrast, most linear phase detectors require at least some analog processing at the full bit rate, limiting process speed and poorly generalizing to multi-phase sampling architectures.

Because of these compelling advantages, the bang-bang loop has become a common design choice for state-of-the CDR designs which are pushing the capability of available IC processes. Fig. 4 surveys CDR designs presented from 1988 to 2001 at the International Solid State Circuits Conference. Designs are plotted by year of presentation against each design's ratio of link speed to effective f_T. The majority of current designs utilize a combination of multiphase sampling structures and bang-bang PLLs. In addition, all CDRs operating at data rates greater than 0.4 f_T are bang-bang designs.

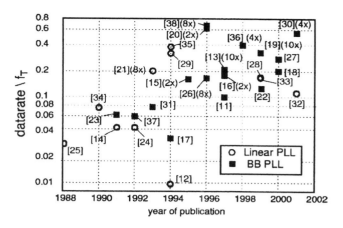

Fig. 4. CDR PLL designs over time. The ratio of link speed to effective process transit frequency is plotted vs year of publication. Multi-phase BB PLLs predominate as data rate approaches the process transit frequency limit. (The number of retiming phases used in each design is given in parentheses.)

II. FIRST-ORDER LOOP DYNAMICS

Unfortunately, transition-sampling flip-flop-based phase detectors can provide only binary (early/late) or ternary (early/late + hold) phase information. This amounts to a hard non-linearity in the loop structure, leading to an oscillatory steady-state and rendering the circuit unanalyzable with standard linear PLL theory. Precise loop behavior can be simulated efficiently with time-step simulators, but this is cumbersome to use for routine design. Fortunately, simple approximate closed-form expressions can be derived for performance parameters of interest, such as loop jitter generation, recovered clock spectrum, and jitter tracking performance as a function of various design parameters.

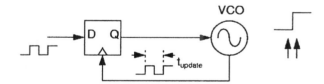

Fig. 5. A simple bang-bang loop using a flip-flop for a phase detector to lock onto square-wave input.

A simple BB PLL is shown in Fig. 5. A flip-flop is used as a phase detector to lock onto a square wave input signal. Depending on whether the VCO phase samples slightly before or after the rising edge of the input square wave, the flip-flop output is either low or high, adjusting the VCO period in such a way as to move the sampling phase error back towards zero. The dynamics of such a binary-quantized loop are equivalent to a data-driven phase detector operating on alternating 0,1 data with 100% transition density, or a master-transition based loop similar to that shown in Fig. 2. For simplicity, we assume that a valid binary phase determination can be made at every timestep. The consequence of random data and the introduction of a ternary hold mode are considered in a later section.

The first-order BB PLL of Fig. 5 can be rendered into a block diagram for analysis as shown in Fig. 6. The loop phase error

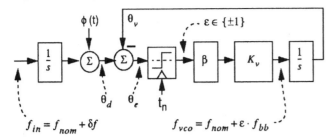

Fig. 6. Block diagram of first order loop showing definition of signal names.

$\theta_e(t_n)$, is defined as the difference between the data phase $\theta_d(t_n)$ and the VCO phase $\theta_v(t_n)$ at the nth sampling time t_n. For convenience, phase is measured with respect to an ideal clock source running at f_{nom}.

The frequency of the incoming data signal differs from the VCO center frequency by δf, and has a zero mean phase jitter of $\phi(t)$. In other words, the data can be considered to have been generated by a pattern generator clocked on the rising edges of the jittered clock signal $\sin[2\pi(f_{nom} + \delta f)t + \phi(t)]$. The data phase $\theta_d(t_n)$ is then $2\pi\delta f t_n + \phi(t_n)$.

The phase detector binary-quantizes the loop phase error at each sampling time to give $\varepsilon_n = \text{sign}[\theta_e(t_n)]$. (Note: In the case of a ternary data-driven phase detector, ε_n may be set to 0 when it is not possible to make a determination of phase error due to consecutive identical bits in the data stream. The consequence of this "hold" state is treated in a later section). The error signal drives the VCO through an attenuator β, to produce a change in frequency of $f_{bb} = \beta K_{vco}$. From time t_n until time t_{n+1}, the VCO operates at one of the two frequencies given by $f_{nom} + \varepsilon_n f_{bb}$.

Because the VCO frequency changes on each cycle, the system has non-uniform sampling times. The time of phase sample $t_{n+1} = t_n + 1/(f_{nom} + \varepsilon_n f_{bb})$. In a typical CDR, f_{bb} is on the order of 0.1% of f_{nom}, so that an analysis assuming uniform time steps of $t_{update} = 1/f_{nom}$ is sufficiently accurate for most purposes. However, for loop analyses requiring exact charge pump balance, such as wide-range loop pull-in without a

frequency detector, these non-uniform sampling times must be accounted for.

With the uniform time step approximation, the VCO phase changes up or down (or "walks off") by $\theta_{bb} = 2\pi(f_{bb}/f_{nom})$ radians during each update period.

In summary, the first order loop obeys a simple set of discrete time difference equations:

$$\theta_d(t_n) = \theta_d(0) + 2\pi\delta f t_n + \phi(t_n) \qquad (1)$$

$$\theta_v(t_{n+1}) = \theta_v(t_n) + \varepsilon_n\theta_{bb} \qquad (2)$$

$$\varepsilon_n = \text{sign}[\theta_d(t_n) - \theta_v(t_n)] \qquad (3)$$

As long as the VCO frequency step brackets the input signal frequency error, the loop will remain phase locked. Assuming $\phi(t)$ small, the lock range is: $-f_{bb} < \delta f < f_{bb}$. The loop generates an excess hunting jitter with a peak-to-peak value of two bang-bang phase steps $J_{pp} = 4\pi(f_{bb}/f_{nom})$.

For the loop to be locked, the average VCO frequency must equal the average data frequency. The phase detector duty cycle C, must satisfy the relation

$$\delta f = C(f_{bb}) + (1-C)(-f_{bb}).$$

The value of C is then given by

$$C = \left(\frac{1}{2} + \frac{\delta f}{2f_{bb}}\right).$$

The phase detector duty cycle, and therefore its average output voltage are proportional to the loop frequency error. Fig. 7 shows a simulated loop with a range of input frequencies. The loop is

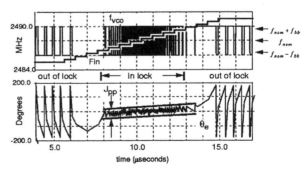

Fig. 7. Simulated response of first-order PLL to a range of input frequencies.

"locked" whenever the input frequency is bracketed by the two VCO frequencies. The rapid alternation between frequencies

36

slightly too high and slightly too low creates a bounded hunting jitter (J_{pp}).

The derivative of the input data phase deviation, $d[\phi(t)]/dt$, adds to the frequency error that must be tolerated by the loop. Assuming $\delta f = 0$, then for $\phi(t) = A\sin(2\pi f_{mod}t)$, the maximum amplitude A of phase modulation at frequency f_{mod} before onset of slew-rate limiting is $|f_{bb}|/f_{mod}$. Fig. 8. demon-

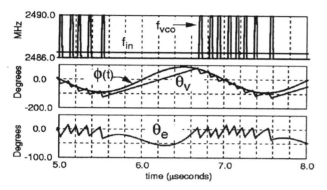

Fig. 8. Simulated response of first-order PLL to sinusoidal input jitter just slightly beyond the tracking capability of the loop.

strates the loop at the onset of jitter-induced slew-rate limiting. Although the average input frequency lies within the lock range of the loop, the added sinusoidal jitter causes the instantaneous input frequency deviation to exceed $\pm f_{bb}$. The loop stops toggling and goes into slew rate limiting, leading to a transient phase error.

A. Summary of First-Order Loop

The first-order bang-bang loop has only one degree of freedom. Jitter generation, lock range, and jitter tolerance are all inconveniently controlled by one parameter, f_{bb}. This situation can be improved by using a second control loop to dynamically adjust the nominal VCO frequency f_{nom} to be equal to the incoming data frequency. Because the phase detector duty cycle is proportional to the loop frequency error, this dynamic centering of VCO frequency can be accomplished by adjusting the VCO center frequency in a feedback loop to drive the phase detector duty cycle C to 50%. This decouples the lock range from jitter tolerance and jitter generation, giving more design freedom.

III. SECOND-ORDER LOOP DYNAMICS

To extend the loop tracking range independent of the jitter generation, an extra integrator is added between the phase detector and the VCO as in Fig. 9. Since the first-order loop dynamic produces a phase detector duty cycle proportional to the loop frequency error, this added integrator can be viewed as an automatic means for keeping the first-order portion of the loop properly cen-

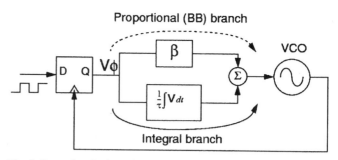

Fig. 9. Second-order bang-bang loop schematic.

tered on the average incoming data frequency. If certain assumptions are met, as described later, we can consider the system to be composed of two non-interacting loops. These are the loops labeled "bang-bang branch" and "integral branch." If the center frequency control loop is slow enough, the resulting loop behavior will be very similar to a simple first-order loop, but with an extended frequency lock range.

A. Stability Factor

To preserve the desirable qualities of the first order loop, it is critical that the phase change due to the proportional branch dominate over the phase change from the integral branch.

The loop phase change in one update time due to the proportional connection is $\Delta\theta_{bb} = \beta V_{\phi}K_v t_{update}$. The phase change due to the integral branch is $\Delta\theta_{int} = V_{\phi}K_v t_{update}^2/(2\tau)$. The ratio of these two is the stability factor of the loop

$$\xi \equiv \frac{\Delta\theta_{proportional}}{\Delta\theta_{integral}} = \frac{2\beta\tau}{t_{update}}.$$

The reader should be careful not to confuse the bang-bang loop stability factor ξ with the linear loop damping factor ζ [5].

The discrete time difference equations for the second-order loop can be written as

$$\theta_d(t_n) = \theta_d(0) + 2\pi\delta f t_n + \phi(t_n) \tag{1}$$

$$\theta_v(t_{n+1}) = \theta_v(t_n) + \theta_{bb}\left(\varepsilon_n + \frac{\varepsilon_n}{\xi} + \frac{2}{\xi}\sum_0^n \varepsilon_n\right) \tag{4}$$

$$\varepsilon_n = \text{sign}[\theta_d(t_n) - \theta_v(t_n)] \tag{3}$$

From this, it can be seen that the second-order loop has two degrees of freedom, the loop phase step θ_{bb} (or equivalently, the loop frequency step f_{bb}) and the stability factor ξ. The added loop integrator extends the frequency tracking range, leaving θ_{bb} free to control jitter tolerance and jitter generation.

B. Simulations of Second-Order Loop

Fig. 10 shows two block diagrams for the second-order loop. The upper diagram is a straightforward translation of the schematic in Fig. 9. The lower diagram is a topological re-arrangement

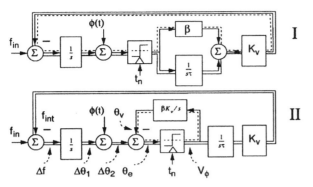

Fig. 10. Two equivalent second-order bang-bang loop block diagrams. The proportional phase-control signal flow is highlighted with a dashed line, and the integral frequency-control loop with a solid line.

which places an inner first-order phase tracking loop inside an outer frequency tracking loop. If one writes the transfer function from the output of the non-linear quantizer block back to the input of the quantizer, it can be shown that both diagrams are exactly equivalent. Some of the signals in the second diagram do not correspond to actual physical variables in the circuit, but they are helpful in understanding the operation of the loop.

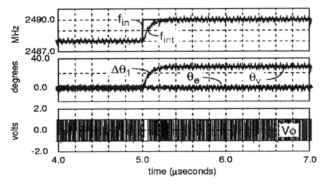

Fig. 11. Second-order loop response to instantaneous frequency step smaller than f_{bb}.

Fig. 11 shows the second-order loop responding to a step change in input frequency f_{in}, producing a slow response f_{int} in the outer integral loop. The resulting phase error $\Delta\theta_1$ is tracked by the inner bang-bang loop θ_v to produce the final sampler phase error θ_e. Notice that, unlike linear PLLs, if the power-supply noise-induced VCO frequency modulation is lim-

ited to $\pm f_{bb}$, then there is *no* jitter accumulation or phase transient at the sampling flip-flop.

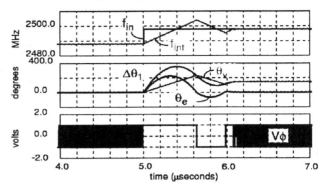

Fig. 12. Second-order loop response to instantaneous frequency step larger than f_{bb}.

Fig. 12 is a simulation in which the input frequency step is bigger than f_{bb}, so the loop goes into slew rate limiting, leading to a transient phase error θ_e at the sampler.

C. Response to Phase Step

For a normalized transient phase step of $\Delta = \theta_{step}/\theta_{bb}$, a first-order loop relocks in Δ update times. The total time for relocking is then $\theta_{step}/(2\pi f_{bb})$.

During the relocking transient of the second-order loop, the loop integrator overshoots the correct steady-state VCO tune voltage. This causes a quadratic overshoot in the phase trajectory.

Fig. 13 shows the second-order phase step response with ξ as a parameter. Up to the first zero crossing, the phase trajectory is given by

$$\frac{\theta(t)}{\theta_{bb}} = \Delta - \left(n + \frac{n^2}{\xi}\right),$$

with $n = t/t_{update}$. The time of the first zero crossing approaches Δ as $\xi \to \infty$, consistent with a first-order loop. In general, the second-order loop is quicker to reach zero phase error than the first-order loop, but pays for this with an oscillatory overshoot. As a conservative rule of thumb, the magnitude of the oscillatory transient of a second-order step response can be considered bounded by the simple linear transient of the first-order loop. The time required to reach steady state, given a step of Δ is always less than or equal to Δ timesteps, independent of ξ.

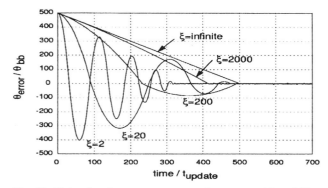

Fig. 13. Noise-free loop response to a phase step with stability factor ξ as a parameter.

IV. SLOPE OVERLOAD

Many systems, such as SONET, specify jitter tolerance in the form of a sinusoidal jitter at various frequencies.

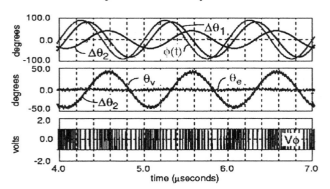

Fig. 14. Second-order loop response to sinusoidal input jitter.

Fig. 14 shows the loop response with a sinusoidal input phase jitter $\phi(t)$. The outer integral loop tracks the input jitter at $\Delta\theta_1$ with a slight phase lag. The resulting phase error $\Delta\theta_2$ is tracked by the inner bang-bang loop θ_v to produce the final sampler phase error θ_e. The duty-cycle of the PD output V_ϕ varies with the slope of $\Delta\theta_2$ which is proportional to the instantaneous frequency error of the outer loop.

In Fig. 15, the phase modulation is increased until the instantaneous frequency error exceeds the inner loop's ability to track. Slew-rate limiting produces a tracking error at the sampler θ_e. A CDR would normally be designed such that slewing would never occur for any valid signal allowed by a particular standard. The next two sections develop an analytic expression for slope over-

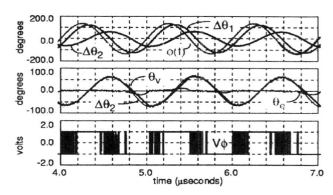

Fig. 15. Second-order loop response to large sinusoidal input jitter.

load so that a loop can be easily designed to never slew for signals meeting a typical frequency-domain jitter tolerance specification.

A. Delta-Sigma Analogy

Before developing an analytic equation for slope overload, it is helpful to introduce a further rearrangement of block diagram II from Fig. 10. Fig. 16 transforms the loop by pulling two integra-

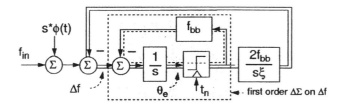

Fig. 16. Redrawing of the loop to show inner $\Delta\Sigma$ inner modulator operating on the loop frequency error.

tors through the last summing node prior to the quantizer. The update time interval is set to 1. The definition for bang-bang frequency step $f_{bb} = \beta K_v V_\phi$, and stability factor $\xi = 2\beta\tau/t_{update}$ are also substituted in.

The shaded area in Fig. 16 shows how the proportional feedback loop can be thought of as an inner $\Delta\Sigma$ modulator producing a phase detector duty cycle proportional to the VCO frequency error [6],[7].

Fig. 17 summarizes an analysis of the first order delta-sigma (after [8]). When the loop is not in slew rate limiting, or in a periodic limit-cycle, the quantizer (e.g., PD) can be replaced with a unity gain element and a noise source $Q(z)$ with the same $A\sin(2\pi t f t/t_{update})/(2\pi f t/t_{update})$ noise characteristics as a random binary bitstream. Both these constraints are met in practice as the VCO phase noise is sufficient to eliminate any deterministic limit cycles, and the loop is designed to never slew rate limit on any conforming input signal. This insight is critical as

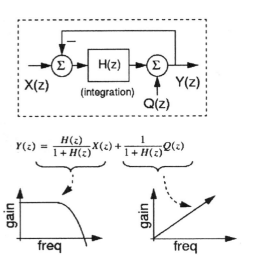

$$Y(z) = \frac{H(z)}{1+H(z)}X(z) + \frac{1}{1+H(z)}Q(z)$$

Fig. 17. Simplified analysis of delta-sigma circuit.

it allows linear analysis to be applied whenever the bang-bang loop is not in slew rate limiting.

With the $\Delta\Sigma$ substitution, the inner loop becomes a wide-band unity-gain block as seen from the viewpoint of the outer integral frequency control loop. The noise in the delta-sigma core is first-order frequency shaped towards high frequencies. However, when the frequency noise is converted to phase noise, the shaping is lost and the noise becomes flat.

B. Expression for Slope Overload

A closed-form analysis of slope overload can now be derived. Referring to Fig. 16, the system slews when $|\Delta F| > f_{bb}$. Assuming no slew rate limiting, we can use the results from the $\Delta\Sigma$ analysis to justify replacing the loop quantizer with a unity gain element. The maximum input phase jitter in UI as a function of frequency, $\sigma_j^{max}(s)$, normalized to θ_{bb} can then be calculated using Laplace transforms.

We want to find an input excitation $F(s)$, for which $|\Delta F| = f_{bb}$ at all frequencies. The inner $\Delta\Sigma$ of Fig. 16 has a linearized transfer function of $1/(s + f_{bb})$. Using standard feedback loop theory, the expression for ΔF can then be written as

$$\Delta F = \frac{F(s)}{1 + \left(\frac{2f_{bb}}{s\xi}\right)\left(\frac{1}{s + f_{bb}}\right)}.$$

Setting $\Delta F = f_{bb}$, and normalizing the equation by letting f_{bb} and $t_{update} = 1$, we can solve for $F(s)/s$ to get the maximum normalized input phase as a function of normalized frequency

$$\frac{\sigma_j^{max}(s)}{\theta_{bb}} = \left(\left(s^2 + s + \frac{2}{\xi}\right)/(s^3 + s^2)\right).$$

This is a curious bootstrapped analysis, in that it assumes a lack of slewing to justify the linearization which permits the computation of the onset of slew rate limiting.

Fig. 18 shows a good agreement between this expression and simulated loop performance in which slewing is defined as a contiguous sequence of ten or more identical phase-error indications. This expression can be used to design a loop for a given jitter tol-

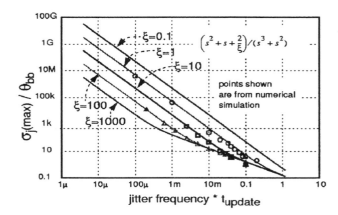

Fig. 18. Normalized amplitude of sinusoidal jitter just sufficient to cause slope overload as a function of normalized jitter frequency and with ξ as a parameter.

erance. The tolerance plots are single-pole slope for high ξ and high jitter frequency, becoming double-pole at lower frequencies and small ξ. At high frequencies, all of the curves become asymptotic to the single-pole tolerance of a first-order bang-bang PLL. The operating region below each of these curves is where the $\Delta\Sigma$ approximation is valid, and where a linear loop analysis is justified.

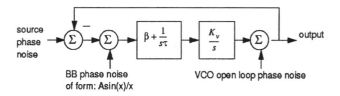

Fig. 19. Loop redrawn replacing phase detector with unity gain element and additive quantization noise.

V. JITTER GENERATION

With these insights, it is possible to accurately predict the loop jitter generation in the frequency domain. Fig. 19 is a redrawing of the loop replacing the phase detector by a unity gain element, and an additive noise source. The forward loop gain is

$$H(s) = \frac{K_{vco}}{s}\left(\beta + \frac{1}{s\tau}\right).$$

From this can be calculated two transfer functions: the lowpass seen by both the source phase noise and the PD noise to the output, $A(s) = 1/[1 + H(s)]$, and the high-pass transfer function from VCO phase noise to the output, $B(s) = H(s)/[1 + H(s)]$. As shown in Fig. 20, with a source phase noise P(s), a PD phase noise Q(s), and a VCO phase noise R(s), the total loop jitter generation spectrum becomes the RMS combination of each of the three weighted terms $J(s) = \sqrt{(PA)^2 + (QA)^2 + (RB)^2}$. The source phase noise is

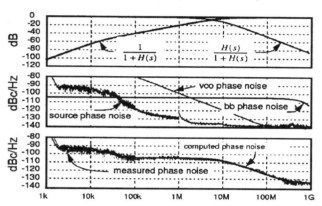

Fig. 20. Example computation of loop jitter generation spectrum with parameters from [11].

generally taken to be the spectrum of the clock driving the data source or BERT, or in the case of a clock multiplying circuit, the spectrum of the reference clock corrected by 20 times the log of the loop frequency multiplication ratio.

The phase noise power is given by

$$S_{RMS} = \int_{0}^{w_{max}} J^2(s)d\omega.$$

The RMS jitter in unit intervals is then

$$J_{RMS} = \text{atan}(\sqrt{S_{RMS}})/\pi .$$

It should be noted that the linearized loop model is only suitable for computation of the jitter spectrum but not for computing the actual sampling point phase error or other time-domain transient response. The linearized response only covers the dynamics of the outer frequency tracking loop, but does not capture the extra tracking of the internal nonlinear $\Delta\Sigma$ core.

VI. GAUSSIAN INPUT NOISE

Fig. 21 is a plot of output jitter vs input jitter with ξ as a

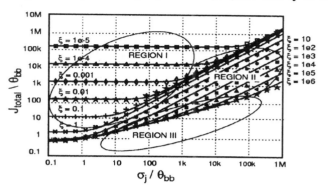

Fig. 21. Normalized output jitter vs input jitter sigma with ξ as a parameter. Simulation is for a non-tristated loop, with square wave data input, 10^8 timesteps per point, and ignoring phase wrapping.

parameter. For convenience, all jitter sigmas are normalized to θ_{bb}, the loop phase step size. The total loop output jitter can be approximated by three regions of operation: $J_{total} \approx J_{idle} + J_{linear} + J_{walk}$. In Region I, the output jitter is independent of input jitter σ_j. This occurs when the self-generated hunting jitter exceeds the input jitter. The RMS jitter in this region is empirically determined to be well approximated by $J_{idle} \approx 0.6 + (1.65/\xi)$. In Region II, the output jitter is proportional to the input jitter. This occurs when the input jitter is so high that, for a given ξ, the bang-bang dynamic is unable to control the second-order portion of the loop. This leads to large quadratic trajectories in the phase domain, causing the loop phase to "hunt" towards the limits of the input jitter distribution. As the loop phase nears the limits of the input jitter distribution, the bang-bang hunting has more effect on stabilizing the second-order loop. In this region, the output jitter is proportional to the input jitter: $J_{lin} \approx 2\sigma_j/(1 + \sqrt{\xi})$. In Region III, the output RMS jitter J_{walk} is approximately equal to $0.7 \cdot \sqrt{\sigma_j}$. This surprising result says that loops with large ξ have output jitter which grows as the square root of the input jitter. Contrast this with a linear PLL which simply low-pass filters the input jitter and thus has an output jitter which grows linearly with the input jitter.

An approximate analysis of loop jitter can shed light on this curious square-root dependence of output jitter on input jitter. Assume a zero-mean input jitter distribution with a sigma σ_j. Using a linearized approximation to the standard probability distribution function, the probability of getting an "early" phase error indication for small loop phase deviation $\Delta\theta$, is approximately

$$p_e \approx \frac{1}{2} - \frac{\Delta\theta}{\sigma_j\sqrt{2\pi}}.$$

The expected phase change in the loop after one update time is

$$\theta_{bb}((1-p_e)-p_e) = \frac{2\Delta\theta}{\sigma_j\sqrt{2\pi}}\theta_{bb}$$

The discrete time equation for the average evolution of loop phase under the condition of a small input phase error can then be expressed as

$$\Delta\theta(t_{n+1}) = \left(1 - \frac{2\theta_{bb}}{\sigma_j\sqrt{2\pi}}\right)\Delta\theta(t_n),$$

This equation has the same form as a discrete time approximation to the capacitor voltage in an RC lowpass filter. By analogy, when time is expressed in units of loop update times, any transient phase error in the bang-bang loop can then be said to decay to zero with a time constant of $\tau = \sigma_j\sqrt{2\pi}/(2\theta_{bb})$.

This "lowpass" loop characteristic is being driven by random energy from the early/late phase detector output. A related problem is the computation of the baseline wander voltage generated by passing a random NRZ data stream through a coupling capacitor. It can be shown that the sigma on the capacitor voltage is given by $\sigma_{BLW} = V_{pp}\sqrt{t_{bit}/(8\tau)}$. Extending this analogy to the loop, we can consider the output of the phase detector as a 50% duty-cycle random NRZ data stream. Given that the output from each "bit" must cause a loop phase change of θ_{bb}, we can compute that the effective V_{pp} to satisfy our loop difference equation must be $\sqrt{2\pi}\sigma_j$. We can then compute the loop jitter by using the analogous baseline wander expression with the effective loop V_{pp} and τ. The result is

$$\sigma_{bb} = \sqrt{\frac{\theta_{bb}\sigma_j\sqrt{2\pi}}{8}} \approx 0.79\sqrt{\theta_{bb}\sigma_j},$$

which is consistent with empirical analysis of simulation results.

One further insight into this behavior is offered. The second-order loop drives the phase detector output to a steady-state 50% duty-cycle. In this condition, the loop phase splits the input jitter distribution into equal early and late halves. This means that the bang-bang loop phase is servoed to the *median* of the input jitter distribution rather than to the *mean* as would be the case with a linear loop. Because of this, the bang-bang loop makes a constant modest correction in response to large jitter outliers, rather than the proportionally large overcompensation of a linear loop. This insight supports the idea that the bang-bang loop jitter should only be sub-linearly affected by the magnitude of the input jitter.

VII. DATA-DRIVEN PHASE DETECTORS

Unless the data contains a guaranteed periodic transition, the CDR will be required to lock onto random transitions embedded in the data stream. The effects of runlength and transition density on loop performance must then be considered. The effect of these two data attributes is dependent on the type of phase detector used. Most modern codes use some variation of Alexander's phase detector [9] shown in Fig. 22. Two matched flip-flops form the

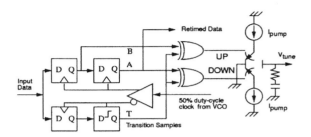

Fig. 22. Modified form of Alexander's ternary-quantized phase detector for NRZ data along with a typical charge pump for driving the VCO tuning input.

front-end of Alexander's phase detector, with the first flip-flop driven on the rising edge of the 50% duty-cycle clock, and the second flip-flop driven on the falling edge of the same clock. (Using a fully-differential monolithic ring-oscillator, it is possible to achieve a very precise 50% duty-cycle clock source). When the loop is locked, the rising-edge retiming flip-flop samples the center of each data bit and produces a retimed data bit at (A) and the following retimed bit at (B). The falling-edge flip-flop functions as a phase detector by sampling the transition (T) between the data bits (A,B). To improve the circuit's operating speed, the (T) sample is delayed an extra half bit time by a latch so that the logic on (A,T,B) has a full bit time for resolution.

The transition sample is then compared to the surrounding data bits to determine whether the clock sampling phase is early or late to derive a binary-quantized (bang-bang) or ternary phase error indication. A truth-table for the logic in Fig. 22 is given in Table 1.

TABLE 1. Truth table for logic in Fig. 22.

State	A	T	B	UP	DOWN	Meaning
0	0	0	0	0	0	hold
1	0	0	1	0	1	early
2	0	1	0	1	1	hold
3	0	1	1	1	0	late
4	1	0	0	1	0	late
5	1	0	1	1	1	hold
6	1	1	0	0	1	early
7	1	1	1	0	0	hold

The states 2 and 5 in Table 1 correspond to the normally impossible condition of sampling a "1" midway between two "0" bits. A custom truth table can use these states to detect either a high bit-error-rate condition [10], a VCO running grossly too slowly (eg: lump these states into the "late" condition), or taken as an indication that a link has locked onto its own VCO crosstalk, perhaps by amplification of power supply noise by pick up from a high-gain optical transimpedance amplifier [11].

Since the mid-bit samples (A,B) straddle the (T) transition sample, it is also possible to detect the lack of a transition. This condition corresponds to states 0 and 7 in Table 1. This information can be used to create an extra ternary hold-state in the PD output, causing the charge pump to hold its value during long runlengths. Both binary and ternary PDs will be discussed in turn, along with their implications on loop performance.

A. Run-length and Latency

Binary phase detectors have no hold state, so the PD continues to put out the last valid phase error indication during long data runlengths. In this situation, the loop idling jitter will be multiplied from the expected value by the maximum runlength of the data. For example, an 8B/10B code has a maximum code runlength of 5 and will have a peak jitter walk-off five times the value of that computed for a "10" repetitive data pattern. The average RMS jitter will be a function of the runlength distributions of each particular code. There is also a trade-off in effective stability factor as a repetitive pattern such as "11110000" will be equivalent to a loop with an effective update time 4 times larger than the expected $t_{update} = 1/f_{nom}$. Since the stability factor is inversely dependent on update time, it is possible for binary PDs to become unstable with data patterns containing very long runs due to the delay in timely phase-error feedback.

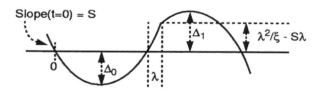

Fig. 23. Setup for computing onset of loop instability with latency λ.

Fig. 23 shows the loop phase trajectory during an acquisition transient. At t=0, the loop crosses zero phase error with $d\phi/dt = S$. From this we can compute an overshoot Δ_o. When the loop phase again crosses zero phase error, the phase detector is late in responding by a time λ. This time is a combination of runlength, latency in the phase detector logic, and high-order poles in the VCO tuning characteristic.

Due to the loop latency λ, the loop overshoots zero phase by $\lambda^2/\xi - S\lambda$ before the "braking" effect of the proportional branch starts to act. The onset of catastrophic instability occurs

when $\Delta_1 > \Delta_0$, for this implies exponential growth of the acquisition transient. The convergence is guaranteed whenever $\xi > 2\lambda$.

Although usable for tightly constrained block codes such as 8b/10B, binary phase detectors are essentially unusable for codes such as 10Gb Ethernet 64b/66b or SONET which can have very long runlengths of up to 66 or 80 bits, respectively.

B. Ternary Phase-Detector

The 3-state, or ternary phase detector provides superior jitter performance for data with long runs [12]. Ternary PDs neither charge nor discharge the loop filter during long runs causing the loop to hold the current estimate of the data frequency. Such loops effectively "stop time" during long runs.

If the charge pump does not have a hold-mode, it is possible to emulate a ternary loop, with some loss of performance, by continuously toggling the phase-detector output to approximately maintain the current charge pump voltage during long runs.

The peak idling jitter for ternary loops is unchanged from the simple 100% transition density analysis. The RMS jitter will be reduced by the average transition density. Because the loop phase cannot change during hold mode, the jitter tolerance will be derated by the average transition density. This can easily be taken into account by increasing θ_{bb} appropriately for the characteristics of the code to be used.

C. VCO Tuning Bandwidth

The previous analyses all assumed an infinite VCO tuning bandwidth for the proportional tuning input. A VCO time-constant τ_{vco}, can slightly reduce hunting jitter if it is small compared to the loop update time.

Timeconstants larger than the loop update time prevent the loop from reversing phase slope within an update period and lengthen the loop limit cycle. If the extra pole is thought of as an extra latency $2\tau_{vco}$, then the result of the previous section can be used to give an approximate bound on loop stability. To avoid divergence: $\tau_{vco} < \xi t_{update}/4$. Comparison with simulation verifies this equation as a conservative limit on τ_{vco}.

However, it cannot be recommended to flirt with this boundary. Unless one meticulously checks performance by numerical simulation, it is safest to design the VCO to essentially respond fully in one update time. This is usually very easy to achieve in ring-oscillators and possible with some care using low-Q LC VCOs.

VIII. Conclusion

Bang-bang CDR circuits have the unique advantages of inherent sampling phase alignment, adaptability to multi-phase sampling structures, and operation at the highest speed at which a process can make a working flip-flop. Approximate equations for loop jitter, recovered clock spectrum, and jitter tracking performance as a function of various design parameters have been derived. The median-tracking property of the bang-bang loop resulting in an output jitter equal to the square root of the input jitter has been presented.

ACKNOWLEDGMENT

The author is grateful to the contributions of Birdy Amrutur, Bill Brown, John Corcoran, Craig Corsetto, Dave DiPietro, Brian Donoghue, Jeff Galloway, Andrew Grzegorek, Tom Hornak, Jim Horner, Tom Knotts, Benny Lai, Adolf Leiter, Bill McFarland, Charles Moore, Rasmus Nordby, Cheryl Owen, Pat Petruno, Kent Springer, Guenter Steinbach, Hugh Wallace, Bin Wu, J.T. Wu, and Chu Yen for technical discussions and helpful insights into bang-bang loop behavior.

REFERENCES

[1]   C. B. Armitage, "SAW Filter Retiming in the AT&T 432 Mb/s Lightwave Regenerator," in *Conference Proceedings: AT&T Bell Labs*, pp. 102-103, Sept. 3-6, 1984.

[2]   C. R. Hogge, Jr., "A Self Correcting Clock Recovery Circuit," *IEEE Transactions on Electron Devices*, vol. ED-32, no. 12, pp. 2704-2706, Dec. 1985.

[3]   J. Tani, Crandall, D., Corcoran, J. Hornak, T., "Parallel Interface ICs for 120Mb/s Fiber Optic Links," in *ISSCC Digest of Technical Papers*, pp. 190-191,390, Feb. 1987.

[4]   R. C. Walker, T. Hornak, C. Yen and K. H. Springer, "A Chipset for Gigabit Rate Data Communication," in *Proceedings of the 1989 Bipolar Circuits and Technology Meeting*, pp. 288-290 September 18-19 1989.

[5]   F. Gardner, *Phaselock Techniques*, New York: John Wiley & Sons, 1979, pp. 8-14.

[6]   I. Galton, "Higher-order Delta-Sigma Frequency-to-Digital Conversion," in *Proceedings of IEEE International Symposium on Circuits and Systems, pp.* 441-444, May 30 - June 2, 1994.

[7]   I. Galton, "Analog-Input Digital Phase-Locked Loops for Precise Frequency and Phase Demodulation," *Transactions on Circuits and Systems-II: Analog and Digital Signal Processing, vol.* 42, no. 10, pp. 621-630, Oct. 1995.

[8]   M. W. Hauser, "Principles of Oversampling A/D Conversion," *J. Audio Eng. So. vol 39*, no. 1/2, pp 3-26, Jan./ Feb. 1991.

[9]   J. D. H. Alexander, "Clock Recovery from Random Binary Signals," *Electronics Letters,* vol. 11, no. 22, pp. 541-542, Oct. 1975.

[10]   J. Hauenschild, D. Friedrich, J. Herrle, J. Krug, "A Two-Chip Receiver for Short Haul Links up to 3.5Gb/s with PIN-Preamp Module and CDR-DMUX," in *ISSCC Digest of Technical Papers*, pp. 308-309,452, Feb. 1996.

[11]   R. C. Walker, C. Stout and C. Yen, "A 2.488Gb/s Si-Bipolar Clock and Data Recovery IC with Robust Loss of Signal Detection," in *ISSCC Digest of Technical Papers* pp. 246-247,466, Feb. 1997.

[12]   N. Ishihara and Y. Akazawa, "A Monolithic 156 Mb/s Clock and Data Recovery PLL Circuit Using the Sample-and-Hold Technique," *IEEE Journal of Solid-State Circuits,* vol. 29, no. 12, pp. 1566-1571, Dec. 1994.

[13]   D. Chen, and M. O. Baker, "A 1.25 Gb/s, 460mW CMOS Transceiver for Serial Data Communication," in *ISSCC Digest of Technical Papers*, pp. 242- 243,465 Feb. 1997.

[14]   L. DeVito, J. Newton, R. Goughwell, J. Bulzacchelli and F.Benkley, "A 52MHz and 155 MHz Clock-Recovery PLL," in *ISSCC Digest of Technical Papers*, pp. 142-143,306, Feb. 1991.

[15]   J. F. Ewen, A. X. Widmer, M. Soyuer, K. R. Wrenner, B. Parker and H. A. Ainspan, "Single-Chip 1062Mbaud CMOS Transceiver for Serial Data Communication," in *ISSCC Digest of Technical Papers*, pp. 32-33,336, Feb. 1995.

[16]   A. Fiedler, R. Mactaggart, J. Welch and S. Krishnan, "A 1.0625Gbps Transceiver with 2x-Oversampling and Transmit Signal Pre-Emphasis," in *ISSCC Digest of Technical Papers, pp. 238-239,464,* Feb. 1997.

[17]   B. Guo, A. Hsu, Y. Wang and J. Kubinec, "125Mb/s CMOS All-Digital Data Transceiver Using Synchronous Uniform Sampling," in *ISSCC Digest of Technical Papers*, pp. 112-113, Feb. 1994.

[18]   Y. M. Greshishchev, P. Schvan, J. L. Showell, M. Xu, J. J. Ojha and J. E. Rogers, "A Fully Integrated SiGe Receiver IC for 10-Gb/s Data Rate," *IEEE Journal of Solid State Circuits*, vol. 35, no. 12, pp. 1949-1957, Dec. 2000.

[19]   R. Gu, J. M. Tran, H. Lin, A. Yee and M. Izzard, "A 0.5-3.5Gb/s Low-Power Low-Jitter Serial Data CMOS Transceiver," in *ISSCC Digest of Technical Papers*, pp. 352-353,478, Feb. 1999.

[20]   J. Hauenschild, C. Dorshcky, T. W. Mohrenfels and R. Seitz, "A 10Gb/s BiCMOS Clock and Data Recovery 1:4-Demultiplexer in a Standard Plastic Package with External VCO," in *ISSCC Digest of Technical Papers*, pp. 202-203,445, Feb. 1996.

[21]   T. He, and P. Gray, "A Monolithic 480 Mb/s AGC/Decision/Clock Recovery Circuit in 1.2 um CMOS," *IEEE Journal of Solid State Circuits*, vol. 28, no. 12, pp. 1314-1320, Dec. 1993.

[22]   P. Larsson, "A 2-1600MHz 1.2-2.5V CMOS Clock-Recovery PLL with Feedback Phase-Selection and Averaging Phase-Interpolation for Jitter Reduction," in *ISSCC Digest of Technical Papers*, pp. 356-357, Feb. 1999.

[23] B. Lai, and R. C. Walker, "A Monolithic 622Mb/s Clock Extraction Data Retiming Circuit," in *ISSCC Digest of Technical Papers,* pp. 144,145, Feb. 1991.

[24] T. H. Lee, and J. F. Bulzacchelli, "A 155MHz Clock Recovery Delay- and Phase-Locked Loop," *IEEE Journal of Solid State Circuits* vol. 27, no. 12, pp. 1736-1746, Dec. 1992.

[25] R. H. Leonowich, and J. M. Steininger, "A 45-MHz CMOS phase/frequency-locked loop timing recovery circuit," in *ISSCC Digest of Technical Papers*, pp. 14-15,278-279, Feb. 1988.

[26] I. Lee, C. Yoo, W. Kim, S. Chai and W. Song, "A 622Mb/s CMOS Clock Recovery PLL with Time- Interleaved Phase Detector Array," in *ISSCC Digest of Technical Papers*, pp. 198-199,444, Feb. 1996.

[27] M. Meghelli, B. Parker, H. Ainspan and M. Soyuer, "A SiGe BiCMOS 3.3V Clock and Data Recovery Circuit for 10Gb/s Serial Transmission Systems," in *ISSCC Digest of Technical Papers*, pp. 56-57, Feb. 2000.

[28] T. Morikawa, M. Soda, S. Shiori, T. Hashimoto, F. Sato and K. Emura, "A SiGe Single-Chip 3.3V Receiver IC for 10Gb/s Optical Communication System," in *ISSCC Digest of Technical Papers*, pp. 380-381,481, Feb. 1999.

[29] A. Pottbacker, and U. Langmann, "An 8GHz Silicon Bipolar Clock-Recovery and Data-Regenerator IC," *IEEE Journal of Solid State Circuits* vol. 29, no. 12, pp. 1572-1576, Dec. 1994.

[30] M. Reinhold, C. Dorschky, F. Pullela, E. Rose, P. Mayer, P. Paschke, Y. Baeyens, J. Mattia and F. Kunz, "A Fully-Integrated 40Gb/s Clock and Data Recovery / 1:4 DEMUX IC in SiGe Technology," in *ISSCC Digest of Technical Papers,* pp. 84-85,435, Feb. 2001.

[31] M. Soyuer, and H. A. Ainspan, "A Monolithic 2.3 Gb/s 100mW Clock and Data Recovery Circuit," in *ISSCC Digest of Technical Papers*, pp. 158-159,282, Feb. 1993.

[32] S. Ueno, K. Watanabe, T. Kato, T. Shinohara, K. Mikami, T. Hashimoto, A. Takai, K. Washio, R. Takeyar and T. Harada, "A Single-Chip 10Gb/s Transceiver LSI using SiGe SOI/BiCMOS," in *ISSCC Digest of Technical Papers,* pp. 82-83,435, Feb. 2001.

[33] H. Wang, and R. Nottenburg, "A 1Gb/s CMOS Clock and Data Recovery Circuit," in *ISSCC Digest of Technical Papers*, pp. 354-355,477, Feb. 1999.

[34] P. Wallace, R. Bayruns, J. Smith, T. Laverick and R. Shuster, "A GaAs 1.5Gb/s Clock Recovery and Data Retiming Circuit," in *ISSCC Digest of Technical Papers*, pp. 192-193, Feb. 1990.

[35] Z. Wang, M. Berroth, J. Seibel, P. Hofmann, A. Hulsmann, Kohler, B. Raynor and J. Schneider, "19GHz Monolithic Integrated Clock Recovery Using PLL and 0.3um Gate-Length Quantum-Well HEMTs," in *ISSCC Digest of Technical Papers*, pp. 118-119, Feb. 1994,

[36] R. C. Walker, K. Hsieh, T. A. Knotts and C. Yen, "A 10Gb/s Si-Bipolar TX/RX Chipset for Computer Data Transmission," in *ISSCC Digest of Technical Papers,* pp. 302-303,450, Feb. 1998.

[37] R. C. Walker, J. Wu, C. Stout, B. Lai, C. Yen, T. Hornak and P. Petruno, "A 2-Chip 1.5Gb/s Bus-Oriented Serial Link Interface," in *ISSCC Digest of Technical Papers,* pp. 226-227,291, Feb. 1992.

[38] C. K. Yang, and M. A. Horowitz, "0.8um CMOS 2.5Gb/s Oversampled Receiver for Serial Links," *IEEE Journal of Solid State Circuits* vol. 31, no. 12, pp. 20150-2023, Dec. 1996.

Richard Walker was born in San Rafael CA, in 1960. He received the B.S. degree in Engineering and Applied Science from the California Institute of Technology in 1982, and an M.S. degree in Computer Science from California State University, Chico, CA in 1992. Rick joined Agilent Laboratories (formerly Hewlett-Packard Laboratories) in 1981, where he is currently a Principal Project Engineer. Since that time, he has worked in the areas of broadband-cable modem design, solid-state laser characterization, phase-locked-loop theory, linecode design, and gigabit-rate serial data transmission. He holds 15 U.S. patents.

Predicting the Phase Noise and Jitter of PLL-Based Frequency Synthesizers

Kenneth S. Kundert

Abstract — **Two methodologies are presented for predicting the phase noise and jitter of a PLL-based frequency synthesizer using simulation that are both accurate and efficient. The methodologies begin by characterizing the noise behavior of the blocks that make up the PLL using transistor-level RF simulation. For each block, the phase noise or jitter is extracted and applied to a model for the entire PLL.**

I. INTRODUCTION

Phase-locked loops (PLLs) are used to implement a variety of timing related functions, such as frequency synthesis, clock and data recovery, and clock de-skewing. Any jitter or phase noise in the output of the PLL used in these applications generally degrades the performance margins of the system in which it resides and so is of great concern to the designers of such systems. Jitter and phase noise are different ways of referring to an undesired variation in the timing of events at the output of the PLL. They are difficult to predict with traditional circuit simulators because the PLL generates repetitive switching events as an essential part of its operation, and the noise performance must be evaluated in the presence of this large-signal behavior. SPICE is useless in this situation as it can only predict the noise in circuits that have a quiescent (time-invariant) operating point. In PLLs the operating point is at best periodic, and is sometimes chaotic. Recently a new class of circuit simulators has been introduced that are capable of predicting the noise behavior about a periodic operating point [1]. SpectreRF is the most popular of this class of simulators and, because of the algorithms used in its implementation, is likely to be the best suited for this application [2]. These simulators can be used to predict the noise performance of PLLs. The ideas presented in this paper allow those simulators to be applied even to those PLLs that have chaotic operating points.

A. Frequency Synthesis

The focus of this paper is frequency synthesis. The block diagram of a PLL operating as a frequency synthesizer is shown in Figure 1 [3]. It consists of a reference oscillator (OSC), a phase/frequency detector (PFD), a charge pump (CP), a loop filter (LF), a voltage-controlled oscillator (VCO), and two

Ken Kundert is with Cadence Design Systems, San Jose, California, *kundert@cadence.com*.

frequency dividers (FDs). The PLL is a feedback loop that, when in lock, forces f_{fb} to be equal to f_{ref}. Given an input frequency f_{in}, the frequency at the output of the PLL is

$$f_{out} = \frac{N}{M} f_{in} \qquad (1)$$

where M is the divide ratio of the input frequency divider, and N is the divide ratio of the feedback divider. By choosing the frequency divide ratios and the input frequency appropriately, the synthesizer generates an output signal at the desired frequency that inherits much of the stability of the input oscillator. In RF transceivers, this architecture is commonly used to generate the local oscillator (LO) at a programmable frequency that tunes the transceiver to the desired channel by adjusting the value of N.

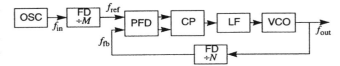

Fig. 1. The block diagram of a frequency synthesizer.

B. Direct Simulation

In many circumstances, SpectreRF[†] can be directly applied to predict the noise performance of a PLL. To make this possible, the PLL must at a minimum have a periodic steady state solution. This rules out systems such as bang-bang clock and data recovery circuits and fractional-N synthesizers because they behave in a chaotic way by design. It also rules out any PLL that is implemented with a phase detector that has a dead zone. A dead zone has the effect of opening the loop and letting the phase drift seemingly at random when the phase of the reference and the output of the voltage-controlled oscillator (VCO) are close. This gives these PLLs a chaotic nature.

To perform a noise analysis, SpectreRF must first compute the steady-state solution of the circuit with its periodic steady state (PSS) analysis. If the PLL does not have a periodic solution, as the cases described above do not, then it will not converge. There is an easy test that can be run to determine if a circuit has a periodic steady-state solution. Simply perform a transient analysis until the PLL approaches steady state and

---

† Spectre is a registered trademark of Cadence Design Systems.

then observe the VCO control voltage. If this signal consists of frequency components at integer multiples of the reference frequency, then the PLL has a periodic solution. If there are other components, it does not. Sometimes it can be difficult to identify the undesirable components if the components associated with the reference frequency are large. In this case, use the strobing feature of Spectre's transient analysis to eliminate all components at frequencies that are multiples of the reference frequency. Do so by strobing at the reference frequency. In this case, if the VCO control voltage varies in any significant way the PLL does not have a periodic solution.

If the PLL has a periodic solution, then in concept it is always possible to apply SpectreRF directly to perform a noise analysis. However, in some cases it may not be practical to do so. The time required for SpectreRF to compute the noise of a PLL is proportional to the number of circuit equations needed to represent the PLL in the simulator times the number of time points needed to accurately render a single period of the solution times the number of frequencies at which the noise is desired. When applying SpectreRF to frequency synthesizers with large divide ratios, the number of time points needed to render a period can become problematic. Experience shows that divide ratios greater than ten are often not practical to simulate. Of course, this varies with the size of the PLL.

For PLLs that are candidates for direct simulation using SpectreRF, simply configure the simulator to perform a PSS analysis followed by a periodic noise (PNoise) analysis. The period of the PSS analysis should be set to be the same as the reference frequency as defined in Figure 1. The PSS stabilization time (tstab) should be set long enough to allow the PLL to reach lock. This process was successfully followed on a frequency synthesizer with a divide ratio of 40 that contained 2500 transistors, though it required several hours for the complete simulation [4].

C. When Direct Simulation Fails

The challenge still remains, how does one predict the phase noise and jitter of PLLs that do not fit the constraints that enable direct simulation? The remainder of this paper attempts to answer that question for frequency synthesizers, though the techniques presented are general and can be applied to other types of PLLs by anyone who is sufficiently determined.

D. Monte Carlo-Based Methods

Demir proposed an approach for simulating PLLs whereby a PLL is described using behavioral models simulated at a high level [5, 6]. The models are written such that they include jitter in an efficient way. He also devised a simulation algorithm based on solving a set of nonlinear stochastic differential equations that is capable of characterizing the circuit-level noise behavior of blocks that make up a PLL [6, 7]. Finally, he gave formulas that can be used to convert the results of the noise simulations on the individual blocks into values for the jitter parameters for the corresponding behavioral models [8]. Once everything is ready, simulation of the PLL occurs with the blocks of the PLL being described with behavioral models that exhibit jitter. The actual jitter or phase noise statistics are observed during this simulation. Generally tens to hundreds of thousands of cycles are simulated, but the models are efficient so the time required for the simulation is reasonable. This approach allows prediction of PLL jitter behavior once the noise behavior of the blocks has been characterized. However, it requires the use of an experimental simulator that is not readily available to characterize the jitter of the blocks.

In an earlier series of papers [9, 10], the relevant ideas of Demir were adapted to allow use of a commercial simulator, Spectre [11], and an industry standard modeling language, Verilog-A[†] [12]. These ideas are further refined in the later half of this paper.

E. Predicting Noise in PLLs

There are two different approaches to modeling noise in PLLs. One approach is to formulate the models in terms of the phase of the signals, producing what are referred to as phase-domain models. In the simplest case, these models are linear and analyzed easily in the frequency domain, making it simple to use the model to predict phase noise, even in the presence of flicker noise or other noise sources that are difficult to model in the time domain. Phase-domain models are described in the first half of this paper.

The process of predicting the phase noise of a PLL using phase-domain models involves:

1. Using SpectreRF to predict the noise of the individual blocks that make up the PLL.
2. Building high-level behavioral models of each of the blocks that exhibit phase noise.
3. Assembling the blocks into a model of the PLL.
4. Simulating the PLL to find the phase noise of the overall system.

The other approach formulates the models in terms of voltage, which are referred to as voltage-domain models. The advantage of voltage-domain models is that they can be refined to implementation. In other words, as the design process transitions to being more of a verification process, the abstract behavioral models initially used can be replaced with detailed gate- or transistor-level models in order to verify the PLL as implemented.

A voltage-domain model is strongly nonlinear and never has a quiescent operating point, making it incompatible with a SPICE-like noise analysis. Often such models have a periodic operating point and so can be analyzed with small-signal RF noise analysis (SpectreRF), but it is also common for that not to be the case. For example, a fractional-N synthesizer does

---

† Verilog is a registered trademark of Cadence Design Systems licensed to Accellera.

not have a periodic operating point. Occasionally, the circuit is sensitive enough that the noise affects the large-signal behavior of the PLL, such as with bang-bang clock-and-data recovery PLLs, which invalidates any use of small-signal noise analysis.

Modeling large-signal noise in a voltage-domain model as a voltage or a current is problematic. Such signals are very small and continuously and very rapidly varying. Extremely tight tolerances and small time steps are required to accurately resolve such signals with simulation. To overcome these problems, the noise is instead represented using the effect it has on the timing of the transitions within the PLL. In other words, the noise is added to circuit in the form of jitter. In this case there is no need for either small time steps or tight tolerances.

The process of predicting the jitter of a PLL with voltage-domain models involves:

1. Using SpectreRF to predict the noise of the individual blocks that make up the PLL.
2. Converting the noise of the block to jitter.
3. Building high-level behavioral models of each of the blocks that exhibit jitter.
4. Assembling the blocks into a model of the PLL.
5. Simulating the PLL to find the jitter of the overall system.

The simple linear phase-domain model described in the first part of this paper, and the nonlinear voltage-domain model described in the second part, represent the two ends of a continuum of models. Generally, the phase-domain models are considerably more efficient, but the voltage-domain models do a better job of capturing the details of the behavior of the loop, details such as the signal capture and escape processes. The phase-domain models can be made more general by making them nonlinear and by analyzing them in the time domain. It is common to use such models with fractional-N synthesizers. Conversely, simplifications can be made to the voltage-domain models to make them more efficient. It is even possible to use both voltage- and phase-domain models for different parts of the same loop. One might do so to retain as much efficiency as possible while allowing part of the design to be refined to implementation level. In general it is best to understand both approaches well, and use ideas from both to construct the most appropriate approach for your particular situation.

II. PHASE-DOMAIN MODEL

It is widely understood that simulating PLLs is expensive because the period of the VCO is almost always very short relative to the time required to reach lock. This is particularly true with frequency synthesizers, especially those with large multiplication factors. The problem is that a circuit simulator must use at least 10-20 time points for every period of the VCO for accurate rendering, and the lock process often involves hundreds or thousands of cycles at the input to the phase detector. With large divide ratios, this can translate to hundreds of thousands of cycles of the VCO. Thus, the number of time points needed for a single simulation could range into the millions.

This is all true when simulating the PLL in terms of voltages and currents. When doing so, one is said to be using *voltage-domain models*. However, that is not the only option available. It is also possible to formulate models based on the phase of the signals. In this case, one would be using *phase-domain models*. The high frequency variations associated with the voltage-domain models are not present in phase-domain models, and so simulations are considerably faster. In addition, when in lock the phase-domain-based models generally have constant-valued operating points, which simplifies small-signal analysis, making it easier to study the closed-loop dynamics and noise performance of the PLL using either AC or noise analysis.

A linear phase-domain model of a frequency synthesizer is shown in Figure 2. Such a model is suitable for modeling the behavior of the PLL to small perturbations when the PLL is in lock as long as you do not need to know the exact waveforms and instead are interested in how small perturbations affect the phase of the output. This is exactly what is needed to predict the phase noise performance of the PLL.

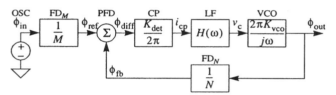

Fig. 2. Linear time-invariant phase-domain model of the synthesizer shown in Figure 1.

The derivation of the model begins with the identification of those signals that are best represented by their phase. Many blocks have large repetitive input signals with their outputs being primarily sensitive to the phase of their inputs. It is the signals that drive these blocks that are represented as phase. They are identified using a ϕ variable in Figure 2. Notice that this includes all signals except those at the inputs of the LF and VCO.

The models of the individual blocks will be derived by assuming that the signals associated with each of the phase variables is a pulse train. Though generally the case, it is not a requirement. It simply serves to make it easier to extract the models. Define $\Pi(t_0, \tau, T)$ to be a periodic pulse train where one of the pulses starts at t_0 and the pulses have duration τ and period T as shown in Figure 3. This signal transitions between 0 and 1 if τ is positive, and between 0 and -1 if τ is negative. The phase of this signal is defined to be $\phi = 2\pi t_0/T$. In many cases, the duration of the pulses is of no interest, in which case $\Pi(t_0, T)$ is used as a short hand. This occurs because the input that the signal is driving is edge triggered. For simplicity, we assume that such inputs are sensitive to the rising edges of the signal, that t_0 specifies the time of a rising edge, and that the signal is transitioning between 0 and 1.

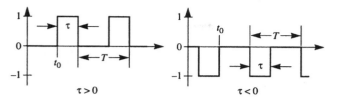

Fig. 3. The pulse train waveform represented by $\Pi(t_0, \tau, T)$.

The input source produces a signal $v_{in} = \Pi(t_0, T)$. Since this is the input, t_0 is arbitrary. As such, we are free to set its phase ϕ to any value we like.

Given a signal $v_i = \Pi(t_0, T)$ a frequency divider will produce an output signal $v_o = \Pi(t_0, NT)$ where N is the divide ratio. The phase of the input is $\phi_i = 2\pi t_0/T$ and the phase of the output is $\phi_o = 2\pi t_0/(NT)$ and so the phase transfer characteristic of a divider is

$$\phi_o = \phi_i/N. \qquad (2)$$

There are many different types of phase detectors that can be used, each requiring a somewhat different model. Consider a simple phase-frequency detector combined with a charge pump [13]. In this case, the detector takes two inputs, $v_1 = \Pi(t_1, T)$ and $v_0 = \Pi(t_0, T)$ and produces an output $i_{cp} = I_{max}\Pi(t_0, t_1-t_0, T)$ where I_{max} is the maximum output current of the charge pump. The output of the charge pump immediately passes through a low pass filter that is designed to suppress signals at frequencies of $1/T$ and above, so in most cases the pulse nature of this signal can be ignored in favor of its average value, $\langle i_{cp} \rangle$. Thus, the transfer characteristic of the combined PFD/CP is

$$\langle i_{cp} \rangle = I_{max}\frac{t_1-t_0}{T} = I_{max}\frac{\phi_1-\phi_0}{2\pi} = \frac{K_{det}}{2\pi}(\phi_1 - \phi_0) \quad (3)$$

where $K_{det} = I_{max}$. Of course, this is only valid for $|\phi_1 - \phi_2| < 2\pi$ at the most. The behavior outside this range depends strongly on the type of phase detector used [3]. Even within this range, the phase detector may be better modeled with a nonlinear transfer characteristic. For example, there can be a flat spot in the transfer characteristics near 0 if the detector has a dead zone. However it is generally not productive to model the dead zone in a phase-domain model.[†]

---

† This phase-domain model is a continuous-time model that ignores the sampling nature of the PFD. A dead zone interacts with the sampling nature of the PFD to create a chaotic limit cycle behavior that is not modeled with the phase-domain model. This chaotic behavior creates a substantial amount of jitter, and for this reason, most modern phase detectors are designed such that they do not exhibit dead zones.

The model of (3) is a continuous-time approximation to what is inherently a discrete-time process. The phase detector does not continuously monitor the phase difference between its two input signals, rather it outputs one pulse per cycle whose width is proportional to the phase difference. Using a continuous time approximation is generally acceptable if the bandwidth of the loop filter is much less than f_{ref} (generally less than $f_{ref}/10$ is sufficient). In practical PLLs this is almost always the case. It is possible to develop a detailed phase-domain PFD model that includes the discrete-time effects, but it would run more slowly and the resulting phase-domain model of the PLL would not have a quiescent operating point, which makes it more difficult to analyze.

The voltage-controlled oscillator, or VCO, converts its input voltage to an output frequency, and the relationship between input voltage and output frequency can be represented as

$$f_{out} = F(v_c) \qquad (4)$$

The mapping from voltage to frequency is designed to be linear, so a first-order model is often sufficient,

$$f_{out} = K_{vco}v_c. \qquad (5)$$

It is the output phase that is needed in a phase-domain model,

$$\phi_{out}(t) = 2\pi \int K_{vco}v_c(t)dt \qquad (6)$$

or in the frequency domain,

$$\phi_{out}(\omega) = \frac{2\pi K_{vco}}{j\omega}v_c(\omega). \qquad (7)$$

A. Small-Signal Stability

This completes the derivation of the phase-domain models for each of the blocks. Now the full model is used to help predict the small-signal behavior of the PLL. Start by using Figure 2 to write a relationship for its loop gain. Start by defining

$$G_{fwd} = \frac{\phi_{out}}{\phi_{diff}} = \frac{K_{det}}{2\pi}H(\omega)\frac{2\pi K_{vco}}{j\omega} = \frac{K_{det}K_{vco}H(\omega)}{j\omega} \quad (8)$$

to be the forward gain,

$$G_{rev} = \frac{\phi_{fb}}{\phi_{out}} = \frac{1}{N} \qquad (9)$$

to be the feedback factor, and

$$T = G_{fwd}G_{rev} = \frac{K_{det}K_{vco}H(\omega)}{j\omega N} \qquad (10)$$

to be the loop gain. The loop gain is used to explore the small-signal stability of the loop. In particular, the phase margin is an important stability metric. It is the negative of the difference between the phase shift of the loop at unity gain and 180°, the phase shift that makes the loop unstable. It should be no less than 45° [14]. When concerned about phase noise or jitter, the phase margin is typically 60° or more to reduce peaking in the closed-loop gain, which results in excess phase noise.

B. Noise Transfer Functions

In Figure 4 various sources of noise have been added. These noise sources can represent either the noise created by the blocks due to intrinsic noise sources (thermal, shot, and flicker noise sources), or the noise coupled into the blocks from external sources, such as from the power supplies, the substrate, etc. Most are sources of phase noise, and denoted

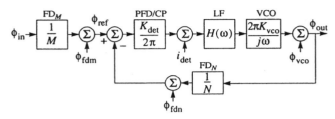

Fig. 4. Linear time-invariant phase-domain model of the synthesizer shown in Figure 2 with representative noise sources added. The ϕ's represent various sources of noise.

ϕ_{in}, ϕ_{fdm}, ϕ_{fdn}, and ϕ_{vco}, because the circuit is only sensitive to phase at the point where the noise is injected. The one exception is the noise produced by the PFD/CP, which in this case is considered to be a current, and denoted i_{det}.

Then the transfer functions from the various noise sources to the output are

$$G_{ref} = \frac{\phi_{out}}{\phi_{ref}} = \frac{G_{fwd}}{1-T} = \frac{NG_{fwd}}{N - G_{fwd}}, \qquad (11)$$

$$G_{vco} = \frac{\phi_{out}}{\phi_{vco}} = \frac{1}{1-T} = \frac{N}{N - G_{fwd}}, \qquad (12)$$

$$G_{in} = \frac{\phi_{out}}{\phi_{in}} = \frac{1}{M}\frac{G_{fwd}}{1-T} = \frac{1}{M}\frac{N}{N - G_{fwd}}, \qquad (13)$$

and by inspection,

$$G_{fdn} = \frac{\phi_{out}}{\phi_{fdn}} = -G_{ref}, \qquad (14)$$

$$G_{fdm} = \frac{\phi_{out}}{\phi_{fdm}} = -G_{ref}, \qquad (15)$$

$$G_{det} = \frac{\phi_{out}}{i_{det}} = \frac{2\pi G_{ref}}{K_{det}}. \qquad (16)$$

On this last transfer function, we have simply referred i_{det} to the input by dividing through by the gain of the phase detector.

These transfer functions allow certain overall characteristics of phase noise in PLLs to be identified. As $\omega \to \infty$, $G_{fwd} \to 0$ because of the VCO and the low-pass filter, and so G_{ref}, G_{det}, G_{fdm}, G_{fdn}, $G_{in} \to 0$ and $G_{vco} \to 1$. At high frequencies, the noise of the PLL is that of the VCO. Clearly this must be so because the low-pass LF blocks any feedback at high frequencies.

As $\omega \to 0$, $G_{fwd} \to \infty$ because of the $1/(j\omega)$ term from the VCO. So at DC, G_{ref}, G_{fdm}, $G_{fdn} \to N$, $G_{in} \to N/M$ and $G_{vco} \to 0$. At low frequencies, the noise of the PLL is contributed by the OSC, PFD/CP, FD$_M$ and FD$_N$, and the noise from the VCO is diminished by the gain of the loop.

Consider further the asymptotic behavior of the loop and the VCO noise at low offset frequencies ($\omega \to 0$). Oscillator phase noise in the VCO results in the power spectral density $S_{\phi vco}$ being proportional to $1/\omega^2$, or $S_{\phi vco} \sim 1/\omega^2$ (neglecting flicker noise). If the LF is chosen such that $H(\omega) \sim 1$, then $G_{fwd} \sim 1/\omega$, and contribution from the VCO to the output noise power, $G_{vco}^2 S_{\phi vco}$, is finite and nonzero. If the LF is chosen such that $H(\omega) \sim 1/\omega$, as it typically is when a true charge pump is employed, then $G_{fwd} \sim 1/\omega^2$ and the noise contribution to the output from the VCO goes to zero at low frequencies.

C. Noise Model

One predicts the phase noise exhibited by a PLL by building and applying the model shown in Figure 4. The first step in doing so is to find the various model parameters, including the level of the noise sources, which generally involves either direct measurement or simulating the various blocks with an RF simulator, such as SpectreRF. Use periodic noise (or PNoise) analysis to predict the output noise that results from stochastic noise sources contained within the blocks using simulation. Use a periodic AC or periodic transfer function (PAC or PXF) to compute the perturbation at the output of a block due to noise sources outside the block, such as on supplies.

Once the model parameters are known, it is simply a matter of computing the output phase noise of the PLL by applying the equations in Section II-B to compute the contributions to ϕ_{out} from every source and summing the results. Be careful to account for correlations in the noise sources. If the noise sources are perfectly correlated, as they might be if the ultimate source of noise is in the supplies or substrate, then use a direct sum. If the sources produce completely uncorrelated noise, as they would when the ultimate source of noise is random processes within the devices, use a root-mean-square sum.

Alternatively, one could build a Verilog-A model and use simulation to determine the result. The top-level of such a model is shown in Listing 1. It employs noisy phase-domain models for each of the blocks. These models are given in Listings 3-7 and are described in detail in the next few sections (III-VI). In this example, the noise sources are coded into the models, but the noise parameters are not set at the top level to simplify the model. To predict the phase noise performance of the loop in lock, simply specify these parameters in Listing 1 and perform a noise analysis. To determine the effect of injected noise, first refer the noise to the output of one of the blocks, and then add a source into the netlist of Listing 1 at the appropriate place and perform an AC analysis.

Listing 1 — Phase-domain model for a PLL configured as a frequency synthesizer.

```
`include "discipline.h"

module pll(out);
output out;
phase out;
parameter integer m = 1 from [1:inf];    // input divide ratio
parameter real Kdet = 1 from (0:inf);    // detector gain
parameter real Kvco = 1 from (0:inf);    // VCO gain
parameter real c1 = 1n from (0:inf);     // Loop filter C1
parameter real c2 = 200p from (0:inf);   // Loop filter C2
parameter real r = 10K from (0:inf);     // Loop filter R
parameter integer n = 1 from [1:inf];    // fb divide ratio
phase in, ref, fb;
electrical c;

oscillator OSC(in);
divider #(.ratio(m)) FDm(in, ref);
phaseDetector #(.gain(Kdet)) PD(ref, fb, c);
loopFilter #(.c1(c1), .c2(c2), .r(r)) LF(c);
vco #(.gain(Kvco)) VCO(c, out);
divider #(.ratio(n)) FDn(out, fb);

endmodule
```

Listings 1 and 3-7 have phase signals, and there is no phase discipline in the standard set of disciplines provided by Verilog-A or Verilog-AMS in *discipline.h*. There are several different resolutions for this problem. Probably the best solution is to simply add such a discipline, given in Listing 2, either to *discipline.h* as assumed here or to a separate file that is included as needed. Alternatively, one could use the *rotational* discipline. It is a conservative discipline that includes torque as a flow nature, and so is overkill in this situation. Finally, one could simply use either the electrical or the voltage discipline. Scaling for voltage in volts and phase in radians is similar, and so it will work fine except that the units will be reported incorrectly. Using the rotational discipline would require that all references to the phase discipline be changed to rotational in the appropriate listings. Using either the electrical or voltage discipline would require that both the name of the disciplines be changed from phase to either electrical or voltage, and the name of the access functions be changed from *Theta* to *V*.

Listing 2 — Signal flow discipline definition for phase signals (the nature *Angle* is defined in *discipline.h*).

```
`include "discipline.h"

discipline phase
    potential Angle;
enddiscipline
```

III. OSCILLATORS

Oscillators are responsible for most of the noise at the output of the majority of well-designed frequency synthesizers. This is because oscillators inherently tend to amplify noise found near their oscillation frequency and any of its harmonics. The reason for this behavior is covered next, followed by a description of how to characterize and model the noise in an oscillator. The origins of oscillator phase noise are described in a conceptual way here. For a detailed description, see the papers by Käertner or Demir et al [15, 16, 17].

A. Oscillator Phase Noise

Nonlinear oscillators naturally produce high levels of phase noise. To see why, consider the trajectory of a fully autonomous oscillator's stable periodic orbit in state space. In steady state, the trajectory is a stable limit cycle, v. Now consider perturbing the oscillator with an impulse and assume that the deviation in the response due to the perturbation is Δv, as shown in Figure 5. Separate Δv into amplitude and phase variations,

$$\Delta v(t) = [1 + \alpha(t)]v\left(t + \frac{\phi(t)}{2\pi f_o}\right) - v(t). \quad (17)$$

where v represents the unperturbed T-periodic output voltage of the oscillator, α represents the variation in amplitude, ϕ is the variation in phase, and $f_o = 1/T$ is the oscillation frequency.

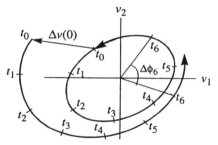

Fig. 5. The trajectory of an oscillator shown in state space with and without a perturbation Δv. By observing the time stamps (t_0, ..., t_6) one can see that the deviation in amplitude dissipates while the deviation in phase does not.

Since the oscillation is stable and the duration of the disturbance is finite, the deviation in amplitude eventually decays away and the oscillator returns to its stable orbit ($\alpha(t) \to 0$ as $t \to \infty$). In effect, there is a restoring force that tends to act against amplitude noise. This restoring force is a natural consequence of the nonlinear nature of the oscillator that acts to suppresses amplitude variations.

The oscillator is autonomous, and so any time-shifted version of the solution is also a solution. Once the phase has shifted due to a perturbation, the oscillator continues on as if never disturbed except for the shift in the phase of the oscillation. There is no restoring force on the phase and so phase deviations accumulate. A single perturbation causes the phase to permanently shift ($\phi(t) \to \Delta\phi$ as $t \to \infty$). If we neglect any short term time constants, it can be inferred that the impulse response of the phase deviation $\phi(t)$ can be approximated with a unit step $s(t)$. The phase shift over time for an arbitrary input disturbance u is

$$\phi(t) \sim \int_{-\infty}^{\infty} s(t - \tau)u(\tau)d\tau = \int_{-\infty}^{t} u(\tau)d\tau, \qquad (18)$$

or the power spectral density (PSD) of the phase is

$$S_{\phi}(\Delta f) \sim \frac{S_u(\Delta f)}{(2\pi\Delta f)^2} \qquad (19)$$

This shows that in all oscillators the response to any form of perturbation, including noise, is amplified and appears mainly in the phase. The amplification increases as the frequency of the perturbation approaches the frequency of oscillation in proportion to $1/\Delta f$ (or $1/\Delta f^2$ in power).

Notice that there is only one degree of freedom — the phase of the oscillator as a whole. There is no restoring force when the phase of all signals associated with the oscillator shift together, however there would be a restoring force if the phase of signals shifted relative to each other. This observation is significant in oscillators with multiple outputs, such as quadrature or ring oscillators. The dominant phase variations appear identically in all outputs, whereas relative phase variations between the outputs are naturally suppressed by the oscillator or added by subsequent circuitry and so tend to be much smaller [8].

B. Characterizing Oscillator Phase Noise

Above it was shown that oscillators tend to convert perturbations from any source into a phase variation at their output whose magnitude varies with $1/\Delta f$ (or $1/\Delta f^2$ in power). Now assume that the perturbation is from device noise in the form of white and flicker stochastic processes. The oscillator's response will be characterized first in terms of the phase noise S_{ϕ}, and then because phase noise is not easily measured, in terms of the normalized voltage noise $\mathcal{L}$. The result will be a small set of easily extracted parameters that completely describe the response of the oscillator to white and flicker noise sources. These parameters are used when modeling the oscillator.

Assume that the perturbation consists of white and flicker noise and so has the form

$$S_u(\Delta f) \sim 1 + \frac{f_c}{\Delta f}. \qquad (20)$$

Then from (19) the response will take the form

$$S_{\phi}(\Delta f) = n\left(\frac{1}{\Delta f^2} + \frac{f_c}{\Delta f^3}\right), \qquad (21)$$

where the factor of $(2\pi)^2$ in the denominator of (19) has been absorbed into n, the constant of proportionality. Thus, the response of the oscillator to white and flicker noise sources is characterized using just two parameters, n and f_c, where n is the portion of S_{ϕ} attributable to the white noise sources alone

at $\Delta f = 1$ Hz and f_c is the flicker noise corner frequency. As shown in Figure 6, n is extracted by simply extrapolating to 1 Hz from a frequency where the noise from the white sources dominates.

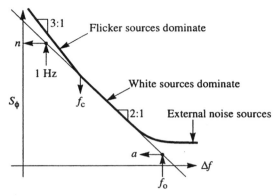

Fig. 6. Extracting the noise parameters, n, a, and f_c, for an oscillator. The parameter a is an alternative to n where $n = af_o^2$. It is used later. The graph is plotted on a log-log scale.

S_{ϕ} is not directly observable and often difficult to find, so now S_{ϕ} is related to $\mathcal{L}$, the power spectral density of the output voltage noise S_v normalized by the power in the fundamental tone. S_v is directly available from either measurement with a spectrum analyzer or from RF simulators, and $\mathcal{L}$ is defined as

$$\mathcal{L}(\Delta f) = \frac{S_v(f_o + \Delta f)}{2|V_1|^2}, \qquad (22)$$

where V_1 is the fundamental Fourier coefficient of v, the output signal. It satisfies

$$v(t) = \sum_{k=-\infty}^{\infty} V_k e^{j2\pi k f_o t}. \qquad (23)$$

In (41) of [15], Demir et al shows that for a free-running oscillator perturbed only by white noise sources[†]

$$\mathcal{L}(\Delta f) = \frac{1}{2}\frac{n}{n^2\pi^2 + \Delta f^2}, \qquad (24)$$

which is a Lorentzian process with corner frequency of

$$f_{corner} = n\pi \ll f_o. \qquad (25)$$

At frequencies above the corner,

$$\mathcal{L}(\Delta f) = \frac{n}{2\Delta f^2} = \frac{1}{2}S_{\phi}(\Delta f), \qquad (26)$$

which agrees with Vendelin [18].

Use (21) to extract f_c. Then use both (21) and (26) to determine n by choosing Δf well above the flicker noise corner frequency, f_c, and the corner frequency of (25), f_{corner}, to avoid ambiguity and well below f_o to avoid the noise from other sources that occur at these frequencies.

---

† Demir uses c rather than n, where $n = cf_o^2$.

C. Phase-Domain Models for the Oscillators

The phase-domain models for the reference and voltage-controlled oscillators are given in Listings 3 and 4. The VCO model is based on (6). Perhaps the only thing that needs to be explained is the way that phase noise is modeled in the oscillators. Verilog-AMS provides the *flicker_noise* function for modeling flicker noise, which has a power spectral density proportional to $1/f^{\alpha}$ with α typically being close to 1. However, Verilog-AMS does not limit α to being close to one, making this function well suited to modeling oscillator phase noise, for which α is 2 in the white-phase noise region and close to 3 in the flicker-phase noise region (at frequencies below the flicker noise corner frequency). Alternatively, one could dispense with the noise parameters and use the *noise_table* function in lieu of the *flicker_noise* functions to use the measured noise results directly. The "wpn" and "fpn"

Listing 3 — Phase-domain oscillator noise model.

```
`include "discipline.h"

module oscillator(out);
output out;
phase out;
parameter real n = 0 from [0:inf);
    // white output phase noise at 1 Hz (rad²/Hz)
parameter real fc = 0 from [0:inf);
    // flicker noise corner frequency (Hz)

analog begin
    Theta(out) <+ flicker_noise(n, 2, "wpn")
                 + flicker_noise(n*fc, 3, "fpn");
end
endmodule
```

Listing 4 — Phase-domain VCO noise model.

```
`include "discipline.h"
`include "constants.h"

module vco(in, out);
input in; output out;
voltage in;
phase out;
parameter real gain = 1 from (0:inf);
    // transfer gain, Kvco (Hz/V)
parameter real n = 0 from [0:inf);
    // white output phase noise at 1 Hz (rad²/Hz)
parameter real fc = 0 from [0:inf);
    // flicker noise corner frequency (Hz)

analog begin
    Theta(out) <+ 2*`M_PI*gain*idt(V(in));
    Theta(out) <+ flicker_noise(n, 2, "wpn")
                 + flicker_noise(n*fc, 3, "fpn");
end
endmodule
```

---

strings passed to the noise functions are labels for the noise sources. They are optional and can be chosen arbitrarily,

though they should not contain any white space. *wpn* was chosen to represent white phase noise and *fpn* stands for flicker phase noise.

When interested in the effect of signals coupled into the oscillator through the supplies or the substrate, one would compute the transfer function from the interfering source to the phase output of the oscillator using either a PAC or PXF analysis. Again, one would simply assume that the perturbation in the output of the oscillator is completely in the phase, which is true except at very high offset frequencies. One then employs (12) and (13) to predict the response at the output of the PLL.

IV. LOOP FILTER

Even in the phase-domain model for the PLL, the loop filter remains in the voltage domain and is represented with a full circuit-level model, as shown in Listing 5. As such, the noise behavior of the filter is naturally included in the phase-domain model without any special effort assuming that the noise is properly included in the resistor model.

Listing 5 — Loop filter model.

```
`include "discipline.h"

module loopFilter(n);
electrical n;
ground gnd;†
parameter real c1 = 1n from (0:inf);
parameter real c2 = 200p from (0:inf);
parameter real r = 10K from (0:inf);
electrical int;

capacitor #(.c(c1)) C1(n, gnd);
capacitor #(.c(c2)) C2(n, int);
resistor #(.r(r)) R(int, gnd);

endmodule
```

---

† The *ground* statement is not currently supported in Cadence's Verilog-A implementation, so instead ground is explicitly passed into the module.

V. PHASE DETECTOR AND CHARGE PUMP

As with the VCO, the noise of the PFD/CP as needed by the phase-domain model is found directly with simulation. Simply drive the block with a representative periodic signal, perform a PNoise analysis, and measure the output noise current. In this case, a representative signal would be one that produced periodic switching at the output. This is necessary to capture the noise present during the switching process. Generally the noise appears as in Figure 7, in which case the noise is parameterized with n and f_c. n is the noise power density at frequencies above the flicker noise corner frequency, f_c, and below the noise bandwidth of the circuit.

The phase-domain model for the PFD/CP is given in Listing 6. It is based on (3). Alternatively, as before one could

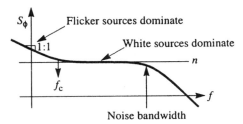

Fig. 7. Extracting the noise parameters, n and f_c, for the PFD/CP. The graph is plotted on a log-log scale.

use the *noise_table* function in lieu of the *white_noise* and *flicker_noise* functions to use the measured noise results directly.

Listing 6 — Phase-domain phase detector noise model.

```
`include "discipline.h"
`include "constants.h"

module phaseDetector(pin, nin, out);
input pin, nin; output out;
phase pin, nin;
electrical out;
parameter real gain = 1 from (0:inf);
    // transfer gain (A/cycle)
parameter real n = 0 from [0:inf);
    // white output current noise (A²/Hz)
parameter real fc = 0 from [0:inf);
    // flicker noise corner frequency (Hz)

analog begin
    I(out) <+ gain * Theta(pin,nin) / (2*`M_PI);
    I(out) <+ white_noise(n, "wpn")
             + flicker_noise(n*fc, 1, "fpn");
end
endmodule
```

VI. FREQUENCY DIVIDERS

There are several reasons why the process of extracting the noise produced by the frequency dividers is more complicated than that needed for other blocks. First, the phase noise is needed and, as of the time when this document was written, SpectreRF reports on the total noise and does not yet make the phase noise available separately. Secondly, the frequency dividers are always followed by some form of edge-sensitive thresholding circuit, in this case the PFD, which implies that the overall noise behavior of the PLL is only influenced by the noise produced by the divider at the time when the threshold is being crossed in the proper direction. The noise produced by the frequency divider is cyclostationary, meaning that the noise power varies over time. Thus, it is important to analyze the noise behavior of the divider carefully. The second issue is discussed first.

A. *Cyclostationary Noise.*

Formally, the term cyclostationary implies that the autocorrelation function of a stochastic process varies with t in a periodic fashion [19, 20], which in practice is associated with a periodic variation in the noise power of a signal. In general, the noise produced by all of the nonlinear blocks in a PLL is strongly cyclostationary. To understand why, consider the noise produced by a logic circuit, such as the inverter shown in Figure 8. The noise at the output of the inverter, n_{out}, comes from different sources depending on the phase of the output signal, v_{out}. When the output is high, the output is insensitive to small changes on the input. The transistor M_P is on and the noise at the output is predominantly due to the thermal noise from its channel. This is region A in the figure. When the output is low, the situation is reversed and most of the output noise is due to the thermal noise from the channel of M_N. This is region B. When the output is transitioning, thermal noise from both M_P and M_N contribute to the output. In addition, the output is sensitive to small changes in the input. In fact, any noise at the input is amplified before reaching the output. Thus, noise from the input tends to dominate over the thermal noise from the channels of M_P and M_N in this region. Noise at the input includes noise from the previous stage and noise from both devices in the form of flicker noise and thermal noise from gate resistance. This is region C in the figure.

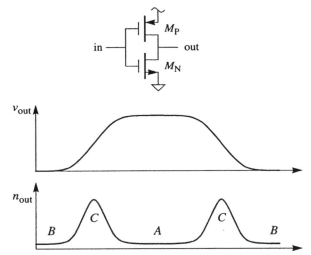

Fig. 8. Noise produced by an inverter (n_{out}) as a function of the output signal (v_{out}). In region A the noise is dominated by the thermal noise of M_P in region B its dominated by the thermal noise of M_N, and in region C the output noise includes the thermal noise from both devices as well as the amplified noise from the input.

The challenge in estimating the effect of noise passing through a threshold is the difficulty in estimating the noise at the point where the threshold is crossed. There are several different ways of estimating the effect of this noise, but the simplest is to use the strobed noise feature of SpectreRF.[†] When the strobed noise feature is active, the noise produced by the circuit is periodically sampled to create a discrete-time random sequence, as shown in Figure 9. SpectreRF then computes the power-spectral density of the sequence. The sample time would be adjusted to coincide with the desired threshold crossings. Since the T-periodic cyclostationary noise process is sampled every T seconds, the resulting noise process is stationary. Furthermore, the noise present at times other than at the sample points is completely ignored.

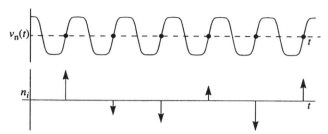

Fig. 9. Strobed noise. The lower waveform is a highly magnified view of the noise present at the strobe points in v_n, which are chosen to coincide with the threshold crossings in v.

B. Converting to Phase Noise

The act of converting the noise from a continuous-time process to a discrete-time process by sampling at the threshold crossings makes the conversion into phase noise easier. If v_n is the continuous-time noisy response, and v is the noise-free response (response with the noise sources turned off), then[†]

$$n_i = v_n(iT) - v(iT). \qquad (27)$$

Then if v_n is noisy because it is corrupted with a phase noise process ϕ, then

$$v_n(t) = v\left(t + \frac{\phi(t)}{2\pi f_0}\right). \qquad (28)$$

Assume the phase noise ϕ is small and linearize v using a Taylor series expansion

$$v_n(t) \cong v(t) + \frac{dv(t)}{dt}\frac{\phi(t)}{2\pi f_0} \qquad (29)$$

and

$$n_i \cong v(iT) + \frac{dv(iT)}{dt}\frac{\phi(iT)}{2\pi f_0} - v(iT) = \frac{dv(iT)}{dt}\frac{\phi(iT)}{2\pi f_0}. \qquad (30)$$

Finally, ϕ_i can be found from n_i using

$$\phi_i = 2\pi f_0 n_i / \frac{dv(iT)}{dt}. \qquad (31)$$

[†] The strobed-noise feature of SpectreRF is also referred to as its time-domain noise feature.

[†] It is assumed that the sequence n_i is formed by sampling the noise at iT, which implies that the threshold crossings also occur at iT. In practice, the crossings will occur at some time offset from iT. That offset is ignored. It is done without loss of generality with the understanding that the functions v and v_n can always be reformulated to account for the offset.

v is T periodic, which makes $dv(iT)/dt$ a constant, and so

$$S_\phi(f) = \left[2\pi f_0 / \frac{dv(iT)}{dt}\right]^2 S_n(f). \qquad (32)$$

where $S_n(f)$ and $S_\phi(f)$ are the power spectral densities of the n_i and ϕ_i sequences.

C. Phase-Domain Model for Dividers

To extract the phase noise of a divider, drive the divider with a representative periodic input signal and perform a PSS analysis to determine the threshold crossing times and the slew rate (dv/dt) at these times. Then use SpectreRF's strobed PNoise analysis to compute $S_n(f)$. When running PNoise analysis, assure that the *maxsidebands* parameter is set sufficiently large to capture all significant noise folding. A large value will slow the simulation. To reduce the number of sidebands needed, use T as small as possible. $S_\phi(f)$ is then computed from (32). Generally the noise appears as in Figure 10. Notice that the noise is periodic in f with period $1/T$ because n is a discrete-time sequence with period T. The parameters n and f_c for the divider are extracted as illustrated. The high frequency roll-off is generally ignored because it occurs above the frequency range of interest.

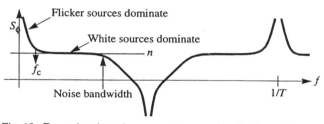

Fig. 10. Extracting the noise parameters, n and f_c, for the divider.

With ripple counters, one usually only characterizes one stage at a time and combines the phase noise from each stage by assuming that the noise in each stage is independent (true for device noise, would not be true for noise coupling into the divider from external sources). The variation due to phase noise accumulates, however it is necessary to account for the increasing period of the signals at each stage along the ripple counter. Consider an intermediate stage of a K-stage ripple counter. The total phase noise at the output of the ripple counter that results due to the phase noise $S_{\phi k}$ at the output of stage k is $(T_K/T_k)^2 S_{\phi k}$. So the total phase noise at the output of the ripple counter is

$$S_{\phi\,\text{out}} = T_K^2 \sum_{k=0}^{K} \frac{S_{\phi k}}{T_k^2} \qquad (33)$$

where $S_{\phi 0}$ and T_0 are the phase noise and signal period at the input to the first stage of the ripple counter.

With undesired variations in the supplies or in the substrate the resulting phase noise in each stage would be correlated, so one would need to compute the transfer function from the signal source to the phase noise of each stage and combine in a vector sum.

Unlike in ripple counters, phase noise does not accumulate with each stage in synchronous counters. Phase noise at the output of a synchronous counter is independent of the number of stages and consists only of the noise of its clock along with the noise of the last stage.

The phase-domain model for the divider, based on (2), is given in Listing 7. As before, one could use the *noise_table* function in lieu of the *white_noise* and *flicker_noise* functions to use the measured noise results directly.

Listing 7 — Phase-domain divider noise model.

```
`include "discipline.h"

module divider(in, out);
input in; output out;
phase in, out;
parameter real ratio = 1 from (0:inf);// divide ratio
parameter real n = 0 from [0:inf);
        // white output phase noise (rads²/Hz)
parameter real fc = 0 from [0:inf);
        // flicker noise corner frequency (Hz)

analog begin
    Theta(out) <+ Theta(in) / ratio;
    Theta(out) <+ white_noise(n, "wpn")
                    + flicker_noise(n*fc, 1, "fpn");
end
endmodule
```

VII. FRACTIONAL-N SYNTHESIS

One of the drawbacks of a traditional frequency synthesizer, also known as an integer-N frequency synthesizer, is that the output frequency is constrained to be N times the reference frequency. If the output frequency is to be adjusted by changing N, which is constrained by the divider to be an integer, then the output frequency resolution is equal to the reference frequency. If fine frequency resolution is desired, then the reference frequency must be small. This in turn limits the loop bandwidth as set by the loop filter, which must be at least 10 times smaller than the reference frequency to prevent signal components at the reference frequency from reaching the input of the VCO and modulating the output frequency, creating spurs or sidebands at an offset equal to the reference frequency and its harmonics. A low loop bandwidth is undesirable because it limits the response time of the synthesizer to changes in N. In addition, the loop acts to suppress the phase noise in the VCO at offset frequencies within its bandwidth, so reducing the loop bandwidth acts to increase the total phase noise at the output of the VCO.

The constraint on the loop bandwidth imposed by the required frequency resolution is eliminated if the divide ratio N is not limited to be an integer. This is the idea behind fractional-N synthesis. In practice, one cannot directly implement a frequency divider that implements non-integer divide ratio except in a few very restrictive cases, so instead a divider that is capable of switching between two integer divide ratios is used, and one rapidly alternates between the two values in such a way that the time-average is equal to the desired non-integer divide ratio [13]. A block diagram for a fractional-N synthesizer is shown in Figure 11. Divide ratios of N and $N + 1$ are used, where N is the first integer below the desired divide ratio, and $N + 1$ is the first integer above. For example, if the desired divide ratio is 16.25, then one would alternate between the ratios of 16 and 17, with the ratio of 16 being used 75% of the time. Early attempts at fractional-N synthesis alternated between integer divide ratios in a repetitive manner, which resulted in noticeable spurs in the VCO output spectrum. More recently, $\Delta\Sigma$ modulators have been used to generate a random sequence with the desired duty cycle to control the multi-modulus dividers [21]. This has the effect of trading off the spurs for an increased noise floor, however the $\Delta\Sigma$ modulator can be designed so that most of the power in its output sequence is at frequencies that are above the loop bandwidth, and so are largely rejected by the loop.

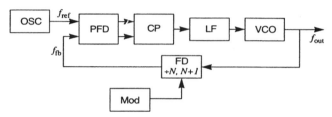

Fig. 11. The block diagram of a fractional-N frequency synthesizer.

The phase-domain small-signal model for the combination of a fractional-N divider and a $\Delta\Sigma$ modulator is given in Listing 8. It uses the *noise_table* function to construct a simple piece-wise linear approximation of the noise produced in an nth order $\Delta\Sigma$ modulator that is parameterized with the low frequency noise generated by the modulator, along with the corner frequency and the order.

VIII. JITTER

The signals at the input and output of a PLL are often binary signals, as are many of the signals within the PLL. The noise on binary signals is commonly characterized in terms of jitter.

Jitter is an undesired perturbation or uncertainty in the timing of events. Generally, the events of interest are the transitions in a signal. One models jitter in a signal by starting with a noise-free signal v and displacing time with a stochastic process j. The noisy signal becomes

$$v_n(t) = v(t + j(t)) \qquad (34)$$

with j assumed to be a zero-mean process and v assumed to be a T-periodic function. j has units of seconds and can be interpreted as a noise in time. Alternatively, it can be reformulated as a noise is phase, or phase noise, using

Listing 8 — Phase-domain fractional-N divider model.

```
`include "discipline.h"

module divider(in, out);
input in; output out;
phase in, out;
parameter real ratio = 1 from (0:inf);// divide ratio
parameter real n = 0 from [0:inf);
        // white output phase noise (rads²/Hz)
parameter real fc = 0 from [0:inf);
        // flicker noise corner frequency (Hz)
parameter real bw = 1 from (0:inf);// ΔΣ mod bandwidth
parameter integer order = 1 from (0:9);// ΔΣ mod order
parameter real fmax = 10*bw from (bw:inf);
        // maximum frequency of concern

analog begin
    Theta(out) <+ Theta(in) / (ratio + noise_table([
            0,       n,
            bw,      n,
            fmax,    n*pow((fmax/bw),order)
        ], "dsn"));
end
endmodule
```

$$\phi(t) = 2\pi f_o j(t), \qquad (35)$$

where $f_o = 1/T$ and

$$v_n(t) = v\left(t + \frac{\phi(t)}{2\pi f_o}\right). \qquad (36)$$

A. Jitter Metrics

Define $\{t_i\}$ as the sequence of times for positive-going zero crossings, henceforth referred to as *transitions*, that occur in v_n. The various jitter metrics characterize the statistics of this sequence.

The simplest metric is the *edge-to-edge jitter*, J_{ee}, which is the variation in the delay between a triggering event and a response event. When measuring edge-to-edge jitter, a clean jitter-free input is assumed, and so the edge-to-edge jitter J_{ee} is

$$J_{ee}(i) = \sqrt{\mathrm{var}(t_i)}. \qquad (37)$$

Edge-to-edge jitter assumes an input signal, and so is only defined for driven systems. It is an input-referred jitter metric, meaning that the jitter measurement is referenced to a point on a noise-free input signal, so the reference point is fixed. No such signal exists in autonomous systems. The remaining jitter metrics are suitable for both driven and autonomous systems. They gain this generality by being self-referred, meaning that the reference point is on the noisy signal for which the jitter is being measured. These metrics tend to be a bit more complicated because the reference point is noisy, which acts to increase the measured jitter.

Edge-to-edge jitter is also a scalar jitter metric, and it does not convey any information about the correlation of the jitter

between transitions. The next metric characterizes the correlations between transitions as a function of how far the transitions are separated in time.

Define $J_k(i)$ to be the standard deviation of $t_{i+k} - t_i$,

$$J_k(i) = \sqrt{\mathrm{var}(t_{i+k} - t_i)}. \qquad (38)$$

$J_k(i)$ is referred to as *k-cycle jitter* or *long-term jitter*[†]. It is a measure of the uncertainty in the length of k cycles and has units of time. J_1, the standard deviation of the length of a single period, is often referred to as the *period jitter*, and it denoted J, where $J = J_1$.

Another important jitter metric is *cycle-to-cycle jitter*. Define $T_i = t_{i+1} - t_i$ to be the period of cycle i. Then the cycle-to-cycle jitter J_{cc} is

$$J_{cc}(i) = \sqrt{\mathrm{var}(T_{i+1} - T_i)}. \qquad (39)$$

Cycle-to-cycle jitter is like edge-to-edge jitter in that it is a scalar jitter metric that does not contain information about the correlation in the jitter between distant transitions. However, it differs in that it is a measure of short-term jitter that is relatively insensitive to long-term jitter [22]. As such, cycle-to-cycle jitter is the only jitter metric that is suitable for use when flicker noise is present. All other metrics are unbounded in the presence of flicker noise.

If $j(t)$ is either stationary or T-cyclostationary, then $\{t_i\}$ is stationary, meaning that these metrics do not vary with i, and so $J_{ee}(i)$, $J_k(i)$, and $J_{cc}(i)$ can be shortened to J_{ee}, J_k, and J_{cc}.

These jitter metrics are illustrated in Figure 12.

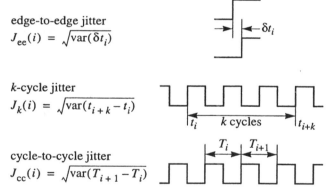

edge-to-edge jitter
$$J_{ee}(i) = \sqrt{\mathrm{var}(\delta t_i)}$$

k-cycle jitter
$$J_k(i) = \sqrt{\mathrm{var}(t_{i+k} - t_i)}$$

cycle-to-cycle jitter
$$J_{cc}(i) = \sqrt{\mathrm{var}(T_{i+1} - T_i)}$$

Fig. 12. The various jitter metrics.

B. Types of Jitter

The type of jitter produced in PLLs can be classified as being from one of two canonical forms. Blocks such as the PFD, CP, and FD are driven, meaning that a transition at their output is a direct result of a transition at their input. The jitter

---

† Some people distinguish between k-cycle jitter and long-term jitter by defining the long-term jitter J_∞ as being the k-cycle jitter J_k as $k \to \infty$.

exhibited by these blocks is referred to as *synchronous jitter*, it is a variation in the delay between when the input is received and the output is produced. Blocks such as the OSC and VCO are autonomous. They generate output transitions not as a result of transitions at their inputs, but rather as a result of the previous output transition. The jitter produced by these blocks is referred to as *accumulating jitter*, it is a variation in the delay between an output transition and the subsequent output transition. Table I previews the basic characteristics of these two types of jitter. The formulas for jitter given in this table are derived in the next two sections.

TABLE I: THE TWO CANONICAL FORMS OF JITTER.

Jitter Type	Circuit Type	Period Jitter
synchronous	driven (PFD/CP, FD)	$J = \dfrac{\sqrt{\mathrm{var}(n_v(t_c))}}{dv(t_c)/dt}$
accumulating	autonomous (OSC, VCO)	$J = \sqrt{aT}$

IX. SYNCHRONOUS JITTER

Synchronous jitter is exhibited by driven systems. In the PLL, the PFD/CP and FDs exhibit synchronous jitter. In these components, an output event occurs as a direct result of, and some time after, an input event. It is an undesired fluctuation in the delay between the input and the output events. If the input is a periodic sequence of transitions, then the frequency of the output signal is exactly that of the input, but the phase of the output signal fluctuates with respect to that of the input. The jitter appears as a modulation of the phase of the output, which is why it is sometimes referred to as phase modulated or PM jitter.

Let η be a stationary or T-cyclostationary process, then

$$j_{\mathrm{sync}}(t) = \eta(t) \tag{40}$$

$$v_n(t) = v(t + j_{\mathrm{sync}}(t)) \tag{41}$$

exhibits synchronous jitter. If η is further restricted to be a white Gaussian stationary or T-cyclostationary process, then $v_n(t)$ exhibits *simple synchronous jitter*. The essential characteristic of simple synchronous jitter is that the jitter in each event is independent or uncorrelated from the others, and (35) shows that it corresponds to white phase noise. Driven circuits exhibit simple synchronous jitter if they are broadband and if the noise sources are white, Gaussian and small. The sources are considered small if the circuit responds linearly to the noise, even though at the same time the circuit may be responding nonlinearly to the periodic drive signal.

For systems that exhibit simple synchronous jitter, from (37),

$$J_{ee}(i) = \sqrt{\mathrm{var}(j_{\mathrm{sync}}(t_i))}. \tag{42}$$

Similarly, from (38),

$$J_k(i) = \sqrt{\mathrm{var}(t_{i+k} - t_i)}, \tag{43}$$

$$J_k(i) = \sqrt{\mathrm{var}([(i+k)T + j_{\mathrm{sync}}(t_{i+k})] - [iT + j_{\mathrm{sync}}(t_i)])}, \tag{44}$$

$$J_k(i) = \sqrt{2\,\mathrm{var}(j_{\mathrm{sync}}(t_i))}. \tag{45}$$

$$J_k(i) = \sqrt{2} J_{ee}(i). \tag{46}$$

Since $j_{\mathrm{sync}}(t)$ is T-cyclostationary $j_{\mathrm{sync}} = j_{\mathrm{sync}}(t_i)$ is independent of i, and so is J_{ee} and J_k. The factor of $\sqrt{2}$ in (46) stems from the length of an interval including the independent variation from two transitions. From (46), J_k is independent of k, and so

$$J_k = J \quad \text{for} \quad k = 1, 2, \dots m. \tag{47}$$

Using similar arguments, one can show that with simple synchronous jitter,

$$J_{cc} = J, \tag{48}$$

Generally, the jitter produced by the PFD/CP and FDs is well approximated by simple synchronous jitter if one can neglect flicker noise.

A. Extracting Synchronous Jitter

The jitter in driven blocks, such as the PFD/CP or FDs, occurs because of an interaction between noise present in the blocks and the thresholds that are inherent to logic circuits.

In systems where signals are continuous valued, an event is usually defined as a signal crossing a threshold in a particular direction. The threshold crossings of a noiseless periodic signal, $v(t)$, are precisely evenly spaced. However, when noise is added to the signal, $v_n(t) = v(t) + n_v(t)$, each threshold crossing is displaced slightly. Thus, a threshold converts additive noise to synchronous jitter.

The amount of displacement in time is determined by the amplitude of the noise signal, $n_v(t)$ and the slew rate of the periodic signal, $dv(t_c)/dt$, as the threshold is crossed, as shown in Figure 13 [23]. If the noise n_v is stationary, then

$$\mathrm{var}(j_{\mathrm{sync}}(t_c)) \cong \frac{\mathrm{var}(n_v)}{[dv(t_c)/dt]^2} \tag{49}$$

where t_c is the time of a threshold crossing in v (assuming the noise is small).

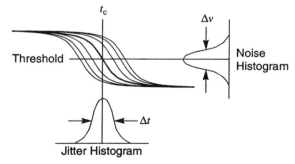

Fig. 13. How a threshold converts noise into jitter.

Generally n_v is not stationary, but cyclostationary (refer back to Section VI-A). It is only important to know when the noisy periodic signal $v_n(t)$ crosses the threshold, so the statistics of n_v are only significant at the time when $v_n(t)$ crosses the threshold,

$$\text{var}(j_{\text{sync}}(t_c)) = \frac{\text{var}(n_v(t_c))}{[dv(t_c)/dt]^2}. \tag{50}$$

The jitter is computed from (42) using (49) or (50),

$$J_{ee} = \frac{\sqrt{\text{var}(n_v(t_c))}}{dv(t_c)/dt}. \tag{51}$$

To compute $\text{var}(n_v(t_c))$, one starts by driving the circuit with a representative periodic signal, and then sampling $v(t)$ at intervals of T to form the ergodic sequence $\{v(t_i)\}$ where $t_i = t_c$ for some i. Then the variance is computed by computing the power spectral density for the sequence by integrating from $f = -f_0/2$ to $f_0/2$. Recall that the noise is periodic in f with period $f_0 = 1/T$ because n is a discrete-time sequence with rate T.

In practice, this is done by using the strobed noise capability of SpectreRF† to compute the power spectral density of the sequence. When the strobed noise feature is active, the noise produced by the circuit is periodically sampled to create a discrete-time random sequence, as shown in Figure 9. SpectreRF then computes the power-spectral density of the sequence. The sample time should be adjusted to coincide with the desired threshold crossings. Since the T-periodic cyclostationary noise process is sampled every T seconds, the resulting noise process is stationary. Furthermore, the noise present at times other than at the sample points is completely ignored.

1) Extracting the Jitter of Dividers: To extract the jitter of a divider, drive the divider with a representative periodic input signal and perform a PSS analysis to determine the threshold crossing times and the slew rate (dv/dt) at these times. Then use SpectreRF's strobed PNoise analysis to compute $S_n(f)$. The sample point should be set to coincide with the point where the output signal crosses the threshold of the subsequent stage (the phase detector) in the appropriate direction. When running PNoise analysis, assure that the *maxsidebands* parameter is set sufficiently large to capture all significant noise folding. A large value will slow the simulation. To reduce the number of sidebands needed, use T as small as possible. SpectreRF computes the power spectral density, which is integrated to compute the total noise at the sample points,

$$\text{var}(n_v(t_c)) = \int_{-f_0/2}^{f_0/2} S_{n_v}(f, t_c)df. \tag{52}$$

Then J_{ee} is computed from (51).

---

† The strobed-noise feature of SpectreRF is also referred to as its time-domain noise feature.

With ripple counters, one usually only characterizes one stage at a time. The total jitter due to noise in the ripple counter is then computed by assuming that the jitter in each stage is independent (again, this is true for device noise, but not for noise coupling into the divider from external sources) and taking the square-root of the sum of the square of the jitter on each stage.

Unlike in ripple counters, jitter does not accumulate with synchronous counters. Jitter in a synchronous counter is independent of the number of stages and consists only of the jitter of its clock along with the jitter of the last stage.

2) Extracting the Jitter of the Phase Detector: The PFD/CP is not followed by a threshold. Rather, it feeds into the LF, which is sensitive to the noise emitted by the CP at all times, not just during transitions. This argues that the noise of the PFD/CP be modeled as a continuous noise current. However, as mentioned earlier, doing so is problematic for simulators and would require very tight tolerances and small time steps. So instead, the noise of the PFD/CP is referred back to its inputs. The inputs of the PFD/CP are edge triggered, so the noise can be referred back as jitter.

To extract the input-referred jitter of a PFD/CP, drive both inputs with periodic signals with offset phase so that the PFD/CP produces a representative output. Use SpectreRF's PNoise analysis to compute the output noise over the total bandwidth of the PFD/CP (in this case, use the conventional noise analysis rather than the strobed noise analysis). Choose the frequency range of the analysis so that the total noise at frequencies outside the range is negligible. Thus, the noise should be at least 40 dB down and dropping at the highest frequency simulated. Integrate the noise over frequency and apply Wiener-Khinchin Theorem [24] to determine

$$\text{var}(n) = \int_{-\infty}^{\infty} S_n(f)df, \tag{53}$$

the total output noise current squared [19]. Then either calculate or measure the effective gain of the PFD/CP, K_{det}. Scale the gain so that it has the units of amperes per second. Then divide the total output noise current by the gain and account for there being two transitions per cycle to distribute the noise over to determine the input-referred jitter for the PFD/CP,

$$J_{ee_{\text{PFD/CP}}} = \frac{T}{2\pi K_{\text{det}}} \sqrt{\frac{\text{var}(n)}{2}}. \tag{54}$$

As before, when running PNoise analysis, assure that the *maxsidebands* parameter is set sufficiently large to capture all significant noise folding. A large value will slow the simulation. To reduce the number of sidebands needed, use T as small as possible.

X. ACCUMULATING JITTER

Accumulating jitter is exhibited by autonomous systems, such as oscillators, that generate a stream of spontaneous output transitions. In the PLL, the OSC and VCO exhibit accumulating jitter. Accumulating jitter is characterized by an undesired variation in the time since the previous output transition, thus the uncertainty of when a transition occurs accumulates with every transition. Compared with a jitter free signal, the frequency of a signal exhibiting accumulating jitter fluctuates randomly, and the phase drifts without bound. Thus, the jitter appears as a modulation of the frequency of the output, which is why it is sometimes referred to as frequency modulated or FM jitter.

Again assume that η be a stationary or T-cyclostationary process, then

$$j_{\text{acc}}(t) = \int_0^t \eta(\tau)d\tau \tag{55}$$

$$v_{\text{n}}(t) = v(t + j_{\text{acc}}(t)) \tag{56}$$

exhibits accumulating jitter. While η is cyclostationary and so has bounded variance, (55) shows that the variance of j_{acc}, and hence the phase difference between $v(t)$ and $v_{\text{n}}(t)$, is unbounded.

If η is further restricted to be a white Gaussian stationary or T-cyclostationary random process, then $v_{\text{n}}(t)$ exhibits *simple accumulating jitter*. In this case, the process $\{j_{\text{acc}}(iT)\}$ that results from sampling j_{acc} every T seconds is a discrete Wiener process and the phase difference between $v(iT)$ and $v_{\text{n}}(iT)$ is a random walk [19]. As shown next, simple accumulating jitter corresponds to oscillator phase noise that results from white noise sources.

The essential characteristic of simple accumulating jitter is that the incremental jitter that accumulates over each cycle is independent or uncorrelated. Autonomous circuits exhibit simple accumulating jitter if they are broadband and if the noise sources are white, Gaussian and small. The sources are considered small if the circuit responds linearly to the noise, though at the same time the circuit may be responding nonlinearly to the oscillation signal. An autonomous circuit is considered broadband if there are no secondary resonant responses close in frequency to the primary resonance.[†]

For systems that exhibit simple accumulating jitter, each transition is relative to the previous transition, and the variation in the length of each period is independent, so the variance in the time of each transition accumulates,

$$J_k = \sqrt{k}J \text{ for } k = 0, 1, 2, \dots, \tag{57}$$

where

$$J = \sqrt{\text{var}(j_{\text{acc}}(t_i + T)) - \text{var}(j_{\text{acc}}(t_i))}. \tag{58}$$

[†] Oscillators are strongly nonlinear circuits undergoing large periodic variations, and so signals within the oscillator freely mix up and down in frequency by integer multiples of the oscillation frequency. For this reason, any low frequency time constants or resonances in supply or bias lines would effectively act like close-in secondary resonances. In fact, this is the most likely cause of such phenomenon.

Similarly,

$$J_{\text{cc}} = \sqrt{2}J. \tag{59}$$

Generally, the jitter produced by the OSC and VCO are well approximated by simple accumulating jitter if one can neglect flicker noise.

A. Extracting Accumulating Jitter

The jitter in autonomous blocks, such as the OSC or VCO, is almost completely due to oscillator phase noise. Oscillator phase noise is a variation in the phase of the oscillator as it proceeds along its limit cycle.

In order to determine the period jitter J of $v_{\text{n}}(t)$ for a noisy oscillator, assume that it exhibits simple accumulating jitter so that η in (55) is a white Gaussian T-cyclostationary noise process (this excludes flicker noise) with a power spectral density of

$$S_\eta(f) = a, \tag{60}$$

and an autocorrelation function of

$$R_\eta(t_1, t_2) = a\delta(t_1 - t_2), \tag{61}$$

where δ is a Kronecker delta function. Then

$$j_{\text{acc}}(t) = \int_0^t \eta_T(\tau)d\tau \tag{62}$$

is a Wiener process [19], which has an autocorrelation function of

$$R_{j_{\text{acc}}}(t_1, t_2) = a\min(t_1, t_2). \tag{63}$$

The period jitter is the standard deviation of the variation in one period, and so

$$J^2 = \text{var}(j_{\text{acc}}(t + T) - j_{\text{acc}}(t)). \tag{64}$$

$$J^2 = \text{E}[(j_{\text{acc}}(t + T) - j_{\text{acc}}(t))^2] \tag{65}$$

$$J^2 = \text{E}[j_{\text{acc}}(t + T)^2 - 2j_{\text{acc}}(t + T)j_{\text{acc}}(t) + j_{\text{acc}}(t)^2] \tag{66}$$

$$J^2 = \text{E}[j_{\text{acc}}(t + T)^2] - 2\text{E}[j_{\text{acc}}(t + T)j_{\text{acc}}(t)] + \text{E}[j_{\text{acc}}(t)^2] \tag{67}$$

$$J^2 = R_{j_{\text{acc}}}(t + T, t + T) - 2R_{j_{\text{acc}}}(t + T, t) + R_{j_{\text{acc}}}(t, t) \tag{68}$$

$$J^2 = a(t + T) - 2at + at \tag{69}$$

$$J = \sqrt{aT} \tag{70}$$

We now have a way of relating the jitter of the oscillator to the PSD of η. However, η is not measurable, so instead the jitter is related to the phase noise S_ϕ. To do so, consider simple accumulating jitter written in terms of phase,

$$\phi_{\text{acc}}(t) = 2\pi f_{\text{o}}j_{\text{acc}}(t) = 2\pi f_{\text{o}}\int_0^t \eta(\tau)d\tau, \tag{71}$$

where $f_{\text{o}} = 1/T$. From (60) and (71) the PSD of ϕ_{acc} is

$$S_{\phi_{\text{acc}}}(\Delta f) = a\frac{(2\pi f_{\text{o}})^2}{(2\pi\Delta f)^2} = \frac{af_{\text{o}}^2}{\Delta f^2}. \tag{72}$$

From (26)

$$\mathcal{L}(\Delta f) = \frac{1}{2} S_{\phi_{\text{acc}}}(\Delta f) = \frac{a f_o^2}{2 \Delta f^2}, \tag{73}$$

$$a = 2\mathcal{L}(\Delta f)\frac{\Delta f^2}{f_o^2}. \tag{74}$$

Determine a by choosing Δf well above the corner frequency, f_{corner}, to avoid ambiguity and well below f_o to avoid the noise from other sources that occur at these frequencies.

1) Example: To compute the jitter of an oscillator, an RF simulator such as SpectreRF is used to find $\mathcal{L}$ and f_o of the oscillator. Given these, a is found with (74), J is found with (70) and J_k is found with (57). This procedure is demonstrated for the oscillator shown in Figure 14. This is a very low noise oscillator designed in 0.35μ CMOS by of Rael and Abidi [25]. The frequency of oscillation is 1.1 GHz and the resonator has a loaded Q of 6.

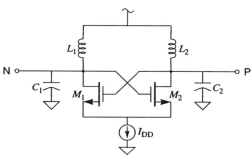

Fig. 14. Differential LC oscillator.

The procedure starts by using an RF simulator such as SpectreRF to compute the normalized phase noise $\mathcal{L}$. Its PNoise analysis is used, with the *maxsidebands* parameter set to at least 10 to adequately account for noise folding within the oscillator.[†] In this case, $\mathcal{L} = -110$ dBc at 100 kHz offset from the carrier. Apply (74) to compute a from $\mathcal{L}$, where $\mathcal{L}(\Delta f) = 10^{-11}$, $\Delta f = 100$ kHz, and $f_o = 1.1$ GHz,

$$a = 2 \cdot 10^{-11}\left(\frac{10^5}{1.1\times 10^9}\right)^2 = 165.3\times 10^{-21}. \tag{75}$$

The period jitter J is then computed from (70),

$$J = \sqrt{a T} = \sqrt{\frac{a}{f_o}} = \sqrt{\frac{165.3 \times 10^{-21}}{1.1 \text{ GHz}}} = 12.3 \text{ fs}. \tag{76}$$

In this example, the noise was extracted for the VCO alone. In practice, the LF is generally combined with the VCO before extracting the noise so that the noise of the LF is accounted for.

[†] At one point it was mistakenly suggested in the documentation for SpectreRF that *maxsidebands* should be set to 0 for oscillators. This causes SpectreRF to ignore all noise folding and results in a significant underestimation of the total noise.

XI. JITTER OF A PLL

If a PLL synthesizer is constructed from blocks that exhibit simple synchronous and accumulating jitter, then the jitter behavior of the PLL is relatively easy to estimate [26]. Assume that the PLL has a closed-loop bandwidth of f_L, and that $\tau_L = 1/2\pi f_L$, then for k such that $kT \ll \tau_L$, jitter from the VCO dominates and the PLL exhibits simple accumulating jitter equal to that produced by the VCO. Similarly, at large k (low frequencies), the PLL exhibits simple accumulating jitter equal to that produced by the OSC. Between these two extremes, the PLL exhibits simple synchronous jitter. The amount of which depends on the characteristics of the loop and the level of synchronous jitter exhibited by the FDs and the PFD/CP. The behavior of such a PLL is shown in Figure 15.

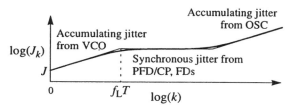

Fig. 15. Long-term jitter (J_k) for an idealized PLL as a function of the number of cycles.

XII. MODELING A PLL WITH JITTER

The basic behavioral models for the blocks that make up a PLL are well known and so will not be discussed here in any depth [27, 28]. Instead, only the techniques for adding jitter to the models are discussed.

Jitter is modeled in an AHDL by dithering the time at which events occur. This is efficient because it does not create any additional activity, rather it simply changes the time when existing activity occurs. Thus, models with jitter can run as efficiently as those without.

A. Modeling Driven Blocks

A feature of Verilog-A allows especially simple modeling of synchronous jitter. The *transition()* function, which is used to model signal transitions between discrete levels, provides a delay argument that can be dithered on every transition. The delay argument must not be negative, so a fixed delay that is greater than the maximum expected deviation of the jitter must be included. This approach is suitable for any model that exhibits synchronous jitter and generates discrete-valued outputs. It is used in the Verilog-A divider module shown in Listing 9, which models synchronous jitter with (41) where j_{sync} is a stationary white discrete-time Gaussian random process. It is also used in Listing 10, which models a simple PFD/CP.

1) Frequency Divider Model: The model, given in Listing 9, operates by counting input transitions. This is done in the

Listing 9 — Frequency divider that models synchronous jitter.

```
`include "discipline.h"

module divider (out, in);

input in; output out; electrical in, out;

parameter real Vlo=–1, Vhi=1;
parameter integer ratio=2 from [2:inf);
parameter integer dir=1 from [–1:1] exclude 0;
        // dir=1 for positive edge trigger
        // dir=–1 for negative edge trigger
parameter real tt=1n from (0:inf);
parameter real td=0 from (0:inf);
parameter real jitter=0 from [0:td/5);// edge-to-edge jitter
parameter real ttol=1p from (0:td/5);// ttol << jitter

integer count, n, seed;
real dt;

analog begin
    @(initial_step) seed = –311;

    @(cross(V(in) – (Vhi + Vlo)/2, dir, ttol)) begin
        // count input transitions
        count = count + 1;
        if (count >= ratio)
            count = 0;
        n = (2*count >= ratio);
        // add jitter
        dt = jitter*$dist_normal(seed,0,1);
    end

    V(out) <+ transition(n ? Vhi : Vlo, td+dt, tt);
end
endmodule
```

Listing 10 — PFD/CP model with synchronous jitter.

```
`include "discipline.h"

module pfd_cp (out, ref, vco);

input ref, vco; output out; electrical ref, vco, out;

parameter real Iout=100u;
parameter integer dir=1 from [–1:1] exclude 0;
        // dir=1 for positive edge trigger
        // dir=–1 for negative edge trigger
parameter real tt=1n from (0:inf);
parameter real td=0 from (0:inf);
parameter real jitter=0 from [0:td/5);// edge-to-edge jitter
parameter real ttol=1p from (0:td/5);// ttol << jitter

integer state, seed;
real dt;

analog begin
    @(initial_step) seed = 716;

    @(cross(V(ref), dir, ttol)) begin
        if (state > –1) state = state – 1;
        dt = jitter*$dist_normal(seed,0,1);
    end

    @(cross(V(vco), dir, ttol)) begin
        if (state < 1) state = state + 1;
        dt = jitter*$dist_normal(seed,0,1);
    end

    I(out) <+ transition(Iout*state, td + dt, tt);
end
endmodule
```

@*cross* block. The cross function triggers the @ block at the precise moment when its first argument crosses zero in the direction specified by the second argument. Thus, the @ block is triggered when the input crosses the threshold in the user specified direction. The body of the @ block increments the count, resets it to zero when it reaches ratio, then determines if count is above or below its midpoint (n is zero if the count is below the midpoint). It also generates a new random dither dT that is used later. Outside the @ block is code that executes continuously. It processes n to create the output. The value of the ?: operator is Vhi if n is 1 and Vlo if n is 0. Finally, the *transition* function adds a finite transition time of tt and a delay of td + dt. The finite transition time removes the discontinuities from the signal that could cause problems for the simulator. The jitter is embodied in dt, which varies randomly from transition to transition. To avoid negative delays, td must always be larger than dt. This model expects jitter to be specified as J_{ee}, as computed with (51).

2) PFD/CP Model: The model for a phase/frequency detector combined with a charge pump is given in Listing 10. It implements a finite-state machine with a three-level output, –1, 0 and +1. On every transition of the VCO input in direction dir, the output is incremented. On every transition of the reference input in the direction dir, the output is decremented. If both the VCO and reference inputs are at the same frequency, then the average value of the output is proportional to the phase difference between the two, with the average being negative if the reference transition leads the VCO transition and positive otherwise [3]. As before, the time of the output transitions is randomly dithered by dt to model jitter. The output is modeled as an ideal current source and a finite transition time provides a simple model of the dead band in the CP.

B. Modeling Accumulating Jitter

1) OSC Model: The delay argument of the *transition()* function cannot be used to model accumulating jitter because of the accumulating nature of this type of jitter. When modeling a fixed frequency oscillator, the *timer()* function is used as shown in Listing 11. At every output transition, the next transition is scheduled using the *timer()* function to be $T/K + J\delta/\sqrt{K}$ in the future, where δ is a unit-variance zero-mean random process and K is the number of output transitions per period. Typically, $K = 2$.

C. VCO Model

A VCO generates a sine or square wave whose frequency is proportional to the input signal level. VCO models, given in

Listing 11 — Fixed frequency oscillator with accumulating jitter.

```
`include "discipline.h"

module osc (out);

output out; electrical out;

parameter real freq=1 from (0:inf);
parameter real Vlo=-1, Vhi=1;
parameter real tt=0.01/freq from (0:inf);
parameter real jitter=0 from [0:0.1/freq);// period jitter

integer n, seed;
real next, dT;

analog begin
    @(initial_step) begin
        seed = 286;
        next = 0.5/freq + $abstime;
    end

    @(timer(next)) begin
        n = !n;
        dT = jitter*$dist_normal(seed,0,1);
        next = next + 0.5/freq + 0.707*dT;
    end

    V(out) <+ transition(n ? Vhi : Vlo, 0, tt);
end
endmodule
```

Listings 12 and 13, are constructed using three serial operations, as shown in Figure 16. First, the input signal is scaled to compute the desired output frequency. Then, the frequency is integrated to compute the output phase. Finally, the phase is used to generate the desired output signal. The phase is computed with *idtmod*, a function that provides integration followed by a modulus operation. This serves to keep the phase bounded, which prevents a loss of numerical precision that would otherwise occur when the phase became large after a long period of time. Output transitions are generated when the phase passes $-\pi/2$ and $\pi/2$.

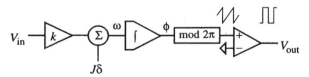

Fig. 16. Block diagram of VCO behavioral model that includes jitter.

The jitter is modeled as a random variation in the frequency of the VCO. However, the jitter is specified as a variation in the period, thus it is necessary to relate the variation in the period to the variation in the frequency. Assume that without jitter, the period is divided into K equal intervals of duration $\tau = T/K = 1/Kf_o$. The frequency deviation will be updated every interval and held constant during the intervals. With jitter, the duration of an interval is

$$\tau_i = \tau + \Delta\tau_i. \tag{77}$$

$\Delta\tau$ is a random variable with variance

$$\text{var}(\Delta\tau) = \frac{\text{var}(T)}{K} = \frac{J^2}{K}. \tag{78}$$

Therefore,

$$\Delta\tau_i = \frac{J\delta_i}{\sqrt{K}} \tag{79}$$

where δ is a zero-mean unit-variance Gaussian random process. The dithered frequency is

$$f_i = \frac{1}{K}\left(\frac{1}{\tau + \Delta\tau_i}\right) = \frac{\frac{1}{K\tau}}{1 + \frac{\Delta\tau_i}{\tau}} = \frac{f_c}{1 + K\Delta\tau_i f_c} \tag{80}$$

Let $\Delta T_i = K\Delta\tau_i$, then

$$f_i = \frac{f_c}{1 + \Delta T_i f_c}. \tag{81}$$

Finally $\text{var}(\tau_i) = J^2/K$, so $\Delta\tau_i = J\delta_i/\sqrt{K}$ and $\Delta T_i = \sqrt{K}J\delta_i$.

The @*cross* statement is used to determine the exact time when the phase crosses the thresholds, indicating the beginning of a new interval. At this point, a new random trial δ_i is generated.

The final model given in Listing 12. This model can be easily modified to fit other needs. Converting it to a model that generates sine waves rather than square waves simply requires replacing the last two lines with one that computes and outputs the sine of the phase. When doing so, consider reducing the number of jitter updates to one per period, in which case the factor of 1.414 should be changed to 1.

Listing 13 is a Verilog-A model for a quadrature VCO that exhibits accumulating jitter. It is an example of how to model an oscillator with multiple outputs so that the jitter on the outputs is properly correlated.

D. Efficiency of the Models

Conceptually, a model that includes jitter should be just as efficient as one that does not because jitter does not increase the activity of the models, it only affects the timing of particular events. However, if jitter causes two events that would normally occur at the same time to be displaced so that they are no longer coincident, then a circuit simulator will have to use more time points to resolve the distinct events and so will run more slowly. For this reason, it is desirable to combine jitter sources to the degree possible.

To make the HDL models even faster, rewrite them in either Verilog-HDL or Verilog-AMS. Be sure to set the time resolution to be sufficiently small to prevent the discrete nature of time in these simulators from adding an appreciable amount of jitter.

1) Including Synchronous Jitter into OSC: One can combine the output-referred noise of FD_M and FD_N and the input-

Listing 12 — VCO model that includes accumulating jitter.

```
`include "discipline.h"
`include "constants.h"

module vco (out, in);

input in; output out; electrical out, in;

parameter real Vmin=0;
parameter real Vmax=Vmin+1 from (Vmin:inf);
parameter real Fmin=1 from (0:inf);
parameter real Fmax=2*Fmin from (Fmin:inf);
parameter real Vlo=-1, Vhi=1;
parameter real tt=0.01/Fmax from (0:inf);
parameter real jitter=0 from [0:0.25/Fmax);// period jitter
parameter real ttol=1u/Fmax from (0:1/Fmax);

real freq, phase, dT;
integer n, seed;

analog begin
    @(initial_step) seed = -561;

    // compute the freq from the input voltage
    freq = (V(in) - Vmin)*(Fmax - Fmin) / (Vmax - Vmin)
        + Fmin;

    // bound the frequency (this is optional)
    if (freq > Fmax) freq = Fmax;
    if (freq < Fmin) freq = Fmin;

    // add the phase noise
    freq = freq/(1 + dT*freq);

    // phase is the integral of the freq modulo 2π
    phase = 2*`M_PI*idtmod(freq, 0.0, 1.0, -0.5);

    // update jitter twice per period
    // 1.414=sqrt(K), K=2 jitter updates/period
    @(cross(phase + `M_PI/2, +1, ttol) or
        cross(phase - `M_PI/2, +1, ttol)) begin
        dT = 1.414*jitter*$dist_normal(seed,0, 1);
        n = (phase >= -`M_PI/2) && (phase < `M_PI/2);
    end

    // generate the output
    V(out) <+ transition(n ? Vhi : Vlo, 0, tt);
end
endmodule
```

referred noise of the PFD/CP with the output noise of OSC. A modified fixed-frequency oscillator model that supports two jitter parameters and the divide ratio M is given in Listing 14 (more on the effect of the divide ratio on jitter in the next section). The accJitter parameter is used to model the accumulating jitter of the reference oscillator, and the syncJitter parameter is used to model the synchronous jitter of FD_M, FD_N and PFD/CP. Synchronous jitter is modeled in the oscillator without using a nonzero delay in the transition function. This is a more efficient approach because it avoids generating two unnecessary events per period. To get full benefit from this optimization, a modified PFD/CP given in Listing 15 is used. This model runs more efficiently by removing support for jitter and the td parameter.

Listing 13 — Quadrature Differential VCO model that includes accumulating jitter.

```
`include "discipline.h"
`include "constants.h"

module quadVco (PIout,NIout, PQout,NQout, Pin,Nin);

electrical PIout, NIout, PQout, NQout, Pin, Nin;
output PIout, NIout, PQout, NQout;
input Pin, Nin;

parameter real Vmin=0;
parameter real Vmax=Vmin+1 from (Vmin:inf);
parameter real Fmin=1 from (0:inf);
parameter real Fmax=2*Fmin from (Fmin:inf);
parameter real Vlo=-1, Vhi=1;
parameter real jitter=0 from [0:0.25/Fmax);// period jitter
parameter real ttol=1u/Fmax from (0:1/Fmax);
parameter real tt=0.01/Fmax;

real freq, phase, dT;
integer i, q, seed;

analog begin
    @(initial_step) seed = 133;

    // compute the freq from the input voltage
    freq = (V(Pin,Nin) - Vmin) * (Fmax - Fmin) / (Vmax - Vmin)
        + Fmin;

    // bound the frequency (this is optional)
    if (freq > Fmax) freq = Fmax;
    if (freq < Fmin) freq = Fmin;

    // add the phase noise
    freq = freq/(1 + dT*freq);

    // phase is the integral of the freq modulo 2π
    phase = 2*`M_PI*idtmod(freq, 0.0, 1.0, -0.5);

    // update jitter where phase crosses π/2
    // 2=sqrt(K), K=4 jitter updates per period
    @(cross(phase - 3*`M_PI/4, +1, ttol) or
        cross(phase - `M_PI/4, +1, ttol) or
        cross(phase + `M_PI/4, +1, ttol) or
        cross(phase + 3*`M_PI/4, +1, ttol)) begin
        dT = 2*jitter*$dist_normal(seed,0,1);
        i = (phase >= -3*`M_PI/4) && (phase < `M_PI/4);
        q = (phase >= -`M_PI/4) && (phase < 3*`M_PI/4);
    end

    // generate the I and Q outputs
    V(PIout) <+ transition(i ? Vhi : Vlo, 0, tt);
    V(NIout) <+ transition(i ? Vlo : Vhi, 0, tt);
    V(PQout) <+ transition(q ? Vhi : Vlo, 0, tt);
    V(NQout) <+ transition(q ? Vlo : Vhi, 0, tt);
end
endmodule
```

2) *Merging the VCO and FD_N:* If the output of the VCO is not used to drive circuitry external to the synthesizer, if the divider exhibits simple synchronous jitter, and if the VCO exhibits simple accumulating jitter, then it is possible to include the frequency division aspect of the FD_N as part of the

Listing 14 — Fixed-frequency oscillator with accumulating and synchronous jitter.

```
`include "discipline.h"

module osc (out);

output out; electrical out;

parameter real freq=1 from (0:inf);
parameter real ratio=1 from (0:inf);
parameter real Vlo=-1, Vhi=1;
parameter real tt=0.01*ratio/freq from (0:inf);
parameter real accJitter=0 from [0:0.1/freq); // period jitter
parameter real syncJitter=0 from [0:0.1*ratio/freq);
        // edge-to-edge jitter

integer n, accSeed, syncSeed;
real next, dT, dt, accSD, syncSD;

analog begin
    @(initial_step) begin
        accSeed = 286;
        syncSeed = -459;
        accSD = accJitter*sqrt(ratio/2);
        syncSD = syncJitter;
        next = 0.5/freq + $abstime;
    end

    @(timer(next + dt)) begin
        n = !n;
        dT = accSD*$dist_normal(accSeed,0,1);
        dt = syncSD*$dist_normal(syncSeed,0,1);
        next = next + 0.5*ratio/freq + dT;
    end

    V(out) <+ transition(n ? Vhi : Vlo, 0, tt);
end
endmodule
```

VCO by simply adjusting the VCO gain and jitter. If the divide ratio of FD_N is large, the simulation runs much faster because the high VCO output frequency is never generated. The Verilog-A model for the merged VCO and FD_N is given in Listing 16. It also includes code for generating a logfile containing the length of each period. The logfile is used in Section XIII when determining S_{VCO}, the power spectral density of the phase of the VCO output.

Recall that the synchronous jitter of FD_M and FD_N has already been included as part of OSC, so the divider model incorporated into the VCO is noiseless and the jitter at the output of the noiseless divider results only from the VCO jitter. Since the divider outputs one pulse for every N pulses at its input, the variance in the output period is the sum of the variance in N input periods. Thus, the period jitter at the output, J_{FD}, is $\sqrt{N}$ times larger than the period jitter at the input, J_{VCO}, or

$$J_{FD} = \sqrt{N} J_{VCO}. \tag{82}$$

Listing 15 — PFD/CP without jitter.

```
`include "discipline.h"

module pfd_cp (out, ref, vco);

input ref, vco; output out; electrical ref, vco, out;

parameter real Iout=100u;
parameter integer dir=1 from [-1:1] exclude 0;
        // dir = 1 for positive edge trigger
        // dir = -1 for negative edge trigger
parameter real tt=1n from (0:inf);
parameter real ttol=1p from (0:inf);

integer state;

analog begin
    @(cross(V(ref), dir, ttol)) begin
        if (state > -1) state = state - 1;
    end
    @(cross(V(vco), dir, ttol)) begin
        if (state < 1) state = state + 1;
    end

    I(out) <+ transition(Iout * state, 0, tt);
end
endmodule
```

Thus, to merge the divider into the VCO, the VCO gain must be reduced by a factor of N, the period jitter increased by a factor of $\sqrt{N}$, and the divider model removed.

After simulation, it is necessary to refer the computed results, which are from the output of the divider, to the output of VCO, which is the true output of the PLL. The period jitter at the output of the VCO, J_{VCO}, can be computed with (82).

To determine the effect of the divider on $S_\phi(\omega)$, square both sides of (82) and apply (70)

$$a_{VCO} T_{VCO} = \frac{a_{FD} T_{FD}}{N}. \tag{83}$$

$T_{VCO} = T_{FD} / N$, and so

$$a_{VCO} = a_{FD} \tag{84}$$

From (72),

$$S_{VCO} \frac{f^2}{f_{VCO}^2} = S_{FD} \frac{f^2}{f_{FD}^2} \tag{85}$$

Finally, $f_{VCO} = N f_{FD}$, and so

$$S_{VCO} = N^2 S_{FD}. \tag{86}$$

Once FD_N is incorporated into the VCO, the VCO output signal is no longer observable, however the characteristics of the VCO output are easily derived from (82) and (86), which are summarized in Table II.

It is interesting to note that while the frequency at the output of FD_N is N times smaller than at the output of the VCO, except for scaling in the amplitude, the spectrum of the noise close to the fundamental is to a first degree unaffected by the presence of FD_N. In particular, the width of the noise spec-

Listing 16 — VCO with FD$_N$.

```
`include "discipline.h"

module vco (out, in);

input in; output out; electrical out, in;

parameter real Vmin=0;
parameter real Vmax=Vmin+1 from (Vmin:inf);
parameter real Fmin=1 from (0:inf);
parameter real Fmax=2*Fmin from (Fmin:inf);
parameter real ratio=1 from (0:inf);
parameter real Vlo=-1, Vhi=1;
parameter real tt=0.01*ratio/Fmax from (0:inf);
parameter real jitter=0 from [0:0.25*ratio/Fmax);
                        // VCO period jitter
parameter real ttol=1u*ratio/Fmax from (0:ratio/Fmax);
parameter real outStart=inf from (1/Fmin:inf);

real freq, phase, dT, delta, prev, Vout;
integer n, seed, fp;

analog begin
    @(initial_step) begin
        seed = -561;
        delta = jitter * sqrt(2*ratio);
        fp = $fopen("periods.m");
        Vout = Vlo;
    end

    // compute the freq from the input voltage
    freq = (V(in) - Vmin)*(Fmax - Fmin) / (Vmax - Vmin)
        + Fmin;

    // bound the frequency (this is optional)
    if (freq > Fmax) freq = Fmax;
    if (freq < Fmin) freq = Fmin;

    // apply the frequency divider, add the phase noise
    freq = (freq / ratio)/(1 + dT * freq / ratio);

    // phase is the integral of the freq modulo 1
    phase = idtmod(freq, 0.0, 1.0, -0.5);

    // update jitter twice per period
    @(cross(phase - 0.25, +1, ttol)) begin
        dT = delta * $dist_normal(seed, 0, 1);
        Vout = Vhi;
    end
    @(cross(phase + 0.25, +1, ttol)) begin
        dT = delta * $dist_normal(seed, 0, 1);
        Vout = Vlo;
        if ($abstime >= outStart) $fstrobe( fp, "%0.10e",
                                        $abstime - prev);
        prev = $abstime;
    end
    V(out) <+ transition(Vout, 0, tt);
end
endmodule
```

trum is unaffected by FD$_N$. This is extremely fortuitous, because it means that the number of cycles we need to simulate is independent of the divide ratio N. Thus, large divide ratios do not affect the total simulation time.

TABLE II: CHARACTERISTICS OF VCO OUTPUT RELATIVE TO THE OUTPUT OF FD$_N$ ASSUMING THE VCO EXHIBITS SIMPLE ACCUMULATING JITTER AND THE FD$_N$ IS NOISE FREE.

Frequency	Jitter	Phase Noise
$f_{VCO} = N f_{FD}$	$J_{VCO} = \dfrac{J_{FD}}{\sqrt{N}}$	$S_{\phi_{VCO}} = N^2 S_{\phi_{FD}}$

To understand why FD$_N$ does not affect the width of the noise spectrum, recall that while we started with a jitter that varied continuously with time, $j(t)$ in (34), for either efficiency or modeling reasons we eventually sampled it to end up with a discrete-time version. The act of sampling the jitter causes the spectrum of the jitter to be replicated at the multiples of the sampling frequency, which adds aliasing. This aliasing is visible, but not obvious, at high frequencies in Figure 18. However, especially with accumulating jitter, the phase noise amplitude at low frequencies is much larger than the aliased noise, and so the close-in noise spectrum is largely unaffected by the sampling. The effect of FD$_N$ is to decimate the sampled jitter by a factor of N, which is equivalent to sampling the jitter signal, $j(t)$, at the original sample frequency divided by N. Thus, the replication is at a lower frequency, the amplitude is lower, and the aliasing is greater, but the spectrum is otherwise unaffected.

XIII. SIMULATION AND ANALYSIS

The synthesizer is simulated using the netlist from Listing 18 and the Verilog-A descriptions in Listings 14-16, modifying them as necessary to fit the actual circuit. The simulation should cover an interval long enough to allow accurate Fourier analysis at the lowest frequency of interest (F_{min}). With deterministic signals, it is sufficient to simulate for K cycles after the PLL settles if $F_{min} = 1/(TK)$. However, for these signals, which are stochastic, it is best to simulate for $10K$ to $100K$ cycles to allow for enough averaging to reduce the uncertainty in the result.

One should not simply apply an FFT to the output signal of the VCO/FD$_N$ to determine $\mathcal{L}(\Delta f)$ for the PLL. The result would be quite inaccurate because the FFT samples the waveform at evenly spaced points, and so misses the jitter of the transitions. Instead, $\mathcal{L}(\Delta f)$ can be measured with Spectre's Fourier Analyzer, which uses a unique algorithm that does accurately resolve the jitter [11]. However, it is slow if many frequencies are needed and so is not well suited to this application.

Unlike $\mathcal{L}(\Delta f)$, $S_\phi(\Delta f)$ can be computed efficiently. The Verilog-A code for the VCO/FD$_N$ given in Listing 16 writes the length of each period to an output file named *periods.m*. Writing the periods to the file begins after an initial delay, specified using outStart, to allow the PLL to reach steady state. This file is then processed by Matlab from MathWorks using the script shown in Listing 17. This script computes $S_\phi(\Delta f)$, the power spectral density of ϕ, using Welch's method [28]. The frequency range is from $f_{out}/2$ to $f_{out}/nfft$. The script com-

---

Listing 17 — Matlab script used for computing $S_\phi(\Delta f)$. These results must be further processed using Table II to map them to the output of the VCO.

```
% Process period data to compute S phi (Δf)
echo off;
nfft=512;        % should be power of two
winLength=nfft;
overlap=nfft/2;
winNBW=1.5;      % Noise bandwidth given in bins

% Load the data from the file generated by the VCO
load periods.m;

% output estimates of period and jitter
T=mean(periods);
J=std(periods);
maxdT = max(abs(periods–T))/T;
fprintf('T = %.3gs, F = %.3gHz\n',T, 1/T);
fprintf('Jabs = %.3gs, Jrel = %.2g%%\n', J, 100*J/T);
fprintf('max dT = %.2g%%\n', 100*maxdT);
fprintf('periods = %d, nfft = %d\n', length(periods), nfft);

% compute the cumulative phase of each transition
phases=2*pi*cumsum(periods)/T;

% compute power spectral density of phase
[Sphi,f]=psd(phases,nfft,1/T,winLength,overlap,'linear');

% correct for scaling in PSD due to FFT and window
Sphi=winNBW*Sphi/nfft;

% plot the results (except at DC)
K = length(f);
semilogx(f(2:K),10*log10(Sphi(2:K)));
title('Power Spectral Density of VCO Phase');
xlabel('Frequency (Hz)');
ylabel('S phi (dB/Hz)');
rbw = winNBW/(T*nfft);
RBW=sprintf('Resolution Bandwidth = %.0f Hz (%.0f dB)',
            rbw, 10*log10(rbw));
imtext(0.5,0.07, RBW);
```

---

putes $S_\phi(\Delta f)$ with a resolution bandwidth of rbw.[†] Normally, $S_\phi(\Delta f)$ is given with a unity resolution bandwidth. To compensate for a non-unity resolution bandwidth, broadband signals such as the noise should be divided by rbw. Signals with bandwidth less than rbw, such as the spurs generated by leakage in the CP, should not be scaled. The script processes the output of VCO/FD_N. The results of the script must be further processed using the equations in Table II to remove the effect of FD_N.

[†] The Hanning window used in the psd() function has a resolution bandwidth of 1.5 bins [29]. Assuming broadband signals, Matlab divides by 1.5 inside psd() to compensate. In order to resolve narrow-band signals, the factor of 1.5 is removed by the script, and instead included in the reported resolution bandwidth.

XIV. EXAMPLE

These ideas were applied to model and simulate a PLL acting as a frequency synthesizer. A synthesizer was chosen with f_{ref} = 25 MHz, f_{out} = 2 GHz, and a channel spacing of 200 kHz. As such, $M = 125$ and $N = 10,000$.

The noise of OSC is –95 dBc/Hz at 100 kHz. Applying (74) to compute a, where $L(\Delta f) = 316 \times 10^{-12}$, $\Delta f = 100$ kHz, and $f_o = 25$ MHz, gives $a = 10^{-14}$. The period jitter J is then computed from (70), giving $J = 20$ ps.

The noise of VCO is –48 dBc/Hz at 100 kHz. Applying (74) and (70) with $L(\Delta f) = 1.59 \times 10^{-5}$, $\Delta f = 100$ kHz, and $f_o = 2$ GHz, gives $a = 7.9 \times 10^{-14}$ and an period jitter of $J = 6.3$ ps.

The period jitter of the PFD/CP and FDs was found to be 2 ns. The FDs were included into the oscillators, which suppresses the high frequency signals at the input and output of the synthesizer. The netlist is shown in Listing 18. The results (compensated for non-unity resolution bandwidth (–28 dB) and for the suppression of the dividers (80 dB)) are shown in Figures 17-20. The simulation took 7.5 minutes for 450k time-points on a HP 9000/735. The use of a large number of time points was motivated by the desire to reduce the level of uncertainty in the results. The period jitter in the PLL was found to be 9.8 ps at the output of the VCO.

Listing 18 — Spectre netlist for PLL synthesizer.

```
// PLL-based frequency synthesizer that models jitter
simulator lang=spectre

ahdl_include "osc.va"      // Listing 14
ahdl_include "pfd_cp.va"   // Listing 15
ahdl_include "vco.va"      // Listing 16

Osc  (in)       osc       freq=25MHz ratio=125 \
                          accJitter=20ps syncJitter=2ns
PFD  (err in fb) pfd_cp   Iout=500ua
C1   (err c)    capacitor c=3.125nF
R    (c 0)      resistor  r=10k
C2   (c 0)      capacitor c=625pF
VCO  (fb err)   vco       Fmin=1GHz Fmax=3GHz \
                          Vmin=–4 Vmax=4 ratio=10000 \
                          jitter=6ps outStart=10ms

JitterSim       tran      stop=60ms
```

The low-pass filter LF blocks all high frequency signals from reaching the VCO, so the noise of the phase lock loop at high frequencies is the same as the noise generated by the open-loop VCO alone. At low frequencies, the loop gain acts to stabilize the phase of the VCO, and the noise of the PLL is dom-

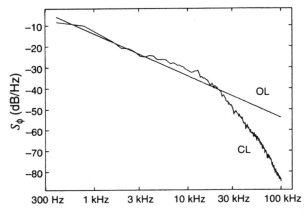

Fig. 17. Noise of the closed-loop PLL at the output of the VCO when only the reference oscillator exhibits jitter (CL) versus the noise of the reference oscillator mapped up to the VCO frequency when operated open loop (OL).

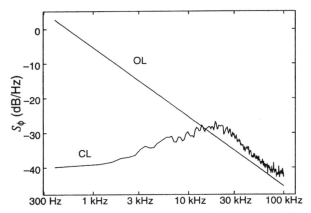

Fig. 18. Noise of the closed-loop PLL at the output of the VCO when only the VCO exhibits jitter (CL) versus the noise of the VCO when operated open loop (OL).

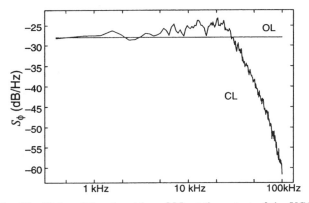

Fig. 19. Noise of the closed-loop PLL at the output of the VCO when only the PFD/CP, FD_M, and FD_N exhibit jitter (CL) versus the noise of these components mapped up to the VCO frequency when operated open loop (OL).

inated by the phase noise of the OSC. There is some contribution from the VCO, but it is diminished by the gain of

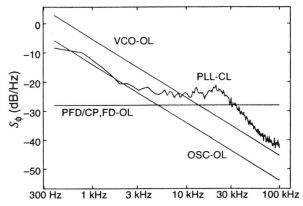

Fig. 20. Closed-loop PLL noise performance compared to the open-loop noise performance of the individual components that make up the PLL. The achieved noise is slightly larger than what is expected from the components due to peaking in the response of the PLL.

the loop. In this example, noise at the middle frequencies is dominated by the synchronous jitter generated by the PFD/CO and FDs. The measured results agree qualitatively with the expected results. The predicted noise is higher than one would expect solely from the open-loop behavior of each block because of peaking in the response of the PLL from 5 kHz to 50 kHz. For this reason, PLLs used in synthesizers where jitter is important are usually overdamped.

XV. CONCLUSION

A methodology for modeling and simulating the phase noise and jitter performance of phase-locked loops was presented. The simulation is done at the behavioral level, and so is efficient enough to be applied in a wide variety of applications. The behavioral models are calibrated from circuit-level noise simulations, and so the high-level simulations are accurate. Behavioral models were presented in the Verilog-A language, however these same ideas can be used to develop behavioral models in purely event-driven languages such as Verilog-HDL and Verilog-AMS. This methodology is flexible enough to be used in a broad range of applications where phase noise and jitter is important.

REFERENCES

[1] Ken Kundert. "Introduction to RF simulation and its application." *Journal of Solid-State Circuits,* vol. 34, no. 9, September 1999.

[2] Cadence Design Systems. "SpectreRF simulation option." *www.cadence.com/datasheets/spectrerf.html.*

[3] F. Gardner. *Phaselock Techniques.* John Wiley & Sons, 1979.

[4] D. Yee, C. Doan, D. Sobel, B. Limketkai, S. Alalusi, and R. Brodersen. "A 2-GHz low-power single-chip CMOS receiver for WCDMA applications." *Proceedings of the European Solid-State Circuits Conference,* Sept. 2000.

[5] A. Demir, E. Liu, A. Sangiovanni-Vincentelli, and I. Vassiliou. "Behavioral simulation techniques for phase/delay-locked systems." *Proceedings of the IEEE Custom Integrated Circuits Conference*, pp. 453-456, May 1994.

[6] A. Demir, E. Liu, and A. Sangiovanni-Vincentelli. "Time-domain non-Monte-Carlo noise simulation for nonlinear dynamic circuits with arbitrary excitations." *IEEE Transactions on Computer-Aided Design of Integrated Circuits and Systems*, vol. 15, no. 5, pp. 493-505, May 1996.

[7] A. Demir, A. Sangiovanni-Vincentelli. "Simulation and modeling of phase noise in open-loop oscillators." *Proceedings of the IEEE Custom Integrated Circuits Conference*, pp. 445-456, May 1996.

[8] A. Demir, A. Sangiovanni-Vincentelli. *Analysis and Simulation of Noise in Nonlinear Electronic Circuits and Systems*. Kluwer Academic Publishers, 1997.

[9] Ken Kundert. "Modeling and simulation of jitter in phase-locked loops." In *Analog Circuit Design: RF Analog-to-Digital Converters; Sensor and Actuator Interfaces; Low-Noise Oscillators, PLLs and Synthesizers*, Rudy J. van de Plassche, Johan H. Huijsing, Willy M.C. Sansen, Kluwer Academic Publishers, November 1997.

[10] Ken Kundert. "Modeling and simulation of jitter in PLL frequency synthesizers." Available from *www.designers-guide.com*.

[11] Kenneth S. Kundert. *The Designer's Guide to SPICE and Spectre*. Kluwer Academic Publishers, 1995.

[12] *Verilog-A Language Reference Manual: Analog Extensions to Verilog-HDL*, version 1.0. Open Verilog International, 1996. Available from *www.eda.org/verilog-ams*.

[13] Ulrich L. Rohde. *Digital PLL Frequency Synthesizers*. Prentice-Hall, Inc., 1983.

[14] Paul R. Gray and Robert G. Meyer. *Analysis and Design of Analog Integrated Circuits*. John Wiley & Sons, 1992.

[15] A. Demir, A. Mehrotra, and J. Roychowdhury. "Phase noise in oscillators: a unifying theory and numerical methods for characterization." *IEEE Transactions on Circuits and Systems I: Fundamental Theory and Applications*, vol. 47, no. 5, May 2000, pp. 655 -674.

[16] F. Käertner. "Determination of the correlation spectrum of oscillators with low noise." *IEEE Transactions on Microwave Theory and Techniques*, vol. 37, no. 1, pp. 90-101, Jan. 1989.

[17] F. X. Käertner. "Analysis of white and $f^{-\alpha}$ noise in oscillators." *International Journal of Circuit Theory and Applications*, vol. 18, pp. 485–519, 1990.

[18] G. Vendelin, A. Pavio, U. Rohde. *Microwave Circuit Design*. J. Wiley & Sons, 1990.

[19] W. Gardner. *Introduction to Random Processes: With Applications to Signals and Systems*. McGraw-Hill, 1989.

[20] Joel Phillips and Ken Kundert. "Noise in mixers, oscillators, samplers, and logic: an introduction to cyclostationary noise." *Proceedings of the IEEE Custom Integrated Circuits Conference*, CICC 2000. The paper and presentation are both available from *www.designers-guide.com*.

[21] T. A. D. Riley, M. A. Copeland, and T. A. Kwasniewski. "Delta-sigma modulation in fractional-N frequency synthesis." *IEEE Journal of Solid-State Circuits*, vol. 28 no. 5, May 1993, pp. 553 -559

[22] Frank Herzel and Behzad Razavi. "A study of oscillator jitter due to supply and substrate noise." *IEEE Transactions on Circuits and Systems – II: Analog and Digital Signal Processing*, vol. 46. no. 1, Jan. 1999, pp. 56-62.

[23] T. C. Weigandt, B. Kim, and P. R. Gray. "Jitter in ring oscillators." *1994 IEEE International Symposium on Circuits and Systems* (ISCAS-94), vol. 4, 1994, pp. 27-30.

[24] A. Papoulis. *Probability, Random Variables, and Stochastic Processes*. McGraw-Hill, 1991.

[25] J. J. Rael and A. A. Abidi. "Physical processes of phase noise in differential LC oscillators." *Proceedings of the IEEE Custom Integrated Circuits Conference*, CICC 2000.

[26] J. McNeill. "Jitter in Ring Oscillators." *IEEE Journal of Solid-State Circuits*, vol. 32, no. 6, June 1997.

[27] H. Chang, E. Charbon, U. Choudhury, A. Demir, E. Felt, E. Liu, E. Malavasi, A. Sangiovanni-Vincentelli, and I. Vassiliou. *A Top-Down Constraint-Driven Methodology for Analog Integrated Circuits*. Kluwer Academic Publishers, 1997.

[28] A. Oppenheim, R. Schafer. *Digital Signal Processing*. Prentice-Hall, 1975.

[29] F. Harris. "On the use of windows for harmonic analysis with the discrete Fourier transform." *Proceedings of the IEEE*, vol. 66, no. 1, January 1978.

PART II

Devices

Physics-Based Closed-Form Inductance Expression for Compact Modeling of Integrated Spiral Inductors

Snezana Jenei, Bart K. J. C. Nauwelaers, and Stefaan Decoutere

Abstract—A closed-form inductance expression for compact modeling of integrated inductors is presented. The expression is more accurate than previously published closed formulas. Moreover, due to its physics-based nature, it is scalable. That is demonstrated by comparison with the measured inductance for a complete set of inductors with different layout parameters.

Index Terms—Closed-form expression, compact inductor model, inductance, integrated inductor, physics-based formula.

I. INTRODUCTION

IN ORDER to facilitate the implementation of integrated inductors [1]–[3], [6], [7], a compact scalable physical model that accurately predicts inductor behavior with different layout and different technology parameters would be a useful tool for the RF IC design and optimization. Using a lumped network representation of spiral inductors based on a limited set of library inductors offers the advantage of validated models, but it limits flexibility of the designer for using intermediate values of inductance. These intermediate values are required when the area constraints request the layout of an inductor with slightly different outer radius and/or when gradient algorithms are applied for the optimization. Recently developed inductor simulators [1]–[3] can calculate inductance precisely. However, the applied method is the Greenhouse algorithm [4]. Although very accurate, this method employs a number of summation steps, depending of the number of interacting segments, over all combinations of parallel segments. Published closed-form expressions for square spiral inductors [6], [7] are simple, but not accurate enough or not scalable. These are non-physical expressions, obtained by fitting with a large number of fitting factors introduced in order to overcome a non-optimal choice of the fitting function. Therefore, an accurate physics-based closed-form expression for the inductance as a function of geometrical parameters, which can be easily applied for the initial inductor design and can be directly implemented in the RF circuit simulator would be a valuable tool.

II. PHYSICS-BASED CLOSED-FORM INDUCTANCE CALCULATION

The starting point for the derivation of our formula is a common point with the precise analytical inductor simulators

Manuscript received March 7, 2001; revised August 7, 2001.

The authors are with IMEC, 3000 Leuven, Belgium (e-mail: Snezana.Jenei@imec.be).

Publisher Item Identifier S 0018-9200(02)00128-2.

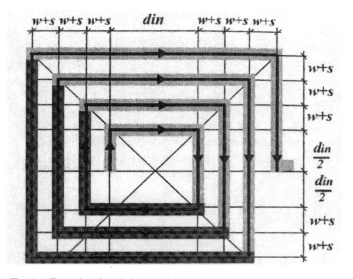

Fig. 1. Example of an inductor with $n = 3.5$ turns, inner diameter d_{in}, segment width w and spacing between segments s. The length of the segments increases with steps $w + s$; the increase occurs every two segments. The total length of the inductor is l. There are, in the average, n segments on every four sides of an inductor, therefore the length of an average segment is $l/(4n)$.

[1]–[3]: decomposition of an inductor into segments. Our formula is derived for the most important practical case: for an inductor laid out with a hollow center of an arbitrary inner diameter d_{in} and with a regular increase in the length of subsequent segments, as in Fig. 1. For such a symmetrical structure, there is a linear relation between the total length l and the inner diameter d_{in}:

$$l = (4n + 1)d_{in} + (4N_i + 1)N_i(w + s) \qquad (1)$$

where n is the number of turns, N_i is the integer part of n, w is the metal width, and s is the spacing between segments. Therefore, the lateral structure of an inductor is fully determined by four parameters: n, w, s, and either l or d_{in}, while the three-dimensional definition of an inductor includes an additional, technology parameter—thickness of the metal t.

The total inductance of an inductor consists of the inductor self-inductance (L_{self}) and of the remaining part, which comes from the total negative (M^-) and the total positive (M^+) mutual inductances that include all negative and all positive interactions between all segments. For straight segments, Naumman's formula for self- and mutual inductance between filaments can be used taking Grover [5] coefficients into account for the real

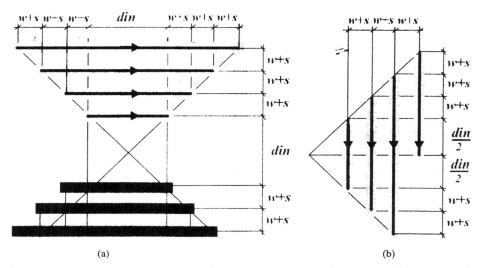

(a)   (b)

Fig. 2. (a) Segments on the opposite sides of the square inductor contribute to the negative mutual inductance. (b) Segments on the same side of the square inductor contribute to the positive mutual inductance.

cross section of segments ($w \times t$) and space (s) between them. The difference between our approach and the Greenhouse algorithm is that we have based our inductance calculation on average segment interaction, rather than summing one by one individual segment interactions.

A straight segment of the length l_{seg} has the self-inductance [5]

$$L_{\text{self seg}} = \frac{\mu_0}{2\pi} l_{\text{seg}} \left(\ln \frac{2l_{\text{seg}}}{w+t} + 0.5 \right). \qquad (2)$$

The self-inductance of the square inductor can be expressed as the sum of $4n$ self-inductances (2) of segments with the average length $l_{\text{AV seg square}} = l/4n$, i.e.,

$$L_{\text{self}} = \frac{\mu_0}{2\pi} l \left(\ln \frac{l}{n(w+t)} - 0.2 \right). \qquad (3)$$

The expressions for mutual inductances benefit in simplicity from the symmetry properties of the square geometry. As presented in Fig. 2(a), anti-parallel segments contribute to the negative mutual inductance. The sum of all interactions in Fig. 2(a) can be approximated by an equivalent of $2n^2$ average interactions—between segments of an average length at an average distance. Since the average distance between segments on the opposite sides of a square inductor is equal to the average segment length, the total negative mutual inductance can be expressed as a very simple function of the total length l and number of turns n as

$$M^- = 2 \cdot 2 \cdot n^2 \left(\frac{\mu_0}{2\pi} \frac{l}{4n} 0.47 \right) = 0.47 \frac{\mu_0}{2\pi} ln. \qquad (4)$$

The third constitutive element of the total inductance is the total positive mutual inductance, to which contribute interactions between parallel segments on the same side of a square [in Fig. 2(b)]. The average distance d^+ for the constituting factor of positive mutual inductance can be calculated by closed formula as

$$d^+ = (w+s) \frac{(3n - 2N_i - 1)(N_i + 1)}{3(2n - N_i - 1)} \qquad (5)$$

where Ni is the integer part of n (number of turns). The total positive mutual inductance is

$$M^+ = \frac{\mu_0}{2\pi} l(n-1) \left(\ln \left(\sqrt{1 + \left(\frac{l}{4nd^+} \right)^2} + \frac{l}{4nd^+} \right) \right.$$
$$\left. - \sqrt{1 + \left(\frac{4nd^+}{l} \right)^2} + \frac{4nd^+}{l} \right). \qquad (6)$$

Thus, the total inductance L of a square spiral inductor is

$$L = \frac{\mu_0}{2\pi} l \left(\ln \frac{l}{n(w+t)} - 0.2 - 0.47 \cdot n + (n-1) \right.$$
$$\cdot \left(\ln \left(\sqrt{1 + \left(\frac{l}{4nd^+} \right)^2} + \frac{l}{4nd^+} \right) \right.$$
$$\left. \left. - \sqrt{1 + \left(\frac{4nd^+}{l} \right)^2} + \frac{4nd^+}{l} \right) \right). \qquad (7)$$

It should be noted that there are no unphysical fitting factors in (7). Parameters in (7) are just the geometry parameters of the inductor: the total length of the inductor l, the number of turns n, the width of segments w, the thickness of the metal t, and the average distance d^+ given by (5). Expression (7) is a closed-form equation for the total inductance. Any circuit simulator can readily evaluate (7). In addition, the expression has the advantage that it indicates to the designer how the relative contributions of self, positive, and negative mutual inductance are related to the geometrical parameters. Also, the expression is scalable and, moreover, valid even for inductors with incomplete inner or/and outer turn(s). This is clearly an advantage over the expressions in [6] and [7], where diameter parameters are not well defined for spirals with "half-turn" turns.

Formula (7) is derived for square inductors, but it can be simply extended for any polygonal spiral. For example, it is possible to use (4) and (6) for mutual components of the total in-

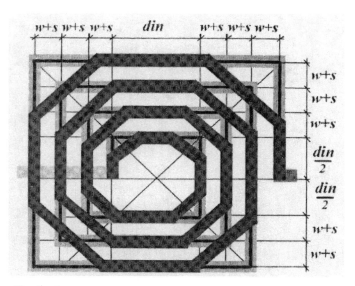

Fig. 3. Octagonal symmetrical spiral and its projection on the square spiral sharing the same vertical and horizontal segments.

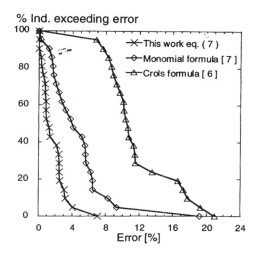

Fig. 4. Number of inductors (in %) exceeding the error (which is defined as the relative difference between measured and calculated inductance in %) for our formula and formulas proposed earlier in [6] and [7] for our set of square inductors.

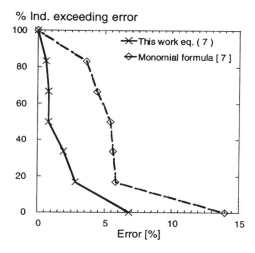

Fig. 5. Number of inductors (in %) exceeding the error for our formula and formula in [7] for the set of inductors from the literature [2].

ductance for octagonal spiral inductors, if the "effective square spiral length"

$$l_{\text{eff square}} = 4n l_{\text{AV seg square}} = 4n \left(\frac{l_{\text{oct}}}{8n} + 2 \cdot \frac{l_{\text{oct}}}{8n} \cos 45° \right)$$

$$= l_{\text{oct}} \left(\frac{1 + \sqrt{2}}{2} \right) = 1.21 \cdot l_{\text{oct}} \qquad (8)$$

is used for the total square inductor length l in (4) and (6), since 45° inclined segments can be decomposed into their vertical and horizontal contributing components (as in Fig. 3). The third part of the total inductance, self-inductance, can be derived as the sum of $8n$ self-inductances of straight segments (2) with an average length $l_{\text{oct}}/8n$. Thus, the total inductance of the octagonal spiral with the length l_{oct} is

$$L = \frac{\mu_0}{2\pi} l_{\text{oct}} \left(\ln \frac{l_{\text{oct}}}{n(w+t)} - 0.9 \right) + \frac{\mu_0}{2\pi} 1.21 l_{\text{oct}}$$

$$\cdot \left(-0.47n + (n-1) \right.$$

$$\cdot \left(\ln \left(\sqrt{1 + \left(\frac{1.21 l_{\text{oct}}}{4nd^+} \right)^2} + \frac{1.21 l_{\text{oct}}}{4nd^+} \right) \right.$$

$$\left. \left. - \sqrt{1 + \left(\frac{4nd^+}{1.21 l_{\text{oct}}} \right)^2} + \frac{4nd^+}{1.21 l_{\text{oct}}} \right) \right). \qquad (9)$$

III. EXPERIMENTAL VALIDATION

We evaluated the accuracy of our expression (7) and other closed form expressions [6], [7] by comparison with measurements. Another way to evaluate the formula accuracy would be the application of numerical, three-dimensional (3D) simulators. Since current 3D simulators can hardly handle the complexity of realistic integrated inductor structure, the usual prac-

tical way to apply them effectively is to assume a number of simplifications and approximations. That is why we are referring rather to the measurements than to the simulations at this moment.

Measured data are inductance values for our set of inductors with different layout, fabricated in 0.35-μm BiCMOS technology, as well as published measured inductance values, for which detailed layout can be found in the literature. In Fig. 4 is presented the comparison of the accuracy of our formula (7), Crols' formula [6], and the monomial formula [7], which is the most accurate closed-form expression published up to date. Data calculated by these three expressions are compared with our measurements, covering the range from 2 nH up to 25 nH. In the case of our formula, the number of very accurately predicted inductances is the highest (the curve is the closest to the y-axis) and, also, the maximum difference between the measured and calculated data (intersect with x-axis) is the smallest. Inductance measurement introduces by itself a non-negligible error and calculations by our formula are mostly within mea-

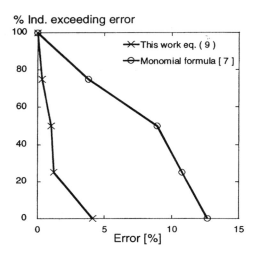

Fig. 6. Number of inductors (in %) exceeding the error for our formula and formula in [7] for the set of octagonal inductors.

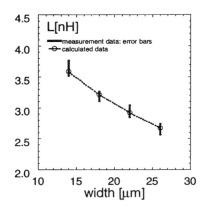

Fig. 7. Calculated and measured data for a different width of inductor segments, while the rest of parameters are constant (number of turns $n = 3.5$, spacing $s = 2$ μm and the total length $l = 2700$ μm).

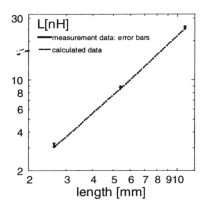

Fig. 8. Calculated and measured data for a different total length of inductor (different number of turns n), while the rest of parameters are constant ($w = 22$ μm, $s = 2$ μm and inner diameter $d_{in} = 105$ μm).

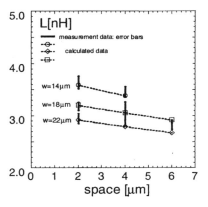

Fig. 9. Calculated and measured data for a different space between inductor segments; parameter is segment width, while the length $l = 2700$ μm and inner radius $d_{in} = 105$ μm.

surement tolerances. Maximum error that is made using our expression is in the worst case less than 8% and in average even much smaller. Similar conclusions can be drawn for the set of measurement data collected in the literature [2] (Fig. 5). Calculated values are compared with the measurements on our set of octagonal inductors and the comparison in accuracy of our formula and monomial expression is given in Fig. 6.

Scalability of the formula (7) is clearly demonstrated in Figs. 7–9. In the designed set of inductors one parameter is varied, while the others are kept constant. All inductance variation with width, length and space is correctly predicted by (7) within the inductance measurement error.

strated by comparison with measurements. The same conclusion can be drawn from comparison with our own measurements and measurement data available in the literature. Scalability of the proposed expression is also clearly shown. All inductance variations by varying geometry parameters are correctly predicted within the experimental error.

IV. CONCLUSION

A closed form inductance expression compatible with circuit simulators has been proposed and validated. Formula derivation is based on a decomposition into segments and an average of the segment interaction. The accuracy, higher than for the previously published closed-form expressions, has been demon-

REFERENCES

[1] Y. Koutsoyannopoulos et al., "A generic CAD model for arbitrary shaped and multilayer integrated inductors on silicon substrates," in Proc. ESSDERC, 1997, pp. 320–323.
[2] J. R. Long and M. A. Copeland, "The modeling, characterization and designed monolithic inductors for silicon RF ICs," IEEE J. Solid-State Circuits, vol. 32, pp. 357–369, Mar. 1997.
[3] A. M. Niknejad and R. G. Meyer, "Analysis, design and optimization of spiral inductors and transformers for Si RF ICs," IEEE J. Solid-State Circuits, vol. 33, pp. 1470–1481, Oct. 1998.
[4] H. M. Greenhouse, "Design of planar rectangular microelectronic inductors," IEEE Trans. Parts, Hyb. Packag., vol. 10, pp. 101–109, 1974.
[5] F. W. Grover, Inductance Calculations. New York: Dover, 1946.
[6] J. Crols et al., "An analytical model of planar inductors on lowely doped silicon substrates for high frequency analog design up to 3 GHz," in Symp. VLSI Circuits Dig. Tech. Papers, 1996, pp. 28–29.
[7] S. S. Mohan, M. M. Hershenson, S. Boyd, and T. H. Lee, "Simple accurate expressions for planar spiral inductances," IEEE J. Solid-State Circuits, vol. 34, pp. 1419–1424, Oct. 1999.

The Modeling, Characterization, and Design of Monolithic Inductors for Silicon RF IC's

John R. Long, *Member, IEEE*, and Miles A. Copeland, *Fellow, IEEE*

Abstract—The results of a comprehensive investigation into the characteristics and optimization of inductors fabricated with the top-level metal of a submicron silicon VLSI process are presented. A computer program which extracts a physics-based model of microstrip components that is suitable for circuit (SPICE) simulation has been used to evaluate the effect of variations in metallization, layout geometry, and substrate parameters upon monolithic inductor performance. Three-dimensional (3-D) numerical simulations and experimental measurements of inductors were also used to benchmark the model accuracy. It is shown in this work that low inductor Q is primarily due to the restrictions imposed by the thin interconnect metallization available in most very large scale integration (VLSI) technologies, and that computer optimization of the inductor layout can be used to achieve a 50% improvement in component Q-factor over unoptimized designs.

Index Terms— Computer-aided design, HF analog integrated circuits, MMIC's, modeling, monolithic inductors, RFIC's, silicon integrated circuit technology.

I. INTRODUCTION

RADIO frequency (RF) circuits fabricated in monolithic microwave integrated circuit technologies (such as GaAs MMIC) make extensive use of on-chip transmission lines to realize an inductance, the inductor being a key component in many high-performance narrowband circuit designs. Silicon IC technologies have rarely been used for analog applications in the radio and microwave range of frequencies, primarily because transmission line structures perform poorly on the semiconducting substrates used to manufacture silicon IC's [1]. In order to exploit the capabilities offered by a monolithic inductance, the limitations imposed by silicon technology upon the component performance must be accurately modeled and characterized. The ability to optimize and refine silicon circuit designs incorporating on-chip inductors has been identified by others [2] as lacking in the present state of the design art.

A computationally efficient, scalable, lumped-element model which can be applied to an arbitrary configuration of microstrip transmission lines, such as the monolithic spiral inductor [3]–[5], will be described in this paper. An efficient and scalable inductor model can be used to rapidly optimize the electrical performance of an inductor for an RF IC applica-

Manuscript received August 15, 1996; revised October 28, 1996. This work was supported by Micronet, the Natural Sciences and Engineering Research Council of Canada (NSERC), and the Telecommunications Research Institute of Ontario (TRIO).

J. R. Long is with the Department of Electrical and Computer Engineering, University of Toronto, Toronto, ON, M5S 3G4 Canada.

M. A. Copeland is with the Department of Electronics, Carleton University, Ottawa, ON, Canada.

Publisher Item Identifier S 0018-9200(97)01294-8.

tion. The accuracy of the lumped element model is evaluated for variations in metallization thickness, layout geometry, and substrate parameters, using a combination of three-dimensional (3-D) numerical simulations and experimental measurements. The results of this investigation highlight areas where process improvements and parameter optimization can be applied to maximize the performance of RF circuits using silicon-based inductor designs. Some design guidelines for maximizing the inductor performance in similar technologies will also be described.

II. THE MONOLITHIC INDUCTOR

Resonant-tuned (LC) circuits offer numerous benefits to the designer of high-frequency circuits. Operation at a low supply voltage, simplified impedance matching between stages, and low dissipation for reduced circuit noise are just a few of the properties of LC circuits that can be exploited to achieve a higher level of performance from a given fabrication technology. However, an on-chip inductance is required for the realization of LC networks for these purposes. At radio and microwave frequencies, a purely passive inductor is often preferable to synthesis of an inductive reactance with an active circuit. Passive components introduce less noise, consume less power, and have a wider bandwidth and linear operating range than their electronic equivalents, such as the gyrator.

Passive inductors can be implemented on-chip using transmission lines. The input impedance (Z_{in}) of a short section of transmission line which is terminated in a short circuit can be written as follows:

$$Z_{in} = Z_0 \cdot \gamma l = (r + j\omega L)l \qquad (1)$$

where Z_0 is the characteristic impedance and γ is the propagation constant for a transmission line of length, l. Here it is assumed that the physical length of the line is shorter than one-tenth of a wavelength at the desired frequency of operation, that is, the line appears to be "electrically short." From (1), the input impedance will be either resistive or inductive, depending upon the value of the resistance/unit length, r, compared to the inductance/unit length, L, of the transmission line. For low resistivity metals such as the interconnect metallization used on an integrated circuit chip, the input impedance can be made to appear predominantly inductive. The ratio of the series inductance to shunt capacitance per unit length defines the characteristic impedance (Z_0) of the transmission line. Maximizing the characteristic impedance (e.g., using a narrow metal line) will reduce the length of line that is required to realize a given inductance.

The microstrip transmission line structure on a silicon IC consists of a metal strip above a conducting (or ground) plane, with the substrate and intermetal oxide layers sandwiched between the two conductors. For dimensions typically encountered in a commercial IC fabrication process (metal linewidths between 5–50 μm on a 350-μm thick substrate), the characteristic impedance of a microstrip line ranges from approximately 100–200 Ω. The substrate behavior depends upon both the resistivity and the frequency of the propagating wave. However, the substrate tends to behave as a lossy dielectric in modern silicon very large scale integration (VLSI) processes [6], where resistivities are typically in the 1–100 Ω-cm range and operation is in the GHz range of frequencies.

The largest inductive reactance that can be realized on-chip is determined from the following equation:

$$\text{Inductive reactance} < 2\pi Z_{0, \text{max}} \bullet \frac{l}{\lambda} = 40\pi \ \Omega \qquad (2)$$

where l is the transmission line length and λ is the guided wavelength. A short transmission line (i.e., less than one-tenth of a wavelength long) is assumed here, with a maximum characteristic impedance of 200 Ω. The inductive reactance from (2) corresponds to an inductance of 20 nH at a frequency of 1 GHz. This is close to the upper limit that can be realized monolithically in a standard silicon process. The smallest inductance value for most practical circuit applications in the 1–3 GHz range of frequencies is on the order of 1 nH. The losses introduced by the semiconducting substrate in a silicon IC technology restrict the Q-factor to less than ten for most commercial IC processes.

In addition to the rectangular geometry used here, circular and octagonal geometries are also widely used to implement microstrip spiral inductors. Although an improvement in Q-factor of up to 10% is possible using a circular rather than square design, the circular spiral consumes greater chip area, and introduces difficulties in the efficient generation of photomasks [7], [8]. In addition, the interaction between the magnetic field components on adjacent sides of a nonrectangular spiral inductor adds extra complexity to the development of a circuit model. Thus, only the rectangular spiral inductor will be considered in the following discussion.

The spiral inductor structure, while compact and more space efficient than an equivalent straight wire inductor, is more difficult to analyze. The accurate characterization of this structure at microwave frequencies requires an analysis of the fringing fields, parasitics, ground plane effects, and most importantly for silicon IC design, an analysis of the effect of the conductive substrate on the component performance. These effects cannot be fully analyzed to yield closed-form expressions which adequately predict the inductor behavior, and hence, numerical analysis is required in order to determine the parameters of the inductor's electrical model.

III. THE MONOLITHIC INDUCTOR IN SILICON TECHNOLOGY

The performance of rectangular spiral microstrip inductors in a 0.8-μm silicon BiCMOS technology was reported by Nguyen and Meyer in 1990 [10]. The inductors were fabricated using a microstrip transmission line wound as a square spiral, and their electrical behavior was modeled adequately by a lumped inductance with a few additional parasitic (RC) components. Some simple LC filtering and signal processing circuits using these inductors were also demonstrated [11], [12], although the performance of the circuits was limited by the low Q of the inductors, which was less than five. Silicon RF integrated circuits incorporating monolithic inductors for product applications have subsequently been developed and demonstrated by a number of manufacturers [13]–[15].

The characteristics of inductors fabricated in various silicon technologies have been studied and reported in the literature. The main motivation behind these studies has been improvement in the inductor quality through a modification of the metallization scheme and/or the properties of the underlying substrate. The methods which have been employed to date have included: thicker metallization [16], stacking of metal layers in a multilevel metal process [17], thicker intermetal oxide [18], fabrication using high-resistivity silicon substrates [16], and the selective removal of silicon from beneath the inductor structure by chemical etching [19]. Justification for these experimental investigations has come from a heuristic assessment of the parameters limiting inductor performance or from numerical simulations of the inductor's electrical characteristics using commercially available electromagnetic field solvers.

IV. SILICON MONOLITHIC INDUCTOR MODEL

A circuit model which describes the electrical behavior of the monolithic inductor at RF and microwave frequencies is required for the computer simulation and optimization of tuned circuits fabricated in silicon IC technologies. Modeling of spiral inductors on silicon has been limited to reports of numerical simulation results [20] and/or parameter fitting of lumped-element equivalent circuits to measured data [16], [17]. Electrical circuit models for an inductor that are derived in this way cannot be scaled to reflect changes in the layout or fabrication technology.

Numerical simulators are now commercially available [21] which compute the electromagnetic field distribution of planar-type conductor configurations in three dimensions. These simulators can extract the circuit parameters (i.e., RLCG parameters) of the spiral inductor from the field solutions for use in a circuit simulator such as SPICE. However, a disadvantage of 3-D numerical simulation is that some sophistication on behalf of the user is required in order to get meaningful results. Another drawback is the limitation that processing speed and memory size place upon the structures which can be analyzed within a reasonable amount of time. Optimization of the Q-factor in a planar structure requires closely spaced microstrip lines for tight coupling of the magnetic field and wide microstrip lines in order to reduce the Ohmic losses in the microstrip. Hence, a large ratio of linewidth to line spacing is necessary, and this requires a substantial amount of computer memory and computation time in order to determine the circuit model parameters if a commercially available field solver package is employed.

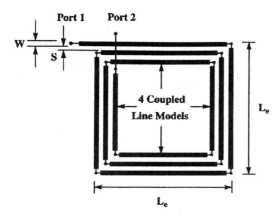

Port 1 Port 2

4 Coupled
Line Models

L_e

L_e

Fig. 1. Simplified microstrip spiral inductor layout (three turns).

TABLE I
MICROSTRIP LINE PARAMETERS [24]

Parameter	Value
Substrate resistivity, ρ	10 Ohm-cm
Substrate thickness, b_2	380 μm
Silicon dielectric constant	11.7
Oxide thickness, b_1	5 μm
Oxide dielectric constant	3.9
Metal resistivity	0.03 Ohm-μm

A. Scalable Inductor Model

A scalable inductor circuit model allows the circuit designer the flexibility to tailor the inductor design for a given RF application. Scalability implies that the electrical circuit parameters of the inductor model can be extracted from geometric and technological parameter specifications. A fully scalable model will allow any inductance value to be designed for use in an RF circuit. It also simplifies integration of the model into a typical CAD framework. The ability to generate inductor parameters on-demand eliminates the need to develop an inductor design library, which can be costly to generate and maintain.

Ohmic losses in the conductive substrate must be accounted for in the scalable model. These losses can be related to the three possible modes for the propagation of signals on transmission lines fabricated in the SiO_2/Si system: the skin-effect, slow-wave, and quasi-TEM modes. Lumped-element equivalent circuits for each propagating mode were proposed by Hasegawa [6], where the shunt parasitic components of the microstrip line can be represented (in general) by a combination of two capacitors and a resistor. This simplifying approximation can be applied to the spiral inductor as shown in Fig. 1, where the inductor is shown as a collection of short microstrip transmission line sections. Each transmission line section is physically short in length, and is therefore adequately described by a lumped-element model that includes series elements to model the inductance and resistance per unit length and shunt elements to model the substrate parasitics and losses. These lumped element models are then joined serially to model the entire spiral structure. The electric and magnetic coupling between parallel conducting strips must be accounted for in the model; however, the weak coupling between orthogonal strips is neglected in order to reduce the model complexity.

This segmented approach to microstrip inductor modeling was originated by Greenhouse [22], refined by others [23], and then extended to an arbitrary configuration of orthogonal microstrip lines by Rabjohn [3]. The flexibility and computational efficiency afforded by this approach have been adopted for the scalable inductor model developed in this work. The model capabilities have been extended to include the Ohmic losses in the conductive substrate.

A computer program, GEMCAP2 [3], [4], was used to extract the electrical parameters for a lumped-element circuit model from the inductor's specifications. The parameters of the lumped-element model for each microstrip line are computed from the layout geometry, using the substrate and metallization properties. Table I lists the technological parameters for the BiCMOS process [24] used to fabricate the inductors described in this paper. The self and mutual inductances for all parallel line segments are calculated from closed-form expressions, where the nonzero metallization thickness is incorporated in the self and mutual inductance calculations using the geometric mean distance of the conductor cross section [25].

The self-inductance, L (in nH), of a straight conductor with a rectangular cross section can be calculated from Grover's formulation for the mutual inductance between two current carrying filaments [26]. The self and mutual capacitances are computed using a two-dimensional (2-D) numerical technique developed for coupled microstrip lines [27]. The shunt resistance of the semiconducting layer can then be estimated directly from the quasistatic capacitance, C_{Si} [6]. Dissipation of the mutual capacitances is neglected because the microstrip lines are closely spaced. The frequency dependent resistance, r_{sk}, is approximated from closed form expressions [28] to form a complete lumped-element equivalent circuit representation.

As an example of the GEMCAP2 analysis technique, a circuit model can be derived for the spiral inductor illustrated in Fig. 1. The physical layout of the microstrip spiral is first partitioned into groups of multiple coupled lines for analysis; one group per side of the rectangular layout as shown in the figure. The parameters of a lumped-element π-equivalent circuit are then extracted for the individual microstrip lines in each group. Four such π-equivalents are required for a single turn of the spiral, as illustrated in Fig. 2, as well as lumped capacitors to model mutual capacitive coupling between the lines (represented by capacitors C_m in Fig. 2). Dependent current sources account for the mutual magnetic coupling between parallel strips in a group of coupled lines.

As the number of turns of the spiral increases, more lumped element sections are added to account for the additional coupled lines within each group. For an eight-turn spiral inductor with $N = 8$ microstrip lines per side, for example, there would be $4N$ or 32 lumped-element sections in total, along with the additional interconnecting elements to model the mutual capacitance between strips as in Fig. 2.

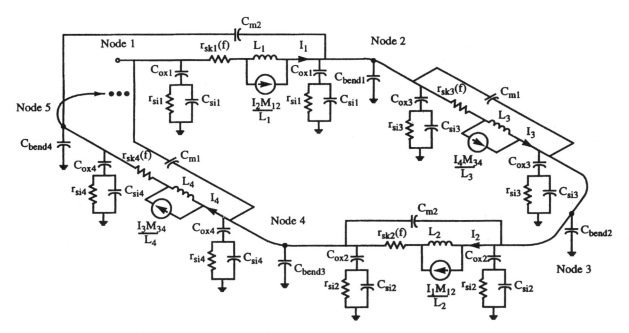

Fig. 2. Lumped-element circuit model for one turn of a spiral inductor.

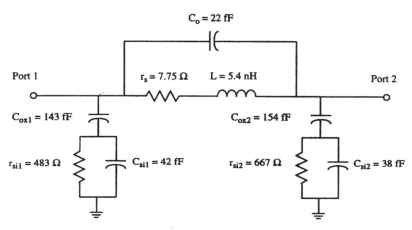

Fig. 3. Compact lumped-element circuit model for a spiral inductor ($N_t = 4.5$, $W = 10\,\mu m$, $S = 5\,\mu m$).

There are a number of other parasitics and higher-order effects which should be taken into consideration. Interwinding capacitance to the conductor used to contact the center of the spiral (usually an "underpass" in a multilevel metal process) can be modeled using lumped capacitors between the external terminal and the appropriate lumped-element sections. These capacitors are not shown in Fig. 2. Current crowding at the corners of the rectangular spiral adds parasitic inductance and capacitance which is accounted for by a connection of lumped elements L_{bend} and C_{bend} at each corner node (note that only C_{bend} is shown in Fig. 2). For frequencies in the low GHz range, this effect is quite small and is often neglected [29]. Coefficients are also added to the dissipative components of the model to account for the dependence of the inductor losses upon operating temperature.

The scalable circuit model can be used directly in a time domain (e.g., SPICE transient analysis) or frequency domain (e.g., Touchstone) circuit simulation along with other active and passive RF circuit elements. The complete model would normally be reduced to a compact model or S-parameter representation for faster optimization of a complex RF circuit.

B. Compact Inductor Model

A simplified or compact version of the fully scalable inductor model (i.e., a model with the minimum possible number of circuit elements) is required for hand calculations and to facilitate optimization of more complex RF circuit structures. For efficient optimization, the compact model should be easily derived from the complex scalable model structure.

A single π-type lumped-element section (as shown in Fig. 3) has been used by others as a compact model to fit experimental measurements of silicon monolithic inductors [10], [16], [17]. The L, C, and r parameters of the compact model can be established through a combination of parameter identification and fitting with the aid of a computer-driven optimizer.

Frlan [30] has shown that the compact model can be estimated directly from the parameters of the scalable model. Using this technique, the series resistance and inductance of the single π-model (r_s and L in Fig. 3) are obtained by simply summing the inductance and resistance of each individual microstrip section connected in series. The parasitic capacitances of the individual sections can be similarly summed into a single lumped capacitance for the compact model, with one shunt arm of the π-model representing the outer windings and the other arm modeling the inner winding parasitics. Resistances r_{Si} in the compact model are computed from the shunt capacitance, C_{Si}, making the complete parameter set consistent. From this initial estimate, it is possible to closely match to the electrical characteristics of the scalable model by refinement of the parameters using a computer optimizer. This approach is simple to implement with most commercial simulators that have the capability to optimize circuit parameters (e.g., HSPICE or Touchstone), and it requires little additional computation time to refine the small number of parameters in a single lumped-element section. The component values for the compact model shown in Fig. 3 were derived by estimation and fitting the scalable (GEMCAP2) model of a 4.5-turn spiral inductor. The shunt parasitics are not symmetric (i.e., C_{ox1} is not equal to C_{ox2}), which is a consequence of the asymmetry inherent in the spiral inductor layout. This model is simple enough to be useful for hand calculations in a circuit and it fits the GEMCAP2 model up to the first self-resonant frequency of the inductor. However, the parameters of the compact model cannot be easily adjusted for slight changes in the inductor design because of the nonphysical nature of this simple model.

C. Q-Factor Extraction and Computation from a Compact Model

The inductor Q-factor is used as a figure of merit to compare the performance of the spiral inductors studied in this work. The Q factor is defined by the ratio of the inductive reactance to the total dissipation, $(\omega L_S)/r_T$, for an equivalent circuit consisting of inductance L_S in series with a resistor (representing the total dissipation), r_T [31]. A measurement or simulation of the impedance parameters for a monolithic inductor will normally include the effect of capacitive parasitics, and therefore the inductive reactance and the total dissipation must be properly identified in order to determine the component Q. For a one-port, the Q-factor is often estimated by taking the ratio of the imaginary and real components of the one-port impedance. While this approximation is valid at low frequencies (less than 500 MHz), a significant error is caused by the parasitic capacitances of a spiral inductor as the frequency increases. This error is avoided by computing the Q-factor directly from the parameters of the compact model.

The procedure for compact model extraction from the fully scalable inductor model was outlined in the previous section of this paper. For experimental data, the compact model parameters (see Fig. 3) are determined by first estimating the inductance, L, and series resistance, r_s, from the impedance measured between Ports 1 and 2 at low frequency (i.e., where

the parasitic capacitances have little effect), and then fitting the remaining parameters of the model using a computer-driven optimizer. Once the compact model parameters have been determined, the inductive reactance, the total dissipation, and the Q-factor are easily computed. For example, the Q-factor of an inductor connected as a one-port (i.e., with Port 2 in Fig. 3 grounded) can be estimated from the following equation:

$$Q\text{-factor} \approx \frac{\omega L}{r_s + \dfrac{\left(\dfrac{\omega}{\omega_{ox}}\right)^4 \cdot r_{\text{Si1}}}{1 + (\omega C_{ox1} r_{\text{Si1}})^2}} \qquad (3)$$

where ω_{ox} is the oxide resonant frequency defined by inductance L and capacitance C_{ox1}. Here the parasitic capacitances C_o and C_{Si} of the compact model have been assumed negligible in order to simplify the resulting expression. The two terms in the denominator represent the sources of dissipation in the monolithic inductor, which consist of the resistance of the metal lines (r_s) and the dissipation added by the conductive substrate.

Some insight into the relationships between circuit parameters and the Q-factor can be gained through examination of (3). It can be seen that decreasing C_{ox} through the use of a thicker oxide layer will increase both the oxide resonant frequency and the substrate corner frequency, $f_{sub} = 1/(2\pi C_{ox1} r_{\text{Si1}})$, resulting in an improvement in the inductor Q-factor. Equation (3) also predicts that an increase in the substrate resistivity will cause a drop in the Q factor for frequencies much less than f_{sub} and an improvement in Q-factor for frequencies much greater than f_{sub}. However, it should be realized that the lumped equivalent model for the inductor shown in Fig. 3 is invalid when highly conductive substrates are used (less than approximately 1 Ω-cm), and hence (3) can only be used to approximate the inductor Q over a limited range of substrate resistivities.

V. MODEL VALIDATION

In order to verify the accuracy of the scalable model, the behavior of a spiral inductor predicted by the lumped-element model (GEMCAP2 simulator) was compared with a 3-D electromagnetic field simulation. The 3-D numerical simulations were performed using a full-wave simulator for planar structures [21]. A 4.5-turn square spiral 5 nH inductor with a 10-μm wide line and a 5-μm line spacing was selected for this comparison. The ratio of line width to line spacing was kept small to ensure that the memory requirements and simulation time for the 3-D numerical computations were not excessive. It would be possible to optimize the inductor design for the best possible circuit performance, such as optimization of the inductor Q-factor, but this is not necessary for model validation.

The substrate and metallization parameters from the BiCMOS process from Table I were used to obtain the simulation results shown in Fig. 4. The inner terminal of the spiral was connected to ground and the impedance of the resulting one-port network as a function of frequency was computed in both cases. Excellent agreement is seen between the real

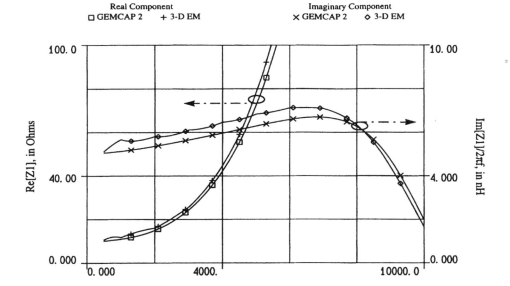

Real Component   Imaginary Component
□ GEMCAP 2   + 3-D EM   × GEMCAP 2   ◇ 3-D EM

Fig. 4.   Simulated one-port impedances for a 5 nH inductor ($N_t = 4.5$, $W = 10\,\mu$m, $S = 5\,\mu$m).

and imaginary parts of the one-port impedance simulated using numerical simulation (3-D EM) and the lumped-element model (GEMCAP2) almost up to the self-resonant frequency (i.e., where the imaginary part of the impedance is zero). Both simulations predict a low-frequency inductance (which is given by the reactive portion of the impedance divided by the radian frequency) close to 5 nH with a self-resonant frequency of approximately 10 GHz. The consistent difference in the reactive component of the impedance is due in part to the fact that the finite thickness of the conducting strip is not accounted for in the numerical simulations. An increase in the conductor thickness causes the inductance, and hence the inductive reactance, to decrease. At low frequencies, the resistive component of the impedance is close to the dc resistance of the inductor, which is about 8.5 Ω. As the frequency increases, the resistive component of the loss increases greatly due to losses in the substrate and skin effect in the metal conductors.

Design of an inductor to fit a particular application would require extensive simulation work in order to determine the number of turns, line width, and line spacing necessary to achieve the best performance within a given technology. Extraction of a lumped element component model is faster than other simulation methods, making this type of optimization more practical. In this particular example, the computation of the lumped-element model parameters was performed in less than 2 min on a SUN SPARC-2 work station. By comparison, 3-D EM simulation of the one-port impedance required approximately 5 min per frequency point on the same work station (equipped with 64 MBytes of RAM), or 50 min for 11 data points between 0.5–10 GHz. It should be noted that a large ratio of line width to line spacing is desirable in order to maximize the coupling between adjacent microstrip lines in the spiral inductor. The time required for a 3-D numerical simulation increases with the ratio of line width

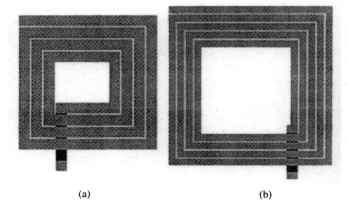

(a)   (b)

Fig. 5.   Inductor layouts used for characterization experiments.

to line spacing, while computation time for the lumped-element model is unaffected by the choice of line width or line spacing. This fact, combined with the large advantage in computational efficiency for even simple structures, lends a considerable advantage in both speed and flexibility to the lumped-element modeling approach when compared to 3-D numerical simulation.

VI. EXPERIMENTAL MEASUREMENT

The effect of layout geometry, metallization thickness, and substrate parameters upon the inductor performance has been investigated experimentally. The number of turns of the rectangular spiral, the metal line width, and the metal spacing were each varied independently in order to determine the effect of the physical layout upon the inductor performance. In addition, the effect of changes in the oxide thickness and the silicon substrate resistivity were also explored experimentally. A 3.5-turn spiral inductor with a nominal inductance of 1800 pH (as shown in part (a) of Fig. 5) was used for the metal and oxide thickness experiments, while the 4.5-turn 5 nH

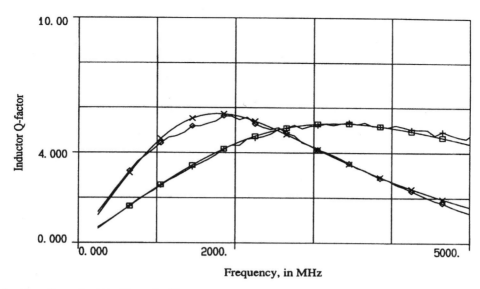

□ 5um sim.   + 5um meas.   ◇ 15um meas.   ✕ 15um sim.

Fig. 6. Q-factor as a function of metal width ($L = 5$ nH).

inductor (shown in part (b) of the same figure) was used to characterize the effect of layout and substrate variations upon the inductor performance. It should be noted that both of these designs exploit the large line width to line spacing ratio that is possible in modern silicon VLSI technologies and would require excessively long simulation times if 3-D numerical simulation were used to predict the inductor behavior. The results of the experimental measurements are useful as a benchmark to confirm the predictions of the lumped-element (i.e., GEMCAP2) computer model and also to provide insight into the possible optimization of different aspects of the inductor performance, such as the Q-factor.

The experimental data was collected from on-wafer measurements of the two-port S-parameters for each inductor test structure using coaxial RF probes. The measured data in each case was fitted to a compact model for the inductor, as described in Section IV-C, and the one-port impedance parameters (or the Q-factor) computed from the resulting parameter set by grounding Port 2 of the inductor compact model (refer to Fig. 3).

A. Inductor Layout (Metal Width, Spacing, and Number of Turns)

The measured Q-factor of a rectangular spiral inductor and the performance predicted by the lumped-element computer model for conductor widths of 5 and 15 μm are shown in Fig. 6. A layer of top-level aluminum 1 μm thick was used to fabricate a 4.5-turn spiral with an inductance of 5 nH in each case. The outside dimensions of the inductor (L_e in Fig. 1) were increased in order to maintain a constant inductance value in both cases.

At frequencies well below the peak in the inductor Q, the shunt parasitics of the spiral inductor have little effect and consequently the inductive reactance and inductor Q increase with frequency. However, as the operating frequency contin-

ues to rise, the dissipation of energy in the semiconducting substrate and the ac resistance of the metallization begin to increase faster than the inductive reactance. Thus, the Q-factor peaks, and then decreases. The larger surface area of the inductor with wide conductor metallization results in higher parasitic capacitances, which lowers the inductor self-resonant frequency and increases the substrate dissipation. This confirms the behavior predicted by (3). In addition, the distribution of current across the conducting strip is nonuniform as a result of the skin effect, causing the ac resistance of the metallization to increase at higher frequencies. As the strip widens, the penalty in ac resistance due to the skin effect will increase at a given frequency [28]. Thus, the peak of the inductor Q-factor shifts to a lower frequency as the conductor width increases. However, this can be used to advantage in the optimization of a given inductor for the maximum Q within a desired range of frequencies.

The effect of the spacing between conductor lines on inductor Q-factor has also been investigated. Narrow line spacings were found to increase the magnetic coupling between windings, which causes an increase in the inductance and the Q-factor for a given layout area as shown in Fig. 7. For the thin metallization used to fabricate on-chip inductors, interwinding capacitance was found to have a negligible effect upon the performance as shown in the figure. However, there is a significant improvement in the Q-factor when the minimum spacing is used.

The relationship between the number of turns of the spiral and the peak Q-factor for a given inductance was also investigated experimentally, and the results are listed in Table II. The outer dimensions of the spiral were varied in order to realize an inductance of 5 nH for each design. The line width and line spacing were kept constant at 10 and 1.5 μm, respectively. The drop in inductor Q as the number of turns increases is related to the distance (or gap) between opposite sides at the center

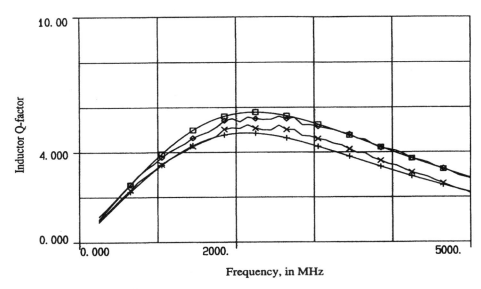

Fig. 7. *Q*-factor as a function of metal spacing ($L = 5$ nH).

TABLE II

NUMBER OF TURNS VERSUS THE PEAK INDUCTOR *Q*-FACTOR FOR A CONSTANT INDUCTANCE OF 5 nH ($W = 10\,\mu$m, $S = 1.5\,\mu$m)

Number of Turns	Peak Q-factor	Outer Length, L_e	Gap Between Opposite Sides	Total Length of Spiral
3.5	5.8	255 μm	177 μm	3.02 mm
4.5	5.7	216 μm	115 μm	2.98 mm
5.5	5.6	199 μm	75 μm	3.00 mm
6.5	5.3	191 μm	45 μm	3.06 mm
7.5	5.0	190 μm	20 μm	3.12 mm

of the spiral. As the number of turns increases, the spacing between opposite sides of the spiral shrinks, causing a drop in inductance because of negative mutual coupling. Thus, more metal is required to maintain a constant inductance value of 5 nH, and the total length of the spiral begins to increase as seen from the data listed in the table. Increasing the number of turns reduces the outer dimension of the spiral (L_e) and conserves chip area; however, some space is required at the center of the spiral in order to realize the highest *Q*-factor.

B. Metallization Thickness

Thin metallization in most VLSI processes limits the quality factor of microstrip inductors, because energy is dissipated by the finite resistivity of the metallization as well as in the conductive substrate. The quality of the spiral inductor, or *Q*-factor, could therefore be improved by increasing the conductor thickness. The measured quality factor for a 1.8-nH spiral inductor fabricated with 1–3 μm thick metal layers is plotted in Fig. 8. This inductor consists of 3.5 turns of 15-μm wide top-level metal in the BiCMOS process with a line spacing of 1.5 μm. The measured behavior of the inductor

fabricated with thicker metal shows close agreement with the predictions of the lumped-element circuit model. The *Q*-factor initially rises with frequency as the reactive component of the impedance increases, peaks, and then decreases due to the increasing dissipation of energy at higher frequencies. The inductor *Q*-factor at 3 GHz improves from five to ten when the thickness of the aluminum metal layer is increased from 1 to 3 μm. The improvement in inductor *Q* is less than a factor of three (as would be expected from resistance considerations) because the inductance is inversely proportional to the thickness of the conductor. Hence, the inductance decreases with increasing metal thickness, lowering the *Q*. The frequency dependence of the conductor resistance, or the skin effect, is more pronounced as the thickness of the metallization is increased, which also contributes to higher dissipation and a lower *Q*-factor at RF. The measured behavior of the inductor fabricated with thicker metal also showed close agreement with the predictions of the lumped-element circuit model. These results indicate that inductors suitable for many RF IC applications could be fabricated in production silicon technologies, if a low resistivity metal of adequate thickness were available.

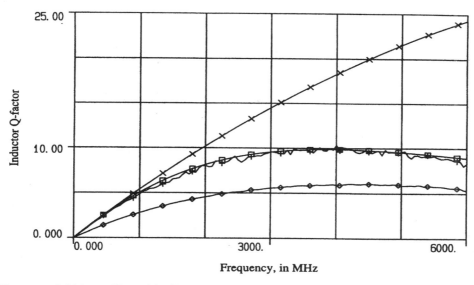

Fig. 8. Q-factor for different metal thickness ($L = 1.8$ nH).

The Q-factor predicted by GEMCAP2 for the same planar spiral on a lossless substrate (Ideal sub.) is also plotted in the Fig. 8 for comparison. The slight downward curvature is the result of capacitive parasitics which cause self-resonance. In addition to ac losses in the metallization, the data shows that the substrate conductivity also introduces a significant degradation in the component Q. This demonstrates that the effect of the substrate conduction on the inductor Q-factor is significant at higher frequencies, but also that operation close to the ideal curve is obtained for frequencies below approximately 1.5 GHz. These results also indicate that inductors with a Q-factor of ten, which is suitable for many RF IC applications, could be fabricated in silicon production technologies if a low resistivity metal of adequate thickness were available.

C. Substrate Effects

The ability to predict the substrate effects on the inductor performance is key to the design and realization of inductors with acceptable performance in silicon technology. When substrate effects are properly accounted for in a circuit simulation, the inductor design can be optimized to minimize these losses and improve circuit performance. In production silicon technologies, an insulating oxide layer separates the top level metal and the semiconducting silicon wafer. The thickness of the oxide layer is an important parameter which influences the inductor performance. The measured effect of changes in the oxide thickness upon the component Q factor is illustrated in Fig. 9, for the 1.8 nH inductor discussed previously (refer to Fig. 8, metal thickness of 3 μm). A thicker oxide layer reduces the parasitic capacitance of the structure, which improves the inductor self-resonant frequency. This is seen in Fig. 9 as a broadening of the inductor Q-factor with frequency as the oxide thickness increases. Performance closer to that predicted for an ideal (i.e., nonconductive) substrate can be achieved when the oxide thickness is maximized.

The measured behavior of the 5 nH spiral inductor of Fig. 6 is shown in Fig. 10 for two different substrate conductivities (10 and 0.01 Ω-cm). A layer of aluminum 1 μm thick and 10 μm wide was used to fabricate the 4.5-turn spiral in both cases. At low frequencies, the real or resistive component of the impedance (Re $[Z_1]$) is close to the dc resistance of the inductor, which is about 8.5 Ω. As the frequency increases, the resistive component increases, primarily due to dissipation in the conductive substrate. As seen in the figure, the dissipation of the inductor fabricated on a low resistivity substrate (0.01 Ω-cm) is substantially greater than for the substrate resistivity normally used in the BiCMOS process (10 Ω-cm). As a result, the peak Q-factor is reduced by a factor of two because of the increased dissipation, which makes the inductor unsuitable for most RF circuit applications.

In Fig. 10, the inductance, which is given by the reactive portion of the measured impedance (Im $[Z_1]$) divided by the frequency, is 5 nH at low frequencies in each case. The effect of parasitic capacitance between the metal and substrate is larger when a highly doped substrate is used to fabricate the inductor, which leads to a lower self-resonant frequency. The reactance of the inductor fabricated on the 0.01 Ω-cm substrate rises to a peak at 4 GHz and then falls rapidly as the inductor approaches self-resonance. However, the low frequency inductance is unaffected. Current flow in the substrate beneath the spiral would cause negative mutual coupling and thus a reduction in the low frequency inductance of the spiral. These results indicate that the substrate current that is induced by the magnetic field is small.

VII. COMPONENT TOLERANCES

It is important to characterize the effects of processing and temperature variations upon the parameters of a monolithic component in order to estimate the yield of working circuits that meet the design specifications. Parameter variations due

□ Ideal sub. + 1um oxide ◇ 3um oxide × 5um oxide

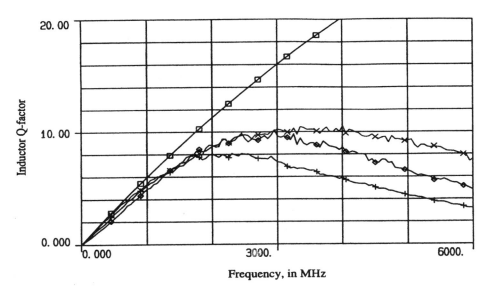

Fig. 9. *Q*-factor as a function of oxide thickness ($L = 1.8$ nH).

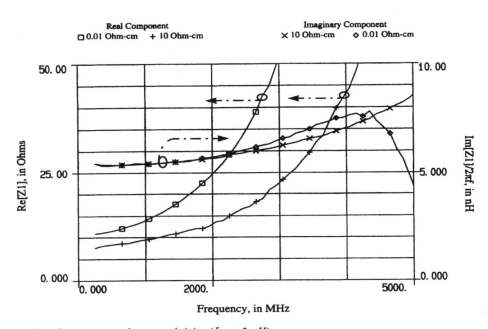

Fig. 10. Measured one-port impedance versus substrate resistivity ($L = 5$ nH).

to fluctuations in the ambient temperature should also be characterized and modeled so that circuit operation over a wide temperature range can be ensured during the design phase. Thus, the processing and temperature induced tolerances in the behavior of the monolithic inductor and transformer are of interest to the RF circuit designer.

The metal lines used in the inductor or transformer layout have dimensions on the order of microns, however, these dimensions are defined photolithographically to within one-tenth of a micron in a submicron IC process. The inductive reactance of a monolithic inductor is set by metal line width and line spacing, and thus the inductance is insensitive to the small lithographic variations in modern silicon IC processes. Simulations predict that the tolerance on the self and mutual inductances of the inductors described in this study will be less than 3% for a ± 0.2 μm change in the line width and line spacing. A larger and more subtle tolerance is introduced by variations in the intermetal oxide thickness and changes in substrate resistivity, both of which affect the parasitic capacitances of the inductor structure. However, simulations also predict that the variation in self-resonant frequency will be less than 5% for a ± 1 μm change in oxide thickness and a ± 50% change in substrate resistivity from the nominal BiCMOS process parameters listed in Table I. These tolerances are far less than those typically encountered for other passive monolithic components, such as resistors and capacitors.

Other sources of variation are the metallization and silicon substrate resistivities, both of which depend upon temperature. The aluminum–copper alloy typically used in silicon VLSI technologies (98% aluminum) has a temperature coefficient of +0.44%/°C [32], which can cause a large increase in the metal losses with increasing temperature. Other interconnect metals, such as gold, also exhibit a strong positive temperature coefficient (0.4%/°C). The shift in metal resistance with temperature directly affects the component quality factor [as seen from (3)], and it must be carefully considered in the design and optimization of an RF circuit.

The substrate resistivity also has a positive temperature coefficient, which is approximately +0.7%/°C [33]. As seen from (3), an increase in substrate resistivity with temperature can cause a degradation in the Q-factor at frequencies much less than the substrate corner frequency ($f_{sub} = 1/2\pi C_{ox} r_{Si}$), and an improvement in Q-factor for frequencies higher than f_{sub}. Choosing to operate in a frequency band close to the peak Q reduces the effect (over temperature) of the substrate resistivity upon the inductor performance.

VIII. INDUCTOR DESIGN AND OPTIMIZATION

The design of a spiral inductor in silicon VLSI technologies involves a complex tradeoff between the various layout and technological parameters, as demonstrated by the experimental results presented in the previous section of this paper. The following set of guidelines or "design rules" summarize the results of this study.

1) Maintain a space of at least five line widths (i.e., $5W$, or further if possible) between the outer turn of the spiral and any surrounding metal features. Parasitic electric and magnetic coupling between conductors is inversely proportional to the separation. Maintaining sufficient space between the inductor and its surroundings will keep unwanted parasitic effects from disturbing the inductor's electrical characteristics.

2) The magnetic coupling between adjacent metal lines is maximized by using the closest spacing (S) between lines that is allowed by the technology. The additional interwinding capacitance from tighter coupling of the electric field between adjacent conductors has only a slight impact on performance, given that metal thickness in most VLSI technologies is usually less than 3 μm. Tight coupling of the magnetic field maximizes the Q-factor and reduces the chip area for a given inductor layout.

3) A strip width of between 10–15 μm is close to the optimum for most inductor designs fabricated with the technological parameters listed in Table I. Increasing the conductor width causes a downward shift in the peak Q-factor and also makes the Q-factor more sensitive to changes in operating frequency. This tradeoff requires extensive computer simulation in order to establish the optimum strip width for a given technology and inductance value.

4) Magnetic flux must be allowed to pass through the center of the spiral. This ensures that negative mutual coupling between opposite sides of the inductor does not significantly affect the inductance and the Q-factor. Thus, the four groups of coupled lines which form the sides of the inductor must be spaced sufficiently apart. Following the previous guideline, a space of greater than five line widths (i.e., $5W$) is recommended.

5) The oxide layer which isolates the metal conductors from the silicon substrate should be kept as thick as possible in order to minimize shunt parasitics and dissipation.

6) The inductor Q is most sensitive to the thickness and resistivity of the metal layer used in fabrication. Although the metal thickness is fixed in production technologies, metal layers in a multilevel metal process can be connected in parallel (metal stacking) to reduce Ohmic losses. This technique has demonstrated a 20% improvement in Q-factor when compared to inductors fabricated using only top metal [17] in some technologies. However, there is a tradeoff between the improvement in metal thickness and the reduction in oxide thickness as a result of metal stacking.

The results of this study have shown that a silicon technology with a substrate resistivity in the 10 Ω-cm range and a 5-μm thick oxide between metal and substrate can be used to fabricate inductors with acceptable Q-factors in the 1–3 GHz frequency range. The metallization thickness has a large effect upon the inductor quality factor, and a Q-factor of five is achievable with 1-μm thick aluminum top metal. However, a Q-factor of ten can be achieved if the metal thickness is increased to 3 μm.

A substantial improvement in inductor performance can also be realized through careful optimization of other aspects of the inductor layout using the lumped-element computer model. A plot of a Q-factor surface (at 1.9 GHz) that is defined by variations in the conductor width and the number of turns of the spiral is shown in Fig. 11. The substrate parameters for the BiCMOS technology given in Table I were used for the simulations, with a 1.5-μm line spacing. The points corresponding to an inductance of 5 nH are plotted onto the surface with an asterisk. These points indicate that an optimum geometry can be identified which maximizes the inductor Q-factor for a given inductance. The spiral geometry which yields the highest Q-factor at 1.9 GHz is also easily identified from this figure. The 3-D plot is a powerful design aid which can be used to visualize the effects of the many parameters which influence the inductor performance. A fast and efficient computer-based model is required in order to generate these types of design aids.

IX. CONCLUSIONS

A scalable lumped-element computer model has been described which adequately models the performance of inductors fabricated in a silicon VLSI process for a wide range of layout geometries. Both the measured and simulated results

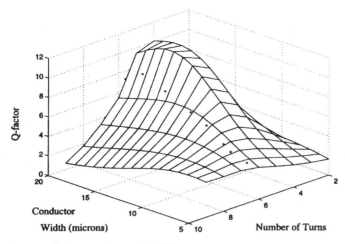

Fig. 11. Q-factor surface at 1.9 GHz.

indicate that the quality factor of the inductor is dominated by losses in the metallization at lower frequency and can be improved through the use of thicker metal in fabrication. Losses introduced by the conductive substrate tend to dominate performance at higher frequencies. However, these losses do not significantly degrade the performance in the low GHz frequency range, making these devices potentially useful in many RF IC applications. Substrate losses can be minimized by proper selection of the conductor width, spacing, and metal thickness for a given set of substrate parameters. The lumped element modeling approach is computationally efficient, fully scalable, and can be used to develop optimized inductor and transformer designs for various RF circuit applications.

ACKNOWLEDGMENT

The authors appreciate fabrication support and services provided by Nortel Semiconductor and Nortel Technology Ltd. Special thanks to A. Veluswami for his assistance in the preparation of graphics for this manuscript.

REFERENCES

[1] T. M. Hyltin, "Microstrip transmission on semiconductor dielectrics," *IEEE Trans. Microwave Theory Tech.*, vol. MTT-13, pp. 777–781, Nov. 1965.

[2] P. R. Gray and R. G. Meyer, "Future directions in silicon IC's for RF personal communications," in *Proc. Custom Integrated Circuits Conf.*, Santa Clara, CA, 1995, pp. 83–89.

[3] G. G. Rabjohn, "Monolithic microwave transformers," M.Eng. thesis, Carleton University, Apr. 1991.

[4] J. R. Long and M. A. Copeland, "Modeling of monolithic inductors and transformers for silicon RFIC design," in *Proc. IEEE MTT-S Int. Symp. Tech. Wireless Appl.*, Vancouver, Canada, Feb. 1995, pp. 129–134.

[5] ——, "Modeling, characterization and design of monolithic inductors for silicon RFIC's," in *Proc. Custom Integrated Circuits Conf.*, San Diego, CA, 1996, pp. 185–188.

[6] H. Hasegawa, M. Furukawa, and H. Yanai, "Properties of microstriplines on Si–SiO$_2$ system," *IEEE Trans. Microwave Theory Tech.*, vol. MTT-19, pp. 869–881, Nov. 1971.

[7] S. Chaki, S. Aono, N. Andoh, Y. Sasaki, N. Tanino, and O. Ishihara, "Experimental study on spiral inductors," in *Proc. IEEE Microwave Symp. Dig. MTT-S*, Orlando, FL, 1995, pp. 753–756.

[8] I. Bahl and P. Bhartia, *Microwave Solid State Circuit Design.* New York, NY: Wiley, 1988, ch. 2.

[9] R. A. Pucel, "Design considerations for monolithic microwave circuits," *IEEE Trans. Microwave Theory Tech.*, vol. MTT-29, pp. 513–534, June 1981.

[10] N. M. Nguyen and R. G. Meyer, "Si IC-compatible inductors and LC passive filters," *IEEE J. Solid-State Circuits*, vol. 25, pp. 1028–1031, Aug. 1990.

[11] ——, "A silicon bipolar monolithic RF bandpass amplifier," *IEEE J. Solid-State Circuits*, vol. 27, pp. 123–127, Jan. 1992.

[12] ——, "A 1.8 GHz LC voltage-controlled oscillator," *IEEE J. Solid-State Circuits*, vol. 27, pp. 444–450, Mar. 1992.

[13] K. Negus, B. Koupal, J. Wholey, K. Carter, D. Millicker, C. Snapp, and N. Marion, "Highly integrated transmitter RFIC with monolithic narrowband tuning for digital cellular handsets," in *Proc. Int. Solid-State Circuits Conf.*, San Francisco, CA, 1994, pp. 38–39.

[14] T. D. Stetzler, I. G. Post, J. H. Havens, and M. Koyama, "A 2.7–4.5 V single chip GSM transceiver RF integrated circuit," *IEEE J. Solid-State Circuits*, vol. 30, pp. 1421–1429, Dec. 1995.

[15] C. Marshall *et al.*, "A 2.7 V GSM transceiver IC with on-chip filtering," in *Proc. Int. Solid-State Circuits Conf.*, San Francisco, CA, 1995, pp. 148–149.

[16] K. B. Ashby, I. A. Koullias, W. C. Finley, J. J. Bastek, and S. Moinian, "High Q inductors for wireless applications in a complementary silicon bipolar process," *IEEE J. Solid-State Circuits*, vol. 31, pp. 4–9, Jan. 1996.

[17] J. N. Burghartz, M. Souyer, and K. A. Jenkins, "Microwave inductors and capacitors in standard multilevel interconnect silicon technology," *IEEE Trans. Microwave Theory Tech.*, vol. 44, pp. 100–104, Jan. 1996.

[18] L. E. Larson, M. Case, S. Rosenbaum, D. Rensch, *et al.*, "Si/SiGe HBT technology for low-cost monolithic microwave integrated circuits," in *Proc. Int. Solid-State Circuits Conf.*, San Francisco, CA, 1996, pp. 80–81.

[19] J. Y.-C. Chang, A. A. Abidi, and M. Gaitan, "Large suspended inductors on silicon and their use in a 2-μm CMOS RF amplifier," *IEEE Electron Device Lett.*, vol. 14, pp. 246–248, May 1993.

[20] D. Lovelace, N. Camillieri, and G. Kannell, "Silicon MMIC inductor modeling for high volume, low cost applications," *Microwave J.*, pp. 60–71, Aug. 1994.

[21] *EM-Sonnet Software User's Manual*, Version 2.3, Sonnet Software Inc., Apr. 10, 1992.

[22] H. M. Greenhouse, "Design of planar rectangular microelectronic inductors," *IEEE Trans. Parts, Hybrids, Packaging*, vol. PHP-10, pp. 101–109, June 1974.

[23] D. Krafesik and D. Dawson, "A closed-form expression for representing the distributed nature of the spiral inductor," in *Proc. IEEE-MTT Monolithic Circuits Symp. Dig.*, 1986, pp. 87–91.

[24] R. Hadaway *et al.*, "A sub-micron BiCMOS technology for telecommunications," *J. Microelectronic Eng.*, vol. 15, pp. 513–516, 1991.

[25] J. C. Maxwell, *A Treatise on Electricity and Magnetism, Parts III and IV.* New York, NY: Dover, 1st ed. 1873, 3rd ed. 1891, reprinted 1954.

[26] F. W. Grover, *Inductance Calculations.* Princeton, NJ: Van Nostrand, 1946, reprinted by Dover Publications, New York, NY, 1954.

[27] D. Kammler, "Calculation of characteristic admittances and coupling coefficients for strip transmission lines," *IEEE Trans. Microwave Theory Tech.*, vol. MTT-16, pp. 925–937, Nov. 1968.

[28] E. Pettenpaul, H. Kapusta, A. Weisgerber, H. Mampe, J. Luginsland, and I. Wolff, "CAD models of lumped elements on GaAs up to 18 GHz," *IEEE Trans. Microwave Theory Tech.*, vol. MTT-36, pp. 294–304, Feb. 1988.

[29] R. J. P. Douville and D. S. James, "Experimental study of symmetric microstrip bends and their compensation," *IEEE Trans. Microwave Theory Tech.*, vol. MTT-26, pp. 175–181, Mar. 1978.

[30] E. Frlan, "Miniature hybrid microwave integrated circuit passive component analysis using computer-aided design techniques," M.Eng. thesis, Carleton University, 1989.

[31] W. H. Hayt and J. E. Kemmerly, *Engineering Circuit Analysis*, 3rd ed. New York, NY: McGraw-Hill, 1978.

[32] D. R. Lide, Ed., *CRC Handbook of Chemistry and Physics.* New York, NY: CRC, 1996, pp. 12-46–12-49.

[33] N. D. Arora, J. R. Hauser, and D. J. Roulston, "Electron and hole mobilities in Si as a function of concentration and temperature," *IEEE Trans. Electron Devices*, vol. ED-29, pp. 292–295, 1982.

Analysis, Design, and Optimization of Spiral Inductors and Transformers for Si RF IC's

Ali M. Niknejad, *Student Member, IEEE*, and Robert G. Meyer, *Fellow, IEEE*

Abstract—Silicon integrated circuit spiral inductors and transformers are analyzed using electromagnetic analysis. With appropriate approximations, the calculations are reduced to electrostatic and magnetostatic calculations. The important effects of substrate loss are included in the analysis. Classic circuit analysis and network analysis techniques are used to derive two-port parameters from the circuits. From two-port measurements, low-order, frequency-independent lumped circuits are used to model the physical behavior over a broad-frequency range. The analysis is applied to traditional square and polygon inductors and transformer structures as well as to multilayer metal structures and coupled inductors. A custom computer-aided-design tool called ASITIC is described, which is used for the analysis, design, and optimization of these structures. Measurements taken over a frequency range from 100 MHz to 5 GHz show good agreement with theory.

Index Terms— Monolithic inductors and transformers, optimization of Si inductors and transformers, spiral inductors and transformers.

I. INTRODUCTION

SILICON integrated circuits (IC's) are finding wide application in the gigahertz frequency range. Modern bipolar (Bi), complementary metal–oxide–semiconductor (CMOS), and BiCMOS processes provide high f_T transistors, allowing Si radio-frequency (RF) IC's to compete with GaAs IC's in the important low-gigahertz frequency ranges. However, the lossy Si substrate makes the design of high Q reactive components difficult. Despite this difficulty, the low cost of Si IC fabrication over GaAs IC fabrication and the potential for integration with baseband circuits makes Si the process of choice in many RF IC applications.

The demands placed on portable wireless communication equipment include low cost, low supply voltage, low power dissipation, low noise, high frequency of operation, and low distortion. These design requirements cannot be met satisfactorily in many cases without the use of RF inductors. Hence, there is a great incentive to design, optimize, and model spiral inductors fabricated on Si substrates. This topic is addressed in this paper.

Since the introduction of Si spiral inductors [1], many authors have reported higher performance inductors on Si substrates, primarily utilizing advances in processing technology. This has included higher conductivity metal layers to reduce

Manuscript received May 27, 1997; revised November 17, 1997. This work was supported by the U.S. Army Research Office under Grant DAAH04-93-G-0200.

The authors are with the Department of Electrical Engineering and Computer Science, University of California, Berkeley, CA 94720-1772 USA.

Publisher Item Identifier S 0018-9200(98)06998-4.

the loss resistance of the inductor [2], use of multimetal layers to increase the effective thickness of the spiral inductor and thereby reduce loss, the connection of multimetal layer spirals in series to reduce the area of the inductors [3], [4], low-loss substrates to reduce losses in the substrate at high frequency [5], and thick oxide or floating inductors to isolate the inductor from the lossy substrate [6]. Little, though, has been written on the analysis and optimization of these structures.

In [7], an analysis approach is presented where an equivalent circuit for each segment of the spiral is calculated and the inductor is considered as an interconnection of such segments. The approach is limited, though, as many important effects are not included. For instance, nonuniform current distribution due to skin and proximity effects within each segment is not considered. In addition, the impedance to substrate is calculated using a two-dimensional approach, making it difficult to apply to arbitrary structures or to coupled inductors.

Most past researchers have used measurement results on previously built inductors to construct models. While this technique is most practical, it does not allow the possibility of optimization, nor does it allow the circuit designer freedom to choose parameters such as inductance, resistance, capacitance, and Q. Alternatively, researchers have used commercial three-dimensional electromagnetic simulators [8], [9] to design and analyze inductors and transformers. While this approach is accurate, it can be computationally very expensive and time consuming. This prevents the designer from performing optimization. In this paper, we present an accurate and computationally efficient approach to overcome some of these difficulties.

II. EQUIVALENT CIRCUIT FORMULATION

A. Electromagnetic Formulation

Consider a typical spiral inductor or transformer. It consists of series and parallel interconnection of metal segments. Applying Maxwell's equations to the conductive portions of such a structure, we obtain [10]

$$\int E_0 \cdot dl - \int \frac{J}{\sigma} \cdot dl - \int \frac{\partial A}{\partial t} \cdot dl - \int \nabla \phi \cdot dl = 0 \quad (1)$$

where E_0 is the applied field, J is the current density, σ is the conductivity of the metal segment, A is the magnetic vector potential, and ϕ is the electric scalar potential. The first term of the above equation arises from the applied field, the second term represents the internal impedance of the segments, the

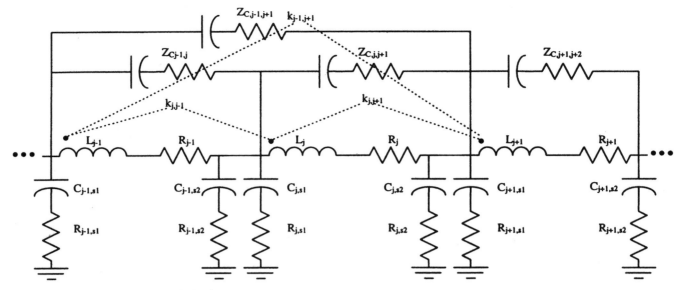

Fig. 1. Equivalent circuit model of spiral segments.

third term represents external inductance, and the fourth term is the capacitive term.

The magnetic and electric potentials are integrals over the charges and currents in the circuit. In the time-periodic case, we have

$$\phi(x, y, z) = \int_v \frac{\rho(x', y', z')e^{-jkR}}{4\pi\varepsilon R} dV' \qquad (2)$$

$$A(x, y, z) = \mu \int_v \frac{J(x', y', z')e^{jkR} dV'}{4\pi R}. \qquad (3)$$

The above retarded potentials can be simplified to static potentials since at frequencies of interest, the exponential term is nearly unity. As they stand, though, the above equations are impractical since they involve integrating over all charges and currents, including those that flow through the substrate. If, however, we replace the R^{-1} term involving the free-space Green function with the appropriate Green functions over a multilayer substrate, the volume of integration reduces to metal segments containing the charge and currents. In [11], we derive the electrostatic Green function over a multilayer conductive substrate that can be used in (2).[1]

By expanding A and ϕ in (1) using (2) and (3) and by considering the currents and charges in each segment separately, we obtain a system of linear equations [12]. Each equation has terms representing the internal impedance of each segment, as well as the mutual magnetic and electrical coupling between each segment and every other segment, and finally terms representing the magnetic and electric coupling to the substrate.

B. Circuit Equation Formulation

From a circuit point of view, (1) can be interpreted as shown in Fig. 1. To derive the two-port parameters of such a structure, we must solve the circuit equations by selecting

an appropriate number of independent equations, using, for instance, nodal or loop analysis. Let us consider the topology of the network in relation to the spiral. For a typical p-sided spiral of n turns, there are $s = n \times p$ segments. Each segment has one branch associated with the series inductance and resistance and one branch associated with the substrate impedance. Assuming a dense capacitive coupling impedance matrix, this adds an additional $s(s-1)/2$ branches of lossy capacitors from segment to segment. Hence, the total number of branches $B = (s+1)(s+2)/2$. The total number of nodes N in the circuit is simply one plus the number of segments, or $s+1$. Consequently, the total number of fundamental loops L in the circuit equals $L = B - N \sim O(s^2)$.

Although loop equations are a more natural formulation, due to the mutual inductive nature of the circuit, the number of required equations is one order of magnitude larger than the number of node equations. Hence, nodal analysis is the method of choice. Writing the nodal equations at node j results in

$$\frac{V_j}{Z_{j0}^C} + \sum_{k \neq j} \frac{V_j}{Z_{jk}^C} - \sum_{k \neq j} \frac{V_k}{Z_{jk}^C} + I_j - I_{j-1} = 0. \qquad (4)$$

In the above equation, V_j is the voltage at node j, I_j is the current flowing into segment j, and Z_{jk}^C is the lossy capacitive impedance coupling segment j and k. The above equations involve the voltages and currents in the segments, similar to modified nodal analysis. The voltage across each segment is related to the currents flowing in the segments by

$$V_j - V_{j+1} = \sum_k Z_{jk}^L I_k \qquad (5)$$

where Z_{jk}^L is an entry of the symmetric inductance matrix Z^L. The diagonal terms represent the internal impedance of each segment, whereas the off-diagonal terms represent the magnetic coupling between segments. The above system of equations yields an invertible complex frequency-dependent matrix. Since there are $2s$ equations in (4) and (5), the system can be solved directly using Gaussian elimination with

---

[1] The magnetostatic Green function can be derived in an analogous way. The computation of the inductance matrix over a lossy conductive substrate is the subject of another paper.

pivoting or LU factorization, in time $O((2s)^3)$. However, the matrix naturally partitions into four square matrices as

$$\begin{bmatrix} Y^C & D^T \\ D & -Z^L \end{bmatrix} \begin{bmatrix} V \\ I \end{bmatrix} = \begin{bmatrix} I_s \\ V_s \end{bmatrix}. \tag{6}$$

Y^C is a matrix with elements

$$[Y_{ij}^C] = \frac{-1}{Z_{ij}^C} \quad [Y_{ii}^C] = \sum_k \frac{1}{Z_{ik}^C}. \tag{7}$$

D is an upper triangular band matrix with its diagonal entries as 1 and with superdiagonal entries as -1. Rewriting (6), we have

$$(Y^C + D^T Y^L D)V = I_s + D^T Y^L V_s \tag{8}$$
$$I = Y^L (DV - V_s) \tag{9}$$

where $Y^L = (Z^L)^{-1}$ is the inverse of the inductance matrix. As we will see later, Y^L is known *a priori,* and no explicit matrix inversion needs to be performed. In addition, due to the simple structure of D, the matrix products involving D and its transpose can be done in order $O(s^2)$. Hence, in forming (8) and (9), no explicit matrix products need to be formed. Thus, we can solve (8) by using an LU decomposition of $(Y^C + D^T Y^L D)$ and then solve (9) directly for I. Therefore, the solution of (6) is dominated by one LU decomposition of an $s \times s$ matrix, which can be performed in time $O(s^3)$. This is more than eight times faster than solving (6) directly.

C. Inductance Matrix

The inductance matrix Z^L must be computed to solve (6). At low frequency, this matrix may be computed easily since the self- and mutual inductance of each segment can be approximated in closed form by using the geometric mean distance (GMD) approximation. Grover [13] provides formulas for such calculations for fairly arbitrary configurations. In [14], an explicit formula appears for the GMD between two finite thickness parallel rectangular cross sections. In a square spiral, all segments are parallel and these formulas suffice. However, for polygon spirals or for arbitrary structures, we need the mutual inductance between nonparallel segments situated in general in three-dimensional space. While the exact formula for filaments placed in these configurations is known in closed form, the GMD concept is difficult to generalize in these cases, as the integrals are not easily evaluated. In addition, at high frequency, these approximations fail altogether as the current distributions become nonuniform and the GMD approximation fails.

The nonuniformity in the current distribution is due not only to the skin effect but also to proximity effects of neighboring current segments. Hence, although many past authors [7] have analyzed and developed closed-form equations for current constriction in an isolated conductor, these results are not directly applicable to the spiral inductor due to proximity effects.

In [14], a nice technique is developed to handle this problem without abandoning the closed-form equations available. The current distribution in the conductor can be approximated in a step-wise fashion by breaking up each segment into subsegments of uniform current distribution. Although this technique is general and accurate, it is computationally expensive. Consider a typical ten-sided spiral with ten turns as an example. Such a spiral has 100 segments, and if we were to divide each segment further into five parallel segments, this would involve 500 segments and a corresponding 500×500 matrix inversion. To alleviate the demand on computational resources, we will assume that the segment current distribution is only influenced by a small set of neighboring segments. For a typical planar spiral, this might involve only two neighboring segments.

We rederive the results presented in [14] with this approximation in mind. For a set of N segments, let the (i, j) segment represent the jth subsegment of segment i and let N_i denote the number of such subsegments in segment i. Let I_{ij} be the current in the (i, j) segment and let I_i be the total current in segment i. With this notation, the voltage across the (i, j) segment is given by

$$V_{ij} = \sum_{k=0}^{N} \sum_{m=1}^{N_k} (r_{ij} \delta_{ki} \delta_{jm} + j\omega L_{ij,km}) I_{km}. \tag{10}$$

In the above equation, r_{ij} is the resistance of the (i, j) conductor, δ_{ij} is the Kroniker delta function, and $L_{ij,km}$ is the mutual inductance between subsegments (i, j) and (k, m). Let C_i denote the set of segments in the neighborhood of segment i, or the set of segments that influence the current distribution of segment i [15]. Equation (10) may be broken up as follows:

$$V_{ij} = V_{ij}^L + V_{ij}^G = \sum_{k \in C_i} \sum_{m=1}^{N_k} Z_{ij,km} I_{km} + \sum_{k \notin C_i} j\omega L_{ik} I_k. \tag{11}$$

Inverting the first term corresponding to the local voltage at segment (i, j), we obtain

$$I_{ij} = \sum_{k \in C_i} \sum_{m=1}^{N_k} Y_{ij,km} V_{km}^L. \tag{12}$$

The total current in segment i is given by

$$I_i = \sum_{j=1}^{N_i} I_{ij} = \sum_{k \in C_i} \sum_{j=1}^{N_i} \sum_{m=1}^{N_k} Y_{ij,km} V_{km}^L. \tag{13}$$

If we make the reasonable assumption that the voltage across each subsegment (i, j) is independent of the index j (as was done in [14]), then (13) becomes

$$I_i = \sum_{k \in C_i} V_{km}^L \sum_{j=1}^{N_i} \sum_{m=1}^{N_k} Y_{ij,km} = \sum_{k \in C_i} V_{km}^L Y_{ik}^L. \tag{14}$$

Equation (11) may be rewritten as

$$V_i = \sum_{k \in C_i} Z_{ik}^L I_k + \sum_{k \notin C_i} j\omega L_{ik} I_k. \tag{15}$$

Repeating for all i, one generates the desired matrix equation. The above technique involves one matrix inversion going from (11) to (12) and another matrix inversion of reduced order going from (14) to (15). The latter matrix inversion can be neglected since the matrix is small. Assuming that matrix inversion is the most computationally intensive operation—and that it is done $O(s^3)$, whereas the original technique presented in [14] is $O[(s \times r)^3]$, where s is the total number of segments and r is the average number of subsegments—this technique is $O[s \times (q \times r)^3]$, where q is one plus the average number of neighbors for each segment. Hence, this technique is s^2/q^3 faster and demands s^2/q^2 less memory to store the matrix. For the example given earlier, a ten-sided, ten-turn spiral with two neighbors and five subsegments per segment, this means that this technique is 370 times faster and more than 1000 times less memory intensive.

D. Capacitance Matrix

Techniques for the calculation of the capacitance matrix are well developed and have wide applications. However, most approaches consider the capacitance matrix of conductors in free space, or over an ideal ground plane. In [11], we present an efficient technique that is directly applicable to the case of a conductive substrate.

With this approach, the true three-dimensional capacitance matrix of the device segments can be extracted and used in (6). Various effects can be included in the capacitance matrix calculation, such as the presence of substrate taps and substrate shields [16]. The important effect of substrate coupling between structures residing on the same substrate can also be included. For more details, refer to [11].

III. ASITIC: A COMPUTER-AIDED-DESIGN TOOL FOR INDUCTOR/TRANSFORMER DESIGN, ANALYSIS, AND OPTIMIZATION

Much is known about the optimization of the technological parameters of a process for optimal inductor performance [7]. Thicker or higher conductivity metal improves the quality factor Q at low frequencies, whereas a higher resistivity substrate and thicker oxide help to isolate the device from the substrate at high frequencies. The optimization of the geometry of inductors and transformers, though, is more difficult. Even for simple structures, such as square spirals, there are several parameters to optimize, including the area of the spiral, the metal width and spacing, and the number of turns. Not much can be said in general since the optimal geometry depends on the frequency of operation.

For instance, at low frequency, one usually uses the minimum spacing available to maximize magnetic coupling, but at high frequency, proximity effects and magnetic coupling favor a larger value of spacing [8]. This is illustrated in Fig. 2, where we plot the resistance of a typical spiral using two values of spacing. Similar considerations apply to the area of the spiral and the number of turns. At lower frequencies, large areas allow wider metal widths to meet a given inductance at lower values of series resistance and therefore higher Q. But at high frequencies, the Q is dominated by the substrate and smaller areas are favorable. This is doubly true for highly conductive substrates that suffer from eddy-current losses at high frequency [8].

With the introduction of multimetal structures, such as shunt and series connected spirals, the situation is even more

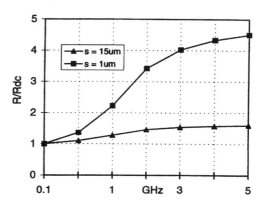

Fig. 2.   Normalized spiral resistance as a function of frequency.

TABLE I
APPROXIMATE PROCESS PARAMETERS

Metal 2	$R_{sh} = 33 m\Omega/sq$	$t = 1.27\mu m$	$C_{sub} = 14 aF/\mu m^2$
Metal 1	$R_{sh} = 50 m\Omega/sq$	$t = 1.00\mu m$	$C_{sub} = 21 aF/\mu m^2$
Metal 0	$R_{sh} = 100 m\Omega/sq$	$t = 0.40\mu m$	$C_{sub} = 105 aF/\mu m^2$
Buried Layer	$\rho = 0.085\Omega$-cm	$t = 1\mu m$	p^+ Si
Bulk Substrate	$\rho = 20\Omega$-cm	$t = 675\mu m$	p^- Si

complicated. Thus, the optimization of such structures must be done carefully on a case-by-case basis. To aid this process, a custom tool has been developed. The techniques presented in this paper have been collected into Analysis and Simulation of Inductors and Transformers for IC's (ASITIC),[2] a user-friendly computer-aided-design (CAD) tool designed to aid the RF circuit designer in the designing, optimizing, and modeling of the spiral inductor and transformers. The tool is flexible, allowing the user to trade off between speed and accuracy. For example, ASITIC can be used to quickly search the parameters space of an inductor optimization problem.

By working with the entire metal layout of the circuit at hand, ASITIC allows general magnetic and substrate coupling to be analyzed between different parts of the circuit. Parasitic metal extraction may also be performed, including the effects of the substrate, and the resulting SPICE file may be included in more accurate simulations. The final design of a spiral may be exported for final layout; this feature is convenient in the layout of complex geometry spirals (nonsquare spirals), since their hand layout is time consuming, especially when changes are made to the geometry of the spiral.

IV. MEASUREMENT RESULTS

A. Description of Process/Measurement Setup

All spiral inductors were fabricated in Philips Semiconductors' Qubic2 BiCMOS process. Approximate process parameters are given in Table I. Using on-chip cascade probes, the average two-port s-parameters of the spirals are measured from 100 MHz to 5 GHz. Since the measurement setup holds wafers in place using vacuum suction, the back plane of the substrate is effectively floating. The substrate surface is grounded by placing substrate taps near the spirals.

[2] See http://www.eecs.berkeley.edu/~niknejad.

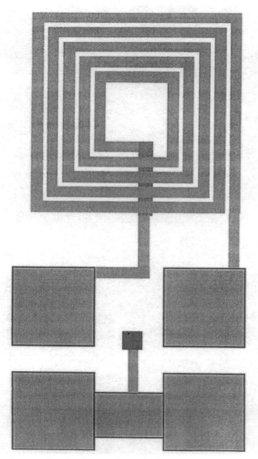

Fig. 3. Layout of spiral and probe pads for measurement.

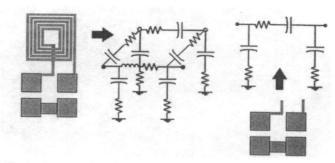

Fig. 4. Layout and equivalent circuit models of spiral and pads used for calibration procedure.

Pad capacitance is deembedded by subtracting out the open-circuit structure y-parameters from the spiral y-parameters. The calibration procedure we use is the standard approach and is reported widely. But there is an intrinsic problem with this calibration approach that can render the measurement results inaccurate. Fig. 3 shows the layout of a typical spiral setup for measurement. To deembed the large substrate capacitance and resistance of the pad structure, the open-pad two-port parameters are also measured, as shown in Fig. 4. In the same figures, we show a schematic of the pad parasitics along with the spiral equivalent circuit. Clearly, the open-circuit pad structure does not contain the parasitic coupling between the spiral structure and the pads. In addition, the pad-to-pad coupling is different in the two cases shown, since the spiral itself tends to shield the pads from one another. Hence, the pad parasitics are not correctly deembedded from the measurement results.

To show this effect, we simulate a sample inductor with and without pads. Next, we deembed the pad parasitics from the simulation results with pads (just as one would in measurement), and the resulting extracted series inductance, resistance, and Q factor are shown in Fig. 5(a) and (b). The inductance value is in error but in both cases increases with frequency; the series resistance values, though, deviate dramatically as a function of frequency. Notice, too, that the trend comes out incorrect, increasing as a function of frequency instead

of decreasing. Also shown is the resulting error in the Q factor. In addition, this error can change the location of the peak Q-factor, resulting in suboptimal designs. Not shown are the equivalent extracted substrate impedance values. The measurement errors can produce dramatic differences here too, changing the substrate resistance by a factor of two. The substrate capacitance can actually change sign due to resonance phenomena. In the following sections, in order to match simulation results to measurements, we simulated our structures under the same conditions as they were measured.

To solve these problems, we can of course always minimize the parasitic coupling between the device under test and the pads by physically moving the structure away from the pads and by placing isolating substrate taps between the structures. Nevertheless, this requires large areas to be consumed on the chip. While this calibration problem occurs for a BiCMOS substrate, simulation also indicates some problems on an epitaxial CMOS substrate. The magnitude of the error, though, is much smaller. For instance, the series L and R change much less, and only the substrate impedance changes drastically due to the calibration step.

B. Extraction of Circuit Parameters from Measurement Data

At each frequency of interest, the equivalent pi-circuit of Fig. 6(a) can be extracted from y-parameters. By taking real and imaginary parts of Y_{12}, equivalent inductance and resistance are extracted. Similarly, input and output shunt capacitance and resistance to substrate may be extracted by taking real and imaginary parts of $Y_{11} + Y_{12}$ and $Y_{22} + Y_{12}$. Since the circuit is passive, it is also reciprocal, so that $Y_{12} = Y_{21}$. This extraction procedure is widely used and is simple to implement since the electrical parameters of the spiral are uniquely determined at each frequency. The drawback, though, is that the extracted inductance is not physical and contains the effects of the capacitive coupling, which tends to boost the inductance as a function of frequency, similar to a parallel tank. The extraction technique presented in [17] is superior in this regard.

In this paper, we use the standard extraction procedure to compare our simulation results to measurements. However, for our final device models, we use frequency-independent circuits such as shown in Fig. 6(b), which contain parasitic elements so that our extraction of inductance is more physical. On the other hand, there is no unique way to assign the parameters

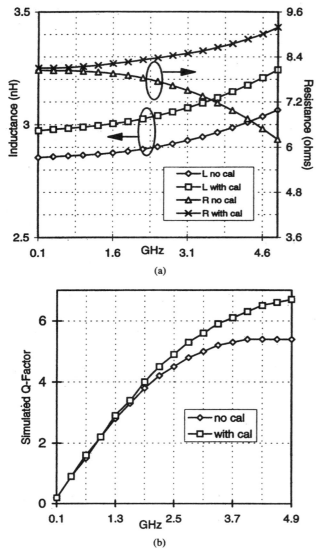

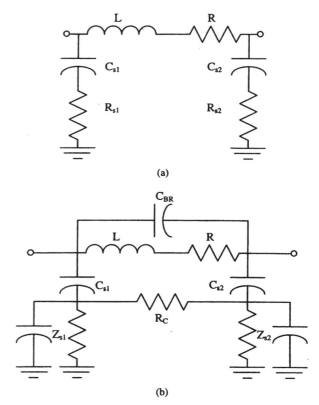

Fig. 6. (a) Traditional spiral inductor model. (b) Modified spiral inductor model.

Fig. 5. (a) Extracted inductance and resistance of spiral inductor with and without calibration procedure. (b) Extracted quality factor Q with and without calibration procedure.

values of our models since we use optimization to fit a low-order frequency-independent model to a distributed structure over a broad frequency range.

The traditional approach of Q extraction involves computing

$$Q_1 = \frac{-\text{Im}(Y_{11})}{\text{Re}(Y_{11})}. \tag{16}$$

The above definition has the awkward property that the Q is zero at self-resonance. Since inductors are usually operated far from self-resonance, this does not present too many problems. But in some applications, the inductor is used as a resonant tank close to self-resonance. In such cases, it is more appropriate to define the Q using a 3-dB bandwidth definition

$$Q = \frac{\omega_0}{\Delta\omega_{3\text{dB}}}. \tag{17}$$

Equally applicable, one may use the rate of change of phase at resonance

$$Q = \frac{\omega_0}{2}\left.\frac{d\phi}{d\omega}\right|_{\omega_0}. \tag{18}$$

The above equations are derived using second-order resonant circuits. For higher order circuits, perhaps the most general definition is based on ratio of energy stored in the circuit to energy dissipated per cycle, or

$$Q = \left.\frac{W_{\text{stored}}}{W_{\text{diss}}}\right|_{\text{per cycle}}. \tag{19}$$

The best approach to defining Q should be application dependent. Our approach to Q extraction is based on (18).

At each frequency of interest, an ideal capacitor is inserted in shunt with the inductor with admittance equal to the imaginary part of Y_{11}. The resulting admittance becomes

$$Y'(\omega) = j\omega C + Y_{11}$$
$$C = -\frac{\text{Im}[Y_{11}(\omega_0)]}{\omega_0}. \tag{20}$$

This capacitance will resonate the device at the frequency of interest ω_0. By examining the rate of change of phase, one can find the equivalent Q

$$\left.\frac{d\phi}{d\omega}\right|_{\omega_0} = \frac{2Q}{\omega_0} = \frac{\angle Y'(\omega_0 + \delta\omega) - \angle Y'(\omega_0 - \delta\omega)}{2\delta\omega}. \tag{21}$$

To illustrate this approach, in Fig. 7 we plot the Q of a typical spiral inductor (square spiral with nine turns of

94

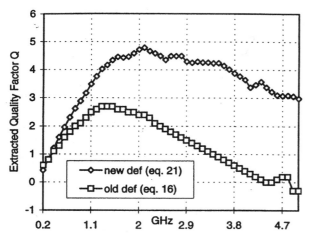

Fig. 7. Extracted quality factor Q of a nine-turn spiral by using (16) and (21).

Fig. 8. Layout of eight-turn square spiral.

dimensions given in the next section) using the definition present in (21). In the same plot, we show the traditional definition of Q, using (16). As we approach self-resonance, the two definitions deviate greatly. Notice also that the peak value of Q occurs at different frequencies for the two definitions.

In the following sections on measurement results, the data are measured at room temperature. It is important to calculate Q at the frequency and temperature of interest [18]. At low frequencies, the Q temperature coefficient (TC) will depend on the metal TC, whereas at high frequency, the Q will change due to the TC of the substrate resistance.

C. Square Spiral Inductors

Since square spirals are the most common inductors in Si RF IC's, we begin by comparing our simulation results of the previous section with measurement results. Many square spirals were fabricated, and measurement results were compared to simulation. The spirals have the following geometry: constant inner area $A_i = 44\ \mu m$, constant width $w = 7\ \mu m$, and constant spacing $s = 5\ \mu m$, with turns $n = 5\text{--}10$. A chip layout of the eight-turn spiral appears in Fig. 8. For example, Fig. 9 shows the measured and simulated s-parameters for the eight-turn spiral. As can be seen from the figure, simulation s-parameters match the measured s-parameters well.

Fig. 10(a) plots the extracted series inductance from both measurement data and simulation data for the five- and eight-turn spirals. Good agreement is found for all spirals. For all spirals, the inductance is an increasing function of frequency. In reality, the physical inductance decreases as a function of frequency due to the current crowding at the edges of the conductors, which leads to a decrease in internal inductance.[3] The extracted increase in L is due primarily to the coupling capacitance, which boosts the effective L as a function of frequency. For the smaller spirals, the same coupling capacitance tends to increase R as well, but the effect of the shunt substrate resistance is to lower the effective series R. For large

[3] For a highly conductive substrate, the eddy currents generated in the bulk substrate also lead to a decrease in inductance as a function of frequency, as these currents partially cancel the magnetic field generated by the device.

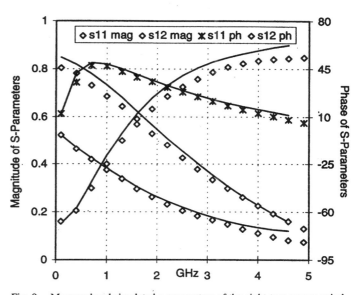

Fig. 9. Measured and simulated s-parameters of the eight-turn square spiral.

spirals in which substrate injection occurs at a lower frequency, the behavior of R is a decreasing function of frequency. For smaller spirals, the substrate injection is minimal due to a small substrate capacitance, and the effect of the coupling capacitance boosts R as a function of frequency.

Fig. 10(b) shows the Q-factor for the five- and eight-turn spirals. Here, we used the phase definition of (21) to compute Q. Good agreement is found between measurements and simulations. For the smaller five-turn spiral, the Q is an almost linear function of frequency, demonstrating that the substrate has negligible effects on the spiral, and the Q is approximately given by $\omega L R^{-1}$. The larger eight-turn spiral, though, shows more complicated behavior due to the substrate. At low frequencies when the substrate impedance is large, simple linear behavior is observed. At high frequencies, when the substrate impedance is smaller than the inductive/resistive impedance of the spiral, the substrate loss dominates and the Q

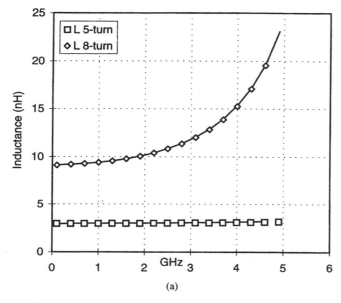

(a)

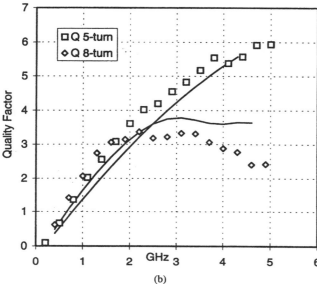

(b)

Fig. 10. (a) Extracted inductance of square spirals from measurements and simulations. (b) Extracted Q-factor of square spirals from measurements and simulations.

TABLE II
SIMULATED AND MEASURED MODELS OF SQUARE SPIRALS

Model Parameter	Meas 5-turn	Sim 5-turn	Meas 8-turn	Sim 8-turn
L (nH)	2.93	2.97	8.78	8.55
R (Ω)	8.02	8.23	17.50	16.99
C_{s1} (fF)	93.5	174.4	176.9	152.6
C_{s1} (fF)	210.7	341.8	656.6	179.9
R_{s1} (Ω)	1.35k	377	527.6	290.2
R_{s2} (Ω)	1.06k	472	587.0	314.7
C_{BR} (fF)	22.4	88.9	63.4	106.2
Σ error^2	.06	0.30	0.37	0.52

TABLE III
POLYGON SPIRAL GEOMETRY

Spirals	L4	L6	L7	L8	L9	L11
Radius R (μm)	105	105	105	105	105	105
Width W (μm)	8.2	7.5	7.5	7.5	7.5	6.5
Space S (μm)	3	3	3	3	3	3
Turns N	6.75	8	8	8	8	7.75
No. of sides	12	12	12	12	12	12
Metal (s)	M2	M2	1+2 via$^{(+)}$	1+2 via$^{(-)}$	1+2+3	M2

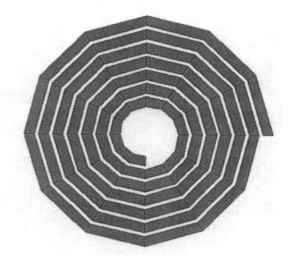

Fig. 11. Layout of polygon spiral inductor L4.

is a decreasing function of frequency. In the frequency range of interest, the Q is a combination of the above-mentioned effects and the Q peaks at some frequency f_{QMAX}. Notice that our analysis predicts this frequency well. This is important since we can design inductors that have peak Q at the frequency of interest.

We model the inductors using the equivalent circuits of Fig. 6(b). Table II compares our results, including the total error involved in the modeling. The error is derived by computing the relative error between the two-port measured s-parameters and the model two-port s-parameters. The error is cumulative over 16 points in the frequency range. Two models were generated, one based on the actual measurement results and the other based solely on simulation results (the simulation model is not a curve-fitted model of the measurements but rather a curve fit to the simulations).

D. Polygon Spiral Inductors

Since polygon spirals with more than four sides have higher Q than square spirals (for the same area), it is advantageous to use these structures. Many spirals are measured and compared with simulation. The geometry of the fabricated spirals is shown in Table III. The chip layout of a sample spiral inductor L4 appears in Fig. 11. S-parameters are plotted for the L4 spiral in Fig. 12, and again good agreement is found between simulation and measurement. Extracted series inductance and Q for the L4 and L11 spirals are plotted in Fig. 13(a) and (b). The inductors are also modeled using the equivalent circuit of Fig. 6(b), and Table IV summarizes the model parameters based on measurement and simulation.

E. Multimetal Spiral Inductors

To improve the low-frequency Q of the spiral, we can place metal layers in shunt to reduce the series resistance of the

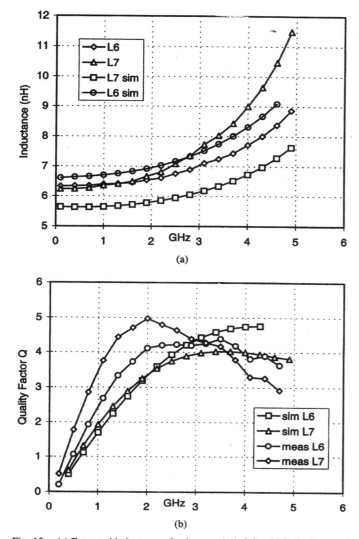

(a)

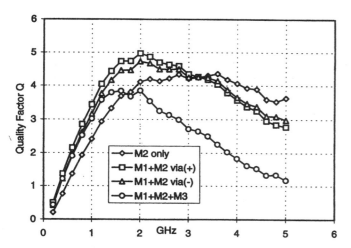

(b)

Fig. 15.   (a) Extracted inductance of polygon spirals L6 and L7. (b) Extracted quality factor of polygon spirals L6 and L7.

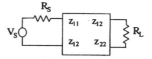

Fig. 16.   Measured Q-factor enhancement of multimetal shunt connected spirals.

F. Coupled Spiral Inductors

Many RF IC designs incorporate several spiral inductors on the same die. Since these structures are physically large,

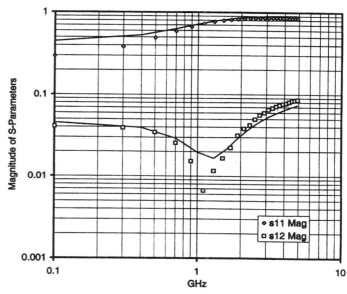

Fig. 17.   Measured and simulated s-parameters of the coupled spirals.

Fig. 18.   Circuit for calculating the power delivered to a resistive load from a reciprocal two-port network.

substrate coupling can be a significant problem. For instance, in any amplification stage, the substrate coupling can act as parasitic feedback, lowering the gain and possibly causing oscillations to occur. Hence, it is very important to model the substrate coupling.

The s-parameters of two eight-turn square spirals separated by a distance of 100 μm were simulated and measured. Fig. 17 shows the magnitude of the measured and simulated s-parameters. As can be seen from the figure, simulation results predict the coupling behavior accurately, such as the minimum S_{21}. To gain further insights into the coupling, we plot the power isolation from one spiral to the other using the following equations. For the arbitrary passive two-port shown in Fig. 18, the ratio of the power delivered to a resistive load R_L through the two-port can be shown to be

$$\frac{P_L}{P_{\text{in}}} = \frac{R_L \|G_I\|^2}{\text{Re}[Z_{\text{in}}]}. \tag{22}$$

In the above equation, G_I is the current gain through the two-port and Z_{in} is the impedance looking into the two-port from the source side

$$G_I = \frac{z_{12}}{R_L + z_{12}} \tag{23}$$

$$Z_{\text{in}} = z_{11} - \frac{z_{12}^2}{R_L + z_{22}}. \tag{24}$$

Using the above equations, we plot the measured and simulated power isolation for the coupled inductors in Fig. 19, where a 50-Ω load resistance is used in the above equations. Clearly, there are two frequencies where the isolation is

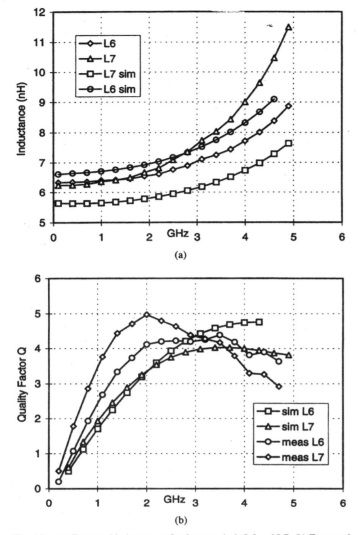

(a)

(b)

Fig. 15. (a) Extracted inductance of polygon spirals L6 and L7. (b) Extracted quality factor of polygon spirals L6 and L7.

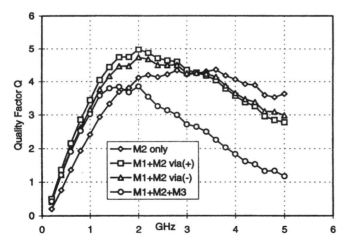

Fig. 16. Measured Q-factor enhancement of multimetal shunt connected spirals.

F. Coupled Spiral Inductors

Many RF IC designs incorporate several spiral inductors on the same die. Since these structures are physically large,

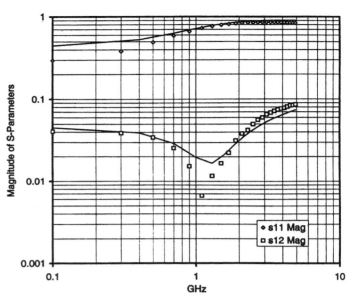

Fig. 17. Measured and simulated s-parameters of the coupled spirals.

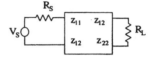

Fig. 18. Circuit for calculating the power delivered to a resistive load from a reciprocal two-port network.

substrate coupling can be a significant problem. For instance, in any amplification stage, the substrate coupling can act as parasitic feedback, lowering the gain and possibly causing oscillations to occur. Hence, it is very important to model the substrate coupling.

The s-parameters of two eight-turn square spirals separated by a distance of 100 μm were simulated and measured. Fig. 17 shows the magnitude of the measured and simulated s-parameters. As can be seen from the figure, simulation results predict the coupling behavior accurately, such as the minimum S_{21}. To gain further insights into the coupling, we plot the power isolation from one spiral to the other using the following equations. For the arbitrary passive two-port shown in Fig. 18, the ratio of the power delivered to a resistive load R_L through the two-port can be shown to be

$$\frac{P_L}{P_{\text{in}}} = \frac{R_L \|G_I\|^2}{\text{Re}[Z_{\text{in}}]}. \quad (22)$$

In the above equation, G_I is the current gain through the two-port and Z_{in} is the impedance looking into the two-port from the source side

$$G_I = \frac{z_{12}}{R_L + z_{12}} \quad (23)$$

$$Z_{\text{in}} = z_{11} - \frac{z_{12}^2}{R_L + z_{22}}. \quad (24)$$

Using the above equations, we plot the measured and simulated power isolation for the coupled inductors in Fig. 19, where a 50-Ω load resistance is used in the above equations. Clearly, there are two frequencies where the isolation is

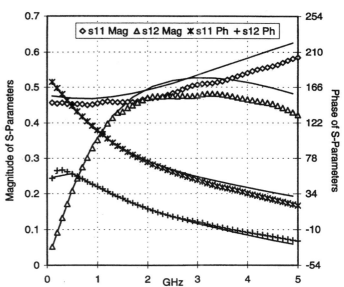

Fig. 21. Measured and simulated s-parameters of planar transformer.

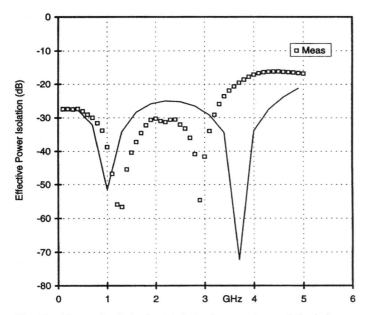

Fig. 19. Measured and simulated isolation between the coupled spirals.

Fig. 20. Layout of planar transformer.

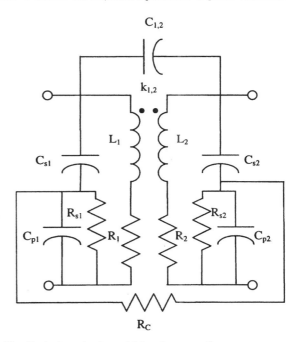

Fig. 22. Equivalent circuit model for planar transformer.

maximum. These frequencies depend on the geometrical layout of the spirals, and this gives the designer the powerful option of placing the spirals in locations to maximize isolation.

G. Planar Transformers

To test the simulation accuracy for transformers, several planar transformers were fabricated and measured. An example transformer structure is shown in Fig. 20. The transformer is made of two interwound spirals each of five turns of 7-μm-wide metal with a spacing of 17 μm. Measured and simulated s-parameters are shown in Fig. 21. Again, good agreement is found between simulation and measurement. We can gain further insight into the measurement results by using (22) to plot the measured loss of the transformer as a function of load

resistance. One can find the optimal value of load resistance at the frequency of interest to minimize losses.

The dynamics of the transformer can be captured in a compact model similar to the coupled inductors, as shown in Fig. 22. The k factor for planar inductors is about 0.7–0.8. Substrate coupling is again modeled with R_C. While the circuit of Fig. 22 is physically based, one can also derive the equivalent circuit, which contains an ideal transformer at the core with parasitic elements [7], [19]. The model parameters from measurement and simulation appear in Table V.

V. CONCLUSION

In this paper, we presented techniques to analyze, model, and optimize spiral inductors and transformers on the Si

TABLE V
SIMULATED AND MEASURED MODELS OF PLANAR TRANSFORMER

Transformer Model Parameter	Measured	Simulated
L (nH)	4.88	4.63
R (Ω)	17.58	16.91
C_s (fF)	124	123
C_p (fF)	430	418
R_s (Ω)	243	252
R_C (Ω)	1.00k	796
$C_{1,2}$ (fF)	53	59
$k_{1,2}$	.703	0.715
Σ error^2	0.79	0.04

substrate. The techniques are accurate, taking into account substrate coupling, current constriction, and proximity effects. The analysis is also fast and efficient, making it suitable for computer optimization. Furthermore, the analysis is general and appropriate for analyzing any arbitrary arrangement of conductors, such as multimetal spirals. A custom CAD tool called ASITIC was developed that incorporates the algorithms discussed in this paper. ASITIC was used to analyze a wide variety of test structures such as square spirals, polygon spirals, coupled spirals, and transformers. The test spirals were also fabricated, and measurement results compared well to simulation. Compact models for the various devices were presented that model the device dynamics over a wide frequency range.

ACKNOWLEDGMENT

The authors thank W. Mack, J. Eisenstadt, and Y. Nguyen of Philips Semiconductors for their help and support in fabricating and measuring the spirals. Also, the authors thank R. Gharpurey of Texas Instruments for his valuable insights.

REFERENCES

[1] N. M. Nguyen and R. G. Meyer, "Si IC-compatible inductors and LC passive filters," *IEEE J. Solid-State Circuits*, vol. 27, pp. 1028–1031, Aug. 1990.
[2] K. B. Ashby, W. C. Finley, J. J. Bastek, S. Moinian, and I. A. Koullias, "High *Q* inductors for wireless applications in a complementary silicon bipolar process," in *Proc. Bipolar and BiCMOS Circuits and Technology Meeting*, Minneapolis, MN, 1994, pp. 179–182.
[3] J. N. Burghartz, M. Soyuer, and K. Jenkins, "Microwave inductors and capacitors in standard multilevel interconnect silicon technology," *IEEE Trans. Microwave Theory Tech.*, vol. 44, pp. 100–103, Jan. 1996.
[4] R. B. Merrill, T. W. Lee, H. You, R. Rasmussen, and L. A. Moberly, "Optimization of high *Q* integrated inductors for multi-level metal CMOS," *IEDM*, 1995, pp. 38.7.1–38.7.3.
[5] L. Zu, Y. Lu, R. C. Frye, M. Y. Law, S. Chen, D. Kossiva, J. Lin, and K. L. Tai, "High *Q*-factor inductors integrated on MCM Si substrates," *IEEE Trans. Comp., Packag., Manufact. Technol. B*, vol. 19, pp. 635–643, Aug. 1996.
[6] J. Y.-C. Chang and A. A. Abidi, "Large suspended inductors on silicon and their use in a 2-μm CMOS RF amplifier," *IEEE Electron Device Lett.*, vol. 14, pp. 246–248, 1993.
[7] J. R. Long and M. A. Copeland, "The modeling, characterization, and design of monolithic inductors for silicon RF IC's," *IEEE J. Solid-State Circuits*, vol. 32, pp. 357–369, Mar. 1997.
[8] J. Craninckx and M. Steyaert, "A 1.8-GHz low-phase-noise CMOS VCO using optimized hollow spiral inductors," *IEEE J. Solid-State Circuits*, vol. 32, pp. 736–745, May 1997.
[9] D. Lovelace and N. Camilleri, "Silicon MMIC inductor modeling for high volume, low cost applications," *Microwave J.*, Aug. 1994, pp. 60–71.
[10] S. Ramo, J. R. Whinnery, and T. Van Duzer, *Fields and Waves In Communication Electronics*, 3rd ed. New York: Wiley, 1994, pp. 324–330.
[11] A. Niknejad, R. Gharpurey, and R. G. Meyer, "Numerically stable green function for modeling and analysis of substrate coupling in integrated circuits," *IEEE Trans. Computer-Aided Design*, vol. 17, pp. 305–315, Apr. 1998.
[12] A. E. Ruehli and H. Heeb, "Circuit models for three-dimensional geometries including dielectrics," *IEEE Trans. Microwave Theory Tech.*, vol. 40, pp. 1507–1516, July 1992.
[13] F. W. Grover, *Inductance Calculations*. Princeton, NJ: Van Nostrand, 1946 (also New York: Dover, 1954).
[14] W. T. Weeks, L. L. Wu, M. F. McAllister, and A. Singh, "Resistive and inductive skin effect in rectangular conductors," *IBM J. Res. Develop.*, vol. 23, pp. 652–660, Nov. 1979.
[15] R. Gharpurey, private communication.
[16] C. P. Yue and S. S. Wong, "On-chip spiral inductors with patterned ground shields for Si-based RF IC's," in *Symp. VLSI Circuits Dig.*, June 1997, pp. 85–86.
[17] C. P. Yue, C. Ryu, J. Lau, T. H. Lee, and S. S. Wong, "A physical model for planar spiral inductors on silicon," in *Int. Electron Devices Meeting Tech. Dig.*, Dec. 1996, pp. 155–158.
[18] R. Groves, K. Stein, D. Harame, and D. Jadus, "Temperature dependence of *Q* in spiral inductors fabricated in a silicon-germanium/BiCMOS technology," in *Proc. 1996 Bipolar/BiCMOS Circuits and Technology Meeting*, New York, 1996, pp. 153–156.
[19] D. O. Pederson and K. Mayaram, *Analog Integrated Circuits for Communications*. Norwell, MA: Kluwer, 1991, pp. 183–184.

Ali M. Niknejad (S'92) was born in Tehran, Iran, on July 29, 1972. He received the B.S.E.E. degree from the University of California, Los Angeles, in 1994 and the master's degree in electrical engineering from the University of California, Berkeley, in 1997, where he is now pursuing the Ph.D. degree.

Mr. Niknejad has held several internship positions in the electronics industry. In the summer of 1994, he was with the Hughes Aircraft Advanced Circuit Technology Center, Torrance, CA, investigating wide-band current feedback op-amp topologies. During the summer of 1996, he was with Texas Instruments, Dallas, TX, where he studied substrate coupling. During the summer of 1997, he was with Lucent Technologies (Bell Labs), Murray Hill, NJ, where he investigated power amplifier topologies for wireless applications. His current research interests include high-frequency electronic circuit design, modeling of passive devices and substrates coupling, digital wireless communication systems, numerical methods in electromagnetics, and radio-frequency computer-aided design.

Robert G. Meyer (S'64–M'68–SM'74–F'81) was born in Melbourne, Australia, on July 21, 1942. He received the B.E., M.Eng.Sci., and Ph.D. degrees in electrical engineering from the University of Melbourne in 1963, 1965, and 1968, respectively.

In 1968, he was an Assistant Lecturer in electrical engineering at the University of Melbourne. Since 1968, he has been with in the Department of Electrical Engineering and Computer Sciences, University of California, Berkeley, where he is now a Professor. His current research interests are high-frequency analog integrated-circuit design and device fabrication. He has been a Consultant on electronic circuit design for numerous companies in the electronics industry. He is a coauthor of the book *Analysis and Design of Analog Integrated Circuits*, (New York: Wiley, 1993), and is editor of the book, *Integrated Circuit Operational Amplifiers* (New York: IEEE Press, 1978).

Dr. Meyer was President of the IEEE Solid-State Circuits Council. He was an Associate Editor of the IEEE JOURNAL OF SOLID-STATE CIRCUITS and IEEE TRANSACTIONS ON CIRCUITS AND SYSTEMS.

Stacked Inductors and Transformers in CMOS Technology

Alireza Zolfaghari, *Student Member, IEEE*, Andrew Chan, *Student Member, IEEE*, and Behzad Razavi, *Member, IEEE*

Abstract—A modification of stacked spiral inductors increases the self-resonance frequency by 100% with no additional processing steps, yielding values of 5 to 266 nH and self-resonance frequencies of 11.2 to 0.5 GHz. Closed-form expressions predicting the self-resonance frequency with less than 5% error have also been developed. Stacked transformers are also introduced that achieve voltage gains of 1.8 to 3 at multigigahertz frequencies. The structures have been fabricated in standard digital CMOS technologies with four and five metal layers.

Index Terms—Inductors, oscillators, quality factor, RF circuits, self-resonance frequency, stacked spirals, transformers, tuned amplifiers.

I. INTRODUCTION

MONOLITHIC inductors have found extensive usage in RF CMOS circuits. Despite their relatively low quality factor (Q) such inductors still prove useful in providing gain with minimal voltage headroom and operating as resonators in oscillators. Monolithic transformers have also appeared in CMOS technology [1], allowing new circuit configurations.

This paper introduces a modification of stacked inductors that increases the self-resonance frequency f_{SR} by as much as 100%, a result predicted by a closed-form expression that has been developed for f_{SR}. Structures built in several generations of standard digital CMOS technologies exhibit substantial reduction of the parasitic capacitance with the technique applied, achieving self-resonance frequencies exceeding 10 GHz for values as high as 5 nH. The modification allows increasingly larger inductance values or higher self-resonance as the number of metal layers increases in each new generation of the technology.

The paper also presents a new stacked transformer that achieves nominal voltage or current gains from 2 to 4. Fabricated prototypes display voltage gains as high as 3 in the gigahertz range, encouraging new circuit topologies for low-voltage operation.

Section II reviews the definitions of Q. Section III provides the motivation for high-value inductors and summarizes the properties of stacked inductors. Section IV deals with the theoretical derivation of the self-resonance frequency of such inductors and Section V exploits the results to propose the modification. Section VI presents the stacked transformers and describes a distributed circuit model used to analyze their behavior. Section VII summarizes the experimental results.

Manuscript received August 11, 2000; revised January 8, 2001. This work was supported in part by the Defense Advanced Research Projects Agency, SRC, Lucent Technologies, and Nokia.

The authors are with the Department of Electrical Engineering, University of California, Los Angeles, CA 90095-1594 USA (e-mail: razavi@icsl.ucla.edu).

Publisher Item Identifier S 0018-9200(01)02586-0.

II. DEFINITIONS OF THE QUALITY FACTOR

Several definitions have been proposed for the quality factor. Among these, the most fundamental is

$$Q = 2\pi \cdot \frac{\text{energy stored}}{\text{energy loss in one oscillation cycle}}. \tag{1}$$

The above definition does not specify what stores or dissipates the energy. However, for an inductor, only the energy stored in the magnetic field is of interest. Therefore, the energy stored is equal to the difference between peak magnetic and electric energies.

If an inductor is modeled by a simple parallel RLC tank, it can be shown that [2]

$$
\begin{aligned}
Q &= 2\pi \cdot \frac{\text{peak magnetic energy} - \text{peak electric energy}}{\text{energy loss in one oscillation cycle}} \\
&= \frac{R_p}{L\omega} \cdot \left[1 - \left(\frac{\omega}{\omega_0}\right)^2 \right] \\
&= \frac{\text{Im}(Z)}{\text{Re}(Z)}
\end{aligned} \tag{2}
$$

where R_p and L are the equivalent parallel resistance and inductance, respectively, ω_0 is the resonance frequency, and Z is the impedance seen at one terminal of the inductor while the other is grounded. Although definition (2) has been extensively used, it is only applicable to the frequencies below the resonance because it falls to zero at the self-resonance frequency.

On the other hand, if only the magnetic energy is considered, then (1) reduces to

$$
\begin{aligned}
Q &= 2\pi \cdot \frac{\text{peak magnetic energy}}{\text{energy loss in one oscillation cycle}} \\
&= \frac{R_p}{L\omega}.
\end{aligned} \tag{3}
$$

Definition (3) has two advantages over (2). First, it can be used over a wider frequency range. Second, it can more explicitly express R_p. It should be noted that at low frequencies, the Q's obtained by (2) and (3) are quite close because the energy stored in the electric field is much smaller than that stored in the magnetic field.

III. LARGE INDUCTORS WITH HIGH SELF-RESONANCE FREQUENCIES

Inductors are extensively used in tuned amplifiers and mixers with high intermediate frequencies (IFs) (Fig. 1). In these applications, to maximize the gain (or conversion gain), the equiva-

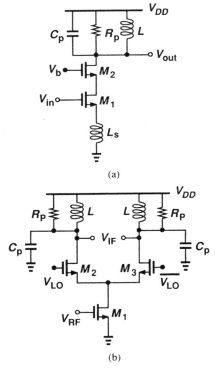

(a)

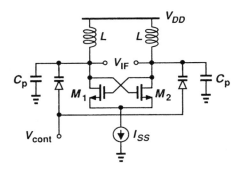

(b)

Fig. 1.   (a) Low-noise amplifier and (b) mixer with high IF.

Fig. 2.   Representative VCO.

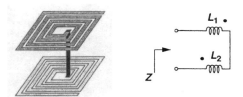

Fig. 3.   Two-layer inductor.

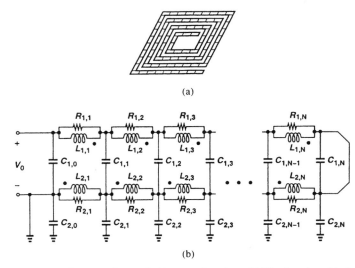

(a)

(b)

Fig. 4.   (a) Decomposing a spiral into equal sections. (b) Distributed model of a two-layer inductor.

and later used in CMOS technology as well [4]. From the circuit model of Fig. 3, it can be seen that the input impedance of this structure is

$$Z = j\omega(L_1 + L_2 + 2M) \qquad (5)$$

where L_1 and L_2 are the self-inductance of the spirals and M is the mutual inductance between the two. In a stacked inductor, the two spirals are identical ($L_1 = L_2 = L$) and the mutual coupling between the two layers is quite strong ($M \approx \sqrt{L_1 L_2} = L$). The total inductance is therefore increased by nearly a factor of 4. Similarly, for an n-layer inductor the total inductance is nominally equal to n^2 times that of one spiral. With the availability of more than five metal layers in modern CMOS technologies, stacking can provide increasingly larger values in a small area.

IV. DERIVATION OF SELF-RESONANCE FREQUENCY

Stacked structures typically exhibit a single resonance frequency. Thus, they can be modeled by a lumped RLC tank with $f_{\text{SR}} = (2\pi \sqrt{L_{eq} C_{eq}})^{-1}$, where L_{eq} and C_{eq} are the equivalent inductance and capacitance of the structure, respectively. While the equivalent inductance can be obtained by various empirical expressions [5], [6], Greenhouse's method [7], or electromagnetic field solvers [8], no method has been proposed to calculate the equivalent capacitance. We derive an expression for the capacitance in this section.

For f_{SR} calculations, we decompose each spiral into equal sections as shown in Fig. 4(a) such that all sections have the same inductance and parasitic capacitance to the substrate or the other spiral. This decomposition yields the distributed model illustrated in Fig. 4(b). In this circuit, inductive elements $L_{i,j}$'s

lent parallel resistance of the inductor (R_p) must be maximized. From definition (3) of the Q, R_p can be expressed as

$$R_p = Q \cdot L\omega. \qquad (4)$$

Therefore, to maximize R_p, the *product* of Q and L must be maximized. Since the Q of on-chip inductors in CMOS technology is quite limited, it is reasonable to seek methods of achieving high inductance values with high self-resonance frequencies and a moderate silicon area.

If a method of reducing the parasitic capacitance C_p of inductors is devised, it also improves the performance of voltage-controlled oscillators (VCOs). In the topology of Fig. 2, for example, reduction of C_p directly translates to a wider tuning range because the varactor diodes can contribute more variable capacitance. Simulations indicate that the inductor modification introduced in this paper increases the tuning range of a 900-MHz CMOS VCO from 4.2% to 23% for a $2\times$ varactor capacitance range.

A candidate for compact high-value inductors is the stacked structure of Fig. 3, originally introduced in GaAs technology [3]

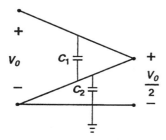

Fig. 5. Voltage profile across each capacitor.

represent the inductance of each section in Fig. 4(a) and they are all mutually coupled. The capacitance between the two layers is modeled by capacitors $C_{1,j}$ and that between the bottom layer and the substrate by capacitors $C_{2,j}$. To include the finite Q of the structure, all sources of loss are lumped into parallel resistor elements $R_{i,j}$. Also, we neglect trace-to-trace capacitances of each spiral. The validity of these assumptions will be explained later.

The simple circuit model of Fig. 4(b) still does not easily lend itself to current and voltage equations. However, we can use the physical definition of resonance. The resonance frequency can be viewed as the frequency at which the peak magnetic and electric energies are equal. In other words, if we calculate the total electric energy stored in the structure for a given peak voltage V_0 and equate that to $C_{eq}V_0^2/2$, then we can obtain C_{eq}.

To derive the electric energy stored in the capacitors, we first compute the voltage profile across the uniformly distributed capacitance of the structure. Assuming perfect coupling between every two inductors in Fig. 4, we express the voltage across each as

$$V_{L,l,m} = \sum_{k=1}^{2} \sum_{n=1}^{N} j\omega I_{k,n} L_{k,n}, \qquad (6)$$

where $I_{l,m}$ is the current through $L_{l,m}$ and N is the number of the sections in the distributed model. Equation (6) reveals that all inductors sustain *equal* voltages. Therefore, for a given applied voltage V_0, we have

$$V_{L,l,m} = \frac{V_0}{2N}. \qquad (7)$$

From (6) and (7), it follows that the voltage varies linearly from V_0 to 0 across the distributed capacitance C_1 and from 0 to $V_0/2$ across C_2 (from left to right in Fig. 5).

Having determined the voltage variation, we write the electric energy stored in the mth element, $C_{1,m}$, as

$$E_{e,C_{1,m}} = \frac{1}{2} C_{1,m}[(V_0 - mV_{L,l,m}) - mV_{L,l,m}]^2. \qquad (8)$$

The total electric energy in C_1 is therefore equal to

$$E_{e,C_1} = \frac{1}{2} \sum_{m=0}^{N} C_{1,m}(V_0 - 2mV_{L,l,m})^2. \qquad (9)$$

As mentioned earlier, all sections are identical, i.e., $C_{1,m} = C_1/(N+1)$, and if we substitute (7) in (9), define a new variable $x = m/N$, and let N go to infinity, then we obtain

$$E_{e,C_1} = \frac{1}{2} C_1 V_0^2 \int_0^1 (1-x)^2 \, dx \qquad (10)$$

$$= \frac{1}{2} \cdot \frac{C_1}{3} V_0^2. \qquad (11)$$

The above equation states that if the voltage across a distributed capacitor changes linearly from zero to a maximum value V_0, then the equivalent capacitance is 1/3 of the total capacitance. Since C_2 sustains a maximum voltage of $V_0/2$, its electric energy is equal to

$$E_{e,C_2} = \frac{1}{2} \cdot \frac{C_2}{3} \cdot \left(\frac{V_0}{2}\right)^2 \qquad (12)$$

$$= \frac{1}{2} \cdot \frac{C_2}{12} V_0^2. \qquad (13)$$

From (11) and (13), the total electric energy stored in the inductor is

$$E_e = E_{e,C_1} + E_{e,C_2} \qquad (14)$$

$$= \frac{1}{2} \cdot \frac{4C_1 + C_2}{12} V_0^2 \qquad (15)$$

yielding the equivalent capacitance as

$$C_{eq} = \frac{1}{12}(4C_1 + C_2). \qquad (16)$$

The foregoing method can be applied to a stack of multiple spirals as well. For an inductor with n stacked spirals, (6) suggests that the voltage is equally divided among the spirals. Therefore, interlayer capacitances sustain a maximum voltage of $2V_0/n$, whereas the bottom-layer capacitance sustains V_0/n. Now, using the result of (11) and adding the electric energy of all layers, we have

$$E_e = \frac{1}{2} \sum_{i=1}^{n-1} \frac{C_i}{3} \left(\frac{2V_0}{n}\right)^2 + \frac{1}{2} \cdot \frac{C_n}{3} \left(\frac{V_0}{n}\right)^2 \qquad (17)$$

$$= \frac{1}{2} \cdot \frac{4\sum_{i=1}^{n-1} C_i + C_n}{3n^2} V_0^2 \qquad (18)$$

and hence

$$C_{eq} = \frac{1}{3n^2} \left(4 \sum_{i=1}^{n-1} C_i + C_n\right). \qquad (19)$$

The simplified model used to derive the equivalent capacitance is slightly different from the exact physical model of a stacked inductor. The following three issues must be considered.

1) We have assumed that all inductors in the distributed model are perfectly coupled. However, the coupling between orthogonal segments of a spiral or different spirals is very small. Nonetheless, if we assume that the inductor elements that are on top of each other are strongly coupled, then they sustain equal voltages. Therefore, the total voltage is still equally divided among the spirals. Furthermore, since each spiral is composed of a few groups of coupled inductors, the linear voltage profile is a reasonable approximation. To verify the last statement, a two-turn single spiral has been simulated. The spiral has been divided into 20 sections (twelve sections for the outer turn and eight sections for the inner turn). Then, inductor elements in the same segment and parallel adjacent segments are strongly coupled while there is no magnetic coupling between other segments (orthogonal and parallel segments with opposite current direction).

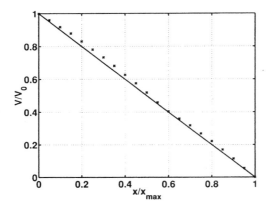

Fig. 6. Simulated voltage profile of a single spiral.

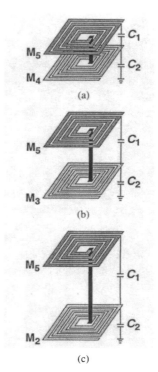

Fig. 7. Modification of two-layer stacked inductors.

Fig. 6 shows the voltage profile for this structure. As seen in this figure, the actual profile is relatively close to the linear approximation.

2) We have neglected the electric energy stored in the trace-to-trace capacitance C_{TT} (the capacitance between two adjacent turns in the same layer). Supported by the experimental results in Section VII, this assumption can be justified by two observations. First, the width of the metal segments is typically much greater than the metal thickness. Therefore, even for a small spacing between the segments, C_{TT} is usually smaller than the interlayer capacitance. Second, the adjacent turns in the same spiral sustain a small voltage difference. Noting that the electric energy is proportional to the square of voltage, we conclude that the effect of C_{TT} is negligible.

3) Presenting all of the loss mechanisms by parallel resistors in the distributed model introduces little error in the calculation of the self-resonance frequency. For metal resistance and magnetic coupling to the substrate, parallel resistors are a good model if $Q^2 \gg 1$.

It is important to note that measurements indicate that (19) provides a reasonable approximation for f_{SR} of a *single* spiral as well, though the focus of the paper is on stacked spirals.

V. MODIFICATION OF STACKED INDUCTORS

For a two-layer inductor, (16) reveals that the interlayer capacitance C_1 impacts the resonance frequency four times as much as the bottom-layer capacitance C_2. In addition, for two adjacent metal layers, C_1 is several times greater than C_2. Therefore, it is plausible to move the spirals farther from each other so as to achieve a higher self-resonance frequency. For example, in a typical CMOS technology with five metal layers, $C_{M_5-M_4} \cong 40$ aF/μm^2 and $C_{M_4-sub} \cong 6$ aF/μm^2, whereas $C_{M_5-M_3} \cong 14$ aF/μm^2 and $C_{M_3-sub} \cong 9$ aF/μm^2. It follows that for the structure of Fig. 7(a), $C_{eq,a} \approx 14$ aF/μm^2, whereas for Fig. 7(b), $C_{eq,b} \approx 5.4$ aF/μm^2, an almost three-fold reduction.

Equation (16) proves very useful in estimating the performance of various stack combinations. For example, it predicts that the structure of Fig. 7(c) has an equivalent capacitance $C_{eq,c} \approx 4$ aF/μm^2 because $C_{M_5-M_2} \cong 9$ aF/μm^2 and $C_{M_2-sub} \cong 12$ aF/μm^2. In other words, the self-resonance

frequency of the inductor in Fig. 7(c) is almost twice that of the inductor in Fig. 7(a).

Note that the value of the inductance remains relatively constant because the lateral dimensions are nearly two orders of magnitude greater than the vertical dimensions. By the same token, the loss through the substrate remains unchanged. Both of these claims are confirmed by measurements (Section VII).

The idea of moving stacked spirals away from each other so as to increase f_{SR} can be applied to multiple layers as well. For example, the structure of Fig. 8(a) can be modified as depicted in Fig. 8(b), thereby raising f_{SR} by 50%.

VI. STACKED TRANSFORMERS

Monolithic transformers producing voltage or current gain can serve as interstage elements if the signals do not travel off chip, i.e., if power gain is not important. Such transformers can also perform single-ended-to-differential and differential-to-single-ended conversion.

A particularly useful example is depicted in Fig. 9, where a transformer having current gain is placed in the current path of an active mixer. Here, the RF current produced by M_1 is amplified by T_1 before it is commutated to the output by M_2 and M_3. The current gain lowers the noise contributed by M_2 and M_3 and it is obtained with no power, linearity, or voltage headroom penalty.

Fig. 10(a) shows the 1-to-2 transformer structure. The primary is formed as a single spiral in metal 4 and the secondary as two series spirals in metal 3 and metal 5. The performance of the transformer is determined by the inductance and series resistance of each spiral and the magnetic and capacitive coupling between the primary and the secondary. To minimize the capacitive coupling, the primary turns are offset with respect to

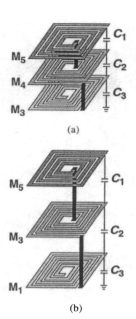

(a)

(b)

Fig. 8. Three-layer stacked inductor modification.

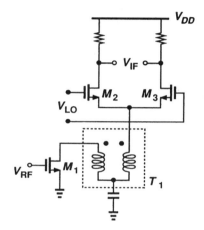

Fig. 9. Example of using a transformer to boost current in an active mixer.

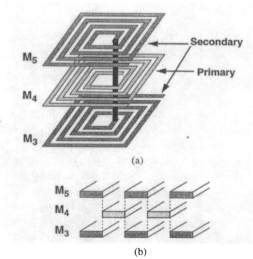

(a)

(b)

Fig. 10. Transformer structure.

the secondary turns as illustrated in Fig. 10(b). Thus, the capacitance arises only from the fringe electric field lines. The number

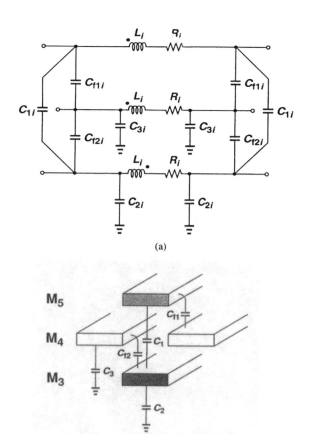

(a)

(b)

Fig. 11. Transformer model.

of turns in each spiral also impacts the voltage (or current) gain at a desired frequency because it entails a tradeoff between the series resistance and the amount of magnetic flux enclosed by the primary and the secondary. For single-ended-to-differential conversion, two of the structures in Fig. 10(a) can be cross-coupled so as to achieve symmetry.

To design the transformer for specific requirements, a circuit model is necessary. Fig. 11 illustrates one section of the distributed model developed for the 1-to-2 transformer. The segments L_i and R_i represent a finite element of each spiral, C_f's denote the fringe capacitances, C_1 models the capacitance between M_5 and M_3, and C_2 and C_3 are the capacitances between the substrate and M_3 and M_4, respectively. The values of L_i and R_i are derived assuming a uniformly distributed model and a Q of 3 for each inductor. The capacitance values are obtained from the foundry interconnect data. Fig. 12 depicts the simulated voltage gain of two transformers, one consisting of eight-turn spirals with 7-μm-wide metal lines and the others consisting of four-turn and three-turn spirals with 9-μm-wide metal lines.

Unlike stacked inductors, whose resonance frequency is not affected by the inductor loss, the transfer characteristics and voltage gain of the transformer depend on the quality factor of the spirals. In this simulation, a Q of 3 has been used for each winding. As Fig. 12, for the eight-turn transformer, capacitive coupling between the spirals is so large that it does not allow the voltage gain to exceed one, while for the four-turn and three-turn transformers we expect a gain of about 1.8 in the

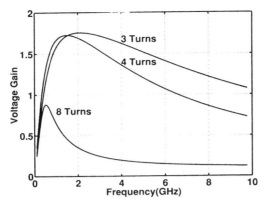

Fig. 12. Simulated voltage gain of the transformers.

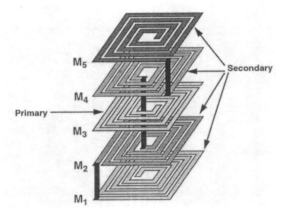

Fig. 13. 1-to-4 transformer structure.

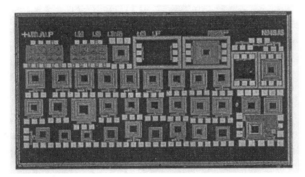

Fig. 14. Die photo.

TABLE I
MEASURED INDUCTORS IN 0.25-μm TECHNOLOGY (LINEWIDTH = 9 μm, LINE SPACING = 0.72 μm)

Inductor	Metal Layers	L (nH)	Measured f_{SR} (GHz)	Calculated f_{SR} (GHz)	Number of Turns
$L_1(240\mu m)^2$	5,4	45	0.92	0.96	7
$L_2(240\mu m)^2$	5,3	45	1.5	1.53	7
$L_3(240\mu m)^2$	5,2	45	1.8	1.79	7
$L_4(240\mu m)^2$	5,4,3	100	0.7	0.7	7
$L_5(240\mu m)^2$	5,3,1	100	1.0	1.0	7
$L_6(200\mu m)^2$	5,3,2	50	1.5	1.46	5
$L_7(200\mu m)^2$	5,2,1	48	1.5	1.54	5

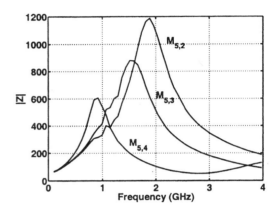

Fig. 15. Measured inductor characteristics.

TABLE II
HIGH-VALUE INDUCTORS IN 0.25-μm TECHNOLOGY (LINEWIDTH = 9 μm, LINE SPACING = 0.72 μm, NUMBER OF TURNS FOR EACH SPIRAL = 7)

Inductor Size	Metal Layers	L (nH)	Measured f_{SR}(GHz)	Calculated f_{SR}(GHz)
$(240\mu m)^2$	5,4	45	0.92	0.96
$(240\mu m)^2$	5,4,3	100	0.7	0.7
$(240\mu m)^2$	5,4,3,2	180	0.55	0.58
$(240\mu m)^2$	5,4,3,2,1	266	0.47	0.49

vicinity of 2 GHz. Note that if the secondary is driven by a current source and the short-circuit current of the primary is measured, the same characteristics are observed.

The concept of stacked transformer can be applied to more layers of metal to achieve higher voltage gains. Fig. 13 shows a stacked transformer with a nominal gain of 4. In this structure, M_3 forms the primary and the rest of the metal layers are used for the secondary.

VII. EXPERIMENTAL RESULTS

A large number of structures have been fabricated in several CMOS technologies with no additional processing steps. Fig. 14 is a die photograph of the devices built in a 0.25-μm process with five metal layers. Calibration structures are also included to de-embed pad parasitics.

Table I shows the measured characteristics of some inductors fabricated in the 0.25-μm process. The Q at self-resonance is approximately equal to 3. As expected from Fig. 7, inductors L_1, L_2, and L_3, with two layers of metal, demonstrate a steady increase in f_{SR} as the bottom spiral is moved away from the top one. Fig. 15 plots the measured impedance of these inductors as a function of frequency, revealing a twofold increase in f_{SR}. For the three-layer inductors (L_4 and L_5 in Table I), proper choice of metal layers can considerably increase f_{SR}. To show how accurately (19) predicts the f_{SR}, calculated values are included as well. The error is less than 5%.

Table II shows how adding the number of metal layers can increase the inductance value. In this table, all inductors have the same dimensions but incorporate a different number of layers.

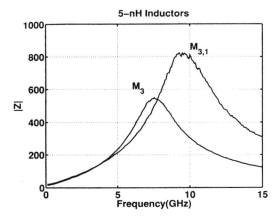

Fig. 16. Comparison of one-layer and two-layer structures for a given value of inductance.

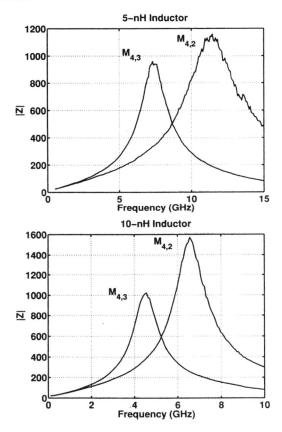

Fig. 17. Measured inductors in 0.4-μm technology.

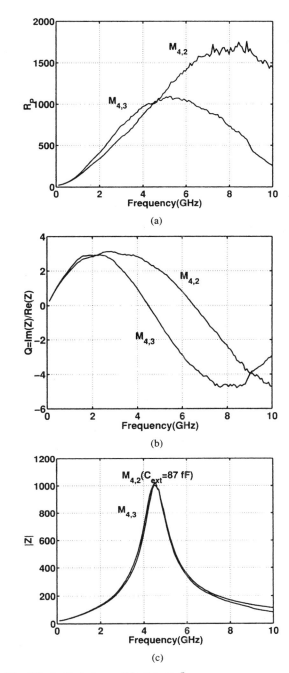

Fig. 18. Effect of inductor modification on Q.

Using five layers of metal yields an inductance value of 266 nH in an area of $(240 \ \mu m)^2$. Accommodating such high values in a small area makes these inductors attractive for integrating voltage regulators and dc–dc converters monolithically.

Stacking inductors can also be useful even for small values. Fig. 16 shows two 5-nH inductors fabricated in a 0.6-μm technology with three layers of metal. The two inductors were designed for the same inductance and nearly equal Q's. The plots in Fig. 16(b) show that the stacked structure has a higher f_{SR} because it occupies less area.

In Fig. 17, some other measured results for two pairs of 5-nH and 10-nH inductors in a 0.4-μm technology (with four layers of metal) are presented. In this case, the self-resonance frequency

increases by 50% with the proposed modification. The Q at self-resonance is between 3 and 5 for the four cases. Note that for the 5-nH inductor resonating at 11.2 GHz, the skin effect is quite significant. Measured and calculated values of f_{SR} [from (19)] differ by less than 4%.

As mentioned before, with the proposed modification, the inductance remains relatively constant because the lateral dimensions are nearly two orders of magnitude greater than the vertical dimensions. This is indeed evident from the slope of $|Z|$ at low frequencies, which is equal to $2\pi L$ (Figs. 15 and 17).

The effect of the proposed modification on the Q is also studied. For the two 10-nH inductors of Fig. 17, we can derive the parallel resistance R_p as a function of frequency [Fig. 18(a)]. If the Q is defined as in (3), then the two inductors have equal

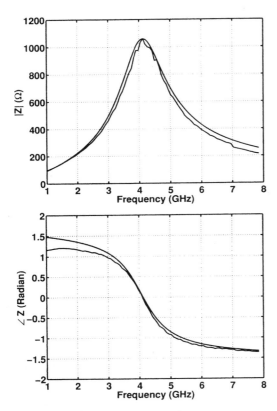

Fig. 19.   Simulation and measurement comparison.

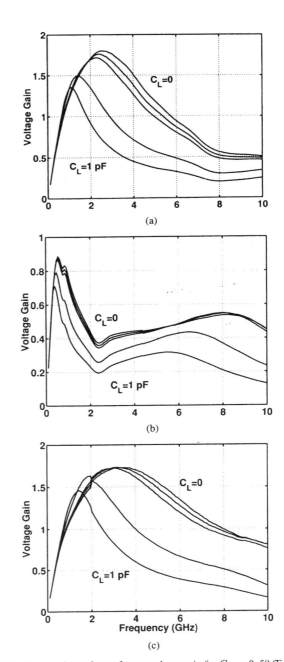

Fig. 20.   Measured 1-to-2 transformer voltage gain for $C_L = 0$, 50 fF, 100 fF, 500 fF, 1 pF. (a) Four turns. (b) Eight turns. (c) Three turns.

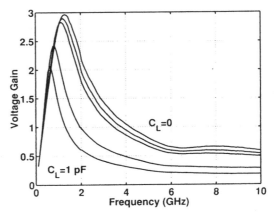

Fig. 21.   Measured 1-to-4 transformer voltage gain for $C_L = 0$, 50 fF, 100 fF, 500 fF, 1 pF.

Q's around 5 GHz, and if (2) is used, the Q's are even closer for frequencies below the resonance [Fig. 18(b)].

Perhaps a fairer comparison is to assume each of the inductors is used in a circuit tuned to a given frequency (e.g., as in a VCO). We then add enough capacitance to the modified structure so that it resonates at the same frequency as the conventional one. Fig. 18(c) shows that the two inductors have the same selectivity and hence the same Q, while the modified structure can sustain an additional capacitance of 87 fF for operation at 4.5 GHz.

To simulate the behavior of an inductor, we can use the distributed circuit of Fig. 3 with a finite number of sections (e.g., 10). However, measured results indicate that for tuned applications, stacked inductors can be even modeled by a simple parallel RLC tank. Fig. 19 compares the simulation results of a parallel RLC tank and the measured characteristics. Here, the equivalent capacitance obtained from (16) and the measured value of the parallel resistance at the resonance frequency are used. These plots suggest that the magnitudes are nearly equal for a wide range and the phases are close for about $\pm 10\%$ around resonance.

Several 1-to-2 transformers have been fabricated in a 0.25-μm technology. Fig. 20 plots the measured voltage gains as a function of frequency. The measured behavior is reasonably close to the simulation results using the distributed model. The four-turn transformer achieves a voltage gain of 1.8 at 2.4 GHz and the three-turn transformer has nearly the same voltage gain over a wider frequency range. The plot also illustrates the effect of capacitive loading on the secondary (calculated using the measured S-parameters), suggesting that capacitances as high as 100 fF have negligible impact on the gain.

Fig. 21 shows the voltage gain of the 1-to-4 transformer of Fig. 13. This transformer is made of three-turn spirals with

9-μm metal lines. The transformer achieves a voltage gain of 3 (9.5 dB) around 1.5 GHz. The short-circuit gain (from secondary to primary) exhibits identical characteristics.

REFERENCES

[1] J. J. Zhou and D. J. Allstot, "Monolithic transformers and their application in a differential CMOS RF low-noise amplifier," *IEEE J. Solid-State Circuits*, vol. 33, pp. 2020–2027, Dec. 1998.

[2] C. P. Yue and S. S. Wong, "On-chip spiral inductors with patterned ground shields for Si-based RF ICs," *IEEE J. Solid-State Circuits*, vol. 33, pp. 743–752, May 1998.

[3] M. W. Green *et al.*, "Miniature multilayer spiral inductors for GaAs MMICs," in *GaAs IC Symp.*, 1989, pp. 303–306.

[4] R. B. Merril *et al.*, "Optimization of high-Q inductors for multilevel metal CMOS," in *Proc. IEDM*, Dec. 1995, pp. 38.7.1–38.7.4.

[5] J. Crols *et al.*, "An analytical model of planer inductors on lowly doped silicon substrates for high-frequency analog design up to 3 GHz," in *Dig. VLSI Circuits Symp.*, June 1996, pp. 28–29.

[6] S. Mohan *et al.*, "Simple accurate expressions for planar spiral inductors," *IEEE J. Solid-State Circuits*, vol. 34, pp. 1419–1424, Oct. 1999.

[7] H. M. Greenhouse, "Design of planar rectangular microelectronic inductors," *IEEE Trans. Parts, Hybrids, Packag.*, vol. PHP-10, pp. 101–109, June 1974.

[8] A. M. Niknejad and R. G. Meyer, "Analysis, design, and optimization of spiral inductors and transformers for Si RF ICs," *IEEE J. Solid-State Circuits*, vol. 33, pp. 1470–1481, Oct. 1998.

[9] B. Razavi, "CMOS technology characterization for analog and RF design," *IEEE J. Solid-State Circuits*, vol. 34, pp. 268–276, Mar. 1999.

[10] J. N. Burghartz *et al.*, "RF circuit design aspects of spiral inductors on silicon," *IEEE J. Solid-State Circuits*, vol. 33, pp. 2028–2034, Dec. 1998.

[11] J. R. Long and M. A. Copeland, "The modeling, characterization, and design of monolithic inductors for silicon RF ICs," *IEEE J. Solid-State Circuits*, vol. 32, pp. 357–369, Mar. 1997.

[12] W. B. Kuhn and N. K. Yanduru, "Spiral inductor substrate loss modeling in silicon RF ICs," *Microwave J.*, pp. 66–81, Mar. 1999.

Andrew Chan (S'98) received the B.S. and M.S. degrees in electrical engineering from the University of California, Los Angeles, in 1998 and 2000, respectively. His master's work was on a low-power frequency synthesizer for a low-power RF transceiver. He also had summer internships at Agilent Technologies and TRW Space & Defense.

Behzad Razavi (S'87–M'90) received the B.Sc. degree in electrical engineering from Sharif University of Technology, Tehran, Iran, in 1985 and the M.Sc. and Ph.D. degrees in electrical engineering from Stanford University, Stanford, CA, in 1988 and 1992, respectively.

He was with AT&T Bell Laboratories, Holmdel, NJ, and subsequently Hewlett-Packard Laboratories, Palo Alto, CA. Since September 1996, he has been an Associate Professor of electrical engineering at the University of California, Los Angeles. His current research includes wireless transceivers, frequency synthesizers, phase-locking and clock recovery for high-speed data communications, and data converters. He was an Adjunct Professor at Princeton University, Princeton, NJ, from 1992 to 1994, and at Stanford University in 1995. He is a member of the Technical Program Committees of the Symposium on VLSI Circuits and the International Solid-State Circuits Conference (ISSCC), in which he is the chair of the Analog Subcommittee. He is an IEEE Distinguished Lecturer and the author of *Principles of Data Conversion System Design* (New York: IEEE Press, 1995), *RF Microelectronics* (Englewood Cliffs, NJ: Prentice-Hall, 1998), and *Design of Analog CMOS Integrated Circuits* (New York: McGraw-Hill, 2000), and the editor of *Monolithic Phase-Locked Loops and Clock Recovery Circuits* (New York: IEEE Press, 1996).

Dr. Razavi received the Beatrice Winner Award for Editorial Excellence at the 1994 ISSCC, the Best Paper Award at the 1994 European Solid-State Circuits Conference, the Best Panel Award at the 1995 and 1997 ISSCC, the TRW Innovative Teaching Award in 1997, and the Best Paper Award at the IEEE Custom Integrated Circuits Conference in 1998. He has also served as Guest Editor and Associate Editor of the IEEE JOURNAL OF SOLID-STATE CIRCUITS and IEEE TRANSACTIONS ON CIRCUITS AND SYSTEMS and *International Journal of High Speed Electronics*.

Alireza Zolfaghari (S'99) was born in Tehran, Iran, on December 23, 1971. He received the B.S. and M.S. degrees in electrical engineering from Sharif University of Technology, Tehran, in 1994 and 1996, respectively. He is currently working toward the Ph.D. degree at the University of California, Los Angeles.

In 1998 he was with TIMA Laboratory, Grenoble, France. His interests include analog and RF circuits for wireless communications.

Estimation Methods for Quality Factors of Inductors Fabricated in Silicon Integrated Circuit Process Technologies

Kenneth O

Abstract— By examining uses of quality (Q) factors for inductors in silicon integrated circuit design, new methods for estimating quality factors are proposed. These methods extract Q factors by numerically adding a capacitor in parallel to measured y_{11} data of an inductor, and by computing the frequency stability factor and 3-dB bandwidth at the resonant frequency of the resulting network. These parameters are then converted to effective quality factors using relationships for simple parallel RLC circuits. By sweeping the numerically added capacitance value, effective quality factors at varying frequencies are computed. These new techniques, in addition to being more relevant for circuit design, provide physically reasonable estimates all the way up to the self-resonant frequencies of inductors. At moderate to high frequencies, the commonly used Q definition [$-\mathrm{Im}(y_{11})/\mathrm{Re}(y_{11})$] can significantly underestimate and can even give unreasonable results. Data obtained using the new methods suggest that quality factors remain high and integrated inductors remain useful all the way up to their self-resonant frequencies, contrary to the behavior obtained using $-\mathrm{Im}(y_{11})/\mathrm{Re}(y_{11})$. These indicate that the commonly used technique can lead to improper use and optimization of integrated inductors.

Index Terms—Integrated inductors, quality factor, silicon IC's.

I. INTRODUCTION

WITH the emergence of RF and microwave applications for silicon integrated circuits, integration of spiral inductors with reasonable characteristics has become an urgent need. In particular, quality (Q) factors of integrated inductors in silicon IC's are typically low, and the understanding and optimization of them have received intense attention [1]–[11]. Despite these efforts, the meaning of reported Q factors is still in a state of confusion. This paper examines existing methods for estimating Q factors and their limitations. To overcome these limitations, new methods, more relevant to circuit design, and useful up to the self-resonant frequencies of the inductors, are proposed. These methods extract quality factors as a function of frequency by numerically adding a capacitor with varying values in parallel to measured y_{11} data of an inductor, and by computing the frequency (phase) stability factors and 3-dB bandwidths at resonant frequencies of the resulting networks. The frequency stability factors and 3-dB bandwidths are then converted to effective quality factors using relationships for simple parallel RLC circuits. Quality factors obtained using these methods are compared to those obtained with the commonly used method based

Manuscript received October 27, 1997; revised March 5, 1998. This work was supported by Rockwell International Corp.

The author is with the Department of Electrical and Computer Engineering, University of Florida, Gainesville, FL 32611 (e-mail: kko@tec.ufl.edu).

Publisher Item Identifier S 0018-9200(98)05527-9.

on $-\mathrm{Im}(y_{11})/\mathrm{Re}(y_{11})$ [1], [3]–[9], [11]. The results of this comparative study suggest that the commonly used method underestimates Q factors, and can lead to improper use and optimization of integrated inductors.

II. DEFINITIONS AND MEASUREMENTS OF QUALITY FACTORS

There are at least two widely used definitions for Q factors. The first and probably the most fundamental definition is based on the maximum energy storage and average power dissipation (P_{diss}) [12] which for simplicity, is referred as Q_{EMAX} in this paper

$$Q_{EMAX} = \frac{\omega W_{\max}}{P_{diss}}. \tag{1}$$

The ω is the radian frequency, and $W_{\max}$ is the maximum total electrical and magnetic energies stored in the system. Unfortunately, as will be discussed later in this section, accurately estimating Q_{EMAX} is difficult.

The most widely used Q definition, which is referred as Q_{CONV} in this paper, is the ratio of a negative of the imaginary part of y_{11} and the real part of $y_{11}[-\mathrm{Im}(y_{11})/\mathrm{Re}(y_{11})]$ [1]. The y_{11} data are obtained by converting measured S-parameters of inductors. A common equivalent circuit for modeling integrated inductors in silicon IC processes [2] is shown in an inset of Fig. 1(a). The y_{11} data are utilized for the quality factor computations because they are the admittances seen looking into port 1 while port 2 is shorted to ground. This is a common configuration in which the inductors are used in amplifiers and oscillators.

Using simple network theory, it can be shown that

$$Q_{CONV} = -\left[\frac{\mathrm{Im}(y_{11})}{\mathrm{Re}(y_{11})}\right] = \frac{2\omega(|\overline{W}_m| - |\overline{W}_e|)}{P_{diss}} \tag{2}$$

where $|\overline{W}_m|$ and $|\overline{W}_e|$ are the average stored magnetic and electrical energies in the system [13]. When port 2 of the equivalent circuit model in Fig. 1(a) is shorted to ground, $|\overline{W}_m|$ and $|\overline{W}_e|$ are energies stored in the inductor L and capacitor C_{p1}, respectively. This Q definition involves the difference between the average stored magnetic and electrical energies rather than the maximum total energy storage. When the average magnetic energy storage is much greater than the electrical storage, this ratio approaches Q_{EMAX}. For the general case as well as for silicon integrated inductors with typically large shunt capacitance to the substrate, thus, significant electrical energy storage, Q_{CONV} can deviate from Q_{EMAX} by a large amount.

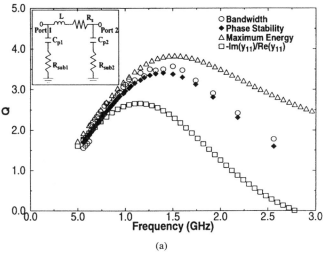

(a)

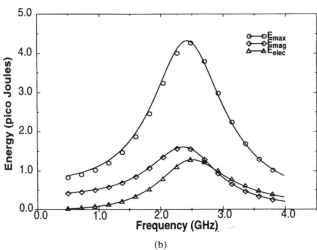

(b)

Fig. 1. (a) Quality factor versus frequency plots of a 15-nH inductor. An equivalent circuit commonly used to model integrated inductors in silicon IC's [2] is shown in an inset. The quality factors are extracted utilizing the commonly used measurement technique, one based on the maximum energy storage using the equivalent circuit model, and newly proposed methods. (b) The average magnetic and electrical energy storages, and the maximum energy storage as a function of frequency for the 15-nH inductor.

TABLE I
MODEL PARAMETERS FOR 15-nH AND 5-nH INDUCTORS

	Inductor 1	Inductor 2
L	15.2 nH	4.8 nH
R_s	28 Ω	10.7 Ω
C_{P1}	0.26 pF	0.44 pF
C_{P2}	0.26 pF	0.42 pF
R_{sub1}	104 Ω	373 Ω
R_{sub2}	120 Ω	351 Ω
# of turns	8, Metal 2	6.5, Metal 1 & 2
Inductor Trace Width & Space	10 μm (W) 5.0 μm (S)	10 μm (W) 2.0 μm (S)
Outer Area	310 x 310 μm²	140 x 140μm²

Fig. 1(a) shows Q versus frequency plots of a 15-nH inductor extracted with the conventional technique (Q_{CONV}), calculated based on the maximum energy storage using the equivalent circuit model (Q_{EMAX}), and using the newly proposed methods discussed in Section III. Modeling parameters for the inductor as well as its geometry data are given in Table I. The inductors in Table I are fabricated on a 20-Ω-cm p-type substrate [6]. The Q_{CONV} and Q_{EMAX} are computed by applying a sine wave to port 1 of the equivalent circuit and by computing the voltage across the capacitor (C_{p1}) and current flowing through the inductor (L) as a function of time. The energies are computed as a function of time using the energy formulas for capacitors and inductors [14]. The energies are averaged over time to obtain $|\overline{W}_m|$ and $|\overline{W}_e|$. These time-dependent energies are also added to compute the total energy storage as a function of time, which is in turn used to extract the maximum energy storage (W_{max}). There are significant differences between the Q_{CONV} and Q_{EMAX}.

Fig. 1(b) shows the average stored magnetic and electrical energies as well as the maximum energy storage for the 15-nH inductor as a function of frequency. At low frequencies, the total energy is dominated by the magnetic energy and Q factors estimated using $[-\text{Im}(y_{11})/\text{Re}(y_{11})]$ (Q_{CONV}) are close to Q_{EMAX}. However, as frequency is increased, the electrical energy storage increases, and the difference between the average stored magnetic and electrical energies decreases, which in turn increases the difference between Q_{CONV} and Q_{EMAX}. As a matter of fact, since the self-resonant frequency of an inductor occurs near a frequency where the difference between the average stored magnetic and electrical energy is zero, or where the imaginary part of y_{11} is equal to zero, from (2), Q factors extracted using $-\text{Im}(y_{11})/\text{Re}(y_{11})$ become zero near the self-resonant frequency. This result, of course, is physically unreasonable. The quality factor should not be zero at the self-resonant frequency.

Lastly, it was stated earlier that accurately estimating Q_{EMAX} is difficult. This is due to problems with representing parasitic as well as inductive element(s) in the model, and extracting their values from measurements. Related difficulties include the distributed nature of the model elements and their frequency dependence. The equivalent circuit model at best is rough. These problems are exacerbated by the fact that the model parameters are extracted using y-parameters which depend on the difference between the average magnetic and electrical energies rather than on the maximum total energy storage. Hence, the use of an extracted equivalent model for computation of the maximum energy storage (and Q_{EMAX}) is prone to errors.

III. EFFECTIVE QUALITY FACTORS OF INTEGRATED INDUCTORS

Before going further, an examination of reasons behind interests for quality factors is in order. For microwave and RF applications, their importance arises from the fact that, for matching networks, Q factors are related to loss, while for bandpass filters, they are related to the 3-dB bandwidths [15]. Quality factors are also related to the phase or frequency stability, and phase noise (through the bandwidth) of oscillators using parallel RLC circuits. The frequency stability factor of an oscillator (S_F) is

$$S_F = -\omega_o \frac{d\phi}{d\omega}\bigg|_{\omega=\omega_o} = -\omega_o \frac{d}{d\omega}\left[\text{atan}\left(\frac{\text{Im}(y_{11})}{\text{Re}(y_{11})}\right)\right]\bigg|_{\omega=\omega_o}$$ (3)

where ω_o is the resonant frequency [16]. Actual parameters of interest are the loss, bandwidths of resonant circuits, and frequency stability factors of oscillators.

Since the bandwidth and frequency stability factor are defined only at the resonant frequency, measured y_{11} data can be used to extract them at the self-resonant frequency of an inductor. These, however, are not typically of great interest because using an inductor in circuits requires adding parasitic or intentional capacitances and resistances in parallel with the inductor. Once a capacitor is added, the resonant frequency is lowered from the inductor self-resonant frequency, and the bandwidth and stability factor are also changed. Unfortunately, these cannot be obtained directly from the measured y_{11} data. An obvious way to estimate these is to fabricate a set of test structures consisting of an inductor and a capacitor connected in parallel with varying capacitance values, and to characterize the structures at their respective resonant frequencies. This of course requires large numbers of test structures and measurements.

Luckily, the fabrication and measurements of the test structures are not necessary. Instead, a bandwidth and a frequency stability factor at a resonant frequency different from the inductor self-resonant frequency can be obtained by numerically adding a capacitor ($C_{num}\omega j$) in parallel to the measured y_{11} data, and by computing the parameters at the resonant frequency of the resulting RLC circuit. The resulting bandwidth and frequency stability factor represent the smallest bandwidth and the highest stability factor which can be achieved using a given inductor at the resonant frequency, since in real circuits, adding a capacitor cannot be accomplished without adding resistance or increasing the loss. To compute these parameters over a range of frequencies of interest, the resonant frequency must be swept, which can be accomplished by sweeping the added capacitance value.

The quality factor is still a useful figure of merit that provides means for quickly estimating circuit performance. The 3-dB bandwidth ($\Delta\omega$) of simple RLC circuits [15], and the frequency (phase) stability factor of oscillators [16] using simple parallel RLC circuits, are related to the quality factor by

$$Q = \frac{\omega_o}{\Delta\omega} = Q_{BW}$$ (4)

$$Q = \frac{S_F}{2} = Q_{PS}.$$ (5)

As seen in the equivalent circuit model of Fig. 1(a), the RLC circuit resulting from numerically adding a capacitor to the equivalent circuit is not a simple RLC circuit. Despite this, using (4) and (5), effective quality factors Q_{BW} (based on the bandwidth) and Q_{PS} (based on the phase stability) can be defined. It should be emphasized that these are effective values. In addition, as mentioned earlier, integrated inductors fabricated in silicon technologies in general have low Q factors, and this makes the validity of the bandwidth-based Q factor definition questionable [15]. However, as figures of

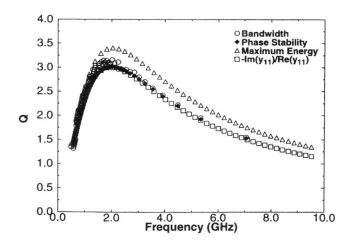

Fig. 2.   Quality factor versus frequency plots of a 5-nH inductor.

merit for estimating and comparing usefulness of inductors, these effective values are still quite useful. Because of their close ties to circuit applications, these effective quality factors should also be more useful and relevant figures of merit than Q_{CONV}. An important feature of the effective Q extractions is that they do not require an equivalent circuit model. This eliminates the modeling and model parameter extraction errors. The extraction routines are implemented in Matlab [17], and the routines are around 30 lines long and straightforward.

IV. DISCUSSION

In order to compare Q_{BW} and Q_{PS} to $Q_{CONV}[-\text{Im}(y_{11})/\text{Re}(y_{11})]$ as well as to Q_{EMAX} (based on the maximum energy storage consideration), Q_{BW} and Q_{PS} are extracted using the equivalent circuit model parameters given in Table I. Figs. 1(a) and 2 show these quality factors for the 15-nH and 5-nH inductor, respectively. In Fig. 1(a), Q_{CONV} deviates significantly from Q_{EMAX}, while Q_{CONV} deviates relatively little (10–15%) for the 5-nH inductor in Fig. 2. For the 15-nH inductor, Q_{CONV} also deviates significantly from Q_{BW} and Q_{PS}. Q_{BW} and Q_{PS} differ from Q_{EMAX} by 10–35% between 1.5 and 2.5 GHz, which is significantly less than the 40–80% deviation of Q_{CONV}. The smaller differences among Q_{CONV}, Q_{BW}, Q_{PS}, and Q_{EMAX} for the 5-nH inductor are due to the fact that the average magnetic energy storage is substantially larger than the electrical energy storage in the examined frequency range. There also exist some differences between Q_{BW} and Q_{PS}, although they are small. Figs. 1(a) and 2 clearly illustrate that Q_{BW} and Q_{PS} better estimate Q_{EMAX}, although significant differences can still exist. For designing filters, tuned amplifiers, and oscillators, Q_{BW} and Q_{PS} should be completely adequate.

Fig. 3(a) and (b) shows plots of Q_{CONV}, Q_{BW}, and Q_{PS} for 2.6- and 12-nH inductors. These quality factors are extracted directly from measured S-parameters rather than using extracted model parameters. For both inductors, at low frequencies, quality factors based on different techniques are approximately the same. However, at high frequencies, Q_{BW} and Q_{PS} are consistently higher than Q_{CONV}. Q_{CONV} can become zero or even negative, while Q_{BW} and Q_{PS} remain

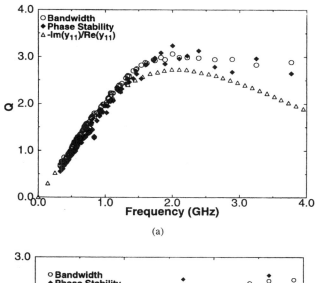

(a)

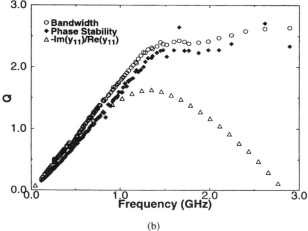

(b)

Fig. 3. (a) and (b) Plots of Q_{CONV}, Q_{BW}, and Q_{PS} versus frequency for 2.6- and 12-nH inductors. Quality factors are extracted directly from measured S-parameter data.

parallel with measured y_{11} data, and extractions of the 3-dB bandwidths and frequency stability factors. Implementations of extraction routines require simple numerical manipulations of measured S-parameters. At low frequencies, quality factors from the new and conventional $[-\mathrm{Im}(y_{11})/\mathrm{Re}(y_{11})]$ methods are approximately the same. On the other hand, at moderate to high frequencies, the conventional method can significantly underestimate quality factors which is caused by the fact that the imaginary part of y_{11} is related to the difference between the average magnetic and electrical energy storage rather than the maximum total energy storage. The conventional Q definition can lead to improper use and optimization of integrated inductors.

REFERENCES

[1] K. B. Ashby, I. A. Koullias, W. C. Finley, J. J. Bastek, and S. Moinian, "High Q inductors for wireless applications in a complementary silicon bipolar process," *IEEE J. Solid-State Circuits*, vol. 31, pp. 4–9, Jan. 1996.
[2] N. M. Nguyen and R. G. Meyer, "Si IC-compatible inductors and LC passive filters," *IEEE J. Solid-State Circuits*, vol. 25, pp. 1028–1031, 1990.
[3] J. N. Burghartz, M. Soyuer, K. Jenkins, and M. Hulvey, "High-Q inductors in standard silicon integrated technology and its application to an integrated RF power amplifier," in *IEDM Tech. Dig.*, Washington, DC, 1995, pp. 1015–1017.
[4] J. N. Burghartz, M. Soyuer, and K. Jenkins, "Integrated RF and microwave components in BiCMOS technology," *IEEE Trans. Electron Devices*, vol. 43, pp. 1559–1570, Sept. 1996.
[5] ———, "Microwave inductors and capacitors in standard multilevel interconnect silicon technology," *IEEE Trans. Microwave Theory Tech.*, vol. 44, pp. 100–104, Jan. 1996.
[6] K. K. O, P. Garone, C. Tsai, G. Dawe, B. Scharf, T. Tewksbury, C. Kermarrec, and J. Yasaitis, "A low cost and low power silicon npn bipolar process with NMOS transistors (ADRF) for RF and microwave applications," *IEEE Trans. Electron Devices*, vol. 42, pp. 1831–1840, Oct. 1995.
[7] K. Kim and K. K. O, "Characteristics of an integrated spiral inductor with an underlying n-well," *IEEE Trans. Electron Devices*, vol. 44, pp. 1565–1567, Sept. 1997.
[8] J. N. Burghartz, M. Soyuer, K. Jenkins, M. Kies, M. Dolan, K. Stein, J. Malinowski, and D. Harame, "Integrated RF components in a SiGe bipolar technology," *IEEE J. Solid-State Circuits*, vol. 32, pp. 1440–1445, Sept. 1997.
[9] D. Harame, L. Larson, M. Case, S. Kovacic, S. Voinigescu, T. Tewksbury, D. Nguyen-Ngoc, K. Stein, J. Cressler, S.-J. Jeng, J. Malinowski, R. Groves, E. Eld, D. Sunderland, D. Rensch, M. Gilbert, K. Schonenberg, D. Ahlgren, S. Rosenbaum, J. Glenn, and B. Meyerson, "SiGe HBT technology: Device and application issues," in *IEDM Tech. Dig.*, Washington, DC, 1995, pp. 731–734.
[10] R. Groves, D. L. Harame, and D. Jadus, "Temperature dependence of Q in spiral inductors fabricated in a silicon germanium/BiCMOS technology," in *Proc. 1996 BCTM*, Minneapolis, MN, pp. 153–156.
[11] R. B. Merrill, T. W. Lee, H. You, R. Rasmussen, and L. A. Moberly, "Optimization of high Q integrated inductors for multi-level metal CMOS," in *IEDM Tech. Dig.*, Washington DC, 1995, pp. 983–986.
[12] D. Fink and H. Beaty, *Standard Handbook for Electrical Engineers*, 11th ed. New York: McGraw-Hill, 1978, pp. 1–20.
[13] C. Desoer and E. Kuh, *Basic Circuit Theory*. New York: McGraw-Hill, 1969, p. 402.
[14] C. Johnk, *Engineering Electromagnetic Field and Waves*. New York: Wiley, 1975.
[15] C. Desoer and E. Kuh, *Basic Circuit Theory*. New York: McGraw-Hill, 1969, p. 316.
[16] J. Smith, *Modern Communication Circuits*. New York: McGraw-Hill, 1986, pp. 251–252.
[17] Matlab, Mathworks Inc, Natick, MA 01760.

well above the zero all the way up to the self-resonant frequencies of the inductors. This, of course, is physically more reasonable. For instance, the 12-nH inductor in Fig. 3(b) has a self-resonant frequency of 2.9 GHz. At this frequency, Q_{BW}, Q_{PS}, and Q_{CONV} are 2.6, 2.4, and 0, respectively. In addition, frequency dependence of Q_{BW} and Q_{PS} at high frequencies is less than that of Q_{CONV}. These data indicate that inductors remain useful all the way up to their resonant frequencies, which is contrary to that suggested by quality factors extracted using the conventional technique. The Q_{PS} data are more scattered because their extraction involves computations of derivatives.

V. SUMMARY

Through an examination of uses for quality factors, new methods for estimating quality factors of integrated inductors for RF and microwave applications are proposed. These methods involve a numerical addition of varying capacitance in

A Q-FACTOR ENHANCEMENT TECHNIQUE FOR MMIC INDUCTORS

Mina Danesh, John R. Long, R. A. Hadaway[1] and D. L. Harame[2]

University of Toronto
Toronto, Ontario, Canada

[1] Nortel Technology, Ottawa, Ontario, Canada
[2] IBM Microelectronics, Burlington, VT, USA

Abstract - An increase of 50% in the peak Q-factor and a wider operating bandwidth for monolithic inductors is achieved by exciting a microstrip structure differentially. Conventional excitation of a 8 nH spiral inductor fabricated in a production silicon IC technology resulted in a peak (measured) Q-factor of 6.6 at 1.6 GHz, while the differential connection showed a maximum Q-factor of 9.7 at 2.5 GHz. These experimental results compared favorably with the behaviour predicted from simulation.

I. INTRODUCTION

Microstrip inductors have been used extensively in radio frequency (RF) and monolithic microwave integrated circuits (MMIC). The quality factor (Q) of microstrip structures is limited by the series resistance of the metallization and, in the case of silicon technology, losses in the conductive substrate (typically 1 to 10 Ω-cm). The Q-factor is typically less than 10 for a metal-insulator-semiconductor (MIS) structure fabricated in a production silicon IC technology [1, 2]. It has been demonstrated that the quality factor can be improved by optimizing both the physical layout and structure of the inductor. Thicker metallization and stacking of metal layers reduces the conductor dissipation, thereby improving the Q [3]. Design guidelines have been proposed to optimize the geometric parameters of microstrip spiral inductors, such as strip width, spacing between the strips, and the gap between groups of coupled lines on opposing sides [1]. Losses in the semiconducting substrate may be reduced through the use of higher resistivity material [4], however, this is incompatible with current silicon device technology (e.g., CMOS). An alternative is to remove the underlying silicon by selective etching of the substrate or by applying a thin membrane beneath the inductor [5, 6], but this requires additional processing steps and there is a loss of mechanical strength when the underlying silicon is removed.

Microstrip spiral inductors are normally driven "single-ended", that is, the driving source is connected to one terminal of the spiral while the other end is grounded. Differential circuit topologies are common in integrated circuits, and consequently an alternative method that is practical for integrated circuits is to excite the spiral inductor differentially, using a source connected between the two ends of the microstrip spiral. The peak Q-factor shows a significant increase under differential excitation, and this high Q value is maintained over a broader bandwidth.

II. THEORY

A transmission line can be approximated over a range of frequencies by a lumped element equivalent circuit model. For a microstrip line fabricated in silicon technology, an appropriate equivalent circuit is shown in Fig. 1, where L is the total inductance of the line and r is the series resistance due to conductor losses and dissipation arising from current flowing in the silicon substrate. The shunt parasitics result from a combination of capacitances, due to an insulating layer of silicon dioxide (C_{ox}) and the underlying substrate (C_{si}), and substrate dissipation (R_{si}) [7]. For the spiral inductor, additional components are required to represent mutual magnetic and electric coupling between adjacent lines [1].

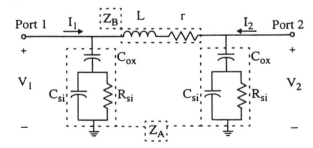

Fig. 1. Microstrip line equivalent circuit.

For single-ended excitation, Port 2 in Fig. 1 is grounded and the inductor is connected as a one-port. The input impedance at Port 1 (in this case Z_{SE}) becomes a parallel combination of two components: one due to the inductance and series dissipation (L and r in Fig. 1), the other due to the shunt R-C parasitic elements, as illustrated in Fig. 2. For a differential excitation, where the signal is applied between the two ports (i.e., between Port1 and Port2), the input impedance (Z_D) is due to the parallel combination of $2Z_A$ and Z_B, where impedances Z_A and Z_B are defined in Fig. 1. Since the substrate parasitics are connected together via the ground plane, the two shunt elements, are now in series (i.e., $Z_A + Z_A$), resulting in the equivalent circuit shown in Fig. 3.

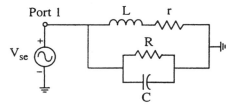

Fig. 2. Single-ended excitation model.

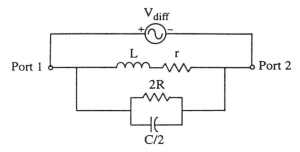

Fig. 3. Differential excitation model.

The quality factor of the inductor below the first resonant frequency is defined by

$$Q = \frac{2\pi f L}{Re[Z_{input}]} \tag{1}$$

where Z_{input} is the series equivalent input impedance. At lower frequencies, the input impedance in either the shunt or differential connections is approximately the same, but as the frequency increases, the substrate parasitics, C and R, come into play. For the case of differential excitation, these parasitics have a higher impedance at a given frequency than in the single-ended connection, as seen from comparison of Figs. 2 and 3. This reduces the real part and increases the reactive component of the input impedance. Therefore, the inductor Q-factor (from eq. 1) is improved when driven differentially, and moreover, a wider operating bandwidth can be achieved.

III. ANALYSIS

A 5-turn square spiral inductor was fabricated and tested in order to verify these predictions. A cross-sectional view of a portion of the structure is illustrated in Fig. 4, and the substrate and metal properties for the fabrication process are listed in Table 1. The outer dimension, A, as shown in Fig. 5, is 250 μm. The inductor consists mainly of topmetal (M3) which is

8 μm wide and the spacing between conductors is 2.8 μm (w and s in Fig. 4, respectively). The relatively narrow conductor width and spacing results in higher magnetic coupling between microstrip lines and lower capacitive parasitics to the substrate. The inner gap between groups of coupled lines, G, is approximately 150 μm, which minimizes negative mutual coupling on opposite sides of the spiral.

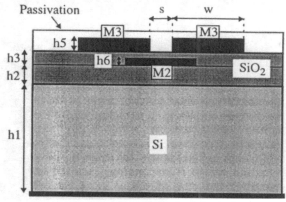

Fig. 4. Cross-sectional view of the inductor.

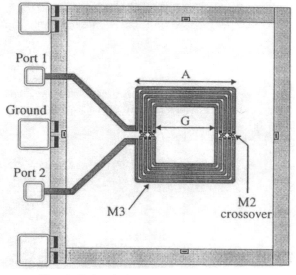

Fig. 5. Spiral inductor test structure layout.

Table 1. Substrate and metal parameters.

Parameter	Value
Oxide relative permittivity	$\varepsilon_r = 3.9$
Oxide thickness over M2	h3 - h6 = 1.3 μm
Oxide thickness below M2	h2 = 3.61 μm
Silicon relative permittivity	$\varepsilon_r = 11.7$
Silicon resistivity	$\rho = 15\ \Omega$–cm
Silicon thickness	h1 = 200 μm
Topmetal M3 resistivity	$\rho = 0.031\ \Omega$–μm
M3 thickness	h5 = 2.07 μm
M2 thickness	h6 = 0.84 μm

IV. RESULTS AND DISCUSSION

Three-dimensional electromagnetic simulation (using HP-EEsof's Momentum) and experimental measurements were obtained for the 2-port spiral inductor. The results for the single-ended configuration were derived by grounding one of the ports (either Port1 or Port2) since the structure is symmetric. The response due to a differential excitation can be derived from 2-port S-parameter measurements using the relationship

$$S_D = S_{11} - S_{21} \qquad (2)$$

The input impedances are determined from the 1-port S-parameters or the 2-port impedance matrix, as:

$$Z_{SE} = Z_{11} - \frac{Z_{12} \cdot Z_{21}}{Z_{22}} = Z_0\left(\frac{1 + S_{11}}{1 - S_{11}}\right) \qquad (3)$$

and

$$Z_D = Z_{11} + Z_{22} - Z_{12} - Z_{21} = 2 \cdot Z_0\left(\frac{1 + S_D}{1 - S_D}\right) \qquad (4)$$

where Z_0 is the system impedance (50 Ω).

Comparisons between experimental measurements and simulations for the input impedance and Q-factor (as defined by eq. 1) are shown in Figs. 6, 7 and 8. The series resistance at low frequencies (i.e., f < 0.5 GHz), as defined by r from the lumped element model of Figs. 2 and 3, was measured to be 7.3 Ω, while the simulation predicts 7.9 Ω. Measured and simulated low frequency inductances are 7.7 nH and 8.3 nH, respectively. At lower frequencies, the difference in Q between differential and single-ended excitations is not significant (<1%). This is because the shunt capacitive parasitic components do not affect the low frequency input impedance and hence, the two cases can be represented by a series L-r model. However, as the frequency increases, the difference between the input impedances becomes substantial; Z_D is much lower than Z_{SE} by an increasingly greater factor. This is caused by the lower substrate parasitics present in the differentially excited case, as previously described. The difference between Q-factors in the differential and single-ended cases, as shown in Fig. 8, illustrates this point. The peak in the Q-factor is a result of the shunt parasitics resonating with the inductance. Lower parasitics for differential excitation result in a higher peak Q-factor and broadening of the Q peak, when compared to the conventional single-ended connection.

For the single-ended excitation, the peak Q-factor is 6.6 at 1.6 GHz from both measurement and simulation. However, for a differential excitation, the resulting peak Q occurs at a higher frequency (2.5 GHz), with a value of 9.7 from measurement and 10.3 from simulation. This is a 47% increase in the measured peak Q between the differential and single-ended cases that can be realized without modification to the fabrication process. Achieving a comparable Q value in the single-ended connection would require an increase of approximately twice the top-metal thickness.

At frequencies beyond the peak, an increase of greater than 50% can be achieved. It should be noted that Q values for the differential case, because they are greater in magnitude, are much more sensitive to slight variations in the measured or simulated input impedance. Thus, the relative effect of an error in either the measurement or simulation near the peak Q for the differential case is more pronounced. A slight increase in the topmetal thickness could account for part of the discrepancy between the measured and simulated performance. This would be consistent with the lower series resistance and inductance observed from the measurements when compared to the simulated values. The measured data shown is from a representative sample; several measurements were performed which gave the same results within a ±5% variation.

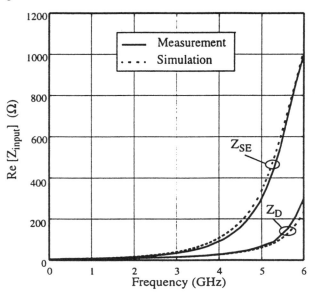

Fig. 6. Resistive Z_{SE} and Z_D.

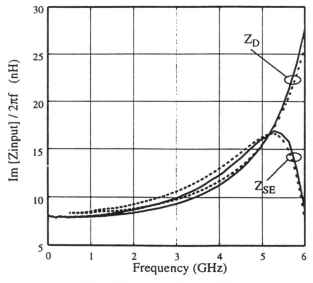

Fig. 7. Inductive Z_{SE} and Z_D.

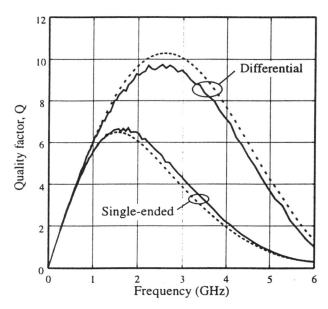

Fig. 8. Measured and simulated Q-factor.

Table 2 lists the parameters of a lumped-element model (see Fig. 9) that was fit over a broadband of frequency (500-6000MHz) for the inductor from the simulated data for both the single-ended and differential connections. The resistive parasitic element is 2.8 times higher for differential excitation, and the shunt capacitance is a small fraction of the single-ended case (C1: 38%, C2: 70%). The modified values for the lumped element models differ from the equivalent circuits shown in Figs. 2 and 3 due to the distributed nature of the actual inductor, which cannot be modeled accurately by a single lumped-element section for both one and two-port configurations.

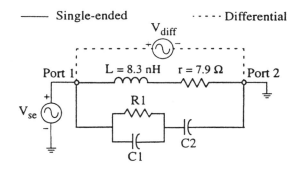

Fig. 9. Equivalent circuit model for both configurations.

Table 2. Lumped element parameter fitting.

Excitation	C1 (fF)	C2 (fF)	R1 (Ω)
Single-ended	135	162	358
Differential	95	62	1000

V. CONCLUSION

Higher peak Q-factor values and a wider operating bandwidth for monolithic inductors can be achieved by exciting a microstrip structure differentially. This could be exploited in many MMIC circuit applications, such as oscillators, mixers and amplifiers, where high-quality components are required. Measured and simulated results presented in this paper have validated the prediction of lower substrate parasitics that were made from a simplified lumped element model. The improvement in performance has been demonstrated for an inductor fabricated in a production silicon IC technology, however, these same results would apply to microstrip inductors fabricated on other substrates, such as GaAs.

ACKNOWLEDGEMENTS

This work was supported by Micronet and the Natural Sciences and Engineering Research Council of Canada (NSERC).

REFERENCES

[1] J. R. Long and M. A. Copeland, "The Modeling, Characterization, and Design of Monolithic Inductors for Silicon RF ICs", *IEEE J. Solid-State Circuits*, Vol. 32, No. 3, March 1997, pp. 357-369.

[2] K. B. Ashby, I. A. Koullias, W. C. Finley, J. J. Bastek and S. Moinian, "High Q Inductors for Wireless Applications in a Complementary Silicon Bipolar Process", *IEEE J. Solid-State Circuits*, Vol. 31, No. 1, January 1996, pp. 4-9.

[3] J. N. Burghartz, M. Soyuer, K. A. Jenkins, M. Kies, M. Dolan, K. J. Stein, J. Malinowski and D. L. Harame, "Integrated RF Components in a SiGe Bipolar Technology", *IEEE J. Solid-State Circuits*, Vol. 32, No. 9, September 1997, pp. 1440-1445.

[4] M. Park, S. Lee, H. K. Yu, J. G. Koo and K. S. Nam, "High Q CMOS-Compatible Microwave Inductors using Double-Metal Interconnection Silicon Technology", *IEEE Microwave and Guided Wave Letters*, Vol. 7, No. 2, February 1997, pp. 45-47.

[5] J. Y.-C. Chang, A. A. Abidi and M. Gaitan, "Large Suspended Inductors on Silicon and Their Use in a 2-μm CMOS RF Amplifier", *IEEE Electron Device Letters*, Vol. 14, No. 5, May 1993, pp. 246-248.

[6] A. C. Reyes, S. M. El-Ghazali, S. J. Dorn, M. Dydyk, D. K. Schroder and H. Patterson, "Coplanar Waveguides and Microwave Inductors on Silicon Substrates", *IEEE Trans. Microwave Theory Tech.*, Vol. 43, No. 9, September 1995, pp. 2016-2022.

[7] H. Hasegawa, M. Furukawa and H. Yanai, "Properties of Microstrip Line on Si-SiO2 System", *IEEE Trans. Microwave Theory Tech.*, Vol. 19, No. 11, November 1971, pp. 869-881.

On-Chip Spiral Inductors with Patterned Ground Shields for Si-Based RF IC's

C. Patrick Yue, *Student Member, IEEE,* and S. Simon Wong, *Senior Member, IEEE*

Abstract— This paper presents a patterned ground shield inserted between an on-chip spiral inductor and silicon substrate. The patterned ground shield can be realized in standard silicon technologies without additional processing steps. The impacts of shield resistance and pattern on inductance, parasitic resistances and capacitances, and quality factor are studied extensively. Experimental results show that a polysilicon patterned ground shield achieves the most improvement. At 1–2 GHz, the addition of the shield increases the inductor quality factor up to 33% and reduces the substrate coupling between two adjacent inductors by as much as 25 dB. We also demonstrate that the quality factor of a 2-GHz LC tank can be nearly doubled with a shielded inductor.

Index Terms— Inductor, inductor model, patterned ground shield, quality factor, self-resonance, substrate loss, substrate noise coupling.

I. INTRODUCTION

RECENTLY, interest in on-chip spiral inductors has surged with the growing demand for radio frequency integrated circuits (RF IC's) [1]. For silicon-based RF IC's, the inductor quality factor (Q) degrades at high frequencies due to energy dissipation in the semiconducting substrate [2]. Noise coupling via the substrate at gigahertz frequencies has been reported [3]. As inductors occupy substantial chip area, they can potentially be the source and receptor of detrimental noise coupling. Furthermore, the physical phenomena behind the substrate effects are complicated to characterize. Therefore, decoupling the inductor from the substrate can enhance the overall performance: increase Q, improve isolation, and simplify modeling.

Some approaches have been proposed to address the substrate issues; however, they are accompanied by drawbacks. Ashby *et al.* [4] suggested the use of high-resistivity (150–200 $\Omega\cdot$cm) silicon substrate to mimic the low-loss semi-insulating GaAs substrate, but this is an uncommon option for current silicon technologies. Chang *et al.* [5] demonstrated that etching a pit in the silicon substrate under the inductors can remove the substrate effects. However, the etch adds extra processing cost, and is not readily available. Moreover, it raises reliability concerns such as packaging yield and long-term mechanical stability. For low-cost integration of inductors, the solution to substrate problems should avoid increasing process complexity.

Manuscript received September 1997; revised December 3, 1997. This work was supported in part by the Center for Integrated Systems Industrial Sponsors and by the National Science Foundation under Contract MIP-9313701.

The authors are with the Center for Integrated Systems, Stanford University, Stanford, CA 94305 USA (e-mail: yuechik@haydn.Stanford.EDU).

Publisher Item Identifier S 0018-9200(98)02230-6.

In this paper, we present a patterned ground shield, which is compatible with standard silicon technologies, to reduce the unwanted substrate effects. To provide some background, Section II presents a discussion on the fundamental definitions of an inductor Q and an LC tank Q. Next, a physical model for spiral inductors on silicon is described. The magnetic energy storage and loss mechanisms in an on-chip inductor are discussed. Based on this insight, it is shown that energy loss can be reduced by shielding the electric field of the inductor from the silicon substrate. Then, the drawbacks of a solid ground shield are analyzed. This leads to the design of a patterned ground shield. Design guidelines for parameters such as shield pattern and resistance are given. In Section III, experiment design, on-wafer testing technique, and parasitic extraction procedure are presented. Experimental results are then reported to study the effects of shield resistance and pattern on inductance, parasitic resistances and capacitances, and inductor Q. Next, the improvement in Q of a 2-GHz LC tank using a shielded inductor is illustrated. A study of the noise coupling between two adjacent inductors and the efficiency of the ground shield for isolation are also presented. Lastly, Section IV gives some conclusions.

II. DESIGN CONSIDERATIONS

A. Definitions of Quality Factor

The quality of an inductor is measured by its Q which is defined as [6]

$$Q = 2\pi \cdot \frac{\text{energy stored}}{\text{energy loss in one oscillation cycle}}. \quad (1)$$

Interestingly, (1) also defines the Q of an LC tank. The definition in (1) is fundamental in the sense that it does not specify what stores or dissipates the energy. The subtle distinction between an inductor Q and an LC tank Q lies in the intended form of energy storage. For an inductor, only the energy stored in the magnetic field is of interest. Any energy stored in the inductor's electric field, because of some inevitable parasitic capacitances in a real inductor, is counterproductive. Hence, Q is proportional to the net magnetic energy stored, which is equal to the *difference* between the peak magnetic and electric energies. An inductor is at self-resonance when the peak magnetic and electric energies are equal. Therefore, Q vanishes to zero at the self-resonant frequency. Above the self-resonant frequency, no net magnetic energy is available from an inductor to any external circuit. In contrast, for an LC tank, the energy stored is the *sum* of the average magnetic and electric energies. Since the energy stored in a (lossless) LC tank is constant and oscillates between magnetic and electric forms, it is also equal to the peak magnetic energy, or the peak electric energy. The rate of the oscillation process is the tank's resonant frequency at which Q is defined. For a lossless LC tank, Q is infinite.

To illustrate the distinction between these two cases, consider a simple parallel RLC circuit first as an inductor model, then as an LC tank model. The expressions for the energies and the resonant frequency ω_0 are:

$$E_{\text{peak magnetic}} = \frac{V_0^2}{2\omega^2 L} \tag{2}$$

$$E_{\text{peak electric}} = \frac{V_0^2 C}{2} \tag{3}$$

$$E_{\text{loss in one oscillation cycle}} = \frac{2\pi}{\omega} \cdot \frac{V_0^2}{2R} \tag{4}$$

$$E_{\text{average magnetic}} = \frac{V_0^2}{4\omega^2 L} \tag{5}$$

$$E_{\text{average electric}} = \frac{V_0^2 C}{4} \tag{6}$$

and

$$\omega_0 = \frac{1}{\sqrt{LC}} \tag{7}$$

where V_0 denotes the peak voltage across the circuit terminals. In terms of an inductor model, C is regarded as the parasitic capacitance of the inductor. The inductor Q is shown in (8), found at the bottom of the page, which equals zero at $\omega = \omega_0$, and is less than zero beyond ω_0. It is worthwhile to mention that the result in (8) can also be obtained using the ratio of the imaginary to the real part of the circuit impedance. The circuit impedance is inductive below ω_0 and capacitive above ω_0. In terms of an LC tank model, C is regarded as the tank capacitance of the LC tank. The tank Q is defined at ω_0 and is expressed in (9), shown at the bottom of the page. The same result can also be derived using a more well-known relationship: the ratio of the resonant frequency to the -3-dB bandwidth.

Both Q definitions discussed are of importance, and their applications are determined by the intended function in a circuit. When evaluating the quality of an on-chip inductor

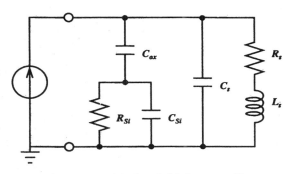

Fig. 1. Lumped physical model of a spiral inductor on silicon.

as a single element, the definition in (8) is more appropriate. In Section III, when LC tanks are studied, the definition in (9) will be used.

B. Understanding of Substrate Effects

The physical model of an inductor on silicon with one port and the substrate grounded is shown in Fig. 1 [2]. An on-chip inductor is physically a three-port element including the substrate. The one-port connection shown in Fig. 1 avoids unnecessary complexity in the following analysis and at the same time preserves the inductor characteristics. In the model, the series branch consists of L_s, R_s, and C_s. L_s represents the spiral inductance which can be computed using the Greenhouse method [7]. R_s is the metal series resistance whose behavior at radio frequency (RF) is governed by the eddy current effect. This resistance symbolizes the energy losses due to the skin effect in the spiral interconnect structure, as well as the induced eddy current in any conductive media close to the inductor. The series feedforward capacitance C_s accounts for the capacitance due to the overlaps between the spiral and the center-tap underpass [8]. The effect of the interturn fringing capacitance is usually small because the adjacent turns are almost equipotential and therefore it is neglected in our model. The overlap capacitance is more significant because of the relatively large potential difference between the spiral and the center-tap underpass. The parasitics in the shunt branch are modeled by C_{ox}, C_{Si}, and R_{Si}. C_{ox} represents the oxide capacitance between the spiral and the substrate. The silicon substrate capacitance and resistance are modeled by C_{Si} and

$$Q_{\text{inductor}} = 2\pi \cdot \frac{\text{peak magnetic energy} - \text{peak electric energy}}{\text{energy loss in one oscillation cycle}} = \frac{R}{\omega L} \cdot \left[1 - \left(\frac{\omega}{\omega_0}\right)^2\right] \tag{8}$$

$$
\begin{aligned}
Q_{\text{tank}} &= 2\pi \cdot \frac{\text{average magnetic energy} + \text{average electric energy}}{\text{energy loss in one oscillation cycle}}\Bigg|_{\omega=\omega_0} \\
&= 2\pi \cdot \frac{\text{peak magnetic energy}}{\text{energy loss in one oscillation cycle}}\Bigg|_{\omega=\omega_0} = \frac{R}{\omega_0 L} \\
&= 2\pi \cdot \frac{\text{peak electric energy}}{\text{energy loss in one oscillation cycle}}\Bigg|_{\omega=\omega_0} = \omega_0 RC \\
&= \frac{R}{\sqrt{L/C}}
\end{aligned}
\tag{9}
$$

R_{Si}, respectively, [9], [10]. The ohmic loss in R_{Si} signifies the energy dissipation in the silicon substrate.

In Fig. 2, the combined impedance of C_{ox}, C_{Si}, and R_{Si} is substituted by R_p and C_p, which are therefore frequency dependent, while L_s, R_s, and C_s remain unchanged as in Fig. 1. The reason for this substitution is twofold: it facilitates the analysis of R_p's effect on Q and the extraction of the shunt parasitics from measured S parameters (see Fig. 8). In terms of the circuit elements in Fig. 2, the energies can be expressed as

$$E_{\text{peak magnetic}} = \frac{V_0^2 L_s}{2 \cdot [(\omega L_s)^2 + R_s^2]} \qquad (10)$$

$$E_{\text{peak electric}} = \frac{V_0^2 (C_s + C_p)}{2} \qquad (11)$$

and

$$E_{\text{loss in one oscillation cycle}} = \frac{2\pi}{\omega} \cdot \frac{V_0^2}{2} \\ \cdot \left[\frac{1}{R_p} + \frac{R_s}{(\omega L_s)^2 + R_s^2} \right] \qquad (12)$$

where

$$R_p = \frac{1}{\omega^2 C_{\mathrm{ox}}^2 R_{\mathrm{Si}}} + \frac{R_{\mathrm{Si}}(C_{\mathrm{ox}} + C_{\mathrm{Si}})^2}{C_{\mathrm{ox}}^2} \qquad (13)$$

$$C_p = C_{\mathrm{ox}} \cdot \frac{1 + \omega^2 (C_{\mathrm{ox}} + C_{\mathrm{Si}}) C_{\mathrm{Si}} R_{\mathrm{Si}}^2}{1 + \omega^2 (C_{\mathrm{ox}} + C_{\mathrm{Si}})^2 R_{\mathrm{Si}}^2} \qquad (14)$$

and V_0 denotes the peak voltage across the inductor terminals. The inductor Q can be derived by substituting (10)–(12) into (8):

$$Q = \frac{\omega L_s}{R_s} \cdot \frac{R_p}{R_p + [(\omega L_s / R_s)^2 + 1]R_s} \\ \cdot \left[1 - \frac{R_s^2 (C_s + C_p)}{L_s} - \omega^2 L_s (C_s + C_p) \right] \\ = \frac{\omega L_s}{R_s} \cdot \text{substrate loss factor} \cdot \text{self-resonance factor} \qquad (15)$$

where $\omega L_s / R_s$ accounts for the magnetic energy stored and the ohmic loss in the series resistance. The second term in (15) is the substrate loss factor representing the energy dissipated in the semiconducting silicon substrate. The last term is the self-resonance factor describing the reduction in Q due to the increase in the peak electric energy with frequency and the vanishing of Q at the self-resonant frequency. Hence, the self-resonant frequency can be solved by equating the last term in (15) to zero.

Fig. 3 shows the measured frequency behavior of Q and the degradation factors for a typical on-chip inductor. At 1 GHz, the measured element set $\{L_s, R_s, C_s, C_p, R_p\}$ is equal to $\{8.2~\text{nH}, 13.4~\Omega, 26~\text{fF}, 102.7~\text{fF}, 1.7~\text{k}\Omega\}$. A detailed comparison between modeled and measured values for a wide variety of spiral inductors can be found in [2]. In Fig. 3(a) at low frequencies, Q is well described by $\omega L_s / R_s$ when both degradation factors have values close to unity. As frequency increases, the degradation factors decrease from unity, as shown in Fig. 3(b). This illustrates that the reduction of Q

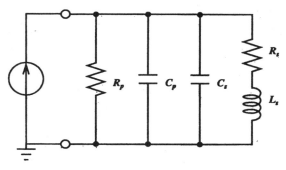

Fig. 2. Equivalent model with the combined impedance of C_{ox}, C_{Si}, and R_{Si} in Fig. 1 substituted by R_p and C_p.

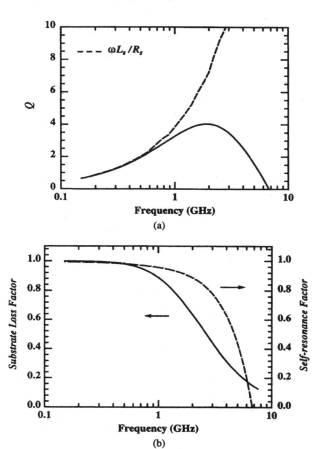

Fig. 3. Typical frequency behavior of (a) Q and (b) the degradation factors.

at high frequencies is a combined effect of substrate loss and self-resonance. In particular, the substrate loss alone causes 10–30% reduction from $\omega L_s / R_s$ at 1–2 GHz. Physically, the substrate loss stems from the penetration of electric field into the silicon. As the potential drop in the semiconductor, i.e., across R_{Si} in Fig. 1, increases with frequency, the energy dissipation in the substrate becomes more severe.

From (15), it can be seen that the substrate loss factor approaches unity as R_p approaches infinity. In other words, by increasing R_p to infinity, we can reduce the substrate loss. From (13), it can be shown that R_p approaches infinity as R_{Si} goes to zero or infinity. This is an important observation because it implies that Q can be improved by making the silicon substrate either a short or an open, thereby eliminating energy dissipation. Using high-resistivity silicon or etching away the silicon is equivalent to making the substrate an open circuit. In this paper, we explored the option of shorting the substrate to eliminate the loss. The approach is to insert a ground plane to block the inductor electric field from entering the silicon.

C. Drawback of Solid Ground Shields

The effectiveness of solid ground shield for reducing silicon parasitics has been reported [11], [12]. Rofougaran *et al.* used metal one as ground shields for metal-two bond pads to improve the input impedance matching of a low-noise amplifier fabricated in a CMOS process. Tsukahara *et al.* used a similar technique with a polysilicon layer as ground shields for metal–insulator–metal capacitors in a bipolar process. The polysilicon ground shields eliminated the silicon parasitics associated with the bottom plate of the capacitors. At 1 GHz, 30-dB reduction in substrate crosstalk was reported.

A solid conductive ground shield can be inserted between the inductor and the substrate to provide a short to ground. This is equivalent to placing a small resistance in parallel with C_{Si} and R_{Si} of the circuit model in Fig. 1. Physically, the electric field of the inductor is terminated before reaching the silicon substrate. One of the serious drawbacks with this approach is that the solid ground shield also disturbs the inductor's magnetic field. According to Lenz's law, image current, also known as loop current, will be induced in the solid ground shield by the magnetic field of the spiral inductor. The image current in the solid ground shield will flow in a direction opposite to that of the current in the spiral. The resulting negative mutual coupling between the currents reduces the magnetic field, and thus the overall inductance.

Using an equivalent circuit model, one can treat the inductor with the ground shield as a transformer. In Fig. 4, the primary and secondary circuits represent the spiral and the solid ground shield, respectively. The induced current flowing in the secondary inductor will impose a counter electromotive force on the primary inductor. This effect can be accounted for by adding a reflected impedance Z_r in series with the impedance of the primary circuit [13]. Z_r can be expressed in terms of the mutual inductance M and the series impedance of the secondary circuit as

$$Z_r = \frac{(\omega M)^2}{R_2 + j\omega L_2}. \tag{16}$$

Therefore, the input impedance seen by the source is

$$Z_{in} = R_1 + j\omega L_1 + Z_r. \tag{17}$$

Note that the imaginary part of Z_r is negative, which signifies the reduction in the overall inductance. Also of importance is the increase in the overall resistance due to the real part of Z_r, which denotes the additional energy loss due to the ground shield conductor. From (16) and (17), one can easily show that the effect of Z_r on Z_{in} diminishes as R_2 approaches infinity. An infinite R_2 can be achieved by inserting features in the ground shield that oppose the flow of the image current.

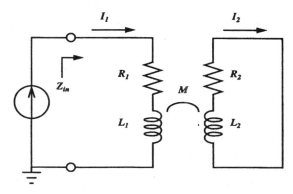

Fig. 4. Circuit model for illustrating the effects of negative mutual coupling between a spiral inductor and a solid ground shield.

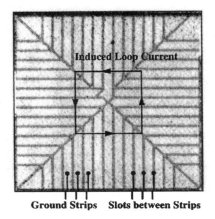

Fig. 5. Close-up photo of the patterned ground shield.

D. Design of Patterned Ground Shields

To increase the resistance to the image current, the ground shield is patterned with slots orthogonal to the spiral as illustrated in Fig. 5. The slots act as an open circuit to cut off the path of the induced loop current. The slots should be sufficiently narrow such that the vertical electric field cannot leak through the patterned ground shield into the underlying silicon substrate. With the slots etched away, the ground strips serve as the termination for the electric field. The ground strips are merged together around the four outer edges of the spiral. The separation between the merged area and the edges is not critical. However, it is crucial that the merged area does not form a closed ring around the spiral since it can potentially support unwanted loop current. The shield should be strapped with the top layer metal to provide a low-impedance path to ground. The general rule is to prevent negative mutual coupling while minimizing the impedance to ground.

The shield resistance is another critical design parameter. The purpose of the patterned ground shield is to provide a good short to ground for the electric field. Since the finite shield resistance contributes to energy loss of the inductor, it must be kept minimal. Specifically, by keeping the shield resistance small compared to the reactance of the oxide capacitance, the voltage drop that can develop across the shield resistance is small. As a result, the energy loss due to the shield resistance is insignificant compared to other losses. A typical on-chip spiral inductor has parasitic oxide capacitance between 0.25–1 pF depending on the size and the oxide thickness. The corresponding reactance due to the oxide capacitance at 1–2 GHz is on the order of 100 Ω and, hence, shield resistance of a few ohms is sufficiently small not to cause any noticeable loss.

As the magnetic field passes through the patterned ground shield, its intensity is weakened due to the skin effect [14]. This directly causes a decrease in the inductance since the magnetic flux is lessened in the space occupied by the ground shield layer. To avoid this attenuation, the shield must be significantly thinner than the skin depth at the frequency of interest. For example, the skin depth of aluminum at 2 GHz is approximately 2 μm, which is only 3–4 times the typical metal-one thickness. This implies that using a typical metal-one layer for the shield may result in reduction of the magnetic field intensity and, hence, the inductance.

III. EXPERIMENTAL RESULTS

A. Experiment Design

In Fig. 6, the test structures are shown for the inductors studied in this work: (a) no ground shield (NGS); (b) solid ground shield (SGS); and (c) patterned ground shield (PGS). Each spiral is fabricated using 2-μm-thick aluminum with 12 mΩ/$\square$ sheet resistance. A 1-μm-thick underpass is used to contact the center of the spiral. The spiral and the ground shield are separated by 5.2 μm of oxide. The ground shield is separated from the silicon substrate by 0.4 μm of oxide. The inductors are fabricated on 10–20 $\Omega \cdot$ cm bulk silicon substrates. Each inductor has seven turns, 15-μm line width, and 5-μm line space. The outer dimension of the spirals is 300 μm. The spiral layout is optimized for the unshielded inductor to achieve maximum Q at about 1.5 GHz. The same layout is used for the shielded inductors to demonstrate the general advantage of inserting the PGS beneath an inductor without deliberate optimization. This implies that further improvement for the shielded inductor is attainable with layout optimization accounting for the parasitics of the shield.

To investigate the effect of shield pattern, ground shields with different slot widths (1.5 and 2.5 μm) and pitches (5 and 20 μm) are fabricated. To study the effect of shield resistance, 0.5-μm aluminum (64 mΩ/$\square$) and 0.5-μm doped polysilicon (12 Ω/$\square$) are used to implement the shield. The polysilicon sheet resistance is chosen to be similar to that of MOSFET gates or BJT emitters. In technologies with silicided gate or emitter, the sheet resistance of the polysilicon layer can be as low as a few ohms per square, which is more suitable for our purpose. Nevertheless, the measured results will reveal that the doped polysilicon is conductive enough not to cause any observable loss.

Noise coupling between inductors is also studied. Crosstalk was measured between two adjacent unshielded inductors on substrate with different resistivities. The test structure is shown in Fig. 7. Each inductor has one end grounded, and the metal ground rings surrounding the inductors are not connected. The efficiency of the ground shield for isolation is evaluated using

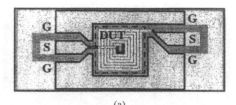
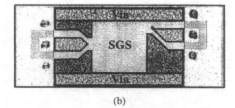

(a)

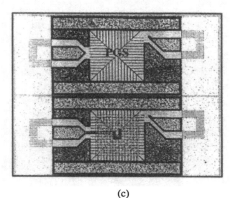

(b)

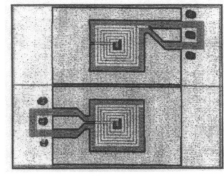

(c)

Fig. 6. Die photos of ground–signal–ground (GSG) test structure and the inductors: (a) spiral inductor with no ground shield (NGS), (b) solid ground shield (SGS) shown without spiral, and (c) patterned ground shield shown without and with spiral.

Fig. 7. Two-port test structure for measuring crosstalk via substrate between two adjacent inductors (shown with unshielded inductors).

the same test structure with shields inserted underneath the inductors.

B. Testing and Extraction Techniques

On-wafer testing was performed with an HP8720B Network Analyzer and Cascade Microtech coplanar ground–signal–ground (GSG) probes. During measurements, the substrate was grounded from the wafer back side through the testing chuck. The shunt parasitics of the test structure were de-embedded using open calibration structures fabricated next to the device under test (DUT). Two-port S parameters were measured, instead of a one-port parameter, to allow extraction of the inductance and other parasitics without curve fitting. The extraction procedure is summarized in Fig. 8. From the de-embedded S parameters, the complex propagation constant and characteristic impedance are computed. Then, the lumped elements in the series and shunt branches of the inductor model (the model from Fig. 2 in its two-port configuration) are solved using the relationships shown in the bottom block of Fig. 8. To extract L_s, R_s, and C_s from the real and imaginary parts of the measured series impedance, some assumptions about L_s and C_s need to be made. L_s and R_s are subject to skin effect, which governs the magnetic field intensity and current density in the conductor at high frequencies [14]. As frequency increases, the penetration of the magnetic the field into the conductor is attenuated, which causes a reduction in the magnetic flux internal to the conductor. However, L_s does not decrease significantly with increasing frequency because it is predominantly determined by the magnetic flux external to the conductor. Thus, L_s can be approximated as constant with frequency. The skin effect on R_s is much more pronounced because R_s is directly affected by the nonuniform current distribution in the conductor. C_s is considered independent of frequency since it represents the metal-to-metal overlap capacitance between the spiral and the center tap. At low frequencies, the reactance is dominated by ωL_s because ωL_s is much greater than $1/\omega C_s$. C_s is extracted using the low-frequency L_s value and the resonant frequency of the series branch. Then, with C_s held constant, L_s and R_s are solved using the real and imaginary parts of the series impedance at each measurement frequency. In the shunt branch, R_p and C_p can be extracted readily from the real and imaginary parts of the shunt admittance, respectively. The extraction technique described has been confirmed with experimental and published data of inductors having different geometric and process parameters [2].

C. Results and Discussion

In Fig. 9(a), measurement results for the effect of aluminum ground shields on L_s are plotted. Two inductors with NGS on 11 and 19 $\Omega \cdot$ cm substrates are included for comparison. The extracted L_s's are about 8 nH: the slight decrease with frequency justifies the assumption that L_s is almost frequency invariant. Furthermore, no noticeable difference in the L_s's is observed for the two cases, confirming that the magnetic fields of the inductors do not interact strongly with the substrates. The extracted C_s's are 18 fF: both inductors have the same C_s since the layout and process parameters are identical except for the substrate resistivity. In the shielded inductors, however, L_s can no longer be assumed as frequency invariant due to the induced loop current and attenuation of the magnetic flux in the shield layer. The extraction of L_s, consequently, is more difficult. In contrast, it is reasonable to expect C_s to remain the same with the introduction of the shield. Therefore, L_s of the shielded inductors are extracted with C_s equal to 18 fF. For the inductor with SGS, the extracted L_s decreases significantly as the frequency increases. This is caused by the negative mutual coupling between the spiral and the SGS,

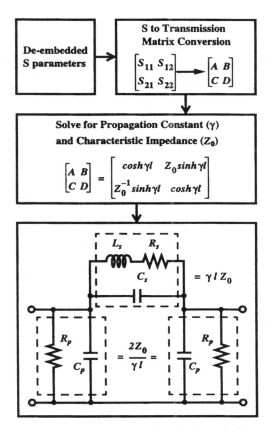

Fig. 8. Parameter extraction procedure for the lumped elements in the inductor model shown in Fig. 2 (l is the overall length of a spiral inductor).

as explained in Section II-C. With the PGS, most of the inductance is recovered, which confirms the effectiveness of the slot pattern for stopping the image current. Close inspection reveals that the inductance for the PGS case is lower than the two NGS cases, and the difference increases with frequency. This suggests that aluminum is too conductive to be optimal as the ground shield layer. In Fig. 9(b), the extracted R_s of the inductors with NGS increases with frequency due to the skin effect of the spiral conductor. The SGS case has a significantly higher R_s due to the image current. On the other hand, the inductor with PGS has the same R_s as the inductors with NGS because there is no image current. For the shunt parasitics shown in Fig. 9(c)–(d), the two NGS cases show a strong frequency dependence. The frequency behaviors of C_p and R_p are governed by C_{ox}, C_{Si}, and R_{Si}. At low frequencies, the electric field terminates at the oxide–Si interface, and C_p is primarily determined by C_{ox}. Since almost all electric energy is stored within the oxide layer along the spiral, little conduction current flows in the silicon substrate, and thus R_p is large. As frequency increases, the electric field starts to penetrate into the silicon substrate, which reduces C_p because of the series connection of oxide and silicon substrate capacitances. The roll-off in R_p signifies increasing energy dissipation in the silicon substrate. For the shielded inductors, C_p's are determined by the oxide capacitance between the spiral and the ground shield, which is slightly higher than the unshielded cases because of a thinner oxide. R_p's of the shielded inductors are very large, indicating that there

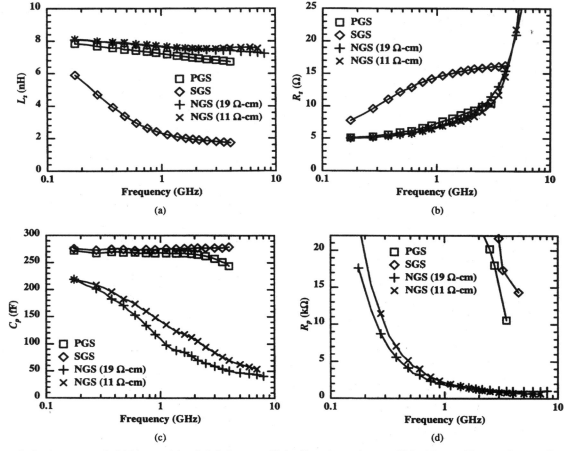

Fig. 9. Effect of aluminum ground shields on: (a) spiral inductance (L_s), (b) series resistance (R_s), (c) parasitic capacitance (C_p), and (d) parasitic resistance (R_p).

is essentially no energy loss in the ground shields. Although lower C_p's for the NGS cases would seem more desirable, they imply the existence of the lossy R_p's. It will be shown that eliminating the substrate loss, i.e., making R_p approach infinity, is more important for improving the inductor Q. That is, the PGS eliminates the lossy frequency-variant capacitance with a slightly larger lossless frequency-invariant one.

In Fig. 10(a)–(d), the measured results for inductors with polysilicon ground shields are plotted against the same unshielded inductors. In the SGS case, the image current starts to build up above 1 GHz. Although it does not lead to noticeable reduction in L_s, it causes R_s to increase more rapidly than the NGS cases. On the other hand, the polysilicon PGS does not deteriorate L_s or R_s, and terminates the inductor's electric field to provide the desired shielding from the substrate. For both aluminum and polysilicon PGS's, the measurement results show no variation for the different slot widths and pitches.

Figs. 11 and 12 show the effects of aluminum and polysilicon ground shields on Q. The inductor with aluminum SGS has the lowest Q because of its lowest L_s and highest R_s. In Fig. 12, the polysilicon SGS yields a Q similar to those of the NGS cases, indicating that it is resistive enough to prevent most of the image current from flowing. Finally, the polysilicon PGS, which combines the appropriate sheet resistance and

pattern, yields the most improvement in Q (ranges from 10 to 33%) between 1–2 GHz. Note that the inclusion of the ground shields increases C_p, which causes a fast roll-off in Q above the peak-Q frequency and a reduction in the self-resonant frequency. Comparison between the inductor parameters for the NGS (11 $\Omega \cdot$ cm) and polysilicon PGS cases is shown in Table I. The results at 2 GHz are compared to emphasize the relative importance of the degradation mechanisms near the peak-Q frequency. In particular, the unshielded inductor suffers greatly from substrate loss with nearly 50% reduction from $\omega L_s / R_s$. Although the shielded inductor has a lower self-resonance factor, it is almost free of substrate loss. The overall effect is a 33% improvement in Q at 2 GHz with the addition of polysilicon PGS. Further optimization of the shielded inductor layout to decrease the self-resonance factor and to increase the Q is possible.

In RF circuits, an inductor is often used to form an LC tank. Fig. 13 plots the frequency behavior of the tank impedance for two 2-GHz LC tanks to demonstrate the impact of the 8-nH inductor with polysilicon PGS on the tank quality factor, Q_{tank}. The tuning capacitances for the shielded and unshielded cases are 0.5 and 0.7 pF, respectively, to account for the difference in the inductors' parasitic capacitance. As mentioned in Section II-A, Q_{tank} can be determined by ratio of the resonant frequency, at which the tank impedance is

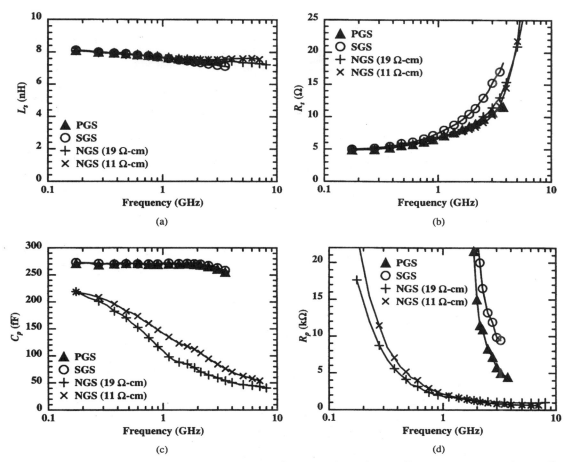

Fig. 10. Effect of polysilicon ground shields on: (a) spiral inductance (L_s), (b) series resistance (R_s), (c) parasitic capacitance (C_p), and (d) parasitic resistance (R_p).

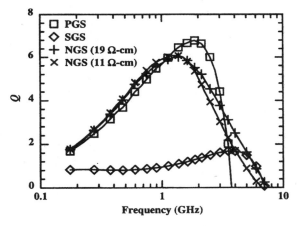

Fig. 11. Effect of aluminum ground shields on Q.

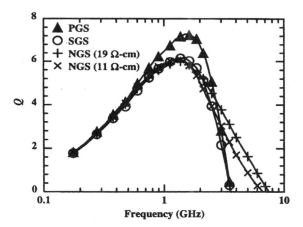

Fig. 12. Effect of polysilicon ground shields on Q.

maximum, to the -3-dB bandwidth. Even though the parasitic capacitances of both inductors are incorporated as part of the tank capacitance, the tank with the unshielded inductor suffers from a lossy R_p. As a result, Q_{tank} is improved from 6.0 for the tank with the unshielded inductor to 10.2 for the one with the shielded inductor. It is important to note that Q_{tank} exceeds the inductor Q for both inductors at 2 GHz (see Table I). This can be attributed to the fact that the reduction of the inductors' Q caused by their parasitic capacitances becomes irrelevant as the capacitances are "absorbed" by the LC tanks.

Substrate noise coupling between two adjacent inductors is measured by the magnitude of the transmission coefficient $|S_{21}|$. Fig. 14 shows that for the unshielded inductors, the one on a more conductive substrate ($11\ \Omega \cdot \text{cm}$) has stronger coupling due to the higher substrate admittance. The peaks in $|S_{21}|$ for the NGS cases correspond to the onset of significant electric field penetration into the silicon substrate, and hence more coupling. In contrast, the inductors shielded by the polysilicon PGS's show significantly better isolation, up to 25 dB, at gigahertz frequencies. It should be noted that, like any

TABLE I
COMPARISON OF MEASURED INDUCTOR PARAMETERS FOR THE
NGS (11 Ω · cm) AND POLYSILICON PGS CASES AT 2 GHz

	NGS	Polysilicon PGS
L_s (nH)	7.5	7.4
R_s (Ω)	8.2	8.5
C_s (fF)	18.0	18.0
C_p (fF)	108.1	268.2
R_p (kΩ)	1.2	15.0
$\omega L_s/R_s$	11.5	10.9
Substrate Loss Factor	0.52	0.94
Self-resonance factor	0.85	0.66
Q	5.08	6.76
Self-resonant Frequency (GHz)	6.8	3.6

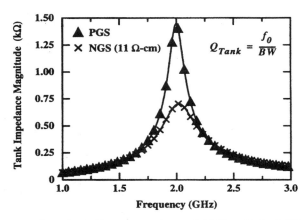

Fig. 13.   Effect of polysilicon patterned ground shield on Q of a 2-GHz LC tank.

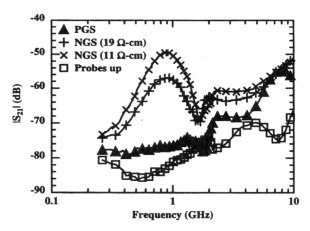

Fig. 14.   Effect of polysilicon patterned ground shield on substrate coupling between two adjacent inductors.

other isolation structure, such as a guard ring, the efficiency of the PGS is highly dependent on the integrity of the ground connection. Designers often need to make a tradeoff between the desired isolation level and the chip area that is required for a low-impedance ground.

IV. CONCLUSIONS

On-chip spiral inductors with patterned ground shields are presented. The parasitic effects of an inductor on silicon are analyzed with the aid of a physical model. A patterned ground shield is devised to eliminate the silicon parasitics of the on-chip spiral inductor. The effects of shield resistance and pattern are studied both theoretically and experimentally. Measurement results confirmed that a patterned ground shield improves the Q and isolation of an on-chip inductor. Furthermore, with the addition of the ground shield, an inductor's characteristics are less dependent on substrate variation, and hence are easier to model. The implementation of the ground shield is compatible with standard silicon IC technology. The experimental results presented in this work are exclusively based on lightly doped (10–20 Ω · cm) substrates. Given the increasing interest in CMOS RF IC's, investigation on the effects of heavily doped (10–20 mΩ·cm) substrates on shielded inductors are underway, and will be reported in the near future.

ACKNOWLEDGMENT

The authors would like to thank the Stanford Nanofabrication Facility staff for their assistance in processing.

REFERENCES

[1] P. R. Gray and R. G. Meyer, "Future directions in silicon IC's for RF personal communications," in *Proc. IEEE 1995 Custom Integrated Circuits Conf.*, May 1995, pp. 83–90.
[2] C. P. Yue, C. Ryu, J. Lau, T. H. Lee, and S. S. Wong, "A physical model for planar spiral inductors on silicon," in *Int. Electron Devices Meet. Tech. Dig.*, Dec. 1996, pp. 155–158.
[3] M. Pfost, H.-M. Rein, and T. Holzwarth, "Modeling substrate effects in the design of high speed Si-bipolar IC's," *IEEE J. Solid-State Circuits*, vol. 31, pp. 1493–1501, Oct. 1996.
[4] K. B. Ashby, I. A. Koullias, W. C. Finley, J. J. Bastek, and S. Moinian, "High Q inductors for wireless applications in a complementary silicon bipolar process," *IEEE J. Solid-State Circuits*, vol. 31, pp. 4–9, Jan. 1996.
[5] J. Y.-C. Chang, A. A. Abidi, and M. Gaitan, "Large suspended inductors on silicon and their use in a 2-mm CMOS RF amplifier," *IEEE Electron Device Lett.*, vol. 14, pp. 246–248, May 1993.
[6] H. G. Booker, *Energy in Electromagnetism.* London/New York: Peter Peregrinus (on behalf of the IEE), 1982.
[7] H. M. Greenhouse, "Design of planar rectangular microelectronic inductors," *IEEE Trans. Parts, Hybrids, Packaging*, vol. PHP-10, pp. 101–109, June 1974.
[8] L. Wiemer and R. H. Jansen, "Determination of coupling capacitance of underpasses, air bridges and crossings in MIC's and MMIC's," *Electron. Lett.*, vol. 23, pp. 344–346, Mar. 1987.
[9] I. T. Ho and S. K. Mullick, "Analysis of transmission lines on integrated-circuit chips," *IEEE J. Solid-State Circuits*, vol. SC-2, pp. 201–208, Dec. 1967.
[10] H. Hasegawa, M. Furukawa, and H. Yanai, "Properties of microstrip line on Si–SiO₂ system," *IEEE Trans. Microwave Theory Tech.*, vol. MTT-19, pp. 869–881, Nov. 1971.
[11] A. Rofougaran, J. Y.-C. Chang, M. Rofougaran, and A. A. Abidi, "A 1 GHz CMOS RF front-end IC for a direct-conversion wireless receiver," *IEEE J. Solid-State Circuits*, vol. 31, pp. 880–889, July 1996.
[12] T. Tsukahara and M. Ishikawa, "A 2 GHz 60 dB dynamic-range Si logarithmic/limiting amplifier with low-phase deviations," in *Int. Solid-State Circuits Conf. Dig. Tech. Papers*, Feb. 1997, pp. 82–83.
[13] P. H. Young, *Electronic Communication Techniques.* New Jersey: Macmillan, 1994.
[14] H. A. Wheeler, "Formulas for the skin effect," *Proc. IRE*, vol. 30, pp. 412–424, Sept. 1942.

The Effects of a Ground Shield on the Characteristics and Performance of Spiral Inductors

Seong-Mo Yim, Tong Chen, and Kenneth K. O

Abstract—The frequency dependence of the model parameters of patterned ground shield (PGS) inductors in large part is explained as a consequence of modeling a distributed system with a lumped model. The effects of PGS shape and material on inductor characteristics have been examined and explained. There is an optimum area for a PGS to maximize Q. Using an n^+ buried/n-well PGS, the peak Q is improved by $\sim 25\%$ from that of an inductor without a PGS while only slightly changing L and C_p in comparison to inductors with other PGSs. Having a PGS does not significantly improve isolation between adjacent inductors when isolation is limited by magnetic coupling since a PGS is specifically designed to limit termination of magnetic fields.

Index Terms—Distributed system, frequency dependence, isolation, spiral inductor.

I. INTRODUCTION

AN ON-CHIP spiral inductor is one of the critical components for implementing radio-frequency integrated circuits such as a low-noise amplifier, a voltage-controlled oscillator, and an impedance matching network [1]–[3]. One of the most important parameters of spiral inductors is the quality factor Q, which is mainly limited by the loss due to inductor metal resistance, substrate resistance, and that associated with induced eddy current below the inductor metal trace [4].

Focusing on the effects of R_{sub}, when R_{sub} in the commonly used equivalent circuit model shown in Fig. 1(b) is increased to infinity, the current through R_{sub} becomes zero, and power dissipation in R_{sub} becomes zero and Q is increased. When R_{sub} is decreased to zero, power dissipation in R_{sub} becomes zero again, and Q is increased [5]. The use of a patterned ground shield (PGS) between inductor metal trace and substrate increases Q [6], [7] by reducing R_{sub} while not inducing significant eddy current which can significantly reduce Q. The patterned sections of a shield are tied together using a metal connection. It is important to make sure that a loop is not formed in this metal connection.

In this paper, the effects of ground shield shape and material which can be formed in a silicon bipolar process are reported. The extracted inductance model parameter L, shown in

Manuscript received January 12, 2001; revised October 5, 2001. This work was supported by a grant from Conexant Systems.

The authors are with the Department of Electrical and Computer Engineering, Silicon Microwave Integrated Circuits and Systems Research Group, University of Florida, Gainesville, FL 32611 USA (e-mail: smyim@tec.ufl.edu).

Publisher Item Identifier S 0018-9200(02)00668-6.

Fig. 1(d), exhibits a stronger frequency dependence for PGS inductors. This frequency dependence of L and the frequency dependence of the series resistance model parameter R_S are explained as a consequence of modeling a distributed system using a lumped element model. Additionally, contrary to a previous report [6], it is shown that isolation characteristics between adjacent inductors are not significantly improved by the addition of a PGS, when isolation is limited by magnetic coupling [5].

II. EXPERIMENT

Fig. 1(a) shows the 3.25-turn 4.8-nH inductor structure mostly used in this study. The metal width and space are 10 and 6 μm. The metal thickness is 3.0 μm and separated from a 20-Ω-cm substrate by ~ 3.0 μm. The outer dimension of the inductor is 250 μm $\times$ 250 μm. Under this inductor structure, a wide variety of patterned ground shields with varying shapes and formed with metal 1, silicided polysilicon, and n^+ buried layer has been placed, and its impact has been investigated. The model parameters of the inductors extracted from measurements are listed in Table I.

III. CHARACTERISTICS OF INDUCTORS

Fig. 1(a) also shows the control inductor (N9) without a PGS, N10 and N12 inductors with a four-piece poly-PGS and an eight-piece poly-PGS which has larger gaps or a reduced PGS area. Fig. 1(c)–(g) shows Q_{bw} [8] and the extracted model parameters for the inductors. Adding the PGS improves the peak Q_{bw} by $\sim 25\%$. The R_{sub} of N10 and N12 are 11 and 7 Ω, respectively, while that for the control inductor is 55 Ω. Another prominent difference between the control and PGS inductors is that inductance of the PGS inductors decreases faster with frequency. Over 4 GHz, the L of N10 and N12 is decreased by $\sim 17\%$ versus 12% for the control. On the other hand, R_s of the control inductor changes more rapidly with frequency and becomes negative at high frequencies. To explain these sometime seemingly nonphysical results, the impact of modeling a distributed system using the simple lumped model is investigated.

IV. DISTRIBUTED MODEL OF INDUCTORS

A spiral inductor is a type of transmission lines [9]–[11] and can be modeled as sections of magnetically coupled transmission

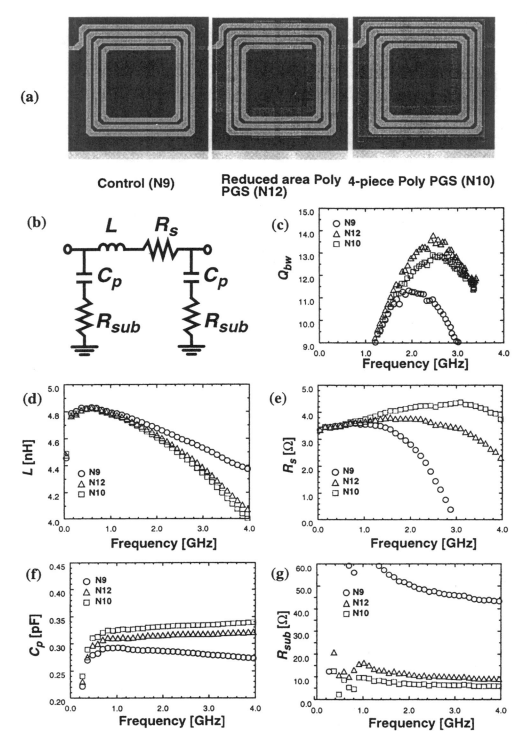

Fig. 1. (a) Inductors with a different ground shield area. (b) Commonly used equivalent circuit model for an inductor. (c) Q_{bw}. (d) L. (e) R_s. (f) C_p. (g) R_{sub} versus frequency.

lines. In reality, the substrate eddy current effect, skin effect, proximity effect, their associated distributed effects, and others should also be included. These make physics-based modeling of spiral inductors difficult. As a first-order attempt to study the impact of modeling a distributed system using a simple lumped element model shown in Fig. 1(b), model parameters for a section of a distributed N-section model have been generated by scaling down L and R_s by N, and C_p by $(N+1)/2$, and by scaling up R_{sub} by $(N+1)/2$ [Fig. 2(a)]. When $N = 1$, these distributed model parameters are the same as the lumped model parameters. In real spiral inductors, the inductors are not symmetric and these sections are not equally distributed. Therefore, the model parameters of these sections are not necessarily the same. However, for the purpose of

128

TABLE I
MEASURED PARAMETERS OF SPIRAL INDUCTORS

	N9	N10	N12	N17	N5	N6	N15	M1	M2	M3
L [nH]	4.8	4.8	4.8	4.8	4.8	4.8	4.8	5.3	5.3	5.3
R_s [Ω]	3.3	3.3	3.3	3.3	3.3	3.3	3.3	18.5	18.5	18.5
C_p [fF]	290	325	310	340	330	300	420	950	380	320
R_{sub} [Ω]	55	11	7	6	7	9	6	6	6	6
Q_{bw} @max.	11.3	12.9	13.8	13.9	14.3	13.5	11.5	---	---	---

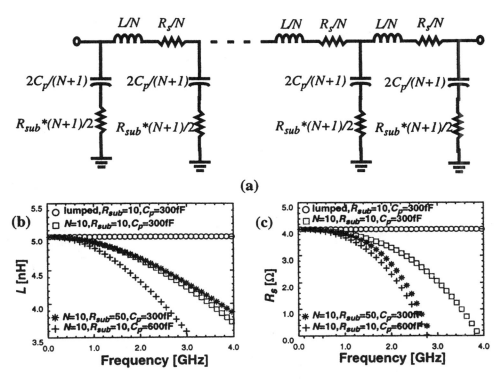

Fig. 2. (a) Distributed line model. (b) Re-extracted L. (c) Re-extracted R_s when the original lumped R_s and L are 4.2 Ω and 5.1 nH.

examining the impact of modeling a distributed system using a lumped model, this approximation should be acceptable. Treating this distributed model as a lumped system, L and R_s are re-extracted and shown in Fig. 2(b) and (c). At low frequencies, as it should be, the extracted L and R_s are the same for both cases. However, when a distributed system is treated as a lumped system, extracted L and R_s decrease with frequency like the measured results shown in Fig. 1. The number of sections was ten. It was found that increasing the number beyond ten had small effects on the re-extracted model parameters for the inductors used in this study. In general, the number of sections N should be chosen that the percentage (%) error (1) should be small.

$$\% \text{ error} = \frac{\coth(\gamma \cdot x) - (N+1) \cdot \frac{\alpha^{2N+1} + \beta^{2N+1}}{\alpha^{2N+2} - \beta^{2N+2}}}{\coth(\gamma \cdot x)} \times 100. \quad (1)$$

In this expression, x is the length of a spiral inductor, and ω is the frequency in radian. $\gamma \cdot x$ is expressed in (2), shown at the bottom of the next page. α and β in (1) are

$$\alpha = \sqrt{N \cdot (N+1) + \left(\gamma \cdot \frac{x}{2}\right)^2} + \left(\gamma \cdot \frac{x}{2}\right)$$

$$\beta = \sqrt{N \cdot (N+1) + \left(\gamma \cdot \frac{x}{2}\right)^2} - \left(\gamma \cdot \frac{x}{2}\right).$$

For this study, inductors with R_{sub} of 10 and 50 Ω, and C_p [Fig. 1(b)] of 300 and 600 fF are evaluated. The frequency dependence of L increases as R_{sub} is decreased and as C_p is increased, though the dependence on R_{sub} is weaker. As seen in Fig. 1, C_p of the PGS inductors is larger than that of the control inductor (N9) while R_{sub} is significantly lower. These can easily explain the stronger frequency dependence of L for the PGS inductors. The re-extracted R_s decreases faster with frequency if R_{sub} and C_p are increased, and it can become negative. These

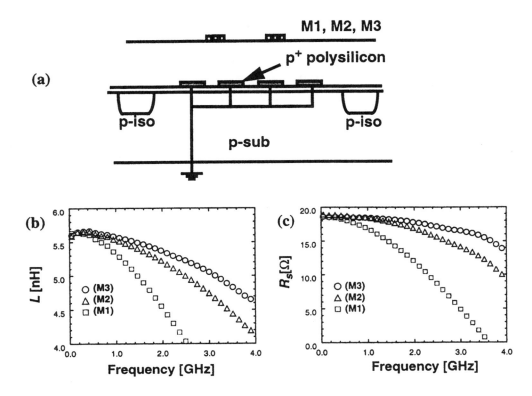

Fig. 3. (a) Inductors formed with a metal-1, metal-2, or metal-3 while using the same polysilicon PGS. (b) L. (c) R_s versus frequency.

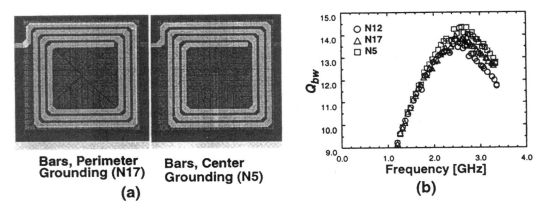

Fig. 4. (a) Inductors with a reduced area poly PGS and with a different current collecting method. (b) Q_{bw} versus frequency.

can once again explain the measured behaviors of R_s shown in Fig. 1(e). For the inductors considered in this investigation, doubling C_p from 300 fF has the same effect as increasing R_{sub} by a factor of 5 from 10 Ω. For N10 and N12 inductors, when the PGSs are added, R_{sub} decreases by almost a factor of 5, while at the maximum, C_p increases by ~20%. Because of this, the R_{sub} effect dominates and the frequency dependence of R_s decreases when a PGS is added.

These also explain the stronger frequency dependence of L and R_s for an inductor fabricated with the metal-1 layer on a polysilicon PGS with higher C_p compared to those fabricated with the metal-2 and metal-3 layer using the same polysilicon PGS (Fig. 3). As the separation between the PGS and inductor metal trace is decreased, C_p is increased, thus the frequency dependence of L and R_s is increased. These effects of modeling a distributed system with a lumped model, however, cannot ex-

$$\gamma \cdot x = \sqrt{(R_s + j\omega L) \cdot \left(\frac{2 \cdot \omega^2 \cdot C_p^2 \cdot R_{\mathrm{sub}}}{1 + \omega^2 \cdot C_p^2 \cdot R_{\mathrm{sub}}^2} + j\omega \cdot \frac{2 \cdot C_p}{1 + \omega^2 \cdot C_p^2 \cdot R_{\mathrm{sub}}^2} \right)} \qquad (2)$$

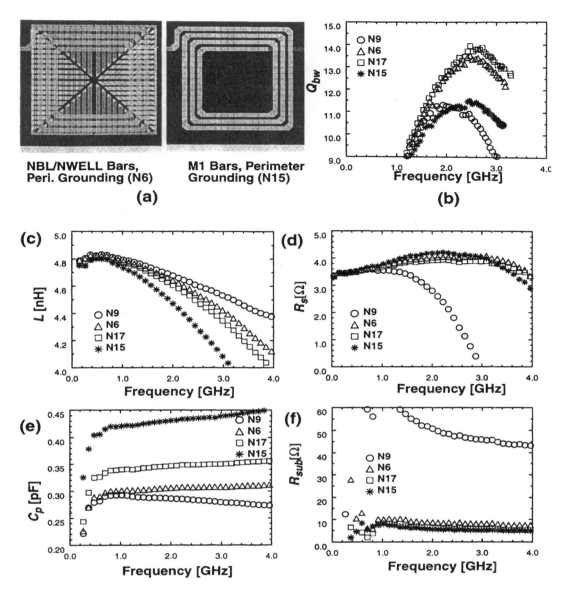

Fig. 5. (a) Inductors with a different ground shield area. (b) Q_{bw}. (c) L. (d) R_s. (e) C_p. (f) R_{sub} versus frequency.

plain the increase in R_s for PGS inductors with frequency at lower frequencies seen in Fig. 1(e). This is typically attributed to the skin effect and eddy current loss, which have not been included in the models.

V. EFFECTS OF GROUND SHIELD SHAPE

Referring back to Fig. 1, the Q of the N12 inductor (with a reduced PGS ground shield area) is slightly higher (around 1) than that for N10 with a four-piece PGS. The C_p of N12 is decreased by ~5% (~15 fF) from that for N10. The fact that the PGS area of N12 is around 60% of that of N10 increases R_{sub} of N12 to 11 Ω from 7 Ω. The net effect of these was to make N12 have slightly lower frequency dependent L and slightly higher frequency dependent R_s than that for N10. Based on these, it is clear that the PGS shape also affect the frequency dependence of L and R_s.

At moderate to high frequencies, R_s of N12 is lower than that for N10. This explains the slightly higher peak Q for N12. Despite the fact that R_s of N9 is lower, because of the higher R_{sub}, the peak Q is lower for N9. Therefore, the reduced area inductor (N12) has the highest Q among N9, N12, and N10. If the area of PGS is reduced too much, R_{sub} will eventually become too high and Q should decrease. As seen for the control inductor with a zero PGS area, because of its highest R_{sub}, it has the lowest Q among the three inductors. These mean there must be an optimum area for PGS to maximize Q. Among the three, N12 is the best design.

Fig. 4 compares Q of inductors with a PGS shape from [6] (N17) to that of N12. Additionally, a PGS structure in which the PGS ground current is extracted from the center has been evaluated (N5). Since the ground current can be significant, there can be nonnegligible magnetic energy storage and inductance associated with the connections for extracting the current. N5

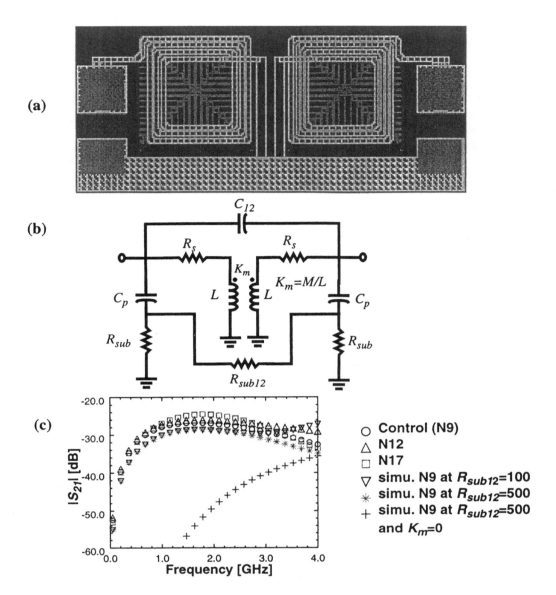

Fig. 6. (a) Inductor isolation test structure. (b) Lumped equivalent circuit model for the isolation test structure. (c) Measured $|S_{21}|$ for the isolation structures using N9, N12, N17, and simulated $|S_{21}|$ for the structure using N12 at $R_{\mathrm{sub12}} = 100, 500\ \Omega$ and for that using N9 at $R_{\mathrm{sub12}} = 500\ \Omega$ with $K_m = 0$.

and N17 have a poly-PGS with 11.6-μm-wide bars separated by 1.6 μm. The peak Q of N5 is slightly higher than that of N12 and N17. Although the differences are small, N5 has the highest Q among the inductors with varying PGS shapes considered in this study.

VI. Effects of Ground Shield Materials

Fig. 5 shows an inductor with a PGS formed using an n^+ buried layer (nbl) and an n-well layer (N6), and another inductor with a metal-1 PGS (N15). The sheet resistance of the buried layer is ~25 Ω/□. The width and space of the nbl shields are 8 and 4 μm, respectively. The width and space of the bars for the metal-1 PGS (N15) are 11.6 and 1.6 μm, respectively. C_p's of N6 and N17 are ~5 and ~25% higher than that for the control inductor (N9). C_p of N15 is the highest and around 1.6 times C_p for N9. Among N6, N17, and N15, the frequency dependence of L is the weakest for N6 which has the largest separation between the inductor metal trace and PGS or the lowest C_p. The frequency dependence is the highest for N15 which has the smallest separation or the highest C_p. This is once again consistent with the earlier discussion on frequency dependence of L resulting from modeling a distributed system with a lumped model. N17 has slightly higher peak Q than N6. N15, however, has the lowest R_{sub}, because its inductance decreases more rapidly with frequency due to higher C_p, the peak Q for N15 is significantly lower than that for N6 and N17. In this bipolar process, by using a PGS with an n^+ buried/n-well layer combination, it is possible to improve the peak Q by ~25% with small changes in the characteristics of C_p and L. If a larger C_p can be tolerated, the peak Q can be improved slightly more by using a polysilicon PGS.

VII. Isolation Between Adjacent Inductors

It has been suggested that having a PGS can improve isolation between adjacent inductors by ~30 dB [6]. To investigate the dependence of isolation on the PGS shape, test structures have been implemented. Fig. 6(a) shows an isolation structure formed with N17. One of the terminals of each inductor is grounded. The second terminal of each inductor is connected to pads and S_{21} between these two pads are measured. The inductors are separated by ~85 μm. The structure like this can be found in a differential LC-tuned voltage-controlled oscillator (VCO). Fig. 6(c) shows $|S_{21}|$ of the isolation structures. Contrary to the previous report [6], the isolation characteristics are not significantly affected by the presence of a PGS and PGS shape. The PGS by design does not significantly perturb the magnetic field generated by the inductor. This means that if the signal coupling between the two inductors is magnetic in nature, then isolation characteristics should not be significantly affected by the presence of a PGS.

To study this, an equivalent circuit model for the isolation structure is constructed and shown in Fig. 6(b). Fig. 6(c) also shows simulated isolation characteristics with and without including the magnetic coupling [$K_m = M/L = 0.037$; see Fig. 6(b)]. In addition to the inductor model parameters discussed earlier, R_{sub12} and C_{12} are included to model signal coupling between the two control inductors (N9) through substrate and the capacitance between two inductors, respectively. C_{12} and its impact are usually small. The model values were $L = 4.8$ nH, $R_s = 3.6$ Ω, $C_p = 300$ fF, $R_{sub} = 50$ Ω, which are comparable to those of the control inductor. The coupling coefficient, K_m was computed using the layout and Greenhouse formula [12]. The simulated isolation degrades by ~20 dB or more at 1.8 GHz, when the magnetic coupling is included.

Fig. 6(c) also shows the dependence of $|S_{21}|$ on R_{sub12} (100, or 500 Ω). When $R_{sub} = 50$ Ω, which is close to that for the control inductor, if R_{sub12} is much greater than 100 Ω, it does not affect $|S_{21}|$. Between 3 and 4 GHz, when R_{sub12} (100 Ω) is comparable to R_{sub}, $|S_{21}|$ increases with frequency. The fact that this rise of $|S_{21}|$ is not seen in the measurements suggests that the R_{sub12} is larger than 100 Ω for the control inductor. The impact of R_{sub12} is more complicated for PGS inductors because of the presence of an oxide layer between the PGS and substrate. At the same time, the R_{sub12} effect should be smaller, since R_{sub} is much smaller and the oxide layer increases the impedance and reduces the substrate coupling.

If the effects of coupling through R_{sub12} and C_{12} are neglected (i.e., $R_s = 0$, $R_{sub} = 0$, $L \gg M$, and $C_{12} = 0$), then the transducer gain or $|S_{21}|^2$ can be estimated as shown in the equation at the bottom of the page, where Z_O is the 50-Ω characteristic impedance. When the frequency goes to 0 and when it becomes high, the expression is reduced to

$$\text{As } \omega \to 0 \text{ Hz, } |S_{21}| \approx \frac{2 \cdot M}{Z_O} \cdot \omega$$

$$\text{As } \omega \to \text{high, } |S_{21}| \approx \frac{2 \cdot M}{Z_O} \cdot \left(\frac{1}{L \cdot C_p}\right)^2 \cdot \left(\frac{1}{\omega}\right)^3 .$$

At low frequencies, as frequency is increased, the coupling increases linearly because of the mutual inductance effect, while at high frequencies, the coupling decreases with frequency because C_p at the port 2 shunts the port to ground. These result in a peak as seen in the measured $|S_{21}|$ plots. The difference between the simulated and the measured $|S_{21}|$ is around 2 dB for the control inductor, which is a good agreement given that all the frequency dependence of the model parameters are neglected. These indeed suggest that isolation is limited by the magnetic coupling and in this limit, having a PGS does not significantly improve isolation between two adjacent inductors.

VIII. Conclusion

The frequency dependence of the model parameters for 3.25-turn 4.8-nH PGS spiral inductors can be largely explained as a consequence of modeling a distributed system using a lumped model. If a distributed system is treated as a lumped system, re-extracted L and R_s decrease as frequency increases. The frequency dependency of L increases as C_p is increased and decreases a little as R_{sub} is increased. The frequency dependency of R_s increases if C_p and R_{sub} are increased. PGS inductors have higher frequency dependency of L and lower frequency dependency of R_s than the control inductors because of their higher C_p and lower R_{sub}. Due to these, Q of PGS inductors is strongly influenced by its distributed nature, and this should be factored in during the design process in order to increase Q. Additionally, there is an optimum area for PGS to maximize Q. Because the connections for ground current extraction can carry significant current, the magnetic energy storage and inductance associated with the ground connection cannot be neglected in designing inductors with a PGS. In the bipolar process utilized for this study, using a PGS formed with an n$^+$ buried/n-well layer combination, it is possible to improve the peak Q by ~25% with small changes in the characteristics of C_p and L. Isolation between two adjacent inductors separated by ~85 μm is limited by magnetic coupling and in this limit, having a PGS does not significantly improve isolation.

$$|S_{21}|^2 \approx \frac{(2 \cdot \omega \cdot M \cdot Z_O)^2}{\left\{\omega^4 \cdot L^2 \cdot C_p^2 \cdot Z_O^2 - \omega^2 \cdot (L^2 - 2 \cdot L \cdot C_p \cdot Z_O^2) + Z_o^2\right\}^2 + (2 \cdot \omega \cdot L \cdot Z_O)^2 \cdot \left\{\omega^2 \cdot L \cdot C_p - 1\right\}^2}$$

Acknowledgment

The authors are grateful to Dr. J. Zheng, P. N. Sherman, and P. Kempf of Conexant Systems for their help.

References

[1] Y.-C. Ho, K.-H. Kim, B. A. Floyd, C. Wann, Y. Taur, I. Lagnado, and K. K. O, "4- and 13-GHz tuned amplifiers implemented in a 0.1-μm CMOS technology on SOI, SOS, and bulk substrates," *IEEE J. Solid-State Circuits*, vol. 33, pp. 2066–2073, Dec. 1998.

[2] C.-M. Hung and K. K. O, "A 1.24-GHz monolithic CMOS VCO with phase noise of −137 dBc/Hz at a 3-MHz offset," *IEEE Microwave and Guided Wave Lett.*, vol. 9, pp. 111–113, Mar. 1999.

[3] A. Hajimiri and T. H. Lee, "Design issues in CMOS differential LC oscillators," *IEEE J. Solid-State Circuits*, vol. 34, pp. 717–724, May 1999.

[4] B. Razavi, "Challenges in the design of frequency synthesizers for wireless applications," in *Proc. CICC*, May 1997, pp. 395–402.

[5] S.-M. Yim, T. Chen, and K. K. O, "The effects of a ground shield on spiral inductors fabricated in a silicon bipolar technology," in *Proc. Bipolar/BiCMOS Circuits and Technology Meeting (BCTM)*, Sept. 2000, pp. 157–160.

[6] C. P. Yue and S. S. Wong, "On-chip spiral inductors with patterned ground shields for SI-based RF ICs," *IEEE J. Solid-State Circuits*, vol. 33, pp. 734–752, May 1998.

[7] T. Chen, K. Kim, and K. K. O, "Application of a new circuit design oriented Q extraction technique to inductors in silicon IC's," in *IEDM Tech. Dig.*, 1998, pp. 527–530.

[8] K. K. O, "Estimation methods for quality factors of inductors fabricated in silicon integrated circuit process technologies," *IEEE J. Solid-State Circuits*, vol. 33, pp. 1249–1252, Aug. 1998.

[9] J. R. Long and M. A. Copeland, "The modeling, characterization, and design of monolithic inductors for silicon RF IC's," *IEEE J. Solid-State Circuits*, vol. 32, pp. 357–369, Mar. 1997.

[10] Y. Eo and W. R. Eisenstadt, "High-Speed VLSI interconnect modeling based on S-parameter measurements," *IEEE Trans. Comp., Hybrids, Manufact. Technol.*, vol. 16, pp. 555–562, Aug. 1993.

[11] R. Groves, D. L. Harame, and D. Jadus, "Temperature dependence of Q and inductance in spiral inductors fabricated in a silicon-germanium/BiCMOS technology," *IEEE J. Solid-State Circuits*, vol. 32, pp. 1455–1459, Sept. 1997.

[12] H. M. Greenhouse, "Design of planar rectangular microelectronic inductors," *IEEE Trans. Parts, Hybrids, Packag.*, vol. PHP-10, pp. 101–109, June 1974.

Temperature Dependence of Q and Inductance in Spiral Inductors Fabricated in a Silicon-Germanium/BiCMOS Technology

Rob Groves, David L. Harame, and Dale Jadus

Abstract— **The behavior of on-chip, planar, spiral inductors fabricated over a conductive silicon substrate has been characterized over the temperature range from −55°C to +125°C. Quality factor (Q) was observed to decrease with increasing temperature at low frequency and increase with increasing temperature at high frequency. Inductance was seen to vary little over the temperature and frequency range. A SPICE model that incorporated the temperature dependence of the inductor's parasitics was presented and shown to give excellent agreement with measured data over the full temperature and frequency range.**

Index Terms— **Microwave devices, modeling, spiral inductors, temperature.**

I. INTRODUCTION

THE fabrication of high Q inductors on a silicon substrate allows for a high degree of integration in analog and analog-digital mixed signal applications. The emerging silicon-germanium technology [1] provides high-frequency analog capability on silicon. Designing oscillators, filters, amplifiers, and other analog circuitry requires a thorough knowledge of the characteristics of inductors fabricated on a silicon substrate. The inductance of a planar spiral is dictated by: the number of turns, line-to-line space, line width, line thickness, and total area. The behavior of inductance with differing geometries is well understood [2]. The inductance of a metal spiral is not expected to vary much with temperature, as it is dominated by the physical dimensions of the spiral. The only factor expected to affect the inductance over temperature is the decreasing penetration of the magnetic field into the conductor, due to skin depth, with increasing metal conductivity. This reduced penetration will decrease the self inductance, which is a small contributor to the overall inductance. The quality factor (Q) of an inductor is a measure of the energy storing capability of the device. Q is expected to vary with the resistive and capacitive parasitics of the inductor, increasing as the parallel impedance gets larger and decreasing as the series resistance gets larger. The sharpness of a resonant circuit's frequency response is directly related to the Q of the passive devices present in the circuit. Designing resonant and matching circuits to operate over a mil-spec temperature range requires a thorough knowledge of the variation of Q and inductance with temperature.

Manuscript received January 29, 1997; revised April 25, 1997.

R. Groves and D. Jadus are with the IBM Microelectronic Division, Hopewell Junction, NY 12533-6531 USA.

D. L. Harame was with the IBM Microelectronic Division, Hopewell Junction, NY 12533-6531 USA. He is now with IBM, Essex Junction, VT 05452 USA.

Publisher Item Identifier S 0018-9200(97)06039-3.

This paper presents measured Q and inductance data over the temperature range from −55°C to +125°C for a 10.5-nH inductor with a peak Q of 5.8 (at 1 GHz/25°C). The inductor's self-resonant frequency is 4 GHz with the output shorted to ground. A knowledge of the variation of Q with temperature allows the circuit designer to design for worst-case temperature conditions found in real world applications.

II. EXPERIMENTAL SETUP

The test structure was a square, spiral, six-turn 10.5-nH inductor utilizing 2-μm thick aluminum metallization over a 15 Ω-cm resistivity P-type substrate. Line width was 16 μm with a line-to-line spacing of 10 μm, as illustrated in the plan view of Fig. 1. This technology provided a 5.7-μm dielectric separation between the inductor and the substrate consisting of the interlayer SiO_2 and shallow trench isolation (STI) below the first metal level. A cross-sectional schematic of the test structure can be seen in Fig. 2. The lumped element model used to simulate the inductor is shown in Fig. 3. This model utilizes the circuit topology suggested by Yue and Ashby [3], [4] and is intended to model the inductor at frequencies at or below the self-resonance point. The various parasitic elements are subdivided to emphasize their physical basis.

The model is comprised of an inductance ($L1$) and its associated parasitics. The inductor has resistive losses and capacitive parasitics as indicated in Fig. 3. Values for the following model elements were determined from the geometry of the device: $L1$ (calculated using Greenhouse [2]), R_{m3}, R_{m2}, C_{m3m2}, C_{m3ox}, and C_{m2ox}. The remaining parasitics (R_{subin}, R_{subout}, C_{subin}, C_{subout}) were determined from the substrate conductance and capacitance values extracted using the method suggested by Yue, Eo, and Eisenstadt in [3], [5], and [6]. This method extracts the complex propagation constant (γ) and characteristic impedance (Z_0) of the inductor from measured S-parameters. From these transmission line parameters (γ and Z_0), the substrate conductivity (G) and capacitance (C) can be extracted (see Fig. 4). The temperature coefficients for the following parasitic elements were included in the model: R_{m3}, R_{m2}, C_{m3ox}, C_{m2ox}, R_{subin}, R_{subout}, C_{subin}, and C_{subout}. The metal resistance temperature coefficients (for R_{m3} and R_{m2}) were determined from dc Kelvin measurements. The metal-to-substrate capacitance temperature coefficients (for C_{m3ox} and C_{m2ox}) were determined from low frequency capacitance measurements. The temperature coefficients of the substrate resistance and capacitance (R_{subin}, R_{subout}, C_{subin}, C_{subout}) were determined from the measured S-parameters [3], [5], [6].

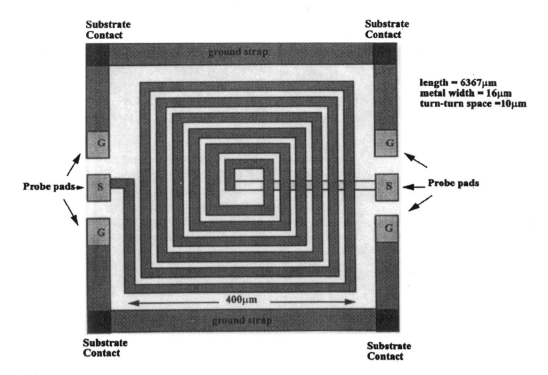

length = 6367μm
metal width = 16μm
turn-turn space = 10μm

400μm

Fig. 1. Plan view of spiral inductor.

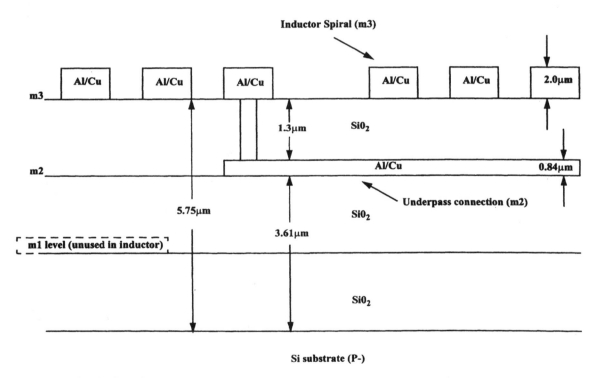

Fig. 2. Inductor cross-sectional schematic.

III. RESULTS AND DISCUSSION

The inductor S-parameters were measured using a Cascade Microtech Summit series automatic water prober with a temperature chamber. Temperature control was provided by a Temptronics 8-in temperature chuck. Condensation was controlled by a constant nitrogen purge at low temperatures. This setup, along with Cascade air coplanar ground-signal-ground 40-GHz wafer probes, proved to be very convenient and reliable over the temperature range from −55°C to +125°C. The measured S-parameters were corrected using the Y-parameter subtraction method [7] to remove the parasitic effects of the probe pads.

136

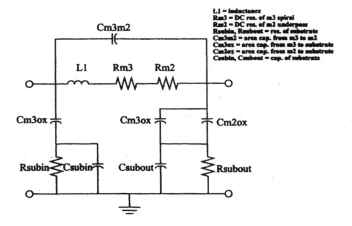

L1 = inductance
Rm3 = DC res. of m3 spiral
Rm2 = DC res. of m2 underpass
Rsubin, Rsubout = res. of substrate
Cm3m2 = area cap. from m3 to m2
Cm3ox = area cap. from m3 to substrate
Cm2ox = area cap. from m2 to substrate
Csubin, Csubout = cap. of substrate

Fig. 3.   Inductor model topology.

Lossy Transmission Line Parameters

$$\gamma = \sqrt{(Z_{series})(Y_{parallel})} \qquad\qquad Z = \sqrt{\frac{Z_{series}}{Y_{parallel}}}$$

$$G = Re\{\gamma/Z_0\} \quad C = (Im\{\gamma/Z_0\})/\omega \quad R = Re\{\gamma Z_0\} \quad L = (Im\{\gamma Z_0\})/\omega$$

Lossy Transmission Line

$$Y_{parallel} = G + j\omega C$$

$$Z_{series} = R + j\omega L$$

Inductor Viewed as Lossy Transmission Line

$$\gamma_{parallel} = \frac{G_{sub} + j\omega C_{sub}}{\left(1 + \dfrac{C_{sub}}{C_{ox}}\right) - j\dfrac{G_{sub}}{\omega C_{ox}}}$$

$$\gamma_{parallel} \cong \frac{1}{1 + \dfrac{C_{sub}}{C_{ox}}}(G + j\omega C_{sub})\Big|_{\omega \gg G_{sub}/C_{ox}}$$

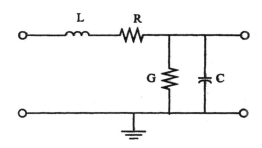

$$Z_{series} = \frac{R + j\omega L}{(1 - \omega^2 L C_{m3m2}) + j\omega R C_{m3m2}}$$

$$Z_{series} \cong R + j\omega L\Big|_{\omega \ll \frac{1}{\sqrt{LC_{m3m2}}}}$$

Fig. 4.   Transmission line model for a spiral inductor.

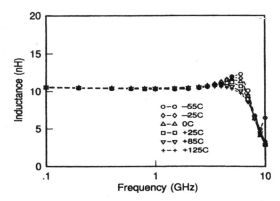

Fig. 5. Measured inductance versus frequency (−55°C to +125°C).

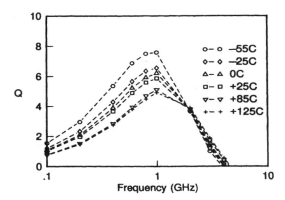

Fig. 6. Measured Q versus frequency (−55°C to +125°C).

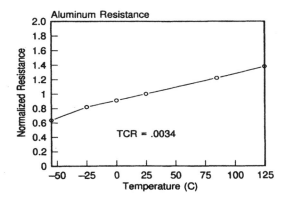

Fig. 7. Normalized metal resistance versus temperature.

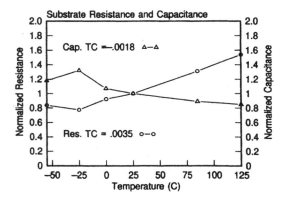

Fig. 8. Normalized substrate resistance and capacitance versus temperature.

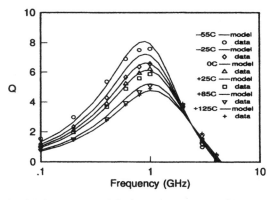

Fig. 9. Modeled and measured Q versus frequency (−55°C to +125°C).

In order to extract the inductance as a function of frequency from the transmission line parameters γ and Z_0, the effects of the feedthrough capacitance (C_{m3m2}) were removed from the measured network parameters by subtracting the capacitance's Y-parameters. The capacitance (C_{m3m2}) was extracted from the measured network parameters using

$$C_{m3m2} = \frac{1}{\omega_{\text{res}}^2 L}(F), \qquad \omega_{\text{res}} = \text{resonant frequency of } Y_{21}. \tag{1}$$

The inductance was then extracted from the corrected transmission line parameters [5], [6]

$$\text{Inductance} = \frac{\text{imag}\{\gamma Z_0\} \cdot l}{\omega}(H), \quad l = \text{length of spiral}. \tag{2}$$

The variation of the extracted inductance over the temperature and frequency range can be seen in Fig. 5.

Q was calculated from the measured data using the short circuit input admittance (Y_{11}), which provides the admittance of the inductor with port 2 shorted

$$Q = \frac{\text{Reactance}}{\text{Resistance}} = \frac{Y_{11} \text{ imag}}{Y_{11} \text{ real}}. \tag{3}$$

In our paper presented at the BCTM in 1996 [8], we explained the error associated with estimating Q using Y_{11} as opposed to the energy-based definition for Q. We calculated a 5% discrepancy, at the point where Q peaks, between the impedance- and

energy-based methods of calculating Q. The energy method yielded a higher value. The configuration with port 2 shorted yields the worst-case value of Q, since Q varies directly with the inductor's port 2 load impedance. Lower values of port 2 load impedance allow a higher current to flow in the inductor causing a higher power loss in the series resistance. Fig. 6 shows the measured Q versus frequency over the range of temperatures.

Q can be seen to decrease with increasing temperature below 2 GHz and increase with increasing temperature above 2 GHz. The variation in extracted inductance with temperature is very small over the entire frequency range. Above the self-resonant frequency the extracted inductance falls off rapidly.

This nonphysical behavior indicates the limitations of the compact lumped element model for frequencies above self-resonance.

Two primary mechanisms are at work in controlling Q over temperature. The aluminum metallization, which is in series with the inductance, has a positive temperature coefficient, which increases resistance and thus the power loss in the inductor as the temperature increases. This serves to decrease Q. The substrate resistivity, which is in parallel with the inductance, also has a positive temperature coefficient. This decreases the power loss in the substrate as the temperature increases (the current is shunted away from the substrate and into the spiral with increasing substrate resistance), serving to increase Q. The resistivity of the aluminum metallization ($m2$ and $m3$) was measured over the temperature range (see Fig. 7) and the temperature coefficient of resistance (TCR) value that was the closest linear fit was used in the model. The substrate resistance and capacitance values were extracted from the measured S-parameters [3], [5], [6], and their temperature coefficients (see Fig. 8) were approximated as linear and used in the model.

At low frequencies (below 2 GHz), the primary power loss in the inductor is in the series resistance of the aluminum. This results from the capacitance to substrate presenting a high impedance to the signal present in the inductor, causing almost all of the current to flow in the inductor metallization. At high frequencies (above 2 GHz), where the capacitance to substrate presents a lower impedance path, and the feedthrough capacitance increasingly shunts out the metal series resistance, the power lost in the substrate begins to dominate. Fig. 6 reveals that the variation of Q with temperature below 2 GHz is dominated by the TCR of the aluminum spiral (Q decreases with increasing temperature), while at frequencies above 2 GHz the variation of Q with temperature is reversed (Q increases with increasing temperature), indicating that the substrate TCR dominates the determination of Q at high frequencies.

Fig. 9 shows a model versus data comparison of Q over the frequency and temperature range. There is excellent agreement between the model and data right up to the self-resonant frequency. In our previous work on temperature dependence [8], we showed a relatively large deviation between model and data above 2 GHz. By extracting the temperature coefficients of the substrate resistance and capacitance at high frequency (using the method suggested by Yue, Eo, and Eisenstadt in [3], [5], and [6]), we were able to model the high frequency behavior of the temperature dependence more accurately.

IV. CONCLUSIONS

The Q and inductance of a planar inductor fabricated over a conductive silicon substrate have been measured over a wide temperature range ($-55°C$ to $+125°C$). The behavior of Q as a function of frequency has been modeled by incorporating the TCR's of the aluminum metallization and silicon substrate and the temperature coefficients of the metal to substrate and substrate capacitances. The inductance was found to vary little with temperature over the full frequency range as expected. The Q was found to decrease with increasing temperature at frequencies up to and including the point where Q peaks. At high frequency, Q was seen to increase with increasing temperature up to the self-resonance point.

REFERENCES

[1] D. L. Harame et al., "Si/SiGe epitaxial-base transistors—Part I: materials, physics, and circuits" and "Si/SiGe epitaxial-base transistors Part II: process integration and analog applications," IEEE Trans. Electron Devices, vol. 42, pp. 455–482, Mar. 1995.

[2] H. M. Greenhouse, "Design of planar rectangular microelectronic inductors," IEEE Trans. Parts, Hybrids, Packaging, vol. PHP-10, pp. 101–109, June 1974.

[3] C. P. Yue et al., "A physical model for spiral inductors on silicon," in 1996 IEDM Proc., pp. 155–158.

[4] K. B. Ashby et al., "High Q inductors for wireless applications in a complementary silicon bipolar process," in 1994 BCTM Proc., pp. 179–182.

[5] Y. Eo and W. R. Eisenstadt, "High-speed VLSI interconnect modeling based on S-parameter measurements," IEEE Trans. Hybrids, Manuf. Technol., vol. 16 pp. 555–562, Aug. 1993.

[6] W. R. Eisenstadt and Y. Eo, "S-parameter based IC interconnect transmission line characterization," IEEE Trans. Comp., Hybrids, Manuf. Technol., vol. 15, pp. 483–490, Aug. 1992.

[7] P. J. Van Wijnen et al., "A new straight forward calibration and correction procedure for "on-wafer" high frequency S-parameter measurements (45 MHz–18 GHz)," in 1987 BCTM Proc., pp. 70–73.

[8] R. Groves et al., "Temperature dependence of Q in spiral inductors fabricated in a silicon-germanium/BiCMOS technology," in 1996 BCTM Proc., pp. 153–156.

Substrate Noise Coupling Through Planar Spiral Inductor

Alan L. L. Pun, *Student Member, IEEE*, Tony Yeung, *Student Member, IEEE*, Jack Lau, *Member, IEEE*, François J. R. Clément, *Member, IEEE*, and David K. Su, *Member, IEEE*

Abstract—While previous studies on substrate coupling focused mostly on noise induced through drain-bulk capacitance, substrate coupling from planar spiral inductors at radio frequency (RF) via the oxide capacitance has not been reported. This paper presents the experimental and simulation results of substrate noise induced through planar inductors. Experimental and simulation results reveal that isolation between inductor and noise sensor is less than -30 dB at 1 GHz. Separation by distance reduces coupling by less than 2 dB in most practical cases. Practical examples reveal an obstacle in integrating RF tuned-gain amplifier with sensitive RF receiver circuits on the same die. Simulation results indicate that hollow inductors have advantages not only in having a higher self-resonant frequency, but also in reducing substrate noise as compared to conventional inductors. The effectiveness of using broken guard ring in reducing inductor induced substrate noise is also examined.

Index Terms—Coupling, cross-talk, guard ring, hollow inductor, isolation, RF power amplifier, spiral inductor, substrate noise.

I. INTRODUCTION

IN THE PAST few years, we have witnessed an increasing interest in integrating inductors in analog radio frequency (RF) chips [1]. Thanks to great advances in multilayer metal processing and a higher frequency requirement, we have been able to integrate inductors of reasonable qualities. The trend to attain higher integration for lower cost is global and incessant. In a relatively small place like Hong Kong, which has a population of over 6 million, the number of cellular phones has exceeded 2 million in the year 1997 and continues to grow rapidly.

An unfortunate byproduct of higher integration is a higher susceptibility to interference. One such interference is substrate noise. Most of the reports [2]–[4] on substrate noise have been done at lower frequencies or have been done on nonepitaxial wafers. In addition, the studies were mainly done on noise generation via junction capacitance. In this paper, we focus on the potential of substrate noise induction through spiral inductors.

Manuscript received August 19, 1997; revised December 5, 1997. This work was supported in part by the Hong Kong Government Research Grant Council.

A. L. L. Pun was with the Department of Electrical and Electronic Engineering, The Hong Kong University of Science & Techhnology, Hong Kong, China. He is now with Pencom Semiconductor, San Jose, CA USA.

T. Yeung and J. Lau are with the Department of Electrical and Electronic Engineering, The Hong Kong University of Science & Techhnology, Hong Kong, China.

F. J. R. Clément is with the Swiss Federal Institute of Technology, Lausanne, Switzerland and Snake Technologies, France.

D. K. Lu is with Hewlett-Packard Laboratories, Palo Alto, CA USA.

Publisher Item Identifier S 0018-9200(98)03507-0.

In today's RF circuits, inductors of value in the range of nanohenrys are found [5]–[7]. Typical applications of inductors include impedance transformation, inductive degeneration, and frequency-tuned loading. The basic inductance value can be determined either analytically, based on a popular theorem proposed by Greenhouse [8], or numerically, based on some electromagnetic (EM) finite-element simulators. Typically, the size of the inductors is on the order of a few hundred micrometers by a few hundred micrometers.

For a triple-layer metal process, the inductors usually reside on roughly 3–4 μm of oxide. As a result, an inductor will have a capacitance on the order of picofarads between itself and the substrate. Perturbation on the inductors will transduce noisy current in the substrate. Intuitively, the substrate noise problem will be more severe in this case as compared to that in a mixed-signal integrated circuit. In the first place, integrated inductors are required to operate in the gigahertz range, while most digital circuits operate at an order of magnitude lower in frequency. Second, in a mixed-signal case, the substrate noise is induced through the drain-bulk junction capacitance. In a typical submicron technology with a substrate doping of 4×10^{16} cm^{-3}, the drain-bulk capacitance is about 0.7 fF/μm^2. Therefore, it will take a very wide digital transistor to deliver the same amount of substrate noise. The difference is more prominent if one considers the fact that wide transistors are mostly used as drivers in digital VLSI and will not be running at high speed.

Section II describes the test chip and the experimental setup in detail. We also introduce a simulation technique in Section III, which can model the substrate effect and generate a lumped model for noise simulation. Section IV provides some results to verify the proposition of inductor-induced substrate noise. In Section V, we provide data on practical substrate noise impact by using the inductors as a tuned-gain stage in RF amplifier. Various methods in reducing substrate noise are also discussed in Section VI.

II. EXPERIMENTAL SETUP

A 2 mm × 2 mm test chip was fabricated in a 0.8-μm triple-layer metal CMOS n-well technology to study the substrate noise coupling effect from spiral inductors. Five three-turn inductors, each of size 350 μm × 350 μm using top-layer metal, are shown in Fig. 1. In addition, ten P^+ diffusion substrate contacts at different locations act as noise sensor for measuring noise. Each of these P^+ diffusion contacts is 50 μm × 50 μm. Dummy contacts without P^+ diffusion are used to calibrate out the nonsubstrate noise.

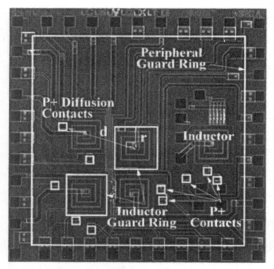

Fig. 1. Microphotograph of the test chip.

Process:

MOSIS 0.8um 3-Layer Metal CMOS Nwell with Heavily Doped Bulk

Size:

2mm x 2mm

d:

P+ Diffusion Contacts to the center of Inductor

r:

Guard Ring to the center of Inductor

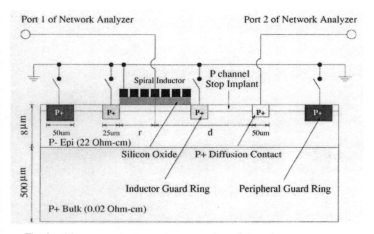

Fig. 2. Measurement setup and cross section of the substrate.

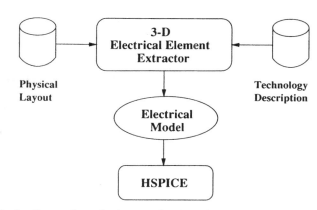

Fig. 3. Process flow of substrate noise simulation tools.

The cross section of the basic test structure and the measurement setup are shown in Fig. 2. There is a roughly 1-μm-thick P^+ channel stop implant of 1 $\Omega \cdot$ cm resistivity near the surface of the chip. The P^- region with a resistivity of about 22 $\Omega \cdot$ cm is found transversing 8 μm into the wafer. As we transverse deeper into the wafer, the resistivity becomes less because of higher doping concentration. Eventually, the resistivity drops to 0.02 $\Omega \cdot$ cm and flattens out.

One of the test chips is mounted on a two-sided 31-mil-thin copper board using conductive epoxy to tie the backside of substrate to ground. In real IC, perfect grounding is not possible because of lead inductance. To study the worst case condition, another chip is mounted using nonconductive epoxy. Measurement with and without substrate tied to ground via P^+ guard rings are both performed. The top side of the PCB has 50-Ω lines connected to SMA connectors, while the bottom side acts as a ground plane. Unused area of the PCB is connected to the ground plane via through-holes to avoid board-level noise.

In the experiment, S-parameters were measured using an HP8753D two-port network analyzer. One end of the inductor was connected to port 1 and another end was connected to

the ground. P^+ diffusion contacts were connected to port 2 of the analyzer. As we excite the inductor, the S_{21} parameter indicates the isolation between the inductor (noise source) and P^+ substrate contact (sensor). Before any measurements were taken, standard TRL calibration was done.

Studies in the past [3] have indicated that P^+ diffusion guard rings are the most effective shield against substrate noise in a conventional CMOS technology. We are interested in studying the efficiency of these guard rings in our setup. Besides the peripheral guard ring that surrounds the die, a second set of guard ring surrounds two of the inductors. These guard rings are placed 175 μm from the center of the inductors. The inductor guard rings are a 25-μm-wide diffusion region with a dedicated biasing pad. The P^+ diffusion ring surrounding the chip is 50 μm wide and is connected to eight bonding pads. For ease of study, the distance between P^+ diffusion contact and the center of inductor is defined as d. r is defined as the distance from the edge of inductor guard ring to the center of inductor.

III. SIMULATION METHODOLOGY

Precise modeling of the substrate noise is often difficult due to the distributive nature of the substrate and the particulars of a specific layout. In order to have a better understanding of the substrate noise propagation, a two-step approach is used to simulate the substrate noise. As it will be shown later, the establishment of the methodology enables us to explore various noise reduction schemes.

A process flow of the substrate noise simulation is shown in Fig. 3. In the first step, physical layout is processed through a three-dimensional electrical element extractor [9]. The program is based on the finite difference method.

The layout defines the surface mesh while the vertical doping profiles of the technology determines the mesh from the top to the bottom of the wafer. The technology description includes information such as mean doping, as well as vertical and lateral resistivity for each combination of the drawn layers and for each subdivision of the mesh. Information of the doping profile of the MOSIS 0.8-μm CMOS technology with epitaxial layer on top of a heavily doped bulk was obtained using spreading resistance analysis. The result shown in Fig. 4 is the resistivity and carrier concentration along the depth of the substrate.

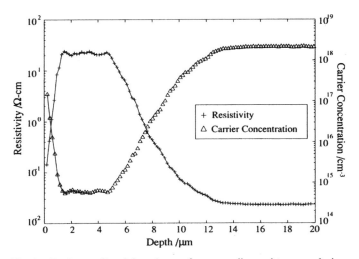

Fig. 4. Doping profile of the substrate from spreading resistance analysis.

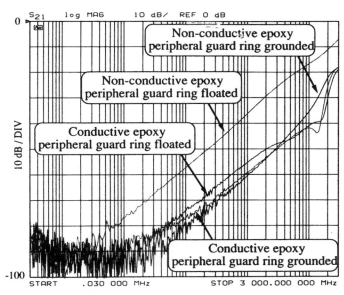

Fig. 5. Measurement of S_{21} versus frequency at $d = 1200$ μm.

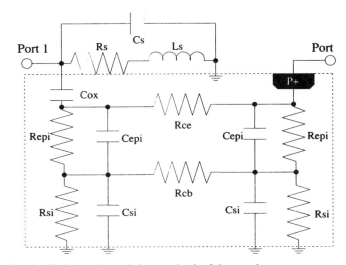

Fig. 6. Equivalent lumped element circuit of the experiment setup.

In the second step of the process, the extracted electrical model is passed to HSPICE through which noise simulation can be carried out.

The two-step approach has been used to study lower-frequency substrate noise on a uniform lightly doped substrate in a 2-μm CMOS technology in an earlier work [4]. It provides an easy and fast way to understand the substrate noise propagation. The program is valid for signals with periods greater than the RC time constant of the substrate. The inverse of the RC time constant is greater than 10 GHz. Details will be explained in the Appendix.

IV. MEASUREMENT RESULTS

A. Effect of Operating Frequency

S_{21} parameters from the inductor to the P^+ diffusion contacts are measured and simulated. The distance between the edge of the inductor and the diffusion contact is 1200 μm. The large distance is chosen to guarantee the substrate noise has no distance dependence.

The frequency is swept from 30 kHz to 3 GHz. Data are collected in two different configurations. In the first case, all bonding pads of the peripheral guard ring were connected to ground. In the second case, all bonding pads of the peripheral guard ring were floated without any connection. In both cases, we use either conductive or nonconductive epoxy for substrate backside contact. So, there is a total of four experiments.

1) Low Frequency: Fig. 5 shows the measured S_{21} for the four different configurations. The worst case occurs with nonconductive epoxy and guard ring floated. With either guard ring grounded or backside tied to grounded, the coupling reduces by some 25 dB at frequency below 1 GHz.

It can be observed that noise coupling increases by 20 dB per decade of frequency, indicating that the coupling is mostly due to ωC_{ox}. While the exact details of the distributive nature of the substrate are hard to picture, a rough equivalent circuit can elucidate the idea (Fig. 6) [1], [10]. In the test, C_{ox} helps create a dominant zero. The value of C_{ox} in this case is about 1 pF for a 4-μm-thick oxide—far greater than that contributed by the substrate capacitance C_{epi} and C_{si}. In order to verify

this fact, after extracting the substrate resistive mesh, we backcalculated the associated substrate capacitance for each resistor grid. We resimulated with and without the substrate capacitance and observed no change in the substrate noise.

2) High Frequency: At frequency above 1 GHz, the S_{21} increases greater than 20 dB per decade of frequency, indicating there is another pole besides ωC_{ox}. This effect can be understood by comparing measured and simulated results. For ease of comparison, the measurement results are reproduced in Fig. 7 along with the simulation results for peripheral guard ring grounded and floated with nonconductive epoxy. In the simulation, we model the bond wire connected to the guard ring as a 2-nH inductor. As frequency increases above 1 GHz, the guard rings become less effective in reducing noise due to the parasitic inductance in the setup for the case of grounded peripheral guard ring. However, no parasitic inductance is involved for the case of peripheral guard ring floated with nonconductive epoxy, and thus S_{21} continues to increase by 20 dB per decade for all frequency range.

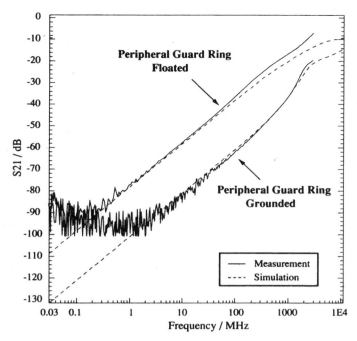

Fig. 7. Measurement and simulation of S_{21} versus frequency at $d = 1200$ μm for peripheral guard ring floated and grounded with nonconductive epoxy.

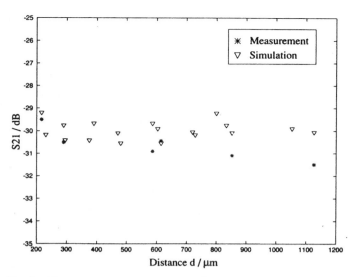

Fig. 9. Measurement and simulation of S_{21} versus distance d with the peripheral guard ring floated and frequency = 500 MHz.

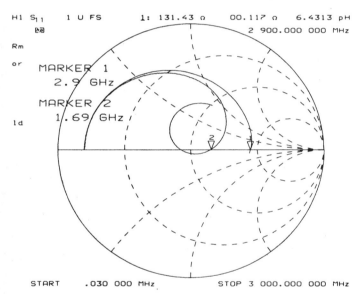

Fig. 8. Measured inductor self-resonant frequency for peripheral guard ring floated (marker 1) and grounded (marker 2).

It should be pointed out that the inductor self-resonant frequency[1] drops greater than 40% when the guard ring is grounded because the effective capacitance to ground is bigger. Fig. 8 shows the measured S_{11} which indicates the inductor self-resonant frequencies are 2.9 and 1.69 GHz for peripheral guard ring floated and grounded, respectively.

B. Effect of Physical Separation

The distance away from the inductor is in general irrelevant, as illustrated in Fig. 9. The variation of S_{21} is less than 2 dB

[1] The self-resonant frequency means the lowest frequency such that the inductance drops to zero on the Smith chart.

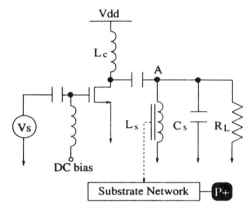

Fig. 10. Schematic of power amplifier for inductor (L_s) substrate noise analysis.

as distance separation changes from 200 μm to 1200 μm. The size of the inductors is quite large and certainly greater than four times the distance of the thickness of epitaxial layer. In an earlier work [2], it has been shown that substrate noise at a distance less than four times the epitaxial thickness depends on the actual distance, beyond which the epitaxial layer becomes an equal-potential node. As an interesting note, for lightly doped wafers, the dependence on distance arises when the separation is less than the wafer thickness.

V. DESIGN CONSIDERATIONS

The inductor substrate model derived above can be used to predict the inductor induced noise coupling from a narrow-band tuned amplifier, a key building block for RF power amplifier is shown in Fig. 10.

In the simulation, the size of the NMOS is chosen to be $\frac{W}{L} = 333$. The L_c is a 3.3-μH off-chip inductor acting as an RF choke. The output load R_L is 50 Ω. The on-chip inductor $L_s = 7$ nH resonates with capacitor $C_s = 10$ pF at 600 MHz. The L_s substrate is connected to the P^+ contact via the substrate equivalent network.

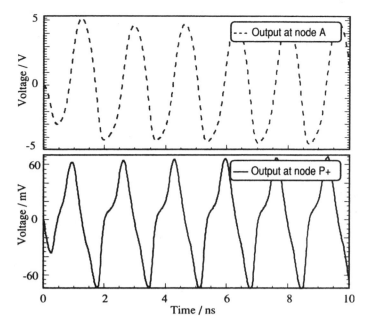

Fig. 11. Transient analysis of noise coupling from the power amplifier inductor (L_s) to P^+ at $d = 300\ \mu$m.

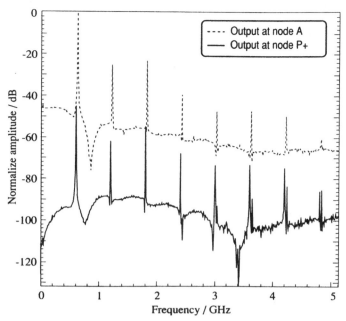

Fig. 12. FFT analysis of noise coupling from the power amplifier inductor (L_s) to P^+ at $d = 300\ \mu$m.

With a power supply Vdd of 5 V, an amplifier in class-B operation can have a maximum output voltage as large as 10 V. This large output voltage across inductor L_s can induce a significant amount of noise in the substrate. Transient simulation results of noise coupling from L_s to a P^+ substrate contact at $d = 300\ \mu$m are shown in Fig. 11. The noise voltage at the P^+ diffusion contact is about 130 mVpp with a 45° of phase shift. The corresponding fast Fourier transform (FFT) analysis shown in Fig. 12 indicates an isolation of −40 dB at the fundamental frequency.

Fig. 13 shows the noise coupling from inductor L_s to the P^+ substrate contact increases at higher harmonic frequencies. These results show that inductor-induced noise can be a major obstacle to integrating an RF power amplifier with sensitive RF receiver circuits on the same die.

VI. VARIOUS METHODS IN REDUCING INDUCTOR-INDUCED SUBSTRATE NOISE

A. Conventional Inductor

The amount of substrate noise that can be reduced by a guard ring surrounding the inductor depends on the distance from the inductor. One can view the inductor substrate capacitance as a massively parallel array of capacitors. Due to the large inductor size and the four times epi thickness requirement, an optimal guard ring placement for minimum noise can be determined. Unfortunately, as the guard ring encroaches into the inductor, the mirror effect degrades the inductance value.

An EM field simulator [11] is used to detect the inductance reduction as a function of guard ring distance. An inductor of five turns with an outer dimension of 180 μm × 180 μm, a line width of 15 μm, a turn-to-turn spacing of 1 μm, and metal (aluminum) thickness of 1.2 μm is simulated to be about 2 nH

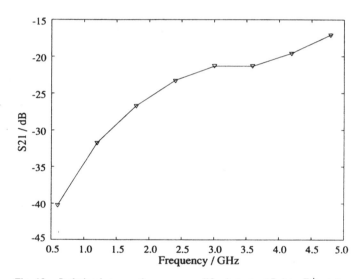

Fig. 13. Isolation between the power amplifier inductor (L_s) to P^+ at $d = 300\ \mu$m.

under free space. The corresponding capacitance C_{ox} is about 0.28 pF and the self-resonant frequency is 6.7 GHz. Fig. 14 shows the simulation results of inductance, together with S_{21} versus r under the conditions of $d = 300\ \mu$m and frequency of 1 GHz, where r is defined as the distance from the center of the inductor to the edge of the guard ring, and d is the distance from the P^+ diffusion contact to the center of the inductor. With the guard ring grounded, the inductance value decreases dramatically as the guard ring moves inward into the inductor. The effect of the guard ring becomes void as it approaches the center of the inductor. Both the inductance value and noise coupling show no change. As the guard ring is placed far away from the inductor, the mirror effect is almost nonexistent and the inductance reduces by only 10%. Yet, the

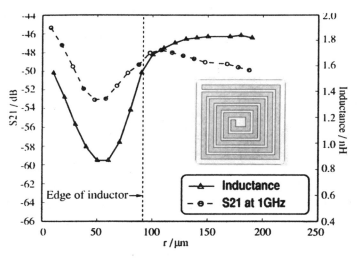

Fig. 14. Simulation results of inductance and S_{21} for solid inductor at $d = 300$ μm. Five-turns 2-nH inductor of size 180 μm × 180 μm.

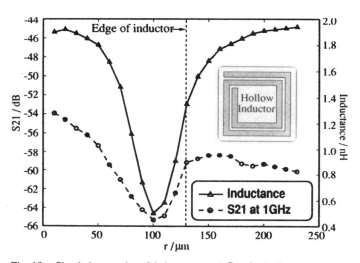

Fig. 15. Simulation results of inductance and S_{21} for hollow inductor at $d = 300$ μm. Two-turns 2-nH inductor of size 260 μm × 260 μm.

noise can be reduced to a reasonable level with the guard ring grounded.

B. Hollow Inductor

It has been understood that the magnitude of coupled noise is directly proportional to the value of oxide capacitance. However, we cannot change the oxide thickness as a way to reduce C_{ox} for a given fabrication process.

A hollow inductor has been proven [12] to have a higher quality factor and self-resonant frequency for the same value of inductance as compared to conventional inductor. We are interested to know if the hollow inductor has any advantage as far as substrate noise is concerned. A two-turn hollow inductor with an outer dimension of 260 μm × 260 μm is simulated to be about 2 nH—the same as that in the conventional inductor. The other dimensions of inductor are the same as the conventional inductor. The corresponding capacitance C_{ox} is about 0.127 pF which yields a self-resonant frequency of

TABLE I
COMPARISONS OF CONVENTIONAL AND HOLLOW INDUCTOR

	Conventional	Hollow
Original inductance	2nH	2nH
Outer dimension	180μm	260μm
Number of turns	5	2
Capacitance C_{ox}	0.28pF	0.127pF
$f_{self-resonance}$	6.7GHz	10GHz
Maximum S_{21}	-45.3dB	-54dB
Minimum S_{21}	-53dB	-65dB
Minimum inductance	0.9nH	0.5nH

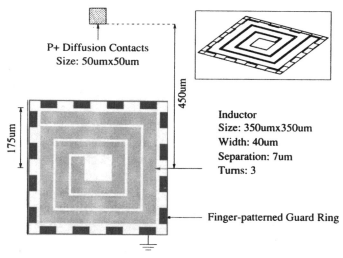

Fig. 16. Broken inductor guard ring at $r = 175$ μm.

about 10 GHz. Results illustrated in Fig. 15 indicate that noise is about 7–12 dB less in the hollow case as compared to that in the conventional one. However, there are tradeoffs for the minimum inductance drops to about 25% of the original value.

A comparison between a conventional inductor and the hollow inductor is tabulated in Table I. The capacitance C_{ox} of the conventional inductor is bigger than the hollow one with the same value of original inductance. Thus, higher self-resonant frequency can be achieved for the hollow inductor. The smaller C_{ox} also implies smaller substrate noise in the hollow inductor. The smaller minimum inductance happened in the hollow case when the guard ring is placed underneath the inductor.

C. Discontinuous Guard Ring

As the guard ring is placed near the inductor, the mirror effect becomes a serious problem as it reduces the original designed inductance value dramatically. In order to reduce the eddy current that flows in the substrate, a broken guard ring is employed. Fig. 16 shows the broken guard ring setup with a width of 25 μm placed near the edge of the inductor. The P^+ substrate contact is set at a distance of 450 μm from the center of inductor. The original inductance value is about 2 nH. Simulation results shown in Fig. 17 reveal that the half-broken guard ring (50% area) reduces noise significantly while the inductance drops by only 8%.

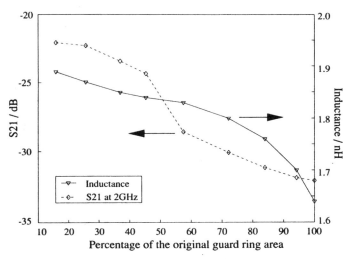

Fig. 17. Simulation results of inductance and S_{21} for broken guard ring, which is specified as area % of the complete ring.

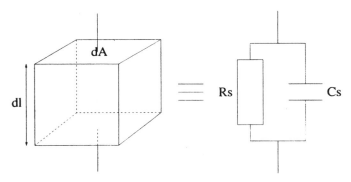

Fig. 18. Model for a piece of homogeneous substrate.

VII. CONCLUSION

The experimental results and simulation verifications provide an understanding of the noise coupling effect of planar spiral inductor in a heavily doped silicon substrate. It would be a major obstacle to integrate an RF power amplifier with a sensitive RF receiver circuit on the same die. Simulation reveals that a hollow inductor has advantages in reducing inductor induced substrate noise as compared to conventional inductor. A broken guard ring can help reduce the substrate eddy current induced by the inductor and hence reduce noise while maintaining the inductance value.

APPENDIX

Capacitive and resistive effects occur throughout the substrate. Inside a doped semiconductor, the conductivity σ is given by $\sigma = q(n\mu_n + p\mu_p)$, where q is the electronic charge, μ_n and μ_p represent the mobility of n- and p-carriers, and n and p stand for the n- and p-carrier density, respectively. Therefore, the current density is given by (1), where $\vec{E}$ is define as the electric field

$$J = \sigma \vec{E} = q(n\mu_n + p\mu_p)\vec{E}. \qquad (1)$$

The μ_n and μ_p parameters vary as functions of the total semiconductor doping and temperature.

In the frequency domain, the equivalent admittance Y_s for a piece of substrate as illustrated in Fig. 18 is given by

$$Y_s = \frac{1 + sR_sC_s}{R_s} = \frac{1 + j\omega T_s}{R_s} \qquad (2)$$

where T_s is defined as the substrate time constant

$$T_s = R_sC_s = \frac{\rho_s dl}{dA} \cdot \frac{\varepsilon_s dA}{dl} = \frac{\varepsilon_0 \varepsilon_{si}}{q(n\mu_n + p\mu_p)}. \qquad (3)$$

For low frequencies, substrate resistance R_s is more important, and the associated capacitance C_s can be neglected. As frequency ω increases, the capacitive effect rises to become equal to the resistive effect at the cutoff frequency f_0 defined by

$$R_s = \frac{1}{\omega_0 C_s} \Longrightarrow f_0 = \frac{1}{2\pi T_s} = \frac{q(n\mu_n + p\mu_p)}{2\pi\varepsilon_0\varepsilon_{si}}. \qquad (4)$$

The minimum f_0 is attained for a lightly doped p-type substrate because mobility is lower for holes than for electrons. The extreme case of a normal initial carrier concentration of 10^{15} cm^{-3} yields a minimum cutoff frequency given by

$$f_{0_{min}} = \frac{q\mu_p p}{2\pi\varepsilon_0\varepsilon_{si}}. \qquad (5)$$

The minimum cutoff frequency is greater than 10 GHz for $\mu_p = 480$ cm^2/Vs at temperature of 300 K. At higher doping concentration, such as that in the highly conductive substrate in an epi-process ($p = 10^{18}$ cm^{-3}), the cutoff frequency is an order of magnitude higher.

Therefore, the capacitance of the substrate can be neglected in the lower RF range for substrates doped homogeneously.

ACKNOWLEDGMENT

The authors would like to thank their colleagues and technical staffs in the Department of the Electrical and Electronic Engineering at The Hong Kong University of Science & Techhnology for their kind help, invaluable suggestions, and technical assistance. In particular, they thank L. Tsui, F. Hui, and Z. Chen in the Consumer Media Laboratory; S. F. Luk in the VLSI Laboratory; P. H. Yin and C. Y. Wong in the Wireless Communication Laboratory; P. Chan in the Analog Laboratory; and A. Ng in the Device Characteristics Laboratory. A special acknowledgment is extended to Y. K. Leung and Prof. S. Wong at the Center for Integrated Systems at Stanford University for providing spreading resistance analysis. The useful comments of the reviewers have also been noted.

REFERENCES

[1] K. B. Ashby, I. A. Koullias, W. C. Finley, J. J. Bastek, and S. Moinian, "High Q inductor for wireless applications in a complementary silicon bipolar process," *IEEE J. Solid-State Circuits*, vol. 31, pp. 4–9, Jan. 1996.

[2] D. K. Su, M. J. Loinaz, S. Masui, and B. A. Wooley, "Experimental results and modeling techniques for substrate noise in mixed-signal integrated circuits," *IEEE J. Solid-State Circuits*, vol. 28, pp. 420–430, Apr. 1993.

[3] K. Joarder, "A simple approach to modeling cross-talk in integrated circuits," *IEEE J. Solid-State Circuits*, vol. 29, pp. 1212–1219, Oct. 1994.

[4] T. Blalack, J. Lau, F. J. R. Clement, and B. A. Wooley, "Experimental results and modeling of noise coupling in a lightly doped substrate," in *Proc. IEEE Int. Electron Devices Meet.*, 1996.

[5] D. K. Shaeffer and T. H. Lee, "A 1.5 V, 1.5-GHz CMOS low noise amplifier," *IEEE J. Solid-State Circuits*, vol. 32, pp. 745–759, May 1997.

[6] N. M. Nguyen and R. G. Meyer, "A Si bipolar monolithic RF bandpass amplifier," *IEEE J. Solid-State Circuits*, vol. 27, pp. 123–127, Jan. 1992.

[7] M. Soyuer, K. A. Jenkins, J. N. Burghartz, H. A. Ainspan, F. J. Canora, S. Ponnapalli, J. F. Ewen, and W. E. Pence, "A 2.4 GHz silicon bipolar oscillator with integrated resonator," *IEEE J. Solid-State Circuits*, vol. 31, pp. 268–270, Feb. 1996.

[8] H. M. Greenhouse, "Design of planar rectangular microelectronic inductors," *IEEE Trans. Parts, Hybrids, Packaging*, vol. 10, pp. 101–109, June 1974.

[9] F. J. R. Clement, E. Zysman, M. Kayal, and M. Declercq, "LAYIN: LAYout inspection CAD tool dedicated to the parasitic coupling effects through the substrate of integrated circuits," *Power and Timing Modeling Optimization and Simulation*, 1995.

[10] C. P. Yue, C. Ryu, J. Lau, T. H. Lee, and S. S. Wong, "A physical model for planar spiral inductors on silicon," in *Proc. IEEE Int. Electron Devices Meet.*, 1996.

[11] _____, *Maxwell Quick 3D Parameter Extractor User's Reference*, Ansoft Corp.

[12] J. Craninckx and M. Steyaert, "A 1.8 GHz low-phase-noise CMOS VCO using optimized hollow spiral inductors," *IEEE J. Solid-State Circuits*, vol. 32, pp. 736–744, May 1997.

Jack Lau (S'89–M'90) received the B.S. and M.S. degrees in electrical engineering from the University of California at Berkeley. He received the Ph.D. degree at the Hong Kong University of Science & Technology in 1994. His M.S. thesis involved the design and analysis of integrated pin drivers for in-circuit testing.

While at Berkeley, he worked at Hewlett-Packard as a Technical Support Intern and at Schlumberger as an Automatic Test Equipment Engineer. After receiving the M.S. degree, he worked as a CPU Design Engineer at Integrated Information Technology Santa Clara, CA. His Ph.D. thesis at the HKUST is on integrated silicon magnetic sensor. From 1995 to 96, he was a Visiting Scholar at Center for Integrated Systems at Stanford University researching in RF CMOS circuits and substrate noise coupling issues. His research interests are in analog high-frequency IC design for wireless applications and integrated magnetic sensor.

Dr. Lau is a recipient of the UC Berkeley Alumni Scholarship, Cray Research Scholarship, and Unisys Scholarship. He is also the recipient of the 1995 School of Engineering Faculty Teaching Appreciation Award. He is also a member of Eta Kappa Nu, and Tau Beta Pi.

Alan L. L. Pun (S'97) was born in Fukien, China, in 1973. He received the B.Eng. degree in electronic engineering and the M.Phil. degree in electrical and electronic engineering from The Hong Kong University of Science & Techhnology (HKUST) in 1996 and 1998, respectively.

He was involved in the design and implementation of a 900-MHz digital wireless phone with frequency hopping spread spectrum for CDMA application in 1995–1996. From 1996 to 1998, he was a Research Assistant at Consumer Media Laboratory. He then conducted research in high-speed CMOS RF integrated circuits in the Wireless Communication Laboratory at HKUST. Currently, he is a design engineer at Pericom Semiconductor, San Jose, CA. His research interests are in analog high-frequency IC design for wireless applications, wireless communication, and CDMA spread spectrum. His current research is on substrate noise cross-talk and RF integrated circuits.

François J. R. Clément (M'96) was born in Fribourg, Switzerland, on June 20, 1965. He received the B.S. and Ph.D. degrees in electrical engineering from the Swiss Federal Institute of Technology in Lausanne (EPFL), Switzerland, in 1991 and 1995, respectively.

In 1996, he was a visiting scholar with Profs. R. W. Dutton and B. A. Wooley at Stanford University. Currently, he serves as a Research Associate with the Swiss Federal Institute of Technology, and as Layin Product Manager with Snake Technologies, France. His research interests include integrated circuit fabrication and physical design. In particular, he has recently worked on the integration of high-voltage devices in standard 5-V CMOS technologies, and parasitic coupling effects through the substrate of integrated circuits.

Tony Yeung (S'97) was born in Hong Kong, China, in 1973. He received the B.Eng. degree of electronic engineering from The Hong Kong University of Science & Techhnology (HKUST) in 1996.

He was involved in the design and implementation of a 900-MHz digital wireless phone with frequency hopping spread spectrum for CDMA application in 1995–1996. Currently, he is engaged in research toward the M.Phil. degree in electrical and electronic engineering at HKUST. He is researching in high-speed CMOS RF integrated circuits in the Wireless Communication Laboratory at HKUST. His research interests are in analog high-frequency IC design for wireless applications, wireless communication, and CDMA spread spectrum. His current research is on substrate noise cross-talk and RF power amplifier circuits.

David K. Su (S'81–M'94) was born in Kuching, Malaysia. He received the B.S. and M.E. degrees in electrical engineering from the University of Tennessee, Knoxville, in 1982 and 1985, and the Ph.D. degree in electrical engineering from Stanford University in 1994.

From 1985 to 1989 he worked as an IC Design Engineer at Hewlett-Packard Company (HP), Corvallis, OR, and Singapore where he designed full-custom and semicustom application-specific integrated circuits. Since 1994, he has been a member of technical staff at the High Speed Electronics Department of HP Laboratories in Palo Alto, CA, where he designs CMOS RF circuits for wireless communications. His current research interests include the design of analog, RF, and mixed-signal integrated circuits for data conversions and communications.

Dr. Su is a member of Tau Beta Pi, Eta Kappa Nu, and Phi Eta Sigma.

Design of High-Q Varactors for Low-Power Wireless Applications Using a Standard CMOS Process

Alain-Serge Porret, *Student Member, IEEE*, Thierry Melly, *Student Member, IEEE*, Christian C. Enz, *Member, IEEE*, and Eric A. Vittoz, *Fellow, IEEE*

Abstract—New applications such as wireless integrated network sensors (WINS) require radio-frequency transceivers consuming very little power compared to usual mainstream applications, while still working in the ultra-high-frequency range. For this kind of applications, the LC-tank-based local oscillator remains a significant contributor to the overall receiver power consumption. This statement motivates the development of good on-chip varactors available in a standard process.

This paper describes and compares the available solutions to realize high-Q, highly tunable varactors in a standard digital CMOS submicrometer process. On this basis, quality factors in excess of 100 at 1 GHz, for a tuning ratio reaching two, have been measured using a 0.5-μm process.

Index Terms—Low-power design, oscillators, RF CMOS, varactors, voltage-controlled oscillators, wireless communication.

I. INTRODUCTION

WIRELESS integrated network sensors (WINS) systems require low-cost micropower transceivers operating in the industrial, scientific, medical (ISM) ultra-high-frequency (UHF) bands (between 370 and 916 MHz) [1], [2]. Although low-power RF circuits have already been implemented in bipolar technology [3], cost and integration considerations motivate the development of full CMOS solutions [4] operating at very low power levels, usually a few milliwatts in receive mode. Also, power consumption and volume requirements makes single battery cell operation highly desirable, ultimately demanding circuits able to operate with a 1-V supply or even less. Such tight constraints motivate a strict limitation of the number of radio-frequency (RF) nodes, usually leading to the choice of a zero- or low-intermediate-frequency (IF) architecture [5]. For this kind of solution, the power consumption of the front end is almost restricted to the low-noise amplifier and the frequency synthesizer. Therefore, great care should be taken with the voltage-controlled oscillator (VCO) design, and more specifically with the LC-tank, which is the focus of this paper. Note that, for many different reasons, a differential implementation of the oscillator is usually chosen [1]–[7]. This will be assumed hereafter.

In the past few years, much effort has been spent on realizing on-chip inductance having reasonable losses. However, quality factors higher than ten may hardly be expected using standard technologies [4], [6] so that off-chip inductors with $Q_L > 40$ should be used for low-power applications.

Assuming that the output swing of an LC-tank VCO is set by the mixer requirements,[1] the power consumption of the oscillator is proportional to the total capacitance of the tank C_T and inversely proportional to the global quality factor Q_T. Moreover, a minimum frequency excursion $x = (\omega_{\max} - \omega_{\min})/(\omega_{\max} + \omega_{\min})$ is required to allow for channel selection, temperature, and parameter scattering compensation. This constraint sets the minimum size of the varactor C_V that should be added to the fixed capacitance C_F. This additional capacitance increases power consumption and should therefore be limited by using varactors with a high tuning range (maximum to minimum capacitance ratio, $\xi = C_{V,\max}/C_{V,\min}$). Unfortunately, high ξ ratios are difficult to conciliate with low-voltage operation. The minimum capacitance of the varactor is approximately given by

$$\zeta = \frac{\bar{C}_V}{C_F} \approx \left(1 - 2x \cdot \frac{\xi + 1}{\xi - 1}\right)^{-1} - 1$$

with

$$\bar{C}_V = \frac{C_{V,\min} + C_{V,\max}}{2} \quad \text{and} \quad x \ll 1. \tag{1}$$

As ξ can hardly be higher than two and x is about 10% for practical applications, parameter ζ is usually larger than one.

The overall tank quality factor is determined by the respective quality factors of the inductor Q_L, the fixed capacitor Q_F, and the varactor Q_V

$$\frac{C_T}{Q_T} = \frac{C_T}{Q_L} + \frac{C_V}{Q_V} + \frac{C_F}{Q_F} \quad \text{with} \quad C_T = C_F + C_V. \tag{2}$$

Last, the power consumption behavior can be expressed as

$$P \propto \frac{\omega_0 C_T}{Q_T} = \omega_0 C_F \cdot \left(\frac{1 + \zeta}{Q_L} + \frac{1}{Q_F} + \frac{\zeta}{Q_V}\right) \tag{3}$$

$$\frac{P}{P_{\min}} = 1 + \zeta \cdot \frac{1 + Q_L/Q_V}{1 + Q_L/Q_F}$$

with

$$P_{\min} = P|_{\zeta=0} \tag{4}$$

Manuscript received July 16, 1999; revised October 18, 1999. This work was supported in part by the Centre Suisse d'Electronique et de Microtechnique (CSEM) and the Swiss Priority Program in Micro and Nano System Technology (MINAST).

The authors are with the Swiss Federal Institute of Technology, Lausanne CH-1015, Switzerland.

Publisher Item Identifier S 0018-9200(00)00541-2.

[1]This is usually the case since the phase noise requirement of such systems is greatly relaxed compared to mainstream applications such as GSM.

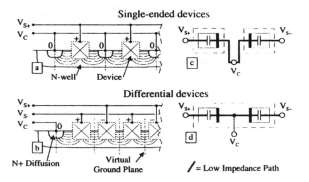

Fig. 1. Principle of "differential" varactors. (a) and (b) Structure of usual, single-ended and proposed, differential devices placed in an N-well. The thin curves symbolize electric field lines. (c) and (d) Connection of both types of devices for differential operation. Elements with a thick black line are voltage dependent.

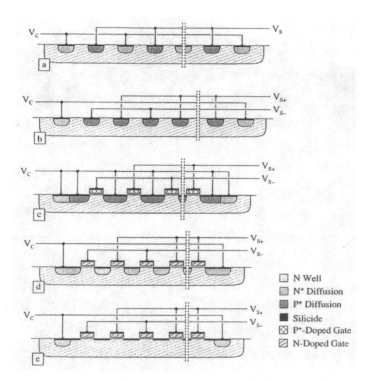

Fig. 2. Structure of different varactors. (a) and (b) "Single-ended" and "differential" diodes. Differential MOS capacitors optimized for (c) inversion mode and (d), (e) accumulation mode.

where P is the nominal power consumption and ω_0 the working frequency. For instance, with $Q_L = Q_F = 40$ and $\zeta = 1.5$, the ratio $P/P_{\min}$ varies between 2 and 2.5 when Q_V goes from 100 down to 40. This simple example clearly illustrates the benefit of using very high Q varactor structures like those that will be discussed in the next sections.

From these considerations, it can be stated that the varactors should simultaneously offer:

- a high quality factor;
- a control voltage range compatible with the supply voltage, ultimately 1 V for single battery cell operation;
- a good tunability over the available control voltage range;
- a small silicon area, to reduce cost;
- a reasonably uniform capacitance variation over the available control voltage range, to make the phase-locked-loop design easier.

Lately, different solutions to this problem have been proposed. Some of them are nonstandard as they involve low-loss switches [7], which are inconsistent with low-voltage operation, or micromechanical devices [8]. Although some other authors focused on standard CMOS [9]–[11], specific solutions addressing micropower transceivers have still not been studied systematically.

II. VARACTOR STRUCTURES AVAILABLE IN CMOS

A. General Considerations

Two classes of devices have to be considered: junctions and MOS capacitors, the latter being operated either in inversion or in accumulation mode. In all cases, the devices should be placed in separate wells in order to be able to use the well potential as the tuning voltage. The two types of devices are therefore p^+/n^- diodes and PMOS capacitors. Consequently, electrical losses are expected to occur mostly in the relatively low-doped n-well.

So far, no solution taking advantage of the differential nature of the signal generated by a balanced LC oscillator has been proposed. Nevertheless, this simple consideration leads to unexpectedly good performances, whether in terms of quality factor, tunability, or silicon area. Moreover, this solution is applicable to both of the above-mentioned types of devices.

As shown in Fig. 1(a), the minimization of the length of the current lines in the well leads to interleave devices and ac grounds. However, Fig. 1(b) clearly shows that, for symmetry reasons, the path to the small-signal ground can still be shortened if devices with opposite polarity alternate. Moreover, this layout increases density and minimizes the number of critical low-impedance interconnections [Fig. 1(c) and (d)]. These paths are critical because their series resistance should be much smaller than $1/\omega_0 C_V \cdot Q_V$ (i.e., even less than 1 Ω for common designs) in order not to deteriorate the resulting quality factor.

B. "Varicap" Diodes

Varactors have been traditionally realized as reverse-biased diodes [11]. Usual single-ended and novel differential p^+/n^- well structures can be found in Fig. 2(a) and (b). An electrical model for this kind of device is given in Fig. 4(a), where C_D is the junction depletion capacitance controlled by the dc voltage $V_{\text{Ctrl}} = V_S - V_C$, V_S and V_C being the tank and control electrode potentials, respectively. C_0 represents additional constant capacitance due to interconnections, and g_D is the diode conductance. G_B is the N-well conductance of the lateral path from the junction to the small-signal ground. The latter can either be an N+ diffusion, for the single-ended case, or a virtual ground plane, for the balanced topologies [see Fig. 1(a) and (b)]. C_D and g_D can be approximated from usually available technological parameters (see Appendix I-A).

If g_D is kept negligible, as it should be during normal operation, the quality factor is proportional to G_B, which is assumed to be voltage independent (Appendix I-A). For a standard comb layout (Fig. 3(a)), the quality factor and the specific capacitance (i.e., the capacitance per total occupied silicon area)

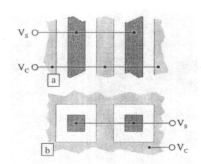

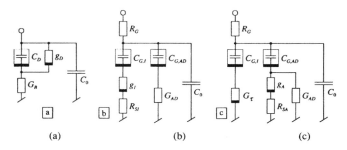

Fig. 3. (a) Comb versus (b) concentric diode structure (single-ended varactor).

Fig. 4. Small-signal equivalent circuit for different varactor types: (a) junction, (b) inversion-mode MOS, and (c) accumulation-mode MOS. Elements with a thick black line are voltage controlled. For "differential" structures, the equivalence presupposes that the feeding is symmetric.

are approximately doubled by using a differential structure. For single-ended devices, a concentric layout can be used [Fig. 3(b)] to significantly lower G_B, at the expense of a very small resulting specific capacitance.

It should be noted that the practical capacitance variation is poor because the steepest change is obtained by forward-biasing the diode. This is limited by the degradation of the quality factor due to parallel conductance g_D and signal clamping. Moreover, for an oscillator swing in excess of a few $U_T = k \cdot T/q_e$, the capacitance excursion is limited by averaging effects over the swing range.

C. Inversion-Mode MOSFET's

For some time, MOSFET capacitor behavior has been successfully exploited between strong inversion and depletion [7], [9]–[11]. The differential structure is still available [Fig. 2(c)], but the improvement is limited because P$^+$ diffusions have still to be inserted between devices, in order to provide enough minority carriers to create the channel. However, thanks to electrical symmetry, source/drain contact and wire resistances are no more critical and diffusions do not have to be contacted all along the device width, allowing to make them narrower. For the same reason, the density of bulk contacts can be reduced, even for depletion-mode operation. Consequently, balanced differential devices are still somewhat better and denser than the single-ended one.

The achievable performances are mainly limited by two factors.

1) In weak and moderate inversion region, the quality factor is restricted by the channel resistance.
2) The effective tunability is limited by the overlap capacitance between the gates and the LDD regions. More-

over, most of the capacitance variation occurs in a narrow range, theoretically a few U_T.

The equivalent circuit is shown in Fig. 4(b). $C_{G,I}$ and $C_{G,\mathrm{AD}}$ are the MOS capacitance related to minority and majority carrier, respectively (see Appendix I-B for a more comprehensive description of the model). C_0 is related to the fixed parasitic capacitance, including interconnections and overlaps. g_I is the equivalent inversion layer conductance, which is related to the ON-resistance of the MOSFET, $g_I = 12/R_{\mathrm{ON}}$. Factor 12 is due to the distributed nature of the channel. R_{SI} accounts for series resistances in the diffusion regions. R_G is the equivalent gate series resistance, which is kept low by the silicide layer. As previously, assuming the depletion width is negligible, the bulk conductance G_{AD} can be considered as constant.

When the channel is formed, it should be noted that both R_{ON} and $C_{G,I}$ increase proportionally to the effective channel length L_{eff}, so that the quality factor is inversely proportional to L_{eff}^2. Therefore, the length should be kept minimum.

The typical behavior of this model is sketched in the leftmost part of Fig. 5. Region 5 corresponds to strong accumulation. The capacitance is saturated to C_{OX} and the quality factor is mainly dependent on G_{AD}. Regions 4 and 3 correspond to the weak accumulation and depletion region, where Q is high because the capacitance is low. In region 2, the inversion layer is developing and its conductivity is still low while the capacitance rapidly increases, producing the dip in the Q curve. Last, in region 1, the channel is deeply inverted, so that the capacitance is constant and the quality factor is mostly limited by R_G and R_{SI}.

D. Accumulation-Mode MOSFET's

The use of PMOS capacitors in accumulation mode is relatively uncommon, although it exhibits very good characteristics. Fig. 2(d) and (e) shows the new proposed structure. The underlying idea is that no minority carrier source is necessary in accumulation mode. Therefore, gates with inverse ac polarity can be placed very close to each other, either without any interpolated structure [Fig. 2(e)] or with interleaved N+ diffusions [Fig. 2(d)]. The first solution limits the parasitic capacitance by avoiding overlaps, whereas the second results in a smaller resistance in series with the accumulation layer. Because of the intrinsically higher electron mobility, both solutions still lead to devices with good characteristics.

At this point, two phenomena should be pointed out. First, the gate of a functional PMOS transistor is heavily P-type doped by the source/drain implants. As this process step must be omitted [Fig. 2(e)], or even replaced by an opposite N-type implant step [Fig. 2(d)], the equivalent threshold voltage of the MOSFET structure will be shifted toward higher absolute value. This is usually advantageous, since it centers the control voltage range near 0 V, allowing easier operation at low supply voltage. Note that, especially if no implant is done [Fig. 2(e)], the final polysilicon doping may be relatively low, depending on the process used, and results in a partial gate depletion, which would tend to reduce the effective capacitance variation. This effect could be very strong for deep-submicrometer processes. Second, in accumulation mode, the surface conductivity of the whole device will be significantly higher than the N-well underneath, and almost all current will flow near the surface. This will still partly

150

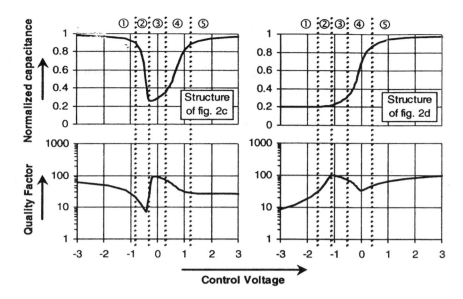

Fig. 5. Schematic plot of the specific capacitance and quality factor qualitative behavior for both inversion and accumulation mode MOS devices, with $C_0 = 0$. The different regions of operation are denoted by the circled numbers.

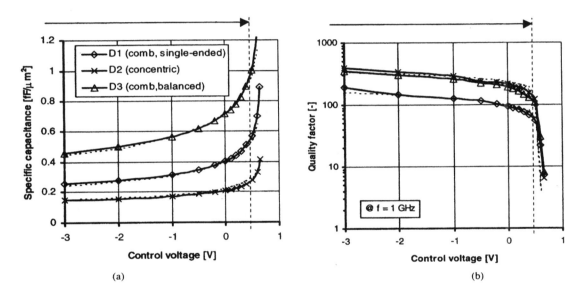

(a)   (b)

Fig. 6. Measurement of the specific capacitance (i.e., the capacitance per total occupied die area) and quality factor as a function of the bias voltage, at 1 GHz, for junction varactors with different shapes. D1 correspond to a comb-shaped diode, D2 to a concentric diode, and D3 to a "differential" diode. The horizontal arrows indicate the useful range of control voltage. The dashed curves refer to the model of Appendix A.

hold if no interstitial diffusions are used [Fig. 2(e)] because most technologies used for RF circuits are salicided, and hence, the interstices between the gates will receive highly conductive silicide.

The equivalent circuit is shown in Fig. 4(c). It is similar to Fig. 4(b), but the inversion layer is not surrounded by P+ diffusions, so that minority carrier density can only change by a generation/recombination process, which is much slower than the operating frequency. A value of conductance G_τ much smaller than $\omega \cdot C_{G,I}$, accounts for this phenomenon. When the inversion layer is formed, $C_{G,I}$ may almost be left out of the equivalent circuit. The equivalent accumulation layer conductance g_A is defined in a similar way as g_I. The lateral resistance R_{SA} results from the series resistance in the well [Fig. 2(e)] or in the LDD re-

gions [Fig. 2(d)]. Again, G_{AD} account for the in-depth conductance of the well. The derivation of expressions for these parameters can be found in Appendix I-B. Clearly, minimum-length devices will again maximize Q.

The rightmost part of Fig. 5 illustrates the behavior of the model. Starting from region 5, where the device is in strong accumulation, the accumulation layer is vanishing, and its conductance decreases in region 4. This may lead to a dip in the quality factor curve, which is similar to the weak inversion dip of the previous section. The inversion layer forming in region 2 is only connected to the ground through G_τ. Therefore, it has a small effect on the capacitance but still severely reduces Q. In region 1, G_τ continues to rise, increasingly degrading the quality factor.

151

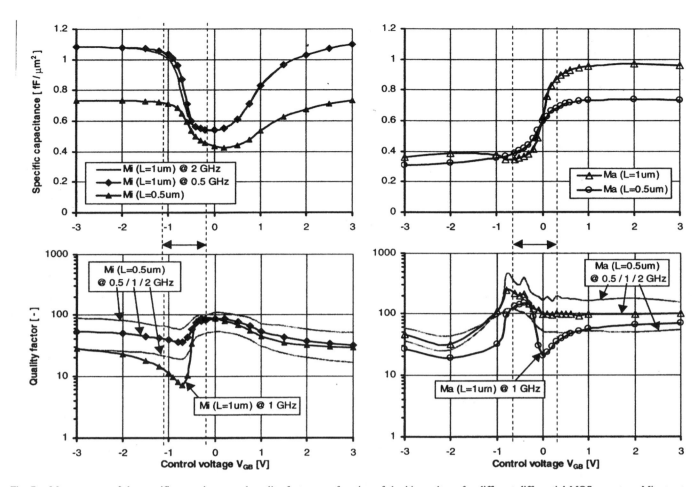

Fig. 7. Measurement of the specific capacitance and quality factor as a function of the bias voltage for different differential MOS varactors. Mi: structures optimized for inversion mode, 1- and 0.5-μm length. Ma: structures optimized for accumulation mode, again 1- and 0.5-μm length. The dashed lines and horizontal arrows indicate the useful range of control voltage.

III. EXPERIMENTAL RESULTS

A. Measurement Setup

All previously discussed varactors have been realized in a standard 0.5-μm digital CMOS process: single-ended comb diodes (D1, Section II-B), single-ended concentric diodes (D2), differential comb diodes (D3), two differential inversion-mode MOSFET's with 1- and 0.5-μm gate length (Mi, Section II-C), and two differential accumulation-mode MOSFET's, with 1- and 0.5-μm gate length as well (Ma, Section II-D). Two additional dummy structures (short/open) were added for deembedding purposes. The measurements were performed from 300 MHz to 13 GHz by using a HP 8719D 13.5-GHz network analyzer and a Cascade prober equipped with two 40-GHz Cascade Microtech Air Coplanar RF probes. For differential devices, a two-port analysis was performed (the two ports being V_{S+} and V_{S-}), and the complete S-parameter matrix was measured. From this information, the equivalent S parameter for the single-ended structure was computed.

Great care has to be taken for these measurements, since at low frequencies ($f < 2$ GHz), the losses are very small and therefore extremely difficult to extract. Therefore, the structures were sized for maximum precision at higher frequencies (3–5 GHz). Consequently, the loss curves become very noisy under 1 or 2 GHz. To avoid this, the quality factor at low fre-

quencies was extrapolated by assuming that the equivalent circuit resistors and conductances of Fig. 4(a)–(c) were independent of frequency. This latter assumption appeared to be consistent with all measurements up to 5 GHz.

B. Measurement Results

Measurements of the seven devices are summarized in Figs. 6 and 7. As expected, all extracted capacitance values were found to be independent of frequency up to more than 5 GHz, except for the 1-μm-long inversion-mode MOSFET (Fig. 7, top-left plot), which is already slightly affected at 2 GHz by non-quasi-static effects. For all devices, the quality factor is almost inversely proportional to frequency in the 300-MHz–3-GHz range of interest. Note that the specific capacitance plotted in these figures is normalized to the total silicon area occupied by the device, corresponding approximately to the area of the well including the structure.

The plots of Fig. 6 show that all reverse-biased diodes have a quality factor higher than 100 at 1 GHz. The sharp fall in the curve for positive control voltage is due to the parallel conductance. It limits the available capacitance variation ξ down to a factor smaller than two, even for signal swings as small as 100-mV peak and for control swing higher than 3 V. The standard comb diodes have the lowest quality factor, the differential

152
152

TABLE I
COMPARISON OF THE MEASURED CHARACTERISTICS OF DIFFERENT STRUCTURES

Device type, Connection type, Layout	Control voltage swing	Tunability ξ	Central specific capacitance $\sqrt{C'_{V,min} \cdot C'_{V,max}}$	Slope non-uniformity $\dfrac{max(dC'_V / dV_{Ctrl})}{min(dC'_V / dV_{Ctrl})}$	Min. quality factor over the useful tuning range, at 1GHz
Junction (D1), single-ended, comb	0.9 V (−0.5 to 0.4 V) 2.4 V (−2.0 to 0.4 V)	1.46 1.83	0.42 fF/μm² 0.39 fF/μm²	8.0 16.3	69
Junction (D2), single-ended, concentric	0.9 V (−0.5 to 0.4 V) 2.4 V (−2.0 to 0.4 V)	1.37 1.61	0.22 fF/μm² 0.20 fF/μm²	8.2 16.7	130
Junction (D3), differential, comb	0.9 V (−0.5 to 0.4 V) 2.4 V (−2.0 to 0.4 V)	1.45 1.82	0.74 fF/μm² 0.70 fF/μm²	8.2 16.8	155
Inv.-mode MOS (Mi, L = 1μm), differential, comb	0.6V (−1.0 to −0.4V)	1.83	0.76 fF/μm²	9	7
Inv.-mode MOS (Mi, L = 0.5μm), differential, comb	0.8V (−1.0V to −0.2V)	1.56	0.57 fF/μm²	4.6	36
Accu.-mode MOS (Ma, L = 1μm), differential, comb	0.8V (−0.4V to 0.4V)	2.45	0.58 fF/μm²	9.4	20
Accu.-mode MOS (Ma, L = 0.5μm), differential, comb	1V (−0.5V to 0.5V)	1.76	0.53 fF/μm²	4.0	95

structure being about twice better. As expected, the concentric one is the best. Concerning tunability and specific capacitance, the differential solution is clearly the best, while the concentric structure is rather poor, mainly because of a small $C_{V,min}/C_0$ ratio.

The leftmost plots of Fig. 7 show the behavior of MOSFET capacitors optimized for inversion mode. The minimum-length structure achieves a quality factor always better than 40 at 1 GHz, for a capacitance variation $\xi \approx 1.6$ and a control voltage swing of 0.8 V. Performances are limited simultaneously by the channel resistance in weak inversion mode (see the dip in the quality factor plot) and by the excessive overlap capacitance contribution.

Last, the rightmost charts show devices exploiting charge accumulation [Fig. 2(d)]. In order to avoid the quality factor dip in "weak accumulation," the structure with the shortest gate length (0.5 μm) should be used. Due to the structure compactness and the high carrier mobility, the achieved quality factor is higher than 200, 100, or 50 over the useful tuning range, at 0.5, 1 or 2 GHz, respectively. The tunability ξ is about 1.8 for a 1-V control voltage swing. Moreover, unlike diodes, this kind of device experiences no risk of signal clamping. These results satisfy the requirements of the aimed applications.

All these results, together with some complementary figures, are summarized in Table I.

IV. CONCLUSION

Although different topologies are available to realize electrically tunable capacitance in a standard CMOS process, it has been shown that improved overall characteristics can be achieved by taking advantage of the differential signals provided by common balanced LC-tank oscillators.

The use of relevant layout techniques allowed to design varactors well suited for low-power wireless applications operating in the 300 MHz–3 GHz range. Using a 0.5-μm technology, the measured quality factors are higher than 100 at 1 GHz, while the

effective capacitance can be controlled between 0.35 and 0.7 fF per square micrometer of occupied silicon area, for a control voltage varying between −1 and +0.5 V. Moreover, improvements are still to be expected with the reduction of feature size resulting from current technology downscaling.

Last, although a compact model fitting all modes of operation of a MOS varactor is difficult to obtain, mainly because of the doping nonuniformity, lumped element equivalent circuits have been proposed, which allow one to predict the behavior of the different structures with sufficient accuracy.

APPENDIX I
MODELS

A. Junction Varactors

In the electrical model of Fig. 4(a), C_D and g_D can be approximated from usually available technological parameters (typewriter-style symbols) [12]

$$C_D = \frac{CJ \cdot A_{eq}}{\left(1 - \frac{V_{ctrl}}{PB}\right)^{MJ}} + \frac{CJSW \cdot P_{eq}}{\left(1 - \frac{V_{ctrl}}{PBSW}\right)^{MJSW}} \quad (5)$$

$$g_D = \frac{JS \cdot A_{eq} + JSW \cdot P_{eq}}{N \cdot U_T} \cdot e^{\frac{-V_{ctrl}}{N \cdot U_T}} \quad (6)$$

where U_T is the thermodynamic voltage, while A_{eq} and P_{eq} are the equivalent diode area and perimeter. Assuming that g_D is negligible, the quality factor is proportional to G_B, which is nearly constant as long as the width of the depletion regions is small relative to the mean current line length. It can be estimated (~±30%) if the well doping profile is known and by making judicious approximations on the current line shape. The dashed lines of Fig. 6 results from these expressions and are almost indistinguishable from the measured curves, ensuring the relevance and accuracy of the model.

For a differential device, the equivalence of Fig. 4 is valid only if the RF signal is perfectly balanced. In case of unbalance, the additional elements of Fig. 8 should be considered because node

W is no longer an RF ground. The well-substrate capacitor C_W is associated with a conductance G_W, which tends to degrade the global Q. However, for a slight unbalance, not higher than 10%, this effect is negligible because C_W is relatively small (the doping levels are low on both side of the junction, so that the depletion region is wide) and G_W is reasonably high (carriers are electrons). In order to avoid any losses in the control voltage conductance G_C, the control terminal should be left at a sufficiently high impedance level. Note that these considerations still hold for MOS differential devices.

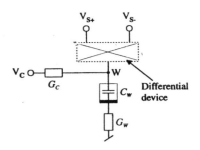

Fig. 8. Effect of the well-substrate capacitance and control voltage input conductance for unbalanced RF signals.

B. MOS Transistor Varactors

The full behavior of the MOS channel is derived from the physical MOS capacitor model [13], [14]. The specific gate capacitance C'_G is given in implicit form, as a function of the surface potential[2] Ψ_S

$$C'_S = \gamma C'_{\mathrm{OX}} \cdot \underbrace{\frac{(1 - e^{-\Psi_S/U_T}) \cdot e^{2\Phi_F/U_T}}{2 \cdot F(\Psi_S)}}_{C'_I}$$

$$+ \gamma C'_{\mathrm{OX}} \cdot \underbrace{\frac{e^{\Psi_S/U_T} - 1}{2 \cdot F(\Psi_S)}}_{C'_{\mathrm{AD}}} \qquad (7)$$

$$C'_G = \frac{C'_S \cdot C'_{\mathrm{OX}}}{C'_S + C'_{\mathrm{OX}}} = \underbrace{\frac{C'_I \cdot C'_{\mathrm{OX}}}{C'_S + C'_{\mathrm{OX}}}}_{C'_{G,I}} + \underbrace{\frac{C'_{\mathrm{AD}} \cdot C'_{\mathrm{OX}}}{C'_S + C'_{\mathrm{OX}}}}_{C'_{G,\mathrm{AD}}} \qquad (8)$$

$$V_{\mathrm{GB}} = \Psi_S + V_{\mathrm{FB}} + \gamma \cdot F(\Psi_S) \qquad (9)$$

with (10) and (11), shown at the bottom of the page, where C'_S is the total specific capacitance of the silicon region under the thin oxide, which can be split in a contribution from the minority carriers (inversion, C'_I) and another from the majority carriers (accumulation and depletion, C'_{AD}). The gate capacitance $C'_G = C'_{G,I} + C'_{G,\mathrm{AD}}$ is equal to C'_S in series with the thin oxide capacitance C'_{OX} and can be split similarly to C'_S. N_B is an equivalent well doping level, including the effect of nonuniform doping along the channel. V_{GB} is the gate potential referred to the well, V_{FB} is the flat-band voltage, ϕ_F is the well Fermi potential, n_i is the intrinsic carrier density of silicon, ε_S is the dielectric constant of silicon, and q_e is the electron charge. If polydepletion is not negligible, when $\psi_S > 0$ for an n-doped gate or when $\psi_S < 0$ for a p-doped gate, (8) and (9) should be replaced by

$$C'_G = \frac{C'_S \cdot C'_{\mathrm{OX,pd}}}{C'_S + C'_{\mathrm{OX,pd}}}$$

$$\text{with } C'_{\mathrm{OX,pd}} = \frac{C'_{\mathrm{OX}}}{1 + \frac{2\gamma}{\gamma_G^2}|F(\Psi_S)|}, \quad \gamma_G = \frac{\sqrt{2\varepsilon_S q_e N_G}}{C'_{\mathrm{OX}}} \qquad (12)$$

$$V_{\mathrm{GB}} = \Psi_S + V_{\mathrm{FB}} + \gamma \cdot F(\Psi_S) + \operatorname{sign} \Psi_S \cdot \left(\frac{\gamma}{\gamma_G} F(\Psi_S)\right)^2 \qquad (13)$$

where N_G is the equivalent gate doping level and assuming total depletion in the gate. The equivalent circuit of the inversion- and accumulation-mode devices is shown in Fig. 4(b) and (c). $C_{G,I}$ and $C_{G,\mathrm{AD}}$ are in series with the equivalent channel conductance g_I and g_A, which are associated with both type of carrier. The latter can be computed from the channel surface conductivity σ_I and σ_A

$$\sigma_I = \mu_p \cdot |Q'_I| \text{ with } |Q'_I| = \begin{cases} \int_{\Psi_S}^{0} C'_I(\Psi) \cdot d\Psi, & \Psi_S < 0 \\ 0, & \Psi_S \geq 0 \end{cases} \qquad (14)$$

$$\sigma_A = \mu_n \cdot |Q'_A| \text{ with } |Q'_A| = \begin{cases} \int_{0}^{\Psi_S} C'_{\mathrm{AD}}(\Psi) \cdot d\Psi, & \Psi_S > 0 \\ 0, & \Psi_S \leq 0 \end{cases} \qquad (15)$$

$$g'_I = \frac{\sigma_I}{12 \cdot L_{\mathrm{eff}}^2} \text{ and } g'_A = \frac{\sigma_A}{12 \cdot L_{\mathrm{eff}}^2} \qquad (16)$$

where Q'_I and Q'_A are the surface charge densities of the inversion and accumulation layers, μ_n and μ_p are the respective mobilities of electrons and holes; g'_I and g'_A are defined per unit of channel active area, and the factor 12 in their expression assumes that the channel is connected at both ends and takes into

[2]The sign convention adopted here is valid only for a P-type MOS. In this case, when the transistor is in inversion mode, ψ_S and V_{GB} are negative.

$$\gamma = \frac{\sqrt{2\varepsilon_S q_e N_B}}{C'_{\mathrm{OX}}}, \quad \phi_F = -U_T \ln \frac{N_B}{n_i} \qquad (10)$$

$$F(\Psi) = \operatorname{sign} \Psi \cdot \sqrt{U_T \cdot e^{2\Phi_F/U_T} \cdot (e^{-\Psi/U_T} - 1) + U_T \cdot (e^{\Psi/U_T} - 1) - \Psi \cdot (1 - e^{2\Phi_F/U_T})} \qquad (11)$$

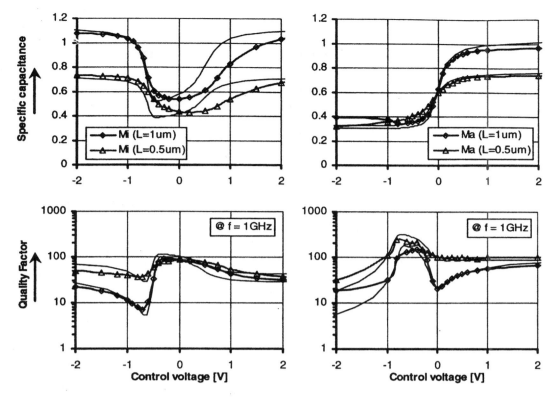

Fig. 9. Comparison between measurements (thick grayed curves with symbols) and the simplified model (thin curves) of Appendix B.

account its distributed nature. These relations presuppose that the quality factor associated with the channel is sufficiently high or, equivalently, that the quasi-static assumption is valid. This condition is required to use equivalent lumped elements as in Fig. 4.

R_{SI} and R_{SA} account mostly for the diffusion series resistance. G_{DA} is the bulk conductance underneath the channel and can be estimated similarly to G_B in Appendix I-A. G_τ is the conductance associated with the recombination of minority carriers and is very small compared to $\omega \cdot C_{G,I}$, because the carrier lifetime is much larger than the signal period. The value of G_τ is difficult to infer from theoretical considerations, but by similarity with the reverse-biased diode discussed in [13], it is assumed to be proportional to $|Q_I| \cdot \sqrt{\omega}$. The exact value is not critical, as it only appears for the accumulation-mode device, outside the normal operating range.

Fig. 9 shows the comparison between the model and the measured curves. Because some parameters such as G_{AD} or G_τ are difficult to evaluate, some discrepancies from measurements remain. The main disagreement is visible in the top-left graph, where the capacitance of the inversion-mode structure is plotted. The variation due to the building of the inversion layer is clearly steeper in the model, especially for the shorter channel device. This difference is input into the severe nonuniformity of the doping level through the channel, mostly as a result of the overlaps with the LDD diffusions. The same effect is assumed to delay and stretch the forming of the accumulation layer for positive control voltage. However, apart from this phenomenon and in spite of its simplicity, this model is found to explain the behavior of the different type of varactors reasonably well.

ACKNOWLEDGMENT

The authors would like to thank F. Théodoloz, Dr. M. Salese, Prof. P. Fazan, and Prof. M. Kayal for helpful technical discussion. They are equally indebted to Dr. D. Bouvet for his expertise in CMOS processes.

REFERENCES

[1] T. Melly, A.-S. Porret, C. C. Enz, and M. Kayal, "A 1.3 V low-power 430 MHz front-end using a standard digital CMOS process," in *Proc. IEEE Custom Integrated Circuit Conf.*, Santa Clara, CA, May 1998, pp. 503–506.
[2] G. Asada, M. Dong, T. S. Lin, F. Newberg, G. Pottie, and W. J. Kaiser, "Wireless integrated network sensors: Low power systems on a chip," in *Proc. Eur. Solid-State Circuits Conf.*, Sept. 1998, pp. 9–16.
[3] M. Pardoen, J. Gerrits, and V. von Kaenel, "A 0.9 V, 1.2 mA, 200 MHz BiCMOS single-chip narrow-band FM receiver," in *Proc. ISSCC*, 1996, pp. 348–349.
[4] A. Rofougaran, J. Y.-C. Chang, M. Rofougaran, and A. Abidi, "A 1 GHz CMOS RF front-end IC for direct-conversion wireless receiver," *IEEE J. Solid-State Circuits*, vol. 31, pp. 880–889, July 1996.
[5] J. Crols and M. S. J. Steyaert, "Low-IF topologies for high-performance analog front ends of fully integrated receivers," *IEEE Trans. Circuits Syst. II*, vol. 45, pp. 269–282, March 1998.
[6] J. R. Long and M. A. Copeland, "The modeling, characterization, and design of monolithic inductors for silicon RF IC's," *IEEE J. Solid-State Circuits*, vol. 32, pp. 357–369, Mar. 1997.
[7] A. Kral, F. Behbahani, and A. A. Abidi, "RF CMOS oscillators with switched tuning," in *Proc. IEEE Custom Integrated Circuit Conf.*, Santa Clara, CA, May 1998, pp. 503–506.
[8] C. T.-C. Nguyen, "Micromechanical devices for wireless communications," in *Proc. IEEE Int. Workshop Micro Electro Mechanical Systems*, Jan. 1998, pp. 1–7.
[9] C.-M. Hung, Y. C. Ho, I.-C. Wu, and K. O. , "High-Q capacitors implemented in a CMOS process for low-power wireless applications," *IEEE J. Solid-State Circuits*, vol. 46, pp. 505–511, May 1998.

[10] R. Castello, P. Erratico, S. Manzini, and F. Svelto, "A ±30% tuning range varactor compatible with future scaled technologies," in *VLSI Symp. Circuits Dig. Tech. Papers*, June 1998, pp. 34–35.

[11] P. Andreani, "A comparison between two 1.8 GHz CMOS VCO's tuned by different varactors," in *Proc. 22nd Eur. Solid-State Circ. Conf. (ESSCIRC)*, The Hague, the Netherlands, Oct. 1998, pp. 380–383.

[12] M. Bucher, C. Lallement, C. C. Enz, F. Théodoloz, and F. Krummenacher. (1998, July) The EPFL-EKV MOSFET equations for simulation version 2.6, revision II, EPFL. [Online]. Available: http://legwww.epfl.ch/ekv/model.html

[13] S. M. Sze, *Physics of Semiconductor Devices*. New York: Wiley, 1981.

[14] Y. P. Tsividis, *Operation and Modeling of the MOS Transistor*, 2nd ed. New York: McGraw Hill, 1999.

Alain-Serge Porret (S'97) was born in Saint-Aubin, Switzerland, in 1971. He received the M.S. degree in electrical engineering from the Swiss Federal Institute of Technology (EPFL), Lausanne, in 1996. He is currently pursuing the Ph.D. degree at the Electronics Laboratories, EPFL.

His current interests are in low-power, fully integrated, CMOS RF circuits, with an emphasis on oscillators and frequency synthesizers. His dissertation is on circuit design for wireless integrated network sensors.

Thierry Melly was born in Ayer, Switzerland, on June 12, 1970. He received the M.S. degree in electrical engineering from the Swiss Federal Institute of Technology (EPFL), Lausanne, in 1996. He is currently pursuing the Ph.D. degree at the Electronics Laboratories, EPFL.

His dissertation is on the implementation of low-power and low-voltage circuits for UHF-band transceivers in CMOS technology.

Christian C. Enz (M'84) received the M.S. and Ph.D. degrees in electrical engineering from the Swiss Federal Institute of Technology (EPFL), Lausanne, in 1984 and 1989, respectively.

From 1984 to 1989, he was a Research Assistant with EPFL, working in the field of micropower analog CMOS integrated circuits (IC) design. In 1989, he was one of the Founders of Smart Silicon Systems S.A. (S3), where he developed several low-noise and low-power IC's, mainly for high-energy physics application. From 1992 to 1997, he was an Assistant Professor at EPFL, working in the field of low-power analog CMOS and BiCMOS IC design and device modeling. From 1997 to 1999, he was Principal Senior Engineer at Conexant (formerly Rockwell Semiconductor Systems), Newport Beach, CA, where he was responsible for the modeling and characterization of MOS transistors for the design of RF CMOS circuits. In 1999, he joined the Swiss Center for Electronics and Microtechnology (CSEM), where he was heading the RF and Analog IC design group and recently was promoted to Head of the Advanced Microelectronics Department. He is also an Adjunct Professor at EPFL. His technical interests and expertise are in the field of very low-power analog and RF IC design and MOS transistor modeling. He is the author or co-author of more than 60 scientific papers and has contributed to numerous conference presentations and advanced engineering courses.

Eric A. Vittoz (A'63–M'72–SM'87–F'89) received the M.S. and Ph.D. degrees in electrical engineering from the Swiss Federal Institute of Technology (EPFL), Lausanne, in 1961 and 1969, respectively.

He joined the Centre Electronique Horloger S.A. (CEH), Neuchâtel, Switzerland, in 1962, where he participated in the development of the first prototypes of electronic watches. In 1971, he was appointed Vice-Director of CEH, supervising advanced developments in electronic watches and other micropower systems. In 1984, he took the responsibility of the Circuits and Systems Research Division of the Swiss Center for Electronics and Microtechnology (CSEM) in Neuchâtel, where he was Executive Vice-President, Integrated Circuits and Systems, from 1991 to 1997, and Advanced Microelectronics later on, after the spinoff of the industrial part of these activities as XEMICS SA. He is now partially retired from CSEM, with the position of Chief Scientist. His field of personal research interest and activity is the design of low-power analog CMOS circuits, with emphasis on their application to advanced perceptive processing. Since 1975, he has been lecturing and supervising undergraduate and graduate student projects in analog circuit design at EPFL, where he became a Professor in 1982. He has published more than 120 papers and has received 26 patents.

On the Use of MOS Varactors in RF VCO's

Pietro Andreani and Sven Mattisson

Abstract—This paper presents two 1.8-GHz CMOS voltage-controlled oscillators (VCO's), tuned by an inversion-mode MOS varactor and an accumulation-mode MOS varactor, respectively. Both VCO's show a lower power consumption and a lower phase noise than a reference VCO tuned by a more commonly used diode varactor. The best overall performance is displayed by the accumulation-mode MOS varactor VCO.

The VCO's were implemented in a standard 0.6-μm CMOS process.

Index Terms—CMOS, radio frequency, varactors, VCO's.

I. INTRODUCTION

THE STUDY OF voltage-controlled oscillators (VCO's) has in the past few years attracted a great deal of attention in the solid-state community, leading both to a deeper understanding of theoretical issues [1]–[5], and to a considerable advancement in the state-of-the-art of the design of integrated VCO's [6]–[8]. The implementation of monolithic VCO's in standard CMOS technologies is justly felt as one of the major challenges that must be overcome in the design of integrated radio-frequency (RF) CMOS transceivers. Recently, a frequency synthesizer almost meeting the very demanding phase-noise specifications of the DCS-1800 communication standard has been presented [9]. The cited work shows that the poor quality factor Q of the reverse-biased diode varactor used to tune the VCO prevents compliance with the phase-noise specifications over the whole tuning range. In fact, even in BiCMOS processes, where the base-collector junction of the bipolar transistor can be exploited, diode varactors with Q's lower than ten have been measured in the lower gigahertz region [10].[1]

Since the diode varactor leaves much to be desired, it is interesting to investigate the behavior of another device, readily available in any CMOS process, having a capacitance value tunable by a voltage: the MOS transistor itself. VCO's based on MOS varactors have been presented in [12]–[15], while both MOS and diode varactors have been used in [16]. A VCO tuned solely via the back-gate voltage of the active pMOS transistors

of the VCO, thus disposing of dedicated varactor devices, has been demonstrated [17]. The tuning range for this VCO is, however, only about 3%.

A previous comparison [18] between two RF CMOS VCO's tuned by a diode varactor and a pMOS varactor, respectively, showed that the diode varactor was still the best choice for the varactors and technology used (a standard 0.8-μm CMOS process).

This work presents the experimental results of the use of two different MOS varactors as tuning components for RF VCO's, and it shows that both varactors allow for a better overall VCO performance than the more commonly used diode varactor.

II. MOS VARACTOR

It is well known that an MOS transistor with drain, source, and bulk (D, S, B) connected together realizes an MOS capacitor with capacitance value dependent on the voltage V_{BG} between B and gate (G). In the case of a pMOS capacitor, an inversion channel with mobile holes builds up for $V_{BG} > |V_T|$, where $|V_T|$ is the threshold voltage of the transistor. The condition $V_{BG} \gg |V_T|$ guarantees that the MOS capacitor works in the strong inversion region, the region where the MOS device shows a transistors behavior. On the other hand, for some voltage $V_G > V_B$, the MOS device enters the accumulation region, where the voltage at the interface between gate oxide and semiconductor is positive and high enough to allow electrons to move freely. Thus, in both strong inversion and accumulation region the value of the MOS capacitance C_{mos} is equal to $C_{ox} = \epsilon_{ox} S / t_{ox}$, where S and t_{ox} are the transistor channel area and the oxide thickness, respectively.

Three more regions can be distinguished for intermediate values of V_{BG}: moderate inversion, weak inversion, and depletion [19]. In these regions there are few or very few mobile charge carriers at the gate oxide interface, which causes a decrease of the capacitance C_{mos} of the MOS device (i.e., C_{mos} is less than C_{ox}). C_{mos} can now be modeled as C_{ox} in series with the parallel of capacitances C_b and C_i. C_b accounts for the modulation of the depletion region (made of ionized donor atoms) below the oxide, while C_i is related to the variation of the number of holes at the gate oxide interface. If C_b (C_i) dominates, the MOS device is working in the depletion (moderate inversion) region; if neither capacitance dominates, the MOS device is working in the weak inversion region. The behavior of C_{mos} versus V_{BG} is reproduced qualitatively in Fig. 1.

Manuscript received July 20, 1999; revised December 10, 1999.

P. Andreani is with the Department of Applied Electronics, Lund University, SE-221 00 Lund, Sweden (e-mail: pietro.andreani@tde.lth.se).

S. Mattisson is with Ericsson Mobile Communications AB, Mobile Phones & Terminals, SE-221 83 Lund, Sweden (e-mail: sven.mattisson@ecs.ericsson.se).

Publisher Item Identifier S 0018-9200(00)04460-7.

[1]Better results have been obtained in a customized SiGe process [11].

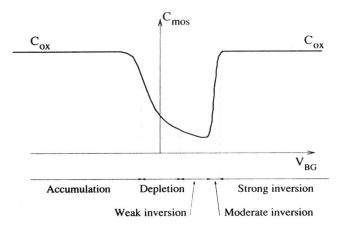

Fig. 1. Tuning characteristics for the pMOS capacitor with B ≡ D ≡ S.

A. Parasitic Resistance

It is easy to show [15] that a good approximation for the parasitic resistance associated with the pMOS channel working in strong inversion is given by

$$R_{\mathrm{mos}} = \frac{L}{12k_p W(V_{\mathrm{BG}} - |V_{\mathrm{T}}|)} \qquad (1)$$

where W, L, and k_p are the width, length, and gain factor of the pMOS transistor. According to (1), R_{mos} increases when V_{BG} approaches $|V_T|$, and is infinite when V_{BG} is equal to $|V_{\mathrm{T}}|$. This is, of course, a gross simplification introduced by the very simple MOS model to which (1) is related; in reality, R_{mos} does increase, but remains finite throughout the whole moderate inversion region, when the concentration of holes at the oxide interface decreases steadily, but C_i is still (much) larger than C_b. The movement of the majority charge carriers in the strong and moderate inversion regions is as indicated in Fig. 2 (solid lines). When the MOS device enters the weak inversion region, however, modulation of the (physical) depletion region is as important as hole injection ($C_i \approx C_b$), to become the dominant effect in the depletion region ($C_i \ll C_b$). In this case, the parasitic resistance is associated with the resistive losses of electrons moving from the bulk contact to the interface between bulk and depletion region (Fig. 2, dashed lines). This is an interesting fact, since electrons have a mobility approximately three times higher than holes. There is, then, the possibility that the parasitic resistance decreases in the depletion region, even compared to that in the strong inversion region.

III. INVERSION-MODE AND ACCUMULATION-MODE MOS CAPACITORS

It should be noted that Fig. 1 shows the C_{mos}-V_{BG} characteristics for a very small signal superimposed on the bias voltage V_{BG}. If the signal at the transistor gate is large (as it happens when the MOS capacitance is used in a VCO), then the instantaneous value of C_{mos} changes throughout the signal period. The average value of C_{mos} over a signal period is of course still a function of V_{BG}, but the tuning capability of the circuit is impaired by the nonmonotonicity of C_{mos}.

One way to obtain an almost monotonic function for C_{mos} is by ensuring that the transistor does not enter the accumulation

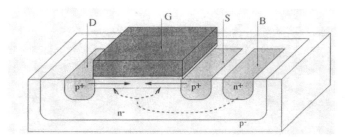

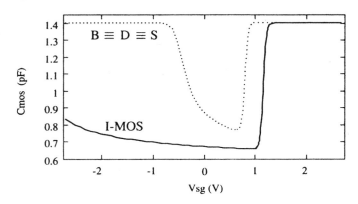

Fig. 2. Charge carrier path for the MOS capacitor working in the strong and moderate inversion regions (solid lines), and in the depletion and accumulation regions (dashed lines).

Fig. 3. Tuning characteristics for the inversion-mode MOS capacitor.

region for a very wide range of values of V_G. This is accomplished by removing the connection between D-S and B, and connecting B to the highest dc-voltage available in the circuit (i.e., the power supply V_{dd}). Fig. 3 shows the comparison between the simulated C_{mos}-V_{SG} characteristics of two equally sized pMOS capacitors designed in the CMOS process used in this work, with $V_{\mathrm{dd}} = 2.7$ V. It is clear that the tuning range of the pMOS capacitor with $V_B = V_{\mathrm{dd}}$ is much wider[2] than for the pMOS capacitor with B ≡ D ≡ S, since the former capacitor is working in the strong, moderate, or weak inversion region only, and never enters the accumulation region (in the following, we will refer to this topology as the inversion-mode MOS, I-MOS for brevity). It should be noted that an equally large tuning range can be obtained with an nMOS capacitor with floating D-S and grounded B, which is readily available in a p-substrate CMOS process [12], [14]. This capacitor has the further advantage of a lower parasitic resistance than the pMOS varactor, but has the drawback of being more sensitive to substrate-induced noise, since it cannot be implemented in a separate p-well.[3]

A more attractive alternative is the use of the pMOS device in the depletion and accumulation regions only [20]–[22]. This solution allows for the implementation of a MOS capacitor with large tuning range, together with a (much) lower parasitic resistance,[4] according to the analysis of the preceding section. To

[2]Strong inversion appears for higher values of V_{SG} because of the substrate effect, which raises the effective value of $|V_T|$ when $V_B > V_S$.

[3]The pMOS varactor is of course also sensitive to substrate noise through the capacitive coupling between substrate and n-well.

[4]Q for this capacitor is proportional to L^{-2} when the device is working in the accumulation region [21], and to L^{-1} in the depletion region [23]. According to [21], Q shows a local minimum at the threshold between the depletion and inversion regions. Such a minimum is undetected in the present work (see Section IV-A).

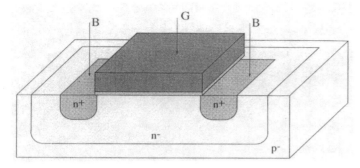

Fig. 4. Accumulation-mode MOS capacitor.

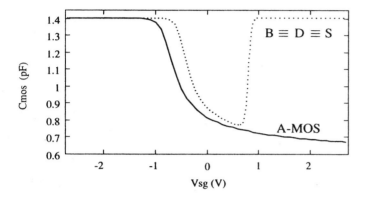

Fig. 5. Tuning characteristics for the accumulation-mode MOS capacitor.

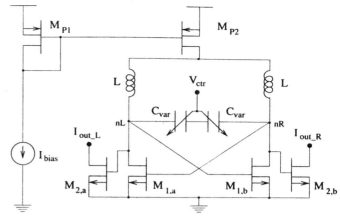

Fig. 6. Schematic view of the VCO.

obtain an accumulation-mode MOS capacitor (to be referred to hereafter as the A-MOS), we must make sure that the formation of the strong, moderate, and weak inversion regions is inhibited, which requires the suppression of any injection of holes in the MOS channel. This, in turn, can be accomplished with the removal of the D–S diffusions (p$^+$-doped) from the MOS device (since we are dealing with circuits working at radio frequencies, we do not need worrying about thermal generation of hole-electron pairs). At the same time, we can implement the bulk contacts (n$^+$) in the place left by D–S, as shown in Fig. 4, which minimizes the parasitic n-well resistance of the device. The tuning characteristics of the A-MOS capacitor is shown in Fig. 5, compared with that of the pMOS capacitor with B ≡ D ≡ S. The mismatch between the two curves in the accumulation region is caused by the difficulty of simulating the A-MOS device with the usual four-terminal transistor model. It is worth noting that we did not introduce any extra process step in the design of the A-MOS capacitor, but on the other hand this device is not supported (even less characterized) by silicon foundries, which means that the designer has to be prepared for some trial-and-error design loops.

According to Figs. 3 and 5, the ratio C_{max}/C_{min} for both I-MOS and A-MOS devices is a little higher than two. Such a rather low value is caused by the deleterious presence of the overlap capacitances between G and D–S, which make up for more than half the value of C_{min}. The intrinsic C_{max}/C_{min} ratio, however, is higher than six, indicating that (much) smaller varactors that those used in this work can be employed, for a specified VCO tuning range, if the length of the varactors can

be increased (at the expense of a lower Q, or by migrating to a CMOS process with shorter minimum transistor length). As a comparison, the diode varactor used here has a C_{max}/C_{min} ratio of about 1.6, close to what has been obtained with other technologies [10], [11], [24]. As a matter of fact, C_{max}/C_{min} can be increased further by biasing the diode closer to forward-biasing. The quality factor, however, drops very quickly as forward-biasing is approached (see e.g. [24]). This fact impairs the diode varactor large-signal tuning capability as well, since the diode should never become forward-biased under the whole signal period. No such limitation is introduced by an MOS varactor.

IV. VCO Design and Measurement Results

Fig. 6 shows the schematic view of the implemented VCO's. This is a standard symmetric CMOS VCO architecture, where transistor M_{P2} provides the bias current, $M_{1,a}$ and $M_{1,b}$ form the negative resistance compensating the VCO losses, and C_{var} are the varactors (transistors $M_{2,a}$ and $M_{2,b}$ allow the VCO signals to be collected off-chip as a drain currents).

Since the objective of the work was to make the behavior of the VCO's sensitive to the respective varactor, it was necessary to design the VCO's in a way that guarantees that the parasitic resistance of the varactors dominates the performances of the VCO's. This condition ruled out the use of monolithic spiral inductors[5] in favor of on-chip bond wire inductors, which possess a high Q-value [6]. The two bond wires are 3 mm long and 0.25 mm apart, corresponding to two inductances of approximate value 3.2 nH and a coupling factor of about 0.4.

Three different VCO's (tuned by a diode varactor, an I-MOS varactor, and an A-MOS varactor, respectively; see Fig. 7) were tailored in order to present the same center frequency and tuning range. Table I presents the dimensions for the components of the three VCO's. All layouts were drawn in a comb-like fashion. The unit transistor is 5 μm wide, in order to keep the gate resistance low (no silicide step was available in the process); the unit diode is 50 μm wide. All varactors were designed for maximum Q (i.e., the lengths of the devices are the minimum lengths allowed by the process).

---

[5]Q's for such inductors are quite modest, especially when the inductors are implemented on a low–resistivity substrate [25].

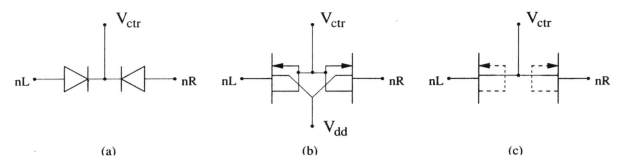

Fig. 7. (a) Diode varactors. (b) I-MOS varactors. (c) A-MOS varactors.

TABLE I
DIMENSIONS (IN μm) OF THE VCO COMPONENTS

Varactor	Varactor Dimensions	$M_{1,a}, M_{1,b}$	$M_{2,a}, M_{2,b}$	M_{P2}
Diode	1900×0.8	500×0.6	60×0.6	2400×0.6
I-MOS	1000×0.6	800×0.6	60×0.6	2400×0.6
A-MOS	1000×0.6	800×0.6	60×0.6	2400×0.6

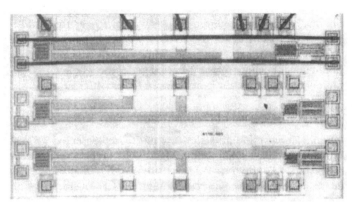

Fig. 8. Die photo of the VCO's. Top: Diode varactor VCO (with bonded inductors). Middle: A-MOS varactor VCO. Bottom: I-MOS varactor VCO.

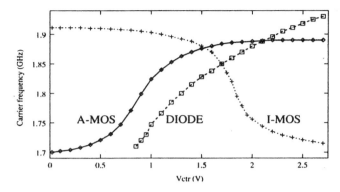

Fig. 9. Tuning characteristics of the diode varactor VCO, I-MOS varactor VCO, and A-MOS varactor VCO.

TABLE II
FREQUENCY BEHAVIOR OF THE VCOs

Varactor	$f_L - f_H$ (GHz)	f_c (GHz)	Tuning Range
Diode	1.73−1.93	1.83	10.9%
I-MOS	1.71−1.91	1.81	11.0%
A-MOS	1.70−1.89	1.80	10.6%

The VCO's were fabricated in a standard double-metal n-well, 0.6-μm CMOS process with a low resistivity p-substrate (≈ 10 m$\Omega \cdot$ cm). A chip photograph of the implemented VCO's is shown in Fig. 8.

All measurements were performed with a power supply voltage $V_{dd} = 2.7$ V. The tuning characteristics of the VCO's are shown in Fig. 9 and summarized in Table II. Both center frequency (≈ 1.8 GHz) and tuning range ($\approx 11\%$) are the same for all VCO, which is a necessary condition for the comparison of the relative varactors to be meaningful. However, the linear portion of the tuning characteristics of both MOS varactor VCO's is somewhat reduced, compared to that of the diode varactor VCO.

A particularly interesting measurement is the minimum current consumption that ensures oscillations in the VCO's. This measurement gives a rough indication of the (relative) parasitic resistance in the LC-tank. It also gives an indication, all other things being equal, of the phase noise shown by the VCO's (because both power consumption and phase noise are proportional to the losses in the LC-tank [1]). The minimum current

consumption for the VCO's is plotted in Fig. 10 versus the oscillation frequency. The current consumption for the A-MOS VCO (0.87 mA $< I_{dd} < 0.99$ mA) is less than half the current consumption for the diode VCO (1.80 mA $< I_{dd} < 2.24$ mA) and about 60% less than for the I-MOS VCO (1.45 mA $< I_{dd} < 1.65$ mA).

A. Varactor Losses

Following [1], the internal losses of the VCO resonator must be compensated by a transconductance G_M (supplied by transistors $M_{1,a}, M_{1,b}$) with value

$$G_M = 1/R_p + (R_l + R_c)(\omega C)^2 \qquad (2)$$

where R_p is the parasitic resistance in parallel to the LC-tank, and R_l and R_c are the parasitic resistances of L and C, respectively. It is well known that the oscillation frequency ω (rad/s) is given by

$$\omega = \frac{1}{\sqrt{LC}} \qquad (3)$$

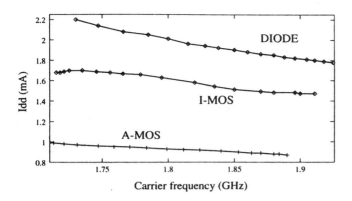

Fig. 10. Minimum current consumption for the diode varactor VCO, I-MOS varactor VCO, and A-MOS varactor VCO.

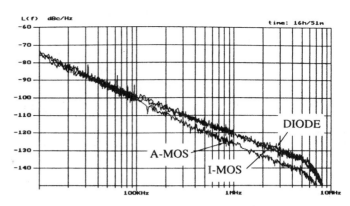

Fig. 11. Phase noise for the diode varactor VCO, I-MOS varactor VCO, and A-MOS varactor VCO at 1.8-GHz carrier frequency.

and we can rewrite (2) as

$$G_M = 1/R_p + \frac{R_l}{L^2}\frac{1}{\omega^2} + \frac{R_c}{L^2}\frac{1}{\omega^2}. \qquad (4)$$

Since $M_{1,a}$, $M_{1,b}$ are working in the moderate inversion region (which can be easily checked through simulations), G_M is proportional to the current flowing in $M_{1,a}$. $M_{1,b}$, and (4) can be written as

$$I_{dd} \propto 1/R_p + \frac{R_l}{L^2}\frac{1}{\omega^2} + \frac{R_c}{L^2}\frac{1}{\omega^2}. \qquad (5)$$

Assuming that the losses in the capacitor are dominating, (5) is simplified as

$$I_{dd} \propto \frac{R_c}{L^2}\frac{1}{\omega^2}. \qquad (6)$$

R_c can, in turn, be divided into two components: a resistance R'_c, independent of the value of C, accounting for the losses in the polysilicon transistor gates and contacts, and a resistance R''_c dependent on the value of C. Since R'_c can be (and has been) minimized through appropriate layout techniques, (6) becomes

$$I_{dd}(\omega) \propto \frac{R''_c(\omega)}{L^2}\frac{1}{\omega^2} \qquad (7)$$

where $R''_c(C)$ has been made dependent on ω through (3).

Referring to Fig. 10, the $I_{dd,min} - f$ characteristics for the A-MOS VCO and diode VCO can be (roughly) recovered by posing $R''_{c,A-MOS}(\omega) = k_A \cdot \omega$ and $R''_{c,diode}(\omega) = k_d$ in (7), respectively. Thus, the parasitic resistance of the A-MOS varactor grows linearly with the frequency throughout the tuning range,[6] while the parasitic resistance of the diode varactor can be considered constant in the whole tuning range. In the case of the I-MOS VCO, an equally simple expression for $R''_{c,I-MOS}(\omega)$ can not be found. It is, however, clear from Fig. 10 that $R''_{c,I-MOS}(\omega)$ reaches a maximum for $f \approx$ 1.72–1.73 GHz, when the I-MOS varactor is working at the interface between the strong and moderate inversion regions.[7]

[6]As we pointed out before, this result is at variance with what shown in [21]

[7]Quality factors for diode. I-MOS, and A-MOS varactors implemented in a 0.5-μm CMOS process have been measured in [24]. In the cited work, diode and A-MOS varactors have very high Q's (> 100), while the I-MOS Q is lower than both.

TABLE III
PHASE NOISE BEHAVIOR OF THE VCO'S

Varactor	I_{dd} (mA)	Phase noise @ 100kHz, 600kHz, 3MHz offset from carrier (dBc/Hz)
Diode	4.5	-100, -116, -130
I-MOS	3.8	-100, -117, -132
A-MOS	2.7	-101, -121, -137

B. Phase Noise

Phase noise measurements where performed with a Europtest PN9711 Phase Noise Test System and the delay line technique (as described in [18]). The error range for the phase noise measurements is specified to be ±2 dB. Again, the performances of the VCO's have been compared, making sure that the buffers of the VCO's deliver the same power (≈ -15 dBm) to the test system.[8] Fig. 11 shows the phase noise of the VCO's at 1.8 GHz carrier frequency (the roll-off at offset frequencies higher than 5 MHz is an artifact of the measurement system), and Table III presents the phase noise at three offset frequencies from the carrier. At offset frequencies lower than 100 kHz the phase noise behavior of the three VCO's is approximately the same. This can be attributed to the rather high $1/f$ noise introduced by the VCO architecture of Fig. 6. This $1/f$ noise could be lowered by adopting a VCO design with an extra pair of cross-coupled pMOS transistors between the current source M_{P2} and the inductors [3], [26]. At higher offset frequencies, however, the phase noise of the VCO's with MOS varactors becomes lower than for the diode varactor VCO. The improvement is only marginal in the case of the I-MOS VCO, but large for the A-MOS VCO.[9]

[8]Actually, this output signal has to be amplified some 15 dB for the test system to work correctly.

[9]Disregarding $1/f$ noise, and using the theory developed in [1], we can estimate the expected phase noise difference Δ_{PN} between two VCO's with equivalent resonator resistance $R_{e,1}$ and $R_{e,2}$, respectively, as $\Delta_{PN} = R_{e,1}/R_{e,2}$, where $R_e (\omega C)^2 = G_M$. Equations (4), (5), together with the data from Fig. 10 at 1.8 GHz, yield $\Delta_{PN} \approx 4$ dB between diode VCO and A-MOS VCO, and $\Delta_{PN} \approx 1$ dB between diode VCO and I-MOS VCO.

V. Conclusions

Three VCO's, tuned by an inversion-mode MOS varactor, an accumulation-mode MOS varactor, and a diode varactor, respectively, have been implemented in a standard 0.6-μm CMOS process and tested. The VCO's have (with very good approximation) the same center frequency (1.8 GHz) and the same tuning range (11%). Measurement results show that the performance of both MOS varactor VCO's is superior to that of the diode varactor VCO. In particular, the accumulation-mode MOS varactor VCO clearly displays the lowest power consumption and the lowest phase noise at large offset frequencies from the carrier. The advantages offered by the MOS varactors are likely to increase when more modern CMOS technologies are adopted.

Acknowledgment

The authors would like to thank P. Petersson, M.Sc., at Ericsson Radio Systems, Kista, Sweden, for providing the phase noise measurement environment and for helping performing the phase noise measurements, and Prof. L. Sundström at the Dept. of Applied Electronics, Lund University, Sweden, for insightful discussions on technical matters.

References

[1] J. Craninckx and M. Steyaert, "Low-noise voltage-controlled oscillators using enhanced *LC*-tanks," *IEEE Trans. Circuits Syst. II*, vol. 42, pp. 794–804, Dec. 1995.

[2] B. Razavi, "A study of phase noise in CMOS oscillators," *IEEE J. Solid-State Circuits*, vol. 31, pp. 331–343, Mar. 1996.

[3] A. Hajimiri and T. H. Lee, "A general theory of phase noise in electrical oscillators," *IEEE J. Solid-State Circuits*, vol. 33, pp. 179–194, Feb. 1998.

[4] S. L. J. Gierkink, E. A. M. Klumperink, T. J. Ikkink, and A. J. M. van Tuijl, "Reduction of intrinsic $1/f$ device noise in a CMOS ring oscillator," in *Proc. ESSCIRC'98*, Sept. 1998, pp. 272–275.

[5] A. Hajimiri and T. H. Lee, "Jitter and phase noise in ring oscillators," *IEEE J. Solid-State Circuits*, vol. 34, pp. 790–804, June 1999.

[6] J. Craninckx and M. Steyaert, "A 1.8-GHz CMOS low-phase-noise voltage-controlled oscillator with prescaler," *IEEE J. Solid-State Circuits*, vol. 30, pp. 1474–1482, Dec. 1995.

[7] ——, "A 1.8-GHz low-phase noise CMOS VCO using optimized hollow spiral inductors," *IEEE J. Solid-State Circuits*, vol. 32, pp. 736–744, May 1997.

[8] M. Zannoth, B. Kolb, J. Fenk, and R. Weigel, "A fully integrated VCO at 2 GHz," *IEEE J. Solid-State Circuits*, vol. 33, pp. 1987–1991, Dec. 1998.

[9] M. Steyaert and J. Craninckx, "A fully integrated CMOS DCS-1800 frequency synthesizer," *IEEE J. Solid-State Circuits*, vol. 33, pp. 2054–2064, Dec. 1998.

[10] J. N. Burghartz, M. Soyuer, and K. A. Jenkins, "Integrated RF and microwave components in BiCMOS technology," *IEEE Trans. Electron Devices*, vol. 43, pp. 1559–1570, Sept. 1996.

[11] J. N. Burghartz, M. Soyuer, K. A. Jenkins, M. Kies, M. Dolan, K. J. Stein, J. Malinowski, and D. L. Harame, "Integrated RF components in a SiGe bipolar technology," *IEEE J. Solid-State Circuits*, vol. 32, pp. 1440–1445, Sept. 1997.

[12] T. I. Ahrens, A. Hajimiri, and T. H. Lee, "A 1.6 GHz 0.5 mW CMOS LC low phase noise VCO using bond wire inductance," in *1st Int. Workshop Design of Mixed-Mode Integrated Circuits and Applications*, 1997, July, pp. 69–71.

[13] A. Kral, F. Behbahani, and A. A. Abidi, "RF-CMOS oscillators with switched tuning," in *Proc. CICC'98*, May 1998, pp. 555–558.

[14] T. I. Ahrens and T. H. Lee, "A 1.4 GHz 3 mW CMOS LC low phase noise VCO using tapped bond wire inductance," in *Int. Symp. Low-Power Electronics and Design*, 1998, Aug., pp. 16–19.

[15] P. Andreani and S. Mattisson, "A 2.4-GHz CMOS monolithic VCO based on an MOS varactor," in *Proc. ISCAS'99*, vol. II, May/June 1999, pp. 557–560.

[16] W. Wong, F. Hui, Z. Chen, K. Shen, J. Lau, and J. Wide tuning range inversion-mode gated varactor and its application on a 2-GHz VCO, *1999 Symp. VLSI Circuit Dig. Tech. Papers*, 1999, June, pp. 53–54.

[17] H. Wang, "A 9.8 GHz back-gate tuned VCO in 0.35 μm CMOS," in *ISSCC Dig. Tech. Papers*, 1999, Feb., pp. 406–407.

[18] P. Andreani, "A comparison between two 1.8 GHz CMOS VCO's tuned by different varactors," in *Proc. ESSCIRC'98*, Sept. 1998, pp. 380–383.

[19] Y. P. Tsividis, *Operation and Modeling of the MOS Transistor*. New York, NY: McGraw-Hill, 1987, ch. 2.

[20] S. Mattisson and A. Litwin, "Swedish Patent <Patent title?>," Application No. 9 703 295-7, 1997.

[21] T. Soorapanth, C. P. Yue, D. R. Shaeffer, T. H. Lee, and S. S. Wong, "Analysis and optimization of accumulation-mode varactor for RF ICs," in *1998 Symp. VLSI Circuit Dig. Tech. Papers*, 1998, June, pp. 32–33.

[22] R. Castello, P. Erratico, S. Manzini, and F. Svelto, "A ±30% tuning range varactor compatible with future scaled technologies," in *1998 Symp. VLSI Circuit Dig. Tech. Papers*, 1998, June, pp. 34–35.

[23] F. Svelto, P. Erratico, S. Manzini, and R. Castello, "A metal-oxide-semiconductor varactor," *IEEE Electron Device Lett.*, vol. 20, pp. 164–166, Apr. 1999.

[24] A.-S. Porret, T. Melly, and C. C. Enz, "Design of high-Q varactors for low-power wireless applications using a standard CMOS process," in *Proc. CICC'99*, May 1999, pp. 641–644.

[25] J. Lee, A. Kral, A. A. Abidi, and N. G. Alexopoulos, "Design of spiral inductors on silicon substrates with a fast simulator," in *Proc. ESSCIRC'98*, Sept. 1998, pp. 328–331.

[26] A. Hajimiri and T. H. Lee, "Design issues in CMOS differential LC oscillators," *IEEE J. Solid-State Circuits*, vol. 34, pp. 717–724, May 1999.

PART III

Phase Noise and Jitter

Low-Noise Voltage-Controlled Oscillators Using Enhanced *LC*-Tanks

Jan Craninckx, *Student Member, IEEE*, and Michiel Steyaert, *Senior Member, IEEE*

Abstract—Frequency synthesizers used in modern telecommunication systems, such as cellular telephones, need to have very low phase noise. Therefore, in the design of high performance frequency synthesizers using Phase Locked Loops (PLL), the Voltage-Controlled Oscillator (VCO) has become a key issue. The trend towards monolithic integration poses some major challenges. This paper discusses the phase noise aspects of *LC*-tuned oscillators. A general formula is developed, based on the concepts of effective resistance and capacitance. The formula also applies for oscillators using active inductors. From these results the importance of an inductor with very low series resistance is apparent. To circumvent the technological limits given by an inductor's series resistance, a presented enhanced *LC*-tank can be used to make a trade-off between noise and power.

I. INTRODUCTION

MODERN TELECOMMUNICATION systems, such as cellular telephones, have caused an uprise in RF integrated circuits. Several receiver building blocks can already be integrated on a single die. These include mixers [1], [2], Low Noise Amplifiers (LNA) [3], etc. An element that has not been realized monolithically yet, at least not with sufficient performance, is the frequency synthesizer generating the LO signal. The toughest specification to achieve is the phase noise spec.

A sinusoidal signal with frequency ω_0 can be written as:

$$V(t) = A \cdot \sin\left[\omega_0 t + \phi(t)\right]. \tag{1}$$

If $\phi(t)$ equals zero, this signal is a pure sine wave and its spectrum is a single line. On the other hand, if $\phi(t)$ is a random process, it will introduce jitter on the total phase, which in turn creates *phase noise* in $V(t)$. Since frequency is the derivative of phase, the frequency of $V(t)$ will also have jitter. So the spectrum of $V(t)$ will also have some power at frequencies very close to the center frequency. The phase noise on a signal is usually characterized in terms of single sided spectral noise density ($\mathcal{L}$) in units of decibel carrier per Hertz (dBc/Hz): (see (2) at the bottom of the next page).

Due to the very narrow channel spacings used in cellular telecommunication networks, such as the European Global System for Mobile communications (GSM) system, extremely low phase noise levels are required. At 10 kHz from the carrier, e.g., a single sided spectral noise density of -100 dBc/Hz is

Manuscript received June 10, 1994; revised January 24, 1995. The work of J. Craninckx was supported by a fellowship of the NFWO (National Fund for Scientific Research). This paper was recommended by Associate Editor D. A. Johns.

The authors are with ESAT-MICAS, Katholieke Universiteit Leuven, B-3001 Heverlee, Belgium.

IEEE Log Number 9415303.

specified. At the moment, this is achieved with an *LC*-tuned oscillator with an external inductor. Indeed, because of the high frequencies, this is the only type of oscillator that will be able to generate a signal with a very pure spectrum. There are ring oscillators also operating at several GHz in Si Bipolar technologies [4], [5], or in CMOS in the 900 MHz range [6], [7]. But in general the phase noise of *LC*-tuned oscillators is expected to be much better because they can use the bandpass characteristic of the *LC*-tank to reduce the phase noise. Ring oscillators suffer from switching effects, can introduce noise in the power supply, and have therefore a worse phase noise than *LC*-tuned oscillators.

This paper starts with a short explanation of the techniques used to analyze the effect of the noise sources present in the oscillator. In Sections III and IV the theoretical calculations resulting in the expressions for the several noise contributions are performed for the basic oscillator type. A distinction is made between oscillators employing active or passive inductors. Two important new concepts are introduced, the effective resistance and the effective capacitance. Section V describes the advantages of crystal oscillators. Based on this, special types of *LC*-tanks are developed that allow a trade-off between noise and power. The last section summarizes the several design options that can be taken when designing an *LC*-tuned oscillator.

II. OSCILLATOR THEORY

An *LC*-tuned oscillator is basically a feedback network as shown in Fig. 1(a). Oscillation will occur at the frequency at which the loop transfer function $\beta A(s)$ is exactly one. This is known as the Barkhausen criterion [8]. The oscillation frequency can easily be found, because the imaginary part of $\beta A(s)$ has to be exactly zero. Oscillation will occur at the frequency where the impedance of the *LC*-tank becomes infinite.

The *LC*-tank of the oscillator shown in Fig. 1(b) includes all parasitic resistances. A series resistance R_c is associated with the capacitor C, and a series resistance R_l is associated with the inductor L. The output resistance of the transconductor, and the parallel resistances across C and L, are represented by R_p.

A restriction made in this paper is the assumption of a linear circuit. This means the amplitude of the oscillation is limited such that the amplifier operates in its linear region. In that case, baseband noise does not contribute to the phase noise. Limiting the oscillation amplitude can be done with an Automatic Gain Control (AGC) circuit. The amplitude can also be limited by

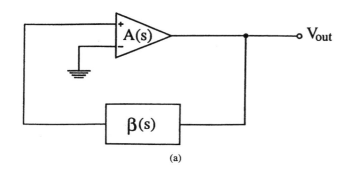

(a)

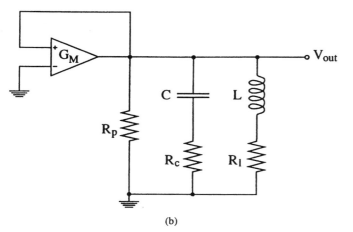

(b)

Fig. 1. (a) General feedback network. (b) An LC-tuned oscillator as a feedback circuit.

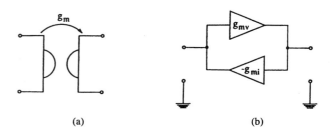

(a)   (b)

Fig. 2. Active inductor: (a) gyrator symbol; (b) gyrator implementation; (c) simulation of an inductor; (d) noise.

2.2 Active Inductors

An often-used way to integrate an inductor is to simulate its operation with active elements. These active elements modify the impedance of a capacitor such that the input impedance of the circuit resembles that of an inductor. In general, a gyrator and a capacitor are required. Fig. 2(a) shows the symbol of a gyrator. The most often-used implementation, i.e., the antiparallel connection of two transconductors [9], is shown in Fig. 2(b).

As shown in Fig. 2(c), the impedance of a capacitor seen through a gyrator looks inductive. It can easily be calculated that the effective inductance value is given by:

$$L_{\text{eff}} = \frac{C_L}{g_{mi} \cdot g_{mv}}. \tag{4}$$

The frequency capability of these active implementations of inductors can be quite high. A GaAs implementation of a floating active inductor [10] can operate up to half the cut-off frequency of the transistors. Therefore, a submicron CMOS version of this circuit is expected to be able to operate at 1 or 2 GHz.

The biggest problem associated with active inductors is their noise contribution to the circuit. Noise is generated by both transconductors in the gyrator. The subscripts i and v refer to whether they generate a current noise source (i) or a voltage noise source (v). Fig. 2(d) shows the equivalent noise sources in an active inductor. The values of these equivalent noise sources can easily be calculated to be [9]:

$$\begin{aligned} di_L^2 &= g_{mi}^2 \cdot dv_{gmi}^2 \\ dv_L^2 &= dv_{gmv}^2. \end{aligned} \tag{5}$$

This expression can be further clarified by substituting an expression for the transconductor equivalent input noise source

$$dv_{gm}^2 = 4kT \cdot \frac{F}{g_m} \cdot df. \tag{6}$$

The factor F represents a noise factor, and is dependent on the particular implementation of the transconductor. Equation (5) now reduces to

$$di_L^2 = 4kT \cdot F_i \cdot g_{mi} \cdot df$$

nonlinear effects, e.g., clipping at the power supplies. In that case, the nonlinearities will ensure that there is some sort of average or effective amplification such that $\beta A(s)$ becomes equal to one. However, in a nonlinear circuit, baseband noise can be upconverted to the oscillation frequency due to the nonlinear effects. Under the assumption made here, only the effect of thermal noise has to be examined because other noise sources, such as $1/f$ noise, have a negligible noise power at the resonance frequency.

2.1 Passive Inductors

The key element in the realization of a good on-chip LC-tank is the inductor. Capacitors are readily available in all IC technologies. But a standard IC technology that allows the realization of a high-quality inductor does not exist, so some tricks have to be used. These techniques severely decrease the performance of the integrated inductor, e.g., limited operating frequency due to parasitic capacitance or low quality factors. The quality factor (Q) of an element is defined as

$$Q = 2\pi \cdot \frac{\text{Peak energy stored}}{\text{Energy loss/cycle}}. \tag{3}$$

$$\mathcal{L}\{\Delta\omega\} = 10 \cdot \log\left(\frac{\text{power in } a \text{ 1 Hz bandwidth at frequency } \omega_0 + \Delta\omega}{\text{carrier power}}\right) \quad [\text{dBc/Hz}]. \tag{2}$$

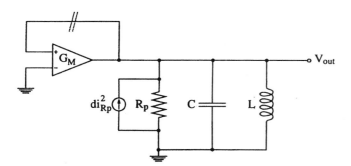

Fig. 3. Basic oscillator with parallel resistor R_p.

$$dv_L^2 = 4kT \cdot \frac{F_v}{g_{mv}} \cdot df. \qquad (7)$$

These formulas explain the subscripts used for the transconductances g_{mi} and g_{mv}, since their associated noise sources are responsible for either an equivalent current noise source, or an equivalent voltage noise source in the active inductor. In Section IV it will be shown that these noise sources have a very bad effect on the phase noise of an LC-tuned oscillator. Therefore, passive implementations of integrated inductors are necessary.

III. PHASE NOISE OF A BASIC OSCILLATOR WITH PASSIVE INDUCTOR

3.1 Parallel Resistance R_p

In Fig. 3, the basic oscillator circuit of Fig. 1(b) is redrawn, with R_p as the only parasitic element. The resistor current noise source di_{Rp}^2 is also included. As a first step, the oscillation frequency has to be calculated. Therefore the loop transfer function is derived by cutting the loop as shown by the two dotted lines in Fig. 3. The result is:

$$T_{\text{loop},R_p}(s) = G_M \cdot \frac{sL}{1 + s\dfrac{L}{R_p} + s^2 LC}. \qquad (8)$$

The imaginary part of the loop transfer function is equal to

$$\Im\{T_{\text{loop},R_p}(\omega)\} = G_M \cdot \frac{\omega L \cdot (1 - \omega^2 LC)}{(1 - \omega^2 LC)^2 + \omega^2 \cdot \left(\dfrac{L}{R_p}\right)^2} \qquad (9)$$

and is zero for

$$\omega_0 = \frac{1}{\sqrt{L \cdot C}}. \qquad (10)$$

This frequency ω_0 is the frequency at which the circuit will oscillate. Now the transconductance G_M, necessary to have a loop transfer function exactly equal to one, can be calculated. This value represents the power necessary to maintain the oscillation in the presence of R_p. It is given by

$$G_{M,R_p} \doteq \frac{G_M}{T_{\text{loop},R_p}(\omega_0)} = \frac{1}{R_p}. \qquad (11)$$

Then the transfer function from the current noise source di_{Rp}^2 to the output voltage V_{out} in closed loop operation must

be calculated. This results in

$$T_{\text{noise},R_p}^2(s) \doteq \frac{dV_{\text{out}}^2}{di_{Rp}^2}(s)$$
$$= \left[\frac{sL}{1 - sL \cdot (G_M - G_p) + s^2 LC}\right]^2. \qquad (12)$$

The notation G_p represents the inverse of the resistance R_p. To calculate the phase noise of the oscillator, we have to evaluate this function at a frequency $\omega_0 + \Delta\omega$ that is slightly offset to the center frequency. At the center frequency ω_0 itself, the noise transfer function is infinite, which is the basic reason for oscillation. To simplify the calculation, we will evaluate the inverse of the noise transfer function, H_{noise,R_p}. So we define

$$H_{\text{noise},R_p}(\omega_0 + \Delta\omega) \doteq \frac{1}{T_{\text{noise},R_p}(\omega_0 + \Delta\omega)}. \qquad (13)$$

We approximate this function with a linearization around the center frequency:

$$H_{\text{noise},R_p}(\omega_0 + \Delta\omega) \approx H_{\text{noise},R_p}(\omega_0) + \frac{dH_{\text{noise},R_p}}{d\omega}(\omega_0) \cdot \Delta\omega. \qquad (14)$$

The first term in this expression, $H_{\text{noise},R_p}(\omega_0)$, is equal to $G_p - G_M$ and is thus zero for $G_M = G_{M,R_p}$. The second term is equal to

$$\frac{dH_{\text{noise},R_p}}{d\omega}(\omega_0) \cdot \Delta\omega = \omega_0 \cdot \frac{dH_{\text{noise},R_p}}{d\omega}(\omega_0) \cdot \left(\frac{\Delta\omega}{\omega_0}\right)$$
$$= 2j \cdot \sqrt{\frac{C}{L}} \cdot \left(\frac{\Delta\omega}{\omega_0}\right). \qquad (15)$$

The transfer function from the current noise source associated with R_p to the output is thus given by

$$T_{\text{noise},R_p}^2(\omega_0 + \Delta\omega) \approx \left|\frac{1}{2j} \cdot \sqrt{\frac{L}{C}}\right|^2 \cdot \left(\frac{\omega_0}{\Delta\omega}\right)^2$$
$$= \frac{1}{4 \cdot (\omega_0 C)^2} \cdot \left(\frac{\omega_0}{\Delta\omega}\right)^2. \qquad (16)$$

So the noise density at frequencies very close to the center frequency will be

$$dV_{\text{out}}^2(\omega_0 + \Delta\omega) = T_{\text{noise},R_p}^2(\omega_0 + \Delta\omega) \times di_{Rp}^2$$
$$\approx \frac{1}{4 \cdot (\omega_0 C)^2} \left(\frac{\omega_0}{\Delta\omega}\right)^2 \times \frac{4kT}{R_p} \cdot df$$
$$= kT \cdot \frac{1}{R_p \cdot (\omega_0 C)^2} \cdot \left(\frac{\omega_0}{\Delta\omega}\right)^2 \cdot df. \qquad (17)$$

Finally, we rewrite this formula with the argument enclosed in curly brackets instead of parentheses. This is just a short notation, since $\{\Delta\omega\}$ has the same meaning as $(\omega_0 + \Delta\omega)$. We also add the subscript R_p, to distinguish the several contributions to the output noise.

$$dV_{\text{out},R_p}^2\{\Delta\omega\} = kT \cdot \frac{1}{R_p \cdot (\omega_0 C)^2} \cdot \left(\frac{\omega_0}{\Delta\omega}\right)^2 \cdot df. \qquad (18)$$

This equation was verified by SPICE circuit simulations in the frequency domain. All other contributions to the phase noise will be written in a format similar to (18).

The noise described here is actually amplified sideband noise. It can be split up into an amplitude modulation (AM) and a phase modulation (PM) component [11]. If the oscillator employs an AGC circuit, the AM component will be removed for frequency offsets smaller than the AGC loop bandwidth. So the absolute value of the noise will be half of the value given by (18). An oscillator with a hard limiter can be regarded as having an AGC with infinite bandwidth, so the two times reduction should be valid for all frequency offsets. In practical circuits this phase noise reduction factor will be somewhere between 1 and 2. We will perform a worst-case analysis, so we will use a factor of 1 in the rest of this paper.

3.2 Other Parasitic Resistors

The noise generated by the other two parasitic resistors can be calculated in an analogous way. The noise contributions of R_l and R_c are given by:

$$dV^2_{\text{out},R_l}\{\Delta\omega\} = kT \cdot R_l \cdot \left(\frac{\omega_0}{\Delta\omega}\right)^2 \cdot df$$

$$dV^2_{\text{out},R_c}\{\Delta\omega\} = kT \cdot R_c \cdot \left(\frac{\omega_0}{\Delta\omega}\right)^2 \cdot df. \quad (19)$$

The power needed to maintain the oscillation in the presence of R_l or R_c is given by:

$$G_{M,R_l} = R_l \cdot (\omega_0 C)^2$$

$$G_{M,R_c} = R_c \cdot (\omega_0 C)^2. \quad (20)$$

3.3 Effective Resistance

Now we can introduce a very important term in the evaluation of LC-tuned oscillators: the *effective resistance*. The phase noise due to the parasitic resistances can be summarized with the following equations:

$$R_{\text{eff}} = R_c + R_l + \frac{1}{R_p \cdot (\omega_0 C)^2} \quad (21)$$

$$G_M = R_{\text{eff}} \cdot (\omega_0 C)^2 \quad (22)$$

$$dV^2_{\text{out},R}\{\Delta\omega\} = kT \cdot R_{\text{eff}} \cdot \left(\frac{\omega_0}{\Delta\omega}\right)^2 \cdot df. \quad (23)$$

So all parasitic resistances can be reduced to one equivalent resistance R_{eff}, which is the equivalent of all series resistances in the loop. This resistance can be used to compare different realizations and implementations of LC-tuned oscillators, since it determines the necessary power as well as the phase noise.

The Q-factor of the LC-tank can also be expressed as a function of this effective resistance. There are three contributions to the Q-factor, one from each resistor. When combining these as done in (24), it is seen that the Q-factor is inversely proportional to R_{eff}:

$$
\begin{aligned}
Q &= \frac{1}{\dfrac{1}{Q_{R_p}} + \dfrac{1}{Q_{R_l}} + \dfrac{1}{Q_{R_c}}} \\[2mm]
&= \frac{1}{\dfrac{1}{R_p(\omega_0 C)} + R_l(\omega_0 C) + R_c(\omega_0 C)} \\[2mm]
&= \frac{1}{R_{\text{eff}}(\omega_0 C)}.
\end{aligned} \quad (24)
$$

3.4 Active Element G_M

The active element in the oscillator also introduces noise. It can be modeled by an output current source $di^2_{G_M}$ equal to

$$di^2_{G_M} = 4kT \cdot F_{G_M} \cdot G_M \cdot df \quad (25)$$

with G_M given by (22) and F_{G_M} the noise factor of the amplifier used. Since this noise source is the same as the one for a parallel resistance R_p, the noise transfer function is given by (16). So the output noise due to the active element is given by

$$
\begin{aligned}
dV^2_{\text{out},G_M}(\omega_0 + \Delta\omega) &= T^2_{\text{noise},R_p}(\omega_0 + \Delta\omega) \cdot di^2_{G_M} \\[2mm]
&\approx kT \cdot \frac{1}{(\omega_0 C)^2} \cdot F_{G_M} \\[2mm]
&\quad \cdot G_M \left(\frac{\omega_0}{\Delta\omega}\right)^2 \cdot df. \quad (26)
\end{aligned}
$$

If we substitute the value of G_M given by (22) this evaluates to

$$dV^2_{\text{out},G_M}\{\Delta\omega\} = kT \cdot R_{\text{eff}} \cdot F_{G_M} \cdot \left(\frac{\omega_0}{\Delta\omega}\right)^2 \cdot df. \quad (27)$$

In a real circuit the transconductance used will be higher than theoretically needed. This can be due to a phase shift in the transconductor, or a safety margin that is used in the design of the oscillator. Sometimes only a part of the output signal is fed back into the amplifier, so the transconductance must be equally higher to make enough negative resistance. We will include this fact in our equations by multiplying the noise with a factor α, representing the amount of noise the actual noisy amplifier generates in excess of an ideal noisy amplifier. To simplify the notation, we define a factor A as being equal to $\alpha \cdot F_{G_M}$. So (27) becomes:

$$dV^2_{\text{out},G_M}\{\Delta\omega\} = kT \cdot R_{\text{eff}} \cdot A \cdot \left(\frac{\omega_0}{\Delta\omega}\right)^2 \cdot df. \quad (28)$$

3.5 Conclusion

The results of the previous sections can be summarized in the following equations:

$$dV^2_{\text{out}}\{\Delta\omega\} = kT \cdot R_{\text{eff}} \cdot [1 + A] \cdot \left(\frac{\omega_0}{\Delta\omega}\right)^2 \cdot df \quad (29)$$

$$\omega_0 = \frac{1}{\sqrt{LC}} \quad (30)$$

$$R_{\text{eff}} = R_l + R_c + \frac{1}{R_p(\omega_0 C)^2} \quad (31)$$

$$A = \alpha \cdot F_{G_M} \quad (32)$$

$$G_M = R_{\text{eff}} \cdot (\omega_0 C)^2. \quad (33)$$

These results were easily verified by circuit simulations. The phase noise can be split up into two parts. The first one is the contribution of the resistances of the LC-tank. It is represented by the factor $R_{\text{eff}} \cdot 1$. The second one is due to the active element used and is represented by the term $R_{\text{eff}} \cdot A$. It is also proportional to R_{eff} because the power needed to maintain the oscillation is determined by this effective resistance. A is usually equal to or larger than 1.

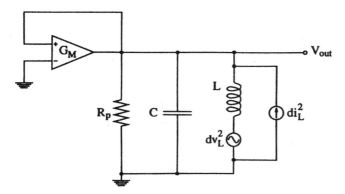

Fig. 4. Basic oscillator with noisy inductor.

The formula to be used for the single sided spectral noise density is:

$$
\begin{aligned}
\mathcal{L}\{\Delta f\} &= \frac{\int_{f_0+\Delta f-(1/2)}^{f_0+\Delta f+(1/2)} dV_{\text{out}}^2}{\text{output power}} \\
&= \frac{kT \cdot R_{\text{eff}} \cdot [1+A] \cdot \left(\frac{\omega_0}{\Delta\omega}\right)^2}{V_{\text{ampl}}^2/2}.
\end{aligned} \tag{34}
$$

This expression can be seen as the inverse of a Signal-to-Noise Ratio (SNR). A very efficient way to reduce the phase noise is enlarging the oscillation amplitude and thus making the signal larger than the noise. There is a limitation in doing this, since the maximum voltage swing at the input of the amplifier G_M is limited by the IC-technology.

The most important parameter in the design of an LC-tuned oscillator is thus the effective resistance R_{eff}. It determines the phase noise of the oscillator as well as the necessary power. To get an idea of the feasibility of low-noise oscillators, we can calculate the maximum resistance allowed to give -100 dBc/Hz at 10 kHz offset from a 1 GHz carrier with oscillation amplitude of 1 V:

$$
R_{\text{eff}} = \frac{10^{-100\,\text{dBc/Hz}/10} \cdot \frac{(1\,\text{V})^2}{2}}{0.41\,10^{-20} \cdot [1+1] \cdot \left(\frac{1\,\text{GHz}}{10\,\text{kHz}}\right)^2} = 0.6\,\Omega. \tag{35}
$$

The power needed depends completely on the value of C. If an inductor of 6 nH is used, the capacitance value is 4 pF. This yields a transconductance of 380 μS. Equations (29)–(33) clearly illustrate the need for an inductor with very low series resistance.

IV. PHASE NOISE OF A BASIC OSCILLATOR WITH ACTIVE INDUCTOR

Noise originating from an active inductor can be modeled by a noise current source di_L^2 and a noise voltage source dv_L^2. To investigate the effect of these two noise sources associated with an active inductor, Fig. 4 will be used. Only the parallel resistor R_p is included in the calculations, but the extension to all parasitic resistors can easily be made by substituting R_p with $1/R_{\text{eff}}(\omega C)^2$.

4.1 Inductor Noise Sources

The current and voltage noise sources have a value given by (7). Using the same techniques as in the previous section, the output noise at frequencies close to the carrier due to the active inductor noise sources can be calculated. This results in:

$$
dV_{\text{out},Li}^2(\omega_0+\Delta\omega) = kT \cdot \frac{1}{(\omega_0 C)^2} \cdot F_{Li} \cdot g_{mi} \cdot \left(\frac{\omega_0}{\Delta\omega}\right)^2 \cdot df
$$

$$
dV_{\text{out},Lv}^2(\omega_0+\Delta\omega) = kT \cdot \frac{F_{Lv}}{g_{mv}} \cdot \left(\frac{\omega_0}{\Delta\omega}\right)^2 \cdot df. \tag{36}
$$

4.2 Total Noise

The total noise is calculated by combining (29) and (36).

$$
dV_{\text{out}}^2\{\Delta\omega\} = kT \cdot \left[R_{\text{eff}}(1+A) + \frac{1}{\omega_0 C} \right.
$$

$$
\left. \cdot \underbrace{\left(\frac{F_{Li} \cdot g_{mi}}{\omega_0 C} + \frac{F_{Lv}}{g_{mv}}(\omega_0 C) \right)}_{N} \left(\frac{\omega_0}{\Delta\omega}\right)^2 \right] \cdot df. \tag{37}
$$

To evaluate this formula, we must now find an expression for the factor N. This can easily be done by substituting the value of the simulated inductance into the expression for ω_0. The result is:

$$
\begin{aligned}
N &= \frac{F_{Li} \cdot g_{mi}}{\omega_0 C} + \frac{F_{Lv}}{g_{mv}}(\omega_0 C) \\
&= F_{Li} \cdot \sqrt{\frac{g_{mi}}{g_{mv}}} \cdot \sqrt{\frac{C_L}{C}} + F_{Lv} \cdot \sqrt{\frac{g_{mi}}{g_{mv}}} \cdot \sqrt{\frac{C}{C_L}} \\
&= \sqrt{\frac{g_{mi}}{g_{mv}}} \cdot \left[F_{Li}\sqrt{\frac{C_L}{C}} + F_{Lv}\sqrt{\frac{C}{C_L}} \right]. \tag{38}
\end{aligned}
$$

Since the noise has to be minimized, we need to search the conditions for minimal N. It seems this can be done by making g_{mv} much larger than g_{mi}. But in that case the voltage swing across the capacitor C_L becomes much larger than the voltage swing across C. If we have already maximized the input voltage of the amplifier G_M up to the limits given by the technology, the voltage across C_L cannot be made higher. So g_{mv} has to be chosen equal to g_{mi}.

To analyze the second factor in the expression for N, we assume F_{Li} and F_{Lv} are both equal to F_L. This is not exactly true, but deviations will be small in most practical active inductors. The conditions for minimum N are then found to be $C = C_L$, in which case N is equal to $2 \cdot F_L$, which is sometimes simplified to $N = 2$, assuming $F_L = 1$.

One last thing that can be done to clarify (37), is to incorporate the formula for the Q-factor given by (24). All results can now be summarized in the following equations:

$$
dV_{\text{out}}^2\{\Delta\omega\} = kT \cdot R_{\text{eff}} \cdot [1+A+N \cdot Q] \cdot \left(\frac{\omega_0}{\Delta\omega}\right)^2 \cdot df \tag{39}
$$

$$\omega_0 = \frac{1}{\sqrt{LC}} = \sqrt{\frac{g_{mi} \cdot g_{mv}}{C \cdot C_L}} \qquad (40)$$

$$R_{\text{eff}} = R_l + R_c + \frac{1}{R_p(\omega_0 C)^2} \qquad (41)$$

$$A = \alpha \cdot F_{G_M} \qquad (42)$$

$$Q = \frac{1}{R_{\text{eff}}(\omega_0 C)} \qquad (43)$$

$$N = \sqrt{\frac{g_{mi}}{g_{mv}}} \cdot \left[F_{Li}\sqrt{\frac{C_L}{C}} + F_{Lv}\sqrt{\frac{C}{C_L}} \right] \qquad (44)$$

$$G_M = R_{\text{eff}} \cdot (\omega_0 C)^2 + g_{mi} + g_{mv}. \qquad (45)$$

Assuming we have designed for the optimal conditions where C_L equals C, where g_{mi} and g_{mv} are both equal to g_{mL}, and if F_{Li} and F_{Lv} are both equal to 1, some of the above equations can be rewritten:

$$dV_{\text{out}}^2\{\Delta\omega\} = kT \cdot R_{\text{eff}} \cdot [1 + A + 2Q] \cdot \left(\frac{\omega_0}{\Delta\omega}\right)^2 \cdot df$$

$$G_M = R_{\text{eff}} \cdot (\omega_0 C)^2 \cdot [1 + 2Q]. \qquad (46)$$

4.3 Conclusion

Equations (39)–(46) allow us to evaluate LC-tuned oscillators using active inductors. The dominant term in the noise expression is the term $2Q$. So at first sight we must create a *low-Q LC*-tank. This is true, but this low Q-factor may not be achieved by increasing R_{eff}, but by increasing the capacitor C. This result doesn't surprise us, since large capacitors usually help to reduce the noise. An expected disadvantage is the proportional increase in necessary power, which is shown in (46). So we can state that, for LC-tanks with Q much larger than one,

$$dV_{\text{out}}^2\{\Delta\omega\} \approx \frac{2kT}{\omega_0 C} \cdot \left(\frac{\omega_0}{\Delta\omega}\right)^2 \cdot df$$

$$G_M \approx 2\omega_0 C. \qquad (47)$$

We can also calculate the capacitance to achieve a phase noise of -100 dBc/Hz at 10 kHz offset from a 1 GHz carrier with oscillation amplitude of 1 V:

$$C = \frac{\dfrac{2 \times 0.41 \, 10^{-20}}{2\pi \times 1\,\text{GHz}} \cdot \left(\dfrac{1\,\text{GHz}}{10\,\text{kHz}}\right)^2}{10^{-100\,\text{dBc/Hz}/10}} = 130\,\text{pF}. \qquad (48)$$

The transconductance associated with this capacitance is 1.6 S. These numbers illustrate the excessive cost of low-noise oscillators with active inductors.

4.4 Comparison with Bandpass Filters

Expression (46) can be compared with results obtained for the total integrated noise of a bandpass filter (BPF). The analogy is not a surprise, since an LC-tuned oscillator consists of a BPF with feedback. Fig. 5(a) shows a general circuit diagram of a BPF.

References [12] and [13] have calculated the total integrated noise at the output of the BPF. It is given by

$$v_{rms}^2 = \frac{kT}{C} \cdot [1 + A + 2Q]. \qquad (49)$$

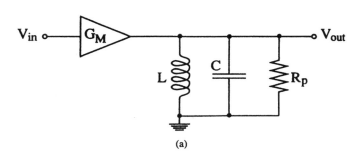

(a)

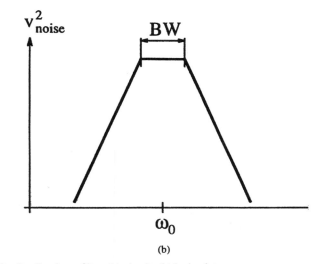

(b)

Fig. 5. Bandpass filter: (a) circuit; (b) bode plot.

Fig. 5(b) shows a plot of the magnitude of the power gain versus frequency. Since the BPF has a bandwidth $BW = 1/R_p C$ to first order, the noise density at the center frequency is

$$dV_{\text{out}}^2(\omega_0) = \frac{v_{rms}^2}{BW} = kT \cdot R_p \cdot [1 + A + 2Q] \cdot df. \qquad (50)$$

This noise level is constant for frequencies ranging from $\omega_0 - BW/2$ to $\omega_0 + BW/2$. At these two corner frequencies the noise power level rolls off with a slope of 20 dB/dec. At a frequency offset $\Delta\omega$ from ω_0 the noise density is thus

$$dV_{\text{out}}^2(\omega_0 + \Delta\omega) = dV_{\text{out}}^2(\omega_0) \cdot \left(\frac{BW}{\Delta\omega}\right)^2 \cdot df$$

$$= kT \cdot R_p \cdot [1 + A + 2Q] \cdot \left(\frac{1/R_p C}{\Delta\omega}\right)^2 \cdot df$$

$$= kT \cdot \frac{R_p}{(R_p \cdot \omega_0 C)^2} \cdot [1 + A + 2Q]$$

$$\cdot \left(\frac{\omega_0}{\Delta\omega}\right)^2 \cdot df. \qquad (51)$$

This reduces to (46) if R_p is replaced with its equivalent R_{eff}.

V. Phase Noise in Crystal Oscillators

Crystal oscillators are known for their excellent phase noise behavior. In [14] a number of examples are given, e.g., -137 dBc/Hz at 10 Hz from a 5 MHz carrier. Suppose the output amplitude is 1 V, the effective resistance can be calculated

$$G_{M,R_c} = R_c \cdot (\omega_0 C_c)^2. \tag{53}$$

5.2 Effective Resistance and Capacitance

Equation (52) shows there can be a large reduction in phase noise because of the factor $(C_s/C_s + C_c)^2$. As in Section 3.3, we can define an effective resistance R_{eff} and combine the previously found results in the following equations:

$$R_{\text{eff}} = \left[R_s + R_c + \frac{1}{R_p \cdot (\omega_0 C_c)^2} \right] \\ \cdot \left(\frac{C_s}{C_s + C_c} \right)^2 \tag{54}$$

$$dV_{\text{out},R}^2\{\Delta\omega\} = kT \cdot R_{\text{eff}} \cdot \left(\frac{\omega_0}{\Delta\omega} \right)^2 \cdot df. \tag{55}$$

We can also define an *effective capacitance* C_{eff}, and derive a general formula for the power consumption analog to (22).

$$C_{\text{eff}} = C_c \cdot \frac{C_s + C_c}{C_s} \tag{56}$$

$$G_M = R_{\text{eff}} \cdot (\omega_0 C_{\text{eff}})^2. \tag{57}$$

Using these two definitions, simple formulas for the phase noise due to the amplifier G_M and an active implementation of the inductor L_s can be derived. We can also write a simple formula for the Q-factor of the simulated crystal. It agrees with the normal expression of (43), but now the effective capacitance C_{eff} is used instead of C.

$$Q_{Xtal} = \frac{1}{R_{\text{eff}} \cdot (\omega_0 C_{\text{eff}})}. \tag{58}$$

So every LC-tank is completely characterized by its effective resistance and capacitance.

5.3 Active Element G_M

Analysis of this noise source can be done as in Section 3.4. The result is:

$$dV_{\text{out}}^2(\omega_0 + \Delta\omega) = kT \cdot R_{\text{eff}} \cdot A \cdot \left(\frac{\omega_0}{\Delta\omega} \right)^2 \cdot df. \tag{59}$$

5.4 Noisy Inductor L_s

We can create the LC-tank of Fig. 6 also using an active inductor. The noise sources of this inductor will create output noise. Its value is given by:

$$dV_{\text{out},L}^2\{\Delta\omega\} = \left[\frac{1}{2} \cdot \frac{1}{\omega_0 C_c} \cdot \frac{\omega_0}{\Delta\omega} \right]^2 \times 4kT \cdot F_{Li} \cdot g_{mi} \cdot df \\ + \left[\frac{1}{2} \cdot \frac{C_s}{C_s + C_c} \cdot \frac{\omega_0}{\Delta\omega} \right]^2 \times 4kT \cdot \frac{F_{Lv}}{g_{mv}} \cdot df. \tag{60}$$

We can attempt to write this formula in the general form of (37) (see (61) at the bottom of the next page).

The expression for N can be simplified to:

$$N = \sqrt{\frac{g_{mi}}{g_{mv}}} \cdot \left[F_{Li} \cdot \sqrt{\frac{C_L \cdot (C_s + C_c)}{C_s C_c}} \\ + F_{Lv} \cdot \sqrt{\frac{C_s C_c}{C_L \cdot (C_s + C_c)}} \right]. \tag{62}$$

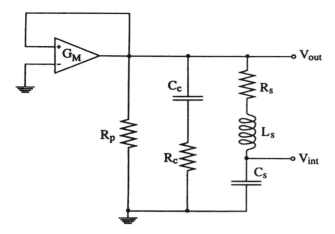

Fig. 6. General model of a simulated crystal oscillator.

according to (29). The result is $10~\mu\Omega$. Since this number is much smaller than the damping resistor of the crystal, another formula must apply to crystal oscillators.

Crystal oscillators can be regarded as being a special type of LC-tuned oscillators. In essence, the crystal is just a load having an infinite impedance at its parallel resonant frequency. So we can simulate a crystal oscillator by employing two capacitors and an inductor. That way we can make a completely integrated "crystal oscillator," without any external component. Fig. 6 shows the model we will examine in this section. The basic crystal consists of C_c, C_s and L_s. Parasitic elements are modeled as follows. R_s is the combined series resistance of C_s and L_s, and R_c is the series resistance of C_c. Parallel resistance of C_c and the output impedance of the amplifier are included in R_p. If the parallel resistances of C_s and L_s must be included, they can be recalculated to an equivalent R_s.

Now the several phase noise contributions will be calculated in the same way as was done in the Sections III and IV.

5.1 Parasitic Resistances

For each parasitic resistance in the LC-tank, the effect of the noise it generates on the output noise can again be calculated. The results are the following:

$$dV_{\text{out},R_p}^2\{\Delta\omega\} = kT \cdot \frac{1}{R_p \cdot (\omega_0 C_c)^2} \cdot \left(\frac{C_s}{C_s + C_c} \right)^2 \\ \cdot \left(\frac{\omega_0}{\Delta\omega} \right)^2 \cdot df$$

$$dV_{\text{out},R_s}^2\{\Delta\omega\} = kT \cdot R_s \cdot \left(\frac{C_s}{C_s + C_c} \right)^2 \cdot \left(\frac{\omega_0}{\Delta\omega} \right)^2 \cdot df$$

$$dV_{\text{out},R_c}^2\{\Delta\omega\} = kT \cdot R_c \cdot \left(\frac{C_s}{C_s + C_c} \right)^2 \cdot \left(\frac{\omega_0}{\Delta\omega} \right)^2 \cdot df \tag{52}$$

with ω_0 equal to $1/\sqrt{C_s L} \cdot \sqrt{C_s + C_c/C_c}$. Each resistor also implies a certain amount of power to compensates for its losses. The necessary transconductances are given by:

$$G_{M,R_p} = \frac{1}{R_p}$$

$$G_{M,R_s} = R_s \cdot (\omega_0 C_c)^2$$

Including the definition of the Q-factor of the crystal, the phase noise originating from the active inductor can be written as:

$$dV^2_{out,L}\{\Delta\omega\} = kT \cdot R_{eff} \cdot [N \cdot Q] \cdot \left(\frac{\omega_0}{\Delta\omega}\right)^2 \cdot df \quad (63)$$

$$N = \sqrt{\frac{g_{mi}}{g_{mv}}} \cdot \left[F_{Li} \cdot \beta + F_{Lv} \cdot \frac{1}{\beta}\right] \quad (64)$$

$$\beta = \sqrt{\frac{C_L \cdot (C_s + C_c)}{C_s C_c}}. \quad (65)$$

5.5 Conclusion

We can summarize the results of the previous sections in the following equations:

$$dV^2_{out}\{\Delta\omega\} = kT \cdot R_{eff} \cdot [1 + A + N \cdot Q] \cdot \left(\frac{\omega_0}{\Delta\omega}\right)^2 \cdot df \quad (66)$$

$$\omega_0 = \frac{1}{\sqrt{C_s L}} \cdot \sqrt{\frac{C_s + C_c}{C_c}}$$

$$= \sqrt{\frac{g_{mi} \cdot g_{mv} \cdot (C_s + C_c)}{C_s C_c C_L}} \quad (67)$$

$$R_{eff} = \left[R_s + R_c + \frac{1}{R_p \cdot (\omega_0 C_c)^2}\right] \cdot \left(\frac{C_s}{C_s + C_c}\right)^2 \quad (68)$$

$$C_{eff} = C_c \cdot \frac{C_s + C_c}{C_s} \quad (69)$$

$$A = \alpha \cdot F_{G_M} \quad (70)$$

$$Q = \frac{1}{R_{eff}(\omega_0 C_{eff})} \quad (71)$$

$$N = \sqrt{\frac{g_{mi}}{g_{mv}}} \cdot \left[F_{Li} \cdot \beta + F_{Lv} \cdot \frac{1}{\beta}\right] \quad (72)$$

$$\beta = \sqrt{\frac{C_L \cdot (C_s + C_c)}{C_s C_c}} \quad (73)$$

$$G_M = R_{eff} \cdot (\omega_0 C_{eff})^2 + g_{mi} + g_{mv}. \quad (74)$$

Compared to the standard LC-tank used in Sections III and IV, there is an improvement in phase noise equal to $(C_s/C_s + C_c)^2$. For a real crystal, the ratio between C_c and C_s can be in the order of 10^3, so a 60 dB reduction of the phase noise can be achieved. That explains the good phase noise characteristics of crystal oscillators.

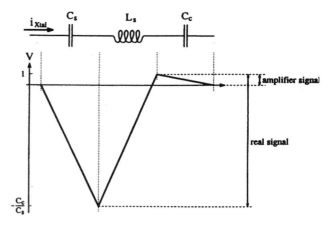

Fig. 7. Internal voltages in a crystal.

There is an easy way to see where this improvement comes from. Referring to Fig. 6, and neglecting the series resistance R_s, the voltage amplitude at the internal node (i.e., between L_s and C_s) equals

$$V_{int} = V_{out} \cdot \frac{\frac{1}{j\omega C_s}}{\frac{1}{j\omega C_s} + j\omega L} = V_{out} \cdot \frac{1}{1 - \omega^2 L C_s}. \quad (75)$$

At resonance frequency ω_0 this simplifies to

$$V_{int} = -V_{out} \cdot \frac{C_c}{C_s}. \quad (76)$$

So the signal at the internal node has a very high amplitude, which results in a large signal-to-noise ratio and thus low phase noise. That is the trick used in crystal oscillators: at an internal node in the equivalent circuit the signal is very large (in the order of kilovolts), so the SNR can also be high. This high voltage does not harm the amplifier, because it only senses the external voltage. And the crystal is only *modeled* by this LC-tank, so there is actually nowhere in the circuit such a large voltage present.

Fig. 7 clarifies this principle. Suppose a current i_{Xtal} flows in the loop of the crystal's LC-tank. The capacitor C_s has a large negative reactance, so afterwards the voltage is very low. The inductor L_s has a positive reactance that is slightly bigger than the reactance of C_s, so the resulting voltage is positive. Finally there is the small negative reactance of C_c. The sum of all the voltages is zero, since at parallel resonance the reactances of the two branches are of opposite sign and equal value.

Basically, it is the sum of all falling (or rising) voltages that represent the factor in phase noise that was gained by the use

$$dV^2_{out,L}\{\Delta\omega\} = \frac{kT}{\omega_0 C_{eff}} \cdot \left(\underbrace{\frac{F_{Li} \cdot g_{mi}}{\omega_0 \frac{C_c^2}{C_{eff}}} + \frac{F_{Lv}}{g_{mv}} \cdot \left(\frac{C_s}{C_s + C_c}\right)^2 \cdot (\omega_0 C_{eff})}_{N}\right) \cdot \left(\frac{\omega_0}{\Delta\omega}\right)^2 \cdot df. \quad (61)$$

of crystal instead of a standard LC-tank. We can call this total voltage the *effective signal* of the oscillator. For a crystal the effective signal equals

$$V_{\text{eff}} = V_{\text{across } C_s} + V_{\text{across } C_s} = \frac{C_c}{C_s} \cdot V_{\text{out}} + 1 \cdot V_{\text{out}}$$

$$= \frac{C_c + C_s}{C_s} \cdot V_{\text{out}}. \qquad (77)$$

So the power of the effective signal is indeed a factor $(C_s/C_s + C_c)^2$ larger than the power of the output signal. The spectral purity of the oscillator output is determined by the ratio of the noise power inside the loop to the effective signal power.

We can substitute the standard LC-tank of our monolithic oscillator with a crystal-like LC-tank. When using a passive inductor, C_c must be made much larger than C_s, in order to benefit from the large internal signal. We will also assume that the effect of the parallel resistance R_p is negligible compared to the effect from the two series resistances R_s and R_c. Then the formulas applying to a crystal-like oscillator with a passive inductor are the following:

$$dV_{\text{out}}^2\{\Delta\omega\} \approx kT \cdot (R_s + R_c) \cdot \left(\frac{C_s}{C_c}\right)^2 \cdot [1 + A]$$

$$\cdot \left(\frac{\omega_0}{\Delta\omega}\right)^2 \cdot df$$

$$G_M \approx (R_s + R_c) \cdot (\omega_0 C_c)^2. \qquad (78)$$

So the power-noise product is constant: power increases quadratically with C_c, and noise decreases quadratically with C_c. The oscillation frequency is mainly determined by C_s. Although the voltages are large, these LC-tanks can be used in IC-technology. The small series capacitance can be made of metal/metal, and can withstand up to 100 V. The large capacitor only sees a small voltage, so the choice between metal/metal, poly/poly or poly/n^+ is free, only depending on the area needed.

A remarkable conclusion is that, for a given inductance, the higher the required oscillation frequency, the easier it is to make the oscillator. For constant L_s and constant C_c/C_s ratio, the effective capacitance decreases quadratically with the oscillation frequency. Assuming the series resistance is mainly determined by R_s, the power needed *decreases* quadratically with ω_0.

When using an active inductor, the voltage swing at the internal node has to be limited. If the output voltage of the oscillator is already maximum, the C_c/C_s ratio cannot be larger than one. So we will make $C_c = C_s = 2C$. So each capacitor has to be twice the size of the capacitor used in an oscillator with a standard LC-tank. The effective capacitance C_{eff} is then equal to $4C$.

To minimize the phase noise we must minimize the factor N in expression (66). For the same reasons as for a standard LC-tanks, we must make g_{mi} and g_{mv} equal to each other. Assume F_{Li} and F_{Lv} are both equal to F_L, than N is minimal

for $\beta = 1$. So the optimum value for C_L is

$$C_L = \frac{C_s C_c}{C_s + C_c} = C. \qquad (79)$$

The power needed for the inductor is then equal to

$$G_{ML} = g_{mi} + g_{mv} = 2\omega_0 C. \qquad (80)$$

For high-Q inductors the phase noise and the power consumption are mainly determined by the active inductor. So the following formulae apply:

$$dV_{\text{out}}^2\{\Delta\omega\} \approx \frac{kT}{2\omega_0 C} \cdot \left(\frac{\omega_0}{\Delta\omega}\right)^2 \cdot df$$

$$G_M \approx 2\omega_0 C. \qquad (81)$$

Compared to the approximate (47) for a standard LC-tank, the phase noise is reduced by a factor of 4 (= 6 dB). The power consumption has not increased. We can again calculate the capacitance to achieve a phase noise of -100 dBc/Hz at 10 kHz offset from a 1 GHz carrier with oscillation amplitude of 1 V:

$$C = \frac{\dfrac{0.41\ 10^{-20}}{2 \times 2\pi \times 1\,\text{GHz}} \cdot \left(\dfrac{1\,\text{GHz}}{10\,\text{kHz}}\right)^2}{10^{-100\,\text{dBc/Hz}/10}} = 33\,\text{pF}. \qquad (82)$$

This value is 4 times lower than for standard LC-tanks. But since two capacitors of twice this value are needed, plus the capacitor for the active inductor, total capacitance is 163 pF. So the total capacitance is now 62.5% of the two large capacitors needed for the standard LC-tank. The transconductance needed now is 400 mS.

So for active inductors the crystal-like LC-tank allows a reduction in power consumption by a factor of 4 for the same noise performance. One has to keep in mind, however, that the active inductor now has to be a floating one, instead of the grounded inductor that was used in the standard LC-tank.

VI. Enhanced LC-Tanks

Now that we have found a way to enhance the phase noise by using special LC-tanks, we can derive other interesting structures. Referring to Fig. 7, efficient ways to generate very large *effective signals* are investigated. What matters is not the difference between the most positive voltage and the most negative voltage, but the sum of all rising (or falling) voltages. The crystal-like LC-tank was a first example.

One big disadvantage of using a large C_c/C_s ratio is the high voltages that are generated. They do not harm the operation of the circuit directly, since there is no electrical connection. But capacitive coupling might disturb other circuits on the IC. Also the large magnetic field associated with the inductors might not be negligible.

An LC-tank that does not have such large voltage swings is shown in Fig. 8(a). The internal voltages are shown in Fig. 8(b). So all nodes in the circuit have the same signal amplitude. That is why we call this tank a low-voltage enhanced LC-tank.

173

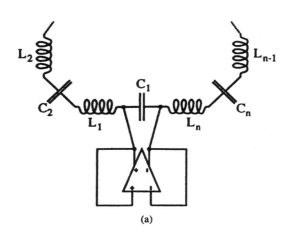

Fig. 8. Oscillator with n inductors and capacitors: (a) circuit; (b) voltage swings.

The effective signal is now n times the external voltage swing. The area needed is n times the area of a standard LC-tank.

The case of passive inductors will be described first. Suppose the series connection of one inductor and one capacitor has resistance R. Then the following formulas apply:

$$
\begin{aligned}
L_s &= n \cdot L \\
C_s &= \frac{C}{n-1} \\
C_c &= C \\
\omega_0 &= \frac{1}{\sqrt{LC}} \\
C_{\text{eff}} &= n \cdot C \\
R_{\text{eff}} &= n \cdot R \cdot \left(\frac{1}{n-1} \right)^2 \approx \frac{R}{n}.
\end{aligned} \tag{83}
$$

This leads to the following expressions for the phase noise and power consumption:

$$
\begin{aligned}
dV_{\text{out}}^2 \{ \Delta \omega \} &\approx \frac{kT \cdot R}{n} \cdot [1 + A] \cdot \left(\frac{\omega_0}{\Delta \omega} \right)^2 \cdot df \\
G_M &\approx n \cdot R \cdot (\omega_0 C)^2.
\end{aligned} \tag{84}
$$

This shows the perfect trade-off that can be made: phase noise decreases proportionally with n, power and area increase proportionally with n. A similar trade-off can be made for active inductors. Here the formulas for phase noise, power and area are:

$$
\begin{aligned}
dV_{\text{out}}^2 \{ \Delta \omega \} &= \frac{2kT}{n \omega_0 C} \cdot N \cdot \left(\frac{\omega_0}{\Delta \omega} \right)^2 \cdot df \\
G_M &= 2n \cdot \omega_0 C.
\end{aligned} \tag{85}
$$

So phase noise decreases proportionally with n, and power and area increase proportionally with n.

VII. CONCLUSION

In this paper a general notation for the phase noise of LC-tuned oscillators was derived, based on the calculation of transfer functions. The results were verified by SPICE simulations. Two important terms were introduced, the effective resistance R_{eff} and the effective capacitance C_{eff}. A distinction must be made between a passive and an active implementation of the inductor.

- In oscillators using passive inductors the phase noise is completely determined by the parasitic series resistances in the loop. Enlarging the capacitance has no influence on the phase noise but increases the power consumption quadratically with capacitance. So inductors as large as possible with minimum series resistance must be constructed.
- In oscillators using active inductors the general rule of decreasing noise by increasing capacitance does work. Phase noise and power are proportional to capacitance.

Several possible improvements of the standard LC-tank with one inductor and one capacitor were studied. Starting point was the intrinsic good noise behavior of crystal oscillators. These structures can be designed in the same way as standard oscillators, using the concepts of effective resistance and capacitance. Again, active and passive inductors require different analysis.

- The phase noise of oscillators with passive inductors can be reduced by using a technique similar to that inherently present in a crystal LC-tank, where the low phase noise originates from the large signal generated on the internal node of the circuit. Passive elements can withstand these large voltages, and the amplifier only sees the small output voltage. A good trade-off can be made between noise and power.
- Oscillators with active inductors cannot withstand extremely large voltages. The internal voltage is as large as the output voltage, but since it has opposite sign, the effective signal is twice as large. So for the same phase noise, the power consumption of oscillators with enhanced LC-tanks is four times lower than that of oscillators with an active inductor that employ a standard LC-tank configuration. Low-phase-noise LC-tuned oscillators using active inductors should always use a crystal-like LC-tank when floating active inductors are available.

Some other LC-tanks were studied, whose main advantage was to reduce the internal voltage swing while holding the noise constant. This should limit the coupling between the oscillator and other circuits integrated on the same die. The most general case is the LC-tank consisting of n inductors and n capacitors. Phase noise decreases proportionally with n, but power and area increase proportionally with n. So a trade-off can be made, but practical limits will put a maximum on the value of n. The area of an on-chip spiral inductor is usually in the order of $250 \times 250 \ \mu \text{m}^2$ [15], or even larger if the inductor is designed for lower series resistance. So a large value of n will lead to a very large silicon area.

Using the concepts derived in this paper, several kinds of LC-tanks can be developed and analyzed, resulting in an optimum solution for cellular-telephone applications.

REFERENCES

[1] A. A. Abidi, "900-MHz downconversion mixer," in *Proc. 1993 Euro. Solid-State Circuits Conf.*, Sevilla, Spain, Sept. 1993, pp. 210–213.

[2] J. Crols and M. Steyaert, "A full CMOS 1.5-GHz highly linear broadband downconversion mixer," in *Proc. 1994 Euro. Solid-State Circuits Conf.*, Ulm, Germany, Sept. 1994, pp. 248–251.

[3] J. Y.-C. Chang and A. A. Abidi, "A 750-MHz RF amplifier in 2-μm CMOS," in *Tech. Dig. 1992 Symp. VLSI Circuits*, May 1993, pp. 111–112.

[4] B. Razavi and J. Sung, "A 6-GHz 60-mW BiCMOS phase locked loop with 2-V supply," in *ISSCC Dig. Tech. Papers*, Feb. 1994, pp. 114–115.

[5] A. Pottbacker and U. Langmann, "An 8-GHz silicon bipolar clock recovery and data-regenerator IC," in *ISSCC Dig. Tech. Papers*, Feb. 1994, pp. 116–117.

[6] M. Thamsirianunt and T. A. Kwasniewski, "A 1.2 μm CMOS implementation of a low-power 900-MHz mobile radio frequency synthesizer," in *Proc. IEEE 1994 Custom Integrat. Circuits Conf.*, May 1994, pp. 383–386.

[7] J. Min, A. Rofourgan, H. Samueli, and A. A. Abidi, "An all-CMOS architecture for a low-power frequency-hopped 900-MHz spread spectrum transceiver," in *Proc. IEEE 1994 Custom Integrat. Circuits Conf.*, May 1994, pp. 379–382.

[8] A. Grebene, *Bipolar and MOS Analog Integrated Circuit Design*. New York: Wiley, 1984, ch. 11.

[9] Y.-T. Wang and A. A. Abidi, "CMOS active filter design at very high frequencies," *IEEE J. Solid-State Circuits*, vol. 25, pp. 1562–1574, Dec. 1990.

[10] G. F. Zhang and J. L. Gautier, "Broad-band, lossless monolithic microwave active floating inductor," *IEEE Microwave Guided Wave Lett.*, vol. 3, pp. 98–100, Apr. 1993.

[11] D. Haigh and J. Everard Ed., *GaAs Technology and its Impact on Circuits and Systems*. London: Peter Peregrinus, 1989, ch. 8.

[12] A. A. Abidi, "Noise in active resonators and the availiable dynamic range," *IEEE Trans. Circuits Syst.—I*, vol. 39, pp. 296–299, Apr. 1992.

[13] J. Crols, *Integration of Receivers for Phase- and Frequentie Modulated Systems*, IWONL Annual Report 1994 (in Dutch), 1993.

[14] A. Laundrie, "Crystal oscillators continue to set stability standards," *Microwaves & RF*, Dec. 1992, pp. 140–147.

[15] N. M. Nguyen and R. G. Meyer, "Si IC-compatible inductors and LC passive filters," *IEEE J. Solid-State Circuits*, vol. 25, pp. 1028–1031, Aug. 1990.

Jan Craninckx (S'95) was born in Oostende, Belgium, in 1969. He received the M.S. degree in electrical and mechanical engineering in 1992 from the Katholieke Universiteit Leuven, Belgium.

Currently, he is a research assistant at the ESAT-MICAS Laboratories of the Katholieke Universiteit Leuven. He is working toward the Ph.D. degree on high-frequency low-noise integrated frequency synthesizers. His research interest are high-frequency integrated circuits for telecommunications.

Michiel Steyaert (S'85–A'89–SM'92) was born in Aalst, Belgium, in 1959. He received the M.S. degree in electrical-mechanical engineering and the Ph.D. degree in electronics from the Katholieke Universiteit Leuven, Heverlee, Belgium in 1983 and 1987, respectively.

From 1983 to 1986, he rceived a IWONL Fellowship (Belgian National Foundation for Industrial Research) which allowed him to work as a Research Assistant at the Laboratory ESAT at the Katholieke Universiteit Leuven. In 1987, he was responsible for several industrial projects in the field of analog micropower circuits at the Laboratory ESAT as an IWONL Project Researcher. In 1988 he was a Visiting Assistant Professor at the University of California, Los Angeles. In 1989 he was appointed as a NFWO Research Associate, and since 1992, a NFWO Senior Research Associate at the Laboratory ESAT, Katholieke Universiteit Leuven, where he has been an Associate Professor since 1990. His current research interests are in high-frequency analog integrated circuits for telecommunications and integrated circuits for biomedical purposes.

Professor Steyaert received the 1990 European Solid-State Circuits Conference Best Paper Award, and the 1991 NFWO Alcatel-Bell-Telephone Award for innovative work in integrated circuits for telecommunications.

A Study of Phase Noise in CMOS Oscillators

Behzad Razavi, *Member, IEEE*

Abstract—This paper presents a study of phase noise in two inductorless CMOS oscillators. First-order analysis of a linear oscillatory system leads to a noise shaping function and a new definition of Q. A linear model of CMOS ring oscillators is used to calculate their phase noise, and three phase noise phenomena, namely, additive noise, high-frequency multiplicative noise, and low-frequency multiplicative noise, are identified and formulated. Based on the same concepts, a CMOS relaxation oscillator is also analyzed. Issues and techniques related to simulation of noise in the time domain are described, and two prototypes fabricated in a 0.5-μm CMOS technology are used to investigate the accuracy of the theoretical predictions. Compared with the measured results, the calculated phase noise values of a 2-GHz ring oscillator and a 900-MHz relaxation oscillator at 5 MHz offset have an error of approximately 4 dB.

I. INTRODUCTION

VOLTAGE-CONTROLLED oscillators (VCO's) are an integral part of phase-locked loops, clock recovery circuits, and frequency synthesizers. Random fluctuations in the output frequency of VCO's, expressed in terms of jitter and phase noise, have a direct impact on the timing accuracy where phase alignment is required and on the signal-to-noise ratio where frequency translation is performed. In particular, RF oscillators employed in wireless tranceivers must meet stringent phase noise requirements, typically mandating the use of passive LC tanks with a high quality factor (Q). However, the trend toward large-scale integration and low cost makes it desirable to implement oscillators monolithically. The paucity of literature on noise in such oscillators together with a lack of experimental verification of underlying theories has motivated this work.

This paper provides a study of phase noise in two inductorless CMOS VCO's. Following a first-order analysis of a linear oscillatory system and introducing a new definition of Q, we employ a linearized model of ring oscillators to obtain an estimate of their noise behavior. We also describe the limitations of the model, identify three mechanisms leading to phase noise, and use the same concepts to analyze a CMOS relaxation oscillator. In contrast to previous studies where time-domain jitter has been investigated [1], [2], our analysis is performed in the frequency domain to directly determine the phase noise. Experimental results obtained from a 2-GHz ring oscillator and a 900-MHz relaxation oscillator indicate that, despite many simplifying approximations, lack of accurate MOS models for RF operation, and the use of simple noise

Manuscript received October 30, 1995; revised December 17, 1995.
The author was with AT&T Bell Laboratories, Holmdel, NJ 07733 USA. He is now with Hewlett-Packard Laboratories, Palo Alto, CA 94304 USA.
Publisher Item Identifier S 0018-9200(96)02456-0.

models, the analytical approach can predict the phase noise with approximately 4 to 6 dB of error.

The next section of this paper describes the effect of phase noise in wireless communications. In Section III, the concept of Q is investigated and in Section IV it is generalized through the analysis of a feedback oscillatory system. The resulting equations are then used in Section V to formulate the phase noise of ring oscillators with the aid of a linearized model. In Section VI, nonlinear effects are considered and three mechanisms of noise generation are described, and in Section VII, a CMOS relaxation oscillator is analyzed. In Section VIII, simulation issues and techniques are presented, and in Section IX the experimental results measured on the two prototypes are summarized.

II. PHASE NOISE IN WIRELESS COMMUNICATIONS

Phase noise is usually characterized in the frequency domain. For an ideal oscillator operating at ω_0, the spectrum assumes the shape of an impulse, whereas for an actual oscillator, the spectrum exhibits "skirts" around the center or "carrier" frequency (Fig. 1). To quantify phase noise, we consider a unit bandwidth at an offset $\Delta\omega$ with respect to ω_0, calculate the noise power in this bandwidth, and divide the result by the carrier power.

To understand the importance of phase noise in wireless communications, consider a generic transceiver as depicted in Fig. 2, where the receiver consists of a low-noise amplifier, a band-pass filter, and a downconversion mixer, and the transmitter comprises an upconversion mixer, a band-pass filter, and a power amplifier. The local oscillator (LO) providing the carrier signal for both mixers is embedded in a frequency synthesizer. If the LO output contains phase noise, both the downconverted and upconverted signals are corrupted. This is illustrated in Fig. 3(a) and (b) for the receive and transmit paths, respectively.

Referring to Fig. 3(a), we note that in the ideal case, the signal band of interest is convolved with an impulse and thus translated to a lower (and a higher) frequency with no change in its shape. In reality, however, the wanted signal may be accompanied by a large interferer in an adjacent channel, and the local oscillator exhibits finite phase noise. When the two signals are mixed with the LO output, the downconverted band consists of two overlapping spectra, with the wanted signal suffering from significant noise due to tail of the interferer. This effect is called "reciprocal mixing."

Shown in Fig. 3(b), the effect of phase noise on the transmit path is slightly different. Suppose a noiseless receiver is to

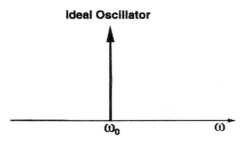

Ideal Oscillator

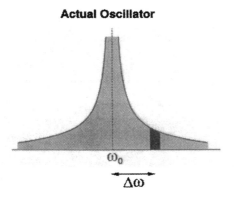

Actual Oscillator

Fig. 1. Phase noise in an oscillator.

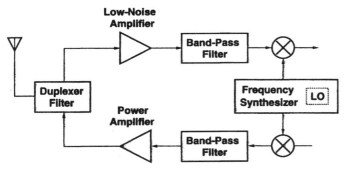

Fig. 2. Generic wireless transceiver.

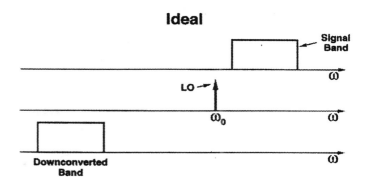

Ideal

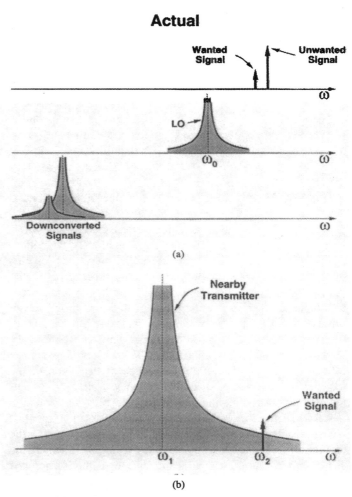

(a)

(b)

Fig. 3. Effect of phase noise on (a) receive and (b) transmit paths.

detect a weak signal at ω_2 while a powerful, nearby tranmitter generates a signal at ω_1 with substantial phase noise. Then, the wanted signal is corrupted by the phase noise tail of the transmitter.

The important point here is that the difference between ω_1 and ω_2 can be as small as a few tens of kilohertz while each of these frequencies is around 900 MHz or 1.9 GHz. Therefore, the output spectrum of the LO must be extremely sharp. In the North American Digital Cellular (NADC) IS54 system, the phase noise power per unit bandwidth must be about 115 dB below the carrier power (i.e., -115 dBc/Hz) at an offset of 60 kHz.

Such stringent requirements can be met through the use of LC oscillators. Fig. 4 shows an example where a transconductance amplifier (G_m) with positive feedback establishes a negative resistance to cancel the loss in the tank and a varactor diode provides frequency tuning capability. This circuit has a number of drawbacks for monolithic implementation. First, both the control and the output signals are single-ended,

making the circuit sensitive to supply and substrate noise. Second, the required inductor (and varactor) Q is typically greater than 20, prohibiting the use of low-Q integrated inductors. Third, monolithic varactors also suffer from large series resistance and hence a low Q. Fourth, since the LO signal inevitably appears on bond wires connecting to (or operating as) the inductor, there may be significant coupling of this signal to the front end ("LO leakage"), an undesirable effect especially in homodyne architectures [3].

Ring oscillators, on the other hand, require no external components and can be realized in fully differential form, but

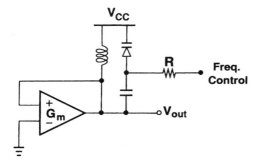

Fig. 4.   LC oscillator.

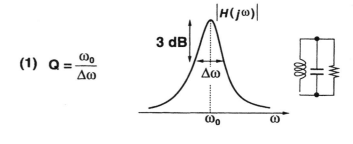

$$(1) \quad Q = \frac{\omega_0}{\Delta \omega}$$

$$(2) \quad Q = 2\pi \frac{\text{Energy Stored}}{\text{Energy Dissipated per Cycle}}$$

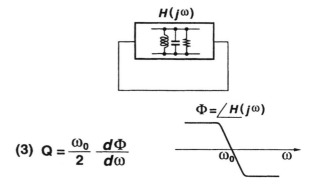

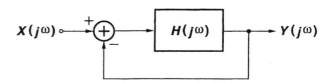

$$(3) \quad Q = \frac{\omega_0}{2} \frac{d\Phi}{d\omega}$$

Fig. 5.   Common definitions of Q.

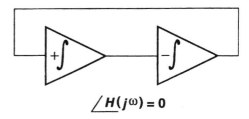

$$\angle H(j\omega) = 0$$

Fig. 6.   Two-integrator oscillator.

their phase noise tends to be high because they lack passive resonant elements.

III. DEFINITIONS OF Q

The quality factor, Q, is usually defined within the context of second-order systems with (damped) oscillatory behavior. Illustrated in Fig. 5 are three common definitions of Q. For an RLC circuit, Q is defined as the ratio of the center frequency and the two-sided −3-dB bandwidth. However, if the inductor is removed, this definition cannot be applied. A more general definition is: 2π times the ratio of the stored energy and the dissipated energy per cycle, and can be measured by applying a step input and observing the decay of oscillations at the output. Again, if the circuit has no oscillatory behavior (e.g., contains no inductors), it is difficult to define "the energy dissipated per cycle." In a third definition, an LC oscillator is considered as a feedback system and the phase of the *open-loop* transfer function is examined at resonance. For a simple LC circuit such as that in Fig. 4, it can be easily shown that the Q of the tank is equal to $0.5\omega_0 \, d\Phi/d\omega$, where ω_0 is the resonance frequency and $d\Phi/d\omega$ denotes the slope of the phase of the transfer function with respect to frequency. Called the "open-loop Q" herein, this definition has an interesting interpretation if we recall that for steady oscillations, the total phase shift around the loop must be precisely 360°. Now, suppose the oscillation frequency slightly deviates from ω_0. Then, if the phase slope is large, a significant change in the phase shift arises, violating the condition of oscillation and forcing the frequency to return to ω_0. In other words, the *open-loop Q* is a measure of how much the *closed-loop* system opposes variations in the frequency of oscillation. This concept proves useful in our subsequent analyses.

While the third definition of Q seems particularlly well-suited to oscillators, it does fail in certain cases. As an example, consider the two-integrator oscillator of Fig. 6, where the open-loop transfer function is simply

$$H(s) = -\left(\frac{\omega_0}{s}\right)^2 \tag{1}$$

yielding $\Phi = \angle H(s = j\omega) = 0$, and $Q = 0$. Since this circuit does indeed oscillate, this definition of Q is not useful here.

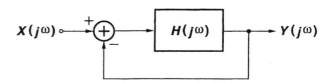

Fig. 7.   Linear oscillatory system.

IV. LINEAR OSCILLATORY SYSTEM

Oscillator circuits in general entail "compressive" nonlinearity, fundamentally because the oscillation amplitude is not defined in a linear system. When a circuit begins to oscillate, the amplitude continues to grow until it is limited by some other mechanism. In typical configurations, the open-loop gain of the circuit drops at sufficiently large signal swings, thereby preventing further growth of the amplitude.

In this paper, we begin the analysis with a linear model. This approach is justified as follows. Suppose an oscillator employs strong automatic level control (ALC) such that its oscillation amplitude remains small, making the linear approximation

178

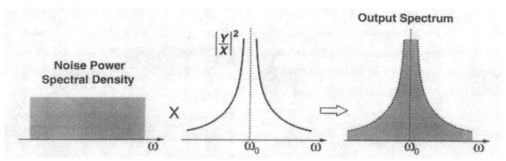

Fig. 8. Noise shaping in oscillators.

valid. Since the ALC can be relatively slow, the circuit parameters can be considered time-invariant for a large number of cycles. Now, let us gradually weaken the effect of ALC so that the oscillator experiences increasingly more "self-limiting." Intuitively, we expect that the linear model yields reasonable accuracy for soft amplitude limiting and becomes gradually less accurate as the ALC is removed. Thus, the choice of this model depends on the *error* that it entails in predicting the response of the actual oscillator to various sources of noise, an issue that can be checked by simulation (Section VIII). While adequate for the cases considered here, this approximation must be carefully examined for other types of oscillators.

To analyze phase noise, we treat an oscillator as a feedback system and consider each noise source as an input (Fig. 7). The phase noise observed at the output is a function of: 1) sources of noise in the circuit and 2) how much the feedback system rejects (or amplifies) various noise components. The system oscillates at $\omega = \omega_0$ if the transfer function

$$\frac{Y}{X}(j\omega) = \frac{H(j\omega)}{1 + H(j\omega)} \qquad (2)$$

goes to infinity at this frequency, i.e., if $H(j\omega_0) = -1$. For frequencies close to the carrier, $\omega = \omega_0 + \Delta\omega$, the open-loop transfer function can be approximated as

$$H(j\omega) \approx H(j\omega_0) + \Delta\omega \frac{dH}{d\omega} \qquad (3)$$

and the noise tranfer function is

$$\frac{Y}{X}[j(\omega_0 + \Delta\omega)] = \frac{H(j\omega_0) + \Delta\omega \dfrac{dH}{d\omega}}{1 + H(j\omega_0) + \Delta\omega \dfrac{dH}{d\omega}}. \qquad (4)$$

Since $H(j\omega_0) = -1$ and for most practical cases $|\Delta\omega \, dH/d\omega| \ll 1$, (4) reduces to

$$\frac{Y}{X}[j(\omega_0 + \Delta\omega)] \approx \frac{-1}{\Delta\omega \dfrac{dH}{d\omega}}. \qquad (5)$$

This equation indicates that a noise component at $\omega = \omega_0 + \Delta\omega$ is multiplied by $-(\Delta\omega \, dH/d\omega)^{-1}$ when it appears at the output of the oscillator. In other words, the noise power spectral density is shaped by

$$\left|\frac{Y}{X}[j(\omega_0 + \Delta\omega)]\right|^2 = \frac{1}{(\Delta\omega)^2 \left|\dfrac{dH}{d\omega}\right|^2}. \qquad (6)$$

This is illustrated in Fig. 8. As we will see later, (6) assumes a simple form for ring oscillators.

To gain more insight, let $H(j\omega) = A(\omega)\exp[j\Phi(\omega)]$, and hence

$$\frac{dH}{d\omega} = \left(\frac{dA}{d\omega} + jA\frac{d\Phi}{d\omega}\right)\exp(j\Phi). \qquad (7)$$

Since for $\omega \approx \omega_0, A \approx 1$, (6) can be written as

$$\left|\frac{Y}{X}[j(\omega_0 + \Delta\omega)]\right|^2 = \frac{1}{(\Delta\omega)^2\left[\left(\dfrac{dA}{d\omega}\right)^2 + \left(\dfrac{d\Phi}{d\omega}\right)^2\right]}. \qquad (8)$$

We define the open-loop Q as

$$Q = \frac{\omega_0}{2}\sqrt{\left(\frac{dA}{d\omega}\right)^2 + \left(\frac{d\Phi}{d\omega}\right)^2}. \qquad (9)$$

Combining (8) and (9) yields

$$\left|\frac{Y}{X}[j(\omega_0 + \Delta\omega)]\right|^2 = \frac{1}{4Q^2}\left(\frac{\omega_0}{\Delta\omega}\right)^2 \qquad (10)$$

a familiar form previously derived for simple LC oscillators [4]. It is interesting to note that in an LC tank at resonance, $dA/d\omega = 0$ and (9) reduces to the third definition of Q given in Section III. In the two-integrator oscillator, on the other hand, $dA/d\omega = 2/\omega_0, d\Phi/d\omega = 0$, and $Q = 1$. Thus, the proposed definition of Q applies to most cases of interest.

To complete the discussion, we also consider the case shown in Fig. 9, where $H_1(j\omega)H_2(j\omega) = H(j\omega)$. Therefore, $Y(j\omega)/X(j\omega)$ is given by (5). For, $Y_1(j\omega)/X(j\omega)$, we have

$$\frac{Y_1}{X}(j\omega) = \frac{H_1(j\omega)}{1 + H(j\omega)} \qquad (11)$$

giving the following noise shaping function:

$$\left|\frac{Y_1}{X}[j(\omega_0 + \Delta\omega)]\right|^2 = \frac{|H_1|^2}{(\Delta\omega)^2\left|\dfrac{dH}{d\omega}\right|^2}. \qquad (12)$$

179

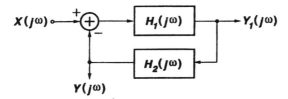

Fig. 9. Oscillatory system with nonunity-gain feedback.

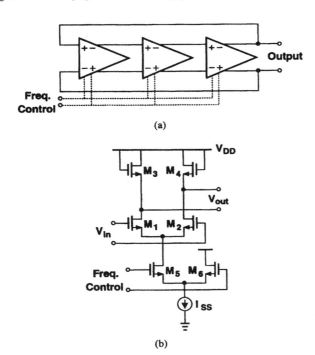

(a)

(b)

Fig. 10. CMOS VCO: (a) block diagram and (b) implementation of one stage.

V. CMOS RING OSCILLATOR

Submicron CMOS technologies have demonstrated potential for high-speed phase-locked systems [5], raising the possibility of designing fully integrated RF CMOS frequency synthesizers. Fig. 10 shows a three-stage ring oscillator wherein both the signal path and the control path are differential to achieve high common-mode rejection.

To calculate the phase noise, we model the signal path in the VCO with a linearized (single-ended) circuit (Fig. 11). As mentioned in Section IV, the linear approximation allows a first-order analysis of the topologies considered in this paper, but its accuracy must be checked if other oscillators are of interest. In Fig. 11, R and C represent the output resistance and the load capacitance of each stage, respectively, ($R \approx 1/g_{m3} = 1/g_{m4}$), and $G_m R$ is the gain required for steady oscillations. The noise of each differential pair and its load devices are modeled as current sources I_{n1}-I_{n3}, injected onto nodes 1–3, respectively. Before calculating the noise transfer function, we note that the circuit of Fig. 11 oscillates if, at ω_0, each stage has unity voltage gain and 120° of phase shift. Writing the open-loop transfer function and imposing these two conditions, we have $\omega_0 = \sqrt{3}/(RC)$ and $G_m R = 2$. The

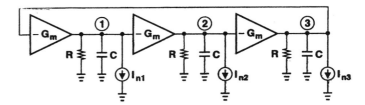

Fig. 11. Linearized model of CMOS VCO.

open-loop transfer function is thus given by

$$H(j\omega) = \frac{-8}{\left(1 + j\sqrt{3}\,\dfrac{\omega}{\omega_0}\right)^3}. \tag{13}$$

Therefore, $|dA/d\omega| = 9/(4\omega_0)$ and $|d\Phi/d\omega| = 3\sqrt{3}/(4\omega_0)$. It follows from (6) or (10) that if a noise current I_{n1} is injected onto node 1 in the oscillator of Fig. 11, then its power spectrum is shaped by

$$\left|\frac{V_1}{I_{n1}}[j(\omega_0 + \Delta\omega)]\right|^2 = \frac{R^2}{27}\left(\frac{\omega_0}{\Delta\omega}\right)^2. \tag{14}$$

This equation is the key to predicting various phase noise components in the ring oscillator.

VI. ADDITIVE AND MULTIPLICATIVE NOISE

Modeling the ring oscillator of Fig. 10 with the linearized circuit of Fig. 11 entails a number of issues. First, while the stages in Fig. 10 turn off for part of the period, the linearized model exhibits no such behavior, presenting constant values for the components in Fig. 11. Second, the model does not predict mixing or modulation effects that result from nonlinearities. Third, the noise of the devices in the signal path has a "cyclostationary" behavior, i.e., periodically varying statistics, because the bias conditions are periodic functions of time. In this section, we address these issues, first identifying three types of noise: additive, high-frequency multiplicative, and low-frequency multiplicative.

A. Additive Noise

Additive noise consists of components that are directly added to the output as shown in Fig. 7 and formulated by (6) and (14).

To calculate the additive phase noise in Fig. 10 with the aid of (14), we note that for $\omega \approx \omega_0$ the voltage gain in each stage is close to unity. (Simulations of the actual CMOS oscillator indicate that for $\omega_0 = 2\pi \times 970$ MHz and noise injected at $\omega - \omega_0 = 2\pi \times 10$ MHz onto one node, the components observed at the three nodes differ in magnitude by less than 0.1 dB.) Therefore, the total output phase noise power density due to I_{n1}-I_{n3} is

$$|V_{1\text{tot}}[j(\omega_0 + \Delta\omega)]|^2 = \frac{R^2}{9}\left(\frac{\omega_0}{\Delta\omega}\right)^2 \overline{I_n^2} \tag{15}$$

where it is assumed $\overline{I_{n1}^2} = \overline{I_{n2}^2} = \overline{I_{n3}^2} = \overline{I_n^2}$. For the differential stage of Fig. 10, the thermal noise current per unit bandwidth

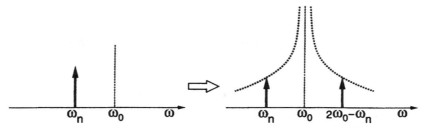

Fig. 12. High-frequency multiplicative noise.

is equal to $\overline{I_n^2} = 8kT(g_{m1} + g_{m3})/3 \approx 8kT/R$. Thus,

$$|V_{1\text{tot}}[j(\omega_0 + \Delta\omega)]|^2 = 8kT\frac{R}{9}\left(\frac{\omega_0}{\Delta\omega}\right)^2. \qquad (16)$$

In this derivation, the thermal drain noise current of MOS devices is assumed equal to $\overline{i_n^2} = 4kT(2g_m/3)$. For short-channel devices, however, the noise may be higher [6]. Using a charge-based model in our simulation tool, we estimate the factor to be 0.873 rather than 2/3. In reality, hot-electron effects further raise this value.

Additive phase noise is predicted by the linearized model with high accuracy if the stages in the ring operate linearly for most of the period. In a three-stage CMOS oscillator designed for the RF range, the differential stages are in the linear region for about 90% of the period. Therefore, the linearized model emulates the CMOS oscillator with reasonable accuracy. However, as the number of stages increases or if each stage entails more nonlinearity, the error in the linear approximation may increase.

Since additive noise is shaped according to (16), its effect is significant only for components close to the carrier frequency.

B. High-Frequency Multiplicative Noise

The nonlinearity in the differential stages of Fig. 10, especially as they turn off, causes noise components to be multiplied by the carrier (and by each other). If the input/output characteristic of each stage is expressed as $V_{\text{out}} = \alpha_1 V_{\text{in}} + \alpha_2 V_{\text{in}}^2 + \alpha_3 V_{\text{in}}^3$, then for an input consisting of the carrier and a noise component, e.g., $V_{\text{in}}(t) = A_0 \cos\omega_0 t + A_n \cos\omega_n t$, the output exhibits the following important terms:

$$V_{\text{out1}}(t) \propto \alpha_2 A_0 A_n \cos(\omega_0 \pm \omega_n)t$$
$$V_{\text{out2}}(t) \propto \alpha_3 A_0 A_n^2 \cos(\omega_0 - 2\omega_n)t$$
$$V_{\text{out3}}(t) \propto \alpha_3 A_0^2 A_n \cos(2\omega_0 - \omega_n)t.$$

Note that $V_{\text{out1}}(t)$ appears in band if ω_n is small, i.e., if it is a *low-frequency* component, but in a fully differential configuration, $V_{\text{out1}}(t) = 0$ because $\alpha_2 = 0$. Also, $V_{\text{out2}}(t)$ is negligible because $A_n \ll A_0$, leaving $V_{\text{out3}}(t)$ as the only significant cross-product.

This simplified one-stage analysis predicts the *frequency* of the components in response to injected noise, but not their *magnitude*. When noise is injected into the oscillator, the magnitude of the observed response at ω_n and $2\omega_0 - \omega_n$ depends on the noise shaping properties of the feedback

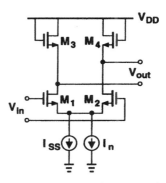

Fig. 13. Frequency modulation due to tail current noise.

oscillatory system. Simulations indicate that for the oscillator topologies considered here, these two components have approximately equal magnitudes. Thus, the nonlinearity folds all the noise components below ω_0 to the region above and vice versa, effectively doubling the noise power predicted by (6). Such components are significant if they are close to ω_0 and are herein called high-frequency multiplicative noise. This phenomenon is illustrated in Fig. 12. (Note that a component at $3\omega_0 + \Delta\omega$ is also translated to $\omega_0 + \Delta\omega$, but its magnitude is negligible.)

This effect can also be viewed as sampling of the noise by the differential pairs, especially if each stage experiences hard switching. As each differential pair switches twice in every period, a noise component at ω_n is translated to $2\omega_0 \pm \omega_n$. Note that for highly nonlinear stages, the Taylor expansion considered above may need to include higher order terms.

C. Low-Frequency Multiplicative Noise

Since the frequency of oscillation in Fig. 10 is a function of the tail current in each differential pair, noise components in this current modulate the frequency, thereby contributing phase noise [classical frequency modulation (FM)]. Depicted in Fig. 13, this effect can be significant because, in CMOS oscillators, ω_0 must be adjustable by more than $\pm 20\%$ to compensate for process variations, thus making the frequency quite sensitive to noise in the tail current. This mechanism is illustrated in Fig. 14.

To quantify this phenomenon, we find the sensitivity or "gain" of the VCO, defined as $K_{\text{VCO}} = d\omega_{\text{out}}/dI_{SS}$ in Fig. 13, and use a simple approximation. If the noise per unit bandwidth in I_{SS} is represented as a sinusoid with the same

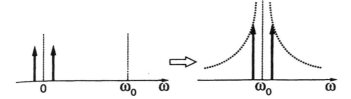

Fig. 14.   Low-frequency multiplicative noise.

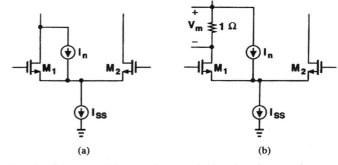

(a)   (b)

Fig. 15.   Gain stage with (a) stationary and (b) cyclostationary noise.

power: $I_m \cos \omega_m t$, then the output signal of the oscillator can be written as

$$v_{\text{out}}(t) = A_0 \cos \left(\omega_0 t + K_{\text{VCO}} \int I_m \cos \omega_m t \, dt \right) \quad (17)$$

$$= A_0 \cos \left(\omega_0 t + \frac{K_{\text{VCO}}}{\omega_m} I_m \sin \omega_m t \right). \quad (18)$$

For $K_{\text{VCO}} I_m / \omega_m \ll 1$ radian ("narrowband FM")

$$v_{\text{out}}(t) \approx A_0 \cos \omega_0 t + \frac{A_0 I_m K_{\text{VCO}}}{2\omega_m}$$
$$\cdot [\cos(\omega_0 + \omega_m)t - \cos(\omega_0 - \omega_m)t]. \quad (19)$$

Thus, the ratio of each sideband amplitude to the carrier amplitude is equal to $I_m K_{\text{VCO}}/(2\omega_m)$, i.e.,

$$|V_n|^2 \text{(with respect to carrier)} = \frac{1}{4} \left(\frac{K_{\text{VCO}}}{\omega_m} \right)^2 I_m^2. \quad (20)$$

Since K_{VCO} can be easily evaluated in simulation or measurement, (20) is readily calculated.

It is seen that modulation of the carrier brings the low frequency noise components of the tail current to the band around ω_0. Thus, flicker noise in I_n becomes particularly important.

In the differential stage of Fig. 3(b), two sources of low-frequency multiplicative noise can be identified: noise in I_{SS} and noise in M_5 and M_6. For comparable device size, these two sources are of the same order and must be both taken into account.

D. Cyclostationary Noise Sources

As mentioned previously, the devices in the signal path exhibit cyclostationary noise behavior, requiring the use of periodically varying noise statistics in analysis and simulations. To check the accuracy of the stationary noise approximation, we perform a simple, first-order simulation on the two cases depicted in Fig. 15. In Fig. 15(a), a sinusoidal current source with an amplitude of 2 nA is connected between the drain and source of M_1 to represent its noise with the assumption that M_1 carries half of I_{SS}. In Fig. 15(b), the current source is also a sinusoid, but its amplitude is a function of the drain current of M_1. Since MOS thermal noise current (in the saturation region) is proportional to $\sqrt{g_m}$, we use a nonlinear dependent source in SPICE [7] as $I_n(t) = \alpha \sqrt[4]{V_m(t)} \sin \omega_n t$, where $\omega_n = 2\pi \times 980$ MHz. The factor α is chosen such that $I_n(t) = 2$ nA $\times \sin \omega_n t$ when $V_m(t) = 1 \, \Omega \times I_{SS}/2$ (balanced

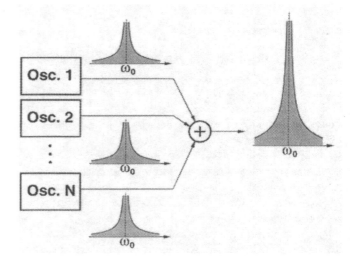

Fig. 16.   Addition of output voltages of N oscillators.

condition). Simulations indicate that the sideband magnitudes in the two cases differ by less than 0.5 dB.

It is important to note that this result may not be accurate for other types of oscillators.

E. Power-Noise Trade-Off

As with other analog circuits, oscillators exhibit a trade-off between power dissipation and noise. Intuitively, we note that if the output voltages of N identical oscillators are added in phase (Fig. 16), then the total carrier power is multiplied by N^2, whereas the noise power increases by N (assuming noise sources of different oscillators are uncorrelated). Thus, the phase noise (relative to the carrier) decreases by a factor N at the cost of a proportional increase in power dissipation.

Using the equations developed above, we can also formulate this trade-off. For example, from (16), since $G_m R \approx 2$, we have

$$|V_{ntot}|^2 = 8kT \frac{2}{9G_m} \left(\frac{\omega_0}{\Delta \omega} \right)^2. \quad (21)$$

To reduce the total noise power by N, G_m must increase by the same factor. For any active device, this can be accomplished by increasing the width and the bias current by N. (To maintain the same frequency of oscillation, the load resistor is reduced by N.) Therefore, for a constant supply voltage, the power dissipation scales up by N.

TABLE I
COMPARISON OF THREE-STAGE AND FOUR-STAGE RING OSCILLATORS

	3-Stage VCO	4-Stage VCO
Minimum Required DC Gain	2	$\sqrt{2}$
Noise Shaping Function	$\dfrac{R^2}{27}(\dfrac{\omega_0}{\Delta\omega})^2$	$\dfrac{R^2}{16}(\dfrac{\omega_0}{\Delta\omega})^2$
Open-Loop Q	$\dfrac{3\sqrt{3}}{4}\ (\approx 1.3)$	$\sqrt{2}\ (\approx 1.4)$
Total Additive Noise	$8kT\dfrac{R}{9}(\dfrac{\omega_0}{\Delta\omega})^2$	$8kT\dfrac{R(1+\sqrt{2})}{12}(\dfrac{\omega_0}{\Delta\omega})^2$
Power Dissipation	1.8 mW	3.6 mW

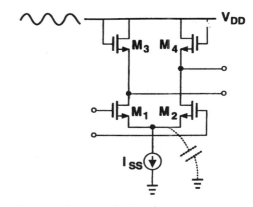

Fig. 17. Substrate and supply noise in gain stage.

F. Three-Stage Versus Four-Stage Oscillators

The choice of number of stages in a ring oscillator to minimize the phase noise has often been disputed. With the above formulations, it is possible to compare rings with different number of stages (so long as the approximations remain valid). For the cases of interest in RF applications, we consider three-stage and four-stage oscillators designed to operate at the same frequency. Thus, the four-stage oscillator incorporates smaller impedance levels and dissipates more power. Table I compares various aspects of the two circuits. We make three important observations. 1) Simulations show that if the four-stage oscillator is to operate at the same speed as the three-stage VCO, the value of R in the former must be approximately 60% of that in the latter. 2) The Q's of the two VCO's (10) are roughly equal. 3) The total additive thermal noise of the two VCO's is about the same, because the four-stage topology has more sources of noise, but with lower magnitudes.

From these rough calculations, we draw two conclusions. First, the phase noise depends on not only the Q, but the number and magnitude of sources of noise in the circuit. Second, four-stage VCO's have no significant advantage over three-stage VCO's, except for providing quadrature outputs.

G. Supply and Substrate Noise

Even though the gain stage of Fig. 10 is designed as a differential circuit, it nonetheless suffers from some sensitivity to supply and substrate noise (Fig. 17). Two phenomena account for this. First, device mismatches degrade the symmetry of the circuit. Second, the total capacitance at the common source of the differential pair (i.e., the source junction capacitance of M_1 and M_2 and the capacitance associated with the tail current source) converts the supply and substrate noise to current, thereby modulating the delay of the gain stage. Simulations indicate that even if the tail current source has a high dc output impedance, a 1-mV$_{pp}$ supply noise component at 10 MHz generates sidebands 60 dB below the carrier at $\omega_0 \pm (2\pi \times 10$ MHz).

VII. CMOS RELAXATION OSCILLATOR

In this section, we apply the analysis methodology described thus far to a CMOS relaxation oscillator [Fig. 18(a)]. When designed to operate at 900 MHz, this circuit hardly "relaxes" and the signals at the drain and source of M_1 and M_2 are close to sinusoids. Thus, the linear model of Fig. 7 is a plausible choice. To utilize our previous results, we assume the signals at the sources of M_1 and M_2 are fully differential[1] and redraw the circuit as in Fig. 18(b), identifying it as a two-stage ring with capacitive degeneration ($C_A = 2C$). The total capacitance seen at the drain of M_1 and M_2 is modeled with C_1 and C_2, respectively. (This is also an approximation because the input impedance of each stage is not purely capacitive.) It can be easily shown that the open-loop transfer function is

$$H(s) = \left[\frac{-g_m RC_A s}{(g_m + C_A s)(RC_D s + 1)} \right]^2 \quad (22)$$

where $C_1 = C_2 = C_D$ and g_m denotes the transconductance of each transistor. For the circuit to oscillate at ω_0, $H(j\omega_0) = 1$, and each stage must have a phase shift of 180°, with 90° contributed by each zero and the remaining 90° by the two poles at $-g_m/C_A$ and $-1/(RC_D)$. It follows from the second condition that

$$\omega_0^2 = \frac{g_m}{RC_A C_D} \quad (23)$$

i.e., ω_0 is the geometric mean of the poles at the drain and source of each transistor. Combining this result with the first condition, we obtain

$$g_m R = \frac{C_A}{C_A - C_D}. \quad (24)$$

After lengthy calculations, we have

$$\left| \frac{dH}{d\omega} \right| = \frac{4}{RC_A \omega_0^2} \quad (25)$$

and

$$Q^2 = 4\left(1 - \frac{C_D}{C_A}\right)\frac{C_D}{C_A}. \quad (26)$$

[1] This assumption is justified by decomposing C into two series capacitors, each one of value $2C$, and monitoring the midpoint voltage. The common-mode swing at this node is approximatley 18 dB below the differential swings at the source of M_1 and M_2.

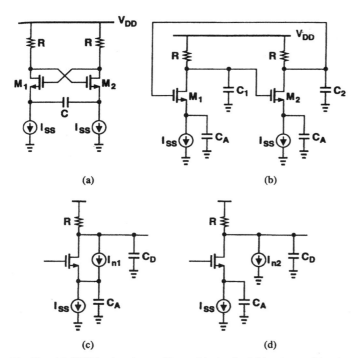

(a)

(b)

(c)

(d)

Fig. 18. (a) CMOS relaxation oscillator, (b) circuit of (a) redrawn, (c) noise current of one transistor, and (d) transformed noise current.

For $C_D = 0.5C_A, Q$ reaches its maximum value—unity. In other words, the maximum Q occurs if the (floating) timing capacitor is equal to the load capacitance. The noise shaping function is therefore equal to $(\omega_0/\Delta\omega)^2/4$.

Since the drain-source noise current of M_1 and M_2 appears between two internal nodes of the circuit [Fig. 18(c)], the transformation shown in Fig. 18(d) can be applied to allow the use of our previous derivations. It can be shown that

$$I_{n2} = \frac{C_A s}{g_m + C_A s} I_{n1} \qquad (27)$$

and the total additive thermal noise observed at each drain is

$$\overline{V_n^2} = \frac{10}{3} kTR \left(\frac{\omega_0}{\Delta\omega}\right)^2. \qquad (28)$$

This power must be doubled to account for high-frequency multiplicative noise.

VIII. SIMULATION RESULTS

A. Simulation Issues

The time-varying nature of oscillators prohibits the use of the standard small-signal ac analysis available in SPICE and other similar programs. Therefore, simulations must be performed in the time domain. As a first attempt, one may generate a pseudo-random noise with known distribution, introduce it into the circuit as a SPICE piecewise linear waveform, run a transient analysis for a relatively large number of oscillation periods, write the output as a series of points equally spaced in time, and compute the fast Fourier transform (FFT) of the output. The result of one such attempt is shown in Fig. 19. It is important to note that 1) many coherent sidebands

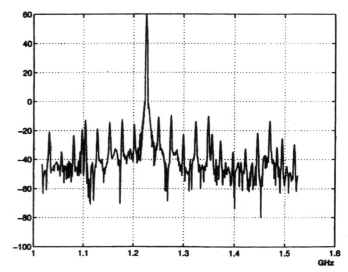

Fig. 19. Simulated oscillator spectrum with injected white noise.

appear in the spectrum even though the injected noise is white, and 2) the magnitude of the sidebands does not directly scale with the magnitude of the injected noise!

To understand the cause of this behavior, consider a much simpler case, illustrated in Fig. 20. In Fig. 20(a), a sinusoid at 1 GHz is applied across a 1-kΩ resistor, and a long transient simulation followed by interpolation and FFT is used to obtain the depicted spectrum. (The finite width results from the finite length of the data record and the "arches" are attributed to windowing effects.) Now, as shown in Fig. 20(b), we add a 30-MHz squarewave with 2 ns transition time and proceed as before. Note that the two circuits share only the ground node. In this case, however, the spectrum of the 1-GHz sinusoid exhibits coherent sidebands with 15 MHz spacing! Observed in AT&T's internal simulator (ADVICE), HSPICE, and Cadence SPICE, this effect is attributed to the additional points that the program must calculate at each edge of the squarewave, leading to errors in subsequent interpolation.

Fortunately, this phenomenon does not occur if only sinusoids are used in simulations.

B. Oscillator Simulations

In order to compute the response of oscillators to each noise source, we approximate the noise per unit bandwidth at frequency ω_n with an impulse (a sinusoid) of the same power at that frequency. As shown in Fig. 21, the "sinusoidal noise" is injected at various points in the circuit and the output spectrum is observed. This approach is justified by the fact that random Gaussian noise can be expressed as a Fourier series of sinusoids with random phase [8], [9]. Since only one sinusoid is injected in each simulation, the interaction among noise components themselves is assumed negligible, a reasonable approximation because if two noise components at, say, −60 dB are multiplied, the product is at −120 dB.

In the simulations, the oscillators were designed for a center frequency of approximately 970 MHz. Each circuit and its linearized models were simulated in the time domain in steps

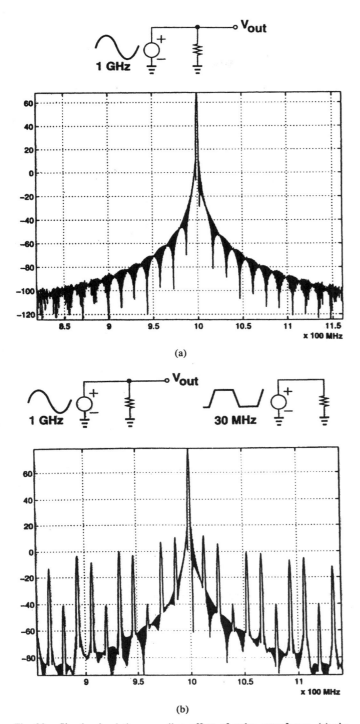

(a)

(b)

Fig. 20. Simple simulation revealing effect of pulse waveforms, (a) single sinusoidal source and (b) sinusoidal source along with a square wave generator.

of 30 ps for 8 μs, and the output was processed in MATLAB to obtain the spectrum. Since simulations of the linear model yield identical results to the equations derived above, we will not distinguish between the two hereafter.

Shown in Fig. 22 are the output spectra of the linear model and actual circuit of a three-stage oscillator in 0.5-μm CMOS technology with a 2-nA$_p$ 980-MHz sinusoidal current injected into the signal path (the drain of one of the differential pairs).

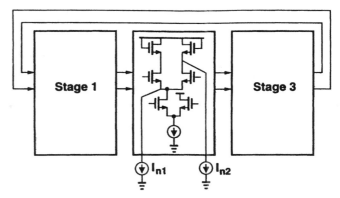

Fig. 21. Simulated configuration.

The vertical axis represents $10 \log V_{\text{rms}}^2$. Note that the observed magnitude of the 980-MHz component differs by less than 0.2 dB in the two cases, indicating that the linearized model is indeed an accurate representation. As explained in Section VI-B, the 960-MHz component originates from third-order mixing of the carrier and the 980-MHz component and essentially doubles the phase noise.

In order to investigate the limitation of the linear model, the oscillator was made progressively more nonlinear. Shown in Fig. 23 is the output spectra of a four-stage CMOS oscillator, revealing approximately 1 dB of error in the prediction by the linear model. The error gradually increases with the number of stages in the ring and reaches nearly 6 dB for an eight-stage oscillator.

For bipolar ring oscillators (differential pairs with no emitter followers), simulations reveal an error of approximately 2 dB for three stages and 7 dB for four stages in the ring.

IX. EXPERIMENTAL RESULTS

A. Measurements

Two different oscillator configurations have been fabricated in a 0.5-μm CMOS technology to compare the predictions in this paper with measured results. Note that there are three sets of results: theoretical calculations based on linear models but including multiplicative noise, simulated predictions based on the actual CMOS oscillators, and measured values.

The first circuit is a 2.2-GHz three-stage ring oscillator. Fig. 24 shows one stage of the circuit along with the measured device parameters. The sensitivity of the output frequency to the tail current of each stage is about 0.43 MHz/μA. The measured spectrum is depicted in Fig. 25(a) and (b) with two different horizontal scales. Due to lack of data on the flicker noise of the process, we consider only thermal noise at relatively large frequency offsets, namely, 1 MHz and 5 MHz.

It is important to note that low-frequency flicker noise causes the center of the spectrum to fluctuate constantly. Thus, as the resolution bandwidth (RBW) of the spectrum analyzer is reduced [from 1 MHz in Fig. 25(a) to 100 kHz in Fig. 25(b)], the carrier power is subject to more averaging and appears to decrease. To maintain consistency with calculations, in which

185

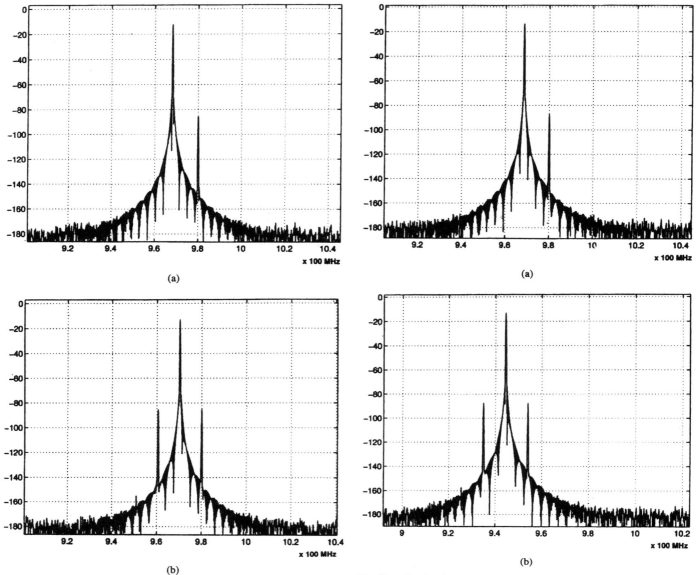

Fig. 22. Simulated output spectra of (a) linear model and (b) actual circuit of a three-stage CMOS oscillator.

Fig. 23. Simulated output spectra of (a) linear model and (b) actual circuit of a four-stage CMOS oscillator.

the phase noise is normalized to a *constant* carrier power, this power (i.e., the output amplitude) is measured using an oscilloscope.

The noise calculation proceeds as follows. First, find the additive noise power in (16), and double the result to account for third-order mixing (high-frequency multiplicative noise). Next, calculate the low-frequency multiplicative noise from (20) for one stage and multiply the result by three. We assume (from simulations) that the internal differential voltage swing is equal to 1 V_{pp} (0.353 V_{rms}) and the drain noise current of MOSFET's is given by $\overline{i_n^2} = 4kT(0.863g_m)$. For $\Delta\omega = 2\pi \times 1$ MHz, calculations yield

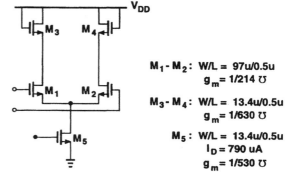

M_1-M_2: W/L = 97u/0.5u
g_m= 1/214 ℧

M_3-M_4: W/L = 13.4u/0.5u
g_m= 1/630 ℧

M_5: W/L = 13.4u/0.5u
I_D = 790 uA
g_m= 1/530 ℧

Fig. 24. Gain stage used in 2-GHz CMOS oscillator.

high-frequency multiplicative noise = -100.1 dBc/Hz   (29)

low-frequency multiplicative noise = -106.3 dBc/Hz   (30)

total normalized phase noise = -99.2 dBc/Hz.   (31)

Simulations of the actual CMOS oscillator predict the total noise to be -98.1 dBc/Hz. From Fig. 25(b), with the carrier power of Fig. 25(a), the phase noise is approximately equal to -94 dBc/Hz.

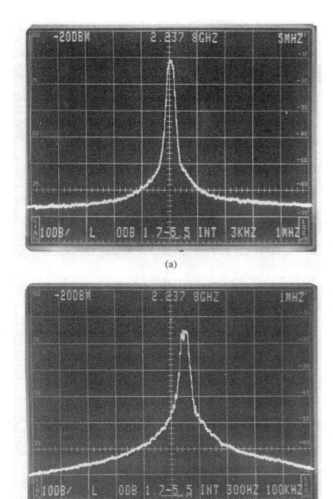

(a)

(b)

Fig. 25. Measured output spectrum of ring oscillator (10 dB/div. vertical scale). (a) 5 MHz/div. horizontal scale and 1 MHz resolution bandwidth, (b) 1 MHz horizontal scale and 100 kHz resolution bandwidth.

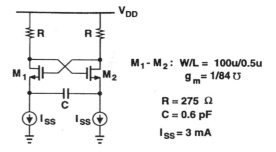

Fig. 26. Relaxation oscillator parameters.

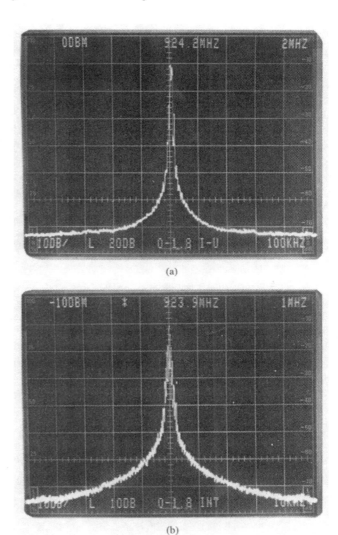

(a)

(b)

Fig. 27. Measured output spectrum of relaxation oscillator (10 dB/div. vertical scale). (a) 2 MHz/div. horizontal scale and 100 kHz resolution bandwidth and (b) 1 MHz horizontal scale and 10 kHz resolution bandwidth.

Similarly, for $\Delta\omega = 2\pi \times 5$ MHz, calculations yield

$$\text{high-frequency multiplicative noise} = -114.0 \text{ dBc/Hz} \quad (32)$$
$$\text{low-frequency multiplicative noise} = -120.2 \text{ dBc/Hz} \quad (33)$$
$$\text{total normalized phase noise} = -113.1 \text{ dBc/Hz} \quad (34)$$

and simulations predict -112.4 dBc/Hz, while Fig. 25(a) indicates a phase noise of -109 dBc/Hz. Note that these values correspond to a center frequency of 2.2 GHz and should be lowered by approximately 8 dB for 900 MHz operation, as shown in (9).

The second circuit is a 920-MHz relaxation oscillator, depicted in Fig. 26. The measured spectra are shown in Fig. 27. Since simulations indicate that the low-frequency multiplicative noise is negligible in this implementation, we consider only the thermal noise in the signal path. For $\Delta\omega = 2\pi \times 1$ MHz, calculations yield a relative phase noise of -105 dBc/Hz, simulations predict -98 dB, and the spectrum in Fig. 27 gives -102 dBc/Hz. For $\Delta\omega = 2\pi \times 5$ MHz, the calculated and simulated results are -119 dBc/Hz and

-120 dBc/Hz, respectively, while the measured value is -115 dBc/Hz.

B. Discussion

Using the above measured data points and assuming a noise shaping function as in (10) with a linear noise-power trade-off (Fig. 16), we can make a number of observations.

How much can the phase noise be lowered by scaling device dimensions? If the gate oxide of MOSFET's is reduced indefinitely, their transconductance becomes relatively independent of their dimensions, approaching roughly that of bipolar transistors. Thus, in the gain stage of Fig. 24 the transconductance of M_1 and M_2 (for $I_{D1} = I_{D2} = 395 \ \mu A$) would go from $(214 \ \Omega)^{-1}$ to $(66 \ \Omega)^{-1}$. Scaling down the load resistance proportionally and assuming a constant oscillation frequency, we can therefore lower the phase noise by $10\log(214/66) \approx 5$ dB. For the relaxation oscillator, on the other hand, the improvement is about 10 dB. These are, of course, greatly simplified calculations, but they provide an estimate of the maximum improvement expected from technology scaling. In reality, short-channel effects, finite thickness of the inversion layer, and velocity saturation further limit the transconductance that can be achieved for a given bias current.

It is also instructive to compare the measured phase noise of the above ring oscillator with that of a 900-MHz three-stage CMOS ring oscillator reported in [10]. The latter employs single-ended CMOS inverters with rail-to-rail swings in a 1.2-μm technology and achieves a phase noise of -83 dBc/Hz at 100 kHz offset while dissipating 7.4 mW from a 5-V supply.

Assuming that

$$\text{Relative Phase Noise} \propto \left(\frac{\omega_0}{\Delta\omega}\right)^2 \frac{1}{V_{\text{swing}}^2} \frac{1}{I_{DD}} \qquad (35)$$

where V_{swing} denotes the internal voltage swing and I_{DD} is the total supply current, we can utilize the measured phase noise of one oscillator to roughly estimate that of the other. With the parameters of the 2.2-GHz oscillator and accounting for different voltage swings and supply currents, we obtain a phase noise of approximately -93 dBc/Hz at 100 kHz offset for the 900-MHz oscillator in [10]. The 10 dB discrepancy is attributed to the difference in the minimum channel length, $1/f$ noise at 100 kHz, and the fact that the two circuits incorporate different gain stages.

ACKNOWLEDGMENT

The author wishes to thank V. Gopinathan for many illuminating discussions and T. Aytur for providing the relaxation oscillator simulation and measurement results.

REFERENCES

[1] A. A. Abidi and R. G. Meyer, "Noise in relaxation oscillators," *IEEE J. Solid-State Circuits*, vol. SC-18, pp. 794–802, Dec. 1983.
[2] T. C. Weigandt, B. Kim, and P. R. Gray, "Analysis of timing jitter in cmos ring oscillators," in *Proc. ISCAS*, June 1994.
[3] A. A. Abidi, "Direct conversion radio tranceivers for digital communications," in *ISSCC Dig. Tech. Papers*, Feb. 1995, pp. 186–187.
[4] D. B. Leeson, "A simple model of feedback oscillator noise spectrum," *Proc. IEEE*, pp. 329–330, Feb. 1966.
[5] B. Razavi, K. F. Lee, and R.-H. Yan, "Design of high-speed low-power frequency dividers and phase-locked loops in deep submicron CMOS," *IEEE J. Solid-State Circuits*, vol. 30, pp. 101–109, Feb. 1995.
[6] Y. P. Tsividis, *Operation and Modeling of the MOS Transistor*. New York: McGraw-Hill, 1987.
[7] J. A. Connelly and P. Choi, *Macromodeling with SPICE*. Englewood Cliffs, NJ: Prentice-Hall, 1992.
[8] S. O. Rice, "Mathematical analysis of random noise," *Bell System Tech. J.*, pp. 282–332, July 1944, and pp. 46–156, Jan. 1945.
[9] P. Bolcato *et al.*, "A new and efficient transient noise analysis technique for simulation of CCD image sensors or particle detectors," in *Proc. CICC*, 1993, pp. 14.8.1–14.8.4.
[10] T. Kwasniewski, *et al.*, "Inductorless oscillator design for personal communications devices—A 1.2 μm CMOS process case study," in *Proc. CICC*, May 1995, pp. 327–330.

Behzad Razavi (S'87–M'91) received the B.Sc. degree in electrical engineering from Tehran (Sharif) University of Technology, Tehran, Iran, in 1985, and the M.Sc. and Ph.D. degrees in electrical engineering from Stanford University, Stanford, CA, in 1988 and 1991, respectively.

From 1992 to 1996, he was a Member of Technical Staff at AT&T Bell Laboratories, Holmdel, NJ, where his research involved integrated circuit design for communication systems. He is now with Hewlett-Packard Laboratories, Palo Alto, CA. His current interests include wireless transceivers, data conversion, clock recovery, frequency synthesis, and low-voltage low-power circuits. He has been a Visiting Lecturer at Princeton University, Princeton, NJ, and Stanford University. He is also a member of the Technical Program Committee of the International Solid-State Circuits Conference. He has served as Guest Editor to the IEEE JOURNAL OF SOLID-STATE CIRCUITS and *International Journal of High Speed Electronics* and is currently an Associate Editor of JSSC. He is the author of the book *Principles of Data Conversion System Design* (IEEE Press, 1995), and editor of *Monolothic Phase-Locked Loops and Clock Recovery Circuits* (IEEE Press, 1996).

Dr. Razavi received the Beatrice Winner Award for Editorial Excellence at the 1994 ISSCC, the best paper award at the 1994 European Solid-State Circuits Conference, and the best panel award at the 1995 ISSCC.

A General Theory of Phase Noise in Electrical Oscillators

Ali Hajimiri, *Student Member, IEEE*, and Thomas H. Lee, *Member, IEEE*

Abstract— A general model is introduced which is capable of making accurate, quantitative predictions about the phase noise of different types of electrical oscillators by acknowledging the true periodically time-varying nature of all oscillators. This new approach also elucidates several previously unknown design criteria for reducing close-in phase noise by identifying the mechanisms by which intrinsic device noise and external noise sources contribute to the total phase noise. In particular, it explains the details of how $1/f$ noise in a device upconverts into close-in phase noise and identifies methods to suppress this upconversion. The theory also naturally accommodates cyclostationary noise sources, leading to additional important design insights. The model reduces to previously available phase noise models as special cases. Excellent agreement among theory, simulations, and measurements is observed.

Index Terms—Jitter, oscillator noise, oscillators, oscillator stability, phase jitter, phase locked loops, phase noise, voltage controlled oscillators.

I. INTRODUCTION

THE recent exponential growth in wireless communication has increased the demand for more available channels in mobile communication applications. In turn, this demand has imposed more stringent requirements on the phase noise of local oscillators. Even in the digital world, phase noise in the guise of jitter is important. Clock jitter directly affects timing margins and hence limits system performance.

Phase and frequency fluctuations have therefore been the subject of numerous studies [1]–[9]. Although many models have been developed for different types of oscillators, each of these models makes restrictive assumptions applicable only to a limited class of oscillators. Most of these models are based on a linear time invariant (LTI) system assumption and suffer from not considering the complete mechanism by which electrical noise sources, such as device noise, become phase noise. In particular, they take an empirical approach in describing the upconversion of low frequency noise sources, such as $1/f$ noise, into close-in phase noise. These models are also reduced-order models and are therefore incapable of making accurate predictions about phase noise in long ring oscillators, or in oscillators that contain essential singularities, such as delay elements.

Manuscript received December 17, 1996; revised July 9, 1997.

The authors are with the Center for Integrated Systems, Stanford University, Stanford, CA 94305-4070 USA.

Publisher Item Identifier S 0018-9200(98)00716-1.

Since any oscillator is a periodically time-varying system, its time-varying nature must be taken into account to permit accurate modeling of phase noise. Unlike models that assume linearity and time-invariance, the time-variant model presented here is capable of proper assessment of the effects on phase noise of both stationary and even of cyclostationary noise sources.

Noise sources in the circuit can be divided into two groups, namely, device noise and interference. Thermal, shot, and flicker noise are examples of the former, while substrate and supply noise are in the latter group. This model explains the exact mechanism by which spurious sources, random or deterministic, are converted into phase and amplitude variations, and includes previous models as special limiting cases.

This time-variant model makes explicit predictions of the relationship between waveform shape and $1/f$ noise upconversion. Contrary to widely held beliefs, it will be shown that the $1/f^3$ corner in the phase noise spectrum is *smaller* than $1/f$ noise corner of the oscillator's components by a factor determined by the symmetry properties of the waveform. This result is particularly important in CMOS RF applications because it shows that the effect of inferior $1/f$ device noise can be reduced by proper design.

Section II is a brief introduction to some of the existing phase noise models. Section III introduces the time-variant model through an impulse response approach for the excess phase of an oscillator. It also shows the mechanism by which noise at different frequencies can become phase noise and expresses with a simple relation the sideband power due to an arbitrary source (random or deterministic). It continues with explaining how this approach naturally lends itself to the analysis of cyclostationary noise sources. It also introduces a general method to calculate the total phase noise of an oscillator with multiple nodes and multiple noise sources, and how this method can help designers to spot the dominant source of phase noise degradation in the circuit. It concludes with a demonstration of how the presented model reduces to existing models as special cases. Section IV gives new design implications arising from this theory in the form of guidelines for low phase noise design. Section V concludes with experimental results supporting the theory.

II. BRIEF REVIEW OF EXISTING MODELS AND DEFINITIONS

The output of an ideal sinusoidal oscillator may be expressed as $V_{\text{out}}(t) = A \cos[\omega_0 t + \phi]$, where A is the amplitude,

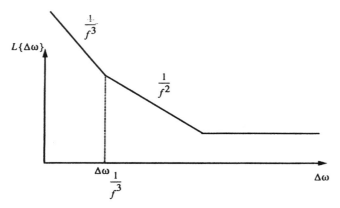

Fig. 1. Typical plot of the phase noise of an oscillator versus offset from carrier.

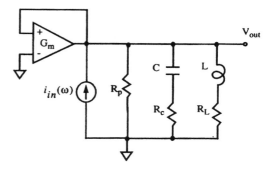

Fig. 2. A typical RLC oscillator.

ω_0 is the frequency, and ϕ is an arbitrary, fixed phase reference. Therefore, the spectrum of an ideal oscillator with no random fluctuations is a pair of impulses at $\pm\omega_0$. In a practical oscillator, however, the output is more generally given by

$$V_{\text{out}}(t) = A(t) \cdot f[\omega_0 t + \phi(t)] \qquad (1)$$

where $\phi(t)$ and $A(t)$ are now functions of time and f is a periodic function with period 2π. As a consequence of the fluctuations represented by $\phi(t)$ and $A(t)$, the spectrum of a practical oscillator has sidebands close to the frequency of oscillation, ω_0.

There are many ways of quantifying these fluctuations (a comprehensive review of different standards and measurement methods is given in [4]). A signal's short-term instabilities are usually characterized in terms of the single sideband noise spectral density. It has units of decibels below the carrier per hertz (dBc/Hz) and is defined as

$$\mathcal{L}_{\text{total}}\{\Delta\omega\} = 10 \cdot \log\left[\frac{\mathcal{P}_{\text{sideband}}(\omega_0 + \Delta\omega, \ 1 \ \text{Hz})}{\mathcal{P}_{\text{carrier}}}\right] \qquad (2)$$

where $\mathcal{P}_{\text{sideband}}(\omega_0 + \Delta\omega, \ 1 \ \text{Hz})$ represents the single sideband power at a frequency offset of $\Delta\omega$ from the carrier with a measurement bandwidth of 1 Hz. Note that the above definition includes the effect of both amplitude and phase fluctuations, $A(t)$ and $\phi(t)$.

The advantage of this parameter is its ease of measurement. Its disadvantage is that it shows the sum of both amplitude and phase variations; it does not show them separately. However, it is important to know the amplitude and phase noise separately because they behave differently in the circuit. For instance, the effect of amplitude noise is reduced by amplitude limiting mechanism and can be practically eliminated by the application of a limiter to the output signal, while the phase noise cannot be reduced in the same manner. Therefore, in most applications, $\mathcal{L}_{\text{total}}\{\Delta\omega\}$ is dominated by its phase portion, $\mathcal{L}_{\text{phase}}\{\Delta\omega\}$, known as phase noise, which we will simply denote as $\mathcal{L}\{\Delta\omega\}$.

The semi-empirical model proposed in [1]–[3], known also as the Leeson–Cutler phase noise model, is based on an LTI assumption for tuned tank oscillators. It predicts the following behavior for $\mathcal{L}\{\Delta\omega\}$:

$$\mathcal{L}\{\Delta\omega\} = 10 \cdot \log\left\{\frac{2FkT}{P_s} \cdot \left[1 + \left(\frac{\omega_0}{2Q_L\Delta\omega}\right)^2\right] \cdot \left(1 + \frac{\Delta\omega_{1/f^3}}{|\Delta\omega|}\right)\right\} \qquad (3)$$

where F is an empirical parameter (often called the "device excess noise number"), k is Boltzmann's constant, T is the absolute temperature, P_s is the average power dissipated in the resistive part of the tank, ω_0 is the oscillation frequency, Q_L is the effective quality factor of the tank with all the loadings in place (also known as loaded Q), $\Delta\omega$ is the offset from the carrier and $\Delta\omega_{1/f}^3$ is the frequency of the corner between the $1/f^3$ and $1/f^2$ regions, as shown in the sideband spectrum of Fig. 1. The behavior in the $1/f^2$ region can be obtained by applying a transfer function approach as follows. The impedance of a parallel RLC, for $\Delta\omega \ll \omega_0$, is easily calculated to be

$$Z(\omega_0 + \Delta\omega) \approx \frac{1}{G_L} \cdot \frac{1}{1 + j2Q_L\dfrac{\Delta\omega}{\omega_0}} \qquad (4)$$

where G_L is the parallel parasitic conductance of the tank. For steady-state oscillation, the equation $G_m R_L = 1$ should be satisfied. Therefore, for a parallel current source, the closed-loop transfer function of the oscillator shown in Fig. 2 is given by the imaginary part of the impedance

$$H(\Delta\omega) = \frac{v_{\text{out}}(\omega_0 + \Delta\omega)}{i_{\text{in}}(\omega_0 + \Delta\omega)} = -j\frac{1}{G_L} \cdot \frac{\omega_0}{2Q_L\Delta\omega}. \qquad (5)$$

The total equivalent parallel resistance of the tank has an equivalent mean square noise current density of $\overline{i_n^2}/\Delta f = 4kTG_L$. In addition, active device noise usually contributes a significant portion of the total noise in the oscillator. It is traditional to combine all the noise sources into one effective noise source, expressed in terms of the resistor noise with

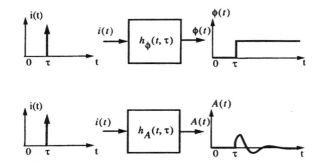

Fig. 3. Phase and amplitude impulse response model.

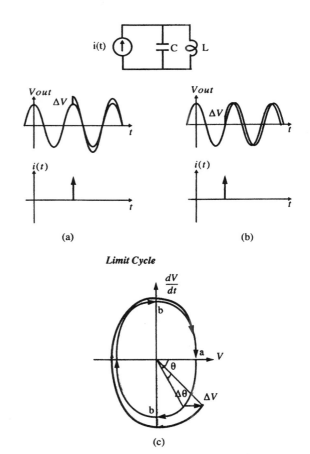

Fig. 4. (a) Impulse injected at the peak, (b) impulse injected at the zero crossing, and (c) effect of nonlinearity on amplitude and phase of the oscillator in state-space.

a multiplicative factor, F, known as the device excess noise number. The equivalent mean square noise current density can therefore be expressed as $\overline{i_n^2}/\Delta f = 4FkTG_L$. Unfortunately, it is generally difficult to calculate F *a priori*. One important reason is that much of the noise in a practical oscillator arises from periodically varying processes and is therefore cyclostationary. Hence, as mentioned in [3], F and $\omega_{1/f}^3$ are usually used as *a posteriori* fitting parameters on measured data.

Using the above effective noise current power, the phase noise in the $1/f^2$ region of the spectrum can be calculated as

$$
\begin{aligned}
\mathcal{L}\{\Delta\omega\} &= 10 \cdot \log\left(\frac{\overline{v_{\text{noise}}^2}}{v_{\text{sig}}^2}\right) \\
&= 10 \cdot \log\left[\frac{\frac{1}{2}\cdot|H(\Delta\omega)|^2\cdot\overline{i_n^2}/\Delta f}{\frac{1}{2}\cdot V_{\text{max}}^2}\right] \\
&= 10 \cdot \log\left[\frac{2FkT}{P_s}\cdot\left(\frac{\omega_0}{2Q\Delta\omega}\right)^2\right].
\end{aligned}
\tag{6}
$$

Note that the factor of 1/2 arises from neglecting the contribution of amplitude noise. Although the expression for the noise in the $1/f^2$ region is thus easily obtained, the expression for the $1/f^3$ portion of the phase noise is completely empirical. As such, the common assumption that the $1/f^3$ corner of the phase noise is the same as the $1/f$ corner of device flicker noise has no theoretical basis.

The above approach may be extended by identifying the individual noise sources in the tuned tank oscillator of Fig. 2 [8]. An LTI approach is used and there is an embedded assumption of no amplitude limiting, contrary to most practical cases. For the RLC circuit of Fig. 2, [8] predicts the following:

$$
\mathcal{L}\{\Delta\omega\} = 10 \cdot \log\left[\frac{kT \cdot \mathcal{R}_{\text{eff}}[1+\mathcal{A}]\cdot\left(\frac{\omega_0}{\Delta\omega}\right)^2}{\mathcal{V}_{\text{max}}^2/2}\right]
\tag{7}
$$

where $\mathcal{A}$ is yet another empirical fitting parameter, and $\mathcal{R}_{\text{eff}}$ is the effective series resistance, given by

$$
\mathcal{R}_{\text{eff}} = R_L + R_C + \frac{1}{R_p(C\omega_0)^2}
\tag{8}
$$

where R_L, R_C, R_p, and C are shown in Fig. 2. Note that it is still not clear how to calculate $\mathcal{A}$ from circuit parameters. Hence, this approach represents no fundamental improvement over the method outlined in [3].

III. MODELING OF PHASE NOISE

A. Impulse Response Model for Excess Phase

An oscillator can be modeled as a system with n inputs (each associated with one noise source) and two outputs that are the instantaneous amplitude and excess phase of the oscillator, $A(t)$ and $\phi(t)$, as defined by (1). Noise inputs to this system are in the form of current sources injecting into circuit nodes and voltage sources in series with circuit branches. For each input source, both systems can be viewed as single-input, single-output systems. The time and frequency-domain fluctuations of $A(t)$ and $\phi(t)$ can be studied by characterizing the behavior of two equivalent systems shown in Fig. 3.

Note that both systems shown in Fig. 3 are time variant. Consider the specific example of an ideal parallel LC oscillator shown in Fig. 4. If we inject a current impulse $i(t)$ as shown, the amplitude and phase of the oscillator will have responses similar to that shown in Fig. 4(a) and (b). The instantaneous voltage change ΔV is given by

$$
\Delta V = \frac{\Delta q}{C_{\text{tot}}}
\tag{9}
$$

where Δq is the total injected charge due to the current impulse and C_{tot} is the total capacitance at that node. Note that the current impulse will change only the voltage across the

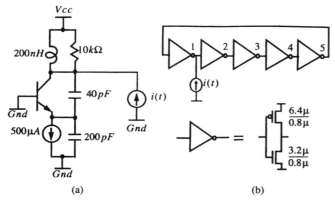

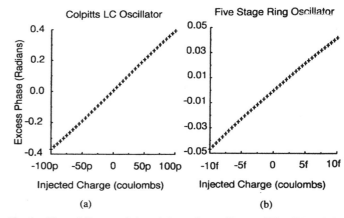

Fig. 5. (a) A typical Colpitts oscillator and (b) a five-stage minimum size ring oscillator.

Fig. 6. Phase shift versus injected charge for oscillators of Fig. 5(a) and (b).

capacitor and will not affect the current through the inductor. It can be seen from Fig. 4 that the resultant change in $A(t)$ and $\phi(t)$ is time dependent. In particular, if the impulse is applied at the peak of the voltage across the capacitor, there will be no phase shift and only an amplitude change will result, as shown in Fig. 4(a). On the other hand, if this impulse is applied at the zero crossing, it has the maximum effect on the excess phase $\phi(t)$ and the minimum effect on the amplitude, as depicted in Fig. 4(b). This time dependence can also be observed in the state-space trajectory shown in Fig. 4(c). Applying an impulse at the peak is equivalent to a sudden jump in voltage at point a, which results in no phase change and changes only the amplitude, while applying an impulse at point b results only in a phase change without affecting the amplitude. An impulse applied sometime between these two extremes will result in both amplitude and phase changes.

There is an important difference between the phase and amplitude responses of any real oscillator, because some form of amplitude limiting mechanism is essential for stable oscillatory action. The effect of this limiting mechanism is pictured as a closed trajectory in the state-space portrait of the oscillator shown in Fig. 4(c). The system state will finally approach this trajectory, called a limit cycle, irrespective of its starting point [10]–[12]. Both an explicit automatic gain control (AGC) and the intrinsic nonlinearity of the devices act similarly to produce a stable limit cycle. However, any fluctuation in the phase of the oscillation persists indefinitely, with a current noise impulse resulting in a step change in phase, as shown in Fig. 3. It is important to note that regardless of how small the injected charge, the oscillator remains time variant.

Having established the essential time-variant nature of the systems of Fig. 3, we now show that they may be treated as linear for all practical purposes, so that their impulse responses $h_\phi(t, \tau)$ and $h_A(t, \tau)$ will characterize them completely.

The linearity assumption can be verified by injecting impulses with different areas (charges) and measuring the resultant phase change. This is done in the SPICE simulations of the 62-MHz Colpitts oscillator shown in Fig. 5(a) and the five-stage 1.01-GHz, 0.8-μm CMOS inverter chain ring oscillator shown in Fig. 5(b). The results are shown in Fig. 6(a) and (b), respectively. The impulse is applied close to a zero crossing,

where it has the maximum effect on phase. As can be seen, the current-phase relation is linear for values of charge up to 10% of the total charge on the effective capacitance of the node of interest. Also note that the effective injected charges due to actual noise and interference sources in practical circuits are several orders of magnitude smaller than the amounts of charge injected in Fig. 6. Thus, the assumption of linearity is well satisfied in all practical oscillators.

It is critical to note that the current-to-phase transfer function is practically linear even though the active elements may have strongly nonlinear voltage-current behavior. However, the nonlinearity of the circuit elements defines the shape of the limit cycle and has an important influence on phase noise that will be accounted for shortly.

We have thus far demonstrated linearity, with the amount of excess phase proportional to the ratio of the injected charge to the maximum charge swing across the capacitor on the node, i.e., $\Delta q/q_{\max}$. Furthermore, as discussed earlier, the impulse response for the first system of Fig. 3 is a step whose amplitude depends periodically on the time τ when the impulse is injected. Therefore, the unit impulse response for excess phase can be expressed as

$$h_\phi(t, \tau) = \frac{\Gamma(\omega_0 \tau)}{q_{\max}} u(t - \tau) \qquad (10)$$

where $q_{\max}$ is the maximum charge displacement across the capacitor on the node and $u(t)$ is the unit step. We call $\Gamma(x)$ the *impulse sensitivity function* (ISF). It is a dimensionless, frequency- and amplitude-independent periodic function with period 2π which describes how much phase shift results from applying a unit impulse at time $t = \tau$. To illustrate its significance, the ISF's together with the oscillation waveforms for a typical LC and ring oscillator are shown in Fig. 7. As is shown in the Appendix, $\Gamma(x)$ is a function of the waveform or, equivalently, the shape of the limit cycle which, in turn, is governed by the nonlinearity and the topology of the oscillator.

Given the ISF, the output excess phase $\phi(t)$ can be calculated using the superposition integral

$$\phi(t) = \int_{-\infty}^{\infty} h_\phi(t, \tau) i(\tau) \, d\tau = \frac{1}{q_{\max}} \int_{-\infty}^{t} \Gamma(\omega_0 \tau) i(\tau) \, d\tau \qquad (11)$$

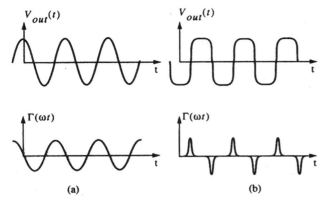

Fig. 7. Waveforms and ISF's for (a) a typical *LC* oscillator and (b) a typical ring oscillator.

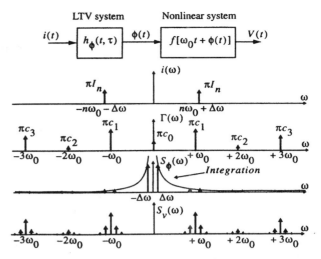

Fig. 8. Conversion of the noise around integer multiples of the oscillation frequency into phase noise.

where $i(t)$ represents the input noise current injected into the node of interest. Since the ISF is periodic, it can be expanded in a Fourier series

$$\Gamma(\omega_0 \tau) = \frac{c_0}{2} + \sum_{n=1}^{\infty} c_n \cos(n\omega_0 \tau + \theta_n) \quad (12)$$

where the coefficients c_n are real-valued coefficients, and θ_n is the phase of the nth harmonic. As will be seen later, θ_n is not important for random input noise and is thus neglected here. Using the above expansion for $\Gamma(\omega_0 \tau)$ in the superposition integral, and exchanging the order of summation and integration, we obtain

$$\phi(t) = \frac{1}{q_{max}} \left[\frac{c_0}{2} \int_{-\infty}^{t} i(\tau) \, d\tau \right. \\ \left. + \sum_{n=1}^{\infty} c_n \int_{-\infty}^{t} i(\tau) \cos(n\omega_0 \tau) \, d\tau \right]. \quad (13)$$

Equation (13) allows computation of $\phi(t)$ for an arbitrary input current $i(t)$ injected into any circuit node, once the various Fourier coefficients of the ISF have been found.

As an illustrative special case, suppose that we inject a low frequency sinusoidal perturbation current $i(t)$ into the node of interest at a frequency of $\Delta\omega \ll \omega_0$

$$i(t) = I_0 \cos(\Delta\omega t) \quad (14)$$

where I_0 is the maximum amplitude of $i(t)$. The arguments of all the integrals in (13) are at frequencies higher than ω_0 and are significantly attenuated by the averaging nature of the integration, except the term arising from the first integral, which involves c_0. Therefore, the only significant term in $\phi(t)$ will be

$$\phi(t) \approx \frac{I_0 c_0}{2 q_{max}} \int_{-\infty}^{t} \cos(\Delta\omega \tau) d\tau = \frac{I_0 c_0 \sin(\Delta\omega t)}{2 q_{max} \Delta\omega}. \quad (15)$$

As a result, there will be two impulses at $\pm\Delta\omega$ in the power spectral density of $\phi(t)$, denoted as $S_\phi(\omega)$.

As an important second special case, consider a current at a frequency close to the carrier injected into the node of interest, given by $i(t) = I_1 \cos[(\omega_0 + \Delta\omega)t]$. A process similar to that of the previous case occurs except that the spectrum of $i(t)$

consists of two impulses at $\pm(\omega_0 + \Delta\omega)$ as shown in Fig. 8. This time the only integral in (13) which will have a low frequency argument is for $n = 1$. Therefore $\phi(t)$ is given by

$$\phi(t) \approx \frac{I_1 c_1 \sin(\Delta\omega t)}{2 q_{max} \Delta\omega} \quad (16)$$

which again results in two equal sidebands at $\pm\Delta\omega$ in $S_\phi(\omega)$.

More generally, (13) suggests that applying a current $i(t) = I_n \cos[(n\omega_0 + \Delta\omega)t]$ close to any integer multiple of the oscillation frequency will result in two equal sidebands at $\pm\Delta\omega$ in $S_\phi(\omega)$. Hence, in the general case $\phi(t)$ is given by

$$\phi(t) \approx \frac{I_n c_n \sin(\Delta\omega t)}{2 q_{max} \Delta\omega}. \quad (17)$$

B. Phase-to-Voltage Transformation

So far, we have presented a method for determining how much phase error results from a given current $i(t)$ using (13). Computing the power spectral density (PSD) of the oscillator output voltage $S_v(\omega)$ requires knowledge of how the output voltage relates to the excess phase variations. As shown in Fig. 8, the conversion of device noise current to output voltage may be treated as the result of a cascade of two processes. The first corresponds to a linear time variant (LTV) current-to-phase converter discussed above, while the second is a nonlinear system that represents a phase modulation (PM), which transforms phase to voltage. To obtain the sideband power around the fundamental frequency, the fundamental harmonic of the oscillator output $\cos[\omega_0 t + \phi(t)]$ can be used as the transfer function for the second system in Fig. 8. Note this is a nonlinear transfer function with $\phi(t)$ as the input.

Substituting $\phi(t)$ from (17) into (1) results in a single-tone phase modulation for output voltage, with $\phi(t)$ given by (17). Therefore, an injected current at $n\omega_0 + \Delta\omega$ results in a pair of *equal* sidebands at $\omega_0 \pm \Delta\omega$ with a sideband power relative to the carrier given by

$$P_{SBC}(\Delta\omega) = 10 \cdot \log \left(\frac{I_n c_n}{4 q_{max} \Delta\omega} \right)^2. \quad (18)$$

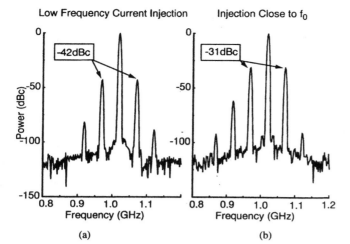

(a)   (b)

Fig. 9. Simulated power spectrum of the output with current injection at (a) $f_m = 50$ MHz and (b) $f_0 + f_m = 1.06$ GHz.

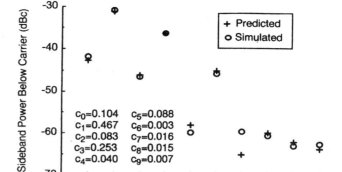

Fig. 10. Simulated and calculated sideband powers for the first ten coefficients.

This process is shown in Fig. 8. Appearance of the frequency deviation $\Delta\omega$ in the denominator of the (18) underscores that the impulse response $h_\phi(t, \tau)$ is a step function and therefore behaves as a time-varying integrator. We will frequently refer to (18) in subsequent sections.

Applying this method of analysis to an arbitrary oscillator, a sinusoidal current injected into one of the oscillator nodes at a frequency $\omega_0 + \Delta\omega$ results in two equal sidebands at $\omega_0 \pm \Delta\omega$, as observed in [9]. Note that it is necessary to use an LTV because an LTI model cannot explain the presence of a pair of equal sidebands close to the carrier arising from sources at frequencies $n\omega_0 + \Delta\omega$, because an LTI system cannot produce any frequencies except those of the input and those associated with the system's poles. Furthermore, the amplitude of the resulting sidebands, as well as their equality, cannot be predicted by conventional intermodulation effects. This failure is to be expected since the intermodulation terms arise from nonlinearity in the voltage (or current) input/output characteristic of active devices of the form $V_{out} = \alpha_1 V_{in} + \alpha_2 V_{in}^2 + \alpha_3 V_{in}^3 + \cdots$. This type of nonlinearity does not directly appear in the phase transfer characteristic and shows itself only indirectly in the ISF.

It is instructive to compare the predictions of (18) with simulation results. A sinusoidal current of 10 μA amplitude at different frequencies was injected into node 1 of the 1.01-GHz ring oscillator of Fig. 5(b). Fig. 9(a) shows the simulated power spectrum of the signal on node 4 for a low frequency input at $f_m = 50$ MHz. This power spectrum is obtained using the fast Fourier transform (FFT) analysis in HSPICE 96.1. It is noteworthy that in this version of HSPICE the simulation artifacts observed in [9] have been properly eliminated by calculation of the values used in the analysis at the exact points of interest. Note that the injected noise is upconverted into two equal sidebands at $f_0 + f_m$ and $f_0 - f_m$, as predicted by (18). Fig. 9(b) shows the effect of injection of a current at $f_0 + f_m = 1.06$ GHz. Again, two equal sidebands are observed at $f_0 + f_m$ and $f_0 - f_m$, also as predicted by (18).

Simulated sideband power for the general case of current injection at $nf_0 + f_m$ can be compared to the predictions of

(18). The ISF for this oscillator is obtained by the simulation method of the Appendix. Here, q_{max} is equal to $C_{eq}V_{swing}$, where C_{eq} is the average capacitance on each node of the circuit and V_{swing} is the maximum swing across it. For this oscillator, $C_{eq} = 26$ fF and $V_{swing} = 5$ V, which results in $q_{max} = 130$ fC. For a sinusoidal injected current of amplitude $I_n = 10$ μA, and an f_m of 50 MHz, Fig. 10 depicts the simulated and predicted sideband powers. As can be seen from the figure, these agree to within 1 dB for the higher power sidebands. The discrepancy in the case of the low power sidebands ($n = 4, 6$–9) arises from numerical noise in the simulations, which represents a greater fractional error at lower sideband power. Overall, there is satisfactory agreement between simulation and the theory of conversion of noise from various frequencies into phase fluctuations.

C. Prediction of Phase Noise Sideband Power

Now we consider the case of a *random* noise current $i_n(t)$ whose power spectral density has both a flat region and a $1/f$ region, as shown in Fig. 11. As can be seen from (18) and the foregoing discussion, noise components located near integer multiples of the oscillation frequency are transformed to low frequency noise sidebands for $S_\phi(\omega)$, which in turn become close-in phase noise in the spectrum of $S_v(\omega)$, as illustrated in Fig. 11. It can be seen that the total $S_\phi(\omega)$ is given by the sum of phase noise contributions from device noise in the vicinity of the integer multiples of ω_0, weighted by the coefficients c_n. This is shown in Fig. 12(a) (logarithmic frequency scale). The resulting single sideband spectral noise density $L\{\Delta\omega\}$ is plotted on a logarithmic scale in Fig. 12(b). The sidebands in the spectrum of $S_\phi(\omega)$, in turn, result in phase noise sidebands in the spectrum of $S_v(\omega)$ through the PM mechanism discuss in the previous subsection. This process is shown in Figs. 11 and 12.

The theory predicts the existence of $1/f^3$, $1/f^2$, and flat regions for the phase noise spectrum. The low-frequency noise sources, such as flicker noise, are weighted by the coefficient c_0 and show a $1/f^3$ dependence on the offset frequency, while

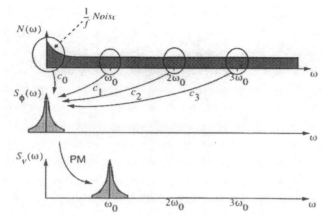

Fig. 11. Conversion of noise to phase fluctuations and phase-noise sidebands.

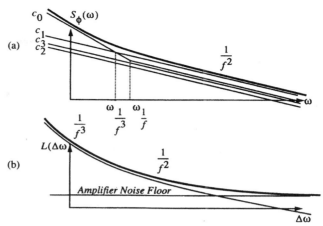

Fig. 12. (a) PSD of $\phi(t)$ and (b) single sideband phase noise power spectrum, $L\{\Delta\omega\}$.

the white noise terms are weighted by other c_n coefficients and give rise to the $1/f^2$ region of phase noise spectrum. It is apparent that if the original noise current $i(t)$ contains $1/f^n$ low frequency noise terms, such as popcorn noise, they can appear in the phase noise spectrum as $1/f^{n+2}$ regions. Finally, the flat noise floor in Fig. 12(b) arises from the white noise floor of the noise sources in the oscillator. The total sideband noise power is the sum of these two as shown by the bold line in the same figure.

To carry out a quantitative analysis of the phase noise sideband power, now consider an input noise current with a white power spectral density $\overline{i_n^2}/\Delta f$. Note that I_n in (18) represents the peak amplitude, hence, $I_n^2/2 = \overline{i_n^2}/\Delta f$ for $\Delta f = 1$ Hz. Based on the foregoing development and (18), the total single sideband phase noise spectral density in dB below the carrier per unit bandwidth due to the source on one node at an offset frequency of $\Delta\omega$ is given by

$$L\{\Delta\omega\} = 10 \cdot \log\left(\frac{\frac{\overline{i_n^2}}{\Delta f}\sum_{n=0}^{\infty} c_n^2}{8q_{max}^2\Delta\omega^2}\right). \quad (19)$$

Now, according to Parseval's relation we have

$$\sum_{n=0}^{\infty} c_n^2 = \frac{1}{\pi}\int_0^{2\pi} |\Gamma(x)|^2\, dx = 2\Gamma_{rms}^2 \quad (20)$$

where Γ_{rms} is the rms value of $\Gamma(x)$. As a result

$$L\{\Delta\omega\} = 10 \cdot \log\left(\frac{\Gamma_{rms}^2}{q_{max}^2}\cdot\frac{\overline{i_n^2}/\Delta f}{4\cdot\Delta\omega^2}\right). \quad (21)$$

This equation represents the phase noise spectrum of an arbitrary oscillator in $1/f^2$ region of the phase noise spectrum. For a voltage noise source in series with an inductor, q_{max} should be replaced with $\Phi_{max} = LI_{swing}$, where Φ_{max} represents the maximum magnetic flux swing in the inductor.

We may now investigate quantitatively the relationship between the device $1/f$ corner and the $1/f^3$ corner of the phase noise. It is important to note that it is by no means

obvious from the foregoing development that the $1/f^3$ corner of the phase noise and the $1/f$ corner of the device noise should be coincident, as is commonly assumed. In fact, from Fig. 12, it should be apparent that the relationship between these two frequencies depends on the specific values of the various coefficients c_n. The device noise in the flicker noise dominated portion of the noise spectrum ($\Delta\omega < \omega_{1/f}$) can be described by

$$\overline{i_{n,1/f}^2} = \overline{i_n^2}\cdot\frac{\omega_{1/f}}{\Delta\omega} \quad (\Delta\omega < \omega_{1/f}) \quad (22)$$

where $\omega_{1/f}$ is the corner frequency of device $1/f$ noise. Equation (22) together with (18) result in the following expression for phase noise in the $1/f^3$ portion of the phase noise spectrum:

$$L\{\Delta\omega\} = 10 \cdot \log\left(\frac{c_0^2}{q_{max}^2}\cdot\frac{\overline{i_n^2}/\Delta f}{8\cdot\Delta\omega^2}\cdot\frac{\omega_{1/f}}{\Delta\omega}\right). \quad (23)$$

The phase noise $1/f^3$ corner, ω_{1/f^3}, is the frequency where the sideband power due to the white noise given by (21) is equal to the sideband power arising from the $1/f$ noise given by (23), as shown in Fig. 12. Solving for ω_{1/f^3} results in the following expression for the $1/f^3$ corner in the phase noise spectrum:

$$\omega_{1/f^3} = \omega_{1/f}\cdot\frac{c_0^2}{2\Gamma_{rms}^2} \approx \omega_{1/f}\cdot\left(\frac{c_0}{c_1}\right)^2. \quad (24)$$

This equation together with (21) describe the phase noise spectrum and are the major results of this section. As can be seen, the $1/f^3$ phase noise corner due to internal noise sources is not equal to the $1/f$ device noise corner, but is *smaller* by a factor equal to $c_0^2/2\Gamma_{rms}^2$. As will be discussed later, c_0 depends on the waveform and can be significantly reduced if certain symmetry properties exist in the waveform of the oscillation. Thus, poor $1/f$ device noise need *not* imply poor close-in phase noise performance.

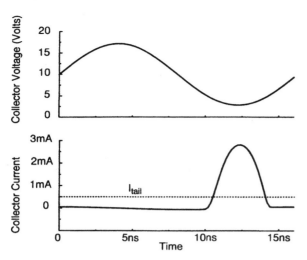

Fig. 13. Collector voltage and collector current of the Colpitts oscillator of Fig. 5(a).

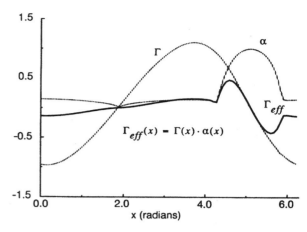

Fig. 14. $\Gamma(x)$, $\Gamma_{\text{eff}}(x)$, and $\alpha(x)$ for the Colpitts oscillator of Fig. 5(a).

D. Cyclostationary Noise Sources

In addition to the periodically time-varying nature of the system itself, another complication is that the statistical properties of some of the random noise sources in the oscillator may change with time in a periodic manner. These sources are referred to as cyclostationary. For instance, the channel noise of a MOS device in an oscillator is cyclostationary because the noise power is modulated by the gate source overdrive which varies with time periodically. There are other noise sources in the circuit whose statistical properties do not depend on time and the operation point of the circuit, and are therefore called stationary. Thermal noise of a resistor is an example of a stationary noise source.

A white cyclostationary noise current $i_n(t)$ can be decomposed as [13]:

$$i_n(t) = i_{n0}(t) \cdot \alpha(\omega_0 t) \tag{25}$$

where $i_n(t)$ is a white cyclostationary process, $i_{n0}(t)$ is a white *stationary* process and $\alpha(\omega t)$ is a deterministic periodic function describing the noise amplitude modulation. We define $\alpha(\omega t)$ to be a normalized function with a maximum value of 1. This way, $\overline{i^2}_{n0}$ is equal to the maximum mean square noise power, $\overline{i_n^2}(t)$, which changes periodically with time. Applying the above expression for $i_n(t)$ to (11), $\phi(t)$ is given by

$$\phi(t) = \int_{-\infty}^{t} i_n(\tau) \frac{\Gamma(\omega_0 \tau)}{q_{\max}} \, d\tau$$
$$= \int_{-\infty}^{t} i_{n0}(\tau) \frac{\alpha(\omega_0 \tau)\Gamma(\omega_0 \tau)}{q_{\max}} \, d\tau. \tag{26}$$

As can be seen, the cyclostationary noise can be treated as a stationary noise applied to a system with an effective ISF given by

$$\Gamma_{\text{eff}}(x) = \Gamma(x) \cdot \alpha(x) \tag{27}$$

where $\alpha(x)$ can be derived easily from device noise characteristics and operating point. Hence, this effective ISF should be used in all subsequent calculations, in particular, calculation of the coefficients c_n.

Note that there is a strong correlation between the cyclostationary noise source and the waveform of the oscillator. The maximum of the noise power always appears at a certain point of the oscillatory waveform, thus the average of the noise may not be a good representation of the noise power.

Consider as one example the Colpitts oscillator of Fig. 5(a). The collector voltage and the collector current of the transistor are shown in Fig. 13. Note that the collector current consists of a short period of large current followed by a quiet interval. The surge of current occurs at the minimum of the voltage across the tank where the ISF is small. Functions $\Gamma(x)$, $\alpha(x)$, and $\Gamma_{\text{eff}}(x)$ for this oscillator are shown in Fig. 14. Note that, in this case, $\Gamma_{\text{eff}}(x)$ is quite different from $\Gamma(x)$, and hence the effect of cyclostationarity is very significant for the LC oscillator and cannot be neglected.

The situation is different in the case of the ring oscillator of Fig. 5(b), because the devices have maximum current during the transition (when $\Gamma(x)$ is at a maximum, i.e., the sensitivity is large) at the same time the noise power is large. Functions $\Gamma(x)$, $\alpha(x)$, and $\Gamma_{\text{eff}}(x)$ for the ring oscillator of Fig. 5(b) are shown in Fig. 15. Note that in the case of the ring oscillator $\Gamma(x)$ and $\Gamma_{\text{eff}}(x)$ are almost identical. This indicates that the cyclostationary properties of the noise are less important in the treatment of the phase noise of ring oscillators. This unfortunate coincidence is one of the reasons why ring oscillators in general have inferior phase noise performance compared to a Colpitts LC oscillator. The other important reason is that ring oscillators dissipate all the stored energy during one cycle.

E. Predicting Output Phase Noise with Multiple Noise Sources

The method of analysis outlined so far has been used to predict how much phase noise is contributed by a single noise source. However, this method may be extended to multiple noise sources and multiple nodes, as individual contributions by the various noise sources may be combined by exploiting superposition. Superposition holds because the first system of Fig. 8 is linear.

196

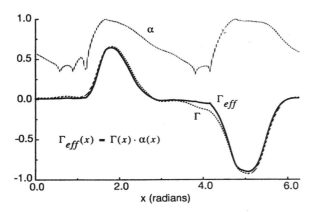

Fig. 15. $\Gamma(x)$, $\Gamma_{\text{eff}}(x)$, and $\alpha(x)$ for the ring oscillator of Fig. 5(b).

The actual method of combining the individual contributions requires attention to any possible correlations that may exist among the noise sources. The complete method for doing so may be appreciated by noting that an oscillator has a current noise source in parallel with each capacitor and a voltage noise source in series with each inductor. The phase noise in the output of such an oscillator is calculated using the following method.

1) Find the equivalent current noise source in parallel with each capacitor and an equivalent voltage source in series with each inductor, keeping track of correlated and noncorrelated portions of the noise sources for use in later steps.

2) Find the transfer characteristic from each source to the output excess phase. This can be done as follows.

  a) Find the ISF for each source, using any of the methods proposed in the Appendix, depending on the required accuracy and simplicity.

  b) Find Γ_{rms} and c_0 (rms and dc values) of the ISF.

3) Use Γ_{rms} and c_0 coefficients and the power spectrum of the input noise sources in (21) and (23) to find the phase noise power resulting from each source.

4) Sum the individual output phase noise powers for uncorrelated sources and square the sum of phase noise rms values for correlated sources to obtain the total noise power below the carrier.

Note that the amount of phase noise contributed by each noise source depends only on the value of the noise power density $\overline{i_n^2}/\Delta f$, the amount of charge swing across the effective capacitor it is injecting into q_{max}, and the steady-state oscillation waveform across the noise source of interest. This observation is important since it allows us to attribute a definite contribution from every noise source to the overall phase noise. Hence, our treatment is both an analysis and design tool, enabling designers to identify the significant contributors to phase noise.

F. Existing Models as Simplified Cases

As asserted earlier, the model proposed here reduces to earlier models if the same simplifying assumptions are made.

In particular, consider the model for *LC* oscillators in [3], as well as the more comprehensive presentation of [8]. Those models assume linear time-invariance, that all noise sources are stationary, that only the noise in the vicinity of ω_0 is important, and that the noise-free waveform is a perfect sinusoid. These assumptions are equivalent to discarding all but the c_1 term in the ISF and setting $c_1 = 1$. As a specific example, consider the oscillator of Fig. 2. The phase noise due solely to the tank parallel resistor R_p can be found by applying the following to (19):

$$\overline{i_n^2}/\Delta f = \frac{4kT}{R_p}$$
$$q_{\text{max}} = C \cdot V_{\text{max}} \qquad (28)$$

where R_p is the parallel resistor, C is the tank capacitor, and V_{max} is the maximum voltage swing across the tank. Equation (19) reduces to

$$\mathcal{L}\{\Delta\omega\} = 10 \cdot \log\left[\frac{1}{2} \cdot \frac{kT}{V_{\text{max}}^2} \cdot \frac{1}{R_p \cdot (C\omega_0)^2} \cdot \left(\frac{\omega_0}{\Delta\omega}\right)^2\right]. \quad (29)$$

Since [8] assumes equal contributions from amplitude and phase portions to $\mathcal{L}_{\text{total}}\{\Delta\omega\}$, the result obtained in [8] is two times larger than the result of (29).

Assuming that the total noise contribution in a parallel tank oscillator can be modeled using an excess noise factor F as in [3], (29) together with (24) result in (6). Note that the generalized approach presented here is capable of calculating the fitting parameters used in (3), (F and $\Delta\omega_{1/f}^3$) in terms of c_n coefficients of ISF and device $1/f$ noise corner, $\omega_{1/f}$.

IV. DESIGN IMPLICATIONS

Several design implications emerge from (18), (21), and (24) that offer important insight for reduction of phase noise in the oscillators. First, they show that increasing the signal charge displacement q_{max} across the capacitor will reduce the phase noise degradation by a given noise source, as has been noted in previous works [5], [6].

In addition, the noise power around integer multiples of the oscillation frequency has a more significant effect on the close-in phase noise than at other frequencies, because these noise components appear as phase noise sidebands in the vicinity of the oscillation frequency, as described by (18). Since the contributions of these noise components are scaled by the Fourier series coefficients c_n of the ISF, the designer should seek to minimize spurious interference in the vicinity of $n\omega_0$ for values of n such that c_n is large.

Criteria for the reduction of phase noise in the $1/f^3$ region are suggested by (24), which shows that the $1/f^3$ corner of the phase noise is proportional to the square of the coefficient c_0. Recalling that c_0 is twice the dc value of the (effective) ISF function, namely

$$c_0 = \frac{1}{\pi} \int_0^{2\pi} \Gamma_{\text{eff}}(x)\, dx \qquad (30)$$

it is clear that it is desirable to minimize the dc value of the ISF. As shown in the Appendix, the value of c_0 is closely related to certain symmetry properties of the oscillation

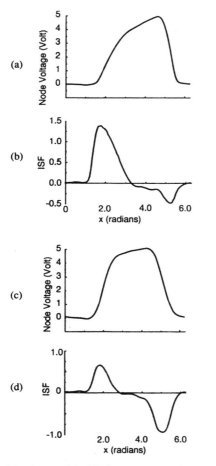

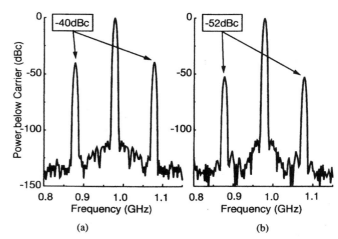

Fig. 17. Simulated power spectrum with current injection at $f_m = 50$ MHz for (a) asymmetrical node and (b) symmetrical node.

Fig. 16. (a) Waveform and (b) ISF for the asymmetrical node. (c) Waveform and (d) ISF for one of the symmetrical nodes.

waveform. One such property concerns the rise and fall times; the ISF will have a large dc value if the rise and fall times of the waveform are significantly different. A limited case of this for odd-symmetric waveforms has been observed [14]. Although odd-symmetric waveforms have small c_0 coefficients, the class of waveforms with small c_0 is not limited to odd-symmetric waveforms.

To illustrate the effect of a rise and fall time asymmetry, consider a purposeful imbalance of pull-up and pull-down rates in one of the inverters in the ring oscillator of Fig. 5(b). This is obtained by halving the channel width W_n of the NMOS device and doubling the width W_p of the PMOS device of one inverter in the ring. The output waveform and corresponding ISF are shown in Fig. 16(a) and (b). As can be seen, the ISF has a large dc value. For comparison, the waveform and ISF at the output of a symmetrical inverter elsewhere in the ring are shown in Fig. 16(c) and (d). From these results, it can be inferred that the close-in phase noise due to low-frequency noise sources should be smaller for the symmetrical output than for the asymmetrical one. To investigate this assertion, the results of two SPICE simulations are shown in Fig. 17. In the first simulation, a sinusoidal current source of amplitude 10 μA at $f_m = 50$ MHz is applied to one of the symmetric nodes of the

oscillator. In the second experiment, the same source is applied to the asymmetric node. As can be seen from the power spectra of the figure, noise injected into the asymmetric node results in sidebands that are 12 dB larger than at the symmetric node.

Note that (30) suggests that upconversion of low frequency noise can be significantly reduced, perhaps even eliminated, by minimizing c_0, at least in principle. Since c_0 depends on the waveform, this observation implies that a proper choice of waveform may yield significant improvements in close-in phase noise. The following experiment explores this concept by changing the ratio of W_p to W_n over some range, while injecting 10 μA of sinusoidal current at 100 MHz into one node. The sideband power below carrier as a function of the W_p to W_n ratio is shown in Fig. 18. The SPICE-simulated sideband power is shown with plus symbols and the sideband power as predicted by (18) is shown by the solid line. As can be seen, close-in phase noise due to upconversion of low-frequency noise can be suppressed by an arbitrary factor, at least in principle. It is important to note, however, that the minimum does not necessarily correspond to equal transconductance ratios, since other waveform properties influence the value of c_0. In fact, the optimum W_p to W_n ratio in this particular example is seen to differ considerably from that used in conventional ring oscillator designs.

The importance of symmetry might lead one to conclude that differential signaling would minimize c_0. Unfortunately, while differential circuits are certainly symmetrical with respect to the desired signals, the differential symmetry *disappears* for the individual noise sources because they are independent of each other. Hence, it is the symmetry of each *half*-circuit that is important, as is demonstrated in the differential ring oscillator of Fig. 19. A sinusoidal current of 100 μA at 50 MHz injected at the drain node of one of the buffer stages results in two equal sidebands, −46 dB below carrier, in the power spectrum of the differential output. Because of the voltage dependent conductance of the load devices, the individual waveform on each output node is not fully symmetrical and consequently, there will be a large

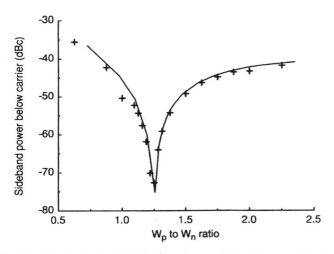

Fig. 18. Simulated and predicted sideband power for low frequency injection versus PMOS to NMOS W/L ratio.

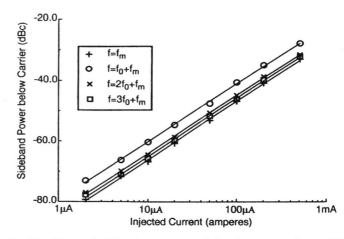

Fig. 20. Measured sideband power versus injected current at $f_m = 100$ kHz, $f_0 + f_m = 5.5$ MHz, $2f_0 + f_m = 10.9$ MHz, $3f_0 + f_m = 16.3$ MHz.

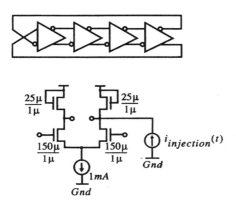

Fig. 19. Four-stage differential ring oscillator.

upconversion of noise to close-in phase noise, even though differential signaling is used.

Since the asymmetry is due to the voltage dependent conductance of the load, reduction of the upconversion might be achieved through the use of a perfectly linear resistive load, because the rising and falling behavior is governed by an RC time constant and makes the individual waveforms more symmetrical. It was first observed in the context of supply noise rejection [15], [16] that using more linear loads can reduce the effect of supply noise on timing jitter. Our treatment shows that it also improves low-frequency noise upconversion into phase noise.

Another symmetry-related property is duty cycle. Since the ISF is waveform-dependent, the duty cycle of a waveform is linked to the duty cycle of the ISF. Non-50% duty cycles generally result in larger c_n for even n. The high-Q tank of an LC oscillator is helpful in this context, since a high Q will produce a more symmetric waveform and hence reduce the upconversion of low-frequency noise.

V. EXPERIMENTAL RESULTS

This section presents experimental verifications of the model to supplement simulation results. The first experiment ex-

amines the linearity of current-to-phase conversion using a five-stage, 5.4-MHz ring oscillator constructed with ordinary CMOS inverters. A sinusoidal current is injected at frequencies $f_m = 100$ kHz, $f_0 + f_m = 5.5$ MHz, $2f_0 + f_m = 10.9$ MHz, and $3f_0 + f_m = 16.3$ MHz, and the sideband powers at $f_0 \pm f_m$ are measured as the magnitude of the injected current is varied. At any amplitude of injected current, the sidebands are equal in amplitude to within the accuracy of the measurement setup (0.2 dB), in complete accordance with the theory. These sideband powers are plotted versus the input injected current in Fig. 20. As can be seen, the transfer function for the input current power to the output sideband power is linear as suggested by (18). The slope of the best fit line is 19.8 dB/decade, which is very close to the predicted slope of 20 dB/decade, since excess phase $\phi(t)$ is proportional to $i(t)$, and hence the sideband power is proportional to I^2, leading to a 20-dB/decade slope. The behavior shown in Fig. 20 verifies that the linearity of (18) holds for injected input currents orders of magnitude larger than typical noise currents.

The second experiment varies the frequency offset from an integer multiple of the oscillation frequency. An input sinusoidal current source of 20 μA (rms) at $f_m, f_0 + f_m$, $2f_0 + f_m$, and $3f_0 + f_m$ is applied to one node and the output is measured at another node. The sideband power is plotted versus f_m in Fig. 21. Note that the slope in all four cases is -20 dB/decade, again in complete accordance with (18).

The third experiment aims at verifying the effect of the coefficients c_n on the sideband power. One of the predictions of the theory is that c_0 is responsible for the upconversion of low frequency noise. As mentioned before, c_0 is a strong function of waveform symmetry at the node into which the current is injected. Noise injected into a node with an asymmetric waveform (created by making one inverter asymmetric in a ring oscillator) would result in a greater increase in sideband power than injection into nodes with more symmetric waveforms. Fig. 22 shows the results of an experiment performed on a five-stage ring oscillator in which one of the stages is modified to have an extra pulldown

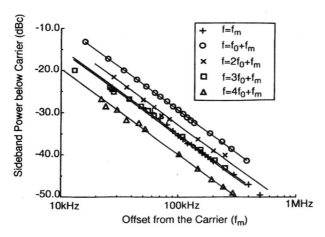

Fig. 21. Measured sideband power versus f_m, for injections in vicinity of multiples of f_0.

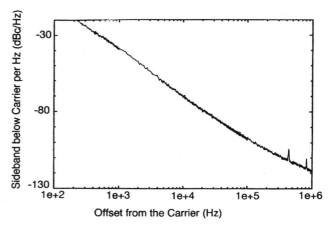

Fig. 23. Phase noise measurements for a five-stage single-ended CMOS ring oscillator. $f_0 = 232$ MHz, 2-μm process technology.

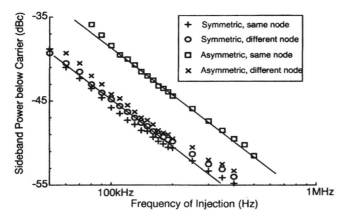

Fig. 22. Power of the sidebands caused by low frequency injection into symmetric and asymmetric nodes of the ring oscillator.

NMOS device. A current of 20 μA (rms) is injected into this asymmetric node with and without the extra pulldown device. For comparison, this experiment is repeated for a symmetric node of the oscillator, before and after this modification. Note that the sideband power is 7 dB larger when noise is injected into the node with the asymmetrical waveform, while the sidebands due to signal injection at the symmetric nodes are essentially unchanged with the modification.

The fourth experiment compares the prediction and measurement of the phase noise for a five-stage single-ended ring oscillator implemented in a 2-μm, 5-V CMOS process running at $f_0 = 232$ MHz. This measurement was performed using a delay-based measurement method and the result is shown in Fig. 23. Distinct $1/f^2$ and $1/f^3$ regions are observed. We first start with a calculation for the $1/f^2$ region. For this process we have a gate oxide thickness of $t_{ox} = 25$ nm and threshold voltages of $V_{TN} = 0.6$ V and $V_{TP} = 0.53$ V. All five inverters are similar with $(W/L)_N = 3\,\mu\text{m}/2\,\mu\text{m}$ and $(W/L)_P = 5\,\mu\text{m}/2\,\mu\text{m}$, and a lateral diffusion of $L_d = 0.1\,\mu$m. Using the process and geometry information, the total capacitance on each node, including parasitics, is calculated to be $C_{\text{total}} = 35.7$ fF. Therefore, $q_{\max} = C_{\text{total}} V_{\text{swing}} =$

179 fC. As discussed in the previous section, noise current injected during a transition has the largest effect. The current noise power at this point is the sum of the current noise powers due to NMOS and PMOS devices. At this bias point, $(\overline{i_n^2}/\Delta f)_{\text{NMOS}} = 4kT\gamma\mu_n C_{ox}(W/L_{\text{eff}})_N(V_{DD}/2 - V_{TN}) = 4.44 \times 10^{-24}$ A^2/Hz and $(i_n^2/\Delta f)_{\text{PMOS}} = 2.19 \times 10^{-24}$ A^2/Hz. Using the methods outlined in the Appendix, it may be shown that $\Gamma_{\text{rms}}^2 \approx 16/N^3$ for ring oscillators. Equation (21) for N identical noise sources then predicts $\mathcal{L}\{\Delta f\} = 10\log(0.84/\Delta f^2)$. At an offset of $\Delta f = 500$ kHz, this equation predicts $\mathcal{L}\{500 \text{ kHz}\} = -114.7$ dBc/Hz, in good agreement with a measurement of -114.5 dBc/Hz. To predict the phase noise in the $1/f^3$ region, it is enough to calculate the $1/f^3$ corner. Measurements on an isolated inverter on the same die show a $1/f$ noise corner frequency of 250 kHz, when its input and output are shorted. The $c_0^2/2\Gamma_{\text{rms}}^2$ ratio is calculated to be 0.3, which predicts a $1/f^3$ corner of 75 kHz, compared to the measured corner of 80 kHz.

The fifth experiment measures the phase noise of an 11-stage ring, running at $f_0 = 115$ MHz implemented on the same die as the previous experiment. The phase noise measurements are shown in Fig. 24. For the inverters in this oscillator, $(W/L)_N = 4\,\mu\text{m}/2\,\mu\text{m}$ and $(W/L)_P = 6\,\mu\text{m}/2\,\mu\text{m}$, which results in a total capacitance of 43.5 fF and $q_{\max} = 217$ fF. The phase noise is calculated in exactly the same manner as the previous experiment and is calculated to be $\mathcal{L}\{\Delta f\} = 10\log(0.152/\Delta f^2)$, or -122.1 dBc/Hz at a 500-kHz offset. The measured phase noise is -122.5 dBc/Hz, again in good agreement with predictions. The $c_0^2/2\Gamma_{\text{rms}}^2$ ratio is calculated to be 0.17 which predicts a $1/f^3$ corner of 43 kHz, while the measured corner is 45 kHz.

The sixth experiment investigates the effect of symmetry on $1/f^3$ region behavior. It involves a seven-stage current-starved, single-ended ring oscillator in which each inverter stage consists of an additional NMOS and PMOS device in series. The gate drives of the added transistors allow independent control of the rise and fall times. Fig. 25 shows the phase noise when the control voltages are adjusted to achieve symmetry versus when they are not. In both cases the control voltages are adjusted to keep the oscillation frequency

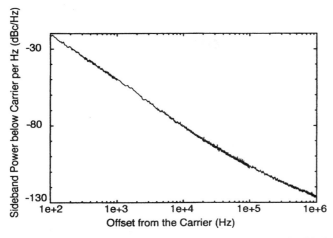

Fig. 24. Phase noise measurements for an 11-stage single-ended CMOS ring oscillator. $f_0 = 115$ MHz, 2-μm process technology.

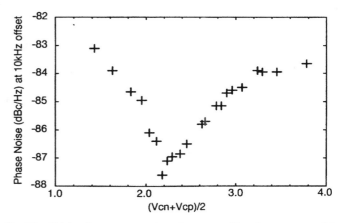

Fig. 26. Sideband power versus the voltage controlling the symmetry of the waveform. Seven-stage current-starved single-ended CMOS VCO. $f_0 = 50$ MHz, 2-μm process technology.

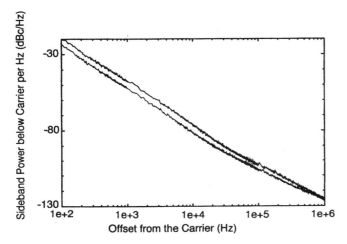

Fig. 25. Effect of symmetry in a seven-stage current-starved single-ended CMOS VCO. $f_0 = 60$ MHz, 2-μm process technology.

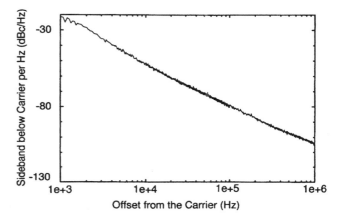

Fig. 27. Phase noise measurements for a four-stage differential CMOS ring oscillator. $f_0 = 200$MHz, 0.5-μm process technology.

constant at 60 MHz. As can be seen, making the waveform more symmetric has a large effect on the phase noise in the $1/f^3$ region without significantly affecting the $1/f^2$ region. Another experiment on the same circuit is shown in Fig. 26, which shows the phase noise power spectrum at a 10 kHz offset versus the symmetry-controlling voltage. For all the data points, the control voltages are adjusted to keep the oscillation frequency at 50 MHz. As can be seen, the phase noise reaches a minimum by adjusting the symmetry properties of the waveform. This reduction is limited by the phase noise in $1/f^2$ region and the mismatch in transistors in different stages, which are controlled by the same control voltages.

The seventh experiment is performed on a four-stage differential ring oscillator, with PMOS loads and NMOS differential stages, implemented in a 0.5-μm CMOS process. Each stage is tapped with an equal-sized buffer. The tail current source has a quiescent current of 108 μA. The total capacitance on each of the differential nodes is calculated to be $C_{total} = 49$ fF and the voltage swing is $V_{swing} = 1.2$ V, which results in $q_{max} = 58.8$ fF. The total channel noise current on each node

is $(\overline{i_n^2}/\Delta f)_{total} = 2.63 \times 10^{-23}$ A^2/Hz. Using these numbers for $N = 4$, the phase noise in the $1/f^2$ region is predicted to be $\mathcal{L}\{\Delta f\} = 10\log(48.1/\Delta f^2)$, or -103.2 dBc/Hz at an offset of 1 MHz, while the measurement in Fig. 27 shows a phase noise of -103.9 dBc/Hz, again in agreement with prediction. Also note that despite differential symmetry, there is a distinct $1/f^3$ region in the phase noise spectrum, because each half circuit is *not symmetrical*.

The eighth experiment investigates cyclostationary effects in the bipolar Colpitts oscillator of Fig. 5(a), where the conduction angle is varied by changing the capacitive divider ratio $n = C_1/(C_1 + C_2)$ while keeping the effective parallel capacitance $C_{eq} = C_1C_2/(C_1 + C_2)$ constant to maintain an f_0 of 100 MHz. As can be seen in Fig. 28, increasing n decreases the conduction angle, and thereby reduces the effective Γ_{rms}, leading to an initial decrease in phase noise. However, the oscillation amplitude is approximately given by $V_{tank} = 2R_{tank}I_{EE}(1 - n)$, and therefore decreases for large values of n. The phase noise ultimately increases for large n as a consequence. There is thus a definite value of n (here, about 0.2) that minimizes the phase noise. This result provides a theoretical basis for the common rule-of-thumb that one should

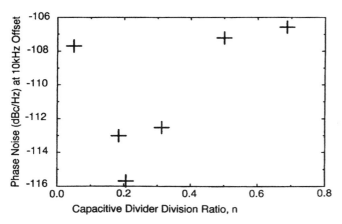

Fig. 28. Sideband power versus capacitive division ratio. Bipolar *LC* Colpitts oscillator $f_0 = 100$ MHz.

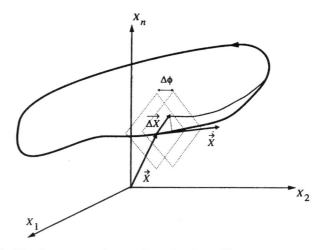

Fig. 29. State-space trajectory of an *n*th-order oscillator.

use C_2/C_1 ratios of about four (corresponding to $n = 0.2$) in Colpitts oscillators [17].

VI. CONCLUSION

This paper has presented a model for phase noise which explains quantitatively the mechanism by which noise sources of all types convert to phase noise. The power of the model derives from its explicit recognition of practical oscillators as time-varying systems. Characterizing an oscillator with the ISF allows a complete description of the noise sensitivity of an oscillator and also allows a natural accommodation of cyclostationary noise sources.

This approach shows that noise located near integer multiples of the oscillation frequency contributes to the total phase noise. The model specifies the contribution of those noise components in terms of waveform properties and circuit parameters, and therefore provides important design insight by identifying *and quantifying* the major sources of phase noise degradation. In particular, it shows that symmetry properties of the oscillator waveform have a significant effect on the upconversion of low frequency noise and, hence, the $1/f^3$ corner of the phase noise can be significantly lower than the $1/f$ device noise corner. This observation is particularly important for MOS devices, whose inferior $1/f$ noise has been thought to preclude their use in high-performance oscillators.

APPENDIX
CALCULATION OF THE IMPULSE SENSITIVITY FUNCTION

In this Appendix we present three different methods to calculate the ISF. The first method is based on direct measurement of the impulse response and calculating $\Gamma(x)$ from it. The second method is based on an analytical state-space approach to find the excess phase change caused by an impulse of current from the oscillation waveforms. The third method is an easy-to-use approximate method.

A. Direct Measurement of Impulse Response

In this method, an impulse is injected at different relative phases of the oscillation waveform and the oscillator simulated

for a few cycles afterwards. By sweeping the impulse injection time across one cycle of the waveform and measuring the resulting time shift Δt, $h_\phi(t, \tau)$ can calculated noting that $\Delta \phi = 2\pi \Delta t/T$, where T is the period of oscillation. Fortunately, many implementations of SPICE have an internal feature to perform the sweep automatically. Since for each impulse one needs to simulate the oscillator for only a few cycles, the simulation executes rapidly. Once $h_\phi(t, \tau)$ is found, the ISF is calculated by multiplication with q_{max}. This method is the most accurate of the three methods presented.

B. Closed-Form Formula for the ISF

An nth-order system can be represented by its trajectory in an n-dimensional state-space. In the case of a stable oscillator, the state of the system, represented by the state vector, $\bar{X}$, periodically traverses a closed trajectory, as shown in Fig. 29. Note that the oscillator does not necessarily traverse the limit cycle with a constant velocity.

In the most general case, the effect of a group of external impulses can be viewed as a perturbation vector $\Delta \bar{X}$ which suddenly changes the state of the system to $\bar{X} + \Delta \bar{X}$. As discussed earlier, amplitude variations eventually die away, but phase variations do not. Application of the perturbation impulse causes a certain change in phase in either a negative or positive direction, depending on the state-vector and the direction of the perturbation. To calculate the equivalent time shift, we first find the projection of the perturbation vector on a unity vector in the direction of motion, i.e., the normalized velocity vector

$$l = \Delta \bar{X} \cdot \frac{\dot{\bar{X}}}{|\dot{\bar{X}}|} \qquad (31)$$

where l is the equivalent displacement along the trajectory, and $\dot{\bar{X}}$ is the first derivative of the state vector. Note the scalar nature of l, which arises from the projection operation. The equivalent time shift is given by the displacement divided by the "speed" $|\dot{\vec{X}}|$

$$\Delta t = \frac{l}{|\dot{\vec{X}}|} = \Delta \vec{X} \cdot \frac{\dot{\vec{X}}}{|\dot{\vec{X}}|^2} \qquad (32)$$

which results in the following equation for excess phase caused by the perturbation:

$$\Delta \phi = 2\pi \frac{\Delta t}{T} = \frac{2\pi}{T} \left(\Delta \vec{X} \cdot \frac{\dot{\vec{X}}}{|\dot{\vec{X}}|^2} \right). \qquad (33)$$

In the specific case where the state variables are node voltages, and an impulse is applied to the ith node, there will be a change in ΔV_i given by (10). Equation (33) then reduces to

$$\Delta \phi_i = \frac{2\pi}{T} \cdot \frac{\Delta q_i}{C_i} \cdot \frac{\dot{v}_i}{|\dot{\vec{v}}|^2} \qquad (34)$$

where $|\dot{\vec{v}}|^2$ is the norm of the first derivative of the waveform vector and $\dot{v}_i$ is the derivative of the ith node voltage. Equation (34), together with the normalized waveform function f defined in (1), result in the following:

$$\Delta \phi = \frac{\Delta q}{q_i} \cdot \frac{f_i'}{|\vec{f}'|^2} \qquad (35)$$

where f_i' represents the derivative of the normalized waveform on node i, hence

$$\Gamma_i(x) = \frac{f_i'}{|\vec{f}'|^2} = \frac{f_i'}{\sum_{j=1}^{n} f_j'^2}. \qquad (36)$$

It can be seen that this expression for the ISF is maximum during transitions (i.e., when the derivative of the waveform function f is maximum), and this maximum value is inversely proportional to the maximum derivative. Hence, waveforms with larger slope show a smaller peak in the ISF function.

In the special case of a second-order system, one can use the normalized waveform f and its derivative as the state variables, resulting in the following expression for the ISF:

$$\Gamma(x) = \frac{f'}{f'^2 + f''^2} \qquad (37)$$

where f'' represents the second derivative of the function f. In the case of an ideal sinusoidal oscillator $f = \cos(x)$, so that $\Gamma(\omega t) = -\sin(\omega t)$, which is consistent with the argument of Section III. This method has the attribute that it computes the ISF from the waveform directly, so that simulation over only one cycle of f is required to obtain all of the necessary information.

C. Calculation of ISF Based on the First Derivative

This method is actually a simplified version of the second approach. In certain cases, the denominator of (36) shows little variation, and can be approximated by a constant. In such a case, the ISF is simply proportional to the derivative of the waveform. A specific example is a ring oscillator with N

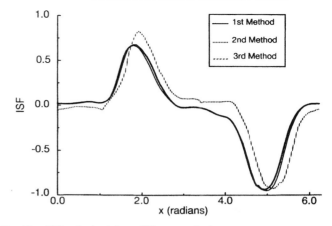

Fig. 30. ISF's obtained from different methods.

identical stages. The denominator may then be approximated by $f'^2_{\max}$

$$\Gamma_i(x) = \frac{f_i'(x)}{f'^2_{\max}}. \qquad (38)$$

Fig. 30 shows the results obtained from this method compared with the more accurate results obtained from methods A and B. Although this method is approximate, it is the easiest to use and allows a designer to rapidly develop important insights into the behavior of an oscillator.

ACKNOWLEDGMENT

The authors would like to thank T. Ahrens, R. Betancourt, R. Farjad-Rad, M. Heshami, S. Mohan, H. Rategh, H. Samavati, D. Shaeffer, A. Shahani, K. Yu, and M. Zargari of Stanford University and Prof. B. Razavi of UCLA for helpful discussions. The authors would also like to thank M. Zargari, R. Betancourt, B. Amruturand, J. Leung, J. Shott, and Stanford Nanofabrication Facility for providing several test chips. They are also grateful to Rockwell Semiconductor for providing access to their phase noise measurement system.

REFERENCES

[1] E. J. Baghdady, R. N. Lincoln, and B. D. Nelin, "Short-term frequency stability: Characterization. theory, and measurement," *Proc. IEEE*, vol. 53, pp. 704–722, July 1965.
[2] L. S. Cutler and C. L. Searle, "Some aspects of the theory and measurement of frequency fluctuations in frequency standards," *Proc. IEEE*, vol. 54, pp. 136–154, Feb. 1966.
[3] D. B. Leeson, "A simple model of feedback oscillator noises spectrum," *Proc. IEEE*, vol. 54, pp. 329–330, Feb. 1966.
[4] J. Rutman, "Characterization of phase and frequency instabilities in precision frequency sources; Fifteen years of progress," *Proc. IEEE*, vol. 66, pp. 1048–1174, Sept. 1978.
[5] A. A. Abidi and R. G. Meyer, "Noise in relaxation oscillators," *IEEE J. Solid-State Circuits*, vol. SC-18, pp. 794–802, Dec. 1983.
[6] T. C. Weigandt, B. Kim, and P. R. Gray, "Analysis of timing jitter in CMOS ring oscillators," in *Proc. ISCAS*, June 1994, vol. 4, pp. 27–30.
[7] J. McNeil, "Jitter in ring oscillators," in *Proc. ISCAS*, June 1994, vol. 6, pp. 201–204.
[8] J. Craninckx and M. Steyaert, "Low-noise voltage controlled oscillators using enhanced LC-tanks," *IEEE Trans. Circuits Syst.–II*, vol. 42, pp. 794–904, Dec. 1995.

[9] B. Razavi, "A study of phase noise in CMOS oscillators," *IEEE J. Solid-State Circuits,* vol. 31, pp. 331–343, Mar. 1996.

[10] B. van der Pol, "The nonlinear theory of electric oscillations," *Proc. IRE,* vol. 22, pp. 1051–1086, Sept. 1934.

[11] N. Minorsky, *Nonlinear Oscillations.* Princeton, NJ: Van Nostrand, 1962.

[12] P. A. Cook, *Nonlinear Dynamical Systems.* New York: Prentice Hall, 1994.

[13] W. A. Gardner, *Cyclostationarity in Communications and Signal Processing.* New York: IEEE Press, 1993.

[14] H. B. Chen, A. van der Ziel, and K. Amberiadis, "Oscillator with odd-symmetrical characteristics eliminates low-frequency noise sidebands," *IEEE Trans. Circuits Syst.,* vol. CAS-31, Sept. 1984.

[15] J. G. Maneatis, "Precise delay generation using coupled oscillators," *IEEE J. Solid-State Circuits,* vol. 28, pp. 1273–1282, Dec. 1993.

[16] C. K. Yang, R. Farjad-Rad, and M. Horowitz, "A 0.6mm CMOS 4Gb/s transceiver with data recovery using oversampling," in *Symp. VLSI Circuits, Dig. Tech. Papers,* June 1997.

[17] D. DeMaw, *Practical RF Design Manual.* Englewood Cliffs, NJ: Prentice-Hall, 1982, p. 46.

Thomas H. Lee (M'83) received the S.B., S.M., Sc.D. degrees from the Massachusetts Institute of Technology (MIT), Cambridge, in 1983, 1985, and 1990, respectively.

He worked for Analog Devices Semiconductor, Wilmington, MA, until 1992, where he designed high-speed clock-recovery PLL's that exhibit zero jitter peaking. He then worked for Rambus Inc., Mountain View, CA, where he designed the phase- and delay-locked loops for 500 MB/s DRAM's. In 1994, he joined the faculty of Stanford University, Stanford, CA, as an Assistant Professor, where he is primarily engaged in research into microwave applications for silicon IC technology, with a focus on CMOS IC's for wireless communications.

Dr. Lee was recently named a recipient of a Packard Foundation Fellowship award and is the author of *The Design of CMOS Radio-Frequence Integrated Circuits* (Cambridge University Press). He has twice received the "Best Paper" award at ISSCC.

Ali Hajimiri (S'95) was born in Mashad, Iran, in 1972. He received the B.S. degree in electronics engineering from Sharif University of Technology in 1994 and the M.S. degree in electrical engineering from Stanford University, Stanford, CA, in 1996, where he is currently engaged in research toward the Ph.D. degree in electrical engineering.

He worked as a Design Engineer for Philips on a BiCMOS chipset for the GSM cellular units from 1993 to 1994. During the summer of 1995, he worked for Sun Microsystems, Sunnyvale, CA, on the UltraSparc microprocessor's cache RAM design methodology. Over the summer of 1997, he worked at Lucent Technologies (Bell-Labs), where he investigated low phase noise integrated oscillators. He holds one European and two U.S. patents.

Mr. Hajimiri is the Bronze medal winner of the 21st International Physics Olympiad, Groningen, Netherlands.

Physical Processes of Phase Noise in Differential LC Oscillators

J. J. Rael and A. A. Abidi

Integrated Circuits & Systems Laboratory
Electrical Engineering Department
University of California
Los Angeles, CA 90095-1594

Introduction

There is an unprecedented interest among circuit designers today to obtain insight into mechanisms of phase noise in LC oscillators. For only with this insight is it possible to optimize oscillator circuits using low-quality integrated resonators to comply with the exacting phase noise specifications of modern wireless systems. Various numerical simulators are now available to assist the circuit designer [1], [2], [3], in some cases accompanied by qualitative interpretations [4]. At present, therefore, the situation of the oscillator designer is similar to the designer of amplifiers who is equipped only with SPICE, but who lacks physical insight and methods for simple yet accurate analysis with which to optimize a circuit.

Over the years, various attempts at phase noise analysis have produced results that are variations on Leeson's classic "heuristic derivation without formal proof" [5], [6]. These analyses are based on a linear model of an LC resonator in steady-state oscillation through application of either feedback or negative conductance. The results confirm Leeson by showing that phase noise is proportional to noise-to-carrier ratio and inversely to the square of resonator quality factor. However, without knowledge of the constant of proportionality, which Leeson leaves as an unspecified noise factor, the actual phase noise cannot be predicted.

It is now well understood that the large-signal periodic switching of a self-limited oscillator [7] underpins this noise factor [8]. At first sight, an accurate noise analysis of an oscillator subject to periodic bias currents appears intractable, however by using sensible approximations Huang has solved this problem for a Colpitts oscillator [9] and obtained good agreement between analysis and measurements of thermally induced phase noise. The mechanisms of flicker noise upconversion, which are important in CMOS oscillators, remain obscure.

In this paper we concentrate on an understanding of the popular differential LC oscillator. We introduce simple models to capture the nonlinear processes that convert voltage or current thermal noise in resistors or transistors into phase noise in the oscillator. The analysis does not require hypothetical elements, such as limiters or amplitude control loops, to fully explain phase noise. A simple expression at the end accurately specifies thermally induced phase noise, and lends substance to Leeson's original hypothesis.

Next, the upconversion of flicker noise into phase noise is traced to mechanisms first identified in the 1930's, but apparently since forgotten. Unlike thermally induced phase noise, which appears as phase modulation sidebands, flicker noise is shown to upconvert by bias-dependent frequency modulation.

The results are validated against SpectreRF simulations and measurements on two differential CMOS oscillators tuned by resonators with very different Q's.

Recognizing Phase Noise

For the purposes of analysis, a noise spectrum is considered as consisting of uncorrelated sinewaves in a 1 Hz bandwidth at any given frequency. Voltage or current noise produces amplitude and phase fluctuations when superimposed on a periodic signal (from now on, a large sinewave $V_0 \sin(2\pi f_0 t)$). This is clearly seen [10] by isolating one sinewave v_n in the noise spectrum, say at a frequency offset $+f_m$ from the sinewave frequency f_0. Figure 1 shows this as a phasor v_n rotating relative to the sinewave phasor V_0, which is then decomposed into two equal collinear phasors at $+f_m$, and two anti-phase conjugate phasors which are assigned a negative relative frequency $-f_m$. Grouping the phasors pairwise as $\pm f_m$, it is seen that one pair modulates the amplitude of the sinewave with time (AM), while the other sweeps its phase (PM). Thus, half of any additive noise on a sinewave produces phase noise, the other half amplitude noise. When $\sin(\omega_0 t)$ is accompanied either by noise sinewave phasors $\pm \sin(\omega_0 + \omega_m)t$, $\mp \sin(\omega_0 - \omega_m)t$ or by $\pm \cos(\omega_0 + \omega_m)t$, $\pm \cos(\omega_0 - \omega_m)t$, then phase noise alone is present.

Simple Model of the Differential Oscillator

This paper treats the well-known tail-current biased differential LC oscillator (Figure 2). In steady state, the differential pair acts as a negative conductance that switches the tail current I_T into the LC resonator. Owing to filtering in the LC circuit, the square wave of current creates a sinusoidal voltage across the resonator of amplitude $(4/\pi)I_T R$. This voltage drives the differential pair into switching, thus sustaining oscillation. In a CMOS oscillator the amplitude may build up to several volts, eventually limited by the supply voltage.

In previous work on noise in mixers [11], we have shown how a simple model of the switching differential pair is sufficient to explain all frequency translations of noise. This model is used here. Suppose that some noise (v_n) accompanies the resonator sinewave. Assuming that a small fraction of the resonator voltage around the zero crossing is enough to fully switch the differential pair, then the noise simply advances or retards the instant of zero crossing (Figure 3(a)). The randomly pulse-width modulated current at the switch output may be decomposed into the original periodic square wave in the absence of noise, superimposed with pulses of constant height but random width (Figure 3(b)). In turn, these pulses may be approximated by a train of impulses at twice the oscillation frequency multiplying the original noise waveform $v_n(t)$ (Figure 3(c)).

Thermally Induced Phase Noise

Resonator Noise

Now consider a current source $i_n \sin((\omega_0 + \omega_m)t + \phi)$ representing noise in the loss conductance of the resonator, where $i_n^2 = 4kT/R$. According to the model above, this modulates the zero crossing instants of the differential pair, producing a current which, in addition to the usual square wave, also consists of current pulses sampling this noise at $2\omega_0$. After sampling, frequency components appear at $\omega_0 \pm \omega_m$, $3\omega_0 \pm \omega_m$, ... However, usually the resonator will filter the 3rd and higher harmonics, leaving $\omega_0 \pm \omega_m$ as the only important terms. These will induce a symmetric voltage response in the resonator, and through feedback arrive at steady state. The steady-state oscillation, in general, is of the form:

$$v_{out} = V_0 \sin \omega_0 t + A \sin(\omega_0 - \omega_m)t + B \cos(\omega_0 - \omega_m)t$$
$$+ C \sin(\omega_0 + \omega_m)t + D \cos(\omega_0 + \omega_m)t$$

and here $A = -C = i_n \times (L\omega_0^2/4\omega_m)$, while $B = D \simeq 0$. The relative signs of A and C prove that the steady-state response to current noise in the resonator's resistor is phase noise in the oscillator. The single-sideband phase noise density is found by the ratio of the sideband power at a given frequency to the power in the fundamental oscillation frequency. Thus, the thermally induced phase noise density due to resonator loss is:

$$\mathcal{L}(\omega_m) = N_1 N_2 \frac{kTR}{V_0^2}\left(\frac{\omega_0}{2Q\omega_m}\right)^2$$

where $N_1 = 2$, the number of loss sources (in the left and right resonators) and $N_2 = 4$ because uncorrelated quadrature noise originating at $\omega_0 \pm \omega_m$ contributes to SSB phase noise at offset ω_m.

Tail Current Noise

The switching action of the differential pair commutates noise in the tail currents like a single-balanced mixer. The noise is translated up and down in frequency, and enters the resonator. The resulting voltage drives the differential pair, the noise components modulating the zero crossing instants. The resulting impulses of current feed back into the resonator. The steady-state solution is found by solving simultaneous equations of a form that anticipates the end result, much like in any feedback circuit.

The single-balanced mixer shows the largest conversion gain around the fundamental switching frequency, $1/3^{rd}$ the current conversion gain around the 3rd harmonic, and so on. Therefore, only mixing by the fundamental at ω_0 is important. Noise originating in the tail current at ω_m upconverts to $\omega_0 \pm \omega_m$. Similarly, noise at $2\omega_0 \pm \omega_m$ downconverts to $\omega_0 \pm \omega_m$.

Analysis shows that the upconversion produces coefficients A=C, B=−D, both of which indicate AM only. It should be noted that AM noise superimposed on the resonator fundamental frequency does not modulate the zero crossings of the switching differential pair, and therefore does not propagate in the feedback loop back into the resonator. However, the downconversion results in phase noise only, with A=−C, and B=D≃0. The phase noise caused by thermal noise originally at $2\omega_0$ is:

$$\mathcal{L}(\omega_m) = \frac{32}{9} \gamma g_m R \frac{kTR}{V_0^2}\left(\frac{\omega_0}{2Q\omega_m}\right)^2$$

where γ is the noise factor of a single FET, classically 2/3. It is important to note that the AM noise resulting from upconversion, if impressed across a varactor at the resonator, will modulate the varactor, thus the oscillation frequency by AM-to-FM conversion [12]. Although the process is different, the resulting sidebands are indistinguishable from PM noise sidebands. Unlike the other mechanisms of phase noise, this effect depends on the varactor characteristics and VCO tuning range and it may be significant only in certain situations.

Differential Pair Noise

Noise originating in the differential pair is unlike the previous two cases. There, only certain parts of the noise spectrum contributed significantly to the total phase noise. White noise in the resonator is filtered at harmonics of the resonant frequency. White noise in the tail current only experiences a significant conversion gain around the second harmonic of the oscillation frequency. However, the simple model says that an impulse train samples white noise in the differential pair, which if true, will cause it to accumulate without bound at any specified offset frequency ω_m.

In reality, any practical differential pair requires a non-zero input voltage excursion to switch, and this is provided by the oscillation waveform across the resonator. Therefore, noise in the differential pair is actually not sampled by impulses, but by time windows of finite width. The window height is proportional to transconductance, and width is set by tail current, and slope of the oscillation waveform at zero crossing. The input-referred noise spectral density of the differential pair is inversely proportional to transconductance. Thus, the narrower the sampling window, that is, the larger the sampling bandwidth, the lower the noise spectral density [11]. Analysis shows that the noise bandwidth product is constant, and produces pure phase noise. After taking into account the accumulation of frequency translations throughout the sampling bandwidth, the following compact yet exact expression is reached:

$$\mathcal{L}(\omega_m) = \frac{32 I_T R \gamma}{\pi V_0} \frac{kTR}{V_0^2}\left(\frac{\omega_0}{2Q\omega_m}\right)^2$$

We note that [8] has arrived at a similar analysis for the first two sources of noise, but was unable to obtain a closed-form expression for this last term.

Proving Leeson's Hypothesis

Leeson originally postulated that thermally induced phase noise in any oscillator takes the form:

$$\mathcal{L}(\omega_m) = \frac{4FkTR}{V_0^2}\left(\frac{\omega_0}{2Q\omega_m}\right)^2$$

where F is an unspecified noise factor. By summing the expressions obtained above for thermally induced phase noise arising from the resonator, differential pair and tail bias current, respectively, for the differential oscillator Leeson's noise factor is:

$$F = 2 + \frac{8\gamma R I_T}{\pi V_0} + \gamma \tfrac{8}{9} g_{mbias} R$$

We emphasize that this simple expression captures all nonlinear effects and frequency translations. At low bias currents while the

206

amplitude of oscillation is smaller than the power supply, the differential pair acts as a pure current switch driving the resonator and $V_0=(4/\pi)RI_T$ [13]. Then the second term comprising F simplifies to 2γ. This means that as tail current increases and assuming $g_{mbias}R$ is held constant, the noise factor remains constant and phase noise improves as V_0^2, that is, as I_T^2. This has been observed by others [13]. However, beyond a critical tail current the amplitude V_0 is pegged constant, limited by supply voltage. Further increases in I_T will cause the differential pair's contribution to noise factor to rise, degrading phase noise proportionally to I_T (Figure 4). Therefore, for least phase noise the tail current should be just enough to drive the amplitude to its maximum possible value.

Flicker Noise Upconversion

Close-in to the oscillation frequency, the slope of the phase noise spectrum in all CMOS VCO's turns from –20 to –30 dB/decade. This is ascribed to the upconversion of flicker noise in FETs. To understand this, let us first see if the analysis above explains this upconversion.

Flicker noise in the tail current source at frequency ω_m indeed upconverts to $\omega_0\pm\omega_m$ and enters the resonator, but as AM, not PM noise. Therefore, in the absence of a high gain varactor to convert AM to FM, flicker noise in the tail current will not appear as phase noise. Next consider flicker noise in the differential pair. The preceding analysis says that this modulates zero crossings, and injects a noise current into the resonator consisting of flicker noise sampled by an impulse train with frequency $2\omega_0$. Thus noise originating at frequency ω_m produces currents at ω_m and at $2\omega_0\pm\omega_m$. Both frequencies are strongly attenuated in the resonator, and neither explains flicker-induced phase noise at $\omega_0\pm\omega_m$. One can only conclude that the mechanisms of flicker noise upconversion are quite different than for thermally induced phase noise.

Fundamental Sources of FM in Oscillators

In 1934, Groszkowski [15] while studying electronic oscillators realized that the steady-state oscillation frequency seldom coincides with the natural frequency of the resonator which tunes the oscillator. He found that the discrepancy arises because the active device in the oscillator, such as the differential pair current switch in the circuit considered here, drives the resonator with a harmonic-rich waveform. The harmonics will flow into the lower impedance capacitor (Figure 5) and upset the exact reactive power balance between the L and the C required for steady state. Now the frequency of oscillation must shift down until the reactive power in the inductor increases to equal the reactive power in the capacitor due to the fundamental and all harmonics. The shift, $\Delta\omega$, is:

$$\frac{\Delta\omega}{\omega_0} = \frac{1}{2Q^2}\sum_{n=2}^{\infty}\frac{n^2(1-n^2)}{(1-n^2)^2+n^2/Q^2}\cdot m_n^2$$

where m_n is the normalized level of the nth harmonic. $\Delta\omega$ is the sum of all negative terms, which means that oscillation frequency slows down with more harmonic content. Now the harmonic content at the output of a periodically switching differential pair is a function of the tail current. In the autonomous oscillator, the drive to the differential pair is also a function of tail current. The sensitivity $\partial\omega/\partial I_T$ is responsible for an "indirect" FM [7] due to flicker noise in I_T.

However, this is not the only mechanism of indirect FM. At RF, active device capacitance is also significant, and it no longer appears as a pure negative resistance to the resonator. For example, the differential pair commutates current flowing in the capacitor C_T at the tail, which presents a negative capacitor (or, equivalently, an inductor in a narrowband sense) at the differential output (Figure 6). This speeds up the oscillation frequency. Flicker noise in the differential pair FETs modulates the duty cycle of commutation, and therefore the effective negative capacitance. Here, too, Groszkowski gives a method of systematic analysis [16], which captures the reactive components in the active devices by measuring the area κ enclosed by hysteresis in the dynamic negative resistance curve.

$$\frac{\Delta\omega}{\omega_0} = -\frac{\kappa}{2Q^2\omega_0 L} + \frac{1}{2Q^2}\sum_{n=2}^{\infty}\frac{n^2(1-n^2)}{(1-n^2)^2+n^2/Q^2}\cdot m_n^2$$

Thus the sensitivity of the reactance to bias current or offset voltage in the differential pair is estimated, which is another means whereby flicker noise modulates the frequency of oscillation.

Validation of Analysis

The phase noise model was validated on two CMOS differential LC oscillators. One oscillator uses a low Q, on-chip inductor, while the other uses off-chip inductors with large Q. Flicker noise is modelled as a bias-independent, gate-referred voltage source [14]. The measured data and SpectreRF simulations are plotted with predictions based on this paper. Excellent agreement (Figure 7) is found across the entire spectrum, which encompasses thermally induced phase noise and upconverted flicker noise.

[1]   K. S. Kundert, "Introduction to RF simulation and its application," *IEEE Journal of Solid State Circuits*, pp. 1298-319, 1999.

[2]   A. Demir, A. Mehrotra, and J. Roychowdhury, "Phase noise in oscillators: a unifying theory and numerical methods for characterisation," in *Design and Automation Conference*, San Francisco, pp. 26-31, 1998.

[3]   B. De Smedt and G. Gielen, "Accurate simulation of phase noise in oscillators," in *European Solid-State Circuits Conference*, pp. 208-11, 1997.

[4]   A. Hajimiri and T. H. Lee, "A general theory of phase noise in electrical oscillators," *IEEE Journal of Solid-State Circuits*, vol. 33, no. 2, pp. 179-94, 1998.

[5]   D. B. Leeson, "A Simple Model of Feedback Oscillator Noise Spectrum," *Proceedings of the IEEE*, vol. 54, pp. 329-330, 1966.

[6]   J. Craninckx and M. Steyaert, "Low-noise voltage-controlled oscillators using enhanced LC-tanks," *IEEE Transactions on Circuits and Systems II: Analog and Digital Signal Processing*, vol. 42, no. 12, pp. 794-804, 1995.

[7]   K. K. Clarke and D. T. Hess, *Communication Circuits: Analysis and Design.* Malabar, FL: Krieger, 1971.

[8]   C. Samori, A. L. Lacaita, F. Villa, and F. Zappa, "Spectrum folding and phase noise in LC tuned oscillators," *IEEE Transactions on Circuits and Systems II: Analog and Digital Signal Processing*, vol. 45, no. 7, pp. 781-90, 1998.

[9]   Q. Huang, "On the exact design of RF oscillators," *CICC Proceedings*, pp. 41-4, 1998.

[10] W. P. Robins, *Phase Noise in Signal Sources.* London: Peter Peregrinus, 1982.

[11] H. Darabi and A. Abidi, "Noise in CMOS Mixers: A Simple Physical Model," *IEEE Journal of Solid State Circuits*, vol. 35, no. 1, in press, 2000.

[12] C. Samori, A. L. Lacaita, A. Zanchi, S. Levantino, and F. Torrisi, "Impact of Indirect Stability on Phase Noise Performance of Fully-Integrated LC Tuned VCOs," in *European Solid-State Circuits Conference*, Duisburg, Germany, pp. 202-205, 1999.

[13] A. Hajimiri and T. H. Lee, "Phase Noise in CMOS Differential LC oscillators," in *Symposium on VLSI Circuits*, Honolulu, HI, pp. 48-51, 1998.

[14] J. Chang, A. A. Abidi, and C. R. Viswanathan, "Flicker Noise in CMOS Transistors from Subthreshold to Strong Inversion at Various Temperatures," *IEEE Transactions on Electron Devices*, vol. 41, pp. 1965-1971, 1994.

[15] J. Groszkowski, "The Interdependence of Frequency Variation and Harmonic Content, and the problem of Constant-Frequency Oscillators," *Proc. of the IRE*, vol. 21, no. 7, pp. 958-981, 1934.

[16] J. Groszkowski, *Frequency of Self-Oscillations.* Oxford: Pergamon Press, 1964.

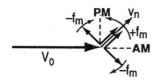

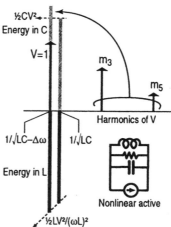

Figure 1. Noise phasor added to a sinewave decomposes into PM and AM sidebands.

Figure 5. Harmonics of oscillating current flow into capacitor, increasing its reactive energy. Steady state frequency shifts down until inductor energy balances.

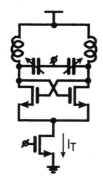

Figure 2. Differential LC oscillator biased by tail current.

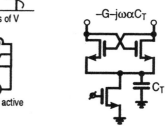

Figure 6. Capacitors associated with active device appear as reactances across the resonator, shifting frequency.

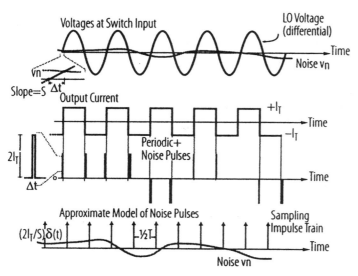

Figure 3. (a) Noise at input of differential pair modulates instants of zero crossing. (b) Output current consists of square wave, plus random noise pulses. (c) Noise pulses modelled as a train of impulses sampling noise waveform.

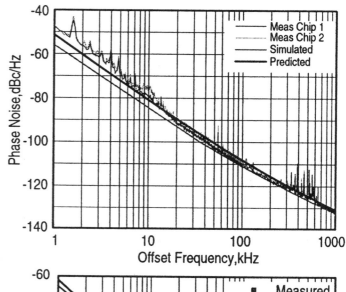

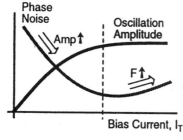

Figure 4. Increasing tail current first causes amplitude to rise, until limited by supply. Phase noise diminishes with rising amplitude, then worsens due to higher noise factor.

Figure 7. Validation of the analysis presented in this paper. Measured phase noise is compared with predictions from analysis, and with SpectreRF simulations. (a) 0.35-μm CMOS 1.1 GHz oscillator using resonator with loaded Q of 6. (b) 0.25-μm CMOS 830 MHz oscillator using discrete inductor with loaded of Q of 25.

Phase Noise in *LC* Oscillators

Konstantin A. Kouznetsov and Robert G. Meyer

Abstract—**Analytical methods for the phase-noise analysis of *LC*-tuned oscillators are presented. The fundamental assumption used in the theoretical model is that an oscillator acts as a large-signal *LC*-tuned amplifier for purposes of noise analysis. This approach allows us to derive closed-form expressions for the close-to-carrier spectral density of the output noise, and to estimate the phase-noise performance of an oscillator from circuit parameters using hand analysis. The emphasis is on an engineering approach intended to facilitate rapid estimation of oscillator phase noise. Theoretical predictions are compared with results of circuit simulations using a nonlinear phase-noise simulator. The analytical results are in good agreement with simulations for weakly nonlinear oscillators. Complete nonlinear simulations are necessary to accurately predict phase noise in oscillators operating in a strongly nonlinear regime. To confirm the validity of the nonlinear phase-noise models implemented in the simulator, simulation results are compared with measurements of phase noise in a practical Colpitts oscillator, where we find good agreement between simulations and measurements.**

Index Terms—**Jitter, noise, oscillator, phase noise, voltage-controlled oscillator.**

I. INTRODUCTION

VOLTAGE-CONTROLLED oscillators (VCO's) are essential building blocks of wireless communication systems, where the spectral quality of the local oscillator signal directly influences the out-of-band interference. Phase noise in oscillators has long been the subject of theoretical and experimental investigation. An early model of phase noise, introduced by Leeson [1], qualitatively described phase-noise spectra in a variety of oscillators. Theoretical derivations of the Leeson formula rely mostly on a linear time-invariant approach to the analysis of noise in oscillators [2]–[4]. Treatments of nonlinear effects on oscillator noise have also been developed [5], [6]. Recently, general theories that allow quantitative predictions of phase noise have been introduced [7]–[9]. These exact approaches naturally lend themselves to computer-aided analysis. A complete nonlinear phase-noise simulator that utilizes approaches similar to those of [7]–[9] has been implemented in a commercial simulation package [10].

In this paper, we explore the validity and limitations of linear noise analysis of *LC*-tuned oscillators. We examine and exploit the assumption that an *LC* feedback oscillator acts as a high-gain, large-signal *LC* amplifier with respect to its noise sources

Manuscript received November 23, 1999; revised April 19, 2000. This work was supported by the Army Research Office under Grant DAAG55-97-1-0340.
K. A. Kouznetsov was with the Department of Electrical Engineering and Computer Science, University of California, Berkeley, CA 94720-1772 USA. He is now with Maxim Integrated Products, Sunnyvale, CA 94086 USA (e-mail: kouznets@mxim.com).
R. G. Meyer is with the Department of Electrical Engineering and Computer Science, University of California, Berkeley, CA 94720-1772 USA.
Publisher Item Identifier S 0018-9200(00)06439-8.

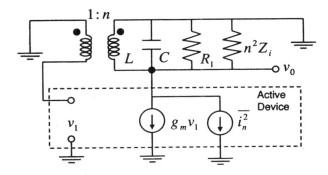

Fig. 1. AC schematic diagram of an *LC*-tuned transformer-coupled oscillator showing a single current noise source referred to the output of the active device.

despite the nonlinear limiting that occurs in steady-state oscillation. This approach allows us to make quantitative estimates of the phase-noise performance of *LC* oscillators with a weak nonlinearity, giving predictions which are in good agreement with nonlinear simulations with the SpectreRF simulator. Our engineering techniques provide insights and quantitative understanding of noise in oscillators and serve as a starting point in a design procedure for practical oscillators that should include complete nonlinear simulations.

II. OSCILLATOR NOISE THEORY

In the analysis of phase noise of an *LC* oscillator, we assume that an oscillator is operating in the steady state at a fundamental frequency f_0 and only wide-sense stationary noise sources are superimposed onto the oscillating waveform. In a nonlinear system noise sidebands at frequencies $kf_0 \pm \delta f$ also contribute to output noise at close-to-carrier frequencies $f_0 \pm \delta f$ via processes of nonlinear mixing; here, k is an integer between $+\infty$ and $-\infty$. For clarity of the discussion that follows, we derive the phase noise of the oscillator at the output frequency $f_0 - \delta f$. We further assume that the oscillator is operating in a near-linear fashion such that noise sidebands close to the carrier provide the dominant contribution to oscillator noise and mixing from other harmonics is suppressed.

Without loss of generality, the circuit diagram of an *LC*-tuned oscillator can be represented as shown in Fig. 1 [11]. The loop gain is assumed to be just less than unity, so that the circuit behaves as a high-gain *LC* positive-feedback amplifier. As the oscillation amplitude builds up, the influence of the negative resistance provided by the active device is reduced (due to soft limiting or AGC action) such that in the steady state, the output spectrum of an oscillator contains a very sharp noise peak at its fundamental frequency, resulting from the oscillator limiting on its own amplified noise.

To determine the noise transfer function from the noise generator $\overline{i_n^2}$ to the oscillator output, we consider effects of noise

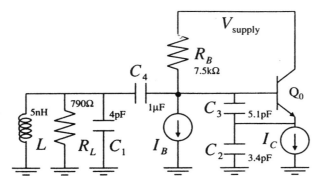

Fig. 2. Circuit diagram of a Colpitts oscillator with the collector of Q_0 ac grounded.

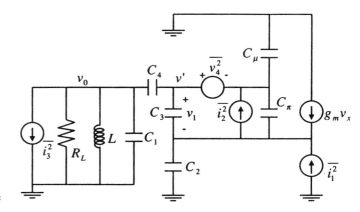

Fig. 3. Small-signal model of the Colpitts oscillator used for the phase noise analysis.

modulation to positive and negative sidebands. As we show in the Appendix, the noise power at $f_0 - \delta f$ will be equally distributed between sidebands at $f_0 - \delta f$ and $-f_0 - \delta f$ leading to the noise transfer function α_0 at $f_0 - \delta f$, which can be written as

$$\alpha_0 = \frac{1}{2} \times \frac{R_L}{2jQ} \left(\frac{f_0}{\delta f} \right). \tag{1}$$

This expression explicitly contains a factor of 1/2 resulting from noise spreading to positive and negative frequencies. Similarly, a noise tone at $-f_0 - \delta f$, which is uncorrelated with the tone at $f_0 - \delta f$, is spread evenly between $-f_0 - \delta f$ and $f_0 - \delta f$ in the output noise spectrum. To determine the total noise at the output at $f_0 - \delta f$, contributions from positive and negative frequencies must be added in quadrature leading to the total noise spectral density at $f_0 - \delta f$, $\overline{v}_t^2$, given by

$$\overline{v}_t^2 = 2 \times \overline{i}_n^2 \times \frac{1}{4} \times \frac{R_L^2}{4Q^2} \left(\frac{f_0}{\delta f} \right)^2. \tag{2}$$

We note that the total output noise of an LC oscillator given by (2) is 1/2 of the noise power of an LC tank with the same Q. This distinction in the noise shaping between an oscillator and a resonant circuit has been pointed out previously in [10].

It is frequently stated that close-to-carrier oscillator noise can be decomposed into AM- and PM-modulated parts with the AM-modulated part being suppressed due to amplitude limiting in the steady state of oscillation. This reasoning often leads to the inclusion of a factor of 1/2 into phase-noise expressions. We perform our calculations in a manner which is similar to the noise analysis of a positive-feedback amplifier, and do not include an additional factor of 1/2 in derivations. Therefore, we use the terms oscillator noise and phase noise interchangeably, meaning total noise at the oscillator output.

III. COLPITTS OSCILLATOR ANALYSIS

To test the validity of our approach to the noise analysis of LC oscillators, we analyze phase noise in a Colpitts oscillator. A circuit diagram of a Colpitts oscillator with the collector of the active device ac grounded is shown in Fig. 2. The active device is a bipolar-junction transistor Q_0 biased by dc current sources I_C and I_B, and resistor R_B. The capacitive transformer formed by capacitors C_2 and C_3 provides the positive feedback

needed to induce oscillation. The coupling capacitor C_4 can be chosen to provide additional transformations from the base of Q_0 to the tank circuit, or a large value of C_4 can be chosen to provide a short circuit between the base of Q_0 and the tank at the oscillation frequency.

We consider the Colpitts oscillator biased at a relatively low value of the collector bias current, $I_C = 0.2$ mA. This value is chosen to set the initial loop gain at 1.5, which is sufficiently close to unity to create a weakly nonlinear oscillator. The small-signal ac circuit used for the noise analysis is shown in Fig. 3. Four noise sources are included in the analysis: collector-current shot noise i_1, base-current shot noise i_2, thermal-current noise in the tank resistance i_3 produced by the resistance $R_L = 790\ \Omega$, and thermal voltage noise in base resistance v_4 produced by the resistance $r_b = 7.6\ \Omega$. Circuit parasitics have not been included for clarity of comparison of analytical results and simulations.

Solving circuit equations, we derive transfer functions from a particular noise source to the output at the frequency $f_0 - \delta f$ in a manner identical to derivations presented in the Appendix. For low values of the collector bias current, the assumption of small g_m, i.e., $g_m \ll |sC_2|,\ |sC_3|$, is valid, in which case, noise transfer functions are found as follows:

$$\alpha_0^{(1)} = \frac{1}{2} \times \frac{C_3}{C_2 + C_3} \times \frac{1}{2j\omega_0 C_{\text{eff}}} \left(\frac{f_0}{\delta f} \right) \tag{3}$$

$$\alpha_0^{(2)} = \frac{1}{2} \times \frac{C_2}{C_2 + C_3} \times \frac{1}{2j\omega_0 C_{\text{eff}}} \left(\frac{f_0}{\delta f} \right) \tag{4}$$

$$\alpha_0^{(3)} = \frac{1}{2} \times \frac{1}{2j\omega_0 C_{\text{eff}}} \left(\frac{f_0}{\delta f} \right) \tag{5}$$

$$\alpha_0^{(4)} = \frac{1}{2} \times \frac{C_2 + C_3}{C_2} \times \frac{1}{2jQ} \left(\frac{f_0}{\delta f} \right) \tag{6}$$

where $\alpha_0^{(1)}, \alpha_0^{(2)}, \alpha_0^{(3)},$ and $\alpha_0^{(4)}$ refer to noise sources $i_1, i_2, i_3,$ and v_4, respectively, $C_{\text{eff}} = C_1 + C_2 C_3/(C_2 + C_3)$ is the effective tank capacitance, and Q is the loaded quality factor of the tank. The phase noise contribution due to the kth noise source is determined by multiplying its noise spectral density by the square of the appropriate transfer function, as follows:

$$PN_0^{(k)} = N_k \left(\alpha_0^{(k)} \right)^2 \tag{7}$$

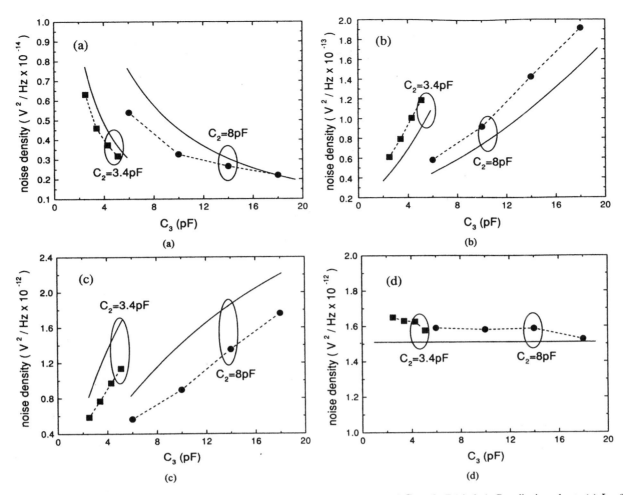

Fig. 4. Simulation results of a weakly nonlinear Colpitts oscillator for $C_2 = 3.4$ pF (squares) and $C_2 = 8$ pF (circles). Contributions due to (a) I_B, (b) r_b, (c) I_c, and (d) R_L are shown; $f_0 = 946$ MHz and $\delta f = 25$ kHz. Ellipses group simulation and analytical curves (solid lines) for indicated values of C_2.

where N_k refers to the spectral density of the kth noise source, $N_1 = 2qI_C, N_2 = 2qI_B, N_3 = 4kT/R_L$, and $N_4 = 4kTr_b$, and the subscript 0 indicates that only the contribution from the 0th harmonic at $f_0 - \delta f$ is calculated. Equations (3)–(7) predict the dependence of phase-noise contributions on circuit parameters such as circuit capacitors, bias currents, and parasitic resistors.

To test theoretical predictions, we compare noise contributions from the 0th harmonic calculated using (3)–(7) with those simulated with the nonlinear phase-noise simulator in SpectreRF in Fig. 4. The phase-noise spectral densities at 25-kHz offset from 946-MHz carrier are plotted as a function of capacitance C_3 for two different values of capacitor C_2. Simulation results for $C_2 = 3.4$ pF and 8 pF are shown as solid squares and solid circles, respectively. Predictions of hand calculations are shown as solid lines in both cases. Results in Fig. 4 demonstrate good quantitative agreement between simulation and analytical results. The theoretical curves correctly capture trends in phase-noise variations and quantitatively lie within 30% error from simulated data.

IV. COMPARISON OF SIMULATIONS AND EXPERIMENT

To confirm the validity of nonlinear models implemented in a simulator, we compare results of simulations and measurements on a hard-driven Colpitts oscillator. The hard-driven case represents the most challenging conditions under which one can test the performance of the simulator. A commercially available Colpitts oscillator circuit [12] was chosen for comparison purposes. Phase-noise simulations were performed with SpectreRF using measured device models and including all on-chip and off-chip parasitics. The frequency of oscillation is set at approximately 946 MHz, and the loaded Q of the off-chip resonator was close to 140. The single-sideband output noise spectrum of one of the production circuits was measured using the single-sweep power spectral density measurement on the spectum analyzer. Measurement results shown in Fig. 5 indicate the total output noise power spectral density, which includes both AM- and PM-modulated components. Simulations with SpectreRF also compute the total output noise spectral density including both AM and PM components. Simulated and measured results are compared in Fig. 5 where we find a very good agreement between measurements and simulations. The flattening out of the phase noise with frequency in measured data at higher frequencies is due to noise in the buffer amplifier, which was not included in simulations.

V. CONCLUSION

In conclusion, we have presented an engineering approach to the analysis of phase noise in LC-tuned oscillators. The basic no-

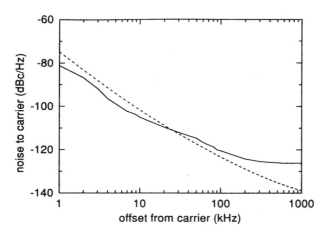

Fig. 5. Comparison of measurement (solid line) and simulation (dashed line) results for the practical Colpitts oscillator; offset is from 946-MHz carrier.

tion that has been explored in our theory is that an oscillator may be treated as a high-gain LC amplifier for phase-noise analysis. The techniques of oscillator noise analysis presented are intuitive and allow quantitative estimates of oscillator noise from circuit parameters. Theoretical predictions are in good agreement with nonlinear simulations using SpectreRF for weakly nonlinear oscillators. Validity of SpectreRF simulation models have been tested by comparing simulations and measurements under the most challenging conditions of a hard-driven oscillator, where we find that simulation and measurement results are in good agreement.

APPENDIX

For an oscillator operating in the steady state at a fundamental frequency f_0, the nonlinear time-varying transconductance $g_m(t)$ can be represented by a sum over multiples of fundamental frequencies, as follows:

$$g_m(t) = \sum_{k=-\infty}^{+\infty} g_m^{(k)} e^{jk2\pi f_0 t} \qquad (8)$$

where k is the number of the harmonic. From Kirchoff's equations at the collector of the active device, we obtain

$$g_m(t)v_1(t) = nv_1(t)/Z_T \qquad (9)$$

where we assumed a linear transformer coupling with a transformation ratio $1 : n$. Substituting (8) into (9) and summing current components at the fundamental frequency f_0, we arrive at the oscillation condition

$$g_{m,\text{eff}} = g_m^{(0)} + g_m^{(2)} = \frac{n}{Z_T(f_0)}. \qquad (10)$$

Next, we consider a noise source i_n at $f_0 - \delta f$ superimposed onto the oscillating waveform as shown in Fig. 1. The nonlinear transconductance $g_m(t)$ modulates this noise source to create noise sidebands at $f_0 - \delta f$ and $-f_0 - \delta f$ in the output waveform. At the input to the active device, we denote rms voltages of the former sideband by $v_1^{(+)}$ and of the latter by $v_1^{(-)}$. Similarly to

(9), we write Kirchoff's equations in the presence of the noise generator i_n for $f_0 - \delta f$ and $-f_0 - \delta f$ as

$$g_m^{(0)} v_1^{(+)} + g_m^{(2)} v_1^{(-)} + i_n = \frac{nv_1^{(+)}}{Z_T(f_0)} \qquad (11)$$

and

$$g_m^{(0)} v_1^{(-)} + g_m^{(2)} v_1^{(+)} = \frac{nv_1^{(-)}}{Z_T(f_0)} \qquad (12)$$

respectively. Solving for $v_1^{(+)} + v_1^{(-)}$ and $v_1^{(+)} - v_1^{(-)}$, we find

$$v_1^{(+)} + v_1^{(-)} = \frac{-i_n}{g_m^{(0)} + g_m^{(2)} - n/Z_T(f_0)} \qquad (13)$$

and

$$v_1^{(+)} - v_1^{(-)} = \frac{-i_n}{g_m^{(0)} - g_m^{(2)} - n/Z_T(f_0)}. \qquad (14)$$

Using the oscillation condition, (10), in (13) and (14), we find that in steady state the right-hand side of (13) peaks very sharply, and is much greater than the right hand side of (14). This leads to $v_1^{(+)} = v_1^{(-)}$, and to the expression for a single-sided transfer function, from i_n to $v_1^{(+)}$, given by

$$v_1^{(+)} = \frac{1}{2} \times \frac{-i_n}{g_{m,\text{eff}}^{(0)} - n/Z_T(f_0)}. \qquad (15)$$

To obtain the output transfer function α_0, we use $v_0^{(+)} = nv_1^{(+)}$ to arrive at

$$\alpha_0 = \frac{1}{2} \times \frac{R_L}{2jQ}\left(\frac{f_0}{\delta f}\right). \qquad (16)$$

This is the noise transfer function from the noise tone at $f_0 - \delta f$ to the output noise of the oscillator at the same frequency. This transfer function explicitly contains a factor of 1/2, which is due to noise translation by $-2f_0$ to the negative frequencies at the output.

ACKNOWLEDGMENT

The authors would like to thank G. Mueller and B. Mack of Maxim Integrated Products, Inc., Sunnyvale, CA, for providing phase-noise measurement results.

REFERENCES

[1] D. B. Leeson, "A simple model for oscillator noise spectrum," *Proc. IEEE*, vol. 54, pp. 329–330, Feb. 1966.
[2] W. P. Robins, *Phase Noise in Signal Sources—Theory and Applications*. London, U.K.: Peregrinus, 1991.
[3] J. Craninckx and M. Steyaert, "Low-noise voltage-controlled oscillators using enhanced *LC*-tanks," *IEEE Trans. Circuits Syst.—II*, vol. 42, pp. 794–804, Dec. 1995.
[4] B. Razavi, "A study of phase noise in CMOS oscillators," *IEEE J. Solid-State Circuits*, vol. 31, pp. 331–343, Mar. 1996.
[5] H. Siweris and B. Schieck, "Analysis of noise upconversion in microwave FET oscillators," *IEEE Trans. Microwave Theory Tech.*, vol. MTT-33, pp. 233–241, Jan. 1985.
[6] C. Samori, A. L. Lacaita, F. Villa, and F. Zappa, "Spectrum folding and phase noise in *LC* tuned oscillators," *IEEE Trans. Circuits Syst.—II*, vol. 45, pp. 781–789, July 1998.
[7] F. Kaertner, "Analysis of white and $f^{-\alpha}$ noise in oscillators," *Int. J. Circuits Theory Appl.*, vol. 18, pp. 485–519, 1990.
[8] A. Demir, "Analysis and Simulation of Noise in Nonlinear Electronic Circuits and Systems," Ph.D. dissertation, Univ. of California, Berkeley, CA, 1997.

[9] A. Hajimiri and T. H. Lee, "A general theory of phase noise in electrical oscillators," *IEEE J. Solid-State Circuits*, vol. 33, pp. 179–194, Feb. 1998.

[10] Oscillator Noise Analysis in SpectreRF: Cadence Design Systems, Inc., to be published.

[11] K. K. Clarke and D. T. Hess, *Communication Circuits: Analysis and Design*. Redding, MA: Addison-Wesley, 1971, ch. 6.

[12] "10 MHz to 1050 MHz Integrated RF Oscillator with Buffered Outputs, Maxim Wireless Analog Design Solutions Guide," Maxim Integrated Products, Inc., Sunnyvale, CA, 1999.

The Effect of Varactor Nonlinearity on the Phase Noise of Completely Integrated VCOs

John W. M. Rogers, *Student Member, IEEE*, José A. Macedo, *Member, IEEE*, and Calvin Plett, *Member, IEEE*

Abstract—This work discusses variations in phase noise over the tuning range of a completely integrated 1.9-GHz differential voltage-controlled oscillator (VCO) fabricated in a 0.5-μm bipolar process with 25-GHz f_t. The design had a phase noise of -103 dBc/Hz at 100 kHz offset at the top of the tuning range, but the noise performance degraded to -96 dBc/Hz at 100 kHz at the bottom of the tuning range. It was determined that nonlinearities of the on-chip varactors, which led to excessively high VCO gain at the bottom of the tuning range, were primarily responsible for this degradation in performance. The VCO has a power output of -5 dBm per side. Calculations predict phase noise with only a small error and provide design insight for minimizing this effect. The oscillator core drew 6.4 mA and the output buffer circuitry drew 6 mA, both from a 3.3-V supply.

Index Terms—Bipolar transistor, circuit theory and design, fully integrated VCO, integrated inductors, phase noise, varactors.

I. Introduction

THE INCREASING demand for portable communications equipment has driven research to produce transceivers at lower cost. This has led to an intense interest in integrating as many components as possible. One high-speed component that has been particularly hard to integrate is the voltage-controlled oscillator (VCO). This is largely due to the poor quality of on-chip passive components in silicon integrated circuit technologies, namely the low quality factor Q of on-chip inductors and the poor linearity of on-chip varactors. In this work, we present a design methodology for a high-power completely integrated Colpitts oscillator. Optimum design of this circuit for best phase-noise performance will be discussed. The effect of poor varactor linearity on phase-noise performance over the complete tuning range will be discussed, an issue not commonly addressed. We will show that with the use of simple theory, this effect can be predicted qualitatively and quantitatively with only a small error.

II. Design Considerations

In this section, the design of a differential Colpitts common-base VCO for low phase-noise performance is discussed. The circuit is shown in its simplest form in Fig. 1. Note that of the Colpitts topologies, either the common-base

Manuscript received December 17, 1999; revised February 13, 2000. This work was supported by the Natural Sciences and Engineering Research Council of Canada (NSERC) and MICRONET.

J. W. M. Rogers is with SiGe Microsystems, Ottawa, ON K2B 8J9, Canada.

J. A. Macedo is with Research In Motion, Kanata, ON K2L 4B6, Canada.

C. Plett is with the Department of Electronics, Carleton University, Ottawa, ON K1S 5B6, Canada.

Publisher Item Identifier S 0018-9200(00)05929-1.

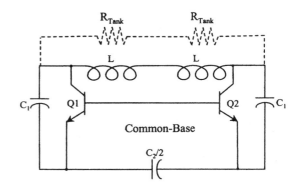

Fig. 1. Differential common-base Colpitts oscillator. Biasing not shown.

or the common-collector is suitable for integrated designs. The common-base was chosen because it is not as widely explored as the common-collector configuration [1]. As well, preliminary simulations showed it had performance equal to or marginally better than the common-collector circuit.

A. Frequency Tuning

The first task in designing an LC oscillator is to set the frequency of oscillation, and hence set the value of the total inductance and capacitance in the circuit. Assuming ideal transistors $Q1$ and $Q2$ the frequency of oscillation is given by

$$\omega_{\text{osc}} = \frac{1}{\sqrt{LC_{\text{tot}}}} \qquad (1)$$

where C_{tot} is the series combination of C_1 and C_2.

To get the unloaded Q as high as possible, as well as increase output swing, it is desirable to make the inductance as large as possible. This is because the losses in the inductor are usually much higher than the losses in the capacitors and because the parallel resistance of the inductor is proportional to the size of the inductor [see (2)]. However, it should be noted that in practice, large monolithic inductors suffer from limited Q. Therefore, in a given technology there will be an optimum size for the inductor.

B. Design of On-Chip Inductors

Inductors prove to be one of the most difficult passive circuit elements to implement on-chip. When they are made in Si technology they suffer from the presence of relatively high-resistance metals and lossy substrates, although much work is being done to improve this [2]–[6]. In order to achieve the best possible Q at the desired frequency, careful layout of the inductor with the help of an existing simulation tool [7], [8] is required.

The parallel resistance that the inductor presents to the tank is given by

$$R_{\text{Tank, L}} = Q_{\text{ind}}\omega_{\text{osc}}L = Q_{\text{ind}}\sqrt{\frac{L}{C_{\text{tot}}}} \qquad (2)$$

The goal of any inductor layout is to design a spiral inductor of specified inductance, and optimize Q for best performance at the frequency of interest. In order to achieve this, careful layout of the structure is required. The resistance of the metal lines causes the inductor to have a high series resistance limiting its performance at low frequencies, while capacitive coupling to the lossy silicon substrate causes the inductor's effective resistance to rise even more at higher frequencies [9], [10]. Large coupling between the inductor and the substrate also causes the structures to have low self-resonance frequencies, placing restrictions on the size of the device that can be built. In order to reduce the series resistance, the line widths should be large; therefore, to obtain a given value of inductance the structure must occupy a greater area, increasing substrate loss. Narrower metal lines mean less substrate loss, but more series resistance. The designer must carefully balance these factors to get the best possible performance at the desired frequency.

Traditionally, on-chip inductors have been square. This is because these have been easier to model than more complicated geometries. A square geometry is by no means optimal however. The presence of the 90° bends adds unnecessary resistance to the structure [11]. Intuitively it is easy to see that as the structure is made more circular the performance will improve. Recent work has also shown that the use of differential inductors can reduce substrate losses and lead to higher Q factors in differential applications [12], [13]. As well, the use of a patterned ground shield underneath the inductor can also lead to improved performance [2]. However, few simulators other than full 3-D electromagnetic simulators are able to predict the performance of such structures so the designer is often left with experimentation as the only possible design approach. For these reasons the inductors used in fabricating the oscillator were made with a square geometry.

In this work a version of GEMCAP [7] was used to simulate inductor performance. Unfortunately, this software could only model square geometries. Despite the fact that the version of the software used did not handle the skin effect or model the underpass of the structure, the values of inductance were simulated very accurately. It also predicted the Q with reliable trends although it tended to be slightly optimistic.

C. Capacitor Ratios

In order to achieve good performance, the use of capacitors is necessary to transform r_e (the dynamic emitter resistance) of transistors $Q1$ and $Q2$ (see Fig. 1) to a higher value. The impedance transformation effectively prevents this typically low impedance from reducing the Q of the oscillator's LC tank. The impedance transformation is given by [14]:

$$R_{\text{Tank}, r_e} = \left(1 + \frac{C_2}{C_1}\right)^2 r_e \qquad (3)$$

where $R_{\text{Tank,re}}$ represents the parallel resistance across the tank due to r_e. Therefore, in order to get the maximum effect of the impedance transformer, it is necessary to make C_2 large and C_1 small. However, if this ratio is made too large then the gain around the loop will drop below one. In addition, the larger this ratio is made the more current must be driven through the transistors to achieve a given output power. This in turn leads to larger noise sources in the tank, thus degrading phase noise. The gain around the loop in the oscillator is given by

$$G = \frac{C_1}{C_1 + C_2}g_m R_{\text{Tank}} = \frac{g_m R_{\text{Tank}}}{1 + \frac{C_2}{C_1}} \qquad (4)$$

where R_{Tank} is approximately equal to the parallel combination of $R_{\text{Tank,L}}$ and $R_{\text{Tank,re}}$. Since the loop gain given in (4) must be greater than one for oscillations to begin, this expression can be solved to give a crude estimate for the minimum bias current required to allow oscillations to start.

$$I_C > V_T \left(1 + \frac{C_2}{C_1}\right) \frac{1}{R_{\text{Tank}}} \qquad (5)$$

where V_T is the thermal voltage. In the limit when C_1 gets very large this expression simplifies to

$$I_C > \frac{V_T}{R_{\text{Tank}}}. \qquad (6)$$

This corresponds to the case where the minimum current is required for start-up of the oscillator. However, this also corresponds to the case where the transistor loss will have the most impact on the Q of the tank and will lead to higher phase noise as shown above.

D. Design and Placement of Varactors

Diodes in the technology used for this work [15] are usually realized using the base–collector junction, as shown in Fig. 2. However, when using this junction there is also a parasitic diode between the collector and the substrate. Observe that the tiedown is typically connected to ground, hence the anode of the parasitic diode is normally grounded. The base–collector junction had a Q that varied from 22.9 to 33.5 with a properly optimized layout as shown in Fig. 2. However, the parasitic junction inherently has a low Q due to the low doping of the substrate. This makes it desirable to remove it from the circuit. Placing two varactors C_{var} in the oscillator as shown in Fig. 3 effectively ac-grounds both sides of the parasitic diodes and removes them from the differential circuit. Note that in this circuit the anode of the varactors is at approximately 1-V dc. Therefore, to prevent forward-biasing them, the control-voltage tuning range will be approximately from 1 to 3.3-V dc. The resistor R_{tune} prevents decoupling both sides of the oscillator. It also provides protection from accidentally forward-biasing the varactors which could otherwise cause large currents to flow.

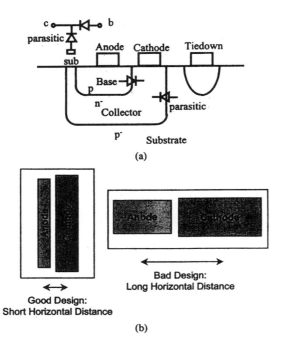

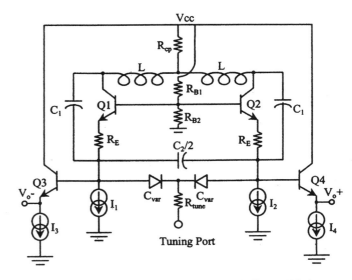

Fig. 3. Final common-base design including output buffers and biasing.

Fig. 2. Integrated varactor with parasitic diode. (a) Cross-sectional view of the diode. (b) Example layout for the varactor.

Unfortunately in bipolar processes the base–collector junction is usually made with constant doping. This leads to a nonlinear C–V characteristic that can be approximated by the following [16]:

$$C(V) = C_o \left(1 - \frac{V}{\psi_o}\right)^{-1/2} \qquad (7)$$

where C_o is the value of the capacitance of the junction with zero volts across it, and ψ_o is the built-in potential of the junction. If available, the use of a junction with a linearly graded doping profile can be shown to lead to increased linearity. The formula for its C–V characteristic is

$$C(V) = C_o \left(1 - \frac{V}{\psi_o}\right)^{-1/3}. \qquad (8)$$

In the limit for large applied reverse bias the varactors can suffer from collector–base punch-through. This will cause the capacitance to become almost constant. At this point, the VCO frequency will cease to vary with changes in the voltage on the control line. As will be shown in Section IV, this occurs in the region of low phase noise.

Equations (1) and (7) could be used to find an expression for the frequency of oscillation as a function of control voltage. However, it is more useful to find such a relationship through simulation and measurement. Nevertheless it should be noted that the shape of the C–V curve described by (7) is nonlinear and will lead to a nonconstant value of K_{VCO} for the oscillator.

E. Addition of Emitter Degeneration

Fig. 3 shows the completed oscillator circuit. The current sources are represented with symbols for simplicity. If sufficient current is driven through the transistors $Q1$ and $Q2$ in the circuit depicted in Fig. 3 such that reasonable output power is achieved, the value of r_e is quite small even after it has been transformed

to the collector. If steps are not taken, this resistance will limit the performance of the design. In order to increase the Q of the tank without sacrificing output power, emitter degeneration R_E is added to the circuit. Degeneration will also further increase the linearity of the transistor amplifiers at the cost of adding some additional noise. This allows the signal levels to increase more before the transistor nonlinearity causes saturation. Care must be taken as the addition of excessive degeneration results in less negative resistance and oscillations will not start. Thus making this resistor too large leads to low output power, and therefore high phase noise. Making R_E small also leads to excessive nonlinear mixing of noise around the carrier that further increases phase noise. Therefore, either simulation or careful analysis must be used to set this value. Simulations showed that 20 Ω was close to optimal.

F. Addition of Output Buffers

The placement of a 50-Ω load either directly across the tank or at the emitter of these circuits such as when testing with a spectrum analyzer would significantly reduce the Q of the circuit. For this reason, output buffers have been added to the circuit. These buffers are emitter followers $Q3$ and $Q4$ with high input impedance and they transform the 50-Ω load into a larger impedance. They were placed at the emitter instead of at the collector in order to take advantage of the impedance transformation provided by the capacitors (C_1, C_2, and C_{var}). This buffer also makes the circuit less sensitive to load pull when driving an active load such as the input to a mixer, and gives increased output power since it can drive 50 Ω without affecting signal amplitude in the oscillator core.

G. Quality Factor (Q) of an Oscillator

The Q of an LC resonator is a figure of merit used for LC tanks and oscillators. It is extremely important because the phase noise produced by the oscillator is a strong function of the Q of the tank. It can be shown that [14]

$$Q = \omega_o R_{\mathrm{Tank}} C_{\mathrm{tot}} = R_{\mathrm{Tank}} \sqrt{\frac{C_{\mathrm{tot}}}{L}} \qquad (9)$$

where R_{Tank} is the total equivalent resistance in parallel with the tank, C_{tot} is the total capacitance of the tank, and L is the inductance. For the oscillator in Fig. 3, R_{Tank} is composed mostly of the parallel resistance of the on-chip inductor (due to its limited Q), and of the transformed dynamic emitter resistance in series with R_E. Thus, for this circuit the resistance in parallel with the tank including the effect of R_E can be approximated by

$$\frac{1}{R_{\text{Tank}}} = \frac{1}{Q_{\text{ind}}\omega_{\text{osc}}L} + \frac{1}{\left((r_e + R_E)\left(1 + \dfrac{C_2 + C_{\text{var}}}{C_1}\right)^2\right)}. \tag{10}$$

Some numeric values using this formula will be given in Section V.

III. OUTPUT POWER OF THE OSCILLATOR

An oscillator relies on the nonlinearity in the transistor amplifier to limit the amplitude of the oscillation. As the voltage swings get larger the transistor starts to behave in a nonlinear manner which causes the loop gain to be reduced until it is exactly unity. At this point the oscillation amplitude stabilizes. If the amplitude of the oscillation is quite small due to a small-signal loop gain (not much more than unity) then transistor cut-off will probably be enough to cause oscillations to stabilize. As the gain is increased and oscillations become more vigorous the transistor will alternate between cut-off and saturation. In this state, the current through the transistor will appear as narrow pulses coinciding with the top of the voltage swing. In this region of operation, a formula is given by [17] to predict the amplitude, as follows:

$$V_{\text{Tank}} \approx 2I_{\text{Bias}}\left(1 - \frac{C_1}{C_1 + C_2 + C_{\text{var}}}\right)(R_{\text{Tank, L}}//R_{\text{load}}) \tag{11}$$

where R_{load} is the resistance presented to the tank by the load. Note that this formula will be slightly less accurate for designs that use degeneration because that will tend to widen the current pulses. For the oscillator in this paper this formula would lead to an output voltage of

$$
\begin{aligned}
V_{\text{out}} &\approx 2A_f I_{\text{Bias}}\left(\frac{C_1}{C_1 + C_2 + C_{\text{var}}}\right)\left(1 - \frac{C_1}{C_1 + C_2 + C_{\text{var}}}\right) \\
&\quad \cdot (R_{\text{Tank, L}}//R_{\text{load}})
\end{aligned} \tag{12}
$$

where A_f is the loss due to the followers. However, it should be noted that oscillations would not grow forever with increasing bias current. Eventually the voltage will reach a hard limit set by one of the supply rails. Further increasing the current will result in a clipped output voltage waveform such that the voltage amplitude of the fundamental frequency is the first harmonic of a square wave. In this region, the formula will become increasingly inaccurate.

IV. PHASE NOISE IN VCOs

When VCO phase-noise performance is reported, it is typically measured at only one point in the tuning range [1]–[21]. However, the phase noise of on-chip designs is not necessarily constant over the tuning range [22]. Unfortunately, on-chip varactors have a nonlinear C–V curve, and this can make the phase noise over the tuning range nonuniform. Any noise on the control line can lead to additional phase noise as shown in [14], however, any noise generated by the VCO at the varactor terminals will also modulate the carrier and create additional phase noise. This term can be added to the well known Leeson's formula [23] to take this additional noise mechanism into account.

$$
L(f_m, K_{\text{VCO}}) = 10 \log\left(\left(\frac{f_o}{2Qf_m}\right)^2\left[\frac{FkT}{2P_s}\left(1 + \frac{f_c}{f_m}\right)\right] + \frac{1}{2}\left(\frac{K_{\text{VCO}}V_m}{2f_m}\right)^2\right) \tag{13}
$$

where

$L(f_m, K_{\text{VCO}})$	phase noise in dBc/Hz;
f_o	frequency of oscillation in Hz;
f_m	frequency offset from the carrier in Hz;
F	noise figure of the transistor amplifier;
k	Boltzmann's constant in J/K;
T	temperature in K;
P_s	RF power produced by the oscillator in W;
f_c	flicker noise corner frequency in Hz;
K_{VCO}	gain of the VCO in Hz/V;
V_m	total amplitude of all low frequency noise sources in V/$\sqrt{\text{Hz}}$.

Since the varactor is nonlinear, K_{VCO} varies over the tuning range. Therefore, in some circumstances, the phase noise in the oscillator can be completely determined by the low-frequency noise. At the bottom of the tuning range where the VCO has a high gain, the low-frequency noise dominates. Conversely at the top of the tuning range where the gain is small, the Leeson's-style noise dominates and determines the phase noise of the oscillator. Thus, the designer must be very careful to minimize these low-frequency noise sources as well as maximizing the Q of the oscillator tank. Note that the varactor can be forward-biased at the top and bottom of its signal swing, but this excess phase noise has nothing to do with forward-biasing the varactor (very little phase noise is generated there regardless) or reducing the varactor's high-frequency Q due to forward-biasing. This same variation can be demonstrated in simulation using an ideal voltage-controlled capacitor if the C–V characteristics match those of a real varactor. Note also that the excess gain term sometimes included in Leeson's formula has been ignored. This term, used to model nonlinearities that can contribute to the noise through nonlinear mixing of higher frequency noise around the carrier, is not significant if the oscillator is designed to have very little excess gain, as is the case here.

V. CALCULATION OF PHASE NOISE

From the preceding discussion it is easy to see how one might go about predicting the phase noise of the VCO. The Q of the oscillator can be calculated using (9) assuming R_{Tank} can be approximated by (10). For this VCO the bias current through $Q1$ and $Q2$ is 3.2 mA, resulting in an r_e of 7.8 Ω and R_E has

TABLE I
CALCULATION OF R_{Tank}

Reverse Bias of the Varactor	C_{var}	$\left(1 + \dfrac{C_2 + C_{\text{var}}}{C_1}\right)^2$	$R_{\text{Tank,re+RE}}$	R_{Tank}	Tank Q
0 Volts	2.9pF	6.91	192.1Ω	99.6Ω	2.68
1 Volt	2.1pF	5.76	160.1Ω	90.2Ω	2.35
2 Volts	1.9pF	5.49	152.6Ω	88.0Ω	2.27
3 Volts	1.8pF	5.35	148.7Ω	86.6Ω	2.22

been selected to be 20 Ω, C_1 is 3.5 pF and C_2 is 2.8 pF. Based on these values, the following table summarizes the effectively transformed emitter resistance seen at the tank ($R_{\text{Tank, re+RE}}$) where the parasitics in $Q1$ and $Q2$ were neglected for simplicity. Also assuming a Q of 5.5 (see Fig. 5) for the 3-nH inductor ($R_{\text{Tank, L}} = 207\,\Omega$ at 1.9 GHz), the total equivalent resistance is obtained using (10). From Table I, it is also easy to see that without R_E, the Q of the oscillator would be drastically reduced.

Note that accounting for the finite varactor Q will not change the results by any measurable amount. Even in the worst-case scenario, the varactor has a parallel resistance of 629 Ω which gets transformed (through the capacitor ratio) to 4.35 kΩ in parallel with the tank. Thus, varactor loss was ignored in this analysis.

Simulations using SPICE will give accurate predictions of the noise figure of the amplifier (we are making the assumption here that using the small-signal noise figure is an accurate method of calculating the transistors noise contribution to the tank) and output power of the circuit. Alternatively, the output power can be estimated from the formula in Section III. V_m can be calculated by summing the noise sources in transistors and resistors present at the terminals of the varactor. Note that this is the noise at f_m modulated to $f_o - f_m$ or $f_o + f_m$. Therefore in calculating the noise one can simplify the circuit knowing that the capacitors are all open circuits and the inductors are short circuits at these frequencies. In minimizing the noise present in the VCO, two things are of paramount importance. First, the current through the transistors must not be made larger than necessary for a given power level, and second, R_{tune} must be kept relatively small. Note that this resistor's purpose is to avoid decoupling both sides of the oscillator and prevent large currents from flowing in case the varactor is forward-biased, however, it does represent a noise source directly applied to the tuning port. If low-frequency noise starts to dominate in the design, it is recommended to reduce the ratio of $C_2 + C_{\text{var}}$ to C_1 at the expense of reducing the Q of the tank in order to reduce the current and therefore reduce the noise at the varactor terminals. K_{VCO} can be predicted provided a model is available for the varactor. Such calculations resulted in the predicted values presented in the next section.

VI. RESULTS

Both the varactor and inductor were tested on-wafer as one-port devices. The collector side of the varactor (see Fig. 2) was grounded so that the parasitic diode was removed form

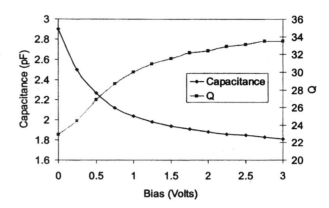

Fig. 4. Measured C–V curve and Q for the varactor used in the VCO.

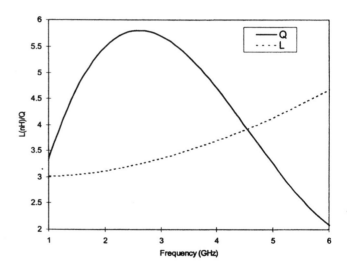

Fig. 5. Measured performance of a 3-nH spiral structure.

the circuit and the effect of only the desired collector–base junction was measured. Fig. 4 shows the C–V curve for the structure. The varactor's capacitance is 2.9 pF for 0-V reverse bias and 1.8 pF for 3-V reverse bias. This gives a capacitor-tuning ratio of 1.6. The Q of the varactor (also shown in Fig. 4) was measured and was greater than 20 over the whole tuning range with this terminal grounded. Note that grounding the base and measuring the Q from the collector side of the junction will show a reduced Q for the two varactors in parallel, usually below five. This further illustrates the importance of the varactor placement in this circuit. The inductor was also measured on-wafer as a one-port test structure. Fig. 5 shows the measured inductor Q and inductance (which was about 5.5 at the frequencies of interest).

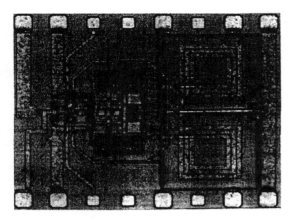

Fig. 6. Microphotograph of the VCO.

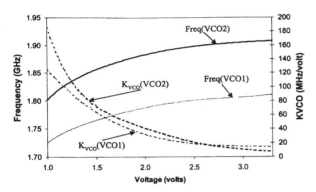

Fig. 7. Frequency versus tuning voltage characteristic for both VCO designs.

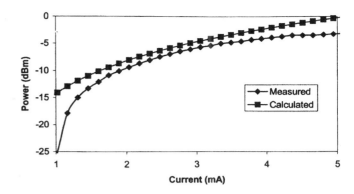

Fig. 8. Output power versus bias current for VCO2. (Both single-ended.)

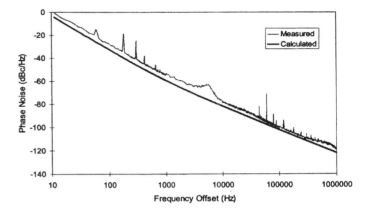

Fig. 9. Phase noise versus frequency offset at 1.9 GHz for VCO2.

A microphotograph of the VCO itself is shown in Fig. 6. The layout was made as symmetric as possible to ensure a true differential circuit. The VCO occupied a chip area of 1.1 mm × 0.8 mm. The inductors were placed apart from the rest of the circuit to ensure good performance, and a guard ring of substrate contacts was placed around them to prevent the VCO signal from traveling through the substrate. Two versions of the VCO were fabricated. The first version of the VCO had $R_{\text{tune}} = 4\,\text{k}\Omega$ and drew 5 mA through each of $Q1$ and $Q2$. The second version of the VCO was reworked to re-optimize device sizes to reduce the low-frequency noise and make the phase noise more uniform over the tuning range. This second design drew 3.2 mA through each of $Q1$ and $Q2$ and had $R_{\text{tune}} = 50\,\Omega$. The capacitor ratio was reduced somewhat in this design to compensate for the reduced gain in the transistors so that the oscillator output power would not change. A third oscillator was fabricated, identical to the first design, but substituted fixed capacitors for the varactors for comparison.

The VCO was measured both by wafer probing and after packaging. It had a tuning range of 90 MHz for the first design and 100 MHz for the second design. All designs had a single-ended output power of −5 dBm at their nominal bias points. The oscillators were operated from a 3.3-V supply. Phase noise measured for both packaged and unpackaged parts differed by less than 3 dB. Plots of frequency versus tuning voltage for both versions of the VCO are shown in Fig. 7. Note that K_{VCO} decreases from over 100 MHz/V down to below 20 MHz/V over the tuning range for both designs. The

output power of the second design was plotted relative to bias current and compared to estimates obtained with the formula presented earlier. Note that there is good agreement provided that the current is not too low or too high. These results are shown in Fig. 8. The phase noise versus frequency offset for the second version of the VCO at 1.9 GHz is shown in Fig. 9. This measurement was done at the top of the tuning range. The phase noise measurements agreed well with values predicted by (13), also shown in Fig. 9. The phase noise was measured under these conditions to be −103 dBc/Hz at 100 kHz offset. The oscillator with no tuning mechanism (without a varactor) was found to oscillate at 1.72 GHz. It was made identical to the first design with the fixed capacitors set to have the same capacitance as the varactors at the bottom of the tuning range. It had a phase noise of −105 dBc/Hz, further demonstrating the importance of the low frequency noise to this circuit.

A plot of phase noise versus tuning voltage for both versions of the oscillator is shown in Fig. 10. This clearly shows that the performance is not uniform over the tuning range. In fact it varies by as much as 10 dB in the first design. The second version of the VCO reduces the range to only 7 dB of variation and gives better overall performance. Calculations using (13) also predicted almost the same variation for both designs. Note that a VCO with constant gain and the same tuning range would have an almost constant phase noise that would rest between these two extremes. This very clearly demonstrates the effect of the varactor on the phase noise. There were of course some slight discrepancies between measured and calculated phase noise. It

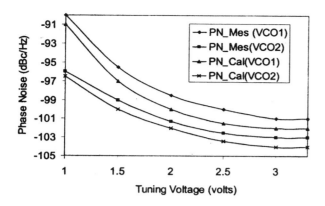

Fig. 10. Measured versus calculated phase noise at 100-kHz offset versus tuning voltage.

should be remembered that oscillators are themselves highly nonlinear and Leeson's formula itself contains approximations including the assumption that the transistor's small-signal noise figure is appropriate. As well, it is difficult to account for all noise sources and transistor parasitic resistance with precise accuracy. Thus, a small error in calculated versus measured results is expected. Nevertheless even using these simple formulas, reasonably accurate results of VCO phase-noise performance can be achieved.

VII. CONCLUSION

VCO phase noise variation over the entire tuning range has been studied. Theoretical calculations and measurements agree very well. A methodology for the design of a completely integrated differential VCO has been presented. A 1.9-GHz VCO was fabricated in a 0.5-μm bipolar technology with a phase noise of -103 dBc/Hz at 100-kHz offset. The analysis in this work shows that the use of linear varactors will lead to more uniform performance levels. Careful consideration of low-frequency noise is necessary in order to optimize the phase-noise performance of the VCO and to ensure that the phase noise is made as uniform as possible over the tuning range.

ACKNOWLEDGMENT

The authors would like to thank Dr. J. Ojha of Nortel Networks' Technology Access and Applications Department for design support and access to technology. The helpful comments of the reviewers were also greatly appreciated.

REFERENCES

[1] L. Dauphinee, M. Copeland, and P. Schvan, "A balanced 1.5-GHz voltage controlled oscillator with an integrated LC resonator," in *IEEE Int. Solid-State Circuits Conf.*, 1997, pp. 390–391.
[2] C. P. Yue and S. S. Wong, "On-chip spiral inductors with patterned ground shields for Si-based RF IC's," *IEEE J. Solid-State Circuits*, vol. 33, pp. 743–752, May 1998.
[3] D. Hisamoto, S. Tanaka, T. Tanimoto, and S. Kimura, "Suspended SOI structure for advanced 0.1-μm CMOS RF devises," *IEEE Trans. Electron Devices*, vol. 45, pp. 1039–1046, May 1998.
[4] D. C. Edelstein and J. N. Burghartz, "Spiral and solenoidal inductor structures on silicon using Cu-damascene interconnects," in *IITC*, 1998, pp. 18–20.
[5] J. Rogers, L. Tan, T. Smy, N. Tait, and N. G. Tarr, "A high Q on-chip Cu inductor post process for Si integrated circuits," in *IITC*, 1999, pp. 239–241.
[6] R. Groves, J. Malinowski, R. Volant, and D. Jadus, "High Q inductors in a SiGe BiCMOS process utilizing a thick metal process add-on module," in *BCTM*, 1999, pp. 149–152.
[7] J. R. Long and M. A. Copeland, "The modeling, characterization, and design of monolithic inductors for silicon RF IC's," *IEEE J. Solid-State Circuits*, vol. 32, pp. 357–369, Mar. 1997.
[8] A. M. Niknejad and R. G. Meyer, "Analysis, design, and optimization of spiral inductors and transformers for Si RF ICs," *IEEE J. Solid-State Circuits*, vol. 33, pp. 1470–1481, Oct. 1998.
[9] J. Craninckx and M. S. J. Steyaert, "A 1.8-GHz low-phase-noise CMOS VCO using optimized hollow spiral inductors," *IEEE J. Solid-State Circuits*, vol. 32, pp. 736–744, May 1997.
[10] W. B. Kuhn and N. K. Yanduru, "Spiral inductor substrate loss modeling in silicon RFICs," *Microwave J.*, pp. 66–81, Mar. 1999.
[11] S. S. Mohan, M. Hershenson, S. P. Boyd, and T.H. Lee, "Simple accurate expressions for planar spiral inductances," *IEEE J. Solid-State Circuits*, vol. 34, pp. 1419–1424, Oct. 1999.
[12] M. Danesh, J. R. Long, R. A. Hadaway, and D. L. Harame, "A Q-factor enhancement technique for MMIC inductors," in *IEEE Radio Frequency Integrated Circuits Symp.*, June 1998, pp. 217–200.
[13] W. B. Kuhn, A. Elshabini-Riad, and F. W. Stephenson, "Center-tapped spiral inductors for monolithic bandpass filters," *Electron. Lett.*, vol. 31, pp. 625–626, Apr. 1995.
[14] B. Razavi, *RF Microelectronics*. Englewood Cliffs, NJ: Prentice Hall, 1998.
[15] S. P. Voinigescu, M. C. Maliepaard, J. L. Showell, G. E. Babcock, D. Marchesan, M. Schroter, P. Schvan, and D. L. Harame, "A scalable high frequency noise model for bipolar transistors with application to optimal transistor sizing for low-noise amplifier design," *IEEE J. Solid-State Circuits*, vol. 32, pp. 1430–1439, Sept. 1997.
[16] P. R. Gray and R. G. Meyer, *Analysis and Design of Analog Integrated Circuits*. New York, NY: Wiley, 1993.
[17] T. H. Lee, *The Design of CMOS Radio Frequency Integrated Circuits*. Cambridge, U.K.: Cambridge University Press, 1998.
[18] J. Craninckx, M. Steyaert, and H. Miyakawa, "A fully integrated spiral-LC CMOS VCO set with prescaler for GSM and DCS-1800 systems," in *IEEE Custom Integrated Circuits Conf.*, 1997, pp. 403–406.
[19] B. Jansen, K. Negus, and D. Lee, "Silicon Bipolar VCO family for 1.1 to 2.2 GHz with fully-integrated tank and tuning circuits," in *IEEE Int. Solid-State Circuits Conf. '97*, pp. 392–393.
[20] M. Zannoth, B. Kolb, J. Fenk, and R. Weigel, "A fully integrated VCO at 2 GHz," *IEEE J. Solid-State Circuits*, vol. 33, pp. 1987–1991, Dec. 1998.
[21] W. Chen and J. Wu, "A 2-V 2-GHz BJT variable frequency oscillator," *IEEE J. Solid-State Circuits*, vol. 33, pp. 1406–1410, Sept. 1998.
[22] J. Craninckx and M. S. J. Steyaert, "A fully integrated CMOS DCS-1800 frequency synthesizer," *IEEE J. Solid-State Circuits*, vol. 33, pp. 2054–2065, Dec. 1998.
[23] D. B. Leeson, "A simple model of feedback oscillator noise spectrum," in *Proc. IEEE*, Feb. 1966, pp. 329–330.

John W. M. Rogers (S'95) was born in Cobourg, ON, Canada. He received the B. Eng. degree in 1997 and the M. Eng. degree in 1999, both in electrical engineering, from Carleton University, Ottawa, ON, Canada. From 1997 to 1999, as part of his Master's degree research he was a Resident Researcher at Nortel Networks' Advanced Technology Access and Applications Group, Ottawa, where he did exploratory work on VCOs for personal communications. During that same period he was involved in the development of a Cu interconnect technology for building high-quality passives for RF applications. He is currently a Resident Researcher with SiGe Microsystems, Ottawa, Canada, while working towards his Ph.D. degree with Carleton University. His research interests are in the areas of RF IC design for wireless applications.

Jitter in Ring Oscillators

John A. McNeill

Abstract—Jitter in ring oscillators is theoretically described, and predictions are experimentally verified. A design procedure is developed in the context of time domain measures of oscillator jitter in a phase-locked loop (PLL). A major contribution is the identification of a design figure of merit κ, which is independent of the number of stages in the ring. This figure of merit is used to relate fundamental circuit-level noise sources (such as thermal and shot noise) to system-level jitter performance. The procedure is applied to a ring oscillator composed of bipolar differential pair delay stages. The theoretical predictions are tested on 155 and 622 MHz clock-recovery PLL's which have been fabricated in a dielectrically isolated, complementary bipolar process. The measured closed-loop jitter is within 10% of the design procedure prediction.

Index Terms—Design methodology, jitter, noise measurement, oscillator noise, oscillator stability, phase jitter, phase-locked loops, phase noise, voltage controlled oscillators.

I. INTRODUCTION

DUE to their speed and ease of integration, ring oscillators are increasingly being used as voltage controlled oscillators (VCO's) in jitter sensitive applications. One example is in clock recovery phase-locked loops (PLL's) for serial data communication [1]–[3]. Other applications that would benefit from the cost and size advantages of a fully integrated low jitter VCO include disk drive clock recovery [4], [5], clock frequency multiplication [6], [7], and oversampling analog-to-digital converters (ADC's) [8], [9].

This paper presents a framework for a theoretical understanding of fundamental limits on jitter performance in ring oscillator VCO's and a design methodology for connecting system-level, closed-loop PLL jitter performance to circuit-level VCO design. Section II begins development of this approach by comparing the ring oscillator to harmonic and relaxation oscillators in the context of noise analysis. Sections III and IV continue development in terms of time domain measures of jitter performance. Section V presents the key equations of the design methodology as applied to a bipolar differential pair delay stage. Section VI gives experimental results.

II. COMPARISON OF OSCILLATOR TYPES

A. Harmonic Oscillator

A harmonic oscillator is characterized by an equivalence to two energy storage elements, operating in resonance, to give a

Manuscript received July 9, 1996; revised December 10, 1996. This work was supported by Analog Devices Semiconductor, Inc.

The author is with Worcester Polytechnic Institute, Worcester, MA 01609 USA.

Publisher Item Identifier S 0018-9200(97)03830-4.

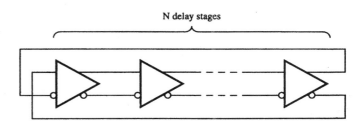

Fig. 1. Typical ring oscillator schematic.

periodic output signal. The actual resonant element might be an *LC* tank or a quartz crystal. Resonant circuit-based VCO's are known to have excellent jitter performance [10], [11]. Unfortunately, the requirement of an off-chip tank or crystal defeats the purpose of integrating the PLL function. Although integrated inductors have been reported in the GHz frequency range [12], these generally have low Q due to resistive losses, and in any case are not practical in the 100 MHz–1 GHz frequency range. Analysis of noise in resonant-based VCO's is well developed in the literature [13]–[15], and design techniques for realizing low jitter performance are relatively well understood. In general, the noise analysis has been approached in the frequency domain, with the high Q of the circuit resonance filtering noise into a narrow band near the fundamental frequency.

B. Multivibrator

A relaxation (multivibrator) oscillator is characterized by an equivalence to one energy storage element, with additional circuitry that senses the element state and controls its excitation to give a periodic output signal [16]. Fully integrated clock recovery PLL's have been described using multivibrator VCO's [1], [17]–[19]. In general, jitter analysis for this type of oscillator has been approached in the time domain [16], [20], [21]. The jitter performance of the multivibrator is known to be worse than the harmonic oscillator, although some design techniques for improving jitter are available [16], [22], [23]. The best jitter performance that has been achieved by a multivibrator is larger than typically desired for fully integrated VCO's.

C. Ring Oscillator

Fig. 1 shows the general ring oscillator investigated in this work: a loop of N delay stages with a wire inversion. The ring will oscillate with a period of $2N$ times the stage delay. Voltage controlled ring oscillators have recently been explored as an alternative to the multivibrator for fully integrated, lower jitter clock recovery PLL's [2], [3], [7], [24]–[31]. Like the multivibrator, a ring oscillator is fully integrable. In addition, some of the empirical results show promise of excellent jitter performance [3]. However, investigation into a theoretical analysis of jitter has only recently begun for bipolar [30], [31] and CMOS [32], [33] ring oscillators. Perhaps one reason that analysis of jitter in ring oscillators has lagged is that the ring does not fit well into either of the harmonic or multivibrator oscillator models. The number of energy storage elements is not as explicit; in fact there are many "energy storage elements" since the ring is composed of multiple stages.

III. METHODS OF QUANTIFYING JITTER

The design technique developed in this paper follows from different methods of measuring jitter in the time domain. Following is a brief description of three relevant time domain measures of jitter. Note that in the closed-loop cases, it is assumed that the VCO is the dominant jitter source.

A. Closed-Loop, Transmit Clock Referenced

For a clock recovery PLL, jitter is usually specified as the standard deviation σ_x of the phase difference between the transmitted clock and the recovered clock. This measurement can be made as shown in Fig. 2(a), using an instrument such as a communications signal analyzer (CSA) [34]. The transmit clock $TCLK$ is used to trigger the CSA; the recovered clock $RCLK$ is observed as the CSA input. In the presence of jitter, a distribution of threshold crossing times is observed as shown in Fig. 2(b). The CSA records a histogram of this distribution; the standard deviation of the distribution is σ_x.

Although this test is a simple indicator of PLL performance, the test provides little information on improving jitter from circuit-level noise sources if σ_x is not satisfactory. This test also requires the PLL to be operating closed-loop. VCO design and simulation would be simplified if we could consider the VCO by itself (open loop), while being able to predict the closed-loop σ_x.

B. Open Loop, Self Referenced

We can also measure the jitter of the VCO on a stand-alone basis as shown in Fig. 2(c). With the VCO free-running at its center frequency, $RCLK$ is used as both the trigger and the input to the CSA. The CSA compares the phase difference between transitions in the clock waveform, separated by an interval ΔT derived from the CSA's internal time base. As in the previous case, the CSA measures the standard deviation of the threshold crossing times $\sigma_{\Delta T(\text{OL})}$.

In this measurement, however, the standard deviation is observed to depend on the measurement interval ΔT. Fig. 2(d) shows a typical plot of $\sigma_{\Delta T(\text{OL})}$ versus ΔT on log–log axes. It can be shown that, for a large class of noise processes [30]–[32], the jitter increases as the square root of the measurement interval

$$\sigma_{\Delta T(\text{OL})}(\Delta T) \approx \kappa \sqrt{\Delta T}. \tag{1}$$

The proportionality constant κ is an important time domain figure of merit which will be used in design to connect open-loop and closed-loop performance, as well as circuit-level and system-level design.

Intuitively, (1) can be understood by considering the jitter over the measurement interval ΔT to be the sum of jitter contributions from many individual stage delays. If these jitter errors are independent, then the standard deviation of the sum increases as the square root of the number of delays being summed.

C. Closed Loop, Self Referenced

The open-loop and closed-loop jitter performance can be related by measuring the "stand-alone" jitter performance of the clock recovered under closed-loop conditions, as shown in Fig. 2(e). When lead-lag compensation is used, the closed-loop transfer function $H(s)$ is that of a second-order system [35]. In clock recovery PLL's, however, it is common to overdamp the loop to avoid peaking in the jitter transfer function [30], and the loop transfer function can be approximated as

$$H(s) = \frac{2\pi f_L}{s + 2\pi f_L} \tag{2}$$

which is a first-order system, where f_L is the loop bandwidth. In this case, the plot of $\sigma_{\Delta T(\text{CL})}$ versus ΔT is of the form shown in Fig. 2(f) [30], [32]. The plot shows two asymptotes which can be understood qualitatively as follows: At short delays, the jitter increases in proportion to the square root of delay, just as in the open-loop case. This is because at time scales shorter than the loop bandwidth time constant, the VCO control voltage cannot change appreciably, and the VCO is essentially running open loop. At longer delays, the phase detector and loop filter are able to sense accumulated phase error due to VCO jitter and adjust the VCO input to bring the VCO phase "back in line" with the transmit clock. At very long delays, the jitter over the measurement interval ΔT is due to the σ_x jitter at the beginning and end of the ΔT time period. Since the jitter errors of clock edges separated by a long delay are uncorrelated, the total jitter is $\sqrt{2}\sigma_x$.

Analysis [30] shows that the two asymptotes intersect at the loop bandwidth time constant $\tau_L = 1/2\pi f_L$. We can use this to solve for the closed-loop σ_x in terms of the open-loop figure of merit κ

$$\sigma_x = \kappa \sqrt{\frac{1}{4\pi f_L}}. \tag{3}$$

If f_L is a free parameter, the closed-loop jitter can be reduced simply by increasing f_L. However, if f_L is fixed (for example, by specification as in SONET [36]), then the only way to reduce closed-loop jitter is to improve the oscillator by reducing the open-loop time domain figure of merit κ.

IV. JITTER INDEPENDENCE OF RING LENGTH

The parameter κ is a link between the open-loop VCO and the closed-loop system-level jitter performance. We will see in Section V that, at the circuit level, it is possible to determine κ for a single delay stage. To complete the design path from circuit level to system level, it is necessary to determine how the circuit-level κ is affected when a number of stages are combined to form a ring oscillator.

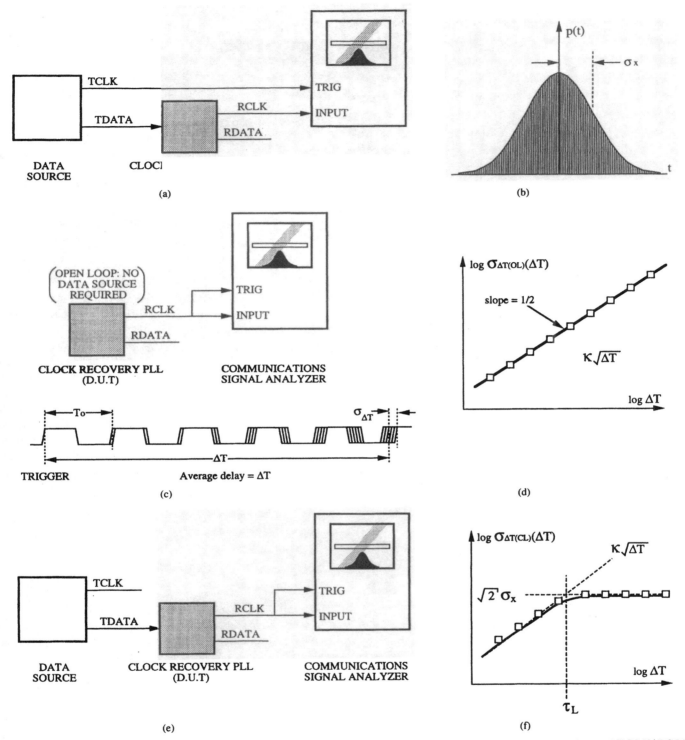

Fig. 2. (a) Measurement technique: time domain, closed loop, and transmit clock referenced. (b) Measurement result: standard deviation of $TCLK/RCLK$ phase. (c) Measurement technique: time domain and open loop. (d) Measurement result: standard deviation versus delay time ΔT. (e) Measurement technique: time domain and closed loop. (f) Measurement result: standard deviation versus delay time ΔT.

An experiment was performed in which ring oscillators of lengths three, four, five, seven, and nine stages were fabricated in a 3-GHz f_T junction-isolated Si bipolar process [31]. The delay element was the gate shown in Fig. 3. The jitter performance for each ring was measured using the open-loop technique described in Section III-B; the results are plotted in Fig. 4. The experimental results in Table I show the least-squares-fit values for κ, as well as the free-running VCO center frequencies.

Fig. 4 shows that the jitter increases roughly as the square root of delay time, consistent with the model of (1). More importantly, the jitter over a given measurement interval is the same regardless of how many stages there are in the ring. Table I shows that the κ values are approximately the same

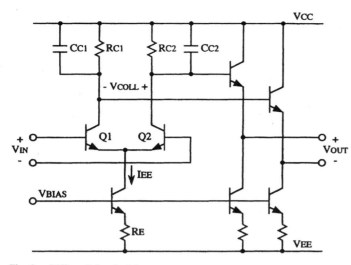

Fig. 3. Differential pair delay gate.

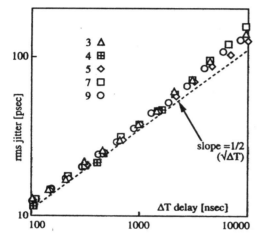

Fig. 4. Jitter versus delay for three-, four-, five-, seven-, and nine-stage rings.

TABLE I
RING EXPERIMENT RESULTS

RING STAGES	κ [E-08√sec]	fo [MHz]
3	4.17	170.1
4	3.56	164.1
5	3.78	102.7
7	3.77	71.9
9	3.94	56.8

regardless of the length of the ring: even as center frequency varies by a 3 : 1 ratio over a range of 56–170 MHz, the value of κ changes by only ±8%. We conclude that the ability of a ring to accurately measure an interval of time depends primarily on the accuracy of its basic delay element as characterized by κ and is essentially independent of the number of stages in the ring. When we have characterized the accuracy of the delay stage in terms of κ, we can predict the jitter for a ring of any length using that stage.

This may seem counterintuitive at first: when more delay stages (and, seemingly, more noise sources) are added, why is the jitter unchanged? The reason can be seen by considering the jitter accumulation process from the "point of view" of the signal transition or "edge" that propagates around the ring. The only delay stage that affects jitter accumulation at a given instant is the stage that is processing the transition. All other gates in the ring are inactive and do not contribute to jitter. Thus, from the standpoint of jitter accumulation, the key measure is the number of gate transitions, not the number of oscillator periods. This is why measures that normalize to the oscillator period are not independent of the number of stages [32].

V. DETERMINING κ EXPRESSIONS FOR JITTER SOURCES

To determine κ at the gate level requires a detailed analysis of each circuit-level noise source and depends on the particular gate used as the delay element in the ring. For design illustration, the simple delay stage shown in Fig. 3 will be analyzed. The input voltage v_{IN} causes differential pair Q_1/Q_2 to steer the tail current I_{EE} to one of the collector loads, R_{C1} or R_{C2}. Capacitors C_{C1} and C_{C2} represent wiring stray, junction, and any explicit capacitances that may be present at the collector node. We begin the analysis by noting that the delay through the gate has two components: the delay through the differential pair (from v_{IN} to v_{COLL}), and the delay through the emitter follower buffers (from v_{COLL} to v_{OUT}). To simplify the analysis and make the results easier to interpret, we will now make some assumptions regarding the gate delay.

i) The gate delay is dominated by the differential pair delay.

ii) The differential pair delay is dominated by the RC time constant of the load.

iii) The differential signal amplitude is much greater than $V_T = kT/q_e$.

iv) The amplitude of the noise is much smaller than the differential signal.

v) All noise sources are white and uncorrelated.

As has been shown in the literature [37], the switching time of a differential pair depends on many factors, so some tradeoff of accuracy is necessary to obtain a simple analytical expression. Generally, as long as assumption iii) holds, the differential pair switches the tail current much faster than the RC time constant of the collector load, and assumption ii) introduces an error less than 20% of delay time. The error due to assumption i) is usually less than 10%. Although the error in delay time due to these assumptions is not insignificant, the assumptions are nevertheless justified since the resulting theory predicts jitter quite well and provides insights for guiding design. The following subsections derive the effective κ for the most significant noise sources in the differential pair delay.

A. Thermal Noise of Collector Load Resistors

For noise analysis, the circuit can be modeled as shown in Fig. 5(a). Thermal noise in the collector load is represented by

224

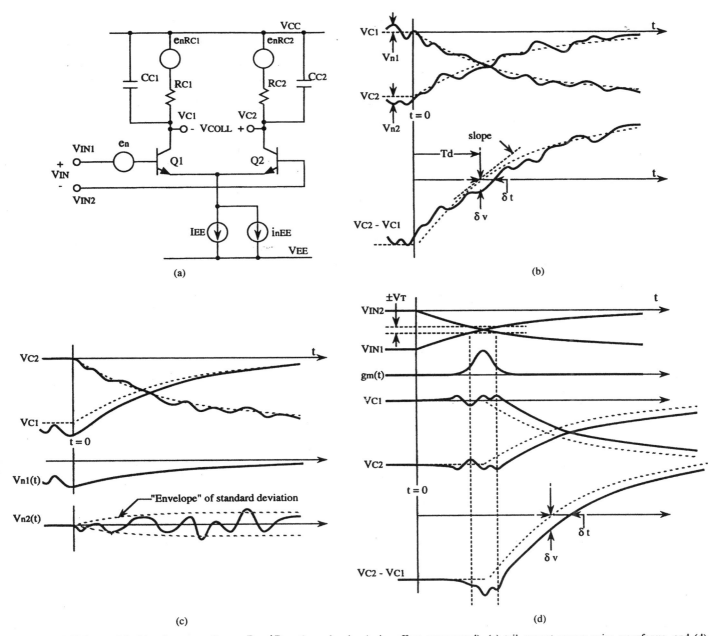

Fig. 5. (a) Noise model, (b) voltage waveforms, R_{C1}/R_{C2} thermal noise (noise effect exaggerated), (c) tail current source noise waveforms, and (d) differential input switching noise waveforms.

$e_{nR_{C1}}$ and $e_{nR_{C2}}$. These sources appear directly at v_{C1} and v_{C2}, but are bandlimited by the $R_{C1}C_{C1}$ and $R_{C2}C_{C2}$ poles. If the differential pair is represented by an ideal switch that is switched at time $t = 0$, the noise-free differential signal is given by

$$v_{COLL}(t) = v_{C2}(t) - v_{C1}(t)$$
$$= I_{EE}R_C \left[1 - 2 \exp \left(\frac{-t}{R_C C_C} \right) \right] \quad (4)$$

with $R_{C1} = R_{C2} = R_C$ and $C_{C1} = C_{C2} = C_C$. The noise-free exponential waveforms are shown as dashed lines in Fig. 5(b). The gate delay T_d is defined as the time when the differential voltage v_{COLL} crosses zero. For the noise-free

waveform, solving (4) for T_d gives

$$T_d = \ln(2) R_C C_C. \quad (5)$$

The slope of the differential signal is given by taking the derivative of (4) with respect to time and evaluating at T_d, giving

$$\left. \frac{dv_{OUT}}{dt} \right|_{T_d} = \frac{I_{EE}}{C_C}. \quad (6)$$

The solid lines in the figure represent the actual collector waveforms, including the exaggerated effect of typical thermal noise waveforms $v_{n1}(t)$ and $v_{n2}(t)$. By superposition, the noise waveforms simply "ride" on the ideal exponential. The result is that, at the time of the ideal differential waveform

225

zero crossing, there is a voltage error δv. This causes a time error in the threshold crossing δt. Since [by assumption iv)] the noise is much less than the exponential signal, then (6) can be used to relate the errors δv and δt, as well as the standard deviations of errors σ_v and σ_t

$$\frac{\sigma_v}{\sigma_t} = \frac{I_{EE}}{C_C}. \tag{7}$$

The standard deviation of the differential voltage error is simply the square root of the sum of the squared (RSS) individual standard deviations σ_{v1} and σ_{v2}. Applying the Johnson noise equation gives the well-known result $\sigma_{v1} = \sigma_{v2} = \sqrt{kT/C_C}$, so the standard deviation of the differential voltage is

$$\sigma_v = \sqrt{\frac{2kT}{C_C}}. \tag{8}$$

Using (7) and (8) gives for the standard deviation of the time error (the jitter)

$$\sigma_t = \sqrt{\frac{2kTC_C}{I_{EE}^2}}. \tag{9}$$

κ for the individual gate is determined by dividing the standard deviation of delay σ_t by the square root of the average delay in (5)

$$\kappa_{R_C} = \frac{\sigma_t}{\sqrt{T_d}}$$
$$= \sqrt{\frac{2}{\ln(2)}} \sqrt{\frac{kT}{I_{EE}^2 R_C}}. \tag{10}$$

κ has dimensions of $\sqrt{\text{sec}}$, and from (10) we see that this comes about by taking the square root of an energy (kT) divided by a power ($I_{EE}^2 R_C$). The rms thermal energy kT represents an uncertainty in the energy of the collector load. $I_{EE}^2 R_C$ represents the dc power dissipation (energy flow) in the collector load. The intuitive meaning of (10) is that κ characterizes the gate's ability to resolve time (jitter) by an energy uncertainty (kT) as a fraction of the energy flow over time ($I_{EE}^2 R_C$).

Since lower κ corresponds to improved jitter, (10) indicates that jitter improves when bias current I_{EE} is increased. This is similar to results that have been reported for differential delay stages in CMOS ring oscillators [32], [33]. Equation (10) also indicates that jitter is improved when the dc power dissipation $I_{EE}^2 R_C$ is increased. This is similar to the results of the noise analyses for harmonic and relaxation oscillators [14], [16].

B. Tail Current Noise

Noise is also present in the tail current of the differential pair. This is represented by noise source i_{nEE}. In this case, the source is switched so the analysis is somewhat more complicated. When v_{IN} is large enough to fully switch the differential pair, the current noise is passed to the output, but is bandlimited by the either the $R_{C1}C_{C1}$ or $R_{C2}C_{C2}$ pole. When v_{IN} is small, the differential pair is approximately balanced, the tail current noise is a common-mode error, and its effect on the differential v_{COLL} is reduced.

The analysis can be simplified by using assumption iii) to idealize the differential pair as switching instantaneously. Fig. 5(c) shows the resulting v_{C1} and v_{C2} waveforms, as well as the superimposed noise waveforms v_{n1} and v_{n2}. Prior to switching, the noise current through R_{C1} causes a noise voltage v_{n1} with standard deviation given by

$$\sigma_{V_{n1}}(t \leq 0) = i_{nEE} R_C \sqrt{\frac{1}{4 R_C C_C}}$$
$$= \frac{i_{nEE}}{2} \sqrt{\frac{R_C}{C_C}}. \tag{11}$$

When switching occurs at $t = 0$, this voltage is sampled on C_{C1}. For $t > 0$, v_{n1} decays exponentially with a time constant of $R_C C_C$, so that

$$\sigma_{V_{n1}}(t > 0) = \frac{i_{nEE}}{2} \sqrt{\frac{R_C}{C_C}} e^{-t/R_C C_C}. \tag{12}$$

For v_{n2}, analysis shows [30] that the standard deviation "builds up" as

$$\sigma_{V_{n2}}(t > 0) = \frac{i_{nEE}}{2} \sqrt{\frac{R_C}{C_C}} \sqrt{1 - e^{-2t/R_C C_C}}. \tag{13}$$

Taking the root sum of (12) and (13) and evaluating at T_d gives the standard deviation of the differential voltage at the zero crossing time as

$$\sigma_v = \frac{i_{nEE}}{2} \sqrt{\frac{R_C}{C_C}}. \tag{14}$$

Using (7) and (14) gives the standard deviation of the time uncertainty as

$$\sigma_t = \frac{1}{2} \sqrt{R_C C_C} \frac{i_{nEE}}{I_{EE}}. \tag{15}$$

Dividing σ_t by the square root of the delay in (5) gives κ

$$\kappa = \frac{1}{2\sqrt{\ln 2}} \frac{i_{nEE}}{I_{EE}}. \tag{16}$$

It is interesting to consider (16) when expressions for i_{nEE} are substituted for shot and thermal noise.

Shot Noise: Substituting the shot noise density $i_{nEE} = \sqrt{2 q_e I_{EE}}$ into (16) gives

$$\kappa = \frac{1}{\sqrt{2 \ln 2}} \sqrt{\frac{q_e}{I_{EE}}}. \tag{17}$$

In this case, the gate's ability to resolve time is characterized by the smallest resolvable unit of charge (q_e) as a fraction of the charge flow over time (I_{EE}).

Thermal Noise: If the tail current source is degenerated, the output noise will be dominated by the thermal noise of the degeneration resistor [38]. Using the thermal noise density $i_{nEE} = \sqrt{4kT/R_E}$ in (16) gives

$$\kappa_{R_E} = \frac{1}{\sqrt{\ln 2}} \sqrt{\frac{kT}{I_{EE}^2 R_E}}. \tag{18}$$

This is similar to (10) in that the gate's ability to resolve time is characterized by the energy uncertainty (kT) as a fraction of the energy flow over time ($I_{EE}^2 R_E$) in the element that determines the current.

Again, in both cases an increase in bias current I_{EE} results in lower κ, which corresponds to improved jitter.

C. Sampling of Input Noise by Switching of Differential Pair

There are also noise sources in series with the inputs of the differential pair, represented by an equivalent e_n in Fig. 5(a). This is due to thermal noise of the Q_1/Q_2 transistor base resistances [39] as well as other wideband noise sources (such as emitter followers in the signal path) going back to v_{COLL} of the preceding stage of the ring. Calculating the jitter effects of these sources is complicated by the fact that the gain from input to output depends on the signal amplitude. Fig. 5(d) shows the input waveforms, the time-dependent transconductance, and the collector voltages v_{C1} and v_{C2}.

The input–output characteristic of a bipolar differential pair is

$$i_{OUT} = I_{EE} \tanh\left(\frac{v_{IN}}{2V_T}\right) \qquad (19)$$

where tanh is the hyperbolic tangent function [39]. The incremental gain is

$$
\begin{aligned}
g_m &= \frac{di_{OUT}}{dv_{IN}} \\
&= \frac{I_{EE}}{2V_T} \operatorname{sech}^2\left(\frac{v_{IN}}{2V_T}\right)
\end{aligned}
\qquad (20)
$$

where sech is the hyperbolic secant.

For input signals that are large compared to V_T, the gain to the output current is small. Thus, the input voltage noise has little effect when the input signal is large. As the input signals cross over during switching, however, the gain rises. During this time, the input voltage noise causes a noise current which is integrated on the collector capacitors. Although the integration is "leaky" due to the discharge path through R_{C1} and R_{C2}, some of the integrated noise still remains when the collector voltages cross approximately T_d later. Assuming all noise sources to be white, and lumped into a single source with density e_n, analysis [30] shows the standard deviation of the differential voltage at the zero crossing is

$$\sigma_v(T_d) = \frac{e_n}{2} \sqrt{\frac{I_{EE}}{3C_C V_T}}. \qquad (21)$$

The time uncertainty obtained by dividing by the slope is

$$\sigma_t = \frac{e_n}{2} \sqrt{\frac{C_C}{3 I_{EE} V_T}}. \qquad (22)$$

Dividing by the square root of T_d gives κ

$$\kappa = \frac{1}{2\sqrt{3}\ln 2} e_n \sqrt{\frac{1}{I_{EE} R_C V_T}}. \qquad (23)$$

Substituting the noise density expression $e_n = \sqrt{4kT r_{bT}}$ (for an equivalent total base resistance r_{bT}) into (23) gives

$$\kappa_{r_{bT}} = \frac{1}{\sqrt{3}\ln 2} \sqrt{\frac{q_e}{I_{EE}} \frac{r_{bT}}{R_C}}. \qquad (24)$$

This is similar to (17) in that the gate's ability to resolve time is characterized by charge (q_e) divided by current (I_{EE}). In this case, the relative magnitude of the total equivalent base resistance r_{bT} and the collector resistance R_C impose an additional scale factor.

D. Noise at VCO Input

For any VCO, white noise at the VCO input will modulate the VCO frequency and add jitter. It can be shown [30] that white noise at the VCO input will give jitter following the κ model. For a white noise density of $e_{n(VCO)}$, κ is given by

$$\kappa_{VCO} = \frac{1}{\sqrt{2}} \frac{K_o}{\omega_o} e_{n(VCO)} \qquad (25)$$

where K_o is the VCO scale factor [rad/V · sec] and ω_o is the VCO center frequency [rad/sec].

E. Other Noise Influences

All of the above-mentioned sources of jitter are an inherent part of the components in the ring. In practice, external influences (such as power supply sensitivity) are often a dominant source of jitter [7], [11], [30]. The "noise floor" set by the sources described above will not be realized unless externally caused jitter can be reduced to a sufficiently low level.

F. Combining κ from Different Sources

Since each κ represents a contribution from an independent noise voltage [by assumption v)], then the κ of all sources together is just the RSS combination of the individual κ terms from (10), (16), (24), and (25).

VI. IMPLICATIONS FOR DESIGN

To design for a desired closed-loop jitter σ_x, the first step is to use (3) to determine the value of κ. Then (10), (16), (24), and (25) can be used in a noise budgeting process to assign contributions of each source to the total κ. For each source, the design equations provide an explicit linkage between system level jitter (as described by κ) and circuit-level design considerations.

For example, in the case of low power design, (10) and (18) set a limit on the best possible jitter that can be achieved for a given dc power dissipation. As another example, (24) shows that for a given equivalent base resistance r_{bT}, there is a link between waveform amplitude $I_{EE} R_C$ and jitter. Thus, in the case of low supply voltage design (with little headroom for large signal swings), we can immediately determine the best possible jitter that could be achieved at a given signal amplitude.

The expressions for different sources of jitter allow the designer to determine which source is the major contributor in a given design. The equations also show the temperature dependence of jitter, which is important since it is possible to compensate by making circuit parameters (such as the tail current I_{EE}) temperature dependent as well.

	Open Loop VCO	155 MHz PLL	622 MHz PLL
Process	Si Bipolar Junction Isolated $f_T = 3$ GHz	Si Bipolar Dielectrically Isolated $f_T = 5$ GHz	Si Bipolar Dielectrically Isolated $f_T = 9$ GHz
R_C	500 Ω	1.9 kΩ	500 Ω
R_E	4.0 kΩ	9.5 kΩ	2.5 kΩ
r_{bT}	1.65 kΩ	4.80 kΩ	1.29 kΩ
I_{EE}	280 μA	122 μA	400 μA
$e_{n(VCO)}$	11 nV/$\sqrt{\text{Hz}}$	95 nV/$\sqrt{\text{Hz}}$	34 nV/$\sqrt{\text{Hz}}$
f_o	164.1 MHz	155.4 MHz	622 MHz
K_o	44.1 MHz/V	70.2 MHz/V	240 MHz/V
f_L	---	228 kHz	330 kHz
κ (predicted)	3.51E-08 $\sqrt{s}$	5.46E-08 $\sqrt{s}$	2.73E-08 $\sqrt{s}$
σ_x (predicted)	---	32.3 ps rms	13.4 ps rms
κ (measured)	3.56E-08 $\sqrt{s}$	6.05E-08 $\sqrt{s}$	2.41E-08 $\sqrt{s}$
σ_x (measured)	---	35.0 ps rms	13.07 ps rms
κ error	-1.3%	-9.7%	+13%
σ_x error	---	-7.8%	+2.7%

The relationship between open-loop (κ) and closed-loop (σ_x) performance expressed in (3) is also useful in evaluation of actual devices. From an open-loop VCO measurement of κ, we can predict what the closed-loop performance should be if limited only by the VCO jitter. Then we can compare this prediction with actual closed-loop measurements to determine if performance is being degraded by jitter coupled from other on-chip circuitry.

VII. EXPERIMENTAL RESULTS

A. Simulation

To test the results of the mathematical techniques developed in Section V, the effects of the individual noise sources in the circuit of Fig. 5(a) were simulated using transient noise sources and a differential pair behavioral model following (19). The simulation environment allowed control over the circuit conditions so that it was possible to isolate the effects of individual noise sources, something that would be difficult if not impossible in a physical circuit.

For each of the noise sources, circuit parameters were varied over an order of magnitude range around design center values. The simulated results [30] showed agreement to within 10% of the predicted κ values, except when one of the assumptions i)–v) of Section V was not met. The only region of significant disagreement was for signals of amplitude $\approx V_T$. This limitation is not encountered in practice since larger signal amplitudes are used to realize lower jitter.

B. Open-Loop Hardware Test

The measured jitter of the three-, four-, five-, seven-, and nine-stage ring oscillators of Section IV can be compared to the prediction of Section V. The design parameters for the ring delay stage circuit are given in the "Open Loop VCO" column of Table II. The predicted value of κ is given by substituting the circuit parameter values into (10), (16), (24), and (25). Combining these in RSS fashion gives $\kappa = 3.51$E-08 $\sqrt{s}$. The dashed line in Fig. 4 shows the predicted $\sigma_{\Delta T(OL)}$ corresponding to this value of κ. Good agreement is seen between this plot and the measured results.

For the four-stage ring, the circuit implementation allowed variation in the I_{EE} tail current. Table III gives the measured results and the predicted κ values. The results in Fig. 6 show good agreement, to within 5%.

C. Closed-Loop Clock Recovery PLL Design

Using the technique of Section V, voltage controlled ring oscillators were designed and fabricated in 155 and 622 MHz clock recovery PLL's [40]. Voltage control of frequency was achieved by taking a linear interpolation of signals at different stages in the ring [41]. Table II gives process information, circuit design values, predicted performance, and measured results for each case. Substituting these into (10), (16), (24), and (25), and combining in RSS fashion gives the predicted value of κ. Using this κ and the loop bandwidth f_L in (3) gives the predicted value of σ_x. As can be seen from Table II, the agreement between the predicted and measured results for κ and σ_x is quite good. Fig. 7 shows the measured closed-loop

TABLE III
MEASURED RESULTS AND PREDICTED κ VERSUS IEE

IEE [μA]	κ components [E-08 √sec]				κ [E-08 √sec]		ERROR [%]
	K_{rbT}	K_{RC}	K_{RE}	K_{VCO}	PREDICTED	MEASURED	
280	3.01	1.75	0.44	0.22	3.51	3.56	- 1.3
470	2.33	1.04	0.26	0.22	2.56	2.57	- 0.3
505	2.24	0.97	0.24	0.22	2.46	2.50	- 1.7
540	2.17	0.91	0.23	0.22	2.36	2.41	- 1.9
570	2.11	0.86	0.22	0.22	2.29	2.37	- 3.3
600	2.06	0.82	0.20	0.22	2.22	2.33	- 4.5

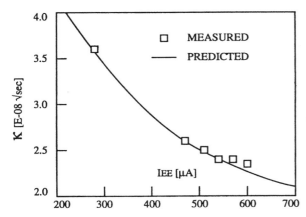

Fig. 6. κ versus tail current IEE.

σ_x of 13.07 ps rms for the 622 MHz clock recovery PLL with a pseudorandom data input.

VIII. CONCLUSION

This paper has developed a methodology to guide design of low-jitter, voltage controlled ring oscillators. The key design parameter is the time domain figure-of-merit κ, which provides the link between circuit-level design and system-level jitter σ_x. (The design technique can also be related to frequency domain measures [30], which is beyond the scope of this paper.) A key insight of this approach is that jitter performance of a ring, as characterized by κ, depends primarily on the individual gate and not on the number of gates in the ring or the ring operating frequency. Explicit expressions were developed to provide a simple, direct means of relating jitter performance to fundamental design parameters. Experimental results at 155 and 622 MHz show that system-level jitter can be predicted to an accuracy of order 10%.

ACKNOWLEDGMENT

The efforts of L. DeVito, A. Gusinov, R. Croughwell, B. Surette, and T. Freitas are greatly appreciated.

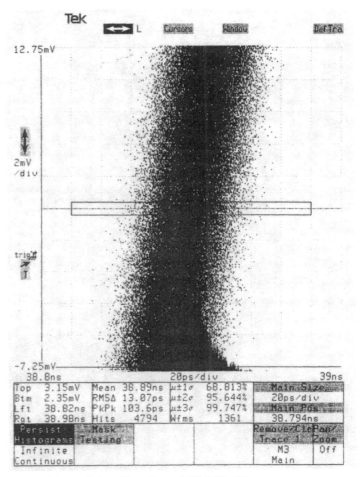

Fig. 7. Measured closed-loop jitter of 622 MHz PLL.

REFERENCES

[1] L. DeVito, J. Newton, R. Croughwell, J. Bulzacchelli, and F. Benkley, "A 52MHz and 155MHz clock-recovery PLL," in *ISSCC Dig. Tech. Papers*, 1991, pp. 142–143.
[2] A. W. Buchwald, K. W. Martin, A. K. Oki, and K. W. Kobayashi, "A 6-GHz integrated phase-locked loop using AlGaAs/GaAs heterojunction bipolar transistors," *IEEE J. Solid-State Circuits*, vol. 27, pp. 1752–1762, Dec. 1992.
[3] B. Lai and R. C. Walker, "A monolithic 622Mb/s clock extraction data retiming circuit," in *ISSCC Dig. Tech. Papers*, Feb. 1991, pp. 144–145.
[4] M. Negahban, R. Behrasi, G. Tsang, H. Abouhossein, and G. Bouchaya, "A two-chip CMOS read channel for hard-disk drives," in *ISSCC Dig. Tech. Papers*, Feb. 1993, pp. 216–217.
[5] W. D. Llewellyn, M. M. H. Wong, G. W. Tietz, and P. A. Tucci, "A 33Mb/s data synchronizing phase-locked-loop circuit," in *ISSCC Dig. Tech. Papers*, Feb. 1988, pp. 12–13.
[6] M. Horowitz, A. Chan, J. Cobrunson, J. Gasbarro, T. Lee, W. Leung, W. Richardson, T. Thrush, and Y. Fujii, "PLL design for a 500Mb/s interface," in *ISSCC Dig. Tech. Papers*, Feb. 1993, pp. 160–161.
[7] I. A. Young, J. K. Greason, J. E. Smith, and K. L. Wong, "A PLL clock generator with 5 to 110MHz lock range for microprocessors," in *ISSCC Dig. Tech. Papers*, Feb. 1992, pp. 50–51.

[8] S. Harris, "The effects of sampling clock jitter on Nyquist sampling analog-to-digital converters, and on oversampling delta-sigma ADC's," *J. Audio Eng. Soc.*, vol. 38, pp. 537–542, July 1990.

[9] _____, "How to achieve optimum performance from delta-sigma A/D and D/A converters," *J. Audio Eng. Soc.*, vol. 41, no. 10, pp. 782–790, Oct. 1993.

[10] R. R. Cordell, J. B. Forney, C. N. Dunn, and W. Garrett, "A 50 MHz phase- and frequency-locked loop," *IEEE J. Solid-State Circuits*, vol. SC-14, pp. 1003–1009, Dec. 1979.

[11] H. Ransijn and P. O'Connor, "A PLL-based 2.5 Gb/s GaAs clock and data regenerator IC," *IEEE J. Solid-State Circuits*, vol. 26, pp. 1345–1353, Oct. 1991.

[12] N. M. Nguyen and R. G. Meyer, "A 1.8 GHz monolithic LC voltage-controlled oscillator," in *ISSCC Dig. Tech. Papers*, 1992, pp. 158–159.

[13] W. A. Edson, "Noise in oscillators," *Proc. IRE*, Aug. 1960, pp. 1454–1466.

[14] M. J. E. Golay, "Monochromaticity and noise in a regenerative electrical oscillator," *Proc. IRE*, pp. 1473–1477, Aug. 1960.

[15] D. B. Leeson, "A simple model of feedback oscillator noise spectrum," *Proc. IEEE*, vol. 54, pp. 329–330, 1966.

[16] C. J. M. Verhoeven, "First order oscillators," Ph.D. dissertation, Delft University, 1990.

[17] K. Kato, T. Sase, H. Sato, I. Ikushima, and S. Kojima, "A low-power 128-MHz VCO for monolithic PLL IC's," *IEEE J. Solid-State Circuits*, vol. SC-23, pp. 474–479, Apr. 1988.

[18] A. Sempel, "A fully integrated HIFI PLL FM demodulator," in *ISSCC Dig. Tech. Papers*, Feb. 1990, pp. 102–103.

[19] M. Souyer and H. A. Ainspan, "A monolithic 2.3Gb/s 100mW clock and data recovery circuit," in *ISSCC Dig. Tech. Papers*, Feb. 1993, pp. 158–159.

[20] A. A. Abidi and R. G. Meyer, "Noise in relaxation oscillators," *IEEE J. Solid-State Circuits*, vol. SC-18, pp. 794–802, Dec. 1983.

[21] B. W. Stuck, "Switching-time jitter statistics for bipolar transistor threshold-crossing detectors," M.S. thesis, Mass. Inst. Technol., 1969.

[22] J. G. Sneep and C. J. M. Verhoeven, "A new low-noise 100-MHz balanced relaxation oscillator," *IEEE J. Solid-State Circuits*, vol. 25, pp. 692–698, June 1990.

[23] C. J. M. Verhoeven, "A high-frequency electronically tunable quadrature oscillator," *IEEE J. Solid-State Circuits*, vol. 27, pp. 1097–1100, July 1992.

[24] M. Banu and A. Dunlop, "A 660Mb/s CMOS clock recovery circuit with instantaneous locking for NRZ data and burst-mode transmission," in *ISSCC Dig. Tech. Papers*, Feb. 1993, pp. 102–103.

[25] M. Banu, "MOS oscillators with multi-decade tuning range and gigahertz maximum speed," *IEEE J. Solid-State Circuits*, vol. SC-23, pp. 1386–1393, Dec. 1988.

[26] S. K. Enam and A. A. Abidi, "A 300MHz CMOS voltage-controlled ring oscillator," *IEEE J. Solid-State Circuits*, vol. 25, pp. 312–315, Feb. 1990.

[27] T. H. Hu and P. R. Gray, "A monolithic 480Mb/s parallel AGC/decision/clock recovery circuit in 1.2μm CMOS," in *ISSCC Dig. Tech. Papers*, Feb. 1993, pp. 98–99.

[28] J. Scott, R. Starke, R. Ramachandran, D. Pietruszynski, S. Bell, K. McClellan, and K. Thompson, "A 16MB/s data detector and timing recovery circuit for token ring LAN," in *ISSCC Dig. Tech. Papers*, Feb. 1989 pp. 150–151.

[29] K. M. Ware, H.-S. Lee, and C. G. Sodini, "A 200MHz CMOS phase-locked loop with dual phase detectors," in *ISSCC Dig. Tech. Papers*, Feb. 1989, pp. 192–193.

[30] J. A. McNeill, "Jitter in ring oscillators," Ph.D. dissertation, Boston University, 1994.

[31] _____, "Jitter in ring oscillators," in *Proc. 1994 ISCAS*, May 1994, vol. 6, pp. 201–204.

[32] T. C. Weigandt, B. Kim, and P. R. Gray, "Analysis of timing jitter in CMOS ring oscillators," in *Proc. 1994 ISCAS*, May 1994, vol. 4, pp. 27–30.

[33] B. Kim, T. C. Weigandt, and P. R. Gray, "PLL/DLL system noise analysis for low jitter clock synthesizer design," in *Proc. 1994 ISCAS*, May 1994, vol. 4, pp. 31–34.

[34] _____, *CSA803 User's Guide*. Beaverton, OR: Tektronix, Inc., 1993.

[35] F. M. Gardner, *Phaselock Techniques*. New York: Wiley, 1979.

[36] _____, "Synchronous optical network (SONET) transport systems—Common generic criteria," *Bellcore Tech. Advisory*, vol. TA-NWT-000253, no. 6, Sept. 1990.

[37] K. M. Sharaf and M. I. Elmasry, "An accurate analytical propagation delay model for high-speed CML bipolar circuits," *IEEE J. Solid-State Circuits*, vol. 29, pp. 31–45, Jan. 1994.

[38] A. Bilotti and E. Mariani, "Noise characteristics of current mirror sinks/sources," *IEEE J. Solid-State Circuits*, vol. SC-10, pp. 516–524, Dec. 1975.

[39] P. R. Gray and R. G. Meyer, *Analysis and Design of Analog Integrated Circuits*. New York: Wiley, 1993.

[40] _____, "AD807 data sheet," Analog Devices Inc., Wilmington, MA.

[41] J. A. McNeill, "Interpolating ring VCO with V-to-f linearity compensation," *Electron. Lett.*, vol. 30, no. 24, pp. 2003–2004, Nov. 1994.

John A. McNeill was born in Syracuse, NY, in 1961. He received the A.B. degree from Dartmouth College, Hanover, NH, in 1983, the M.S. degree from the University of Rochester, Rochester, NY, in 1991, and the Ph.D. degree from Boston University, Boston, MA, in 1994.

From 1983 to 1986 he worked for Analogic Corp., Wakefield, MA, in the area of high-speed, high-resolution analog-to-digital converters and data acquisition systems. In 1986 he joined Adaptive Optics Associates, Cambridge, MA, where he designed low noise interface electronics for charge coupled device (CCD) cameras used in high-speed, wide dynamic range imaging systems. In 1994 he joined the Electrical and Computer Engineering Department faculty of Worcester Polytechnic Institute, Worcester, MA. His teaching and research interests are in the area of analog and mixed signal integrated circuit design.

In 1995, Dr. McNeill was named the Joseph Samuel Satin Distinguished Fellow in Electrical and Computer Engineering.

Jitter and Phase Noise in Ring Oscillators

Ali Hajimiri, Sotirios Limotyrakis, and Thomas H. Lee, *Member, IEEE*

Abstract—A companion analysis of clock jitter and phase noise of single-ended and differential ring oscillators is presented. The impulse sensitivity functions are used to derive expressions for the jitter and phase noise of ring oscillators. The effect of the number of stages, power dissipation, frequency of oscillation, and short-channel effects on the jitter and phase noise of ring oscillators is analyzed. Jitter and phase noise due to substrate and supply noise is discussed, and the effect of symmetry on the upconversion of $1/f$ noise is demonstrated. Several new design insights are given for low jitter/phase-noise design. Good agreement between theory and measurements is observed.

Index Terms—Design methodology, jitter, noise measurement, oscillator noise, oscillator stability, phase jitter, phase-locked loops, phase noise, ring oscillators, voltage-controlled oscillators.

I. INTRODUCTION

DUE to their integrated nature, ring oscillators have become an essential building block in many digital and communication systems. They are used as voltage-controlled oscillators (VCO's) in applications such as clock recovery circuits for serial data communications [1]–[4], disk-drive read channels [5], [6], on-chip clock distribution [7]–[10], and integrated frequency synthesizers [10], [11]. Although they have not found many applications in radio frequency (RF), they can be used for some low-tier RF systems.

Recently, there has been some work on modeling jitter and phase noise in ring oscillators. References [12] and [13] develop models for the clock jitter based on time-domain treatments for MOS and bipolar differential ring oscillators, respectively. Reference [14] proposes a frequency-domain approach to find the phase noise based on an linear time-invariant model for differential ring oscillators with a small number of stages.

In this paper, we develop a parallel treatment of frequency-domain phase noise [15] and time-domain clock jitter for ring oscillators. We apply the phase-noise model presented in [16] to obtain general expressions for jitter and phase noise of the ring oscillators.

The next section briefly reviews the phase-noise model presented in [16]. In Section III, we apply the model to timing jitter and develop an expression for the timing jitter of oscillators, while Section IV provides the derivation of a closed-form expression to calculate the rms value of the impulse sensitivity function (ISF). Section V introduces expressions for jitter and phase noise in single-ended and differential ring oscillators

Manuscript received April 8, 1998; revised November 2, 1998.
A. Hajimiri is with the California Institute of Technology, Pasadena, CA 91125 USA.
S. Limotyrakis and T. H. Lee are with the Center for Integrated Systems, Stanford University, Stanford, CA 94305 USA.
Publisher Item Identifier S 0018-9200(99)04200-6.

in long- and short-channel regimes of operation. Section VI describes the effect of substrate and supply noise as well as the noise due to the tail-current source in differential structures. Section VII explains the design insights obtained from this treatment for low jitter/phase-noise design. Section VIII summarizes the measurement results.

II. PHASE NOISE

The output of a practical oscillator can be written as

$$V_{\text{out}}(t) = A(t) \cdot f[\omega_0 t + \phi(t)] \qquad (1)$$

where the function f is periodic in 2π and $\phi(t)$ and $A(t)$ model fluctuations in amplitude and phase due to internal and external noise sources. The amplitude fluctuations are significantly attenuated by the amplitude limiting mechanism, which is present in any practical stable oscillator and is particularly strong in ring oscillators. Therefore, we will focus on phase variations, which are not quenched by such a restoring mechanism.

As an example, consider the single-ended ring oscillator with a single current source on one of the nodes shown in Fig. 1. Suppose that the current source consists of an impulse of current with area Δq (in coulombs) occurring at time $t = \tau$. This will cause an instantaneous change in the voltage of that node, given by

$$\Delta V = \frac{\Delta q}{C_{\text{node}}} \qquad (2)$$

where C_{node} is the effective capacitance on that node at the time of charge injection. This produces a shift in the transition time. For small ΔV, the change in the phase $\phi(t)$ is proportional to the injected charge

$$\Delta \phi = \Gamma(\omega_0 t) \frac{\Delta V}{V_{\text{swing}}} = \Gamma(\omega_0 t) \frac{\Delta q}{q_{\text{swing}}} \qquad (3)$$

where V_{swing} is the voltage swing across the capacitor and $q_{\text{swing}} = C_{\text{node}} V_{\text{swing}}$. The dimensionless function $\Gamma(\omega_0 t)$ is the time-varying proportionality constant and is periodic in 2π. It is large when a given perturbation causes a large phase shift and small where it has a small effect [16]. Since $\Gamma(x)$ thus represents the sensitivity of every point of the waveform to a perturbation, $\Gamma(x)$ is called the *impulse sensitivity function*.

The time dependence of the ISF can be demonstrated by considering two extreme cases. The first is when the impulse is injected during a transition; this will result in a large phase shift. As the other case, consider injecting an impulse while the output is saturated to either the supply or the ground. This impulse will have a minimal effect on the phase of the oscillator, as shown in Fig. 2.

231

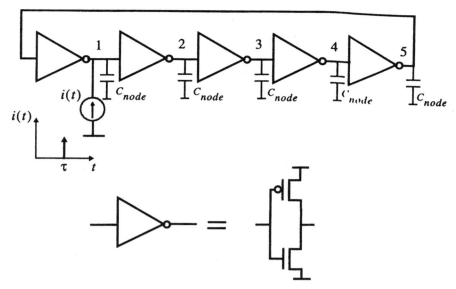

Fig. 1. Five-stage inverter-chain ring oscillator.

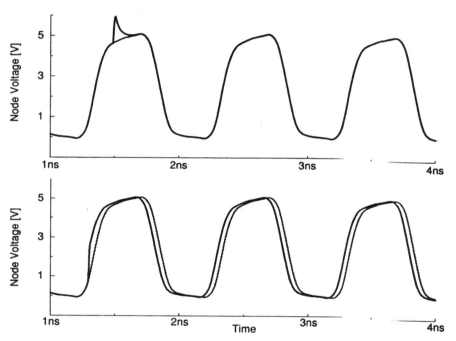

Fig. 2. Effect of impulses injected during transition and peak.

Being interested in its phase $\phi(t)$, we can treat an oscillator as a system that converts voltages and currents to phase. As is evident from the discussion leading to (3), this system is linear for small perturbations. It is also time variant, no matter how small the perturbations are.

Unlike amplitude changes, phase shifts persist indefinitely, since subsequent transitions are shifted by the same amount. Thus, the phase impulse response of an oscillator is a time-varying step. Also note that as long as the introduced change in the voltage due to the current impulse is small, the resultant phase shift is linearly proportional to the injected charge, and hence the transfer function from current to phase is linear.

The unit impulse response of the system is defined as the amount of phase shift per unit current impulse [16]. Based on the foregoing argument, we obtain the following time-dependent impulse response:

$$h_\phi(t, \tau) = \frac{\Gamma(\omega_0\tau)}{q_{max}}u(t - \tau) \quad (4)$$

where $u(t)$ is a unit step.

Knowing the response to an impulse, we can calculate $\phi(t)$ in response to any injected current using the superposition integral

$$\phi(t) = \int_{-\infty}^{\infty} h_\phi(t, \tau)i(\tau)\,d\tau$$
$$= \int_{-\infty}^{t} \frac{\Gamma(\omega_0\tau)}{q_{max}}i(\tau)\,d\tau \quad (5)$$

232

where $i(t)$ represents the noise current injected into the node of interest. Note that the integration arises from the closed-loop nature of the oscillator. The single-sideband phase-noise spectrum due to a white-noise current source is given by [16][1]

$$L\{f_{\text{off}}\} = \frac{\Gamma_{\text{rms}}^2}{8\pi^2 f_{\text{off}}^2} \cdot \frac{\overline{i_n^2}/\Delta f}{q_{\text{max}}^2} \qquad (6)$$

where Γ_{rms} is the rms value of the ISF, $\overline{i_n^2}/\Delta f$ is the single-sideband power spectral density of the noise current source, and f_{off} is the frequency offset from the carrier. In the case of multiple noise sources injecting into the same node, $\overline{i_n^2}/\Delta f$ represents the total current noise due to all the sources and is given by the sum of individual noise power spectral densities [17]. If the noise sources on different nodes are uncorrelated, the waveform (and hence the ISF) of all the nodes are the same except for a phase shift, assuming identical stages. Therefore, the total phase noise due to all N noise sources is N times the value given by (6) (or $2N$ times for a differential ring oscillator).

From (5), it follows that the upconversion of low-frequency noise, such as $1/f$ noise, is governed by the dc value of the ISF. The corner frequency between $1/f^2$ and $1/f^3$ regions in the spectrum of the phase noise is called f_{1/f^3} and is related to the $1/f$ noise corner $f_{1/f}$ through the following equation [16]:

$$f_{1/f^3} = f_{1/f} \cdot \frac{\Gamma_{\text{dc}}^2}{\Gamma_{\text{rms}}^2} \qquad (7)$$

where Γ_{dc} is the dc value of the ISF. Since the height of the positive and negative lobes of the ISF is determined by the slope of the rising and falling edges of the output waveform, respectively, symmetry of the rising and falling edges can reduce Γ_{dc} and hence the upconversion of $1/f$ noise.

III. JITTER

In an ideal oscillator, the spacing between transitions is constant. In practice, however, the transition spacing will be variable. This uncertainty is known as clock jitter and increases with measurement interval ΔT (i.e., the time delay between the reference and the observed transitions), as illustrated in Fig. 3. This variability accumulation (i.e., "jitter accumulation") occurs because any uncertainty in an earlier transition affects all the following transitions, and its effect persists indefinitely. Therefore, the timing uncertainty when ΔT seconds have elapsed is the sum of the uncertainties associated with each transition.

The statistics of the timing jitter depend on the correlations among the noise sources involved. The case of each transition's being affected by independent noise sources has been considered in [12] and [13]. The jitter introduced by each stage is assumed to be totally independent of the jitter introduced by other stages, and therefore the total variance of the jitter is given by the sum of the variances introduced at each stage. For ring oscillators with identical stages, the variance will be given by $m\sigma_s^2$, where m is the number of transitions during ΔT and

[1] A more accurate treatment [17] shows that the phase noise does not grow without bound as f_{off} approaches zero (it becomes flat for small values of f_{off}). However, this makes no practical difference in this discussion.

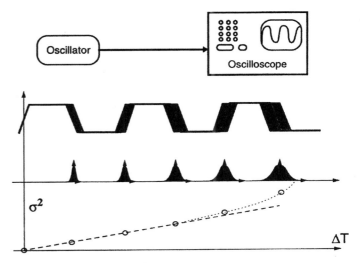

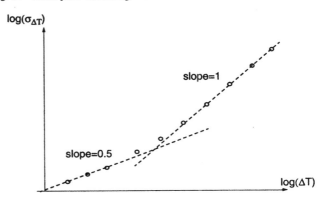

Fig. 3. Clock jitter increasing with time.

Fig. 4. RMS jitter versus measurement time on a log–log plot.

σ_s^2 is the variance of the uncertainty introduced by one stage during one transition. Noting that m is proportional to ΔT, the standard deviation of the jitter after ΔT seconds is [13]

$$\sigma_{\Delta T} = \kappa\sqrt{\Delta T} \qquad (8)$$

where κ is a proportionality constant determined by circuit parameters.

Another instructive special case that is not usually considered is when the noise sources are totally correlated with one another. Substrate and supply noise are examples of such noise sources. Low-frequency noise sources, such as $1/f$ noise, can also result in a correlation between induced jitter on transitions over multiple cycles. In this case, the standard deviations rather than the variances add. Therefore, the standard deviation of the jitter after ΔT seconds is proportional to ΔT

$$\sigma_{\Delta T} = \zeta\Delta T \qquad (9)$$

where ζ is another proportionality constant. Noise sources such as thermal noise of devices are usually modeled as uncorrelated, while substrate and supply-noise sources, as well as low-frequency noise, are approximated as partially or fully correlated sources. In practice, both correlated and uncorrelated sources exist in a circuit, and hence a *log–log* plot of the timing jitter $\sigma_{\Delta T}$ versus the measurement delay ΔT for an open-loop oscillator will demonstrate regions with slopes of 1/2 and 1, as shown in Fig. 4.

233

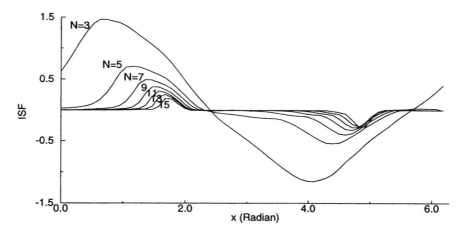

Fig. 5. ISF for ring oscillators of the same frequency with different number of stages.

In most digital applications, it is desirable for $\sigma_{\Delta T}$ to decrease at the same rate as the period T. In practice, we wish to keep constant the ratio of the timing jitter to the period. Therefore, in many applications, phase jitter, defined as

$$\sigma_{\Delta\phi} = 2\pi\frac{\sigma_{\Delta T}}{T} = \omega_0\sigma_{\Delta T} \qquad (10)$$

is a more useful measure.

An expression for $\sigma_{\Delta\phi}$ can be obtained using (5). As shown in Appendix A, for $\Delta T \gg T$ or $\Delta T = nT$ where n is an integer, the phase jitter due to a single white noise source is given by

$$\sigma_{\Delta\phi}^2 = \frac{\Gamma_{\rm rms}^2 \cdot \overline{i_n^2}/\Delta f}{2q_{\rm max}^2}\Delta T. \qquad (11)$$

Using (10) and (11), the proportionality constant κ in (8) is calculated to be

$$\kappa = \frac{\Gamma_{\rm rms}}{q_{\rm max}\omega_0}\sqrt{\frac{1}{2}\frac{\overline{i_n^2}}{\Delta f}}. \qquad (12)$$

IV. CALCULATION OF THE ISF FOR RING OSCILLATORS

To calculate phase noise and jitter using (6) and (12), one needs to know the rms value of the ISF. Although one can always find the ISF through simulation, we obtain a closed-form approximate equation for the rms value of the ISF of ring oscillators, which usually makes such simulations unnecessary.

It is instructive to look at the actual ISF of ring oscillators to gain insight into what constitutes a good approximation. Fig. 5 shows the shape of the ISF for a group of single-ended CMOS ring oscillators. The frequency of oscillation is kept constant (through adjustment of channel length), while the number of stages is varied from 3 to 15 (in odd numbers). To calculate the ISF, a narrow current pulse is injected into one of the nodes of the oscillator, and the resulting phase shift is measured a few cycles later in simulation.

As can be seen, increasing the number of stages reduces the peak value of the ISF. The reason is that the transitions of the normalized waveform become faster for larger N. Since the sensitivity during the transition is inversely proportional to the slope, the peak of the ISF drops. It should be noted that only the peak of the ISF is inversely proportional to the slope, and

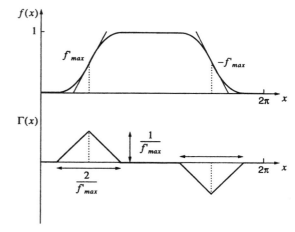

Fig. 6. Approximate waveform and ISF for ring oscillator.

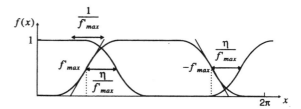

Fig. 7. Relationship between rise time and delay.

this relation should not be generalized to other points in time. Also, the widths of the lobes of the ISF decrease as N becomes larger, since each transition occupies a smaller fraction of the period. Based on these observations, we approximate the ISF as triangular in shape and with symmetric rising and falling edges, as shown in Fig. 6. The case of nonsymmetric rising and falling edges is considered in Appendix B.

The ISF has a maximum of $1/f'_{\rm max}$, where $f'_{\rm max}$ is the maximum slope of the normalized waveform f in (1). Also, the width of the triangles is approximately $2/f'_{\rm max}$, and hence the slopes of the sides of the triangles are ± 1. Therefore, assuming equality of rise and fall times, $\Gamma_{\rm rms}$ can be estimated as

$$\Gamma_{\rm rms}^2 = \frac{1}{2\pi}\int_0^{2\pi}\Gamma^2(x)\,dx = \frac{4}{2\pi}\int_0^{1/f'}x^2\,dx$$
$$= \frac{2}{3\pi}\left(\frac{1}{f'_{\rm max}}\right)^3. \qquad (13)$$

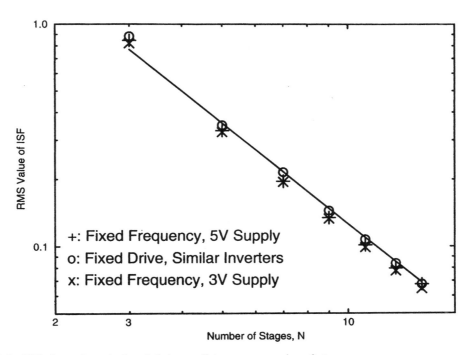

Fig. 8. RMS values of the ISF's for various single-ended ring oscillators versus number of stages.

On the other hand, stage delay is proportional to the rise time

$$\hat{t}_D = \frac{\eta}{f'_{\max}} \quad (14)$$

where $\hat{t}_D$ is the normalized stage delay and η is a proportionality constant, which is typically close to one, as can be seen in Fig. 7.

The period is $2N$ times longer than a single stage delay

$$2\pi = 2N\hat{t}_D = \frac{2N\eta}{f'_{\max}}. \quad (15)$$

Using (13) and (15), the following approximate expression for Γ_{rms} is obtained:

$$\Gamma_{\mathrm{rms}} = \sqrt{\frac{2\pi^2}{3\eta^3} \frac{1}{N^{1.5}}}. \quad (16)$$

Note that the $1/N^{1.5}$ dependence of Γ_{rms} is independent of the value of η. Fig. 8 illustrates Γ_{rms} for the ISF shown in Fig. 5 with plus signs on log–log axes. The solid line shows the line of $\Gamma_{\mathrm{rms}} \approx 4/N^{1.5}$, which is obtained from (16) for $\eta = 0.75$. To verify the generality of (16), we maintain a fixed channel length for all the devices in the inverters while varying the number of stages to allow different frequencies of oscillation. Again, Γ_{rms} is calculated, and is shown in Fig. 8 with circles. We also repeat the first experiment with a different supply voltage (3 V as opposed to 5 V), and the result is shown with crosses. As can be seen, the values of Γ_{rms} are almost identical for these three cases.

It should not be surprising that Γ_{rms} is primarily a function of N because the effect of variations in other parameters, such as $q_{\max}$ and device noise, have already been decoupled from $\Gamma(x)$, and thus the ISF is a unitless, frequency- and amplitude-independent function.

Equation (16) is valid for differential ring oscillators as well, since in its derivation no assumption specific to single-ended oscillators was made. Fig. 9 shows the Γ_{rms} for three sets of differential ring oscillators, with a varying number of stages (4–16). The data shown with plus signs correspond to oscillators in which the total power dissipation and the drain voltage swing are kept constant by scaling the tail-current sources and load resistors as N changes. Members of the second set of oscillators have a fixed total power dissipation and fixed load resistors, which result in variable swings and for whom data are shown with circles. The third case is that of a fixed tail current for each stage and constant load resistors, whose data are illustrated using crosses. Again, in spite of the diverse variations of the frequency and other circuit parameters, the $1/N^{1.5}$ dependency of Γ_{rms} and its independence from other circuit parameters still holds. In the case of a differential ring oscillator, $\Gamma_{\mathrm{rms}} \approx 3/N^{1.5}$, which corresponds to $\eta = 0.9$, is the best fit approximation for Γ_{rms}. This is shown with the solid line in Fig. 9. A similar result can be obtained for bipolar differential ring oscillators.

Although Γ_{rms} decreases as the number of stages increases, one should not prematurely conclude that the phase noise can be reduced using a larger number of stages because the number of noise sources, as well as their magnitudes, also increases for a given total power dissipation and frequency of oscillation.

In the case of asymmetric rising and falling edges, both Γ_{rms} and Γ_{dc} will change. As shown in Appendix B, the $1/f^3$ corner of the phase-noise spectrum is inversely proportional to the number of stages. Therefore, the $1/f^3$ corner can be reduced either by making the transitions more symmetric in terms of rise and fall times or by increasing the number of stages. Although the former always helps, the latter has other implications on the phase noise in the $1/f^2$ region, as will be shown in the following section.

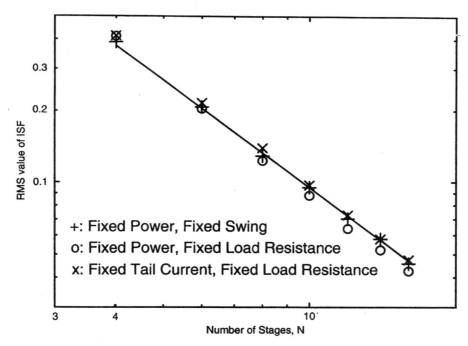

Fig. 9. RMS values of the ISF's for various differential ring oscillators versus number of stages.

V. EXPRESSIONS FOR JITTER AND PHASE NOISE IN RING OSCILLATORS

In this section, we derive expressions for the phase noise and jitter of different types of ring oscillators. Throughout this section, we assume that the symmetry criteria required to minimize Γ_{dc} (and hence the upconversion of $1/f$ noise) are already met and that the jitter and phase noise of the oscillator are dominated by white noise. For CMOS transistors, the drain current noise spectral density is given by

$$\frac{\overline{i_n^2}}{\Delta f} = 4kT\gamma g_{d0} = 4kT\gamma\mu C_{ox}\frac{W}{L}\Delta V \qquad (17)$$

where g_{d0} is the zero-bias drain source conductance, μ is the mobility, C_{ox} is the gate-oxide capacitance per unit area, W and L are the channel width and length of the device, respectively, and ΔV is the gate voltage overdrive. The coefficient γ is 2/3 for long-channel devices in the saturation region and typically two to three times greater for short-channel devices [18]. Equation (17) is valid in both short- and long-channel regimes as long as an appropriate value for γ is used.

A. Single-Ended CMOS Ring Oscillators

We start with a single-ended CMOS ring oscillator with equal-length NMOS and PMOS transistors. Assuming that $V_{TN} = |V_{TP}|$, the maximum total channel noise from NMOS and PMOS devices, when both the input and output are at $V_{DD}/2$, is given by

$$\frac{\overline{i_n^2}}{\Delta f} = \left(\frac{\overline{i_n^2}}{\Delta}\right)_N + \left(\frac{\overline{i_n^2}}{\Delta f}\right)_N = 4kT\gamma\mu_{eff}C_{ox}\frac{W_{eff}}{L}\Delta V \quad (18)$$

where

$$W_{eff} = W_n + W_p \qquad (19)$$

and

$$\mu_{eff} = \frac{\mu_n W_n + \mu_p W_p}{W_n + W_p} \qquad (20)$$

and ΔV is the gate overdrive in the middle of transition, i.e., $\Delta V = (V_{DD}/2) - V_T$.

During one period, each node is charged to q_{max} and then discharged to zero. In an N-stage single-ended ring oscillator, the power dissipation associated with this process is $Nq_{max}V_{DD}f_0$. However, during the transitions, some extra current, known as crowbar current, is drawn from the supply, which does not contribute to charging and discharging the capacitors and goes directly from supply to ground through both transistors. In a symmetric ring oscillator, these two components are approximately equal, and their difference will depend on the ratio of the rise time and stage delay. Therefore, the total power dissipation is approximately given by

$$P = 2\eta N V_{DD}q_{max}f_0. \qquad (21)$$

Assuming $\mu_n W_n = \mu_p W_p$ to make the waveforms symmetric to the first order, we have

$$f_0 = \frac{1}{2Nt_D} = \frac{1}{\eta N(t_r + t_f)} \approx \frac{\mu_{eff}W_{eff}C_{ox}\Delta V^2}{8\eta NLq_{max}} \quad (22)$$

where t_D is the delay of each stage and t_r and t_f are the rise and fall time, respectively, associated with the maximum slope during a transition.

Assuming that the thermal noise sources of the different devices are uncorrelated, and assuming that the waveforms (and hence the ISF) of all the nodes are the same except for a phase shift, the total phase noise due to all N noise sources is

N times the value given by (6). Taking only these inevitable noise sources into account, (6), (16), (18), (21), and (22) result in the following expressions for phase noise and jitter:

$$L\{\Delta f\} \approx \frac{8}{3\eta} \cdot \frac{kT}{P} \cdot \frac{V_{\text{DD}}}{V_{\text{char}}} \cdot \frac{f_0^2}{\Delta f^2} \qquad (23)$$

$$\kappa \approx \sqrt{\frac{8}{3\eta}} \cdot \sqrt{\frac{kT}{P} \cdot \frac{V_{\text{DD}}}{V_{\text{char}}}} \qquad (24)$$

where V_{char} is the *characteristic voltage* of the device. For long-channel mode of operation, it is defined as $V_{\text{char}} = \Delta V / \gamma$. Any extra disturbance, such as substrate and supply noise, or noise contributed by extra circuitry or asymmetry in the waveform will result in a larger number than (23) and (24). Note that lowering threshold voltages reduces the phase noise, in agreement with [12]. Therefore, the minimum achievable phase noise and jitter for a single-ended CMOS ring oscillator, assuming that all symmetry criteria are met, occurs for zero threshold voltage

$$L\{\Delta f\} > \frac{16\gamma}{3\eta} \cdot \frac{kT}{P} \cdot \frac{f_0^2}{\Delta f^2} \qquad (25)$$

$$\kappa > \sqrt{\frac{16\gamma}{3\eta}} \cdot \sqrt{\frac{kT}{P}}. \qquad (26)$$

As can be seen, the minimum phase noise is inversely proportional to the power dissipation and grows quadratically with the oscillation frequency. Further, note the lack of dependence on the number of stages (for a given power dissipation and oscillation frequency). Evidently, the increase in the number of noise sources (and in the maximum power due to the higher transition currents required to run at the same frequency) essentially cancels the effect of decreasing Γ_{rms} as N increases, leading to no net dependence of phase noise on N. This somewhat surprising result may explain the confusion that exists regarding the optimum N, since there is not a strong dependence on the number of stages for single-ended CMOS ring oscillators. Note that (25) and (26) establish the lower bound and therefore should not be used to calculate the phase noise and jitter of an arbitrary oscillator, for which (6) and (12) should be used, respectively.

We may carry out a similar calculation for the short-channel case. For such devices, the drain current may be expressed as

$$I_D = \frac{\mu C_{\text{ox}}}{2} W E_c \Delta V \qquad (27)$$

where E_c is the critical electric field and is defined as the value of electric field resulting in half the carrier velocity expected from low field mobility. Combining (17) with (27), we obtain the following expression for the drain current noise of a MOS device in short channel:

$$\frac{\overline{i_n^2}}{\Delta f} = 8kT \frac{\gamma I_D}{E_C L}. \qquad (28)$$

The frequency of oscillation can be approximated by

$$f_0 = \frac{1}{2Nt_D} = \frac{1}{\eta N(t_r + t_f)}$$
$$= \frac{\mu_{\text{eff}} W_{\text{eff}} C_{\text{ox}} \Delta V^2}{8\eta N L q_{\text{max}}}. \qquad (29)$$

Using (28) and (29), we obtain the same expressions for phase noise and jitter as given by (23) and (24), except for a new V_{char}

$$V_{\text{char}} = \frac{E_c L}{\gamma} \qquad (30)$$

which results in a larger phase noise and jitter than the long-channel case by a factor of $\gamma \Delta V / E_c L$. Again, note the absence of any dependency on the number of stages.

B. Differential CMOS Ring Oscillators

Now consider a differential MOS ring oscillator with resistive load. The total power dissipation is

$$P = N I_{\text{tail}} V_{\text{DD}} \qquad (31)$$

where N is the number of stages, I_{tail} is the tail bias current of the differential pair, and V_{DD} is the supply voltage. The frequency of oscillation can be approximated by

$$f_0 = \frac{1}{2Nt_D} \approx \frac{1}{2\eta N t_r} \approx \frac{I_{\text{tail}}}{2\eta N q_{\text{max}}}. \qquad (32)$$

Surprisingly, tail-current source noise in the vicinity of f_0 does not affect the phase noise. Rather, its low-frequency noise as well as its noise in the vicinity of *even* multiples of the oscillation frequency affect the phase noise. Tail noise in the vicinity of even harmonics can be significantly reduced by a variety of means, such as with a series inductor or a parallel capacitor. As before, the effect of low-frequency noise can be minimized by exploiting symmetry. Therefore, only the noise of the differential transistors and the load are taken into account. The total current noise on each single-ended node is given by

$$\frac{\overline{i_n^2}}{\Delta f} = \left(\frac{\overline{i_n^2}}{\Delta f} \right)_N + \left(\frac{\overline{i_n^2}}{\Delta f} \right)_{\text{Load}}$$
$$= 4kT I_{\text{tail}} \left(\frac{1}{V_{\text{char}}} + \frac{1}{R_L I_{\text{tail}}} \right) \qquad (33)$$

where R_L is the load resistor, $V_{\text{char}} = (V_{\text{GS}} - V_T)/\gamma$ for a balanced stage in the long-channel limit and $V_{\text{char}} = E_c L / \gamma$ in the short-channel regime. The phase noise and jitter due to all $2N$ noise sources is $2N$ times the value given by (6) and (12). Using (16), the expression for the phase noise of the differential MOS ring oscillator is

$$L_{\text{min}}\{\Delta f\} = \frac{8}{3\eta} \cdot N \cdot \frac{kT}{P} \cdot \left(\frac{V_{\text{DD}}}{V_{\text{char}}} + \frac{V_{\text{DD}}}{R_L I_{\text{tail}}} \right) \cdot \frac{f_0^2}{\Delta f^2} \qquad (34)$$

and is given by

$$\kappa_{\text{min}} = \sqrt{\frac{8}{3\eta}} \cdot \sqrt{N \cdot \frac{kT}{P} \cdot \left(\frac{V_{\text{DD}}}{V_{\text{char}}} + \frac{V_{\text{DD}}}{R_L I_{\text{tail}}} \right)}. \qquad (35)$$

Equations (34) and (35) are valid in both long- and short-channel regimes of operation with the right choice of V_{char}.

Note that, in contrast with the single-ended ring oscillator, a differential oscillator does exhibit a phase noise and jitter dependency on the number of stages, with the phase noise degrading as the number of stages increases for a given frequency and power dissipation. This result may be understood as a consequence of the necessary reduction in the charge swing that is required to accommodate a constant frequency of oscillation at a fixed power level as N increases. At the same time, increasing the number of stages at a fixed total power dissipation demands a proportional reduction of tail-current sources, which will reduce the swing, and hence q_{max}, by a factor of $1/N^2$.

C. Bipolar Differential Ring Oscillator

A similar approach allows us to derive the corresponding results for a bipolar differential ring oscillator. In this case, the power dissipation is given by (31) and the oscillation frequency by (32). The total noise current is given by the sum of collector shot noise and load resistor noise

$$\frac{\overline{i_n^2}}{\Delta f} = 2q_e I_C + \frac{4kT}{R} = 4kT I_{tail}\left(\frac{1}{V_{char}} + \frac{1}{R_L I_{tail}}\right) \quad (36)$$

where q_e is the electron charge, $I_C = I_{tail}/2$ is the collector current during the transition, and $V_{char} = 4kT/q_e$. Using these relations, the phase noise and jitter of a bipolar ring oscillator are again given by (34) and (35) with the appropriate choice of V_{char}.

VI. OTHER NOISE SOURCES

Other noise sources, such as tail-current source noise in a differential structure, or substrate and supply noise sources, may play an important role in the jitter and phase noise of ring oscillators. The low-frequency noise of the tail-current source affects phase noise if the symmetry criteria mentioned in Section II are not met by each half circuit. In such cases, the ISF for the tail-current source has a large dc value, which increases the upconversion of low-frequency noise to phase noise. This upconversion is particularly prominent if the tail device has a large $1/f$ noise corner.

Substrate and supply noise are among other important sources of noise. There are two major differences between these noise sources and internal device noise. First, the power spectral density of these sources is usually nonwhite and often demonstrates strong peaks at various frequencies. Even more important is that the substrate and supply noise on different nodes of the ring oscillator have a very strong correlation. This property changes the response of the oscillator to these sources.

To understand the effect of this correlation, let us consider the special case of having equal noise sources on all the nodes of the oscillator. If all the inverters in the oscillator are the same, the ISF for different nodes will only differ in phase by multiples of $2\pi/N$, as shown in Fig. 10. Therefore, the total phase due to all the sources is given by superposition of (5)

$$\phi(t) = \frac{1}{q_{max}} \sum_{n=0}^{N-1} \int_{-\infty}^{t} i(\tau)\Gamma\left(\omega_0\tau + \frac{2\pi n}{N}\right) d\tau$$

$$= \frac{1}{q_{max}} \int_{-\infty}^{t} i(\tau)\left[\sum_{n=0}^{N-1} \Gamma\left(\omega_0\tau + \frac{2\pi n}{N}\right)\right] d\tau. \quad (37)$$

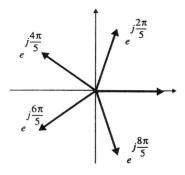

Fig. 10. Phasors for noise contributions from each source.

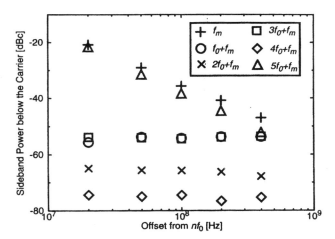

Fig. 11. Sideband power below carrier for equal sources on all five nodes at $nf_0 + f_m$.

Expanding the term in brackets in a Fourier series, we can show that it is zero except at dc and multiples of $N\omega_0$, i.e.,

$$\phi(t) = \frac{N}{q_{max}} \sum_{n=0}^{\infty} c_{(nN)} \int_{-\infty}^{t} i(\tau)\cos(nN\omega_0\tau)\,d\tau \quad (38)$$

where c_i is the ith Fourier coefficient of the ISF. Equation (38) means that for identical sources, only noise in the vicinity of integer multiples of $N\omega_0$ affects the phase.

To verify this effect, sinusoidal currents with an amplitude of 10 μA were injected into all five nodes of the five-stage ring oscillator of Fig. 1 at different offsets from integer multiples of the frequency of oscillation, and the induced sidebands were measured. The measured sideband power with respect to the carrier is plotted in Fig. 11.

As can be seen in Fig. 11, only injection at low frequency and in the vicinity of the fifth harmonic are integrated, and show a -20 dB/dec slope. The effect of injection in the vicinity of harmonics that are not integer multiples of N is much smaller than at the integer ones. Ideally, there should be no sideband induced by the injection in the vicinity of harmonics that are not integer multiples of N; however, as can be seen in Fig. 11, there is some sideband power due to the amplitude response.

Low-frequency noise can also result in correlation between uncertainties introduced during different cycles, as its value does not change significantly over a small number of periods.

Therefore, the uncertainties add up in amplitude rather than power, resulting in a region with a slope of one in the *log–log* plot of jitter even in the absence of external noise sources such as substrate and supply noise.

VII. DESIGN IMPLICATIONS

One can use (23) and (34) to compare the phase-noise performance of single-ended and differential MOS ring oscillators. As can be seen for N stages, the phase noise of the differential ring oscillator is approximately $N[1 + V_{\text{char}}/(R_L I_{\text{tail}})]$ times larger than the phase noise of a single-ended oscillator of equal N, P_{tot} and f_0. Since the minimum N for a regular ring oscillator is three, even a properly designed differential CMOS ring oscillator underperforms its single-ended counterpart, especially for a larger number of stages. This difference is even more pronounced if proper precautions to reduce the noise of the tail current are not taken. However, the differential ring oscillator may still be preferred in IC's because of the lower sensitivity to substrate and supply noise, as well as lower noise injection into other circuits on the same chip. The decision to use differential versus single-ended ring oscillators should be based on both of these considerations.

The common-mode sensitivity problem in a single-ended ring oscillator can be mitigated to some extent by using two identical ring oscillators laid out close to each other that oscillate out of phase because of small coupling inverters [19]. Single-ended configurations may be used in a less noisy environment to achieve better phase-noise performance for a given power dissipation.

As shown in Appendix B, asymmetry of the rising and falling edges degrades phase noise and jitter by increasing the $1/f^3$ corner frequency. Thus, every effort should be taken to make the rising and falling edges symmetric. By properly adjusting the symmetry properties, one can suppress or even eliminate low-frequency-noise upconversion [16]. As shown in [16], differential symmetry is insufficient, and the symmetry of each half circuit is important. One practical method to achieve this symmetry is to use more linear loads, such as resistors or linearized MOS devices. This method reduces the $1/f$ noise upconversion and substrate and supply coupling [20]. Another revealing implication, shown in Appendix A, is the reduction of the $1/f^3$ corner frequency as N increases. Hence for a process with large $1/f$ noise, a larger number of stages may be helpful.

One question that frequently arises in the design of ring oscillators is the optimum number of stages for minimum jitter and phase noise. As seen in (23), for single-ended oscillators, the phase noise and jitter in the $1/f^2$ region is not a strong function of the number of stages for single-ended CMOS ring oscillators. However, if the symmetry criteria are not well satisfied and/or the process has a large $1/f$ noise, a larger N will reduce the jitter. In general, the choice of the number of stages must be made on the basis of several design criteria, such as $1/f$ noise effect, the desired maximum frequency of oscillation, and the influence of external noise sources, such as supply and substrate noise, that may not scale with N.

The jitter and phase noise behavior are different for differential ring oscillators. As (34) suggests, jitter and phase noise increase with an increasing number of stages. Hence if the $1/f$ noise corner is not large, and/or proper symmetry measures have been taken, the minimum number of stages (three or four) should be used to give the best performance. This recommendation holds even if the power dissipation is not a primary issue. It is not fair to argue that burning more power in a larger number of stages allows the achievement of better phase noise, since dissipating the same total power in a smaller number of stages results in better jitter/phase noise as long as it is possible to maximize the total charge swing.

Another insight one can obtain from (34) and (35) is that the jitter of a MOS differential ring oscillator at a given V_{DD}, P, N, and f_0 is smaller than that of a differential bipolar ring oscillator, at least for today's range of circuit and process parameters. As we go to shorter channel lengths, the characteristic voltage for the MOS devices given by (30) becomes smaller, and thus phase noise degrades with scaling. Bipolar ring oscillators do not suffer from this problem.

LC oscillators generally have better phase noise and jitter compared to ring oscillators for two reasons. First, a ring oscillator stores a certain amount of energy in the capacitors during every cycle and then dissipates all the stored energy during the same cycle, while an LC resonator dissipates only $2\pi/Q$ of the total energy stored during one cycle. Thus, for a given power dissipation in steady state, a ring oscillator suffers from a smaller maximum charge swing q_{max}. Second, in a ring oscillator, the device noise is maximum during the transitions, which is the time where the sensitivity, and hence the ISF, is the largest [16].

VIII. EXPERIMENTAL RESULTS

The phase-noise measurements in this section were performed using three different systems: an HP 8563E spectrum analyzer with phase-noise measurement capability, an RDL NTS-1000A phase-noise measurement system, and an HP E5500 phase-noise measurement system. The jitter measurements were performed using a Tektronix CSA 803A communication signal analyzer.

Tables I–III summarize the phase-noise measurements. All the reported phase-noise values are at a 1-MHz offset from the carrier, chosen to achieve the largest dynamic range in the measurement. Table I shows the measurement results for three different inverter-chain ring oscillators. These oscillators are made of the CMOS inverters shown in Fig. 12(a), with no frequency tuning mechanism. The output is taken from one node of the ring through a few stages of tapered inverters. Oscillators number 1 and 2 are fabricated in a 2-μm, 5-V CMOS process, and oscillator number 3 is fabricated in a 0.25-μm, 2.5-V process. The second column shows the number of stages in each of the oscillators. The W/L ratios of the NMOS and PMOS devices, as well as the supply voltages, the total measured supply currents, and the frequencies of oscillation are shown next. The phase-noise prediction using (23) and (6), together with the measured phase noise, are shown in the last three columns.

TABLE I
INVERTER-CHAIN RING OSCILLATORS

Index	N	NMOS W/L μm/μm	PMOS W/L μm/μm	V_{DD}	I_{sup} mA	f_o	Pred. (23) dBc/Hz	Pred. (6) dBc/Hz	Meas. PN dBc/Hz
1	5	3/2	5/2	5.0	0.3	232MHz	-119.9	-117.7	-118.5
2	11	4/2	6/2	5.0	0.5	115MHz	-127.2	-126.4	-126.0
3	19	10/0.25	20/0.25	2.5	10	1.33GHz	-111.8	-113.0	-111.5

TABLE II
CURRENT-STARVED INVERTER-CHAIN RING OSCILLATORS

Index	N	Ninv W/L μm/μm	Pinv W/L μm/μm	Ntail W/L μm/μm	Ptail W/L μm/μm	I_{sup} mA	f_o MHz	Pred. (23) dBc/Hz	Pred. (6) dBc/Hz	Meas. PN dBc/Hz
4	3	35/0.53	70/0.53	28/0.53	56/0.53	2.34	751	-113.8	-116.6	-114.0
5	5	21/0.39	42/0.39	23/0.39	46/0.39	2.51	850	-111.7	-111.9	-112.6
6	7	14/0.36	28/0.36	36.8/0.36	73.5/0.36	2.49	931	-110.5	-110.4	-111.7
7	9	12.6/0.32	25.2/0.32	28/0.32	56/0.32	2.73	932	-110.4	-113.5	-112.5
8	11	10.5/0.32	21/0.32	146/0.32	291/0.32	2.65	869	-110.9	-110.1	-112.2
9	15	9.1/0.28	18.2/0.28	146/0.28	291/0.28	2.8	929	-110.0	-110.7	-112.3
10	17	7.4/0.25	12.6/0.25	25.2/0.28	50.4/0.28	3.8	898	-111.2	-109.4	-112.0
11	19	6.3/0.25	12.6/0.25	56/0.25	112/0.25	3.9	959	-110.6	-110.1	-110.9

As an illustrative example, we will show the details of phase-noise calculations for oscillator number 3. Using (16) to calculate Γ_{rms}, the phase noise can be obtained from (6). We calculate the noise power when the stage is halfway through a transition. At this point, the drain current is simulated to be 3.47 mA. An E_c of 4×10^6 V/m and a γ of 2.5 is used in (28) to obtain a noise power of $\overline{i^2}/\Delta f = 2.87 \times 10^{-22}$ A^2/Hz. The total capacitance on each node is $C_{\text{total}} = 71.8$ fF, and hence $q_{max} = 179.5$ fC. There is one such noise source on each node; therefore, the phase noise is N times the value given by (6), which results in $\mathcal{L}\{1\,\text{MHz}\} = -113.0$ dBc/Hz.

Table II summarizes the data obtained for current-starved ring oscillators with the cell structure shown in Fig. 12(b), all implemented in the same 0.25-μm, 2.5-V process. Ring oscillators with a different number of stages were designed with roughly constant oscillation frequency and total power dissipation. Frequency adjustment is achieved by changing the channel length, while total power dissipation control is performed by changing device width. The W/L ratios of the inverter and the tail NMOS and PMOS devices are shown in Table II. The node $Nbias$ is kept at V_{DD}, while node $Pbias$ is at 0 V. The measured total current dissipation and the frequency of oscillation can be found in columns 7 and 8. Phase-noise calculations based on (23) and (6) are in good

agreement with the measured results. The die photo of the chip containing these oscillators is shown in Fig. 13. The slightly superior phase noise of the three-stage ring oscillator (number 4) can be attributed to lower oscillation frequency and longer channel length (and hence smaller γ).

Table III summarizes the results obtained for differential ring oscillators of various sizes and lengths with the inverter topology shown in Fig. 12(c), covering a large span of frequencies up to 5.5 GHz. All these ring oscillators are implemented in the same 0.25-μm, 2.5-V process, and all the oscillators, except the one marked with N/A, have the tuning circuit shown. The resistors are implemented using an unsilicided polysilicon layer. The main reason to use poly resistors is to reduce $1/f$ noise upconversion by making the waveform on each node closer to the step response of an RC network, which is more symmetrical. The value of these load resistors and the W/L ratios of the differential pair are shown in Table III. A fixed 2.5-V power supply is used, resulting in different total power dissipations. As before, the measured phase noise is in good agreement with the predicted phase noise using (34) and (6). The die photo of oscillator number 26 can be found in Fig. 14.

To illustrate further how one obtains the phase-noise predictions shown in Table III, we elaborate on the phase-noise

TABLE III
DIFFERENTIAL RING OSCILLATORS

Index	N	W/L μm/μm	R_L Ω	I_{tail} mA	P_{tot} mW	f_{max}	Tuning range	Pred. (34) dBc/Hz	Pred. (6) dBc/Hz	Meas. PN dBc/Hz
12	4	4.2/0.25	2k	1	10	2.81GHz	34%	-95.4	-95.5	-95.2
13	4	8.4/0.25	1k	2	20	4.47GHz	42%	-95.1	-94.0	-94.3
14	4	16.8/0.25	500	4	40	3.89GHz	44%	-98.5	-97.2	-97.4
15	4	33.6/0.25	250	8	80	5.43GHz	25%	-98.7	-99.6	-98.5
16	4	8.4/0.25	2k	1	10	2.87GHz	37%	-95.2	-96.6	-93.8
17	4	16.8/0.25	1k	2	20	3.39GHz	45%	-96.7	-97.9	-96.8
18	4	33.6/0.25	500	4	40	5.33GHz	32%	-95.8	-97.2	-95.3
19	4	16.8/0.25	2k	1	10	1.75GHz	73%	-99.5	-97.5	-95.2
20	4	33.6/0.25	1k	2	20	2.24GHz	58%	-100.3	-100.3	-99.0
21	4	33.6/0.25	2k	1	10	1.27GHz	67%	-104.4	-101.8	-100.2
22	4	67.2/0.25	1k	2	20	1.19GHz	76%	-105.8	-102.6	-100.2
23	4	33.6/0.25	2k	1	10	1.53GHz	N/A	-100.6	-98.9	-97.3
24	6	13.4/0.25	3k	0.67	10	859MHz	58%	-103.9	-106.0	-104.3
25	8	6.7/0.25	4k	0.5	10	731MHz	74%	-104.1	-106.3	-106.2
26	12	4.2/0.25	6k	0.33	10	447MHz	52%	-106.6	-110.4	-109.5

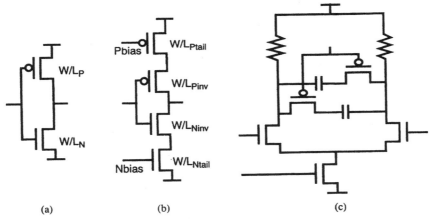

Fig. 12. Inverter stages for (a) inverter-chain ring oscillators, (b) current-starved inverter-chain ring oscillators, and (c) differential ring oscillators.

calculations for oscillator number 12. The noise current due to one of differential pair NMOS devices is given by (28). The total capacitance on each node in the balanced case is $C_{\text{total}} = 41.6$ fF, and the simulated voltage swing is 1.208 V; therefore, $q_{\max} = 50.3$ fC. In the balanced case, this current is half of the tail current, i.e., $I_D = 0.5$ mA, and therefore the noise current of the NMOS device has a single-sideband spectral density of $\overline{i^2}/\Delta f = 4.14 \times 10^{-23}$ A^2/Hz. The thermal noise due to the load resistor is $\overline{i^2}/\Delta f = 8.28 \times 10^{-24}$ A^2/Hz; therefore, the total current noise density is given by $\overline{i^2}/\Delta f = 4.97 \times 10^{-23}$ A^2/Hz. For a differential ring oscillator with N stages, there is one such noise source on each node; therefore, the phase noise is $2N$ times the value given by (6), which results in $\mathcal{L}\{1\,\text{MHz}\} = -95.5$ dBc/Hz. The total power dissipation is $NV_{\text{DD}}I_{\text{tail}} = 10$ mW, and $R_L = 2\,\text{k}\Omega$.

Therefore, with an η of 0.9, (34) predicts a phase noise of $\mathcal{L}\{1\,\text{MHz}\} = -95.4$ dBc/Hz.

Timing jitter for oscillator number 12 can be measured using the setup shown in Fig. 15. The oscillator output is divided into two equal-power outputs using a power splitter. The CSA 803A is not capable of showing the edge it uses to trigger, as there is a 21-ns minimum delay between the triggering transition and the first acquired sample. To be able to look at the triggering edge and perhaps the edges before that, a delay line of approximately 25 ns is inserted in the signal path in front of the sampling head. This way, one may look at the exact edge used to trigger the signal. If the sampling head and the power splitter were noiseless, this edge would show no jitter. However, the power splitter and the sampling head introduce noise onto the signal, which cannot be easily

241

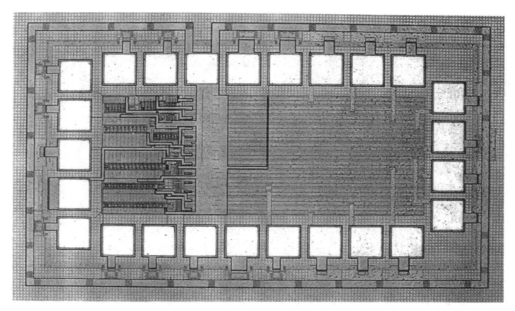

Fig. 13. Die photograph of the current-starved single-ended oscillators.

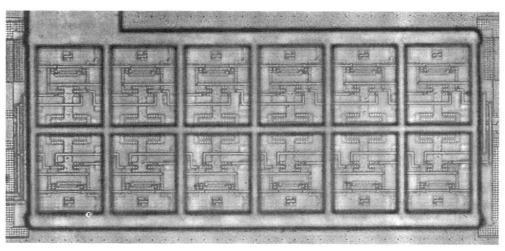

Fig. 14. Die photograph of the 12-stage differential ring oscillator.

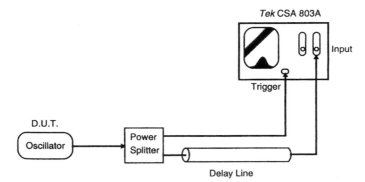

Fig. 15. Timing jitter measurement setup using CSA803A.

distinguished from the device under test (DUT)'s jitter. This extra jitter can be directly measured by looking at the jitter on the triggering edge. This edge can be readily identified since it has lower rms jitter than the transitions before and after it.

The effect of this excess jitter should be subtracted from the jitter due to the DUT. Assuming no correlation between the jitter of the DUT and the sampling head, the equivalent jitter due to the DUT can be estimated by

$$\sigma_{\Delta T, \text{eff}} = \sqrt{\sigma^2_{\Delta T, \text{meas}} - \sigma^2_{\Delta T, \text{min}}} \qquad (39)$$

where $\sigma_{\Delta T, \text{eff}}$ is the effective rms timing jitter, $\sigma_{\Delta T, \text{meas}}$ is the measured rms jitter at a delay ΔT after the triggering edge, and $\sigma_{\Delta T, \text{min}}$ is the jitter on the triggering edge.

Fig. 16 shows the rms jitter versus the measurement delay for oscillator number 12 on a *log–log* plot. The best fit κ for the data shown in Fig. 16 is $\kappa = 6.18 \times 10^{-9}\sqrt{s}$. Equations (12) and (35) result in $\kappa = 5.95 \times 10^{-9}\sqrt{s}$ and $\kappa = 6.07 \times 10^{-9}\sqrt{s}$, respectively. The region of the jitter plot with the slope of one can be attributed to the $1/f$ noise of the devices, as discussed at the end of Section VI.

In a separate experiment, the phase noise of oscillator number 7 is measured for different values of $Nbias$ and

242

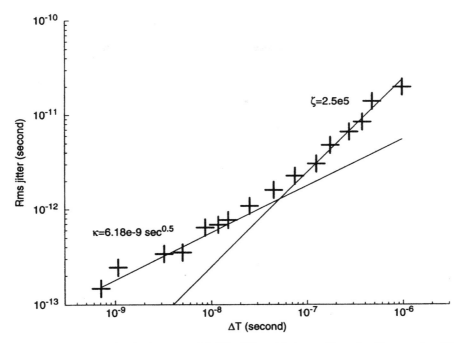

Fig. 16.   RMS jitter versus measurement interval for the four-stage, 2.8-GHz differential ring oscillator (oscillator number 12).

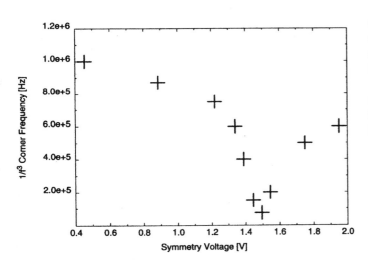

Fig. 17.   Phase noise versus symmetry voltage for oscillator number 7.

PBias. These bias voltages are chosen in such a way as to keep a constant oscillation frequency while changing only the ratio of rise time to fall time. The $1/f^3$ corner of the phase noise is measured for different ratios of the pullup and pulldown currents while keeping the frequency constant. One can observe a sharp reduction in the corner frequency at the point of symmetry in Fig. 17.

IX. CONCLUSION

An analysis of the jitter and phase noise of single-ended and differential ring oscillators was presented. The general noise model, based on the ISF, was applied to the case of ring oscillators, resulting in a closed-form expression for the phase noise and jitter of ring oscillators [(6), (23), (34)]. The model was used to perform a parallel analysis of jitter and phase noise for ring oscillators. The effect of the number of stages

on the phase noise and jitter at a given total power dissipation and frequency of oscillation was shown for single-ended and differential ring oscillators using the general expression for the rms value of the ISF. The upconversion of low-frequency $1/f$ was analyzed showing the effect of waveform asymmetry and the number of stages. New design insights arising from this approach were introduced, and good agreement between theory and measurements was obtained.

APPENDIX A
RELATIONSHIP BETWEEN JITTER AND PHASE NOISE

The phase jitter is

$$\sigma_{\Delta\phi}^2 = E\{\Delta\phi^2\} = E\{[\phi(t + \Delta T) - \phi(t)]^2\} \tag{40}$$

where

$$\Delta\phi = \int_0^{\Delta T} \frac{\Gamma(\omega_0\tau)}{q_{\max}} i(\tau)\, d\tau. \tag{41}$$

Therefore

$$\sigma_{\Delta\phi}^2 = \frac{1}{q_{\max}^2} \int_0^{\Delta T} \int_0^{\Delta T} \Gamma(\omega_0\tau_1)\Gamma(\omega_0\tau_2) \\ \cdot E[i(\tau_1)i(\tau_2)]\, d\tau_1\, d\tau_2. \tag{42}$$

For a white-noise current source, the autocorrelation function is $R_{ii}(t_1, t_2) = (1/2)(\overline{i_n^2}/\Delta f)\delta(t_1 - t_2)$; therefore

$$\sigma_{\Delta\phi}^2 = \frac{1}{2}\frac{\overline{i_n^2}/\Delta f}{q_{\max}^2} \int_0^{\Delta T} \Gamma^2(\omega_0\tau)\, d\tau \tag{43}$$

which is

$$\sigma_{\Delta\phi}^2 = \frac{1}{2}\frac{\overline{i_n^2}/\Delta f}{q_{\max}^2}\Gamma_{\mathrm{rms}}^2\Delta T \quad \text{for} \quad \begin{matrix} \Delta T \gg T \\ \text{or} \\ \Delta T = mT. \end{matrix} \tag{44}$$

Analog and digital designers prefer using phase noise and timing jitter, respectively. The relationship between these two parameters can be obtained by noting that timing jitter is the standard deviation of the timing uncertainty

$$\sigma_{\Delta\phi}^2 = \frac{1}{\omega_0^2} E\{[\phi(t+\Delta T) - \phi(t)]^2\}$$
$$= \frac{E[\phi^2(t)]}{\omega_0^2} + \frac{[\phi^2(t+\Delta T)]}{\omega_0^2} - \frac{E[\phi(t)\phi(t+\Delta T)]}{\omega_0^2} \tag{45}$$

where $E[\cdot]$ represents the expected value. Since the autocorrelation function of $\phi(t)$, $R_\phi(\Delta T)$, is defined as

$$R_\phi(\tau) = E[\phi(t)\phi(t+\Delta T)] \tag{46}$$

the timing jitter in (45) can be written as

$$\sigma_{\Delta\phi}^2 = \frac{2}{\omega_0^2}[R_\phi(0) - R_\phi(\Delta T)]. \tag{47}$$

The relation between the autocorrelation and the power spectrum is given by the Khinchin theorem [21], i.e.,

$$R_\phi(\tau) = \int_{-\infty}^{\infty} S_\phi(f)e^{j2\pi f\tau}\,df \tag{48}$$

where $S_\phi(f)$ represents the power spectrum of $\phi(t)$. Therefore, (47) results in the following relationship between clock jitter and phase noise:

$$\sigma_{\Delta\phi}^2 = \frac{8}{\omega_0^2}\int_0^{\infty} S_\phi(f)\sin^2(\pi f\tau)\,df. \tag{49}$$

It may be useful to know that $S_\phi(f)$ can be approximated by $\mathcal{L}\{\Delta f\}$ for large offsets [22]. As can be seen from the foregoing, the rms timing jitter has less information than the phase-noise spectrum and can be calculated from phase noise using (49). However, unless extra information about the shape of the phase-noise spectrum is known, the inverse is not possible in general.

In the special case where the phase noise is dominated by white noise, $\mathcal{L}\{\Delta f\}$ and κ are given by (6) and (12). Therefore, κ can be expressed in terms of phase noise in the $1/f^2$ region as

$$\kappa = \frac{\Delta f}{f_0} \cdot 10^{-\mathcal{L}\{\Delta f\}/20} \tag{50}$$

where $\mathcal{L}\{\Delta f\}$ is the phase noise measured in the $1/f^2$ region at an offset frequency of Δf and f_0 is the oscillation frequency. Therefore, based on (8), the rms cycle-to-cycle jitter will be given by

$$\sigma_{\text{CTC}} = \frac{f}{f_0^{1.5}} \cdot 10^{-\mathcal{L}\{\Delta f\}/20}. \tag{51}$$

Note that for (50) and (51) to be valid, the phase noise at Δf should be in the $1/f^2$ region.

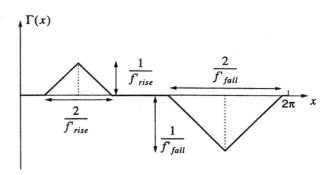

Fig. 18. Approximate waveform and the ISF for asymmetric rising and falling edges.

APPENDIX B
NONSYMMETRIC RISING AND FALLING EDGES

We approximate the ISF in this Appendix by the function depicted in Fig. 18. The rms value of the ISF is

$$\Gamma_{\text{rms}}^2 = \frac{1}{2\pi}\int_0^{2\pi}\Gamma^2(x)\,dx$$
$$= \frac{1}{\pi}\left[\int_0^{1/f'_{\text{rise}}} x^2\,dx + \int_0^{1/f'_{\text{fall}}} x^2\,dx\right]$$
$$= \frac{1}{3\pi}\left(\frac{1}{f'_{\text{rise}}}\right)^3(1+A^3) \tag{52}$$

where f'_{rise} and f'_{fall} are the maximum slope during the rising and falling edge, respectively, and A represents the asymmetry of the waveform and is defined as

$$A \equiv \frac{f'_{\text{rise}}}{f'_{\text{fall}}} \tag{53}$$

noting that

$$2\pi = \eta N\left(\frac{1}{f'_{\text{rise}}} + \frac{1}{f'_{\text{fall}}}\right) = \frac{\eta N}{f'_{\text{rise}}}(1+A). \tag{54}$$

Combining (52) and (54) results in the following:

$$\Gamma_{\text{rms}}^2 = \frac{2\pi^2}{3\eta^3}\frac{1}{N^3}\left[4\frac{1+A^3}{(1+A)^3}\right] \tag{55}$$

which reduces to (16) in the special case of $A = 1$, i.e., symmetric rising and falling edges. The dc value of the ISF, Γ_{dc}, can be calculated from Fig. 18 in a similar manner and is given by

$$\Gamma_{\text{dc}} = \frac{2\pi}{\eta^2}\frac{1}{N^2}\left(\frac{1-A}{1+A}\right). \tag{56}$$

Using (7), the $1/f^3$ corner is given by

$$f_{1/f^3} = f_{1/f} \cdot \frac{3}{2\eta N} \cdot \frac{(1-A)^2}{(1-A+A^2)}. \tag{57}$$

As can be seen for a constant rise-to-fall ratio, the $1/f^3$ corner decreases inversely with the number of stages; therefore, ring oscillators with a smaller number of stages will have a larger $1/f^3$ noise corner. As a special case, if the rise and fall time are symmetric, $A = 1$, and the $1/f^3$ corner approaches zero.

ACKNOWLEDGMENT

The authors would like to thank M. A. Horowitz, G. Nasserbakht, A. Ong, C. K. Yang, B. A. Wooley, and M. Zargari for helpful discussions and support. They would further like to thank Texas Instruments, Inc., and Stanford Nano-Fabrication facilities for fabrication of the oscillators.

REFERENCES

[1] L. DeVito, J. Newton, R. Croughwell, J. Bulzacchelli, and F. Benkley, "A 52 and 155 MHz clock-recovery PLL," in *ISSCC Dig. Tech. Papers*, Feb. 1991, pp. 142–143.

[2] A. W. Buchwald, K. W. Martin, A. K. Oki, and K. W. Kobayashi, "A 6-GHz integrated phase-locked loop using AlCaAs/Ga/As heterojunction bipolar transistors," *IEEE J. Solid-State Circuits*, vol. 27, pp. 1752–1762, Dec. 1992.

[3] B. Lai and R. C. Walker, "A monolithic 622 Mb/s clock extraction data retiming circuit," in *ISSCC Dig. Tech. Papers*, Feb. 1993, pp. 144–144.

[4] R. Farjad-Rad, C. K. Yang, M. Horowitz, and T. H. Lee, "A 0.4 mm CMOS 10 Gb/s 4-PAM pre-emphasis serial link transmitter," in *Symp. VLSI Circuits Dig. Tech Papers*, June 1998, pp. 198–199.

[5] W. D. Llewellyn, M. M. H. Wong, G. W. Tietz, and P. A. Tucci, "A 33 Mbi/s data synchronizing phase-locked loop circuit," in *ISSCC Dig. Tech. Papers*, Feb. 1988, pp. 12–13.

[6] M. Negahban, R. Behrasi, G. Tsang, H. Abouhossein, and G. Bouchaya, "A two-chip CMOS read channel for hard-disk drives," in *ISSCC Dig. Tech. Papers*, Feb. 1993, pp. 216–217.

[7] M. G. Johnson and E. L. Hudson, "A variable delay line PLL for CPU-coprocessor synchronization," *IEEE J. Solid-State Circuits*, vol. 23, pp. 1218–1223, Oct. 1988.

[8] I. A. Young, J. K. Greason, and K. L. Wong, "A PLL clock generator with 5–110 MHz of lock range for microprocessors," *IEEE J. Solid-State Circuits*, vol. 27, pp. 1599–1607, Nov. 1992.

[9] J. Alvarez, H. Sanchez, G. Gerosa, and R. Countryman, "A wide-bandwidth low-voltage PLL for PowerPC™ microprocessors," *IEEE J. Solid-State Circuits*, vol. 30, pp. 383–391, Apr. 1995.

[10] I. A. Young, J. K. Greason, J. E. Smith, and K. L. Wong, "A PLL clock generator with 5–110 MHz lock range for microprocessors," in *ISSCC Dig. Tech. Papers*, Feb. 1992, pp. 50–51.

[11] M. Horowitz, A. Chen, J. Cobrunson, J. Gasbarro, T. Lee, W. Leung, W. Richardson, T. Thrush, and Y. Fujii, "PLL design for a 500 Mb/s interface," in *ISSCC Dig. Tech. Papers*, Feb. 1993, pp. 160–161.

[12] T. C. Weigandt, B. Kim, and P. R. Gray, "Analysis of timing jitter in CMOS ring oscillators," in *Proc. ISCAS*, June 1994.

[13] J. McNeill, "Jitter in ring oscillators," *IEEE J. Solid-State Circuits*, vol. 32, pp. 870–879, June 1997.

[14] B. Razavi, "A study of phase noise in CMOS oscillators," *IEEE J. Solid-State Circuits*, vol. 31, pp. 331–343, Mar. 1996.

[15] A. Hajimiri, S. Limotyrakis, and T. H. Lee, "Phase noise in multigigahertz CMOS ring oscillators," in *Proc. Custom Integrated Circuits Conf.*, May 1998, pp. 49–52.

[16] A. Hajimiri and T. H. Lee, "A general theory of phase noise in electrical oscillators," *IEEE J. Solid-State Circuits*, vol. 33, pp. 179–194, Feb. 1998.

[17] ———, *The Design of Low Noise Oscillators*. Boston, MA: Kluwer Academic, 1999.

[18] A. A. Abidi, "High-frequency noise measurements of FET's with small dimensions," *IEEE Trans. Electron Devices*, vol. ED-33, pp. 1801–1805, Nov. 1986.

[19] T. Kwasniewski, M. Abou-Seido, A. Bouchet, F. Gaussorgues, and J. Zimmerman, "Inductorless oscillator design for personal communications devices—A 1.2 μm CMOS process case study," in *Proc. CICC*, May 1995, pp. 327–330.

[20] J. G. Maneatis and M. A. Horowitz, "Precise delay generation using coupled oscillators," *IEEE J. Solid-State Circuits*, vol. 28, pp. 1273–1282, Dec. 1993.

[21] W. A. Gardner, *Introduction to Random Processes*. New York: McGraw-Hill, 1990.

[22] W. F. Egan, *Frequency Synthesis by Phase Lock*. New York: Wiley, 1981.

Ali Hajimiri received the B.S. degree in electronics engineering from Sharif University of Technology, Tehran, Iran, in 1994 and the M.S. and Ph.D. degrees in electrical engineering from Stanford University, Stanford, CA, in 1996 and 1998, respectively.

He was a Design Engineer with Philips, where he worked on a BiCMOS chipset for GSM cellular units from 1993 to 1994. During the summer of 1995, he was with Sun Microsystems, where he worked on the UltraSparc microprocessor's cache RAM design methodology. During the summer of 1997, he was with Lucent Technologies (Bell Labs), where he investigated low-phase-noise integrated oscillators. In 1998, he joined the Faculty of the California Institute of Technology, Pasadena, as an Assistant Professor. His research interests are high-speed and RF integrated circuits. He is coauthor of *The Design of Low Noise Oscillators* (Boston, MA: Kluwer Academic, 1999).

Dr. Hajimiri was the Bronze Medal Winner of the 21st International Physics Olympiad, Groningen, the Netherlands. He was a corecipient of the International Solid-State Circuits Conference 1998 Jack Kilby Outstanding Paper Award.

Sotirios Limotyrakis was born in Athens, Greece, in 1971. He received the B.S. degree in electrical engineering from the National Technical University of Athens in 1995 and the M.S. degree in electrical engineering from Stanford University, Stanford, CA, in 1997, where he currently is pursuing the Ph.D. degree.

In the summer of 1993, he was with K.D.D. Corp., Saitama R&D Labs, Japan, where he worked on the design of communication protocols. During the summers of 1996 and 1997, he was with the Texas Instruments Inc. R&D Center, Dallas, TX, where he focused on LNA, low-phase-noise oscillator design, and GSM mobile unit transmit path architectures. His current research interests include the design of mixed-signal circuits for high-speed data conversion and broad-band communications.

Mr. Limotyrakis received the W. Burgess Dempster Memorial Fellowship from the School of Engineering, Stanford University, in 1995.

Thomas H. Lee (S'87–M'87), for a photograph and biography, see p. 585 of the May 1999 issue of this JOURNAL.

A Study of Oscillator Jitter Due to Supply and Substrate Noise

Frank Herzel and Behzad Razavi

Abstract—This paper investigates the timing jitter of single-ended and differential CMOS ring oscillators due to supply and substrate noise. We calculate the jitter resulting from supply and substrate noise, show that the concept of frequency modulation can be applied, and derive relationships that express different types of jitter in terms of the sensitivity of the oscillation frequency to the supply or substrate voltage. Using examples based on measured results, we show that thermal jitter is typically negligible compared to supply- and substrate-induced jitter in high-speed digital systems. We also discuss the dependence of the jitter of differential CMOS ring oscillators on transistor gate width, power consumption, and the number of stages.

Index Terms—Jitter, oscillator, phase-locked loops, supply noise.

I. INTRODUCTION

High-speed digital circuits such as microprocessors and memories employ phase locking at the board-chip interface to suppress timing skews between the on-chip clock and the system clock [1]–[3]. Fabricated on the same substrate as the rest of the circuit, the phase-locked loop (PLL) must typically operate from the global supply and ground busses, thus experiencing both substrate and supply noise. The noise manifests itself as jitter at the output of the PLL, primarily through various mechanisms in the voltage-controlled oscillator (VCO). As exemplified by measured results reported in the literature, we show that the contribution of device electronic noise to jitter is typically much less than that due to supply and substrate noise.

This paper describes the effect of supply and substrate noise on the performance of single-ended and differential ring oscillators, providing insights that prove useful in the design of other types of oscillators as well. Section II summarizes the oscillators studied in this work and Section III defines various types of jitter. Sections IV and V quantify the jitter due to thermal noise in the oscillation loop and frequency-modulating noise, respectively. Sections VI and VII apply the developed results to the analysis of supply and substrate noise, and Section VIII presents the dependence of jitter upon parameters such as device size, the number of stages, and power dissipation.

II. RING OSCILLATORS UNDER INVESTIGATION

In this paper, we investigate both single-ended ring oscillators (SERO's) and differential ring oscillators (DRO's). The latter are much more important in digital circuit applications, since DRO's are less affected by supply and substrate noise. The circuit topologies are shown in Fig. 1 for the SERO and in Fig. 2 for the DRO.

Manuscript received October 1, 1997; revised August 2, 1998. This paper was recommended by Associate Editor B. H. Leung.

F. Herzel was with the Electrical Engineering Department, University of California at Los Angeles, Los Angeles, CA 90095, USA, on leave from the Institute for Semiconductor Physics, Frankfurt, Oder, Germany.

B. Razavi is with the Electrical Engineering Department, University of California, Los Angeles, CA 90095 USA.

Publisher Item Identifier S 1057-7130(99)01471-8.

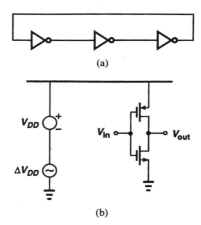

Fig. 1. Single-ended ring oscillator: (a) block diagram and (b) implementation of one stage.

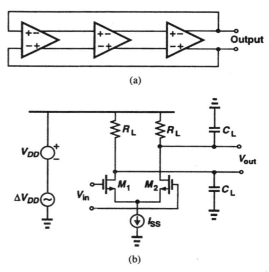

Fig. 2. Differential ring oscillator: (a) block diagram and (b) implementation of one stage.

The simulations were performed with the SPICE parameters of a 0.6-μm CMOS technology. We employed the minimum gate length throughout the paper. Furthermore, unless indicated otherwise, we use the following parameters for the differential stage: $W = 80$ μm, $R_L = 1$ kΩ, $I_{SS} = 1$ mA, $C_L = 0$, $V_{DD} = 3$ V. The rms value of ΔV_{DD} was chosen to be 71 mV, corresponding to a peak amplitude of 100 mV for a sinusoidal perturbation.

III. DEFINITIONS OF JITTER

We consider the output voltage $V_{out}(t)$ of an oscillator in the steady state. The time point of the nth minus-to-plus zero crossing of $V_{out}(t)$ is referred to as t_n. The nth period is then defined as $T_n = t_{n+1} - t_n$. For an ideal oscillator, this time difference is independent of n, but in reality it varies with n as a result of noise in the circuit. This results in a deviation $\Delta T_n = T_n - \bar{T}$ from the mean period $\bar{T}$. The quantity ΔT_n is an indication of jitter.

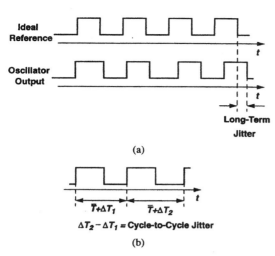

(a)

$\overline{T}+\Delta T_1$ $\overline{T}+\Delta T_2$

$\Delta T_2 - \Delta T_1 =$ Cycle-to-Cycle Jitter

(b)

Fig. 3.   Illustration of (a) long-term jitter and (b) cycle-to-cycle jitter.

More specifically, absolute jitter or long-term jitter

$$\Delta T_{\text{abs}}(N) = \sum_{n=1}^{N} \Delta T_n \qquad (1)$$

is often used to quantify the jitter of phase-locked loops. Modeling the total phase error with respect to an ideal oscillator [Fig. 3(a)], absolute jitter is nonetheless illsuited to describing the performance of *oscillators* because, as shown later, the variance of ΔT_{abs} diverges with time.

A better figure of merit for oscillators is cycle jitter, defined as the rms value of the timing error ΔT_n[1]

$$\Delta T_c = \lim_{N \to \infty} \sqrt{\frac{1}{N} \sum_{n=1}^{N} \Delta T_n^2}. \qquad (2)$$

Cycle jitter describes the magnitude of the period fluctuations, but it contains no information about the dynamics.

The third type of jitter considered here is cycle-to-cycle jitter [Fig. 3(b)] given by

$$\Delta T_{\text{cc}} = \lim_{N \to \infty} \sqrt{\frac{1}{N} \sum_{n=1}^{N} (T_{n+1} - T_n)^2} \qquad (3)$$

representing the rms difference between two consecutive periods.

Note the difference between the cycle jitter and the cycle-to-cycle jitter: the former compares the oscillation period with the *mean* period and the latter compares the period with the *preceding* period. Hence, in contrast to cycle jitter, cycle-to-cycle jitter describes the short-term dynamics of the period. The long-term dynamics, on the other hand, are not characterized by cycle-to-cycle jitter. For example, if $1/f$ noise modulates the frequency slowly, ΔT_{cc} does not reflect the result accurately. With respect to the zero crossings, the cycle-to-cycle jitter is a double-differential quantity in that three zero crossings of the output voltage are related to each other. As discussed in Section V, this results in a completely different dependence on the modulation frequency than for the cycle jitter.

We should note that an oscillator embedded in a phase-locked loop periodically receives correction pulses from the phase detector and charge pump, and hence its long-term jitter strongly depends on the PLL dynamics. Thus, for the analysis of a free-running oscillator, cycle jitter and cycle-to-cycle jitter are more meaningful, particularly

[1] In this paper, we use a time average definition of jitter which is equivalent to the stochastic average if and only if the process ΔT^2 is ergodic.

because the latter type hardly changes when the oscillator is placed in the loop.

A more general quantification of the jitter is possible by means of the steady-state autocorrelation function (ACF) defined as

$$C_{\Delta T}(m) = \lim_{N \to \infty} \frac{1}{N} \sum_{n=1}^{N} (\Delta T_{n+m} \Delta T_n). \qquad (4)$$

To obtain an intuitive understanding of this quantity, we insert (4) with $m = 0$ in (2), obtaining

$$\Delta T_c^2 = C_{\Delta T}(0). \qquad (5)$$

Equation (5) states that the ACF with zero argument is the squared cycle jitter. For a nonzero argument, the ACF decreases with increasing m, finally approaching zero for $m \to \infty$. This indicates that the timing error ΔT_n has a finite memory. In order to express the cycle-to-cycle jitter by the ACF, we rewrite (3) as

$$\Delta T_{\text{cc}}^2 = \lim_{N \to \infty} \frac{1}{N} \sum_{n=1}^{N} (\Delta T_{n+1} - \Delta T_n)^2$$
$$= 2C_{\Delta T}(0) - 2C_{\Delta T}(1). \qquad (6)$$

This expression will be used for an analytical calculation of the cycle-to-cycle jitter in Section V.

IV. JITTER DUE TO DEVICE ELECTRONIC NOISE

The electronic noise of the devices in an oscillator loop leads to phase noise and jitter [5], [7], [8]. Our objective is to express jitter in terms of phase noise and vice versa. These relationships are useful as they relate two measurable quantities.

In this paper, we neglect the effect of $1/f$ noise because it introduces only slow phase variations in the oscillator. Such variations are suppressed by the large loop bandwidth of PLL's used in today's digital systems.

As derived in the Appendix, for white noise sources in the oscillator, the single-sideband phase noise S_ϕ (phase noise with respect to the carrier) can be expressed in terms of the cycle-to-cycle jitter according to

$$S_\phi(\omega) = \frac{(\omega_0^3/4\pi)\Delta T_{\text{cc}}^2}{(\omega - \omega_0)^2 + (\omega_0^3/8\pi)^2 \Delta T_{\text{cc}}^4} \approx \frac{(\omega_0^3/4\pi)\Delta T_{\text{cc}}^2}{(\omega - \omega_0)^2} \qquad (7)$$

where ω_0 is the oscillation frequency and $\omega - \omega_0$ is the offset frequency. The Appendix also shows that the cycle-to-cycle jitter can be deduced from the phase noise according to

$$\Delta T_{\text{cc}}^2 \approx \frac{4\pi}{\omega_0^3} S_\phi(\omega)(\omega - \omega_0)^2. \qquad (8)$$

To obtain an estimate of the thermal jitter, we consider the differential CMOS ring oscillator in [5]. For the 2.2-GHz oscillator with a phase noise of -94 dBc/Hz at 1 MHz offset, we obtain from (8) a thermal cycle-to-cycle jitter of 0.3 ps, i.e., less than $0.3°$. Similar values are obtained for the 900-MHz CMOS ring oscillators reported in [6]. In most timing applications, such small values are negligible with respect to other sources of random jitter.

The thermal absolute jitter is proportional to the square root of the measurement interval Δt. As derived in the Appendix, the absolute jitter is given by

$$\Delta T_{\text{abs}} = \sqrt{\frac{f_0}{2}} \Delta T_{\text{cc}} \sqrt{\Delta t}. \qquad (9)$$

In [9], the rms value of absolute jitter has been divided by the square root of the measurement time to obtain a time-independent figure of merit. This is not possible for supply and substrate noise, since (7)–(9) are derived for white noise in the feedback loop. Supply and substrate noise, however, are generally not white.

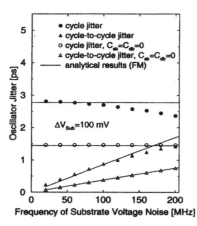

Fig. 9. Cycle jitter and cycle-to-cycle jitter of the DRO versus substrate voltage noise frequency. Solid lines represent the quasi-static FM expressions. The empty symbols show the jitter with the drain-bulk and source-bulk capacitances set to zero.

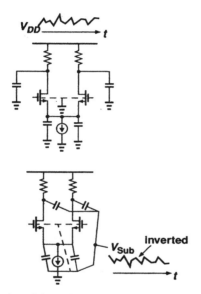

Fig. 10. Illustration of the equivalence of supply and substrate noise for the DRO.

In this section, we study jitter as a function of three parameters: transistor gate width, power dissipation, and the number of stages. To make meaningful comparisons, the circuit is modified in each case such that the frequency of oscillation remains constant. These parameters also affect the thermal jitter to some extent, but, considering the vastly different designs reported in [5] and [6], we note that this type of jitter still remains negligible.

A. Effect of Transistor Gate Width

The differential three-stage ring oscillator of Fig. 2 begins to oscillate for $W \geq 30 \ \mu m$.

Fig. 11 shows the effect of the gate width on the jitter, where the oscillation frequency is kept constant by adjusting C_L in Fig. 2. The jitter reaches a minimum for $W \approx 80 \ \mu m$. For large W, the value of C_L must be reduced so as to maintain the same oscillation frequency, yielding a larger voltage-dependent fraction due to drain and source junctions of each device and hence a higher sensitivity to noise.

B. Effect of Power Consumption

The jitter resulting from device electronic noise generally exhibits an inverse dependence upon the oscillator power dissipation [5], [10].

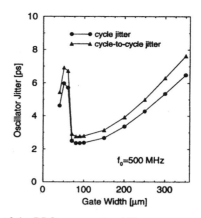

Fig. 11. Jitter of the DRO versus gate width.

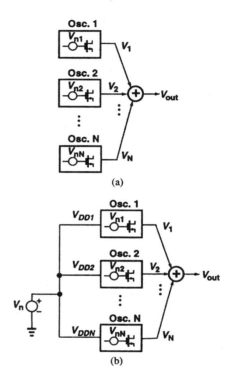

(a)

(b)

Fig. 12. Illustration of the relationship between power consumption and noise for (a) device electronic noise and (b) supply noise.

By contrast, the effect of supply and substrate noise on the jitter of a given oscillator topology is relatively independent of the power drain. This can be understood with the aid of the conceptual illustrations in Fig. 12, where the output voltages of N identical oscillators are added in phase. In Fig. 12(a), only the device electronic noise is considered [5]. Since the noise in each oscillator is uncorrelated, the output noise voltage is $\sqrt{N}$ times that of each oscillator, whereas the output signal voltage is $N \times V_j$. In Fig. 12(b), on the other hand, all oscillators are disturbed by the same noise source, thus exhibiting completely correlated noise. That is, both the noise voltage and the signal voltage are increased by a factor of N.

To confirm the above observation, the gate width and tail current were decreased while the load resistance was increased proportionally. Table I shows that the jitter is quite constant.

C. Effect of Number of Stages

In applications where the required oscillation frequency is considerably lower than the maximum speed of the technology, a ring

248

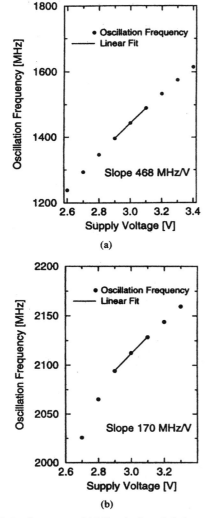

(a)

(b)

Fig. 6. Oscillation frequency of (a) the single-ended ring oscillator and (b) the differential ring oscillator as a function of static supply voltage.

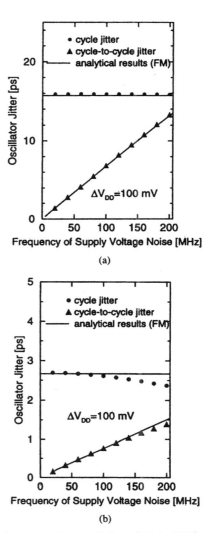

(a)

(b)

Fig. 7. Cycle jitter and cycle-to-cycle jitter of (a) the SERO and (b) the DRO as a function of supply voltage noise frequency. The solid lines represent the quasi-static FM expressions.

In the following, we investigate the jitter due to sinusoidal supply voltage perturbations. The calculation of the jitter consists of the following steps: interpolation of the voltage waveform to find the zero crossings; calculation of the periods T_n and subtraction of the mean period $\bar{T}$ to obtain ΔT_n; and calculation of the cycle jitter (2) and the cycle-to-cycle jitter (3) by performing time averaging.

We should also mention that simulations indicate that jitter has a relatively linear dependence on the noise amplitude for supply variations as large as a few hundred millivolts.

Fig. 7 plots the analytical and simulated cycle and cycle-to-cycle jitter of single-ended and differential ring oscillators. As can be seen, the analytical results of Section V predict the jitter with reasonable accuracy.

VII. JITTER DUE TO SUBSTRATE NOISE

Substrate noise can be treated in the same fashion as supply noise. For the numerical simulation of substrate noise, the bulk terminal of the transistors is driven by a noise source (Fig. 8). Fig. 9 shows the calculated jitter of the DRO as a function of the noise frequency. Comparison with Fig. 7 indicates that for the DRO, a supply voltage perturbation is almost equivalent to a substrate voltage perturbation of opposite sign. To understand this, note from Fig. 10 that, with an ideal tail current source, a change of ΔV in V_{DD} is equivalent to a change

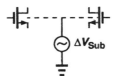

Fig. 8. Substrate noise modeling.

of $-\Delta V$ in V_{sub}. Simulations confirm that the static sensitivity K_0 is indeed equal for supply and substrate noise, apart from the sign. Fig. 9 also demonstrates that the quasi-static FM approach is suited to describing the jitter introduced by substrate noise. Furthermore, it suggests that a substantial fraction of the jitter results from the voltage dependence of C_{db} and C_{sb}.

VIII. OSCILLATOR DESIGN FOR LOW JITTER

The simulation results presented thus far indicate the superior performance of differential oscillators with respect to single-ended topologies. Nonetheless, even differential configurations have a wide design space; device size, voltage swings, power dissipation, and the number of stages in a ring oscillator influence the overall sensitivity to supply and substrate noise.

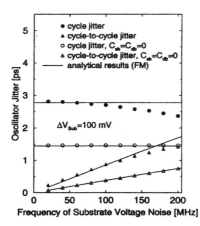

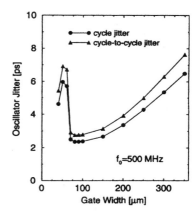

Fig. 9. Cycle jitter and cycle-to-cycle jitter of the DRO versus substrate voltage noise frequency. Solid lines represent the quasi-static FM expressions. The empty symbols show the jitter with the drain-bulk and source-bulk capacitances set to zero.

Fig. 11. Jitter of the DRO versus gate width.

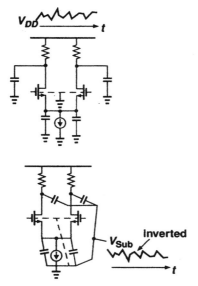

Fig. 10. Illustration of the equivalence of supply and substrate noise for the DRO.

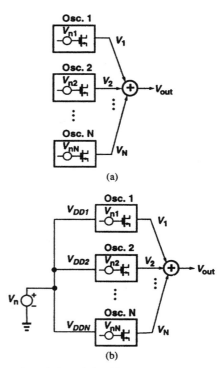

Fig. 12. Illustration of the relationship between power consumption and noise for (a) device electronic noise and (b) supply noise.

In this section, we study jitter as a function of three parameters: transistor gate width, power dissipation, and the number of stages. To make meaningful comparisons, the circuit is modified in each case such that the frequency of oscillation remains constant. These parameters also affect the thermal jitter to some extent, but, considering the vastly different designs reported in [5] and [6], we note that this type of jitter still remains negligible.

A. Effect of Transistor Gate Width

The differential three-stage ring oscillator of Fig. 2 begins to oscillate for $W \geq 30 \ \mu m$.

Fig. 11 shows the effect of the gate width on the jitter, where the oscillation frequency is kept constant by adjusting C_L in Fig. 2. The jitter reaches a minimum for $W \approx 80 \ \mu m$. For large W, the value of C_L must be reduced so as to maintain the same oscillation frequency, yielding a larger voltage-dependent fraction due to drain and source junctions of each device and hence a higher sensitivity to noise.

B. Effect of Power Consumption

The jitter resulting from device electronic noise generally exhibits an inverse dependence upon the oscillator power dissipation [5], [10].

By contrast, the effect of supply and substrate noise on the jitter of a given oscillator topology is relatively independent of the power drain. This can be understood with the aid of the conceptual illustrations in Fig. 12, where the output voltages of N identical oscillators are added in phase. In Fig. 12(a), only the device electronic noise is considered [5]. Since the noise in each oscillator is uncorrelated, the output noise voltage is $\sqrt{N}$ times that of each oscillator, whereas the output signal voltage is $N \times V_j$. In Fig. 12(b), on the other hand, all oscillators are disturbed by the same noise source, thus exhibiting completely correlated noise. That is, both the noise voltage and the signal voltage are increased by a factor of N.

To confirm the above observation, the gate width and tail current were decreased while the load resistance was increased proportionally. Table I shows that the jitter is quite constant.

C. Effect of Number of Stages

In applications where the required oscillation frequency is considerably lower than the maximum speed of the technology, a ring

TABLE I
IMPACT OF POWER CONSUMPTION

I_{ss} [mA]	R_L	W [μm]	ΔT_c [ps]	ΔT_{cc} [ps]
4	250	320	2.51	1.49
2	500	160	2.51	1.50
1.33	750	106	2.49	1.49
1	1000	80	2.49	1.48
0.8	1250	64	2.49	1.51
0.67	1500	53	2.50	1.49

TABLE II
THREE-STAGE VERSUS SIX-STAGE OSCILLATOR

n	W [μm]	C_L [fF]	R_L [Ω]	ΔT_c [ps]	ΔT_{cc} [ps]
3	40	535	1000	4.62	5.44
3	60	540	1000	5.71	6.72
3	80	535	1000	2.35	2.74
3	100	515	1000	2.38	2.80
3	150	420	1100	2.65	3.12
3	200	320	1200	3.37	3.97
3	250	220	1300	4.43	5.21
3	358	0	1500	7.58	8.76
6	40	265	550	8.09	9.48
6	60	260	530	3.96	4.59
6	80	235	530	3.85	4.52
6	100	200	530	4.25	5.00
6	150	110	530	5.33	6.27
6	207	0	530	6.78	7.98

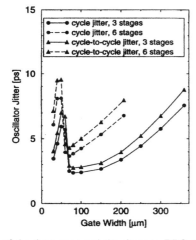

Fig. 13. Jitter of the three-stage and the six-stage DRO versus gate width for an oscillation frequency of 500 MHz.

oscillator may incorporate more than three stages. Thus, the optimum number of stages with respect to the jitter is of interest.

Shown in Table II and plotted in Fig. 13 is the jitter of three-stage and six-stage oscillators designed for a frequency of 500 MHz with constant tail current and voltage swings. We note that the minimum values of cycle jitter and cycle-to-cycle jitter are smaller in a three-stage topology. This is because for the three-stage oscillator, the reduction of the oscillation frequency to the desired value is obtained by means of the fixed capacitances C_L rather than by the voltage-dependent capacitances of the transistors. Hence, a smaller fraction of the total load capacitance is subject to variations with supply and substrate noise.

The addition of a fixed capacitor to each stage nonetheless entails the issue of substrate noise coupling to the bottom plate of the capacitor. In order to minimize this effect, a grounded shield must isolate the capacitor from the substrate, as illustrated in Fig. 14. Here,

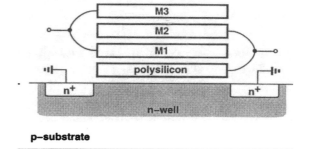

Fig. 14. Grounded shield used under the capacitor to block substrate noise.

an n-well, grounded by a low-resistance n^+ ring, is placed under the capacitor so as to block the noise produced in the substrate.

IX. CONCLUSION

We have investigated the timing jitter in oscillators subject to supply and substrate noise. For digital timing applications, the effect of supply and substrate noise on the jitter is typically much more pronounced than that of thermal noise. For supply and substrate noise, we have derived analytical relationships between the cycle-to-cycle jitter and the low-frequency sensitivity of the oscillation frequency to supply or substrate noise. These relationships have been verified by means of numerical calculations for single-ended and differential CMOS ring oscillators. For differential ring oscillators, we have investigated the dependence of the jitter on the transistor gate width, power consumption, and the number of stages. As a special result, we have found that in applications where the required oscillation frequency is lower than the maximum speed of the technology, a three-stage ring oscillator with additional load capacitances gives the lowest jitter.

APPENDIX
JITTER AND PHASE NOISE DUE TO THERMAL AND SHOT NOISE

The output voltage of an oscillator can be written as

$$V(t) = V_0 \cos[\omega_0 t + \varphi(t)] \tag{18}$$

where V_0 is the amplitude, ω_0 is the oscillation frequency, and $\phi(t)$ is the slowly varying excess phase. The excess frequency is

$$\Delta\omega(t) = \frac{d}{dt}\phi(t) \tag{19}$$

and hence

$$\phi(t) = \int_0^t \Delta\omega(u)\,du + \phi(0). \tag{20}$$

Thermal and shot noise may be considered as white noise since their cutoff frequencies are typically much higher than the oscillation frequency. White noise in the feedback loop of the oscillator results in phase diffusion, a phenomenon described by a Wiener process [11]. Extensive investigations of phase noise indicate that white noise sources in all types of oscillators give rise to a phase noise power spectrum proportional to $1/(\Delta\omega)^2$, where $\Delta\omega$ is the offset frequency with respect to the carrier frequency [4], [5]. This trend is valid for offset frequencies as high as several percent of the carrier frequency. Thus, *frequency* noise, $\Delta\omega(t)$, can be assumed white in such a band. The autocorrelation of $\Delta\omega(t)$ is given by

$$\overline{\Delta\omega(t+\tau)\Delta\omega(t)} = 2D_o\delta(\tau) \tag{21}$$

where D_ϕ is the diffusivity and $\delta(\tau)$ the Delta function. The probability density of $\phi(t)$ represents a Gaussian distribution centered at $\phi(0)$ with the variance

$$\sigma_\phi^2 = 2D_\phi t. \tag{22}$$

As evident from (22), the variance diverges with time. The autocorrelation of $V(t)$ is known [12] and reads

$$\langle V(t+\tau)V(t)\rangle = \frac{V_0^2}{2}\exp(-D_\phi|\tau|)\cos(\omega_0\tau). \tag{23}$$

Performing the Fourier transformation, we obtain the one-sided power spectral density

$$S_V(\omega) = V_0^2 \frac{D_\phi}{(\omega-\omega_0)^2 + D_\phi^2}. \tag{24}$$

This quantity is often normalized to $V_0^2/2$ and referred to as relative phase noise with respect to the carrier [5] or as single-sideband phase noise [8], given by

$$S_\phi(\omega) = \frac{2D_\phi}{(\omega-\omega_0)^2 + D_\phi^2}. \tag{25}$$

For $\omega - \omega_0 \gg D_\phi$, we obtain from (25)

$$S_\phi(\omega) \approx \frac{2D_\phi}{(\omega-\omega_0)^2}. \tag{26}$$

Next, we will relate the cycle-to-cycle jitter and the single-sideband phase noise to each other. Note that the stationary Wiener process has no memory and the increments in different time intervals are statistically independent [11]. Therefore, the rms mean increment of the excess phase $\phi(t)$ within one cycle, i.e., the cycle jitter of the phase, equals the increment of $\phi(t)$ between $t=0$ and $t=\bar{T}$. Thus, from (22), we obtain the phase cycle jitter as

$$\Delta\phi_c = \sqrt{2D_\phi\bar{T}}. \tag{27}$$

The excess phase change during the nth cycle is referred to as $\Delta\phi_n$. The nth oscillation period is defined by the relation

$$2\pi f_0 T_n = 2\pi + \Delta\phi_n. \tag{28}$$

For the deviation of the nth period T_n from the mean period $\bar{T} = 1/f_0$, we then find

$$\Delta T_n = \frac{\Delta\phi_n}{2\pi f_0} = \Delta\phi_n \frac{\bar{T}}{2\pi}. \tag{29}$$

Hence, the cycle jitter ΔT_c of the period during one cycle is related to $\Delta\phi_c$ according to

$$\Delta T_c = \Delta\phi_c \frac{\bar{T}}{2\pi}. \tag{30}$$

For white noise sources, two successive periods are uncorrelated. Since cycle-to-cycle jitter represents the difference between two periods, the variance of cycle-to-cycle jitter is twice as large as the variance of one period, yielding

$$\Delta T_{cc} = \sqrt{2}\Delta T_c. \tag{31}$$

Combining (27), (30), and (31), we obtain

$$\Delta T_{cc}^2 = \frac{8\pi}{\omega_0^3}D_\phi \tag{32}$$

with

$$\omega_0 = \frac{2\pi}{\bar{T}}. \tag{33}$$

The cycle-to-cycle jitter can now be expressed in terms of the single-sideband phase noise by inserting (32) in (26) to give

$$\Delta T_{cc}^2 \approx \frac{4\pi}{\omega_0^3}S_\phi(\omega)(\omega-\omega_0)^2. \tag{34}$$

On the other hand, the phase noise can be expressed by the cycle-to-cycle jitter by inserting (32) in (25), yielding

$$S_\phi = \frac{(\omega_0^3/4\pi)\Delta T_{cc}^2}{(\omega-\omega_0)^2 + (\omega_0^3/8\pi)^2 \Delta T_{cc}^4}. \tag{35}$$

A similar expression has been derived for ring oscillators in [10] and reads in our notation

$$S_\phi = \frac{f_0^3 \Delta T_c^2}{(f-f_0)^2} \tag{36}$$

where $f_0 = \omega_0/2\pi$. Equation (36) turns out to be a special case of (35) for $\omega - \omega_0 \gg D_\phi$.

The absolute jitter increases proportionally to the square root of the measurement interval Δt as evident from (22). Hence, the absolute phase jitter is

$$\Delta\phi_{abs} = \sqrt{2D_\phi\Delta t} = \kappa\sqrt{\Delta t}. \tag{37}$$

Using (32), the proportionality constant κ can be related to the cycle-to-cycle jitter according to

$$\kappa = \sqrt{2D_\phi} = \sqrt{2}\pi f_0^{3/2}\Delta T_{cc}. \tag{38}$$

REFERENCES

[1] I. A. Young, J. K. Greason, and K. L. Wong, "A PLL clock generator with 5 to 110 MHz of lock range for microprocessors," *IEEE J. Solid-State Circuits*, vol. 27, pp. 1599–1607, Nov. 1992.

[2] J. Alvarez, H. Sanchez, G. Gerosa, and R. Countryman, "A wide-bandwidth low-voltage PLL for powerPC™ microprocessors," *IEEE J. Solid-State Circuits*, vol. 30, pp. 383–391, Apr. 1995.

[3] R. Bhagwan and A. Rogers, "A 1 GHz dual-loop microprocessor PLL with instant frequency shifting," in *IEEE Proc. ISSCC*, San Francisco, CA, Feb. 1997, pp. 336–337.

[4] D. B. Leeson, "A simple model of feedback oscillator noise spectrum," in *Proc. IEEE*, pp. 329–330, Feb. 1966.

[5] B. Razavi, "A study of phase noise in CMOS oscillators," *IEEE J. Solid-State Circuits*, vol. 31, pp. 331–343, Mar. 1996.

[6] T. Kwasniewski *et al.*, "Inductorless oscillator design for personal communications devices—A 1.2 μm CMOS process case study," in *Proc. CICC*, May 1995, pp. 327–330.

[7] F. X. Kärtner, "Analysis of white and $f^{-\alpha}$ noise in oscillators," *Int. J. Circuits Theory, Appl.*, vol. 18, pp. 485–519, Sept. 1990.

[8] W. Anzill and P. Russer, "A general method to simulate noise in oscillators based on frequency domain techniques," *IEEE Trans. Microwave Theory Tech.*, vol. 41, pp. 2256–2263, Dec. 1993.

[9] J. A. McNeill, "Jitter in ring oscillators," *IEEE J. Solid-State Circuits*, vol. 32, pp. 870–879, June 1997.

[10] T. C. Weigandt, B. Kim, and P. R. Gray, "Analysis of timing jitter in CMOS ring oscillators," in *Proc. IEEE Int. Symp. Circuits and Systems (ISCAS'94)*, London, U.K., June 1994, vol. 4, pp. 27–30.

[11] C. W. Gardiner, *Handbook of Stochastic Methods.* Berlin: Springer-Verlag, 1983.

[12] R. L. Stratonovich, *Topics in the Theory of Random Noise.* New York: Gordon and Breach, 1967.

Measurements and Analysis of PLL Jitter Caused by Digital Switching Noise

Patrik Larsson

Abstract—When integrating analog and digital circuits onto a mixed-mode chip, power supply noise coupling is a major limitation on the performance of the analog circuitry. Several techniques exist for reducing the noise coupling, of which one of the cheapest is separating the power supply distribution networks for the analog and digital circuits. Noise coupling from a digital noise-generating circuit through the power supply/substrate into an analog phase-locked loop (PLL) is analyzed for three different power supply schemes. The main mechanisms for noise coupling are identified by comparing different PLLs and varying their bandwidths. It is found that the main cause of jitter strongly depends on the power supply configuration of the PLL. Measurements were done on mixed-mode designs in a standard 0.25-μm digital CMOS process with a low-resistivity substrate. The same circuits were also implemented with triple-well processing for comparisons.

Index Terms—Jitter, phase-locked loop, supply noise.

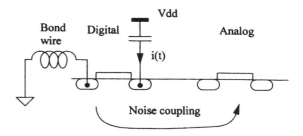

Fig. 1. Noise generated by digital logic couples through the substrate to an analog circuit.

I. INTRODUCTION

THE continuing scaling of CMOS process technologies enables higher degree of integration, which in the last few years has strongly contributed to rapid growth in mixed-mode designs. Integrating analog building blocks, such as AD/DA converters, oscillators, and phase-locked loops (PLLs), on large-scale digital chips enables single-chip solutions which reduces system cost. However, when combining an analog chip and a digital chip into one mixed-mode design, a particular area of concern is on-chip noise coupling from the digital to the analog circuitry which do not exist in either of the two original chips.

The principle of substrate noise coupling is shown in Fig. 1. An MOS transistor in the digital section of the chip turns on and discharges a capacitive load generating a brief current pulse in the V_{ss} network. The current pulse is forced through the inductive bonding wire in the V_{ss} path and generates a voltage bounce on V_{ss}. This noise couples through the resistive substrate or through a shared supply network into the analog section of the chip. The amount of noise reaching the analog circuitry is proportional to the inductance and the amplitude of the current spike, but inversely proportional to the impedance of the connecting path. Here it is assumed that noise coupling due to resistance/inductance/capacitance of the on-chip power distribution network can be avoided by proper layout.

There are several techniques attempting to reduce the noise injected into the analog circuitry (see refs. in [1]–[3]). Each of the techniques focuses on one of the three components above

Manuscript received November 28, 2000; revised April 14, 2001.
The author was with Bell Laboratories, Holmdel, NJ 07733 USA. He is now at Union, NJ 07083 USA.
Publisher Item Identifier S 0018-9200(01)05789-4.

(inductance of power supply path, amplitude of current pulse, impedance of connecting path), but often several techniques are combined to achieve a better noise reduction. From a designer's perspective, the easiest approach is to select a low-inductance package or a package with high pin-count such that many pins can be dedicated for V_{dd}/V_{ss}. For smaller circuits, this might be a viable approach, but it is prohibitively costly for large circuits and the technique is not suitable in future scaled designs [2]. The current spikes generated by standard CMOS logic can be greatly reduced using other logic styles. However, low-noise logic families are based on current steering methods that has constant power supply current with a power consumption penalty. Noise coupling can also be reduced by increasing the impedance between the digital and analog circuitry. This can be done by separating the on-chip power distribution networks for the analog and digital circuits. At additional cost, this method can be enhanced by nonstandard processing technology such as SOI, triple-well [4]–[6], or high-resistivity substrate.

The following material analyzes the efficiency of using separate power supply distribution networks in an attempt to reduce noise coupling into a phase-locked loop (PLL), which is the most common analog block in mixed-mode designs of today. The main coupling mechanism is identified for three different supply schemes. Measurements from 0.25-μm CMOS circuits fabricated in a standard digital CMOS process with low-resistivity substrate are compared to results from circuits processed in a triple-well technology.

II. CIRCUIT CONFIGURATION

The three different power supply distribution schemes in Fig. 2 will be studied in the following. In the first case, the digital circuitry and the analog PLL share both V_{dd} and V_{ss}. Since the switching noise appears on both V_{dd} and V_{ss} with opposite phase [3], we can expect large noise coupling resulting in large PLL jitter. In the second case, a separate V_{dd} is used for the analog PLL. Intuitively, we would expect a noise reduction

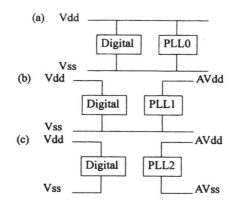

Fig. 2. Three power supply schemes under investigation. (a) PLL0: common V_{dd} and V_{ss}, (b) PLL1: separate analog V_{dd}, (c) PLL2: separate analog V_{dd} and V_{ss}.

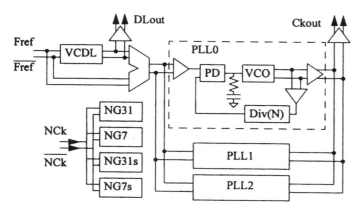

Fig. 3. Chip with noise generators and PLLs. Power supply distribution not shown for clarity.

TABLE I
POWER SUPPLY DISTRIBUTION

	Vdd	Vss	Dec. Cap.
NG7, NG31	Vdd0	Vss0	260pF
NG7s, NG31s	Vdds	Vsss	260pF
PLL0	Vdd0	Vss0	40pF
PLL1	Vdd1	Vss0	40pF
PLL2	Vdd2	Vss2	40pF

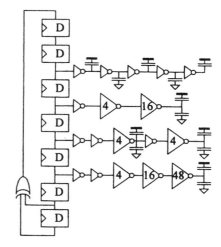

Fig. 4. Noise generator NG7 with various internal loads to emulate random logic.

by a factor of two, since half of the noise (the noise coupled through V_{dd}) is eliminated. With large amount of decoupling capacitance between V_{ss} and AV_{dd}, the noise would turn into common-mode noise that cannot disturb the PLL, ideally giving no jitter. However, even with ideal capacitive AV_{dd}-V_{ss} coupling, resonance effects where AV_{dd} and V_{ss} resonates at opposite phase will turn the common-mode noise into differential-mode noise [3]. In the third case, the PLL is supplied with both separate V_{dd} and V_{ss}. With infinite-impedance substrate, this would completely eliminate any noise coupling, but in the case of standard low-resistivity substrate, there is still some coupling. In all three cases, the local V_{ss} is used for substrate contacts. The three power supply schemes are compared in a single chip containing several PLLs and noise generators (NG) as shown in the chip block diagram in Fig. 3. Table I summarizes the power supply distribution schemes and the amount of decoupling capacitance used in each circuit.

Digital noise-generating circuits are realized by pseudo-random bit sequence generators NG7 and NG31, which are 7 and 31 bit shiftregisters with feedback XOR functions, respectively. They are clocked by the differential external NCk that is locally turned into a single-ended clock. Each noise generator can be individually enabled by external control and has various internal loads. To achieve realistic power spectrum densities of the generated noise, the switching of random logic was emulated by different size buffers having different logic

depth and different capacitive loads as shown for NG7 in Fig. 4. The measured spectral density of the generated supply noise was very similar to the spectral density of a random NRZ bit sequence without any strong tones. Due to the limited repetition length of NG7, its spectrum consisted of 127 discrete tones as expected, but the tones of the NG31 were so closely spaced that they could not be distinguished on a spectrum analyzer. The large internal capacitive loads of the noise generators can optionally be driven directly by NCk, emulating a heavily pipelined circuit generating strong tones in the power spectrum at harmonics of the clock rate. Power consumption measurements indicated that NG7 and NG31 had a load capacitance of 100 pF and 280 pF, respectively. NG7s/NG31s in Fig. 3 have separate supplies as indicated in Table I. A single bond wire with approximately 4–6 nH inductance on a 44-pin TQFP package was reserved for each supply listed in Table I.

All PLLs are identical except for their power supply schemes and each contains a differential VCO with wide tuning range, a fully programmable divider and a phase detector (PD) with low logic depth similar to those in [7]. Their analog components are built according to the substrate referencing technique [8]. The reference clock, F_{ref} in Fig. 3, is distributed differentially and locally turned into a single-ended signal before driving the phase detector. The reference clock conditionally goes through an on-chip voltage controlled delay line (VCDL), whose output is available externally. The VCDL is used when setting the PLL bandwidth. To achieve a desired bandwidth, a sine wave with the same frequency as the desired bandwidth is applied to the delay-modulating input of the VCDL and the jitter at the VCDL output is measured. The PLL charge-pump current is tuned until the peak-to-peak PLL output jitter is half of the VCDL output jitter. At this operating point, the PLL 3 dB-bandwidth is equal to the sine wave frequency. After the bandwidth has been set, the VCDL is bypassed with the multiplexer in Fig. 3 and the reference clock directly feeds the PLLs, such that the VCDL does not contribute to the measured jitter.

All the following jitter measurements are based on the setup in Fig. 5 [9] that measures the tracking jitter with respect to the reference clock. The jitter of the RF source itself was measured as 3 ps RMS, which is negligible compared to the following measurement results.

The measured jitter at 4-MHz PLL bandwidth is plotted in Fig. 6, which shows the RMS jitter as function of the NCk frequency when NG31 was enabled. In this case, the reference clock was 100 MHz and the division ratio in the PLLs was set to 8, such that the VCOs were running at 800 MHz. These numbers are also used in all measurements presented below unless otherwise stated. Note that some of the markers in Fig. 6 are shown off the lines as will be explained later. The following sections will determine the main source of noise coupling by analyzing the jitter of each PLL in detail.

III. POSSIBLE SOURCES OF JITTER

For the chip in Fig. 3, there are several possibilities of noise injection into the PLLs. These possible jitter sources are shown in the PLL model in Fig. 7 and are described as follows.

- F_{ref} input: The reference input is routed as a differential signal to a differential-to-single-ended converter. Due to differences in inductive coupling in the bonding wires and capacitive coupling in the package and layout, the differential routing is not ideal and can exhibit differential mode disturbance. The local differential pair in each PLL is also a source of noise coupling, since mismatch causes non-ideal power supply rejection. The noise generated by the reference signal source is also a part of this term.
- PD: Power supply noise can cause delay variations (=phase noise) in the phase detector.
- Filter: Noise coupled into the filter that stores the VCO control voltage causes FM modulation that gives jitter at the VCO output.
- VCO: Power supply variations of the VCO changes the delay of the delay stages inside the VCO.
- Divider: Similar to the PD; power supply noise on the divider causes delay variations.
- Buffer: Same mechanism as for the PD and divider.

From the chip schematic in Fig. 3, we can see that the output buffer noise is common for all PLLs. Since PLL1 has significantly less jitter than the other PLLs, the main source of jitter in PLL0 and PLL2 cannot be the output buffer. Assuming that the noise coupling due to mismatch in the F_{ref} differential converter at the PLL input does not vary drastically between the PLLs, the same argument as for the output buffer also holds for the F_{ref} input. This was also verified by measuring several chips, that can be expected to have different mismatch.

More detailed jitter measurements for PLL1 are presented in Fig. 8. The two upper curves were obtained when the PLL bandwidth was 1 MHz, while the two lower curves are for a bandwidth of 4 MHz. Increasing the bandwidth above 4 MHz

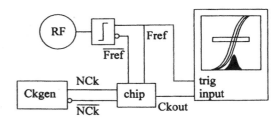

Fig. 5. Test setup for measuring tracking jitter in the presence of on-chip switching noise.

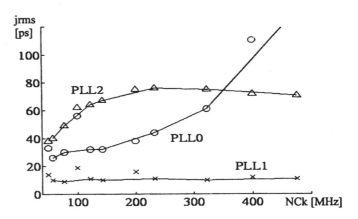

Fig. 6. Jitter induced by NG31 into PLLs having a bandwidth of 4 MHz.

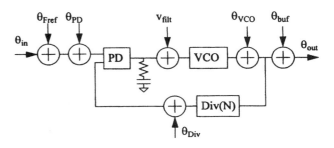

Fig. 7. Possible jitter sources of a PLL.

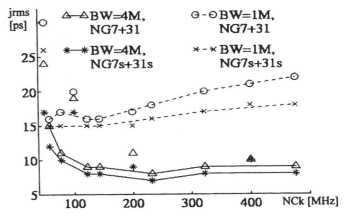

Fig. 8. Jitter of PLL1 when NG7 and NG31 are active and have a strong clock component.

further reduced the jitter. Again, the output buffer noise can be discarded as the major source of noise, since it would be independent of the PLL bandwidth. When changing the amplitude of the noise-generating input clock, NCk, the jitter changed by less

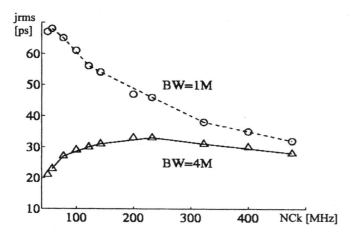

Fig. 9. Jitter induced by noise generator NG7 into PLL2 with a bandwidth of 1 and 4 MHz.

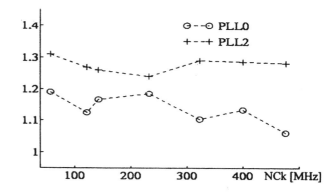

Fig. 10. The ratio of the jitter caused by NG7 at 600 and 800 MHz with a PLL bandwidth of 1 MHz.

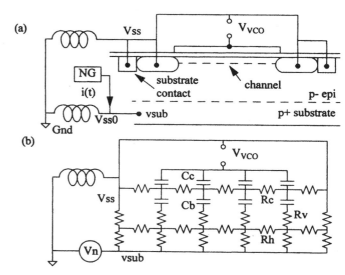

Fig. 11. (a) Physical and (b) electrical model of MOS device used as loop filter capacitor. V_{ss}: local V_{ss}. G_{nd}: external g_{nd}, v_{sub}: substrate noise, v_{VCO}: filter voltage fed to VCO, R_c: channel resistance, R_h: horizontal epi resistance, R_v: vertical epi resistance, C_c: gate-to-channel capacitance, C_b: channel-to-bulk capacitance.

than 1 ps RMS indicating that the capacitive/inductive portion of the F_{ref} noise source is negligible. Based on this reasoning, it will be assumed in the following discussion that neither F_{ref} or the output buffer is the main source of jitter.

When the noise generators were disabled the jitter was 12–14 and 5–6 ps for 1 and 4 MHz bandwidth, respectively. This indicates the lower limit when supply noise is absent. The curves in Fig. 8 are close to these numbers proving that PLL1 has good noise performance. The data for the NG7s and NG31s are slightly lower than NG7/NG31 indicating that some improvement can still be obtained using separate supplies for the digital section even though a low-resistivity substrate is used.

IV. SEPARATE V_{DD} AND V_{SS} (PLL2)

The first approach to distinguish between the possible jitter sources is to change the PLL bandwidth. The noise transfer function from the PD and the divider has a lowpass filter shape such that the resulting output jitter will be a lowpass filtered version of the divider/PD phase noise. A PLL whose jitter is dominated by the PD/divider noise would therefore exhibit less jitter when reducing its bandwidth. Fig. 9 shows that the jitter for PLL2 is reduced for a higher bandwidth, indicating that the main source of jitter is noise injection either into the loop filter or the VCO which has a highpass noise transfer function.

To distinguish between filter noise and VCO noise, we look into an ideal model of a VCO. The output phase of a VCO is

$$\theta = \int \omega \, dt = K_{vco} \int v_{in}(t) \, dt \qquad (1)$$

where v_{in} is the VCO control voltage and K_{vco} is the voltage-to-frequency transfer constant of the VCO. From simulation of the oscillator, K_{vco} was extracted to approximately 1 GHz/V and as later explained in connection with Fig. 12, a realistic estimate of the VCO input noise is a 20-mV pulse with a 1 ns rise/falltime and a duration of 1 ns. This gives a phase advancement of

$$\Delta\theta = 1e^9 \cdot 2\pi \cdot 20e^{-3} \cdot \left(1 + \frac{1+1}{2}\right) e^{-9} = 0.25 \text{ rad} \qquad (2)$$

which translates into 50 ps at a VCO frequency of 800 MHz. Note that the same noise pulse would give a phase error of

67 ps (33% more) when the VCO runs at 600 MHz. With this in mind, the jitter of PLL2 was measured at a VCO frequency of 600 MHz by changing F_{ref} from 100 to 75 MHz, keeping the PLL bandwidth and damping constant. The result is plotted in Fig. 10 which indicates that PLL2 fits well with the hypothesis of noise injection into the loop filter.

The large difference in jitter performance between PLL1 and PLL2 in Fig. 6 indicates that there is a significant difference between their loop filters. The filters are realized by the same components and the difference appears only when modeling parasitic substrate resistances as in Fig. 11, which shows the filter capacitor implemented with an NMOS transistor. The noise generator (NG) produces a current $i(t)$ that is transferred into a voltage noise (V_n) by the impedance of the inductive bond wire connecting V_{ss0} to the external G_{nd}. Global V_{ss} (V_{ss0}) and the node below the epi layer underneath the MOS device (v_{sub}) can be assumed shorted since many substrate contacts are used inside the NG blocks and the substrate has low resistivity in this particular process. Since PLL1 uses global V_{ss}, the nodes V_{ss} and v_{sub} in Fig. 11(a) are shorted. Thus, any noise appearing in the substrate will directly couple to each node of the filter model

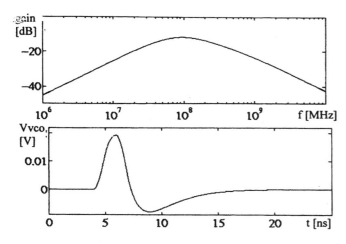

Fig. 12. (a) AC and (b) transient response to substrate noise of filter capacitor.

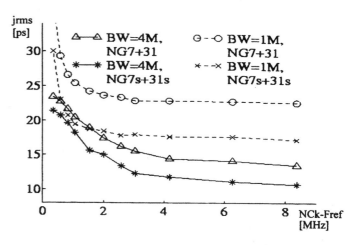

Fig. 13. Jitter of PLL1 as function of the frequency offset between NCk and F_{ref}.

to the right in Fig. 11(b) such that the VCO will receive a constant v_{VCO} avoiding any jitter. However, the filter of PLL2 does not have the nodes V_{ss} and v_{sub} shorted such that noise injected into the substrate node of PLL2 will cause the gate of the NMOS device to fluctuate differently from V_{ss}. Therefore, v_{VCO} varies with time and modulates the VCO frequency.

The ac response of the filter model in Fig. 11(b) is plotted in Fig. 12 indicating a maximum noise coupling of -10 dB at 100 MHz. A 100-mV substrate noise pulse with 1 ns rise/fall time and a duration of 1 ns gives the voltage pulse in Fig. 12 on the loop filter. Referring back to (2), this noise pulse would give a jitter of around 50 ps.

From the above reasoning, it is evident that it is advantageous to reduce the impedance between the local V_{ss} and v_{sub} (or more accurate: the node beneath the MOS channel). However, it is actually the ratio R_{local}/R_{dig} which is important, where R_{local} is the impedance between local V_{ss} and v_{sub}, and R_{dig} is the impedance between V_{ss} for the digital circuitry and v_{sub}. One alternative to reducing R_{local} is then to increase R_{dig}, which can be done by using a process with high-resistivity substrate and placing the digital circuitry physically far from the analog circuits. Yet another alternative to increase R_{dig} is to use triple-well technology as described below.

V. SHARED V_{DD} AND V_{SS} (PLL0)

Increasing the bandwidth of PLL0 also reduced its jitter, so the main source of noise is again either noise injected into the filter or directly into the VCO. By careful layout, any capacitive coupling from V_{dd} into the loop filter was avoided, such that PLL0 has essentially an identical filter as PLL1 and should therefore not have the same filter noise coupling as PLL2. Furthermore, the different noise characteristics shown in Fig. 6 indicate that there is a different noise mechanism in PLL0 compared to PLL2. Another indication of PLL0 not being limited by filter noise is Fig. 10 showing a noise ratio different from the expected 33%. The only option left is that the jitter of PLL0 is dominated by power supply noise directly injected into the VCO.

If a VCO has nonideal power supply rejection, it is likely that the delay of each delay cell is linearly proportional to a change in V_{dd}. This would result in an identical equation as

(1) and filter noise cannot be differentiated from VCO noise. However, a noise pulse ac coupled (capacitive coupling) from the power supplies inside the delay cells in a VCO only causes a fixed phase shift as opposed to a continuous phase drift. So for ac coupled noise, the phase noise would not depend on the VCO frequency. From this reasoning, Fig. 10 indicates that the VCO in PLL0 has both ac coupled noise and nonideal power supply rejection. At high noise frequencies, the ac coupling is dominating such that the curve in Fig. 10 approaches unity, but for lower noise frequencies where the curve approaches 1.33, nonideal power supply rejection is dominating. This also explains the different noise behavior of PLL0 in Fig. 6. Since a dominant source of jitter is capacitive coupling, the jitter increases with the noise frequency. The noise coupling in PLL2 discussed above, has a bandpass characteristic as demonstrated in Fig. 12(a), which gives a different behavior in Fig. 6.

VI. SEPARATE V_{DD}, SHARED V_{SS} (PLL1)

To find the main cause of jitter in PLL1, it is observed that certain frequencies of NCk (50, 100 and 200 MHz) generates larger jitter in Fig. 6 and 8. These frequencies are harmonics of F_{ref} for which the markers are not joined with the curves as briefly mentioned previously. A more detailed measurement near a frequency where NCk and F_{ref} are harmonically related reveals the graph in Fig. 13. The jitter increases significantly when the frequency difference between F_{ref} and NCk drops below the PLL bandwidth of 1 and 4 MHz, respectively. Since low-frequency coupling into the VCO or filter would be suppressed by the PLL feedback, these curves show that the noise coupling is through the PD or divider. It is concluded that the periodic power supply noise causes delay variation in the divider or PD that mixes F_{ref} into a low-frequency beat note which is tracked by the PLL. This was further verified with a spectrum analyzer, which show a tone close to the VCO frequency.

The jitter outside the PLL bandwidth is still higher than the case without supply noise, which was 14 and 6 ps for 1 and 4 MHz bandwidth, respectively. The right hand side of Fig. 13 shows that the noise outside the bandwidth drops for higher bandwidths indicating that the secondary noise source is either

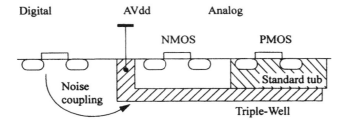

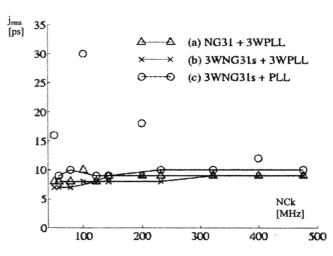

Fig. 14. Triple-well processing provides a buried well that breaks the resistive noise coupling path.

Fig. 15. Jitter induced by NG31 into PLLs having a bandwidth of 4 MHz. Prefix 3W indicates that the block resides in a triple-well.

through the filter, the VCO, or a combination of the two. Fig. 8 shows that for large frequency offsets, the jitter is close to the noise floor set by device noise. The most likely secondary jitter source is therefore through the filter and the steady decrease in jitter for larger frequency offsets (>PLL bandwidth) is caused by the integrating property of the VCO which limits its modulation bandwidth.

None of the measurements can identify if the PD or the divider is the main source of jitter. However, since the PD has two identical paths, these should respond reasonably equal to supply noise. Therefore, the single-ended divider implementation is the most likely jitter source.

The enlarged jitter at harmonics has a direct consequence on the intended application. For a microprocessor clock generator, where the digital circuits are driven by the PLL, the noise generating clock is identical to the VCO frequency. In this case, there will be no mixing and the jitter performance is set by the noise floor indicated in Fig. 6. In case of a multi-channel clock recovery device, where the incoming data streams may have been generated by different crystals, the jitter enlargement at closely spaced frequencies must be considered.

VII. TRIPLE-WELL PROCESSING

Triple-well technology [4]–[6] is a relatively cheap process enhancement which can be made compatible with standard CMOS processing. It requires an additional implant layer, which is the buried well in Fig. 14 that extends the standard PMOS tub underneath the NMOS devices. This well breaks the resistive path from the digital noise source into the analog circuits, indicating that it can work as a noise blocking feature when using separate V_{dd} and V_{ss} networks for the digital and analog circuits. However, there is still a finite impedance between the digital and analog sections, since the triple-well has capacitive coupling to the substrate. The cross section in Fig. 14 shows a triple-well beneath the analog circuit, but it can also be located under the digital circuitry or both.

The chip described in Fig. 3 was processed in a triple-well technology enabling measurements of three different triple-well configurations on a single chip. For a PLL with separate supplies residing in a triple-well, the jitter in Fig. 15(a) was obtained when the NG31 noise generator was active. The jitter of PLL2 in Fig. 6 caused by substrate noise coupling into the filter as discussed in Section IV is not present for the triple-well PLL. Furthermore, no noise peaks can be observed at harmonics of the noise clock in Fig. 15(a), indicating that the PD/divider noise coupling also is eliminated.

A buried well was added to the noise generator NG31s in Fig. 3 that has separate supply. The jitter of the PLL with a triple-well is shown in Fig. 15(b), indicating that the additional benefit in using triple-well beneath both the analog and digital circuits is minor. However, this cannot be a general statement, since the jitter when using triple-well only under PLL2 is very close to the limit of 5–6 ps that was observed without any noise source.

When activating NG31s having a triple-well and observing the jitter of a PLL without triple-well, there is still significant coupling causing large jitter at harmonics of the noise clock frequency as shown in Fig. 15(c). From this, we conclude that it is more effective to place the PLL in a triple-well than the digital circuitry. The main difference between these two schemes is the effective area of the noise blocking triple-well. The area of the PLL was about 200×250 μm while the area of the noise generator was 1100×700 μm. The area is a direct measure of the capacitive coupling from the substrate to the triple-well. With larger area, the impedance is lower and therefore placing the triple-well under the noise generator is less efficient in blocking the noise.

VIII. CONCLUSION

The jitter of three identical PLLs with different power supply configurations have been characterized. The PLLs were built in a standard low-resistivity substrate process. It was found that power supply noise introduced by on-chip digital switching circuits has a different coupling mechanism into each of the three PLLs.

For a PLL that shares V_{dd} and V_{ss} with the digital circuitry, the main jitter source is supply coupling into the VCO. Using separate V_{dd} and V_{ss} for a PLL causes substrate noise to couple into the loop filter node. The reason for this coupling is parasitic resistances in the epi layer below the MOS transistor used as filter capacitor.

A PLL with separate V_{dd} but sharing V_{ss} with the digital circuitry exhibits far less jitter than the other PLLs. The main cause of jitter in this case is delay variations in the feedback divider that mixes the PLL reference frequency into a low-frequency beat note. If this beat note frequency is lower than the PLL bandwidth, the PLL tracks the beat note. This occurs only when a harmonic of the clock of the noise generating digital circuitry is close to the reference clock driving the PLL.

The same PLL was also fabricated in a triple-well process. It is shown that both the noise coupling through the filter and the divider noise at harmonics are eliminated. The importance of keeping the triple-well area small to reduce capacitive coupling was also demonstrated.

REFERENCES

[1] H. B. Bakoglu, *Circuits, Interconnections, and Packaging for VLSI.* Reading, MA: Addison-Wesley, 1990.
[2] P. Larsson, "Power supply noise in future ICs: A crystal ball reading," in *Proc. IEEE CICC*, 1999, pp. 467–474.
[3] ——, "*di/dt* noise in CMOS integrated circuits," *Analog Integrated Circuits and Signal Processing*, no. 1/2, pp. 113–130, Sept. 1997.
[4] Y. Okazaki *et al.*, "Characteristics of a new isolated p-well structure using thin epitaxy over the buried layer and trench isolation," *IEEE Trans. Electron Devices*, vol. 39, pp. 2758–2764, Dec. 1992.
[5] R. B. Merrill, W. M. Young, and K. Brehmer, "Effect of substrate material on crosstalk in mixed analog/digital integrated circuits," in *Proc. IEEE IEDM*, 1994, pp. 433–436.
[6] K. Joardar, "Substrate crosstalk in BiCMOS mixed mode integrated circuits," *Solid-State Electron.*, vol. 39, no. 4, pp. 511–516, Apr. 1996.
[7] P. Larsson, "A 2–1600-MHz CMOS clock recovery PLL with low-V_{dd} capability," *IEEE J. Solid-State Circuits*, vol. 34, pp. 1951–1960, Dec. 1999.
[8] D. J. Allstot and W. C. Black Jr., "A substrate-referenced data-conversion architecture," *IEEE Trans. Circuits Syst.*, vol. 38, pp. 1212–1217, Oct. 1991.
[9] J. A. McNeill, "Jitter in ring oscillators," *IEEE J. Solid-State Circuits*, vol. 32, pp. 870–879, June 1997.

Patrik Larsson received the Ph.D. degree in 1995 on signal integrity and high-speed CMOS circuit design.

He was with Bell Laboratories for five years, where he worked on low-power digital filtering, blind equalization, clock recovery PLLs for Gb/s communication, and *di/dt* noise.

On-Chip Measurement of the Jitter Transfer Function of Charge-Pump Phase-Locked Loops

Benoît R. Veillette, *Student Member, IEEE*, and Gordon W. Roberts, *Member, IEEE*

Abstract— An all-digital technique for the measurement of the jitter transfer function of charge-pump phase-locked loops (PLL's) is introduced. Input jitter may be generated using one of two methods. Both rely on delta–sigma modulation to shape the unavoidable quantization noise to high frequencies. This noise is filtered by the low-pass characteristic of the device and has little impact on the test results. For an input–output response measurement, the output jitter is compared against a threshold. As the stimulus generation and output analysis circuits are digital, do not require calibration, and demand a small area overhead, this jitter transfer function measurement scheme may be placed on the die to adaptively tune a PLL after fabrication. The technique can also implement built-in self-test (BIST) for the characterization or manufacture test of PLL's. The validity of the scheme was verified experimentally with off-the-shelf components.

Index Terms—Mixed analog-digital integrated circuits, phase-locked loops, self-testing, semiconductor device testing, sigma–delta modulation.

I. INTRODUCTION

PHASE-LOCKED loops (PLL's) operating on digital signals are fundamental in microelectronic systems. Indeed, they realize essential functions such as clock distributions and clock recovery. They can thus be found in large digital circuits such as microprocessors and on mixed-signal IC's for digital communications. These devices process digital signals where the phase information is contained in the transition times. They usually make use of sequential phase detectors which require charge-pumps and are thus called charge-pump PLL's [1]. The specifications for these PLL's are extremely aggressive, especially when embedded in high-speed digital communication systems. Process variations can adversely affect circuit performance and result in low yield. To increase the number of good parts, some of the components can be trimmed. This is, however, a very expensive process. Furthermore, aging or different operating conditions can later affect circuits such that they no longer meet specifications. The solution is to allow the PLL to self-calibrate. With this property, the expensive trimming stage can be avoided and changes in the operating conditions can be tracked. However, the challenge now is to integrate on-chip a characterization scheme. This involves two functional blocks: a stimulus source and a response analyzer.

Manuscript received July 15, 1997; revised October 20, 1997. This work was supported by NSERC and the Micronet, a Canadian federal network of centers of excellence dealing with microelectronic devices, circuits, and systems for ultra large scale integration.

The authors are with Microelectronics and Computer Systems Laboratory, McGill University, Montréal, PQ H3A 2A7, Canada.

Publisher Item Identifier S 0018-9200(98)01021-X.

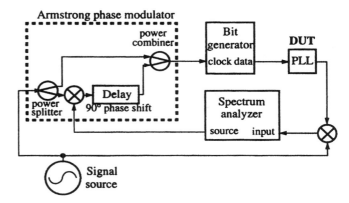

Fig. 1. Jitter transfer function test setup.

These circuits should not necessitate calibration themselves. Furthermore, the silicon area overhead should be small.

Fig. 1 shows a typical test setup for the measurement of the jitter transfer function in a laboratory [2]. A signal source generates a high-frequency carrier which is phase modulated with the source output of a spectrum analyzer using an Armstrong phase modulator [3]. This jittery signal is fed to the clock input of a data generator. The recovered signal from the PLL is down-modulated and observed on the spectrum analyzer. It can be seen that this procedure requires precision analog signal sources and instruments. Looking for shortcuts, engineers are tempted to break the task and measure components independently [4] to infer the device characteristics. However, the nature of the phase-locked loop renders this solution unattractive as the tight feedback and the sensitivity of some nodes to parasitics makes it difficult to relate the values of components to the PLL behavior. A significant level of accuracy may only be achieved by measuring the system as a whole.

In this paper we propose to verify charge-pump PLL's characteristics using mostly synchronous digital circuits. It implies that the stimulus signals can only change at the clock edges and that the output signals may only be sampled at the same clock edges. This would seem like an unbearable constraint as jitter, both created and measured, is quantized to the test clock period. The results of any test would thus be severely limited in precision. However, this hurdle can be overcome using low-pass delta–sigma ($\Delta\Sigma$) modulation [5]. This technique can encode high quality signals, in this case a sinewave, on one or a few bits. The quantization noise introduced in the operation is shaped to high frequencies. Since PLL's are low-pass, high-frequency jitter is filtered

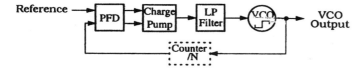

Fig. 2. Digital phase locked loop functional block diagram.

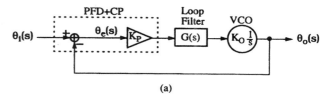

(a)

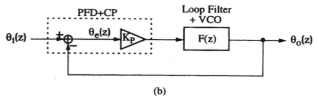

(b)

Fig. 3. Models of charge-pump phase-locked loop: (a) continuous time and (b) discrete time.

out. The output will thus exhibit a sufficiently high SNR to be considered a pure sinewave. Therefore, the input high-frequency jitter noise does not affect test results. This filtering principle was demonstrated for a voice codec, another low-pass circuit [6].

Two methods will be presented to create jitter. The first one is the digital modulation of the edges of a reference signal using a higher frequency clock. The alternative is the injection of a sinusoidal signal in the loop through a second charge pump. The second method requires a lower clock frequency but the first one is nonintrusive. Contrary to the usual measurement procedure, a fixed output jitter is selected and the amplitude of the input jitter at a given frequency is varied. Ultimately, the test signal amplitude that results in the selected output jitter is obtained and used to compute the jitter transfer function. It is important to note that while both jitter creation schemes are digital, the jitter loop injection method does not require a test clock frequency larger than the PLL operating frequency. Nevertheless, as will be shown, the test accuracy can be increased with the digital phase modulation and a higher clock frequency.

The outline of the paper is the following. First, an overview of the PLL will be provided in Section II to establish the notation. In Sections III and IV, the two methods for stimulating the device will be explained. Section V will study signal generation using delta–sigma modulation. The analysis of the output signal as well as the test methodology will be the topics of Section VI. Section VII will briefly look at the issue of accuracy. Experimental setup and results will be discussed in Sections VIII and IX, respectively. Overhead will be addressed in Section X. Finally, conclusions will be drawn.

II. PHASE-LOCKED LOOP OVERVIEW

A block diagram of a simple charge pump PLL is shown in Fig. 2. To the left, a sequential phase detector compares the transitions of the reference input and voltage-controlled oscillator (VCO) output signals. A two-output phase-frequency detector (PFD) is illustrated, but other types of sequential phase detectors may also be employed. The output of the PFD can be any of three logic states, and thus a charge pump is required for digital-to-analog conversion. The charge pump can be of the current or voltage type. The jitter signal injection technique of Section V relies on current charge pumps, but it could also be adapted for a voltage charge pump as long as the filter allows summation of voltages. Referring back to Fig. 2, the low-pass filter removes short term variations and shapes the PLL characteristics. The VCO in turn generates a square wave whose frequency depends on the level of its analog input. A counter may be inserted in the feedback loop to lock the VCO clock to a lower frequency reference signal.

The continuous-time linear model of this PLL in steady-state operation is illustrated in Fig. 3(a). The variables $\theta_i(s)$ and $\theta_o(s)$ are the phase of the reference signal and the VCO output signal, respectively. It should be understood that while these variables represent jitter in a signal, the other variables in the circuit stand for either voltage or current. The phase detector converts phase to one of the two analog quantities while the VCO performs the complementary operation. In Fig. 3(a), the parameter K_P is the composite gain of the phase detector and charge-pump circuits expressed in A/rad or V/rad depending on the type of charge pump. The transfer function of the loop filter is denoted $G(s)$. The gain of the VCO is labeled K_O and is stated in rad/V. If a counter is present, then its effect is lumped into this constant by dividing the VCO intrinsic gain by the counter length (N in Fig. 2). With this operation, a counter in the PLL becomes transparent for the proposed measurement method.

The continuous-time model allows satisfactory predictions of the PLL behavior. On the other hand, the phase of digital signals is contained in the signal transitions and is thus better represented as a discrete-time sequence. Therefore, the analysis of PLL operating on digital signals should be performed using difference equations or z-transform tools. Indeed, it has been shown that the discrete-time model is more accurate, especially at high jitter frequency [7]. This model is shown in Fig. 3(b). The closed-loop equation governing the operation of the PLL is

$$H(z) = \frac{\theta_o(z)}{\theta_i(z)} = \frac{K_P \cdot F(z)}{1 + K_P \cdot F(z)}. \tag{1}$$

In this equation, the discrete-time transfer function $F(z)$ is the impulse invariant transform of the series combination of the loop filter and VCO transfer functions ($K_O G(s)/s$). The function $H(z)$ is labeled the jitter transfer function. Many PLL specifications can be extracted from this transfer function such as the jitter bandwidth and the jitter peaking. The latter measure corresponds to the maximum value of $|H(z)|$.

Process variations can significantly alter the charge pump and VCO gains as well as the filter passive components values. The method we propose can be used to automatically trim components for compliance with a desired $H(z)$. Alternatively, it can be used to screen devices by comparing the measured magnitude of $H(z)$ against a mask.

III. INPUT DIGITAL PHASE MODULATION

The first method that we will consider to generate the test stimulus is the digital phase modulation of the PLL input. Sinusoidal jitter of arbitrary frequency and amplitude can be generated by modifying the edges of the reference signal digitally. A test clock frequency which is a multiple of the reference signal frequency is a prerequisite. This method therefore may not be suitable for high-frequency PLL's. The principle is to control the instantaneous phase of a 1-b digital signal by delaying it by an amount set by a multibit digital signal. Fig. 4 illustrates a circuit that can realize this operation. The input signal to the delay cells string is a pulse whose period is an integer multiple of the test clock period. The digital signal generator output, denoted $d(z)$, is updated at every PLL cycle and delays the input signal by a variable amount. Fig. 5 shows a typical waveform at the output of this circuit for a test clock (f_S) eight times the reference signal frequency (f_R). The dotted lines represent the test clock edges with the thick ones used as zero jitter marks. The PLL input signal jitter can thus be expressed as

$$\theta_i(z) = d(z)\frac{2\pi f_R}{f_S}\ (\text{rad}). \qquad (2)$$

Obviously, the number of inputs to the multiplexer is limited by the silicon area available but also by the ratio of the frequency of the clock operating the delay cells to the frequency of the reference signal. The limited number of delay cells is likely to make the jitter of the input signal $\theta_i(z)$ coarsely quantized such that its SNR is unacceptably low. However, PLL's are frequency selective with respect to jitter. Indeed, the jitter transfer function is low-pass and the device filters high-frequency jitter. When incorporating a $\Delta\Sigma$ modulator in the signal generator, quantization noise can be shaped to high frequencies [5]. Therefore, the encoded signal from a low-pass $\Delta\Sigma$ modulator, such as $d(z)$ in Fig. 5, contains a high-quality low-frequency sinewave and high-frequency quantization noise as illustrated in Fig. 6(a). While $d(z)$ is a multibit signal as it controls an $N \times 1$ multiplexer, $\Delta\Sigma$ modulation can ultimately encode signals on a single bit. Because the PLL is low-pass and is designed to reject high-frequency components in the loop, the quantization noise will be suppressed, leaving only the sinusoid in the output jitter as shown in Fig. 6(b). Section V will examine how such a signal can be generated.

IV. SIGNAL INJECTION IN THE LOOP

The second option for the generation of jitter is the injection of a test signal at the input of the loop filter as shown in Fig. 7. This signal source, represented here by the variable $\theta_x(z)$, is injected through a second charge pump with gain K_X. It should be understood that this signal source is not a jittery digital signal but an analog signal embedded in a 1-b digital signal encoded using a $\Delta\Sigma$ modulator. However, this signal, when referred back to the input, is equivalent to input

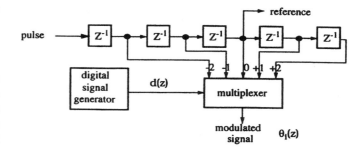

Fig. 4. Digital phase modulation circuit.

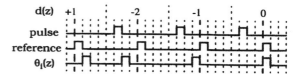

Fig. 5. Typical digital phase modulated waveform.

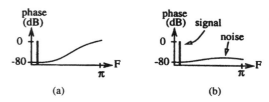

Fig. 6. Phase spectrum: (a) input signal jitter and (b) output signal jitter.

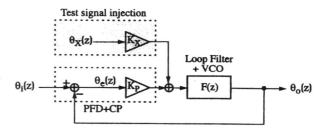

Fig. 7. Discrete-time model of phase-locked loop modified for signal injection.

jitter. The PLL reference signal meanwhile is a square wave and therefore the input jitter $\theta_i(z)$ will be zero. This setup will be used to evaluate the characteristics of the PLL through the measure of its $\theta_o(z)/\theta_x(z)$ transfer function. Examining the model of Fig. 7, it can be seen that this transfer function will be equal to

$$\frac{\theta_o(z)}{\theta_x(z)} = \frac{K_X}{K_P} \cdot \frac{K_P \cdot F(z)}{1 + K_P \cdot F(z)}. \qquad (3)$$

It is thus equivalent within a multiplicative constant to the jitter transfer function $\theta_o(z)/\theta_i(z)$. For a spectral test such as the jitter transfer function, the test signal is a sinewave encoded into a single bit. The quantization noise is concentrated at high frequencies and is filtered out as explained in the previous section. It is important to note that the clock period for signal injection must be an integer multiple of the reference signal period to prevent aliasing of the quantization noise back in the PLL passband. This condition implies that the signal

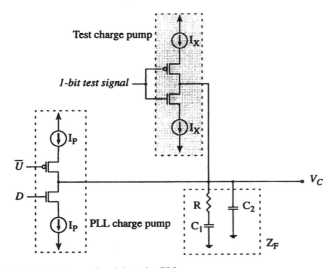

Fig. 8. Injecting a signal into the PLL.

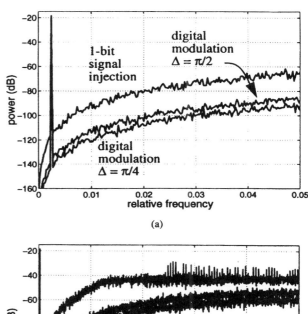

(a)

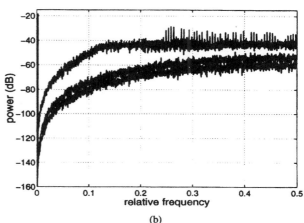

(b)

Fig. 9. Spectrum of test jitter signals: (a) signal band and (b) Nyquist interval.

injection frequency cannot be higher than the reference signal frequency. Converting the 1-b digital stream to an analog signal and summing it with the output of the phase detector is quite simple. A second current charge pump is placed in parallel with the phase detector charge pump and both outputs are connected together, forming a current summing node as shown in Fig. 8. The accuracy of this analog-to-digital conversion is a function of the matching of the two current sources I_P and I_X, typically ranging between 0.1–1% in monolithic form. In this schematic, the impedance Z_F implements the loop filter and V_C is the controlling voltage of the VCO.

V. NOISE-SHAPED SINUSOID GENERATION

The basis of the jitter generation methods described in the previous two sections is low-pass delta–sigma modulation. This technique allows the encoding of a bandlimited signal represented by a large number of bits onto a very small number of bits. An error is created by this quantization operation but it is filtered such that it appears mostly at high frequencies. This feat is realized by feeding back the quantization error to a filter and summing it with the input before the next quantization. Using delta–sigma modulation, arbitrary signals, such as a sinusoid, can be represented by a square wave (1-b signal) with the difference being composed mostly of high-frequency noise.

Fig. 9 shows the spectrum of a sinusoid for jitter generation coded using different quantizers. Two curves show the power density spectrum of multibit signals for the digital phase modulation method. They were generated for different test clock ratios (f_S/f_R) and thus different quantization steps (Δ). The last curve shows a 1-b signal for the signal injection method. The same second-order low-pass $\Delta\Sigma$ modulator was used to produce each curve, only the quantizer was changed. The input signal amplitude was kept the same in each case. As explained previously, the signal is located at low frequencies where the noise power is small. It can also be seen that the quality of the signal improves with the number of quantization steps (smaller step size Δ). This is evident by the drift in

the curve downward in Fig. 9 as Δ decreases. Three possible methods, illustrated in Fig. 10, may be employed on-chip to generate any of the signals of Fig. 9. The selection of one of these for implementation is based on the available resources in the system, silicon area, PLL speed, and required accuracy. Each method is described below.

A. Direct Frequency Synthesis

The most straightforward and versatile signal generation scheme is a ROM-based digital frequency synthesizer [8] followed by a low-pass $\Delta\Sigma$ modulator. The output of the $\Delta\Sigma$ modulator may be a 1-b signal or a multibit signal [9] as required by the jitter creation technique used. However, the direct frequency synthesis solution requires a large silicon area and is thus usually not a good choice for built-in self-test (BIST). However, if a large block of RAM is present in the system, then it could be enrolled to store data for signal generation during the PLL test phase.

B. Delta–Sigma Oscillator

An alternative is to use a circuit called a low-pass $\Delta\Sigma$ oscillator [10]. The circuit, illustrated in Fig. 10(b), is a digital resonator where the frequency setting multiplier has been replaced by the combination of a $\Delta\Sigma$ modulator and

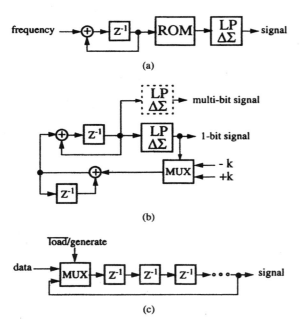

(a)

(b)

(c)

Fig. 10. Generating delta–sigma encoded sinusoids: (a) direct digital frequency synthesis, (b) delta–sigma oscillator, and (c) fixed-length periodic byte stream.

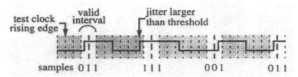

Fig. 11. Comparing waveform against jitter threshold.

a multiplexer. To avoid large multiplexers, the output of the $\Delta\Sigma$ oscillator is usually a single bit. Therefore, a second $\Delta\Sigma$ modulator may be required for multibit signal generation. However, as the structure is similar to the first one except for the quantizer, hardware can be time-shared. Because a ROM or a multiplier is not required, the silicon implementation of $\Delta\Sigma$ oscillators is very efficient. However, the presence of a nonlinear block in the feedback loop makes this device difficult to predict using a linear model. Off-line simulations are required to achieve maximum accuracy. Nevertheless, $\Delta\Sigma$ oscillators can be implemented quickly and may possibly use available on-chip computing resources such as a digital signal processor (DSP).

C. Fixed-Length Bit Streams

The last method consists in generating a data stream from a software sinewave generator and low-pass $\Delta\Sigma$ modulator and then selecting a subset of this stream. This subset is stored in memory and the resulting data stream is then repeated [11] as illustrated in Fig. 10(c). One can view this approach as a special case of the ROM-based digital frequency synthesis scheme presented above. This method is particularly useful for single-bit signals as they can be represented using a very small number of bits, on the order of a hundred. The downside is that signal quality will vary widely with different subsets of the same length of a given $\Delta\Sigma$ modulator output if care is not

taken. Some form of optimization is necessary to obtain good results [12]. Also, a different data stream is required for each frequency and amplitude desired. Nonetheless, considerable speed can be achieved with this technique, and the overhead is the smallest of the three methods.

VI. EVALUATION OF THE OUTPUT JITTER

The exact gauging of the jitter response $\theta_o(z)$ is rather difficult. However, a measure that can be made with good accuracy is the point when $\theta_o(z)$ reaches a predetermined value or threshold. The edges of the test clock, assumed to be jitter free, will be used to establish this threshold. Fig. 11 illustrates how this can be accomplished. The dashed lines represent the rising edge of the test clock with the bold ones indicating the rising transition of the reference clock. The output signal is sampled at positive test clock edges until two adjacent samples yield a zero followed by a one, indicating a rising edge of the signal. If this rising edge occurred in intervals immediately before or after the reference test clock edge, then jitter is below threshold; otherwise, an error is generated. Over a time interval, the number of errors are counted and a bit error rate (BER) measure can be obtained. This averaging is done to prevent glitches and noise signals from significantly affecting the final result. In Fig. 11, a ratio of the test clock frequency over the reference signal frequency of eight allows a minimum value for the threshold of $\pi/4$. Larger values could also be used for the threshold by allowing more test clock cycles in the valid interval.

The previous method requires a test clock frequency at least three times the reference frequency. Indeed, three periods of the test clock for a PLL period is the limiting case as two of the clock periods must compose the valid interval, leaving one to catch jitter above threshold. A different scheme, however, can be used that compares jitter against a π rad threshold using a 50% duty cycle test clock of the same frequency as the reference clock. Fig. 12(a) illustrates how it can be implemented with a few gates and registers. The threshold is verified by sampling a data signal with both the reference clock, used at the input of the PLL, and the recovered clock at the VCO output. This data signal will toggle between one and zero at the reference signal falling edges, resulting in a frequency half that of the reference clock. When the output jitter exceeds π rad, sampling errors will occur with the VCO output clock because it will sample a different data than the reference clock. This is shown in Fig. 12(b) where an error occurs when the jitter goes from 0.9 to 1.1 π. This circuit is somewhat similar to the circuit typically used for a jitter tolerance test [2].

A single frequency point test is thus performed as follows: an input jitter ($\theta_i(z)$ or $\theta_x(z)$) is applied to the PLL and its amplitude is modified until the maximum value resulting in output jitter below threshold is found. The fastest procedure for obtaining $\theta_i(z)$ or $\theta_x(z)$ is to use a binary search algorithm. The initial amplitude (A_0) is zero and the initial increment (Δ_0) is 0.5. A test is performed with amplitude $A_0 + \Delta_0$. If the VCO output jitter is lower than the threshold, then the next amplitude A_1 will be $A_0 + \Delta_0$. Otherwise, the

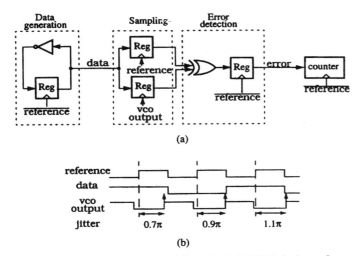

Fig. 12. (a) Circuit to evaluate π rad jitter threshold. (b) Typical waveforms.

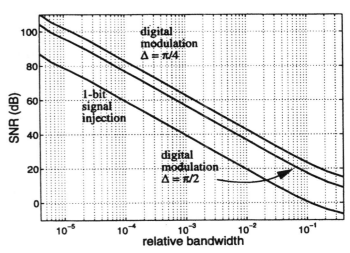

Fig. 13. Signal-to-noise ratio of the output jitter versus the PLL relative bandwidth.

amplitude remains the same ($A_1 = A_0$). The increment is then divided by two and the procedure is repeated until the desired accuracy is achieved. The uncertainty associated with the signal amplitude following a k step measure will be $\Delta_k/2$ which equals $2^{-(k+1)}$. However, arbitrary accuracy cannot be achieved as noise sources are always present.

VII. ACCURACY

The accuracy of the measured results will depend on many factors. Foremost is the residual input signal jitter quantization noise which passes through the PLL. The jitter creation clock frequency along with the $\Delta\Sigma$ modulator noise transfer function, the number of quantization bits, and the PLL bandwidth will influence the effective SNR of the measured jitter at the output. It is interesting to note that for each frequency point on the jitter transfer function, the SNR of the measured output jitter will remain constant. This is because, under a white noise assumption for the quantizer error, the noise present in the PLL input signal jitter does not vary for a selected $\Delta\Sigma$ modulator implementation. It is only a function of the quantizer step and does not change with the signal amplitude and frequency. Neglecting internal noise sources, the PLL output jitter noise power density is a result of the input jitter noise power density shaped by the jitter transfer function. It can thus be seen that the PLL output jitter noise power is independent of the parameters of the test sinusoid. Furthermore, as the threshold is fixed, the PLL output sinewave jitter amplitude must also be constant. In the measurement procedure, the input jitter sinewave amplitude is varied to account for the response of the PLL at a given frequency. Consequently, for a given PLL, the SNR of the measured output jitter is fully determined from the quantizer step size for the input signal jitter and the jitter threshold at the output. Fig. 13 shows the theoretical SNR of the output jitter with respect to the ratio of the cutoff frequency of a second-order PLL over the PLL lock frequency, referred to here as the PLL relative bandwidth. A curve is displayed for each of the three input jitter quantization granularities. A second-order $\Delta\Sigma$ modulator is used to generate the input signal, and the output jitter amplitude is held constant at π. For

reference, digital data communication systems such as SONET mandate a relative loop bandwidth of about 0.1%.

Also of concern is the amount of jitter present in the test clock. However, the clock signal should be generated by a tester and, provided a sound floorplan, this source of error should be negligible. A more significant problem is the jitter, both static and random, generated internally. It will add to the jitter from the input signal and thus modify the effective output jitter threshold. Its effect can only be reduced by using a jitter threshold much larger than the jitter noise. Alternatively, it could be accounted for and subtracted from the final results. Finally, for the signal injection method, the matching of the two charge pumps is obviously a cause of errors.

VIII. EXPERIMENTAL SETUP

A test setup, whose schematic is shown in Fig. 14, was implemented on a breadboard using off-the-shelf components. The device under test (DUT) is centered around the VCO from a 74HC4046 monolithic phase-locked loop. However, because this IC uses a voltage charge pump followed by a passive filter and since its phase detector could not be separated from this block, an XC4010 FPGA was programmed to implement the classical phase-frequency detector [13]. The charge pump is built out of discrete NPN and PNP transistors, a resistor, and analog switches from a 74HC4066. A circuit is also required to maintain the transistors inside their linear region when the switches are open, but it is not shown here. The PLL was operated at 100 kHz as parasitics of the board and the time constant of some components do not allow for a higher frequency. However, the measurement scheme should be extendable to much higher frequency as the test circuits are similar in nature to the PLL components.

Both digital signals for jitter creation ($d(z)$ and osc) are generated by a low-pass $\Delta\Sigma$ oscillator programmed on the same field-programmable gate array (FPGA). It uses 24-b buses to achieve a tunability of 55 parts per million of its programmable clock. A 3-b quantizer in addition to the standard 1-b quantizer makes the signal generator capable of

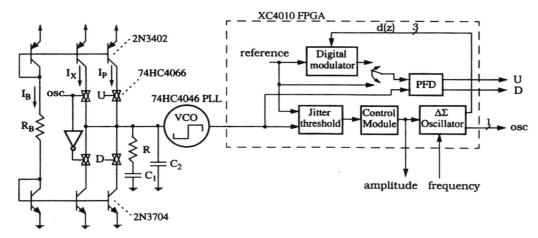

Fig. 14. Experimental setup.

multibit output [refer to Fig. 10(b)] for the purpose of digital phase modulation. It should be noted that apart from the quantizer, $\Delta\Sigma$ modulator circuitry is not duplicated as it will be operated in time-shared mode at double speed for multibit operation.

The input to the PLL can be set to accommodate both jitter generation methods. For the loop jitter injection scheme, a 100-kHz square wave is presented to the input of the phase detector. The input can also be the same signal phase modulated with the help of an 800-kHz test clock. Eight jitter steps are therefore possible, resulting in a $\pi/4$ quantization.

Various jitter threshold circuits are also implemented on the FPGA. The π jitter threshold circuit of Fig. 12(a) will be employed in conjunction with the loop jitter injection. On the other hand, thresholds of $\pi/2$ and $\pi/4$ are implemented for the digital phase modulation method, making use of the higher frequency test clock.

For each test, a warm-up stage of 2^{14} data cycles is executed to remove transients before a 2^{16} data cycle test stage is performed. The error threshold is set to 64, corresponding to a BER of 10^{-3}. A control module built around a finite state machine selects the amplitude of the input jitter for the ensuing test according to the output of the jitter threshold circuit, using the binary search algorithm. At each frequency point, the amplitude is resolved to an accuracy of 15 b within 13 s. The entire digital circuitry for all the experiments requires 81% of the resources of an XC4010 FPGA. This experimental setup is connected to a workstation through I/O modules to allow a driving software to set the low-pass $\Delta\Sigma$ oscillator frequency as well as read the amplitude.

IX. EXPERIMENTAL RESULTS

The jitter transfer function measurement was carried out for both the jitter injection and the digital phase modulation techniques on two different PLL configurations with different bandwidths and damping values. Table I summarizes the main parameters of these experiments. The same current amplitude was used for both charge pumps ($I_x = I_p$). The transfer functions are presented in the continuous-time domain as this is more typical of what can be found in industry.

TABLE I
EXPERIMENT PARAMETERS

	Experiment 1	Experiment 2
R	1.0 KΩ	2.2 KΩ
C_1	2.0 μF	5.1 μF
C_2	100 pF	480 pF
I_X	47 μA	47 μA
K_P	7.5 μA/rad	7.5 μA/rad
K_O	175 krad/V	175 krad/V

The results of the experiments on the first configuration are shown in Fig. 15. A measured jitter transfer function for each jitter generation method is displayed. The dotted line represents the theoretical jitter transfer function as predicted from the direct measurement of the components. The phase modulation scheme used a $\pi/2$ threshold for this experiment. The curve shows a 0.4-dB offset which can be attributed mostly to the static jitter of the PLL. For the other jitter creation scheme, the signal injection clock was chosen to be 50 kHz, that is half the PLL rate, in order to demonstrate the flexibility in selecting this parameter. The offset is larger, possibly because of mismatch between the two charge pumps realized out of discrete transistors. The first two columns of Table II summarize the features of the curves after removal of the offsets. Both methods yield similar results for the PLL bandwidth and jitter peaking. The theoretical predictions are slightly off, most probably because of the parasitics of the setup which were not accounted for in the calculations.

The jitter transfer functions measured in the second experiment are shown in Fig. 16. This PLL exhibits a larger bandwidth and is more damped. It can be seen that the curve obtained here with the jitter injection technique is of lesser quality. This came about because the larger bandwidth yields a lower output jitter SNR. From the graph of Fig. 13, it can be seen that this SNR is barely over 20 dB. On the other hand, the digital phase modulation still shows a smooth curve because of the 3-b quantization which results in lower jitter

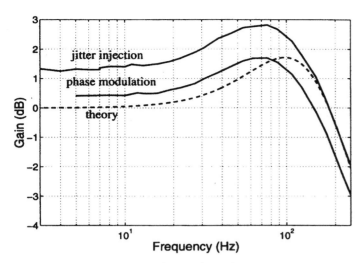

Fig. 15. Jitter transfer function for experiment 1.

TABLE II
RESULTS SUMMARY

	$R = 1.0 \ K\Omega, \ C_l = 2.0 \ \mu F$		$R = 2.2 \ K\Omega, \ C_l = 5.1 \ \mu F$	
	3 dB (Hz)	j.p. (dB)	3 dB (Hz)	j.p. (dB)
signal injection	238	1.49	435	0.42
phase modulation	234	1.29	395	0.24
theory	283	1.71	466	0.22

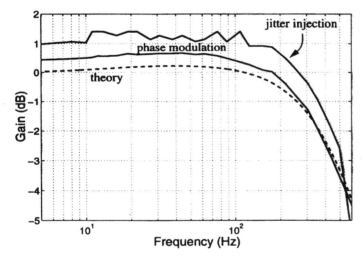

Fig. 16. Jitter transfer function for experiment 2.

noise levels at the output. Again, the meaningful parameters are summarized in the two right-most columns of Table II.

X. IMPLEMENTATION

For any integrated measurement scheme, the area overhead is obviously a major concern. While the digital portion of the experimental setup uses a large portion of the FPGA, a much more compact implementation is possible. Indeed, a $\Delta\Sigma$ oscillator was selected as the signal generator because of its versatility as a complete jitter transfer function was

sought. However, in many applications, a smaller number of signal frequencies and amplitudes are necessary and a fixed-length periodic bit stream could generate the signals for the cost a few kilobits of RAM. For example, to verify that the bandwidth of the PLL is smaller than some value, only one test is required. Specific values for overhead or measurement time depend heavily on the PLL application. Moreover, one should not have a dogmatic stance about overhead as the addition of the on-chip measurement circuits add value to a system. The economic gains from a BIST may far outweigh the cost of extra silicon.

XI. CONCLUSIONS

We have presented a PLL jitter transfer function measurement technique which is entirely digital except for the possible addition of a charge pump. The technique is suitable for on-chip measurement since it does not require trimming and the silicon overhead is small. Two methods were introduced for the creation of jitter, allowing tradeoffs between test clock frequency on one side and loading, complexity, and accuracy on the other side. Experimental results were presented which suggest this scheme could be successfully implemented on silicon.

ACKNOWLEDGMENT

The authors acknowledge the suggestions of B. Gerson from PMC Sierra.

REFERENCES

[1] F. M. Gardner, "Charge-pump phase-lock loops," *IEEE Trans. Commun.*, vol. COMM-28, pp. 1849–1857, Nov. 1980.
[2] L. DeVito, "A versatile clock recovery architecture and monolithic implementation," in *Monolithic Phase-Locked Loops and Clock Recovery Circuits: Theory and Design*, B. Razavi, Ed. New York: IEEE Press, 1996.
[3] E. H. Armstrong, "A method of reducing disturbances in radio signaling by a system of frequency modulation," in *Proc. IRE*, May 1936, vol. 24, no. 5, pp. 689–740.
[4] P. Goteti, G. Devarayanadurg, and M. Soma, "DFT for embedded charge-pump PLL systems incorporating IEEE 1149.1," in *Proc. IEEE 1997 CICC*, Santa Clara, CA, May 1997, pp. 210–213.
[5] M. W. Hauser, "Principles of oversampling A/D conversion," *J. Audio Eng. Soc.*, vol. 39, nos. 1/2, pp. 3–26, Jan./Feb. 1991.
[6] M. F. Toner and G. W. Roberts, "A BIST scheme for an SNR, gain tracking, and frequency response test of sigma–delta ADC," *IEEE Trans. Circuits Syst.-II*, vol. 41, pp. 1–15, Jan. 1995.
[7] J. P. Hein and J. W. Scott, "z-domain model for discrete-time PLL's," *IEEE Trans. Circuits Syst.*, vol. 35, pp. 1393–1400, Nov. 1988.
[8] J. Tierney, C. M. Rader, and B. Gold, "A digital frequency synthesizer," *IEEE Trans. Audio Electroacoustic*, vol. 19, pp. 48–57, 1971.
[9] J. G. Kenney and L. R. Carley, "Design of multi-bit noise shaping data converter," *Analog Integrated Circuits and Signal Processing J.*, May 1993, vol. 3, no. 3, pp. 259–272.
[10] A. K. Lu, G. W. Roberts, and D. A. Johns, "A high-quality analog oscillator using oversampling D/A conversion techniques," *IEEE Trans. Circuits Syst.-II*, vol. 41, pp. 437–444, July 1994.
[11] E. M. Hawrysh and G. W. Roberts, "An integration of memory-based analog signal generation into current DFT architectures," in *Proc. 1996 ITC*, Washington, DC, Oct. 1996, pp. 528–537.
[12] B. Dufort and G. W. Roberts, "Signal generation using periodic single and multi bit sigma–delta modulated streams," in *Proc. IEEE 1997 ITC*, Washington, DC, Nov. 1997, pp. 396–405.
[13] C. A. Sharpe, "A 3-state phase detector can improve your next PLL design," *EDN Mag.*, pp. 55–59, Sept. 1976.

PART IV

Building Blocks

A Low-Noise, Low-Power VCO with Automatic Amplitude Control for Wireless Applications

Mihai A. Margarit, Joo Leong (Julian) Tham, *Member, IEEE,*
Robert G. Meyer, *Fellow, IEEE,* and M. Jamal Deen, *Senior Member, IEEE*

Abstract— Voltage-controlled oscillators (VCO's) used in portable wireless communications applications, such as cellular telephony, are required to achieve low phase-noise levels while consuming minimal power. This paper presents the design challenges of a monolithic VCO with automatic amplitude control, which operates in the 300 MHz to 1.2 GHz frequency range using different external resonators. The VCO phase-noise level is −106 dBc/Hz at 100-KHz offset from an 800-MHz carrier, and it consumes 1.6 mA from a 2.7-V power supply.

An extensive phase-noise analysis is employed for this VCO design in order to identify the most important noise sources in the circuit and to find the optimum tradeoff between noise performance and power consumption.

Index Terms—Phase noise, voltage-controlled oscillator (VCO), wireless applications.

I. INTRODUCTION

THE remarkable growth in telecommunication systems, such as cellular telephony, demands continuous efforts toward the improvement of radio-frequency (RF) circuit performance at ever increasing levels of integration. Complete transceiver solutions that integrate low-noise amplifiers (LNA's), mixers, voltage-controlled oscillators (VCO's), and transmit modulators already exist. Moreover, the stringent noise and spurious emissions requirements for cellular communications systems, such as GSM, DCS, and PCS, need to be achieved with even lower power-consumption levels.

This paper describes the analysis and implementation of a monolithic VCO with automatic amplitude control (AAC), which is part of a one-chip transceiver dedicated for dual-band cellular systems [1]. The VCO is capable of operating from 300 MHz to 1.2 GHz using different resonators. The measured phase-noise level is −106 dBc/Hz at 100-kHz offset from an 800-MHz carrier.

This paper begins with a presentation of the VCO and the AAC circuits, with emphasis on the critical aspects of the design. Section III presents the phase-noise analysis of the circuit done in order to help identify the important noise sources in the circuit. Based on these results, the optimized

Manuscript received April 13, 1998; revised February 3, 1999.
M. A. Margarit is with Rockwell Semiconductor Systems, Newport Beach, CA 92658 USA and the School of Engineering Science, Simon Fraser University, Vancouver, BC V5A 1S6 Canada.
J. L. Tham was with Rockwell Semiconductor Systems, Newport Beach, CA 92658 USA. He is now with Maxim Integrated Products, Sunnyvale, CA 94086 USA.
R. G. Meyer is with the Department of Electrical Engineering and Computer Sciences, University of California, Berkeley, CA 94720 USA.
M. J. Deen is with the School of Engineering Science, Simon Fraser University, Vancouver BC V5A 1S6 Canada.
Publisher Item Identifier S 0018-9200(99)04203-1.

design is discussed. In Section IV, the experimental data are compared with simulations. The last section summarizes the conclusions drawn from this analysis.

II. DESIGN OF A HIGH-FREQUENCY VCO WITH AUTOMATIC AMPLITUDE CONTROL

A. Practical Design Considerations

A simplified block diagram of a VCO with integrated active circuitry and an external parallel resonator is presented in Fig. 1(a). The actual resonator is composed of the inductor L_1 and varactor CR_1. R_p models the losses in the inductor, C_b the parasitic capacitances of the inductor and the board, L_2 the inductances of the bond wire and package lead, and C_p the parasitic capacitance of the transconductor. The transconductor G_m, along with positive feedback capacitors C_1 and C_2, provides the negative resistance needed to compensate for the losses in the tank. When the oscillator reaches steady state, the loop gain equals one and the following relationship holds:

$$\frac{G_m}{n} \cdot R_p = 1 \tag{1}$$

where

$$n = \frac{C_1 + C_2}{C_1}$$

and G_m is the large-signal transconductance.

To ensure proper startup of the oscillator, the small-signal transconductance is typically chosen to be three to four times larger than the large-signal G_m. This condition imposes a design restriction on the feedback ratio n. As will be discussed later in this paper, the value of n for optimum noise performance is different than that needed for proper startup. One approach to ensure proper oscillation startup while maintaining the optimum feedback ratio n for noise performance is to implement an oscillator with AAC, which forces the transconductance to the value needed to provide constant oscillation amplitude.

The circuit in Fig. 1(a) has more than one resonance mode, as becomes evident when the equation for the total admittance of the circuit Y_t seen by the transconductor is written as shown in (2) at the bottom of the next page. Assuming that $sL_v \ll (1/sC_v)$, the fundamental parallel mode is given by $s \cdot L_1 \| (1/s \cdot (C_v + C_b + C_{p1}))$. There is a second spurious mode associated with $(1/s \cdot C_{p1}) \| [s \cdot L_2 + s \cdot L_1 \| (1/s \cdot (C_v + C_b))]$, which is due to the parasitic bondwire and package inductance L_2. The addition of resistor R_s shown in Fig. 1(b) damps the spurious oscillation mode and

271

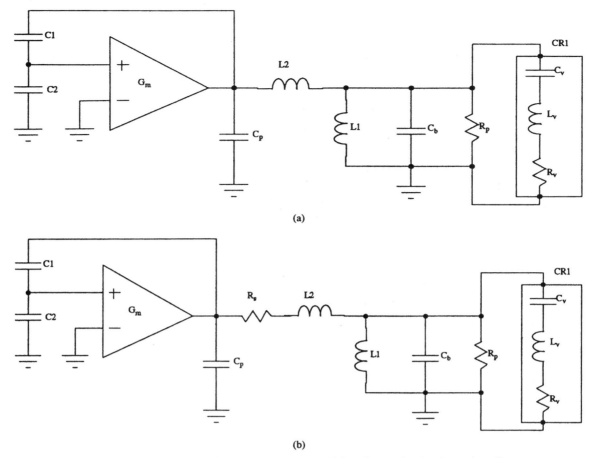

Fig. 1. (a) Simplified diagram of a voltage-controlled oscillator. (b) Voltage-controlled oscillator with damping resistor R_s.

has negligible effect on the fundamental mode [2]. However, care needs to be taken in the design, since too large a value of R_s degrades the noise performance of the oscillator. A different solution for damping the second mode has been reported recently [3]. However, this solution requires more assembly steps.

B. Design of the VCO with AAC

The schematic of a differential VCO with AAC is shown in Fig. 2. The design considerations discussed above are implemented here. L_1 and L_2 in Fig. 2 represent the parasitic bond-wire inductances. R_1 and R_2 are used to damp the spurious oscillation mode. Capacitors C_1, C_2 and C_3, C_4 provide positive feedback with ratio as shown in (1), where $C_1 = C_3$ and $C_2 = C_4$. To provide a wide-frequency range of

operation, the value of these capacitors has to be minimized. However, their minimum value is limited by requiring that $(1/s \cdot C_2), (1/s \cdot C_4) \ll r_\pi$ in order to minimize the excess phase shift at lower operating frequencies.

The AAC function is performed by the following blocks in Fig. 2: a high-frequency rectifier, a low-pass filter, a differential-to-single-ended amplifier, and a voltage reference. The output from the high-frequency rectifier is low-pass filtered and provides the inverting input of the differential amplifier a dc signal proportional to the oscillation amplitude. This signal is compared with the reference voltage. With the use of replica bias circuits for the voltage reference and the rectifier, the two dc voltages track very well over process and temperature. R_{14} provides negative feedback for the differential amplifier in order to maintain constant

$$Y_t = s \cdot C_{p1} + \cfrac{1}{s \cdot L_2 + \cfrac{1}{s \cdot C_b + \cfrac{1}{s \cdot L_1} + \cfrac{1}{R_p} + \cfrac{1}{s \cdot L_v + R_v + \cfrac{1}{s \cdot C_v}}}} \qquad (2)$$

where

$$C_{p1} = C_p + \frac{C_1 \cdot C_2}{C_1 + C_2}$$

272

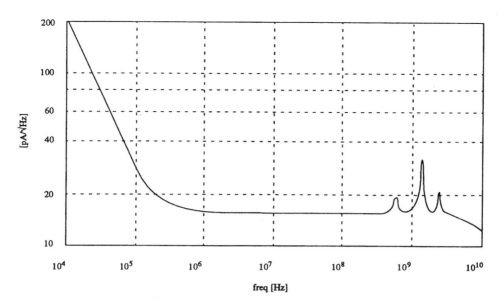

Fig. 5. Tail-current noise spectrum.

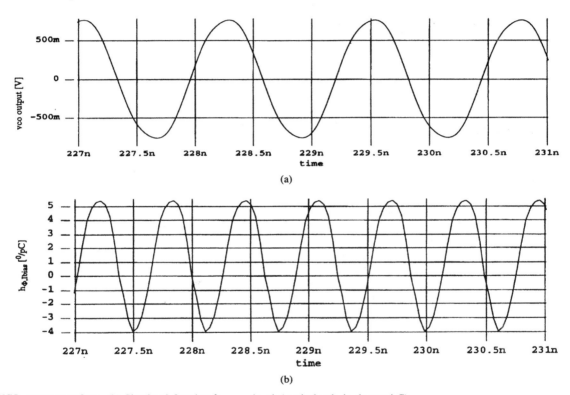

(a)

(b)

Fig. 6. (a) VCO output waveform. (b) Simulated function $h_{\Phi, I_{\text{bias}}}(t, \tau)$ (vertical axis in degrees/pC).

across the nodes "out" and "outb" (Fig. 2), and it has the value $2 \cdot R_{\text{eq}} = 720 \ \Omega$. The on-chip resistors R_1 and R_2 have the minimum value required to damp the parasitic oscillation. The degradation in phase noise at 100-kHz frequency offset from an 800-MHz carrier due to these resistors is no more than 0.5 dB. The spectrum for $\overline{i_{\text{bias}}^2}$ is simulated using small-signal SPICE analysis and is shown in Fig. 5.

The behavioral test generator was used to inject 64 pulses as shown in Fig. 3(a). The VCO was run at 800 MHz while the test generator was applied. The number of samples per period was chosen to be large enough to minimize aliasing effects.

The excess phase induced by each pulse $h_{\Phi}(t, \tau)$ generates a periodic function with respect to the launch phase of the pulses.

Fig. 6 shows the function $h_{\Phi, I_{\text{bias}}}(t, \tau)$ for current pulses injected at the tail of the emitter-coupled pair. This function has a periodicity that is half the oscillation period. To obtain more meaningful information on the phase sensitivity for perturbations in the tail current of the VCO, the function $h_{\Phi, I_{\text{bias}}}(t, \tau)$ is plotted together with the oscillation output waveform. It can be seen that for perturbations injected around the zero crossings and the peaks of the oscillation, the phase sensitivity is close to zero and reaches its maxima

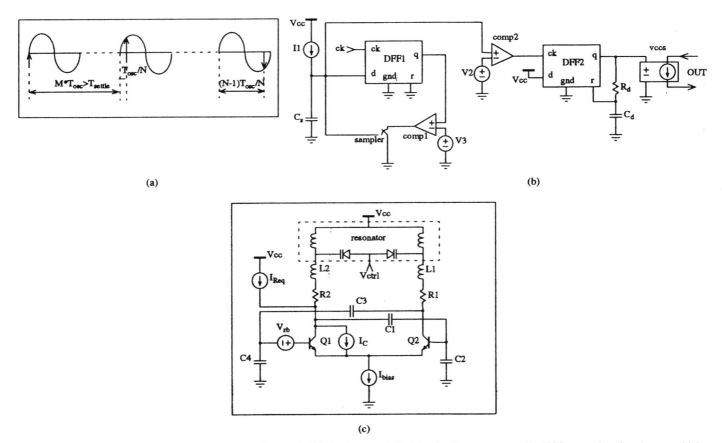

(a)

(b)

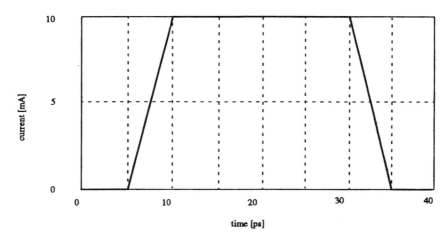

Fig. 3. (a) Sequence of pulses used to excite the oscillator. (b) Block diagram of the behavioral test generator. (c) VCO test points for phase sensitivity to injected charge.

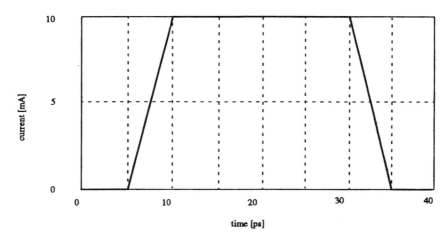

Fig. 4. Impulse shape.

is almost constant from DC to 25 GHz, which is sufficient for this design that uses a bipolar process with $f_t = 25$ GHz. Fig. 3(c) shows the injection points into the oscillator. Current source I_{bias} is injected to simulate the effect of the noise generated by the bias generator and the AAC circuit $\overline{i^2_{\text{bias}}}$. Current source I_c simulates the effect of the collector shot noise of transistor Q_1, $\overline{i^2_c}$; voltage source V_{rb} the effect of the noise voltage generated by r_b, $\overline{v^2_{r_b}}$; and current source $I_{R_{\text{eq}}}$ the effect of the resistive losses in the resonator $\overline{i^2_{R_{\text{eq}}}}$. The noise densities are as follows:

$$\overline{i^2_c}/\Delta f = 2q \cdot i_{c,\max} \tag{4}$$

$$\overline{v^2_{r_b}}/\Delta f = 4kTr_b \tag{5}$$

$$\overline{i^2_{R_{\text{eq}}}}/\Delta f = \frac{4kT}{R_{\text{eq}}}. \tag{6}$$

The collector shot noise of transistor Q_1 is evaluated at the peak of the collector current ($i_{c,\max} = 1.6$ mA). The cyclostationarity of $\overline{i^2_c}$, which is due to the time-varying nature of the collector current of Q_1, will be considered further in this analysis. The base resistance of Q_1, r_b is 8.6 Ω. The losses in the resonant circuit are expressed by the equivalent resistance R_{eq}. The total resistive loss in the resonator is calculated

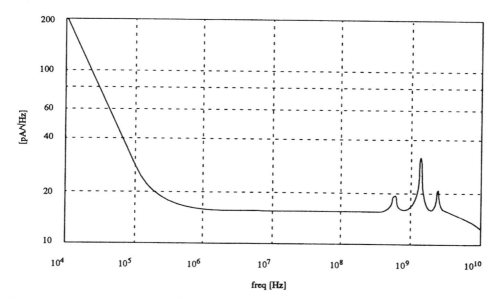

Fig. 5. Tail-current noise spectrum.

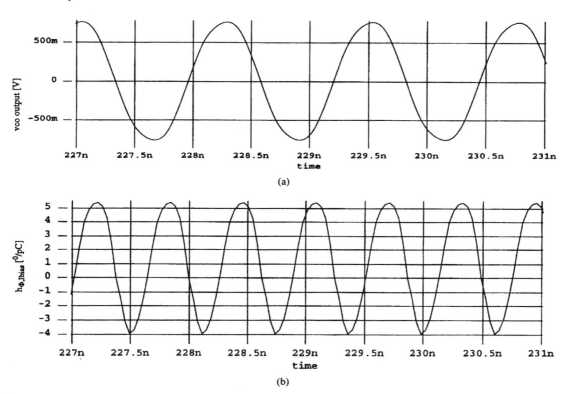

Fig. 6. (a) VCO output waveform. (b) Simulated function $h_{\Phi, I_{\text{bias}}}(t, \tau)$ (vertical axis in degrees/pC).

across the nodes "out" and "outb" (Fig. 2), and it has the value $2 \cdot R_{\text{eq}} = 720 \ \Omega$. The on-chip resistors R_1 and R_2 have the minimum value required to damp the parasitic oscillation. The degradation in phase noise at 100-kHz frequency offset from an 800-MHz carrier due to these resistors is no more than 0.5 dB. The spectrum for $\overline{i_{\text{bias}}^2}$ is simulated using small-signal SPICE analysis and is shown in Fig. 5.

The behavioral test generator was used to inject 64 pulses as shown in Fig. 3(a). The VCO was run at 800 MHz while the test generator was applied. The number of samples per period was chosen to be large enough to minimize aliasing effects.

The excess phase induced by each pulse $h_{\Phi}(t, \tau)$ generates a periodic function with respect to the launch phase of the pulses.

Fig. 6 shows the function $h_{\Phi, I_{\text{bias}}}(t, \tau)$ for current pulses injected at the tail of the emitter-coupled pair. This function has a periodicity that is half the oscillation period. To obtain more meaningful information on the phase sensitivity for perturbations in the tail current of the VCO, the function $h_{\Phi, I_{\text{bias}}}(t, \tau)$ is plotted together with the oscillation output waveform. It can be seen that for perturbations injected around the zero crossings and the peaks of the oscillation, the phase sensitivity is close to zero and reaches its maxima

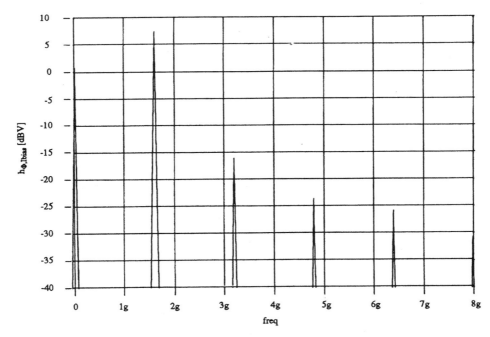

Fig. 7. Frequency spectrum of $h_{\Phi, I_{\text{bias}}}(t, \tau)$ (vertical axis in dBV).

for $(2i - 1/4) \cdot T_{\text{osc}}$, where $i = 0, 1, \cdots, n.$ The spectrum of $h_{\Phi, I_{\text{bias}}}(t, \tau)$ is shown in Fig. 7. As expected, there are harmonics at multiples of double the oscillation frequency. The harmonics mix with the noise around these frequencies and contribute to the total phase noise. For calculation of the phase-noise contribution from the tail current, the Fourier coefficients of $h_{\Phi, I_{\text{bias}}}(t, \tau), C_k$, need to be computed. The phase noise contributed by the tail current is given by the sum of the squares of the phase errors from each harmonic of $h_{\Phi, I_{\text{bias}}}(t, \tau)$ [4]

$$L_{I\text{-tail}}(\Delta\omega) = \frac{\sum (C_k)^2 \left[\overline{i_{\text{bias}}^2(k, \Delta\omega)}/(\Delta f)\right]}{4 \cdot (\Delta\omega)^2} \quad (7)$$

where $\overline{i_{\text{bias}}^2(k, \Delta\omega)}/(\Delta f)$ is the noise density of the bias circuitry at offset from the kth harmonic of the oscillation frequency. In this analysis, the summation is performed over the first five harmonics. Higher order harmonics have insignificant contribution to the phase noise.

Of particular interest is the dc component, coefficient C_0. The corner frequency where the $1/f^3$ region of the phase noise intersects the $1/f^2$ region is given by [4]

$$\omega_C = \omega_{1/f} \cdot \left(\frac{C_0}{C_1}\right)^2 \quad (8)$$

where $\omega_{1/f}$ is the $1/f$ corner frequency of the tail-current noise. Since $\omega_{1/f}$ is 200 kHz, (8) predicts a value for ω_C of 3 kHz, which is in agreement with the phase-noise measurements. To lower ω_C, the corner frequency of the low-pass filter, set by $R_{12} - C_7$, can be decreased by increasing C_7.

A similar analysis is carried out for the collector shot noise, the base resistance thermal noise of Q_1, and for thermal noise $\overline{i_{R_{\text{eq}}}^2}$. Thermal noise generated by the base biasing resistors, R_4 and R_5 in Fig. 2, is shunted by the relatively low impedance seen across capacitors C_2 and C_4, respectively, at

the frequencies of interest, and is therefore neglected. While the noise spectral density of the base resistance and the parallel resistance of the resonator are assumed to be stationary, the effect of cyclostationarity needs to be considered for the collector shot noise. For this, the unit impulse response function $h_{\Phi, I_c}(t, \tau)$ is multiplied by the collector current waveform and normalized to the collector peak current. To account for the cyclostationarity of the collector shot noise, an effective periodic function $h_{\Phi\text{eff}}(t, \tau)$ is defined as

$$h_{\Phi\text{eff}}(t, \tau) = \frac{h_{\Phi, I_c}(t, \tau) \cdot i_c(\tau)}{i_{c,\max}}. \quad (9)$$

In the above equation, $i_{c,\max}$ is the same as in (4). The functions $h_{\Phi\text{eff}}(t, \tau)$ and $h_{\Phi, I_c}(t, \tau)$ are shown in Fig. 8. It can be seen that the effect of cyclostationarity is to reduce the collector shot noise contribution. $h_{\Phi\text{eff}}(t, \tau)$ has a period equal to the oscillation period. Again, in order to see the effect of this noise source, the oscillation output waveform and the collector current of Q_1 are plotted in the same figure. The function $h_{\Phi\text{eff}}(t, \tau)$ reaches its maximum when the collector current is close to the peak, and it reaches the minimum when i_c is around zero. The plot of Fig. 8(b) shows the collector current's going negative for a small fraction of the oscillation period. This is due to the displacement current flowing through the collector-base capacitance C_μ. The internal collector current, which generates collector shot noise, does not go negative. This displacement current, however, is a negligible error in the noise estimation when the total collector current is considered. The Fourier components of $h_{\Phi\text{eff}}(t, \tau)$ in Fig. 9 show that the collector shot noise is mixed mostly with the first and second harmonics of the oscillation frequency to contribute to the total phase noise. However, the collector shot noise of Q_1 is the dominant noise contributor (as will be seen below), and an optimization based on the feedback ratio n was performed. To do this, the ratio n was varied, and the rms value of $h_{\Phi\text{eff}}(t, \tau)$

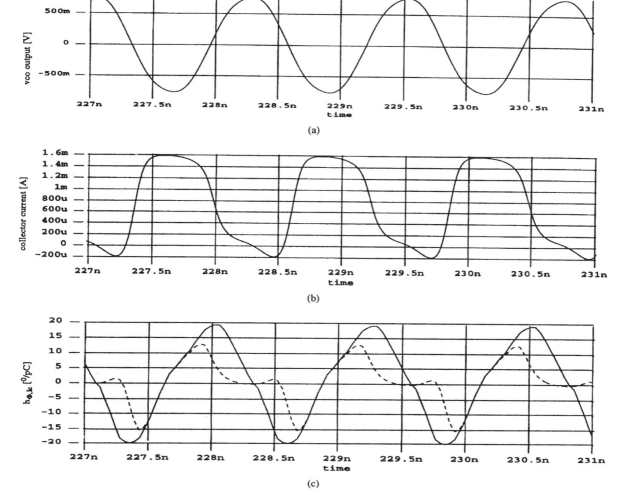

Fig. 8. (a) VCO output waveform. (b) Collector current of Q_1. (c) Simulated functions $h_{\Phi,I_c}(t,\tau)$ (continuous line) and $h_{\Phi\text{eff}}(t,\tau)$ (dashed line). Vertical axis in degrees/pC.

was calculated using Parseval's relation

$$2 \cdot (h_{\Phi\text{eff,rms}}(t,\tau))^2 = \sum C_k^2. \tag{10}$$

The simulations show that if the feedback ratio n is too low, $h_{\Phi\text{eff}}(t,\tau)$ increases, and so does the contribution to the total phase noise. This is due to the fact that the base-emitter junctions of Q_1 and Q_2 are overdriven. Although the function $h_{\Phi,I_c}(t,\tau)$ remains unchanged, the collector current approaches the form of a square wave, leading to an increase in $h_{\Phi\text{eff}}(t,\tau)$. For very large values of n, the AAC circuit forces more bias current into the VCO, and both the collector shot noise of Q_1 [(4)] and the tail noise current [Fig. (5)] increase. Although the phase-noise level does not degrade for values of n higher than six, this comes at the expense of increased power consumption. The phase-noise level achieved for a given power consumption can be used as a figure of merit (FoM), which is defined as

$$\text{FoM} = \frac{(S/N)\,[\text{dBc/Hz}]}{P_{\text{dc}}[\text{mW}]}. \tag{11}$$

Both the phase-noise level and the figure of merit as a function of the feedback ratio n are shown in Fig. 10. It can be seen that

the optimum feedback ratio for this design is around 4.5. Using the optimum feedback ratio, the small-signal loop gain is

$$T_l = \frac{G_m \cdot R_{\text{eq}}}{n} = 2.5 \tag{12}$$

where $I_{\text{bias}} = 1.6$ mA.

This shows that with the use of the AAC circuit, phase-noise optimization (through choosing the optimal feedback ratio) can be accomplished independently while ensuring proper oscillator startup.

A similar analysis was carried out for the thermal noise generated by the base resistance of transistor Q_1 and the equivalent resistance of the resonant tank. Table I gives the contributions of various noise sources to the total phase noise of the VCO at 100-kHz offset from an 800-MHz carrier. These values are given as noise-to-signal ratios. The contribution of the base resistance thermal noise is simulated to be less than 5% of the total phase noise. The noise contributed by the resonator is more significant (the loaded Q of the resonant tank is simulated to have a value of 16), with the dominant noise contributor being the collector shot noise.

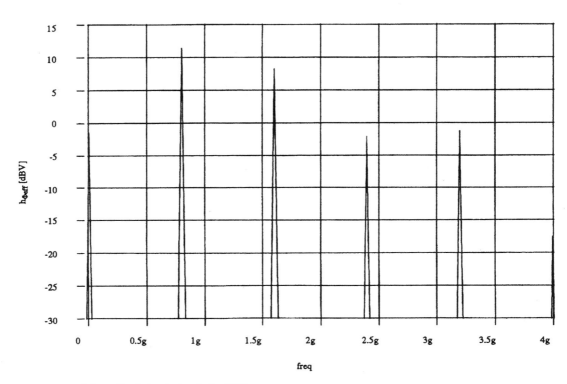

Fig. 9. Frequency spectrum of $h_{\Phi\text{eff}}(t, \tau)$ (vertical axis in dBV).

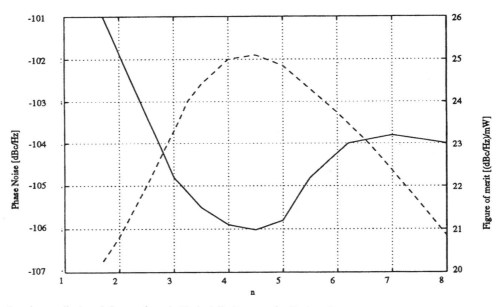

Fig. 10. Phase noise (continuous line) and figure of merit (dashed line) versus feedback ratio n.

TABLE I
NOISE CONTRIBUTIONS AT 100-kHz OFFSET FROM AN 800-MHz CARRIER

	N/S	% of total
Collector shot noise $\overline{i_c^2}/\Delta f$	$2 \cdot 0.816 \cdot 10^{-11}$	68%
Noise from tank losses $\overline{i_{R_{eq}}^2}/\Delta f$	$2 \cdot 0.256 \cdot 10^{-11}$	21%
Tail current noise $\overline{i_{bias}^2}/\Delta f$	$0.168 \cdot 10^{-11}$	7%
Noise from base resistance $\overline{v_{r_b}^2}/\Delta f$	$2 \cdot 0.05 \cdot 10^{-11}$	4%

Although the noise contribution from the tail current is not important at this offset frequency, it becomes the major noise source at offset frequencies less than 3 kHz.

The factor of two for some of the noise sources in Table I accounts for noise sources that are considered twice due to the circuit symmetry [5]. The sum of these values gives a noise-to-signal ratio of -106.2 dBc/Hz.

A phase-noise simulation was also carried out using Spectre-eRF as a comparison. It can be seen in Fig. 11 that the results are within 2 dB in the $1/f^2$ region of the spectrum. SpectreRF predicts a phase-noise level of -108 dBc/Hz at 100-kHz offset.

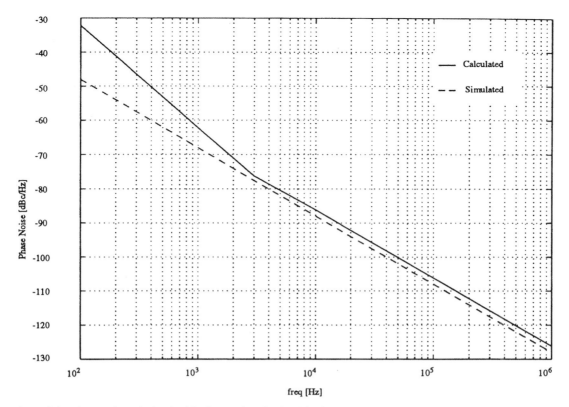

Fig. 11. Comparison of the phase noise calculated with phase noise simulated in SpectreRF.

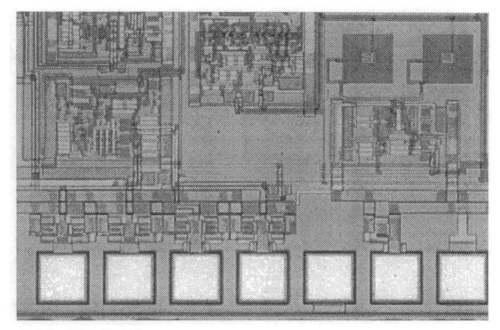

Fig. 12. Microphotograph of the VCO with AAC.

However, the current version of SpectreRF did not predict the $1/f^3$ region of the spectrum, which was calculated in the above analysis and confirmed by measurements.

IV. MEASUREMENT RESULTS

The VCO with AAC was fabricated in a bipolar process with $f_t = 25$ GHz. The microphotograph of the circuit is shown in Fig. 12. The VCO can operate from 300 MHz up to 1.2 GHz with different resonators. The output spectrum is shown in Fig. 13. It can be seen that the second harmonic is -43 dBc and the third harmonic is -58 dBc. The effect of quasi-linear operation of the VCO made possible by the AAC loop is to reduce the level of harmonic distortion. For a given resonator, the VCO with AAC reduces the third-order harmonic distortion by 6 dB when compared to a VCO without AAC. The phase noise is -106 dBc/Hz at 100-

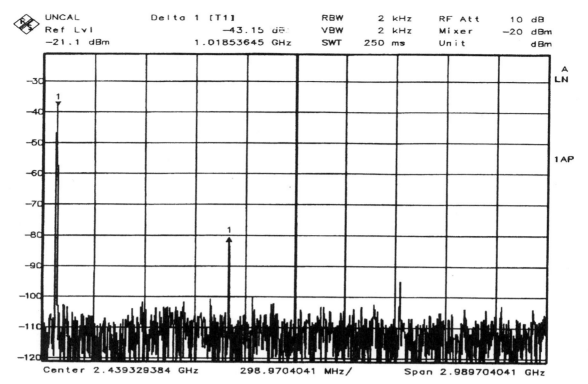

UNCAL   Delta 1 [T1]   RBW   2 kHz   RF Att   10 dB
Ref Lvl   -43.15 dB   VBW   2 kHz   Mixer   -20 dBm
-21.1 dBm   1.01853645 GHz   SWT 250 ms   Unit   dBm

Center 2.439329384 GHz   298.9704041 MHz/   Span 2.989704041 GHz

Fig. 13.   VCO output spectrum.

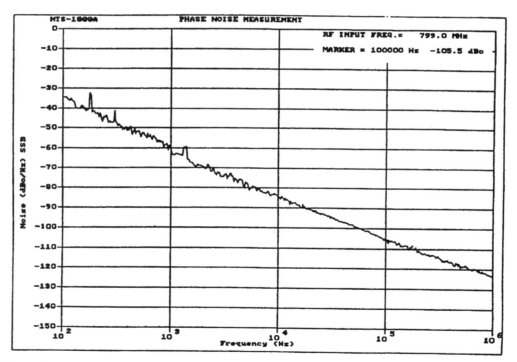

Fig. 14.   Measured phase noise.

kHz offset for a carrier frequency of 800 MHz, which is in agreement with the analysis presented in this paper (Fig. 14). The VCO core consumes 1.6 mA from a 2.7-V power supply. The remaining circuits used for the AAC (rectifier, voltage reference, and amplifier) consume 0.25 mA. If better phase-noise performance is desired, only the current consumption in the VCO core needs to be increased, while the consumption of the AAC circuits remains unchanged.

V. CONCLUSIONS

In this paper, the possibilities of developing a low-noise, low-power VCO with capabilities for wireless applications have been explored. An automatic amplitude control circuit was implemented, which allows the choice of the optimum oscillator feedback ratio for noise performance without being constrained by startup considerations. At the same time, the automatic amplitude control allows proper VCO operation for a wide range of the resonator quality factor. A novel method was used to study the phase-noise performance of the VCO. The method predicts results that are close to the measurements and allows the designer to obtain detailed information about the processes that contribute to oscillator phase noise.

ACKNOWLEDGMENT

The authors would like to thank Dr. C. Hull and R. Magoon for helpful discussions.

REFERENCES

[1] J. L. Tham, M. Margarit, B. Pregardier, C. Hull, and F. Carr, "A 2.7 V 900 MHz/1.9 GHz dual-band transceiver IC for digital wireless communication," in *Proc. CICC*, 1998, p. 559.
[2] J. L. Tham, "Integrated radio frequency LC voltage-controlled oscillators," College of Engineering, University of California, Berkeley, Electronics Research Laboratory Memo., 1995.
[3] P. Davis, P. Smith, E. Campbell, J. Lin, K. Gross, G. Bath, Y. Low, M. Lau, Y. Degani, J. Gregus, R. Frye, and K. Tai, "Si-on-Si integration of a GSM transceiver with VCO resonator," in *Proc. ISSCC 1998*, vol. 41, Feb. 1998, p. 248.
[4] A. Hajimiri and T. H. Lee, "A general theory of phase noise in electrical oscillators," *IEEE J. Solid-State Circuits*, vol. 33, pp. 179–194, Feb. 1998.
[5] C. D. Hull and R. G. Meyer, "A systematic approach to the analysis of noise in mixers," *IEEE Trans. Circuits Syst. I*, vol. 40, pp. 909–919, Dec. 1993.

Mihai A. Margarit received the Dipl.Ing. degree in electrical engineering from the "Politehnica" University Bucharest, Romania, in 1984. He currently is pursuing the Ph.D. degree in electrical engineering at Simon Fraser University, Burnaby, B.C., Canada.

Since 1984, he has worked in analog circuit design for the National Institute for Microelectronics, Bucharest, the Fraunhofer Institute, Erlangen, Germany, and Simon Fraser University, Vancouver, Canada. He is currently with Rockwell Semiconductor Systems, Newport Beach, CA, where he is a Senior Design Engineer working on high-frequency circuits for wireless communication applications.

Joo Leong (Julian) Tham (S'88–M'96) received the B.S. degree in electrical engineering (with highest honors) from the University of California, Santa Barbara, and the M.S. degree in electrical engineering from the University of California, Berkeley.

He has worked at Raytheon and Trimble Navigation. His previous work includes autocalibration systems and global positioning system receivers. From 1993 to 1999, he was with Rockwell Semiconductor Systems, Newport Beach, CA, where he was a Principal Design Engineer and Manager working on radio-frequency integrated circuits for wireless communication applications. He currently is with Maxim Integrated Products, Sunnyvale, CA. His current interests are in the areas of high-frequency circuit design and integrated transceiver architectures.

Mr. Tham is a member of Eta Kappa Nu, Tau Beta Pi, and the Golden Key Honor Society. He was named Rockwell Semiconductor Systems Engineer of the Year in 1995.

Robert G. Meyer (S'64–M'68–SM'74–F'81) was born in Melbourne, Australia, on July 21, 1942. He received the B.E., M.Eng.Sci., and Ph.D. degrees in electrical engineering from the University of Melbourne in 1963, 1965, and 1968, respectively.

In 1968, he was an Assistant Lecturer in electrical engineering at the University of Melbourne. Since September 1968, he has been with the Department of Electrical Engineering and Computer Sciences, University of California, Berkeley, where he is now a Professor. His current research interests are high-frequency analog integrated-circuit design and device fabrication. He has been a Consultant on electronic circuit design for numerous companies in the electronics industry. He is a coauthor of *Analysis and Design of Analog Integrated Circuits* (New York: Wiley, 1993) and Editor of *Integrated Circuit Operational Amplifiers* (New York: IEEE Press, 1978).

Dr. Meyer was President of the IEEE Solid-State Circuits Council and was an Associate Editor of the IEEE JOURNAL OF SOLID-STATE CIRCUITS and of the IEEE TRANSACTIONS ON CIRCUITS AND SYSTEMS.

M. Jamal Deen (S'81–M'86–SM'92) was born in Georgetown, Guyana. He received the B.Sc. degree in physics and mathematics from the University of Guyana in 1978 and the M.S. and Ph.D. degrees in electrical engineering and applied physics from Case Western Reserve University, Cleveland, OH, in 1982 and 1985, respectively.

From 1978 to 1980, he was an Instructor of physics at the University of Guyana. From 1980 to 1983, he was a Research Assistant at Case Western Reserve University. He was a Research Engineer (1983–1985) and an Assistant Professor (1985–1986) at Lehigh University, Bethlehem, PA. In 1986, he joined the School of Engineering Science, Simon Fraser University, Vancouver, BC, Canada, as an Assistant Professor and since 1993 has been a full Professor. He was a Visiting Scientist at the Herzberg Institute of Astrophysics, National Research Council, Ottawa, Ont., Canada, in summer 1986, and he spent his sabbatical leave as a Visiting Scientist at Northern Telecom, Ottawa, in 1992–1993. He was also a Guest Professor in the Faculty of Electrical Engineering, Delft University of Technology, The Netherlands, in summer 1997 and a CNRS scientist at the Physics of Semiconductor Devices Laboratory, Grenoble, France, in summer 1998. His current research interests include integrated devices and circuits; device physics, modeling, and characterization; and low-power, low-noise, high-frequency circuits.

Dr. Deen is a member of Eta Kappa Nu, the American Physical Society, and the Electrochemical Society. He was a Fulbright-Laspau Scholar from 1980 to 1982, an American Vacuum Society Scholar from 1983 to 1984, and an NSERC Senior Industrial Fellow in 1993.

A Fully Integrated VCO at 2 GHz

Markus Zannoth, Bernd Kolb, Joseph Fenk, and Robert Weigel, *Senior Member, IEEE*

Abstract—A fully integrated voltage-controlled oscillator at a frequency of 2 GHz with low phase noise has been implemented in a standard bipolar process with a f_t of 25 GHz. The design is based on an LC-resonator with vertical-coupled inductors. Only two metal layers have been used. The supply voltage of the oscillator is 2.7 V. The phase noise is only −136 dB/Hz at 4.7-MHz frequency offset. A tuning range of 150 MHz is achieved with integrated tuning diodes.

Index Terms— Bipolar, fully integrated VCO, noise requirement for cordless phones.

I. INTRODUCTION

WITH the fast growth of the wireless application market, there is a growing need for smaller designs and higher levels of integration for the reduction of costs and size. Because of the very poor performance of integrated resonators on silicon IC's, local RF oscillators are difficult to integrate with regard to the phase noise requirements. At the moment, external resonators are used with external hyperabrupt tuning diodes. The integrated tuning diodes have low performance because it is not possible to produce a hyperabrupt pn-junction in our standard bipolar process. The limiting factor is the inductor of the resonator, which can only archive quality factors Q of about four. While the performance of mobile telecommunication standards like global system for mobile communication (GSM) or digital communication system (DSC-1800) requires such high noise performance, which cannot be reached in our technology with the use of full integration of the local oscillator, the requirements for cordless phones like for the Digital European Cordless Telecommunications (DECT) standard seem to be achievable. In a DECT system, the critical point concerning oscillator phase noise is the emissions due to modulation. There the emitted power at the output in the third adjacent channel is specified to be smaller than 20 nW, which equals −47 dBm [1]. With a maximum output power of +25 dBm, there is a difference of 72 dB. The specified measurement filter to get this power has a bandwidth of 1 MHz. So the noise requirement at this point becomes −132 dB/Hz, which results from: $L = P_{\text{noise}} - P_{\text{out}} - 10\log(BW) = -47$ dBm − 25 dBm − 60(dB/Hz) = −132(dB/Hz). This is the difference between the noise level and the output-power normalized to 1 Hz. The frequency offset for this specification is the start frequency of the measurement filter. As this is in the third

Manuscript received April 10, 1998; revised July, 17, 1998.
M. Zannoth, B. Kolb, and J. Fenk are with Siemens AG, Muenchen D-81541, Germany.
R. Weigel is with the University of Linz, Institute for Communication and Information Engineering, Linz A-4040 Austria.
Publisher Item Identifier S 0018-9200(98)08854-4.

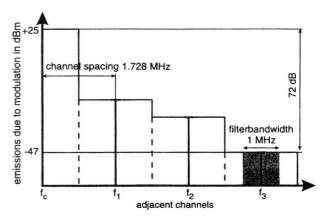

Fig. 1. DECT specification.

adjacent channel, three times the channel spacing of 1.728 MHz has to be taken (see Fig. 1). The filter is centered in the channel, half of the filter-bandwidth of 1 MHz has to be subtracted to get the offset frequency

$$f_{\text{offset}} = 3 * 1.728 \text{ MHz} - \frac{1 \text{ MHz}}{2}$$
$$= 4.684 \text{ MHz} \approx 4.7 \text{ MHz.}$$

Normalized to a 100-kHz frequency offset, the requirement is −98.6 dB/Hz, if the phase noise has a constant slope of −20 dB/decade, as assumed by [3]. These are the requirements for the main transmitter-voltage-controlled oscillator (TX-VCO), when the following blocks are not dominating in noise. This is indeed fulfilled.

With fully integrated oscillators these requirements seem to be possible to realize. This paper presents a fully integrated oscillator, which achieves the required specification. It uses an LC oscillator consisting of integrated tuning diodes and integrated vertically coupled inductors. In this design only two metal layers in a standard bipolar process are used.

II. OSCILLATOR WITH COUPLED INDUCTORS

The phase noise in oscillators depends on the quality factor of the resonator, the noise figure of the amplifier creating a negative resistance, and the energy in the resonator [3], [4]. For low phase noise, passive resonators are chosen. With active inductors, noise is added by the active devices that cannot be compensated by increasing the quality factor. For best integration and reproducibility spiral inductors are used, as described in [5]. The quality factor of the resonator is mainly limited by the inductors. Here it is limited to four, as only two metal layers are available. One metal layer has a thickness of 0.8 μm. With limited chip area for the inductor and the use of

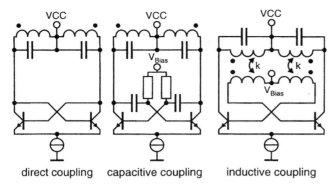

direct coupling   capacitive coupling   inductive coupling

Fig. 2. Possibilities of feedback.

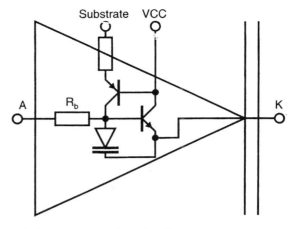

Fig. 3. Equivalent circuit of the tuning diode.

vertically coupled inductors, a quality factor of only four was achievable without changing technology parameters.

The simplest way to reduce phase noise is increasing the resonator energy by applying higher voltages to the resonator. In this design an emitter-coupled pair with cross feedback is used as a negative resistance, which is responsible for undamping the resonator. The limit of the maximum oscillation amplitude depends on the feedback. There are three ways of feeding the output-signal back to the input (see Fig. 2). The easiest way is direct coupling, where no biasing network is needed and very low power consumption can be achieved. Using direct coupling the voltage across the resonator is limited by the base-collector diode of the transistors. When forward biased, this diode inserts additional damping and current noise to the resonator causing increased phase noise. With capacitive coupling [6] this can be avoided. Here no resistive element is inserted into the feedback. With capacitive feedback a phase noise of -100 dB/Hz at 100 kHz could be achieved by [6], having quality factors of eight. The disadvantage is the need of a high-impedance biasing network at the transistors base. This biasing network can be realized by noisy resistors or by large inductors that cost a lot of chip-space. If resistors are used, uncorrelated noise is introduced to both halfwaves of the signal, when the oscillator acts in its linear region. This noise is nearly negligible, when a low Q resonator is used. In our case the impedance at the input of the feedback amplifier is about 500 Ω. This impedance consists of the feedback capacitor and the tank impedance at resonance. The Bias resistor in parallel is about 4 $k\Omega$, and so the effect of adding noise is not very dominant. With inductive coupling the bias current can be fed through the inductor. This allows connecting a low-impedance biasing network which can be made of a voltage source. The advantage of connecting a voltage source directly to the circuit is the absence of resistive elements that cause white noise, which would be converted to phase noise by the nonlinear elements. Every DC path can be blocked carefully against emissions from the supplies without any resistive element. The maximum voltage at the resonator can be adjusted by the biasing voltage so that the base-collector diode is not the limiting element. Now the amplitude of the swing is limited by the base emitter diode of the transistors and the limitation of the current source. The energy in the resonator can be increased and so the phase noise is reduced. Now the maximum voltage is not one diode-voltage, as in the

case of direct coupling. In the presented design the resonator voltage reaches a value of 3 V_{PP}.

The limiting elements for the maximum voltage of the resonator are now the two serial-connected tuning diodes. To decrease their voltage without reducing the resonator-energy, a capacitor is added in series at the cost of tuning range. This capacitor is also responsible for getting a linear tuning characteristic. To provide the DC-path for the tuning diodes, resistors are connected in parallel to the coupling capacitances. These resistors are negligible in sight of reducing the quality factor because they have a large value of 1 $k\Omega$, which is much larger than the capacitances impedance of 40 Ω (see Fig. 6). The quality factor of the capacitance is about 24, which is at the same range as the varactor. These quality factors are negligible high relative to that of the inductor.

The inductors are produced as symmetrical quadratic spirals. At our standard bipolar process only two metal layers could be used to create vertically coupled inductors. The crosses are made in the gap between two metal lines (see Fig. 7). The cost of this technique is the wide gap between the lines, which causes an increment of the size and parasitic effects like series resistance and substrate capacitance. The quality factor is as low as four. This is caused by the technology, where the metal layers have a poor conductivity and high capacitances to the medium-doped substrate. In this design an inductor of 2.7 nH was used. Its series resistance is 4.2 Ω. The coupling factor was estimated to 0.85. The values of the equivalent circuit (see Fig. 8) where first calculated by algorithms from [7] and then fitted to measurement. The coupling capacitor was estimated from the plate capacitance of the two metal layers.

For tuning, the base-emitter diode of a transistor is used (see Fig. 3). This has the disadvantage of a high series resistance (base resistance) of 2.6 Ω and a relatively low capacitance variation by a factor of 1.75 applying a voltage difference of 2.7 V (see Fig. 4). However, this represents the only way to create a tuning diode without changing the standard bipolar process, where no hyperabrupt pn-junctions are available. The Q of this varactor was simulated to be about 25 (see Fig. 5) when it is calculated from $1/(jwRC)$. The base-collector diode could not be used for tuning, because it does not have such a large capacitance variation.

283

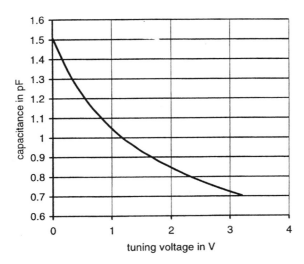

g. 4. Capacitance of the tuning diode.

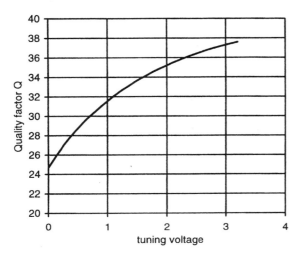

Fig. 5. Quality factor of the tuning diode.

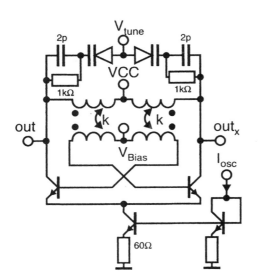

'ig. 6. VCO with coupled inductors.

To get the signal into a 50-Ω measurement system a buffer is added (see Fig. 9). As the signal is taken directly from the

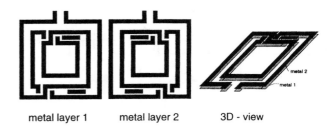

metal layer 1   metal layer 2   3D - view

Fig. 7. Simplified layout of the inductor.

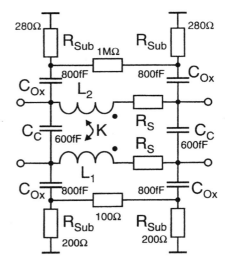

Fig. 8. Equivalent circuit of the inductor.

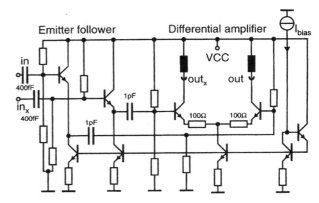

Fig. 9. Output buffer.

resonator, a high impedance is required for minimizing the effects of the load. The signal is fed through small coupling capacitances (400 fF) to emitter-followers that provide this high input impedance. This first stage drives a differential amplifier with open collector outputs. A balun can be connected that transforms the differential signal to a single ended one that can be connected to 50 Ω. The current-consumption of the amplifier is about 6 mA. Its output power is about −8 dBm at 50 Ω, which is enough for noise measurements.

III. RESULTS

The measured phase noise of the oscillator can be calculated from expressions by Leeson [3] and has a slope of −20 dB/Dec

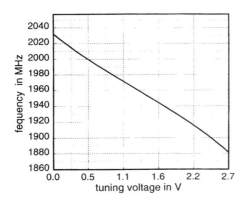

Fig. 10. Measured tuning characteristic.

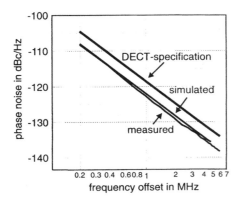

Fig. 11. Measured and simulated phase noise.

measured at offset-frequencies between 100 kHz and 50 MHz, which represent the measurement limits. If the phase noise is calculated from

$$\mathcal{L}\{\Delta\omega\} = \frac{kTR_{\text{eff}}[1+A]\left(\frac{\omega_0}{\Delta\omega}\right)^2}{P_{\text{signal}}} \quad [5]$$

we get a value of -135 dB/Hz at an offset frequency of 4.7 MHz. The effective series resistance is about 16 Ω. It is taken from the series resistances of the inductor, the tuning diodes, and the coupling capacitances. The amplitude of the oscillation was simulated to be 1.5 V_{Peak} (3 V_{PP}) at a resonance frequency of 2 GHz. The noise factor of the amplifier was supposed to be about two. The expressions from [3] give a approximation for the expected noise. For better calculation, nonlinear effects have to be considered. In [12], such calculations as the correlation between waveform and phase noise are shown.

The simulation with spectre RF shows nearly the same values as the measured ones. In Fig. 11 the simulated and measured noise are shown up to an offset frequency of 6 MHz. In the simulation the equivalent circuits are taken from above. The simulator uses nonlinear methods [13], [14] to calculate noise. It gives nearly the same results as the measurement and the small signal approximation from above.

The measurement shows that the free running oscillator has a resonant frequency of 1.96 GHz, a phase noise of -136

TABLE I
SUMMARY OF THE VCO CHARACTERISTICS

Supply voltage	2.7 V
Current of oscillator core	12 mA
Current of amplifier	6 mA
Output power	-8 dBm
Center frequency	1.96 GHz
Tuning constant (KVCO)	-55 MHz/V
Tuning range (0...2.7 V)	150 MHz
Phase noise @ 100 kHz	-102 dBc/Hz
Phase noise @ 4.7 MHz	-136 dBc/Hz
Suppression of harmonics	23 dB
Size of one inductor	215 μm $\times$ 215 μm
Chip size without Pads	600 μm $\times$ 600 μm

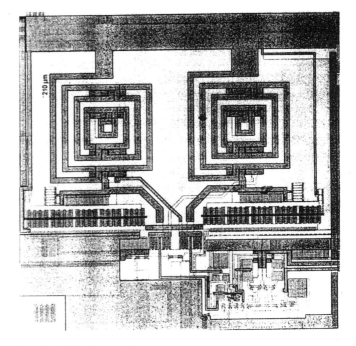

Fig. 12. Chip photograph.

dB/Hz at 4.7 MHz offset frequency, and a VCO constant of -55 MHz/V (see Figs. 10 and 11). The phase noise is measured in the middle of the tuning range. With a thicker metal layer (1.2-μm aluminum instead of 0.8 μm) a phase noise of even -137 dB/Hz can be achieved. This improvement is achieved by reducing the series resistance of the inductor and increasing its quality factor. The tuning characteristic is nearly linear (see Fig. 10) and the noise performance varies only less than 1 dB at the whole tuning range from 0 V–2.7 V. The linearity of this characteristic is improved by the series capacitance (Fig. 6) added to the varactor diode. The current through the oscillator core is about 12 mA; the supply voltage is 2.7 V. At tuning voltages above the supply voltage the varactor diodes get forward biased and so the frequency stays nearly constant and the noise rises.

This occurs because of the reduction of the quality factor and the introduction of additional current noise due to the forward-biased diodes.

IV. CONCLUSION

A fully integrated bipolar VCO is realized (see Fig. 12) that achieves a measured phase noise of -136 dB/Hz at 4.7 MHz. The oscillator has a linear tuning characteristic with a tuning range of 150 MHz at a center frequency of 1.96 GHz. Further characteristics are given in Table I.

In this design two metal layers are used to build vertically coupled integrated inductors. These have quality factors of about four. Integrated varactor diodes are implemented by using base-emitter diodes of transistors. With this design the noise requirements of the DECT-specification of -132 dB/Hz at 4.7 MHz frequency offset are achieved with a margin of 4 dB. The output power is -8 dBm at 50 Ω, with a center frequency of 1.95 GHz. For the use of this oscillator in a DECT product, the varactor-capacitance will be increased until the required center frequency of 1.88 GHz is reached. The design has been realized in standard high-volume bipolar process with an f_T of 25 GHz.

REFERENCES

[1] ETSI, *Digital European Cordless Telecommunications (DECT) Common Interface, Part 2: Physical Layer*, Oct. 1992.
[2] L. L. Larson, *RF and Microwave Circuit Design for Wireless Communications*. Boston: Artech House, 1996.
[3] B. D. Leeson, "A simple model of feedback oscillator noise spectrum," *Proc. Lett. IEEE*, pp. 329–330, Feb. 1966.
[4] G. Sauvage, "Phase noise in oscillators: A mathematical analysis of Leeson's model," *IEEE Trans. Instrum. Meas.*, vol. IM-26, pp. 408–410, Dec. 1977.
[5] J. Craninckx and M. S. J. Steyaert, "A 1.8-GHz low-phase-noise CMOS VCO using optimized hollow spiral inductors," *IEEE J. Solid-State Circuits*, vol. 32, pp. 736–744, May 1997.
[6] G. Palmisano, M. Paparo, F. Torrisi, and P. Vita, "Noise in fully integrated PLL's," in *Proc. 6th Workshop Advances in Analog Circuit Design AACD'97*, Como, Italy, pp. 1–19.
[7] J. Crols, P. Kinget, J. Craninckx, and M. Steyaert, "An analytical model of planar inductors on lowly doped silicon substrates for high frequency analog design up to 3 GHz," in *IEEE Symp. VLSI Circuit Dig. Tech. Papers*, 1996, pp. 28–29.
[8] J. N. Burghartz, M. Soyuer, and K. A. Jenkins, "Microwave inductors and capacitors in standard multilevel interconnect silicon technology," *IEEE Trans. Microwave Theory Tech.*, vol. 44, pp. 100–104, Jan. 1996.
[9] L. Dauphinee, M. Copeland, and P. Schvan, "A balanced 1.5 GHz voltage controlled oscillator with an integrated LC resonator," in *Proc. ISSCC'97, Session 23, Analog Techniques*, pp. 390–391.
[10] I. B. Jansen, K. Negus, and D. Lee, "Silicon bipolar VCO family for 1.1 to 2.2 GHz with fully-integrated tank and tuning circuits," in *Proc. ISSCC'97, Session 23, Analog Techniques*, p. 392.
[11] B. Razavi, "A 1.8 GHz CMOS voltage—Controlled oscillator," in *Proc. ISSCC'97, Session 23, Analog Techniques*, pp. 388–389.
[12] K. A. Hajimiriand and T. H. Lee, "A general theory of phase noise in electrical oscillators," *IEEE J. Solid-State Circuits*, vol. 33, pp. 179–194, Feb. 1998.
[13] CADENCE, *Oscillator Noise Analysis in SpectreRF*, application note to SpectreRF, 1998.
[14] F. X Kärtner, "Untersuchung des Rauschverhaltens von Oszillatoren," Ph. D. dissertation, Tech. Univ. Munich, Munich, Germany, 1988.

Markus Zannoth was born in Munich, Germany, in 1971. He received the Dipl. Ing. degree in electrical engineering in 1996 from the Technical University of Munich, Munich, Germany. Since 1996, he has been working towards the Dr.Ing. degree at Siemens AG and the Technical University of Munich.

His doctoral research is on integrated oscillators.

Bernd Kolb was born in 1972. He studied electrical engineering with an emphasis on telecommunication techniques at the Georg-Simon-Ohm-Polytechnic Nuremberg. There, he received the Dipl.Ing. (FH) degree in 1995.

He joined the Siemens High Frequency IC Department in 1995. Since then, he has worked in the field of oscillators, frequency dividers, and vector modulators. He has focused on designing highly integrated transmitter IC's for mobile communication. He is now with Lucent Network Systems GmbH Nuremberg, Germany, where he designs high-frequency parts of base station for mobile communication.

Joseph Fenk received the diploma in electronics from the Technical University of Munich, Munich, Germany, in 1968.

He is responsible for product definition and project management of communications RF-integrated circuits at Siemens Components, Inc., Integrated Circuit Division. After joining Siemens in 1968, he worked as a Development Engineer on high-frequency components in the Discrete Components Group, developing transmitters, aerial and tuner transistors, FET's, and Varactor and PIN diodes. In 1976, he joined the Integrated Circuits Group as a Design Engineer for consumer products. He has been engaged in the development of integrated circuits for infrared preamplifiers, prescalers, IF-amplifiers/demodulators for FM-radio and satellite-TV, mixer/oscillators FM radio, TV-and SAT-TV, and TV UHF/VHF modulator IC's, as well as circuits for narrowband FM mobile radio. He holds more than 50 patents relating to IC and system design and has presented technical papers at numerous industry conferences and forums.

Robert Weigel (S'88–M'89–SM'95) was born in Ebermannstadt, Germany, in 1956. In 1989, he received the Dr.Ing. degree, and in 1992 the Dr.Ing.habil degree, both in electrical engineering from the Technical University of Munich, Munich, Germany.

From 1982 to 1988, he was a Research Assistant, from 1988 to 1994. he was a Senior Research Engineer, and from 1988 to 1996, he was a Professor at the Technical University of Munich. In the winter of 1994–1995, he was a Guest Professor at the Technical University of Vienna, Vienna, Austria. Since 1996, he has been Head of the Institute for Communication and Information Engineering at the University of Linz, Austria. He has been engaged in research and development on microwave theory and techniques, integrated optics, high-temperature superconductivity, surface acoustic wave (SAW) technology, and digital and microwave communication systems. In these fields, he has published more than 120 papers and has given more than 90 international presentations. His work includes European research projects and international journals.

Dr. Weigel is a senior member of the IEEE Microwave Theory and Techniques and the Ultrasonics, Ferroelectrics, and Frequency Control Societies. He is also a member of the Institute for Systems and Components of the Electromagnetics Academy, the Informationstechnishe Gesellschaft (ITG) in the Verband Deutscher Elekrotechniker (VDE), and the Society of Photo-Opticals Instrumentation Engineers (SPIE). In 1993 he was a co-recipient of the MIOP-award.

Tail Current Noise Suppression in RF CMOS VCOs

Pietro Andreani, *Member, IEEE*, and Henrik Sjöland, *Member, IEEE*,

Abstract—This paper presents the experimental results of two different techniques, inductive degeneration and capacitive filtering, for reducing the phase noise in tail-biased RF CMOS voltage-controlled oscillators (VCOs). Both techniques prevent the low-frequency tail current noise from being converted into phase noise. The techniques are applied to two distinct VCO designs, showing that the largest phase noise reduction (up to 6–7 dB at 3-MHz offset frequency from the carrier) is achieved via inductive degeneration. Capacitive filtering, however, also substantially reduces the phase noise at high offset frequencies and may therefore become a valid alternative to inductive degeneration, as discrete capacitors are of more common use than discrete inductors.

Index Terms—CMOS, phase noise, RF, VCO.

I. INTRODUCTION

THE THEORY and practice of the negative resistance *LC*-tank voltage-controlled oscillator (VCO) has recently made significant progress. Not long ago, the design of the tail transistor was considered uncritical, because of the linear circuit approximation often used in the VCO analysis. However, it was eventually recognized [1], [2] that the tail transistor may have a large impact on the generation of phase noise, often being the largest contributor. This is because the high-frequency tail current noise at twice the oscillation frequency is down-converted into phase noise by the hard switching oscillator. Subsequently, it has been shown [3] that, to a first approximation, the phase noise contribution from the switches is independent of their transconductance, and a closed-form expression has been given for the phase noise generated by high-frequency noise sources in the VCO.

The phase noise performance of the VCO is also affected by the low-frequency noise sources present in the VCO, of which the largest contribution is usually again coming from the tail current. A number of effects have been identified, such as the AM-to-PM conversion in the nonlinearities of the varactor [4], the modulation of the tail capacitance [3], the Groszkowski effect [3], the modulation of the bias point [5], and possibly other mechanisms. There is also some evidence that different effects have different weights in different VCO implementations. For the sake of completeness, we note that in bipolar junction transistors (BJT) oscillators yet another conversion mechanism has been identified: the phase-delay modulation effect [6].

The elucidation of the mechanisms of phase noise generation has been quickly followed by countermeasures tending to re-

Manuscript received July 19, 2001; revised October 5, 2001.

P. Andreani was with the Department of Electroscience, Lund University, SE-221 00 Lund, Sweden. He is now with the Center for Physical Electronics, Orsted DTU, Technical University of Denmark, DK-2800 Kgs. Lyngby, Denmark (e-mail: Pietro.Andreani@oersted.dtu.dk).

H. Sjöland is with the Department of Electroscience, Lund University, SE-221 00 Lund, Sweden (e-mail: Henrik.Sjoland@es.lth.se).

Publisher Item Identifier S 0018-9200(02)01691-8.

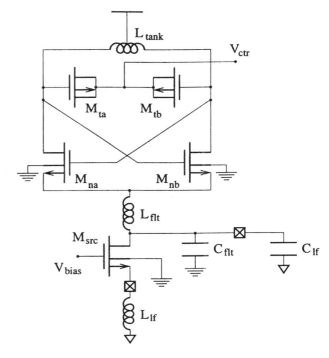

Fig. 1. Schematic view of VCO1.

duce the generation of phase noise as much as possible, either by eliminating the noise sources, or by preventing them from contributing to phase noise (the highest credit should also be given to those analog circuit simulators capable of simulating the noise performance in highly nonlinear circuits).

Thus, Hegazi *et al.* [7] have presented an evolution of the basic topology of the VCO, where the high-frequency tail current noise is removed by an on-chip lowpass *LC* filter (L_{flt} and C_{flt} in Fig. 1) placed between the tail transistor M_{src} and the switch transistors M_{na} and M_{nb}. The low-frequency noise from M_{src} is not removed, but its effects are minimized by reducing the size of the CMOS varactor, which, in that case, is the main responsible for the conversion of the low-frequency tail current noise into phase noise. In order to achieve a sufficiently wide tuning range, capacitors from an array are switched on and off in parallel with the *LC* tank. This approach results in a CMOS VCO with a superior figure of merit, at the cost of a more complicated tuning scheme for the VCO, and with the need of high-quality on-chip capacitors.

In this paper, we present two different techniques, inductive degeneration and capacitive filtering, which suppress the low-frequency tail current noise as well. Since both high-frequency and low-frequency tail current noise are now prevented from flowing into the oscillator core, the nonlinearities in the *LC* tank are no longer critical. As a result, any varactor can be adopted, avoiding the increased complexity of switched tuning.

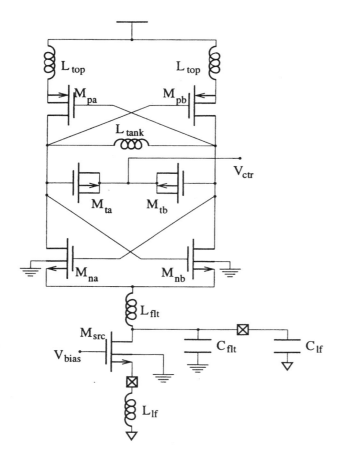

Fig. 2. Schematic view of VCO2.

It is worth noting that the $1/f$ noise of M_{src} could in principle be reduced at pleasure by making M_{src} very large; its low-frequency white noise, however, can be reduced only by reducing its g_m. For a constant tail current, this implies an increase of the transistor over-drive, and consequently an increase of the minimum drain–source voltage for which M_{src} still acts as a current source. The maximum voltage amplitude of the oscillations would then decrease, with a net increase of the phase noise as a result.

II. LOW-FREQUENCY NOISE SUPPRESSION

In order to filter off the tail current noise at low frequencies, at least one large reactive component is needed. Two possibilities exist: inductor or capacitor. Such a component, due to the value of the reactance involved, must be discrete.

In the first alternative [8], [9], an off-chip inductor is placed between the source of the tail transistor M_{src} and ground (L_{lf} in Figs. 1 and 2). The inductor degenerates M_{src}, and the power of the transistor noise current is reduced by the factor $|1 + j\, g_m\, \omega\, L_{fl}|^2$, where g_m is the transconductance of M_{src}. The technique suppresses the noise in a frequency band which is limited upwards by the parasitic parallel capacitance of the inductor, and downwards by the inductance value (the larger the inductance, the lower the frequency limit). For a g_m of 50 mS, and an L_{lf} of at least 30 μH, the noise reduction begins at about 100 kHz. Thus, standard off-the-shelf inductors with values in

the range 10–100 μH provide noise suppression over the offset frequencies of interest in GSM (depending on the technology used, and on what offset frequencies are of interest, the value of L_{lf} can be reduced, or must be increased). The inductance value, the quality factor, and the self-resonance frequency (10 MHz or higher for the inductors tested) are not critical; furthermore, unlike an external resonator, the off-chip inductor carries no high-frequency signals, so the package parasitics and PCB layout are uncritical as well.[1]

The second way of reducing the phase noise of the VCOs, capacitive filtering [10], is to use a large off-chip capacitor C_{lf} in parallel with C_{flt} (Figs. 1 and 2), shunting also the low-frequency noise generated by M_{src} to ground.[2] This choice is motivated by the fact that a capacitor between 10 and 100 nF is a less bulky, more ideal, and cheaper component than an inductor around 100 μH, the value adopted in this work. This approach, however, is less robust than inductive degeneration, since it creates a low impedance path to ground for the low-frequency noise of the switching transistors through the large capacitance C_{lf}, which may well counteract the beneficial effects on the tail current noise.

It should be noted that the use of a single capacitor in parallel with the tail transistor has been proposed in [2]. In the cited work, however, the role of the capacitor was to prevent the up-conversion of the low-frequency tail noise into phase noise and not to remove the low-frequency tail noise itself. If this approach has been shown to help reducing the corner frequency of the up-converted $1/f$ noise [11], it has the drawback of lowering the high-frequency impedance at the drain of M_{src}, which in this case is directly connected to the sources of the switch transistors. This node should be kept at a high impedance level for even harmonics of the oscillation frequency, in order not to degrade the quality factor of the oscillator [7]. Inductor L_{flt}, in parallel with the parasitic capacitance at the sources of the switch transistors, restores such a high impedance level by resonating the node at the (most important) second harmonic.

The low-frequency tail noise suppression techniques have been tested on both the single-ended nMOS oscillator in Fig. 1, and the push–pull oscillator in Fig. 2, with a very significant reduction of phase noise in both cases. In the following, we will refer to the VCO with only one switch pair as VCO1 (Fig. 1), and to the VCO with two switch pairs as VCO2 (Fig. 2).

A. More Noise Considerations

All simulations were performed with SpectreRF and the BSIM3 MOS model, version 3.2.2 [12]. A shortcoming of this model is that it neglects the effects of the induced gate noise; however, this noise contribution is negligible as long as the oscillation frequency is much lower than the transit frequency of the transistor. This condition is satisfied in the experiments presented here. An advantage of BSIMv3 is the possibility to

---

[1]It should be remembered that external components have the disadvantage that they may pick up some external noise and inject it into the VCO. The large inductor, however, can also reject such an external noise, as is evident from the measurement results in Fig. 12.

[2]Strictly speaking, C_{lf} and C_{flt} are in parallel only at low frequencies, since both bond wire and package introduce large inductive impedances at radio frequencies.

choose between two different models for the transistor thermal channel noise. The first is the Spice2 noise model, given by

$$S_{i_{ds}} = 4\,k_B\,T\,\gamma\,(g_m + g_{mb} + g_d) \qquad (1)$$

where k_B is Boltzmann's constant, T is the absolute temperature, and g_m, g_{mb}, and g_d are the transconductance, the bulk transconductance, and the output channel conductance of the transistor, respectively. The value of γ is unity when the transistor is working in the deep linear region and decreases to 2/3 when the transistor is working in saturation. The second noise model is physical and charge-based [13]

$$S_{i_{ds}} = 4\,k_B\,T\,\frac{\mu_{\mathrm{eff}}}{L_{\mathrm{eff}}^2}\,|Q_{\mathrm{inv}}| \qquad (2)$$

where μ_{eff} is the effective mobility of the charge carriers, L_{eff} is the effective length of the transistor, and Q_{inv} is the inversion charge layer in the transistor channel.

Equation (1) is the well-known long-channel approximation for the channel thermal noise, except for the presence of g_{mb} and g_d, which increases the noise level by approximately 20% when the transistor is working in saturation. In fact, this expression gives a more accurate noise estimate in our case, since the exact long-channel approximation should account for the presence of the bulk effect along the transistor channel [13]. For the CMOS process used here, the presence of g_{mb} simulates rather well the noise increase due to the bulk effect.

Equation (1) underestimates the actual channel noise when short-channel effects are important, such as the velocity saturation of the charge carriers in presence of high longitudinal channel fields. Equation (2), however, does capture the noise increase caused by velocity saturation (it does not take into account any hot electron phenomena, though), and the difference in noise level given by (1) and (2), respectively, can be taken as a measure of how much the transistor behavior has departed from the ideal long-channel regime. Accordingly, the phase noise in both VCO1 and VCO2 was simulated employing both noise models. Surprisingly, the contribution of the transistors to the overall phase noise was a few percent less when (2) was used, which indicated that the long-channel approximation is in fact applicable. Although unexpected, this result can be readily understood by analyzing the waveforms (Fig. 3) of the drain current I_d (scaled up by a factor 100 for readability), the gate–drain voltage V_{gs}, and the drain–source voltage V_{ds} for M_{na} (or M_{nb}) in VCO1 (the same qualitative analysis applies to VCO2 as well). It is clear that almost all current flowing from M_{na} into the LC tank is delivered in the presence of a very low V_{ds}. In such conditions, the electron velocity in the channel is far from its saturation limit, and the transistor is indeed working in the long-channel regime, at least as long as its contribution to the phase noise is considered, even though a (deep) submicrometer process has been used.

We conclude by remarking that BSIM3 has two different models for the $1/f$ noise, too. All simulations results presented here were obtained with the $1/f$ noise model yielding the largest $1/f$ noise.

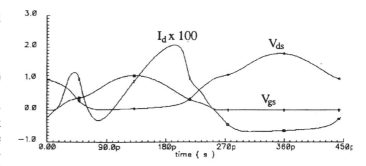

Fig. 3. Signal waveforms for transistor M_{na} in VCO1.

TABLE I
DIMENSIONS AND VALUES FOR THE ON-CHIP VCO COMPONENTS

Transistors (W×L in μm)		Reactors	
VCO1			
M_{src}	2000×1.0	L_{tank}	$\approx 2.3\,\mathrm{nH}$
M_{na}, M_{nb}	500×0.35	Q_{tank}	≈ 9 at $2.2\,\mathrm{GHz}$
M_{ta}, M_{tb}	1000×0.35	L_{flt}, C_{flt}	$\approx 3.0\,\mathrm{nH}$, $10\,\mathrm{pF}$
VCO2			
M_{src}	2000×1.0	L_{tank}	$\approx 2.3\,\mathrm{nH}$
M_{na}, M_{nb}	300×0.35	Q_{tank}	≈ 8 at $1.8\,\mathrm{GHz}$
M_{pa}, M_{pb}	800×0.35	L_{flt}, C_{flt}	$\approx 3.0\,\mathrm{nH}$, $10\,\mathrm{pF}$
M_{ta}, M_{tb}	1600×0.35	L_{top}	$\approx 0.7\,\mathrm{nH}$

III. VCO DESIGN AND SIMULATION RESULTS

Both VCO1 and VCO2 have been designed with the on-chip noise filter L_{flt}–C_{flt}. In the case of VCO2, an on-chip inductive degeneration of the pMOS switches has been added (L_{top} in Fig. 2). It should be noted that, according to simulations, the use of a single L_{top} in series with the common source of both pMOS switches results in a lower phase noise reduction. However, there is an asymmetry between the nMOS and the pMOS switches of VCO2: the nMOS switches see a high low-frequency impedance at their sources, which rejects much of their low-frequency noise; this is not true of the pMOS switches, and therefore it is expected that they contribute more to phase noise than the nMOS switches. This is confirmed by simulations. Both VCOs employ the same monolithic tank inductor L_{tank}, while the varactors are pMOS devices working in the accumulation and depletion regions only. Table I shows dimensions and features for the various components of the VCOs.

In all simulations, a supply voltage of 1.4 V and a supply current of 9 mA were used for VCO1, whereas the values 2.0 V and 6 mA were used for VCO2. The effectiveness of the different noise reduction techniques for VCO1 can be seen in Fig. 4, which shows four different phase noise simulations for VCO1 at the (simulated) minimum oscillation frequency. One simulation is for the "plain" VCO design, with neither on-chip LC filter, nor off-chip inductor (or off-chip capacitor). This is nothing but the standard VCO implementation (case "a" in Table II). Case "b" is the VCO with on-chip filter, but no off-chip inductor; in case "c" there is no on-chip filter, but an off-chip inductor of 100 μH. Finally, case "d" is when both on-chip filter and

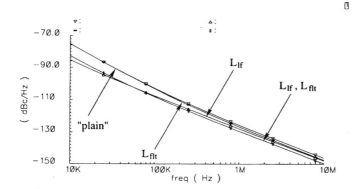

Fig. 4. Phase noise simulations for VCO1.

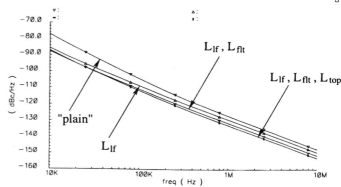

Fig. 5. Phase noise simulations for VCO2.

TABLE II
SIMULATED PHASE NOISE FOR VCO1 AT 3-MHz OFFSET FREQUENCY FOR THE
MINIMUM CARRIER FREQUENCY. THE FOUR SIMULATIONS CASES ARE: (A)
L_{lf} = NO, ON-CHIP FILTER = NO; (B) L_{lf} = NO, ON-CHIP FILTER = YES; (C)
L_{lf} = YES, ON-CHIP FILTER = NO; (D) L_{lf} = YES, ON-CHIP FILTER = YES

f_c = 2.0 GHz	a	b	c	d
Phase noise (dBc/Hz)	-136.3	-137.3	-138.3	-139.3
% R_{tank}	33.7	40.6	49.6	62.2
% $M_{na}+M_{nb}$ (white)	20.5	21.6	30.3	33.3
% $M_{na}+M_{nb}$ (1/f)	0.5	0.5	0.6	0.7
% M_{src} (white)	40.6	31.5	17.1	0.0
% M_{src} (1/f)	3.2	3.4	0.0	0.0
% Other sources	1.5	2.4	2.4	3.8

TABLE III
SIMULATED PHASE NOISE FOR VCO2 AT 3-MHz OFFSET FREQUENCY FOR THE
MINIMUM CARRIER FREQUENCY. THE FOUR SIMULATIONS CASES ARE: (A)
L_{lf} = NO, ON-CHIP FILTER = NO, L_{top} = NO; (B) L_{lf} = YES, ON-CHIP
FILTER = NO, L_{top} = NO; (C) L_{lf} = YES, ON-CHIP FILTER = YES, L_{top} =
NO; (D) L_{lf} = YES, ON-CHIP FILTER = YES, L_{top} = YES

f_c = 1.7 GHz	a	b	c	d
Phase noise (dBc/Hz)	-137.0	-138.2	-139.4	-140.3
% R_{tank}	26.9	35.8	50.2	56.5
% $M_{na}+M_{nb}$ (white)	24.6	32.7	13.4	15.8
% $M_{na}+M_{nb}$ (1/f)	0.0	0.0	0.2	0.2
% $M_{pa}+M_{pb}$ (white)	20.2	26.7	32.3	22.5
% $M_{pa}+M_{pb}$ (1/f)	0.0	0.0	0.1	0.3
% M_{src} (white)	24.3	2.2	0.1	0.1
% M_{src} (1/f)	1.9	0.0	0.0	0.0
% Other sources	2.1	2.6	3.7	4.6

off-chip inductor are used. Fig. 4 clearly shows that the advantages yielded by the inductive degeneration increase with lower offset frequencies, where the $1/f$ noise of the tail transistor totally dominates the phase noise. Table II presents in detail the phase noise contributions of the different components of VCO1, at the offset frequency of 3 MHz, for the four different simulation cases. The overall phase noise reduction is 3 dB, of which 2 dB are due to the inductive degeneration and 1 dB is due to the on-chip LC filter. Even more interesting is that the combination of on-chip filtering and off-chip inductive degeneration completely removes the tail current noise, which is otherwise the single largest phase noise source, as is manifested in Table II. When the tail current noise is removed, the phase noise is mostly generated by the resonator tank itself, the switch transistor pair contributing roughly half as much (at high enough offset frequencies). Thus, the noise factor of VCO1 is approximately 1.5.

Four different phase noise simulations for VCO2 are presented in Fig. 5, again at the (simulated) minimum oscillation frequency. Table III shows the phase noise contributions of the different components of VCO2, at the offset frequency of 3 MHz, for the four different simulation cases. The overall phase noise reduction is 3.3 dB, of which 1.2 dB is due to the inductive degeneration. The three techniques combined result in a noise factor for VCO2 of approximately 1.8. It is worth noting that, although in these simulations the effect of the inductive degeneration is comparable to that of the on-chip noise reduction techniques, across the whole tuning range of

the VCOs the inductive degeneration yields on average the largest phase noise reduction. This is especially true in the case of the highest oscillation frequencies (see, e.g., [8] and the measurement results).

As previously stated, the use of the capacitive filtering technique should in any case result in a less robust design than when inductive degeneration is adopted, since C_{lf} offers a low impedance way to ground for the low-frequency noise (both white and $1/f$) generated by the switch transistors, while this noise is otherwise rejected by the high impedance presented at low frequencies by the drain of M_{src}. Thus, when the value of C_{lf} increases, the contribution of M_{src} to the phase noise decreases, while that of the switch transistors increases. The actual amount of increase or decrease of the different noise contributions will determine whether the use of C_{lf} is advantageous or not. Since the current theory of phase noise generation does not provide a closed-form expression for the phase noise caused by the presence of C_{lf}, we have to rely on simulations to evaluate the phase noise performances of the different VCO designs.

The results of the phase noise simulations are quite unexpected (and, considering the measurement results, partially erroneous). Figs. 6 and 7 show the phase noise of VCO1 and VCO2, respectively, when either C_{lf} or L_{lf} are employed, for the highest value of the carrier frequency. As previously explained, with this

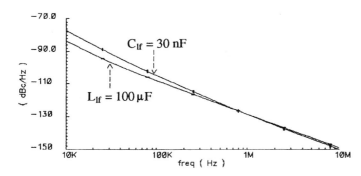

Fig. 6. Phase noise simulations for VCO1, when either inductive degeneration or capacitive filtering are used.

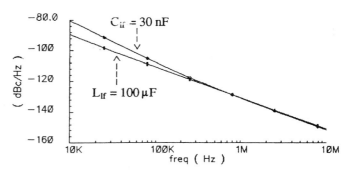

Fig. 7. Phase noise simulations for VCO2, when either inductive degeneration or capacitive filtering are used.

choice for the carrier frequency the effects of the tail current noise are most dominant. For offset frequencies higher than approximately 400 kHz, the two approaches lead to quantitatively equivalent results, and in fact C_{lf} yields in the case of VCO1 a 1-dB lower phase noise at 3-MHz offset frequency. The list of the contributions of the individual components in the VCO to the phase noise shows that, while the noise contribution of the switch transistors does in fact increase when C_{lf} is used, the contribution from the tank losses decreases. As an example, in VCO1 at 3-MHz offset frequency the noise of the switch transistors increases by 20% compared to when L_{lf} is used, but the noise from the tank decreases by as much as 70%. It is not clear why this happens, but it can be checked that the signal waveforms in the VCO are identical when either C_{lf} or L_{lf} are used (as they should, since L_{flt} decouples the tank circuit from the tail circuit).

Finally, it was checked that the phase noise displayed by the VCOs is largely insensitive of the actual value of C_{lf}, for capacitance values higher than approximately 10 nF.

IV. MEASUREMENT RESULTS

The VCOs were fabricated in a standard 0.35-μm CMOS process (three metal layers with resistivity 70 mΩ/$\square$, 70 mΩ/$\square$, and 40 mΩ/$\square$, respectively; gate resistivity 9 Ω/$\square$, substrate resistivity 8 Ω·cm; no analog features except low-Q poly–poly capacitors).

The die photographs of VCO1 and VCO2 are shown in Figs. 8 and 9, respectively. The same values for supply voltages and supply currents were used when taking measurements as when performing simulations: 1.4 V and 9 mA for VCO1, and 2.0 V and 6 mA for VCO2, respectively.[3] The frequency of VCO1 could be tuned between 1.96 and 2.36 GHz, and VCO2 could be tuned between 1.69 and 1.96 GHz, giving a tuning range of 18% and 15%, respectively. The phase noise of the two oscillators were measured at several different carrier frequencies. In Fig. 10, the phase noise at 600-kHz and 3-MHz offset frequency versus carrier frequency is shown for VCO1, both without and with inductive degeneration ($L_{lf} = 100\ \mu$H). The same plots for VCO2 are shown in Fig. 11. As expected from the simulation results, the largest benefit of the technique is at the extremes of the tuning interval, especially at the highest frequencies. Since at

[3]The results presented here are slightly different from those found in [8] and [9]. While the abnormal peaking in the phase noise plots has now disappeared, a phase noise level 1–2 dB higher has been measured.

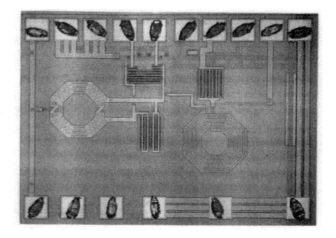

Fig. 8. Die photograph of VCO1 (dimensions: 1.10 mm × 0.85 mm).

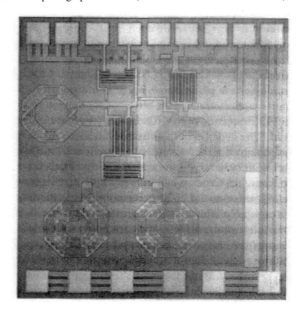

Fig. 9. Die photograph of VCO2 (dimensions: 1.10 mm × 1.15 mm).

these extreme frequencies the varactors do not have any tuning capability and are therefore quite linear, it seems that in these experiments the major culprit for the low-frequency tail current noise conversion into phase noise are the remaining (fixed) capacitances of the LC tank, mostly made of the gate and drain capacitances of the MOS switches. Such capacitances are of course very nonlinear across a whole oscillation period.

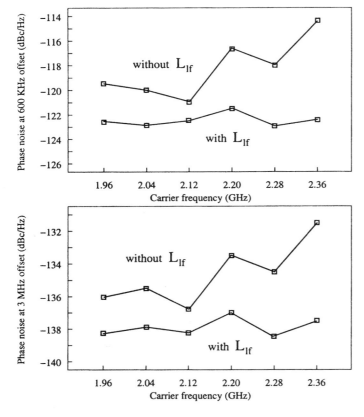

Fig. 10. Measured phase noise for VCO1 at 600-kHz (top) and 3-MHz (bottom) offset frequency.

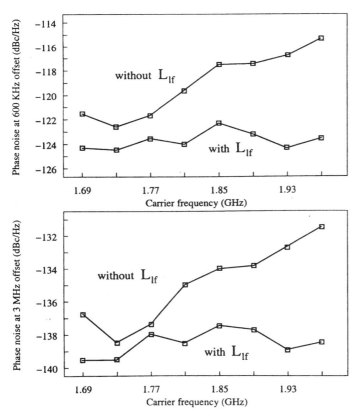

Fig. 11. Measured phase noise for VCO2 at 600-kHz (top) and 3-MHz (bottom) offset frequency.

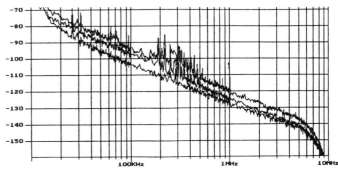

Fig. 12. Phase noise measurements for VCO1, with no phase noise reduction techniques (highest curve), capacitive filtering (middle curve), and inductive degeneration (lowest curve).

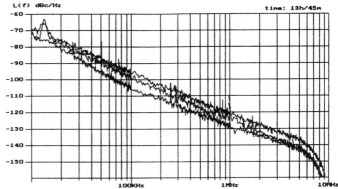

Fig. 13. Phase noise measurements for VCO2, with no phase noise reduction techniques (highest curve), capacitive filtering (middle curve), and inductive degeneration (lowest curve).

Figs. 12 and 13 show the phase noise measurements for VCO1 and VCO2, respectively, when the carrier frequency is set to the highest possible value, where the effects of the tail current noise are most dominant. Three different measurements are shown in each figure: one when neither C_{lf} nor L_{lf} are used (highest curve), one when a C_{lf} of 33 nF is used (middle curve), and one when an L_{lf} of 100 μH is used (lowest curve). We will refer to these curves as plain phase noise, C_{lf} phase noise, and L_{lf} phase noise, respectively. It is obvious that the C_{lf} phase noise performance does not completely live up to the expectations generated by the simulations results (Figs. 6 and 7), since the C_{lf} phase noise is higher than the L_{lf} phase noise for all offset frequencies. The C_{lf} phase noise approaches the plain phase noise at lower offset frequencies, while it approaches the L_{lf} phase noise at higher offset frequencies (as previously noted, L_{lf} can have the additional beneficial effect of greatly damping the interference from external signals, see Fig. 12). Thus, whether C_{lf} leads to a substantial improvement of the phase noise performance is related to what the offset frequencies of interest are. If offset frequencies lower than approximately 300 kHz are not important, the phase noise decrease yielded by the use of C_{lf} is large, even though not as large as when inductive degeneration is adopted. Table IV shows that the difference between plain phase noise and C_{lf} phase noise is much larger than the difference between C_{lf} phase noise and L_{lf} phase noise already at an offset frequency of 600 kHz, and even larger at 3 MHz.

TABLE IV
MEASURED PHASE NOISE DATA (IN dBc/Hz) FOR BOTH VCOs

Offset frequency	$L_{lf} = 0$ $C_{lf} = 0$	$L_{lf} = 0$ $C_{lf} = 30\,\text{nF}$	$L_{lf} = 100\,\mu\text{H}$ $C_{lf} = 0$
VCO1			
100 kHz	-94.0	-96.5	-103.5
600 kHz	-114.0	-119.0	-121.5
3 MHz	-131.0	-136.0	-137.5
VCO2			
100 kHz	-95.5	-98.0	-105.5
600 kHz	-116.0	-120.0	-123.5
3 MHz	-131.5	-136.5	-138.5

It is also important to emphasize that the VCOs presented here have in no way been optimized for use together with the capacitive filtering technique. Therefore, it is reasonable to expect at least some performance improvements to be gained with dedicated VCO designs.

V. CONCLUSION

This paper has presented simulation and measurement results of two different approaches to phase noise reduction in RF CMOS VCOs, achieved through the removal of the low-frequency tail current noise. It was shown that, although capacitive filtering can be effective in some cases, inductive degeneration is in general a better and more robust choice. When the low-frequency tail current noise is prevented from flowing into the oscillator core, any nonlinear mechanism capable of converting such a low-frequency noise into phase noise becomes uncritical; as a consequence, even varactors displaying a large AM-to-PM conversion may be employed without penalties. Inductive degeneration, together with additional on-chip filtering, makes the otherwise dominating noise contribution of the tail current source totally negligible. As a result, VCOs exhibiting mediocre phase noise behavior are turned into high-performance ones.

ACKNOWLEDGMENT

The authors would like to thank P. Petersson, at Ericsson Radio Systems in Kista, Sweden, and Dr. T. Mattsson, at Ericsson Mobile Platforms in Lund, Sweden, for helping performing the phase noise measurements; and Dr. C. Samori, at the Department of Electronics, University of Milan, Italy, and Dr. A. Zanchi, at Texas Instruments, Inc., Dallas, TX, for valuable discussions on technical matters.

REFERENCES

[1] C. Samori, A. L. Lacaita, F. Villa, and F. Zappa, "Spectrum folding and phase noise in *LC* tuned oscillators," *IEEE Trans. Circuits Syst. II*, vol. 45, pp. 781–790, July 1998.

[2] A. Hajimiri and T. H. Lee, "Design issues in CMOS differential *LC* oscillators," *IEEE J. Solid-State Circuits*, vol. 34, pp. 717–724, May 1999.

[3] J. J. Rael and A. Abidi, "Physical processes of phase noise in differential *LC* oscillators," in *Proc. CICC 2000*, May 2000, pp. 569–572.

[4] C. Samori, A. L. Lacaita, A. Zanchi, S. Levantino, and F. Torrisi, "Impact of indirect stability on phase noise performance of fully-integrated *LC* tuned VCOs," in *Proc. ESSCIRC '99*, Sept. 1999, pp. 202–205.

[5] C. Samori, S. Levantino, and V. Boccuzzi, "A −94dBc/Hz@100kHz, fully-integrated, 5-GHz, CMOS VCO with 18% tuning range for Bluetooth applications," in *Proc. CICC '01*, May 2001, pp. 201–204.

[6] C. Samori, A. L. Lacaita, A. Zanchi, S. Levantino, and G. Calì, "Phase noise degradation at high oscillation amplitudes in *LC*-tuned VCO's," *IEEE J. Solid-State Circuits*, vol. 35, pp. 96–99, Jan. 2000.

[7] E. Hegazi, H. Sjöland, and A. Abidi, "A filtering technique to lower oscillator phase noise," in *Proc. ISSCC 2001*, Feb. 2001, pp. 364–365.

[8] P. Andreani and H. Sjöland, "A 2.2 GHz CMOS VCO with inductive degeneration noise suppression," in *Proc. CICC '01*, May 2001, pp. 197–200.

[9] ——, "A 1.8GHz CMOS VCO with reduced phase noise," in *Proc. 2001 Symp. VLSI Circuits*, June 2001, pp. 121–122.

[10] P. Andreani, "Phase noise reduction in RF CMOS VCO's via capacitive filtering," in *Proc. NORCHIP 2001*, Nov. 2001, pp. 21–27.

[11] B. De Muer, M. Borremans, M. Steyaert, and G. Li Puma, "A 2-GHz low-phase-noise integrated *LC*-VCO set with flicker-noise upconversion minimization," *IEEE J. Solid-State Circuits*, vol. 35, pp. 1034–1038, July 2000.

[12] W. Liu *et al.*, *BSIM3v3.2.2 MOSFET Model, User's Manual*. Berkeley, CA: Dept. of Electrical Engineering and Computer Sciences, University of California, 1999.

[13] Y. Tsividis, *Operation and Modeling of the MOS Transistor*, 2nd ed. New York: McGraw-Hill, 1999, ch. 8.

Pietro Andreani (M'99) received the M.S.E.E. degree from the University of Pisa, Italy, in 1988, and the Ph.D. degree from Lund University, Sweden, in 1999.

He joined the Department of Applied Electronics, Lund University, in 1990, where he contributed to the development of software tools for digital ASIC design. After working at the Department of Applied Electronics, University of Pisa, as a CMOS IC Designer during 1994, he rejoined the Department of Applied Electronics (now Department of Electroscience), Lund University, as an Associate Professor. Since November 2001, he has been a Professor in the Center for Physical Electronics at the Technical University of Denmark, Lyngby, Denmark.

Henrik Sjöland (M'98) received the M.Sc. degree in electrical engineering and the Ph.D. degree in applied electronics both from Lund University, Lund, Sweden, in 1994 and 1997, respectively.

He is currently an Associate Professor at Lund University and his research interests include the design and analysis of analog integrated circuits, feedback amplifiers, and RF CMOS. He spent one year visiting at the Abidi group at the University of California at Los Angeles as a Fulbright postdoc in 1999. He is also the author of *Highly Linear Integrated Wideband Amplifiers* (Boston, MA: Kluwer, 1999).

Low-Power Low-Phase-Noise Differentially Tuned Quadrature VCO Design in Standard CMOS

Marc Tiebout, *Member, IEEE*

Abstract—This paper describes the design and optimization of voltage controlled oscillators with quadrature outputs. Systematic design of fully integrated *LC*-VCOs with a high inductance tank leads to a cross-coupled double core *LC*-VCO as the optimal solution in terms of power consumption. Futhermore, a novel fully differential frequency tuning concept is introduced to ease high integration. The concepts are verified with a 0.25-μm standard CMOS fully integrated quadrature voltage-controlled oscillator (VCO) for zero- or low-IF DCS1800, DECT, or GSM receivers. At 2.5-V power supply voltage and a total power dissipation of 20 mW, the quadrature VCO features a worst-case phase noise of −143 dBc/Hz at 3-MHz frequency offset over the tuning range. The oscillator is tuned from 1.71 to 1.99 GHz through a differential nMOS/pMOS varactor input.

Index Terms—Differential tuning, LC-tank, phase noise, quadrature, VCO.

I. INTRODUCTION

MOBILES in telecommunications systems such as DCS1800 or GSM900 offer from generation to generation higher performances and longer standby times at a lower cost. Multimode multiband capability, supporting, e.g., DCS1800, GSM900 and DECT, and in the near future UMTS, becomes very important. Receiver architectures like zero-IF or low-IF offer flexibility and low cost through high integration. The tough combination of the very low phase noise specifications with very low power consumption (battery operation) pushes designers to use *LC*-VCOs. A great research effort has been invested in the design of integrated voltage-controlled oscillators (VCOs) using integrated or external resonators, but as their power consumption is still unacceptable, today's mobile phones commonly use external *LC*-VCO modules [1], [2]. As in low-IF [3] or zero-IF transceivers, quadrature signals (0° and 90°) are needed for *I/Q*-(de)modulation, it is important to offer quadrature generation at a minimal power consumption. This work aims at the overall optimized design of integrated VCOs providing quadrature outputs and fulfilling the phase-noise specifications for GSM and DCS1800 at a power consumption comparable to, or even lower than, external VCO modules.

The paper is organized into seven sections. Section II covers systematic *LC*-VCO design for low power and low phase noise. Section III discusses fully integrated design. The quadrature generation issues are discussed in Section IV. Section V treats the novel differential tuning. Section VI presents the prototype design and measurements, followed by the conclusion in Section VII.

II. *LC*-VCO DESIGN

A. *LC*-VCO Basics

A general *LC*-VCO can be symbolized as in Fig. 1. The oscillator consists of an inductor L and a capacitor C, building a parallel resonance tank, and an active element $-R$, compensating the losses of the inductor (R_L in Fig. 1) and the losses of the capacitor (R_C in Fig. 1). As the capacitance C is proportional to a tuning input voltage, the circuit results in a VCO with angular center frequency

$$\omega_c = \frac{1}{\sqrt{LC}}. \tag{1}$$

The capacitor C in Fig. 1 not only consists of a variable capacitor to tune the oscillator, but it also includes the parasitic or fixed capacitances of the inductor, the active elements, and the load (output driver, mixer, prescaler, etc.). VCO designs for mobile telecommunications demand the extremely tough combination of low phase noise and low power, explaining the avalanche of publications on this topic. Many publications [4]–[6], [3], [7], [8] treat the phase noise mechanisms in bipolar and MOS VCOs. As the systematic tank design used in this work is quite different to the approach used in recently published very low phase noise VCO design [9], the next section recapitulates the energy consideration theorem and Leeson's empirical phase-noise expression. These expressions are unconditionally valid for a large signal oscillator and lead to the optimized design of Section VI.

B. Design for Low Power

Recalculating capacitor and inductor losses to one single resistor, the *LC*-tank resonator simplifies to Fig. 2. Using the energy conservation theorem, the maximal energy stored in the inductor must equal the maximal energy stored in the capacitor

$$\frac{CV_{\text{peak}}^2}{2} = \frac{LI_{\text{peak}}^2}{2} \tag{2}$$

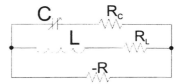

Fig. 1. Basic *LC*-VCO.

Manuscript received November 27, 2000; revised January 29, 2001.

The author is with Infineon Technologies AG, Department of Wireless Products, Technology and Innovation, D-81609 Munich, Germany (e-mail: marc.tiebout@infineon.com).

Publisher Item Identifier S 0018-9200(01)04523-1.

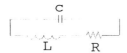

Fig. 2. Basic *LC* resonator tank.

with V_{peak} the peak amplitude voltage across the capacitor and I_{peak} the peak amplitude current through the inductor. Now we can calculate the loss in the tank as

$$P_{loss} = RI_{peak}^2 = C\frac{R}{L}V_{peak}^2 \qquad (3)$$

or with (1)

$$P_{loss} = RC^2\omega_c^2 V_{peak}^2 = \frac{R}{L^2\omega_c^2}V_{peak}^2. \qquad (4)$$

As the power consumption of a VCO must compensate at least the losses in the tank, these equations lead to some interesting conclusions for the power consumption of any *LC*-VCO.

1) It is not surprising that power loss decreases linearly for lower series resistances in the resonance tank. These equations demonstrate that for some given unavoidable series resistance in the coil, there is still the degree of freedom to increase the inductance in order to decrease the power loss.
2) Normally the frequency of oscillation is specified and cannot be changed. In this case, (4) clearly shows that power consumption decreases *quadratically*, if the tank inductance can be increased. Measurements (also simulations) on several fully integrated *LC*-VCOs, using the topology of Fig. 1, show a decreasing power consumption when tuned at higher frequencies, as predicted by (3).

C. Design for Low Phase Noise

Already in 1966, Leeson [10] published the following heuristic expression for the phase noise of an *LC*-VCO:

$$S_{SSB} = F\frac{kT}{2P_{sig}}\frac{\omega_c^2}{Q^2\Delta\omega^2} \qquad (5)$$

where Q is the loaded quality factor of the tank, $\Delta\omega = 2\pi\Delta f$ is the angular frequency offset, and F is called the device noise excess factor or simply noise factor. The equation was verified in [7], providing insight in the noise factor F. Equation (5) shows that one obvious way to reduce phase noise is to increase $P_{sig} \propto V_{peak}^2$. As Q goes in quadratically, (5) also clearly shows that the most effective way to lower phase noise is to use an *LC*-tank with a higher Q. Conventional inductor $Q = -Im(y_{11})/Re(y_{11})$ is useless as it equals to zero at tank resonance. With Barkhausen oscillation criterion, the phase stability definition for Q appears to be the most appropriate for the oscillator application (see [11] for a very thorough explanation and comparison of Q definitions). The phase stability quality factor is defined as

$$Q_{PS} = -\frac{\omega_c}{2}\frac{d\phi}{d\omega}\bigg|_{\omega=\omega_c}. \qquad (6)$$

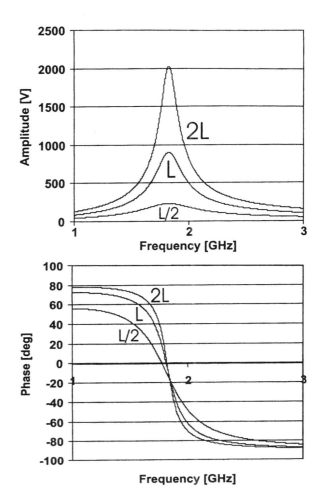

Fig. 3. *LC*-tank bode plot.

Keeping in mind that $\phi = 0$ at ω_c, Q_{PS} for the *LC*-tank of Fig. 2 can easily be simplified to

$$Q_{PS} = \frac{1}{R}\sqrt{\frac{L}{C}} \qquad (7)$$

or with (1)

$$Q_{PS} = \frac{L}{R}\omega_c. \qquad (8)$$

To provide a less mathematic insight, Fig. 3 shows the amplitude and phase plot for an *LC*-tank, keeping constant resonance frequency for different L/C ratios. A higher L/C ratio results in the phase domain into a steeper phase rolloff, or into a tank rejecting stronger any phase deviation. The amplitude representation shows, that higher L/C ratios correspond to smaller band-pass filters, or more selective filters. In practice, an upper limit for the L/C ratio is set by the required frequency tuning range. The minimal tuning range is determined by the combination of the specified frequency bands with the tolerances of the VCO components.

Substituting (8) into (5)

$$S_{SSB} = F\frac{kT}{2P_{sig}}\frac{R^2}{L^2\Delta\omega^2} = F\frac{kT}{V_{peak}^2}\frac{R^3}{L^2\Delta\omega^2} \qquad (9)$$

295

TABLE I
LOW-POWER LOW-PHASE-NOISE OPTIMIZATION SUMMARY

	low power	low phasenoise
L	maximize	maximize
C	minimize	minimize
R	minimize	minimize
Amplitude	minimize	maximize

it is remarkable that phase noise is not dependent on ω_c, or phase noise over tuning range should be constant, if V_{peak} can be kept constant through an amplitude control mechanism. Equation (9) leads to a very different tank, compared to [9] stating that phase noise only depends on the effective tank resistance. Equation (9) clearly shows that, despite the unavoidable high series resistances in standard CMOS processes, phase noise still can be optimized.

The conclusions of this section are summarized in Table I.

III. FULLY INTEGRATED DESIGN

A. Optimized LC-Tank

From the previous section, it is clear that an LC-tank with maximal L/R and L/C ratio is needed. The best available inductors are external inductors. These external components feature very high Qs, but their use is limited by the bondwires, the package and pad parasitics, and the electrostatic discharge (ESD) protection networks. Especially when aiming at highly integrated zero-IF receivers, board coupling becomes a very critical issue. For reasons of manufacturability and cost, a fully integrated solution will be the first choice, if power consumption can be reduced to an acceptable level (e.g., to the power consumption of external modules [1]). As the main goal of this work is a VCO with quadrature outputs, the complete power of VCO and interfacing and quadrature generation circuitry must be compared, which is the topic of Section IV.

The power consumption of a fully integrated solution can be reduced through an optimization of the fully integrated tank using the following design options:

- *MOS varactor.* MOS transistors are used as varactors [12] because of their high C_{max}/C_{min}. This enables an LC-tank with higher L/C ratio. For maximal varactor Q, a minimal length of 0.25 μm should be preferred to minimize the resistive paths in the channel. However, to decrease the effect of the fixed overlap gate–source and gate–drain capacitances and to increase C_{max}/C_{min}, MOS varactor length is set to 0.35 μm. A multifingered folded layout is used for minimal gate resistance or maximal varactor Q.

- *high winding count.* With n the number of turns in a coil, the coil inductance $L \propto n^2$ and the coil series resistance $R_s \propto n$, so a higher n will improve the R^3/L^2 ratio from (9) by n. An upper boundary for n comes from the fact that the winding-to-winding capacitance also increases with n.

- *differential coil.* The use of one differential coil, instead of two single coils [13], exploits the coupling factor to increase the inductance, leading to a higher L/C ratio. As the differential coil is used in a balanced configuration,

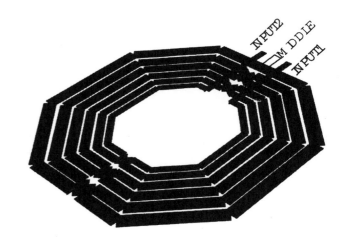

Fig. 4. Coil layout.

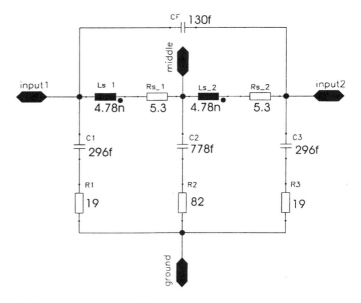

Fig. 5. Lumped circuit model for the coil.

the middle capacitance of the lumped model (Fig. 5) is cancelled out. This effectively increases the differential self-resonance frequency and the differential Q_{PS}.

- *special coil layout.* The differential coil is further optimized through a special coil layout as shown in Fig. 4. The middle tap of the coil is the outer winding in the layout. It is laid out wider to reduce the ohmic resistance of this winding, without any capacitive penalty as its capacitance is at common mode. This increases L/R largely as the outer winding is the longest one. The inner windings are thinner, this decreases the capacitive load at the RF nodes of the oscillator increasing L/C and the differential self-resonance frequency.

The main drawback of this complex coil layout is that the accuracy of the lumped model needed to design the oscillator to the specified frequency and tuning range only can be extracted out of measurements. To avoid two silicon runs, the coil from

296

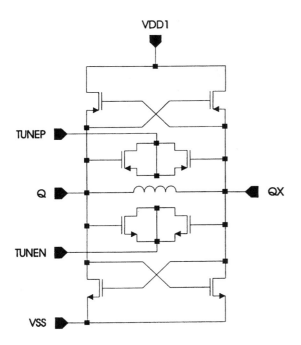

Fig. 6.   Current reusing VCO topology.

[14] was reused. The 1.3-GHz coil was scaled to 1.8-GHz, through a linear geometry downscaling of the layout. The components of the lumped model were adapted according their physical meaning. The size of the resulting coil for 1.8-GHz applications is 260 μm by 260 μm. The coil was separately characterized using s-parameter structures. The lumped model extracted from the measurements is presented in Fig. 5. As the resimulation of the lumped model and the measured s-parameters match quasi-perfectly from 100-MHz to 3-GHz, the resistive losses dominate in this coil design, and no skin effect or eddy current losses had to be modeled. In balanced mode, Q_{PS} is about 8 in the 1.8-GHz frequency range.

B. Design of the Active Part

For the realization of the negative resistance, the combination of nMOS and pMOS transistors, as can be found, e.g., in [6], was chosen as it reuses the dc current. Furthermore, the current source was omitted to maximize the signal swing (Fig. 6). Omitting the current source has a few more advantages. It eliminates an important phase-noise source [7]. As all VCO-core transistors are put in a gigahertz-switching bias condition [15], flicker noise terms apparently are reduced by about 10 dB comparing measurement and simulation for several measured designs using this topology. The main disadvantage of omitting the current source is the increased sensitivity to the power supply (higher frequency pushing). This effect can be reduced by the integration of a voltage regulator. This topology can be redrawn to the combination of two digital inverters and an LC-tank. Other possible disadvantages of this very digital structure are the increased spectral impurity of the oscillator signal and the less symmetrical waveforms [6], resulting in an increased upconversion of flicker noise. Both effects can be damped by the band-pass characteristic of a well-designed resonance tank.

Finally, this topology (with or without current source) guarantees to limit all gate voltages to the supply voltage. In contrast

to many VCO designs using a coil at power supply to obtain a VCO swing of twice the power supply voltage, this topology allows a reliable operation over many years within the process limits and inherently avoids degeneration effects due to, e.g., hot-electron effects.

IV. QUADRATURE GENERATION

Three design options are available to generate quadrature signals.

1) Combination of VCO, polyphase-filter (or R-C C-R filter), and output buffers (or limiters) as used in, e.g., [16], [17].
2) VCO at double frequency followed by master–slave flipflops.
3) Two cross-coupled VCOs as proposed in [18].

The first option needs four output buffers or limiters consuming a lot of power. If buffers are inserted between VCO and filters, even more power is needed; if the filters are directly connected to the VCO tank, tank capacitance is increased, leading to higher power consumption (4) and worse phase noise (9). Furthermore, a lot of chip area is needed, as the filters need good matching.

The second option has the smallest area. This option needs a VCO designed at double frequency, which should not consume more power, as a higher Q_{PS} for integrated tanks at higher frequencies is achievable (Inductance scales down linearly when sizing down an integrated coil, coil capacitance scales down quadratically, so L/C improves). However, the master–slave flipflops, which have to be designed for the doubled frequency, consume too much power in current widely used CMOS technologies (0.25 μm). As soon as next technology generations (0.18 μm, 0.12 μm) become available, this could change very rapidly. If primary design concern is low cost or small area, then this solution clearly must be preferred, as the VCO designed at double frequency features a smaller coil and the area of the master–slave flipflops in submicron CMOS is negligible.

The third option comes at the cost of double VCO area. This option outperforms the other solutions in terms of power consumption, as soon as a well-designed VCO core consumes less power than the four output buffers or limiters of the first option, or as soon as the VCO core consumes less power than the master–slave flipflops designed for double frequency needed to realize the second option. The two-core solution, furthermore, provides a very high voltage swing, which eases the design of prescaler and mixer circuits connected to the VCO. This consideration can also be extended to VCOs using external high-quality inductors. When using external inductors, a lot of current must be spent to amplify the VCO signal to drive the polyphase filters (external VCO at nominal frequency) or to drive the flipflops (external VCO at double frequency).

V. MOS VARACTOR TUNING

The nMOS varactors have a very steep capacitance over voltage characteristic, as shown in Fig. 7. Through gate folding, a very high measured Q at 2 GHz over gate and source/drain voltage was obtained (Fig. 8). As varactor Q is much higher than inductor Q, the MOS varactors can be considered as ideal

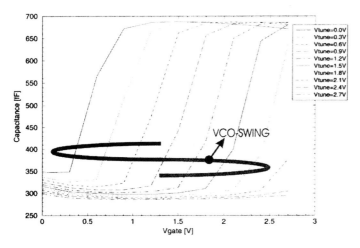

Fig. 7. Measured nMOS varactor CV characteristic.

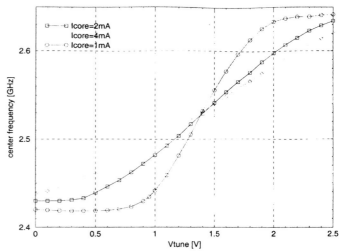

Fig. 9. Frequency tuning for different oscillator amplitudes.

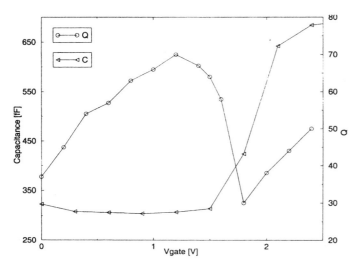

Fig. 8. Measured nMOS varactor CV & Q @ 2 GHz for $V_{tune} = 1.2$ V.

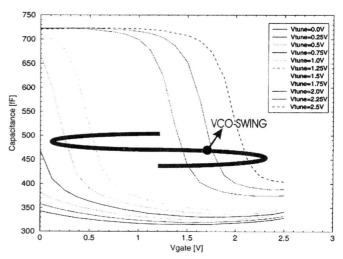

Fig. 10. Measured pMOS varactor characteristic.

capacitors. To linearize the oscillator tuning characteristic, the varactors are directly coupled to the large signal swing over the inductors. The frequency is indirectly set through the backgate voltage V_{tune} of the MOS transistors (Fig. 7 or 11). As shown in Fig. 7, the steep curve can be shifted through the tuning voltage V_{tune} on the backgates (labels as in Fig. 11). As the oscillator has a very large signal swing (nearly full power supply or 2.5 V), the effective capacitance of the varactor is averaged over each period. The resulting capacitance varies linearly with V_{tune} in a range defined by the oscillation amplitude, a picture of a typical resulting frequency characteristic is given in Fig. 9. The capacitance variation over each oscillation period results in harmonic distortion of the oscillator sine. Although this distortion is partially rejected by the high Q of the LC-tank, it is probably a source of increased flicker noise upconversion. Also, pMOS transistors can be used as varactors (measured data in Fig. 10). VCO gain (K_{vco}, [MHz/V]) and VCO gain linear region depend on the signal swing of the oscillator and the ratio C_{min}/C_{max} of the varactor (Fig. 9). C_{max} is obtained in strong inversion mode and depends on the oxide thickness t_{ox} of the MOS gates; it is identical for nMOS and pMOS, $C_{max-pMOS} = C_{max-nMOS}$. Varactor C_{min}

is obtained in weak inversion mode [19], and depends on the complex channel doping profile, including LDD and HALO implants. In the submicron CMOS process used here, the ratio $C_{min-pMOS}/C_{min-nMOS} \cong 0.85 \cong 1$ (the number is process dependent). Finally, the signal swing, of course, is identical for pMOS and nMOS varactors. So a parallel connection of nMOS and pMOS varactors enables in first order a (quasi-) differentially tuned VCO with equal but opposite signed gain K_{vco} to nMOS and pMOS tuning input.

VI. PROTOTYPE DESIGN AND MEASUREMENTS

As for the 0.25-μm CMOS process available for this design, the cross-coupled quadrature VCO definitely outperforms the other quadrature solutions in terms of power consumption. It was combined with the differential tuning concept leading to the final schematic of the prototype IC, presented in Fig. 11. The 1.5-Ωcm low-ohmic substrate nonepi standard CMOS process offers four thin aluminum metal layers. This is the worst-case scenario for coil design, comparing with bipolar, BiCMOS, or GaAs RF technologies. In the integrated inductor, the three top

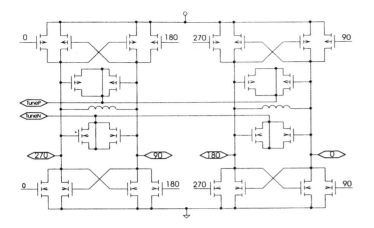

Fig. 11.   Quadrature VCO schematic.

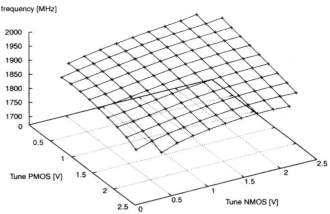

Fig. 13.   VCO tuning range.

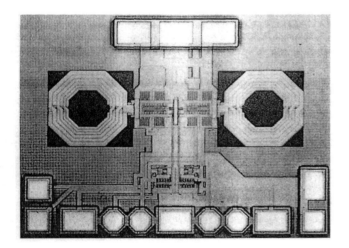

Fig. 12.   Chip photo of VCO.

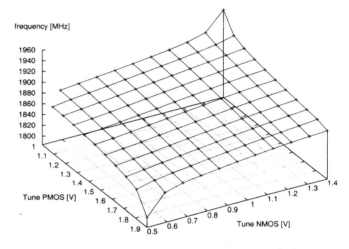

Fig. 14.   Zoomed VCO tuning range.

metals were put in parallel to reduce the series resistances. For maximal speed and minimal tank capacitive load, all transistors (not including the varactors) lengths were set to minimal length of 0.25 μm. Excessive white noise [20]–[23] could become a strong argument in more advanced submicron processes to choose lengths greater than minimal.

The widths of the MOS transistors of the cores were dimensioned to obtain sine waveforms of maximal amplitude and of maximal symmetry, using ordinary SPICE transient simulations. Also by means of transient simulations, optimal core cross-coupling width was set to one third of the width of the core transistors. If the cross coupling is made too weak, a two-tone oscillation is possible; if it is made too strong, power is wasted and extra parasitics load the tanks. Due to the high Q of the tank, all MOS widths and currents can be kept small. The chip photo in Fig. 12 shows the perfectly symmetrical layout. Testchip die size is 1500 μm by 700 μm.

The 3-D tuning characteristic of the VCO is presented in Figs. 13 and 14. As the slope is very linear, it is well suited for a frequency synthesizer realization (PLL). Phase noise was measured using Europtest PN9000 equipment (delay line method); the result is presented in Fig. 15. A quadrature accuracy of ~ 3° was measured. This number is mainly due to limited accuracy

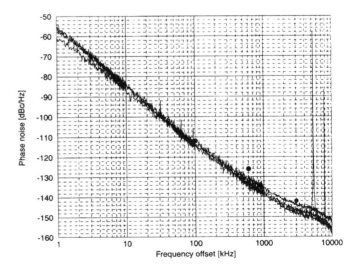

Fig. 15.   Measured phase noise over frequency tuning at 2.5-V power supply, 7.8-mA core current. Dots indicate GSM requirements.

of the measurements, e.g., a bondwire length mismatch of only 0.2 mm ($\cong$0.2 nH) leads to ~ 1° phase mismatch. Table II gives a summary of the measured performance.

TABLE II
MEASURED QUADRATURE VCO PERFORMANCE SUMMARY

Center frequency	1.72GHz-1.99GHz
Tuning range	280MHz (17%) differentially
Phase noise	< -143dBc/Hz@3MHz
Quadrature mismatch	$< 5°$
Power	20mW (7.8mA @ 2.5V)
Technology	standard 0.25μm CMOS
Area	1.1mm^2 (incl. pads)

TABLE III
PERFORMANCE OF SOME RECENTLY PUBLISHED FULLY INTEGRATED
VCOs, PHASE NOISE IS RECALCULATED TO 1.8 GHz AT 3 MHz OFFSET,
FOM AS DEFINED IN (10)

VCO	Tech. [μm]	Power [mW]	Phasenoise [dBc/Hz]	FOM [dBc/Hz]
[24]	0.7	6	-130	-177.8
[25]	0.25	24	-132	-174.1
[26]	0.8	66	-132	-171.1
[9]	0.25	32.4	-141	-181.5
(this)	0.25	20	-143	-185.5

A normalized phase noise has been defined as a figure of merit (FOM) for oscillators:

$$\text{FOM} = S_{\text{SSB}} \left(\frac{\Delta f}{f_0} \right)^2 P_{\text{vco}}/\text{mW} \qquad (10)$$

where P_{vco} is the total VCO power consumption. This results in a FOM of -185.5 dBc/Hz for this design. In Table III, some recently published VCOs are listed. It shows that this design has a state-of-the-art phase-noise performance at an extremely low power dissipation, especially when considering that all other designs need additional power to generate quadrature outputs.

VII. CONCLUSION

The design problem to minimize the overall power consumption of a VCO with quadrature outputs is solved using two cross-coupled fully integrated high-inductance VCO cores. A prototype quadrature LC-VCO for 1.8 GHz was designed in 0.25-μm standard digital CMOS. A differential tuning range of 280 MHz was obtained through the use of nMOS and pMOS varactors. Measured worst-case phase noise is -143 dBc/Hz at 3 MHz. This VCO fulfills GSM and DCS1800 receive phase noise requirements at a power consumption of only 20 mW.

REFERENCES

[1] Murata VCO Data Sheets, MQ or MQE [Online]. Available: http://www.murata.com
[2] AVX VCO Data Sheets [Online]. Available: http://www.avx-corp.com/docs/Catalogs/yk-vco.pdf
[3] M. Steyaert and J. Craninckx, *Wireless CMOS Frequency Synthesiser Design*. London, U.K.: Kluwer, 1998.
[4] B. Razavi, "A study of phase noise in CMOS oscillators," *IEEE J. Solid-State Circuits*, vol. 31, pp. 331–343, May 1996.
[5] A. Hajimiri and T. H. Lee, "A general theory of phase noise in electrical oscillators," *IEEE J. Solid-State Circuits*, vol. 33, pp. 179–194, Feb. 1998.
[6] ——, "Design issues in CMOS differential LC oscillators," *IEEE J. Solid-State Circuits*, vol. 34, pp. 717–724, Feb. 1999.
[7] J. J. Rael and A. A. Abidi. "Physical processes of phase noise in differential LC-oscillators," in *Proc. IEEE Custom Integrated Circuits Conf. (CICC)*, Orlando, FL, May 2000, pp. 569–572.
[8] Q. Huang, "Phase noise to carrier ratio in LC oscillators," *IEEE Trans. Circuits Syst. I*, vol. 47, pp. 965–980, July 2000.
[9] M. Borremans, B. De Muer, N. Itoh, and M. Steyaert, "A 1.8-GHz highly tunable low-phase-noise CMOS VCO," in *Proc. IEEE Custom Integrated Circuits Conf. (CICC)*. Orlando, FL, May 2000, pp. 585–588.
[10] D. B. Leeson, "A simple model of feedback oscillator noise spectrum," *Proc. IEEE*, pp. 329–330, Feb. 1966.
[11] K. O, "Estimation methods for quality factors of inductors fabricated in silicon integrated circuit process technologies," *IEEE J. Solid-State Circuits*, vol. 33, pp. 1249–1252. Aug. 1998.
[12] P. Andreani, "A comparison between two 1.8-GHz CMOS VCOs tuned by different varactors," in *Proc. 24th Eur. Solid-State Circuits Conf. (ESSCIRC)*, The Hague, The Netherlands, Sept. 1998, pp. 380–383.
[13] M. Danesh, J. R. Long, R. A. Hadaway, and D. L. Harame, "A Q-factor enhancement technique for MMIC Inductors," in *IEEE Radio Frequency Integrated Circuits Symp.*, June 1998, p. 217.
[14] M. Tiebout, "A fully integrated 1.3-GHz VCO for GSM in 0.25-μm standard CMOS with a phase noise of -142 dBc/Hz at 3-MHz offset," in *Proc. 30th Eur. Microwave Conf.*. Paris, France, Oct. 2000, pp. 140–143.
[15] E. A. M. Klumperink, S. L. J. Gierkink, A. P. van der Wel, and B. Nauta, "Reducing MOSFET 1/f noise and power consumption by switched biasing," *IEEE J. Solid-State Circuits*, vol. 35, pp. 994–1001, July 2000.
[16] J. Craninckx and M. S. J. Steyaert, "A fully integrated CMOS DCS-1800 frequency synthesizer," *IEEE J. Solid-State Circuits*, vol. 33, pp. 2054–2065, Dec. 1998.
[17] M. Borremans and M. Steyaert, "A CMOS 2-V quadrature direct up-converter chip for DCS-1800 integration," in *Proc. 26th Eur. Solid-State Circuits Conf. (ESSCIRC)*, Stockholm, Sweden, Sept. 2000, pp. 88–91.
[18] M. Rofougaran, A. Rofougaran, J. Rael, and A. A. Abidi, "A 900-MHz CMOS LC-oscillator with quadrature outputs," in *Proc. IEEE Int. Solid-State Circuits Conf.*, New York. NY, 1996, p. 392.
[19] A.-S. Porret *et al.*, "Design of high-Q varactors for low-power wireless applications," *IEEE J. Solid-State Circuits*, vol. 35, pp. 337–345, Mar. 2000.
[20] R. P. Jindal, "Hot-electron effects on channel thermal noise in file-line nMOS field-effect transistors," *IEEE Trans. Electron Devices*, vol. ED-33, pp. 1395–1397, Sept. 1986.
[21] A. Birbas, D. Triantis, and D. Kondis, "Thermal noise modeling for short channel MOSFETs," *IEEE Trans. Electron Devices*, vol. ED-43, pp. 2069–2075, Nov. 1996.
[22] S. Tedja *et al.*, "Analytical and experimental studies of thermal noise in MOSFETs," *IEEE Trans. Electron Devices*, vol. ED-41, pp. 2069–2075, Nov. 1994.
[23] P. Klein, "An analytical thermal noise model of deep submicron MOSFETs for circuit simulation with emphasis on the BSIM3v3 SPICE model," in *Proc. ESSDERC*, 1998, pp. 460–463.
[24] J. Craninckx and M. Steyaert, "A 1.8-GHz low-phase-noise CMOS VCO using optimized hollow spiral inductors," *IEEE J. Solid-State Circuits*, vol. 32, pp. 736–744, May 1997.
[25] B. De Muer, N. Itoh, and M. S. Steyaert, "Low supply voltage fully integrated CMOS VCO with three terminals spiral inductor," in *Proc. 25th European Solid-State Circuits Conf.*, Duisburg, Germany, Sept. 1999, pp. 207–209.
[26] C.-M. Hung and O. Kenneth, "A 1.24-GHz monolithic CMOS VCO with phase noise of -137 dBc/Hz at a 3-MHz offset," *IEEE Microwave and Guided Wave Lett.*, vol. 9, pp. 331–343, Mar. 1999.

Marc Tiebout (S'90–M'93) was born in Asse, Belgium, in 1969. He received the M.S. degree in electrical and mechanical engineering in 1992 from the Katholieke Universiteit Leuven, Belgium.

After his studies, he joined Siemens AG, Corporate Research and Development, Microelectronics, Munich, Germany. He is currently with Infineon Technologies AG, Department of Wireless Products, Technology and Innovation, working on RFCMOS circuits and transceivers for wireless communications.

Analysis and Design of an Optimally Coupled 5-GHz Quadrature *LC* Oscillator

Johan van der Tang, *Member, IEEE*, Pepijn van de Ven, Dieter Kasperkovitz, and
Arthur van Roermund, *Senior Member, IEEE*

Abstract—A 5-GHz quadrature *LC* oscillator has been realized,
in which the two *LC* stages are coupled with phase shifters. Analysis on the behavioral level shows that an N-stage *LC* oscillator
is optimally coupled when each stage is connected with phase
shifters providing $\pm 180°/N$ phase shift. Simulation of the 5-GHz
two-stage quadrature *LC* oscillator reveals a 4.3-dB reduction in
phase noise compared to a quadrature *LC* oscillator without phase
shifters. Measurements of the 5-GHz quadrature *LC* oscillator,
made in a 30-GHz f_T process, show a phase noise lower than
-113 dBc/Hz, with a resonator quality factor of only 4 and an
oscillator core power dissipation of 21.2 mW.

Index Terms—Analog integrated circuits, *LC* oscillators, modeling, optimization, phase noise, quadrature generation.

I. INTRODUCTION

MODERN receiver architectures, such as the zero-IF receiver and the low-IF receiver, allow a high degree of integration and are therefore often utilized in wireless transceiver designs [1]. In order to avoid loss of information, these architectures normally have an in-phase (I) and quadrature (Q) signal processing path. Usually the received signal is split after the LNA and multiplied with a quadrature (I/Q) signal source. Examples of other receiver architectures that require I/Q signals are image-reject receivers such as the Hartley and Weaver architecture [2]. Quadrature signals may also be needed at the transmit side of a wireless transceiver. In direct-conversion transmitters, baseband I and Q data streams are multiplied with a quadrature carrier signal, added, and transmitted. In this way, modulation and upconversion are performed in the same circuit [3]. In receiver front ends for optical transmission, I/Q signals are used in advanced data clock recovery (DCR) architectures. The function of the DCR circuit is to extract the clock information from the serial data stream which is transmitted over the optical fiber. Most integrated DCR circuits are phase-locked loop (PLL) based and require a frequency aid, which brings the oscillator in the PLL within the acquisition range of the PLL. If an I/Q oscillator is used, integrated DCR architectures can be constructed with a built-in frequency discriminator [4]. In this case, no expensive external reference frequency such as a crystal oscillator signal is needed to bring the oscillator within the DCR acquisition range. Another interesting application of I/Q oscillators in the optical transceiver field is half-rate DCR architectures [5].

A half-rate DCR architecture employs I/Q signals to be able to operate at half the frequency of the incoming data rate.

All of the mentioned architectures have in common that they require a signal source which provides I/Q output signals. These I/Q signals can be generated in many ways. If phase noise requirements are not stringent, an even-stage ring oscillator is often most convenient [6], [7]. However, in many applications, for example, telecommunication front ends, given a realistic power budget, the phase-noise specification can only be achieved with *LC* oscillators. An even-stage *LC* oscillator delivers, like even-stage ring oscillators, I/Q signals "correct-by-construction" [8]–[13]. It is constructed with identical stages, and the quadrature relation has its roots in the phase condition for oscillation (Barkhausen's phase criterion). In practice, the quadrature relation is not perfect, but limited by device mismatches, and is dependent on the layout symmetry. Calibration techniques can be employed to improve phase and amplitude matching [14]–[16].

This paper addresses the issue of optimum coupling of *LC* oscillators. When quadrature generation by means of an even-stage *LC* oscillator is chosen for an application, the design question of the optimum way of constructing a quadrature *LC* oscillator immediately arises. Optimum, in this context, meaning minimum $\mathcal{L}(f_m)$, which is the phase noise relative to the carrier at offset frequency f_m and has the units (dBc/Hz).

In order to answer the posed design question, an N-stage *LC* oscillator is investigated in this paper. Linear behavioral modeling is used to derive first-order expressions for the quality factor Q_N and the phase noise $\mathcal{L}_N(f_m)$ of an N-stage *LC* oscillator. Analysis of Q_N and $\mathcal{L}_N(f_m)$ shows that coupling *LC* oscillators with phase shifters can improve $\mathcal{L}_N(f_m)$. Based on this insight, a 5-GHz quadrature *LC* oscillator with phase shifters is realized, which can be used in half-rate DCR architectures in optical receivers [5].

In Section II, the conventional way of coupling two oscillator stages to form a quadrature oscillator is described. Section III introduces a behavioral model of an N-stage *LC* oscillator, which includes phase shifters. Linear analysis of this model will lead to the formulation of Q_N and $\mathcal{L}_N(f_m)$ for an N-stage *LC* oscillator. Section IV discusses the implications of the presented theory and concludes that phase shifters can be used to optimize $\mathcal{L}_N(f_m)$ of coupled *LC* oscillators. An optimally coupled 5-GHz quadrature oscillator employing phase shifters is described in Section V. The measurement results of this oscillator are discussed in Section VI.

II. CONVENTIONAL COUPLING OF *LC* OSCILLATORS

The simplest I/Q *LC* oscillator consists of two identical single-phase *LC* oscillators. Normally, these *LC* oscillators

Manuscript received August, 2001; revised December, 2001.
J. van der Tang and A. van Roermund are with the Eindhoven University of Technology, Mixed-Signal Microelectronics Group, 5600 MB Eindhoven, The Netherlands (e-mail: j.d.v.d.tang@tue.nl; a.h.m.v.roermund@tue.nl).
P. van de Ven and D. Kasperkovitz are with the Philips Research Laboratories, Integrated Transceivers Group, 5656 AA Eindhoven, The Netherlands (e-mail: pepijn.van.de.ven@philips.com; dkasperk@itom.nl).
Publisher Item Identifier S 0018-9200(02)03663-6.

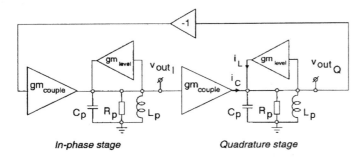

Fig. 1. Behavioral model of a quadrature LC oscillator with conventional coupling.

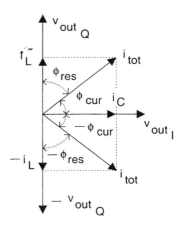

Fig. 2. Combined phasor diagram of the output voltages $\vec{v}_{\text{out}_I}$ and $\vec{v}_{\text{out}_Q}$ and the currents in the quadrature stage, $\vec{i}_L$ and $\vec{i}_C$.

are coupled with transconductances which force the two single-phase LC oscillators to oscillate in quadrature [8]–[13]. The behavioral model of an I/Q LC oscillator which uses this conventional coupling method is shown in Fig. 1. The I/Q oscillator has two identical stages, an in-phase and a quadrature stage, and has an inversion in the feedback path. In each stage, an LC oscillator is present, implemented with a parallel resonance circuit (the parallel circuit of C_p, R_p, and L_p) and transconductance gm_{level}. Transconductance gm_{couple} couples the two stages. The quality factor Q_p of the resonator at resonance frequency $\omega_o = (\sqrt{C_p L_p})^{-1}$ is

$$Q_p = Rp\sqrt{\frac{C_p}{L_p}} = \frac{\omega_o}{2}\left|\frac{\delta\phi}{\delta\omega}\right| \qquad (1)$$

where $\delta\phi/\delta\omega$ is the slope of the phase characteristic of the resonator.

The principle of operation of the conventionally coupled oscillator is explained by Fig. 2. Assuming a quadrature oscillation mode exists, output voltages (see Fig. 1) $\vec{v}_{\text{out}_I}$ and $\vec{v}_{\text{out}_Q}$ will be in quadrature. Output voltage $\vec{v}_{\text{out}_Q}$ may lead or lag $\vec{v}_{\text{out}_I}$ by 90°, hence, the phasor $-\vec{v}_{\text{out}_Q}$ is also drawn. The phasors of the currents in the quadrature stage, $\vec{i}_L = gm_{\text{level}} \cdot \vec{v}_{\text{out}_Q}$ and $\vec{i}_C = gm_{\text{couple}} \cdot \vec{v}_{\text{out}_I}$ are shown as well in Fig. 2[1]. Based on symmetry, the currents in the in-phase stage will be identical in amplitude and phase relation, and we can restrict our analysis to one stage. The sum of $\vec{i}_C$ and $\vec{i}_L$ can be written as $\vec{i}_t = i_{\text{tot}} e^{j\phi_{\text{cur}}}$ with $i_{\text{tot}} = |\vec{i}_L + \vec{i}_C|$ and $\tan(\phi_{\text{cur}}) = |\vec{i}_L|/|\vec{i}_C|$. The current $\vec{i}_t$ is injected into the resonator which provides a phase shift ϕ_{res}. For the behavioral model in Fig. 1, it can be shown that $|\vec{v}_{\text{out}_Q}| = i_{\text{tot}} R_p \cos(\phi_{\text{res}})$. The resonator phase shift $|\phi_{\text{res}}|$ must be $90° - |\phi_{\text{cur}}|$ for quadrature oscillation.

As illustrated in Fig. 2, $\phi_{\text{cur}} + \phi_{\text{res}}$ must be either +90° or −90° for quadrature operation. In practice, this ambiguity of several solutions for quadrature oscillation[2] is solved because a practical resonator, unlike the theoretical parallel resonator in Fig. 1, is asymmetric. This asymmetry provides a unique solution to the oscillation conditions, where the phase condition for quadrature oscillation is met and the loop gain is the highest

[17]. From the preceding discussion, it will be clear that, in conventionally coupled I/Q LC oscillators, the resonator phase shift ϕ_{res} is nonzero. For example, if i_C and i_L have the same amplitude, $\phi_{\text{cur}} = \pm45°$ and ϕ_{res} will also be equal to $\pm45°$. At zero resonator phase shift, the quality factor of a parallel resonator is maximum and equal to Q_p [see (1)]. Therefore, intuitively, $\phi_{\text{res}} = 0°$ seems like a good operating point for maximum quality factor and thus minimum $\mathcal{L}(f_m)$ in quadrature LC oscillators. In the next section, the influence of ϕ_{res} on the quality factor and $\mathcal{L}(f_m)$ of multiphase LC oscillators will be investigated.

III. PHASE NOISE IN N-STAGE LC OSCILLATORS

In Fig. 3, the behavioral model of an N-stage LC oscillator is shown. Compared to the quadrature oscillator in Fig. 1, there are three major differences. First, the model in Fig. 3 is an N-stage LC oscillator model and obviously reduces to a quadrature LC oscillator for $N = 2$. Second, in each of the N stages, a noise source i_n is present, which models the total noise generated in a stage. Third, a phase shifter with a fixed phase shift ϕ_p is inserted in each stage.

The linear time-invariant model in Fig. 3 will be used to calculate $\mathcal{L}(f_m)$ of an N-stage LC oscillator. A linear approach needs justification, given the availability of recent phase-noise theories, which include the time-variant and nonlinear nature of a practical oscillator [18]–[20]. The main reason for a linear approach is its simplicity, while still providing valuable design insights. A linear approach to phase noise in an oscillator can be quantitatively accurate within a few decibels for oscillators which are operating in the linear or weakly nonlinear region [21], [22]. For the large class of oscillators operating in the strongly nonlinear region, phase-noise expressions derived using linear analysis cannot be used for accurate quantitative predictions. However, linear analysis is still of value for this class, as it reveals the dependencies of the phase-noise performance on design parameters.

The phase-noise-to-carrier ratio of an N-stage LC oscillator is calculated using the linear modeling method outlined in [22]. At system level, this model describes how white noise is shaped within a linear feedback system into phase noise. Application of

---

[1]In a high-frequency I/Q oscillator parasitic shift will cause $\vec{i}_C$ and $\vec{i}_L$ to lag $\vec{v}_{\text{out}_I}$ and $\vec{v}_{\text{out}_Q}$, respectively. Also, the quadrature relation between the currents may not be present in a practical circuit, due to nonidentical parasitic phase shift and other second-order effects. However, this does not alter the principle of operation, and for brevity, an in-depth discussion of nonidealities is omitted.

[2]For a practical resonator, there can be more than two solutions, where the oscillation conditions for quadrature operation are met [17].

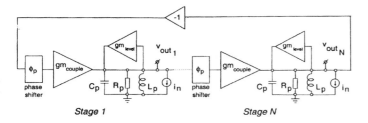

Fig. 3. Behavioral model of an N-stage LC oscillator with phase shifters.

this model on the N-stage LC oscillator in Fig. 3 results in the following expression:

$$\mathcal{L}_N(f_m) = 10 \log \left(\frac{1}{2} \cdot \frac{1}{4Q_N^2} \left(\frac{f_{osc}}{f_m} \right)^2 \cdot \frac{N \cdot \overline{i_n^2}}{i_{carrier}^2} \right) \quad (2)$$

where f_{osc} is the carrier frequency $\omega_{osc}/2\pi$, $i_{carrier}^2$ is the squared rms carrier current, and Q_N is defined as

$$Q_{N|\omega=\omega_{osc}} = N \cdot \frac{\omega_{osc}}{2}$$
$$\cdot \sqrt{\left(|H(j\omega)|^{N-1} \cdot \frac{\delta |H(j\omega)|}{\delta\omega} \right)^2 + \left(\frac{\delta \arg (H(j\omega))}{\delta\omega} \right)^2} \quad (3)$$

in which $H(j\omega)$ is the transfer function of one stage of the N-stage LC oscillator in Fig. 3. In order to gain insight into the behavior of Q_N as a function of N and the circuit parameters in one LC stage, ω_{osc}, $|H(j\omega)|$, and the derivatives of $|H(j\omega)|$ and $\arg(H(j\omega))$ need to be calculated. Calculation of these parameters, substitution in (3), and simplification leads to a good approximation of Q_N:

$$Q_N(\phi_{res}) \approx N \cdot Q_p \cdot \cos(\phi_{res}) \quad (4)$$

which has an error less than 4% for ϕ_{res} ranging from 0° to 80° with respect to the more complicated exact expression for Q_N reported in [23].

IV. Optimum Coupling of LC Oscillators

In this section, the implications of the theory in Section III will be discussed. In Section IV-A, the design question posed in the introduction concerning the optimum way to couple LC oscillators is answered. To implement optimally coupled quadrature oscillators, phase shifters are needed. Two simple phase shifters are described in Section IV-B.

A. Design Implications of Q_N and $\mathcal{L}_N(f_m)$

The quality factor Q_N in (4) is maximum for $\phi_{res} = 0°$. For $\phi_{res} = 0°$, the quality factor of an N-stage LC oscillator is simply NQ_p. For an I/Q LC oscillator ($N = 2$), Q_N is plotted in Fig. 4 versus $|\phi_{res}|$ for three values of Q_p. Close to zero resonator phase shift, Q_N stays close to its optimum of $2\,Q_p$. However, for large values of $|\phi_{res}|$, Q_N decreases rapidly. At $|\phi_{res}| = 60°$, Q_N is halved and at $|\phi_{res}| = 76°$, approximately only 1/4 of the maximum quality factor remains.

Reduction of Q_N results in $\mathcal{L}_N(f_m)$ degradation. In Fig. 4(b), the degradation of $\mathcal{L}_N(f_m)$ versus $|\phi_{res}|$ is plotted. In a practical LC oscillator, $\mathcal{L}_N(f_m)$ can be proportional to $1/Q_N^3$ or

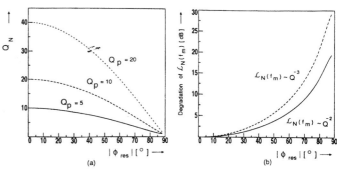

Fig. 4. (a) Quality factor Q_N versus the resonator phase shift, for Q_p is 5, 10, and 20, and $N = 2$. (b) Degradation of the $\mathcal{L}_N(f_m)$ versus the resonator phase shift, for $N = 2$ and $Q_p = 5$.

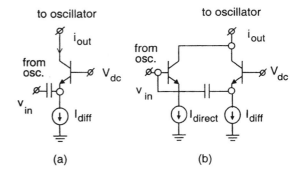

Fig. 5. (a) Phase shifter with fixed phase shift. (b) Tunable phase shifter.

$1/Q_N^2$ [19], and the degradation of $\mathcal{L}_N(f_m)$ for both situations is plotted in Fig. 4(b), for $Q_p = 5$ and $N = 2$. At a resonator phase shift of 0° ... 30°, the degradation of $\mathcal{L}_N(f_m)$ is only a few decibels. However, for $|\phi_{res}| = 80°$, for example, the degradation in $\mathcal{L}_N(f_m)$ is 14 and 21 dB for quadratic and cubic dependency on Q_N, respectively. The degradation of $\mathcal{L}_N(f_m)$ will be higher for larger values of Q_p. Fig. 4 points out that the stages in an N-stage LC oscillator are optimally coupled if ϕ_{res} is made zero. Zero resonator phase shift is obtained if the phase shifters in Fig. 3 have a phase shift $\phi_p = \pm 180°/N$. For this phase shift, the output currents of the transconductance gm_{couple} and transconductance gm_{level} both have the same phase as the desired voltage phase in a stage. In other words, $\phi_p = \pm 180°/N$ makes the phase shift ϕ_{cur} in each stage effectively $\pm 180°/N$. Therefore, the phase condition for quadrature oscillation will be met for $\phi_{res} = 0$.

Expression (2) with $Q_N = NQ_p$ substituted, shows that increasing the number of stages improves $\mathcal{L}_N(f_m)$ with $10 \log(N)$ dB. However, an increased number of stages also means increased power consumption. When normalizing for power dissipation, $\mathcal{L}_N(f_m)$ is independent of the number of stages.

B. Two Simple Phase Shifters

In a two-stage I/Q LC oscillator, minimum $\mathcal{L}_N(f_m)$ will be obtained with a phase shift $|\phi_p|$ of 90°. As a ±90° phase shift for sinusoidal signals is similar to an integrating or differentiating action, it is possible to implement the phase shifters based on an integrator or a differentiator. Fig. 5(a) shows the circuit implementation of a differentiator. The circuit not only provides the

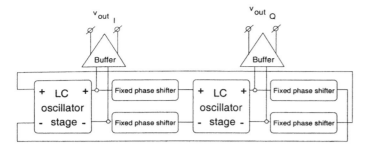

Fig. 6. Block diagram of the I/Q oscillator with phase shifters.

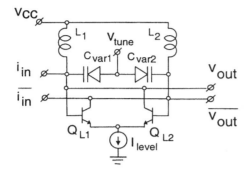

Fig. 7. Circuit diagram of an *LC* oscillator stage.

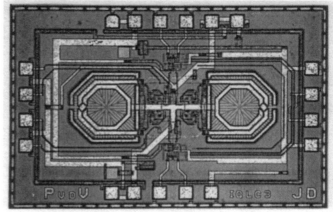

Fig. 8. Micrograph of the 5-GHz quadrature *LC* oscillator.

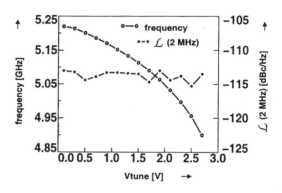

Fig. 9. Frequency and $\mathcal{L}(2 \text{ MHz})$ versus the tuning voltage V_{tune}.

required phase shift, but also provides isolation between the oscillator stages. The current i_{out} flowing into the collector shows a $-90°$ phase shift compared to the input voltage v_{in} up to a certain frequency. Above this frequency, the phase will deviate from $-90°$ due to additional poles in the transistor. This effect can be compensated by adding a direct path between v_{in} and i_{out} as is shown in Fig. 5(b). In principle, any phase shift can be made by proper dimensioning of the currents I_{direct} and I_{diff} of the direct path and the differentiating path, respectively. At high frequencies, however, parasitic effects will limit the range of the phase variation.

V. A 5-GHz QUADRATURE *LC* OSCILLATOR WITH PHASE SHIFTERS

The phase noise improvement due to adding phase shifters in a multiphase *LC* oscillator has been investigated for a two-stage *LC* oscillator at high frequencies. A 5-GHz I/Q *LC* oscillator has been realized which is based on the architecture in Fig. 3. This oscillator can be used in half-rate DCR architectures which were mentioned in Section I. With a quadrature oscillator running at 5 GHz, a 10-Gb/s half-rate DCR can be constructed [5].

A block diagram of the realized I/Q *LC* oscillator is shown in Fig. 6. The phase shifter from Fig. 5(a) is used to couple the two *LC* oscillator stages. The cross-coupled wires implement the inverter in the behavioral model. Fig. 7 shows the circuit diagram of one *LC* oscillator stage. The I/Q *LC* oscillator is implemented in a BiCMOS process with a 30-GHz cut-off frequency (f_T) [24]. The inductors L_1 and L_2 in Fig. 7 are implemented as one balanced coil with a center tap which is connected to V_{CC}. Constraints in design time led to reuse of a well-characterized resonator (the parallel circuit of L_1, L_2, C_{var1}, and C_{var2}). At 5 GHz, the quality factor Q_p of this resonator is only 4, because the resonator was optimized for lower frequencies. Furthermore,

its large parasitic capacitance reduced the effective tuning range of the p-n-type varactors. Obviously, a resonator with higher Q_p and lower parasitics will improve $\mathcal{L}(f_m)$ as well as the tuning range. The transistors Q_{L1} and Q_{L2} implement gm_{level}. Supply voltage V_{CC} was set to 2.7 V and I_{level} was set to 1.44 mA.

In order to obtain a value for the $\mathcal{L}(f_m)$ of the I/Q *LC* oscillator, the circuit simulator SpectreRF was used. At 5 GHz, the simulated $\mathcal{L}(2 \text{ MHz})$ is about -114 dBc/Hz. This figure was compared with $\mathcal{L}(f_m)$ of the same design but coupled without phase shifters. Phase-noise simulations show a degradation of 4.3 dB at 5 GHz compared to the implementation of Fig. 6. Even at 5 GHz and with a moderate resonator quality factor Q_p, the proposed architecture with phase shifters has improved performance in comparison to conventionally coupled I/Q *LC* oscillators.

VI. EXPERIMENTAL RESULTS

The oscillator presented in Section V was designed and fabricated in a 30-GHz f_T BiCMOS process [23]. The active chip area is 1450 μm × 2280 μm and the VCO core dissipation is 21.2 mW with a supply voltage of 2.7 V. Fig. 8 shows a micrograph of the I/Q *LC* oscillator with phase shifters.

Measurement results of the oscillation frequency and $\mathcal{L}(2 \text{ MHz})$ versus varactor tuning voltage V_{tune} are shown in Fig. 9. The $\mathcal{L}(2 \text{ MHz})$ is better than -113 dBc/Hz over the complete tuning range and the phase noise simulation of -114 dBc/Hz matches well with the measurements. Measurements of $\mathcal{L}(2 \text{ MHz})$ and the tuning range were performed using

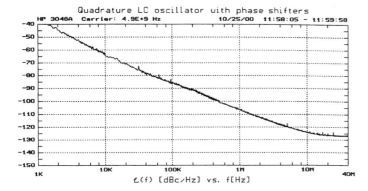

Fig. 10. $\mathcal{L}(f_m)$ versus the offset frequency f_m. The carrier frequency is 4.9 GHz.

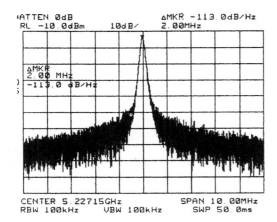

Fig. 11. Power spectrum of the oscillator at 5.22 GHz.

a spectrum analyzer. The $\mathcal{L}(f_m)$ measurements were verified with a phase-noise sideband measurement setup based on the HP 3048. Fig. 10 shows the measured phase-noise sideband of the I/Q LC oscillator at an oscillation frequency of 4.9 GHz. The power spectrum of the oscillator running at its highest frequency of 5.22 GHz is shown in Fig. 11.

VII. CONCLUSION

Quadrature LC oscillators can be used for low-power low-phase-noise quadrature signal generation. Coupling of N LC oscillator stages by means of phase shifters having a phase shift of $\pm 180°/N$ implements zero resonator phase shift and therefore optimum coupling. Two-stage quadrature LC oscillators are optimally coupled if the phase shifters which couple the stages have $\pm 90°$ phase shift. Based on this insight, a 5-GHz quadrature LC oscillator with phase shifters has been realized in a 30-GHz f_T BiCMOS process. Compared to a similar design without phase shifters, a 4.3-dB improvement in $\mathcal{L}(f_m)$ was simulated at 5 GHz. Measured tuning range was 4.91 to 5.22 GHz. The $\mathcal{L}(2\ \text{MHz})$ is better than -113 dBc/Hz with only 21.2 mW core dissipation and a resonator quality factor of 4.

ACKNOWLEDGMENT

This work is the result of a cooperation between Eindhoven University of Technology and Philips Research Laboratories, Eindhoven. The authors would like to express their gratitude toward Philips Research for IC fabrication and providing measurement facilities. The authors would also like to thank Dr. C. Vaucher from Philips Research for participating in many fruitful discussions.

REFERENCES

[1] J. Fenk, "Highly integrated RF-ICs for GSM, DECT, and UMTS systems, a status review and development trends," in *Proc. Eur. Solid State Circuits Conf. (ESSCIRC)*, 1999, pp. 11–14.
[2] B. Razavi, *RF Microelectronics*. Upper Saddle River, NJ: Prentice Hall, 1998.
[3] ——, "RF transmitter architectures and circuits," in *Proc. IEEE Custom Integrated Circuits Conf. (CICC)*, 1999, pp. 197–204.
[4] A. Pottbacker and U. Langmann, "An 8-GHz silicon bipolar clock-recovery and data-regenerator IC," in *IEEE Int. Solid State Circuits Conf. (ISSCC) Dig. Tech. Papers*, 1994, pp. 116–117.
[5] J. Hauenschild, "A plastic packaged 10 Gb/s BiCMOS clock and data recovering 1:4-demultiplexer with external VCO," *IEEE J. Solid-State Circuits*, vol. 31, pp. 2056–2059, Dec. 1996.
[6] J. Van der Tang and D. Kasperkovitz, "A 0.9–2.2-GHz monolithic quadrature mixer oscillator for direct-conversion satellite receivers," in *IEEE Int. Solid State Circuits Conf. (ISSCC) Dig. Tech. Papers*, 1997, pp. 88–89.
[7] J. Van der Tang *et al.*, "A 9.8–11.5-GHz I/Q ring oscillator for optical receivers," in *Proc. IEEE Custom Integrated Circuits Conf. (CICC)*, 2001, pp. 323–326.
[8] A. Rofougaran *et al.*, "A 900 MHz CMOS *LC*-oscillator with quadrature outputs," in *IEEE Int. Solid State Circuits Conf. (ISSCC) Dig. Tech. Papers*, 1996, pp. 392–393.
[9] B. Razavi, "A 1.8-GHz CMOS voltage-controlled oscillator," in *IEEE Int. Solid State Circuits Conf. (ISSCC) Dig. Tech. Papers*, 1997, pp. 388–389.
[10] T. Wakimoto and S. Konaka, "A 1.9-GHz Si bipolar quadrature VCO with fully integrated *LC* tank," in *Symp. VLSI Circuits Dig. Tech. Papers*, 1998, pp. 30–31.
[11] C. Lo and H. Luong, "2-V 900 MHz quadrature coupled *LC* oscillators with improved amplitude and phase matchings," in *Proc. IEEE Int. Symp. Circuits and Systems (ISCAS)*, vol. 2, 1999, pp. 585–588.
[12] T. Lui, "1.5-V 10–12.5-GHz integrated CMOS oscillators," in *Symp. VLSI Circuits Dig. Tech. Papers*, 1999, pp. 55–56.
[13] J. Kim and B. Kim, "A low-phase-noise CMOS *LC* oscillator with a ring structure," in *IEEE Int. Solid State Circuits Conf. (ISSCC) Dig. Tech. Papers*, 2000, pp. 430–431.
[14] S. Navid *et al.*, "Level-locked loop, a technique for broadband quadrature signal generation," in *Proc. Custom Integrated Circuits Conf. (CICC)*, 1997, pp. 411–414.
[15] D. Lovelace and J. Durec, "A self calibrating quadrature generator with wide frequency range," in *IEEE Radio Frequency Integrated Circuits Symp.*, 1997, pp. 147–150.
[16] C. Park *et al.*, "A 1.8-GHz self-calibrating phase-locked loop with precise I/Q matching," *IEEE J. Solid-State Circuits*, vol. 36, pp. 777–783, May 2001.
[17] A. Rofougaran *et al.*, "A single-chip 900-MHz spread-spectrum wireless transceiver in 1-μm CMOS—Part I: Architecture and transmitter design," *IEEE J. Solid-State Circuits*, vol. 33, Apr. 1998.
[18] A. Hajimiri *et al.*, "A general theory of phase noise in electrical oscillators," *IEEE J. Solid-State Circuits*, vol. 33, pp. 179–194, Feb. 1998.
[19] Q. Huang, "Phase noise to carrier ratio in *LC* oscillators," *IEEE Trans. Circuits Syst. I*, vol. 47, pp. 965–980, July 2000.
[20] J. Rael and A. Abidi, "Physical processes of phase noise in differential *LC* oscillators," in *Proc. Custom Integrated Circuits Conf. (CICC)*, 2000, pp. 569–572.
[21] J. van der Tang and D. Kasperkovitz, "Fast phase-noise analysis method for noise optimization of oscillators," in *Proc. Eur. Solid State Circuits Conf. (ESSCIRC)*, 1998, pp. 504–507.
[22] B. Razavi, "A study of phase noise in CMOS oscillators," *IEEE J. Solid-State Circuits*, vol. 31, pp. 331–343, Mar. 1996.
[23] P. van de Ven *et al.*, "An optimally coupled 5-GHz quadrature *LC* oscillator," in *Symp. VLSI Circuits*, 2001, pp. 115–118.
[24] A. Pruijmboom *et al.*, "QUBiC3: A 0.5 μm BiCMOS production technology, with $f_T = 30$ GHz, $f_{max} = 60$ GHz and high-quality passive components for wireless telecommunication applications," in *Bipolar/BiCMOS Circuits and Technology Meeting Dig. Tech. Papers*, 1998, pp. 120–123.

A 1.57-GHz Fully Integrated Very Low-Phase-Noise Quadrature VCO

Peter Vancorenland and Michiel S. J. Steyaert, *Senior Member, IEEE*

Abstract—A very low-phase-noise quadrature voltage-controlled oscillator is presented, featuring an inherently better figure of merit than existing architectures. Through an improved circuit schematic and a special layout technique, the phase noise of the circuit can be lowered. The circuit draws 15 mA from a 2-V supply. The phase noise is -133.5 dBc/Hz at 600 kHz and the tuning range is 24% wide at a center frequency of 1.57 GHz.

Index Terms—CMOS, quadrature oscillator, RF, wireless.

I. INTRODUCTION

THE EXPLODING telecommunications market pushes commercial RF circuits to integration in standard low-cost digital CMOS processes. Although most commercial designs employ bipolar technology, CMOS is a promising candidate for full integration of RF front ends, due to its low production cost. Much research has been done to integrate RF building blocks in CMOS, but some obstacles still exist. A major challenge in the design of CMOS single-chip transceiver systems is the integration of a low-power low-noise voltage-controlled oscillator (VCO) with quadrature outputs. This can be done by using a polyphase filter to generate the quadrature signals [1] or with a VCO with four outputs [2]–[4].

In this paper, a fully integrated quadrature *LC* VCO is presented that implements a novel quadrature voltage-controlled oscillator (QVCO) architecture featuring very low-phase-noise performance.

II. ARCHITECTURE

A QVCO can be created by locking two differential VCOs onto each other as described in [2]. A schematic diagram of this circuit is depicted in Fig. 1. Here, the four output nodes I_p, Q_p, I_n, and Q_n are depicted. They have the same amplitude and frequency, but their phases are 0°, 90°, 180°, and 270° shifted, respectively. The triangles are the inverters, implemented by nMOS transistors.

The α inverters are cross-coupled between two inverting nodes. These transistors will generate a negative resistance on these nodes, compensating the positive resistances from the circuit parasitics, thus ensuring oscillator startup.

The β inverters generate a cross-coupling between the two resonating differential VCOs. These transistors will lock both VCOs onto each other, forcing them to resonate in quadrature phases.

The inductors in the resonant *LRC* tank of a quadrature oscillator can be connected together in different ways [2], [4].

Manuscript received August, 2001; revised December, 2001.
The authors are with the Katholieke Universiteit Leuven, Dept. Elektrotechniek, ESAT-MICAS, B-3001 Heverlee, Belgium (e-mail: vanco@esat.kuleuven.ac.be; steyaert@esat.kuleuven.ac.be).
Publisher Item Identifier S 0018-9200(02)03665-X.

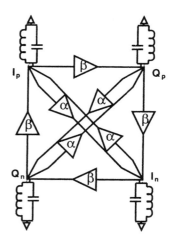

Fig. 1. Block schematic of the quadrature voltage-controlled oscillator of [2].

Three different connection types are shown in Fig. 2, where Fig. 2(a) represents the architecture from [2], Fig. 2(c) from [4], and Fig. 2(b) is the presented architecture. For each of these connections, the inherent performance of these QVCOs can be calculated.

These architectures can be compared by means of their figures of merit (FOM), which are carrier-to-noise ratio (CNR)/power ratios normalized for center frequency, offset frequency, and tuning range [5].

All the terms involved in the calculation of the FOM of the circuits can be calculated from the ac behavior of the equivalent small-signal schematic [6]. Table I summarizes the calculated results. The formulas herein are derived under the approximations of a high-ohmic substrate (high substrate resistance, negligible effect of the parasitic capacitance of the inductor). From this, it can clearly be seen that the connection of the coils in this architecture gives rise to an inherent improvement of the FOM of 3 dB better as compared to both of the other existing architectures.

III. CIRCUIT DESIGN

The β transistors from Fig. 1 are required to cross couple both differential VCOs. Increasing their relative transconductance improves the quality of the phase lock, but will increase the total power consumption, since they do not generate any negative resistance necessary to keep the oscillation running and will reduce the oscillation amplitude and degrade the quality factor of the resonance tank [7].

The active current in the tank is a combination of contributions from the α and β transistors, whose currents are 90° shifted in phase. The combination of both currents is out of phase with the required tank current for optimal $\delta\phi/(\delta f)$ (where ϕ is the phase of the tank and f the working frequency), hence the optimal quality factor of the tank will not be reached [8].

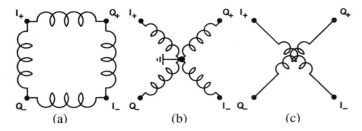

(a)   (b)   (c)

Fig. 2. Different types of coil connections for a quadrature LC oscillator.

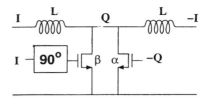

Fig. 3. Phase shift at gate of the β transistor.

TABLE I
COMPARISON OF THIS ARCHITECTURE WITH THE OTHER PUBLISHED ARCHITECTURES FROM [2] AND [4] FOR VCOs AT THE SAME FREQUENCY USING THE SAME COIL

Architecture	This	[2]	[4]	
CNR	$\dfrac{32\Delta\omega^2 I_p^2}{\omega_0^2 L^2}$	$\dfrac{16\Delta\omega^2 I_p^2}{\omega_0^2 L^2}$	$\dfrac{16\Delta\omega^2 I_p^2}{\omega_0^2 L^2}$	
ω_0	$\sqrt{\dfrac{2}{LC}}$	$\sqrt{\dfrac{1}{LC}}$	$\sqrt{\dfrac{2}{LC}}$	Hz
$\omega_{0,Diff}$	$\sqrt{2}\omega_0$	ω_0	∞	Hz
g_m	$\dfrac{2}{\omega_0(1+Q)}$	$\dfrac{1}{\omega_0(1+Q)}$	$\dfrac{2}{\omega_0(1+Q)}$	S
C	$\dfrac{2}{\omega_0^2 L}$	$\dfrac{1}{\omega_0^2 L}$	$\dfrac{2}{\omega_0^2 L}$	F
FOM [5]	0	-3	-3	dB

The phase noise as given by Leeson's formula [9] is inversely proportional to the signal power and the quality factor of the resonance tank. Increasing the transconductance of the β transistors will thus decrease the phase-noise performance for this architecture.

On the other hand, minimizing the β transconductance will decrease the maximum attainable image rejection ratio (IMRR) in a quadrature mixer system since the phase lock between the oscillators will be weaker.

In order to get good phase-noise performance and good image rejection, a new cross-coupling circuitry is needed which can lock both VCOs and which only generates currents in the correct 90° shifted phase.

This circuit can be obtained by shifting the voltage at the gate of the β transistor over an extra 90°, as depicted in Fig. 3. The extra benefit of this approach is that, since the currents at the drain of the β transistor are in phase with the currents at the drain of the α transistor, they will contribute to the negative resistance enabling a reduction of the α transconductances, while still achieving the minimal negative resistance requirement, thus reducing the power consumption.

One quarter of the circuit including the α and β transistors is given in Fig. 4, where M_3 and M_4 are the β transistors. The 90° phase shift is performed here in two steps. First, there is a 45° shift since the gates of M_3 and M_4 are connected to signals that are each 90° shifted in phase. Second, the current is shifted another 45° by the capacitor C_β that is connected between the sources of M_3 and M_4.

The bulk transconductance and the C_{sb} of the transistors, however, partially reduce the desired phase-shift effect to about 70°–80° instead of 90°. This can corrected by adding an extra correction circuit consisting of transistors M_1 and M_2 that inject current to compensate for the 10°–20° loss in phase shift. Although adding this circuit has a drawback on the power con-

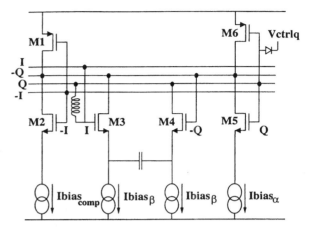

Fig. 4. Phase-compensated circuit. M_5 and M_6 are α transistors, M_3 and M_4 are phase-compensated β transistors, M_1 and M_2 are correction transistors.

sumption, it is small compared to the power reduction of the phase-compensated circuit.

The diodes used to tune the frequency are p+/n junction diodes with the cathodes connected to the control voltage. The full circuit schematic is given in Fig. 5.

IV. COIL DESIGN AND LAYOUT

The layout of the inductors core is given in Fig. 6(a). This core consists of four symmetrical coils [5] which are laid out quadrisymmetrically. They are interconnected with leads of the same width as the coil windings. The current flow direction is given for the left and the right coil. On the left side, the current flows from Q_- toward I_+, and from I_- toward Q_+ on the right side. This means that inside the coil, all the currents flow in the same circular direction around the center of the coil, but in the interconnection leads, this is no longer the case, since for a current in these leads, there is a current on the other side of the coils core which flows in the inverse circular direction around the center of the coils core. These currents have no net increase in the inductance of the LC tank, so the only parasitic effect of the leads is their resistance.

With this layout, one can neglect the magnetic effects of the leads and must only take their resistance into account. This allows the VCO coils to be optimized separately as a single coil, instead of optimizing the whole core including four single coils, their leads, and all of their (magnetic) interactions. This will tremendously reduce the optimization times and complexity, while not degrading the performance.

Since the layout is made fully quadrisymmetrical, there are two leads connecting each of the nodes I_p, Q_p, I_n, and Q_n to the rest of the circuitry, so the effective resistance of the coil–circuit

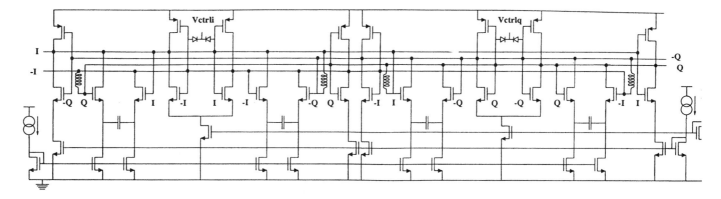

Fig. 5. Full circuit schematic.

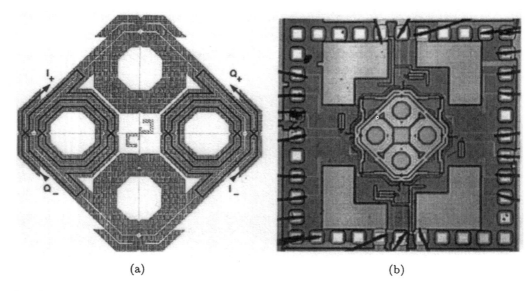

(a)   (b)

Fig. 6. (a) Current flows in the quadrisymmetrical coil. (b) Chip photograph.

connection is half the resistance of one lead. Since these are considerably smaller than the total length of the coil windings, their effect on the phase noise is minimized.

The optimization of the single coils is done with a finite-element simulator in an optimization loop [10]. The optimal coil inductance was 2.27 nH for a parasitic series resistance of only 1.15 Ω. The quality factor of the tank is 20.

V. MEASUREMENTS

The chip was processed in a 0.25-μm technology. A chip photograph is shown in Fig. 6(b). The chip was laid out on a 3 mm × 3 mm area, and the diameter of the coil's core is 900 μm. The largest part of the area is filled up by decoupling capacitance.

The chip was bonded on a ceramic substrate and measured. The current consumption is 15 mA from the 2-V supply. The tuning range is 330 MHz from 1.36 to 1.69 GHz, or 24%. The variation of the frequency with the control voltage is given in Fig. 7(a). The phase noise was measured using a Europtest PN9000 phase-noise measurement system. The measurement was performed on a single buffered output of the QVCO.

The variation of the quadrature phase noise with the frequency is given in Fig. 8(a). From this figure, one can see that

TABLE II
PERFORMANCE SUMMARY OF THE QUADRATURE VCO

Frequency	1.57	GHz
Phase noise	-133.5	dBc/Hz
Tuning Range	24	%
Supply	2	V
Current	15	mA
Technology	0.25μ	CMOS

the minimal phase noise of the QVCO is −133.5 dBc/Hz at 600 kHz offset from the carrier.

The normalized phase noise has been defined as an FOM [11]

$$\text{FOM} = 10 \log \left(\left(\frac{\omega_0}{\Delta\omega}\right)^2 \frac{1}{L\{\Delta\omega\}P} \right). \quad (1)$$

The FOM for this circuit is 187.1. This can be compared to other QVCO designs [2]–[4], [7], [12], [13]. In Fig. 8(b), the FOM is plotted versus the normalized tuning range to include all important VCO specifications, as explained in [5]. The designs in the upper-right corner of this figure have the best performance. It can clearly be seen that this design outperforms previously published QVCOs.

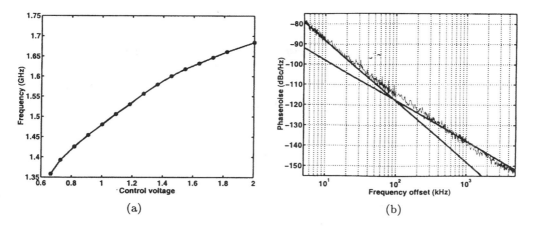

(a)   (b)

Fig. 7.   (a) Oscillation frequency versus control voltage. (b) Phase noise at 1.61 GHz.

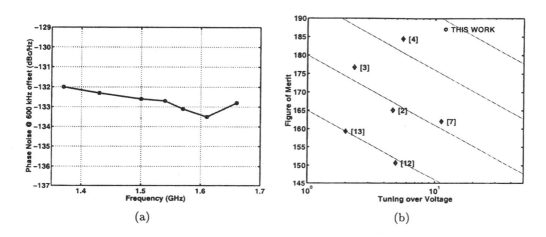

(a)   (b)

Fig. 8.   Phase noise at 600 kHz versus frequency. (b) Figure of merit versus relative tuning range.

VI. CONCLUSION

A very low-phase-noise QVCO has been presented. The FOM of this new architecture is inherently better than the existing architectures. Through an improved circuit schematic and a special layout technique, the phase noise of the circuit can be lowered to amongst the lowest published. The circuit draws 15 mA from a 2-V supply. The phase noise is −133.5 dBc/Hz at 600 kHz, and the tuning range is 24% wide at a center frequency of 1.57 GHz, making it suitable for GPS applications. The performance of the VCO is summarized in Table II.

ACKNOWLEDGMENT

The authors would like to thank Kawasaki Microelectronics for the funding of the test chip.

REFERENCES

[1] Steyaert et al., "A 2-V CMOS cellular tranceiver front end," in *IEEE Int. Solid State Circuits Conf. (ISSCC) Dig. Tech. Papers*, Feb. 2000. Session TA8.3.
[2] A. Rofougaran, J. Real, M. Rofougaran, and A. Abidi, "A 900 MHz CMOS LC oscillator with quadrature outputs," in *IEEE Int. Solid State Circuits Conf. (ISSCC) Dig. Tech. Papers*, 1996, pp. 392–393.
[3] T. Wakimoto et al., "A 1.9-GHz Si bipolar quadrature VCO with fully integrated LC tank," in *Proc. VLSI Symp.*, 1998.
[4] M. Tiebout, "A differentially tuned 1.73 GHz–1.99 GHz quadrature CMOS VCO for DECT, DCS1800 and GSM900 with a phase noise over tuning range between −128 dBc/Hz and −137 dBc/Hz at 600-kHz offset," in *Proc. Eur. Solid State Circuits Conf. (ESSCIRC)*, 2000, pp. 444–447.
[5] B. De Muer et al., "A 1.8-GHz highly tunable low-phase-noise CMOS VCO," in *Proc. IEEE Custom Integrated Circuits Conf. (CICC)*, 2000, pp. 585–588.
[6] P. Vancorenland et al., IEEE Trans. Circuits Syst., submitted for publication.
[7] T. Liu, "A 6.5-GHz monolithic CMOS voltage-controlled oscillator," in *IEEE Int. Solid State Circuits Conf. (ISSCC) Dig. Tech. Papers*, Feb. 1999, pp. 404–405.
[8] B. Razavi, "A study of phase noise in CMOS oscillators," *IEEE J. Solid-State Circuits*, vol. 31, pp. 331–343, Mar. 1996.
[9] D. B. Leeson, "A simple model of feedback oscillator noise spectrum," *Proc. IEEE*, pp. 329–330, Feb. 1966.
[10] B. De Muer et al., "A simulator optimizer for the design of very low-phase-noise CMOS LC oscillators," in *Proc. ICECS*, 1999, pp. 1157–1561.
[11] P. Kinget, *Integrated GHz Voltage Controlled Oscillators.* Norwell, MA: Kluwer, 1999, pp. 353–381.
[12] C. Lam et al., "A 2.6/5.2-GHz CMOS voltage-controlled oscillator," in *IEEE Int. Solid State Circuits Conf. (ISSCC) Dig. Tech. Papers*, 1999, pp. 402–403.
[13] B. Razavi, "A 1.8-GHz CMOS voltage-controlled oscillator," in *IEEE Int. Solid State Circuits Conf. (ISSCC) Dig. Tech. Papers*, Feb. 1997, pp. 388–389.

A Low-Phase-Noise 5GHz Quadrature CMOS VCO using Common-Mode Inductive Coupling

Sander L.J. Gierkink   Salvatore Levantino   Robert C. Frye   Vito Boccuzzi

Agere Systems
600 Mountain Ave.
Murray Hill NJ 07974 USA
E-mail:sanderg@agere.com

Abstract

A new concept for quadrature coupling of LC oscillators is introduced and demonstrated on a 5GHz CMOS VCO. It uses injection-locking through common-mode inductive coupling to enforce quadrature. The technique provides quadrature over a wide tuning range without introducing any phase noise- or power consumption increase. The realized VCO is tunable between 4.6GHz and 5.2GHz and measures a phase noise lower than −124dBc/Hz at 1MHz offset over the entire tuning range. The circuit draws 8.75mA from a 2.5V supply.

1. Introduction

The wireless LAN market is expected to continue its exponential growth in the coming years. In this market, the 5GHz 54Mb/s 802.11a standard is slowly replacing the 2.4GHz 11Mb/s 802.11b standard. The development of single-chip 802.11a CMOS solutions is desirable to enable implementations at lowest cost. The full integration of transceivers implies the use of low-IF or zero-IF architectures that require quadrature signals for I/Q-(de)modulation.

Several techniques exist to generate quadrature. A VCO running at the double frequency can be divided by two to give quadrature. This solution shows poor quadrature accuracy, as it requires an accurate 50% duty cycle VCO. A VCO followed by a polyphase filter gives quadrature [1]; however, it requires buffers that increase the power consumption considerably. Alternatively, two separate oscillators can be forced to run in quadrature by using coupling transistors [2]. This approach suffers from a trade-off between accurate quadrature and low phase noise. Moreover, the coupling transistors increase the power consumption. To circumvent the phase noise penalty, additional 90° phase shifters can be placed at the gates of the coupling transistors [3-5]. However, the increase in power consumption remains. Recently, an alternative quadrature topology has been proposed, where the negative resistance transistors are cascoded by the coupling transistors [6]. Although this approach gives

low phase noise and does not increase power consumption, the technique is not well suited for implementation of widely tunable oscillators in the 5GHz range. This is because the coupling transistors have to be about five times larger than the negative resistance transistors [6], thus loading the oscillator with large parasitic capacitors that limit the tuning range.

The solutions so far proposed suffer from an increase in phase noise and/or an increase in power consumption or they result in a limited tuning range, when used at high frequencies of oscillation.

This paper presents a fully integrated 5GHz quadrature CMOS VCO that uses a new technique to generate quadrature over a wide tuning range without suffering from an increase in power consumption and phase noise. The technique uses injection-locking through common-mode inductive coupling to enforce the quadrature relation between two oscillators. The proposed oscillator is implemented in a 0.25-μm CMOS process.

2. Quadrature by common mode coupling

The schematic view of the proposed quadrature oscillator is shown in Figure 1. It consists of two separate oscillators whose common-mode 2nd-order harmonics are coupled by the inductor pair L_5, L_6.

The quadrature behavior can be explained by first focussing on the waveforms occurring in one of the two oscillators, shown in Figure 2. Due to the varactors, the two single-ended output waveforms are distorted. The varactors are accumulation-depletion pMOS devices: they exhibit a step-like C(V) curve with maximum capacitance at high gate-to-well voltages. Therefore, when the single-ended waveform voltage is high, the associated varactor capacitance is high and the waveform "slows down" and flattens. When the voltage is low, the capacitance is low and the waveform "speeds up" and sharpens. This gives rise to a 2nd order harmonic that shows up as a common-mode signal. The minima of the 2nd order harmonic align in phase with the minima of the fundamental [7].

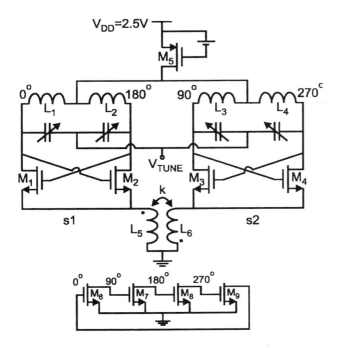

Figure 1. Schematic of the quadrature VCO.

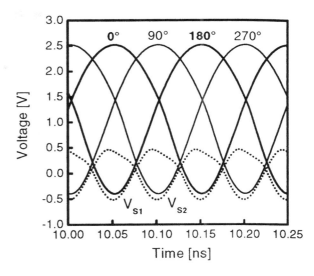

Figure 2. Simulated waveforms of the VCO.

The inductors L_1 and L_2 in Figure 1 are deliberately not combined into a single symmetrical inductor (the same holds for L_3, L_4) but rather each inductor is implemented separately. A symmetrical inductor exhibits a larger quality factor when driven differentially [8]. This would result in a larger suppression of the common mode 2nd harmonic, which is instead necessary to maximize quadrature coupling in this topology.

Being a common mode signal, the 2nd-order harmonic does not appear at the differential output of the two separate oscillators. However, it is present in the common-mode current flowing through transistors $M_{1..4}$. By anti-phase coupling the common-mode currents of the two oscillators, the respective output waveforms at frequency f_0 are forced to run in quadrature. This anti-phase coupling is implemented by means of the coupled inductors L_5 and L_6. Figure 2 shows the four simulated quadrature output waveforms, along with the source voltages V_{S1} and V_{S2}.

In order to maximize the oscillation amplitude, the minima of V_{S1} and V_{S2} need to align with the minima of the output waveforms, as depicted in Figure 2. This condition is satisfied if the impedance at the sources is real at frequency $2f_{osc}$. Therefore, the tail inductor value is chosen to give resonance at $2f_{osc}$.

Figure 2 also shows that both V_{GS} and V_{DS} of transistors $M_{1..4}$ become nearly zero during portions of the oscillation period. Thus, $M_{1..4}$ are periodically pushed into deep triode at the waveform minima and they can conduct current only close to the zero crossings of the waveforms. Since the current pulses of M_1 and M_2 (and of $M_{3,4}$) are $180°$ out of phase, their sum runs at $2f_{osc}$. Its maxima align with the zero-crossings of the fundamental and thus it is in phase with the 2nd harmonic coming from

the varactors. This provides an additional contribution to the 2nd harmonic that helps to maintain a strong quadrature coupling.

Transistors $M_{6..9}$ in Figure 1 are minimum-size devices that are added to give directivity to the quadrature phases. Without them, the oscillator would have no preference for either $+90°$ or $-90°$ phase difference. The current flowing through $M_{6..9}$ is negligible compared to the current in the transistors $M_{1..4}$, as the ratio in their (W/L)'s is more than one hundred.

3. Ring-based quadrature coupling versus injection-locked coupling

In the following, we compare the fundamental differences between ring-based quadrature VCO's and the new coupling scheme based on injection locking.

In quadrature LC oscillators based on a ring structure, the frequency of oscillation is not necessarily coincident with the resonance frequency of the individual tanks. The Barkhausen criterion applied to a conventional four-stage ring oscillator implies that the phase shift across each stage of the ring is $90°$. However, the maximum phase shift between voltage and current that a resonator can provide is $\pm 90°$; this condition only occurs at frequencies where the impedance drops to zero and the phase diagram is flat. Therefore, phase shifters must be included in the loop.

In the two-stage quadrature oscillator topology proposed in [2], a $45°$ phase shift per stage is obtained by summing two quadrature signals. Another $180°$ phase shift is obtained by a simple sign-inversion. As a result, the oscillation frequency is forced to deviate from the tank resonance frequency, enabling each resonator to

Figure 3. Chip photograph.

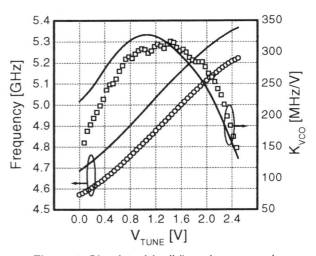

Figure 4. Simulated (solid) and measured (dots) tuning curves and their derivatives.

provide 45° of phase shift. However, the tank is now no longer operating at the frequency where the impedance is maximum and the phase characteristic is steepest. Consequently, the oscillation amplitude and the phase stability are reduced and the phase noise increases.

This problem has been addressed in [3,4,5] by applying additional phase shift, such that the tanks operate at zero phase shift, i.e. at their resonance frequency. However, the additional phase shifters increase power consumption and design complexity, and can potentially introduce extra noise.

In the proposed VCO, two oscillators are coupled through reciprocal injection locking. This mechanism only enforces the quadrature phase relation between the two oscillators and does not require the oscillation frequency to deviate from the tank resonance. Consequently, the coupling does not reduce the phase stability of each individual oscillator and no phase noise increase is seen. Moreover, since the quadrature coupling is established by means of coupled inductors rather than by transistors, the coupling devices introduce no significant extra sources of noise.

The coupling inductors L_5 and L_6 have additional beneficial effects on phase noise whilst not increasing the power consumption. They allow the oscillation waveforms to reach values below the negative supply rail (see Figure 2). Thus, the oscillation amplitude is maximized and the $1/f^2$-phase noise is reduced. In addition, the coupling inductors enable the V_{GS} and V_{DS} of $M_{1..4}$ to periodically reach zero-value. This helps to reduce their contribution to $1/f^3$- phase noise [9].

Injection locking techniques are often considered not enough reliable. They feature a limited capture range that is difficult to predict in practical cases. However, here, a sufficient capture range is only required to overcome potential mismatches between the two tanks.

4. Implementation

The proposed quadrature VCO is realized in Agere Systems' 0.25-μm CMOS process. The inductors are laid out in the three top metal layers of this five Al-metal layer process. An in-house electro-magnetic simulator (IES³) is used to model all inductors.

Figure 3 shows a photograph of the chip. The four inductors $L_{1..4}$ of the oscillator core are in the center of the figure. They have a simulated inductance of 1.8nH and a quality factor of 9 at 5GHz. The coupled inductors $L_{5,6}$ are laid out as a center tapped symmetrical inductor (see right hand side of Figure 3). This inductor is intentionally placed relatively further from the core, to minimize parasitic coupling to the other inductors. Its connecting leads are also included in the electro-magnetic simulation. The two coupled inductors have a simulated inductance of 0.62nH each and a coupling coefficient k of 0.55 at 10GHz.

The layout is highly symmetrical, not only with respect to the signal paths, but also with respect to the ground return paths. All ground currents combine in one point in the center of the oscillator layout and go from there through a single path to the ground bondpads. The supply is de-coupled by means of on-chip capacitors.

5. Experimental Results

The simulated and measured tuning curves are shown in Figure 4. The shapes of the two curves are nearly identical. The measured tuning curve is only offset by about 100MHz with respect to the simulated one. This is most likely due to inaccuracies in the estimated stray capacitances. The oscillator is tunable between 4.57 and 5.21GHz (13% tuning range). The maximum and minimum values of K_{VCO} are within a ±35% range of the average value of 233MHz/V. The relatively constant K_{VCO} is due to the large signal amplitude, which effectively averages the steep C(V)-curve of the varactor

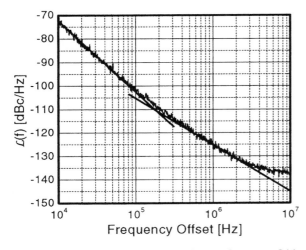

Figure 5. Measured phase noise at f_{osc}=4.88GHz.

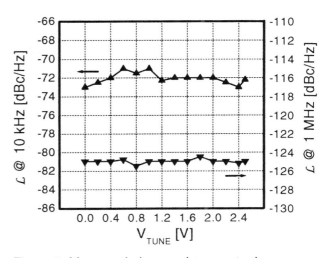

Figure 6. Measured phase noise over tuning range.

[7]. A constant K_{VCO} is beneficial for PLL design, since it gives a constant loop gain and thus does not require dynamic adjustment of the charge pump current.

Figure 5 shows the phase noise at f_{osc}=4.88GHz. It measures –125dBc/Hz at 1MHz offset from the carrier. At offset frequencies larger than about 2MHz, the measured phase noise is limited to –140dBc/Hz by the noise floor of the measurement setup.

Figure 6 shows the measured phase noise along the entire tuning range, for offset frequencies of 10kHz (flicker-noise dominated) and 1MHz (white-noise dominated). Both the $1/f^3$- and $1/f^2$-phase noise are remarkably constant over the tuning range. The worst-case noise levels are -71dBc/Hz at 10kHz and -124.5dBc/Hz at 1MHz. Table 1 shows a comparison of the achieved figure of merit (FOM) [10] to that of other published quadrature oscillators. All FOMs are calculated from the worst-case phase noise.

Table 1. Comparison with previous art

Ref.	f_{osc} [GHz]	P [mW]	FOM
[3]	1.88-1.98	27	178
[4]	4.91-5.23	21	168
[5]	1.36-1.66	30	181
[11]	1.77-1.99	20	185
[6]	1.64-1.97	50	178
This work	**4.60-5.20**	**22**	**185**

6. Conclusion

A new quadrature-coupling concept has been introduced. In the proposed scheme, two oscillators are injection-locked in quadrature by means of common-mode inductive coupling. This gives quadrature over a wide tuning range, without introducing any phase noise- or power consumption increase. The advantages over conventional ring-based quadrature oscillators have been discussed. A 0.25μm-CMOS 5GHz quadrature VCO demonstrates the proposed concept, featuring a phase noise lower than -124dBc/Hz at 1MHz offset over the 4.6-5.2GHz tuning range at 22mW power consumption.

7. References

[1] J. Crols and M. Steyaert, "A Fully Integrated 900MHz CMOS Double Quadrature Downconverter," *ISSCC Dig. of Technical Papers*, pp.136-137, Feb. 1995.

[2] A. Rofougaran et al., "A Single-Chip 900-MHz Spread-Spectrum Wireless Transceiver in 1-μm CMOS—Part I: Architecture and Transmitter Design," *IEEE J. of Solid-State Circuits*, vol.33, pp.515-534, April 1998.

[3] A.M. ElSayed and M.I. Elmasry, "Low-Phase-Noise *LC* Quadrature VCO Using Coupled Tank Resonators in a Ring Structure," *IEEE J. of Solid-State Circuits*, vol.36, no.4, pp.701-705, April 2001.

[4] P. van de Ven et al., "An optimally coupled 5GHz quadrature LC oscillator," *2001 Symposium on VLSI Circuits Dig. of Tech. Papers*, pp.115-118, June 2001.

[5] P. Vancorenland and M. Steyaert, "A 1.57 GHz Fully Integrated Very Low Phase Noise Quadrature VCO," *2001 Symposium on VLSI Circuits Dig. of Tech. Papers*, pp.111-114, June 2001.

[6] P. Andreani, "A Low-Phase-Noise Low-Phase-Error 1.8GHz Quadrature CMOS VCO," *ISSCC Dig. Of Tech. Papers*, pp.290-291, Feb. 2002.

[7] S. Levantino et al., "Frequency Dependence on Bias Current in 5-GHz CMOS VCOs': Impact on Tuning Range and Flicker Noise Up-conversion," to be published in *IEEE J. of Solid-State Circuits*.

[8] M. Danesh and J.R. Long, "Differentially Driven Symmetric Microstrip Inductors," *IEEE Trans. on Microwave Theory and Tech.*, vol.50, pp.332-340, Jan. 2002.

[9] E.A.M. Klumperink et al, "Reducing MOSFET 1/f Noise and Power Consumption by Switched Biasing", *IEEE J. of Solid-State Circuits*, Vol.35, pp.994-1001, July 2000.

[10] P. Kinget, "Integrated GHz Voltage Controlled Oscillators," in *Analog Circuit Design: (X)DSL and Other Communication Systems: RF MOST Models: Integrated Filters and Oscillators*, W. Sansen, J. Huijsing, and R. van de Plassche, Eds., Kluwer, Boston, MA, 1999, pp. 353-381.

[11] M. Tiebout, "Low-Power Low-Phase-Noise Differentially Tuned Quadrature VCO Design in Standard CMOS," *IEEE J. of Solid-State Circuits*, vol.36, July 2001, pp.1018-1024.

An Integrated 10/5GHz Injection-locked Quadrature LC VCO in a 0.18μm digital CMOS process

A. Ravi[1,2], K. Soumyanath[1], L. R. Carley[2], R. Bishop[1]

[1]Communication and Interconnect Technology, Intel Labs, Hillsboro OR 97124   [2]Carnegie Mellon University, Pittsburgh PA15213

Abstract

We present a 5GHz quadrature LC voltage controlled oscillator based on super-harmonic injection locking to an on-chip free-running 10GHz master oscillator. The oscillators are fabricated in a 0.18μm, low voltage digital CMOS process with a lossy substrate (ρ ~20mohm-cm) and thin (1.6μm), high resistivity M6. All the oscillators use fully integrated, low Q (~4) spiral inductors. The 10GHz master oscillator achieves a tuning range of 1.4GHz, and a phase noise of -118dBc/Hz at a 1MHz offset. We measured the phase noise of the quadrature VCO to be 6dBc/Hz lower than that of the free-running master oscillator. The oscillator core draws 14mA from a 1.6V supply. An image reject receiver built using the quadrature signals provides nearly 40dB of image rejection, confirming better than 1⁰ of quadrature matching. An on-chip frequency locked loop is shown to guarantee capture and accurate quadrature generation in the presence of an additional, artificially introduced, 10% mismatch in the natural frequencies.

1. Introduction

Accurate In-phase (I) and Quadrature (Q) signals are required in many wireless transceiver architectures. Traditionally, the required signals have been generated using poly-phase filters [1], [2], coupled oscillators [3], [4] or by a digital frequency divider from an oscillator at twice the desired frequency. The broadband nature and duty cycle sensitivity of digital dividers leads to degradation of phase noise performance and higher power consumption. Injection locked oscillators [5], [6], [7] can be used as narrowband frequency dividers, and are therefore, more suitable for generating quadrature signals with better spectral purity at lower power. In this paper, we propose a master oscillator at 10GHz and super-harmonic injection locking for I/Q generation at 5GHz. This method has the additional advantage of providing immunity to frequency pulling from on-chip, 5Ghz power amplifiers. Finally, tuning schemes for enhanced mismatch immunity can be implemented without disturbing the dynamics of a PLL that incorporates the master VCO. A lossy substrate and resistive metallizations are typical of a digital process that has been optimised for high density. Although such a process presents significant challenges to VCO design, system considerations demand their use.

2. Architecture

Fig. 1 shows a simplified schematic of the injection locked quadrature oscillator scheme. The master and slave oscillators are full complementary LC cross-coupled differential VCOs [8]. The better half-circuit symmetry of the circuit (compared to NMOS only versions) reduces the up-conversion of flicker noise. The lossy substrate and

This work was performed while the first author was with Communications and Interconnect Technology, Intel Labs, Hillsboro OR 97124.

resistive M6 restrict inductor Q to ~4, at the 0.5nH design value.

The master oscillator in the 10GHz band is buffered and AC coupled into the common mode node of the slave oscillators. The two slaves are tuned to the 5GHz band, with a common control voltage that is shared with the master. This ensures that the natural frequencies of the master and slaves are approximately equal. Though we have used MiM capacitors ($1fF/\mu^2$) for coupling the master and slaves; the value required is of the order of 100fF, and can be realized efficiently using capacitors such as vertical wall arrays. We use the same value for the inductor in the master and slave oscillators but the slave varactors are four times those of the master. The matching between the master and slave oscillators does not have to be exact for them to injection lock. The incident power into the slaves determines the injection locking range and thus the maximum allowable mismatch (Fig. 2). As long as the mismatch between the free running frequencies of the master and slaves is less than the lock range, the slaves oscillate at half the frequency of the master oscillator and as a result of the inherent frequency division in super-harmonic injection locking, maintain a relative phase shift of 90 degrees.

The phase response of injection locked oscillators can be modelled as that of a first order phase-locked loop (PLL). Within the injection lock range, the output phase noise is 6dB lower than that of the master oscillator. At higher frequencies the phase noise is determined entirely by the slave oscillators. Since the lock range is a function of the amplitude of the injection signal, and the master oscillator is expected to be noisier to begin with (it operates at a higher frequency), the lock range can be optimised for best phase noise performance. However, reducing the amplitude of the injected signal increases the sensitivity to process mismatch. Mismatch effects can be compensated using an on-chip frequency locked loop (FLL) described in Section 5. We have built a second flavour of the quadrature oscillator incorporating this scheme.

3. MOS Varactor

The tuning scheme we present does not use diodes or accumulation mode capacitors. The former is not compatible with low voltage techniques and the latter is not available naturally in a low cost digital process. The varactor is shown in (Fig. 4(a)). We have achieved a large tuning range (~1.4GHz) by using a depletion/inversion varactor ($L_{gate} > L_{min}$), which maintains monotonicity across large signal swings. At the maximum control voltage, the channel will have disappeared and the capacitance is dominated by the overlap capacitance, resulting in $C_{min} = W \cdot C_{ov}$. At minimum control voltage, the presence of the inversion layer results in $C_{max} = W \cdot L \cdot C_{ox}$. Hence by using an $L_{gate} > L_{min}$, we achieve the required large C_{max}/C_{min}

ratio (Fig 4(b)). We have chosen L_{gate} to ensure that the varactor Q remains much higher than that of the inductor.

4. Measurements

The outputs of the oscillator are measured on-wafer using open drain buffers to drive the 50ohm probe loads. Fig. 5(a) shows a die shot of the injection locked quadrature oscillator. The die in Fig. 5(b) consists of the same oscillator with the additional, mismatch compensation circuit of Fig. 3.

4.1. Frequency Tuning Range

Fig. 6 compares the measured injection-locked oscillator frequency tuning range to that of a free-running slave oscillator. The injection locked tuning range is 16% (4.5GHz – 5.3GHz) corresponding to a rail-to-rail control voltage variation. The quadrature oscillator core, consisting of the master oscillator and the two slaves, draws 14mA of current from a 1.6V supply.

4.2. Phase Noise

Fig. 7 is the output spectrum of the master VCO at 9.76GHz. A low bandwidth (~1kHz), off-chip AFC loop built around a Foster-Seeley discriminator stabilizes the VCO frequency and allows the phase noise to be measured on a spectrum analyser. The phase noise is –118.2dBc/Hz at a 1MHz offset from the carrier. The crossover from $1/f^3$ noise to $1/f^2$ noise occurs at a 500kHz offset from the carrier (Fig. 8). The phase noise of the quadrature VCO is measured to be 6dBc/Hz lower at a 1MHz offset, matching theory.

4.3. Figure of Merit

Fig. 9 compares the performance of the quadrature VCO reported here with other recently published VCOs. The Figure of Merit (FOM) [10] is calculated using:

$$\text{FOM} = -\mathcal{L}\{f_{off}\} + 20\log(f_c/f_{off}) - 10\log(Q^2 \cdot P_{diss})$$

where $\mathcal{L}\{f_{off}\}$ is the phase noise at the frequency offset f_{off} from the carrier frequency f_c and P_{diss} is the DC power dissipation. This FOM is normalized for the quality factor, Q, of the inductors.

4.4. Quadrature Accuracy

We have used three different techniques to determine the quadrature phase accuracy of the VCO. The external AFC loop prevents the oscillator output frequency from drifting during the measurement.

Fig. 10(a) shows the simplest technique. The oscillator outputs are mixed down to a lower frequency and the phase shift is measured in the time domain. During calibration, an RF signal is split into the two paths, and the external DC phase modulation input (shown as ϕ_{in} (cal) in the figure) on one of the two phase locked signal sources (used for down-conversion) is adjusted until the low frequency outputs line up in phase. Amplitude imbalances can also be corrected at the down-conversion signal sources. Upon replacing the calibration signal with the VCO outputs, we are able to resolve the phase accuracy to be better than 5^0.

The second technique uses the network analyser in the tuned receiver mode (Fig. 10(b)). In this mode the phase shift between the two ports can be measured. During calibration the phase offset is adjusted to null out the residual imbalance. After switching to the oscillator outputs, we measured a quadrature accuracy of about 1^0.

Finally, we use an image rejection Weaver mixer shown in Fig. 10(c). The external DC phase modulation on the IF mixer LO input (ϕ_{in} (cal) in the figure) is adjusted to calibrate out off-chip fixturing imbalances. For the measurement, the signal to be de-modulated is set to 105MHz above the RF LO while the image is set to 90MHz below it. The Weaver mixer provides about 40dB of image rejection (Fig. 11), corresponding to a measured quadrature phase accuracy of about 1^0. This is at the resolution floor of our fixturing, indicating that better than 1^0 of phase accuracy can be measured with improved calibration.

5. FLL based Mismatch Compensation

The flavour incorporating the FLL of Fig. 3 has been used to study its effect on mismatch immunity. The varactors of the slave oscillators are partitioned into two sections. The first (typically larger) section tracks the control voltage of the master oscillator in regular operation. XOR gates detect the phase difference between the two slaves and this signal is filtered and applied as feedback correction to the other section of the slave varactors. This circuit is a frequency locked loop, with the distinction that for this application frequency discrimination can be performed through a phase detector.

In the second test chip we split the slave varactors 3:1 between the two sections. Independent control voltages for the master and the larger section of the slave varactor allow us to introduce a controlled amount of mismatch and determine the point of failure for injection locking. The FLL circuit doubles the effective injection locking range (from 500MHz to 1GHz). Alternatively, this can be viewed as guaranteeing capture and quadrature accuracy in the presence of an additional 10% mismatch in the natural frequencies of the master/slave system.

6. Conclusions

We have described a completely integrated 10/5GHz super-harmonic injection locked LC quadrature VCO in a digital CMOS process. The tuning range is 16% and the phase noise is 6dB better than the –118.2dBc/Hz at a 1MHz offset, of the free-running on-chip 10GHz master oscillator. The oscillator core consisting of the master and the two slaves draws 14mA from a 1.6V supply. We have also presented three different approaches to measuring the phase difference between the two channels of the oscillator. The in-phase and quadrature outputs are accurate to better than nearly 1^0. These results are obtained with resistive metallization and on substrates that are nearly three orders of magnitude more conductive (ρ of the order of 20mohm-cm) than most previously reported oscillators.

7. Acknowledgements

The authors would like to thank Brad Bloechel, Greg Dermer, Dinesh Somasekhar, Gaurab Banerjee, Shekhar Borkar and Steve Pawlowski for their contributions. We also acknowledge Al Fazio, Greg Atwood, Derchang Kau and California Technology & Manufacturing, Intel Corp. for process support.

8. References

[1] B. Razavi, *RF Microelectronics*, Prentice Hall, 1998, pp. 138-146.
[2] F. Behbahani et al, JSSC vol. 36, No. 6, pp. 873-887.
[3] A. Rofougaran et al, 1996 ISSCC Dig., pp. 392-393.

[4] T. P. Liu, 1999 ISSCC Dig., pp. 404-405.
[5] R. Adler, Proc. IRE vol. 34, pp. 351-357.
[6] H. Rategh et al, JSSC vol. 34, No. 6, pp. 813-821.
[7] I. Schmideg, Proc. IEEE, Aug 1971, pp. 1250-1251.
[8] A. Hajimiri et al, JSSC vol. 34, No. 5, pp. 717-724.
[9] P. Vancorenland et al, VLSI Circ. Symp, pp. 111-114.
[10] P. van de Ven et al, VLSI Circ. Symp, pp. 115-118.
[11] A. L. Coban et al, VLSI Circ. Symp, pp. 119-122.
[12] M. Tiebout, Proc. ESSCIRC 2000.
[13] C. Lam et al, 1999 ISSCC Dig., pp. 402-403.
[14] B. Razavi, 1997 ISSCC Dig., pp. 388-389.

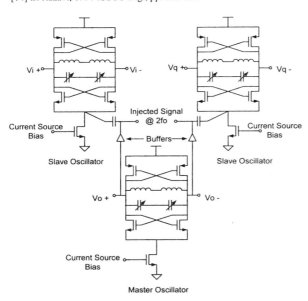

Figure 1. Injection Locked Quadrature Oscillator

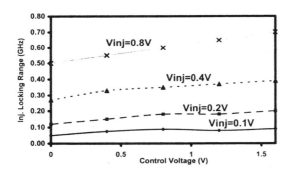

Figure 2. Injection Locking Range as a Function of Injection Signal Amplitude (V$_{inj}$)

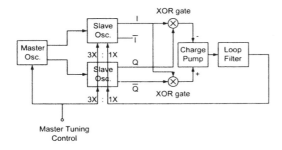

Figure 3. Mismatch Compensation Circuit

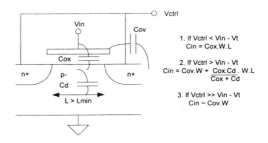

NMOS Varactor

Figure 4(a). MOS Varactor

1. If Vctrl < Vin - Vt
   Cin = Cox.W.L

2. If Vctrl > Vin - Vt
   Cin = Cov.W + $\frac{Cox.Cd}{Cox + Cd}$. W.L

3. If Vctrl >> Vin - Vt
   Cin ~ Cov.W

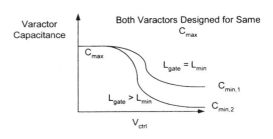

Figure 4(b). MOS Varactor Tuning

Figure 5(a). Die Shot

Figure 5(b). Die with On-chip FLL

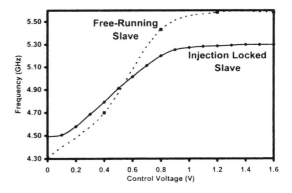

Figure 6. Tuning Characteristics

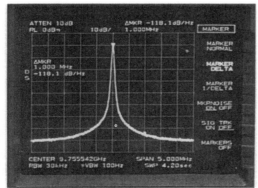

Figure 7. Master Oscillator Spectrum

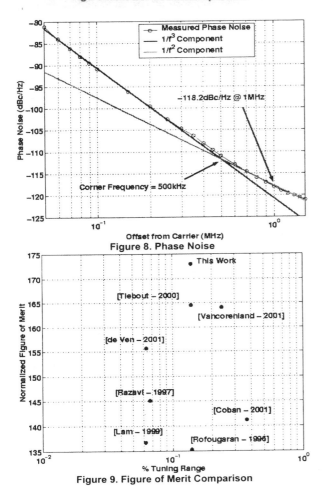

Figure 8. Phase Noise

Figure 9. Figure of Merit Comparison

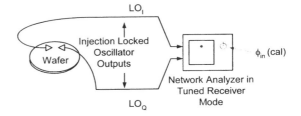

Figure 10(b). Tuned Receiver for Quadrature Accuracy Measurement

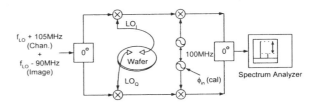

Figure 10(c). Image Reject Receiver for Quadrature Accuracy Measurement

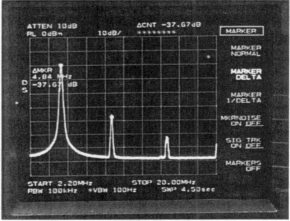

Figure 11. Image Rejection Measurement

Figure 10(a). Time Domain Quadrature Accuracy Measurement

Rotary Traveling-Wave Oscillator Arrays: A New Clock Technology

John Wood, Terence C. Edwards, *Member, IEEE*, and Steve Lipa, *Student Member, IEEE*

Abstract—Rotary traveling-wave oscillators (RTWOs) represent a new transmission-line approach to gigahertz-rate clock generation. Using the inherently stable *LC* characteristics of on-chip VLSI interconnect, the clock distribution network becomes a low-impedance distributed oscillator. The RTWO operates by creating a rotating traveling wave within a closed-loop differential transmission line. Distributed CMOS inverters serve as both transmission-line amplifiers and latches to power the oscillation and ensure rotational lock. Load capacitance is absorbed into the transmission-line constants whereby energy is recirculated giving an adiabatic quality. Unusually for an *LC* oscillator, multiphase (360°) square waves are produced directly. RTWO structures are compact and can be wired together to form rotary oscillator arrays (ROAs) to distribute a phase-locked clock over a large chip. The principle is scalable to very high clock frequencies. Issues related to interconnect and field coupling dominate the design process for RTWOs. Taking precautions to avoid unwanted signal couplings, the rise and fall times of 20 ps, suggested by simulation, may be realized at low power consumption. Experimental results of the 0.25-μm CMOS test chip with 950-MHz and 3.4-GHz rings are presented, indicating 5.5-ps jitter and 34-dB power supply rejection ratio (PSRR). Design errors in the test chip precluded meaningful rise and fall time measurements.

Index Terms—Clocks, MOSFET oscillators, phase-locked oscillators, phased arrays, synchronization, timing circuits, transmission line resonators, traveling-wave amplifiers.

I. INTRODUCTION

CLOCKING at gigahertz rates requires generators with low skew and low jitter to avoid synchronous timing failures. The notion of a "clocking surface" becomes untenable at gigahertz rates [1], frequently mandating that large VLSI chips are subdivided into multiple clock domains and/or utilize skew-tolerant multiphase circuit design techniques [2].

Techniques such as distributed phase-locked loops (PLLs) [3] and delay-locked loops (DLLs) [4] can control systematic skew to within ± 20 ps, but are complex, introduce random skew (i.e., jitter), and have area penalties. H-tree distribution systems, while simple, are difficult to balance and can use upwards of 30% of a chip's total power budget [5]. All these systems are inherently single-phase, induce large amounts of simultaneous switching noise, and can be highly susceptible to this noise.

Researchers have therefore looked to alternative oscillator mechanisms for better phase stability and lower power consumption. Previous transmission-line systems such as salphasic distribution [6], distributed amplifiers [7], and adiabatic *LC* resonant clocks [8] provide only a sinusoidal or semisinusoidal clock, making fast edge rates difficult to achieve.

This paper introduces the rotary traveling-wave oscillator (RTWO); a differential *LC* transmission-line oscillator which produces gigahertz-rate multiphase (360°) square waves with low jitter. Extension of the RTWO to rotary oscillator arrays (ROAs) offers a scalable architecture with the potential for low-power low-skew clock generation over an arbitrary chip area without resorting to clock domains. Simulations predict rise and fall times of 20 ps on a 0.25-μm process and a maximum frequency limited only by the f_T of the integrated circuit technology used.

Experiments show that although the RTWO operates differentially, careful attention is required to guard against magnetic field couplings between the clock conductors and other structures if the potential performance of these oscillators is to be realized.

II. CONCEPT OF THE ROTARY CLOCK OSCILLATOR

A. Fundamentals and Structures

The basic ROA architecture is shown in Fig. 1. A representative multigigahertz rotary clock layout has 25 interconnected RTWO rings placed onto a 7×7 array grid. Each ring consists of a differential line driven by shunt-connected antiparallel inverters distributed around the ring. This arrangement produces a single clock edge in each ring which sweeps around the ring at a frequency dependent on the electrical length of the ring. Pulses are synchronized between rings by hard wiring which forces phase lock.

Fig. 2 illustrates the theory behind the individual RTWO. Fig. 2(a) depicts an open loop of differential transmission line (exhibiting *LC* characteristics) connected to a battery through an ideal switch. When the switch is closed, a voltage wave begins to travel counterclockwise around the loop. Fig. 2(b) shows a similar loop, with the voltage source replaced by a cross-connection of the inner and outer conductors to cause a signal inversion. If there were no losses, a wave could travel on this ring indefinitely, providing a full clock cycle every other rotation of the ring (the Möbius effect).

In real applications, multiple antiparallel inverter pairs are added to the line to overcome losses and give rotation lock. Rings are simple closed loops and oscillation occurs spontaneously upon any noise event. Unbiased, startup can occur in

Manuscript received March 20, 2001; revised June 28, 2001. This work was supported by Multigig Ltd., and also supported in part by the National Science Foundation under Award EIA-31332.

J. Wood is with MultiGig, Ltd., Northampton NN8 1RF, U.K. (e-mail: john.wood@multigig.com).

T. C. Edwards is with Engalco, Huntington, YO32 9NY, U.K. (e-mail: enquiries@engalco.com).

S. Lipa is with the Microelectronics Systems Laboratory, North Carolina State University, Raleigh, NC 27695 USA.

Publisher Item Identifier S 0018-9200(01)08220-8.

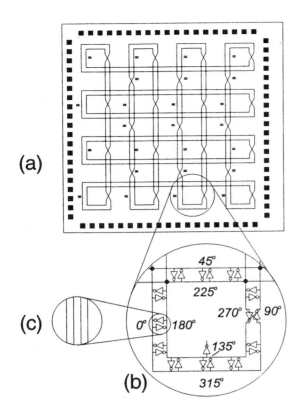

(a)

(c)

(b)

45°
225°
270° 90°
0° 180°
135°
315°

Fig. 1. Basic rotary clock architecture. The = signs denote points with same phase.

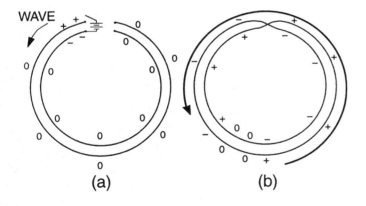

WAVE

(a)

(b)

Fig. 2. Idealized theory underlying the RTWO. (a) Open loop of differential conductors to a battery via a switch. (b) Similar loop but with the voltage source replaced by the inner and outer conductors cross-connected.

either rotational sense—usually in the direction of lowest loss. Deterministic rotation biasing mechanisms are possible, e.g., directional coupler technology or gate displacement [9]. Once a wave becomes established, it takes little power to sustain it, because unlike a ring oscillator, the energy that goes into charging and discharging MOS gate capacitance becomes transmission line energy, which is recirculated in the closed electromagnetic path. This offers potential power savings as losses are not related to $CV^2 f$ but rather to $I^2 R$ dissipation in the conductors where R can be reduced, e.g., by adoption of copper metallization.

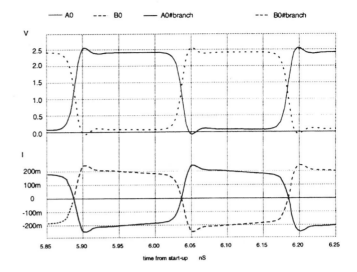

— A0   --- B0   — A0#branch   --- B0#branch

time from start-up   nS

Fig. 3. Waveforms of line voltage and line current for the 3.4-GHz clock simulation example.

B. Waveforms

Fig. 3 shows simulated waveforms of a 3.4-GHz RTWO taken at an arbitrary position on the ring. The design has the following characteristics for reference:

- Conductors: Width = 20 μm
- Pitch = 40 μm
- Ring Length = 3200 μm
- Metallization: 1.75 μm copper
- Loop inductance total = 1.87 nH
- Process: 0.25-μm CMOS
- Nch total width: 2000 μm
- Pch total width: 5000 μm
- Number of inverters: 24 pairs.

Very large distributed transistor widths give substantial capacitive loading to the lines, thus lowering velocity to give a reasonably low clock rate from a compact oscillator structure. In application, up to 75% of this capacitance can come from load capacitance, reducing the size of the drive transistors accordingly.

The upper traces of Fig. 3 show the simulated voltage waveforms on the differential line at points labeled A0, B0. The lower traces show the current in the conductors to be ±200 mA, while the supply current is simulated at 84 mA with ±4.5 mA of ripple. This clearly illustrates that energy is recycled by the basic operation of the RTWO. Just driving the 34 pF of capacitance present would require 275 mA at this frequency (from CVf).

C. Phase Locking

Interconnected rings, as in Fig. 1(a), will run in lockstep, ensuring that the relative phase at all points of an ROA are known. It is possible to use a large array of interconnected rings to distribute a clock signal over a large die area with low clock skew. For example, referring to Fig. 1(a), all the points marked with the equals sign (=) have the same relative phase as that arbitrarily marked as 0°. At any point along the loop, the two signal conductors have waveforms 180° out of phase (two-phase

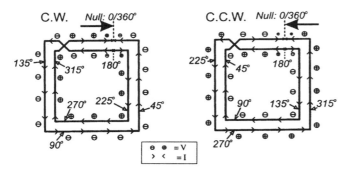

Fig. 4. Voltage, current, and phase relationships versus rotation direction (Poynting's vector).

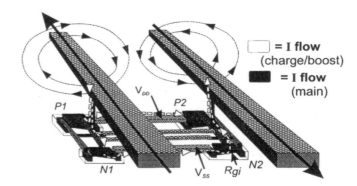

Fig. 5. Three-dimensional view of the structure. The two differential lines are shown, with current flow arrows (main and charge/boost) and encircling H-fields. CMOS transistors are also shown complete with supply voltages (V_{DD} and V_{SS}) and both p- and n-channels.

nonoverlapping clock). A full 360° is measured along the complete closed path of the loop. In principle, an arbitrary number of clock phases can be extracted. Phase advances or retards depend on the direction of rotation, and Fig. 4 shows the current–voltage relationships for clockwise and counterclockwise rotation.

D. Network Rules

Although the square-ring shape is convenient to show diagrammatically, it is only one example of a more general network solution which requires ROAs to conform closely to the following rules.

1) Signal inversion must occur on all (or most) closed paths.
2) Impedance should match at all junctions.
3) Signals should arrive simultaneously at junctions.

From 1) above, any *odd number* of crossovers are allowed on the differential path and regular crossovers forming a braided or "twisted pair" effect can dramatically reduce the unwanted coupling to wires running alongside the differential line.

The differential lines would typically be fabricated on the top metal layer of a CMOS chip where the reverse-scaling trend of VLSI interconnect offers increasingly high performance [10].

E. Fields and Currents

Fig. 5 illustrates a three-dimensional section of the ring structure connected to a pair of CMOS inverters expanded to show the four individual transistors. The main current flow in the differential conductors is shown by solid arrows, the magnetic field surrounding these conductors by dashed loops, and the capacitance charge/signal-boost current flowing through the transistors by dashed lines.

An important feature of differential lines is the existence of a well-defined "go" and "return" path which gives predictable inductance characteristics in contrast to the uncertain return-current path for single-ended clock distribution [11].

Capacitance arises mainly from the transistor gate and depletion capacitance and interconnect capacitance does not dominate.

R_{gi} indicates intrinsic gate resistance, i.e., the ohmic path through which the gate charge flows. The term R_{gi} implies a parasitic gate term, but in reality, most of this resistance is in the series circuit of the channel under the gate electrode. This is shared by the D-S channel, as illustrated by the triangular region (shown with transistors operating in the pinchoff region).

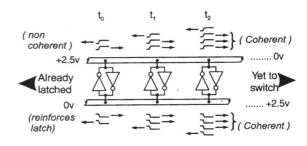

Fig. 6. Expanded view of short sections of the transmission line, including three sets of back-to-back inverters as a wavefront passes.

F. Coherent Amplification, Rotation Locking

Fig. 6 is an expanded view of a short section of transmission line with three sets of back-to-back inverters shown. It is assumed that startup is complete and the rotating wave is sweeping left to right. For this analysis, we view the inverter pairs as discrete latch elements.

Each latch switches in turn as the incident signal, traveling on the low impedance transmission line, overrides the ON resistance of the latch and its previous state. This "clash" of states occurs only at the rotating wavefront and therefore only one region is in this cross-conduction condition at any one time. The transmission-line impedance is of the order of 10 Ω and the differential on-resistance of the inverters is in the 100-Ω–1-kΩ range, depending on how finely they are distributed throughout the structure.

Once switched, each latch contributes for the remainder of the half cycle, adding to the forward-going signal. Coherent buildup of switching events occurs in this forward direction only. An equal amount of energy is launched in the reverse direction, but the latches in that direction cannot be switched further into the state to which they have already switched. The reverse-traveling components simply reduce the amount of drive required from those latches.

Importantly, it is the nonlinear latching action which is responsible for the self-locking of direction (a highly linear amplifier has no such directionality).

To clarify the above statements, Fig. 7 demonstrates how a large CMOS latch responds to an imposed differential signal. The curve trace shows a central differential-amplification region bounded by two absorptive ohmic regions (shaded) corre-

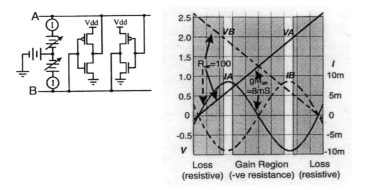

Fig. 7. DC transfer characteristic of two back-to-back inverters to an imposed differential signal.

sponding to the two latched states. Except at the wavefront location where amplification takes place, the ring structures will be terminated ohmically to the supplies.

The four-transistor "full-bridge" circuit minimizes supply current ripple to the cross-conduction period.

G. Frequency and Impedance Relations

In simulation models (and indeed as fabricated), the RTWO transmission line is built up from multiple *RLC* segments, and therefore, these primary line constants must be identified.

Fig. 8(a) is the basic RF macromodel of a short length (*SegLen*) of RTWO line with all significant RF components and parasitics annotated (as per Fig. 5). Suffixes identify per-unit-length *perlen*, lumped *lump* and *total* (or *loop*) values. There are N_{seg} segments connected together, plus a crossover, to produce a closed ring of length *RingLen*.

Fig. 8(b) is a capacitive equivalent circuit for the transistor and load capacitances. AC0 indicates an ac ground point (V_{DD} and V_{SS}).

The differential lumped capacitance C_{lump} of one such segment is given approximately by

$$C_{lump} = C_{ABint} + C_{dg\,N1} + C_{dg\,N2} + C_{dg\,P1} + C_{dg\,P2}$$
$$+ (C_{gs\,N1} + C_{gs\,N2} + C_{db\,N1} + C_{db\,N2} + C_{gs\,P1}$$
$$+ C_{gs\,P2} + C_{db\,P1} + C_{db\,P2} + C_{loadA} + C_{loadB})/4$$

(1)

where

C_{ABint}	interconnect capacitance for the line AB;
C_{dg}	gate overlap and Miller-effect feedback capacitance;
C_{gs}	total channel capacitance;
C_{db}	drain depletion capacitance to bulk (substrate);
C_{load}	load capacitance added to a line.

(Note that the /4 is used to convert the in-parallel "to ground" values into in-series differential values of capacitance.)

C_{ABint} is usually a small part of total capacitance and accurate formulas are available [12] if needed.

To calculate the per-unit-length differential inductance, i.e., accounting for mutual coupling, we use [13], expressed below.

$$L_{perlen} = \left(\frac{\mu_0}{\pi}\right) \log\left\{ \left(\frac{\pi \cdot s}{w + t_c}\right) + 1 \right\}$$

(2)

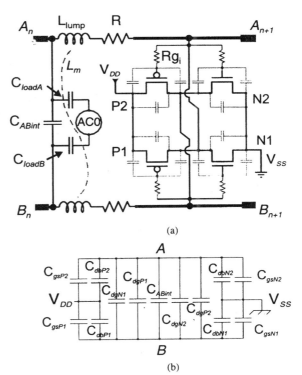

Fig. 8. Development of the rotary clock model. (a) Complete RF circuit. (b) Capacitance circuit.

where

s	conductor separation;
w	conductor width;
t_c	conductor thickness.

The phase velocity is given by

$$v_p = \frac{1}{\sqrt{L_{perlen}C_{eff,perlen}}} \quad \text{where } C_{eff,perlen} = \frac{C_{lump}}{SegLen}.$$

(3)

For heavily loaded RTWO structures, v_p can be as low as 0.03 of c (where c is the free space velocity, i.e., 2.998×10^8 m/s).

The clock frequency f_c is given approximately by

$$f_c = \frac{v_p}{[2 \cdot RingLen]}.$$

(4)

(The $\times 2$ factor arises from the pulse requiring two complete laps for a single cycle.)

Differential characteristic impedance is given by

$$Z_0 = \sqrt{\frac{L_{perlen}}{C_{eff,perlen}}}.$$

(5)

Transmission line characteristics dominate over *RC* characteristics when [14]

$$R_{loop} < 2Z_0.$$

(6)

H. Bandwidth and Power Consumption

Seen from an RF perspective, Fig. 8(a) shows the RTWO to be two push–pull distributed amplifiers folded on top of each other. Distributed amplifiers exhibit very wide bandwidth because parasitic capacitances are "neutralized" by becoming part

TABLE I
CHANGES OF CHARACTERISTICS WITH N_{seg}

Number of Segments $Nseg$	Clock Frequency f_c GHz	Cutoff frequency f_{cutoff} GHz	Rise/Fall ps	Supply Current mA
8	3.25	8.6	50	100
24	3.38	25.9	25	87
72	3.44	77.6	<15	78

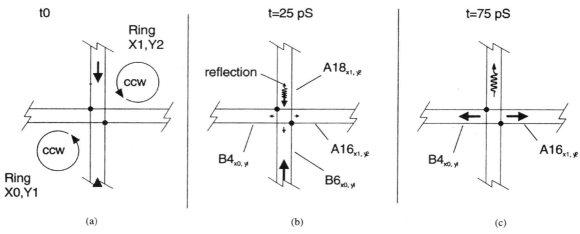

Fig. 9. A four-port junction of two RTWO rings carrying anticlockwise signals, with a noncoincident signal arrival time.

of the transmission-line impedance [15]. Performance is limited by the carrier transit time of the MOSFETs [16], not by the traditional digital inverter propagation time t_{pd}, which is not applicable where gates and drains are driven cooperatively by an imposed low-impedance signal, and where the load capacitance is hidden in the transmission line.

Operation of the RTWO is largely adiabatic when the voltage drop required to charge the capacitances is developed mainly across the inductance:

$$Z(L_{lump}) \gg R_{seg} \qquad (7)$$

and when the intrinsic gate resistance is low relative to the reactance of the gate capacitance.

$$R_{gi} \ll Z(C_{gate}). \qquad (8)$$

RTWO rise and fall times are controllable by setting the cutoff frequency of the transmission lines.

$$f_{cutoff} = \frac{1}{2\pi \sqrt{L_{lump} C_{lump}}}. \qquad (9)$$

Edges become faster and cross-conduction losses are reduced when the structure is more distributed.

Table I lists characteristic changes with N_{seg}, where $L_{lump} = L_{total}/N_{seg}$, $C_{lump} = C_{total}/N_{seg}$ with L_{total}, and C_{total} held constant.

The most significant power loss mechanism for the RTWO is $I^2 R$ power dissipated in the interconnect, given by

$$P_{disp} = \frac{V_{DD}^2}{Z_0^2} R_{loop}. \qquad (10)$$

Most of the remaining losses in Table I are attributed to cross-conduction and parasitic R_{gi} losses. R_{gi} is a real loss mechanism for gigahertz signals, and RTWO rise/fall times can be doubled by this phenomenon. In newer CMOS processes, R_{gi} improves with shorter channel length.

III. MORE DETAILED CONSIDERATIONS

A. Skew Control

Interconnected RTWO loops offer the potential to control skew in spite of relatively large open-loop time-of-flight mismatches. Functionally, phase averaging occurs by pulse combination at the junction of multiple transmission lines. For a four-port junction, the normal operating mode will see two pulses arriving at the junction simultaneously. These two sources will feed two output ports and signal flow will be unimpeded by reflections if impedance is matched. This amounts to a situation similar to that described in [17], [18], although for ROAs, the mechanism is LC transmission-line energy combination, not ohmic combination of CMOS inverter outputs.

Where there exists a time-of-flight mismatch, one pulse arrives at the junction before the other. Fig. 9(a) depicts the operation of a four-port junction between of two interwired but velocity-mismatched RTWO loops. Each of these rings has been divided into segments numbered $0 \ldots 23$ (each as Fig. 8). Four rings are wired together (similar to Fig. 16, shown later). Only the junction of the rings $X0,Y1$ and $X1,Y2$ are considered here; the latter having a higher open-loop operating frequency.

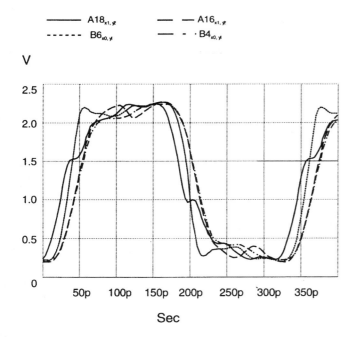

—— A18ₓ₁,y₂   — — A16ₓ₁,y₂
- - - - B6ₓ₀,y₁   — - · B4ₓ₀,y₁

Fig. 10. Waveforms corresponding with Fig. 9.

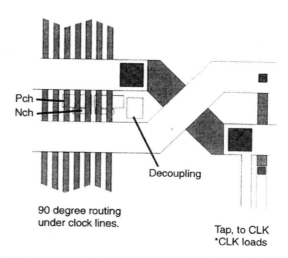

Fig. 11. Segment of chip layout showing 90° routing beneath clock lines and a tap to clock (CLK. *CLK) loads.

From simulation, two pulse-combination effects appear to be present, the simplest of which is the impedance match effect where the first signal to arrive at a junction must try to drive three transmission lines. If all ports have equal impedance, the junction can only reach a quarter of the full signal value and a reflection occurs driving an inverted signal back down the incident port [Fig. 9(b)]. Initially, detrimental effects on signal fidelity arising from this reflection are overcome when the other pulse arrives, whereupon the pulses combine and branch into the output ports, as shown in Fig. 9(c).

The second pulse combination effect is believed to be due to nonlinear MOSFET drain capacitance, which can modulate the velocity of the line. Reflections can drive the MOSFETS from the ohmic state into the low-capacitance pinchoff region, locally increasing velocity.

Quantitative Results From Simulation: Fig. 10 presents the results of a SPICE simulation of the above situation with an extreme condition of velocity mismatch. A +50% variation of T_{ox} oxide thickness is modeled across a small 2.4 × 2.4 mm chip having four interconnected rings. Thick oxide (lower C) devices are on the right side of the chip, giving a 22.5% phase velocity increase relative to the left side.

Looking at these results with reference to Fig. 9 reveals that the first pulse arrives from ring (x_1, y_2) and passes point A18$_{x1,y2}$ at time $t_0 + 25$ ps and begins its rise time. Within this rise time, the leading edge reaches the nearby junction, where negative reflections bounce back to momentarily prevent A18$_{x1,y2}$ passing through the 1.5-V level.

The second pulse arrives from the slower left-hand ring (x_0, y_1), reaching point B6$_{x0,y1}$ at approximately $t_0 + 45$ ps. It then combines with the first pulse at the junction to branch into the two output ports without further reflections.

By $t_0 + 75$ ps, the signals have reached points A16$_{x1,y2}$ and B4$_{x0,y1}$ and are essentially coincident—forward progress of the waves in rings (x_0, y_1) and (x_1, y_2) are now synchronized.

The phase-locking phenomenon occurs at every junction of the array (not just the junction considered here) and twice per oscillation cycle which accounts for the smaller than expected initial skew seen between the rings.

Simulations of typical arrays show that lockup is achieved within a few nanoseconds from powerup after signals settle into the lowest-energy state of coherent mesh.

B. Coupling Issues Related to Layout

The induced magnetic fields from the rotary clock structures can be strong. This is because $\partial I/\partial t$ is relatively high (square waves). The magnetic coupling coefficient, however, depends on the angle between source and victim and falls to zero when the angle becomes 90°.

Fig. 11 illustrates a 90° layout technique to minimize inductive coupling problems. The top metal M5 (running left to right) is used to create the differential RTWO, while orthogonal M4 is used as a routing resource for busses into and out of areas bounded by the clock transmission line.

For capacitive coupling, fast rise and fall times imply high displacement currents and a potentially aggressive noise source. Differential transmission lines tend to mitigate such effects [19], and in Fig. 11, the total capacitive coupling area between each of the transmission-line conductors and any M4 conductor is balanced. If the clock source were ideally differential, no net charge would be coupled to the M4 wires. For the RTWO, distributed inverters force the waveforms to be substantially differential and nonoverlapping, keeping glitches below the sensitivity of a typical gate.

For the five-metal test chip (Section V), a 45% utilization of M4 was used for the 90° routing pattern immediately underneath the RTWO rings. This coverage allows the M4 to act as both a routing resource and as an electrostatic shield similar to [20], preventing electrostatic coupling to signal lines further below. Magnetic fields are not attenuated much by this configuration, because the spaces between the thin perpendicular M4 lines break up the circulating currents which could repel a magnetic field. Substrate magnetic fields [21] are, therefore, to be expected.

Coupling to co-parallel (0°) victim conductors is potentially much more problematic (discussed later in Section IV-C).

C. Tapoff Issues and Stub Loadings

It is possible to "tap into" the ROA structure (Fig. 11) anywhere along its length and extract a locally two-phase signal with known phase relationship to the rest of the network. This signal can then be routed via a fast differential transmission line to other circuits and will generally represent a capacitive stub on the RTWO ring.

For minimum signal distortion, the round-trip time-of-flight τ_p (forward and backward along the stub) must be much less than the rise time τ_r and fall time τ_f of the clock waveform:

$$\tau_p \ll \tau_r, \tau_f. \tag{11}$$

When the above condition is met, the capacitance can be taken as being effectively lumped on the main RTWO ring at the tap point for the purposes of predicting oscillator frequency and ring impedance.

Although not immediately apparent, this condition is achievable in practice due to three factors. The first factor is that the tap line velocity is relatively fast for SiO_2 dielectric. It is approximately $0.5c$, while the main RTWO oscillator ring might be operating at perhaps $0.075c$. The second factor is that the tap length only has to be long enough to reach within a single RTWO ring. The third factor is that it requires two signal rotations on the RTWO to complete a clock cycle. These three factors work together to make the RTWO rings physically small compared to the expected speed-of-light dimensions. The distances to be spanned by the fast tap wires are therefore short enough that transmission-line effects on these lines are unimportant—certainly at the clock fundamental frequency and even at higher harmonics.

This can be illustrated by reference to a specific 3.4-GHz RTWO, 3200 μm long with 20-ps rise/fall times. Within one of these rise or fall periods, a stub transmission line with velocity $0.5c$ is able to communicate a signal over a distance of 3 mm. For a stub length of 400 μm (to reach the center of the ring), this equates to 3.75 round-trip times along the stub.

Fig. 12 shows simulated waveforms with 2 pF of total to-ground capacitance at the end of one such stub. Reflected energy gives rise to the ringing which is evident with this level of capacitance. The line resistance of the stubs must be low to maintain reflective energy conservation.

The ratiometric factors outlined above between ring length, frequency, rise/fall time, and stub lengths are expected to hold as ROAs are scaled to higher frequencies and smaller ring lengths without requiring special stub tuning measures.

Capacitive Loading Limits: Substantial total-chip capacitive loading can be tolerated by the RTWO relative to conventionally resonant systems [8], [22], [23]. However, the loading effects of interconnect, active, and stub capacitances cannot be increased without limit. The consequential lowering of line impedance increases circulating currents until I^2R losses become a concern. Eventually, the impedance becomes so low relative to the loop resistance that the relation (6) cannot be maintained, whereupon oscillation ceases altogether.

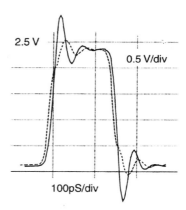

Fig. 12. Signal at either end of a 2-pF total tap loading line.

D. Frequency/Impedance Adjustment

Rewriting (4) in the form below shows that frequency is set only by the total inductance and capacitance of the RTWO loop.

$$f_{\text{osc}} = \frac{1}{2\sqrt{L_{\text{total}}C_{\text{total}}}}. \tag{12}$$

Total loop inductance L_{total} is proportional to *RingLen* and varies strongly as a function of the width and pitch of the top metal differential conductors. This allows a coarse frequency selection through the top-metal mask definition. Unit-to-unit inductance variation is expected to be small because of the good lithographic reproduction of the relatively large clock conductors and the weak sensitivity of inductance to metal thickness variations.

Total capacitance C_{total} for the RTWO is the sum of all lumped capacitances connected to the loop (1). C_{total} tends to be dominated by gate-oxide capacitance (C_{ox}) from the drive FETs and the clock load FETs. C_{ox} is inversely proportional to gate-oxide thickness T_{ox}, which on a modern CMOS SiO_2 is controlled to approximately $\pm5\%$ variation over extended wafer lots [24]. Drain depletion capacitances exist on bulk CMOS where the active transistors connect to the ring.

During the VLSI layout phase, a CAD tool (expected release: Q1 2002) can target a fixed operating frequency. The tool will be able to correct impedance discontinuities caused by lumped load capacitance by the addition of dummy "padding" capacitance elsewhere around the loop, and postcompensate an overly capacitive-loaded clock network by reducing the differential inductances through pitch reduction—hence restoring velocity and thus frequency. Alternatively, at the expense of using more metallization, a new layout with more numerous, shorter length rings could be used. The tool will need to simultaneously solve impedance matching issues [refer to Section II-A, (5)]. By manipulation of both L and C simultaneously, it is possible to control v_p and Z independently, as shown diagrammatically in Fig. 13. For example, velocity v_p can be reduced by increasing both L and C by the same factor to cancel the effect on Z. These adjustments can support arbitrary branch-and-combine networks (at least in theory).

Post fabrication, adding together the sources of variation and given that frequency is related to $\sqrt{C}$ and $\sqrt{L}$, a $\pm5\%$ initial tolerance of operating frequency between parts is expected.

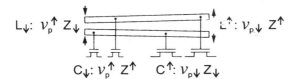

Fig. 13. Differential line with varing trace separations and capacitive inverter loadings indicating the effects of altering several parameters.

Matching within a die should be better, but temperature gradients and transistor size variations as they affect capacitance will lead to phase velocity changes requiring correction by the *Skew Control* mechanism (described in Section III-A).

Temperature can alter frequency through variation of L_{total} and C_{total}. Inductance variation is assumed to be negligible compared to capacitance variation and is not considered. Gate-oxide thickness variation could potentially affect C_{total}, but for SiO_2 dielectric, with properties similar to quartz, this can be ignored. More significant are temperature variations of drain depletion capacitance and of transistor L_{eff}, W_{eff}.

To tune an ROA clock to an exact reference frequency, allowing limited "speed-binning" and reduced internal phase mismatches, closed-loop control of distributed switched capacitors [9] or varactors [25] is envisaged.

E. Active Compensation for Interconnect Losses

Resistive interconnect losses make it difficult to communicate high-frequency clock signals over a large chip without waveshape distortion and attenuation, which impacts on the practicality of reflective energy conservation schemes [6], [22], [23]. The skin effect loss mechanism has been evident in clock tree conductors for some time [26] and is frequency dependent. High-speed H-trees tend to use hierarchical buffers within the trees to maintain amplitude and edge rates.

Active compensation of VLSI differential transmission lines to overcome clock attenuation was shown by Bußmann and Langmann [27] to be applicable to sine-wave signals. Shunt-connected negative impedance convertors (NICs) were used with linear compensation to prevent oscillations.

The distributed inverters used within RTWOs afford active compensation for transmission-line losses, raising the apparent Q of the resonant rings and helping to maintain a uniformly high clock amplitude around the structure.

F. Logic Styles

Two-phase latched logic [28] is the style most compatible with RTWO. It is highly skew tolerant and through dataflow-aware placement [27] offers the potential to exploit the full $360°$ of clock phase to reduce clock-related surging [29], which in future systems could exceed 500 A [30]. Conventional single-phase D-latch designs can be driven where timing improvements through skew scheduling [31] might be possible. A locally four-phase system to support domino logic [2] could be implemented by wrapping two loops of RTWO line around the region to clock. Unfortunately, all of these techniques are beyond the capability of current logic synthesis tools.

A. Approach

To enable rapid "what-if" evaluation of potential RTWO structures, a simulation/visualization program known as Rotary Explorer [32] has been developed. Rotary Explorer is GUI driven and parametrically creates a SPICE deck of macro-models linking to FASTHENRY subcircuits [33] for multipole magnetic analysis of skin, proximity, and LR coupling effects in the time domain. MOSFETs are modeled using BSIM3v3 nonquasi-static model with an external resistor added to model R_{gi} (Fig. 8). The BSIM4 model [34], which properly accounts for R_{gi} as a D-S channel component, was not available.

With the Rotary Explorer program, it is possible to simulate RTWO rings independently or as interlocked X, Y arrays. The effects of tap loads, oxide thickness variations, and magnetically induced "victim" noise can be evaluated.

As a visualization aid, Rotary Explorer gives a "live" display of color-coded SPICE voltages projected onto a scaled image of the ROA structure being simulated. This aids in the intuitive understanding of reflections and how the structure achieves a steady-state phase-locked operation.

B. Results

Two very important performance metrics for any oscillator are its sensitivity to changes in temperature and supply voltage. Simulations of these effects on a nominally 3.34-GHz rotary clock resulted in the data given in Tables II and III.

Supply Induced Jitter: Following on from the above and in light of the RTWO's time-of-flight oscillation mechanism, it is inferred that such voltage sensitivity will also apply to phase modulation versus voltage, i.e., jitter—at least at low supply-noise frequencies. For a single RTWO ring, the power-supply induced jitter ϕ will be related to ΔV and the power-supply rejection ratio (PSRR) by

$$\phi = \Delta V * (\text{PSRR}) \qquad (13)$$

where ΔV, because of the distributed nature of the oscillator, is the mean supply voltage deviation as experienced along the path of an edge as it travels two complete rotations. To improve PSRR, plans are in place to add voltage-dependent capacitance to the structure to give first-order compensation.

From simulations, we see that jitter reduces for multiple ring structures due to averaging effects.

C. Coupling II—Simulated Coupling

The Rotary Explorer program makes it easy to simulate coupled noise between an RTWO ring and user defined victim trace (drawn with the aid of a mouse). Simulated results are shown in Table IV for a 3.4-GHz RTWO configured to have 20 ps rise and fall times, and with geometry as shown in Fig. 14.

Peak coupling magnitude occurs at 60-μm *victim length*. A trace longer than this will see a coupling cancellation effect that approaches zero for each pitch of the braiding it traverses.

Fig. 15 illustrates a notably strong coupled signal waveform at *victim distance* = 17 μm, with no loading on the victim

TABLE II
VARIATIONS WITH TEMPERATURE

Temperature °C	-50	+25	+150
Clock frequency GHz	3.36	3.34	3.31

TABLE III
VARIATIONS WITH DC SUPPLY VOLTAGE V_{DD}

Supply voltage V_{DD}	1.5	2.5	3.5
Clock frequency GHz	3.45	3.34	3.32

TABLE IV
INDUCED NOISE AS A FUNCTION OF VICTIM DISTANCE AND LENGTH

Victim Distance μ	Victim Length μ	Noise (approx) mV
17	30	40
	60	60
	120	20
35	30	20
	60	30
	120	10
70	30	10
	60	15
	120	5

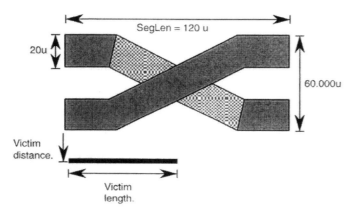

Fig. 14.   Crossover traces, a visualization output from the Rotary Explorer tool.

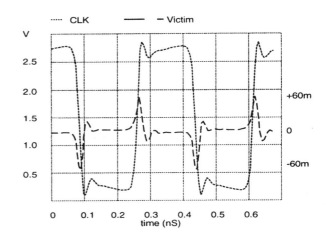

Fig. 15.   Example of notably strong coupled signal waveform.

trace and one end connected to ground. Note the more sensitive noise scale.

The absolute maximum coupling occurs if *victim distance* is allowed to go to zero. In this case, mutual coupling between aggressor and victim is 100% with no cancellation effects from the other differential trace. As a numerical example, it follows that a 2.5-V signal with a rise time of 20 ps on a transmission line with a velocity of $0.072c$ has the 2.5-V gradient over 430 μm of length (Fig. 4 illustrates the concept). Over the 60-μm length discussed above, this equates to 348 mV. Slower edge rates, faster transmission lines, and lower supply voltages reduce this figure proportionally.

Long-range inductive noise coupling from the differential transmission line is expected to be small, since (from a distance) the 'go' and 'return' currents are equal and opposite.

Potential problems exist in short-range magnetic coupling to wiring in the vicinity of the clock lines. Inductance is lowered

by coupling to any highly conductive structure in which eddy currents can flow to decrease and distort the inducing field. Couplings to less conductive circuits such as the substrate give a loss mechanism which can be modeled as a shunt term in the transmission-line equations. LC resonance in the small-scale coupled structures is unlikely because of the high resonant frequencies. All of the coupling mechanisms mentioned are edge-rate dependent, and this can limit the achievable rise and fall times of the RTWO by attenuating the high-frequency signal components.

Full RLC layout extraction is essential in the neighborhood of the clock lines if routing is allowed in these areas. An alternative proposal under investigation is to predefine a VLSI structure combining clock and power distribution into the same grid to give consistent characteristics and shielding.

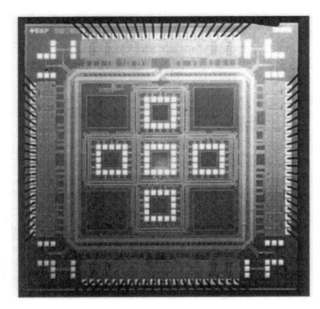

Fig. 16. Die photograph of a prototype chip.

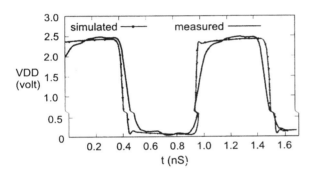

Fig. 17. Measurement versus simulation waveforms for the large 965-MHz ring.

V. SOME EXPERIMENTAL RESULTS

Fig. 16 shows a die photograph of a prototype built using a 0.25-μm 2.5-V CMOS process with 1-μm Al/Cu top metal M5. The conductors are relatively wide in order to minimize resistive losses of the rather thin M5. The available top-metal area consumed by the transmission lines was 15%. A general feature of the RTWO and ROA is that power can be reduced by increasing the metal area devoted to clock generation. The simple substitution of copper metallization could halve the width of the lines for the same power consumption.

The prototype features a large ring independent of four interconnected smaller rings. The 12 000-μm outer ring uses 60-μm conductors on a 120-μm pitch, with 128 62.5-μm/25-μm inverter pairs distributed along its length.

For the large ring, simulations predicted a clock frequency of approximately 925 MHz. Measurements of the actual performance versus simulated with $V_{DD} = 2.5$ V are shown in Fig. 17. The oscillation frequency was 965 MHz. Jitter was measured at 5.5 ps rms using a Tektronix 11 801A oscilloscope with an SD-26 sampling head.

The slower than simulated rise-time discrepancy is believed to be due to the large extrinsic gate electrode resistance on the Pch FETs. At design time, the importance of this parameter was

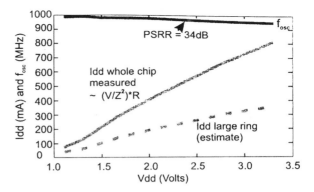

Fig. 18. Clock frequency versus V_{DD} for the large ring and I_{DD} versus V_{DD} for the entire chip with all five rings.

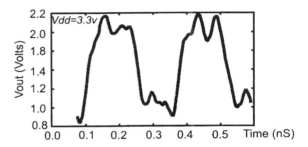

Fig. 19. Measured output on one of the 3.42-GHz rings.

overlooked. Transistors are now laid out according to RF design rules with the gate driven from both sides of the device.

Fig. 18 shows that the oscillation frequency F_{osc} versus V_{DD} is quite flat over a large V_{DD}. We calculate from the measured slope that PSRR is approximately 34 dB for oscillators fabricated on this process. The oscillator was seen to be functional down to 0.8-V supply voltage, although 1.1 V was required to initiate startup.

The test chip incorporates 15 pF of on-chip decoupling capacitance per ring. No off-chip decoupling was required. Effectively, the equivalent of ten single-ended lines each having <10 Ω impedance were active, but simultaneous switching surges are low because of the distributed switching times of the inverters.

The quad of inner rings each have the following characteristics:

- Conductors: Width = 20 μm
- Pitch = 40 μm
- Ring Length = 3200 μm.

Total channel widths are 2000 μm for the Nch FET and 5000 μm for the Pch FET spread over 40 pairs of inverters.

Fig. 19 shows the measured waveform from one of the 3.4-GHz rings. The oscillation frequency is 3.38 GHz versus a simulated frequency of 3.42 GHz. However, the waveshape is disappointingly distorted, the amplitude is low, and even-mode artifacts are visible.

Investigation of the fault identified a 'co-parallel' (0°) inductive coupling problem between the clock signal lines and V_{DD} and V_{SS} supply traces running directly beneath on M3 for the complete loop length. Only when a complete FASTHENRY analysis was performed including these power traces was it apparent that induced current loops (circulating through the decoupling capacitors) were strongly attenuating the rotary signal. In this condition, the latching action (Fig. 7) does not fully develop and the rings support linear amplification of noise signals—hence the problematic multimode action. (This effect was much less severe on the large 965-MHz ring because the V_{DD}/V_{SS} lines were much closer to the magnetically neutral center line of the transmission line). The problem can be mitigated by use of braided transmission lines. (as detailed in Section IV-C).

Analysis of the test chip showed that $90°$ coupling between M5 and the orthogonal thin M4 lines is not a significant problem, making it possible to route power and signals between regions bounded by the rotary clock structures.

VI. Conclusion and Further Work Planned

This paper has described the rotary traveling-wave oscillator (RTWO) and its potential application to gigahertz-rate VLSI clocking. The oscillator is unique for a resonant-style LC-based oscillator in that it produces square waves directly and can be hardwired to form rotary oscillator arrays (ROAs). Being LC-based, the oscillator is stable and jitter is low.

The formulas presented here give practical adiabatic oscillator designs suitable for VLSI fabrication. The structure and operation of the RTWO is fundamentally simple and amenable to analysis. We find that agreement between simulation and measurement is good.

We need to demonstrate *skew control* (believed to be inherent) to fully establish that the simulated performance of multiring ROAs is realizable, and to measure susceptibility to induced high-frequency noise. Further work is planned to establish firm mathematical/analytical foundations for the prediction of both jitter and skew and to determine exact stability criteria for arrayed oscillators. Currently, a test chip using braided transmission line design to minimize coupling and incorporating varactors to control frequency is awaiting packaging and test.

Looking to the future, our simulations predict that the oscillator scales well. On a more modern 0.18-μm copper process, 10.5-GHz square-wave oscillator/distributors should be realizable consuming less than 32 mA per ring using slimmer 10-μm conductors. From simulation, the RTWO also appears to be viable on SOI processes.

Acknowledgment

The authors would like to thank P. Franzon and M. Steer, both of North Carolina State University, for their assistance, and the Raunds and British public library service.

References

[1] E. G. Friedman, *High Performance Clock Distribution Networks*. Boston, MA: Kluwer, 1997.

[2] D. Harris, *Skew Tolerant Circuit Design*. San Mateo, CA: Morgan Kaufmann, 2000.

[3] G. A. Pratt and J. Nguyen, "Distributed synchronous clocking," *IEEE Trans. Parallel Distributed Syst.*, vol. 6, pp. 314–328, Mar. 1995.

[4] S. Tam, S. Rusu, U. N. Desai, R. Kim, J. Zhang, and I. Young, "Clock generation and distribution for the first IA-64 microprocessor," *IEEE J. Solid-State Circuits*, vol. 35, pp. 1545–1552, Nov. 2000.

[5] C. J. Anderson et al., "Physical design of a forth-generation power GHz microprocessor," in *ISSCC 2001 Dig. Tech. Papers*, Feb. 2001, pp. 232–233.

[6] V. L. Chi, "Salphasic distribution of clock signals for synchronous systems," *IEEE Trans. Comput.*, vol. 43, pp. 597–602, May 1994.

[7] B. Kleveland et al., "Monolithic CMOS distributed amplifier and oscillator," in *ISSCC Dig. Tech. Papers*, Feb. 1999, pp. 70–71.

[8] W. Athas, N. Tzartzanis, L. J. Svensson, L. Peterson, H. Li, X. Jiang, P. Wang, and W.-C. Liu, "AC-1: A clock-powered microprocessor," in *Proc. Int. Symp. Low-Power Electronics and Design*, Aug. 1997, [Online] Available: http://www.isi.edu/acmos/people/nestoras/papers/97-08.MontereyAC1.ps.

[9] J. Wood. PCT/GB00/00175. MultiGig Ltd.. [Online]. Available: http://www.delphion.com/cgi-bin/viewpat.cmd/WO000044093A1

[10] B. Kleveland, T. H. Lee, and S. S. Wong, "50-GHz interconnect design in standard silicon technology," presented at the IEEE MTT-S Int. Microwave Symp., Baltimore, MD, June 1998, [Online] Available: http://smirc.stanford.edu/papers/mtts98p-bendik.pdf.

[11] B. Kleveland, X. Qi, L. Madden 1, R. W. Dutton, and S. S. Wong, "Line inductance extraction and modeling in a real chip with power grid," presented at the IEEE IEDM Conf., Washington, D. C., Dec. 1999, [Online] Available: http://gloworm.stanford.edu/tcad/pubs/device/iedm.pdf.

[12] N. Delorme et al., "Inductance and capacitance analytic formulas for VLSI interconnect," *Electron. Lett.*, vol. 32, no. 11, May 23, 1996.

[13] C. S. Walker, *Capacitance, Inductance and Crosstalk Analysis*. Norwood, MA: Artech, 1990, p. 95.

[14] A. Deutsch et al., "Modeling and characterization of long on-chip interconnections for high-performance microprocessors," *IBM J. Res. Develop.*, vol. 39, no. 5, pp. 547–567, Sept. 1995. p. 549.

[15] J. B. Beyer et al., "MESFET distributed amplifier design guidelines," *IEEE Trans. Microwave Theory Tech.*, vol. MTT-32, pp. 268–275, Mar. 1984.

[16] Y. Tsividis, *Operation and Modeling of the MOS Transistor*, 2nd ed. New York: McGraw-Hill, 1999, pp. 339–340.

[17] H. Larsson, "Distributed synchronous clocking using connected ring oscillators," Master's thesis, Computer Systems Engineering Centre for Computer System Architecture, Halmstad Univ., Halmstad, Sweden, Jan. 1997. [Online]. Available: http://www.hh.se/ide/ccaweb/publications/97/distclock/9705.ps.

[18] L. Hall, M. Clements, W. Liu, and G. Bilbro, "Clock distribution using cooperative ring oscillators," in *Proc. IEEE 17th Conf. Advanced Research in VLSI (ARVLSI'97)*, 1997, [Online] Available: http://www.computer.org/proceedings/arvlsi/7913/79130062abs.htm.

[19] T. C. Edwards and M. B. Steer, *Foundations of Interconnect and Microstrip Design*, Chichester, U.K.: Wiley, 2000, ch. 6. sec. 6.11.

[20] C. P. Yue and S. S. Wong, "On-chip spiral inductors with patterned ground shields for Si-based RF ICs," *IEEE J. Solid-State Circuits*, vol. 33, pp. 743–752, May 1998.

[21] C. P. Yue and S. S. Wong, "A study on substrate effects of silicon-based RF passive components," in *MTT-S Int. Microwave Symp. Dig.*, June 1999, pp. 1625–1628.

[22] M. E. Becker and T. F. Knight Jr. Transmission line clock driver. presented at 1999 IEEE Int. Conf. Computer Design. [Online]. Available: http://www.computer.org/proceedings/iccd/0406/04060489abs.htm

[23] P. Zarkesh-Ha and J. D. Meindl, "Asymptotically zero power dissipation Gigahertz clock distribution networks," *IEEE Electrical Performance and Electronic Packaging*, pp. 57–60, Oct. 1999.

[24] K. Bernstein, K. Carrig, C. M. Durham, and P. A. Hansen, *High Speed CMOS Design Styles*. Norwood, MA: Kluwer, 1998, p. 22.

[25] T. Soorapanth, C. P. Yue, D. Shaeffer, T. H. Lee, and S. S. Wong, "Analysis and optimization of accumulation-mode varactor for RF ICs," presented at the Symp. VLSI Circuits, Honolulu, HI, June 11–13, 1998, [Online] Available: http://smirc.stanford.edu/papers/VLSI98p-chet.pdf.

[26] H. B. Bakoglu, J. T. Walker, and J. D. Meindl, "A symmetric clock-distribution tree and optimized high speed interconnections for reduced clock skew in ULSI and WSI circuits," in *IEEE Int. Conf. Computer Design*, Oct. 1986, pp. 118–122.

[27] M. Bußmann and U. Langmann, "Active compensation of interconnect losses for multi-GHz clock distribution networks," *IEEE Trans. Circuits and Syst. II*, vol. 39, pp. 790–798, Nov. 1992.

[28] M. C. Papaefthymiou and K. H. Randall, "Edge-triggering vs. two-phase level-clocking," presented at the 1993 Symp. Research on Integrated Systems, Mar. 1993, [Online] Available: http://www.eecs.umich.edu/~marios/papers/sis93.ps.

[29] L. Benni *et al.*, "Clock skew optimization for peak current reduction," *J. VLSI Signal Processing*, vol. 16, pp. 117–130, 1997.

[30] International Semiconductor Roadmap for Semiconductors (1999). [Online]. Available: http://public.itrs.net/files/1999_SIA_Roadmap/Design.pdf

[31] I. S. Kourtev and E. G. Friedman, *Timing Optimization Through Clock Skew Scheduling*. Boston, MA: Kluwer, 2000.

[32] MultiGig, Ltd. Rotary Explorer. [Online]. Available: http://www.multigig.com/software.htm

[33] M. Kamon, M. J. Tsuk, and J. K. White, "FASTHENRY: A multipole-accelerated 3-D inductance extraction program," *IEEE Trans. Microwave Theory Tech.*, vol. 429, pp. 1750–1758, Sept. 1994.

[34] BSIM Research Group. (2000–2001) The BSIM4 Short-Channel Transistor Model. Univ. of California at Berkeley. [Online]. Available: http://www-device.eecs.berkeley.edu/~bsim3/bsim4.html

Terence C. Edwards (M'89) received the M.Phil. degree in microwaves.

He is the Executive Director of Engalco, a consultancy firm based in the U.K., mainly specializing in signal transmission technologies and the global RF and microwave industry. He researches and takes responsibility for regular releases of Microwaves North America, published 1995, 1998, and 2001. He has authored several publications (including papers published in the IEEE TRANSACTIONS ON MICROWAVE THEORY AND TECHNIQUES), has led management seminars on fiber optics, presented a paper on mobile technologies at the IMAPS Microelectronics Symposium, Philadelphia, PA, October 1997, and has written several articles and books. These include (jointly with Prof. Michael Steer) one recently on MICs (New York: Wiley) and on gigahertz and terahertz technologies (Norwood, MA: Artech, 2000). He is on the editorial advisory board for the *International Journal of Communication Systems*. He regularly consults for both national and overseas companies and is on the prestigious IEE (London) President's List of Consultants.

Mr. Edwards is a Fellow of the Institution of Electrical Engineers (IEE), U.K.

John Wood is the Engineering Director of MultiGig, Ltd., a U.K. technology startup specializing in multi-gigahertz circuit design I.P.

Previously, he has worked as a consultant design engineer on multidomain design projects in mechanical, power electronics, infrared optics, and software development roles. He holds a number of patents which have been licensed for manufacture in the fields of infrared plastic welding and high-speed digital signaling. His technical interests include all areas of engineering design, but particularly electromagnetics, VLSI circuit design, and high-speed analog techniques.

Steve Lipa (S'00) received the B.S. degree in electrical engineering from the University of Virginia, Charlottesville, in 1980, and the M.S. degree in electrical engineering from North Carolina State University, Raleigh, in 1993. He is currently working toward the Ph.D. degree in electrical engineering at North Carolina State University.

He is currently a Research Assistant and Laboratory Manager with the Microelectronics Systems Laboratory at North Carolina State University. He has ten years of experience as an Integrated Circuit Design Engineer, primarily in the design of high-speed digital logic circuits. His current research is in the area of high-speed clock distribution.

35-GHz Static and 48-GHz Dynamic Frequency Divider IC's Using 0.2-μm AlGaAs/GaAs-HEMT's

Zhihao Lao, *Member, IEEE*, Wolfgang Bronner, Andreas Thiede, Michael Schlechtweg, *Member, IEEE*,
Axel Hülsmann, Michaela Rieger-Motzer, Gudrun Kaufel, Brian Raynor, and Martin Sedler

Abstract— Two static and two dynamic frequency dividers based on enhancement and depletion 0.2-μm gate length AlGaAs/GaAs-high electron mobility transistor (HEMT) ($f_T = 60$ and 55 GHz) technology were designed and fabricated. High-speed operations up to 35 GHz for the static frequency dividers and 48 GHz for the dynamic dividers, respectively, have been achieved. The single-ended input and differential outputs to ground simplify many applications. The power consumption is 250 mW for the divide-by-two dividers and 350 mW for the divide-by-four dividers using two supply voltages of 4 and −2.5 V.

Index Terms—AlGaAs/GaAs-HEMT IC's, asymmetrical structure of current switch, dynamic frequency dividers, high-speed circuits, static frequency dividers.

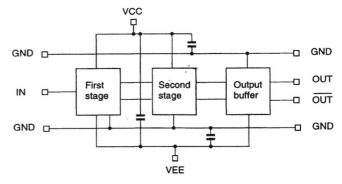

Fig. 1. Block diagram of the 4:1 frequency divider.

I. INTRODUCTION

HIGH-SPEED frequency dividers belong to the key components in measurement equipment, microwave, and satellite communication systems. Therefore, many different high-speed static and dynamic frequency dividers based on various kinds of device technology have been developed. The fastest frequency dividers to date are the AlInAs/GaInAs heterojunction bipolar transistor (HBT) ($f_T = 150$ GHz) static frequency divider operating at 39.5 GHz [1], the AlGaAs/GaAs high electron mobility transistors (HEMT) dynamic frequency divider at 34 GHz [2], and the 51-GHz T-gate AlGaAs/InGaAs MODFET dynamic frequency divider [3], respectively. An operating speed of 30 GHz for a static divider based on Si bipolar technology has been reported [4]. In this paper we present two kinds of static and dynamic frequency dividers, each configured as divide-by-two (2:1) and divide-by-four (4:1) dividers. The four frequency dividers for different applications have been designed in our standard 0.2-μm gate length enhancement (e-mode) and depletion (d-mode) AlGaAs/GaAs-HEMT ($f_T = 60$ and 55 GHz) process technology. The static frequency dividers use master-slave toggle flip-flops operating up to 35 GHz, the fastest realized so far in HEMT technology. By choosing a different gate width for the input transistors, the static dividers have high input sensitivity. The design features of the dynamic frequency dividers eliminated latches compared to the static divider and approaches an operating speed of 48 GHz, which is the fastest

one in AlGaAs/GaAs-HEMT technology. Single-ended input and differential outputs to ground are used in the two circuits to simplify applications.

II. CIRCUIT DESIGN

A. The Static Frequency Dividers

The 2:1 and 4:1 static frequency divider circuits are implemented using source-coupled FET logic (SCFL). The 2:1 static divider consists of a master-slave D flip-flop with output fed back to the data input (MS-TFF). The 4:1 static divider has two stages of divide-by-two MS-TFF in series, as shown in Fig. 1. Fig. 2 shows the circuit schematic of the 2:1 static frequency divider. The frequency divider is designed to operate with a single-ended input to ground without any additional bias by using two supply voltages. Differential signals at internal load resistors are applied to reduce voltage swing and to offer good common-mode suppression. Double-stage source-followers are employed between the D-latches to provide necessary voltage level shifting and to enhance latch speed. The internal logic swing is about 600 mV at the load resistors with a gate current of 4 mA. An inductor of 0.2 nH was placed in series with the load resistor. The value of the inductor is a compromise between operating speed and stability. By introducing this inductive load, the maximum operation frequency is increased by about 5%. A small buffer is inserted between the divider and the output buffer to lighten the load of the divider. Two stages of source-followers with two diodes are applied for level shifting so that the output buffer can provide standard SCFL or emitter coupled logic (ECL) interfaces of the differential outputs to ground.

Manuscript received January 22, 1997; revised April 16, 1997.
The authors are with Fraunhofer-Institute of Applied Solid-State Physics, D-79108 Freiburg, Germany.
Publisher Item Identifier S 0018-9200(97)06312-9.

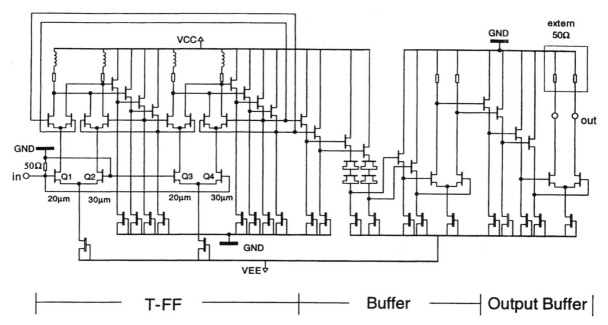

Fig. 2. Schematic of the 2:1 static frequency divider.

The design of the static frequency divider targets not only at high maximum operating frequency, but also at lowest possible low frequency limit for a sinewave input. The low slew rate of a sinewave signal at low frequency represents long transparent time for an MS-TFF circuit and determines the low frequency limit of the MS-TFF circuit. Inserting a clock buffer can increase the sensitivity of a static frequency divider at lower frequencies, but it would also limit the speed performance due to its low 3-dB bandwidth in contrast to maximal operating frequency of the frequency divider. An important characteristic of this design to extend the low frequency limit is the asymmetrical structure of the current switches connected to the clock input. The main variation for the 2:1 static divider (and for the first stage in the 4:1 static divider) is illustrated in Fig. 2. The gate widths of the latch transistors Q_2 and Q_4 for clock signal are larger than that of the track transistors Q_1 and Q_3 so that Q_2 and Q_4 have a lower switching voltage. When the clock goes low and its complement is high, the transistors Q_2 and Q_4 will turn on earlier than Q_3 and Q_1 (not simultaneously as usual). Thus the two upper latches in series of Q_1 and Q_3 are always acquiring data during the latch mode of prior latches and thus the transparent time is reduced. The transistor sizes of the upper latches remain unchanged. Although the maximum operating frequency is just slightly reduced (about 0.5 GHz by simulations compared to the conventional symmetrical structure) due to the larger gate-width of Q_2 and Q_4 for the input signal, the low frequency limit is lowered much more for sinewave input signal.

An open drain buffer without an on-chip line termination was chosen at the low-speed output to keep power dissipation low and to preserve high-speed performance. The high-speed signal input is terminated with an on-chip 50 Ω resistor to reduce signal reflections due to impedance mismatch. The circuit parameters, including the size of each transistor, were optimized to reduce propagation delay time. In particular, all e-mode FET's operate around maximum transconductance and FET widths are chosen based on driving and loading requirements.

B. The Dynamic Frequency Dividers

As seen from Fig. 2, the D-latch is composed of two pairs of current switches connected to the load resistors. The data path is configured with fan-out and fan-in of two at both sides of the double-stage source followers. As is well known, the most effective way to minimize the parasitic capacitances is to reduce the number of fan-out and fan-in. If the clock signal is too fast for the input signals to change during the positive (or negative) clock period, the latches in Fig. 3(a) can be omitted as shown in Fig. 3(b). As a result, the number of fan-out and fan-in of the data path is therefore reduced to merely one. Furthermore, the load of the input clock signal is also half that of the static divider. Similar approaches have been taken in two data decision circuits by weakening the latch function [5], [6]. Simulation indicated that the dynamic frequency divider has a 40% higher maximum operating frequency $f_{\max\_in}$ than a static one and can reach an operating frequency around the transit frequency f_T of the used transistors without consideration of connection line parameters. The dynamic frequency divider has a minimum operating frequency of about $2f_{\max\_in}/3$ for a sinewave input signal. Fig. 4 shows the schematic of the 4:1 dynamic frequency divider which differs from 4:1 static one only in the first 2:1 divider stage.

III. CIRCUIT FABRICATION

The four kinds of frequency dividers were fabricated using our standard enhancement and depletion 0.2-μm gate length

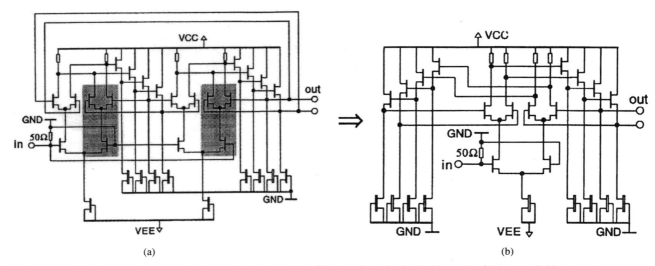

(a)   (b)

Fig. 3. Evolution of the dynamic frequency divider: reducing the width of the transistors in the latching path of the static divider toward zero.

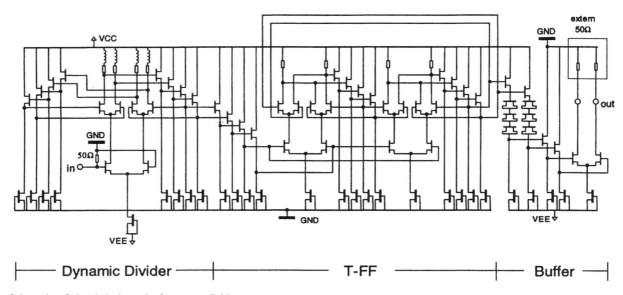

|— Dynamic Divider —|—————— T-FF ——————|— Buffer —|

Fig. 4. Schematic of the 4:1 dynamic frequency divider.

AlGaAs/GaAs-HEMT process [7]. The following mean values for the 0.2-μm gate length enhancement and depletion HEMT's were obtained, respectively: threshold voltages $V_T =$ 0.05 and -0.7 V, maximum transconductance $g_m = 600$ and 400 ms/mm, transit frequencies $f_T = 60$ and 55 GHz, as shown in Table I.

Fig. 5 shows the micrographs of the four frequency dividers. The size of the 2:1 divider chips is 0.7×1 mm^2; the 4:1 divider chips are 1×1 mm^2. Three on-chip metal-insulator-metal (MIM) capacitors for each circuit were placed between the voltage sources and ground to minimize harmonic distortion. Air-bridge inductors were used in series with load resistors. The resistors are made by a thin-film NiCr process and have a sheet resistance of 50 Ω per square. All signal lines are kept as short as possible, particularly the feedback lines in the static dividers, and all signal lines from input pads to the output pads are arranged symmetrically to avoid differences in transit time.

TABLE I
MAIN PARAMETERS OF 0.2-μm GATE-LENGTH HEMT'S

	E-HEMT ($V_{gs} = 0.6$V)	D-HEMT ($V_{gs} = 0$V)
f_T (GHz)	60	55
I_{dsmax} (mA/mm)	200	180
V_T (V)	0.05	-0.7
R_s (Ωmm)	0.6	0.6
g_m (mS/mm)	600	400

IV. CIRCUIT PERFORMANCE

All chips were measured on wafer by using 50-Ω coplanar test probes. The static frequency dividers operate up to 35 GHz

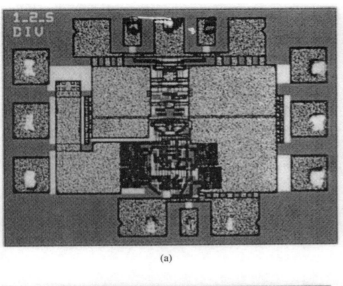

(a)

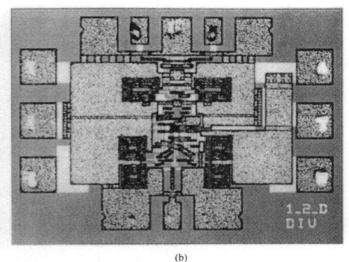

(b)

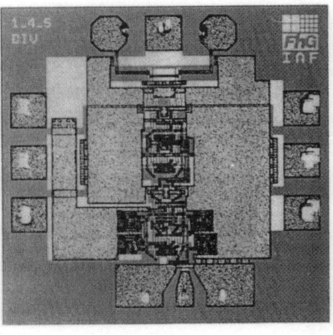

(c)

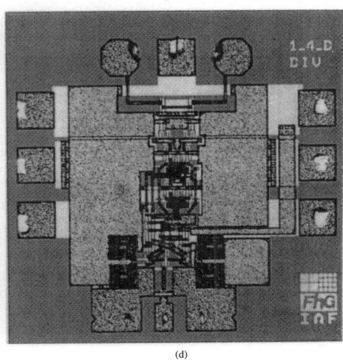

(d)

Fig. 5. Micrographs of the dividers: (a) 2:1 static divider (0.7×1 mm^2), (b) 2:1 dynamic divider (0.7×1 mm^2), (c) 4:1 static divider (1×1 mm^2), and (d) 4:1 dynamic divider (1×1 mm^2).

and the dynamic dividers up to 48 GHz, each driven by a single-ended input signal without any bias. The measured minimum input sensitivity V_{p-p} versus the input frequency f_{in} of the static and the dynamic frequency dividers is plotted in Fig. 6(a) and (b). Above the minimum input voltage reliable operation is guaranteed. The sensitivity curve Fig. 6(a) confirms the effects of applying the asymmetrical structure for the MS-TFF to increase the sensitivity at low frequencies. The minimum input voltage of the static frequency dividers is less than 0.9 V_{p-p} in the frequency range of 2–35 GHz. The input amplitudes of 600 and 700 mV$_{p-p}$ at the maximum operating frequencies are needed for the dynamic and the static dividers, respectively. The resonance frequency is about 23 GHz for the static dividers and 43 GHz for the dynamic dividers, respectively. Fig. 7(a) and (b) shows the measured input and output waveforms of the 2:1 static divider at the input frequency of 35 GHz, and the 4:1 static one at the input frequency of 34 GHz, respectively. In Fig. 7(a) the upper trace is the single-ended 35-GHz input and the lower trace is the 17.5-GHz output signal at one output. Fig. 8(a) and (b) show the measured input and output waveforms of the 2:1 and 4:1 dynamic dividers at the input frequency of 48 GHz. The small time difference of 2 ps of the outputs in Fig. 8(a) is due to the measurement setup. Fig. 9 shows the output

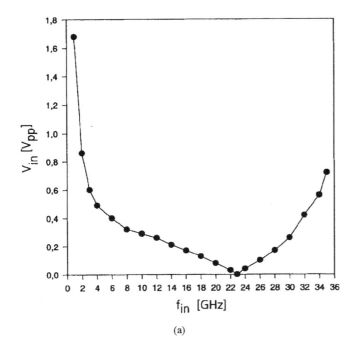

(a)

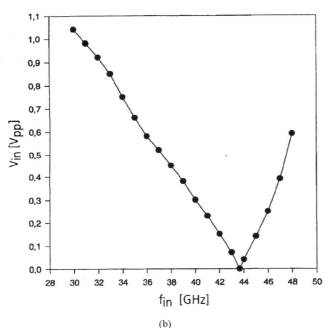

(b)

Fig. 6. Measured sensitivity of input voltage versus input frequency of (a) the static frequency dividers and (b) the dynamic dividers.

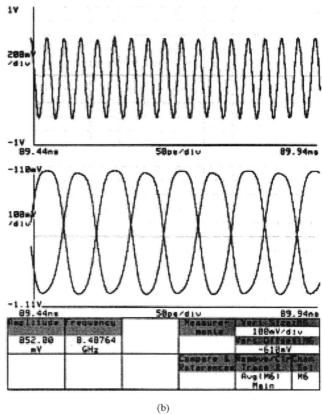

spectrum of the 2:1 dynamic divider at the input frequency of 48 GHz. Note that such spectrum can only be obtained at high frequencies. At low frequencies the output waveform is near square-form and the output spectrum contains high order harmonic components beside the fundamental frequency. More input power is required at lower frequencies because the circuit needs a minimum turn-on/off time for the input signal. Also, the input amplitude should be increased to maintain a required slew rate for a sinewave input signal at lower frequencies, as

Fig. 7. Measured waveforms of the (a) 2:1 static divider at 35 GHz and (b) 4:1 static divider at 34 GHz.

334

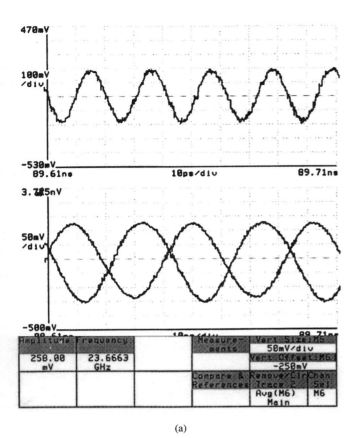

Amplitude	Frequency		Measure-ments	Vert Size:M6 50mV/div
258.00 mV	23.6663 GHz			Vert Offset:M6 -250mV
			Compare & References	Remove:Clr Chan Trace 2 Sel
				Avg(M6) M6 Main

(a)

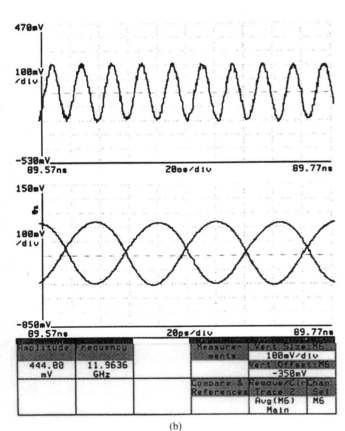

Amplitude	Frequency		Measure-ments	Vert Size:M6 100mV/div
444.00 mV	11.9636 GHz			Vert Offset:M6 -350mV
			Compare & References	Remove:Clr Chan Trace 2 Sel
				Avg(M6) M6 Main

(b)

Fig. 8. Measured waveforms of the (a) 2 : 1 dynamic divider and (b) 4 : 1 one at 48 GHz: upper single-ended input signal; below differential output signals. The input amplitude was damped by 6 dB.

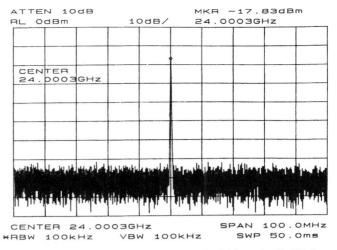

Fig. 9. The output spectrum of the 2 : 1 dynamic divider at the 48 GHz input frequency (horizontal 10 MHz/div., vertical 10 dB/div.).

given by

$$\left. \frac{dV}{dt} \right|_{t=0} = \left. \frac{d}{dt} V_0 \sin \omega t \right|_{t=0}$$
$$= V_0 \omega \cos \omega t |_{t=0}$$
$$= V_0 \omega.$$

The differential output voltage swing is 680 mV$_{p-p}$ for the 2 : 1 dividers and 900 mV$_{p-p}$ for the 4 : 1 dividers. Fig. 10(a) and (b) shows the maximum input frequency distribution of all operational 4 : 1 frequency dividers. The power dissipation is 250 mW for the 2 : 1 dividers and 350 mW for the 4 : 1 dividers, respectively, using two supply voltages of 4 and −2.5 V.

V. Conclusion

The design and performance of static and dynamic frequency dividers has been presented. The static and the dynamic frequency dividers operate reliably up to 35 and 48 GHz, respectively. A minimum input signal amplitude of less than 0.9 V$_{p-p}$ was obtained in the frequency range of 2–35 GHz for the static frequency dividers. By applying a novel design of the dynamic frequency dividers, the maximum operating speed has been improved by 40% as compared to the MS-TFF circuit. The static frequency dividers and the dynamic dividers are the fastest dividers in HEMT technology and in AlGaAs/GaAs-HEMT technology, respectively. The high-speed performance of the four 2 : 1 and 4 : 1 frequency dividers makes them suitable for application in measurement equipment, microwave, and satellite communication systems.

Acknowledgment

The authors thank G. Weimann for encouragement, M. Berroth, now with Technical University of Stuttgart, for many years of continuous support, valuable advice, and encouragement throughout this work, and T. Jakobus for his expert technology management. The authors also thank the entire HEMT processing staff for successful fabrication of these circuits.

number

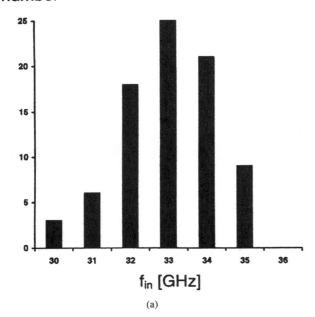

f_{in} [GHz]

(a)

number

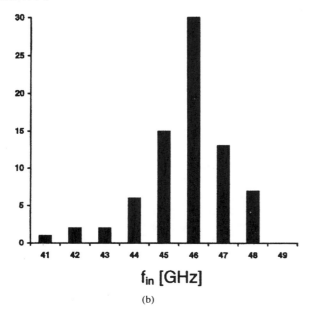

f_{in} [GHz]

(b)

Fig. 10. Maximum input frequency distribution of the (a) 4:1 static frequency divider and (b) 4:1 dynamic divider.

REFERENCES

[1] J. F. Jensen, M. Hafizi, R. A. Metzger, W. E. Stanchina, D. B. Rensch, and Y. K. Allen, "39.5 GHz static frequency divider implemented in AlInAs/GaInAs HBT technology," in *IEEE GaAs IC Symp. Dig.*, 1992, pp. 101–104.

[2] A. Thiede, M. Berroth, U. Nowotny, J. Seibel, R. Bosch, K. Köhler, B. Raynor, and J. Schneider, "An 18–34 GHz dynamic frequency divider based on 0.2 μm AlGaAs/GaAs/AlGaAs quantum-well transistors," in *ISSCC Dig.*, 1993, pp. 176–177.

[3] A. Thiede, P. Tasker, A. Hülsmann, K. Köhler, W. Bronner, M. Schlechtweg, M. Berroth, J. Braunstein, and U. Nowotny, "28–51 GHz dynamic frequency divider based on 0.15 μm T-Gate $Al_{0.2}Ga_{0.8}As/In_{0.25}Ga_{0.75}As$ MODFET's," *Electron. Lett.*, vol. 29, no. 10, pp. 933–934, 1993.

[4] A. Felder, M. Möller, J. Popp, J. Böck, and H.-M. Rein, "46 Gb/s DEMUX, 50 Gb/s MUX, and 30 GHz static frequency divider in silicon bipolar technology," *IEEE J. Solid-State Circuits*, vol. 31, no. 4, pp. 481–486, 1996.

[5] K. Murata, T. Otsuji, E. Sano, M. Ohhata, M. Togashi, and M. Suzuki, "A novel high-speed latching operation flip-flop (HLO-FF) circuit and its application to a 19-Gb/s decision circuit using a 0.2 μm GaAs MESFET," *IEEE J. Solid-State Circuits*, vol. 30, no. 10, pp. 1101–1108, 1995.

[6] T. Otsuji, M. Yoneyama, K. Murata, and E. Sano, "A super-dynamic flip-flop circuit for broadband application up to 24 Gb/s utilizing production-level 0.2-μm GaAs MESFET's," in *IEEE GaAs IC Symp. Dig.*, 1996, pp. 145–148.

[7] A. Hülsmann, G. Kaufel, K. Köhler, B. Raynor, J. Schneider, and T. Jakobus, "E-beam direct-write in a dry-etched recess gate HEMT process for GaAs/AlGaAs circuits," *Jpn. J. Appl. Phys.*, vol. 29, no. 10, pp. 2317–2320, 1990.

Zhihao Lao (M'96), for a photograph and biography, see p. 1392 of the September 1997 issue of this JOURNAL.

Wolfgang Bronner, for a photograph and biography, see p. 1393 of the September 1997 issue of this JOURNAL.

Andreas Thiede, for a photograph and biography, see p. 1393 of the September 1997 issue of this JOURNAL.

Michael Schlechtweg (M'89), for a biography, see p. 1393 of the September 1997 issue of this JOURNAL.

Axel Hülsmann, for a biography, see p. 1393 of the September 1997 issue of this JOURNAL.

Michaela Rieger-Motzer, for a photograph and biography, see p. 1393 of the September 1997 issue of this JOURNAL.

Gudrun Kaufel, for a photograph and biography, see p. 1393 of the September 1997 issue of this JOURNAL.

Brian Raynor, for a photograph and biography, see p. 1393 of the September 1997 issue of this JOURNAL.

Martin Sedler, for a photograph and biography, see p. 1393 of the September 1997 issue of this JOURNAL.

Superharmonic Injection-Locked Frequency Dividers

Hamid R. Rategh, *Student Member, IEEE*, and Thomas H. Lee, *Member, IEEE*

Abstract—Injection-locked oscillators (ILO's) are investigated in a new theoretical approach. A first-order differential equation is derived for the noise dynamics of ILO's. A single-ended injection-locked frequency divider (SILFD) is designed in a 0.5-μm CMOS technology operating at 1.8 GHz with more than 190 MHz locking range while consuming 3 mW of power. A differential injection-locked frequency divider (DILFD) is designed in a 0.5-μm CMOS technology operating at 3 GHz and consuming 0.45 mW, with a 190 MHz locking range. A locking range of 370 MHz is achieved for the DILFD when the power consumption is increased to 1.2 mW.

Index Terms—Analog and digital frequency dividers, injection-locked oscillators, radio-frequency integrated circuits.

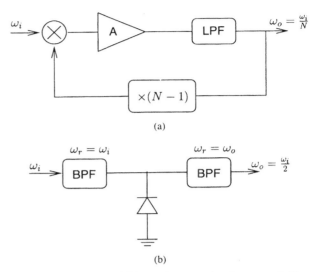

Fig. 1. Analog frequency dividers: (a) regenerative frequency divider and (b) parametric frequency divider.

I. INTRODUCTION

CONVENTIONAL phase-locked loops (PLL's) use frequency dividers in their feedback path to achieve frequency multiplication. Most PLL's designed for wireless systems use flip-flop-based digital frequency dividers. These dividers are wide band and their power consumption increases with the frequency of operation. In frequency synthesizers used in modern wireless systems, frequency dividers consume a large percentage of the total power [2], [8]. Most often, off-chip frequency dividers are used as the first stage in a stack of dividers in high-frequency PLL's [8]. The limitation on power and maximum frequency of operation of conventional digital frequency dividers is associated with the wide-band nature of these dividers. However, since most wireless systems are themselves narrow band, narrow-band analog frequency dividers may be used to reduce power and increase the maximum frequency of operation.

Regenerative frequency dividers [Fig. 1(a)] are the most widely used analog frequency dividers [5]–[7]. Frequency division in such a divider results from combining frequency multiplication in the feedback path with mixing at the input. Regenerative dividers can operate at frequencies higher than flip-flop-based dividers [13]. However, they require many functional blocks to guarantee frequency division [7]. As a result, regenerative frequency dividers are not the best solution for low-power systems.

Parametric frequency dividers [Fig. 1(b)] are another group of analog frequency dividers used in microwave systems [3], [5], [15]. The frequency division principle of a parametric frequency divider relies on exciting a varactor at frequency f and realizing a negative resistance that sustains a loop gain of unity at $f/2$. High Q varactors and inductors are key

Manuscript received October 5, 1998; revised February 22, 1999.
The authors are with the Center for Integrated Systems, Stanford University, Stanford, CA 94305-4070 USA (e-mail: hamid@smirc.stanford.edu).
Publisher Item Identifier S 0018-9200(99)04190-6.

elements in parametric frequency dividers [15]. Since high Q passive elements cannot be implemented in contemporary silicon technologies, parametric dividers are not amenable to integration.

The third group, injection-locked frequency dividers (ILFD's), work by synchronizing an oscillator with an incident signal. Depending upon the ratio of the incident frequency to the oscillation frequency, three classes of injection-locked oscillators (ILO's) may be defined: first-harmonic, subharmonic, and superharmonic ILO's. In a first-harmonic ILO, the oscillation frequency is the same as the fundamental frequency of the incident signal [1], while in a subharmonic ILO, the incident frequency is a subharmonic of the oscillation frequency [4], [9], [14], [20]. Likewise, in a superharmonic ILO, the incident frequency is a harmonic of the oscillation frequency. Uzunoglu *et al.* [16], [17] used synchronous oscillators (SO's) as frequency dividers, without providing a physical model for the frequency division functionality of SO's. The SO proposed in [17] is a nonlinear oscillator with a very large internal gain and a saturated output amplitude (voltage limited). High bias currents are required to provide the large gain and to operate SO's in a voltage-limited amplitude regime. Therefore, SO's are not appropriate for low-power systems. Unlike SO's, superharmonic ILO's can be designed as very low-power frequency dividers [10].

In this paper, we present a new method to calculate the locking range of ILO's. We also introduce two different mechanisms for failure of injection locking. A differential equation is derived that models the noise dynamics of ILO's.

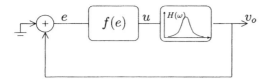

Fig. 2.   Model for a free-running *LC* oscillator.

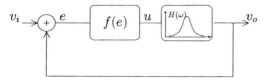

Fig. 3.   Model for an injection-locked oscillator.

Measurements on a single-ended ILFD (SILFD) are compared with simulations. The simulation results of a differential ILFD (DILFD) are reported as well.

II. MODEL FOR INJECTION-LOCKED OSCILLATORS

An *LC* oscillator can be modeled as a nonlinear block $f(e)$, followed by a frequency selective block (e.g., an *RLC* tank) $H(\omega)$, in a positive feedback loop as shown in Fig. 2. The nonlinear block models all the nonlinearities in the oscillator, including any amplitude-limiting mechanism. To have a steady-state oscillation, a loop gain of unity should be maintained. We would like to express the oscillation condition in terms of gain and phase criteria for reasons that will be clear later. The gain condition is satisfied if the output amplitude V_o is the same as the amplitude of $e(t)$ in an open-loop excitation of the system at the oscillation frequency ω_o. The phase condition requires that the excess phase introduced in the loop at $\omega = \omega_o$ be zero.

With an additional external signal (i.e., the incident signal), this same model can be used to model an ILO. This model is shown in Fig. 3. To investigate the injection-locking phenomenon in an ILO, we define

$$v_i(t) = V_i \cos(\omega_i t + \varphi) \tag{1}$$

$$v_o(t) = V_o \cos(\omega_o t) \tag{2}$$

$$u(t) = f(e(t)) = f(v_o(t) + v_i(t)) \tag{3}$$

$$H(\omega) = \frac{H_0}{1 + j2Q\frac{\omega - \omega_r}{\omega_r}} \tag{4}$$

where $v_i(t)$ is the incident signal, $v_o(t)$ is the output signal, φ is the phase difference between those two signals, and ω_r and Q are the resonant frequency and quality factor of the *RLC* tank, respectively. The output of the nonlinear block $u(t)$ may contain various harmonic and intermodulation terms of $v_i(t)$ and $v_o(t)$. As shown in Appendix A, we can write $u(t)$ as

$$u(t) = f(v_i + v_o)$$

$$= \sum_{m=0}^{\infty} \sum_{n=0}^{\infty} K_{m,n} \cos(m\omega_i t + m\varphi) \cos(n\omega_o t) \tag{5}$$

where each $K_{m,n}$ is an intermodulation coefficient of $f(v_i + v_o)$.

We assume that all frequency components of $u(t)$ far from the resonant frequency of the tank are filtered out, so the frequency of the output signal can be written as $\omega_o = \omega_r + \Delta\omega$. Thus, we need only consider intermodulation terms with frequency ω_o, that is, $|m\omega_i - n\omega_o| = \omega_o$. For an Nth-order superharmonic ILO (i.e., $\omega_o = (1/N)\omega_i$), the intermodulation terms with $n = Nm \pm 1$ possess a frequency equal to $1/N$ of the incident frequency. The signal $u_{\omega_o}(t)$, which is the component of $u(t)$ with frequency ω_o, can be written as

$$u_{\omega_o}(t) = K_{0,1}\cos(\omega_o t) + \frac{1}{2}\sum_{m=1}^{\infty} K_{m,Nm\pm 1}\cos(\omega_o t + m\varphi). \tag{6}$$

Using a complex exponential to replace sines and cosines, and applying the oscillation condition, the output signal can be written as

$$v_o = V_o e^{j\omega_o t} = \frac{H_0 e^{j\omega_o t}}{1 + j2Q\frac{\Delta\omega}{\omega_r}}\left[K_{0,1} + \frac{1}{2}\sum_{m=1}^{\infty} K_{m,Nm\pm 1}e^{jm\varphi}\right] \tag{7}$$

or

$$V_o\left(1 + j2Q\frac{\Delta\omega}{\omega_r}\right) = H_0\left[K_{0,1} + \frac{1}{2}\sum_{m=1}^{\infty} K_{m,Nm\pm 1}e^{jm\varphi}\right]. \tag{8}$$

The real and imaginary parts of (8) can be separated as

$$V_o = H_o\left[K_{0,1} + \frac{1}{2}\sum_{m=1}^{\infty} K_{m,Nm\pm 1}\cos(m\varphi)\right] \tag{9}$$

$$2V_o Q\frac{\Delta\omega}{\omega_r} = \frac{H_0}{2}\sum_{m=1}^{\infty} K_{m,Nm\pm 1}\sin(m\varphi). \tag{10}$$

Equations (9) and (10) are the fundamental equations for a superharmonic injection-locked oscillator. The simultaneous solution of these two equations specifies V_o and φ for any incident amplitude V_i and any incident frequency ω_i or, equivalently, for any offset frequency $\Delta\omega = (\omega_i/N) - \omega_r$. Equation (10) can be rearranged as

$$\Delta\omega = \Delta\omega_A\left[\frac{H_0}{2V_i}\sum_{m=1}^{\infty} K_{m,Nm\pm 1}\sin(m\varphi)\right] \tag{11}$$

where $\Delta\omega_A = (\omega_r/2Q)(V_i/V_o)$ is Adler's *locking range* figure of merit [1]. The fundamental equations, (9) and (10), are very general but provide limited intuition. However, as shown in the next section, for the special case of $N = 2$ (i.e., divide-by-two) and a third-order nonlinearity (i.e., $f(e) = a_0 + a_1 e + a_2 e^2 + a_3 e^3$), (9) and (10) can be solved analytically, which allows the development of design insight.

A. Special Case ($N = 2$ and $f(e)$ Is a Third-Order Nonlinear Function)

For the special case of $N = 2$ and $f(e) = a_0 + a_1 e + a_2 e^2 + a_3 e^3$, the only unknown in (10) is the input–output phase difference φ, which means the phase condition can be satisfied independently of the gain condition

$$\sin(\varphi) = \frac{2Q}{H_o a_2 V_i} \frac{\Delta \omega}{\omega_r}$$
$$|\sin(\varphi)| < 1 \Rightarrow \left| \frac{\Delta \omega}{\omega_r} \right| < \left| \frac{H_0 a_2 V_i}{2Q} \right|. \tag{12}$$

On the other hand, satisfying the gain condition and solving (9) results in an expression for the oscillation amplitude

$$V_o = \sqrt{\frac{4}{3} \frac{1}{a_3 H_0} \left[1 - H_0 \left(a_1 + \frac{3}{2} a_3 V_i^2 + a_2 V_i \cos(\varphi) \right) \right]}. \tag{13}$$

As (12) suggests, the locking range can be increased by increasing either H_0/Q or the incident amplitude V_i. Increasing H_0/Q in an LC oscillator is equivalent to using an inductor with a larger value ($H_0/Q = \omega L$). The self-resonant frequency of the inductor puts a limit on the maximum inductor size and effectively limits the locking range by failing to satisfy the phase condition. The increase of the locking range with the incident amplitude is also limited. When the term under the square root in (13) becomes negative, the gain condition fails and limits the locking range. As a result, injection locking fails and the locking range is limited by failure of either the phase condition (phase limited) or the gain condition (gain limited). The effect of each limiting mechanism on the noise performance of an ILO is discussed in more detail in Section V-A.

As mentioned before, the locking range in an ILO is a function of the incident amplitude. So, by injecting the incident signal into a high-impedance node, the required incident power can be reduced significantly. Due to the high impedance of the gate of MOS transistors, MOS transistors are a good candidate for injection-locked oscillators.

The underlying assumption in the derivation of (9) and (10) is that the resonant frequency of the LC tank does not change as the incident frequency changes. However, to achieve a larger tuning range, the free-running oscillation frequency of the ILO can be modified such that it tracks the incident frequency [11], [12].

III. NOISE IN ILO's

To investigate the phase noise performance of an ILO, we first consider the response of a first-harmonic ILO to a deterministic sinusoidal noise. For convenience, the model for an ILO is repeated in Fig. 4 with the noise v_n added to the summing junction. The noise can be either from the incident signal or from the ILO itself. The incident signal, output signal, and sinusoidal noise are represented by their equivalent phasors in Fig. 5 and mathematically defined as

$$v_i(t) = V_i \cos(\omega_o t) \tag{14}$$
$$v_o(t) = V_o \cos(\omega_o t + \varphi) \tag{15}$$
$$v_n(t) = V_n \cos((\omega_o + \omega_n)t + \varphi_n). \tag{16}$$

When the output signal is injection locked to the incident signal in the absence of noise, the input–output phase difference is constant ($\varphi = \varphi_o$). However, when sinusoidal noise

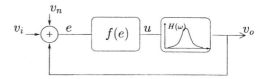

Fig. 4. ILO model used for noise analysis.

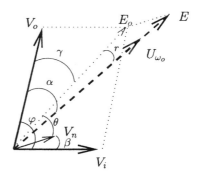

Fig. 5. Phasor representation of signals in Fig. 4.

with an offset frequency ω_n is added to the system, φ is no longer constant and the instantaneous output frequency ω is defined as

$$\omega = \omega_o + \frac{d\varphi}{dt}. \tag{17}$$

It is the variation of φ that generates phase noise in the output signal. As shown in Appendix B, $\frac{d\varphi}{dt}$ can be approximated as

$$\frac{d\varphi}{dt} \simeq -\Delta\omega_o - \frac{1}{A} \left[\frac{V_i}{V_o} \sin(\varphi) - \frac{V_n}{V_o} \cos(\varphi) \sin(\beta) \right] \tag{18}$$

where $\Delta\omega_0$ is the difference between the incident frequency and the free-running frequency, $A = (2Q)/\omega_r$, and $\beta = \omega_n t + \varphi_n$.

The input–output phase difference can be written as

$$\varphi = \varphi_o + \varphi_\epsilon \tag{19}$$

where φ_o is the input–output phase difference in the absence of noise and is a constant [$\Delta\omega_o = -V_i/(AV_o)\sin(\varphi_o)$ from (45)] and φ_ϵ is the time-variant portion of φ. $\varphi_\epsilon \ll 1$ because $V_n \ll V_i \ll V_o$. Hence (18) can be simplified to

$$\frac{d\varphi_\epsilon}{dt} + K\varphi_\epsilon = \frac{V_n}{AV_o} \cos(\varphi_o) \sin(\beta) \tag{20}$$

where

$$K = \frac{V_i}{AV_o} \cos(\varphi_o) - \frac{V_n}{AV_o} \sin(\varphi_o) \sin(\beta). \tag{21}$$

If $\tan(\varphi_o) \ll V_i/V_n$, meaning that the incident frequency is not at the edge of a phase-limited locking range, K can be approximated as

$$K \simeq \frac{V_i}{AV_o} \cos(\varphi_o) \tag{22}$$

339

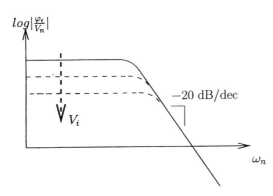

Fig. 6. Noise transfer function of an ILO.

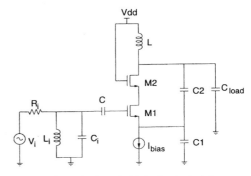

Fig. 7. Schematic of the single-ended injection-locked frequency divider.

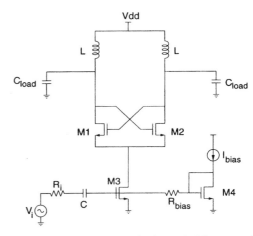

Fig. 8. Schematic of the differential injection-locked frequency divider.

which allows simplification of (20) to a first-order differential equation

$$\frac{d\varphi_\epsilon}{dt} + \left[\frac{V_i}{AV_o}\cos(\varphi_o)\right]\varphi_\epsilon = \left[\frac{V_n}{AV_o}\cos(\varphi_o)\right]\sin(\omega_n t + \varphi_n). \tag{23}$$

The noise transfer function from v_n to the output phase (23) is shown in Fig. 6. From (23) and Fig. 6, it is clear that an ILO has the same noise transfer function as a first-order PLL. The noise from the incident signal is shaped by the low-pass characteristic of the noise transfer function, and the output signal tracks the phase variations of the incident signal within the loop bandwidth $(V_i/(AV_o)\cos(\varphi_o))$. However, unlike a first-order PLL, the loop bandwidth of an ILO is a function of the incident amplitude and is larger for a larger incident amplitude.

The interpretation of the noise transfer function is a little different if the noise comes from the ILO itself. Within the loop bandwidth, the noise from the ILO is suppressed by the ratio of the noise power to the incident power. Outside the loop bandwidth, the noise suppression increases by 20 dB per decade of offset frequency, and a $1/f^2$ phase noise region is observed.

The noise dynamics in a superharmonic ILO are the same as those of a first-harmonic ILO, except $(d\varphi_\epsilon)/(dt)$ is $1/N$ of that in a first-harmonic ILO due to the frequency division operation. So (23) for an Nth-order ILFD can be modified as

$$\frac{d\varphi_\epsilon}{dt} + \left[\frac{V_i}{AV_o}\cos(\varphi_o)\right]\varphi_\epsilon$$
$$= \frac{1}{N}\left[\frac{V_n}{AV_o}\cos(\varphi_o)\right]\sin(\omega_n t + \varphi_n) \tag{24}$$

where φ_o is no longer a simple function of $\Delta\omega_0$ but is determined by solving the superharmonic ILO's fundamental equations, (9) and (10). As the division ratio N increases, the noise rejection increases proportionally. So in a divide-by-two ILFD, the output close-in phase noise is $20\log(2) = 6$ dB lower than that of the incident signal.

IV. CIRCUIT IMPLEMENTATION

In this paper, we propose two different architectures for ILFD's. Fig. 7 shows the schematic of an SILFD. For simplic-

ity, the biasing circuitry is not shown in this figure. A Colpitts oscillator forms the core of the SILFD. The incident signal is injected into the gate of M1. Transistors M1 and M2 are used in cascode, mainly to provide more isolation between the input and output. Transistor M2 is sized to be smaller than M1 by almost a factor of three to reduce the parasitic capacitance at the output node (drain of M2). As a result, a larger inductor can be used to resonate this reduced capacitance. As discussed in Section II-A, using a larger inductor increases the locking range. The power consumption is also reduced due to the increased effective parallel impedance of the LC tank, assuming that tank losses are mainly from the inductor. Last, Li and Ci in the gate of M1 are used to model the LC tank of the preceding LC oscillator. The analogy of this circuit with the model in Fig. 3 can be realized by observing that transistor M1 functions as the summing element for the incident and output signals.

The schematic of a DILFD is shown in Fig. 8. The incident signal is injected into the gate of M3, which delivers the incident signal to the common source connection of M1 and M2. The output signal is fed back to the gates of M1 and M2. The output and incident signals are thus summed across the gates and sources of M1 and M2. The common source connection of M1 and M2, even in the absence of the incident signal, oscillates at twice the frequency of the output signal, which makes this node an appropriate injection node for a divide-by-two operation.

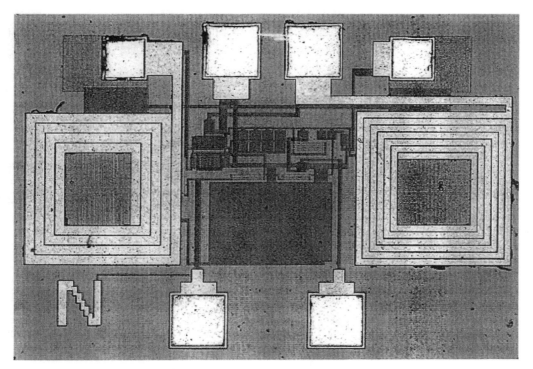

Fig. 9. Die micrograph of the SILFD (0.7×1 mm^2).

V. SIMULATION AND MEASUREMENT RESULTS

A. Single-Ended ILFD

The SILFD shown in Fig. 7 is designed in a 0.5-μm CMOS technology and operates on 2.5 V and a bias current of 1.2 mA. The free-running frequency of oscillation is 920 MHz, and the incident frequency is around 1840 MHz. Both inductors are on-chip spiral inductors with patterned ground shields [18], [19]. The die micrograph of the SILFD is shown in Fig. 9. The total area of the die is 0.7 mm^2 (0.7×1 mm^2).

The oscillation amplitude of the SILFD is plotted in Fig. 10 as a function of the incident frequency for different incident amplitudes. The locking range is determined by the frequency difference between the two ends of each curve. At small incident amplitudes, the locking range is phase limited, as explained in Section II-A, and increases with the incident amplitude. However, for incident amplitudes beyond 300 mV, the locking range is gain limited and shrinks as the incident amplitude increases. Simulated and measured locking range as a function of incident amplitude are shown in Fig. 11. A locking range of more than 190 MHz (11% of the center frequency) is achieved when consuming 3 mW of power. The maximum locking range as a function of bias current is shown in Fig. 12. A locking range of more than 135 MHz is achieved with a bias current as low as 600 μA.

Phase noise measurement results are shown in Fig. 13. The thin solid line in this figure shows the phase noise of the free-running SILFD. The thick solid line is the phase noise of the HP8664A signal generator used as the incident signal. The nonsolid lines are the phase noise measurement of the SILFD when locked to three different incident frequencies, referred to as middle-frequency, phase-limited, and gain-limited curves.

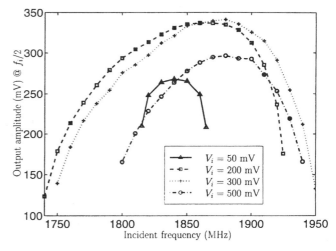

Fig. 10. Oscillation amplitude in the SILFD.

The middle-frequency curve is the output phase noise measured at an incident frequency in the middle of the locking range. The phase- and gain-limited curves are measured when the incident frequency is at the edge of a phase- and gain-limited locking range, respectively.

At low offset frequencies, the divider output phase noise is almost 6 dB lower than the incident phase noise, as is expected from the divide-by-two operation and predicted by (24). However, at higher offset frequencies, the excess noise from the divider increases the output phase noise. The far-out phase noise at the edge of a gain-limited locking range is even worse than the phase noise of the free-running oscillator. The small oscillation amplitude at the edge of a gain-limited

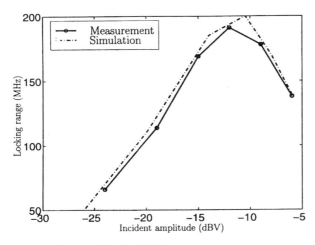

Fig. 11. Locking range for the SILFD.

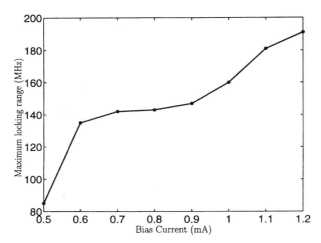

Fig. 12. Locking range as a function of the bias current in the SILFD.

locking range explains this higher phase noise of the ILFD at large offset frequencies.

Despite the large close-in phase noise of the free-running ILFD, the divider phase noise tracks the phase noise of the incident signal for offset frequencies up to 100 kHz. As a result, the ILFD can be designed for very low-power operation without sacrificing the noise performance of the system. Also, very low Q on-chip spiral inductors, with small physical dimensions, can be used in ILFD's.

B. Differential ILFD

A DILFD (Fig. 8) is designed in a 0.5-μm CMOS technology. The supply voltage is 1.5 V and the tail current is nominally 300 μA. The DILFD oscillates at 1.6 GHz in free-running operation, and the incident frequency is in the vicinity of 3.2 GHz. On-chip spiral inductors with a Q of 5.8 are used in this design.

The oscillation amplitude as a function of incident frequency is shown in Fig. 14. Comparing this with Fig. 10, two differences are observed. In Fig. 14, the curves are flatter and the locking range increases monotonically with incident amplitude. These suggest that the locking range in the DILFD is phase limited, unlike the gain-limited locking range in the

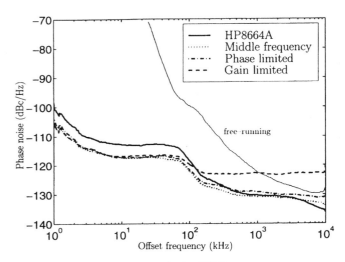

Fig. 13. Phase noise measurement in the SILFD.

SILFD at large incident amplitudes. This can partially be due to the subunity voltage gain of M3 in Fig. 8. As a result, the amplitude of the injected signal at the summing node (the common source connection of M1 and M2) of the DILFD is less than that of the SILFD. Also, the increased tail current in the presence of a large incident signal changes $f(e)$, which can effectively change the phase-limited region of the locking range in DILFD's.

More than 190 MHz of locking range is achieved with only 0.45 mW of power (Fig. 14). By increasing the power to 1.2 mW, the locking range increases to 370 MHz (12% of the center frequency). The DILFD is expected to have a better phase noise than the SILFD over the entire locking range, due to its phase-limited locking range.

The performance of the SILFD and DILFD is summarized in Table I. For comparison purposes, the performance of a conventional frequency divider made out of two back-to-back connected source-coupled-logic (SCL) latches designed in the same technology is also tabulated. The SCL divider operates at about half the frequency of the DILFD and consumes more than four times the power. The SCL divider also fails to operate above 3 GHz. The last column in Table I shows the simulated acquisition time in ILFD's. The acquisition time, which measures how fast an ILFD locks to an incident signal, is inversely proportional to the locking range. Therefore, as long as the locking range is phase limited, increasing the incident amplitude reduces the acquisition time.

C. Noise Transfer Function

To verify the noise dynamics derived in Section III, the SILFD is injection locked to an incident frequency while a second signal is injected at different offset frequencies from the incident frequency. As demonstrated in Fig. 15, two sidebands are generated in the output signal spectrum. The power below carrier of the sidebands is measured at different offset frequencies and is shown in Figs. 16 and 17. In Fig. 16, the incident power P_i is constant and the noise transfer function is measured for three noise power levels P_n.

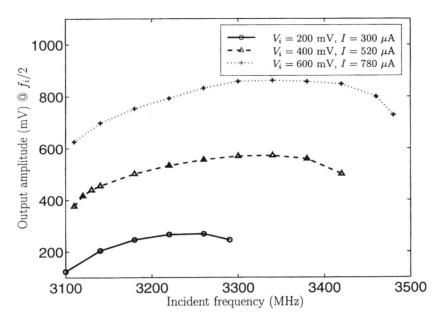

Fig. 14. Oscillation amplitude in the DILFD.

Divider	f(GHz)	V_{dd}	$I(\mu A)$	P(mW)	Δf(MHz)	t_a(ns)
SILFD	1.8	2.5	600	1.5	135	≤ 9
(Measured)			1200	3.0	191	≤ 7
DILFD	3.5	1.5	300	0.45	190	≤ 7
(Simulated)			400	0.60	260	≤ 6
			780	1.17	370	≤ 5
SCL Latch	1.8	2.0	1000	2.0	1800	
(Simulated)	3.0			Failed		

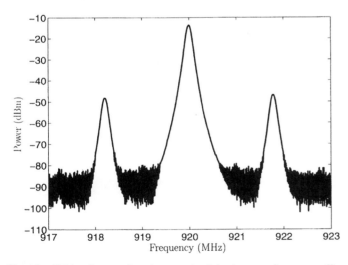

Fig. 15. Sideband generation due to noise injection at a frequency offset from the incident frequency.

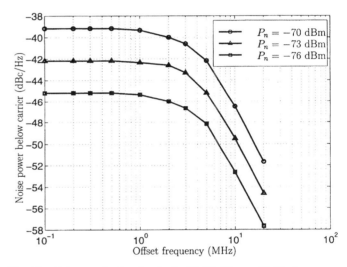

Fig. 16. Noise transfer function in the SILFD ($P_i = -40$ dBm).

noise transfer function measurement results of Figs. 16 and 17 are in very good agreement with (24).

VI. CONCLUSION

A new method is reported for calculating the locking range of injection-locked oscillators. Two different mechanisms for the failure of injection locking are introduced. It is shown mathematically that the noise transfer function of an ILO is the same as that of a first-order PLL. Two novel circuits for single-ended and differential ILFD's are proposed. The measurement results of the SILFD verify the theory of injection locking and the model for the noise dynamics of ILO's. It is shown that ILFD's can operate at frequencies where conventional digital frequency dividers fail and still consume less power than digital frequency dividers operating at lower frequencies (Table I). Unlike digital frequency dividers, the power consumption in

As predicted by (24), reducing the noise power by 3 dB shifts the noise transfer function curve down by the same amount.

The same measurement is repeated for different incident powers while keeping the noise power constant. The results are shown in Fig. 17. When the incident power increases by 3 dB, both the loop bandwidth and the close-in noise rejection increase by 3 dB, while the far-out noise does not change. The

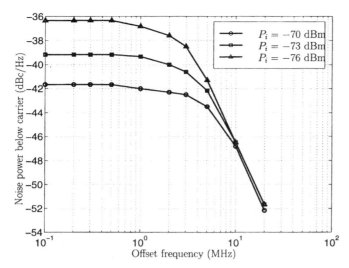

Fig. 17. Noise transfer function in the SILFD ($P_n = -70$ dBm).

an ILFD does not increase linearly with the frequency of operation. Therefore, injection-locked frequency dividers are attractive for digital CMOS frequency dividers, especially for low-power and high-frequency wireless systems.

APPENDIX A

To simplify the proof of (5), we redefine v_i, v_o, and $f(e)$ as

$$v_i = V_i \cos(\beta) \tag{25}$$

$$v_o = V_o \cos(\alpha) \tag{26}$$

$$f(e) = f(v_i + v_o) \tag{27}$$

where $\alpha = \omega_i t + \varphi$ and $\beta = \omega_o t$. Function $f(e)$ is periodic with respect to both α and β. For every β, we can define a periodic function $g(\alpha)$ as

$$g(\alpha) = f(v_o + V_i \cos(\alpha)). \tag{28}$$

Since $g(\alpha + 2\pi) = g(\alpha)$ and $g(-\alpha) = g(\alpha)$, $g(\alpha)$ can be represented by its Fourier series as

$$g(\alpha) = \sum_{m=0}^{\infty} L_m(\beta) \cos(m\alpha) \tag{29}$$

where each $L_m(\beta)$ is a Fourier series coefficient of $g(\alpha)$ and is calculated as

$$L_0(\beta) = \frac{1}{2\pi} \int_0^{2\pi} f(V_o \cos(\beta) + V_i \cos(\alpha)) \, d\alpha \tag{30}$$

$$L_m(\beta) = \frac{1}{\pi} \int_0^{2\pi} f(V_o \cos(\beta) + V_i \cos(\alpha)) \cos(m\alpha) \, d\alpha. \tag{31}$$

Since each L_m is even and periodic with period 2π, it can be represented in terms of its Fourier series as

$$L_m(\beta) = \sum_{n=0}^{\infty} K_{m,n} \cos(n\beta) \tag{32}$$

where

$$K_{m,0} = \frac{1}{2\pi} \int_0^{2\pi} L_m(\beta) \, d\beta \tag{33}$$

$$K_{m,n} = \frac{1}{\pi} \int_0^{2\pi} L_m(\beta) \cos(n\beta) \, d\beta. \tag{34}$$

Now to complete the proof, insert (32) into (29) and replace $g(\alpha)$ by $f(v_i + v_o)$

$$f(v_i + v_o) = \sum_{m=0}^{\infty} \sum_{n=0}^{\infty} K_{m,n} \cos(m\alpha) \cos(n\beta). \tag{35}$$

APPENDIX B

To derive (23), we start by evaluating the excess phase introduced in the loop, excluding the phase added by the frequency selective block in a first-harmonic ILO.

The phasor representation of $e(t)$, E (Figs. 4 and 5), is calculated as the vector sum of V_i, V_o, and V_n. As E experiences the nonlinearities of $f(e)$, new harmonics are generated, but u_{ω_o}, the component of $u(t)$ with the same instantaneous frequency as $e(t)$, stays in phase with $e(t)$. So U_{ω_o}, the phasor representation of u_{ω_o}, and E have the same direction, as shown in Fig. 5. The phase difference introduced between V_o and U_{ω_o} is equal to

$$\alpha = \gamma + r \tag{36}$$

where γ is the phase difference between V_o and E_o (vector sum of V_o and V_i) and r is the phase difference between E_o and U_{ω_o} (Fig. 5). Since $V_n \ll V_i \ll V_o$, we can approximate γ and r as

$$\gamma \simeq \tan(\gamma) = \frac{V_i \sin(\varphi)}{V_o + V_i \cos(\varphi)} \simeq \frac{V_i}{V_o} \sin(\varphi) \tag{37}$$

$$r \simeq \tan(r) \simeq \frac{V_n \sin(\theta)}{E} \simeq \frac{V_n}{V_o + V_i \cos(\varphi)} \sin(\theta) \cos(\gamma) \tag{38}$$

$$r \simeq \frac{V_n}{V_o} \sin(\theta) \tag{39}$$

where

$$\theta = \varphi - \gamma - \beta \simeq \varphi - \beta. \tag{40}$$

To satisfy the phase condition, α should be canceled out by the phase introduced by the RLC tank ($\Delta\phi_{RLC}$). Thus

$$\alpha = \Delta\phi_{RLC} \simeq -\frac{2Q}{\omega_r}(\Delta\omega) = -A\Delta\omega \tag{41}$$

where

$$A = \frac{2Q}{\omega_r} \tag{42}$$

and

$$\Delta\omega = \omega - \omega_r = \omega - (\omega_o - \Delta\omega_0) = \frac{d\varphi}{dt} + \Delta\omega_o \tag{43}$$

where ω is replaced by its equivalent from (17). To calculate $(d\varphi)/(dt)$, we insert (43) and (36) into (41) and rearrange the terms

$$\frac{d\varphi}{dt} = -\Delta\omega_o - \frac{1}{A}(\gamma + r). \tag{44}$$

Equation (44) can be further expanded by replacing γ and r from (37) and (39)

$$\frac{d\varphi}{dt} = -\Delta\omega_o - \frac{1}{A}\left[\frac{V_i}{V_o} \sin(\varphi) + \frac{V_n}{V_o} \sin(\theta)\right]. \tag{45}$$

Now if we replace θ by $\varphi - \beta$ from (40) and expand $\sin(\varphi - \beta)$, (45) can be written as

$$\frac{d\varphi}{dt} = -\Delta\omega_o - \frac{1}{A}\left[\sin(\varphi)\left(\frac{V_i}{V_o} + \frac{V_n}{V_o}\cos(\beta)\right) \right.$$
$$\left. - \frac{V_n}{V_o}\cos(\varphi)\sin(\beta)\right]. \tag{46}$$

Since $V_n \ll V_i$, we can approximate $(d\varphi)/(dt)$ as

$$\frac{d\varphi}{dt} \simeq -\Delta\omega_o - \frac{1}{A}\left[\frac{V_i}{V_o}\sin(\varphi) - \frac{V_n}{V_o}\cos(\varphi)\sin(\beta)\right] \tag{47}$$

which ends our derivation.

ACKNOWLEDGMENT

The authors would like to acknowledge Dr. A. Hajimiri and R. Betancourt for their valuable discussions and comments. They are also grateful to Rockwell Semiconductor for fabricating the SILFD.

REFERENCES

[1] R. Adler, "A study of locking phenomena in oscillators," *Proc. IRE*, vol. 34, pp. 351–357, June 1946.
[2] T. S. Aytur and B. Razavi, "A 2-GHz, 6 mW BiCMOS frequency synthesizer," *IEEE J. Solid-State Circuits*, vol. 30, pp. 1457–1462, Dec. 1995.
[3] I. Bahl and P. Bhartia, *Microwave Solid State Circuit Design*. New York: Wiley, 1988.
[4] A. S. Daryoush, T. Berceli, R. Saedi, P. Herczfeld, and A. Rosen, "Theory of subharmonic synchronization of nonlinear oscillators," in *IEEE MTT-S Dig.*, 1989, pp. 735–738.
[5] M. M. Driscoll, "Phase noise performance of analog frequency dividers," *IEEE Trans. Ultrason., Ferro-Elect., Freq. Contr.*, vol. 37, pp. 295–301, July 1990.
[6] R. G. Harrison, "Theory of regenerative frequency dividers using double-balanced mixers," in *IEEE MTT-S Dig.*, 1989, pp. 459–462.
[7] V. Manassewitsch, *Frequency Synthesizers: Theory and Design*. New York: Wiley, 1987.
[8] C. G. S. Michael, H. Perrott, and T. L. Tewksbury, "A 27-mW CMOS fractional-N synthesizer using digital compensation for 2.5-Mb/s GFSK modulation," *IEEE J. Solid-State Circuits*, vol. 32, pp. 2048–2059, Dec. 1997.
[9] G. R. Poole, "Subharmonic injection locking phenomenon in synchronous oscillators," *Electron. Lett.*, vol. 26, pp. 1748–1750, Oct. 1990.
[10] H. R. Rategh and T. H. Lee, "Superharmonic injection locked oscillators as low power frequency dividers," in *Symp. VLSI Circuits Dig.*, 1998, pp. 132–135.
[11] H. R. Rategh, H. Samavati, and T. H. Lee, "A 5 GHz, 32 mW CMOS frequency synthesizer with an injection locked frequency divider," in *Symp. VLSI Circuits Dig.*, 1999, pp. 12.1.1–12.1.4.
[12] ———, "A 5 GHz, 1 mW CMOS voltage controlled differential injection locked frequency divider," in *CICC Dig.*, 1999, pp. 24.5.1–24.5.4.
[13] B. Razavi, *RF Microelectronics*. Englewood Cliffs, NJ: Prentice-Hall, 1998.
[14] I. Schmideg, "Harmonic synchronization of nonlinear oscillators," *Proc. IEEE*, pp. 1250–1251, Aug. 1971.
[15] G. R. Sloan, "The modeling, analysis, and design of filter-based parametric frequency dividers," *IEEE Trans. Microwave Theory Tech.*, vol. 41, pp. 224–228, Feb. 1993.
[16] V. Uzunoglu, Z. Ma, and M. H. White, "Coherent phase-locked synchronous oscillator (graphical design technique)," *IEEE Trans. Circuits Syst.*, vol. 40, pp. 60–63, Jan. 1993.
[17] V. Uzunoglu and M. H. White, "The synchronous oscillator: A synchronization and tracking network," *IEEE J. Solid-State Circuits*, vol. SC-20, pp. 1214–1226, Dec. 1985.
[18] C. P. Yue, C. Ryu, J. Lau, T. H. Lee, and S. S. Wong, "A physical model for planar spiral inductors on silicon," in *Proc. Int. Electron Devices Meeting*, 1996, pp. 6.5.1–6.5.4.
[19] C. P. Yue and S. S. Wong, "On-chip spiral inductors with patterned ground shields for Si-based RF IC's," in *Symp. VLSI Circuits Dig.*, 1997, pp. 85–86.
[20] X. Zhang, X. Zhou, B. Aliener, and A. S. Daryoush, "A study of subharmonic injection locking for local oscillators," *IEEE Microwave Guided Wave Lett.*, vol. 2, pp. 97–99, Mar. 1992.

Hamid R. Rategh (S'98) was born in Shiraz, Iran, in 1972. He received the B.S. degree in electrical engineering from Sharif University of Technology, Iran, in 1994 and the M.S. degree in biomedical engineering from Case Western Reserve University, Cleveland, OH, in 1996. He currently is pursuing the Ph.D. degree in the Department of Electrical Engineering, Stanford University, Stanford, CA.

During the summer of 1997, he was with Rockwell Semiconductor Systems in Newport Beach, CA, where he was involved in the design of a CMOS dual-band, GSM/DCS1800, direct conversion receiver. His current research interests are in low-power radio-frequency integrated circuits design for high-data-rate wireless local-area network systems. He was a member of the Iranian team in the 21st International Physics Olympiad, Groningen, the Netherlands.

Mr. Rategh received the Stanford Graduate Fellowship in 1997.

Thomas H. Lee (S'87–M'87), for a photograph and biography, see p. 585 of the May 1999 issue of this JOURNAL.

A Family of Low-Power Truly Modular Programmable Dividers in Standard 0.35-μm CMOS Technology

Cicero S. Vaucher, Igor Ferencic, Matthias Locher, Sebastian Sedvallson, Urs Voegeli, and Zhenhua Wang

Abstract—A truly modular and power-scalable architecture for low-power programmable frequency dividers is presented. The architecture was used in the realization of a family of low-power fully programmable divider circuits, which consists of a 17-bit UHF divider, an 18-bit *L*-band divider, and a 12-bit reference divider. Key circuits of the architecture are 2/3 divider cells, which share the same logic and the same circuit implementation. The current consumption of each cell can be determined with a simple power optimization procedure. The implementation of the 2/3 divider cells is presented, the power optimization procedure is described, and the input amplifiers are briefly discussed. The circuits were processed in a standard 0.35 μm bulk CMOS technology, and work with a nominal supply voltage of 2.2 V. The power efficiency of the UHF divider is 0.77 GHz/mW, and of the *L*-band divider, 0.57 GHz/mW. The measured input sensitivity is >10 mVrms for the UHF divider, and >20 mVrms for the *L*-band divider.

Index Terms—CMOS integrated circuits, current-mode logic, frequency synthesizers, phase-locked loop, programmable frequency counter, programmable frequency divider.

I. INTRODUCTION

THE feasibility of RF functions implemented in CMOS technology has been demonstrated by a.o. the work presented in [1]–[3]. They show that the scaling of CMOS technologies to deep submicron has made CMOS a technological option for the low-gigahertz frequency range. However, for CMOS to become a *commercial* option for RF building blocks requires compliance to all trends of the consumer market: miniaturization, low cost, high reliability and long battery lifetime. Bulk CMOS technologies presently available satisfy the low cost and reliability trends by standard design practice. Complying to miniaturization and long battery lifetime, on the other hand, demands CMOS building blocks with low-power dissipation and good electromagnetic compatibility (EMC) characteristics. A critical RF function in this context is the frequency synthesizer, more particularly the programmable frequency divider. The divider consists of logic gates which operate at (or close to) the highest RF frequency. Due to the divider's complexity, high operation frequency normally leads to high power dissipation.

Other crucial aspects of the present-day consumer electronics industry are the short time available for the introduction of new products in the market, and the short product lifetime. On top of that, the lifespan of a given CMOS technology is also short, due to the aggressive scaling of minimum feature sizes. Short time-to-market demands architectures providing easy optimization of power dissipation, fast design time and simple layout work. High reusability, in turn, requires an architecture which provides easy adaptation of the input frequency range and of the maximum and minimum division ratios of existing designs.

The choice of the divider architecture is therefore essential for achieving low-power dissipation, high design flexibility and high reusability of existing building blocks. A modular architecture complies with these requirements, as shall be demonstrated in this paper. The focus of the paper is first on the truly modular architecture and on the implementation of the circuits. Then the power optimization procedure and the design of the input amplifier are briefly discussed. Finally, a collection of measured data and the conclusions are presented.

II. PROGRAMMABLE DIVIDER ARCHITECTURES

A. Architecture Based on a Dual-Modulus Prescaler

Fig. 1 depicts the divider architecture based on a dual-modulus prescaler [4], [5]. The design of the dual-modulus prescaler itself has been extensively treated in the literature [4], [6]–[10]. On the other hand, the architecture of Fig. 1 has some undesirable characteristics. One readily notices the lack of modularity of the concept: besides the dual-modulus prescaler, the architecture requires two additional counters for the generation of a given division ratio. The programmable counters—which are, in fact, fully programmable dividers, albeit not operating at the full RF frequency—represent a substantial load at the output of the dual-modulus prescaler, so that power dissipation is increased. Besides, the additional design and layout effort required for the programmable counters increase the time-to-market of new products. These properties led us to conclude that the dual-modulus-based architecture is not an interesting option for the realization of building blocks with high reusability, high flexibility, and short design time.

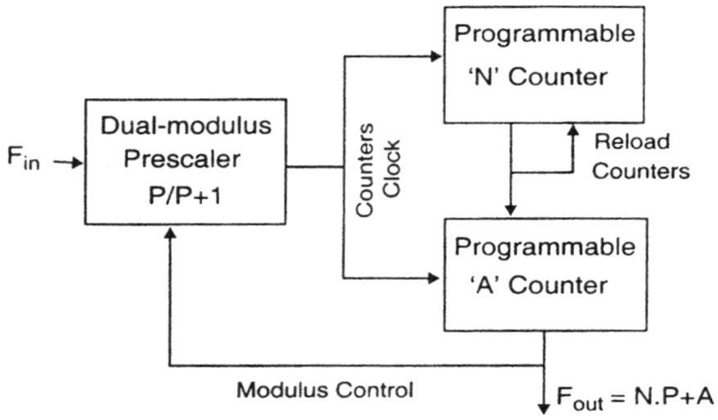

Fig. 1. Fully programmable divider based on a dual-modulus prescaler.

Manuscript received November 16, 1999; revised January 24, 2000.
C. S. Vaucher is with the Philips Research Laboratories, 5656AA Eindhoven, The Netherlands.
I. Ferencic, M. Locher, S. Sedvallson, U. Voegeli, and Z. Wang are with Philips Semiconductors Zurich, 8045 Zurich, Switzerland.
Publisher Item Identifier S 0018-9200(00)03878-6.

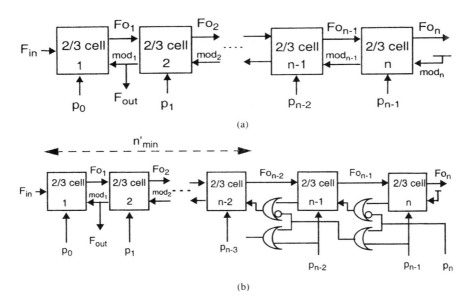

Fig. 2. Programmable prescaler. (a) Basic architecture. (b) With extended division range.

B. Programmable Prescaler Architectures

The "basic" programmable prescaler architecture is depicted in Fig. 2(a). The modular structure consists of a chain of 2/3 divider cells connected like a ripple counter [11]. The structure of Fig. 2(a) is characterized by the absence of long delay loops, as feedback lines are only present between adjacent cells. This "local feedback" enables simple optimization of power dissipation. Another advantage is that the topology of the different cells in the prescaler is the same, therefore facilitating layout work. The architecture of Fig. 2(a) resembles the one presented in [12], which is also based on 2/3 divider cells. Yet there are two fundamental differences. First, in [12] all cells operate at the same (high) current level. Second, the architecture of [12] relies on a common strobe signal shared by all cells. This leads to high power dissipation, because of high requirements on the slope of the strobe signal, in combination with the high load presented by all cells in parallel.

The programmable prescaler operates as follows. Once in a division period, the last cell on the chain generates the signal mod_{n-1}. This signal then propagates "up" the chain, being reclocked by each cell along the way. An active mod signal enables a cell to divide by 3 (once in a division cycle), provided that its programming input p is set to 1. Division by 3 adds one extra period of each cell's input signal to the period of the output signal. Hence, a chain of n 2/3 cells provides an output signal with a period of

$$
\begin{aligned}
T_{\text{out}} &= 2^n \cdot T_{\text{in}} + 2^{n-1} \cdot T_{\text{in}} \cdot p_{n-1} + 2^{n-2} \cdot T_{\text{in}} \cdot p_{n-2} \\
&\quad + \cdots + 2 \cdot T_{\text{in}} \cdot p_1 + T_{\text{in}} \cdot p_0 \\
&= (2^n + 2^{n-1} \cdot p_{n-1} + 2^{n-2} \cdot p_{n-2} \\
&\quad + \cdots + 2 \cdot p_1 + p_0) \times T_{\text{in}}.
\end{aligned}
\tag{1}
$$

In (1), T_{in} is the period of the input signal F_{in}, and $p_0, \cdots, p_{n-1}$ are the binary programming values of the cells 1 to n, respectively. The equation shows that all integer division ratios ranging from 2^n (if all $p_n = 0$) to $2^{n+1} - 1$ (if all $p_n = $

1) can be realized. The division range is thus rather limited, amounting to roughly a factor two between maximum and minimum division ratios.[1] The division range can be extended by combining the prescaler with a set-reset counter [13]. In that case, however, the resulting architecture is no longer modular.

The divider implementation presented in Fig. 2(b) extends the division range of the basic prescaler, whilst maintaining the modularity of the basic architecture [14]. The operation of the new architecture is based on the direct relation between the performed division ratio and the bus programmed division word $p_n, p_{n-1}, \cdots, p_1, p_0$. Let us introduce the concept of *effective length* n' of the chain. It is the number of divider cells that are effectively influencing the division cycle. Deliberately setting the *mod* input of a certain 2/3 cell to the active level overrules the influence of all cells to the right of that cell. The divider chain behaves as if it has been shortened. The required *effective length* n' corresponds to the index of the most significative (and active) bit of the programmed division word. Only a few extra OR gates are required to adapt n' to the programmed division word, as depicted on the right side of Fig. 2.

With the additional logic the division range becomes:

- minimum division ratio: $2^{n'_{\min}}$;
- maximum division ratio: $2^{n+1} - 1$.

We see that the minimum and maximum division ratios can be set independently, by choice of $n'_{\min}$ and n respectively. Subsequent changes in an optimized design can be realized with low risk. A somewhat similar technique, applied to an asynchronous programmable counter, is described in [9].

III. TRULY MODULAR PROGRAMMABLE DIVIDERS FAMILY

A. Realized Circuits

The modular architecture of Fig. 2(b) was applied in the realization of a family of fully programmable frequency dividers.

---

[1]In principle, it is also possible to divide by 3^n, but the gap between this value and the continuous division range makes it useless in standard synthesizer applications.

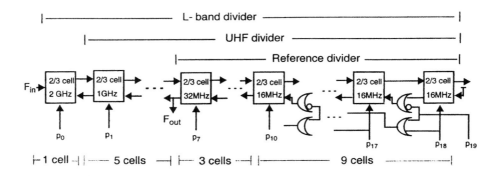

Divider	Minimum division ratio	Maximum division ratio
Reference	8	8191
UHF	256	262143
L - band	512	524287

Fig. 3. Family of truly modular programmable dividers, and corresponding division range of the different implementations.

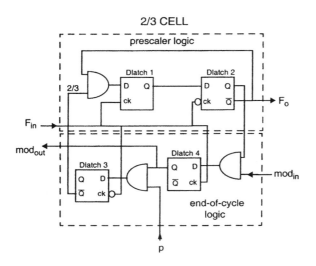

Fig. 4. Functional blocks and logical implementation of a 2/3 divider cell.

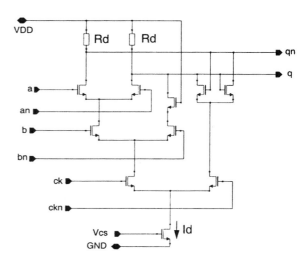

Fig. 5. SCL implementation of an AND gate combined with a latch function.

Three circuits were implemented: an 18-bit L-band divider, a 17-bit UHF divider, and a 12-bit reference divider. The architecture and the division range of the dividers is presented in Fig. 3. The L-band divider was used as the basis for the UHF and for the reference divider. The UHF divider consists of the same circuitry as the L-band divider, except for the first 2/3 cell, which was removed. The reference divider is simply the L-band divider stripped off its six high frequency cells.

B. Logic Implementation of the 2/3 Divider Cells

A 2/3 divider cell comprises two functional blocks, as depicted in Fig. 4. The *prescaler logic* block divides, upon control by the *end-of-cycle logic*, the frequency of the F_{in} input signal either by 2 or by 3, and outputs the divided clock signal to the next cell in the chain. The end-of-cycle logic determines the momentaneous division ratio of the cell, based on the state of the mod_{in} and p signals. The mod_{in} signal becomes active once in a division cycle. At that moment, the state of the p input is checked, and if $p = 1$, the end-of-cycle logic forces the prescaler

to swallow one extra period of the input signal. In other words, the cell divides by 3. If $p = 0$, the cell stays in division by 2 mode. Regardless of the state of the p input, the end-of-cycle logic reclocks the mod_{in} signal, and outputs it to the preceding cell in the chain (mod_{out} signal).

C. Circuit Implementation of the 2/3 Divider Cells

The use of standard rail-to-rail CMOS logic techniques makes the integration of digital functions with sensitive RF signal processing blocks difficult, due to the generation of large supply and substrate disturbances during logic transitions. Source coupled logic (SCL), often referred to as MOS current mode logic (MCML), has better EMC properties, because of the constant supply current and differential voltage switching operation [8]. Besides, SCL has lower power dissipation than rail-to-rail logic, for (very) high input frequencies [15].

The logic functions of the 2/3 cells are implemented with the SCL structure presented in Fig. 5. The logic tree combines an AND gate with a latch function. Three AND_latch circuits are

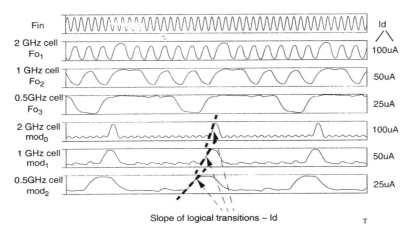

Fig. 6. Transient simulation of optimized L-band divider.

used to implement Dlatch1, Dlatch3, Dlatch4 and the AND gates of the 2/3 cells (see Fig. 4). Therefore, six logic functions are achieved, at the expense of three tail currents only. Dlatch2 is implemented as a "normal" D latch (without the differential pair connected to the b–bn inputs).

The nominal voltage swing is set to 500 mV in the high frequency (and high current) cells, and to 300 mV in the low current cells ($Id \leq 2 \; \mu$A). The voltage is generated by the tail current, set by the current source Id. and by the load resistances Rd.

D. Power Dissipation Optimization

The absence of long delay loops in the architecture of Fig. 2 enables fast and reliable optimization of power dissipation, since simulation runs may be done for clusters of two cells each time.

The critical point in the operation of the programmable prescaler are the divide by 3 actions [11]. There is a maximum delay between the *mod* and the clock signals in a given cell that still allows properly timed division by 3. The maximum delay is $\tau_{\max} = 1.5 * T_{\text{in}}$, where T_{in} is the period of the cell's input signal. The input frequency for each cell is scaled down by the previous one. As a consequence, the maximum allowed delay increases as one moves "down" the chain. As the delay in a cell is inverse proportional to the cell's current consumption (which is a property of current mode logic circuits), the currents in the cells may be scaled down as well.

The results of a transient simulation with the optimized high frequency cells of the L-band divider are presented in Fig. 6. The influence of current consumption on the slope of the digital signals (and hence on the time delay) is clearly observed. Table I presents the tail current and the resistance values of the optimized divider cells. Layout optimization took about three iteration cycles. Transient simulations, including extracted parasitics, showed that layout parasitics caused a decrease of about 30% in the highest operation frequency, when compared to the original simulations.

E. Input Amplifiers

The input amplifier provides the required amplification of the voltage-controlled oscillator (VCO) signal to "digital" levels, determined by the sensitivity specifications and by the divider

TABLE I
SCALING OF CURRENTS IN THE 2/3 DIVIDER CELLS

Cell	Nominal current Id (μA)	Nominal load res. Rd (kΩ)
2 GHz	100	5
1 GHz	50	10
500 MHz	25	20
250 MHz	12.5	40
125 MHz	6.25	80
62 MHz	3	150
32 MHz	2	150
16 MHz	1	300

circuitry. High input sensitivity enables the divider to be directly coupled to a wide range of VCO's, without the need for external (discrete) buffers. In addition, the input amplifier performs other important functions, which are listed below.

- It provides reverse isolation, to prevent the divider activity from "kicking-back" and disturbing the VCO.
- It provides single-ended to differential conversion of the (very often) single-ended VCO signal.
- It enables the VCO to be ac coupled to the divider function, and provides a signal to the first divider cell with the proper dc level.

The required amplification of the UHF amplifier, set by sensitivity requirements (-20 dBm), has been split into two differential stages. Each differential pair operates with 50-μA nominal current, and has load resistances of 14 kΩ. The L-band input amplifier is a scaled version of the UHF input amplifier. The tail currents were doubled, and the drain resistances were halved. The nominal low frequency small signal gain of the UHF amplifier is 26 dB; the gain of the L-band amplifier is 23 dB. Negative feedback from the output node to the input was implemented with 50 kΩ resistances. The feedback provides dc biasing to the first stage, and allows AC coupling of the VCO signal to the first differential pair.

IV. MEASUREMENTS

The control currents for the UHF and L-band dividers can be set externally, through input pins. The input amplifier current is

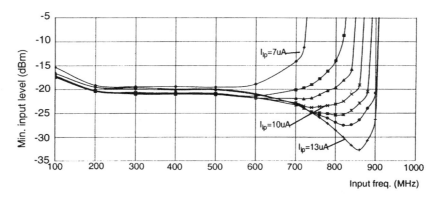

Fig. 7. Sensitivity of the UHF divider, for different divider current settings. Division ratio = 511, nominal current is $I_{lp} = 10\,\mu$A.

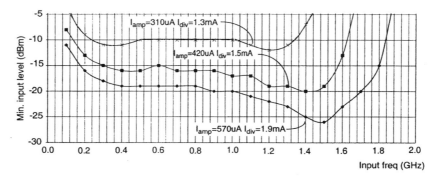

Fig. 8. Sensitivity curves of the L-band divider, for a few divider and amplifier current settings. Division ratio = 1023.

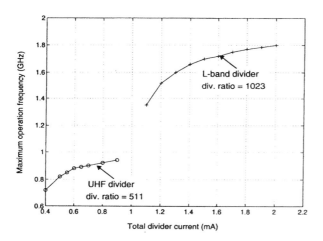

Fig. 9. Maximum operation frequency of the UHF and L-band dividers, as function of divider current consumption (excluding input amplifiers).

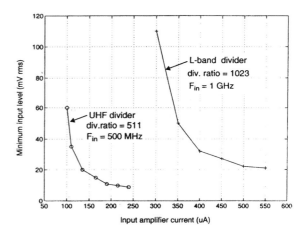

Fig. 10. Minimum input level for correct division of the frequency dividers, as function of the input amplifiers consumption.

controlled by the input current I_{bf}, and the 2/3 divider cell currents by input current I_{lp}. The curves presented in this section were obtained with the nominal supply voltage of 2.2 V, except where otherwise noted.

A. Input Sensitivity and Maximum Operation Frequency

Fig. 7 presents sensitivity curves of UHF divider, for different current settings of the control current I_{lp}. The nominal value of I_{lp} is 10 μA. Fig. 8 shows sensitivity curves of the L-band divider, for different current settings in the divider and input amplifier. Such as the UHF divider, the L-band divider is highly sensitive over a large frequency range. The circuits were driven by a differential input signal, carried over printed circuit board (PCB) strip-lines with a characteristic impedance of 50 Ω. The strip-lines were terminated with discrete resistances of 50 Ω to ground, which were set close to the input leads of the input amplifiers. The maximum operation frequencies of the UHF and L-band dividers, as function of the current consumption, are plotted in Fig. 9. The effect of the input amplifiers current consumption on the input sensitivities is displayed Fig. 10. Setting the UHF amplifier current to 230 μA yields a sensitivity value in excess of 10 mVrms.

The influence of the supply voltage on the maximum operating frequency was found to be small ($\approx 5\%$ for V_{dd} decreased from 2.2 V down to 1.8 V). It is interesting to mention that

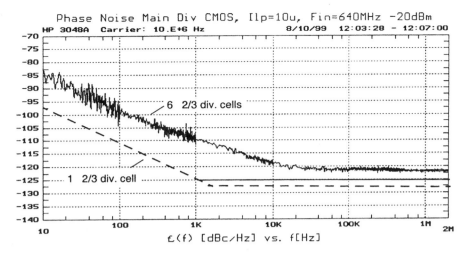

Fig. 11. Phase noise of the reference and UHF divider, measured at 10 MHz. - - - Reference divider ($I_{lp} = 10\ \mu$, $F_{in} = 20$ MHz, -10 dBm). — Reference divider ($I_{lp} = 10\ \mu$, $F_{in} = 20$ MHz, -20 dBm).

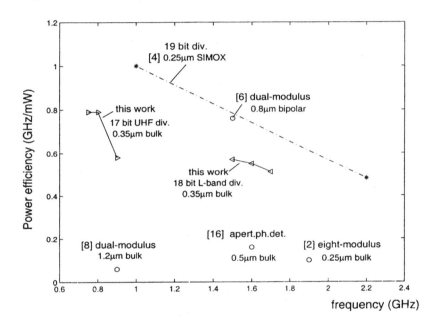

Fig. 12. Comparison of power efficiency (GHz/mW).

MCML circuits have been demonstrated to operate with supply voltages as low as 1.2 V [15], without significant loss of speed.

B. Phase Noise

The phase noise of the UHF and reference dividers was measured with a dedicated phase noise measurement system. We used coherent demodulation techniques (phase-locked loop configuration), and employed a low-noise 10 MHz signal source during the evaluation of the circuits. To facilitate the measurements, we implemented signal taps on the Fo output of certain cells on the divider chain. The UHF divider was provided with a tap on the Fo output of its sixth cell; the reference divider had the output of the first cell tapped.

Fig. 11 presents the phase noise of the UHF divider, with nominal settings for the supply current. The straight lines represent measured phase noise of the reference divider. We see a dependency of the noise floor of the reference divider on the level of the 20-MHz input signal. For the UHF divider, however, no dependency of the noise floor on the level of the input signal at 640 MHz was observed. An increase of 25% in current led to a change in noise floor from -122 dBc/Hz (nominal bias, $I_{lp} = 10\ \mu$A) to -124 dBc/Hz (with $I_{lp} = 12.5\ \mu$A). The noise floor of the reference divider went from -127.5 dBc/Hz down to -130 dBc/Hz, with increased bias. Fig. 11 shows that the high frequency cells of the UHF divider (see Fig. 3) contribute significantly to the phase noise, specially in the "$1/f$ region." An increase of $1/f$ noise of about 15 dB is observed, compared to the noise of the single reference divider's cell.

C. Power Efficiency

Fig. 12 presents the power efficiency of the UHF and L-band dividers, in comparison to recently published data on low-power dividers and tuning systems. Power efficiency is defined here as the ratio of the divider's maximum operation frequency to its power dissipation, with dimensions of GHz/mW. The authors have found that (most of) the dividers presented in the literature do not include an input amplifier. Therefore, only the current consumption of the "core" divider circuits is taken in the calculations. This leads to a fair comparison of the available data.

Refs. [2] and [8] describe prescalers implemented in bulk CMOS technology. Reference [16] proposes a new synthesizer architecture, where the divider is "powered-down" after lock has been achieved. Ref. [14] describes a fully programmable divider implemented in an ultrathin-film 0.25-μm CMOS/SIMOX process. The CMOS/SIMOX divider power efficiency is about 30% higher than the L-band divider's. Our divider, however, is implemented in a standard 0.35-μm bulk technology. The power efficiency of a bipolar dual-modulus prescaler [6] included as well, for technology benchmarking. Its power efficiency is similar to the power efficiency of the CMOS/SIMOX divider. The fully programmable dividers described here demonstrate that architectural choices and optimization procedures can take standard 0.35 μm CMOS to performance levels comparable to more expensive technologies, such as bipolar and CMOS/SIMOX processes.

V. CONCLUSION

This paper presented a truly modular and power-scalable architecture for low-power fully programmable frequency dividers. The flexibility and reusability properties of the architecture were demonstrated with the realization of a family of programmable divider circuits, consisting of the UHF divider (17 bits), the L-band divider (18 bits), and the reference divider (12 bits). The UHF and reference divider were implemented by simple removal of divider cells from the L-band circuitry. The implementation of the 2/3 divider cells was presented, and the power dissipation optimization procedure was described. To cope with EMC considerations, the dividers were implemented in CMOS SCL (current mode logic). The circuits were processed in a standard 0.35-μm bulk CMOS technology, and operate with a nominal supply voltage of 2.2 V. The power efficiency of the UHF divider is 0.77 GHz/mW, and of the L-band divider, 0.57 GHz/mW. The measured input sensitivity, including the input amplifiers, is >10 mVrms for the UHF divider, and >20 mVrms for the L-band divider.

ACKNOWLEDGMENT

The authors wish to thank G. van Veenendaal, of PS-Systems Laboratory, Eindhoven, The Netherlands, for evaluation work done on the programmable dividers. Many thanks go to D. Kasperkovitz and J. de Haas for the support provided during the project.

REFERENCES

[1] S. Wu and B. Razavi, "A 900 MHz/1.8 GHz CMOS receiver for dual-band applications," *IEEE J. Solid-State Circuits*, vol. 33, pp. 2178–2185, Dec. 1998.

[2] J. Craninckx and M. Steyaert, "A fully integrated CMOS DCS-1800 frequency synthesizer," *IEEE J. Solid-State Circuits*, vol. 33, pp. 2054–2065, Dec. 1998.

[3] Q. Huang *et al.*, "GSM transceiver front-end circuits in 0.25-μm CMOS," *IEEE J. Solid-State Circuits*, vol. 34, pp. 292–303, Mar. 1999.

[4] Y. Kado *et al.*, "An ultralow power CMOS/SIMOX programmable counter LSI," *IEEE J. Solid-State Circuits*, vol. 32, pp. 1582–1587, Oct. 1997.

[5] U. L. Rohde, *RF and Microwave Digital Frequency Synthesizers*. New York, NY: Wiley, 1997.

[6] T. Seneff *et al.*, "A sub-1 mA 1.6 GHz silicon bipolar dual modulus prescaler," *IEEE J. Solid-State Circuits*, vol. 29, pp. 1206–1211, Oct. 1994.

[7] J. Craninckx and M. Steyaert, "A 1.75 GHz/3 V dual-modulus divide-by-128/129 prescalar in 0.7 μm CMOS," *IEEE J. Solid-State Circuits*, vol. 31, pp. 890–897, July 1996.

[8] F. Piazza and Q. Huang, "A low power CMOS dual modulus prescaler for frequency synthesizers," *IEICE Trans. Electron.*, vol. E80-C, pp. 314–319, Feb. 1997.

[9] P. Larsson, "High-speed architecture for a programmable frequency divider and a dual-modulus prescaler," *IEEE J. Solid-State Circuits*, vol. 31, pp. 744–748, May 1996.

[10] J. Navarro Soares, Jr. and W. A. M. Van Noije, "A 1.6 GHz dual-modulus prescaler using the extended true-single-phase-clock CMOS circuit technique (E-TSPC)," *IEEE J. Solid-State Circuits*, vol. 34, pp. 97–102, Jan. 1999.

[11] C. S. Vaucher and D. Kasperkovitz, "A wide-band tuning system for fully integrated satellite receivers," *IEEE J. Solid-State Circuits*, vol. 33, pp. 987–998, July 1998.

[12] N. H. Sheng *et al.*, "A high-speed multimodulus HBT prescaler for frequency synthesizer applications," *IEEE J. Solid-State Circuits*, vol. 26, pp. 1362–1367, Oct. 1991.

[13] C. S. Vaucher, "An adaptive PLL tuning system architecture combining high spectral purity and fast settling time," *IEEE J. Solid-State Circuits*, vol. 35, pp. 490–502, Apr. 2000.

[14] C. S. Vaucher and Z. Wang, "A low-power truly modular 1.8 GHz programmable divider in standard CMOS technology," in *Proc. 25th Eur. Solid-State Circuits Conf.*, Sept. 1999, pp. 406–409.

[15] M. Mizuno *et al.*, "A GHz MOS adaptive pipeline technique using MOS current-mode logic," *IEEE J. Solid-State Circuits*, pp. 784–791, June 1996.

[16] A. R. Shahani *et al.*, "Low-power dividerless frequency synthesis using aperture phase detection," *IEEE J. Solid-State Circuits*, vol. 33, pp. 2232–2239, Dec. 1998.

A 1.75-GHz/3-V Dual-Modulus Divide-by-128/129 Prescaler in 0.7-μm CMOS

Jan Craninckx and Michiel S. J. Steyaert, *Senior Member, IEEE*

Abstract— A dual-modulus divide-by-128/129 prescaler has been developed in a 0.7-μm CMOS technology. A new circuit technique enables the limitation of the high-speed section of the prescaler to only one divide-by-two flipflop. In that way, a dual-modulus prescaler with the same speed as an asynchronous divider can be obtained. The measured maximum input frequency of the prescaler is up to 2.65 GHz at 5 V power supply voltage. Running at a power supply of 3 V, the circuit consumes 8 mA at a maximum input frequency of 1.75 GHz.

I. INTRODUCTION

RECENT publications have demonstrated the ever-increasing importance of CMOS RF circuits [1]–[3]. CMOS offers the big advantage of cheap processing and single-chip integration with digital building blocks. But to obtain the high frequencies required in modern telecommunication systems, at a reasonable power consumption, new circuit techniques must be developed.

The frequency synthesizer is one of the major building blocks for integrated transceivers. It requires a lot of effort to achieve the required specs in a standard CMOS process without any external components. Most frequency synthesizers are of the phase-locked loop (PLL) type, as shown in Fig. 1. The only two blocks operating at the full frequency are the voltage-controlled oscillator (VCO) and the prescaler. For reasons of low noise and low power, the VCO must be an oscillator based on the resonance frequency of an LC-tank with passive on-chip inductors [4], [3].

The prescaler must also operate at the full frequency. It divides the VCO output frequency (F_{out}) by a certain ratio to a low-frequency signal (F_{div}). This signal is locked by the PLL onto a very stable reference frequency (F_{ref}). The division ratio must be variable to allow a fast changing of the synthesized frequency. In the configuration of Fig. 1, the modulus is controlled by the overflow bit *of* of the accumulator. The synthesized frequency is then equal to

$$F_{\text{out}} = (N + n) \times F_{\text{ref}}. \tag{1}$$

So frequencies equal to a fractional multiple of the reference frequency can be synthesized, which explains the term *fractional-N frequency synthesizer*. High-speed dual-modulus prescalers are more difficult to construct than frequency dividers with a fixed division ratio, because the extra dual-modulus logic slows down the system [5].

Manuscript received December 4, 1995; revised February 11, 1996.
The authors are with the Departement Elektrotechniek, afd. ESAT-MICAS, Katholieke Universiteit Leuven, B-3001 Heverlee, Belgium.
Publisher Item Identifier S 0018-9200(96)04469-1.

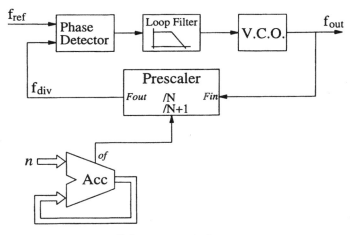

Fig. 1. PLL fractional-N frequency synthesizer.

In this paper, a high-speed CMOS dual-modulus prescaler is presented that does not suffer from this speed degradation. The topology used to obtain this result is explained in the next section. It is based on the 90° phase relationship between the master and the slave section of a Master/Slave (M/S) toggle flipflop. Section III shows the several high-speed building blocks that were designed to obtain a 1.75-GHz input frequency at a supply voltage of only 3 V. Section IV reports the measured performance of the circuit. The last section draws some conclusions.

II. PRESCALER TOPOLOGY

A conventional high-speed dual-modulus prescaler generally consists of a synchronous divide-by-4/5 part and an asynchronous divide-by-32 part as shown in Fig. 2 [5]. The synchronous divider is the only part of the circuit operating at the maximum input frequency. Most of the time, its control signal *Ctrl* is low, so the output frequency $F4$ is determined by the loop over the first two D-flipflops and equals one-fourth of the input frequency. This frequency is divided by 32 in the asynchronous divider to obtain an output frequency F_{out} equal to $F_{\text{in}}/128$. The divide-by-129 operation is enabled by setting the *Mode*-input high. When the outputs of all flipflops of the asynchronous divider are high, i.e., once every period of the output signal F_{out}, the *Ctrl*-signal becomes high. This causes the loop in the synchronous divider to be closed over three flipflops instead of two. This extra delay is equivalent to a divide-by-five operation. So the prescaler divides once by five and 31 times by four, which results in a division by 129.

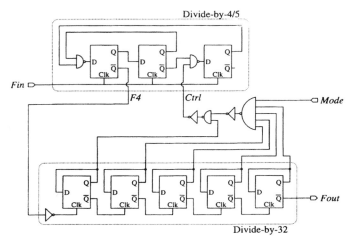

Fig. 2. Conventional dual-modulus prescaler architecture.

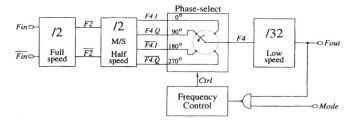

Fig. 3. New dual-modulus prescaler architecture.

The problems with this prescaler topology are, of course, situated in the synchronous divider. It is the only part of the circuit operating at the maximum frequency, the rest of the circuit runs at maximum one-fourth of the input frequency. The synchronous divider contains three high-frequency fully functional D-flipflops. These flipflops are the reason for a lot of power consumption and clock load. Moreover, the NAND-gates in the critical path of the synchronous divider loop will cause a decrease in the maximum input frequency. Clever design can reduce this effect by embedding the NAND-gate into the first stage of the flipflop, but can never eliminate it completely. Therefore, a dual-modulus prescaler will always have a smaller operating speed than an asynchronous frequency divider [5].

The new dual-modulus prescaler topology proposed here solves all these problems. The block diagram is shown in Fig. 3. It consists of a chain of seven pure divide-by-two circuits interrupted by a phase-select block. Since the frequency-limiting first stage is only one toggle-flipflop, input frequencies as high as asynchronous dividers can be obtained. The dual-modulus operation is based on the 90-degrees phase relationship between the outputs of the master and the slave of a Master/Slave (M/S) D-flipflop.

The new prescaler operates as follows. The differential input signal F_{in} is fed to a first high-speed divide-by-two flipflop. This flipflop is the only one operating at the full input frequency. It must not be a fully functional flipflop, but can be optimized for divide-by-two operation. The resulting signal $F2$ is once again divided by a second high-speed divide-by-two flipflop. For the dual-modulus operation, this flipflop must be of the master/slave kind as shown in Fig. 4(a). Four output signals result: the differential output of the master ($F4.I$ and $\overline{F4.I}$) and the differential output of the slave ($F4.Q$ and $\overline{F4.Q}$). When the *Mode*-input is low, the frequency-control block is disabled and its output signal *Ctrl* will be constant. Therefore, the phase-select circuitry simply picks one of its four input signals and connects it to the input of the asynchronous divide-by-32 block. This block is a chain of five divide-by-two flipflops operating at a relatively low speed (maximum one-fourth of the input frequency). The resulting

output frequency is thus a factor of $2 \times 2 \times 32 = 128$ smaller than the input frequency.

The divide-by-129 operation is enabled by setting the *Mode*-input high. The frequency-control block is now working. On every positive edge of the output signal F_{out}, the control signal *Ctrl* will be changed in such a way that the phase-select block will connect $F4$ to the signal that is 90° delayed with respect to the present signal. So when $F4$ is initially connected to $F4.I$, after the rising edge of F_{out}, a connection will be made to $F4.Q$. This is shown in Fig. 4(b). Since the signal $F4.Q$ lags the signal $F4.I$ by 90°, the signal $F4$ will be delayed. The delay is equal to 90° of a signal with period $4 \times T_0$, or equal to T_0. So also the output period is increased with this delay and now equals $128 \times T_0 + T_0$. The prescaler division factor is now 129, and a dual-modulus operation is obtained by using only divide-by-two toggle-flipflops.

This new prescaler architecture is thus inherently faster than a conventional one. Instead of three full-speed D-flipflops that must be able to perform all logical functions, only one full-speed divide-by-two flipflop is needed. No extra logic is needed as in a 4/5-divider. The next section will discuss the several building blocks used in the block diagram of Fig. 3.

III. CIRCUIT DESIGN

A. Full-Speed Divide-by-Two

The frequency-limiting building block in the architecture is of course the first divide-by-two block. The fastest standard CMOS D-flipflops up to now are the dynamic circuits of [6], [5], or the level-triggered latches of [7]. However, they both operate with a 5-V power supply and their speed drops rapidly at lower voltages.

For example, we can make an estimation of the maximum frequency obtainable with the dynamic flipflop of [6] by analyzing the oscillation frequency of a three-inverter ring oscillator. This analogy is shown in Fig. 5. The toggle-flipflop can be regarded as being a three-inverter ring oscillator with some additional control transistors that will regulate the oscillation frequency to a certain value, i.e., half of the input toggle frequency. These control transistors will slow down the circuit, so the maximum input frequency obtainable with the dynamic D-flipflop is less then twice the oscillation frequency of the ring oscillator.

In our 0.7-μm CMOS technology, the oscillation frequency of the ring oscillator is only 1.5 GHz at a 5-V power supply, and only 0.8 GHz at 3 V. This means the maximum input frequency of the toggle-flipflop will be lower than twice

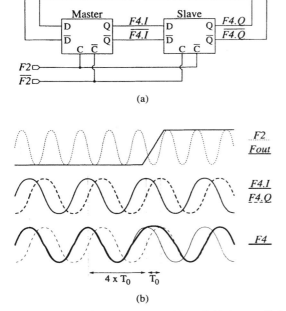

(a)

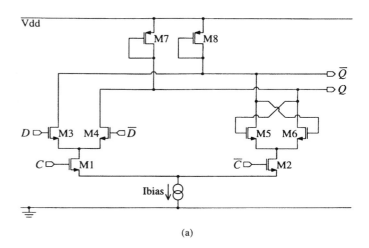

(a)

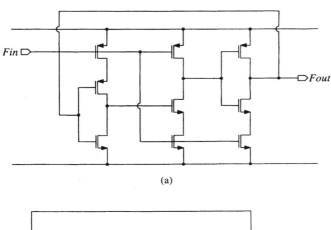

(b)

Fig. 4. Phase-selection principle: (a) Master/Slave divide-by-two flipflop and (b) waveforms.

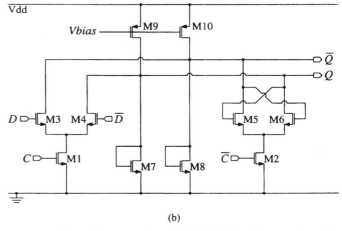

(b)

Fig. 6. (a) CMOS implementation of an ECL flipflop, (b) one section of the high-frequency D-flipflop.

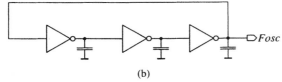

(b)

Fig. 5. Analogy between: (a) dynamic CMOS toggle-flipflop and (b) three-inverter ring oscillator.

this value, or 1.6 GHz at 3 V. To compare this with the results obtained in [6], the simulation is also performed for 1.2-μm design rules. For a 5-V supply, this renders an oscillation frequency 0.8 GHz. So the possible speed enhancement by going from a 1.2-μm process to 0.7 μm is completely eliminated by lowering the supply voltage from 5 V to 3 V.

In this design, the output swing of the flipflop is limited. The circuit is based on a standard M/S ECL D-flipflop. A straightforward CMOS implementation of one section is shown in Fig. 6(a). To increase the maximum toggle frequency, the current source I_{bias} can be omitted, and the source

of transistors $M1$ and $M2$ is connected directly to the ground rail. To drive these transistors sufficiently into and out of saturation, the input signal swing must be large enough. In this design, a 1.5-V_{ptp} input amplitude was used. Since this prescaler can be driven directly by the VCO, this poses no problem because the VCO output swing is maximized for low phase noise [4]. So the circuits exploit the speed enhancement possible by the reduction in voltage swing from input to output. This circuit was already implemented successfully in a 4-V fixed-modulus prescaler [3]. Simulations indicate a speed improvement of 20% over a standard implementation. Since the second divide-by-two operates at only half the frequency, the smaller output swing poses no problems.

To fit this circuit into a 3-V power supply, another measure must be taken. To avoid the V_{GS} voltage drop across the diode load transistors $M7$ and $M8$, they must be folded to ground. They can then be replaced with NMOS transistors, which is another small advantage because of smaller parasitic capacitances. This circuit is shown in Fig. 6(b).

The sizing of this circuit is not straightforward. Because the circuit is biased with the dc level of the VCO signal, care has to be taken to ensure proper operation over all process variations. This was investigated with numerous Monte-Carlo simulations. Therefore, a SPICE model library was developed

TABLE I
(a) Process Parameters. (b) Transistor Sizes of
the Full-Frequency Divide-by-Two Flipflop

	NMOS	PMOS	
V_T	0.75	-0.95	V
μ_0	470	160	cm^2/Vs
T_{OX}	15	15	nm

(a)

	W [μm]	L [μm]		W [μm]	L [μm]
M1	34.0	0.7	M2	12.0	0.7
M3	17.0	0.7	M4	17.0	0.7
M5	6.0	0.7	M6	6.0	0.7
M7	2.2	0.9	M8	2.2	0.9
M9	15.0	0.7	M10	15.0	0.7

(b)

that contained the dependencies of the model parameters on the possible process variations. This library was used the generate statistically random models, which were used to simulate the circuit under varying process conditions. The 3-σ variation of, e.g., the threshold voltage V_T was 150 mV. The model parameters were generated consistent with reality, e.g., the oxide thickness of the NMOS and the PMOS model was kept the same. The operating temperature was also varied.

So the circuit was not sized to obtain a maximum operation speed with typical model parameters, but to be robust for process variations and temperature changes. The operating point was chosen to guarantee a reasonable yield (e.g., 95%) at a somewhat reduced frequency. Of course, the yield under real processing conditions cannot be predicted safely with this procedure, but certainly an improvement over a design with only typical parameter simulations is achieved.

The resulting transistor sizes, together with a summary of the most important process parameters, are shown in Table I. The optimized circuit has a simulated maximum toggle frequency of 2 GHz at 3 V. The output amplitude is approximately 0.7 V_{ptp}.

B. Half-Speed Divide-by-Two

The second divide-by-two circuit must be an M/S toggle flipflop. Its design is based on the first divider, but the bias current source is not omitted. This is necessary to cope with the smaller input amplitude and the higher dc level of its input signal, which is the output signal of the first divider-by-two. Since this divider operates at half the input frequency, the speed enhancement which resulted from omitting the bias current source is no longer necessary. The output amplitude is approximately 0.5 V_{ptp}.

C. Phase-Selection

The phase-select circuitry is shown in Fig. 7. The selection of the correct signal is done in two stages. In a first stage both the in-phase (I) and quadrature (Q) signals are amplified

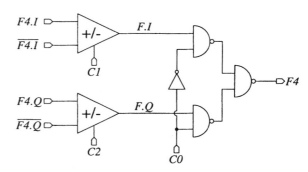

Fig. 7. Phase-selection circuitry.

in a differential-to-single-ended amplifier. This is necessary because the amplitude of the output signals of the second divide-by-two circuit is only 0.5 V_{ptp}. The amplifier can be switched between positive and negative amplification, thereby making a selection between the positive signals ($F4.I$ and $F4.Q$) or the negative signals ($\overline{F4.I}$ and $\overline{F4.Q}$). The control signals $C1$ and $C2$ are used for this.

The circuit schematic of this amplifier is shown in Fig. 8. By changing the polarity of the control signals C and $\overline{C}$, the bottom current mirror can be coupled to the rest of the circuit in a positive or a negative mode. Because of the high dc level of the input signals, this configuration was not possible with a PMOS current mirror directly above the input differential pair. The currents had to be mirrored down first.

A selection between the two remaining signals $F.I$ and $F.Q$ is made with three simple NAND-gates and the control signal $C0$ (see Fig. 7). A very important aspect of this circuit is what happens at the transfer from one selection to another. Uncareful design can cause spikes in the signal $F4$, resulting in an improper division by the divide-by-32 block. A smooth conversion in $F4$ must be guaranteed for all possible variations in processing or temperature and for all input frequencies. This is only important for the NAND-gates controlled by $C0$, because the control signals $C1$ and $C2$ are changed when the second stage of the circuit has selected the other signal (i.e., when $C0$ selects $F.I$ to be connected to $F4$, $C2$ is changed). The transients that occur when switching the amplifiers from one amplification mode to another are therefore not important.

For high input frequencies, the risk of creating spikes doesn't exist, because the NAND-gates do not react fast enough. However, for lower input frequencies, there is a possibility of spikes. A simulation of this is shown in Fig. 9. The following waveforms are shown from top to bottom: the signals $F.I$ and $F.Q$ (notice the 90° delay of $F.Q$), the control signal $C0$ and the resulting signal $F4$. The simulation is performed with fast transistor models and at an operating frequency of 250 MHz, which corresponds to an input frequency of 1 GHz. When $C0$ is low, the high-to-low transitions in $F4$ are determined by the signal $F.I$. After $0 \rightarrow 1$ change of $C0$, the transitions are determined by $F.Q$. This is indicated on the figure. However, if the control signal $C0$ has a very steep slope, $F.I$ can be deselected (and the signal $F4$ goes from high to low) before $F.Q$ is high enough. This will cause a negative spike in $F4$, as shown on the bottom of Fig. 9.

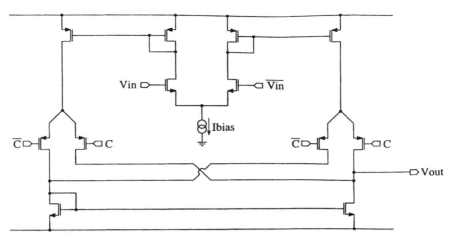

Fig. 8. Switchable amplifier for phase-selection.

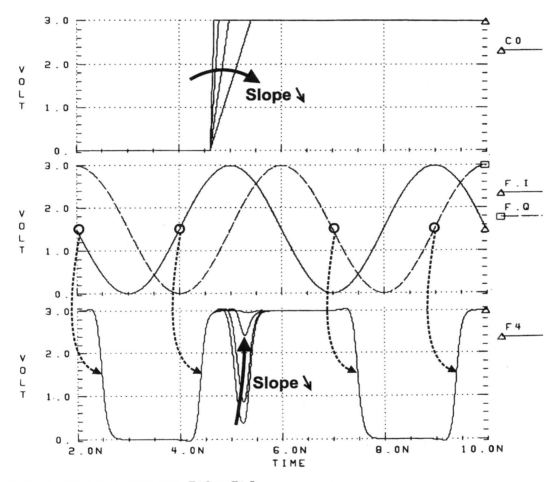

Fig. 9. Risk of spikes in $F4$ at the transition from $F4.I$ to $F4.Q$.

The presence of this effect is dependent on the exact time of arrival and on the slope of the change in $C0$. The spike only occurs for a very critical time of arrival. But since the delay in $C0$ cannot be controlled nor guaranteed, the possibility of having exactly that arrival time cannot be ruled out. One could think of controlling the control signal arrival time by inserting a clocked D-flipflop into the path of $C0$. This flipflop could be clocked by, e.g., the signal $F4$. But even in this case, the delay of the flipflop is not guaranteed, and the risk of spikes is not eliminated.

However, the problem is solved more easily by lowering the slope of $C0$. In Fig. 9 it can be seen that the spike disappears as the slope of $C0$ lowers. So a very small buffer inverter is used to steer the control signals, in order to limit the slope. As for process variations, the range over which the slope can vary without the risk of spikes is very large. So the buffer inverter is

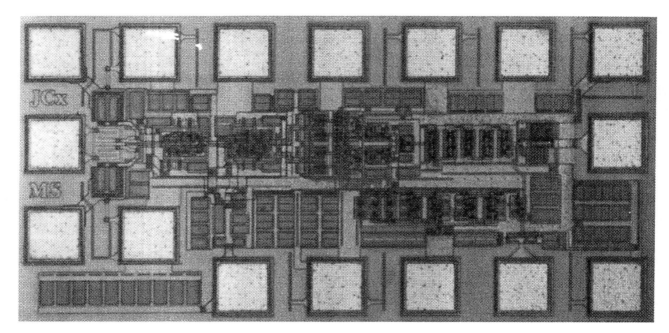

Fig. 10.   Microphotograph of the IC.

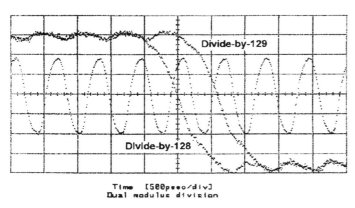

Fig. 11.   Input and output waveforms for both division factors.

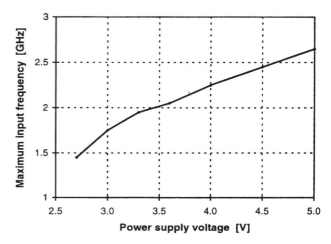

Fig. 12.   Measured maximum input frequency versus supply voltage.

designed such that a smooth transition in $F4$ is guaranteed for all possible arrival times of $C0$ and for all possible transistor processing variations and operating temperatures.

D. Low-Speed Divide-by-32

This divider is an asynchronous divider that consists of a chain of five toggle-flipflops. The maximum operating frequency is one-fourth of the input frequency, so the dynamic flipflop proposed in [6], as already shown in Fig. 5(a), can be used.

IV. MEASUREMENT RESULTS

The circuit has been implemented in a standard 0.7-μm CMOS process. A chip microphotograph is shown in Fig. 10. The die size, including bonding pads, is 1150×550 μm^2. Measurements were made at room temperature with the chip bonded on a ceramic substrate. The maximum input frequency of this technique is limited by the parasitic bondwire inductance and bondpad capacitance. The input signals were terminated with on-chip 200-Ω resistances. This results in a small overshoot in the input transfer characteristic in compari-

son with a 50-Ω termination, but the maximum input frequency is extended. The required differential input signal was easily made by inserting a small delay into one of the two inputs. This delay is not critical, since the applied phase shift (normally 180° for a perfect differential signal) can be varied by more than 50°. In this prototype, the bias current for the high-speed flipflops and the switchable amplifier is supplied externally.

The IC consumes, output buffer not included, 8 mA. This is divided over the several building blocks as follows: full-speed divide-by-two: 2.5 mA; half-speed divide-by-two: 1.5 mA; phase-selection: 2.5 mA; low-speed divide-by-32: 1.5 mA.

The measured maximum input frequency is up to 1.75 GHz at 3 V. The output signal for both division factors, together with the 1.75-GHz input signal, is shown in Fig. 11. It can clearly be seen that the divide-by-129 signal has a high-to-low transition that is delayed by one period of the input signal with respect to the divide-by-128 signal. This measured result is about 10% less than what could be expected from simulations.

358

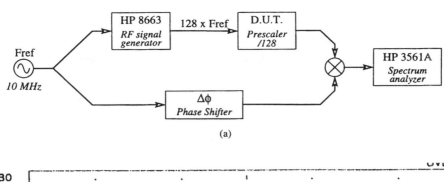

(a)

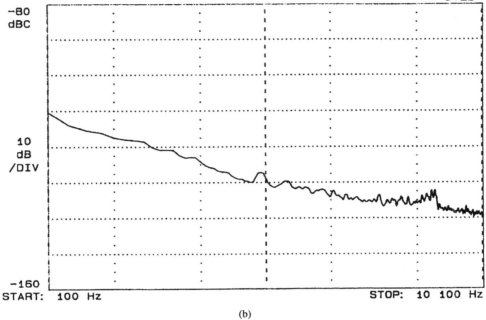

(b)

Fig. 13.   Phase noise measurement. (a) System configuration. (b) Measurement result.

This difference is probably due to insufficient modeling of the high-frequency effects of the transistors.

As for the influence of processing variations, these could not be measured completely since only a limited number of samples were available. However, an indication of the robustness of the circuit is given by the influence of, e.g., the power supply voltage or the externally supplied biasing current on the operating frequency. The maximum input frequency versus supply voltage is shown in Fig. 12. Although this circuit was optimized for 3-V operation, an input frequency as high as 2.65 GHz was measured with a 5-V power supply. This proves that this circuit is capable of operation in future mobile communication circuits, i.e., in the frequency range of 2–2.5 GHz.

The phase noise of this prescaler was measured with the HP3048A phase noise measurement system, with a configuration as shown in Fig. 13(a). The measurement results for a 1.28-GHz input signal are shown in Fig. 13(b). This frequency was chosen because a very stable 10-MHz reference was available. The measured output phase noise is −111 dBc/Hz at 100 Hz offset, −131 dBc/Hz at 1 kHz offset, and flattens for higher offset frequencies to a noise floor of −142 dBc/Hz. This corresponds to an equivalent input phase noise of −141 dBc/Hz at on offset frequency of 600 kHz, which is more than enough for the GSM or DCS1800 systems.

V. CONCLUSION

A high-frequency dual-modulus divide-by-128/129 prescaler has been developed in a standard 0.7-μm CMOS technology. A new circuit topology has reduced the full-frequency part of the prescaler to only one divide-by-two flipflop. Therefore, this prescaler can operate at the same speed as an asynchronous divider.

A new toggle-flipflop was developed, suitable for a 3-V power supply voltage. The maximum input frequency is 1.75 GHz. The power consumption is only 8 mA, and the output phase noise is as low as −131 dBc/Hz at 1 kHz offset.

REFERENCES

[1] J. Min, A. Rofourgan, H. Samueli, and A. A. Abidi, "An all-CMOS architecture for a low-power frequency-hopped 900-MHz spread spectrum transceiver," in *Proc. IEEE 1994 Custom Integrated Circuits Conf.*, May 1994, pp. 379–382.
[2] J. Crols and M. Steyaert, "A fully integrated 900-MHz CMOS double quadrature downconvertor," in *ISSCC Dig. Tech. Papers*, San Francisco, Feb. 1995, pp. 136–137.
[3] J. Craninckx and M. Steyaert, "A 1.8-GHz low-phase-noise voltage controlled oscillator with prescaler," *IEEE J. Solid-State Circuits*, vol. 30, pp. 1474–1482, Dec. 1995.
[4] _____, "Low-noise voltage controlled oscillators using enhanced LC-tanks," *IEEE Trans. Circuits Syst. II: Analog and Digital Signal Processing*, vol. 42, pp. 794–804, Dec. 1995.

[5] R. Rogenmoser, Q. Huang, and F. Piazza, "1.57-GHz asynchronous and 1.4-GHz dual modulus 1.2-μm CMOS prescalers," in *Proc. IEEE 1994 Custom Integrated Circuits Conf.*, May 1994, pp. 16.3.1–4.

[6] R. Rogenmoser, N. Felber, Q. Huang, and W. Fichtner, "A 1.16-GHz dual modulus 1.2-μm CMOS prescaler," in *Proc. IEEE 1993 Custom Integrated Circuits Conf.*, San Diego, May 1993, pp. 27.6.1–4.

[7] N. Foroudi and T. A. Kwasniewski, "CMOS high-speed dual-modulus frequency dividers for RF frequency synthesis," *IEEE J. Solid-State Circuits*, vol. 30, pp. 93–100, Feb. 1995.

Jan Craninckx was born in Oostende, Belgium, in 1969. He received the M.S. degree in electrical and mechanical engineering in 1992 from the Katholieke Universiteit Leuven, Belgium.

Currently, he is a Research Assistant at the ESAT-MICAS laboratories of the Katholieke Universiteit Leuven. He is working toward the Ph.D. degree on high-frequency low-noise integrated frequency synthesizers. For this work, he obtained a fellowship of the NFWO (National Fund for Scientific Research). His research interest are high-frequency integrated circuits for telecommunications.

Michiel S. J. Steyaert (S'85–A'89–SM'92) was born in Aalst, Belgium, in 1959. He received the masters degree in electrical-mechanical engineering and the Ph.D. degree in electronics from the Katholieke Universiteit Leuven, Heverlee, Belgium in 1983 and 1987, respectively.

From 1983 to 1986, he obtained an IWONL fellowship (Belgian National Foundation for Industrial Research) which allowed him to work as a Research Assistant at the Laboratory ESAT at the Katholieke Universiteit Leuven. In 1987, he was responsible for several industrial projects in the field of analog micropower circuits at the Laboratory ESAT as an IWONL Project Researcher. In 1988 he was a Visiting Assistant Professor at the University of California, Los Angeles. In 1989 he was appointed as a NFWO Research Associate, and since 1992, a NFWO Senior Research Associate at the Laboratory ESAT, Katholieke Universiteit Leuven, where he has been an Associate Professor since 1990. His current research interests are in high-performance and high-frequency analog integrated circuits for telecommunications systems and analog signal processing.

Dr. Steyaert received the 1990 European Solid-State Circuits Conference Best Paper Award, and the 1991 NFWO Alcatel-Bell-Telephone Award for innovative work in integrated circuits for telecommunications.

A 1.2 GHz CMOS Dual-Modulus Prescaler Using New Dynamic D-Type Flip-Flops

Byungsoo Chang, Joonbae Park, and Wonchan Kim

Abstract—A 1.2 GHz dual-modulus prescaler IC fabricated with 0.8 μm CMOS technology is presented in this paper. The dual-modulus prescaler includes a synchronous counter (divide-by-4/5) and an asynchronous counter (divide-by-32). A new dynamic D-flip-flop (DFF) is developed for high-speed synchronous counter. The maximum operating frequency of 1.22 GHz with power consumption of 25.5 mW has been measured at 5 V supply voltage.

I. Introduction

A HIGH-SPEED prescaler IC is essential for wide-band frequency synthesizers employing the pulse swallow method. In the pulse swallow phase-locked loop (PLL), the fast-varying output of voltage-controlled oscillator (VCO) is fed to the prescaler directly. Accordingly, VCO and prescaler become limiting factors in determining the operating speed of the PLL. A prescaler operating at several hundred megahertz or above was traditionally built with Si-bipolar or GaAs technology [1], [2]. However, with the scaling down of MOS transistors in their feature size, CMOS prescaler IC's and frequency dividers operating above gigahertz range have been introduced in previous works [3]–[5].

Typically, a dual-modulus prescaler is constructed with two parts—a synchronous counter and an asynchronous counter—to reduce the number of flip-flops operating at high frequency and the power consumption. In this two-stage approach, the synchronous counter is the critical part in determining the speed of a prescaler. The optimization of D-flip-flop (DFF's) in the synchronous counter is essential to increase the operating frequency of the prescaler. The high-speed operation of MOS transistors is limited by their low transconductance. Therefore, new circuit techniques such as dynamic and sequential circuit techniques must be used in designing the synchronous counter. True-single-phase-clocked (TSPC) dynamic DFF's and differential latches are examples [6], [7].

In this paper, a high-speed dual-modulus prescaler (divide-by-128/129) using 0.8 μm CMOS technology is described. We propose a new dynamic DFF adopting ratioed logic technique for the synchronous counter (divide-by-4/5) in the prescaler. The simplified structure of the DFF reduces the effective capacitance of internal and external nodes, which leads to the reduction of the power

Manuscript received June 5, 1995; revised August 28, 1995.
The authors are with the Department of Electronics Engineering, Seoul National University, Shilim-dong, Kwanak-gu, Seoul 151-742, Korea.
Publisher Item Identifier: S 0018-9200(96)03396-3.

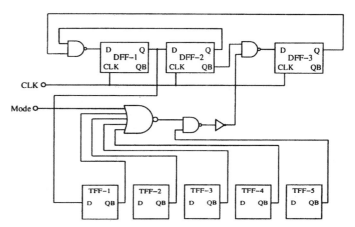

Fig. 1. Block diagram of the dual-modulus prescaler.

consumption as well as the propagation delay. The measured maximum operating frequency and power consumption of the prescaler are 1.22 GHz and 25.5 mW at the supply voltage of 5 V.

II. Circuit Design

Fig. 1 shows a schematic of the designed prescaler IC in this paper. It consists of three DFF's, five toggle-flip-flops (TFF's), input and output circuits, and several gates. The DFF's and the NAND gates form a divide-by-4/5 counter, and the chained TFF's form a divide-by-32 counter. The fractional division ratio is selected according to a mode signal. When the "mode" is one (zero), the division ratio is set to 128 (129). A pseudo-nMOS NOR gate is used for the five input NOR gate to reduce the load capacitance of the TFF's.

A. Synchronous Counter

As mentioned above, the operating speed of prescalers is limited by that of the DFF's. The operating speed of the TSPC DFF in Fig. 2 is severely affected by the large RC delay due to the stacked structures. The effect of transistor sizing is not so evident, because most transistors are drivers and loads at the same time. The propagation delay can be reduced by increasing the size of clocked transistors, but it increases the load capacitance to the clock driver and thus increases the power consumption. A level-triggered latch instead of an edge-triggered flip-flop can be used to achieve a high-speed prescaler [7]. However, the prescaler using such sequential circuits operates at the

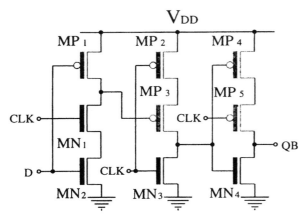

Fig. 2. TSPC DFF proposed by Yuan and Svensson in [6].

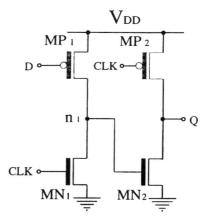

Fig. 3. The circuit schematic of a ratioed latch.

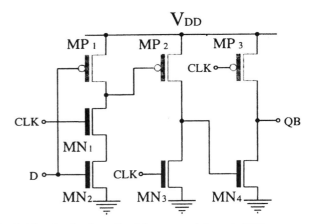

Fig. 4. The circuit schematic of the proposed dynamic DFF.

limited range of input frequency due to the constraint of the clock period.

In high-frequency operating mode, the concept of zero standby power consumption has little meaning because the transition time of a signal takes a considerable portion of a clock period. Therefore, a ratioed logic can replace a ratioless logic without paying much penalty on the power consumption. Fig. 3 shows the schematic of the proposed ratioed latch. In the dynamic DFF shown in Fig. 2, the C^2MOS and precharge stages are necessary to prevent the transparent path from the input D to output Q; hence stacked structures are unavoidable. On the other hand, two cascaded pseudo-nMOS inverters in the ratioed latch replace the N-precharge stage and the N-C^2MOS stage in Fig. 2. The ratioed latch is in the hold mode (evaluation mode) when the clock is high (low).

The hold mode operation is as follows. The W/L ratios of MN_1 and MP_1 are determined so that the voltage of the node n_1 remains below V_{Tn} of MN_2, regardless of the input D, during the high period of the clock. Thus, pull-up or pull-down of the output Q does not happen, because MN_2 and MP_2 remain in cut-off when the clock is high. Therefore, the signal path from the input D to the output Q is not transparent and the latch enters into the hold mode.

When the clock changes its state from high to low, the latch enters into the evaluation mode. If the input D is low, node n_1 is pulled up to V_{DD} only by MP_1. The pull-down strength of MN_2 must be sufficiently larger than the pull-up strength of MP_2 so that the output low voltage $V_{OL,Q}$ is lower than the input low voltage (V_{IL}) of the following gates. If the input D remains high, the node n_1 remains at ground level. The output Q is pulled up to V_{DD} only by MP_2, because the transistor MN_2 remains in cut-off. When the latch in Fig. 3 is in the evaluation mode (i.e., CLK = low), the high-to-low transition of the input D can overwrite the stored data in the output Q. The N-C^2MOS stage must be connected to the input of the latch to prevent the pull-down of the latch input D during the evaluation mode. Therefore, the output of the latch may change its state only one time during a single clock period. Fig. 4 shows the circuit schematic of the proposed

negative edge-triggered DFF. In contrast to the TSPC DFF in Fig. 2, only seven transistors are necessary for the basic flip-flop function, while nine transistors are needed to perform the same function in the case of the TSPC DFF. The number of transistors driven by the clock signal is also reduced, which increases the driving capability of the clock driver and decreases the power consumption substantially.

Toggle-flip-flop (TFF) circuits are simulated to compare the maximum operating frequency of the proposed DFF and the TSPC DFF. Fig. 5 shows the relationships of the maximum input frequency versus the supply voltage. The power consumptions of the TFF's are also shown in Fig. 6. The simulation results confirm the advantage of the proposed DFF over the TSPC DFF. Unlike the static DFF, the proposed DFF has the lower limit on the operating frequency due to the leakage current. The divide-by-two function of the proposed DFF can work even at 10 kHz when simulated at 5 V supply. This bound is low enough for the aimed operating frequency of the DFF's in the dual-modulus prescaler which is higher than several hundred MHz.

B. Asynchronous Counter

As shown in Fig. 1, the output of the synchronous counter is connected to the first TFF in the asynchronous

362

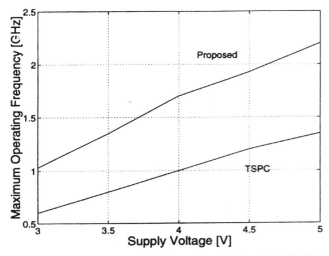

Fig. 5. Simulated operating speed of the proposed DFF and the TSPC DFF.

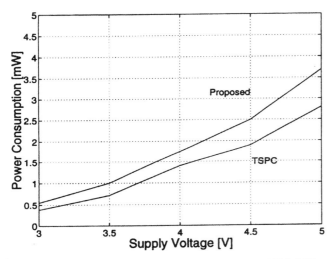

Fig. 6. Power consumption of the proposed DFF and the TSPC DFF.

counter. The operating frequency of the divide-by-32 counter is four or five times lower than that of the divide-by-4/5 counter. Therefore, the optimization of TFF's in this stage is focused on the reduction of power consumption. In the asynchronous divider, where the transition period takes only a small portion in a whole period, the use of ratioed logic is not advantageous because of the static power consumption. To alleviate the load of the synchronous counter, only the first TFF in the divider chain is formed by the proposed DFF. The other TFF's are TSPC DFF's in the ripple counter configuration to diminish the overall power consumption.

C. Input and Output Circuit

The input of the prescaler is supplied from a buffered output of a three-stage ring oscillator integrated with the prescaler to ease measurements. The output buffer is a simple inverter chain to drive the pad and external parasitics.

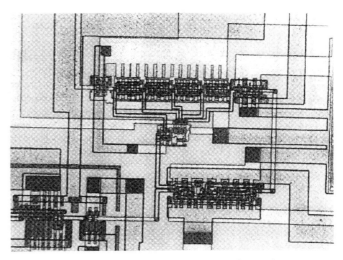

Fig. 7. The chip microphotograph of the fabricated prescaler.

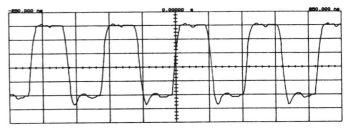

Fig. 8. Measured waveform of the prescaler divided by 128 (f_{in} = 1.22 GHz). (Vertical scale: 1 V/div; horizontal scale: 50 ns/div.)

III. EXPERIMENTAL RESULTS

The dual-modulus prescaler has been fabricated with 0.8 μm n-well CMOS process. The threshold voltages of nMOS transistors and pMOS transistors are about 0.7 V and 1.2 V, respectively. Fig. 7 shows the microphotograph of the fabricated prescaler. The NAND gates in the synchronous counter are merged with the DFF's to enhance the performance. The active area of the synchronous counter is 35 μm $\times$ 140 μm and that of the asynchronous counter is 40 μm $\times$ 220 μm. The maximum operating frequency and the power dissipation of the prescaler are 1.22 GHz and 25.5 mW at 5 V single supply, respectively. The high-speed nature of the prescaler is mainly attributed to the ratioed DFF's exhibiting the properties of small RC-time constant, namely small propagation delay, and small loading to the prescaler input. Figs. 8 and 9 show the oscilloscope traces of the prescaler (f_{in} = 1.22 GHz) output at 5 V supply. The divide-by-4/5 counter, divide-by-32 counter, and the gates consume 62%, 10%, and 28% of the total power consumption according to HSPICE simulation, respectively. The power consumption in the input and the output buffers is not included in the above calculation. Most of the power consumption in the gates are drained from the pseudo-nMOS NOR gate and could be reduced by employing a static NOR gate. Fig. 10 shows the maximum operating frequency as a function of supply voltage, varied from

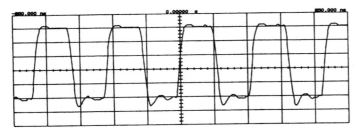

Fig. 9. Measured waveform of the prescaler divided by 129 (f_{in} = 1.22 GHz). (Vertical scale: 1 V/div; horizontal scale: 50 ns/div.)

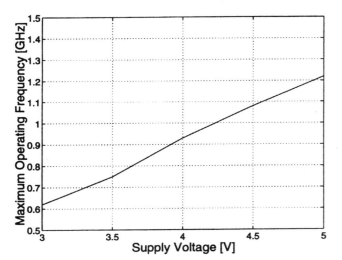

Fig. 10. Measured operating speed of the prescaler as a function of supply voltage.

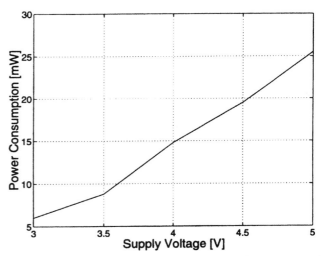

Fig. 11. Measured power consumption of the prescaler versus supply voltage.

nMOS or pMOS transistors. The fabricated prescaler operates up to 1.22 GHz with the power dissipation of 25.5 mW at 5 V supply.

3–5 V. The power dissipation versus supply voltage at maximum operating condition is shown in Fig. 11.

IV. CONCLUSIONS

A high-speed dual-modulus prescaler ($\div 128/129$) has been presented using 0.8 μm CMOS technology. The dynamic latch without stacked devices and pass gates is adopted to the dynamic DFF's for the synchronous divider. The ratioed logic technique used for the latch builds a dynamic data storage not relying on the series gating of

REFERENCES

[1] Y. Yamauchi, O. Nakajima, K. Nagata, and M. Hirayama, "A 15-GHz monolithic two-modulus prescaler," IEEE J. Solid-State Circuits, vol. 26, no. 11, pp. 1632–1636, 1991.
[2] H. P. Singh, R. A. Sadler, W. J. Tanis, and A. N. Schenberg, "GaAs prescalers and counters for Fast-settling frequency synthesizers," IEEE J. Solid-State Circuits, vol. 25, no. 2, pp. 239–245, 1990.
[3] H.-I Cong, J. M. Andrews, D. M. Boulin, S.-C. Fang, S. J. Hillenius, and J. A. Michejda, "A 2-GHz CMOS dual-modulus prescaler IC," in ISSCC Dig. Tech. Papers, Feb. 1988, pp. 138–139.
[4] R. Rogenmoser, Q. Huang, and F. Piazza, "1.57 GHz asynchronous ans 1.4 GHz dual-modulus 1.2 μm CMOS prescalers," in Proc. CICC, May 1994, pp. 16.3.1–16.3.4.
[5] B. Razavi, K. F. Lee, and R.-H. Yan, "A 13.4-GHz CMOS frequency divider," in ISSCC Dig. Tech. Papers, Feb. 1994, pp. 176–177.
[6] J. Yuan and C. Svensson, "High-speed CMOS circuit technique," IEEE J. Solid-State Circuits, vol. 24, no. 2, pp. 62–70, 1989.
[7] N. Foroudi and T. A. Kwasniewski, "CMOS high-speed dual-modulus frequency divider for RF frequency synthesis," IEEE J. Solid-State Circuits, vol. 30, no. 2, pp. 93–100, 1995.

High-Speed Architecture for a Programmable Frequency Divider and a Dual-Modulus Prescaler

Patrik Larsson

Abstract—We present a prescaler architecture that is suitable for high-speed CMOS applications. We apply the architecture to a 4/5 and an 8/9 dual-modulus prescaler and obtain a measured maximum clock frequency of 1.90 GHz in a standard 0.8 μm CMOS bulk process. This is 13% faster than the traditional prescaler architecture keeping the same power consumption. We also apply the key part of the prescaler to a divide-by-N circuit reaching 1.75 GHz. This is three times faster than any previously reported CMOS implementation and comparable to GaAs implementations.

I. INTRODUCTION

ONE of the high-frequency building blocks in a communication system is a frequency synthesizer including a frequency divider. Traditionally, the divider has been realized with a high-speed technology such as bipolar or GaAs. CMOS is the cheapest technology today and seems to be the cheapest alternative for the foreseen future, indicating that CMOS implementations are advantageous.

There are three types of frequency dividers: cascaded divide-by-two stages, dual-modulus prescaler, and programmable divider (also called divide-by-N circuit). Many applications require a programmable division ratio, excluding the use of divide-by-two stages. In this paper, we present a novel architecture for both a dual-modulus prescaler and a divide-by-N circuit.

Some CMOS prescalers have been presented [1]–[3] with a maximum operating frequency not far behind bipolar [4], [5] or GaAs [6], [7] implementations. See [5], [7], and [8] for further references. These circuits have used advanced processing and/or special circuit techniques in combination with a standard frequency divider architecture. In this paper we focus on improving the architecture, and we compare our results with the traditional architecture using the same processing technology and circuit technique.

II. DUAL-MODULUS PRESCALER

All prescalers that we have found are based on a shift register ring with logic similar to Fig. 1(a). This is an 8/9 prescaler [2], which easily can be modified to a 4/5 prescaler [1], [6]. Our approach is to preprocess the

Manuscript received May 4, 1995; revised August 26, 1995.

The author was with the Department of Physics and Measurement Technology, Linköping University, S-581 83 Linköping, Sweden. He is now with AT&T Bell Laboratories, Holmdel, NJ 07733 USA.

Publisher Item Identifier S 0018-9200(96)03397-5.

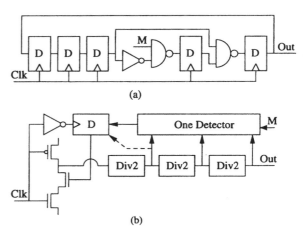

Fig. 1. Dual-modulus 8/9 prescalers. (a) Based on a shift register ring. (b) With preprocessing of the clock signal.

clock signal and then use cascaded divide-by-two stages as sketched in Fig. 1(b). The one detector gives a low output pulse when the M signal and the outputs of all divide-by-two stages are high. This pulse is delayed one clock cycle and synchronized with the inverted Clk signal in the D flip-flop. A low output will prohibit one negative pulse of the Clk signal from reaching the first divide-by-two stage. After canceling one clock pulse, it will take eight clock pulses before the detector output goes low, giving a division ratio of nine. By setting M low, the detector output is always high and the prescaler divides by eight. The circuit operates as an 8/9 prescaler, but can easily be modified to a $2^b/2^b + 1$ prescaler by adding/removing divide-by-two stages and extending/shortening the detector.

Since the rightmost divide-by-two stage reaches one before the other stages, the one detector can combine early arriving information first [9], and the critical path in the detector is the detection of the leftmost bit in the divider chain. In principle, this bit should be sent to the detector as in Fig. 1(b). To enhance speed, the detection of the leftmost bit was done with very simple logic inside the D flip-flop as the dashed line indicates. Since the D flip-flop has a small load of only one transistor, it has roughly the same speed as the first divide-by-two stage. This indicates that the maximum operating frequency of the clock preprocessing prescaler is nearly the same as that of a chain of cascaded divide-by-two stages. The following divide-by-two stages operate at a lower frequency than that of the first stage, which makes it possible to use smaller transistor sizes reducing the load of the first stage. Therefore, transistor sizing is an efficient method for speed enhancement and power reduction of this architecture.

The gating of the clock signal in Fig. 1(b) utilizes dynamic logic since the input of the first divide-by-two stage is floating when *Clk* is high and the output of the *D* flip-flop low. If the circuit is to be clocked at very low speed, dynamic logic is not suitable. We can simply eliminate the dynamic mode by adding a *P* transistor driven by the output of the *D* flip-flop, i.e., the gating of the clock signal is done with a *Nand* gate instead of the three-transistor structure in Fig. 1(b).

We implemented the shift register ring in Fig. 1(a) and the clock preprocessing structure in Fig. 1(b) for both an 8/9 and a 4/5 prescaler. We used the same circuit technique for both prescalers. All transistor widths were limited to 20 μm except for some transistors in the clock buffers and for a few clocked transistors in the ring structure that was sized larger. The prescalers were implemented in a standard 0.8 μm CMOS process for which relevant process parameters are given in Table I. The inverter delay was measured in a five-stage ring oscillator which is characterized in Fig. 2(a). The inverter delay was smaller for a three-stage oscillator with identical delay stages as seen in Fig. 2(b). We believe this is caused by the internal nodes not having full swing in the three-stage oscillator. Each delay stage was an inverter with a 100 μm wide *P*-transistor and a 52 μm wide *N*-transistor with no sharing of source and drain diffusion areas.

Table II shows the measurement results from the two prescaler structures, and their measured maximum operating frequency are plotted as function of power supply voltage for two different chips in Fig. 3(a) and (b). The prescalers work properly for both division ratios at the maximum operating frequency. The maximum operating frequency is comparable to that of [1]–[3] taking into account the different process technologies and flip-flop implementations. To avoid high-speed I/O signals, a divide-by-16 circuit was added at the output of the prescalers. This was implemented with the same dynamic circuit technique as the prescalers. Its minimum output frequency was measured to be lower than 100 kHz, and we conclude that the ring structure can be clocked at a frequency of 100 kHz. In the clock preprocessing architecture, the last stage is clocked at a much lower frequency than the input clock leading to a minimum input frequency of $2^b \times 100$ kHz, where 2^b is the division ratio. The minimum clock frequency limit can be alleviated by using static CMOS for the MSB's operating at lower speed than the high-speed dynamic LSB stages. The figures for power consumption in Table II do not include the clock power that is considerably larger for the shift register ring than the clock preprocessing architecture. Simulations of extracted layout predicted a power consumption of 22–28 μW/MHz for the prescalers including the clock power. We only give approximate results of power consumption since noise coupling between internal VCO's on the chip reduced the accuracy of the power measurements as explained in the following. Several different types of divid-

TABLE I
CMOS PROCESS PARAMETERS

Power supply	Vdd = 5V
Minimum gate length	L = 0.8μm
Gate oxide thickness	t_{ox} = 16nm
Measured inverter delay at Vdd=5V	t_{inv} = 114ps

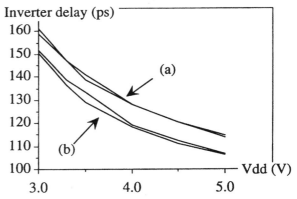

Fig. 2. Measured inverter delay in ring oscillator for two different chips as function of power supply voltage. (a) Five-stage and (b) three-stage oscillator.

ers were laid out on a single chip sharing a single power supply. Each divider had its own individually controlled VCO. The only way of estimating the power consumption of a specific divider was to measure the total power consumption as function of the input frequency of that divider and assume that the power consumption is linearly proportional to the input frequency as is common for CMOS circuits. However, when changing the speed of the VCO feeding the divider for which power consumption was to be estimated, noise coupled from the VCO to other VCO's changing the input frequencies of the other dividers slightly. Therefore, the change in power consumption was not only due to the frequency change of the divider for which we intended to measure the power consumption.

III. AN ASYNCHRONOUS DIVIDE-BY-*N* CIRCUIT

A programmable frequency divider is often built from a counter with load or reset and a detection circuit as in Fig. 4. By using a backward counter, detecting the 0 state and loading the counter with the division ratio, the propagation delay of the detector can be made small [9]. By realizing that the higher order bits (MSB's) of the counter reach the 0 state before the lower order bits (LSB's), we can combine early arriving information first and we only need quick detection of the LSB's. Since the MSB's reach zero earlier than the LSB's, we can also let the toggling of MSB's lag behind the LSB's [10]. Therefore, an asynchronous ripple counter consisting of cascaded programmable divide-by-two stages is sufficient. This avoids the carry propagation that is limiting the speed of synchronous counters. The dividing function is also obtained with a forward counter, but then the counter is loaded with the one's complement of the binary word that represents

TABLE II
MEASURED PRESCALER CHARACTERISTICS

Type of prescaler	Division ratio	Speed (GHz)	P/f @ Vdd=5V (μW/MHz)	Area (μm²)
Shift register ring	4/5	1.68	≈12-16*	70x150
[1, 2 and others]	8/9	1.68	≈12-16*	100x140
Clock preprocessing	4/5	1.90	≈16-20*	70x140
[This work]	8/9	1.90	≈16-20*	80x160

* Approximate measurement excluding input clock power.

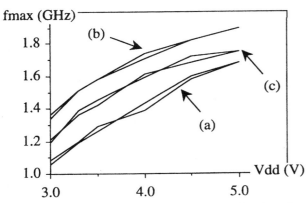

Fig. 3. Measured maximum operating frequency as function of power supply voltage. (a) Traditional dual-modulus prescaler in Fig. 1(a). (b) Dual-modulus prescaler in Fig. 1(b). (c) Divide-by-*N* circuit in Fig. 7.

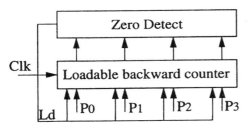

Fig. 4. Programmable frequency divider based on a loadable counter.

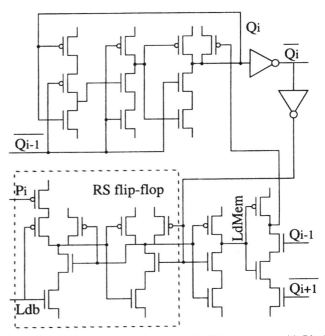

Fig. 5. Transistor schematic of a single bitcell in a counter with Ripple Inhibited Load.

the division ratio. In the following we will discuss a forward counter that detects the 11 . . . 11 state to perform the loading.

A problem for an asynchronous counter is the loading. In a forward counter, the divide-by-two stages consist of a negative edge-triggered flip-flop. After detecting the 1 state, the counter is loaded with a word that contains some zeros. If bit i is loaded with a zero, bit $i + 1$ will toggle since it triggers on negative edge. Therefore, the load signal needs to be applied to all bits simultaneously to prohibit some bits from toggling. Thus, the load signal is global and has high fanout indicating a speed bottleneck.

We remove the speed limiting requirement of having a global load signal with a simple idea; instead of loading some bits of the counter with zeros from the 1 state, we let the counter go one step beyond 11 . . . 11, which is 00 . . . 00. From this state we load some bits with ones. Loading bit i with a 1 will not toggle bit $i + 1$. With this principle, the load signal need not be global, and it can

be buffered as it propagates along the bitslices. First, we create the load signal when all bits reach one. Then, the load signal is propagated along the divider to set a memory bit, LdMem, in each bitcell as indicated in Fig. 5. When the counter is switching from 11 . . . 11 to 00 . . . 00, there will be a ripple signal through all divide-by-two stages. Immediately when a bit has switched from one to zero and the LdMem bit is set, the output is returned to one and LdMem is reset. We call this principle ripple inhibited load. If bit i is to be loaded with a zero, no loading is necessary and therefore the LdMem bit is only set for those bits that are to be loaded with a one. This is realized with the P transistor driven by P_i in Fig. 5. The top half of Fig. 5 is a loadable positive edge-triggered divide-by-two circuit followed by a static inverter.

P_i is the ith bit of the binary word that determines the division ratio. If P_i is low, bit i should be loaded with a one since ones complement of P should be loaded into the counter. If P_i is low and Ldb falls, the LdMem node will go high and it stays high even if Ldb returns to high. When Q_i falls, the RS flip-flop is reset, but LdMem is latched at the value loaded by the Ldb signal. To make sure that Q_i

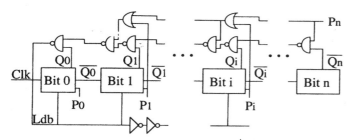

Fig. 6. Block schematic of our divider based on a counter with Ripple Inhibited Load.

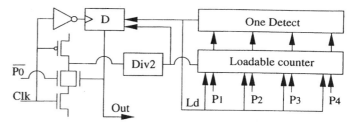

Fig. 7. A divide-by-N circuit with preprocessing of the clock signal.

stays low long enough to toggle the next bit, we introduce a simple handshaking protocol with bit $i + 1$. This is realized by the N-transistor that is driven by the inverse of Q_{i+1}. Bit i is not allowed to load itself with a one until Q_{i+1} has been falling, i.e., the inverse of Q_{i+1} is high.

To simplify the load operation of the toggle stage of bit i, we do not perform the loading until the clock signal of bit i is low, i.e., Q_{i-1} is high. Since we know that the clock signal is low, the loading of bit i can be performed with a single P transistor that pulls up the output of bit i. Some weak transistors were added to the bitcell in Fig. 5 to make sure that noise will not corrupt charge storage on dynamic nodes. This makes the divider very safe at the cost of reduced maximum clock frequency.

The block schematic of our asynchronous programmable divider is shown in Fig. 6. If the division ratio is a small number, e.g., 010 . . . 00, the ripple of (unimportant) higher order bits when going from state 11 . . . 11 to 00 . . . 00 might corrupt the next detection of the 11 . . . 11 state. Therefore, we preprocess the binary word P. If all bits of P to the right of position i are zero, the output of the 1 state detection circuit of bit i (the leftmost NAND gate) is set to high i.e., the real values of $Q_j, j \geq i$ are irrelevant. All bitcells are identical to the transistor schematic in Fig. 5 except for the last stage, which is a simple divide-by-two circuit without load. The reason why the last bit does not need any load is that we combine its output with the P_n value in a NAND gate before sending it to the 1 state detection chain. If P_n is zero, we can ignore the actual value of Q_n and if P_n is high, it should be loaded with a zero (ones complement of one) which need not be done since Q_n is zero after the 11 . . . 11 to 00 . . . 00 transition. Since the load signal does not need to be a global signal applied to all bitslices simultaneously, the Ldb signal is buffered after two bitslices. A more detailed description of this divider can be found in [8].

IV. COMBINING THE PRESCALER AND THE DIVIDE-BY-N CIRCUIT

When examining our prescaler in Fig. 1(b), we see that it consists of a state detection circuit, a ripple chain, and a preprocessing circuit. The ripple chain is a fixed chain without load. By replacing the ripple chain and the state detection circuit with the asynchronous divider presented in the previous section, we get a new divide-by-N circuit

drawn in Fig. 7, which is shown to have considerably higher maximum operating frequency than the asynchronous divider by itself. The output frequency of the preprocessing unit in Fig. 7 is at most half of the input frequency, so the preprocessing unit can be clocked at twice the frequency of the asynchronous divider. The basic operation is as follows. If the division ratio is an odd number ($P_0 = 1$), we cancel one input clock pulse when the load signal is activated. After canceling one pulse, there is an even number of clock pulses before the load signal is activated the next time. Therefore, we can divide the incoming clock signal by two before feeding the asynchronous divider. If the division ratio is an even number ($P_0 = 0$), we divide the incoming clock by two without canceling any clock pulse. Note the irregular connection of P_0 in Fig. 7. This is the least significant bit of the binary word that represents the division ratio and it determines whether to cancel a clock pulse or not when the load signal is active.

We used the 0.8 μm CMOS process characterized in Table I when designing a six-bit frequency divider. We measured a maximum operating frequency of 1.75 GHz at $V_{dd} = 5$ V as indicated in Fig. 3(c), which shows the maximum operating frequency as function of the power supply voltage. The minimum frequency was the same as for the prescalers discussed above. Previously published works are summarized in Table III. Note that the programmable divider here is running faster than the traditional prescalers based on a shift register ring listed in Table II.

Reference [10] is a maximally pipelined synchronous divider with minor transistor sizing leaving most transistors 4 μm wide. The maximum operating frequencies of our clock preprocessing divider and the dividers presented in [10] and [11] are independent of the number of bits in the dividers, while the structures in [7] and [9] have a large fan-out that will be troublesome to drive for dividers with many bits. As for the dual-modulus prescalers, noise coupling degraded the accuracy of the power measurements, so we give only approximate values of the power consumption.

V. CONCLUSIONS

Most high-speed prescalers and frequency dividers that have been reported use an advanced process and/or circuit technique to reach high operating frequency. We focus on architectural developments and introduce a new dual-

TABLE III
PERFORMANCE OF DIVIDE-BY-N CIRCUITS

Ref.	Technology	Vdd	Ratio	Bits	Speed (MHz)	P @ fmax (mW)	Area/bit (μm²)
[7]	0.7μm GaAs	1.0V	≥2	4	1700	36	?
[7]	0.7μm GaAs	1.0V	≥2	6	1400	90	127000
[9]	1μm GaAs	3, -3V	≥3	6	1500	1400	275000
[10]	1μm CMOS	5V	≥2	6	400	80	7600
[11]	2μm BiCMOS	4.5V	≥8	15	≥165*	55	25000
[12]	0.8μm CMOS	5V	≥3	14	500	8	16000
This work	0.8μm CMOS	5V	≥2	6	1750	50-60**	6900

* Simulated with worst case speed parameters.
** Approximate measurements excluding clock power.

modulus prescaler architecture. In a standard $0.8 \ \mu$m CMOS bulk process, we measured a maximum operating frequency of 1.90 GHz for our architecture, compared to 1.68 GHz for the traditional prescaler architecture implemented in the same process.

We combine the key part of the prescaler with a novel asynchronous divider and reach 1.75 GHz in a standard $0.8 \ \mu$m CMOS process. This is three times faster than any previously reported CMOS or BiCMOS divide-by-N circuit and is similar to several reported GaAs circuits.

Dual-modulus prescalers were invented because of the inferior speed of divide-by-N circuits. With our architecture, we show that there is only a small difference in maximum operating frequency between a dual-modulus prescaler and a divide-by-N circuit, indicating that dual-modulus prescalers might be obsolete in the future.

REFERENCES

[1] H. I. Cong, J. M. Andrews, D. M. Boulin, S. C. Fang, S. J. Hillenius, and J. A. Michejda, "Multigigahertz CMOS dual-modulus prescalar IC," *IEEE J. Solid-State Circuits*, vol. 23, pp. 1189–1194, Oct. 1988.

[2] R. Rogenmoser, Q. Huang, and F. Piazza, "1.57 GHz asynchronous and 1.4 GHz dual-modulus 1.2 μm CMOS prescalers," in *Proc. Custom Integrated Circuit Conf. '94*, 1994, pp. 16.3.1–4.

[3] N. Foroudi and T. A. Kwasniewski, "CMOS high-speed dual-modulus frequency divider for RF frequency synthesis," *IEEE J. Solid-State Circuits*, vol. 30, pp. 93–100, Feb. 1995.

[4] T. Aytur and B. Razavi, "A 2GHz, 6mW BiCMOS frequency synthesizer," in *Proc. International Solid-State Circuit Conf.*, 1995, pp. 264–265.

[5] T. Seneff, L. McKay, K. Sakamoto, and N. Tracht, "A sub-1 mA 1.5-GHz silicon bipolar dual modulus prescaler," *IEEE J. Solid-State Circuits*, vol. SC-29, pp. 1206–1211, Oct. 1994.

[6] M. Ohhata, T. Takada, M. Ino, N. Kato, and M. Ida, "A 4.5-GHz GaAs dual-modulus prescaler IC," *IEEE T. Microwave Theory Tech.*, vol. 36, pp. 158–160, Jan. 1988.

[7] H. P. Singh, R. A. Sadler, W. J. Tanis, and A. N. Schenberg, "GaAs Prescalers and counters for fast-settling frequency synthesizers," *IEEE J. Solid-State Circuits*, vol. 25, pp. 239–245, Feb. 1990.

[8] P. Larsson, "Analog Phenomena in Digital Circuits," Ph.D. dissertation no. 376, Linköping Univ., 1995.

[9] M. G. Kane, P. Y. Chan, S. S. Cherensky, and D. C. Fowlsi, "A 1.5-GHz programmable divide-by-N GaAs counter," *IEEE J. Solid-State Circuits*, vol. 23, pp. 480–484, April 1988.

[10] P. Larsson, "A wide-range programmable high-speed CMOS frequency divider," in *Proc. IEEE International Symp. Circuits and Systems*, 1995, pp. 195–198.

[11] C. S. Choy, C. Y. Ho, G. Lunn, B. Lin, and G. Fung, "A BiCMOS programmable frequency divider," *IEEE T. Circuits and Systems-II*, vol. 39, pp. 147–154, Mar. 1992.

[12] J. C. Wu and H. H. Chang, 'A 550 MHz 9.3 mW CMOS frequency divider," in *Proc. IEEE Int. Symp. Circuits and Systems*, 1995, pp. 199–202.

A 1.6-GHz Dual Modulus Prescaler Using the Extended True-Single-Phase-Clock CMOS Circuit Technique (E-TSPC)

J. Navarro Soares, Jr., and W. A. M. Van Noije

Abstract—The implementation of a dual-modulus prescaler (divide by 128/129) using an extension of the true-single-phase-clock (TSPC) technique, the extended TSPC (E-TSPC), is presented. The E-TSPC [1], [2] consists of a set of composition rules for single-phase-clock circuits employing static, dynamic, latch, data-precharged, and NMOS-like CMOS blocks. The composition rules, as well as the CMOS blocks, are described and discussed. The experimental results of the complete dual-modulus prescaler, implemented in a 0.8 μm CMOS process, show a maximum 1.59 GHz operation rate at 5 V with 12.8 mW power consumption. They are compared with the results from other recent implementations showing that the proposed E-TSPC circuit can reach high speed with both smaller area and lower power consumption.

Index Terms— CMOS digital, high-speed circuits, prescalers, single-phase-clock design.

I. INTRODUCTION

FOR MORE than 15 years, CMOS has been the main technology for very-large-scale integration (VLSI) system design. From the beginning to nowadays, several CMOS clock policies have been proposed. The pseudotwo-phase logic was one of the earliest techniques [3]. Later on, two-phase logic structures were proposed. The domino technique [4] associated successfully both two-phase and dynamic CMOS circuits. With the NORA technique [5], [6], an extensive no-race approach for two-phase and dynamic circuits was developed. A single-phase-clock policy was introduced in [7] [the true single-phase-clock (TSPC)]. This technique was subsequently advanced by [8]–[10].

Single-phase-clock policies are superior to the others due to the simplification of the clock distribution. They reduce the wiring costs and the number of clock-signal requirements (no problems with phase overlapping, for instance). Consequently, higher frequencies and simpler designs can be achieved.

Introduced by [1] and [2], the extended true-single-phase-clock CMOS circuit technique (E-TSPC), an extension of the TSPC, consists of composition rules for single-phase circuits using static, dynamic, latch, data-precharged, and NMOS-like blocks. The composition rules enlarge the block-connection possibilities and avoid races; additionally, NMOS-like blocks enhance the technique for high-speed operations.

The design of a dual-modulus prescaler (divide by 128/129) with the E-TSPC in a standard 0.8 μm CMOS process (0.7 μm

effective channel length) is presented. The prescaler implementation purpose is the evaluation of the E-TSPC technique potentialities.

This paper is organized as follow. In Section II, the principal features of the E-TSPC technique, blocks and design rules, are presented. In Section III, some different dual-modulus implementations are analyzed. Experimental results and comparisons are reported in Section IV, and the principal conclusions are drawn in Section V.

II. E-TSPC CIRCUIT BLOCKS AND COMPOSITION RULES

A. Basic CMOS Blocks

An E-TSPC circuit should use any of the blocks: CMOS static block, n-dynamic block [Fig. 1(a)], p-dynamic block [Fig. 1(c)], n-latch block [Fig. 1(e)], p-latch block [Fig. 1(g)], and high (PH) and low (PL) data-precharged blocks (Fig. 2).

In Fig. 1, the clocked transistors of the n- and p-latches are placed close to the power rail, following the suggestion of [11]. This configuration can attain a higher speed but suffers charge-sharing problems. Clocked transistors close to either the power rail or the block output are admissible latch configurations.

In data-precharged blocks [10], some input signals, called precharging inputs or pc-inputs, control the output precharge (see Fig. 2). If all PH block pc-inputs are high, or if all PL block pc-inputs are low, then the PH or PL block is precharged. In this case, the PH block output goes to low, and the PL block output to high. In Fig. 2, the CMOS static block that executes the logic function $d = a \cdot (b + c)$ is drawn, along with all equivalent PH and PL blocks [Fig. 2(b) and (c)]. The pc-inputs of each block are also indicated. The PH and PL blocks that have the output precharged when the clock is low will be called n-Dp blocks; similarly, the PH and PL blocks that have the output precharged when the clock is high will be called p-Dp blocks.

B. Composition Rules

First, the definition of data chains, fundamental to the design rules, is given.

Definition: An *n-data chain* is any noncyclic signal propagation path:

1) containing at least one n-latch, one n-dynamic, or one n-Dp block;

2) starting in a circuit external input, or in the output of a p-latch, p-dynamic, or p-Dp block; when this output is followed by static blocks in the normal data flow, the data chain starts in the output of the last static block;

Manuscript received February 16, 1998; revised May 25, 1998. This work was supported in part by FAPESP and CNPq, Brazil.

The authors are with the LSI/PEE, Escola Politécnica, University of São Paulo, São Paulo, S.P. 05508-900 Brazil (e-mail: navarro@lsi.usp.br; noije@lsi.usp.br).

Publisher Item Identifier S 0018-9200(99)00410-2.

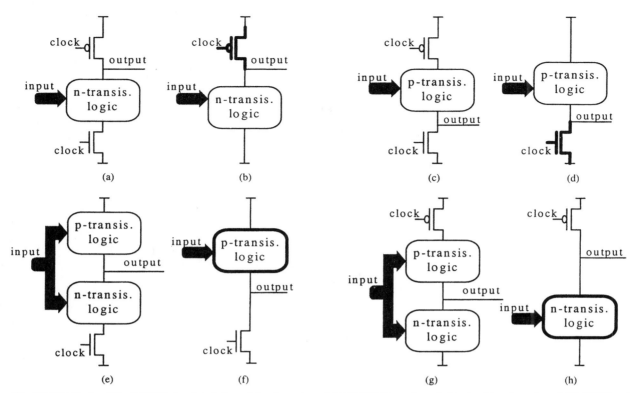

Fig. 1. Construction blocks of the E-TSPC circuit technique: (a) n-dynamic and (b) NMOS-like n-dynamic blocks; (c) p-dynamic and (d) NMOS p-dynamic blocks; (e) n-latch and (f) NMOS-like n-latch blocks; and (g) p-latch and (h) NMOS-like p-latch blocks.

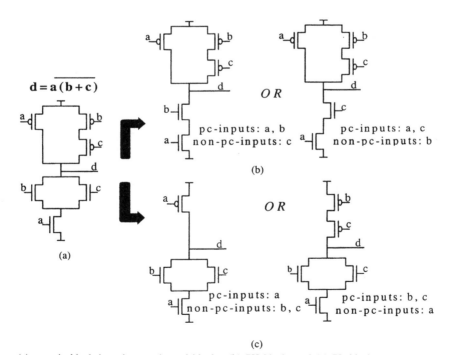

Fig. 2. Transformation from (a) a static block into data-precharged blocks: (b) PH blocks and (c) PL blocks.

3) going through static, n-dynamic, n-Dp, or n-latch blocks;
4) regardless of the number and ordering of the blocks defined above;
5) finishing in a circuit external output, or in the input of the first p-latch, p-dynamic, or p-Dp block.

For the p-data chains, an equivalent definition applies, replacing n with p and vice versa.

When clock is high, n-data chains are in evaluation phase; otherwise, they are in holding phase. P-data chains evaluate when clock is low.

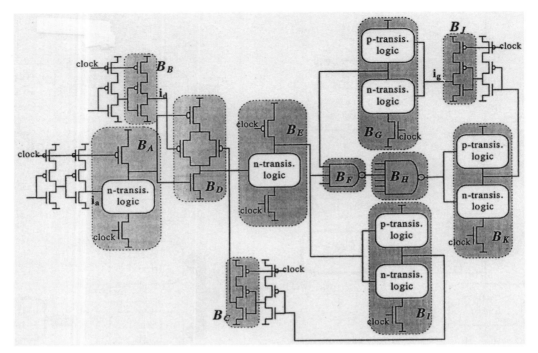

Fig. 3. Example of n-data chains. The blocks mentioned in the text are named and hatched in the figure.

In Fig. 3, part of a circuit schematic is depicted with seven complete n-data chains. Some examples are the data chain starting at input i_a and going through blocks B_A, B_D, B_E, and B_I; the data chain starting at i_d and going through B_D, B_E, B_F, B_H, and B_K; and the data chain starting at i_g and going through B_G, B_F, B_H, and B_K.

Five of the six E-TSPC composition rules are now listed. Their purpose is to ensure the observance of some constraints during the evaluation and holding phases. To simplify the rule statements, the symbol χ will be used to denote n or p in nouns like χ-data chain, χ-dynamic block, etc.

Composition Rule (r_1): The χ-data chain input should be an input of a dynamic block, an input of a latch, or a nonpc-input of a Dp block.

Composition Rule (r_2): A χ-latch must not drive, directly or through static blocks, a χ-dynamic or a χ-Dp block.

Composition Rule (r_3): The number of inversions between:

r_{3a}) any two adjacent dynamic blocks must be *odd*[1];

r_{3b}) any two adjacent Dp-blocks of the same type (PH and PH or PL and PL) must be *odd*;

r_{3c}) any two adjacent Dp-blocks of complementary types must be *even*;

r_{3d}) a PH (PL) block and an adjacent n- (p)-dynamic (or vice versa) in an n- (p)-data chain must be *even*;

r_{3e}) a PL (PH) block and an adjacent n- (p)-dynamic (or vice versa) in an n- (p)-data chain must be *odd*.

(Two blocks are called adjacent if there are only static blocks between them.)

Composition Rule (r_4): Consider the last dynamic block in the χ-data chain (when it exists). The number of inversions (due to any block) from this dynamic block up to at least one χ-latch must be *even*.

[1] Through all the rules, zero inversion will be considered even.

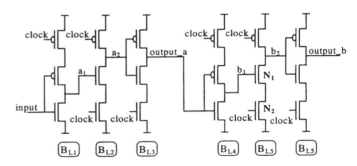

Fig. 4. Two TSPC D-flip-flops connected in series.

Composition Rule (r_5):
The χ-data chain must have one of the following two configurations:

r_{5a}) at least one dynamic block and one latch;

r_{5b}) at least two latches and an *even* number of inversions (latches or static blocks) between them.

It is worth noting that these five composition rules are very similar to the five rules proposed in the NORA technique [6]. In a circuit where all data chains obey the five rules, it can be proved that (six theorems presented in [1] and [2]):

a) all data-precharged blocks are precharged during the holding phase of the data chains to which they belong;

b) the dynamic and the data-precharged blocks are not incorrectly discharged during the evaluation phase;

c) the output of the data-chain last latch is steady during the holding phase of the data chain.

C. Exception Rule

Although the above-described rules are necessary to avoid race problems, typical TSPC systems do not follow some of them. The most common exception is found in connecting two D-flip-flops (D-FF's), as shown in Fig. 4. In such a

TABLE I
CONDITIONS FOR CORRECT OPERATION OF THE NMOS-LIKE BLOCKS

Circuit block	clock=high	clock=low
N-MOS like n-dynamic	no constraint	output must be high (independently of inputs)
N-MOS like p-dynamic	output must be low (independently of inputs)	no constraint
N-MOS like n-latch	output is high if the p-transistor logic is conducting; otherwise, low	no constraint
N-MOS like p-latch	no constraint	output is low if the n-transistor logic is conducting; otherwise, high

configuration, the p-data chains are constituted of only one p-latch block, namely, B_{L1} or B_{L4} (r_5 violation). In consequence, the p-latch output may change during its holding time. A faulty sequence example is depicted below: consider an initial state on which the signals *clock, input,* and *output_a* are low, and both blocks B_{L1} and B_{L4} are evaluating. At the end of the evaluation period, the outputs a_1 and b_1 are high. Subsequently, when the clock goes to high, the other blocks will evaluate. Suppose that a_1 works properly, holding its former value (high). In this case, the node a_2 goes to low, *output_a* goes to high, and b_1 goes to low. As a result, the transistor N_1 is cut, and the final value of node b_2 will depend on the circuit delays.

Commonly, the delay between nodes a_1 and b_1 is long enough to ensure that b_2 is fully discharged through transistors N_1 and N_2; in this case, the second D-FF works properly.

A simple exception rule is added to cover the utilization of the well-established TSPC D-FF's (Fig. 4).

Exception Rule (r_e): Configurations similar to that of Fig. 4, where rules r_4 and r_5 are not obeyed, are accepted if enough delay exists.

The data chains where r_e is applied, to the detriment of r_4 and r_5, do not have a latch with steady output at the holding phase. Since the correct operation of the circuit will depend on the block delays, the *exception rule* should be used with caution.

Considering the connection rules presented in former works [7]–[10], our six proposed rules differ in the following aspects.

a) The "nonlatched domino logic," a timing strategy considered in [10], is not accepted in our proposal.

b) The proposed rules permit a more flexible usage of both data-precharge blocks, due to the distinction between pc and nonpc-inputs, and static logic blocks (static logic is allowed between dynamic and latch blocks). In Fig. 2, where no rule violations occur, several connections not allowed by former work rules are provided, for instance, the connection between blocks B_B and B_D, between B_G and B_F, between B_E and B_F, etc.

D. NMOS-Like Logic Extension

When high speed is also a requirement, restrictions on the use of p-dynamic and p-latch blocks should be imposed. These blocks have at least two p-transistors in series, which may

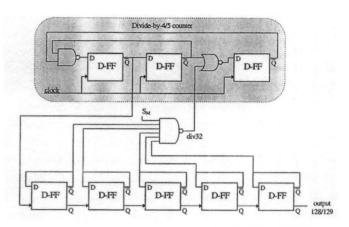

Fig. 5. Schematic of the dual-modulus prescaler (divide by 128/129).

reduce considerably the maximum speed. In such applications, the p-data chains are limited to one block, and most logic operations are handled with n-data chains with limited logic dept. Thus, deep pipelines will be necessary to implement complex and fast logic designs.

NMOS-like dynamic and latch blocks can be used to minimize this difficulty and also to increase the n-data chain speed. They are *ratioed* logic blocks, where the n-transistor section and the p-transistor section may conduct simultaneously. A similar technique was used in [12], but restricted to D-FF's. In Fig. 1, the NMOS-like versions of the dynamic and latch blocks are drawn. To assure a correct operation, these blocks should satisfy the constraints summarized in Table I. The transistor section that must impose the output value, when both sections are conducting, is drawn with bold lines in Fig. 1.

The NMOS-like blocks are faster due to the reduced number of transistors in series, but, unfortunately, they consume more power. In consequence, they should be used only in critical data chains, where the desirable speed has not been reached. Since the connection characteristics do not depend on whether it is a conventional or an NMOS-like block, the composition rules (r_1–r_5 and r_e) are valid and necessary for both; as a result, NMOS-like blocks and conventional blocks can replace one another, and the judicious selection of NMOS-like blocks is made easy.

Summarizing, the static blocks, the n/p-dynamic, the n/p-latch, the PH/PL data-precharged, the NMOS-like blocks, and the composition rules r_1–r_5 and r_e compose the E-TSPC technique.

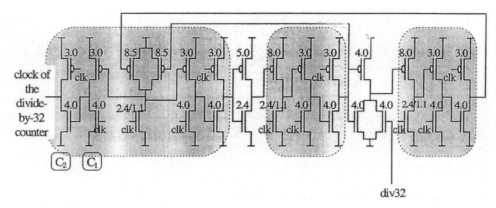

Fig. 6. Transistor schematic of the divide-by-4/5 counter D_{G4}. The transistor width or, when the length is different from 0.8 μm, the transistor width/length, in μm, is also indicated in the figure.

TABLE II
MAXIMUM SPEED AND POWER-CONSUMPTION RESULTS FOR
THE FOUR DESIGNED DIVIDE-BY-4/5 COUNTERS (SPICE
SIMULATIONS, SLOW PARAMETERS, AND $V_{DD} = 5$ V)

Design	Speed (GHz)	Power (μw/MHz)
D_{G1}	0.98	3.27
D_{G2}	1.28	4.45
D_{G3}	1.39	4.85
D_{G4}	1.67	5.62

III. DUAL-MODULUS DESIGN

Dual-modulus prescalers, a circuit with applications in frequency synthesis systems, have been frequently used to compare different high-speed implementations [12] and [13], our current goal. A high-speed dual-modulus prescaler (divide by 128/129) was designed using a standard 0.8 μm CMOS bulk process.

The schematic of the dual-modulus prescaler is depicted in Fig. 5. The circuit inside the cross-hatched box, composed of three D-FF's and two logic gates, forms a divide-by-4/5 counter. The *div32* signal selects if it counts up to four (*div32 = high*) or up to five (*div32 = low*). The five D-FF's at the bottom of the figure form a divide-by-32 counter. The fractional division ratio of the prescaler, 128 or 129, is selected according to the S_M signal.

Four different approaches were applied to draw a layout of the divide-by-4/5 counter, which is the critical high-speed part of the prescaler. The approaches are:

D_{G1}) design with conventional rise edge-triggered TSPC D-FF (Fig. 4);

D_{G2}) design with rise edge-triggered D-FF, and further optimization applying the E-TSPC technique;

D_{G3}) design with a modified fall edge-triggered D-FF [12];

D_{G4}) design with fall edge-triggered D-FF, and further optimization applying the E-TSPC technique.

In Fig. 6, the transistor schematic of the D_{G4} approach, with transistor dimensions, is depicted. The three cross-hatched boxes mark the D-FF's; the first D-FF (left) has a buffered output.

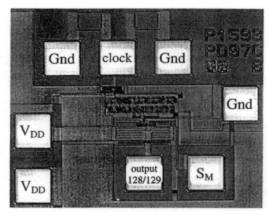

Fig. 7. Photograph of the prescaler test chip.

The maximum speed and the power consumption for each design are shown in Table II. These results were obtained with SPICE simulations from the extracted netlists of the layouts for slow parameters, room temperature, and power supply at 5 V. The comparison of the results exhibits some advantages of the E-TSPC technique. From the D_{G1} to D_{G4} approach, the speed improvement is higher than 70%, and from D_{G3} to D_{G4} is 20%. On the other hand, the power consumption increases 72% from D_{G1} to D_{G4}. As D_{G4} uses only NMOS-like blocks, the latter result is not surprising, and confirms that these blocks should be restricted to critical circuit parts. Since the composition rules favor the replacement of conventional blocks with NMOS-like ones and vice versa, E-TSPC circuits can reach high speed and keep the power consumption low.

To better evaluate the above results, the following notes should be taken into account:

- all approaches use small transistor sizes, usually minimum sizes (as indicated in Fig. 6);

- the Fig. 5 divide-by-4/5 counter schema was slightly modified for each design ($D_{G1}, \cdots, D_{G4}$) to conform with its structure characteristics;

- the NOR configuration of Fig. 6 is similar to an NMOS logic, but the load is now a PMOS transistor. It is faster than the CMOS static NOR and is used in the D_{G1}, D_{G3}, and D_{G4} approaches;

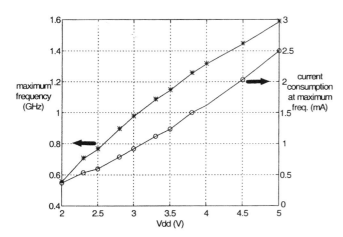

Fig. 8. Measured results for the prescaler maximum frequency (f_{max}), left axis (*), and current consumption at f_{max}, right axis (o), as a function of the power supply.

- C_1 and C_2 blocks, Fig. 6, drive the clock signal to the divide-by-32 counter. All four designs have similar configuration.

IV. EXPERIMENTAL RESULTS

The full prescaler circuit, occupying a 0.0126 mm² area, was formed with the counter D_{G4}. The D-FF's of the 32 asynchronous counter were built with conventional rise edge-triggered TSPC D-FF (Fig. 4). The clock signal from the divide-by-4/5 counter, Fig. 6, is inverted before being sent to the 32 counter. This expedient allows a longer time interval for preparation of the signal *div32*.

The prescaler test chip, whose photograph is shown in Fig. 7, was mounted on an alumina substrate with the chip-on-board technique. A coplanar radio-frequency probe was used to feed the unique prescaler high-speed signal, the clock input.

In Fig. 8, the measured maximum frequency and current consumption as a function of the power supply are shown. Since the used pulse generator has a maximum excursion of 3 V, the circuit real maximum frequencies are expected to be slightly higher than the measured results for power supply above 3 V.

Performance results of this work, of two recently published prescalers using TSPC D-FF's, and of a new prescaler architecture are summarized in Table III. In [13], the prescaler is implemented with rise edge-triggered TSPC D-FF's, which were size optimized to reach maximum speed; in consequence, not only the circuit speed but also the area and power consumption are high. Fall edge-triggered TSPC D-FF's with small-sized transistors and with some NMOS-like blocks are used in [12]. The resulting circuit has a small area and a low power consumption but a reduced maximum operation rate. Our implementation, with the E-TSPC technique and small-sized transistors, provides the smallest area and the lowest power consumption; the speed, in addition, is comparable to [13] and [14].

V. CONCLUSIONS

A complete high-speed dual-modulus prescaler (divide by 128/129) was developed in a 0.8 μm CMOS process. The

TABLE III
AREA, SPEED, AND POWER-CONSUMPTION RESULTS FOR FOUR DIFFERENT PRESCALERS

Prescaler	Technology (μm)	power supply (V)	Area ($10^{-3}mm^2$)	Speed (GHz)	Power (μW/MHz)
[13]	1.0	5	39.1	1.6	31.2
[12]	0.8	5	13.7(*)	1.22	20.9
[14]	0.7	3	632 (**)	1.75	13.7
(this work)	0.8	5	12.6	1.59	8.0

(*) only the divide-by-4/5 and the divide-by-32 counters are considered.
(**) including bonding pads and buffers.

measured circuit attained 1.59 GHz and 8.0 mW/MHz power consumption with 5 V power supply. It can be advantageously compared with other implementations in terms of area and power consumption; in terms of speed, it matches the fastest TSPC prescaler. The studies done during the design reveal that, to take full advantage of the TSPC technique, every possible configuration should be considered. The E-TSPC, being an extension of TSPC, permits exploring a larger number of solutions and, in consequence, finding the best configuration. The dual-modulus prescaler results exhibit some significant improvements produced by the E-TSPC.

REFERENCES

[1] J. Navarro and W. Van Noije, "E-TSPC: Extended true single-phase clock CMOS circuit technique," in *VLSI: Integrated Systems on Silicon, IFIP International Conference on VLSI*, R. Reis and L. Claesen, Eds. London, U.K.: Chapman & Hall, 1997, pp. 165–176.
[2] _____, "E-TSPC: Extended true single-phase-clock CMOS circuit technique for high speed applications," SBMICRO, *J. Solid-State Devices Circuits*, vol. 5, pp. 21–26, July 1997.
[3] N. H. E. Weste and K. Eshraghian, *Principles of CMOS VLSI Design*, 1st ed. Reading, MA: Addison-Wesley, 1985.
[4] R. H. Krambeck, C. M. Lee, and H.-F.S. Law, "High-speed compact circuits with CMOS," *IEEE J. Solid-State Circuits*, vol. SC-17, pp. 614–619, June 1982.
[5] N. F. Gonçalves and H. J. De Man, "NORA: A racefree dynamic CMOS technique for pipelined logic structures," *IEEE J. Solid-State Circuits*, vol. SC-18, pp. 261–266, June 1983.
[6] N. F. Gonçalves, "NORA: A racefree CMOS technique for register transfer systems," Ph.D. dissertation, Katholieke Universiteit Leuven, Leuven, Belgium, 1984.
[7] Y. Ji-ren, I. Karlsson, and C. Svensson, "A true single-phase-clock dynamic CMOS circuit technique," *IEEE J. Solid-State Circuits*, vol. SC-22, pp. 899–901, Oct. 1987.
[8] J. Yuan and C. Svensson, "High speed CMOS circuit technique," *IEEE J. Solid-State Circuits*, vol. 24, pp. 62–70, Feb. 1989.
[9] M. Afghahi and C. Svensson, "A unified single-phase clocking schema for VLSI systems," *IEEE J. Solid-State Circuits*, vol. 25, pp. 225–235, Feb. 1990.
[10] P. Larsson, "Skew safety and logic flexibility in a true single phase clocked system," in *Proc. IEEE ISCAS*, Seattle, WA, May 1995, pp. 941–944.
[11] Q. Huang, "Speed optimization of edge-triggered nine-transistor D-flip-flop for gigahertz single-phase clocks," in *Proc. IEEE ISCAS*, Chicago, IL, May 1993, pp. 2118–2121.
[12] B. Chang, J. Park, and W. Kin, "A 1.2 GHz CMOS dual-modulus prescaler using new dynamic D-type flip-flops," *IEEE J. Solid-State Circuits*, vol. 31, pp. 749–752, May 1996.
[13] Q. Huang and R. Rogenmoser, "Speed optimization of edge-triggered CMOS circuits for gigahertz single-phase clocks," *IEEE J. Solid-State Circuits*, vol. 31, pp. 456–465, Mar. 1996.
[14] J. Craninckx and M. S. J. Steyaert, "A 1.75-GHz/3-V dual-modulus divide-by-128/129 prescaler in 0.7-μm CMOS," *IEEE J. Solid-State Circuits*, vol. 31, pp. 890–897, July 1996.

A Simple Precharged CMOS Phase Frequency Detector

Henrik O. Johansson

Abstract— We propose a simple precharged CMOS phase frequency detector (PFD). The circuit uses 18 transistors and has a simple topology. Therefore, the detector, in a 0.8-μm CMOS process, works up to clock frequencies of 800 MHz according to SPICE simulations on extracted layout. Further, the detector has no dead-zone in the phase characteristic which is important in low jitter applications. The phase and frequency characteristics are presented and comparisons are made to other PFD's. The phase offset of the detector is sensitive to differences of the duty-cycle between the inputs. Mixed-mode simulations are presented of the lock-in procedure for a phase-locked loop (PLL) where the detector is used. Measurements on the detector are presented for a test-chip with a delay-locked loop (DLL) where the phase detection ability of the detector has been verified.

Index Terms— CMOS integrated circuits, delay lock loops, phase detectors, phase lock loops.

I. INTRODUCTION

A part of a phase-locked loop (PLL) is the phase detector (PD) [1]. The PD detects the phase difference between the reference frequency and the controlled slave frequency. Some PD's also detect frequency errors, they are then called phase frequency detectors (PFD's). A PFD is usually built with a state machine with memory elements such as flip-flops [2], [3], Figs. 1 and 2, respectively. We propose a new simple PFD, ncPFD, which uses two nc-stages [4] and six inverters, Fig. 3(a).

A drawback with some phase detectors is a dead zone in the phase characteristic at the equilibrium point. The dead zone generates phase jitter since the control system does not change the control voltage when the phase error is within the dead zone.

In Section II the ncPFD circuit is described. The phase and frequency characteristics are discussed in Sections III and IV, respectively, and comparisons are made to other PFD's. Behavioral mixed-mode simulations are made to check the lock-in properties of the ncPFD detector and these simulations are shown in Section V. Experiments on the phase detection abilities of the ncPFD are presented in Section VI.

II. CIRCUIT

The transistor schematic of the ncPFD is shown in Fig. 3(a). The detector has a 0-rad phase offset. The main part of the circuit is the nc stage [4]. Delays (two inverters) are inserted at the reference and slave inputs in order to remove the dead zone in the phase characteristics around π rad phase error. In Fig. 4, waveforms for the circuit in Fig. 3(a) are shown when the slave input lags the reference input.

Manuscript received March 11, 1997; revised August 21, 1997.

The author is with Electronic Devices, Department of Physics and Measurement Technology, Linköping University, S-58183 Linköping, Sweden.

Publisher Item Identifier S 0018-9200(98)00732-X.

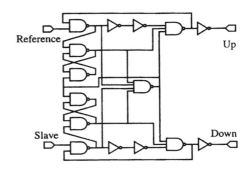

Fig. 1. Conventional phase frequency detector (conPFD) from [2].

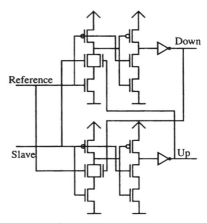

Fig. 2. Precharge type phase frequency detector (ptPFD) from [3].

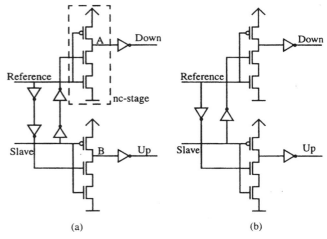

(a)   (b)

Fig. 3. (a) The ncPFD in zero degree phase offset version. (b) Modified version with π rad phase offset.

The detector can easily be modified to one with π-rad phase offset, as shown in Fig. 3(b), where one, or in general an odd number, of inverter(s) are used for the delays.

If the phase detector is used only as a phase detector, i.e., not as a frequency detector, the circuit in Fig. 3(a) can be used as

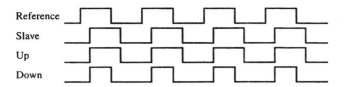

Fig. 4. Waveforms for the case when slave lags after the reference signal. The pulse width of the up signal is larger than for the down signal.

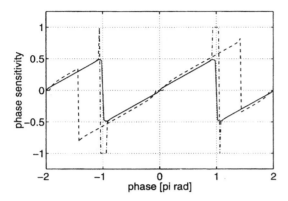

Fig. 5. Phase characteristics of the ncPFD (solid line), conPFD (dashed line), and the ptPFD (dash-dot line) from SPICE level-2 simulations of extracted layout, $V_{DD} = 3.0$ V and $f = 50$ MHz.

a π-rad phase detector by switching the up and down signals. The equilibrium point will then be on the negative slope of the phase characteristics at π rad instead of at the positive slope at zero, Fig. 5. Similarly, the π-rad phase detector, Fig. 3(b), can be modified to a 0-rad phase detector.

III. PHASE CHARACTERISTIC

The phase characteristic of the proposed ncPFD is shown in Fig. 5 together with the characteristics of the conventional PFD (conPFD) of Fig. 1 [2] and the precharge type PFD (ptPFD) shown in Fig. 2 [3]. Unlike the conPFD and ptPFD, there is no dead-zone in the characteristics of the ncPFD. A magnification of the characteristics at zero phase is shown in Fig. 6. The dead zone of the conPFD can be reduced by inserting delay at the output of the four-input-NAND-gate. But if delays are inserted in the feedback signals from the up and down outputs of the ptPFD, the dead zone unfortunately increases.

In an ncPFD, when the PLL is locked, both up and down signals are active. Therefore the phase offset of the PLL depends on the matching between the up and down currents of the charge pump.

All data in this section are based on simulations of extracted layout with SPICE (level-2) when $V_{DD} = 3.0$ V and $f = 50$ MHz unless otherwise stated. The layout was made in a 0.8-μm standard CMOS process and the N and P-transistors are 2.0 and 4.0 μm wide, respectively. The outputs were connected to 4.0 fF capacitors, and the inputs were driven with inverters with a tapering factor of one.

A. Duty-Cycle and Transition-Time Dependence

The output of the ncPFD depends on the pulse-width of the input signals. Hence, the duty cycle will affect the phase

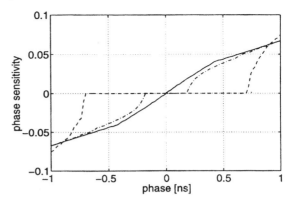

Fig. 6. Magnified phase characteristics at zero phase error of the ncPFD (solid line), conPFD (dashed line), and the ptPFD (dash-dot line) from SPICE level-2 simulations of extracted layout, $V_{DD} = 3.0$ V and $f = 50$ MHz.

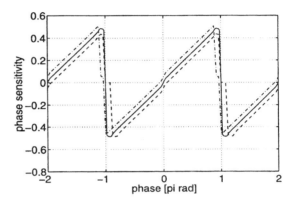

Fig. 7. Phase characteristics for three cases with different duty cycles. The reference input duty cycle is 50% for all cases and the slave input has the duty-cycles 45%, 50%, and 55% for the dashed, solid, and dashed–dotted lines, respectively.

characteristics. The phase characteristics are checked for three different duty cycles, 45, 50, and 55%.

When both the reference and slave have the same duty cycle, the phase offset is not affected. There is a dead zone at π-rad when the duty cycle is less than 50%. A duty cycle of 45% gives a dead zone width of 0.50 rad, 1.6 ns, at π rad. This dead zone may result in a metastable state of the control loop.

When the duty cycle is different for the two inputs, the phase offset will be nonzero, Fig. 7. A duty cycle difference of 5% at 50 MHz, i.e., 1 ns, gives a phase offset of $0.063 * \pi$ rad, i.e., 630 ps.

The phase characteristic of the ncPFD is not affected by variations of the rise and fall times when they are in the range of 300 ps up to 600 ps.

B. Maximum Operation Frequency

A maximum operation frequency definition can be found in [3]. The definition is that the maximum operation frequency is one over the shortest period with correct up and down signals when the inputs have the same frequency and 90° phase difference. This definition is easily applicable on flip-flop-based PFD's where this frequency is easily identified. Unfortunately, the degradation of the performance of the

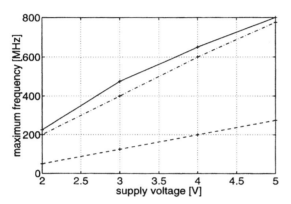

Fig. 8. The width of the dead zones of the ncPFD (solid), ptPFD (dashed), and conventional PFD (dash-dot) as function of frequency. The frequency resolution is 100 MHz and the supply voltage is 5.0 V. The plot is based on SPICE simulations of extracted layout.

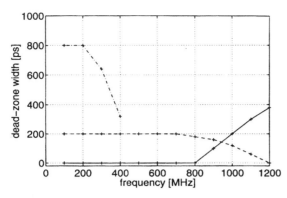

Fig. 9. Maximum frequency as function of supply voltage for the ncPFD (solid line), the ptPFD (dash-dot line), and the conPFD (dashed line). The frequency resolution is 25 MHz. The plot is based on simulations of extracted layout. The layouts are made in a standard 0.8-μm CMOS process.

ncPFD is gradual for increasing frequency and this makes it hard to find a specific frequency where the circuit starts to malfunction.

Therefore, we define the maximum operation frequency to be the frequency where the size of the dead zone starts to deviate significantly from the low-frequency value. This definition gives similar results for the flip-flop-based phase detectors as for the definition in [3], and it is applicable on the ncPFD. An example of how the dead-zone-width varies with the frequency is shown in Fig. 8.

The maximum speeds for different supply voltages are plotted in Fig. 9 for the three PFD's of Figs. 1, 2, and 3(a). As seen, the maximum speed of the ncPFD and the ptPFD are similar and the conPFD is approximately three times slower.

IV. FREQUENCY CHARACTERISTICS

A frequency dependent phase detector always has some kind of memory. For the ncPFD, the memory consists of the two dynamic nodes at the output of the nc-stages. In Fig. 10, the frequency of the slave input is approximately three times higher than the reference input frequency, as a result, the down signal has a higher duty cycle than the up signal. Thus the slave frequency should decrease.

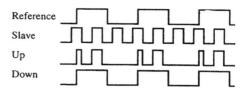

Fig. 10. Waveforms for the case when the slave has a higher frequency than the reference signal. The down signal has higher duty cycle than the up signal.

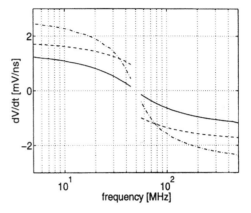

Fig. 11. Frequency sensitivity for the ncPFD (solid), ptPFD (dash-dot), and conPFD (dashed). The plot is based on behavioral simulations with 20 different initial phases for each frequency and the mean-value for each frequency is plotted. The reference frequency is 50 MHz.

The average frequency sensitivities of the ncPFD, ptPFD, and conPFD are shown in Fig. 11. The frequency sensitivity is represented by the rate of change in the control voltage of the loop filter of a PLL when the slave input is driven by a pulse generator with a fixed frequency instead of the voltage-controlled oscillator (VCO) output. Each frequency is simulated 20 times with different initial phases, i.e., skew between the inputs.

The ptPFD has the largest sensitivity, followed by the conPFD, and the ncPFD has the lowest. The sensitivity goes to zero as the slave frequency approaches the reference frequency for both the ncPFD and ptPFD. But for the conPFD, the sensitivity is relatively high even for frequencies close to the reference.

In Fig. 12 the sensitivity for the ncPFD is shown with the mean, minimum, and maximum values from the 20 simulations for each frequency. Note that the behavior of the minimum and maximum values are almost random.

For the ncPFD, the minimum absolute value of the sensitivity is close to zero for certain frequencies, Fig. 12. Actually, the sensitivity is zero for some frequency ratios and phase combinations. This is the case also for the ptPFD but not for the conPFD. The condition for this seems to be that when the frequency ratio of the reference and slave inputs is a rational number and the ratio is in the interval 1/2 to 2, including the limits, the sensitivity is zero for certain initial phases. We have no general proof of the previous statement but, for example, the sensitivity of the ncPFD for $f_{slave} = 4/5 * f_{reference}$ as function of initial phase is shown in Fig. 13. The sensitivity is zero for the phases 0.0, 2.5, and 5.0 ns. This lack of sensitivity may lead to false locking for a PLL in operation. However,

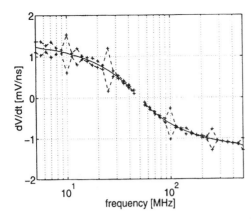

Fig. 12. Frequency sensitivity for the ncPFD for a number of frequencies. The plot is based on behavioral simulations with 20 different initial phases for each frequency. The solid line is the mean value and the "+" symbols are the minimum and maximum values. The reference frequency is 50 MHz.

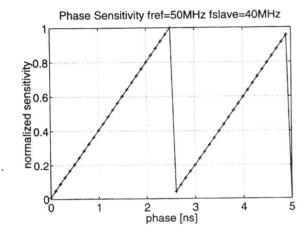

Fig. 13. Frequency sensitivity for the ncPFD when the slave frequency is 4/5 of the reference frequency. For the initial phases of 0.0, 2.5, and 5.0 ns the sensitivity is zero.

this false locking will not be stable, since a small phase change results in a nonzero sensitivity and drives the loop back to lock.

One way to add small phase changes to the simulation is to include phase noise which is always present in an oscillator. When we add phase noise of approximately 300 ps peak-to-peak to the simulations, the normalized minimum sensitivity which was zero will increase to approximately 0.01. The improvement is not significant but the sensitivity will be nonzero and positive for all phases. Hence, false locking is avoided. To further enhance the phase noise during the lock in process, one could use dithering techniques, i.e., add the signal from a noise/signal source to the control voltage of the VCO.

V. BEHAVIORAL MIXED-MODE SIMULATIONS

In order to understand the sensitivity to frequency errors and lock-in properties of the proposed detector, a complete third-order charge pump PLL system was simulated using a multilevel mixed-mode simulator, Lsim [5]. The PFD was represented by a schematic simulated in switch mode. The VCO, phase-noise generator, and charge pump are represented by behavioral models written in the hardware description

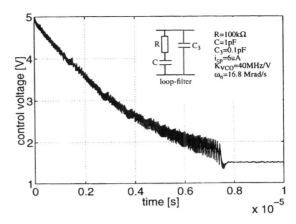

Fig. 14. Lock-in process of a third-order PLL with the ncPFD as phase frequency detector. The loop filter and PLL data are shown in the upper right corner.

language M [6]. The loop filter used ideal R and C models in circuit mode with analog voltages. The loop filter and PLL data are shown as an inset in Fig. 14. A lock-in simulation is shown in Fig. 14. The simulation is done with the presence of 300 ps peak-to-peak phase noise.

Because of the sawtooth-shaped frequency sensitivity of the ncPFD (for a fixed frequency offset and varied initial phase), Fig. 13, and the presence of noise, the lock-in time is not deterministic but random. The lock-in times for 60 simulations have been analyzed. Most simulations show a lock-in time of 7 μs and the largest time is 16 μs. There is no upper limit on the lock-in time. One simulation took approximately 3 cpu-min on a SPARC 10 workstation.

VI. EXPERIMENTS

The phase detection properties of the ncPFD have been verified experimentally with a test chip. The test chip is a line receiver for serial data that utilizes several parallel samplers to receive bit rates of 2.0 Gb/s [7]. The phase detector was used in a delay-locked loop (DLL) which generates control signals for the sampling switches used in the line receiver. The ncPFD, Fig. 3(a), was used as a π-rad phase detector and the delay line was half a wavelength long.

The skew between the reference and slave signals is not possible to measure directly. This quantity has been measured indirectly through measurement error compensation circuits to be about 125 ps at $f = 250$ MHz. Unfortunately, there is no control of how large the measurement error is.

The circuit blocks used to measure the offset are shown in Fig. 15. The two clocks that we want to compare come from the beginning and the end of the delay line. They are fed into two matched inverter chains where the propagation delay for rising and falling edges are matched against process variations [8]. The delay from the multiplexer inputs to the oscilloscope screen for the two signal paths are not matched. Two measurements are done to compensate this. One where the delay line input signal goes uninverted through *Output buffer 1* and one where the same signal goes inverted through the *Output buffer 2*. The measured skew including the measurement error

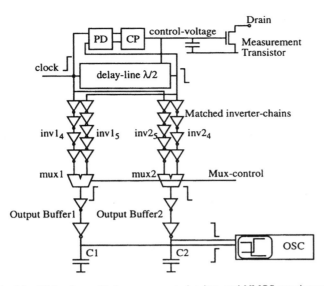

Fig. 15. DLL, phase offset measurement circuitry, and NMOS transistor to access the control voltage.

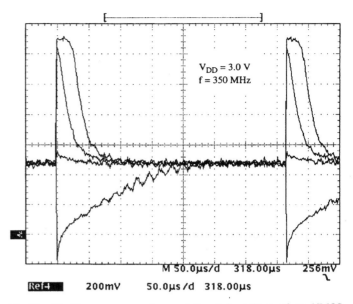

Fig. 16. Oscilloscope screen dump of the drain voltage of an NMOS transistor with external pull-up resistor where the gate is connected to the control voltage. Four different lock-in procedures are shown. The initial control voltages are 0.0, 1.0, 2.0, and 3.0 V for the curves from top to bottom, respectively.

for the measurements will be as follows:

$$
\text{skew}_1 = (\Delta + \text{inv2}_{4\text{fall}} + \text{mux2}_{\text{fall}} + \text{Buf2}_{\text{fall}})
$$
$$
- (\text{inv1}_{4\text{rise}} + \text{mux1}_{\text{rise}} + \text{Buf1}_{\text{rise}}) \quad (1)
$$
$$
\text{skew}_2 = (\Delta + \text{inv1}_{5\text{fall}} + \text{mux1}_{\text{rise}} + \text{Buf1}_{\text{rise}})
$$
$$
- (\text{inv2}_{5\text{rise}} + \text{mux2}_{\text{fall}} + \text{Buf2}_{\text{fall}}) \quad (2)
$$

where Δ is the real skew and $\text{inv2}_{4\text{fall}}$ and $\text{inv1}_{4\text{rise}}$ are the delays through the four inverters' long chains for falling and rising edges through the left and right chain, respectively. Similarly, $\text{inv1}_{5\text{fall}}$ and $\text{inv2}_{5\text{rise}}$ are for the five inverters' long chains. And $\text{mux1}_{\text{rise}}$ and $\text{mux2}_{\text{fall}}$ are the delays through the multiplexers. The $\text{Buf1}_{\text{rise}}$ and $\text{Buf2}_{\text{fall}}$ are the delays through the output-buffers and through the oscilloscope input-channels.

The sum of the skews (1) and (2) is

$$
\text{skew}_1 + \text{skew}_2 = 2\Delta + \text{inv2}_{4\text{fall}} - \text{inv1}_{4\text{rise}}
$$
$$
+ \text{inv1}_{5\text{fall}} - \text{inv2}_{5\text{rise}}. \quad (3)
$$

Note that the expression is independent of the *mux* and *Buf* delays. Hence, theoretically, if the rise and fall delays of the inverter chains are matched properly, there will not be any measurement error.

In Fig. 16 an oscilloscope screen dump with four lock-in procedures is shown. The signal is the drain voltage of an NMOS transistor with an external pull-up resistor and with the gate connected to the control voltage as shown in Fig. 15. The lock-in time is less than 200 μs. Ideally, the control voltage should go monotonically to the equilibrium voltage. Therefore, the beating in the lock-in procedure when the initial control voltage is 3.0 V is unexpected. The reason for this is unknown.

VII. CONCLUSIONS

A new PFD without a dead zone has been proposed. The circuit topology is simple and has no feedback loops. Simulation results indicate that the circuit can operate up to 800 MHz in 0.8-μm CMOS with a 5-V supply. The detector's phase offset depends on the duty cycle of the inputs. Measurements have been performed on the detector when it was used in a DLL as a phase detector and the functionality was verified.

REFERENCES

[1] R. E. Best, *Phase-Locked Loops,* 2nd ed. New York, NY: McGraw-Hill, 1993.
[2] N. H. E. Weste and K. Eshragrian, *Principles of CMOS VLSI Design,* 2nd ed. Reading, MA: Addison Wesley, 1993.
[3] H. Kondoh, H. Notani, T. Yoshimura, H. Shibata, and Y. Matsuda, "A 1.5-V 250-MHz to 3.0-V 622-MHz operation CMOS phase-locked loop with precharge type phase-detector," *IEICE Trans. Electron.,* vol. E78-C, no. 4, pp. 381–388, Apr. 1995.
[4] P. Larsson and C. Svensson, "Skew safety and logic flexibility in a true single phase clocked system," in *Proc. IEEE Int. Symp. Circuits Syst.,* 1995, pp. II:941–944.
[5] Mentor Graphics, *Explorer Lsim User's Manual.* Mentor Graphics Corp., 1992.
[6] Mentor Graphics, *M Language User's Guide.* Mentor Graphics Corp., 1991.
[7] H. O. Johansson, J. Yuan, and C. Svensson, "A 4 Gsamples/s Line-Receiver in 0.8 μm CMOS," in *Proc. Symp. VLSI Circuits,* 1996, pp. 116–117.
[8] M. Shoji, "Elimination of process-dependent clock skew in CMOS VLSI," *IEEE J. Solid-State Circuits,* vol. SC-21, pp. 875–880, Oct. 1986.

PART V

Clock Generation by PLLs and DLLs

A 320 MHz, 1.5 mW @ 1.35 V CMOS PLL for Microprocessor Clock Generation

Vincent von Kaenel, *Member, IEEE*, Daniel Aebischer, Christian Piguet, and Evert Dijkstra

Abstract— This paper describes a low-power microprocessor clock generator based upon a phase-locked loop (PLL). This PLL is fully integrated onto a 2.2-million-transistors microprocessor in a 0.35-μm triple-metal CMOS process without the need for external components. It operates from a supply voltage down to 1 V at a VCO frequency of 320 MHz. The PLL power consumption is lower than 1.2 mW at 1.35 V for the same frequency. The maximum measured cycle-to-cycle jitter is ±150 ps with a square wave superposed to the supply voltage with a peak-to-peak amplitude of 200 mV and rise/fall time of about 30 ps. The input frequency is 3.68 MHz and the PLL internal frequency ranges from 176 MHz up to 574 MHz, which correspond to a multiplication factor of about 100.

I. INTRODUCTION

THIS paper reports on the design of a phase-locked loop (PLL) for on-chip clock generation in a high performance microprocessor (μP) [1]. The μP is targeted for portable computing applications where power consumption must be minimized. The primary motivations for the PLL design are as follows.

First, the PLL is used as a frequency multiplier to generate a high-speed internal clock for the microprocessor from a low-frequency, low-cost, 3.68 MHz quartz oscillator. The multiplication factor is in the range of 100 X. Additionally, the microprocessor requires a precisely controlled 50% duty cycle clock. The PLL incorporates a common technique of generating a clock at twice the microprocessor clock frequency that is then divided by two to achieve an accurate 50% duty cycle. The integration of the PLL and divide-by-two circuitry on-chip eliminates the need to drive the high capacitance board level interconnects with high frequency signals and effectively lowers the power consumption of the system board.

Second, the PLL must allow fast recovery from the μP's idle (low power) mode to normal (full speed) mode. Two design approaches were considered. The first is a PLL that is disabled during idle mode. This approach minimizes power during idle mode but requires a low settling time for the PLL during the transition from idle to normal mode. Alternatively, a low-power PLL design that runs continuously during idle mode was considered. This approach allows instantaneous transition between the modes but requires the power consumption of the PLL to be minimized. This paper describes such a PLL.

Manuscript received May 1, 1996; revised August 5, 1996. This work has been carried out under a contract of Digital Equipment Corporation.

The authors are with CSEM, Swiss Center for Electronics and Microtechnology, Inc., 2007 Neuchâtel, Switzerland.

Publisher Item Identifier S 0018-9200(96)08095-X.

TABLE I
PROCESS PARAMETERS SUMMARY

0.35μm CMOS Parameters summary	
Feature size	0.35μm
Channel length	0.25μm
Threshold voltages	±0.35V
Gate oxide	6.5nm
Substrate	p-epi with n-well

Finally, when looking at target systems for the μP, we found no pressing need for phase locking. Consequently, we removed this as a design criteria and concentrated our efforts on minimizing clock cycle-to-cycle jitter.

The integration of the PLL onto a microprocessor complicates the design as the PLL is exposed to the microprocessor power supply switching noise. This noise induces output clock period variation (jitter). To obtain the highest performance from the microprocessor, the output jitter of the PLL has to be as low as possible. The power consumption of the microprocessor has been further reduced by scaling the supply voltage down to 1.35 V. The microprocessor and the PLL have been implemented in a 0.35-μm CMOS process which features low threshold MOS devices to maintain the speed performance at low supply voltage (Table I).

In summary, the challenge was to design a PLL with a high multiplication factor that combines limited output jitter, low supply voltage, and low power consumption.

II. SYSTEM TRADEOFFS

In our case, the major noise source for the output jitter is the power supply switching noise of the digital circuits (estimated to be 10% of the supply voltage, so the absolute minimum supply voltage is 1.2 V). This noise has a very large bandwidth, even larger than the μP clock frequency. So, the power supply noise rejection (PSNR) of analog circuits inside the PLL must be maximized.

For low-power PLL's, the second jitter source is the intrinsic noise of MOS devices in the VCO. This noise can be reduced by increasing the current flowing into sensitive devices, thus increasing the power consumption. On the other hand, to obtain low-voltage analog circuits, the saturation voltage of MOS devices needs to be reduced by using wider devices. This results in a larger parasitic capacitance between the supply voltage and the internal sensitive analog nodes, which might decrease the PSNR for the same current consumption. This paper describes how these contradictory effects have been

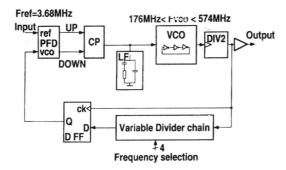

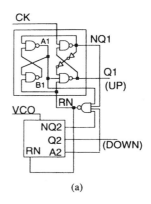

Fig. 1. PLL Architecture.

taken into account to optimize the current consumption with a high PSNR at low-supply voltage [12].

The PLL architecture is depicted on Fig. 1. The reference frequency is delivered externally by a quartz oscillator at 3.68 MHz. This signal is the reference frequency for the phase and frequency detector (PFD). The feedback of internal clock is compared to the reference clock for phase and frequency error. The PFD generates Up or Down pulses to the charge pump circuit (CP) which is followed by the loop filter (LF). This filter stabilizes the PLL. The loop filter voltage controls the voltage controlled oscillator (VCO). The output frequency of the VCO is divided by two in order to deliver a 50% duty cycle clock signal. The output frequency of the VCO can be programmed from a minimum frequency of 176 MHz to a maximum frequency of 574 MHz. The divider ratio can be selected between 16 values ranging from 48 to 156.

III. PLL BLOCKS

A. Phase and Frequency Detector

The PFD uses a dual D-flip-flop (DFF) structure [2], [3]. As the phase difference between the inputs decreases, the pulse width on Up or Down also decreases. To avoid the appearance of a dead zone (range of phase difference where no PFD output is generated), it is usual to have a minimum pulse width on Up and Down even if there are no phase differences between the two inputs. These simultaneous Up and Down signals in the steady state of the PLL create a short circuit in the charge pump which results in a perturbation on the LF voltage, and hence produces jitter. To limit the LF voltage perturbation without having a dead zone, we have reduced the minimum Up and Down pulse widths by reducing the reset delay in the PFD. Moreover, the linearity of the PFD and charge pump is affected by this minimum pulse width on Up and Down when there is no phase difference between the inputs of the PFD. The new PFD has no dead zone and the nonlinearity near the steady state of the PLL is reduced.

Fig. 2(a) shows the conventional NAND-based PFD [10]. It is constructed with two flip-flops which are set when the clock transition goes low. Fig. 2(b) shows the flow graph of one of these flip-flops. The reset signal RN has to be generated as soon as the flip-flop reaches the stable state 1010 ($A1$, $B1$, $NQ1$, $Q1$) from the preceding unstable state 1011. The variable NQ, switching before Q (Up or Down), is used to

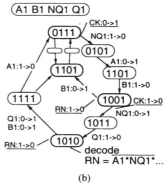

Fig. 2. Conventional NAND-Based PFD with flow graph.

generate the reset signal. We consider now the steady state of the PLL (perfect synchronization between VCO and CK). Up ($Q1$) and Down ($Q2$) rise. Then $NQ1$ and $NQ2$ fall after a gate delay. As $A1$ and $A2$ are already active, the reset is decoded. The assertion of reset causes the Up and Down signals to transition low. So, the minimum pulse width of Up and Down is at least the delay three gates (two inputs NAND, four inputs NAND and three inputs NAND). These simultaneous pulses on Up and Down are a contradictory signal for the charge pump resulting in a perturbation on the loop filter voltage and finally in jitter. Moreover, due to the chosen state assignment, one has to use both internal variables NQ and A to generate properly the reset signal RN. If not, the state 0111 will generate a reset signal when it is not wanted. Neither NQ nor A alone can be used to generate the reset. It results in a NAND gate RN with four inputs ($NQ1$, $A1$, $NQ2$, $A2$). Such a gate is slow, as the reset is achieved with $RN: 1 \rightarrow 0$, through four MOS transistors in series. Designers of this PFD have inserted two inverters between the transitions of NQ and Q in such a way as to create a delay between NQ and Q due to this slow reset gate. The proposed design (Fig. 5) solves this problem while using a fast reset gate with only two MOS device in series.

The design of the new PFD has been performed using an asynchronous race-free design method. A detailed description of this design methodology is presented in [5] and [6]. A basic schematic of such a circuit is shown in Fig. 3. This circuit has the same basic function as the NAND-Based PFD, but the state assignment is different. Two D-flip-flops with $D = 1$ are clocked with the two clock signals which are compared. If CK is active before VCO, the Up output is generated while Down

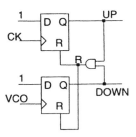

Fig. 3. PFD schematic.

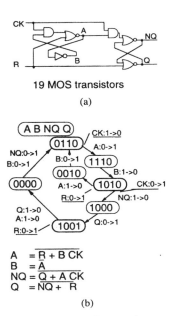

19 MOS transistors

(a)

A = $\overline{R + B\ CK}$
B = $\overline{A}$
NQ = $\overline{Q + A\ CK}$
Q = $\overline{NQ + R}$

(b)

Fig. 4. D-flip-flop with D = 1 with flow graph.

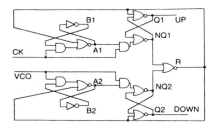

Fig. 5. Phase and frequency detector.

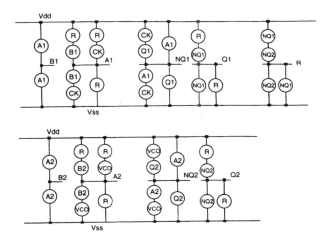

Fig. 6. PFD MOS circuit (MOS devices represented by circles. NMOS devices are between internal nodes (A1, A2, B1, B2, . . .) and VSS, PMOS are between internal nodes and VDD).

is produced if VCO is active before CK. As soon as the two clock signals are simultaneously active, the two D-flip-flops are reset. If CK and VCO are active at the same time, ideally neither Up nor Down has to be set. In such a case, the reset signal R has to be fast enough to minimize the Up and Down activation time when the PLL is in the steady state (perfect synchronization between CK and VCO).

Fig. 4 shows an implementation of a D-flip-flop with D = 1 having a noncritical race. The idea is to minimize the simultaneous activation time of Up and Down. As both A and Q have to go Down, while B and NQ have to switch on, both A and Q can switch simultaneously when the reset is active, producing the simultaneous switching on of B and NQ. As two variables switch simultaneously (A and Q between states 1001 and 0000, 1000 or 0001 are possible intermediate states), there is a race. However, the final stable state (0110) does not depend on the order of switching of A and Q. Hence, the result of the race is always the same (noncritical race) and is layout independent (it does not depend on the value of parasitic capacitances on internal nodes nor of the size of MOS devices).

Based on this D-flip-flop implementation, Fig. 5 shows the phase frequency detector accordingly to Fig. 3. To speed Up the reset even more, a NOR gate producing the reset signal R is controlled by the NQ signals of the D-flip-flops. These NQ signals switch before the Q (Up or Down) outputs between

states 1010 and 1001 (see Fig. 4). As soon as the NQ signal is low, the NOR gate generates a reset signal for the D-flip-flops. As the Up or Down signal is generated after NQ1 or NQ2, the minimum pulse width on Up or Down (Q1, Q2) is the delay of only one gate (two inputs NOR). By a careful sizing of the Q NOR gate and the reset gate, it is possible to have a very low pulse width on Up and Down in the steady state of the PLL. This results in a lower jitter and better linearity near the steady state without dead zone.

Each D-flip-flop contains 19 transistors in a branch-based implementation [8], [9]. Fig. 6 depicts the PFD MOS circuit with 42 transistors. The delay has been reduced compared to the conventional NAND-based PFD circuit [10] due to the fact that the number of serial transistors in branches is limited to two for all gates. It is possible to invert the MOS schematic of Fig. 6 in order to have a NAND gate based PFD. However, no significant speed advantage has been observed between these two version since the maximum number of PMOS in series in each branch remains the same. A careful analysis has been performed on the complete PFD circuit of Fig. 5. This design requires that both flip-flops are simultaneously reset. If one flop-flop resets early it can result in the reset signal returning to R = 0 before the second flip-flop has reset. Such a behavior is inherent to the basic structure of Fig. 3 and can be controlled with careful matching of the two flip-flops

B. Charge Pump

The charge pump circuit (CP) uses a switch structure that cancels the charge injection by using dummy devices (Fig. 7) [6]. Even with low-threshold voltage MOS devices, the static

TABLE II

CALCULATED OR SIMULATED (WITH PARASITIC COUPLING
CAPACITANCES EXTRACTED FROM LAYOUT) CONTRIBUTIONS TO
THE OUTPUT PEAK-TO-PEAK JITTER AT 320 MHz @ 1.35 V

Jitter contributor without supply noise	Jitter (p-p)	unit
White Noise in VCO	±30	ps
Dead zone of PFD	<±10	ps
Leakage on LF and Charge injection	±15	ps
Total Jitter without supply noise	±55	ps
Jitter due to a 0.2V supply jump in 30ps		
VCO induced jitter	±80	ps
Jitter induced by the change of the LF voltage	±10	ps
Total Jitter due to a 0.2V supply jump	±90	ps
Jitter due to a 10mV substrate jump in 30ps		
VCO induced jitter	<±5	ps
Total jitter due to a 10mV substrate jump	±5	ps
Total Jitter(sum of the above contributors)	**±150**	**ps**

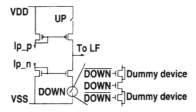

Fig. 7. Charge pump and switch detail.

phase error due to channel leakage current is lower than 4 mrad (2 ps @ 320 MHz). This jitter source is negligible compared to the supply noise contribution (Table II). The specified lock-time (100 μs) made it possible to choose a low PLL natural frequency (100 kHz), thus reducing the ripple voltage on the LF, and hence the output jitter. However, in order that the PLL compensates for the $1/f$ noise in the MOS devices of the VCO, the natural frequency of the loop should not be lower than 100 kHz. The LF realizes a zero and two poles, which results in a third-order system [2]–[4].

C. VCO

As far as the PSNR and low supply voltage operation are concerned, the VCO is the most critical block because its internal noise results directly in jitter. Moreover, low-voltage operation limits the design options. To obtain a fully integrated PLL, a current controlled ring oscillator (CCO) is the basic element of the VCO (Fig. 8). It allows high-frequency operation at low-voltage operation, since no additional capacitances have been used on internal nodes other than the ones created by the inverter devices.

The design procedure for the CCO is described in the following lines. The CCO MOS devices are in the strong inversion region and

$$\beta_n = \beta_p = \beta$$
$$n_n = n_p = n$$
$$V_{Tp} = V_{Tn} = V_T$$
$$I_D = \frac{\beta}{2n}(V_{gs} - V_T)^2 \qquad (1)$$

where β is the gain of the MOS device, n the weak inversion slope, V_T the threshold voltage, and I_D the drain current in

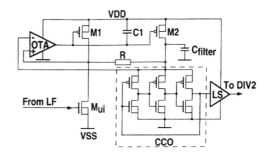

Fig. 8. VCO detailed schematic.

saturation. As this is a low-power, low-voltage CCO, the MOS device are not in velocity saturation. Assuming that only one inverter is active at a time

$$\frac{I_{cco}}{\beta} \propto (V_{cco} - V_T)^2. \qquad (2)$$

The minimum voltage across the CCO is obtained when I_{cco}/β is minimum for a given process. The β is maximum when the length of the MOS device is minimum for a given width. The CCO current can be calculated from the well known formula of consumption in digital circuits $I_{cco} = fC_{eq}V_{cco}$ (3). In our case, the gate and the drain capacitance are much larger than the interconnect capacitance, therefore the equivalent capacitance on each internal node of the CCO C_{eq} is proportional to the width of MOS device, hence proportional to β for a minimum length MOS device. The CCO current over β is also

$$\frac{I_{cco}}{\beta} \propto fV_{cco}. \qquad (3)$$

If the CCO voltage is much larger than the threshold voltage, from (2) and (3) we can say that (for a given β)

$$f \propto \sqrt{I_{cco}} \propto V_{cco}. \qquad (4)$$

Equations (3) and (4) are not sufficient to calculate the CCO MOS devices β value for a given frequency and process. The white noise (thermal noise in MOS devices) affects the cycle-to-cycle jitter of the VCO through two major contributions. First, the voltage V_{cco} across the CCO collects noise coming from the voltage to current converter, assuming the CCO with C_{filter} as an equivalent parallel RC load. This causes a rather low frequency modulation to the VCO period. The second source of jitter results from the CCO internal charging and discharging time uncertainty of the node capacitors by the noisy MOS currents.

To limit the jitter due to the MOS device channel white noise (thermal noise), the current flowing in the CCO must be sufficiently high. The output period variance σ_T decreases when the current flowing in the CCO increases [11]

$$I_{cco} \propto \left(\frac{1}{\sigma_T^2}\right)^a. \qquad (5)$$

Allowing a limited amount of jitter (due to thermal noise) in the CCO and given an operating frequency, the minimum current flowing in the CCO can be calculated. Then, using (2) and (3), the β can be derived. To have a minimum CCO

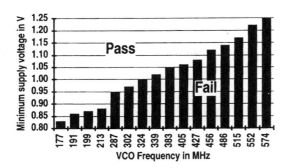

Fig. 9. Measured minimum PLL supply voltage.

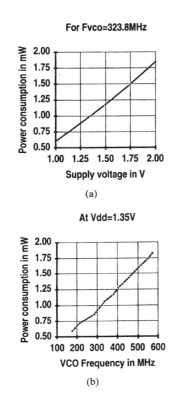

For Fvco=323.8MHz

(a)

At Vdd=1.35V

(b)

Fig. 10. Measured PLL power consumption.

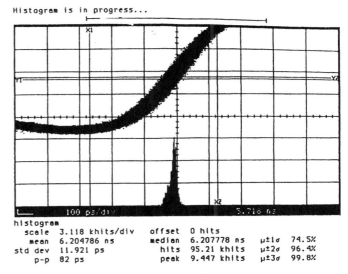

histogram
scale	3.118 khits/div	offset	0 hits
mean	6.204786 ns	median	6.207778 ns	μ±1σ 74.5%
std dev	11.921 ps	hits	95.21 khits	μ±2σ 96.4%
p-p	82 ps	peak	9.447 khits	μ±3σ 99.8%

Fig. 11. Measured jitter without supply noise.

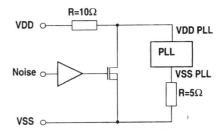

Fig. 12. Integrated supply noise generator.

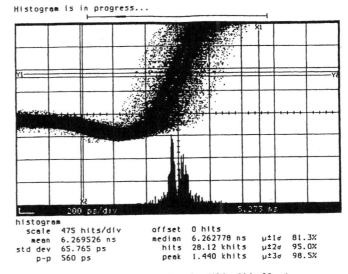

histogram
scale	475 hits/div	offset	0 hits
mean	6.269526 ns	median	6.262778 ns	μ±1σ 81.3%
std dev	65.765 ps	hits	28.12 khits	μ±2σ 95.0%
p-p	560 ps	peak	1.440 khits	μ±3σ 98.5%

Fig. 13. Measured jitter with supply noise (436 mV in 30 ps).

voltage in a given process, a minimum length for MOS devices should be chosen and no additional capacitances should be added on internal nodes of the CCO (C_{eq} minimum).

Then, as the supply voltage is specified, the maximum saturation voltage of the mirror supplying the current to the CCO is determined. It is not possible to use a cascoded current mirror for operation as low as 1.2 V because the saturation voltage of such a mirror is too high for the expected PSNR. Therefore, to ensure a maximum PSNR even at low supply voltages, a new circuit called active cascode is used (Fig. 8). Two characteristics linked to the VCO sensitivity to supply noise are interesting: first, the sensitivity to very low frequency or dc variations on the supply voltage; second, the sensitivity to high frequency variation on the supply voltage. The VCO dc sensitivity to supply voltage is low (1 ps per 25 mV of supply variation for $f_{vco} = 320$ MHz by worst case simulation at a supply voltage of 1.2 V and 1 ps per at least 50 mV in the same condition has been measured) due to the

feedback loop, with an operational transconductance amplifier (OTA) that maintains the voltage equality across the mirror devices ($M1$ and $M2$). The high frequency PSNR is mainly determined by the ratio of the drain parasitic capacitance of the mirror MOS devices and the filtering capacitor (C_{filter}) across the CCO. This last capacitor cannot be increased too much because it introduces a pole in the PLL which can make it unstable. The size of the mirror devices is determined by the maximum allowable saturation voltage. The simulated and

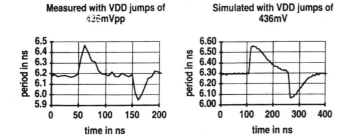

Fig. 14. Cycle-to-cycle period measurements and simulation.

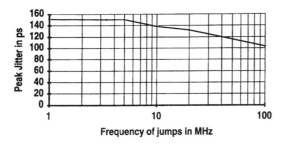

Fig. 15. Measured jitter with supply noise (200 mV).

TABLE III
SIMULATED MAXIMUM CURRENT CONSUMPTION AT 320 MHz AND 1.35 V

Block	I	unit
VCO, current reference	800	µA
Divider chain	240	µA
Divider by two	50	µA
Charge Pump	10	µA
Phase and Frequency Detector	2	µA
Total	1102	µA

measured high frequency PSNR is 1 ps per 2.5 mV of supply variation in closed loop for a VCO output frequency of 320 MHz and with a supply voltage of 1.2 V. When the supply voltage is higher, the PSNR is also higher, the given values are worst case. The stability of this active cascode is ensured by the resistor R between the input and output of the mirror. The relation between current and frequency in this CCO is not linear. To a first approximation, the frequency depends on the square root of the current for a voltage across the CCO much larger than the maximum threshold voltage of P or N-MOS devices (4). To obtain a linear gain for the VCO, the voltage to current converter (V/I) should have a quadratic transfer function. This is achieved by using a M_{ui} in the strong inversion region (Fig. 8).

The level shifter (LS) provides a full signal swing to the first divide-by-two circuit to generate a 50% duty cycle signal for the µP.

D. Dividers

In order to reduce the power consumption, the divider chain has been realized with asynchronous dividers (or ripple counters). The jitter introduced by these dividers is cancelled by a D-flip-flop that resynchronizes the output signal of the dividers with the output frequency of the PLL.

The divider ratio can be selected between 16 values ranging from 48 to 156. In order to have low-power programmable

TABLE IV
PLL PERFORMANCE SUMMARY FOR $F_{vco} = 320$ MHz

	Measured	Calculated
Supply voltage range	1.0 - 2.2V	1.2 - 2.2V
Power consumption (V$_{DD}$=1.5V)	1.2mW	<1.5mW
VCO period sensitivity to supply voltage (V$_{DDmin}$=1.2V)	1ps/52mV	<1ps/25mV
Jitter without supply noise	±41ps	±55ps
Jitter with V$_{DD}$ jumps of 200mVpp, 30ps edge time (V$_{DDmin}$=1.2V)	±150ps	±150ps
Settling time	60µs	<100µs
Area	0.21mm^2	

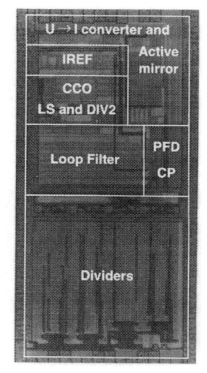

Fig. 16. PLL micrograph.

dividers, we have implemented 16 independent ripple counters that can be selected by a multiplexor and a demultiplexor. The area of these 16 dividers is larger than the area of a single programmable divider, however, the power consumption is lower because of the simplicity of the ripple counter.

IV. MEASUREMENT RESULTS

The minimum operating supply voltage versus the VCO frequency is represented on Fig. 9. For these measurements, the process of the measured PLL was approximately slow for both N and P-MOS devices. So this is a worst case. For the nominal VCO frequency (320 MHz) the minimum supply voltage is 1 V which is well below the minimum specified operating voltage of 1.2 V.

The power consumption versus supply voltage for the nominal frequency (320 MHz) is depicted in Fig. 10(a). At the nominal supply voltage (1.35 V), the power consumption is lower than 1.2 mW. Fig. 10(b) shows the power consumption as a function of the VCO frequency with a supply voltage of 1.35 V. At the maximum frequency, which is 574 MHz, the power consumption is lower than 2 mW.

The jitter without supply noise has been measured and plotted on Fig. 11 for the nominal frequency and supply voltage. This measurements gives ±41 ps of cycle-to-cycle jitter (PLL in closed loop). The worst case calculated value is ±55 ps.

The noise spectrum generated by the digital part is related to the internal speed of the process. Hence, as a model for this noise, we have used a supply voltage square wave with rise and fall times given by the maximum speed of the process and an amplitude given by the simulations of the package response. In our case, we have assumed 200 mV peak-to-peak amplitude of noise and rise and fall times of 30 ps (simulated). This noise model is really more severe than an externally generated sine or square wave on the supply voltage up to the frequency of the VCO because of the wide noise spectrum generated. As it is not possible to generate externally such a noisy supply voltage, we have integrated a noise generator to characterize the PSNR of this PLL. Fig. 12 shows the integrated supply noise generator. A switch controlled by an external signal induces a voltage drop across a serial resistor with the PLL supply voltage. When the switch is off, the serial resistance produce a negligible voltage drop because of the low-power consumption of the PLL. The simulated rise and fall time of the square wave produced by this generator is 30 ps, and in any case related to the maximum speed of the process. These rise and fall times are difficult to measure. However, with an oscilloscope we have observed rise and fall times lower than 1 ns (by on-chip probing) which is lower than one cycle of the CCO. One advantage of this noise model is that it can be used in simulation during the design. The serial resistor and the width of the switch has been designed to produce a voltage jump of at least 200 mV. In our case of process, the voltage drop was around 400 mV. Then, by linear interpolation between the measured jitter without supply noise and the measured jitter with supply noise, we have calculated the jitter for a jump of 200 mV. Improvement in the noise generator made it possible to electrically adjust the amplitude of the square wave to 200 mV. The jitter measurements with these improved test structure show that the linear interpolation gives accurate results. Fig. 13 shows the measured jitter with a repetition frequency of jumps of 1 MHz asynchronous to the reference frequency and to the output frequency, an amplitude of jumps of 436 mV (the supply voltage jumps between 1.2 V and 1.636 V), and for the nominal VCO frequency of 320 MHz. The measured jitter with the PLL in closed loop is ±280 ps which corresponds to ±150 ps for supply voltage jumps of 200 mV. The two peaks on the histogram of Fig. 13 are not due to a high sensitivity to dc supply voltage, but due to the shape of the excitation waveform (noise generator) used on the supply voltage (square wave).

To closely analyze the output jitter due to supply noise, we have used a high sampling rate single-shot oscilloscope (8 GS/s) and a dedicated program that calculates each PLL output period duration by interpolation between the sampled points. The accuracy of this analysis can be as good as ±1 ps RMS [14]. Fig. 14(a) shows the period as function of the time during a period of a square wave on the supply voltage. Fig. 14(b) shows the simulated response of the PLL to the

same supply voltage noise (including all parasitic coupling capacitances extracted from layout). The simulations and the measurements results are close. So by this simulation method it is possible to have an accurate prediction of the measured PSNR of the PLL in closed loop.

The measured jitter for jumps of 200 mV as a function of the repetition frequency of the jumps is depicted on Fig. 15. It is important to note that Fig. 15 is not a Bode plot of the noise response of the PLL because the input signal is a square wave superposed on the supply voltage and the output signal is the jitter. The maximum measured jitter for the nominal output frequency of 160 MHz (nominal VCO frequency of 320 MHz divided by two) with a minimum supply voltage of 1.2 V is ±150 ps.

The current consumption of each subblock is presented in Table III. The PLL area is 0.21 mm² (Fig. 16). The temperature range is from 0°C to 100°C. Even with a better process (0.18-μm CMOS) and for a VCO operating frequency of 200 MHz, the lowest power consumption reported so far is 2 mW for a comparable jitter without supply noise [13]. At the same VCO operating frequency of 200 MHz, our PLL consumes only 1 mW [15]. The performances summary is presented in Table IV.

V. CONCLUSION

The designed circuit shows that it is possible to overcome the issue of PLL short settling time by using a very low power PLL generating the clock signal even in idle mode. So, the recovery time from idle mode to the normal mode is virtually zero.

The resulting PLL power consumption is lower than 1.2 mW @ 1.35 V. The measured output cycle-to-cycle jitter with a square wave of 200 mVpp with 30 ps rise and fall time as a supply voltage noise is ±150 ps. The jitter measured with the integrated noise generator is really close to the simulation result.

The measured PLL minimum supply voltage for a VCO frequency of 320 MHz is 1.0 V with a slow process corner.

The association of very low operating supply voltage and high PSNR has been achieved by using the active cascode in the VCO. The low power consumption has been achieved by a careful analysis of the jitter created by thermal noise in the VCO and choosing the minimum VCO current that allows to reach our goal.

This PLL is integrated on a microprocessor [1] that is now in full production. The PLL met all the design specifications.

ACKNOWLEDGMENT

The collaboration of DEC's StrongArm design team and T. Lee of Stanford University has been extremely valuable.

REFERENCES

[1] James Montanaro *et al.*, "A 160 MHz 32b 0.5 W CMOS RISC microprocessor," in *ISSCC'96*, Feb. 8–10, 1996, San Francisco CA, pp. 214–215.
[2] F. M. Gardner, *Phaselock Techniques* 2nd ed. New York: Wiley, 1979.
[3] ——, "Charge-pump phase-lock loops," *IEEE Trans. Commun.*, vol. COM-28, no. 11, pp. 1849–1858, Nov. 1980.
[4] M. van Pamel, "Analysis of a charge pump PLL's," *IEEE Trans. Commun.*, vol. 42, no. 7, pp. 2490–2498, July 1994.

[5] C. Piguet, "Logic synthesis of race-free asynchronous CMOS circuits," *IEEE J. Solid-State Circuits*, vol. 26, no. 3, pp. 271–380, Mar. 1991.

[6] Vittoz, E. Dijkstra, Shiels, Eds., *Low Power Design: A Collection of CSEM Papers*. Cleveland, OH: Electronic Design Books, A Division of Penton, 1995.

[7] T. Ibaraki and S. Muroga, "Synthesis of networks with a minimal number of negative gates," *IEEE Trans. Comput.*, vol. C-20, no. 1, Jan. 1971.

[8] J.-M. Masgonty *et al.*, "Technology- and power-supply-independent cell library," in *IEEE CICC'91*, May 12–15, 1991, San Diego, CA, USA, Conf. 25.5

[9] C. Piguet *et al.*, "Low-power low-voltage digital CMOS cell design," in *Proc. PATMOS'94*, Oct. 17–19, 1994 Barcelona, Spain, pp. 132–139.

[10] I. A Young *et al.*, "A PLL clock generator with 5 to 100 MHz of lock range for microprocessor," *IEEE J. Solid-State Circuits*, vol. 27, no. 11, pp. 1599–1607, Nov. 1992.

[11] J. A. Mcneil, "Jitter in ring oscillators," Ph.D. dissertation, Boston College of Engineering, 1994

[12] E. A. Vittoz, "Low power design: Ways to approach the limits," Plenary address ISSCC 1994, San Francisco, CA.

[13] M. Mizuno *et al.*, "A 0.18 μm CMOS hot-standby PLL using a noise-immune adaptive gain voltage-controlled oscillator," presented at *ISSCC'95*, Feb. 15–17 1995, San Francisco CA.

[14] M. K. Williams, "A discussion of methods for measuring low-amplitude jitter," presented at *Int. Test Conf.*, Washington DC, 1995.

[15] V. von Kaenel *et al.*, "A 320 MHz, 1.5 mW @ 1.35 V CMOS PLL for microprocessor clock generation," in *ISSCC'96*, Feb. 8–10, 1996, San Francisco CA, pp. 132–133.

Daniel Aebischer was born in Bienne, Switzerland, in 1964. He received the M.S. degree in electrical engineering from the Swiss Federal Institute of Technology (EPFL), Lausanne, Switzerland, in 1991.

In 1991, he joined the Centre Suisse d'Electronique et de Microtechnique S.A., Neuchâtel, Switzerland. He is currently working in the field of low-power crystal oscillators and fully integrated CMOS VCO's.

Christian Piguet, for a photograph and biography, see p. 879 of the July 1996 issue of this JOURNAL.

Evert Dijkstra was born in Heerenveen, The Netherlands, on January 1, 1960. From June 1983 to December 1984, he executed his technical training period and his M.S.E.E. thesis (cum laude) for the Twente University of Technology, The Netherlands, at the Centre Electronique Horloger, Neuchâtel, Switzerland.

After graduation, he joined the Centre Suisse d'Electronique et de Microtechnique (CSEM), Neuchâtel, Switzerland. From February 1988 to September 1988 he was on leave at VLSI Technology, San Jose, CA, for a joint CSEM-VLSI Technology E2PROM development project. He is presently Head of the CSEM IC Design business unit, working on battery operated designs. His personnel technical interests include the design of micropower mixed mode systems, A/D converters, DSP systems, and E2PROM developments.

Mr. Dijkstra served as the Technical Programme Chairman of the 1996 ESSCIRC conference.

Vincent von Kaenel (M'93) was born in Neuchâtel, Switzerland, on March 5, 1964. He received the M.S. degree in electronics with honors form the Ecole d'Ingénieur d'Yverdon, Yverdon, Switzerland, in 1989.

He joined the Centre Suisse d'Electronique et de Microtechnique S.A., Neuchâtel, Switzerland, in 1989. Since then, he has been working on the design of low-voltage, low-power, and high-speed digital circuits, micropower Q-oscillators, and low-voltage, highly integrated FM receivers. He is currently involved in the design of low-power, high-speed PLL's. He is currently on leave from CSEM and consulting at Digital Equipment Corporation in Palo Alto, CA.

A Low Jitter 0.3–165 MHz CMOS PLL Frequency Synthesizer for 3 V/5 V Operation

Howard C. Yang, Lance K. Lee, and Ramon S. Co

Abstract— This paper describes a phase-locked loop (PLL)-based frequency synthesizer. The voltage-controlled oscillator (VCO) utilizing a ring of single-ended current-steering amplifiers (CSA) provides low noise, wide operating frequencies, and operation over a wide range of power supply voltage. A programmable charge pump circuit automatically configures the loop gain and optimizes it over the whole frequency range. The measured PLL frequency ranges are 0.3–165 MHz and 0.3–100 MHz at 5 V and 3 V supplies, respectively (the VCO frequency is twice PLL output). The peak-to-peak jitter is 81 ps (13 ps rms) at 100 MHz. The chip is fabricated with a standard 0.8-μm n-well CMOS process.

Index Terms—CMOS phase-locked loop, current-steering amplifier, current-steering logic, frequency synthesizer, low noise, low voltage VCO.

I. INTRODUCTION

WITH the ever-increasing performance and decreasing price of microprocessors and PC/workstation systems, much more stringent requirements have been placed on the design of system clock synthesizers. Today's high-performance clock synthesizers are often required: 1) to operate over a wide frequency range (high frequency for increased performance and low frequency for power saving) using a crystal oscillator input with constant frequency; 2) to have small phase jitter and frequency variation; 3) to operate from 3 V/5 V V_{dd} supplies for both portable and desktop systems; 4) to have smooth transition between high and low frequencies; and 5) to have integrated loop filter on the chip. To satisfy all these requirements simultaneously, a stable loop over the entire operating frequency range is needed; and a low jitter, wide frequency range, and variable supply voltage-controlled oscillator (VCO) circuit is essential.

Several design techniques that improve the performance of a phase-locked loop (PLL) are presented in this paper. A current D/A converter is implemented to control the PLL bandwidth so that the loop performance is optimized over the operating frequency range of 0.3–165 MHz. The VCO is formed by a ring of single-ended current-steering amplifier (CSA) cells, which were first introduced as a current-steering logic (CSL) family for low noise and low power supply applications [1]. This VCO circuit can operate over a wide frequency range with low phase jitter at variable power supply. To achieve smooth frequency transitions, a pulse width limiting circuit is used to control the pulse width of the phase/frequency detector output.

Manuscript received June 25, 1996; revised October 4, 1996.
H. C. Yang and L. K. Lee are with the Shanghai Belling Microelectronics Manufacturing Co., Ltd., Shanghai 200233, P.R. China.
R. S. Co is with the Kingston Technology Corp., Fountain Valley, CA 92708 USA.
Publisher Item Identifier S 0018-9200(97)02477-3.

In Section II of this paper, the PLL architecture is described with an analysis of the loop stability and loop optimization. The circuit design techniques for the PLL are considered in Section III. The measured results are discussed in Section IV. Finally, conclusions are made in Section V regarding this work.

II. PLL ARCHITECTURE

It is often difficult to design a PLL that can operate over a wide frequency range due to the practical limit of the capacitor size that can be integrated for the loop filter. One method to widen the frequency range is to vary the PLL bandwidth as a function of the desired output frequency. This principle is applied in our design by utilizing a current D/A converter which controls the charge pump current. With this technique, we can also optimize the loop performance, including damping factor and loop gain, over the entire operating frequency range.

A block diagram of a conventional frequency synthesizer is enclosed in the dashed-line box in Fig. 1. The output frequency F_{out} is synthesized as

$$F_{\text{out}} = (M/N)F_{\text{in}}. \tag{1}$$

Using linear approximations, the loop equations of the PLL [2] for stability analysis are

$$K = \frac{(K_o/N)I_pR}{2\pi} \tag{2}$$

$$\zeta = \frac{RC_2}{2}\sqrt{\frac{(K_o/N)I_p}{2\pi C_2}} \tag{3}$$

where K is the loop gain, ζ is the damping factor, K_o is the VCO gain, I_p is the charge pump current, R is the loop filter resistor, and C_2 is the integration capacitor of the loop filter. In order to have an adequate margin of stability, (2) and (3) must satisfy the constraints $K < 2\pi f_{\text{PFD}}$ and $\zeta > 0.5$, where f_{PFD} is the operating frequency of the phase/frequency detector. The former constraint is required to prevent aliasing effects as a result of the Nyquist Criterion, while the latter constraint is required to have a satisfactory transient response. In practice, a loop gain which is ten times less than the phase/frequency detector frequency [2] is more than adequate.

For a clock synthesizer with a frequency range of 0.3–165 MHz, the feedback divider N in the loop equations can vary by more than 20 times. The tolerances of the loop parameters (K_o, I_p, R, and C_2) would also have to be accounted for. These conditions make the constraints for stability margin extremely difficult to satisfy if C_2 is kept within a reasonable size (a few hundred picofarads) for on-chip integration. As shown in Figs. 1 and 2, the proposed

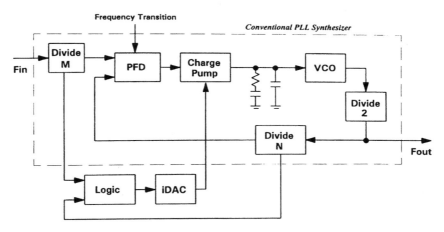

Fig. 1. Block diagram of frequency synthesizer.

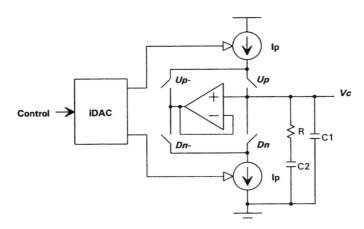

Fig. 2. Programmable charge pump with D/A converter control.

PLL architecture uses a current D/A converter to control the charge pump current, I_p. The product $(K_o/N)I_p$ is optimized for given values of M and N such that the stability margin constraints are satisfied by (2) and (3). Using a decoding table, the optimization is performed by the logic block in Fig. 1 which sets the charge pump current upon examination of M and N. Typically, a loop bandwidth which is ten times less than f_{PFD} for the entire frequency range can be easily achieved using this architecture.

III. CIRCUIT DESIGN TECHNIQUES

A. VCO Circuit

Two types of VCO based on the ring oscillator topology are commonly used in CMOS PLL design: the current-starved inverter based VCO [3]–[5] and the differential-pair based VCO [6]–[8]. In spite of a wide frequency range, the first type of VCO is sensitive to power supply noise. Although an on-chip voltage regulator can be used to reduce the effect of power supply noise [5], it is not effective for operation at high frequency since a voltage regulator inherently has poor ac rejection. Another drawback of using an on-chip regulator is that it reduces the useful power supply range, making it undesirable for low-supply applications. A local feedback

loop on the VCO can also be used for reducing the PLL jitter; however, it is likely to cause glitches or overshoots whenever the frequency transition mode is activated due to complicated feedback loops [8]. Hence, it is not a suitable approach for microprocessor applications, wherein a smooth transition between frequencies is usually required. While the second type of VCO rejects power supply noise well, the frequency range of operation may not be sufficient for some applications. To widen the frequency range of differential-pair-based VCO's, complex MOS resistors [7] can be used at the cost of higher V_{dd} supply and more complex design.

The VCO design in this work utilizes a simple CSA circuit [1]. Fig. 3(a) shows a CSA cell which consists of a current source, I_b, and a pair of NMOS devices. $M1$ is the input device and $M2$ is the load. When V_{in} is high, $M1$ turns on, sinking the bias current I_b, while $M2$ shuts off. Under this condition, the on resistance of $M1$ defines the output low voltage, V_{OL}. When V_{in} is low, $M1$ turns off and I_b is steered to $M2$. Under this condition, the resistance of the diode-connected $M2$ defines the output high voltage, V_{OH}. By varying the bias current, I_b, a current-controlled CSA-based ring oscillator is formed with an output voltage swing of

$$\Delta V = V_{\text{OH}} - V_{\text{OL}} = V_{\text{th}} + \sqrt{\frac{(W/L)_1 - (W/L)_2}{(W/L)_1 \cdot (W/L)_2} \cdot \frac{2I_b}{K'}}. \quad (4)$$

Equation (4) indicates that ΔV (typically between 1 and 2 V) varies with $\sqrt{I_b}$. Thus, the voltage swing of the CSA cell increases correspondingly with frequency. This is a desirable feature because the signal level improves at high frequency when the power supply switching noise becomes worse. Since the voltage swing is limited by the diode-connected $M2$, the current source always operates in the saturation region; consequently, very small switching noise is generated. For an n-well process, the PMOS current source can be guarded by its own well and isolated from the noisy p-substrate. The current source also buffers the output from V_{dd}, thereby reducing the noise injected from V_{dd} to the output. Any ground noise (coupled from other circuitry within the chip) is rejected by the CSA as a common mode noise because both its output and input are referred to the same ground. By referring the charge pump, loop filter, V/I converter, VCO, and other analog

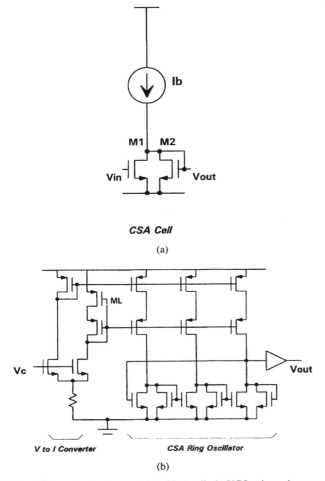

CSA Cell

(a)

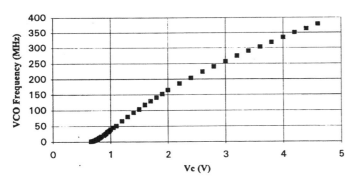

V to I Converter   CSA Ring Oscillator

(b)

Fig. 3. (a) Current steering amplifier (CSA) cell. (b) VCO using a three-stage CSA ring oscillator.

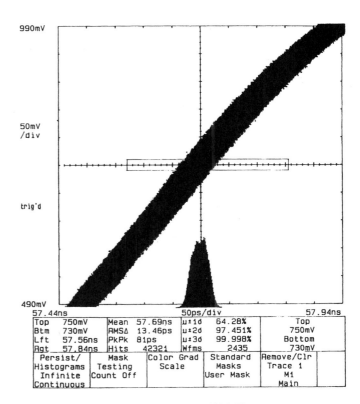

Fig. 5. Histogram of PLL period jitter at 100 MHz.

Top	750mV	Mean	57.69ns	μ±1σ	64.28%	Top	
Btm	730mV	RMSΔ	13.46ps	μ±2σ	97.451%		750mV
Lft	57.56ns	PkPk	81ps	μ±3σ	99.998%	Bottom	
Rgt	57.84ns	Hits	42321	Wfms	2435		730mV
Persist/		Mask		Color Grad		Remove/Clr	
Histograms		Testing		Scale		Trace 1	
Infinite		Count Off				M1	
Continuous						Main	

Fig. 4. Measured VCO performance.

circuits in the PLL to the same ground, i.e., p-substrate, the ground noise can be substantially rejected [9].

Fig. 3(b) shows the VCO circuit using a three-stage CSA ring oscillator. The current sources in the CSA ring oscillator are cascoded with high-swing bias. In the V/I converter, a single degenerated g_m stage provides a first-order linear relationship between the oscillation frequency and the control voltage. ML is forced in the linear region to provide the high-swing cascoded bias for the VCO. This VCO is suitable for low supply voltage operation since it only needs about 2.5 V

supply for proper operation. At $V_{dd} > 2.5$ V, the frequency of the CSA ring oscillator is practically independent of power supply. The maximum useful frequency of the VCO is limited by the saturation voltage (~0.5 V) of the cascoded PMOS current source in the charge-pump circuit.

The measured VCO performance is illustrated in Fig. 4. The frequency range of the VCO was observed from 174 kHz to 378.8 MHz at 5 V supply (the VCO frequency is twice the PLL output frequency). At $V_c \approx 2.3$ V, the VCO frequency achieved 200 MHz indicating that the PLL can operate at 100 MHz for 3 V power supply.

B. Phase/Frequency Detector

A modified dead-zone free, phase/frequency detector (PFD) is used in this design [2]. In the frequency transition mode, a pulse width limiting circuit in the PFD limits the pulse width of the UP and DN signals. Such limited UP/DN pulses provide a finite amount of charge to the integration capacitor of the loop filter in order to slow down the frequency ramp of the VCO. The UP/DN pulses, however, cannot be made arbitrarily narrow. The noise level in the chip may dominate over the minute UP/DN correction pulses causing the PLL not to properly acquire in frequency. The maximum *frequency transition* rate of less than 0.1%, i.e., the difference in period between two consecutive clock cycles, is achieved by using this technique.

C. Charge Pump and Loop Filter

The charge pump shown in Fig. 2 is designed using cascoded current sources with CMOS switches. The amount

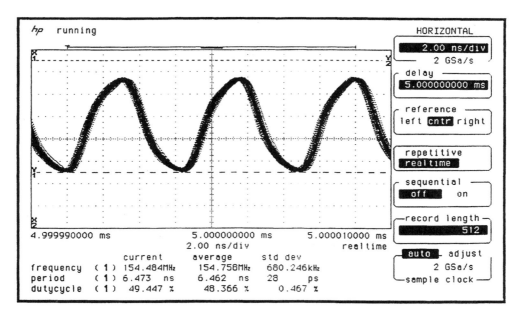

Fig. 6.   Measured PLL output at 155 MHz after 5 ms delay, i.e., after 775 000 clock cycles. The rms jitter (standard deviation) is 28 ps as shown.

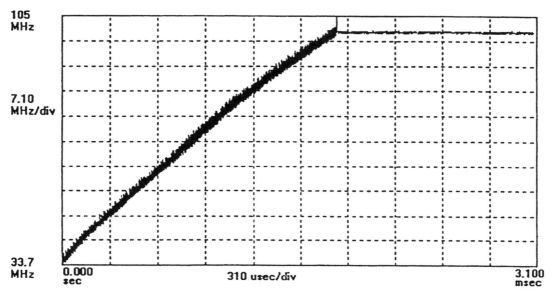

Fig. 7.   Measured frequency transition from 33.3 to 100 MHz.

of the current, typically 10–150 μA, is controlled by the current digital-to-analog converter (DAC). A bandgap current reference circuit is used to compensate the I_p variation over temperature. The opamp in the charge pump circuit reduces the transients caused by the charge transfer [4] as I_p is switched. The capacitors in the loop filter are formed by NMOS devices with the sources and drains connected to ground and the gate connected to the filter output node. The capacitor C2 is about 400 pF.

IV. MEASURED RESULTS

The measured results show that the PLL operates from 0.3 to 165 MHz and 0.3 to 100 MHz at 5 V and 3 V power supplies, respectively. Since the output frequency is the VCO frequency divided by two, a PLL output with 50% duty cycle is guaranteed. Fig. 5 shows the histogram of PLL *period jitter* (also referred to as short-term or clock-to-clock jitter) at 100 MHz after 42 321 hits using a standard 14.318 MHz crystal as the input reference frequency. The peak-to-peak jitter is 81 ps with an rms jitter of 13 ps. The long-term jitter of the PLL was also measured. An rms jitter of 28 ps is observed at 155 MHz with 5 ms delay, i.e., measured after a delay of 775 000 clock cycles from the triggered clock cycle, as shown in Fig. 6. In order to appreciate the noise rejection capability of this PLL, the chip was used to generate the clock for an HP laser printer. During the printing process, which is very noisy electrically and thermally, both the period jitter and the long-term jitter of the clock must be low for good printing quality. A

TABLE I
Summary of Measured PLL Performance

Frequency Range	0.3–165 MHz
Period (Short-Term) Jitter at 100 MHz	13 ps rms, 81 ps peak-to-peak
Long-Term Jitter at 155 MHz	28 ps rms after 5 ms delay
Supply Voltage Range	2.5 V to 7 V
Output Duty Cycle	50%, VCO frequency is twice PLL output
VCO Linearity	2% from 10–200 MHz
VCO Current Consumption	500 μA at 200 MHz
Crosstalk between 2 PLL's	50 dB down

comparison of test results between the use of this PLL and the original crystal clock showed no perceptible difference even under high magnification. Fig. 7 shows a smooth frequency transition of the PLL from 33 to 100 MHz. No frequency glitches and overshoots were observed during the transition time. Since there are two independent PLL's on the same chip, crosstalk between the two PLL's was also measured. The signal coupling between the two PLL's is at least 50 dB down. The measured performance of the PLL is summarized in Table I.

V. CONCLUSION

In this paper, we demonstrated the design of a fully integrated CMOS PLL circuit that achieves wide operating frequency range and low jitters (both short-term and long-term) over a wide range of power supply voltage. The key element in the design that provides all these features is the CSA-based VCO circuit. A programmable current DAC is also used to optimize the loop gain of the PLL. Smooth frequency transition is realized by using a modified PFD with a pulse width limiting circuit. The chip is implemented in a standard 0.8-μm CMOS process.

REFERENCES

[1] D. J. Allstot, G. Liang, and H. C. Yang, "Current-mode logic techniques for CMOS mixed-mode ASIC's," in *Proc. IEEE Custom Integrated Circuits Conf.*, 1991, pp. 25.2.1–25.2.4.
[2] F. M. Gardner, "Charge-pump phase-lock loops," *IEEE Trans. Commun.*, vol. 28, pp. 1849–1858, Nov. 1980.
[3] R. Shariatdoust, K. Nagaraj, M. Saniski, and J. Plany, "A low jitter 5 MHz to 180 MHz clock synthesizer for video graphics," in *Proc. IEEE Custom Integrated Circuits Conf.*, 1992, pp. 24.2.1–25.2.5.
[4] M. G. Johnson and E. L. Hudson, "A variable delay line PLL for CPU-coprocessor synchronization," *IEEE J. Solid-State Circuits*, vol. 23, pp. 1218–1223, Oct. 1988.
[5] K. M. Ware, H.-S. Lee, and C. G. Sodini, "A 200-MHz CMOS phase-locked loop with dual phase detectors," *IEEE J. Solid-State Circuits*, vol. 24, pp. 1560–1568, Dec. 1989.
[6] B. Kim, D. N. Helman, and P. R. Gray, "A 30-MHz hybrid analog/digital clock recovery circuit in 2-μm CMOS," *IEEE J. Solid-State Circuits*, vol. 25, pp. 1385–1394, Oct. 1990.
[7] I. A. Young, J. K. Greason, and K. L. Wong, "A PLL clock generator with 5 to 110 MHz of lock range for microprocessors," *IEEE J. Solid-State Circuits*, vol. 27, pp. 1599–1606, Nov. 1992.
[8] D. Mijuskovic *et al.*, "Cell-based fully integrated CMOS frequency synthesizers," *IEEE J. Solid-State Circuits*, vol. 29, pp. 271–279, Mar. 1994.
[9] D. J. Allstot and W. C. Black Jr., "A substrate-referenced data-conversion architecture," *IEEE Trans. Circuits Syst.*, vol. 38, pp. 1212–1217, Oct. 1991.

Low-Jitter Process-Independent DLL and PLL Based on Self-Biased Techniques

John G. Maneatis

Abstract— Delay-locked loop (DLL) and phase-locked loop (PLL) designs based upon self-biased techniques are presented. The DLL and PLL designs achieve process technology independence, fixed damping factor, fixed bandwidth to operating frequency ratio, broad frequency range, input phase offset cancellation, and, most importantly, low input tracking jitter. Both the damping factor and the bandwidth to operating frequency ratio are determined completely by a ratio of capacitances. Self-biasing avoids the necessity for external biasing, which can require special bandgap bias circuits, by generating all of the internal bias voltages and currents from each other so that the bias levels are completely determined by the operating conditions. Fabricated in a 0.5-μm N-well CMOS gate array process, the PLL achieves an operating frequency range of 0.0025 MHz to 550 MHz and input tracking jitter of 384 ps at 250 MHz with 500 mV of low frequency square wave supply noise.

I. INTRODUCTION

DELAY-LOCKED loops (DLL's) and phase-locked loops (PLL's) are often used in the I/O interfaces of digital integrated circuits in order to hide clock distribution delays and to improve overall system timing. In these applications, DLL's and PLL's must closely track the input clock. However, the rising demand for high-speed I/O has created an increasingly noisy environment in which DLL's and PLL's must function. This noise, typically in the form of supply and substrate noise, tends to cause the output clocks of DLL's and PLL's to jitter from their ideal timing. With a shrinking tolerance for jitter in the decreasing period of the output clock, the design of low jitter DLL's and PLL's has become very challenging.

Achieving low jitter in PLL and DLL designs can be difficult due to a number of design tradeoffs. Consider a typical PLL which is based on a voltage controlled oscillator (VCO). The amount of input tracking jitter produced as a result of supply and substrate noise is directly related to how quickly the PLL can correct the output frequency. To reduce the jitter, the loop bandwidth should be set as high as possible. Unfortunately, the loop bandwidth is affected by many process technology factors and is constrained to be well below the lowest operating frequency for stability [1]. These constraints can cause the PLL to have a narrow operating frequency range and poor jitter performance. Although a typical DLL is based on a delay line and thus simpler from a control perspective, it can have a limited delay range which leads to a set of problems similar to that of the PLL.

This paper describes both a DLL and PLL design based upon self-biased techniques [2]. Self-biasing can remove virtu-

Manuscript received May 3, 1996; revised July 29, 1996.
The author is with Silicon Graphics, Inc., Mountain View, CA 94043.
Publisher Item Identifier S 0018-9200(96)07946-2.

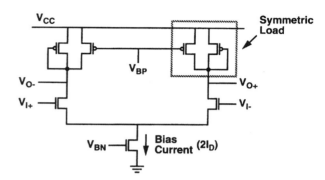

Fig. 1.   Differential buffer delay stage with symmetric loads.

ally all of the process technology and environmental variability that plagues PLL and DLL designs. Self-biasing can provide a bandwidth that tracks the operating frequency. This tracking bandwidth can in turn provide a very broad frequency range, minimized supply and substrate noise induced jitter with a high input tracking bandwidth, and, in general, very robust designs. Other benefits include a fixed damping factor for PLL's and input phase offset cancellation. Both the damping factor and the bandwidth to operating frequency ratio are determined completely by a ratio of capacitances giving effective process technology independence. The key idea behind self-biasing is that it allows circuits to choose the operating bias levels in which they function best. By referencing all bias voltages and currents to other generated bias voltages and currents, the operating bias levels are essentially established by the operating frequency. The need for external biasing, which can require special bandgap bias circuits, is completely avoided.

This paper will begin by reviewing a differential buffer stage design that provides high supply and substrate noise rejection and allows the possibility of self-biasing. The loop architecture for self-biased DLL and PLL designs will be presented in Section III and Section IV, respectively. A number of other loop components are critical to achieving low jitter in DLL and PLL designs. Section V will describe an improved phase-frequency comparator and differential-to-single-ended converter. The paper will also present some experimental results demonstrating the performance of the DLL and PLL designs.

II. DIFFERENTIAL BUFFER STAGE

In order to achieve low jitter operation, DLL and PLL designs require buffer stage designs with low supply and substrate noise sensitivity. The voltage-controlled delay line (VCDL) and the VCO used in the DLL and PLL designs are

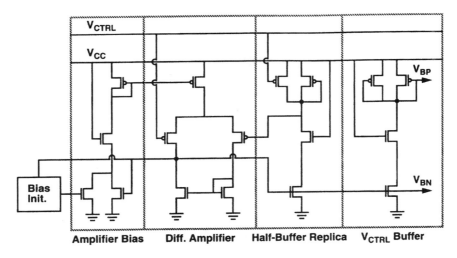

Fig. 2. Replica-feedback current source bias circuit.

based upon the differential buffer delay stages with symmetric loads and replica-feedback biasing [3].

The buffer stage, shown in Fig. 1, contains a source coupled pair with resistive load elements called symmetric loads. Symmetric loads consist of a diode-connected PMOS device in shunt with an equally sized biased PMOS device. The PMOS bias voltage V_{BP} is nominally equal to V_{CTRL}, the control input to the bias generator. Because of this equality, V_{CTRL} will be used instead of V_{BP} in subsequence references to the PMOS bias voltage. V_{CTRL} defines the lower voltage swing limit of the buffer outputs. The buffer delay changes with V_{CTRL} since the effective resistance of the load elements also changes with V_{CTRL}. It has been shown that these load elements lead to good control over delay and high dynamic supply noise rejection. The simple NMOS current source is dynamically biased with V_{BN} to compensate for drain and substrate voltage variations, achieving the effective performance of a cascode current source. However, this current source can provide high static supply and substrate noise rejection without the extra supply voltage required by cascode current sources.

The bias generator, shown in Fig. 2, produces the bias voltages V_{BN} and V_{BP} from V_{CTRL}. Its primary function is to continuously adjust the buffer bias current in order to provide the correct lower swing limit of V_{CTRL} for the buffer stages. In so doing, it establishes a current that is held constant and independent of supply voltage. It accomplishes this task by using a differential amplifier and a half-buffer replica. The amplifier adjusts V_{BN} so that the voltage at the output of the half-buffer replica is equal to V_{CTRL}, the lower swing limit. If the supply voltage changes, the amplifier will adjust to keep the swing and thus the bias current constant. The bandwidth of the bias generator is typically set equal to the operating frequency of the buffer stages so that the bias generator can track all supply and substrate voltage disturbances at frequencies that can affect the DLL and PLL designs. With this bias generator, the buffer stages have been shown to achieve a static supply noise rejection of about 0.25%/V while operating over a broad delay range with low supply voltage requirements that scale with the operating delay. The bias generator also provides

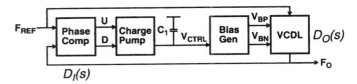

Fig. 3. Typical DLL block diagram (clock distribution omitted).

a buffered version of V_{CTRL} at the V_{BP} output using an additional half-buffer replica. This output isolates V_{CTRL} from potential capacitive coupling in the buffer stages and plays an important role in the self-biased PLL design.

Buffer stages with low supply and substrate noise sensitivity are essential for low jitter DLL and PLL operation. With the foundation of a buffer stage design with low noise sensitivity, the next two sections will consider techniques for self-biasing DLL and PLL designs. Such techniques will provide further reductions in input tracking jitter by allowing the loop bandwidth to be set as close as possible to the operating frequency.

III. SELF-BIASED DELAY-LOCKED LOOP

A self-biased DLL is constructed by taking advantage of the control relationship offered by a typical DLL. A typical DLL is shown in Fig. 3. It is composed of a phase comparator, charge pump, loop filter, bias generator, and voltage controlled delay line. The negative feedback in the loop adjusts the delay through the VCDL by integrating the phase error that results between the periodic reference input and the delay line output. Once in lock, the VCDL will delay the reference input by a fixed amount to form the output such that there is no detected phase error between the reference and the output. The VCDL delay, therefore, must be a multiple of the reference input clock period. With the chip-wide clock distribution network included as part of the VCDL delay, the DLL can be used to rebuffer the input clock signal without adding any effective delay.

A. Closed-Loop Response

The frequency response of the DLL can be analyzed with a continuous time approximation, where the sampling operation

397

the phase comparator is ignored. This approximation holds for bandwidths about a decade or more below the operating frequency. This bandwidth constraint is also required for stability due to the reduced phase margin near the higher order poles that result from the delay around the sampled feedback loop. Because the loop filter integrates the phase error, the DLL has a first-order closed loop response. The response could be formulated in terms of input phase and output phase. However, this set of variables is incompatible with a continuous time analysis since the sampled nature of the system would need to be considered. A better set of variables is input delay and output delay. The output delay is the delay between the reference input and the DLL output or, equivalently, the delay established by the VCDL. The input delay is the delay to which the phase comparator compares the output delay or, equivalently, the phase difference for which the phase comparator and charge pump generate no error signal. The output delay, $D_O(s)$, is related to the input delay, $D_I(s)$, by

$$D_O(s) = (D_I(s) - D_O(s)) \cdot F_{\text{REF}} \cdot \frac{I_{CH}}{sC_1} \cdot K_{DL} \quad (1)$$

where F_{REF} is the reference frequency (Hz), I_{CH} is the charge pump current (A), C_1 is the loop filter capacitance (F), and K_{DL} is the VCDL gain (s/V). The product of the delay difference and the reference frequency is equal to the fraction of the reference period in which the charge pump is activated. The average charge pump output current is equal to this fraction times the peak charge pump current. The output delay is then equal to the product of the average charge pump current, the loop filter transfer function, and the delay line gain. The closed loop response is then given by

$$\frac{D_O(s)}{D_I(s)} = \frac{1}{1 + s/\omega_N} \quad (2)$$

where ω_N, defined as the loop bandwidth (rad/s), is given by

$$\omega_N = I_{CH} \cdot K_{DL} \cdot F_{\text{REF}} \cdot \frac{1}{C_1}. \quad (3)$$

If the charge pump current I_{CH} and the VCDL gain K_{DL} are constant, the loop bandwidth ω_N will track the operating frequency ω_{REF}. However, the parameters I_{CH}, K_{DL}, and C_1 are process technology dependent and will cause the loop bandwidth to vary around the design target. In addition, constant gain VCDL's are typically implemented by interpolating between two closely delayed signals with a weighted sum which leads to a narrow delay range. Linear results are obtained only when the delay spacing is less than the signal edge rate. As the delay spacing increases, the interpolation becomes increasingly nonlinear. The delay range can be extended by using a VCDL with nonlinear delay control at the expense of a tracking loop bandwidth.

B. Bandwidth Tracking

Symmetric load buffer stages can be used to implement the VCDL in order to obtain a broad delay range. Fig. 4 shows their typical delay as a function of control voltage. The delay can change over a very broad range, but it is

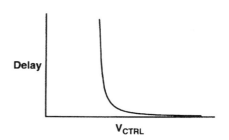

Fig. 4.   Typical symmetric load buffer stage delay as a function of control voltage.

nonlinear with respect to the control voltage. In fact, the delay changes proportionally to $1/(V_{\text{CTRL}} - V_T)$, with a slope K_{DL} proportional to $1/(V_{\text{CTRL}} - V_T)^2$ or $1/I_D$, where I_D is one half of the buffer bias current.

As the operating frequency is reduced, K_{DL} becomes larger, which increases the loop bandwidth relative to the operating frequency. This behavior is undesirable because the stability of the loop is undermined at lower frequencies, which in turn constrains the operating frequency range. Thus, even though nonlinear control over delay allows a VCDL to have a large delay range, stability constraints still lead to a small operating frequency range for the DLL.

The effect on stability for nonlinear control over delay can be corrected by applying self-biased techniques. Suppose that the charge pump current I_{CH} is set equal to the buffer bias current $2 \cdot I_D$. The $1/I_D$ dependence of K_{DL} can then be cancelled out leading to loop bandwidth that tracks the operating frequency without constraining the operating frequency range.

With this solution, the DLL design is completely self-biased as all bias voltages and currents are referenced to other generated bias voltages and currents. The bias generator generates all of the needed biases for the VCDL from V_{CTRL}, and the charge pump uses the current formed by V_{CTRL} to generate corrections to V_{CTRL}. The key difference from typical DLL designs is that no special bandgap bias circuit or the equivalent is needed to establish the charge pump current.

C. Quantitative Analysis

A more detailed analysis will show a very simple result for the relationship between the loop bandwidth and the operating frequency. First, a relationship between VCDL gain and control voltage is needed. Fig. 5 shows a symmetric load and its typical IV characteristics. The buffer bias current is $2 \cdot I_D$. It can be shown that the effective resistance of a symmetric load R_{EFF} is directly proportional to the small signal resistance at the ends of the swing range which is just one over the transconductance g_m for one of the two equally sized devices when biased at V_{BP} or, equivalently, V_{CTRL}. Thus, the buffer delay can be defined as

$$t = R_{\text{EFF}} \cdot C_{\text{EFF}} = \frac{1}{g_m} \cdot C_{\text{EFF}} \quad (4)$$

where C_{EFF} is the effective buffer output capacitance.

Using a half-buffer replica, the bias generator sets the buffer bias current equal to the current through a symmetric load

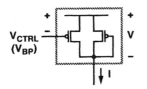 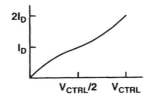

Fig. 5. Typical symmetric load IV characteristics.

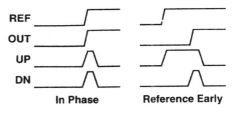

Fig. 6. Phase-frequency comparator waveforms with *UP* and *DN* asserted on every cycle.

with its output voltage at V_{CTRL}. In this case, the two equally sized PMOS devices are both biased at V_{CTRL} and each source half of the buffer bias current. Since these devices typically have greater than minimum channel length and are biased with moderate voltages, a simple quadratic model can be used for the drain current. The drain current for one of the two equally sized devices biased at V_{CTRL} is then given by

$$I_D = \frac{k}{2} \cdot (V_{CTRL} - V_T)^2 \qquad (5)$$

where k is the device transconductance of one of the PMOS devices. Taking the derivative with respect to V_{CTRL}, the transconductance is given by

$$g_m = k \cdot (V_{CTRL} - V_T) = \sqrt{2 \cdot k \cdot I_D}. \qquad (6)$$

The buffer delay is then given by

$$t = \frac{C_{EFF}}{k \cdot (V_{CTRL} - V_T)}. \qquad (7)$$

The delay for an n stage VCDL is given by

$$D = n \cdot t = \frac{C_B}{2 \cdot k \cdot (V_{CTRL} - V_T)} \qquad (8)$$

where C_B is defined as $2 \cdot n \cdot C_{EFF}$ or, equivalently, the total buffer output capacitance for all stages. Taking the derivative with respect to V_{CTRL}, the VCDL gain is given by

$$K_{DL} = \left| \frac{dD}{dV_{CTRL}} \right| = \frac{C_B}{2 \cdot k \cdot (V_{CTRL} - V_T)^2}$$
$$= \frac{C_B}{4 \cdot I_D}. \qquad (9)$$

Thus, the gain is inversely proportional to buffer bias current.

With the relationship for K_{DL} established, the bandwidth to operating frequency ratio can be derived. Let the charge pump current be set to some multiple x of the buffer bias current such that

$$I_{CH} = x \cdot (2 \cdot I_D). \qquad (10)$$

The bandwidth to operating frequency ratio is then given by

$$\frac{\omega_N}{\omega_{REF}} = \frac{1}{\omega_{REF}} \cdot I_{CH} \cdot K_{DL} \cdot F_{REF} \cdot \frac{1}{C_1}$$
$$= \frac{1}{2\pi} \cdot I_{CH} \cdot K_{DL} \cdot \frac{1}{C_1}$$
$$= \frac{1}{2\pi} \cdot x \cdot (2 \cdot I_D) \cdot \frac{C_B}{4 \cdot I_D} \cdot \frac{1}{C_1}$$
$$= \frac{x}{4\pi} \cdot \frac{C_B}{C_1}. \qquad (11)$$

Thus, the bandwidth to operating frequency ratio is constant and completely determined by a ratio of capacitances that can be matched reasonably well in layout, dramatically reducing the process technology sensitivity of the design. In addition, the DLL can operate over the same broad frequency range achievable by a VCO based on the same buffer stages operating open loop.

D. Zero-Offset Charge Pump

The self-biased DLL design requires a charge pump current that will vary several decades over the operating frequency range. At low current levels, small charge offsets can lead to significant phase offsets. In addition, all phase comparators will typically assert their *UP* and *DN* outputs for equal durations on every cycle once the loop is in lock. In order to achieve zero static phase offset, the charge pump must transfer no net charge to the loop filter for these equal duration *UP* and *DN* pulses, which requires that the *UP* and *DN* currents be identical and independent of the charge pump output voltage.

An XOR phase comparator used for quadrature locking separately asserts its *UP* and *DN* outputs twice per cycle for equal durations. The same is true for a phase-frequency comparator (PFC) which is used for in-phase locking. In order to successfully avoid a dead-band region in the PFC as seen by the charge pump, the PFC must assert both *UP* and *DN* outputs on every cycle as shown in Fig. 6. A dead-band region is the range of input phase differences for which the PLL takes no corrective action. Such a dead-band region will result in additional input tracking jitter equal to the magnitude of the dead-band region. This requirement means that for in-phase inputs, the charge pump will see both its *UP* and *DN* inputs asserted for an equal and short period of time. If in-phase PFC inputs produce no *UP* or *DN* pulses, then it will take some finite phase difference before a large enough pulse is produced to turn on the charge pump, which leads to a dead-band region. If the reference is early with both *UP* and *DN* outputs asserted on every cycle, then the difference in duration between the *UP* and *DN* pulses will be equal to the input phase difference.

Self-biasing makes it possible to design a charge pump to have zero static phase offset when both the *UP* and *DN* outputs of the phase comparator are asserted for equal durations on every cycle with in-phase inputs. By constructing the charge pump from the symmetric load buffer stage, it can be guaranteed that the *UP* and *DN* currents for these equal duration pulses completely cancel out and transfer no net charge to the loop filter. A simplified schematic for the zero-offset charge pump is shown in Fig. 7. The charge

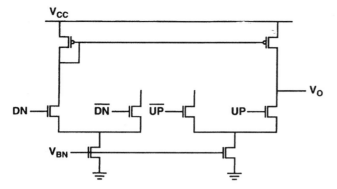

Fig. 7. Simplified schematic of the offset-cancelled charge pump.

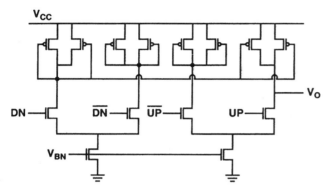

Fig. 8. Complete schematic of the offset-cancelled charge pump with symmetric loads.

pump is composed of two NMOS source coupled pairs each with a separate current source and connected by a current mirror made from symmetric load elements. Charge will be transferred from or to the loop filter connected to the output when the *UP* input or *DN* input, respectively, is switched high.

With both the *UP* and *DN* inputs asserted, the left source-coupled pair will behave like the half-buffer replica in the bias circuit and produce V_{CTRL} at the current mirror node. The PMOS device in the right source coupled pair will have V_{CTRL} at its gate and drain which is connected to the loop filter. This device will then source the exact same buffer bias current that is sunk by the remainder of the source coupled pair. With no net charge transferred to the loop filter, the charge pump will have zero static phase offset.

The complete schematic for the zero-offset charge pump is shown in Fig. 8. The current mirror is constructed from symmetric load elements. Also, the unselected source coupled pair outputs are connected to symmetric load elements to match the voltages at the other outputs. At this point, all of the elements necessary to construct a self-biased DLL have been considered. The following section shows that similar self-biasing techniques can be applied to a phase-locked loop.

IV. SELF-BIASED PHASE-LOCKED LOOP

A self-biased PLL, like the self-biased DLL, is constructed by taking advantage of the control relationship offered by a typical PLL. However, the control relationship and the additions to make it self-biased are more complicated than for

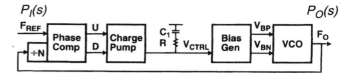

Fig. 9. Typical PLL block diagram (clock distribution omitted).

the DLL. A typical PLL is shown in Fig. 9. It is composed of a phase comparator, charge pump, loop filter, bias generator, and VCO. The key differences from the DLL are that it uses a VCO instead of a delay line and that it requires a resistor in the loop filter for stability. The negative feedback in the loop adjusts the VCO output frequency by integrating the phase error that results between the periodic reference input and the divided VCO output. Once in lock, the VCO will generate an output with a frequency that is N times larger than that for the reference input such that there is no detected phase error between the reference and the divided output. With the chip-wide clock distribution network added in the feedback path to the divider, the PLL can be used to multiply and rebuffer an input clock signal without adding delay.

A. Closed-Loop Response

As with the DLL, the frequency response of the PLL can be analyzed with a continuous time approximation for bandwidths a decade or more below the operating frequency. This bandwidth constraint is also required for stability due to the reduced phase margin near the higher order poles that result from the delay around the sampled feedback loop. Because the loop filter integrates the charge representing the phase error and the VCO integrates the output frequency to form the output phase, the PLL has a second-order closed response. The output phase, $P_O(s)$, is related to the input phase, $P_I(s)$, by

$$P_O(s) = \left(P_I(s) - \frac{P_O(s)}{N}\right) \cdot I_{CH} \cdot \left(R + \frac{1}{sC_1}\right) \cdot K_V \cdot \frac{1}{s} \tag{12}$$

where I_{CH} is the charge pump current (A), R is the loop filter resistor (Ω), C_1 is the loop filter capacitance (F), and K_V is the VCO gain (Hz/V). The closed loop response is then given by

$$\frac{P_O(s)}{P_I(s)} = \left(\frac{1}{N} + \frac{s}{I_{CH} \cdot (R + 1/(sC_1)) \cdot K_V}\right)^{-1}$$
$$= \frac{N \cdot (1 + s \cdot C_1 \cdot R)}{1 + s \cdot C_1 \cdot R + s^2/(I_{CH}/C_1 \cdot K_V/N)} \tag{13}$$

or, equivalently, by

$$\frac{P_O(s)}{P_I(s)} = N \cdot \frac{1 + 2 \cdot \zeta \cdot (s/\omega_N)}{1 + 2 \cdot \zeta \cdot (s/\omega_N) + (s/\omega_N)^2} \tag{14}$$

where ζ, defined as the damping factor, is given by

$$\zeta = \frac{1}{2} \cdot \sqrt{\frac{1}{N} \cdot I_{CH} \cdot K_V \cdot R^2 \cdot C_1} \tag{15}$$

and ω_N, defined as the loop bandwidth (rad/s), is given by

$$\omega_N = \frac{2 \cdot \zeta}{R \cdot C_1}. \tag{16}$$

The loop bandwidth and damping factor characterize the closed-loop response. The PLL is critically damped with a damping factor of one and overdamped with damping factors greater than one.

For a typical PLL, the charge pump current I_{CH}, VCO gain K_V, and loop filter resistance R are constant [4]. These conditions give rise to a constant damping factor and a constant loop bandwidth. A constant bandwidth can constrain the achievement of a wide operating frequency range and low input tracking jitter. A PLL adjusts its output frequency, not its output phase like a DLL. If the frequency is disturbed, the phase error that results from each cycle of the disturbance will accumulate for many cycles until the loop can correct the frequency error. The error will be accumulated for a number of cycles proportional to the operating frequency divided by the loop bandwidth. Thus, ω_N should be positioned as close as possible to ω_{REF} to minimize the total phase error. In addition, ω_N depends on I_{CH}, K_V, R, and C_1, but not on ω_{REF}. All of these parameters have independent variability. However, ω_N must be a decade below the lowest operating frequency for stability. The result is that ω_N must be conservatively set for stability at the lowest operating frequency with worst case process variations, rather than set for optimized jitter performance.

B. Bandwidth Tracking

Ideally, both ζ and ω_N/ω_{REF} should be constant so that there is no limit on the operating frequency range and so that the jitter performance can be improved. I_{CH} could be set equal to the buffer bias current $2 \cdot I_D$ as done in self-biasing the DLL, but this is not sufficient since ζ would change with operating frequency as a result of a square root of I_D dependence. To keep ζ constant with operating frequency, two parameters, such as I_{CH} and R, must vary. I_{CH} can be set equal to the buffer bias current and R can be set to vary inversely proportionally to the square root of the buffer bias current. ζ will then remain constant, but ω_N will be proportional to the square root of the buffer bias current.

To obtain a tracking bandwidth, the VCO operating frequency should have the same dependency on the buffer bias current as the loop bandwidth ω_N. Symmetric load buffer stages can be used to implement the VCO in order to obtain a broad frequency range. Fig. 10 shows their typical frequency as a function of control voltage. The frequency is proportional to $V_{CTRL} - V_T$ or, equivalently, the square root of I_D, and the slope is constant. Thus, K_V is constant and the reference frequency ω_{REF} is proportional to the square root of the buffer bias current. Since both ω_N and ω_{REF} are proportional to the square root of the buffer bias current, the loop bandwidth will track the operating frequency.

C. Feed-Forward Zero

It may seem difficult to obtain a resistor for the loop filter that varies inversely proportionally to the square root of the buffer bias current. However, this resistor can be formed from the small-signal resistance $1/g_m$ for a diode-connected device, where g_m is proportional to the square root of the buffer bias

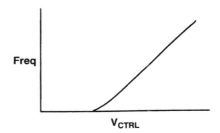

Fig. 10. Typical VCO frequency as a function of control voltage when implemented with symmetric load buffer stages.

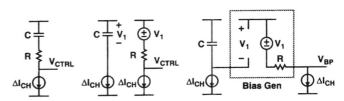

Fig. 11. Transformation of the loop filter for the integration of the loop filter resistor.

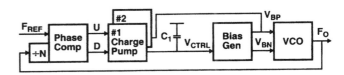

Fig. 12. Complete self-biased PLL block diagram (clock distribution omitted).

current. The integration of such a resistance into the loop filter can be accomplished by applying a transformation to the loop filter as illustrated in Fig. 11.

The loop filter for a PLL is typically a capacitor in series with a resistor that is driven by the charge pump current ΔI_{CH}. The control voltage is then the sum of the voltage drops across the capacitor and resistor. The voltage drops across the capacitor and resistor can be generated separately, as long as the same charge pump current is applied to each of them. The two voltage drops can then be summed to form the control voltage by replicating the voltage across the capacitor with a voltage source placed in series with the resistor.

It just so happens that the bias generator can conveniently implement this voltage source and resistor since it buffers V_{CTRL} to form V_{BP} with a finite output resistance. Referring back to the buffer bias circuit in Fig. 2, it is evident that this resistance is established by a diode-connected symmetric load or, equivalently, a diode-connected PMOS device. Thus, the resistance is equal to $1/g_m$ or inversely proportional to the square root of the buffer bias current. Thus, the self-biased PLL can be completed simply by adding an additional charge pump current [5] to the bias generator's V_{BP} output as shown in Fig. 12. Therefore, as with the DLL, this PLL design is completely self-biased.

D. Quantitative Analysis

As for the DLL, a more detailed analysis will show a very simple result for the damping factor and the relationship between the loop bandwidth and the operating frequency. First, a relationship between VCO gain and control voltage is needed. The oscillation frequency for an n-stage VCO is given by

$$F = \frac{1}{2 \cdot n \cdot t} = \frac{k \cdot (V_{\text{CTRL}} - V_T)}{C_B} = \frac{\sqrt{2 \cdot k \cdot I_D}}{C_B} \quad (17)$$

where C_B is once again defined as $2 \cdot n \cdot C_{\text{EFF}}$. Thus, the oscillation frequency is proportional to the square root of the buffer bias current. Taking the derivative with respect to V_{CTRL}, the VCO gain K_V is given by

$$K_V = \left| \frac{dF}{dV_{\text{CTRL}}} \right| = \frac{k}{C_B} \quad (18)$$

which is independent of the buffer bias current.

With the relationship for K_V established, the damping factor and bandwidth to operating frequency ratio can be derived. Let the charge pump current be set to some multiple x of the buffer bias current such that

$$I_{CH} = x \cdot (2 \cdot I_D). \quad (19)$$

Also, let the diode-connected symmetric load in the bias generator that establishes the loop filter resistance be y times larger than the symmetric loads used in the buffer stages such that

$$R = \frac{y}{2 \cdot g_m} = \frac{y}{\sqrt{8 \cdot k \cdot I_D}}. \quad (20)$$

Substituting in the expressions for I_{CH}, K_V, and R, the damping factor is then given by

$$\begin{aligned}
\zeta &= \frac{1}{2} \cdot \sqrt{\frac{1}{N} \cdot I_{CH} \cdot K_V \cdot R^2 \cdot C_1} \\
&= \frac{1}{2} \cdot \sqrt{\frac{1}{N} \cdot x \cdot (2 \cdot I_D) \cdot \frac{k}{C_B} \cdot \frac{y^2}{8 \cdot k \cdot I_D} \cdot C_1} \\
&= \frac{y}{4} \cdot \sqrt{\frac{x}{N}} \cdot \sqrt{\frac{C_1}{C_B}}. \quad (21)
\end{aligned}$$

Thus, the damping factor is simply a constant times the square root of the ratio of two capacitances. Substituting in the expressions for ζ, R, and F_{REF}, the loop bandwidth to operating frequency ratio is given by

$$\begin{aligned}
\frac{\omega_N}{\omega_{\text{REF}}} &= \frac{1}{2\pi \cdot F_{\text{REF}}} \cdot \frac{2 \cdot \zeta}{R \cdot C_1} \\
&= \frac{1}{2\pi} \cdot \frac{N \cdot C_B}{\sqrt{2 \cdot k \cdot I_D}} \cdot \frac{y}{4} \cdot \sqrt{\frac{x}{N}} \cdot \sqrt{\frac{C_1}{C_B}} \\
&\quad \cdot \frac{\sqrt{8 \cdot k \cdot I_D}}{y} \cdot \frac{2}{C_1} \\
&= \frac{x \cdot N}{2\pi} \cdot \sqrt{\frac{C_B}{C_1}}. \quad (22)
\end{aligned}$$

The loop bandwidth to operating frequency ratio is also a constant times the square root of the ratio of the same two capacitances.

Thus, the loop bandwidth will track operating frequency and, therefore, sets no constraint on the operating frequency range. The PLL can operate over the same broad frequency range achievable by the VCO operating open loop. The only process technology dependence is on a ratio of capacitances that can be matched reasonably well in layout. ω_N and ζ can be aggressively set to minimize jitter accumulation over all operating frequencies.

E. PLL Capture Behavior

A useful artifact of the self-biasing used in the PLL is a nonlinear capture behavior. Typical PLL's will slew toward the final target frequency at roughly a constant rate. A phase-frequency comparator will detect on average a phase error of a half cycle which will cause the charge pump to source or sink on average half of its charge pump current to the loop filter. The resulting change in control voltage is given by

$$\left| \frac{dV_{\text{CTRL}}}{dt} \right| = \frac{1}{2} \cdot \frac{1}{C_1} \cdot I_{CH} \quad (23)$$

which leads to the result

$$V_{\text{CTRL}}(t) = V_{\text{CTRL}}(0) \pm \frac{1}{2 \cdot C_1} \cdot I_{CH} \cdot t \quad (24)$$

and

$$t = |V_{\text{CTRL}}(t) - V_{\text{CTRL}}(0)| \cdot \frac{2 \cdot C_1}{I_{CH}}. \quad (25)$$

For the self-biased PLL, the charge pump current, which is proportional to the buffer bias current, changes with the control voltage and the VCO output frequency. This dependency means that the rate of change in the control voltage or the VCO output frequency will increase when approaching higher frequencies and decrease when approaching lower frequencies. The resulting change in control voltage is given by

$$\left| \frac{dV_{\text{CTRL}}}{dt} \right| = \frac{1}{2} \cdot \frac{1}{C_1} \cdot I_{CH} = \frac{x \cdot k}{4 \cdot C_1} \cdot (V_{\text{CTRL}} - V_T)^2 \quad (26)$$

which leads to the result

$$V_{\text{CTRL}}(t) = \left(\frac{1}{V_{\text{CTRL}}(0) - V_T} \pm \frac{x \cdot k \cdot t}{4 \cdot C_1} \right)^{-1} + V_T \quad (27)$$

and

$$t = \frac{4 \cdot C_1}{x \cdot k} \cdot \left| \frac{1}{V_{\text{CTRL}}(0) - V_T} - \frac{1}{V_{\text{CTRL}}(t) - V_T} \right|. \quad (28)$$

The control voltage as a function of time during capture is plotted in Fig. 13. The result is that the self-biased PLL will slew toward lock at the fastest rate possible using the maximum charge pump current such that the instantaneous loop bandwidth does not exceed that required to lock at the current VCO output frequency. In contrast, a typical PLL will only be able to slew toward lock at a constant rate using a fixed charge pump current such that the loop bandwidth does not exceed that required to lock at the lowest operating frequency of interest. Thus, the self-biased PLL will exhibit much faster locking times when locking from similar or higher operating frequencies. If, however, the self-biased PLL is started at a very low operating frequency, possibly in the low kilohertz range, it will exhibit very slow locking times.

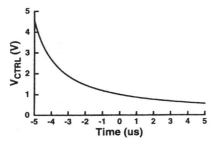

Fig. 13. Self-biased PLL control voltage as a function of time during capture.

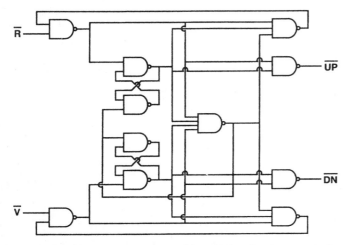

Fig. 14. Phase-frequency comparator with equal short duration output pulses for in-phase inputs.

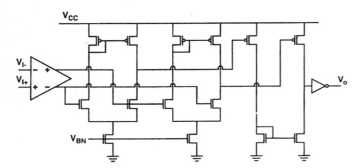

Fig. 15. Differential-to-single-ended converter with 50% duty cycle output.

V. ADDITIONAL LOOP COMPONENTS

In completing the self-biased DLL and PLL designs, some additional loop components are required. These include a phase-frequency comparator and a differential-to-single-ended converter. Although they are not essential to the self-biased techniques, they are important to the overall performance of the self-biased DLL and PLL designs.

A. Phase-Frequency Comparator

Equal and short duration pulses at the UP and DN outputs of the PFC are needed for in-phase inputs in order to eliminate a dead-band in the PFC as seen by the charge pump. Such a dead-band will lead to additional input tracking jitter, as discussed in Section III-D. In order to guarantee equal and

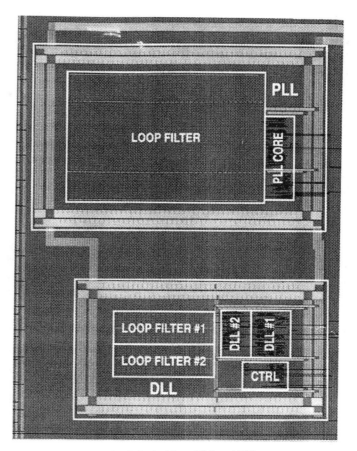

Fig. 16. Die micrograph of the dual-loop DLL and PLL.

short duration pulses, some delay is typically added in the reset path. Because of this added delay, a signal transition at the R or V input will cause the corresponding UP or DN output to be asserted for some short delay before both outputs are reset for the case when the other output was already asserted. However, adding delay in the reset path can reduce the maximum operating frequency of the PFC. The maximum operating frequency is determined by the amount of time required to reset the PFC after receiving the last set of input transitions so that it is ready to detect the next set of input transitions. The phase-frequency comparator shown in Fig. 14, based on a conventional PFC [4], instead adds delay only in the output reset path by forming the outputs without including the reset signal generated by the four input NAND gate. Rather than obtaining the outputs from the three input NAND gates at the far right, the outputs are obtained from copies of the gates with the reset signal input deleted. The outputs are still reset, but through a slower path that includes the two NAND gates which form the SR latches. Since the input reset path is unchanged, the maximum operating frequency is unaffected.

B. Differential-to-Single-Ended Converter

PLL's are typically designed to operate at twice the chip operating frequency so that their outputs can be divided by two in order to guarantee a 50% duty cycle [4]. This practice is particularly wasteful if the delay elements generate differential signals. The maximum operating frequency will be reduced

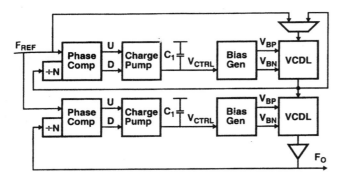

Fig. 17. Dual-loop DLL block diagram.

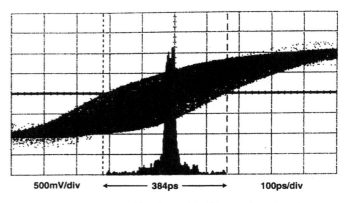

Fig. 18. Measured PLL tracking jitter with 500 mV of supply noise.

by a factor of two and the input tracking jitter performance can be adversely affected. The requirement for a 50% duty cycle can be satisfied without operating the PLL at twice the chip operating frequency if a single-ended output with 50% duty cycle can be obtained from the differential output signal. The differential-to-single-ended converter circuit shown in Fig. 15 can produce such a 50% duty cycle output. It is composed of two opposite phase NMOS differential amplifiers driving two PMOS common-source amplifiers connected by an NMOS current mirror. The two NMOS differential amplifiers are constructed from symmetric load buffer stages using the same NMOS current source bias voltage as the driving buffer stages so that they receive the correct common-mode input voltage level. They provide signal amplification and a dc bias point for the PMOS common-source amplifiers. The PMOS common-source amplifiers provide additional signal amplification and conversion to a single-ended output through the NMOS current mirror. Because the two levels of amplification are differentially balanced with a wide bandwidth, the opposing differential input transitions have equal delay to the output. The limitations of this circuit in converting the differential signal transitions into rising and falling single-ended output transitions at medium and high bias levels are identical to those of a divider in converting single direction transitions into rising and falling single-ended output transitions. However, using this circuit instead of a divider to generate a 50% duty cycle output can substantially relax the design constraints on the VCO for high frequency designs.

VI. EXPERIMENTAL RESULTS

Both the self-biased DLL and PLL designs were fabricated in a 0.5-μm N-well CMOS gate array process. A micrograph of the fabricated designs with a superimposed floor plan is shown in Fig. 16. The loop filter capacitors for both the DLL and PLL are integrated on-chip using PMOS gate array devices. The capacitor arrays also contain an equal number of NMOS devices that are in rows interleaved between the PMOS devices. These NMOS devices are used to make a supply bypass capacitor. The DLL design actually implemented was a dual-loop DLL design [2], [6], as shown in Fig. 17, which contains two cascaded DLL's to allow it to perform frequency multiplication and duty cycle adjustment.

The PLL input-to-output tracking jitter performance with 500 mV of 1 MHz square wave supply noise is illustrated

TABLE I
DUAL-LOOP DLL PERFORMANCE CHARACTERISTICS MEASURED AT
250 MHz WITH 500 mV OF 1 MHz SQUARE WAVE SUPPLY NOISE

Operating frequency range:	0.0025 MHz–400 MHz @ 3.3 V
Minimum supply requirements:	2.45 V, 8.6 mA
Input offset, sensitivity:	112 ps, <100 ps/100 MHz
Tracking jitter, sensitivity:	610 ps, 1165 ps/V (P-P)
Cycle-to-cycle jitter, sensitivity:	262 ps, 430 ps/V (P-P)
Block area:	1.18 mm^2
Technology:	0.5-μm N-well CMOS gate array

TABLE II
PLL PERFORMANCE CHARACTERISTICS MEASURED AT 250 MHz
WITH 500 mV OF 1 MHz SQUARE WAVE SUPPLY NOISE

Operating frequency range:	0.0025 MHz–550 MHz @ 3.3 V
Minimum supply requirements:	2.10 V, 4.4 mA
Input offset, sensitivity:	<25 ps, <10 ps/100 MHz
Tracking jitter, sensitivity:	384 ps, 704 ps/V (P-P)
Cycle-to-cycle jitter, sensitivity:	144 ps, 290 ps/V (P-P)
Block area:	1.91 mm^2
Technology:	0.5-μm N-well CMOS gate array

in Fig. 18. This square wave supply noise has edge transition times less than 10 ns. It is important to note that low frequency square wave supply noise is one of the most extreme jitter tests that can be performed on a PLL. Sine wave supply noise at the loop bandwidth typically leads to much less jitter. The confined central peaks indicate very low static phase offset sensitivity to supply voltage.

Performance characteristics of the dual-loop DLL and PLL are summarized in Table I and Table II, respectively. Both designs have a frequency range spanning five orders of magnitude. This large range should allow a single design to satisfy a variety of operating frequency requirements. The cycle-to-cycle jitter listed is the jitter in the period of the output. The measured jitter, although small, was increased by the gate array implementation of the loop filter capacitors which contain the interleaved rows of unrelated NMOS devices that lead to control voltage coupling to ground. The die area for the dual-loop DLL and PLL was substantially increased by the inefficient implementation of the loop filter capacitors. With custom silicon, the PLL only occupies 0.4 mm^2.

The PLL design performed about 50% better than the dual-loop DLL design. This difference in performance may have resulted from larger capacitive coupling to the DLL loop filter capacitors since they were eight times smaller than the PLL loop filter capacitor.

VII. CONCLUSIONS

Self-biasing greatly simplifies DLL and PLL designs. It eliminates the need for precise currents, eliminates virtually all process technology dependencies, and makes a wide operating frequency range possible. The bandwidth to operating frequency ratio and the PLL damping factor are fixed completely by a ratio of capacitances. The operating frequency range is limited only by the buffer stage design. Self-biasing facilitates the construction of an input offset-cancelled charge pump. Self-biasing also allows a PLL to have the largest possible loop bandwidth over all operating frequencies for minimal jitter accumulation. The phase-frequency comparator design provides equal short duration output pulses for in-phase inputs without reducing its maximum operating frequency. The differential-to-single-ended converter can convert differential input signals into single-ended output signals with 50% duty cycle, avoiding the need for dividing by two. Fabricated in a 0.5-μm N-well CMOS gate array process, the PLL achieves an operating frequency range of 0.0025 MHz to 550 MHz and input tracking jitter of 384 ps at 250 MHz with 500 mV of low frequency square wave supply noise.

REFERENCES

[1] F. Gardner, "Charge-pump phase-lock loops," *IEEE Trans. Commun.*, vol. COM-28, no. 11, pp. 1849–1858, Nov. 1980.
[2] J. Maneatis, "Low-jitter process-independent DLL and PLL based on self-biased techniques," in *ISSCC 1996 Dig. Tech. Papers*, Feb. 1996, pp. 130–131.
[3] J. Maneatis and M. Horowitz, "Precise delay generation using coupled oscillators," *IEEE J. Solid-State Circuits*, vol. 28, no. 12, pp. 1273–1282, Dec. 1993.
[4] I. Young *et al.*, "A PLL clock generator with 5 to 110 MHz of lock range for microprocessors," *IEEE J. Solid-State Circuits*, vol. 27, no. 11, pp. 1599–1607, Nov. 1992.
[5] D. Mijuskovic *et al.*, "Cell-based fully integrated CMOS frequency synthesizers," *IEEE J. Solid-State Circuits*, vol. 29, no. 3, pp. 271–279, Mar. 1994.
[6] A. Waizman, "A delay line loop for frequency synthesis of de-skewed clock," in *ISSCC 1994 Dig. Tech. Papers*, Feb. 1994, pp. 298–299.

John G. Maneatis was born in San Francisco, CA, on November 7, 1965. He received the B.S. degree in electrical engineering and computer science from the University of California, Berkeley, in 1988 and the M.S. and Ph.D. degrees in electrical engineering from Stanford University, Stanford, CA, in 1989 and 1994.

He worked at Hewlett-Packard Laboratories, Palo Alto, CA, during the summer of 1989 on high-speed analog-to-digital conversion and monolithic clock recovery, and at Digital Equipment Corporation Western Research Laboratory, Palo Alto, CA, during the summer of 1990 on CAD tool development and ECL circuit design. While at Stanford University, his research interests included high-performance circuit design for phase-locked loops, microprocessors, and data conversion. Since 1994 he has been a circuit designer at Silicon Graphics, Inc., Mountain View, CA, working in the area of microprocessor design.

Dr. Maneatis is a member of Tau Beta Pi, Eta Kappa Nu, and Phi Beta Kappa and a Registered Professional Electrical Engineer in the State of California.

A Low-Jitter PLL Clock Generator for Microprocessors with Lock Range of 340–612 MHz

David W. Boerstler

Abstract— A fully integrated, phase-locked loop (PLL) clock generator/phase aligner for the POWER3 microprocessor has been designed using a 2.5-V, 0.40-μm digital CMOS6S process. The PLL design supports multiple integer and noninteger frequency multiplication factors for both the processor clock and an L2 cache clock. The fully differential delay-interpolating voltage-controlled oscillator (VCO) is tunable over a frequency range determined by programmable frequency limit settings, enhancing yield and application flexibility. PLL lock range for the maximum VCO frequency range settings is 340–612 MHz. The charge-pump current is programmable for additional control of the PLL loop dynamics. A differential on-chip loop filter with common-mode correction improves noise rejection. Cycle–cycle jitter measurements with the microprocessor actively executing instructions were 10.0 ps rms, 80 ps peak to peak (P-P) measured from the clock tree. Cycle-cycle jitter measured for the processor in a reset state with the clock tree active was 8.4 ps rms, 62 ps P-P. PLL area is 1040 × 640 μm^2. Power dissipation is <100 mW.

Index Terms— Clock generator, clocking, microprocessors, phase-locked loop (PLL).

I. BACKGROUND

THE use of phase-locked loops (PLL) for generating phase-synchronous, frequency-multiplied clocks in microprocessors has been prevalent in industry [1]–[4]. In recent years, the trend toward ever increasing clock frequency has made PLL's even more attractive due to the difficulties in distributing high-frequency clocks through several levels of packaging [5], [6], but the jitter penalty for using a PLL has not kept pace with the rate of reduction in processor cycle time. Until this year,[1] the best reported microprocessor PLL jitter penalties ranged from 82 to 83 ps peak to peak (P-P) for inactive processors [1], [5], and a PLL on a small (600-K transistor) graphics display chip has been reported with 80 ps P-P jitter for a quiet supply at 320 MHz [7]. Many examples of higher jitter PLL designs exist in the literature. Power-supply noise created from the digital switching activity on a microprocessor is recognized as a major source of PLL jitter, and the primary focus of designers has been directed toward reducing this sensitivity.

Manuscript received December 10, 1997; revised August 10, 1998.

The author is with the IBM Research Division, IBM Austin Research Laboratory, Austin, TX 78758 USA.

Publisher Item Identifier S 0018-9200(99)02429-4.

[1]Recent announcements of a 1-GHz microprocessor PLL [12], [13] and a PLL with an on-chip regulator [14] reported jitter of <±9 ps (quiet conditions)/<±36 ps (processor active) and <±10 ps (sinusoidal external noise)/<±20 ps (square wave external noise), respectively.

II. INTRODUCTION

This paper describes a fully integrated PLL-based clock generator/phase aligner used for the POWER3 microprocessor. The microprocessor is fabricated in IBM CMOS6S technology and contains approximately 12 million transistors. With the microprocessor actively executing instructions, this PLL achieved cycle–cycle jitter of 10.0 ps rms, 80 ps P-P in its application environment and 8.4 ps rms, 62 ps P-P with the microprocessor in a reset state with a portion of the clock tree active.

A simplified block diagram of the PLL clock generator is shown in Fig. 1. The external reference or BUSCLK enters a receiver and is divided by two by divider stage M_1 before entering the phase/frequency detector (PFD) as Φ_i. The internal feedback signal Φ_o from divider M_6 is compared to Φ_i by the PFD, which generates an error signal Φ_e, which is used by the charge-pump and filter network to control the voltage-controlled oscillator (VCO). The output frequency of the VCO is divided by M_3 and is used as the main processor clock (PCLK) after passing through four levels of clock buffering in an H-tree clock distribution network. The processor clock is passed through a delay-matching receiver before entering divider M_6, completing the feedback path. Since at equilibrium the inputs of the PFD will be matched in frequency (and phase), the processor-to-bus frequency ratio is equal to the ratio f_{PCLK}/f_{BUS}, which is equal to the ratio M_6/M_1, allowing integer or noninteger frequency synthesis by changing divider ratios. Since the technique does not require clock choppers [2], the duty cycle and phase alignment are relatively insensitive to environment and process tolerances. The output of the VCO is also connected to frequency divider M_5, which is used for the L2 cache clock (L2CLK). Since $f_{VCO} = M_3 \cdot f_{PCLK} = M_5 \cdot F_{L2CLK}$, the processor-to-L2 clock-frequency ratio is also adjustable to integer or noninteger ratios. Other phase-synchronous clocks may be designed in similar fashion, and quadrature or interstitial clocks may be created by a polarity change at the divider input. Using the structure of Fig. 1, the VCO frequency f_{vco} is equal to M_3 times the processor clock frequency f_{PCLK}. For cases when M_3 is even, the processor clock edges are generated from only one VCO clock edge; hence a nearly ideal 50% processor clock duty cycle may be achieved through its independence from the VCO duty cycle.

III. PROCESS TECHNOLOGY

The microprocessor and integral clock generator PLL are fabricated in a five-layer CMOS process with 0.4-μm feature

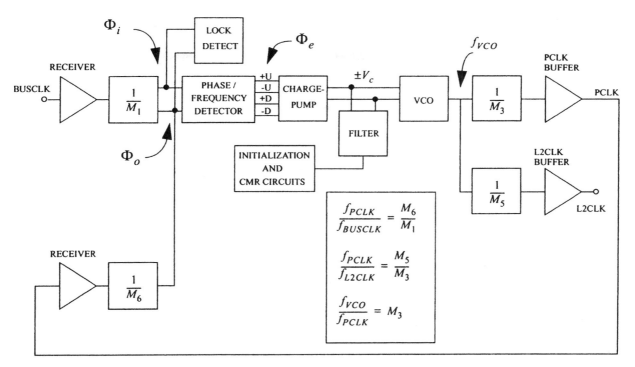

Fig. 1. PLL block diagram.

TABLE I
CMOS6S PROCESS SUMMARY

Minimum lithographic image	0.3 μm
Supply Voltage	2.5 V
N-well / P-epi on P+ bulk substrate	
Leff	0.200 μm
Tox	5.5 nm
Idsat	552 (n) 254 (p) μA/μm
Vt	0.5 V
5 metal levels + local interconnect	
M1 pitch	1.12 μm
M2-4 pitch	1.44 μm
M5 pitch	3.84 μm

sizes. Table I lists some of the relevant attributes of this process technology.

The PLL clock generator is shown in the microprocessor die photograph of Fig. 2(a). The dimensions of the entire PLL are 1040×640 μm^2. It is shown with the major features identified in Fig. 2(b).

IV. PLL Clock Generator Components

A. Phase/Frequency Detector

The digital PFD generates a signal that conveys relative phase and frequency error information about its inputs to the charge pump and filter. The PFD design is based on a three-state machine structure [8], as depicted in Fig. 3(a). From the initial reset state, a rising edge on the Φ_i input will assert the +UP output until the rising edge of Φ_o appears, which deasserts +UP and forces a reset of both flip-flops [Fig. 3(b)].

A rising edge first appearing on Φ_o similarly asserts +DOWN until a rising edge arrives at Φ_i, followed by a subsequent reset. Complementary outputs are generated by the PFD for use in the differential charge-pump stage that follows the PFD. The pulse width of the output varies proportionally with the phase error between the two inputs, except for the dead-zone region as the difference approaches zero. This dead zone exists when the phase error becomes small relative to the combined response time of the PFD, charge pump, and filter circuits. Circuit simulation results show a nominal dead zone of ± 25 ps. Concerns of current mismatch in the charge-pump and filter networks are reduced at the expense of increased dead zone by preventing simultaneous assertions of +UP and +DOWN.

B. Power-Supply Isolation

A separate analog power connection (AVDD) is used for the analog circuits [current reference, charge pump, common-mode rejection (CMR), filter initialization, and VCO circuits] to increase the isolation of the sensitive circuits from the logic-induced switching noise present on the main power supply. To allow the detection of potential defects using conventional testing, the AVDD pin is held low, disabling the analog devices that normally draw dc current. Both on-chip and on-module decoupling is used on AVDD.

C. Reference Circuit

A thermal voltage-referenced current source is used to provide temperature- and supply-independent biasing for the analog circuits in the PLL. The circuit contains an array of P+ diffusions in the N-well connected to form two forward-biased diodes with areas that differ by a factor of ten. When connected

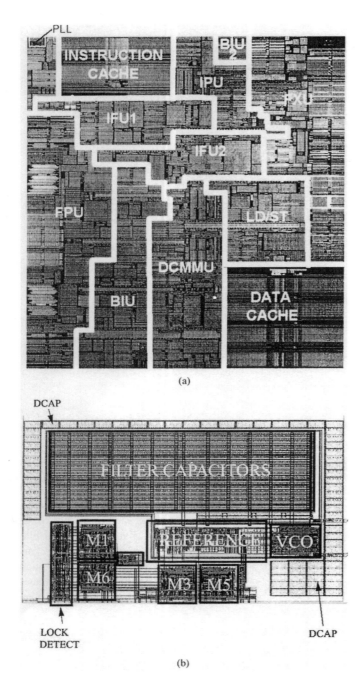

(a)

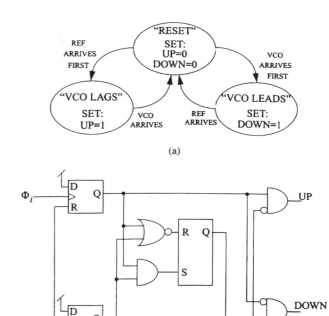

(a)

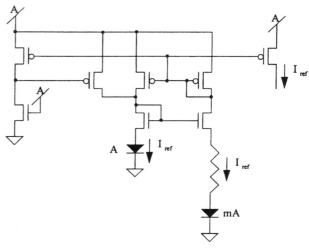

(b)

Fig. 3. (a) PFD state diagram. (b) Phase detector implementation.

Fig. 4. Reference circuit.

Fig. 2. (a) Die photograph of POWER3 microprocessor. (b) PLL layout.

as shown in Fig. 4, the current I_{ref} through each leg has two stable operating points, $I_{ref} = (kT/q)\,(\ln(10)/R) = 10.6\ \mu A$ or $I = 0$. The startup circuit prevents the zero current state from occurring by injecting current into one leg during initial power-on. The resistor is implemented using the precision resistor available in the process, which has a temperature coefficient (TC) of $+2000$ ppm/°C. The positive TC's of the thermal voltage term and the resistor tend to cancel, providing a reference current TC of $+785$ ppm/°C at 85°C. The reference current I_{ref} is used for subsequent generation of reference currents and the PMOS bias voltage V_p through mirroring. Sensitivity to power-supply change is $+1.7\%/V$ for $\pm 20\%$ change on VDD.

D. Process and Temperature Compensation

Variations in L_{eff} due to process are monitored using the circuit shown in Fig. 5(a). All of the current sources are generated directly from the reference circuit current I_{ref}. A constant current is passed through a branch containing short-channel NMOS devices, creating a monitoring voltage $V_{l,n,mon}$, which is sensitive to NMOS device length variations. This voltage is compared to a reference voltage $V_{l,n,ref}$ generated by a constant current through a long-channel NMOS device that is relatively insensitive to length variations. The devices and bias currents used for length sensing are sized so that $V_{l,n,mon}$ and $V_{l,n,ref}$ are equivalent for a nominal L_{eff} process. To minimize temperature sensitivity, the bias currents correspond to the zero-temperature coefficient (0-TC) region of the devices.

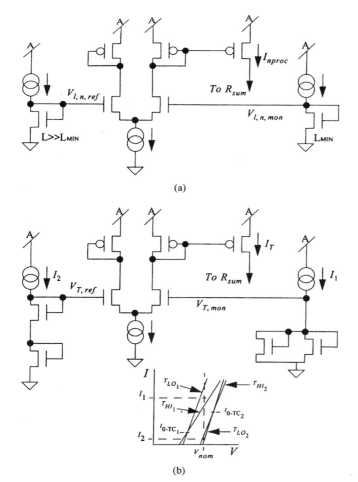

(a)

(b)

Fig. 5. (a) Process compensation circuit. (b) Temperature compensation circuit.

The two voltages are compared using a differential amplifier, which generates a current proportional to the NMOS L_{eff} offset from nominal. This current is mirrored to produce a current I_{nproc} that is injected into a precision resistor R_{sum} used for combining various process monitors to generate a compensating reference voltage. The compensating reference voltage is connected to the active load elements of the VCO, which control the VCO's voltage swing. A current I_{pproc} generated from a similar PMOS circuit also is injected into the resistor.

Weighted combinations of standard bias circuits with differing voltage and temperature coefficients have been used previously to compensate reference circuits for VCO's [9]. In this case, however, temperature was monitored directly by comparing the voltage of two series-connected devices biased by current I_2 below their 0-TC operating point to the voltage of two parallel devices biased by current I_1 significantly above their effective 0-TC point [Fig. 5(b)]. The devices and bias currents are sized so that both branches of the differential amplifier are balanced at V_{nom} for nominal temperature conditions. The inset shows the I–V characteristics as a function of temperature for the series (subscript 2) and parallel (subscript 1) connected devices; the 0-TC points correspond to the crossing point where the current is

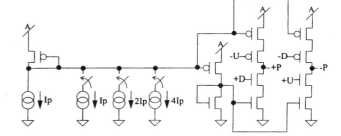

Fig. 6. Charge-pump circuit.

invariant with temperature. The current in one leg of the differential amplifier varies proportionally with temperature and is mirrored and added to the summing junction of the resistor R_{sum}. A constant bias current is also added to the summing junction to establish the correct weighting of the various compensating currents and to correct for the TC of the summing junction resistor.

Using a statistical process model, the process compensation was designed to favor the stabilization of the "best case" side of the distribution over the "worst case" side in anticipation of future process trends. Given the limited range over which a circuit may be practically compensated, the performance for the "best case" devices was not sacrificed at the expense of extensive compensation of the poorest performing devices. For the unsorted population, this approach allowed a reduction in the sensitivity of the VCO to process variability by a factor of 3.6 (55.4–15.2%) over the uncompensated VCO; temperature sensitivity was reduced by a factor of 4.7 (38.6–8.2%).

E. Charge Pump

The reference circuit is used to generate the currents I_p, $2I_p$, and $4I_p$ for use within the charge pump. The peak charge-pump current may be adjusted in 30-μA increments from 30 to 240 μA by scaling the mirror currents as shown in Fig. 6. The error signals $\pm U$ and $\pm D$ generated by the PFD are used to switch the peak current selected. Adjusting the charge pump allows for optimization of the loop characteristics for different divider and VCO settings. Differential outputs $+P$ and $-P$ are included for high CMR in the subsequent analog circuits.

F. Loop Filter

The differential loop filter and initialization circuits are shown in Fig. 7. Currents to and from the charge-pump circuit enter the filter at nodes $+P$, $-P$. The input to the filter contains NMOS transmission-gate clamping devices to limit the maximum filter voltage to $AVDD - V_{\text{tn}+}$, where $V_{\text{tn}+}$ is the NMOS threshold voltage for a large source-bulk voltage. For the CMOS6S process, the clamps prevent the filter voltage from exceeding approximately 1.8 V, eliminating concern for the VCO input stage's shutting off. The filter capacitors are accumulation-mode gate-oxide devices, and are interleaved to improve the matching. Both loop-filter capacitors together occupy an area of approximately 865×280 μm^2 and are approximately 450 pF each. Precision resistors (1.2 KΩ each) are used to produce a zero in the filter transfer function.

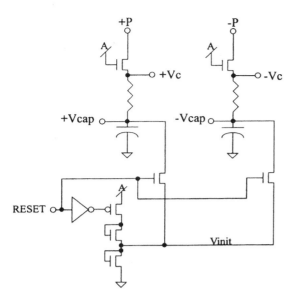

Fig. 7. Loop filter and filter initialization.

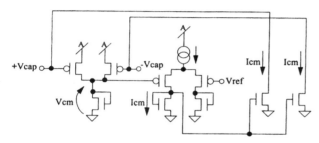

Fig. 8. Common-mode control.

The filter output is connected to the VCO control input at nodes $+V_c, -V_c$. An initialization circuit activated during the initial system power-on-reset is used to precharge the filter capacitors to the nominal common-mode voltages at nodes $+V_{\text{cap}}, -V_{\text{cap}}$.

G. Common-Mode Control

It is possible for common-mode voltages to develop in the filter from leakage, drift, or device mismatch. Since the common-mode voltage can introduce frequency offsets in the VCO or even inhibit operation for extreme cases, the circuit shown in Fig. 8 was used in conjunction with the filter clamps described earlier. The common-mode voltage of the filter is sensed by generating currents proportional to $+V_{\text{cap}}$ and $-V_{\text{cap}}$ and summing them across a load device to produce V_{cm}. A differential amplifier compares V_{cm} to a reference voltage and generates a current I_{cm}, which is proportional to the common-mode voltage. The current I_{cm} is mirrored by two identical current sources, which bleed current from both filter capacitors simultaneously without affecting the differential voltage between them. The maximum drain currents for this structure, which corresponds to the case when both clamps have activated, are approximately 16 μA. For typical cases where the common-mode voltage is below 600 mV, the bleed currents are <1 μA. Stability of the network is assured by heavy dominant-pole compensation.

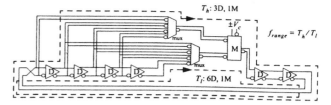

Fig. 9. Voltage-controlled oscillator.

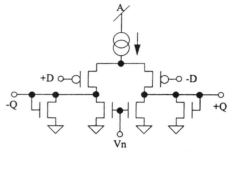

(a)

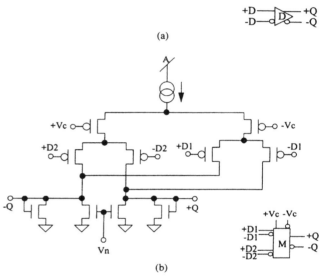

(b)

Fig. 10. (a) Delay element. (b) Mixer circuit.

H. Voltage-Controlled Oscillator

The VCO design is based upon a delay-interpolating ring oscillator structure [9]–[11], as shown in Fig. 9. In contrast to the current-starved and current-modulated VCO's, which are very commonly used for microprocessor clock generators, delay-interpolating VCO's have relatively low-to-moderate VCO gains and are well suited to fully differential control and signal path circuit implementations. The lower VCO gain of the delay-interpolating VCO's produces significantly less jitter due to coupled noise than higher gain structures. The limited operating frequency range for delay-interpolating VCO's, which must be less than 2 : 1 to ensure monotonicity, may be effectively augmented by selecting suitable divider ratios or by adding programmability to the VCO signal paths.

The frequency limits of the VCO are determined by the longest and shortest path delays through the structure. Fig. 9 shows an example high-frequency limit of period T_h composed of three delay units and one mixer unit, and a low-frequency

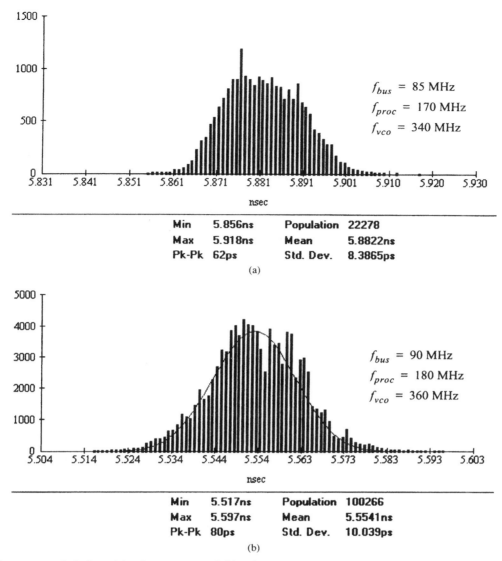

Min	5.856ns	Population	22278
Max	5.918ns	Mean	5.8822ns
Pk-Pk	62ps	Std. Dev.	8.3865ps

(a)

Min	5.517ns	Population	100266
Max	5.597ns	Mean	5.5541ns
Pk-Pk	80ps	Std. Dev.	10.039ps

(b)

Fig. 11. Cycle–cycle processor clock jitter: (a) quiet processor and (b) active processor.

limit of period T_l composed of six delay units and one mixer unit. These frequency limits also affect the VCO gain (for a given mixer design) as well as the center frequency. The frequency limits may be independently controlled using the multiplexers shown in Fig. 9, allowing flexible control of the VCO operating range and greater than ten-to-one adjustment range for VCO gain.

The delay elements and mixer designs are based upon PMOS source-coupled pair differential amplifiers with NMOS load networks [Fig. 10(a) and (b)] which allow voltage-controlled swing adjustment through effective load-line translation by adjusting the voltage V_n. The high impedance provided by the current source improves the supply noise rejection for the source-coupled pair, and the N-well improves the isolation to the p+ bulk substrate noise. The variation of the threshold voltage due to bulk effect is eliminated using bulk-to-source biasing throughout the structure. Sensitivity of the VCO to low repetition rate, 100-mV steps on VDD and AVDD is 0.418 ps/mV. Center-frequency common-mode voltage sensitivity is <3.5% over the full input range dictated by

the common-mode control circuit. Nominal VCO gain for the settings that produce the maximum VCO range is 185 MHz/V. The worst case VCO power dissipation is 30 mW.

I. Dividers and Receivers

Dividers M_1, M_3, M_5, and M_6 (Fig. 1) may be individually programmed and support division by 2, 3, 4, 5, 6, 8, or 10. The dividers are placed in pairs within the layout to improve device matching between M_3 and M_5 and between M_1 and M_6. The receivers shown in Fig. 1 are also placed together and are located near the I/O pad for BUSCLK.

V. PLL MEASUREMENTS

The damping factor, loop gain, and natural frequency of the PLL may be adjusted over a wide range to match the application by changing the charge-pump and VCO gain as described above. System testing was conducted with 90-A peak charge-pump current using the maximum frequency and range on the VCO with a variety of divider settings and BUSCLK

411

frequencies. The processor clock was accessed from the clock tree through a series of inverters. A time-interval measurement (TIM) system was used to measure cycle–cycle period jitter statistics for a number of packaged die representing various process skews. The processor was operated using an array initialization program loop with the fixed-point and floating-point processors active for the "active" processor tests, and was also operated in a "quiet" mode reset state. All tests were performed at room temperature with ambient forced-air cooling. Conventional first-cycle oscilloscope-based jitter measurements were performed periodically and provided P-P jitter results that were consistent with those measured on the TIM system. The external clock was provided by a high-frequency pulse generator, with 7.3 ps rms, 36 ps P-P jitter.

Fig. 11(a) shows a histogram of cycle–cycle period measurements taken with the processor in an inactive reset state but with the clock tree active. The frequencies of the reference clock, processor clock, and VCO are 85, 170, and 340 MHz, respectively, which corresponds to a -3-dB loop bandwidth of 2 MHz. The distribution of samples in the histogram follows a Gaussian distribution with period jitter of 8.4 ps rms, 62 ps P-P. The minimum period measured for this sample size ($n = 22\,278$) was 26.2 ps less than the mean (3.1 sigma away). Assuming that cycle-time failures only occur on the minimum period side, the worst case clock jitter penalty for this system (i.e., a "quiet" processor) is 26.2 ps at 3.1 sigma confidence (or 25.2 ps penalty at 3.0 sigma). Since a peak-to-peak jitter approximately equal to the PFD dead zone can exist for the PLL, the ±25 ps simulated value for the dead zone may be a significant component of the measured jitter.

Fig. 11(b) shows a clock-jitter histogram for the processor executing the array initialization routine for a large population ($n = 100\,266$). A Gaussian curve has been superimposed on the histogram for comparison purposes. The frequencies of the reference clock, processor clock, and VCO are 90, 180, and 360 MHz, respectively. For this system (i.e., an "active" processor), the period jitter has increased to 10.0 ps rms, 80 ps P-P, and the worst case clock-jitter penalty is 37.1 ps at 3.7 sigma confidence (or 30.1 ps at 3.0 sigma). The effective noise penalty for running the array initialization routine is 4.9 ps at 3.0 sigma.

VI. CONCLUSION

This work demonstrates the viability of a low-jitter PLL design approach amenable to high-speed microprocessors. Measured jitter for the design was 8.4 ps rms, 62 ps P-P for quiet conditions and 10.0 ps rms, 80 ps P-P for the processor active. A tunable, moderate-gain VCO with active process and temperature compensation provides high power-supply rejection and low sensitivity to temperature and process variability. A differential design approach maintains noise immunity in both control and signal paths within the analog portions of the PLL.

ACKNOWLEDGMENT

The author wishes to thank J. Peter for layout of the PLL, N. James and H. Casal for the hardware characterization and divider implementation, R. Kodali for circuit simulation and specification, D. Woeste and J. Strom for the divider and lock detector circuits, and S. Dhong and M. Papermaster for their continuous support of this work.

REFERENCES

[1] I. Young, M. Mar, and B. Bhushan, "A 0.35 μm CMOS 3–880 MHz PLL $N/2$ clock multiplier and distribution network with low jitter for microprocessors," in *ISSCC Dig. Tech. Papers*, Feb. 1997, pp. 330–331.

[2] J. Alvarez, H. Sanchez, G. Gerosa, and R. Countryman, "A wide-bandwidth low-voltage PLL for powerPC microprocessors," *IEEE J. Solid-State Circuits*, vol. 30, pp. 383–391, Apr. 1995.

[3] J. Cho, "Digitally-controlled PLL with pulse width detection mechanism for error correction," in *ISSCC Dig. Tech. Papers*, Feb. 1997, pp. 334–335.

[4] I. Young, J. Greason, and K. Wong, "A PLL clock generator with 5–110 MHz of lock range for microprocessors," *IEEE J. Solid-State Circuits*, vol. 27, pp. 1599–1607, Nov. 1992.

[5] V. von Kaenel, D. Aebischer, C. Piguet, and E. Dijkstra, "A 320 MHz, 1.5 mW at 1.35 V CMOS PLL for microprocessor clock generation," in *ISSCC Dig. Tech. Papers*, Feb. 1996, pp. 132–133.

[6] P. E. Gronowski, P. Bannon, M. Bertone, R. Blake-Campos, G. Bouchard, W. Bowhill, D. Carlson, R. Castelino, D. Donchin, R. Fromm, M. Gowan, A. Jain, B. Loughlin, S. Mehta, J. Meyer, R. Mueller, A. Olesin, T. Pham, R. Preston, and P. Rubinfeld, "A 433 MHz 64b quad-issue RISC microprocessor," in *ISSCC Dig. Tech. Papers and Slide Supplement*, Feb. 1996, pp. 222–223.

[7] Z. Zhang, H. Du, and M. Lee, "A 360 MHz 3V CMOS PLL with 1 V peak-to-peak power supply noise tolerance," in *ISSCC Dig. Tech. Papers*, Feb. 1996, pp. 134–135.

[8] D. H. Wolaver, *Phase-Locked Loop Circuit Design*. Englewood Cliffs, NJ: Prentice-Hall, 1991, pp. 59–61.

[9] J. F. Ewen, A. Widmer, M. Soyuer, K. Wrenner, B. Parker, and H. Ainspan, "Single-chip 1062 Mbaud CMOS transceiver for serial data communication," in *ISSCC Dig. Tech. Papers*, Feb. 1995, pp. 32–33.

[10] B. Lai and R. Walker, "A monolithic 622 Mb/s clock extraction and data retiming circuit," in *ISSCC Dig. Tech. Papers*, Feb. 1991, pp. 144–145.

[11] S. K. Enam and A. Abidi, "NMOS IC's for clock and data regeneration in gigabit-per-second optical fiber receivers," *IEEE J. Solid-State Circuits*, vol. 27, pp. 1763–1774, Dec. 1992.

[12] D. W. Boerstler and K. Jenkins, "A phase-locked loop clock generator for a 1 GHz microprocessor," in *Symp. VLSI Circuits Dig. Tech. Papers*, June 1998, pp. 212–213.

[13] J. Silberman, N. Aoki, D. Boerstler, J. Burns, S. Dhong, A. Essbaum, U. Ghoshal, D. Heidel, P. Hofstee, K. Lee, D. Meltzer, H. Ngo, K. Nowka, S. Posluszny, O. Takahashi, I. Vo, and B. Zoric, "A 1.0 GHz single-issue 64b PowerPC integer processor," in *ISSCC Dig. Tech. Papers*, Feb. 1998, pp. 230–231.

[14] V. von Kaenel, D. Aebischer, R. van Dongen, and C. Piguet, "A 600 MHz CMOS PLL microprocessor clock generator with a 1.2 GHz VCO," in *ISSCC Dig. Tech. Papers*, Feb. 1998, pp. 396–397.

David W. Boerstler received the B.S. degree in electrical engineering from the University of Cincinnati, Cincinnati, OH, in 1978 and the M.S. degree in computer engineering and in electrical engineering from Syracuse University, Syracuse, NY, in 1981 and 1985, respectively.

Since joining IBM in 1978, he has held a variety of assignments, including the design of high-frequency PLL's for clock generation and recovery, fiber-optic transceiver and system design, and other analog, digital, and mixed-signal bipolar and CMOS circuit development projects. He currently is a Research Staff Member with the High-Performance VLSI group at the IBM Austin Research Laboratory, Austin, TX. His current research interests include high-frequency synchronization techniques and signaling approaches for high-speed interconnect.

Mr. Boerstler has received IBM Outstanding Technical Achievement Awards for his work on the design of the serializer/deserializer for the ESCON fiber-optic channel products and for the clock-generator design of IBM's 1-GHz PowerPC microprocessor prototype. He has received seven IBM Invention Achievement Awards.

A 960-Mb/s/pin Interface for Skew-Tolerant Bus Using Low Jitter PLL

Sungjoon Kim, *Student Member, IEEE*, Kyeongho Lee, *Student Member, IEEE*, Yongsam Moon, *Student Member, IEEE*, Deog-Kyoon Jeong, *Member, IEEE*, Yunho Choi, and Hyung Kyu Lim, *Member, IEEE*

Abstract—This paper describes an I/O scheme for use in a high-speed bus which eliminates setup and hold time requirements between clock and data by using an oversampling method. The I/O circuit uses a low jitter phase-locked loop (PLL) which suppresses the effect of supply noise. Measured results show peak-to-peak jitter of 150 ps and rms jitter of 15.7 ps on the clock line. Two experimental chips with 4-pin interface have been fabricated with a 0.6-μm CMOS technology, which exhibits the bandwidth of 960 Mb/s per pin.

Index Terms— Skew-tolerant, high speed bus, oversampling, phase locked loop, jitter, CMOS, phase frequency detector, voltage controlled oscillator.

I. INTRODUCTION

AS the speed of high-speed digital systems tends to be limited by the bandwidth of pins, new I/O architectures are gaining momentum over conventional ones. The advent of 64 Mb and 256 Mb DRAM's and faster logic chips also propels the need for high-speed I/O interface while reducing the number of pins and hence the system cost. Synchronous DRAM's increased chip bandwidth up to 220 Mb/s/pin [1]. A revolutionary architecture using delay-locked loops (DLL's) or phase-locked loops (PLL's) was also successful in providing over 500 Mb/s/pin bandwidth [2], [3]. Such a narrow, high-speed bus provides large bandwidth in a small, low pin-count package, but such high-speed bus architectures inevitably require strict phase relationships between clock and data. A phase-tolerant I/O scheme was also developed previously for a point-to-point link [4]. This paper describes an I/O scheme for use in a high-speed bus which eliminates setup and hold time margins by using blind 3× oversampling and data recovery. In the new scheme, the clock line delivers only frequency information. The data receiving circuits extract phase information from the data itself. An 8-b data bus employing this skew insensitive scheme can deliver over 960 MB/s. Two experimental chips with 4-pin interface were fabricated.

In Section II, the chip architecture and the skew-tolerant I/O scheme will be presented. The circuit design techniques for low jitter PLL and other circuits are discussed in Section III.

The chip layout and experimental results are presented in Section IV followed by a conclusion in the final section.

II. SYSTEM ARCHITECTURE

Two chips, bus master and bus slave, were designed. Bus masters in a system bus initiate bus transactions, and slaves respond to the tenured master. For example, a memory controller works as the master chip and a memory with a high-speed interface works as the slave chip. A simplified block diagram of the two chips is shown in Fig. 1. The bus signals are composed of 4-b wide data lines, a clock line, and a reference line. A charge pump PLL multiplies the external clock by two and generates two sets of multiphase clocks for both bit serialization and data oversampling. The relationship between internal 12-phase clocks and external clock is shown in Fig 2. First set of multiphase clocks are 12 multiphase clocks with 30° of phase separation. These 12 clocks are shown in Fig 2(a) as PCK[0] to PCK[11]. These multiphase clocks were laid out to minimize the interference. Fig 2(b) shows the multiphase clock distribution. Ground lines were inserted between each multiphase clock to minimize the interference. When one clock is switching, the adjacent clocks are guaranteed to be in stable state. This configuration minimizes coupling between clocks. The second set of multiphase clocks are four multiphase clocks with 90° of phase separation. This second set of multiphase clocks, TCK[0] to TCK[3], are in phase with PCK[0], PCK[3], PCK[6], PCK[9], respectively. We generate these two separate sets of clocks to equalize loading conditions.

An 8-b parallel data stream is first converted to a 4-b data stream by an internal clock and then serialized with a serialization circuit. The serializer circuit used is the same type of circuit reported in [4]. The only difference is that four phase clocks instead of ten phase clocks of the previous design are used in this design, thereby reducing area and parasitic capacitance at high-speed nodes. The serial stream is driven by a current controlled open-drain output driver. The second set of multiphase clocks, TCK[0] to TCK[3], are used by the transmitter to serialize 4 b of data. Each pin connected to a high-speed bus has 12 oversamplers and a output driver. In [6], 32 clock phases are generated to oversample the incoming data. The decision on the degree of oversampling is a tradeoff between input data phase jitter tolerance, power, and area. If too many clock phases are used per bit period, power consumption and chip area will increase. But low oversampling ratio may affect the tolerance of phase

Manuscript received August 20, 1996; revised December 3, 1996.
S. Kim, K. Lee, Y. Moon, and D.-K. Jeong are with the Inter-University Semiconductor Research Center, Seoul National University, Seoul 151-742, Korea.
Y. Choi and H. Lim are with Samsung Electronics Co., Yongin-City, Kyungki-Do, Korea.
Publisher Item Identifier S 0018-9200(97)02850-3.

413

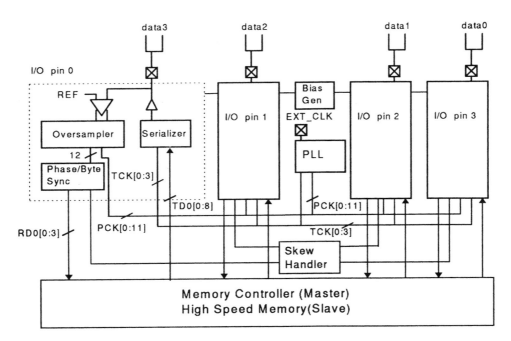

Fig. 1. Simplified block diagram of master and slave chip.

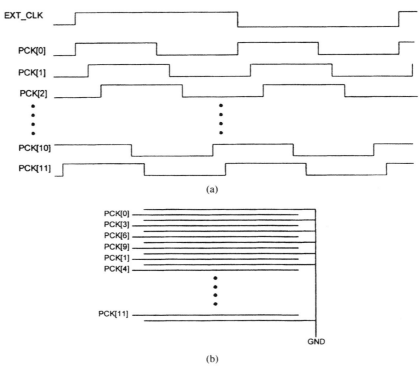

(a)

(b)

Fig. 2. (a) External clock and 12 multiphase clocks relationship. (b) Multiphase clock layout.

jitter on the incoming data. If the phase jitter on the incoming data is low and the PLL has low jitter characteristics, the oversampling ratio can be as low as three [7]. The oversampler oversamples the bus data three times per bit using 12 phase clocks provided by a PLL. To extract correct phase information from the data stream, the high-to-low transition is inserted in each head of a packet on each pin for correct data sampling. The slaves of the bus keep oversampling the bus signals to catch the start of a bus transfer. This process is illustrated in

Fig. 3. The serial input data is sampled at the rising edges of each multiphase clock. The receiver samples the serial data blindly without any constraint on setup and hold time margins. The sampled data is amplified again regeneratively to reduce possible metastability. Fig. 3 shows two high-speed bus signals, bus signal 0 and bus signal 1, with skew between them. When the signal receiver detects the first 1-to-0 transition, it selects the next bit as the first valid data. The third bit after the first valid bit is also selected as valid. It is assumed

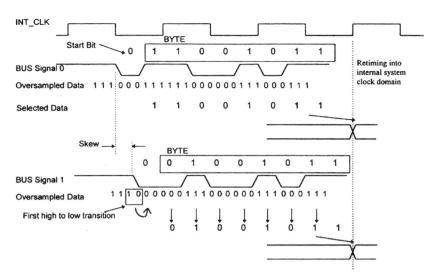

Fig. 3. Skew-insensitive bus operation.

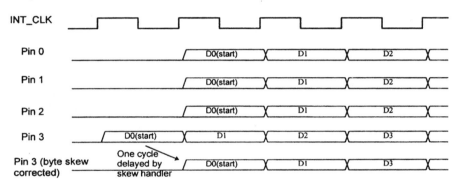

Fig. 4. Byte skew handling operation.

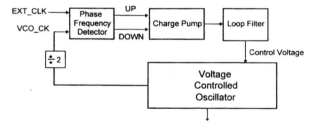

Fig. 5. Functional block diagram of charge pump PLL.

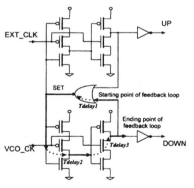

Fig. 7. Implemented phase frequency detector.

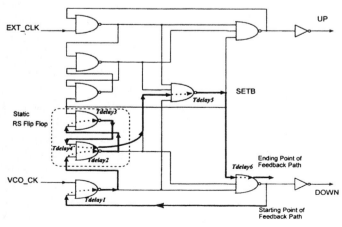

Fig. 6. Conventional phase frequency detector.

that the next oversampled bit after the first 1-to-0 transition was sampled near the center of data eye pattern. Each pin of the data bus tracks the start phase of a data transfer separately. After each pin catches the start of a data transfer, the demultiplexed data of each pin is retimed into a single internal clock domain. Since this process can be done in one clock cycle, the masters can respond quickly as distance from the signal source changes.

Since this scheme allows skew not only in clock line but also among data lines, there is a possibility that some of the demultiplexed parallel data are one internal clock cycle earlier or later than the other demultiplexed data after retiming.

415

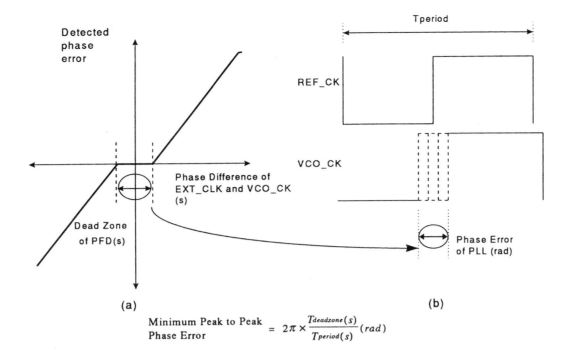

$$\text{Minimum Peak to Peak Phase Error} = 2\pi \times \frac{T_{deadzone}(s)}{T_{period}(s)} (rad)$$

Fig. 8. (a) PFD dead zone and (b) PLL jitter.

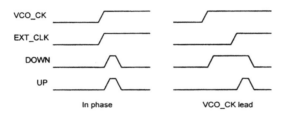

Fig. 9. Voltage controlled oscillator circuit diagram.

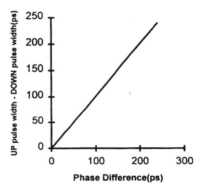

Fig. 10. Simulated UP/DOWN pulse width difference as a function of input phase difference.

The skew handler examines the parallel output of each pin and checks whether every pin is aligned properly. If some of the parallel outputs are not aligned, skew handler delays the parallel outputs which arrived earlier. Fig 4 explains the operation of the interpin skew handler operation.

III. CIRCUIT DESCRIPTION

A. Low Jitter PLL

The performance of the PLL or DLL is one of the limiting factors of the high-speed interface or serial communications. The jitter characteristics become more important especially for such applications that require integration of PLL or DLL with noisy digital circuits. Integration with digital circuits induces noise on the supply rails or on the substrate. Since the charge pump PLL used in this design generates multiple phase clocks to divide one external clock period into many equally spaced intervals, the accuracy and the jitter characteristics become more important.

Fig. 5 shows the functional block diagram of a charge pump PLL clock generator. It consists of a phase frequency detector, charge pump, loop filter, clock divider, and a voltage-controlled oscillator (VCO). With a six-stage differential VCO, 12 clock phases are available to oversample the incoming data and to serialize parallel data into serial bit stream.

One of the critical building blocks of the PLL is the phase frequency detector (PFD). A low precision PFD has a wide dead zone (undetectable phase difference range), which results in increased jitter. The jitter caused by the large dead zone can be reduced by increasing the precision of the phase frequency detector. Fig. 6 shows a conventional implementation of a static PFD [8]. This conventional PFD is an asynchronous state machine. The delay time to reset all internal nodes determines the circuit speed. The critical path of the conventional PFD is shown in bold lines in Fig. 6. The critical path forms a feedback path with six gate delays. The dead-zone occurs when the loop is in a lock mode and the output of the charge pump

416

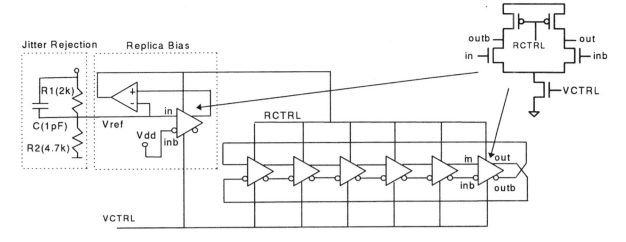

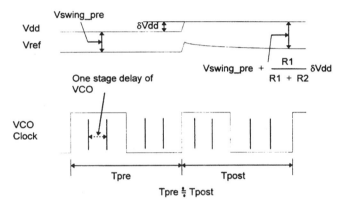

Fig. 11. Voltage controlled oscillator circuit diagram.

does not change for small changes in the input signals at the PFD. Any width of the dead-zone directly translates to jitter in the PLL and must be avoided.

To overcome the speed limitation and to reduce the dead zone, a new dynamic logic style PFD was designed. A similar dynamic comparator was reported before [9]. But our implementation requires fewer number of transistors. Fig. 7 shows the circuit diagram of the PFD. Conventional static logic circuitry was replaced by dynamic logic gates. As a result, the number of transistors in the PFD core is reduced from 44 to 16. The critical path of this PFD is shown also in Fig. 7. The critical path of this PFD is composed of three-gate feedback path. The shortened feedback path delay and dynamic operation allow high precision in the high-frequency operation.

Fig. 8. shows the relation between dead zone of PFD and the phase error of PLL. If the phase difference of EXT clock and VCO clock is smaller than the dead zone, the PFD cannot detect the phase difference. So the phase error signal of PFD will remain zero, resulting in unavoidable phase error between EXT clock and VCO clock. The minimum peak-to-peak phase error caused by this dead zone is

$$\text{Minimum Peak-to-Peak Phase Error} = 2\pi \times \frac{T_{\text{deadzone}}}{T_{\text{period}}}. \quad (1)$$

In order to avoid dead zone, the PFD asserts both UP and DOWN outputs as shown in Fig 9. For in-phase inputs of EXT_CLK and VCO_CK, the charge pump will see both UP and DOWN pulse for the same short period of time. If there is a phase difference between EXT_CLK and VCO_CK, the width of UP and DOWN pulse will be proportional to the phase differences of the inputs. Fig. 10 shows the SPICE simulation result of the UP/DOWN pulse width differences as a function of the input phase differences. The deadzone of the PFD is significantly smaller than the measured maximum PLL jitter.

Several critical parameters of the PLL, such as speed, timing jitter, spectral purity, and power dissipation, strongly depend

Fig. 12. VCO operation for step supply noise.

on the performance of the VCO. So the noise insensitivity of the VCO is very important. The VCO implemented in this design has a simple bias circuit to reject supply step noise. The processor or bus can have intervals when there is heavy circuit activity in switching large amounts of capacitance and intervals when there is very little circuit activity. This will show up as steps or impulses on the power supply of PLL [8]. The actual peak-to-peak jitter in this case becomes dominated by the peaks in the impulse transient noise response. The VCO used in the design is a six-stage differential-type ring oscillator with limited voltage swing and is shown in Fig. 11. Each stage is made up of a differential NMOS pair with variable resistance loads made of PMOS devices operating in the triode region. The bias voltage for the PMOS is generated by a replica bias circuit. The operation of this bias circuit is shown in Fig. 12. The V_{ref} voltage dynamically tracks the supply variations. The replica bias circuit which consists of replica delay cell and an op-amp sets the minimum voltage level of the internal VCO swing to V_{ref}. The V_{ref} signal is generated by two resistors and one capacitor. When the supply rail is quiet, the voltage swing of the internal VCO is V_{dd}-V_{ref}. Let us assume that there is a supply voltage step variation of δV_{dd} at some point. After the

417

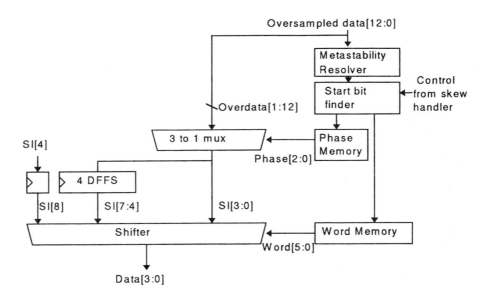

Fig. 13. Phase and byte sync block diagram.

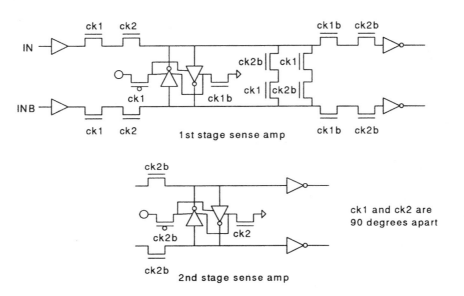

Fig. 14. Sampler circuit diagram.

step change at the supply, the V_{ref} level settles to

$$\frac{R_2}{R_1 + R_2}(V_{dd} + \delta V_{dd}) \qquad (2)$$

with a time constant of

$$\frac{R_1 \cdot R_2 \cdot C}{R_1 + R_2}. \qquad (3)$$

At the instant of supply step change, the voltage difference between V_{dd} and V_{ref} remains the same due to the capacitor at the V_{ref} generator. If V_{dd}-V_{ref} is fixed, the delay cells run a little bit faster due to the supply voltage increase instead of keeping exact constant delay. Since V_{dd}-V_{ref} remains the same temporarily, the delay cells run a little bit faster due

to the increased supply voltage for a short period of time. And the voltage swing at the VCO increases with a time constant determined by R_1, R_2, C, and OPAMP bandwidth and approaches to

$$\frac{R_1}{R_1 + R_2}(V_{dd} + \delta V_{dd}) \qquad (4)$$

which result in the increase of one stage delay. This gives an averaging effect on the VCO delay after the supply step change, making the delay change minimized with supply step change. If we select R_1, R_2, and C values for a minimum average delay change, the effect of supply step change can be nullified. The values we chose for this particular process are $R_1 = 4$ kΩ, $R_2 = 10$ kΩ, and $C = 1.5$ pF.

PLL circuits can be sensitive to noise pickup from the supplies and substrate. So the PLL circuit has a dedicated power and ground pads. Bypass capacitors are included in the layout to stabilize VDD and GND of PLL. Guard rings are used to isolate PLL and other digital parts. The placement of multiphase clocks were carefully chosen to remove possible coupling between clocks.

B. Phase and Byte Sync

Phase and byte sync block at Fig. 1 is shown in Fig 13. It consists of 3-to-1 mux array, metastability resolver, start bit finder, phase memory, word memory, shifter, and D-flip-flops (DFF's). This circuit finds the start bit and decimates the oversampled 12 b and aligns the byte boundary. The oversampled 12 b are sent from the sampler to the metastability resolver. Since the oversampled 12 b are not sampled at the center of the eye, there is a possibility that some of the bits are still at the metastable state. The metastability is practically removed by one more stage of synchronizers in the metastability resolver. The start bit finder receives information from the metastability resolver and selects one of the three phases as a correct phase and also extracts byte align information. The phase and byte align information are stored at the phase and word memory. The 3-to-1 mux array decimates 12 b into 4 b. The shifter at the final stage aligns the byte boundary according to the value of the word memory.

C. Oversampler

The oversampler used in the data receiver is shown in Fig. 14. Each oversampler is a cascaded sense amplifier and uses four clocks for correct, timely sampling. It is very important to reduce the probability of metastability by careful design and layout. The same size is used for both PMOS and NMOS in the core synchronizing amplifier to maximize the loop bandwidth.

IV. EXPERIMENTAL RESULTS

Two prototype chips, master and slave, have been fabricated in a 0.6-μm double-metal CMOS process. Fig. 15 shows the microphotograph of the fabricated master chip. This chips occupies 4100 μm × 4300 μm including pad area. The master chip incorporates a common skew-insensitive I/O macro block, a bus protocol handler, and a self-test circuit for chip and system diagnostics. The common skew-insensitive I/O macro block includes a charge-pump PLL for multiphase generation, oversamplers, I/O buffers, parallel-to-serial converters, and a bias generator for internal use. The core area for the skew-insensitive I/O macro block is 3600 μm × 700 μm for 4-pin interface. The microphotograph of the fabricated slave chip is shown in Fig 16. It has the same die size as the master chip. Many blocks are shared with the master chip. The skew-insensitive I/O macro block and the charge pump PLL are the same as those of the master's. The slave chip includes a small internal fast SRAM to verify correct read/write operations.

The measured charge pump PLL jitter histogram of the master and the slave chips is shown in Fig. 17. Since the two chips use the same PLL, it showed similar jitter performance.

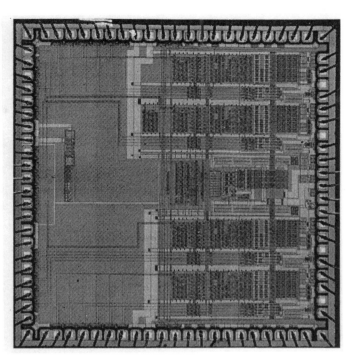

Fig. 15. Microphotograph of master chip.

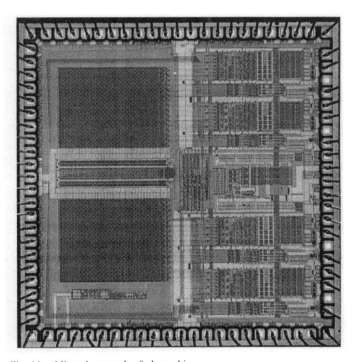

Fig. 16. Microphotograph of slave chip.

The rms jitter is 15.7 ps when the tested chip is active. The peak-to-peak jitter was measured to be less than 150 ps. This PLL jitter characteristic is especially important for multiphase operation.

Fig. 18 shows an output data waveform at 960 Mb/s. The master chip is sending data to the bus according to the predetermined bus protocol. The jitter at the output data is larger than the jitter at the charge pump PLL clock due to the extra modulation effect of supply voltage fluctuation to data

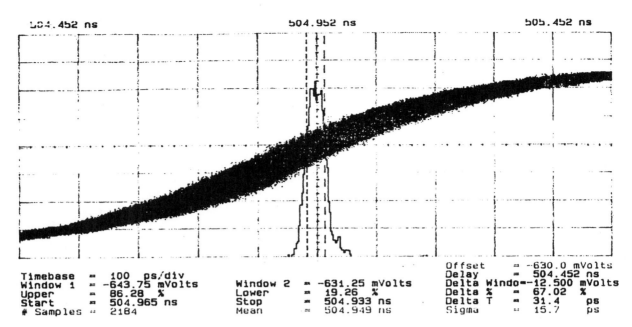

Timebase   =   100 ps/div   Window 2   = -631.25 mVolts   Offset   = -630.0 mVolts
Window 1   = -643.75 mVolts   Lower   =   19.26 %   Delay   = 504.452 ns
Upper   =   86.28 %   Stop   = 504.933 ns   Delta Windo=-12.500 mVolts
Start   = 504.965 ns   Mean   = 504.949 ns   Delta %   =   67.02 %
Samples   =   2184   Delta T   =   31.4   ps
                                                                       Sigma   =   15.7   ps

Fig. 17. PLL jitter histogram.

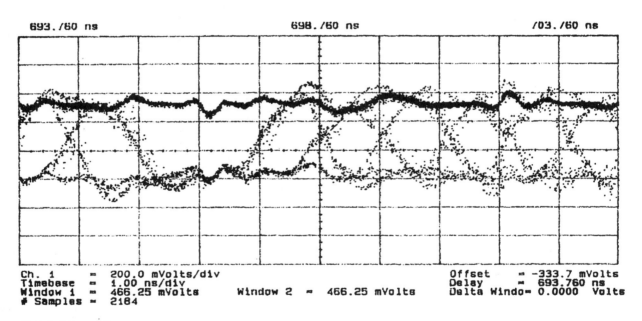

Ch. 1   =   200.0 mVolts/div   Offset   = -333.7 mVolts
Timebase   =   1.00 ns/div   Delay   = 693.760 ns
Window 1   =   466.25 mVolts   Window 2   = 466.25 mVolts   Delta Windo= 0.0000 Volts
Samples   =   2184

Fig. 18. Output data waveform.

output. The speed limit came from several reasons. CMOS driving capability limitation and the signal degradation through chip packaging and printed circuit board (PCB) were among the main factors.

The skew-insensitive receiving operation was also observed. There are four high-speed pins in the prototype chip. We made a PCB with four high-speed impedance controlled bus lines. The length of normal lines is 12 cm. One of the high-speed signal paths was made intentionally longer than the other signals by 10 cm. The 960 Mb/s high-speed serial data was sent into the receiver. The receiver recovers the serial data into 8-b 120-MHz parallel data. Fig. 19 shows 120 MHz recovered parallel data. The upper waveform is from the

pin with a longer trace. The lower waveform is from the normal length pin. Although the two pins have different trace lengths, the chips could receive data without errors. The power dissipation at 960 Mb/s was 0.7 W for the master chip. The chip characteristics is summarized in Table I.

TABLE I
MAIN FEATURES OF THE CHIP

Core Area	3.6 mm × 0.7 mm
Technology	0.6-μm double-metal CMOS
Supply Voltage	3.3 V
Data Rate	960 Mb/s
PLL jitter	15.8 ps rms @ 960 Mb/s
Power	0.7 W fully active

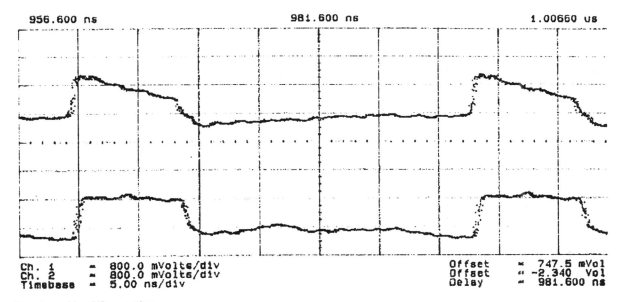

Ch. 1	= 800.0 mVolts/div	Offset	= 747.5 mVol
Ch. 2	= 800.0 mVolts/div	Offset	= -2.340 Vol
Timebase	= 5.00 ns/div	Delay	= 981.600 ns

Fig. 19. Skew-insensitive I/O operation.

V. CONCLUSION

A new high-speed skew-insensitive I/O scheme has been described in this paper. Two chips that incorporated the new I/O scheme using the low jitter PLL technique have been fabricated in a 0.6-μm double-metal CMOS process. Three times oversampling technique relaxed the strict requirement of setup and hold margins of high-speed chip-to-chip interfaces. Newly designed fast phase frequency detector and a high noise immunity VCO circuit improved jitter performance of PLL. The measured PLL rms jitter was 15.7 ps. Accurate multiphase clock generation for oversampling the bus signal was made possible by utilizing the low jitter PLL. By using such techniques, skew-insensitive data transfer was tested. This skew-insensitive I/O scheme is useful for high-speed ASIC-to-memory and ASIC-to-ASIC interfaces. This scheme will become more important as the chip-to-chip data transfer speed goes up.

REFERENCES

[1] M. Horiguchi et al., "An experimental 220 MHz 1 Gb DRAM," in ISSCC 1995 Dig. Tech. Papers, pp. 252–253.
[2] M. Horowitz et al., "PLL design for a 500 MB/s interface," in ISSCC 1993 Dig. Tech. Papers, pp. 160–161.
[3] T. H. Lee et al., "A 2.5 V CMOS delay-locked loop for an 18 Mbit, 500 Megabytes/s DRAM," IEEE J. Solid-State Circuits, vol. 29, pp. 1491–1496, Dec. 1994.
[4] E. Reese et al., "A phase tolerant 3.8 GB/s data-communication router for a multiprocessor supercomputer backplane," in ISSCC 1994 Dig. Tech. Papers, Feb. 1994, pp. 296–297.
[5] S. Kim et al., "A pseudo-synchronous skew-insensitive I/O scheme for high bandwidth memories," in Proc. Symp. VLSI Circuits, June 1994, pp. 41–42.
[6] M. Bazes and R. Ashuri, "A novel CMOS digital clock and data decoder," IEEE J. Solid-State Circuits, vol. 27, pp. 1934–1940, Dec. 1992.
[7] S. Kim et al., "An 800 Mbps multi-channel CMOS serial link with 3× oversampling," in Proc. IEEE Custom Integrated Circuit Conf., 1995, pp. 451–454.
[8] I. Young et al., "A PLL clock generator with 5 to 110 MHz lock range for microprocessors," IEEE J. Solid-State Circuits, vol. 27, pp. 1599–1607, Nov. 1992.
[9] H. Notani et al., "A 622-MHz CMOS phase-locked loop with precharge-type phase frequency detector," in Proc. Symp. VLSI Circuits, June 1994, pp. 129–130.

Sungjoon Kim (S'91) was born in Pusan, Korea, on June 2, 1970. He received the B.S. and M.S. degrees in electronics engineering from Seoul National University in 1992 and 1994, respectively. Since 1994 he has been working toward the Ph.D. degree in the same university.

He spent the summer of 1995 working on the limiting factors of CMOS Gb/s transmission at SUN Microsystems, CA. His research interests include clock and data recovery for high-speed communication and high-speed I/O interface circuits.

Kyeongho Lee (S'92) was born in Seoul, Korea, on August 5, 1969. He received the B.S. and M.S. degrees in electronics engineering from Seoul National University in 1993 and 1995, respectively. He is currently working toward the Ph.D. degree in electronics engineering of the same university.

He is working on various CMOS high-speed circuits for data communication. His research interests include high-speed CMOS interface circuits, high-speed video display system, and PLL systems for Gigabit communication.

Yongsam Moon (S'97) was born in Incheon, Korea, on March 1, 1971. He received the B.S. and M.S. degrees in electronics engineering from Seoul National University in 1994 and 1996, respectively, where he is currently working toward the Ph.D. degree.

He has been working on architectures and CMOS circuits for microprocessors. His current research interests are in clock and data recovery circuits for high-speed data communication.

Active GHz Clock Network Using Distributed PLLs

Vadim Gutnik, *Member, IEEE*, and Anantha P. Chandrakasan, *Member, IEEE*

Abstract—A novel clock network composed of multiple synchronized phase-locked loops is analyzed, implemented, and tested. Undesirable large-signal stable (mode-locked) states dictate the transfer characteristic of the phase detectors; a matrix formulation of the linearized system allows direct calculation of system poles for any desired oscillator configuration. A 16-oscillator 1.3-GHz distributed clock network in 0.35-μm CMOS is presented here.

Index Terms—Clock network, multiple oscillator system, phase-locked loop.

I. INTRODUCTION

THE CLOCK distribution network of a modern microprocessor uses a significant fraction of the total chip power and has substantial impact on the overall performance of the system. For example, the 72-W 600-MHz Alpha processor [1] dissipates 16 W in the global clock distribution, and another 23 W in the local clocks: more than half the power goes to driving the clock net. The clock uncertainty budget for a global clock is 10% of a clock period, which translates to a 10% reduction in maximum operating speed; as argued below, this penalty is likely to increase for currently popular clock architectures.

Most conventional microprocessors use a balanced tree to distribute the clock [1]–[3]. Because the delays to all nodes are nominally equal, trees may be expected to have low skew. However, at gigahertz clock speeds a large fraction of skew and jitter comes from random variations in gate and interconnect delay. The majority of jitter in a clock tree is introduced by buffers and inter-line coupling to the clock wires; a relatively small amount comes from noise in the source oscillator [4]. Therefore, a primary consideration in clock design is matching delay along the clock path.

As clock speed increases, signal delay across a chip becomes comparable to a clock cycle. For example, a 2-cm-long wire in a 0.25-μm process has a delay of 0.86 ns, while the clock might be as high as 1 GHz; scaling to 4 GHz, the same wire (with optimal buffering) will have a delay of approximately 0.43 ns, compared to a clock period of 0.25 ns. In all practical cases a signal that takes longer than a clock cycle to propagate would be pipelined, and hence re-clocked. The fundamental weakness of tree distribution (and networks that depend on tree matching)

is that skew is only relevant between communicating latches, but the clock path is always the length of the chip. Clock speeds increase with gate delay, and processor architectures can exploit both locality of blocks and pipelining to avoid penalty due to long signal paths, but the error in a global clock scales with the total path delay, and is thus a growing fraction of a clock cycle.

In this paper, we consider the effects of static and dynamic mismatch on a few representative clock networks in Section II and propose a distributed generation scheme that needs only local synchronization to generate a global clock. Large and small-signal stability of the proposed network is analyzed in Section III. This clock was implemented on a test chip; circuit details and results are presented in Sections IV and V.

II. MODELING RANDOM SKEW

A. Assumptions

Given sufficiently accurate models, systematic skew can be corrected at design time. Therefore, the primary interest is random zero-mean variations. For the sake of comparing architectures, we make several simplifying assumptions.

1) Delay mismatch, both static and dynamic, is proportional to total delay.
2) Wire RC delay is independent of gate delay (d).
3) The clock period proportional to gate delay.
4) Chip size is independent of gate delay.
5) In 0.25-μm technology, signal delay across a die equals one clock period.

Assumption 1 is inaccurate, but convenient. Mismatch due to gradients scales as delay squared; purely random short-distance mismatch scales as the square root of delay. For the sake of analysis, however, we will assume that uncertainty scales linearly. Assumptions 2, 3, and 4 are approximately true, given historical data: as the geometries scale the resistance increase in clock wires is offset by lower capacitance; processor cycle time is generally on the order of 8–16 gate delays; and chip sizes hover around 15×15 mm^2.

Assumption 5 serves to normalize signal delay, chip size, and clock speed. It is not coincidental that random variation has become a noticeable issue at about the time when cross-die signal delay is comparable to one clock cycle: as a heuristic, 10% of a clock cycle is allocated for unmodeled skew and jitter margin, and delay uncertainty is about 5%–10% of delay. Hence, when delay across a chip is comparable to clock cycle time, random delay is a considerable fraction of the total clock error budget.

B. Tree

To keep internal clock skew low, a tree is generally made deep enough that a tile driven by a single leaf is small compared to the size of the chip [5], [6]. In turn, this means that the path from the

Manuscript received March 24, 2000; revised June 24, 2000. This paper was supported by the MARCO Focused Research Center on Interconnects, which was funded at the Massachusetts Institute of Technology through a subcontract from the Georgia Institute of Technology, and supported in part by a Graduate Fellowship from the Intel Corporation.

V. Gutnik was with M.I.T. Microsystems Technology Lab, Cambridge, MA 02139 USA. He is now with Silicon Laboratories, Austin, TX 78749 USA (e-mail: gutnik@mit.edu).

A. P. Chandrakasan is with M.I.T. Microsystems Technology Lab, Cambridge, MA 02139 USA (e-mail: anantha@mtl.mit.edu).

Publisher Item Identifier S 0018-9200(00)09441-5.

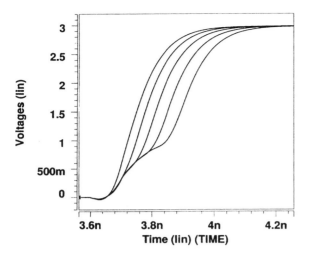

Fig. 1. Simulated edge in a grid with skew to the drivers.

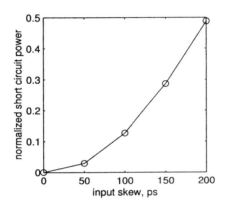

Fig. 2. Short circuit power in a grid vs. input tree skew.

clock source to the load is comparable to the size of the entire die. Because the worst-case skew occurs between two adjacent leaves for which the clock path was completely different, worst case mismatch depends on the entire source-to-leaf delay. And worse, the problem becomes worse with process scaling. Because RC delay does not scale, delay along an optimally buffered line scales only as $\sqrt{d}$; hence the skew as a fraction of the clock period grows as $1/\sqrt{d}$ with falling d.

C. Grid

Modern grids are H-tree-grid hybrids: a short H-tree distributes clock to a few (4 or 16, for example) buffers around a chip, and those buffers drive a clock grid in parallel. Shorting the buffers together helps drive down some of the uncertainty at the cost of increased short-circuit power during switching and somewhat slower edge rates. However, rise time scales linearly with d, so by the same reasoning as applied to the tree scaling arguments, skew as a fraction of rise time will increase with $1/\sqrt{d}$ as gate delay falls. When the tree skew exceeds rise time, short circuit power dissipation increases rapidly, and the clock edges begin to show an unacceptable kink. Fig. 1 shows simulated edge shapes with increasing input skew for a grid driven from a 4-level tree with skews from 0 to 200 ps, and Fig. 2 shows the corresponding short-circuit power dissipation, plotted as a fraction of CV^2f-power for the clock grid.

D. Active Feedback

As is evident from the given examples, most of the skew comes from the initial long-distance distribution of a clock to relatively small loads. A delay-locked loop (DLL) could be adapted to measure and cancel out wire variations, as shown in Fig. 3. If the round-trip delay is tuned to an even number of clock cycles, the wire has nominally 0 delay.

Unfortunately, despite the apparent symmetry, the forward and reverse paths do not match well for two reasons. First, "matched" buffers are physically separated. In Fig. 3, b_1 should match b_7, although it would be physically near b_8. b_1 is not as far away from its matched pair as it might be in a tree, but it will still typically be millimeters away. Second, there is no temporal correlation. The clock signal passes w_1 at a different time than it passes w_7, so any time-dependent variations, including those due to power supply and signal coupling, do not match.

Another approach, proposed by Intel, is shown in Fig. 4 [7]. Here, a DLL matches delays to two half-trees; an obvious generalization, with four DLLs matching quarter-trees is shown in Fig. 5. Static delay variations of some nearest neighbors are canceled out by the DLL to within the precision of the matching of the comparators. The drawback is that some neighboring nodes, as A and B in Fig. 5, are only related through multiple DLLs. A much better result can be obtained by using DLLs that take multiple reference inputs, and adjust output phase to be aligned exactly between the two inputs. The network can then be redrawn somewhat more symmetrically, as Fig. 6. (For clarity, the local tree was not drawn, and the connections to the comparators are abstracted.)

Optimization of the number of tiles is straightforward. Internal skew scales with tile area, so as the number of tiles increases, internal skew falls. However, every boundary between tiles introduces some skew because of mismatch in the phase detector (PD). Hence, as the number of tiles increases, the number of boundaries increases. Fig. 7 shows the optimization curves calculated for this clock metric. As in other clock networks, faster clocks require a more finely grained architecture. Jitter in a DLL network will rise in exactly the same way as it increases in clock trees, and for the same reasons. Skew scales linearly with d because it is comprised of comparator mismatches and delays across each leaf-patch. Note, however, that in a phase-locked loop (PLL) the noise can be expected to scale with d; a PLL network like the one in Fig. 6 would have total clock uncertainty that is a constant fraction of the clock period.

III. STABILITY

We propose a distributed clock network comprised of an array of synchronized PLLs. Independent oscillators generate the clock signal at multiple points ("nodes") across a chip; each oscillator distributes the clock to only to a small section of the chip ("tile") (Fig. 8). PDs at the boundaries between tiles produce error signals that are summed by an amplifier in each tile and used to adjust the frequency of the node oscillator. In general, the network need not be square or regular.

With locally generated clocks, there are no chip-length clock lines to couple in jitter; skew is introduced only by asymmetries in PDs instead of mismatches in physically separated buffers,

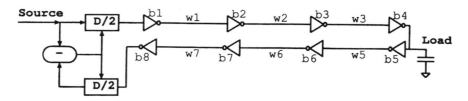

Fig. 3. Low-skew wire with DLL.

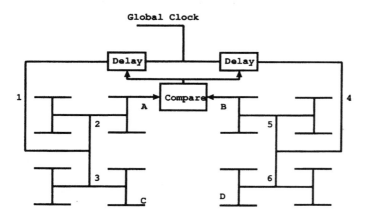

Fig. 4. Matching tree leaves with a DLL.

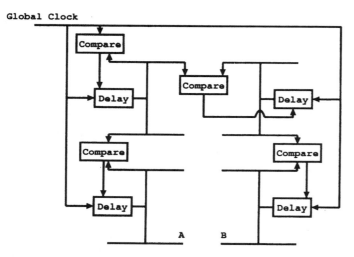

Fig. 5. DLL architecture.

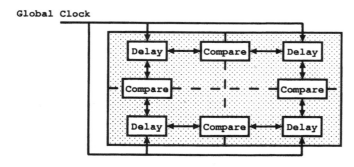

Fig. 6. Multi-input delay cell DLL architecture.

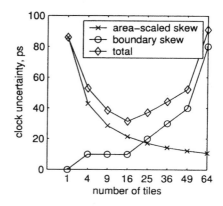

Fig. 7. Tile number optimization.

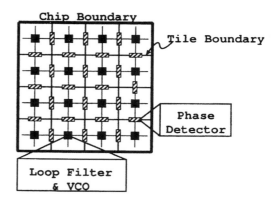

Fig. 8. Distributed clocking network.

and the clock is regenerated at each node, so high-frequency jitter does not accumulate with distance from the clock source. Unlike earlier work on multiple clock domains which suggested the use of multiple independent clocks [8], this approach produces a single fully synchronized clock. The rest of this section examines small and large signal stability of a distributed PLL.

A. Small Signal

In a multiple-oscillator PLL large-signal and small-signal behavior are interrelated. In normal operation, the oscillators are phase-locked, and jitter depends on the network response to noise. Because startup is expected to take a negligibly small fraction of time, the connection of the oscillators is optimized for small-signal behavior rather than to make initial acquisition more efficient. The linearized small-signal behavior, valid when the oscillators are nearly in phase, is analyzed first.

B. General Derivation

The block diagram (Fig. 9) of a multiple-oscillator PLL is essentially identical to the one for a conventional PLL, except that the connections between blocks are vectors instead of individual signals, and the gains and transfer functions are matrices instead of scalars. This means that the PD becomes matrix A_1,

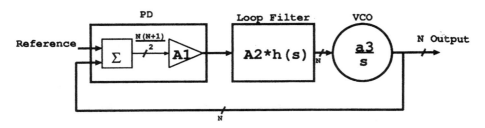

Fig. 9.   Linear system model of a multi-oscillator PLL.

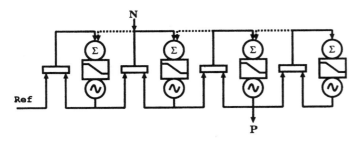

Fig. 10.   One-dimensional PLL array; symmetrical with the dotted-line connections.

of size $N(N+1)/2 \times N$, and the loop filter becomes A_2, a corresponding $N \times N(N+1)/2$ matrix. $G = A_2 A_1$ is an intuitively meaningful $N \times N$ matrix. The network of oscillators is similar to a lumped circuit C with a node for each oscillator and a branch for each connection between pairs of oscillators. Node voltages in C represent oscillator phase, and branch currents represent the error signals on the output of the PD. G is the conductance matrix for C with unity conductance branches. G for a four-oscillator network is shown in (1). Each off-diagonal entry g_{ij} is -1 if there is a PD between node i and node j; g_{ii} is the number of detectors attached to node i.

$$G = \begin{pmatrix} 3 & -1 & -1 & 0 \\ -1 & 2 & 0 & -1 \\ -1 & 0 & 2 & -1 \\ 0 & -1 & -1 & 2 \end{pmatrix}. \qquad (1)$$

DC gain in the loop can be lumped into a_3.

Writing the transfer function in matrix form gives

$$\Phi = [sI + a_3 A_2 A_1 h(s)]^{-1} h(s) a_3 A_2 u \qquad (2)$$

where u is the phase error input to each phase comparator. $u(1)$ is the reference phase, and $u(2) \cdots u(n)$ are the noise contributions from interconnect and PD mismatch.

C. Examples

Matrix A_1 is determined by the geometry of the tiles, and hence will constrained by the placement of clock loads, which for this problem is fixed. Assuming the simplest possible PLL, $h(s) = (s+z)/s$. This leaves A_2, a_3, and z as design variables.

There are still far too many choices to find the general optimum, but a few examples may help guide the search.

1) One-Dimensional Array: A one-dimensional array of oscillators with PDs between neighbors is the simplest generalization of a single PLL. In a perfectly asymmetrical array (call this

system S_1), the output of PLL i is the input to PLL $i+1$, as shown in Fig. 10. S_1 is described by

$$A_1 = \begin{pmatrix} 1 & 0 & 0 & 0 \\ -1 & 1 & 0 & 0 \\ 0 & -1 & 1 & 0 \\ 0 & 0 & -1 & 1 \end{pmatrix} \quad A_{2,1} = \begin{pmatrix} 1 & 0 & 0 & 0 \\ 0 & 1 & 0 & 0 \\ 0 & 0 & 1 & 0 \\ 0 & 0 & 0 & 1 \end{pmatrix}. \qquad (3)$$

This system has multiple poles at the same place where a single-oscillator PLL has single poles.

On the other hand, in a perfectly symmetrical array (call it S_2), the input to each oscillator i is the phase of oscillators $i-1$ and $i+1$ (Fig. 10, with the dotted-line connections). The A_1 matrix is the same because the physical arrangement of nodes is identical, but A_2 changes:

$$A_{2,2} = \begin{pmatrix} 1 & -1 & 0 & 0 \\ 0 & 1 & -1 & 0 \\ 0 & 0 & 1 & -1 \\ 0 & 0 & 0 & 1 \end{pmatrix}. \qquad (4)$$

To achieve the same phase margin in S_2 as in S_1, it is necessary to lower the gain a_3. This can be shown with a geometrical argument: in S_2, when the phase of oscillator i changes by $\Delta\phi$, the change is measured at two PDs, so oscillator i feels twice the feedback that it would have felt in S_1, and at the same time, oscillators $i-1$ and $i+1$ both adjust in the opposite direction, giving four times the effective gain. Hence, the gain must be decreased by a factor of approximately four. Mathematically, the largest eigenvalues of $A_{2,1} A_1$ is 1, but the largest eigenvalue of $A_{2,2} A_1$ is 3.5. Poles of the symmetrical system, solved via (2), are plotted in Fig. 12(a). The key difference between S_1 and S_2 is the systems' response to noise. In both cases, noise at frequencies higher than the unity gain frequency ω_0 are attenuated. For frequencies much lower than ω_0, the response can be calculated via (2). Fig. 11 shows a Bode plot of noise at node P in response to a noise source at node N. Noise performance of S_1 is much worse for intermediate frequencies because there is no feedback so errors propagate forever. In S_2, the feedback limits the influence of preceding stages, and this in turn attenuates noise. For this reason, networks with feedback are preferred, despite the more complicated stability calculation.

2) Two-Dimensional Array: A two-dimensional array is analyzed exactly the same as is a one-dimensional array, except that the gain has to decrease by another factor of two because the center oscillators see four neighbors rather than two. A 16-element array in a 4×4 grid is implemented in this thesis. Its poles are shown in Fig. 12(b).

425

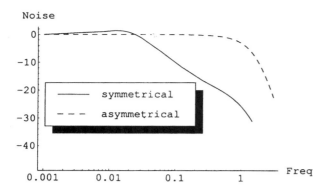

Fig. 11. Comparison of noise responses for symmetrical and asymmetrical networks.

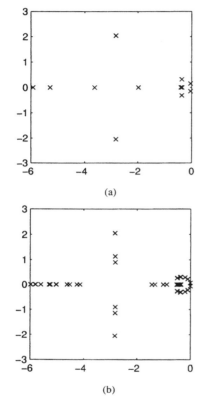

(a)

(b)

Fig. 12. Root locus for 1-D and 2-D PLL arrays. (a) 1-D array. (b) 2-D array.

D. Large Signal: Mode Locking

The analysis of the previous section indicates that fully connected networks should have a better noise response than asymmetrical networks. However, the feedback allows the possibility of undesirable large-signal modes. Consider the matrices for a 2×2 PLL network:

$$A_1 = \begin{pmatrix} -1 & 0 & 0 & 0 \\ 1 & -1 & 0 & 0 \\ 1 & 0 & -1 & 0 \\ 0 & 1 & 0 & -1 \\ 0 & 0 & 1 & -1 \end{pmatrix}$$

$$A_2 = A_1^T = \begin{pmatrix} -1 & 1 & 1 & 0 & 0 \\ 0 & -1 & 0 & 1 & 0 \\ 0 & 0 & -1 & 0 & 1 \\ 0 & 0 & 0 & -1 & -1 \end{pmatrix}. \quad (5)$$

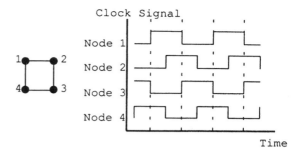

Fig. 13. Mode-locking example.

Because phase is periodic with period 2π, the phase measured at the PDs $\Delta\phi = A_1\phi \bmod 2\pi$. For small ϕ, $(A_1\phi \bmod 2\pi) = A_1\phi$, so the nonlinearity is irrelevant. However, with $\phi = \phi_x = [0, \pi/2, -\pi/2, \pi]^T$

$$A_2(A_1\phi_x \bmod 2\pi) = A_2[0, -\pi/2, \pi/2, -\pi/2, \pi/2]^T = 0 \quad (6)$$

so ϕ_x is a stationary point. This is intuitively easy to see, in reference to Fig. 13: each oscillator leads one neighbor, and lags behind another neighbor by exactly the same amount. The net phase error is zero, so clearly there is no restoring force to drive the phases to 0. Because the nonlinearity does not change for small deviations from ϕ_x, dynamics about ϕ_x are the same as those about 0 and hence this state is stable. The locking of a distributed oscillator to nonzero relative phases has been called *mode-locking* [9]. At startup, each oscillator in a distributed PLL starts at a random phase, so there is a nonzero chance of converging to a mode-locked state. Simulations show that for a network like the one shown here, the system ends mode-locked from $\approx 1/3$ of random initial states. The probability goes up rapidly with the size of the system; a 4×4 array ends up mode-locked well over 99% of the time.

Pratt and Nguyen proved several useful properties about systems in mode-lock [9]. The key result, generalized for non-Cartesian networks, is that *for a system in mode-lock, there must be a phase difference θ between two oscillators such that $\theta \geq 2\pi/n$ where n is the number of nodes in the largest minimal loop in the network and a minimal loop is a loop in the graph that cannot be decomposed into multiple loops*

This result suggests a way to distinguish between mode-locked states and the desired 0-phase state: in mode-lock, there must be at least one branch with a large phase error. If the gain of the PD is designed to be negative for a phase difference larger than θ, then all mode-locked states are made unstable without affecting the in-phase equilibrium. Pratt and Nguyen suggest that XOR PDs preclude mode-lock in a rectangular network of oscillators because the response decreases for phase errors larger than $\pi/2$, [9]. This result follows directly from the result derived above: in a rectangular array, the largest minimal loop has four nodes, so $\theta = 2\pi/4 = \pi/2$. A PD described in the next section, with $\theta < \pi/2$, would be useful in nonrectangular networks, and where more gain near 0 phase is desirable.

IV. IMPLEMENTATION

The distributed clock network generates the clock signal with PLLs at multiple points ("nodes") across a chip, and distributes

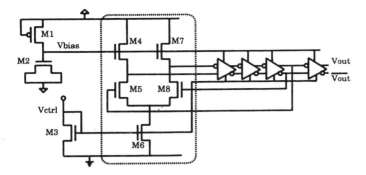

Fig. 14. Ring oscillator schematic

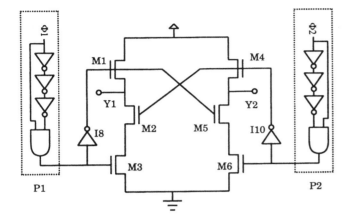

Fig. 15. Phase detector (PD).

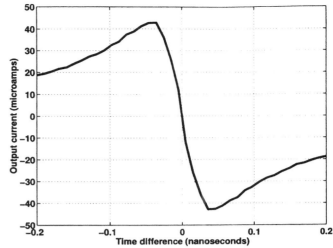

Fig. 16. Simulated phase transfer curve

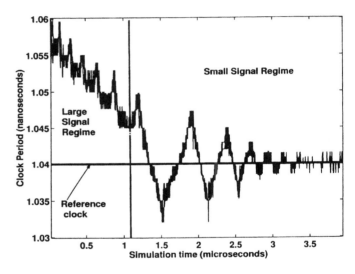

Fig. 17. Locking behavior of the PLL array

each only to a small section of the chip ("tile") (Fig. 8). PDs at the boundaries between tiles produce error signals that are summed by an amplifier in each tile and used to adjust the frequency of the node oscillator

Because the proposed network has many nodes, the power and size constraints on each node are even more stringent than the constraints on a single global PLL. The oscillator, PD, and loop filter of a working demonstration chip, fabricated in a standard 0.35-μm single-poly triple metal process, are considered in turn below.

A. Oscillator

The demonstration chip used an nMOS-loaded differential ring oscillator as a voltage-controlled oscillator (VCO) (Fig. 14). Transistors $M_4 - M_8$ comprise the differential inverter. The differential pair is $M_{5,8}$, the tail current is driven by M_6, and $M_{4,7}$ act as the nMOS load. The nMOS loads allow fast oscillation and shield the output signal from V_{DD} noise. V_{bias} is a low-pass version of V_{DD} generated by subthreshold leakage through PFET M_1; supply noise coupling in through C_{gd} of $M_{4,7}$ is bypassed by M_2. The oscillation frequency is only dependent on the supply voltage through capacitor nonlinearity and the output conductance of $M_{4,7}$, and feedback of the PLL compensates drift of V_{DD} and V_{bias}.

B. Phase Detector (PD)

The PD, shown in Fig. 15, has a sufficient nonlinearity, higher gain at small input phase difference and less high-frequency content than an XOR PD. The core ($M_1 - M_6$) is an nMOS-loaded arbiter which acts as a nonlinear PD. For no input phase difference, the output is balanced. As the phase difference increases from zero, one output will be asserted for the full duration of an input pulse, while the other output will be asserted for only the remainder of the input pulse duration after the first input pulse ends, which is equal to the input phase difference. Thus the detector has very high gain near zero phase error that drops off to zero as the input phase difference approaches the input pulse width (Fig. 16).

The pulse generators P_1 and P_2 enable this arbiter to give frequency-error feedback. If one input is at a higher frequency than the other, its output will be asserted for more input pulses than the other. Because the width of the pulses is independent of input frequency, the average output voltage corresponds to frequency. Unlike a typical phase-frequency detector, however, the strength of the error signal falls to zero as frequency difference goes to 0, so there can be no mode-lock problems, yet large signal frequency (and hence, phase) locking is enhanced. Fig. 17 shows the large signal correction and small signal behavior of the entire array of PLLs as the already internally locked array

427

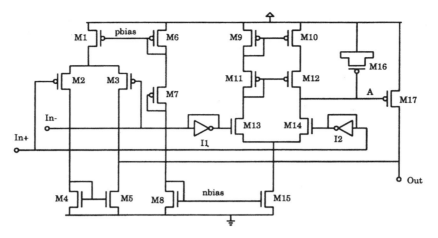

Fig. 18. Loop filter schematic.

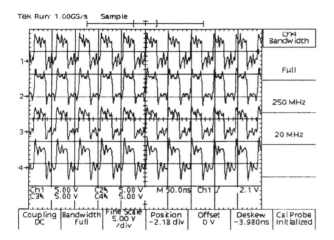

Fig. 19. Frequency-locked divider outputs.

Fig. 20. Micrograph of the 16-oscillator 1.3-GHz chip.

approaches and locks to the reference clock. The detector fits in $30~\mu\text{m} \times 30~\mu\text{m}$.

C. Loop Filter

The loop filter is shown in Fig. 18. $M_1 - M_5$ make up amplifier A_1, while $M_9 - M_{17}$ make up A_2. The differential output currents from the PDs at the edges of each tile are summed at nodes $In+$ and $In-$, and drive both amplifiers. A_1 is a single stage differential pair so it has relatively low gain but a bandwidth limited by g_m/C_{gs}. A_2 has a high-gain cascoded stage driving a common source PFET M_{17}. M_{16} is a large gate capacitor which serves to set the dominant pole of A_2 such that the PLL network is stable. M_{15} is biased at very low current to boost gain and enable a low time constant (as low as 12 kHz) with a $15~\mu\text{m} \times 15~\mu\text{m}$ gate capacitor. The simple design and feed-forward compensation allow the loop filter to fit in only $15~\mu\text{m} \times 45~\mu\text{m}$. Each clock node, consisting of an oscillator and a loop filter, takes just $45~\mu\text{m} \times 45~\mu\text{m}$.

V. RESULTS

A chip was fabricated with a 4×4 array of nodes and PD between nearest neighbors. Counting one node and two PDs the area overhead is approximately $0.0038~\text{mm}^2$ per tile. Another

PD was placed between one of the nodes and the chip clock input to lock the network to an external reference. The output of the 16 oscillators was divided by 64 and driven off chip. At $V_{DD} = 3$ V, the divided outputs were seen to be frequency locked at 17 to 21 MHz, corresponding to oscillator phase lock at 1.1 to 1.3 GHz. An oscilloscope plot of four locked output signals is shown in Fig. 19. Long-term jitter between neighbors is less than 30 ps. Cycle-to-cycle jitter is less than 10 ps. The oscillators, amplifiers and all the biasing draws 130 mA at 3 V. A chip plot is shown in Fig. 20. (The rest of the area on the 3mm × 3mm chip is taken up by test circuits.)

VI. CONCLUSION

Design and measurements on this chip confirm that generating and synchronizing multiple clocks on chip is feasible. Neither the power nor the area overhead of multiple PLLs is substantial compared to the cost of distributing the clock by conventional means. Most importantly, a distributed clock network can take advantage of improved devices by shrinking the size of the cells, lowering the overall skew and jitter, so performance will scale with the speed of devices, rather than with the much slower improvement of on-chip interconnect speed.

REFERENCES

[1] D. W. Bailey and B. J. Benschneider, "Clocking design and analysis for a 600-MHz Alpha microprocessor," *J. Solid State Circuits*, vol. 33, no. 11, pp. 1627–1633, Nov. 1998.

[2] C. F. Webb, "A 400-MHz S/390 microprocessor," in *ISSCC Dig. Tech. Papers*, Feb. 1997, pp. 168–169.

[3] T. Yoshida, "A 2-V 250-MHz multimedia processor," in *ISSCC Dig. Tech. Papers*, Feb. 1997, pp. 266–267.

[4] I. A. Young, M. F. Mar, and B. Bhushan, "A 0.35-μm CMOS 3-880-MHz PLL N/2 clock multiplier and distribution network with low jitter for microprocessors," in *ISSCC Dig. Tech. Papers*, Feb. 1997, pp. 330–331.

[5] H. B. Bakoglu, J. T. Walker, and J. D. Meindl, "A symmetric clock-distribution tree and optimized high-speed interconnections for reduced clock skew in ULSI and WSI circuits," in *IEEE Int. Conf. Computer Design*, NY, Oct. 1986, pp. 118–122.

[6] P. Zarkesh-Ha, T. Mule, and J. D. Meindl, "Characterization and modeling of clock skew with process variations," in *Proc. IEEE 1999 Custom Integrated Circuits Conf.*, pp. 441–444.

[7] G. Geannopoulos and X. Dai, "An adaptive digital deskewing circuit for clock distribution networks," in *ISSCC Dig. Tech. Papers*, Feb. 1998, pp. 400–401.

[8] F. Ançeau, "A synchronous approach for clocking VLSI systems," *J. Solid State Circuits*, vol. SC-17, no. 1, pp. 51–56, Feb. 1982.

[9] G. A. Pratt and J. Nguyen, "Distributed synchronous clocking," *IEEE Trans. Parallel and Distributed Systems*, Mar. 1995.

Vadim Gutnik (M'00) received the B.S. degree in electrical engineering and materials science from the University of California, Berkeley, in 1994, and the S.M. and Ph.D. degrees in electrical engineering from the Massachusetts Institute of Technology, Cambridge, in 1996 and 2000, respectively.

Previous research interests have included micromechanical resonators, and variable-voltage power supplies. He is currently working as a Design Engineer at Silicon Laboratories, Austin, TX.

Dr. Gutnik received an NDSEG fellowship in 1994, and the Intel Foundation Fellowship in 1997.

Anantha P. Chandrakasan (M'95) received the B.S., M.S., and Ph.D. degrees in electrical engineering and computer sciences from the University of California, Berkeley, in 1989, 1990, and 1994, respectively.

Since September, 1994, he has been the Analog Devices Career Development Assistant Professor of electrical engineering at the Massachusetts Institute of Technology, Cambridge. His research interests include the ultra-low-power implementation of custom and programmable digital signal processors, wireless sensors and multimedia devices, emerging technologies, and CAD tools for VLSI. He is a co-author of the book titled *Low Power Digital CMOS Design* (Norwood, MA: Kluwer, 1995). He has served on the technical program committee of various conferences including ISSCC, VLSI Circuits Symposium, DAC, ISLPED, and ICCD. He is the Technical Program Co-Chair for the 1997 International Symposium on Low-Power Electronics and Design and for VLSI Design'98.

He received the National Science Foundation Career Development Award in 1995, the IBM Faculty Development Award in 1995, and the National Semiconductor Faculty Development Award in 1996. He received the IEEE Communications Society 1993 Best Tutorial Paper Award for the *IEEE Communications Magazine* paper titled, "A Portable Multimedia Terminal."

A Low-Noise Fast-Lock Phase-Locked Loop with Adaptive Bandwidth Control

Joonsuk Lee, *Student Member, IEEE*, and Beomsup Kim, *Senior Member, IEEE*

Abstract—This paper presents a salient analog phase-locked loop (PLL) that adaptively controls the loop bandwidth according to the locking status and the phase error amount. When the phase error is large, such as in the locking mode, the PLL increases the loop bandwidth and achieves fast locking. On the other hand, when the phase error is small, this PLL decreases the loop bandwidth and minimizes output jitters. Based on an analog recursive bandwidth control algorithm, the PLL achieves the phase and frequency lock in less than 30 clock cycles without pre-training, and maintains the cycle-to-cycle jitter within 20 ps (peak-to-peak) in the tracking mode. A feed forward-type duty-cycle corrector is designed to keep the 50% duty cycle ratio over all operating frequency range.

Index Terms—Adaptive bandwidth PLL, analog implementation, clock recovery, fast locking time, frequency hopping, gear-shifting algorithm, low jitter, phase-locked loops, time-varying channel.

I. INTRODUCTION

PHASE-LOCKED loops (PLL's) have been widely used in high-speed data communication systems such as Ethernet receivers, disk drive read/write channels, digital mobile receivers, high-speed memory interfaces, and so forth, because PLL's efficiently perform clock recovery or clock generation with relatively low cost. Those PLL's used in the systems are required to generate low-noise or low-jitter clock signals and at the same time need to achieve fast locking.

Conventional analog PLL's in clock recovery applications use a narrow-band loop filter to reduce output jitters at the expense of elongated locking time. In order to improve the locking-time characteristics, digital or hybrid analog/digital PLL's with a loop bandwidth stepping capability have been studied [1], [2]. Since the stepping hardware is implemented with complex digital building blocks, these PLL's usually suffer from high power dissipation, low operating speed and large die size. In order to reduce consuming power and die size, simpler algorithms such as a gear-shifting or a lock-detection algorithm were attempted [3], [4]. The PLL's with such algorithms control the loop bandwidth according to a prestored charge-pump current control sequence in memory during the start-up mode. However, in clock recovery applications such as HDD and DVD, where the channel characteristics vary in time, the prestored control sequence cannot make the PLL's

Manuscript received October 29, 1999; revised February 23, 2000.

J. Lee was with the Boston Design Center, IBM Microelectronics, Lowell, MA 01851 USA. He is now with the Korea Advanced Institute of Science and Technology, Taejon 305-701, Korea.

B. Kim is with the Korea Advanced Institute of Science and Technology, Taejon 305-701, Korea (e-mail: bkim@ee.kaist.ac.kr).

Publisher Item Identifier S 0018-9200(00)06435-0.

respond properly to unpredictable phase fluctuation, instant frequency shift, and time-varying jitter because the sequence was calculated with preknown fixed noise statistics.

Discrete-time PLL's, which are programmed on DSP processors, based on a recursive least squared (RLS) algorithm [5] or the Kalman filter algorithm [6] can respond to such unpredictable jitter variations, but require enormous amount of hardware. The outputs generated from the discrete-time PLL's are in a digital domain, and therefore the discrete-time PLL's require digital-to-analog converters (DAC) and an analog-to-digital converter (ADC) to sample input signals for detection. Slow signal-processing speed of the digital-to-analog conversion in the discrete-time PLL's limits the operating frequency and confines the use of the PLL's to the applications dealing with low-frequency signals like digital wireless base stations.

This paper presents a new analog adaptive PLL (AAPLL) architecture capable of varying the loop bandwidth according to an adaptively updated control sequence under a time-varying noise environment. Since the control sequence is generated from analog signal processing, the PLL operates at several hundred megahertz and can be easily modified to run at gigahertz frequency ranges.

This paper consists of five sections including the present section. Section II describes the AAPLL architecture and the analog adaptive bandwidth-control algorithm. Stability and jitter analysis for the AAPLL are given in Section III. AAPLL locking behaviors are also discussed in this section. Section IV shows the AAPLL IC implementation and measurement results. Finally, a brief summary of this paper is given in Section V.

II. RECURSIVE EQUATION AND ANALOG LOOP BANDWIDTH CONTROLLER

In this section, a recursive bandwidth update algorithm for the analog adaptive controller and its implementation are described.

A. Adaptive Bandwidth Control

As mentioned in the introduction, a common approach to improve the locking speed of a PLL is to use a gear-shifting method for loop bandwidth control. In such a PLL, when fast locking is required, as in the initial frequency/phase acquisition mode, the loop bandwidth of the PLL is expanded by the increased charge-pump current or the phase detector gain [2]–[4]. Zero phase start (ZPS) is also helpful to reduce the phase acquisition time [4], but limited to the case when the initial-frequency locking has been already established. For the case where rapid initial-frequency locking is required, various techniques with a prestored gear-shifting sequence have been studied [4],

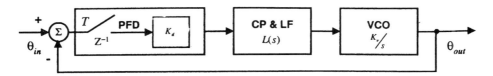

Fig. 1. Linearized model of a CP-PLL.

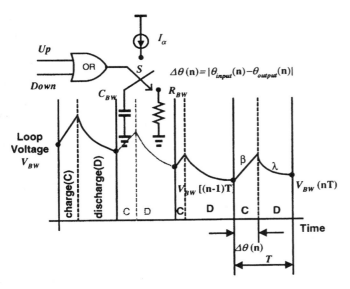

Fig. 2. Conceptual diagram of an analog adaptive controller.

[5]. However, in the case where the channel characteristics vary in time, such as in a disk drive, the prestored gear-shifting sequence is not helpful. Unpredictable phase fluctuation, instant frequency shift, and varying input jitter force such a PLL to use an indefinite wide loop bandwidth in order not to lose the locking. Although a discrete-time adaptive PLL can adjust the bandwidth according to the input noise statistics, it still requires complex hardware and its applications are limited to the low frequency operating systems [5].

A linearized model of a charge-pump PLL (CP-PLL) is shown in Fig. 1. The transfer function for z domain is represented by

$$H(z) = \frac{K_d K_v L'(z) z^{-1}}{1 + K_d K_v L'(z) z^{-1}} \qquad (1)$$

where $L'(z) = T \cdot \mathbf{Z}\{\mathbf{L}^{-1}[L(s)/s]|_{t=nT}\}$. Here K_d and K_v are the phase detector gain given by $I_{CP}/2\pi$ and the voltage-controlled oscillator (VCO) gain given by df/dv, respectively, and $L'(z)$ is the z-transform of the sampled version of $L(s)/s$, where $L(s)$ is the PLL loop filter in s domain given by $R_1 + 1/sC_1$ if a simple passive low-pass filter (R_1, C_1) is assumed to be used as a loop filter. The quantity $K_d K_v R_1 T$ is called the PLL loop gain K_{loop}.

Discrete-time PLL's that have an adaptive stepping capability can control the loop bandwidth by a loop-gain update equation minimizing the RLS error [7]. However, it is difficult to fully implement the update equation used in the discrete-time PLL because it requires a significant amount of die size and power consumption. A simpler but still an effective loop gain update

equation is required and used in the proposed AAPLL. The update equation is given by (2), in terms of the loop gain K_{loop} that is proportional to the loop bandwidth.[1]

$$K_{loop}(n+1) = \lambda \cdot K_{loop}(n) + \alpha \cdot |\theta_{input}(n) - \theta_{output}(n)|. \qquad (2)$$

Here, λ is a forgetting factor that has a positive value close to but less than unity. α is a coefficient that normalizes and converts the absolute value of input–output phase errors from radians to dimensionless numbers. The loop gain $K_{loop}(n+1)$ is calculated by a recursive manner according to (2). When the input–output phase error becomes zero, the forgetting factor λ makes the loop gain converge to zero as the discrete time nT increases. Equation (2) reflects the most recent input–output phase error $|\theta_{input}(n) - \theta_{output}(n)|$ most significantly. This recursive relation is similar to the RLS algorithms commonly used for an estimator [8].

The loop gain $K_{loop}(n+1)$ at time $(n+1)T$ is calculated as the weighted sum of the present loop gain $K_{loop}(n)$ and the absolute value of the present input–output phase error, $|\theta_{input}(n) - \theta_{output}(n)|$ at time nT. The equation indicates that the loop gain, thus the loop bandwidth, rapidly grows when the recent absolute phase errors become large, and greatly improves the PLL loop tracking capability. When the recent absolute phase errors become small, the loop gain shrinks, as does the loop bandwidth, because the first part of (2) dominates. The reduced loop bandwidth improves the PLL's input jitter rejection capability. Therefore, (2) satisfies the necessary loop bandwidth control under the presence of unpredictable jitter variation.

B. Analog Adaptive Bandwidth Control

Equation (2) is achieved by a CP-PLL with a small amount of extra hardware. The second term of (2), an absolute phase error, is obtained from outputs of a phase frequency detector (PFD). Since the PFD and the following charge-pump circuit generate *up/down* current signals proportional to the input–output phase difference, simply combining these *up/down* signals through an OR gate gives the absolute phase error signal $|\Delta\theta(n)| = |\theta_{input}(n) - \theta_{output}(n)|$ at time nT. Fig. 2 shows a conceptual diagram of how the recursive loop gain of the AAPLL is calculated in the controller. The bandwidth voltage V_{BW} becomes the bias voltage of the following current source in the charge-pump circuit and controls the amount of charge-pump currents. The current switch S steers the current proportional to the phase error and increases the voltage across the capacitor C_{BW} by the corresponding amount at a constant rate while the resistor R_{BW} exponentially discharges the capacitor. The resistor and the capacitor realize the first part of (2) with the forgetting factor λ. As

---

[1]Here, the loop bandwidth and the loop gain have the following relationship: $K_{loop} = W_{loop}/f$.

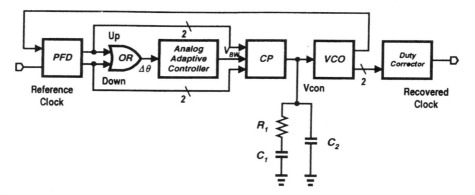

Fig. 3. AAPLL total block diagram.

derived in the Appendix, in the steady state the voltage across the capacitor C_{BW} at time $(n+1)T$ is given by

$$V_{BW}[(n+1)T] = \lambda \cdot V_{BW}[nT] + \beta \cdot |\Delta\theta(n)|. \quad (3)$$

Here, the forgetting factor λ equals $\exp[-T/R_{BW}C_{BW}]$, β is $(I_\alpha/C_{BW}) \cdot (T/2\pi) \cdot \lambda$, and I_α is the amount of the charging current in the controller.

The loop gain is asymptotically proportional to the bandwidth voltage V_{BW} governed by (3) because the charge-pump current is directly controlled by this voltage. It means that the bandwidth of the AAPLL follows (2).

III. CIRCUIT IMPLEMENTATION

This section describes the circuit implementation of the adaptive bandwidth controller, the charge-pump circuit, the VCO, and the duty cycle correction circuit. Fig. 3 shows the overall block diagram of the AAPLL, which modifies a conventional PLL by attaching an analog adaptive band-width-controlling block. Due to the minor change, the AAPLL is easily applicable to various PLL applications and still takes advantage of the full adaptability.

A. Adaptive Bandwidth Controller and Charge Pump

The well-designed PFD is used as a phase-detecting block instead of a mixer, though the input signal frequency is high, in order to achieve a wideband capturing capability. The PFD shown in Fig. 4 consists of two simplified true single-phase clock (TSPC) D-flip–flops and one NOR gate. Since the input frequency of the AAPLL is selected to recover the clock signal in DVD systems, whose clock frequency is about 250 MHz, the PFD should generate *up* and *down* signals at such a high speed. In order to minimize the abnormal operation of the PFD, TSPC D-flip–flops are used as leaf cells since these intrinsic delays are smaller than those of conventional ones.

The adaptive bandwidth controller shown in Figs. 3 and 6 consists of an OR gate and a differential switch, which takes the differential signals from the OR gate. The OR gate sums the *up* and *down* signals generated from the PFD and gives the absolute phase error. The differential switch controls current paths from V_{dd} to the bandwidth capacitor C_{BW} according to the phase difference $|\theta_{input}(n) - \theta_{output}(n)|$ for one clock period. When the phase difference signal is mostly on over a period, such as in the initial-phase acquisition state, the charging rate

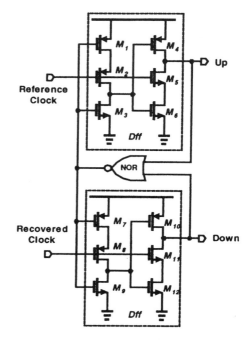

Fig. 4. PFD schematic with simplified TSPC D-flip–flops.

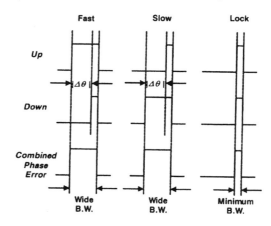

Fig. 5. *Up/Down* and phase-error signal diagram.

of the capacitor C_{BW} exceeds the discharging rate. Hence the capacitor voltage V_{BW} of the capacitor C_{BW} and the pumping current in the charge pump increase. As a result, the phase detector gain increases and so does the loop bandwidth. On the other hand, when the phase error signal is off for the most part

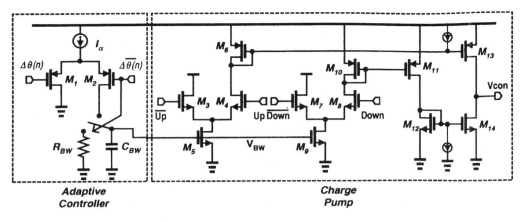

Fig. 6. Adaptive bandwidth controller and CP schematics.

of one period, such as in the tracking state, the discharging rate exceeds the charging rate and the capacitor voltage decreases. Therefore both the phase-detector gain and the loop bandwidth decrease. In the steady-state tracking mode, the AAPLL loop bandwidth can be very narrow because the phase error becomes zero. However, the AAPLL still maintains the minimum loop bandwidth even in such a case because of the *up/down* signals generated from the set/reset type PFD, as shown in Fig. 4. In the zero-phase-error and perfect locking case in Fig. 5, the OR-gated *effective* phase-error signal can still supply currents to the bandwidth capacitor. The statistical variation of the input signal also contributes to maintain this minimum bandwidth. Fig. 5 shows the relation between the phase error and the bandwidth control of the AAPLL.

Fig. 6 shows the circuit diagram of the analog adaptive bandwidth controller with a charge-pump circuit. As mentioned before, the voltage $V_{\rm BW}$ across the capacitor $C_{\rm BW}$ in parallel with a resistor $R_{\rm BW}$ controls the phase-detection gain K_d. In order to control the discharging rate of the capacitor $C_{\rm BW}$, a voltage-controlled resistor (VCR) $R_{\rm BW}$, as shown in Fig. 7, is used. The VCR is designed to have fully linear I–V characteristics for a given power supply range. By adjusting the magnitude of the bias current in the VCR branches, the resistance of $R_{\rm BW}$ is changed and so does the discharging rate of the capacitor $C_{\rm BW}$.

The output node of $C_{\rm BW}$ is connected to the gates of nMOS transistors, and controls the charge-pump current by adjusting the bias point of M_5 and M_9 in Fig. 6. The charge pump consists of two differential input stages, a mirror stage, an output stage, and two small extra current sources. These two small current sources help the rapid turn-on/off operation for MOS', M_{13} and M_{14}. The differential PFD signals drive the charge-pump inputs. When the *down* signal goes high, the current controlled by the voltage of the bandwidth capacitor $C_{\rm BW}$ is drawn from the loop filter. When the *up* signal goes high, the same amount of current is supplied to the loop filter.

B. Voltage-Controlled Oscillator (VCO)

A four-stage VCO as shown in Fig. 8 is used for the AAPLL. The basic delay cell consists of six transistors. The cross-coupled pMOS transistors, M_3 and M_4, guarantee the differential operation of the delay cell without a tail-current bias. Auxiliary pMOS transistors, M_5 and M_6, control the oscillation frequency. Unlike

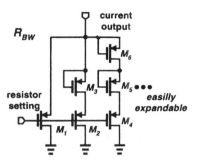

Fig. 7. VCR schematic.

conventional differential VCO's with a current bias, this VCO allows the AAPLL to operate under a single 1.5-V power supply, consuming 1.5 mA. Because the output signal of the VCO swings rail-to-rail, no additional level shifter with a carefully designed replica bias circuit is required to generate CMOS level outputs. The latch, configured with pMOS', M_3 and M_4, sharpens the edge of the output signal so that the added noise has little chance to be converted as jitters. Eventually this latch helps the reduction process of the VCO jitter [9], [10].

C. Duty-Cycle Corrector

Maintaining a 50% duty-cycle ratio for a clock signal is extremely important in most high-speed clock recovery and clock generation applications because several systems, such as double-data rate (DDR) SDRAM's and pipelined microprocessors, use negative transition edges of a clock signal in order to increase total system throughputs. This is often achieved by a VCO running at twice as high as the desired clock frequency, and then dividing the VCO frequency by 2. Other approaches use a feedback-type duty-cycle corrector. Since precise placement of the falling edge between two successive rising edges of the VCO output signal is generally controlled by an additional feedback loop, the duty-cycle correctors require an extra training period to stabilize the feedback loop.

The AAPLL uses a feed forward-type duty-cycle corrector instead of the feedback type in order to eliminate the extra feedback hardware and the training period, as shown in Fig. 9(a). The duty cycle corrector utilizes multiphase signals generated from a multistage differential VCO. The signal $\bar{A}$ in Fig. 9(b) selected from the multiphase signals turns on MOS', M_1 and

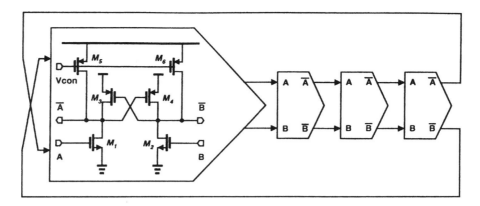

Fig. 8. Four-stage VCO.

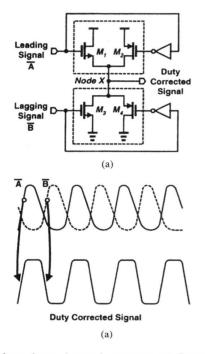

(a)

(a)

Fig. 9. Feed forward-type duty-cycle corrector. (a) Duty-cycle corrector schematic. (b) Conceptual diagram of the correcting operation.

M_2, and charges the output node X of the duty-cycle corrector almost instantaneously, because the discharge path of the node X is already off due to the signal $\bar{B}$. The signal $\bar{B}$, which is also selected from the multiphase signals, is the one whose rising edge is shifted by $180°$ in phase from that of $\bar{A}$. Similarly, the signal $\bar{B}$ rapidly discharges the node X and delivers the desired 50% duty-cycle signal. Since this duty-cycle correction circuit consists of only two transmission gates and two inverters, the silicon area is minimal and the power consumption is negligible. In HSPICE simulation, the proposed duty-cycle corrector keeps the output duty cycle almost perfectly at 50% with the input duty cycles varying from 10 to 90%.

IV. ANALYSIS AND SIMULATION

In this section, the stability of the AAPLL is analyzed for the adaptively generated loop sequence, and behavioral simulation results for fast lock and large jitter reduction are described.

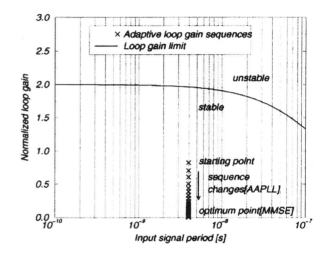

Fig. 10. Stability diagram for loop gain K_{loop}.

A. Stability

Since the AAPLL automatically changes the loop bandwidth, a careful loop stability analysis is required. As mentioned in the previous section, an analog adaptive controller adjusts the phase-detector gain of the CP-PLL. Therefore, stability checking for the PLL for each different phase gain should be accomplished first. A complete stability analysis for the CP-PLL is cumbersome because a PLL operates in both a linear and a nonlinear region. A simplified stability analysis for a second-order CP-PLL [8] is used in this section. When the criterion is extended to include the logic delay effect, it can be expressed as

$$K_{\text{loop}}[n] < \frac{2}{1 + \frac{1}{2} \cdot \frac{T - \tau_d}{\tau}}. \qquad (4)$$

Here, T, τ_d, and τ are the clock period, the logic delay, and the RC time constant of a loop filter respectively. The stability limit for the loop gain K_{loop} of the AAPLL is derived and simulated using this criterion as shown in Fig. 10. The adaptively generated loop gain sequence by the recursive equation is also shown in the same figure to verify the AAPLL stability. The sequence converges to the minimum bandwidth and the amplitude of this bandwidth is almost similar to that derived from the MMSE criterion [4]. Equation (4) can be written to obtain the stability criterion for the bandwidth voltage V_{BW} by solving a MOS I-V

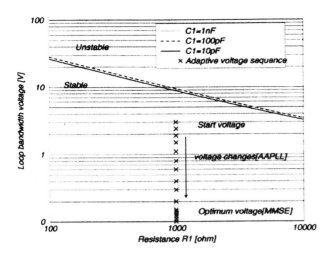

Fig. 11. Stability limit graph for bandwidth voltage $V_{\rm BW}$.

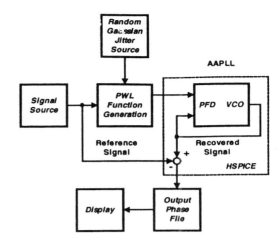

Fig. 12. Simulation setup for the locking behavior measurement.

characteristic equation for M_5, M_9 in Fig. 6 assumiing a saturation condition.

$$V_{\rm BW}[n] < \sqrt{\frac{4\pi}{\left(1 + \frac{1}{2} \cdot \frac{T - \tau_d}{\tau}\right) \cdot (K_N K_V R_1 T)}} + V_T. \quad (5)$$

Here, K_N, K_V, V_T, and R_1 are $(\mu_N C_{\rm ox})/2 \cdot (W/L)_{5,9}$, the VCO gain, the nMOS threshold voltage, and the size of a resistor of the loop filter in Fig. 3, respectively. Here, $(W/L)_{5,9}$ is the size of M_5, M_9. The stability limit for the bandwidth voltage, which is one of the observable values in the measurement setup, is visualized in Fig. 11 for various capacitor C_1 and resistor R_1 values. The figure shows that all the sequences $K_{\rm loop}[n]$ and $V_{\rm BW}[n]$ are within the stable region for the various resistor and capacitor values.

B. Output Jitter

Recently, it was reported that a CP-PLL has an optimum loop bandwidth that generates minimum jitter in the steady state [11]. A clean tone, that is assumed to have only noise floor and no random walking phase noise, is used as a reference signal for the jitter derivation. Because the AAPLL eventually achieves the steady state locking with a clean reference signal like other conventional PLL's, the output cycle-to-cycle jitter of the AAPLL can be calculated by

$$\Delta \tau_{\rm rms} = \left(\delta \tau_{\rm rms1} + \frac{T^2}{2\pi} K_v \Delta V_{\rm rms4}\right) \cdot \sqrt{\frac{1}{2K_{\rm loop}}}$$

$$+ \left(\delta \tau_{\rm rms2} + \frac{T}{2\pi} \frac{\Delta I_{\rm rms3}}{K_d}\right) \cdot \sqrt{\frac{K_{\rm loop}}{2}}. \quad (6)$$

Here, $\delta \tau_{\rm rms1}$, $\delta \tau_{\rm rms2}$, $\Delta I_{\rm rms3}$, and $\Delta V_{\rm rms4}$ are the internal jitter from the VCO, the jitter of the input signal, the rms value of the charge-pump current variation, and the rms value of VCO control voltage noise in the steady state, respectively.

C. Behavioral Simulation of a Locking Feature

Closed-form analysis of locking behaviors for the AAPLL is difficult because of its nonlinear operation. In this paper, a simulation-based approach like the Monte Carlo Method is used instead. The AAPLL is modeled in a SPICE circuit simulator and

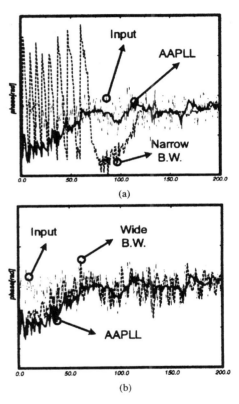

(a)

(b)

Fig. 13. Simulation results for the locking behavior of the AAPLL and conventional ones. (a) Fixed narrow-bandwidth PLL. (b) Fixed wide-bandwidth PLL.

extensively tested by the circuit simulator. Fig. 12 shows the simulation setup for the AAPLL. Fig. 13 compares the simulated locking behavior of the AAPLL with that of a conventional PLL. The bandwidths of the conventional PLL are selected to have two typical values. One is optimized for the initial locking, and the other for the steady-state tracking. The gray line in Fig. 13(a) indicates an incoming signal in the phase domain. The solid line in the figures shows the phase of the AAPLL output signal from initial locking to steady-state tracking. The phase variation of the conventional PLL optimized for steady-state tracking with a narrow bandwidth is shown in the same figure as a dashed line.

Fig. 14. Micrograph of the fabricated AAPLL chip.

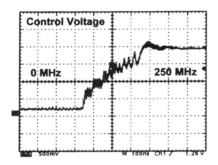

Fig. 15. Control voltage change for a 0–250 MHz frequency input.

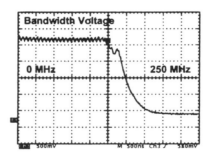

Fig. 16. Loop bandwidth voltage change for a 0–250-MHz frequency input.

In Fig. 13(b), the phase change of the conventional PLL optimized for initial locking is also shown as a dashed line. This simulation result gives several characteristics of the AAPLL. The AAPLL controlled by the recursive algorithm achieves fast lock in the initial locking period, comparable to the speed obtained from a wide-bandwidth PLL because the consecutive error signals rapidly increase the loop bandwidth of the AAPLL. In the steady-state tracking mode, the AAPLL substantially rejects the input jitter due to the narrower loop bandwidth.

V. EXPERIMENTAL RESULTS

The AAPLL is fabricated in a 0.6-μm single-poly triple-metal n-well CMOS process [12]. The die size for the AAPLL is 0.11 mm^2. The total power consumption is less than 15 mW with a single 3-V supply. A microphotograph of the AAPLL

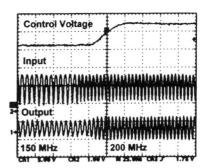

Fig. 17. Experimental results of the locking for a 150–200-MHz input signal.

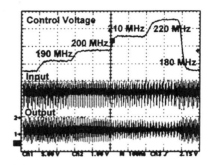

Fig. 18. Experimental results of the locking for a 180–220-MHz input signal by four steps.

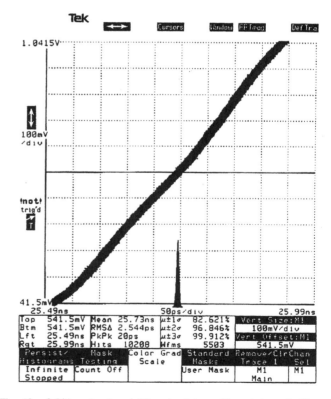

Fig. 19. 2.544-ps (rms) and 20-ps (peak-to-peak) cycle-to-cycle jitter at 250-MHz input.

is shown in Fig. 14. To get the forgetting factor $\lambda = 0.97$, $R_{BW} = 6.6$ kΩ, $C_{BW} = 20$ pF are used. And $I_\alpha = 100$ μA is selected to get $\beta = 3 \cdot 10^{-3}$. The locking-speed measurement is carried out using an abrupt change of the input signal frequency from 0 to 250 MHz. Fig. 15 shows the corresponding

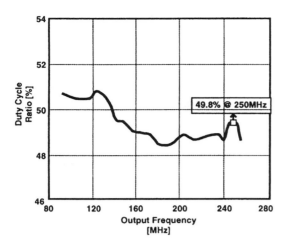

Fig. 20. 50% duty-cycle correction operation over the entire frequency range.

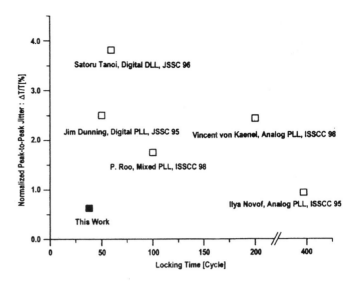

Fig. 22. Comparison between recently reported PLL's and DLL's and this work.

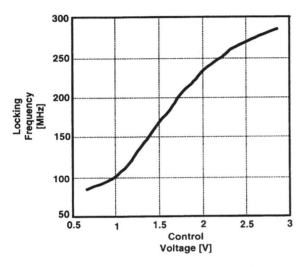

Fig. 21. VCO linearity.

TABLE I
AAPLL CHARACTERISTICS SUMMARY

Technology	0.6-μm *single-poly* 3-level metal n-well Standard CMOS process
AAPLL core size	0.11 mm^2
Measured Cycle-to-Cycle jitter	2.544 ps (RMS) 20 ps (peak-to-peak) at 250-MHz input signal
Power dissipation	15 mW at 3 V 2.3 mW at 1.5 V
Duty cycle ratio	49.8 % at 250-MHz input signal
Locking time	< 30 cycles to lock onto unknown frequency signal < 10 cycles to lock onto frequency-hopping signal

VCO control-voltage variations. In order to measure the locking speed precisely, the running cycles of the output waveform are counted from the frequency triggering point in the initial locking state. The AAPLL requires less than 30 clock cycles for both frequency and phase lock in this case. The voltage of C_{BW}, representing the AAPLL bandwidth, is also measured. Fig. 16 shows the measured voltage and describes the adaptation of the loop bandwidth in the AAPLL. Figs. 17 and 18 show the measured control voltages and the corresponding output signals when the input frequencies vary from 150 to 200 MHz and from 180 to 220 MHz by four steps respectively. In this case, the frequency and the phase locking require less than 10 symbol periods because the frequency steps are much smaller compared to the previous case. The measured cycle-to-cycle jitters of the AAPLL output signal at a 250-MHz input signal are 2.54 ps (rms) and 20 ps (peak-to-peak) as depicted in Fig. 19. This jitter value contains the inherent measurement setup jitter [13].

In order to test the performance of the duty cycle correction circuit, the duty cycle ratio of the output signal is measured with the input signals from 90 to 260 MHz. Fig. 20 shows the measured result of the corresponding duty cycle ratio. This result indicates the feed forward-type duty-cycle corrector maintains 50% duty-cycle ratio within 2% error for the region. Fig. 21 shows a VCO linearity diagram. The VCO gain is about 100 MHz/V at a 250-MHz input frequency. The AAPLL operates from 80 to 290 MHz with a 3-V supply voltage. Fig. 22 compares the normalized peak-to-peak jitter and the lock time of the AAPLL with those of recently reported PLL's and DLL's. Measured characteristics are summarized in Table I.

VI. CONCLUSION

This paper presents the design of a 250-MHz low-jitter fast-lock analog adaptive bandwidth-controlled PLL on a single chip. The chip is implemented in a 0.6-μm standard CMOS process. Simple recursive control logic is proposed to control the bandwidth effectively. The measured locking time is less than 10 cycles in a 10-MHz frequency step and less than 30 cycles from an unknown frequency signal to the 250–MHz signal respectively. The measured output jitters are 2.6 ps (rms) and 20 ps (peak-to-peak). All the components are designed using analog technique and hence the required die size and the power consumption are minimal.

APPENDIX

As shown in Fig. 2, the OR gate gives the control signal for the switch S according to the phase error signal in the bandwidth controller. When the phase error of the signal is high, the controller signal from the OR gate feeds current to the bandwidth capacitor $C_{\rm BW}$ and the voltage across the capacitor increases at a constant rate $I_\alpha/C_{\rm BW}$. As a result, the bandwidth voltage $V_{\rm BW}$ increases proportional to the normalized phase error $\Delta\theta(n)$. After the charging process, the controller signal from the OR gate disconnects the path from the current source and connects to the resistor. So the capacitor $C_{\rm BW}$ discharges through the resistor $R_{\rm BW}$. The switching action occurs every clock cycle period.

The voltage $V_{\rm BW}$ of the bandwidth capacitor at time $(n+1)T$ can be written as

$$V_{\rm BW}[(n+1)T] = \left[V_{\rm BW}[nT] + \frac{I_\alpha}{C_{\rm BW}} \cdot \frac{T}{2\pi}|\Delta\theta(n)|\right]$$
$$\times e^{-\frac{T}{R_{\rm BW}C_{\rm BW}}\left(1-\frac{|\Delta\theta(n)|}{2\pi}\right)} \qquad (7)$$

where $V_{\rm BW}[nT]$ is the voltage of the previous capacitor voltage at time nT. The voltage equation can be simplified to (8).

$$V_{\rm BW}[(n+1)T] = \lambda(n) \cdot V_{\rm BW}[nT] + \beta(n) \cdot |\Delta\theta(n)| \qquad (8)$$

Here $\lambda(n) = \exp[-T/R_{\rm BW}C_{\rm BW}(1-|\Delta\theta(n)|/2\pi]$, $\beta(n) = (I_\alpha/C_{\rm BW}) \cdot (T/2\pi) \cdot \lambda(n)$. In the initial locking mode, the AAPLL does the locking operation based on (8). Once the AAPLL finished the phase and frequency locking, the phase error is far less than 2π. In this case, the forgetting factor and the proportional coefficient can be be replaced by $\lambda = \exp[-T/R_{\rm BW}C_{\rm BW}]$ and $\beta = (I_\alpha/C_{\rm BW}) \cdot (T/2\pi) \cdot \lambda$.

REFERENCES

[1] J. Dunning *et al.*, "An all-digital phase-locked loop with 50-cycle lock time suitable for high-performance microprocessors," *IEEE J. Solid-State Circuits*, vol. 30, pp. 412–422, Apr. 1995.
[2] B. Kim, D. N. Helman, and P. R. Gray, "A 30-MHz hybrid analog/digital clock recovery circuit in 2-μm CMOS," *IEEE J. Sold-State Circuits*, vol. 25, pp. 1385–1394, Dec. 1990.
[3] M. Mizuno *et al.*, "A 0.18 μm CMOS hot-standby phase-locked loop using a noise immune adaptive-gain voltage-controlled oscillator," *ISSCC Dig. Tech. Papers*, pp. 268–269, Feb. 1995.
[4] G. Roh, Y. Lee, and B. Kim, "An optimum phase-acquisition technique for charge-pump phase-locked loops," *IEEE Trans. Circuit Syst. II*, vol. 44, pp. 729–740, Sept. 1997.
[5] B. Chun, Y. Lee, and B. Kim, "Design of variable loop gain of dual-loop DPLL," *IEEE Trans. Commun.*, vol. 45, pp. 1520–1522, Dec. 1997.
[6] P. F. Driessen, "DPLL bit synchronizer with rapid acquisition using adaptive Kalman filtering techniques," *IEEE Trans. Commun.*, vol. 452, pp. 2673–2675, Sept. 1994.
[7] B. Kim, "Dual-loop DPLL gear-shifting algorithm for fast synchronization," *IEEE Trans. Circuits Syst. II*, vol. 44, pp. 577–586, July 1997.
[8] S. Haykin, *Adaptive Filter Theory*. Englewood Cliffs, NJ: Prentice Hall, 1995.
[9] T. C. Weigandt, B. Kim, and P. R. Gray, "Analysis of timing jitter in CMOS ring oscillators," in *Proc. Int. Symp. Circuit and Systems*, vol. 4, London, U.K., June 1994, pp. 27–30.
[10] C. H. Park and B. Kim, "A low-noise 900-MHz VCO in 0.6-μm CMOS," *IEEE J. Solid-State Circuits*, vol. 34, pp. 586–591, May 1999.
[11] K. Lim, C. H. Park, and B. Kim, "Low noise clock synthesizer design using optimal bandwidth," in *Proc. Int. Symp. Circuit and Systems*, Monterey, CA, June 1998, pp. 163–166.
[12] J. Lee and B. Kim, "A 250 MHz low jitter adaptive bandwidth PLL," *ISSCC Dig. Tech. Papers*, pp. 346–347, Feb. 1999.
[13] J. McNeil, "Jitter in ring oscillators," *IEEE J. Solid-State Circuits*, vol. 32, pp. 870–879, June 1997.

Joonsuk Lee (S'99) received the B.S. and M.S. degrees in electrical engineering and computer sciences from Korea Advanced Institute of Science and Technology (KAIST), Taejon, Korea, in 1995 and 1997, respectively. Since 1997 he has been working toward the Ph.D. degree at the same university.

From 1999 to 2000, he was with IBM Microelectronics, Boston, MA, as an Analog and Mixed Signal Designer involved in a high performance sigma–delta ADC/DAC project with Motorola, Lowell, MA. His research interests include PLL/DLL, timing recovery algorithms, high-speed SDRAM interface, and LAN and mixed-mode signal processing technique for telecommunication IC's.

Mr. Lee is the Gold Medal winner of the Human-Tech Thesis Prize from Samsung Electronics Co. Ltd. in 1997, the Gold Medal winner of the Chip Design Contest from LG Semicon Co. Ltd. in 1998, and the Gold Medal winner of the Integrated Design Center (IDEC) Award in 1998.

Beomsup Kim (S'87–M'90–SM'95) received the B.S. and M.S. degrees in electronic engineering from Seoul National University, Seoul, Korea, in 1983 and 1985, respectively, and the Ph.D. degree in electrical engineering and computer sciences from the University of California, Berkeley, in 1990.

From 1986 to 1990, he worked as a Graduate Researcher and Graduate Instructor at Department of Electrical Engineering and Computer Sciences, University of California, Berkeley. From 1990 to 1991, he was with Chips and Technologies, Inc., San Jose, CA, where he was involved in designing high speed-signal processing IC's for disk drive read/write channels. From 1991 to 1993, he was with Philips Research, Palo Alto, CA, where he was conducting research on digital signal processing for video, wireless communication, and disk drive applications. During 1994, he was a Consultant, developing the partial-response maximum likelihood detection scheme of the disk drive read/write channel. In 1994, he became an Assistant Professor with the Department of Electrical Engineering, Korea Advanced Institute of Science and Technology (KAIST), Taejon, Korea, and is currently an Associate Professor. During 1999, he took a sabbatical leave and stayed at Stanford University, Stanford, CA, and also consulted for Marvell Semiconductor Inc., San Jose, CA, on the Gigabit Ethernet and wireless LAN DSP architecture. His research interests include mixed-mode signal processing IC design for telecommunications, disk drive, local area network, high-speed analog IC design, and VLSI system design.

Dr. Kim is a corecipient of the Best Paper Award (1990–1991) for the IEEE JOURNAL OF SOLID-STATE CIRCUITS, and received the Philips Employee Reward in 1992. Between June 1993 and June 1995, he served as an Associate Editor for the IEEE TRANSACTIONS ON CIRCUITS AND SYSTEMS II: ANALOG AND DIGITAL SIGNAL PROCESSING.

A Low-Jitter 125–1250-MHz Process-Independent and Ripple-Poleless 0.18-μm CMOS PLL Based on a Sample–Reset Loop Filter

Adrian Maxim, *Member, IEEE*, Baker Scott, *Associate Member, IEEE*, Edmund M. Schneider, *Member, IEEE*,
Melvin L. Hagge, *Member, IEEE*, Steven Chacko, and Dan Stiurca, *Member, IEEE*

Abstract—This paper describes a low-jitter phase-locked loop (PLL) implemented in a 0.18-μm CMOS process. A sample–reset loop filter architecture is used that averages the oscillator proportional control current which provides the feedforward zero over an entire update period and hence leads to a ripple-free control signal. The ripple-free control current eliminates the need for an additional filtering pole, leading to a nearly 90° phase margin which minimizes input jitter peaking and transient locking overshoot. The PLL damping factor is made insensitive to process variations by making it dependent only upon a bandgap voltage and ratios of circuit elements. This ensures tracking between the natural frequency and the stabilizing zero. The PLL has a frequency range of 125–1250 MHz, frequency resolution better than 500 kHz, and rms jitter less than 0.9% of the oscillator period.

Index Terms—CMOS, frequency synthesizer, jitter, oscillator, phase-locked loop, sample–reset loop filter.

I. INTRODUCTION

PHASE-LOCKED loops (PLLs) in mixed analog–digital VLSI ICs operate in a very noisy environment, often with considerable noise introduced through coupling to the supply and substrate. Low-jitter PLLs require high loop bandwidths to reject the oscillator internal noise and the substrate and supply noise [1]–[4]. Designs that are insensitive to process and environmental conditions lead to well-controlled damping factors so that PLL bandwidth can be maximized [2]. Also, adaptive-bandwidth PLLs are desirable for fast-locking applications [5], and fully differential architectures can be used to provide additional supply and substrate noise immunity [6].

Since the introduction of charge-pump PLLs, numerous architectures for PLL components (phase-frequency detectors, charge-pumps, loop filters, and oscillators) have been proposed. The phase-frequency detector (PFD) can be based on NAND/NOR gates or D flip-flops, and several dead-zone avoidance techniques have been proposed [1]–[6]. Connected to PFDs, charge-pumps can be single ended or differential, with pump control switches that are connected to the drain, gate, or source of the current mirror, or using a current steering technique [7]. Fully differential charge pumps have gained an

increased interest due to their better supply and substrate noise rejection, although they require increased on-chip capacitance and extra circuitry for common-mode feedback [6].

Passive loop filters are popular for their simplicity [1], [2], [4], [5], but the control of their loop time constants lacks flexibility. A wider range of loop time constants and decreased area of on-chip capacitance can be realized by using active loop filters in conjunction with feedforward charge pumps [3], [6].

Ring oscillators can have a wide frequency range and can be single ended or differential, with voltage [1], [5] or current [2]–[4] control. Cross-coupled differential oscillators with no tail current can be used for low voltage designs [5]. Active load differential inverters are often preferred for their high signal-to-noise ratio, whereas diode clamped differential inverters are used in high-speed applications [3], [4].

Many applications require a stable 50% duty cycle PLL output. A common way of achieving a 50% duty cycle is to run the oscillator at twice the output frequency and then divide the output by two [3]. This method can consume additional power, but is very effective for deep-submicron technologies where use of differential comparators is suboptimal because of large device mismatches. More sophisticated methods of duty cycle control have been developed by using feedback or feedforward correction [5].

Most charge-pump PLLs have two major drawbacks. First, the loop filter pole position must be chosen as a compromise between the loop phase margin and the jitter performance. Second, loop damping factor variation with process requires wide margin for the natural frequency to stabilizing zero separation to keep the jitter peaking and the transient overshoot under the specified maximum value.

This paper describes a sample–reset loop filter architecture that averages the proportional control current which provides the feedforward zero over each update period, minimizing the ripple on the oscillator control signal and thus reduces the reference spurs [8]. The PLL damping factor is made insensitive to process variations by making it dependent only upon a bandgap voltage and ratios of circuit elements. This introduces a tracking mechanism between the loop natural frequency and the stabilizing zero.

II. SAMPLE–RESET LOOP FILTER TECHNIQUE

Charge-pump PLLs that have two poles at the origin require a zero to be introduced in the loop for stability. Common methods

Manuscript received April 4, 2001; revised June 26, 2001.
A. Maxim is with Maxim Integrated Products, Austin, TX 78750 USA (e-mail: amaxim@texas.net).
B. Scott is with Channel Technology, Boulder, CO 80302 USA.
E. M. Schneider, M. L. Hagge, S. Chacko, and D. Stiurca are with Cirrus Logic Inc., Austin, TX 78749 USA.
Publisher Item Identifier S 0018-9200(01)08223-3.

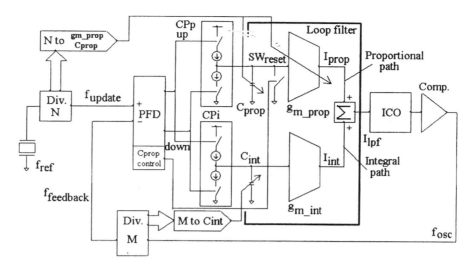

Fig. 1. Sample–reset PLL block diagram.

of adding a zero are introducing a resistor in series with the charge-pump capacitance, or using a feedforward technique. Most charge-pump PLLs use a proportional signal based on the instantaneous phase difference. This signal is characterized by narrow high-amplitude pulses that lead to an abrupt variation of the oscillator control signal and rapid frequency changes that can degrade the PLL's jitter performance. To reduce these effects and hence jitter, the oscillator control signal can be smoothed with a ripple filtering pole, which, unfortunately, degrades the PLL's phase margin.

A PLL's reference input (with frequency f_{ref}) often passes through an input divider (N). The output of the N divider is the input to the PFD. The PFD updates its state based on the frequency at the input of the PFD, thus we will refer to this as the update frequency f_{update} and its corresponding period as the update period T_{update}: $f_{update} = f_{ref}/N$ and $T_{update} = N/f_{ref}$.

Traditional feedforward charge pumps provide a periodic output current equal to I_{cp-p} for the phase difference time period $\Delta\varphi$ and 0 otherwise. The average proportional current per update period $I_{ave} = I_{cp-p} \cdot \Delta\varphi/(2 \cdot \pi)$ determines the position of the stabilizing zero.

The key idea of the sample–reset loop filter architecture is to generate a proportional current that is constant over the entire update period and has a value equal to the average current (I_{ave} as defined in the previous paragraph). This value leads to the same position of the stabilizing zero as in the standard charge-pump PLL, but generates a ripple-free oscillator control current, and thus minimizes the jitter. It can be achieved by first sampling the phase difference for each update period on a capacitor C_{prop} and then injecting a constant control current proportional to the sampled phase difference during the rest of the update period. At the beginning of each update period, a reset must be performed on the sampling capacitance voltage to eliminate the memory of the proportional path. This eliminates an additional pole at the origin that would otherwise make the loop unstable. The reset signal is synchronized with the update fre-

The major advantage of this architecture is staircase-shaped oscillator control, which does not need additional filtering, and leads to a better jitter performance. The architecture provides a type-II second-order PLL that has nearly 90° phase margin with negligible jitter peaking and transient locking overshoot.

Fig. 1 shows a block diagram of the sample–reset PLL architecture. It consists of a reference frequency generator (f_{ref}), an input divider (N), a current controlled oscillator (ICO), a feedback divider (M), a PFD with the associated integral (CPi) and proportional (CPp) charge pumps, and the loop filter. The filter has two signal paths, the integral path consisting of the phase integration capacitance C_{int} and the integral transconductance stage g_{m_int}, and the proportional path that provides the feedforward stabilizing zero with the phase sampling capacitance C_{prop} and the corresponding reset switch SW_{reset}, and the proportional transconductance stage g_{m_prop}.

The single sampling capacitance architecture shown in Fig. 2(a) has two shortcomings. First, the proportional control current still has some ripple due to the fact that during each update period the voltage on C_{prop} samples, holds, and finally resets. Second, the reset needs to be very short (less than a few percent of the period). This results in difficult constraints for the reset circuitry. These problems are solved by separating the sample and reset phases from the holding phase with a double capacitance architecture shown in Fig. 2(b). One capacitor is reset, and then samples the phase difference, while the other capacitor is in hold mode and generates the proportional current. The differential single and double sampling capacitance architectures presented in Fig. 2(c) and (d) result in better jitter performance due to higher supply and substrate noise rejection.

A comparison between the proportional path current used in standard charge-pump PLLs and the sample–reset PLL is shown in Fig. 3. The standard charge-pump PLL proportional path output is based on instantaneous phase difference. The sample–reset proportional path output is based on the average phase difference for each update period.

Open loop gain and therefore damping factor varies with the modulus of the feedback divider (M). A nearly constant

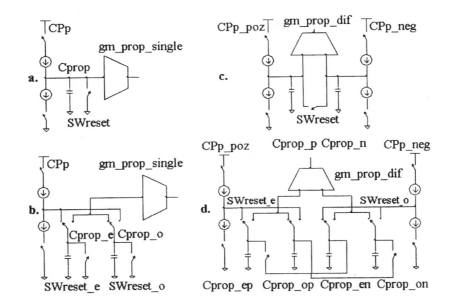

Fig. 2. Single ended/differential and single/double sampling capacitance sample–reset proportional path architectures.

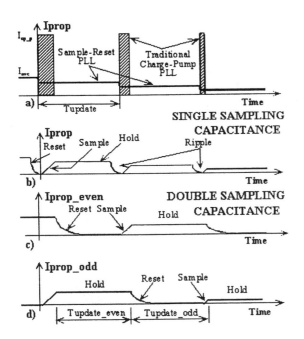

Fig. 3. Sample–reset versus standard PLL proportional current.

C_{int} into the output of the integral path charge pump as the M divider increases. A tradeoff between the amount of C_{int} capacitance switching and the maximum damping factor variation over a given range of M divider modulus must be done. The C_{int} capacitance is tuned dynamically by a digital decoder that converts the M divider value into the appropriate digital control signals to the C_{int} capacitance switches. The proportional path current is averaged over each update period. Changing N changes the update period and therefore the position of the stabilizing zero. The position of the zero is inversely proportional to the update period multiplied by the proportional path gain. The proportional path gain $(g_{m\_\text{prop}}/C_{\text{prop}})$ can be varied as a function of the update frequency to minimize the

variation of the zero frequency. A tunable capacitance C_{prop} combined with a programmable g_{m_prop} transconductance minimizes the area of on-chip capacitance for a given power consumption in the proportional path. A second digital decoder is used to generate the control signals for the proportional path transconductance and capacitance switches.

III. STABILITY OF THE SAMPLE–RESET PLL

The closed-loop small-signal model of the sample–reset PLL is shown in Fig. 4. Key components are the input divider $(1/N)$, PFD $(K_{\text{pfd}} = 1/2\pi)$, the integral $(I_{\text{cp\_i}})$ and proportional $(I_{\text{cp\_p}})$ charge pumps, the loop filter with the integral $(K_{\text{int}}/s = g_{m\_\text{int}}/(C_{\text{int}} \cdot s))$ and proportional $(K_{\text{prop}} = (T_{\text{update}} \cdot g_{m\_\text{prop}})/C_{\text{prop}})$ paths, the current controlled oscillator (K_{ico}/s), and the feedback divider $(1/M)$.

The total current supplied by the sample–reset loop filter I_{lpf} is the sum of the integral I_{int} and proportional I_{prop} path currents:

$$I_{\text{lpf}}(s) = I_{\text{int}}(s) + I_{\text{prop}}(s)$$
$$= \frac{I_{\text{cp\_i}}}{2\pi} \cdot \frac{1}{sC_{\text{int}}} \cdot g_{m\_\text{int}} + \frac{I_{\text{cp\_p}}}{2\pi} \cdot \frac{T_{\text{update}}}{C_{\text{prop}}} \cdot g_{m\_\text{prop}}.$$
$$(1)$$

The open loop transfer function and the natural frequency (square root of the dc gain K_{DC}) are

$$H_{\text{open}}(s) = \frac{I_{\text{cp\_i}}}{2\pi} \cdot \frac{g_{m\_\text{int}}}{sC_{\text{int}}}$$
$$\cdot \left[\frac{I_{\text{cp\_p}}}{I_{\text{cp\_i}}} \cdot \frac{C_{\text{int}}}{C_{\text{prop}}} \cdot \frac{1}{f_{\text{update}}} \cdot \frac{g_{m\_\text{prop}}}{g_{m\_\text{int}}} \cdot s + 1 \right]$$
$$\cdot \frac{K_{\text{ico}}}{s} \cdot \frac{1}{M}$$
$$(2)$$

$$\omega_n = \sqrt{K_{\text{DC}}} = \sqrt{\frac{K_{\text{ico}} \cdot I_{\text{cp\_i}} \cdot g_{m\_\text{int}}}{2\pi \cdot C_{\text{int}} \cdot M}}.$$
$$(3)$$

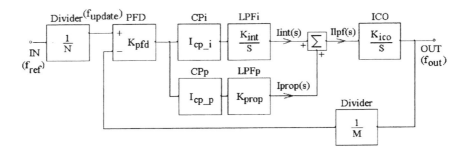

Fig. 4. Sample–reset PLL small-signal model.

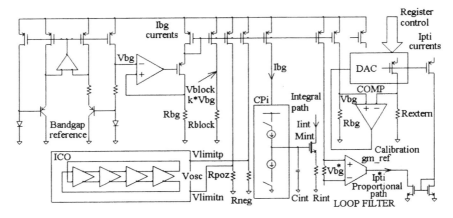

Fig. 5. PLL architecture for process-independent damping factor.

The resulting expression for the stabilizing zero ωz is shown in (4). It is important to note that the zero position is f_{update} times a ratio of currents, times a ratio of capacitances, times a ratio of transconductances, that with proper design and layout will exhibit very low process variation:

$$\omega_z = f_{\text{update}} \cdot \frac{I_{\text{cp\_i}}}{I_{\text{cp\_p}}} \cdot \frac{C_{\text{prop}}}{C_{\text{int}}} \cdot \frac{g_{m\_\text{int}}}{g_{m\_\text{prop}}}. \qquad (4)$$

The damping factor ξ of type-II second-order PLLs is equal to half the natural frequency divided by the stabilizing zero:

$$\xi = \frac{1}{2} \cdot \frac{\omega_n}{\omega_z} = \frac{1}{2} \cdot \sqrt{\frac{K_{\text{ico}} \cdot I_{\text{cp\_i}} \cdot C_{\text{int}}}{2\pi \cdot M \cdot g_{m\_\text{int}}}}$$
$$\cdot \frac{I_{\text{cp\_p}}}{I_{\text{cp\_i}}} \cdot \frac{1}{C_{\text{prop}}} \cdot \frac{g_{m\_\text{prop}}}{f_{\text{update}}}. \qquad (5)$$

A tracking mechanism between the PLL's natural frequency and stabilizing zero ensures process independence of the damping factor. A bandgap voltage V_{bg} is used to generate all the on-chip biasing currents and reference voltages (see Fig. 5). There are two types of bias currents used throughout the chip. The first current, $I_{\text{bg}} = V_{\text{bg}}/R_{\text{bg}}$, is obtained by imposing the bandgap voltage across an internal polysilicon resistor R_{bg}, which results in a current inversely proportional to that resistance. Mirroring the I_{bg} current into resistors ($R_{\text{block}} = K \cdot R_{\text{bg}}$) that are carefully matched with the bandgap resistor R_{bg} provides a technique of replicating a fraction of the process-independent bandgap voltage within the different

circuit blocks in the IC ($V_{\text{block}} = K \cdot R_{\text{bg}} \cdot I_{\text{bg}} = K \cdot V_{\text{bg}}$). The second bias current ($I_{\text{pti}} = V_{\text{bg}}/R_{\text{extern}}$) is obtained by calibrating a temperature-compensated replica of the I_{bg} current to an external high-precision resistor R_{extern}. An onboard digital-to-analog converter (DAC) and control circuit performs the calibration, resulting in a process and temperature independent current I_{pti}.

The most significant contribution to the damping factor process variation [see (5)] is due to the proportional path transconductance g_{m_prop}. This can be canceled out by first generating a reference process-independent transconductance g_{m_ref} and then deriving g_{m_prop} as a weighted version of g_{m_ref}. The g_{m_ref} transconductance is generated using a differential amplifier with its input signal set to a fraction V_{bg}^* of the bandgap voltage, and its output current set to a fraction I_{pti}^* of the process-independent bandgap trimmed current. Feedback is used to tune the tail current, so the resulting transconductance $g_{m\_\text{ref}} = I_{\text{pti}}^*/V_{\text{bg}}^*$ is virtually process independent.

The next major contribution is from the term $I_{\text{cp\_i}}/g_{m\_\text{int}} \approx I_{\text{cp\_i}} \cdot R_{\text{int}}$ that can be canceled out by mirroring I_{bg} to be used as charge-pump currents and carefully matching the integral path resistor R_{int} with the bandgap resistor R_{bg}, as shown in Fig. 5. Thus $I_{\text{cp\_i}} \cdot R_{\text{int}} = V_{\text{bg}} \cdot R_{\text{int}}/R_{\text{bg}}$ is a fraction of the bandgap voltage, and is process independent.

Finally, the supply and process variation of the oscillator gain ($K_{\text{ico}} \alpha 1/V_{\text{osc}}$) is reduced by clamping the amplitude of oscillation V_{osc} to a fraction of the bandgap voltage. The damping levels are given by $V_{\text{limitp}} = R_{\text{poz}} \cdot I_{\text{bg}} = K_1 \cdot V_{\text{bg}}$ and $V_{\text{limitn}} = R_{\text{neg}} \cdot I_{\text{bg}} = K_2 \cdot V_{\text{bg}}$.

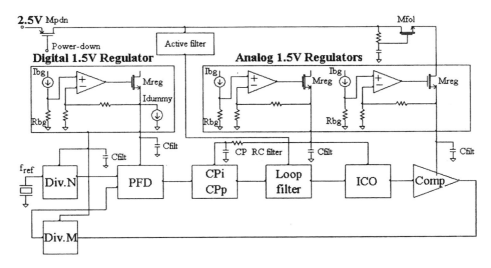

Fig. 6. Power supply circuitry for low-jitter operation.

The natural frequency process variation [see (3)] is given by the integral charge-pump current $I_{cp\_i} = V_{bg}/R_{bg}$ and the integral path transconductance $g_{m\_int} \approx 1/R_{int}$, resulting in a process variation inversely proportional to the polysilicon bandgap resistor ($\omega_n \propto \sqrt{1/R_{bg} \cdot 1/R_{int}} \propto 1/R_{bg}$). The stabilizing zero process variation [see (4)] is mainly influenced by the integral path transconductance ($g_{m\_int} = 1/R_{int}$) which is also inverse to polysilicon resistance (assuming g_{m_prop} is process independent and I_{cp_i}/I_{cp_p} and C_{prop}/C_{int} have very low process dependence). Thus, both the natural frequency and stabilizing zero are process dependent, but they track each other, yielding a process-independent damping factor.

To avoid granularity effects, sampling ratios (S = update frequency f_{update} divided by loop bandwidth B) greater than 10 are required [2]. It is desirable to have sampling ratios with low process variation and invariance to the divider's modulus (M, N). The sampling ratio varies with $\sqrt{M}$ over a given frequency range. Reducing its variation requires the division of the frequency range into multiple subranges with low M_{max}/M_{min} ratios:

$$S = \frac{f_{update}}{B}$$
$$= \frac{f_{out}}{M} \cdot \frac{1}{\sqrt{1 + 2 \cdot \xi^2 + \sqrt{2 + 4 \cdot \xi^2 + 4 \cdot \xi^4}} \cdot f_n}$$
$$= \frac{f_{out}}{func(\xi)} \cdot \sqrt{\frac{2\pi C_{int} \cdot R_{int}}{K_{ico} \cdot I_{cp\_i} \cdot M}}. \tag{6}$$

IV. POWER SUPPLY STRATEGY FOR LOW-JITTER OPERATION

PLLs operating in large mixed analog–digital chips are subject to noise injection from the power supply, ground, and substrate. This degrades the overall jitter performance significantly. Therefore, !ow-jitter operation requires a careful design of the power supply for the PLL building blocks. Fig. 6 is a diagram of the PLL power supply circuitry.

A high power-supply rejection ratio (PSRR) regulator is used to separate the supply of the ICO from the rest of the

PLL [1]–[5]. A further reduction of the supply-injected noise is achieved by separating the drain of the regulator's serial transistor M_{reg} from the rail with a source follower M_{fol}. The 1.5-V supplies are filtered with local bypass capacitors C_{filt}.

The differential to single-ended converter (comparator) needs to generate sharp edges to maintain a precise 50% duty cycle. The sharp edges cause significant supply ripple. Thus, the comparator's supply needs to be separated from the ring oscillator supply.

The charge pump is one of the PLL components that is most sensitive to supply noise. RC filtering is used to isolate the charge pump from the other elements in its supply domain.

The loop filter is biased from the 2.5-V analog chip supply through an active filter. The 2.5-V supply voltage is necessary to provide proper headroom in the loop filter.

An optimum power management methodology in large mixed analog–digital chips requires an ability to power down the PLL. A large area PFET (M_{pdn}) is used as serial switch to cut or pass the supply current to the PLL.

V. SAMPLE–RESET PLL BUILDING BLOCKS

A. Phase-Frequency Detector

Fig. 7 shows the PFD and the block that generates the control signals for the proportional capacitance switches. The PFD uses the standard seven NAND gates architecture [1]–[4]. Generating two synchronous narrow pulses during each update period for both pump-up and pump-down charge-pump signals eliminates the dead zone at small phase differences. A oneshot circuit triggered by the PFD reset signal is used. Pass gates are added at the UP and DOWN outputs to match the propagation time of the inverters from the UPb and DOWNb outputs.

The PFD also generates the control signals for the proportional path switches, which select between the even and odd capacitors and determine the sample, hold, and reset phases. A flag for even and odd update periods is generated by dividing the frequency of the dead-zone avoidance oneshot circuit by two. Nonoverlapping control signals are used to connect the proportional capacitors alternately to the charge pump and the propor-

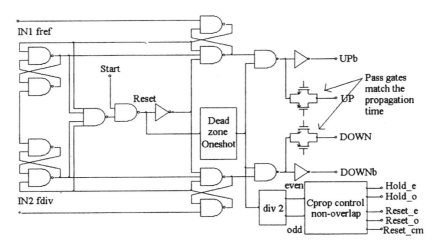

Fig. 7. Phase-frequency detector and C_{prop} switches control block.

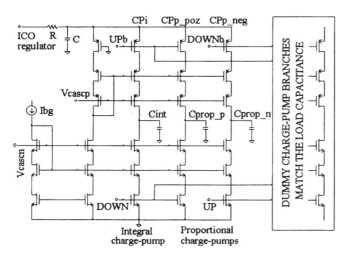

Fig. 8. Integral and proportional charge pumps.

tional transconductance (Hold_e, Hold_o). There are also signals for charge-sharing reset on the proportional capacitor (Reset_e, Reset_o) and a proportional transconductance common-mode reset (Reset_cm).

B. Charge Pumps

Fast-locking PLLs require wide bandwidth. They also require the PFD and charge pumps to operate at high update frequencies. It is important to have low clock feedthrough and charge sharing effects. The source switch architecture of the charge pump is shown in Fig. 8. This architecture ensures high speed since all internal nodes are low impedance. Clock feedthrough is also low because the drains of the switches are separated from the high-impedance output node. Dummy switches are added to compensate for charge sharing. Each PFD output sees the same number of nMOS and pMOS switches, leading to a matched switching time.

C. Sample–Reset Loop Filter

Fig. 9 shows a detailed schematic of the sample–reset loop filter. The single ended integral path is a voltage-to-current

(V–I) converter onto a capacitor C_{int}. The transistor M_{int} and a source degeneration resistor R_{int} has a transconductance $g_{m\_\text{int}} = g_m/(1 + g_m \cdot R_{\text{int}}) \approx 1/R_{\text{int}}$. The transconductance is set by R_{int} and will not depend upon M_{int} transistor g_m. Care was taken to Kelvin connect the C_{int} and R_{int} ground connections.

The proportional path is more susceptible to substrate and supply noise and therefore uses a fully differential architecture. There are three subcircuits: a reference transconductance stage g_{m_ref} that provides a process-independent transconductance, a programmable transconductance stage g_{m_prop}, and a switching block SW$_{\_\text{cprop}}$. The programmable g_m stage generates a binarily weighted replica of the reference transconductance to produce the proportional path control current. The voltage sampled on the capacitor C_{prop} controls the current in the programmable g_m stage. The switching block successively connects the proportional capacitors first to the proportional charge pumps (CP$_{\text{p\_poz}}$, CP$_{\text{p\_neg}}$), then to the programmable transconductance stage, and finally to the reset voltage $V_{\text{cm\_reset}}$. The size of C_{int} and C_{prop} capacitors is chosen so that the contribution of kT/C noise is negligible.

The process-independent reference transconductance stage is a differential pair (M_{rg_m+} and M_{rg_m-}) tuned to a constant g_m by a feedback circuit that regulates its tail current. The differential pair input is a fraction of the bandgap voltage generated locally by injecting a replica of the bandgap current I_{bg} into a resistor R_{in} which matches the bandgap resistor R_{bg}. A trimmed bandgap current I_{pti} is summed into the drains of the differential pair. The resulting transconductance ($g_{m\_\text{ref}} \propto I_{\text{pti}}/V_{\text{bg}}$) is virtually process, temperature, and supply independent.

This circuit has both positive and negative feedback. Connecting a large compensation capacitor C_c to the high impedance output of the differential pair ensures stable operation. The resulting dominant pole is at a very low frequency and the reference g_m circuit has a long settling time. Therefore, the reference stage must be isolated from any major source of ripple, particularly the steps due to the switching of the proportional capacitors.

The tail current of the reference g_m stage is mirrored into several replica g_m stages that are binarily weighted versions

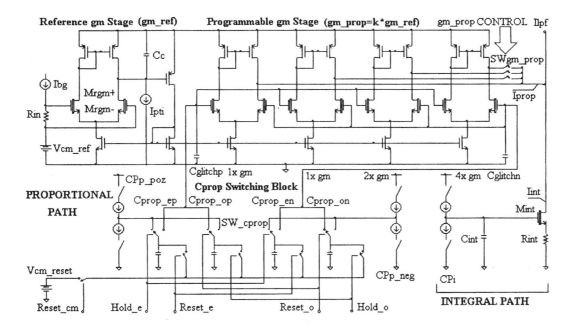

Fig. 9. Current-mode sample–reset loop filter.

of the reference. The currents from the binarily weighted g_m stages are delivered to a set of switches $SW_{g_m\text{-prop}}$. The state of the switches determines the transconductance ($g_{m\text{-prop}} = f(SW_{g_m\text{-prop}}) \cdot g_{m\text{-ref}}$). This variable g_m is used in conjunction with the capacitor C_{prop} to tune the value of the stabilizing zero as a function of the loop update frequency.

The differential architecture of the programmable g_m stage requires two pairs of proportional path capacitors, even ($C_{\text{prop\_ep}}/C_{\text{prop\_en}}$) and odd ($C_{\text{prop\_op}}/C_{\text{prop\_on}}$) pairs, and two proportional charge pumps ($CP_{\text{p\_poz}}$ and $CP_{\text{p\_neg}}$). Differential charging significantly simplifies the reset circuit. Fast reset is achieved through charge sharing between the two capacitors within a pair. $C_{\text{prop\_ep}}$ shares with $C_{\text{prop\_en}}$, and $C_{\text{prop\_op}}$ shares with $C_{\text{prop\_on}}$.

During the even update period, the even capacitor pair is first reset to the common-mode voltage $V_{\text{cm\_reset}}$ and then connected to the proportional charge pumps ($CP_{\text{p\_poz}}$ and $CP_{\text{p\_neg}}$). Then the pair samples the current phase difference. Meanwhile, the odd capacitor pair that has sampled the previous phase difference is connected to the programmable g_m stage to generate a proportional path current. During the odd update period, the odd capacitor pair is first reset and then connected to the proportional charge pumps to sample the phase difference. Meanwhile, the even capacitor pair is connected to the programmable g_m stage [see waveforms in Fig. 3(c), (d)]. Two sets of reset switches are used. The first switches (controlled by Reset_e and Reset_o) connect the positive and negative capacitors within a pair to form a coarse charge-sharing reset. The second switches (controlled by Reset_cm) connect the two shorted capacitors to a reference voltage to prevent common-mode voltage drift.

The matching of the even and odd switches is critical for the proper operation of the sample–reset loop filter. Two small capacitances ($C_{\text{glitchp}}, C_{\text{glitchn}}$) are permanently connected at the input of the programmable g_m stage. These capacitances prevent glitching on the proportional path by keeping the g_m stage from floating when neither the even nor the odd capacitances are connected at the input.

The phase margin of the sample–reset PLL is sensitive to any additional delay in the loop, particularly from the proportional path. Reducing the delay in the proportional path to nearly zero is accomplished by connecting the proportional capacitors to the $g_{m\text{-prop}}$ stage immediately after the sampling phase. Dead-zone avoidance pulses that appear after the phase difference time period ends trigger the hold signals (Hold_e, Hold_o).

D. Controlled Oscillator

Maximizing loop bandwidth and minimizing oscillator phase noise minimizes jitter. Variation of oscillator gain directly influences the loop damping factor [see (5)]. Therefore, a process, temperature, frequency, and supply independent gain leads to a larger potential phase margin and better linearity of the PLL. Fig. 10 shows a diagram of the ICO.

The ring oscillator uses differential inverters to improve supply rejection and reduce substrate noise effects. Symmetric active load provides high oscillation amplitude and therefore good signal-to-noise ratio.

A simplified model of the ICO is a relaxation oscillator where the output capacitor is charged and discharged between two threshold levels by a constant current. The resulting gain is inversely proportional to the capacitance at the output of each ring oscillator stage (C_{out}) and the amplitude of oscillation (V_{osc}): $K_{\text{ico}} \propto 1/(C_{\text{out}} \cdot V_{\text{osc}})$.

A process and supply independent gain is achieved by clamping ring element amplitude to a fraction of the bandgap voltage. The positive and negative clamping voltages are generated locally by injecting two I_{bg}-based currents ($I_{\text{bgp}}, I_{\text{bgm}}$)

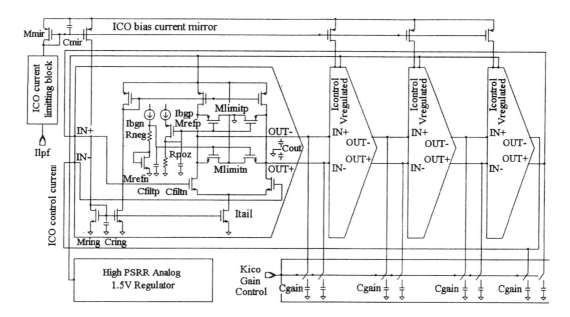

Fig. 10. ICO with process-independent output amplitude.

into two resistors (R_{poz}, R_{mg}) matched to the bandgap resistor. Two diode connected devices (M_{refp}, M_{refn}) are connected in series with R_{poz} and R_{mg}. Careful matching of M_{refp} and M_{refn} to clamping devices M_{limitp} and M_{limitn} help cancel the V_{GS} component in the positive and negative limiting voltages. Soft clamping is used, where the limiting devices are ON for a very small fraction of the oscillator period. Therefore, the process dependent input impedance of the limiting devices has a negligible influence on the oscillator period. Additional filtering at the positive and negative limiting nodes (C_{filtp}, C_{filtn}) significantly improves the ring element PSRR.

The loop has no ripple filtering pole. The most significant high-order parasitic poles are those of the current mirror that biases the four ring elements ($g_m(M_{mir})/C_{mir}$) and of the ring elements ($g_m(M_{ring})/C_{ring}$) (see Fig. 10). Both poles are at least two decades higher than the stabilizing zero and have a negligible effect on loop phase margin.

To improve oscillator linearity and provide wide frequency range, the ICO has several subranges where additional capacitance C_{gain} is gradually switched onto the output of each ring element as needed to reduce frequency. The oscillator is thus required to operate on only a narrow region of a nonlinear $f(I)$ curve, ensuring a fairly constant gain. This increases oscillator linearity but reduces PLL open-loop gain. To compensate for this reduction in gain, the integral path capacitor C_{int} is increased, to keep the $K_{ico} \cdot C_{int}$ term in the damping factor expression constant (5).

During the nonlinear locking process the transient overshoot can drive the oscillator to an output frequency higher than the maximum operating frequency of the divider. This can cause a catastrophic divider failure, and the PLL can fail to lock, and can also fail to recover. To prevent this, a current limiting circuit is introduced into the ICO. The circuit limits the maximum control current so that the ICO frequency cannot exceed the maximum divider frequency.

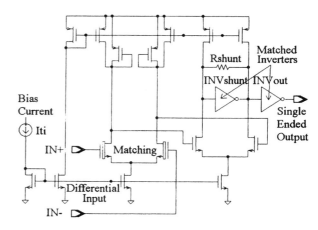

Fig. 11. Differential-to-single-ended converter.

E. Output Comparator

A differential-to-single-ended converter (comparator) must be used to drive the feedback divider. The comparator should not significantly load the ring oscillator.

A critical parameter in many systems is a well-controlled 50% duty cycle over process, temperature, operating frequency, and supply voltage variations. This circuit uses a feedback architecture that closely matches the trip point of the comparator decision stage with that of the output buffer.

Differential comparators in deep-submicron technologies often have duty cycle control limited by the matching of the comparator input devices. Native (zero VT) devices are used due to their better threshold voltage matching [9], which can be explained by their lower channel doping (no threshold adjustment implant).

The differential-to-single-ended converter is shown in Fig. 11. The first stage is a wide-bandwidth diode-load differential-gain stage that presents a very low input capacitance. The decision stage is a differential pair with feedback. Feedback is

TABLE I
SAMPLE–RESET PLL PERFORMANCE SUMMARY

Technology	0.18μm CMOS
Frequency range	125-1250 MHz
Frequency resolution	500 kHz
Analog supply voltage	2.5 V
Digital supply voltage	1.5 V
RMS jitter	Less than 0.9% of oscillator period
Power consumption	90 mW from 2.5V supply
Die area (2 PLLs, supply regulators, and bandgap)	300 μm x 800 μm

accomplished with a resistor R_{shunt} that restricts the output swing to within several hundred millivolts of the output inverter trip point. Additional feedback is implemented by inverter $\text{INV}_{\text{shunt}}$ that is carefully matched to the output inverter INV_{out}. $\text{INV}_{\text{shunt}}$ ensures a precise tracking between the trip point of the decision stage and the output buffer. This produces a well-controlled 50% duty cycle over process, temperature, frequency, and supply voltage variation. To obtain high-speed operation, tail currents of several milliamperes are used.

F. Dividers

The feedback and input variable modulus dividers are designed using a pulse swallower architecture. The speed is increased with a divide by two prescalar followed by a synchronous variable modulus divider. The jitter introduced by the feedback divider is reduced with an output flip-flop synchronized at the oscillator output frequency.

VI. INCREASING FREQUENCY RESOLUTION WITH CASCADED PLL CONFIGURATION

The resolution of the output frequency ($f_{\text{out}} = f_{\text{ref}} \cdot M/N$) is given by the feedback (M) and/or input (N) divider modulus ranges. The M divider is included in the feedback loop and therefore directly influences the loop stability. The N divider, though not in the loop, does influence stability through the update period over which the averaging of the proportional path current is performed. A compromise between the high resolution and the high loop bandwidth is achieved by cascading two PLLs.

In magnetic read-channel applications, there are typically two PLLs, a low-resolution servo PLL and a high-resolution data PLL. Boosting the frequency at the input of the high-resolution data PLL up to the maximum output frequency (1.25 GHz) and then dividing it down by the input N divider ensures a wide range for the N divider while maintaining high update frequencies. Operating at high update frequencies allows a high loop bandwidth which minimizes the jitter introduced by the internal oscillators.

Bandwidths of the two PLLs must be carefully selected to minimize overall output jitter. Input jitter gain is related to the ratio of output and input frequencies. Single PLL and cascaded

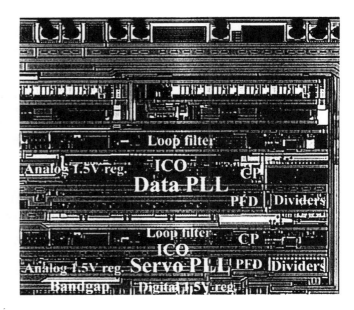

Fig. 12.   0.18-μm CMOS read-channel die photo (PLL detail).

PLL architectures have similar input jitter transfer, if jitter peaking of the two cascaded PLLs does not overlap.

The bandwidth of the high-resolution PLL is designed to be less than or equal to the bandwidth of the low-resolution frequency-boosting PLL. A good choice is to select the two PLL bandwidths to be equal. This choice should not significantly degrade the system jitter transfer performance since the sample–reset architecture provides negligible jitter peaking.

VII. EXPERIMENTAL RESULTS

This sample–reset PLL has been fabricated in a five-metal-layer 0.18-μm CMOS bulk process. Fig. 12 shows a portion of a die photo of the magnetic read-channel chip that uses the sample–reset PLL architecture. Table I summarizes the performance of the sample–reset PLLs.

Fig. 13 shows the simulated transient locking waveforms (integral, proportional, and total oscillator control current) for the sample–reset PLL operating at 1.25 GHz with $B \approx 2$ MHz and $\xi \approx 0.9$ [Fig. 13(a)], in comparison with the waveforms of a standard charge-pump PLL operated at 600 MHz with $B \approx 1$ MHz, natural frequency to stabilizing zero separation of

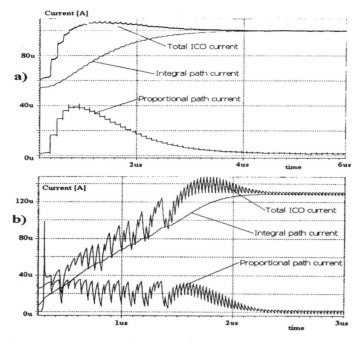

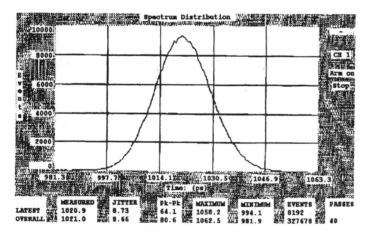

Fig. 13. Transient locking waveforms. (a) Sample–reset PLL. (b) Standard charge-pump PLL.

Fig. 14. Measured output jitter for cascaded PLL configuration.

1.8, and a pole-to-zero separation of 25 ($\xi \approx 0.9$) [Fig. 13(b)] used in a previous part. The sample–reset architecture has eliminated the high-amplitude ripple from the proportional current.

Fig. 14 shows the experimentally measured rms jitter of less than 9 ps, for a 1-GHz output for two PLLs connected in cascade, operating in conjunction with noisy digital circuitry (≈ 100 mV$_{pp}$ supply noise). The measured output jitter of a single sample–reset PLL is comparable with that of two cascaded PLLs. This confirms our initial assumption that the output jitter is dominated by the reference spurs and the supply and substrate injected noise. The jitter improvement is equally contributed by the sample–reset architecture that minimizes the reference spurs and the high PSRR regulators that minimizes the supply-injected noise.

Using native devices with lower substrate transconductance in the signal path and of accumulation capacitors in the loop filter minimizes the substrate injected noise.

The absence of the ripple filtering pole and the high PLL bandwidth (low time constants in the loop filter) lead to smaller on-chip capacitance, reducing the die size of the sample–reset PLL.

VIII. CONCLUSION

A charge-pump PLL with a sample–reset loop filter has been described. This architecture averages the proportional control current of the oscillator over each update period. Time averaging employs an additional charge pump and a sampling capacitor to generate a constant proportional path current for the entire update period. The major advantage of the sample–reset architecture is the ripple-free staircase shape of the proportional control signal. The control signal does not need additional filtering for low-jitter operation. The PLL is type II with a single stabilizing zero and nearly 90° phase margin. As a consequence, the transient locking has no overshoot, and the input jitter peaking is negligible.

The PLL damping factor is made insensitive to process variations by making it dependent only upon a bandgap voltage and ratios of circuit elements. This ensures tracking between the natural frequency and the stabilizing zero. The stable damping factor and the high update frequency lead to the maximization of the loop bandwidth and low settling time with good rejection of the internal oscillator noise.

To improve jitter performance, several high PSRR regulators are used to bias the different PLL building blocks, and *RC* supply filtering is provided for the blocks with high supply-noise sensitivity.

REFERENCES

[1] I. Yang, J. Greason, and K. Wong, "A PLL clock generator with 5 to 110 MHz of lock range for microprocessors," *J. Solid-State Circuits*, vol. 27, pp. 1599–1607, Nov. 1992.

[2] J. Maneatis, "Low-jitter process-independent DLL and PLL based on self-biased techniques," *J. Solid-State Circuits*, vol. 31, pp. 1723–1732, Nov. 1996.

[3] D. Mijuskovic, M. Bayer, T. Chomicz, H. Garg, P. James, P. McEntarfer, and J. Poiter, "Cell-based fully integrated CMOS frequency synthesizer," *J. Solid-State Circuits*, vol. 29, pp. 271–279, Mar. 1994.

[4] I. Novof, J. Austin, R. Kelkar, and D. Strayer, "Fully integrated CMOS phase-locked loop with 15 to 240 MHz locking range and 50 ps jitter," *J. Solid-State Circuits*, vol. 30, pp. 1259–1266, Nov. 1995.

[5] J. Lee and B. Kim, "A low noise fast-lock phase-locked loop with adaptive bandwidth control," *J. Solid-State Circuits*, vol. 35, pp. 1137–1145, Aug. 2000.

[6] K. Lin, L. Tee, and P. Gray, "A 1.4 GHz differential low noise CMOS frequency synthesizer using a wideband PLL architecture," in *ISSCC Dig. Tech. Papers*, San Francisco, CA, Feb. 2000, pp. 147–149.

[7] W. Rhee, "Design of high performance CMOS charge-pumps in phase locked loops," in *Proc. IEEE Int. Symp. Circuits and Systems*, Orlando, FL, May. 1999, pp. II.545–II.548.

[8] A. Maxim, B. Scott, E. Schneider, M. Hagge, S. Chacko, and D. Stiurca, "A low-jitter 125–1250 MHz process-independent 0.18-μm CMOS PLL based on a sample–reset loop filter," in *ISSCC Dig. Tech. Papers*, San Francisco, CA, Feb. 2001, pp. 394–395.

[9] A. Maxim and G. Maxim, "A novel physical based model of deep-sub-micron CMOS transistors mismatch for monte carlo SPICE simulation," in *Proc. IEEE Int. Symp. Circuits and Systems*, Sidney, NSW, Australia, May 2001, pp. V.511–V514.

A Dual-Loop Delay-Locked Loop Using Multiple Voltage-Controlled Delay Lines

Yeon-Jae Jung, Seung-Wook Lee, Daeyun Shim, Wonchan Kim, Changhyun Kim, *Member, IEEE*, and Soo-In Cho

Abstract—This paper describes a dual-loop delay-locked loop (DLL) which overcomes the problem of a limited delay range by using multiple voltage-controlled delay lines (VCDLs). A reference loop generates quadrature clocks, which are then delayed with controllable amounts by four VCDLs and multiplexed to generate the output clock in a main loop. This architecture enables the DLL to emulate the infinite-length VCDL with multiple finite-length VCDLs. The DLL incorporates a replica biasing circuit for low-jitter characteristics and a duty cycle corrector immune to prevalent process mismatches. A test chip has been fabricated using a 0.25-μm CMOS process. At 400 MHz, the peak-to-peak jitter with a quiet 2.5-V supply is 54 ps, and the supply-noise sensitivity is 0.32 ps/mV.

Index Terms—Clock synchronization, delay-locked loop, duty cycle corrector, replica biasing, voltage-controlled delay lines.

I. INTRODUCTION

FOR high-performance microprocessors and memory ICs, the use of phase-locked loops (PLLs) or delay-locked loops (DLLs) is essential to minimize the negative effects caused by skews and jitters of clock signals. In applications where the frequency multiplication is not required, a DLL is a natural choice since it is free from the jitter accumulation problem of an oscillator-based PLL. Conventional DLLs, however, suffer from the problem of their limited delay range since DLLs adjust only the phase, not the frequency.

We propose a new dual-loop DLL architecture that allows unlimited delay range by using multiple voltage-controlled delay lines (VCDLs). In our architecture, the reference loop generates four evenly spaced clocks, which are then delayed with controllable amounts by four VCDLs and multiplexed to generate the output clock in the main loop. The selection and delay control in the main loop permit the DLL to emulate the infinite delay range with a multiple of finite-length VCDLs. Moreover, a fully analog control technique can be applied to exploit the established benefits of conventional DLLs such as low skew and low jitter. To reduce supply-noise sensitivity further, a new low-jitter scheme is employed in a replica biasing circuit, which compensates the delay variation of a delay line against the injected supply noise. Finally, a duty cycle corrector immune to process mismatches is also used.

This paper is arranged as follows. In Section II, following a brief overview of conventional DLLs, the proposed architecture

Manuscript received June 27, 2000; revised October 15, 2000.
Y.-J. Jung, S.-W. Lee, D. Shim, and W. Kim are with the School of Electrical Engineering, Seoul National University, Seoul 151-742, Korea.
C. Kim and S.-I. Cho are with Samsung Electronics Company, Kyungki-do, Korea.
Publisher Item Identifier S 0018-9200(01)03028-1.

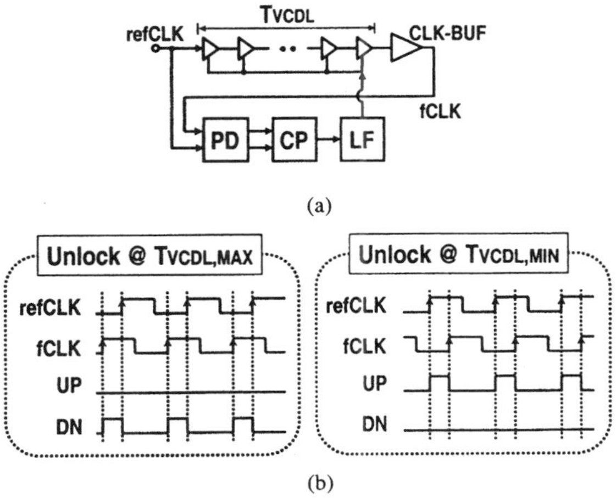

Fig. 1.   (a) Block diagram of a conventional DLL. (b) Lock-failure cases.

is described with design concepts and various building blocks. Section III describes circuits for low-jitter scheme and duty cycle correction. Section IV discusses the prototype chip implementation and shows experimental results. Section V concludes this paper with a summary.

II. ARCHITECTURE

A. Limited Range Problem of Conventional DLLs

A simplified block diagram of a conventional DLL [1] is outlined with its lock-failure cases in Fig. 1. In the normal condition, the DLL forces the output clock ($fCLK$) to be aligned with the input reference clock ($refCLK$) through the negative feedback loop, which comprises a voltage-controlled delay line, a phase detector, a charge pump, and a loop filter. The clock buffer (CLK-BUF) is inserted to provide the chip-wide clock. Although this simple architecture offers many design flexibilities, the main problem in the conventional DLL of Fig. 1(a) is that the delay time of the VCDL (T_{VCDL}) has a minimum and a maximum boundary. Therefore, the DLL has states in which it does not work, as shown in Fig. 1(b). When T_{VCDL} has a maximum delay and the $fCLK$ leads the $refCLK$, DN pulses are generated but the VCDL can not produce any more delay. On the other hand, when T_{VCDL} has a minimum delay and the $fCLK$ lags the $refCLK$, UP pulses are generated but the VCDL cannot reduce any more delay. These lock-failure cases arise from the facts that the range of T_{VCDL} is limited and the initial value of T_{VCDL} is not known at loop startup. An additional loop startup control circuitry may solve this problem and the DLL acquire lock. Unfortunately, the delay time of the clock buffer and following clock distribution tree ($T_{\mathrm{CLK-BUF}}$)

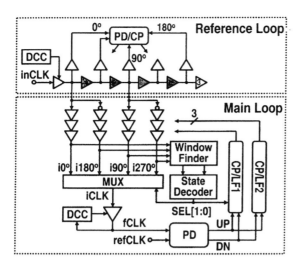

Fig. 2. Block diagram of the proposed dual-loop DLL.

deviates from the value at the simulation stage according to temperature and voltage variations [2]. When the variation of $T_{\text{CLK-BUF}}$ is excessive, the DLL loses the lock and falls into the lock-failure cases in Fig. 1(b).

A DLL relying on quadrature phase mixing [3] has been proposed to overcome the limited range problem of the conventional DLL. The phase mixing technique using quadrature clocks provides unlimited phase shift capability. However, phase mixing uses two small slew-rate clocks to obtain linear results. Therefore, this approach has the disadvantage of the increased dynamic noise sensitivity and jitter. In the semidigital DLL [4], a digitally controlled phase interpolator uses internally generated 30°-spaced clocks through the dual DLL architecture. Although noise sensitivity issues on the phase interpolation could be alleviated by smaller interpolation intervals, inherent digital nature causes dithering around zero phase error due to continuous control-bit updates. A digital DLL architecture with infinite phase capture ranges [5] is also not free from the same dithering problem and requires a large chip area for fine delay control.

B. Proposed Dual-Loop DLL

Fig. 2 shows a block diagram of the proposed dual-loop DLL architecture [6]. This architecture is based on two loops: the reference loop and the main loop. The reference loop is locked at 180° phase shift through the conventional DLL architecture. Since the reference loop VCDL is composed of four main delay cells, each delay cell generates a 45° phase shift at locked condition. All delay cells including delay buffers are differential elements commonly controlled by the output of the charge pump. The delay cell named "3" means three parallel-connected delay cells, so that the load balance between 0° and 180° clock is preserved. The reference loop provides two differential clocks spaced by 90° to the main loop. To cover the entire 360° phase range, clocks from the reference loop are partially inverted and inputted to four sets of VCDL in the main loop. Each main loop VCDL is composed of three delay cells and generates low swing internal clocks-$i0°$, $i90°$, $i180°$, and $i270°$. These clocks experience the analog delay time control by two kinds of four con-

trol voltages generated from two main loop charge pumps. The multiplexer selects one of four clocks as $iCLK$ and this clock feeds the clock buffer whose function is to convert low swing to full CMOS-level as well as provide the chip-wide output clock, $fCLK$. The $fCLK$ drives the phase detector which compares it to the reference clock. The output of the phase detector is used by two charge pumps and four loop filters to control the delay time of each main loop VCDL. Four-to-one clock switching is implemented by the window finder and the state decoder block. The window finder monitors the boundary where the selected $iCLK$ is switched and forces the state decoder to update the two-bit selection code at the switching event. The selection code not only controls the clock selection at the multiplexer but changes the configuration of two charge pumps and four loop filters to accommodate the clock switching. Duty cycle correction (DCC) is employed to remove the duty cycle imperfections of the input clock $inCLK$ and the output clock $fCLK$. Finally, although two input clocks, $inCLK$ and $refCLK$, can be merged into one clock input, lower jitter clock source is preferred as the $inCLK$, if possible, since it determines the jitter characteristics of the whole DLL.

In this architecture, the clock selection scheme enables the output clock to cover the entire phase range (modulo 2π). Furthermore, seamless clock switching is possible by optimizing the main loop VCDL delay control scheme. Moreover, the phase locking is achieved by fully analog control in all loops, so that we can apply low-skew and low-jitter techniques, established in conventional DLLs.

C. Reference Loop Design

The objectiveness of the reference loop is to provide quadrature clocks to the main loop. Since the main loop uses these multiphase clocks as references, the phase distribution in the output clocks should be preserved against a possible harmonic lock. The reference loop phase detector depicted in Fig. 3(a) has the capability to detect and escape up to the second harmonic lock. This design is made of two level-sensitive AND/NAND logic which requires 45° and 90° clocks as well as 0° and 180° clocks. At one period lock, clocks and UP/DN output waveforms are shown in Fig. 3(b). The phase detector asserts their UP and DN outputs for equal duration due to 45° clock in order to avoid a dead-zone problem, although the phase offset of the reference loop gives negligible effects on the offset of the main loop output clock. At the second harmonic lock as shown in Fig. 3(c), the phase detector detects that the loop is in the harmonic lock due to 90° clock and asserts only UP output to escape the harmonic lock. By limiting the delay range of a delay line, there is no possibility of harmonic lock over third since the reference loop is composed only of delay cells with no additional delay elements such as the clock buffer.

D. Main Loop Design

The main loop design is focused on the selection control and delay control of the main loop VCDL to achieve the infinite delay range by using four finite-length VCDLs. Fig. 4(a) shows the conceptual timing diagram of the main loop VCDL selection control. Assuming $i0°$ clock is selected as $iCLK$, the $iCLK$

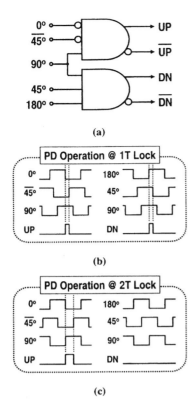

(a)

(b)

(c)

Fig. 3. Reference loop phase detector. (a) Block diagram . (b) Operation at 1 period lock. (c) Operation at 2 period lock.

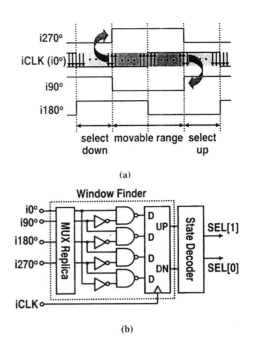

(a)

(b)

Fig. 4. Selection control of the main loop VCDL. (a) Conceptual timing diagram. (b) Block diagram of the control logic.

moves in the movable range according to the output of the main loop phase detector. Other clocks remain fixed at the initial phase relationship spaced by 90°. When the rising edge of the $iCLK$ coincides with that of $i90°$ (or $i270°$) clock, "select up" (or "select down") is generated and then $iCLK$ is changed to $i90°$ (or $i270°$) clock. Now $i90°$ (or $i270°$) clock acts as a new selected clock in a right-shifted (or left-shifted) movable range. Thus, clock switching at the quadrant boundaries can be repeated in this manner, to cover the entire phase range. Fig. 4(b) shows a block diagram of the selection control logic. Since the $iCLK$ passes through the MUX stage, a MUX replica is required for delay matching between the $iCLK$ and all internal clocks. Therefore, clock waveforms in Fig. 4(a) are validated. In the window finder, one inverter–one NAND pair makes the window which is bounded by rising edges of two input clocks. Thus, four windows are generated. Sampled values of these windows by the $iCLK$ enable the window finder to find which window the $iCLK$ belongs to. If the found window is the "select up" or "select down" region, UP or DN signal is generated, respectively. Then, the state decoder updates two-bit selection code to change the $iCLK$ in one clock cycle. Although clock switching occurs immediately after the switching event, there is the possibility of the small delay difference in the $iCLK$ since the rising edge of old $iCLK$ may have a different time position with that of new $iCLK$ after clock switching. This delay difference can be represented as a switching jitter at the lock state.

The delay control of the main loop VCDL should be optimized between two conflicting conditions, delay range and power consumption. More delay cells mean larger delay range but their power consumption is proportional to the number of

required delay cells. Furthermore, a larger delay causes a larger jitter. Intuitively, we apply a single control scheme as shown in Fig. 5(a), where only the $iCLK$ rotates and other clocks remain fixed in phase space. Thus, clock switching occurs at the quadrant boundaries. Unfortunately, since the required delay range is from $-90°$ to $+90°$, this control scheme consumes the same number of delay cells per VCDL as those in the reference loop. In order to reduce the number of required delay cells, a differential delay control scheme is employed. The differential control means that when the $iCLK$ rotates counterclockwise, all other clocks rotate clockwise with their phase relationship fixed. If all clocks move with same speed, the required delay range is from $-45°$ to $+45°$, as shown in Fig. 5(b). However, if the $iCLK$ must rotate in the opposite direction after switching due to the delay fluctuation of the reference clock or the clock buffer, there is the problem of losing the lock since the delay range of a VCDL was already exhausted. In Fig. 5(c), we adopt a differential delay control with $3\times$ speed difference, where the $iCLK$ moves three times faster than other clocks, so that 3/4 of delay cells in the single delay control case satisfy the required delay range, $-67.5°$ to $+67.5°$. Since $3\times$ speed difference provides a shared region in the available delay range of two neighboring clocks, seamless clock switching is possible in any direction without losing the lock with three delay cells per VCDL.

Fig. 6 shows the configuration of the main loop phase detector, charge pumps, and loop filters. Outputs of the phase detector are connected to the charge pump1 (CP1) directly and to the charge pump2 (CP2) with inversion. Thus, if the CP1 generates an increasing control voltage for a VCDL which generates the $iCLK$, the CP2 generates a decreasing control voltage for all other VCDLs. As a result, two substantially identical charge pumps are used for the differential delay control scheme. Three times speed difference is implemented by the fact that the CP1 has one loop filter and the CP2 has three loop filters. In

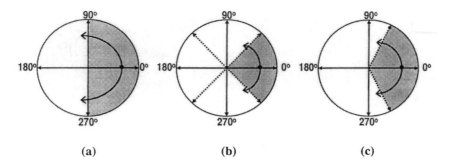

(a)   (b)   (c)

Fig. 5.   Delay control of the main loop VCDL. (a) Single control with other clocks fixed. (b) Differential control with same speed. (c) Differential control with 3× speed difference.

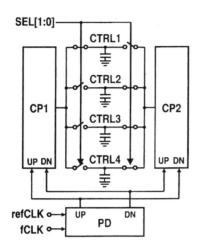

Fig. 6.   Configuration of the main loop phase detector, charge pumps, and loop filters.

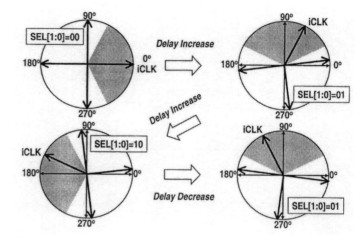

Fig. 7.   Example of the main loop VCDL control procedure.

case of clock switching, the selection code alters the connection between charge pumps and loop filters. Consequently, charge redistribution occurs between three loop filters except a loop filter for the new $iCLK$. This charge redistribution proceeds rapidly since two different voltages converge into one value. The fast VCDL control voltage change prevents possible dithering around the clock switching phase.

Fig. 7 shows one example of the main loop VCDL control procedure starting at the unlock state. Let us assume the $iCLK$ should be near $180°$ in phase space to acquire the lock. Initially, assuming the selection code is "00," $i0°$ clock is selected as the $iCLK$. The $iCLK$ rotates counterclockwise in phase space according to outputs of the phase detector. All clocks excluding the $iCLK$ rotates clockwise with one-third speed compared to that of the $iCLK$. Before the delay range of the VCDL generating the $iCLK$ is reached at a limit, the $iCLK$ is changed to $i90°$ clock. Thus, the selection code is "01." All clocks except the new $iCLK$ settle near their original phase positions with $90°$-phase space by the charge redistribution of loop filters. After clock switching, the $iCLK$ still moves counterclockwise to be switched to $i180°$ clock. Since this "10" state is near the lock state, the DLL can acquire the lock by a minor delay control. However, let us assume the delay time of the $iCLK$ must decrease due to the delay fluctuation of the reference clock or the clock buffer. Similarly in the delay increase case, before a VCDL delay range is exhausted, the $iCLK$ is returned

to $i90°$ clock, "01" state. In result, the proposed DLL covers the entire phase range and remains at the lock state in any direction switching by optimizing the control schemes of multiple VCDLs. Therefore, since this architecture makes it possible to emulate the infinite-length VCDL by using multiple finite-length VCDLs, the DLL overcomes the problem of conventional DLLs, described by the limited delay range and the initial phase relationship constraint.

III. LOW JITTER SCHEME AND DCC

A. Low-Jitter Scheme

The jitter performance of the DLL is degraded by various noise sources, typically in the form of supply and substrate noise in high speed and highly integrated circuits. To reduce the jitter, the loop bandwidth should be set as high as possible but must have an upper limit for stability issues. Thus, low-jitter DLL designs strongly depend on the delay characteristics of a delay line with supply-noise injection. In order to design the delay line with low supply-noise sensitivity, the replica biasing for the delay control must be considered in noisy environment. The replica biasing circuit, which consists of a half-replica of a differential delay cell and an operational amplifier (op-amp), sets the low swing level of the delay cell to the reference voltage, V_{ref}. In the conventional replica biasing, the V_{ref} tracks the supply variation with the same amount. Unfortunately, this is not the optimal solution. The variation of the op-amp gain and

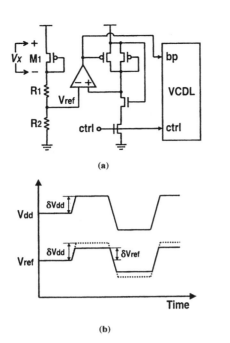

(a)

(b)

Fig. 8. (a) Circuit diagram of a replica biasing. (b) Operation of a reference voltage generator under supply-noise injection.

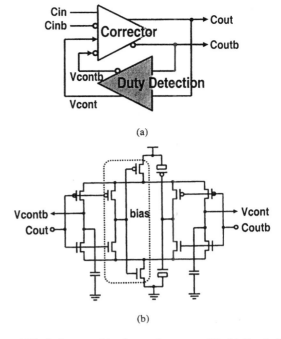

(a)

(b)

Fig. 9. (a) Block diagram of the duty cycle corrector [3]. (b) Circuit diagram of the proposed duty detection stage.

the tail-current source distorts the delay characteristics of the delay cell. The delay equation of this case is described by

$$T_{\text{delay}} = \frac{C_L \cdot (V_{\text{swing}} + \delta V_{\text{swing}})}{I_{\text{tail}} + \delta I_{\text{tail}}} \quad (1)$$

where

T_{delay} delay time of a delay cell;
C_L load capacitance;
V_{swing} swing voltage of the delay cell;
I_{tail} current of the tail-current source.

For a positive supply variation of δV_{dd}, since δI_{tail} is positive and δV_{swing} negative, T_{delay} greatly decreases.

In the design depicted in Fig. 8(a), an additional reference voltage generator is attached to the replica biasing circuit. The reference voltage generator is composed of one transistor and two resistors and generates the reference voltage, V_{ref}, in the nominal supply condition. When there is a supply variation of δV_{dd}, the reference voltage generator produces a predetermined variation of δV_{ref}, which is a reduced swing compared to δV_{dd}, as shown in Fig. 8(b). The reduced swing compensates the delay variation due to the aforementioned variations induced by supply noise. Thus, supply-noise sensitivity can be minimized. For a given δV_{dd} supply noise, the desired δV_{ref} is a function of δV_X across M_1 transistor as follows:

$$(\delta V_{dd} - \delta V_X) \cdot \frac{R_2}{R_1 + R_2} = \delta V_{\text{ref}}. \quad (2)$$

The sensitivity of δV_{ref} over δV_{dd} to process variations should be analyzed to guarantee a reliable operation. For example, the sensitivity to the threshold voltage variation ΔV_T of the transistor M_1 can be obtained by (3), shown at the bottom of the page. In (3), β_p means $\mu C_{\text{ox}}(W/L)$ of transistor M_1. The sensitivity value is in the order of 10^{-3} with a ΔV_T of 100 mV. Similar analyses with other process parameters also show that the predetermined δV_{ref} is kept nearly constant under moderate process variations. This replica biasing circuit is commonly applied to all VCDLs of the reference loop and the main loop to achieve the low-jitter characteristics through the whole DLL.

B. Duty Cycle Corrector

The duty cycle of clock signals within the DLL deviates from its ideal value of 50% due to various asymmetries in signal paths and voltage offsets in an off-chip generated reference clock. For applications in which the timing of both edges of the clock is critical, a duty cycle corrector (DCC) is required to maximize timing margins. A DCC [3] in Fig. 9(a) is configured as the error-voltage feedback with a corrector stage and a duty detection stage. The duty detection stage outputs the differential control voltage (V_{cont}, V_{contb}), which is proportional to the duty cycle error of inputted clocks (C_{out}, C_{outb}). This differential control voltage then effectively introduces offset voltage to clock inputs (C_{in}, C_{inb}) at the corrector stage to correct the duty cycle of output clocks.

$$S_{\Delta V_T}^{\delta V_{\text{ref}}/\delta V_{dd}} = -\frac{1}{\{1 + \beta_p (R_1 + R_2)(V_{dd} - V_X - V_T - \Delta V_T)\}^2 \cdot (V_{dd} - V_X - V_T - \Delta V_T)}. \quad (3)$$

453

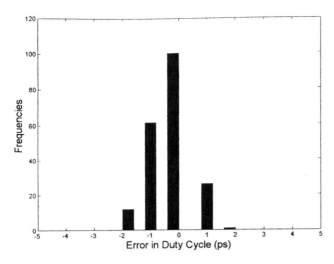

Fig. 10. Simulated mismatch sensitivity characteristics of the DCC with the proposed duty detection stage.

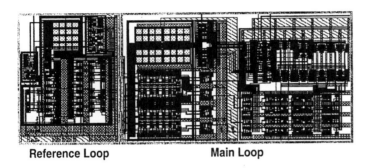

Reference Loop   **Main Loop**

Fig. 11. Prototype chip layout.

As the clock frequency is increased, tighter bound is placed on the performance of the DCC. Even worse, process mismatches between transistors work as a serious error factor in the DCC especially under deep-submicron technology. Although process mismatches plague all devices, special care must be paid to the duty detection stage since near-ideal performance of this stage can remove the duty cycle distortion caused by the mismatches of all other nonideal blocks. The proposed duty detection block is based on two stacked source-coupled pairs configuration, as shown in Fig. 9(b). The source-coupled pair is immune to device mismatches due to its current steering capability, i.e., since for fairly large input signals, the source-coupled pair conducts the current set by the tail-current source through only one branch, various mismatch effects in transistors can be hidden. The common-mode problem of this approach is solved by the transistors in boxed area, comprising the self-biasing technique [7], which enables the output common mode to be dynamically adjusted by input clocks. Two transistors with source and drain tied are added to eliminate the load imbalance caused by the self-biasing circuit. Fig. 10 shows the simulated mismatch sensitivity characteristics of the DCC with the proposed duty detection stage over typical process mismatch parameters, $A_{VT} = \sigma_{\Delta VT} \cdot \sqrt{WL} = 8$ mV $\cdot \mu$m and $A_\beta = \sigma_{\Delta\beta} \cdot \sqrt{WL} = 3\% \cdot \mu$m [8]. Under 50% duty cycle of the input clock, the duty cycle error is less than ± 2 ps, which guarantees a robust operation against process mismatches.

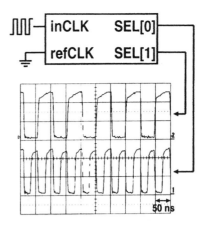

Fig. 12. Selection code waveforms with the $refCLK$ input grounded.

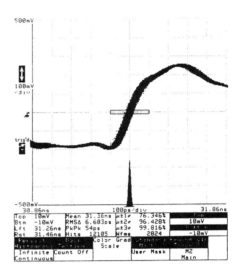

(a)

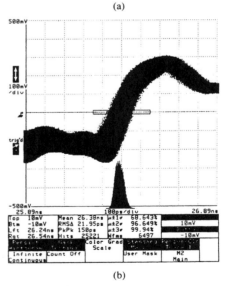

(b)

Fig. 13. Jitter histograms at 400 MHz. (a) Quiet supply. (b) Added 2.5-MHz 300-mV square-wave supply noise.

IV. EXPERIMENTAL RESULTS

The test chip has been fabricated using a 0.25-μm five-metal CMOS process. The threshold voltages in this process are

454

TABLE I
PERFORMANCE CHARACTERISTICS OF THE PROTOTYPE DLL

Process	0.25 μm 5-metal CMOS process
Active Area	0.13 mm²
Supply Voltage	2.5 V
Operating Range	150-600 MHz
Static Phase Error	< 20 ps
Power Dissipation	60 mW @ 400 MHz
Jitter	54 ps pk-to-pk with quiet supply
	150 ps pk-to-pk with 300-mV,
	2.5-MHz supply noise
Supply Sensitivity	0.32 ps/mV @ 400 MHz

achieves 54-ps peak-to-peak jitter and 0.32-ps/mV jitter supply-noise sensitivity.

REFERENCES

[1] M. Johnson and E. Hudson, "A variable delay line PLL for CPU-co-processor synchronization," *IEEE J. Solid-State Circuits*, vol. 23, pp. 1218–1223, Oct. 1988.
[2] T. Yoshimura, Y. Nakase, N. Watanabe, Y. Morooka, Y. Matsuda, M. Kumanoya, and H. Hamano, "A delay-locked loop and 90° phase shifter for 800-Mb/s double data rate memories," in *Symp. VLSI Circuits Dig. Tech. Papers*, June 1998, pp. 66–67.
[3] T. H. Lee, K. S. Donnelly, J. T. C. Ho, J. Zerbe, M. G. Johnson, and T. Ishikawa, "A 2.5-V CMOS delay-locked loop for an 18-Mb 500-Mbyte/s DRAM," *IEEE J. Solid-State Circuits*, vol. 29, pp. 1491–1496, Dec. 1994.
[4] S. Sidiropoulos and M. A. Horowitz, "A semidigital dual delay-locked loop," *IEEE J. Solid-State Circuits*, vol. 32, pp. 1683–1692, Nov. 1997.
[5] K. Minami *et al.*, "A 1-GHz portable digital delay-locked loop with infinite phase capture ranges," in *ISSCC Dig. Tech. Papers*, Feb. 2000, pp. 350–351.
[6] Y.-J. Jung, S.-W. Lee, D. Shim, W. Kim, C.-H. Kim, and S.-I. Cho, "A low-jitter dual-loop DLL using multiple VCDLs with a duty cycle corrector," in *Symp. VLSI Circuits Dig. Tech. Papers*, June 2000, pp. 50–51.
[7] M. Bazes, "Two novel fully complementary self-biased CMOS differential amplifiers," *IEEE J. Solid-State Circuits*, vol. 26, pp. 165–168, Feb. 1991.
[8] M. J. M. Pelgrom, H. P. Tuinhout, and M. Vertregt, "Transistor matching in analog CMOS applications," in *IEDM Dig. Tech. Papers*, Dec. 1998, pp. 915–918.

0.57 V (nMOS) and −0.55 V (pMOS). The gate-oxide thickness is 5.8 nm. Fig. 11 shows the layout of the prototype chip. The active area of the DLL occupies 0.13 mm².

Waveforms depicted in Fig. 12 shows two-bit selection code with the reference clock input grounded, while running the input clock at its nominal frequency of 400 MHz. In this configuration, the main loop phase detector always asserts DN signals. Therefore, the selection code is continuously updated in accordance with sequences of "00," "01," "10," and "11." This means the infinite times rotation of the output clock throughout the full 0°–360° range.

Fig. 13(a) and (b) shows the jitter histograms of the DLL clock output at 400 MHz. Fig. 13(a) shows 6.7 ps RMS and 54 ps peak-to-peak jitter characteristics with a quiet power supply. With a 300-mV 2.5-MHz square-wave supply noise, the peak-to-peak jitter increases to 150 ps, as shown in Fig. 13(b). The ratio of the peak-to-peak jitter to the RMS jitter is well maintained in spite of supply-noise injection. Supply-noise sensitivity is measured to be 0.32 ps/mV.

Table I summarizes the DLL performance characteristics. The DLL operates from 150- to 600- MHz frequency range with a 2.5-V supply. Static phase error between the reference clock and the output clock of the DLL is less than 20 ps. Operating at 400 MHz, the DLL dissipates 60 mW.

V. CONCLUSION

We have described a dual-loop DLL architecture that allows the unlimited delay range by using multiple VCDLs. The reference loop generates four evenly spaced clocks without a possible harmonic lock. Clock selection in the main loop enables the DLL to cover the entire phase range and seamless clock switching is achieved by optimizing the main loop VCDL delay range control. Thus, this architecture can emulate the infinite-length VCDL with multiple finite-length VCDLs. To obtain low supply-noise sensitivity, the low-jitter scheme generates a reduced swing voltage compared to supply noise for the delay compensation of a delay line. Finally, a duty cycle corrector presents a high immunity to process mismatches with the help of two stacked source-coupled pairs configuration. A prototype fabricated using 0.25-μm CMOS technology

Yeon-Jae Jung was born in Korea in 1974. He received the B.S. and M.S. degrees from the School of Electrical Engineering, Seoul National University, Seoul, Korea, in 1997 and 1999, respectively, where he is currently working toward the Ph.D. degree.

He has worked on architectures and CMOS circuits for high-speed I/O interfaces. His current research interests include high-speed CMOS circuits and communication ICs.

Seung-Wook Lee was born in Seoul, Korea, in 1971. He received the B.S. and M.S. degrees in electronics engineering from Seoul National University, Seoul, Korea, in 1995 and 1997, respectively, where he is currently working toward the Ph.D. degree in the School of Electrical Engineering.

His research interests include CMOS RF circuit design and high-speed communication interfaces.

Mr. Lee is the winner of the Bronze Prize of the IC design contest held by the Federation of Korean Industries in 1995.

Daeyun Shim was born in Seoul, Korea, in 1962. He received the B.S., M.S., and Ph.D. degrees in electronics engineering from Seoul National University, Seoul, Korea, in 1985, 1987, and 2000, respectively. His Ph.D. dissertation was related to the design of high-speed locking clock generators.

Since 1987, he has been working on digital video signal processing and ASIC design at Samsung Electronics Corporation. His research interests are video signal processing and compression, high-speed digital circuit design, and high-speed locking systems. He is currently working on DVD-PRML system design.

An All-Analog Multiphase Delay-Locked Loop Using a Replica Delay Line for Wide-Range Operation and Low-Jitter Performance

Yongsam Moon, *Student Member, IEEE*, Jongsang Choi, Kyeongho Lee, *Member, IEEE*,
Deog-Kyoon Jeong, *Member, IEEE*, and Min-Kyu Kim

Abstract—This paper describes an all-analog multiphase delay-locked loop (DLL) architecture that achieves both wide-range operation and low-jitter performance. A replica delay line is attached to a conventional DLL to fully utilize the frequency range of the voltage-controlled delay line. The proposed DLL keeps the same benefits of conventional DLL's such as good jitter performance and multiphase clock generation. The DLL incorporates dynamic phase detectors and triply controlled delay cells with cell-level duty-cycle correction capability to generate equally spaced eight-phase clocks. The chip has been fabricated using a 0.35-μm CMOS process. The peak-to-peak jitter is less than 30 ps over the operating frequency range of 62.5–250 MHz. At 250 MHz, its jitter supply sensitivity is 0.11 ps/mV. It occupies smaller area (0.2 mm²) and dissipates less power (42 mW) than other wide-range DLL's [2]–[7].

Index Terms—Delay-locked loop, duty-cycle correction, dynamic phase detector, multiphase clock generation, replica delay line, triply controlled delay cell.

I. INTRODUCTION

AS THE SPEED performance of VLSI systems increases rapidly, more emphasis is placed on suppressing skew and jitter in the clocks. Phase-locked loops (PLL's) and delay-locked loops (DLL's) have been typically employed in microprocessors, memory interfaces, and communication IC's for the generation of on-chip clocks. However, it becomes increasingly difficult to reduce the clock skew and jitter, whether they are inherent or result from substrate and supply noise, as the clock speed and circuit integration levels are increased.

While the phase error of PLL's is accumulated and persists for a long time in a noisy environment, that of DLL's is not accumulated, and thus, the clock generated from DLL's has lower jitter. Therefore, DLL's offer a good alternative to PLL's in cases where the reference clock comes from a low-jitter source, although their usage is excluded in applications where frequency tracking is required, such as frequency synthesis and clock recovery from an input signal. However, the main problem of conventional DLL's [1] is that they are very difficult to design to

work over process, voltage, and temperature (PVT) variations. Since DLL's adjust only phase, not frequency, the operating frequency range is severely limited. We propose a new DLL architecture that operates in a wide frequency range while keeping the low-jitter performance.

Various wide-range DLL architectures [2]–[7], with similar motivations, have been developed, which can be classified into three categories: analog type [2], digital type [3], [4], and dual-loop type [5]–[7]. While a conventional analog DLL [1] uses a voltage-controlled delay line (VCDL), the wide-range analog DLL [2] uses phase mixers for wide-range operation. However, because of its relatively high analog complexity, the analog DLL requires a process-specific implementation, making it relatively difficult to port across multiple processes [4]. Thus, digital DLL's [3], [4] have been proposed for better process portability. However, skew error and jitter are increased due to continuous change of phase selections among quantized delay times with supply and temperature variations. To overcome these problems, dual-loop architectures have been proposed [5]–[7]. In [5], a PLL is added to make the core DLL lock to a reference frequency, and a phase mixer interpolates two intermediate clocks in the core DLL and produces the final output clock. Or, almost continuous phase is obtained with addition of a fine delay line [6] or a phase interpolator [7] to a digital DLL. However, additional chip area and power consumption of these wide-range DLL's are excessive, and furthermore, their jitter performance gets worse compared with conventional DLL's since the number of delay cells or gates in the clock propagation paths becomes larger.

We propose a new DLL architecture that achieves a large operating range by attaching a replica delay line in parallel with a conventional analog DLL. Since the replica delay line occupies one-fourth the area of the core DLL, it incurs only a small increase in chip area and power consumption. Since the replica delay line is out of the clock propagation path, it does not do any harm on low-jitter performance. While other wide-range DLL's [2]–[7] use phase mixers or phase selections to generate a single output, the proposed DLL uses a similar multistage analog VCDL to what conventional analog DLL's use. Therefore, the proposed DLL can generate multiphase clocks without using excessive amount of hardware. Furthermore, by incorporating a dynamic phase detection circuit and cell-level duty cycle correction method, the multiphase clocks are equally spaced even in high-frequency operations. A prototype DLL designed for eight-phase clocks can

Manuscript received July 20, 1999; revised October 6, 1999.

Y. Moon, J. Choi, and D.-K. Jeong are with the School of Electrical Engineering, Seoul National University, Seoul 151-742 Korea (e-mail: ysmoon@griffin.snu.ac.kr).

K. Lee was with the School of Electrical Engineering, Seoul National University, Seoul 151-742 Korea. He is now with Global Communication Technology Inc., Los Altos, CA 94024 USA.

M.-K. Kim is with Silicon Image, Inc., Sunnyvale, CA 94086 USA.

Publisher Item Identifier S 0018-9200(00)00538-2.

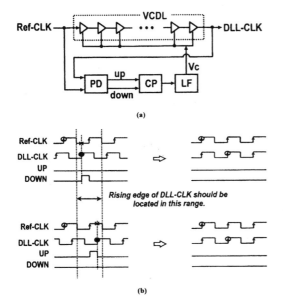

(a)

(b)

Fig. 1. Block diagram of (a) a conventional DLL and (b) a DLL locking operation and operating frequency range limitation.

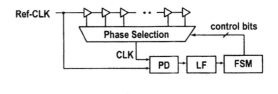

(a)

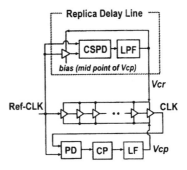

(b)

Fig. 2. Block diagrams of (a) a digital DLL and (b) a dual-loop DLL.

Fig. 3. Block diagram of the proposed analog DLL.

be used in applications such as gigabit serial interfaces [8], [9].

This paper is arranged as follows. Section II describes a conventional analog DLL and includes an analysis of its operational frequency range. This section also overviews other wide-range DLL architectures. In Section III, the proposed architecture is presented with design ideas, issues, and various analyses. Section IV describes various circuits used in the design. Section V discusses the prototype chip implementation and shows experimental results. Section VI concludes this paper with a summary.

II. CONVENTIONAL ARCHITECTURES

A. Range Problem of Conventional DLL's

A simplified block diagram of a conventional DLL [1] is outlined with its operation mechanism in Fig. 1. When the delay time (T_{VCDL}) of the VCDL is initially smaller (or larger) than the period (T_{CLK}) of the reference clock (*Ref-CLK*), the DLL adjusts T_{VCDL} until phase difference disappears in a negative feedback loop, as shown in Fig. 1(b). The phase difference is detected by sampling the reference clock with the rising edge of the output clock (*DLL-CLK*). Depending on the sampled value, a *DOWN* or *UP* pulse is generated. These pulses discharge (or charge) a capacitor in the loop filter, thereby decreasing (or increasing) the control voltage V_C and reducing the phase difference gradually.

However, if the sampling edge of *DLL-CLK* deviates from the lock range indicated in Fig. 1(b), the DLL falls prey to a stuck or a harmonic lock problem. In order to avoid this problem, the minimum ($T_{\text{VCDL·min}}$) of T_{VCDL} should be located between $0.5 \times T_{\text{CLK}}$ and T_{CLK}, and the maximum ($T_{\text{VCDL·min}}$) between T_{CLK} and $1.5 \times T_{\text{CLK}}$. These stuck-free conditions can be expressed as the following inequality:

$$0.5 \times T_{\text{CLK}} < T_{\text{VCDL·min}} < T_{\text{CLK}}$$
$$T_{\text{CLK}} < T_{\text{VCDL·max}} < 1.5 \times T_{\text{CLK}} \quad (1)$$

or, equivalently, in terms of T_{CLK}

$$\text{Max}(T_{\text{VCDL·min}}, 2/3 \times T_{\text{VCDL·max}}) < T_{\text{CLK}}$$
$$< \text{Min}(2 \times T_{\text{VCDL·min}}, T_{\text{VCDL·max}}). \quad (2)$$

The range of stuck-free clock period is determined by inequality (2). If the target clock period satisfies inequality (2), the DLL works without the stuck problem. However, it should be noted that inequality (2) has the maximum range $2/3 \times T_{\text{VCDL·max}} < T_{\text{CLK}} < T_{\text{VCDL·max}}$, when $T_{\text{VCDL·max}} = 2 \times T_{\text{VCDL·min}}$. In addition, if $T_{\text{VCDL·max}} \geq 3 \times T_{\text{VCDL·min}}$, there is no range of T_{CLK} that satisfies inequality (2), and the DLL is prone to the stuck problem. Since the PVT variations of T_{VCDL} can be as much as 2:1 in a typical CMOS process, the stuck-free condition can be satisfied over only a very narrow range of T_{CLK}, and thus a time-consuming and tedious circuit trimming job is required when process migrations are performed across different processes.

B. Digital DLL's and Dual-Loop DLL's

Digital DLL's [3], [4] have been developed to overcome the narrow frequency range problem of conventional analog DLL's. A simplified block diagram of a typical digital DLL [3] is outlined in Fig. 2(a). Multistage delay cells in the VCDL provide

457

fixed and quantized delay times. The finite-state machine selects one clock output with closest phase to the reference clock's by using digital control bits instead of using an analog control voltage.

Therefore, major drawbacks in the digital DLL's are large skew due to quantized delay time and large jitter due to control-bit updates during operation. To increase the resolution and cover a wide delay range, a large delay cell array must be used, and that inevitably increases chip area and power consumption. In order to cope with these problems, Garlepp et al. [4] proposed a phase blending technique in a hierarchical structure for improved phase resolution. However, the inherent problems of digital DLL's are not solved entirely.

Dual-loop DLL's [6], [7] have been proposed to minimize the problems of digital DLL's. A simplified block diagram of architecture proposed in [6] is shown in Fig. 2(b). A fine delay line, which is analog controlled, is attached to a digital DLL in the subsequent stage. In [7], a phase interpolator is cascaded to a digital DLL for unlimited phase capture range. These dual-loop architectures achieve both a wide frequency range and relatively low jitter performance. However, due to digital DLL's inherent nature, jitter histogram of the generated clock shows the superposition of two Gaussian distributions [7] resulting from the control-bit updates. In addition, the overhead of chip area and power consumption is significant.

III. PROPOSED ARCHITECTURE

A. All-Analog DLL Using a Replica Delay Line

Fig. 3 shows a high-level block diagram of the proposed architecture. The delay time ($T_{\rm VCDL}$) of the main analog DLL (core DLL) is primarily controlled by a control voltage Vcr, which is generated from a replica delay line. Another control voltage Vcp fine-tunes $T_{\rm VCDL}$. The replica delay line consists of only one replica delay cell, a current steering phase detector (CSPD), and a low-pass filter (LPF). The replica delay cell is identical to the delay cells in the core DLL. Due to sharing of Vcr, the delay time ($T_{\rm RDC}$) of the replica delay cell is almost equal to the delay time ($T_{\rm DC}$) of each delay cell in the core DLL. They are not exactly the same unless Vcp equals bias. Due to the characteristics of the proposed CSPD [10], $T_{\rm RDC}$ is forced to be one-eighth of $T_{\rm CLK}$. Therefore, $T_{\rm VCDL}$ of the core DLL becomes equal to $T_{\rm CLK}$ when the number of delay cells in the core DLL is eight. With the replica delay line with a wide frequency range, the core DLL's operating frequency bounds will be established, and thus the core DLL will not fall into such a harmonic lock problem as conventional analog DLL's do. With only a negligible increase in chip area and power consumption, the proposed architecture offers many advantages compared with other wide-range DLL's. Since the DLL is analog controlled and the clock path is not extended, the DLL can keep the low-jitter performance of the conventional DLL. In addition, because it uses a multistage analog VCDL, the proposed DLL can generate multiphase clocks.

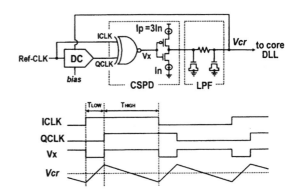

Fig. 4. Configuration and operation of a replica delay line.

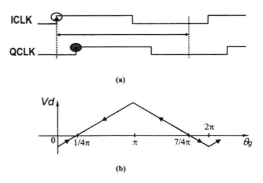

Fig. 5. (a) Delay capture range of the replica delay line and (b) gain curve of CSPD.

B. Delay Capture Range of the Replica Delay Line

Fig. 4 shows the circuit diagram and operation waveforms of the replica delay line. The replica delay line generates a control voltage Vcr to pass to the core DLL. Vcr is used as a reference voltage in the core DLL to lock to the input frequency. The CSPD takes two inputs $ICLK$ and $QCLK$. $ICLK$ is directly connected to Ref-CLK, and $QCLK$ is delayed from Ref-CLK by one delay cell. $T_{\rm RDC}$ is equal to the delay difference between $ICLK$'s and $QCLK$'s rising edges. In the charge pump, the pullup current I_P is tuned to three times the pulldown current I_N ($I_P : I_N = 3 : 1$). When Vx is high, the charge on the filter capacitors will be decreased by $I_N \times T_{\rm high}$, and Vcr will go down. On the other hand, when Vx is low, the charge will be increased by $I_P \times T_{\rm low}$, and Vcr will go up three times faster. When the feedback loop is locked, a stable value of Vcr will be obtained with the relation of $I_P \times T_{\rm low} = I_N \times T_{\rm high}$. Therefore, in the locked state, XNOR output Vx has the low-to-high duration ratio of 1:3 ($T_{\rm low} : T_{\rm high} = 1 : 3$), and the rising edge of $ICLK$ leads that of $QCLK$ by one-eighth of $T_{\rm CLK}$.

Fig. 5(a) shows the capture range of the replica delay line when $I_P : I_N = 3 : 1$. This can be derived from the gain curve of the CSPD shown in Fig. 5(b). If $T_{\rm RDC}$ is smaller than $1/8 \times T_{\rm CLK}$, the change of Vcr, denoted as Vd, will become negative and $T_{\rm RDC}$ will increase. That action is indicated in the gain curve as the corresponding arrow pointing to the right. If $T_{\rm RDC}$ is between $1/8 \times T_{\rm CLK}$ and $7/8 \times T_{\rm CLK}$, Vd will be positive and the corresponding arrow points to the left. Eventually, $T_{\rm RDC}$ will be settled at $1/4 \cdot \pi$, which represents $1/8 \times T_{\rm CLK}$. However, if $T_{\rm RDC}$ is larger than $7/8 \times T_{\rm CLK}$, the settling point will run away and a harmonic lock problem will occur.

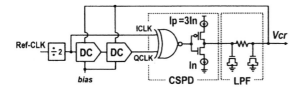

Fig. 6. Replica delay line for high-frequency operation.

The operating conditions explained above, in which the delay line locks correctly, can be summarized in the inequalities as follows:

$$T_{RDC\cdot min} < 1/8 \times T_{CLK}$$
$$1/8 \times T_{CLK} < T_{RDC\cdot max} < 7/8 \times T_{CLK} \qquad (3)$$

or equivalently in terms of T_{CLK}

$$Max\{8 \times T_{RDC\cdot min}, 8/7 \times T_{RDC\cdot max}\} < T_{CLK} < 8 \times T_{RDC\cdot max}. \qquad (4)$$

If the delay range of the controlled delay cell satisfies the relation $T_{RDC\cdot max} < 7 \times T_{RDC\cdot min}$, the DLL will have a frequency range determined by the entire delay range of the delay cell. However, even if we make the delay range wider and satisfy $T_{RDC\cdot max} > 7 \times T_{RDC\cdot min}$ in an effort to increase the frequency range, the lock range is limited to only 7:1.

In some applications where the frequency range must be larger than 7:1, changing the pump-current ratio of the CSPD can make the frequency range wider. For example, with $I_P : I_N = 4 : 1$, the frequency range of 9:1 can be obtained. With $I_P : I_N = 5 : 1$, the frequency range of 11:1 can be obtained.

In high-frequency operations, T_{CLK}, especially T_{low}, may be too short to drive the XNOR gate. So, a divide-by-two circuit and a pair of delay cells are used to slow down the frequency of *Ref-CLK* [11]. The new configuration shown in Fig. 6 is effectively the same as the one in Fig. 4 but offers a more robust operation in the high-frequency operations.

C. Core DLL

Fig. 7 shows a simplified block diagram of the core DLL. It consists of a VCDL, a dynamic phase detector, a charge pump, and a loop filter. The core DLL generates eight-phase clock outputs through eight delay cells (DC's) in the VCDL. The core DLL is the same as a conventional analog DLL except that it has another control voltage Vcr. Vcr from the replica delay line coarsely determines the delay time (T_{VCDL}) of the VCDL so that T_{VCDL} is equal to T_{CLK} in the locked state. In the locked state, the eighth clock output, *CLK7* in Fig. 7, is aligned with *Ref-CLK*.

In high-frequency operations, there may be some static phase mismatch between *CLK7* and *Ref-CLK* due to the long rise/fall times of signal transition edges compared with the period of the clock. So the fine-tuning is required. The dynamic phase detector (PD) in the core DLL generates control signal Vcp, fine-tunes T_{VCDL}, and removes residual phase mismatch so that the rising edge of *Ref-CLK* is exactly aligned with that of *CLK7*.

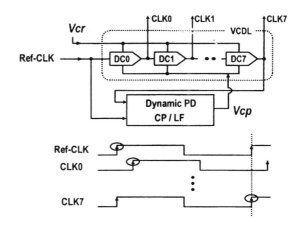

Fig. 7. Block diagram of the core DLL.

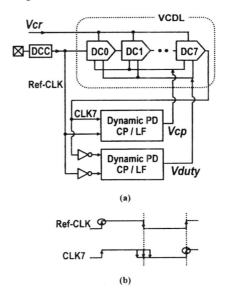

Fig. 8. (a) Core DLL with cell-level duty-cycle correction and (b) rising and falling edge alignment.

D. Cell-Level Duty-Cycle Correction

Fig. 8 shows the core DLL with a cell-level duty-cycle correction mechanism. In high-frequency operations, clock outputs with a short cycle time can be severely distorted as the clock passes through many delay cells. Even if the duty cycle of *Ref-CLK* is 50% at the entrance, that of *CLK7* may deviate significantly from 50%. It causes multiphase clock outputs to have phase error, which could be fatal, especially in high-speed communication applications. A conventional solution is to attach duty-cycle correction circuits to all clock output drivers with the price of added area, increased jitter, and further phase mismatch due to elongated path. So a cell-level duty cycle correction is proposed.

The second phase detector shown in Fig. 8 takes inverted *Ref-CLK* and inverted *CLK7* as its inputs, generating a control signal *Vduty* as the output. It fine-tunes the cell current ratio, and thus aligns the falling edges of *Ref-CLK* and *CLK7*. In the steady state, therefore, both rising and falling edges of *CLK7* and *Ref-CLK* are synchronized in phase, and both clocks have the same duty cycle. It should be noted that the duty-cycle correction circuit (DCC) used right at the input of *Ref-CLK* corrects

459

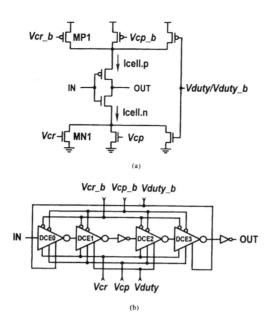

(a)

(b)

Fig. 9. Triply controlled DC. (a) Circuit diagram of a DCE and (b) configuration of a triply controlled DC.

the duty cycle of *Ref-CLK* only. With cell-level duty cycle correction, not only *CLK7* but also the other intermediate clock outputs maintain a 50% duty cycle without any additional circuits.

Although two control voltages *Vcp* and *Vduty* are simultaneously adjusted in the coupled negative feedback loops, the stability is guaranteed by making one of its loops have a sufficiently low bandwidth.

IV. CIRCUIT DESIGN

A. Triply Controlled Delay Cell

According to the noise analyses of [12] and [13], a fast-slewing (short rise/fall time) delay cell with a fully switching capability offers less phase noise. Although offering a full swing output, a shunt-capacitor delay cell [14], with its capacitor, would increase the chip area and power. Therefore, we decided to use the current-starved inverter [15] as a basic controlled delay cell. Since the current-starved inverter does not require a level conversion circuit, which is required for a differential delay cell, it has less chip area and power, although substrate and supply noise might cause detrimental influence.

A triply controlled delay cell is used as the basic delay cell element (DCE). The circuit diagram of the DCE and the configuration of one unit of DC are shown in Fig. 9. Four DCE's and two inverters compose a DC and make its rising/falling delay times symmetric. The delay time (T_{DC}) of the triply controlled delay cell is determined by six control signals: *Vcr*, *Vcr_b*, *Vcp*, *Vcp_b*, *Vduty*, and *Vduty_b*. Of those signals, *Vcr* and *Vcr_b* come from the replica delay line. In the DCE, the sizes of MP1 and MN1 are made larger than the others' so that *Vcr* and *Vcr_b* can control $I_{cell \cdot p}$ and $I_{cell \cdot n}$ primarily. The other control signals, which are generated by the core DLL, make only small adjustments to $I_{cell \cdot p}$ and $I_{cell \cdot n}$ for the fine-tuning of T_{DC}. *Vcp* and *Vcp_b* are used to align the rising edges of *Ref-CLK* and *CLK7*. *Vduty* and *Vduty_b* are responsible for maintaining the correct duty cycle and, thus, aligning the falling edges.

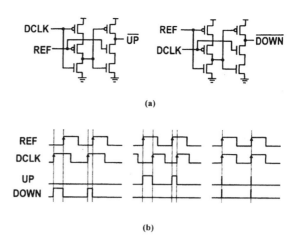

(a)

(b)

Fig. 10. (a) Dynamic phase detector and (b) its operations.

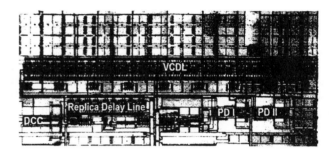

Fig. 11. Prototype chip microphotograph.

Since the high and low levels alternate in an inverter chain, duty-cycle control signals must alternate between *Vduty* and *Vduty_b* as well. Therefore, *Vduty_b* controls DCE0 and DCE3 and *Vduty* controls DCE1 and DCE2, as shown in Fig. 9(b). In the delay circuit, either *Vduty* or *Vduty_b* changes the duty cycle of the clock outputs by adjusting the current ratio of $I_{cell \cdot p}$ to $I_{cell \cdot n}$. With this mechanism, the multiphase clock outputs, $CLK0 \cdots CLK7$, will be duty-cycle corrected and equally spaced. There is no need to attach a DCC circuit in each clock output.

B. Dynamic Phase Detector

Since the tuning precision of the core DLL depends on the characteristics of the phase detector, we propose a new high-precision dynamic phase detector. Fig. 10(a) shows the circuit diagram of the proposed dynamic phase detector, which is improved from the published phase-frequency detector [8] by removing a feedback path and replacing the feedback input with an *REF* and *DCLK* signal. The phase detector can operate with less phase offset at high frequencies due to symmetry of circuit, shallow logic depth of only two gates, and fast operation with a dynamic logic circuit. While the widths of *UP* and *DOWN* pulses are proportional to the phase difference of the inputs as shown in Fig. 10(b), there remains a chain of short pulses in the locked state. These pulses in the locked state serve to reduce the dead zone of the phase detector [8]. However, the accuracy of the phase detector is improved when the pulse duration is shorter. Furthermore, smaller capacitor in the loop filter can be used since the amount of pumped charge is smaller compared

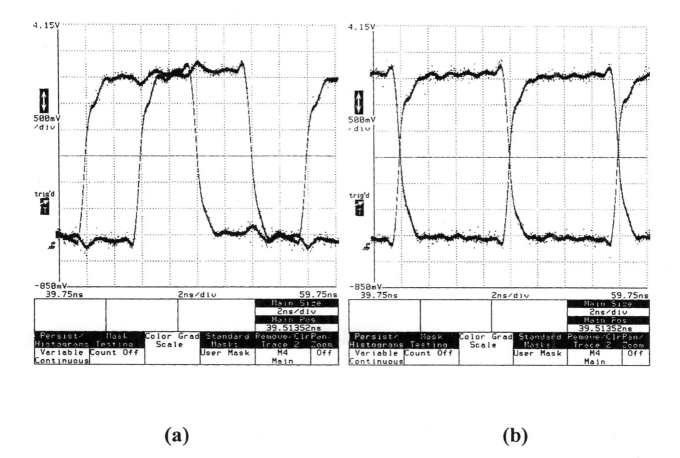

(a)   **(b)**

Fig. 12.   Clock waveforms at 62.5 MHz. (a) *CLK0, CLK2* and (b) *CLK0, CLK4.*

with a conventional "bang-bang" type of phase detector or a proportional phase detector with wider pulse width.

V. EXPERIMENTAL RESULTS

The test chip has been fabricated using a 0.35-μm, N-well, triple-metal CMOS process. The threshold voltages in this process are 0.42 V (NMOS) and −0.22 V (PMOS). The gate-oxide thickness is 75 nm. Fig. 11 shows a microphotograph of the fabricated chip. The chip integrates the DLL with an on-chip decoupling capacitance of 270 pF. The active area of the DLL occupies 0.08 mm² and the decoupling capacitor 0.12 mm². Since the pulse currents of the multiphase clock outputs are interspersed, the ac component of the supply current is present at the eighth harmonic frequencies of the clock. Therefore, the 270-pF on-chip capacitor is adequate to reduce the on-chip supply noise induced by switching of digital circuits.

The prototype chip operates from 62.5 to 250 MHz with a 3.3-V power supply. Fig. 12(a) shows the waveforms of *CLK0* and *CLK2* at 62.5 MHz. These clock outputs are the first and the third clocks, respectively, and have a 90° phase difference. Fig. 12(b) shows the waveforms of *CLK0* and *CLK4*, which are an inversion of each other with a 180° phase difference. Fig. 13(a) and (b) shows the same waveforms at 250 MHz. In spite of some ringing due to capacitance and inductance of the board and measurement instrument, the measurement results

show that the clock outputs are aligned with precise phase relationships of less than 1% error over an operating frequency range from 62.5 to 250 MHz. The delay range of the VCDL is estimated to be between 4 and 16 ns. With minor change of device sizes of the VCDL, the operating frequency range could be extended toward a higher frequency range.

Fig. 14(a) and (b) shows the jitter histograms in the clock output *CLK7*. The frequency of *Ref-CLK* is 250 MHz. Fig. 14(a) shows 4-ps rms and 29-ps peak-to-peak jitter characteristics in a quiet power supply, where only the DLL is activated in the chip. When other digital circuits are turned on, rms and peak-to-peak jitter are increased to 6.4 and 44 ps, respectively, and internal supply noise of about 200 mV is measured. If a 500-mV, 1.1-MHz square wave is injected externally on the power supply, the peak-to-peak jitter increases to 83 ps, as shown in Fig. 14(b). At 250 MHz, jitter supply sensitivity is measured to be only 0.11 ps/mV. Furthermore, from 62.5 to 250 MHz, the clock outputs show almost flat jitter performance. Since the delay range of the VCDL in the core DLL is primarily set by *Vcr* and *Vcr_b*, the gain of the VCDL is nearly flat over a wide range of operating frequency. The jitter performance of the proposed DLL is better than or at least comparable to other wide-range DLL's [2]–[7].

Table I summarizes the DLL performance characteristics. The power dissipation is proportional to the operating frequency. Operating at 250 MHz, the DLL draws 12.6-mA dc from a 3.3-V power supply.

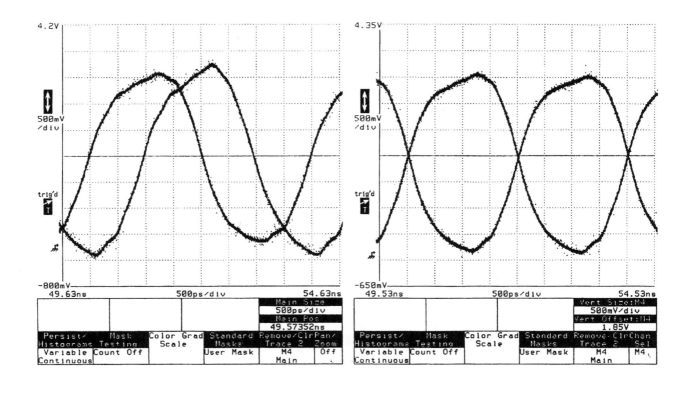

(a)   **(b)**

Fig. 13.   Clock waveforms at 250 MHz. (a) *CLK0, CLK2* and (b) *CLK0, CLK4*.

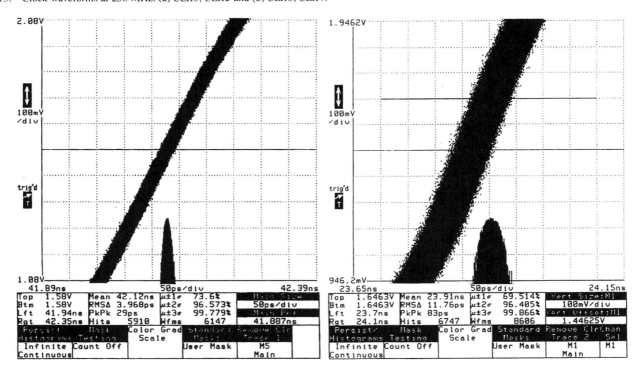

(a)   **(b)**

Fig. 14.   Jitter histograms at 250 MHz in (a) a quiet supply and (b) with added 1.1-MHz, 500-mV square wave noise.

TABLE I
PERFORMANCE CHARACTERISTICS OF
PROTOTYPE CHIP

Process	0.35 μm *N*-well triple-metal CMOS process (Vtn = 0.42 V, Vtp = -0.22 V, Tox = 7.5 nm)
Area	0.08 mm² (active area) 0.2 mm² (with on-chip decoupling-capacitor)
Supply Voltage	3.3 V
Icc	0.042× fclk (MHz) + 2.1 (mA) 12.6 mA @ 250 MHz
Jitter (@ 250 MHz)	4 ps RMS / 29 ps pk-to-pk (with a quiet supply) 6.4 ps RMS / 44 ps pk-to-pk (with digital circuits activated)
Supply sensitivity	0.11 ps/mV (@ 250 MHz)
Phase Offset	< 40 ps

VI. CONCLUSION

By including a replica delay line with a CSPD, the core DLL operates in a wide frequency range from 62.5 to 250 MHz. Since the replica delay line occupies a quarter of the area of the core DLL, the area cost and power consumption of the prototype chip are much smaller than those of other wide-range DLL's [2]–[7]. Both the analog-control scheme and the flat gain of the VCDL offer a low-jitter performance of 4-ps rms and 29-ps peak-to-peak, and a low supply sensitivity of 0.11 ps/mV. The DLL incorporates dynamic phase detectors and triply controlled delay cells with cell-level duty-cycle correction capability in order to generate equally spaced eight-phase clocks.

The DLL can be used not only as an internal clock buffer of microprocessors and memory IC's but also as a multiphase clock generator for gigabit serial interfaces. With a faster VCDL with minor change of device sizes, the DLL will operate at a higher and wider frequency range.

REFERENCES

[1] M. Johnson and E. Hudson, "A variable delay line PLL for CPU-co-processor synchronization," *IEEE J. Solid-State Circuits*, vol. 23, pp. 1218–1223, Oct. 1988.

[2] T. H. Lee, K. S. Donnelly, J. T. C. Ho, J. Zerbe, M. G. Johnson, and T. Ishikawa, "A 2.5 V CMOS delay-locked loop for an 18 Mbit, 500 Megabyte/s DRAM,," *IEEE J. Solid-State Circuits*, vol. 29, pp. 1491–1496, Dec. 1994.

[3] A. Efendovich, Y. Afek, C. Sella, and Z. Bikowsky, "Multifrequency zero-jitter delay-locked loop," *IEEE J. Solid-State Circuits*, vol. 29, pp. 67–70, Jan. 1994.

[4] B. W. Garlepp, K. S. Donnelly, J. Kim, P. S. Chau, J. L. Zerbe, C. Huang, C. V. Tran, C. L. Portmann, D. Stark, Y.-F. Chan, T. H. Leen, and M. A. Horowitz, "A Portable Digital DLL for High-Speed CMOS Interface Circuits," *IEEE J. Solid-State Circuits*, vol. 34, pp. 632–644, May 1999.

[5] S. Tanoi, T. Tanabe, K. Takahashi, S. Miyamoto, and M. Uesugi, "A 250–622 MHz deskew and jitter-suppressed clock buffer using two-loop architecture," *IEEE J. Solid-State Circuits*, vol. 31, pp. 487–493, Apr. 1996.

[6] K. Lee, Y. Moon, and D.-K. Jeong, "Dual loop delay-locked loop,", U.S. patent pending.

[7] S. Sidiropoulos and M. A. Horowitz, "A semi-digital dual delay-locked loop," *IEEE J. Solid-State Circuits*, vol. 32, pp. 1683–1692, Nov. 1997.

[8] S. Kim, K. Lee, Y. Moon, D.-K. Jeong, Y. Choi, and H. K. Lim, "A 960-Mb/s/pin interface for skew-tolerant bus using low jitter PLL," *IEEE J. Solid-State Circuits*, vol. 32, pp. 691–700, May 1997.

[9] D.-L. Chen and M. O. Baker, "A 1.25 Gb/s, 460 mW CMOS transceiver for serial data communication," in *IEEE ISSCC Dig. Tech. Papers*, Feb. 1997, pp. 242–243.

[10] Y. Moon, D. K. Jeong, and G. Kim, "Clock dithering for electromagnetic compliance using spread spectrum phase modulation," in *IEEE ISSCC Dig. Tech. Papers*, Feb. 1999, pp. 186–187.

[11] Y. Moon, J. Choi, K. Lee, D.-K. Jeong, and M.-K. Kim, "A 62.5–250 MHz multi-phase delay-locked loop using a replica delay line with triply controlled delay cells," in *Proc. IEEE Custom Integrated Circuits Conf.*, May 1999, pp. 299–302.

[12] B. Kim, "High speed clock recovery in VLSI using hybrid analog/digital techniques," Ph.D. dissertation, Univ. of California, Berkeley, Memo. UCB/ERL M90/50, June 1990.

[13] A. Hajimiri, S. Limotyrakis, and T. H. Lee, "Jitter and phase noise in ring oscillators," *IEEE J. Solid-State Circuits*, vol. 34, pp. 790–804, June 1999.

[14] M. Bazes, "A novel precision MOS synchronous delay line," *IEEE J. Solid-State Circuits*, vol. SC-20, pp. 1265–1271, Dec. 1985.

[15] D.-K. Jeong, G. Borriello, D. A. Hodges, and R. H. Katz, "Design of PLL-based clock generation circuits," *IEEE J. Solid-State Circuits*, vol. SC-22, pp. 255–261, Apr. 1987.

Yongsam Moon (S'96) was born in Incheon, Korea, on March 1, 1971. He received the B.S. and M.S. degrees in electronics engineering from Seoul National University, Seoul, Korea, in 1994 and 1996, respectively, where he is currently pursuing the Ph.D. degree.

He has been working on architectures and CMOS circuits for microprocessors. His current research interests include clock and data recovery for high-speed communication and high-speed I/O interface circuits.

Jongsang Choi was born in Korea on September 11, 1974. He received the B.S. and M.S. degrees in electronics engineering from Seoul National University, Seoul, Korea, in 1997 and 1999, respectively, where he is currently pursuing the Ph.D. degree.

He has been working on architectures and CMOS circuits for high-speed communication. His current research interests include high-speed CMOS circuits and gigabit network systems.

Kyeongho Lee (S'92–M'00) was born in Seoul, Korea, on August 5, 1969. He received the B.S., M.S., and Ph.D. degrees in electronics engineering from Seoul National University, Seoul, Korea, in 1993, 1995, and 2000, respectively.

Since 2000 he has been with Global Communication Technology, Inc., Los Altos, CA. He is working on various CMOS high-speed circuits for RF communication. His research interests include high-speed CMOS circuits and PLL systems.

Deog-Kyoon Jeong (S'87–M'89) received the B.S. and M.S. degrees in electronics engineering from Seoul National University, Seoul Korea, in 1981 and 1984, respectively., and the Ph.D. degree in electrical engineering and computer sciences from the University of California at Berkeley, Berkeley, CA, in 1989.

From 1989 to 1991, he was with Texas Instruments Incorporated, Dallas, TX,, where he was a Member oif the Technical Staff. He worked on modeling and design of BiCMOS circuits and single-chip implementation of the SPARC architecture. Since 1991, he has been on the Faculty of the School of Electrical Engineering, Seoul National University, Seoul, Korea, as an Associate Professor. His research interests include high-speed circuits, microrocessor architectures, and memory systems.

Min-Kyu Kim was born in Seoul, Korea, in 1965. He received the B.S., M.S., and Ph.D. degrees in electronics engineering from Seoul National University, Seoul, Korea, in 1988, 1990, and 1998, respectively.

From 1995 to 1996, he was with the Electronics and Telecommunications Research Institute, Taejon, Korea, working on the development of high-speed communication IC's for ATM switches. Since 1998, he has been working on high-speed serial link technologies at Silicon Image, Inc., Cupertino, CA. His current interests include circuit design for high-speed communication systems and digital-interface display systems.

A Semidigital Dual Delay-Locked Loop

Stefanos Sidiropoulos, *Student Member, IEEE,* and Mark A. Horowitz, *Senior Member, IEEE*

Abstract—This paper describes a dual delay-locked loop architecture which achieves low jitter, unlimited (modulo 2π) phase shift, and large operating range. The architecture employs a core loop to generate coarsely spaced clocks, which are then used by a peripheral loop to generate the main system clock through phase interpolation. The design of an experimental prototype in a 0.8-μm CMOS technology is described. The prototype achieves an operating range of 80 kHz–400 MHz. At 250 MHz, its peak-to-peak jitter with quiescent supply is 68 ps, and its jitter supply sensitivity is 0.4 ps/mV.

Index Terms—Clock synchronization, delay-locked loops, phase interpolation, phase-locked loops.

I. INTRODUCTION

PHASE-LOCKED loops (PLL's) and delay-locked loops (DLL's) are routinely employed in microprocessor and memory IC's in order to cancel the on-chip clock amplification and buffering delays and improve the I/O timing margins. However, the increasing clock speeds and integration levels of digital circuits create a hostile operating environment for these phase alignment circuits. The supply and substrate noise resulting from the switching of digital circuits affects the PLL or DLL operation and results in output clock jitter which subtracts from the I/O timing margins.

In applications where no clock synthesis is required, DLL's offer an attractive alternative to PLL's due to their better jitter performance, inherent stability, and simpler design. The main disadvantage of conventional DLL's, however, is their limited phase capture range. This paper presents a dual DLL architecture which combines several techniques to achieve unlimited phase capture range, low jitter and static-phase error, and four orders of magnitude operating frequency range. This architecture is based on a cascade of two loops. The core loop generates six clocks evenly spaced by 30° which are then used by the peripheral loop to generate the output clock, under the control of a digital finite state machine (FSM). By using phase interpolation, the dual loop can provide unlimited phase shift without the use of a voltage controlled oscillator (VCO). Using an FSM for phase control offers the advantage of enabling the flexible implementation of complicated phase capture algorithms in the digital domain. Finally, by utilizing self-biased techniques, the loop achieves large operating range and low jitter.

This paper begins with a brief overview of conventional DLL design. After outlining some of the disadvantages of

Manuscript received April 10, 1997; revised June 5, 1997. This work was supported by ARPA under contract DABT63-94-C-0054.

The authors are with the Computer Systems Laboratory, Stanford University, Stanford, CA 94305 USA and with Rambus Inc., Mountain View, CA 94040 USA.

Publisher Item Identifier S 0018-9200(97)08033-5.

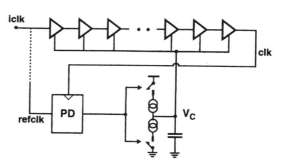

Fig. 1. Block diagram of a conventional DLL.

conventional approaches, Section II presents the dual interpolating DLL architecture. Section III discusses circuit design issues that arose in the prototype implementation of the architecture in a 0.8-μm CMOS technology. Section IV discusses the experimental results, and concluding remarks follow in Section V.

II. ARCHITECTURE

A. Conventional DLL's

A simplified block diagram of a conventional DLL [1] is outlined in Fig. 1. The components are a voltage controlled delay line (VCDL), a phase detector, a charge pump, and a first-order loop filter. The input reference clock drives the delay line which comprises a number of cascaded variable delay buffers. The output clock *clk* drives the loop phase detector (depicted in this example as a conventional flip-flop). The output of the phase detector is integrated by the charge pump and the loop filter capacitor to generate the loop control voltage V_C. The loop negative feedback drives the control voltage to a value that forces a zero phase error between the output clock and the reference clock.

This simple design offers many advantages compared to VCO-based PLL's. Due to frequency acquisition constraints, PLL's usually resort to a specific type of phase detector, the state-machine-based phase frequency detector (PFD). In contrast, DLL's can be easily implemented by using "bang–bang" control—i.e., the control signal of the loop, rather than being proportional to the phase error magnitude, can simply be a binary "up" or "down" indication. Thus, in a "bang–bang" DLL the phase detector can be a replica of the input data receiver resulting in an optimal placement of the sampling clock in the center of the input receiver's sampling uncertainty window. Additionally, since DLL's do not use a VCO, phase errors induced by supply or substrate noise do not accumulate over many clock cycles. This improved noise immunity is

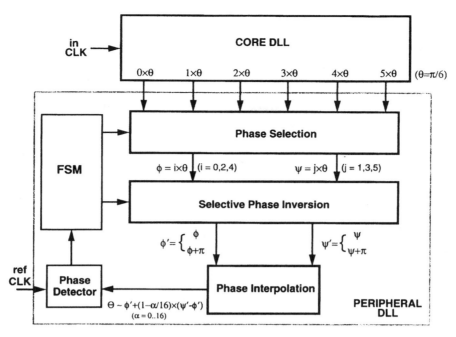

Fig. 2. Dual interpolating DLL architecture.

the main reason for the increased adoption of DLL's in applications that do not require clock synthesis.

The conventional DLL architecture of Fig. 1 suffers from two important disadvantages: clock jitter propagation and limited phase capture range. Since the VCDL simply delays the reference clock by a single clock cycle, the reference clock jitter directly propagates to the output clock. This all-pass filter behavior with respect to the frequency of the jitter of the reference clock results in reduced I/O timing margins, especially in "source-synchronous" interfaces where the reference clock emanates from another noisy digital chip. To overcome this problem, a separate low-jitter differential clock can be used as the input to the delay line. This way the on-chip common-mode noise and the reference clock jitter do not affect the I/O timing margins.

A more important problem is that a VCDL does not have the cycle slipping capability of a VCO. Therefore, at a given operating clock frequency, the DLL can delay its input clock by an amount bounded by a minimum and a maximum delay. As a consequence, extra care must be taken by the designer so that the loop will not enter in a state in which it tries to lock toward a delay which is outside these two limits. A compromising solution is to extend the VCDL range and use an FSM that controls the loop start-up. However, DLL's relying on quadrature phase mixing [2], [3] completely eliminate this problem. This approach is based on the fact that quadrature clocks can be easily generated, given a clock of the correct frequency. The quadrature clocks are then fed to a phase mixer which can produce a clock whose phase can span the complete 0–360° phase interval. This approach eliminates the limited phase range problem of conventional DLL's since it can essentially rotate the output clock phase infinite times providing seamless switching at the quadrant boundaries. The main disadvantage of quadrature mixing is that the output of

the phase mixer is a clock with a slew rate inherently limited by $4 \times V_{sw}/T$, where V_{sw} is the output swing of the phase mixer and T the period of the clock. This slow clock exhibits increased dynamic noise sensitivity, thus degrading the jitter performance of quadrature mixing DLL's.

The approach presented here overcomes this limitation of quadrature mixing DLL's since it generates the output clock by interpolating between smaller 30° phase intervals [5]. Simultaneously, by avoiding the use of a VCO it eliminates the phase error accumulation problem of similar approaches [4].

B. Dual Interpolating DLL

Fig, 2 shows a high-level block diagram of the proposed architecture. This architecture is based on cascading two loops. A conventional first-order core DLL is locked at 180° phase shift. Assuming that the delay line of the core DLL comprises six buffers, their outputs are six clocks which are evenly spaced by 30°. The peripheral digital loop selects a pair of clocks, ϕ and ψ, to interpolate between. Clocks ϕ and ψ can be potentially inverted in order to cover the full 0–360° phase range. The resulting clocks, ϕ' and ψ', drive a digitally controlled interpolator which generates the main clock Θ. The phase of this clock can be any of the N quantized phase steps between the phases of clocks ϕ' and ψ', where $0 \cdots N$ is the interpolation controlling word range.

The output clock Θ of the interpolator drives the phase detector which compares it to the reference clock. The output of the phase detector is used by the FSM to control the phase selection, the selective phase inversion, and the interpolator phase mixing weight. The FSM moves the phase of the clock Θ according to the phase detector output. In the more common case this means just changing the interpolation mixing weight by one. If, however, the interpolator controlling word has reached its minimum or maximum limit, the FSM must change the phase of clock ϕ or ψ to the next appropriate selection. This phase selection change might also involve an inversion of the corresponding clock if the current interpolation interval is adjacent to the 0° or 180° boundary. Since these phase selection changes happen only when the corresponding phase mixing weight is zero, no glitches occur on the output clock. The digital "bang–bang" nature of the control results in dithering around the zero phase error point in the lock condition. The dither amplitude is determined by the interpolator phase step and the delay through the peripheral loop.

In this architecture the output clock phase can be rotated, so no hard limits exist in the loop phase capture range: the loop provides unlimited (modulo 2π) phase shift capability. This property eliminates boundary conditions and phase relationship constraints, common in conventional DLL's. The only requirement is that the DLL input clock and the reference clock are plesiochronous (i.e., their frequency difference is bounded), making this architecture suitable for clock recovery applications. Since the system does not use a VCO, it does not suffer from the phase error accumulation problem of conventional PLL's. Moreover, the input clocks of the phase interpolator are spaced by just 30°, so the output of the phase interpolator does not exhibit the noise sensitivity of the quadrature mixing approach. Finally, the fact that the capture algorithm can be completely implemented in the digital domain gives great flexibility in its implementation as will be discussed in Section III. Although the prototype described in this paper is implemented with an analog core loop, possible implementations of the architecture can use digital control in both loops, further enhancing the system versatility. Moreover, the architecture can be easily extended to use a clock recirculating scheme in the core loop, so that the output clock frequency is a multiple of the input clock [7].

C. Dual-Loop Dynamics

Cascading two loops can compromise the overall system stability and lead to undesired jitter peaking effects. However, as the analysis in this section will show, this dual-loop architecture does not exhibit any jitter peaking irrespective of the dynamics of the two loops. The behavior of the DLL can be analyzed with respect to two types of perturbations: i) input or reference clock delay variations and ii) delay variations resulting from supply and substrate noise. The frequency response of the dual loop can be analyzed by making a continuous time approximation, in which the sampling operation of the phase detectors and the digital nature of the peripheral loop are ignored. This approximation is valid for core and peripheral loop bandwidths at least a decade below the operating frequency. This constraint needs to be satisfied anyway in a DLL in order to eliminate the effects of higher order poles resulting from the delays around loop.

Fig. 3 shows the dual loop linearized model including both the loop clocks $D_{IN}(s)$ and $D_{REF}(s)$, and delay errors introduced by supply or substrate noise $D_N(s)$. Each of the two loops is modeled as a single pole system, in which the input, output, and error variables are delays, similar to the single-loop analysis published in [7]. For example, the output delay of the

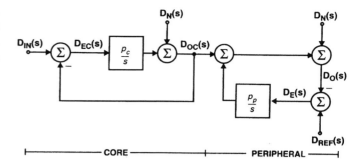

Fig. 3. Linearized dual DLL model.

core loop $D_{OC}(s)$ (in seconds) is the delay established by the core loop delay line, while the input delay $D_{IN}(s)$ is the delay for which the core loop phase detector and charge pump do not generate an error signal. Since the core loop VCDL spans half a clock cycle, $D_{IN}(s)$ is equal to half an input clock period. By using these loop variables, the input-to-output transfer function of the core loop can be easily derived

$$\frac{D_{OC}(s)}{D_{IN}(s)} = \frac{1}{1 + s/p_c} \qquad (1)$$

where p_c (in rads/s) is the pole of the core loop as determined by the charge pump current, the phase detector and delay line gain, and the loop filter capacitor. Similarly, the noise-to-delay error transfer function of the core loop can be shown to be

$$-\frac{D_{EC}(s)}{D_N(s)} = \frac{s/p_c}{1 + s/p_c} \qquad (2)$$

where $D_N(s)$ is the additional delay introduced in the core loop from supply or substrate noise, and $D_{EC}(s)$ is the delay error seen by the core loop phase detector. This transfer function indicates that noise induced delay errors can be tracked up to the loop bandwidth and that the response of the loop to a supply step consists of an initial step followed by a decaying exponential with a time constant equal to $1/p_c$.

Before proceeding to analyze the response of the dual loop, it should be noted that the linearized model of Fig. 3 uses a simplifying assumption. The assumption is that the delay error $D_N(s)$ introduced by supply or substrate variations is identical in both loops and does not depend on the state of the phase selection multiplexers. Since the supply and substrate sensitivity of the peripheral loop depends on the phase selection and will be typically higher due to the presence of the final CMOS system clock buffer, this assumption is not necessarily accurate. However, it does not affect the conclusions drawn below about the stability of the loop, since it only removes a modifying constant, which is equal to the ratio in the delay sensitivities of the two loops. This constant only affects the relative location of the poles and zeros of the resulting transfer function, and, as it will be shown below, the loop is unconditionally stable irrespective of the relation between the individual poles and zeros. Using the model of Fig. 3, it is straightforward to show that the transfer function $D_O(s)/D_{REF}(s)$ of the peripheral loop is identical in form to that of the core loop. This result agrees with intuition since in the dual loop system reference clock perturbations do not

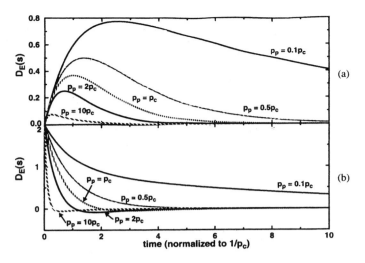

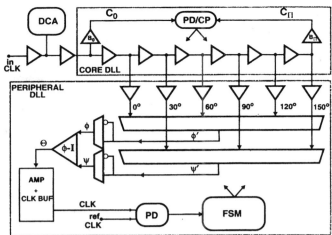

Fig. 5. Dual DLL detailed block diagram.

Fig. 4. Dual loop response to: (a) step change in input clock and (b) supply noise step.

affect the core loop. More interesting is the transfer function of the input clock $D_{\text{IN}}(s)$ to dual-loop error $D_{\text{E}}(s)$ since changes in the period of the input clock will cause both the core and peripheral loop to react. Based on (1) and (2), this transfer function can be shown to be

$$-\frac{D_{\text{E}}(s)}{D_{\text{IN}}(s)} = \frac{s/p_p}{(1 + s/p_c) \cdot (1 + s/p_p)}. \tag{3}$$

This bandpass transfer function exhibits no peaking at any frequency regardless of the relative magnitudes of p_c and p_p. The step response of the system, shown in Fig. 4(a), reveals that unit-step changes in $D_{\text{IN}}(s)$ (i.e., step changes in the input clock period) will initially peak at a less than unity value determined by the ratio of the two poles.[1] Moreover, as the magnitude of p_p increases, the disturbance on the output is reduced since the peripheral loop compensates quickly for disturbances at the output caused by changes of the input clock.

Finally, the transfer function from supply or substrate noise-induced delay errors $D_{\text{N}}(s)$ to the delay error of the dual loop $D_{\text{E}}(s)$ can be derived

$$-\frac{D_{\text{E}}(s)}{D_{\text{N}}(s)} = \frac{(1 + 2s/p_c) \cdot s/p_p}{(1 + s/p_c) \cdot (1 + s/p_p)}. \tag{4}$$

Equation (4) also exhibits no peaking at any frequency since the location of the last zero can never be above that of the poles. The step response of the system is plotted in Fig. 4(b) for various ratios of the core to peripheral pole frequencies. Under all conditions, the initial delay error is equal to twice the injected unity error $D_{\text{N}}(s)$ since this error is added on both loops. When the peripheral loop bandwidth is less than half that of the main loop, there is no overshoot in the dual-loop step response. This result occurs because the core loop compensates for its delay error quickly, while the slower peripheral loop compensates for the output delay

[1] It should be noted that in case in-CLK and ref-CLK are identical or correlated, the resulting transfer function exhibits a low-pass peaked behavior. Nevertheless the resulting peaking is small, exhibiting a maximum of 15% when $p_p = p_c$, while it is less than 5% as long as p_p and p_c are an order of magnitude apart in frequency.

error later. When the pole frequencies of the two loops are very close, the system overshoots since the peripheral loop compensates for the output delay error at approximately the same rate as the peripheral loop. The worst case overshoot of approximately 4.5% of the initial disturbance occurs when the peripheral loop bandwidth is twice that of the core loop. As the peripheral loop bandwidth increases, the overshoot becomes progressively smaller since the peripheral loop corrects for both the peripheral and core delay errors. Subsequently, the influence of the slower core loop correction on the output delay error is compensated by the peripheral loop. Therefore, even in the worst case, the dual loop cascade exhibits only minor overshoot.

III. CIRCUIT DESIGN

A. Overview

A more detailed block diagram of the dual loop is shown in Fig. 5. This design uses a separate local differential clock as the input to the delay line. Although the use of this clock is not inherent in the loop architecture, it minimizes the supply sensitivity in applications such as "source synchronous" interfaces. To minimize the effects of input clock duty cycle imperfections and common-mode mismatches, a duty cycle adjuster (DCA) [2] is employed after the first clock receiving buffer. The 50% duty cycle clock drives the core DLL. The core delay line consists of six differential buffers. An extra pair of buffers B_0, B_{Π} generate two clocks which drive the core loop 180° phase detector. The output of the phase detector controls the charge pump which forces clocks C_0 and C_{Π} to be 180° out of phase. Since all the buffers in the core delay line (including B_0 and B_{Π}) have the same size, all the core VCDL stages have the same fan-out and delay. Therefore, forcing C_0 and C_{Π} to be 180° out of phase will generate six evenly spaced by 30° clocks at the outputs of the core delay line.

The phase selection and phase inversion multiplexers are differential elements controlled by the core loop control voltage. In order to eliminate jitter-sensitive slow clocks, all buffers in the clock path need to have approximately the same

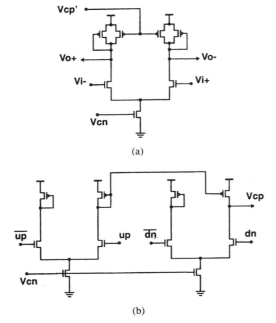

(a)

(b)

Fig. 6. (a) Core loop delay buffer and (b) charge pump.

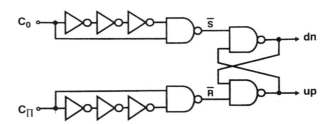

Fig. 7. Core loop phase detector.

bandwidth. For this reason, the phase selection in this design is implemented as a combination of a 3-to-1 and a 2-to-1 multiplexers, instead of a single 6-to-1 differential multiplexer with lower total power. Since the phase selection multiplexer can affect the phase shift of the core delay line through data-dependent loading, the six output clocks are buffered before driving the phase selection multiplexers. This way, changing the multiplexer select does not affect the core delay line phase shift.

The outputs ϕ, ψ of the phase inversion multiplexer drive the phase interpolator which generates the low swing differential clock Θ. This clock is then amplified and buffered through a conventional CMOS inverter chain generating the main clock (CLK). The peripheral loop phase detector [1] compares that clock to the reference clock, generating a binary phase error indication that is then fed to the FSM. The FSM based on the phase detector (PD) output selects phases ϕ, ψ and controls the phase interpolation.

B. Core Loop

To minimize the jitter supply sensitivity, all the delay buffers in the design, from the input clock (in-CLK) to the output of the phase interpolator (Θ), use differential elements with replica feedback biasing [6]. In order to linearize the loop gain and obtain large operating range, the core loop charge pump current is scaled along with the VCDL buffer current as illustrated in Fig. 6 [7]. Voltage V_{cn} is generated through the replica-feedback biasing circuit while V'_{cp} is a buffered version of the charge pump control voltage V_{cp}. In addition to the core VCDL buffers, voltages V'_{cp} and V_{cn} control the differential buffer elements of the peripheral loop. This ensures that all the buffers in the design have approximately equal delays and that the edge rates of the interpolator input clocks (ϕ, ψ) scale with the operating frequency of the loop.

The sensitivity of the dual-loop architecture to the core loop phase offset depends on the particular application. For the case that the dual DLL is used to just generate a clock whose phase is directly controlled by the phase detector output, the phase offset of the core loop does not affect the system phase offset. In this case, the loop operation will not be affected as long as the core loop phase offset is bounded. An absolute core loop offset less than 30° ensures monotonic switching at the 0° and 180° interpolation boundaries, so the interpolating loop functions correctly, albeit with a larger than nominal interpolation phase step. Core loop phase offsets larger than this amount will result in a hysteretic locking behavior at the quadrant boundaries, which will increase the dither jitter if the reference clock phase forces the dual loop to lock at this point.

The dual-loop operation becomes more sensitive to core loop phase offsets in case the designer wants to use this architecture to generate an additional clock that is offset by 90° relative to the reference clock. In such an application, the quadrature clock would be generated by using an extra pair of phase selection and inversion multiplexers whose selects would be offset by three relative to those generating the main clock. This would create a 90° interpolation interval offset, resulting in the required quadrature phase shift. In this case the core loop phase offset would impact the quadrature phase if the selects of the extra multiplexers happen to wrap around the 0° or 180° interpolation interval boundaries.

Even though the prototype does not implement quadrature phase generation, a low offset phase detector and careful matching of the layout were used to ensure uniform spacing of the six clocks. A self-biased DLL requires a linear phase detector. To avoid start-up problems that would result from the use of a conventional state machine PFD [7], the core loop uses the phase detector depicted in Fig. 7. This design comprises an S–R latch augmented with two input pulse generators. The absence of extra state storage in this design eliminates any start-up false locking conditions. Additionally, its symmetric structure and the use of pulse triggering minimize the core loop phase offset.

The core of the phase detector is an S–R latch-based phase detector. The S–R latch ensures a 180° phase shift between the falling edges of its inputs only when the duty cycle of the two input clocks is identical. However, when the duty cycle of the two input clocks is different, this mismatch will propagate as a core loop phase locking offset. This happens because an unbalanced overlap of the two input clocks causes the output of the S–R latch to have a duty cycle deviating from 50%. To compensate for this effect, the S–R latch is

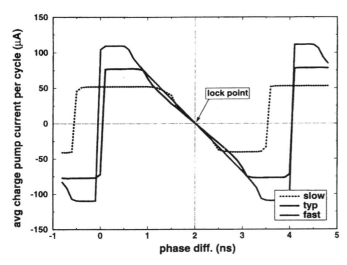

Fig. 8. Phase detector and charge pump simulated transfer function.

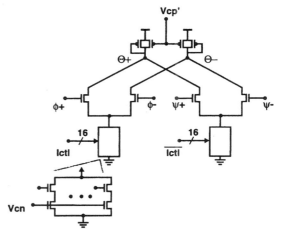

Fig. 9. Phase interpolator (type-I) schematic.

augmented with two pulse generators which propagate a low pulse on the positive edges of the input clocks. Since potential overlaps are minimized, the design can tolerate large duty cycle imperfections and still provide an accurate 180° lock in the core loop.

Fig. 8 shows the simulated transfer characteristics of the phase detector and charge pump over three extreme process and environment conditions. The cycle time of the two input clocks is set at 4 ns, while their duty cycles are mismatched by 0.5 ns such that the duty cycle of C_0 is 37.5% while the duty cycle of clock C_Π is 62.5%. It can be seen that the transfer function is linear and has no offset or dead-band around the 2-ns point where the loop actually locks. However, the combination of input pulsing and duty cycle imperfections results in nonlinear transfer function characteristics at the vicinity of the boundaries of the locking range (i.e., 0 and 4 ns). The only effect of this nonlinearity is that the core loop can exhibit an initial slew-rate limited reduction of its phase error, since the output of the phase detector and charge pump is constant. After the phase error has been reduced, such that the phase detector operates within its linear region, the core loop will exhibit a conventional single-pole response. Harmonic locking problems, common in PLL's using S–R phase detectors, are eliminated in this design since the core loop is reset to its minimum delay at system start-up.

C. Phase Interpolator Design

The most critical circuit in the design of the peripheral digital loop is the phase interpolator. The phase interpolator receives two clocks ϕ, ψ and generates the main clock Θ whose phase is the weighted sum of the two input phases. Essentially, the phase interpolator converts a digital weight code generated from the FSM to the phase of clock Θ. Linearity is not important in the design of this digital-to-phase converter since it is enclosed in the peripheral loop feedback. The important requirement is that the interpolation process is monotonic to ensure that no hysteresis exists in the loop locking characteristics. Additionally, the phase step must be minimized since it determines the loop dither amplitude. In this

case, the interpolation step is 1/16 of the 30° interval resulting in approximately 2° peripheral loop nominal dither. Another important requirement is that the design should provide for seamless interpolation-boundary switching. This means that when the input code is such that the weight on one of the input clocks is zero, this clock should have no influence on the output.

Fig. 9 shows a schematic diagram of the interpolator used in the prototype chip. This design is a dual input differential buffer which uses the same symmetric loads as all the core VCDL buffers and peripheral loop multiplexers. The bias voltages V'_{cp} and V_{cn} are identical with those biasing the rest of the loop, ensuring that its total delay is approximately 30° of the clock period which is the same as the rest of the loop buffers. Therefore, the transition time of the interpolator input clocks is larger than the minimum delay through the interpolator, and the two input transitions overlap. This condition ensures that the interpolator outputs never settle at half of the swing range. The current sources of the two differential pairs are thermometer controlled elements. The thermometer codes are generated by a 16-b long up/down shift register which is controlled by the peripheral loop FSM. By changing the thermometer code, the FSM adjusts in a complementary fashion the currents of the two input differential pairs resulting in a mixing of the two input clock phases. This design (type-I) does not completely satisfy the seamless boundary-switching requirement. Even when the current through one of the differential pairs is zero, the input still influences the output of the interpolator. This influence is due to the capacitive coupling of the gate-to-drain capacitance of the differential pair input transistors.

Fig. 10 shows an alternative design which does not suffer from this problem. In this design (type-II), the interpolator differential pairs consist of unit cell differential pairs. Therefore, when one of the interpolation weight thermometer codes is zero, the corresponding input is completely cut off from the output, eliminating the gate-to-drain coupling capacitance.

Fig. 11 shows the simulated transfer function of the interpolator alternative designs. This simulation includes random (<20 mV) threshold voltage offsets in the thermometer code

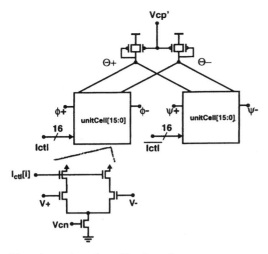

Fig. 10. Phase interpolator (type-II) schematic.

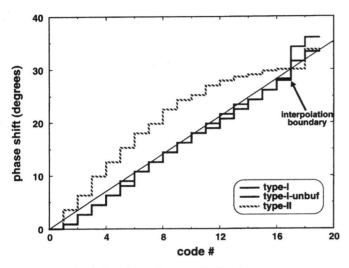

Fig. 11. Simulated phase interpolator transfer function.

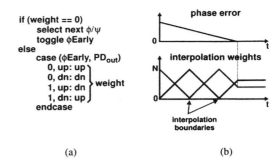

(a)   (b)

Fig. 12. (a) Simplified FSM algorithm and (b) resulting loop behavior.

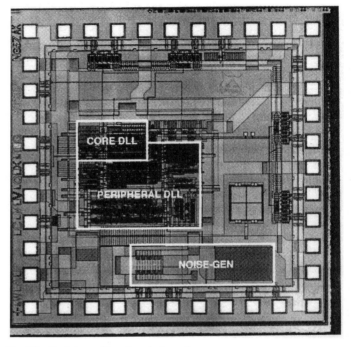

Fig. 13. Prototype chip microphotograph.

current sources. The type-I design exhibits a nominal step of approximately 2°. However, due to the gate-to-drain capacitive coupling effect, the maximum step of 3.8° occurs at the interpolation boundary when the input clock ϕ is switched to the next selection. In the lower power implementation where no buffering is used at the core delay line outputs (type-I-unbuf), the data-dependent loading on the previous stage results on a double phase step at the interpolation interval boundaries. Although the alternative design (type-II) does not exhibit a boundary phase step, it was not used since it occupies more layout area and exhibits more nonlinear characteristics due to data-dependent loading of the previous stage. So in the present implementation, worst-case dithering occurs at the interpolation interval boundaries and has an approximate magnitude of 3.8°.

D. Finite State Machine

A simplified version of the peripheral loop FSM algorithm is outlined in Fig. 12(a). The single state ϕEarly of the FSM indicates the relationship of the two interpolator input clocks.

On every cycle of its operation, the FSM might undertake two actions.

- In the more frequent case of in-range interpolation (i.e., weight $\neq$ 0), the FSM simply increments or decrements the interpolation weight by shifting up or down the interpolator controlling shift register. The direction of the shift is decided based on the phase detector output and the current value of the state ϕEarly.

- If the peripheral loop has run out of range in the current interpolation interval, the FSM seamlessly slides the current interpolation interval by switching phase ϕ or ψ to the next selection. The fact that the interpolation has run out of range in the current interval is simply indicated by a combination of the current value of the state ϕEarly, the most or least significant bit of the thermometer register, and the output of the phase detector. In case the current selection of phase ϕ or ψ is adjacent to the 0 or 180° interpolation interval boundary, switching to the next selection involves toggling the select of the second-stage phase inversion multiplexer.

470

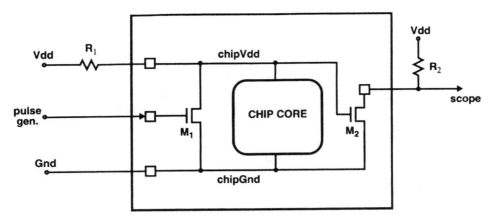

Fig. 14. Noise generation and monitoring circuits.

The loop phase capture behavior resulting from this simple algorithm is illustrated in Fig. 12(b). The phase error decreases at a linear rate until the system achieves lock. Subsequently, the loop dithers around the zero phase error point with a dither magnitude of one phase interpolation interval. This occurs because in this type of "bang–bang" system, the output of the phase detector is just a binary phase error without any indication of the magnitude of the phase error. The complementary interpolation weights slew linearly, changing direction at the interpolation interval boundaries. Once the system finds lock, they either dither by one or they stay constant if the dither point happens to lie on an interval boundary.

The magnitude of the peripheral loop phase dither is determined by the minimum interpolation step and the delay through the feedback loop. In conventional analog "bang–bang" DLL's, the loop delay is largely determined by the delay through the delay line and the clock distribution network. However, this digital implementation has a larger minimum loop delay. The underlying reason is that driving the FSM directly from the phase detector output might lead into metastability problems, especially since the whole loop operation is driving the phase detector to its metastable point of operation. For this reason, in this implementation the output of the phase detector is delayed by three metastability hardened flip-flops. This increases the mean time between failures (MTBF) of the system to a calculated worst case of approximately 100 years, but at the same time increases the peripheral loop delay by three cycles. To compensate for that delay and decrease the loop dither, the FSM logic implements a front-end filter which counts eight continuous phase detector "up" or "down" results before propagating this signal to the core FSM. This causes the FSM to delay its next decision until the results of its previous action have been propagated to the phase detector output and reduces the inherent peripheral loop dither to one phase interpolation interval.

The digital nature of the peripheral loop control enabled the implementation of the FSM to be done through synthesis of a behavioral verilog model followed by a simple standard cell place and route. The FSM behavioral model was verified by simulation in conjunction with a behavioral core loop model. The significance of this automated methodology is that other more complicated algorithms can be implemented requiring minimal effort from the designer. Faster phase acquisition can be obtained by disabling the front end counter/filter and changing the interpolation step by a larger amount while the loop is not in lock. The loop can also implement a periodic phase calibration algorithm. In this case, the FSM is activated initially to drive the loop to zero phase error. Then it is shut down to save power and it is periodically turned on to compensate for slow phase drifts. Since the FSM can run at a frequency slower than that of the system clock, the implementation of different algorithms is not in the system critical path.

IV. EXPERIMENTAL RESULTS

To verify the dual DLL architecture, a chip has been fabricated through MOSIS in the HP CMOS26B process. This is a 1.0-μm drawn process with the channel lengths scaled to 0.8 μm. Although the gate oxide in this process is ~170 Å allowing 5-V operation, the loop design and testing was done with a 3.3-V power supply voltage.

Fig. 13 is a micrograph of the chip. The chip integrates the dual DLL, along with noise injection and monitoring circuits and current-mode differential output buffers. The dual DLL occupies 0.8 mm^2 of silicon area, the majority (~60%) of which is devoted to the peripheral loop logic. This is mainly due to the relatively large standard cell size of the library used in this implementation.

The block labeled NOISE-GEN in Fig. 13 is used to inject and measure on-chip supply noise. Fig. 14 shows a schematic diagram of these circuits. The 1000-μm wide transistor M_1 shorts the on-chip supply rails creating a voltage drop across the off-chip 4-Ω resistor R_1. In order to monitor the droop on the on-chip supply, device M_2 and the external 5-Ω load resistor R_2 form a broadband attenuating buffer which drives the 50-Ω scope. The gain of the buffer is computed during an initial calibration step. The use of these circuits enables the injection and monitoring of fast (<1-ns rise time) steps on the on-chip supply.

The dither jitter of the loop with quiescent on-chip supply varies with the input phase. This occurs because the offset of the interpolator and the phase selection multiplexers change according to the point of lock. Fig. 15 shows the worst-

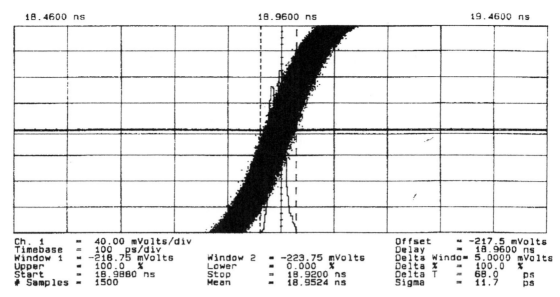

18.4600 ns	18.9600 ns	19.4600 ns

Ch. 1	=	40.00 mVolts/div		Offset	= -217.5 mVolts
Timebase	=	100 ps/div		Delay	= 18.9600 ns
Window 1	=	-218.75 mVolts	Window 2 = -223.75 mVolts	Delta Windo=	5.0000 mVolts
Upper	=	100.0 %	Lower = 0.000 %	Delta %	= 100.0 %
Start	=	18.9880 ns	Stop = 18.9200 ns	Delta T	= 68.0 ps
# Samples	=	1500	Mean = 18.9524 ns	Sigma	= 11.7 ps

Fig. 15. Jitter histogram with quiet supply.

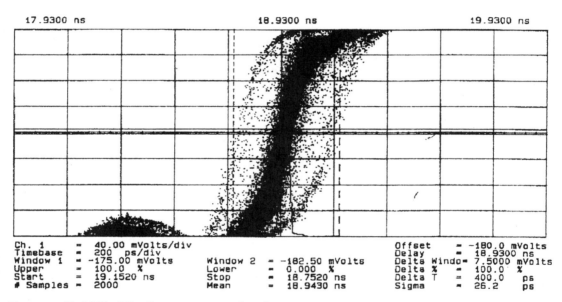

17.9300 ns	18.9300 ns	19.9300 ns

Ch. 1	=	40.00 mVolts/div		Offset	= -180.0 mVolts
Timebase	=	200 ps/div		Delay	= 18.9300 ns
Window 1	=	-175.00 mVolts	Window 2 = -182.50 mVolts	Delta Windo=	7.5000 mVolts
Upper	=	100.0 %	Lower = 0.000 %	Delta %	= 100.0 %
Start	=	19.1520 ns	Stop = 18.7520 ns	Delta T	= 400.0 ps
# Samples	=	2000	Mean = 18.9430 ns	Sigma	= 26.2 ps

Fig. 16. Jitter histogram with 1-MHz 750-mV square wave supply noise.

case jitter (68 ps) with quiescent supply. The jitter histogram consists of the superposition of two Gaussian distributions resulting from the switching of the peripheral loop between two adjacent interpolation intervals. The distance between the peaks of the two superimposed distributions is about 40 ps, which is in fair agreement with the simulation results. With the noise generation circuits injecting a 750-mV 1-MHz square wave on the chip supply, the peak-to-peak jitter increases to 400 ps (Fig. 16). It should be noted that simulation results indicate that approximately 50% of this jitter is not inherent to the loop, but is due to the supply sensitivity of the succeeding static CMOS clock buffer and off-chip driver.

Fig. 17 illustrates the linearity of the interpolation process in the peripheral loop. The figure shows the histogram of the output clock with the peripheral loop FSM continuously rotating that clock. The histogram was generated by keeping the reference clock to a constant voltage while the input clock ran at its nominal frequency of 250 MHz. The histogram valleys correspond to the interpolation interval boundaries. The spacing of the valleys is within 10% of their nominal 333-ps distance, indicating good matching of the delays of the core loop buffers. The absence of one valley at the 180° interpolation boundary indicates a slight offset in the core loop. The fact that the magnitude of the highest peak of the histogram is smaller than the magnitude of the deepest valley indicates that the interpolator achieves the 4-b target linearity (the 4-b linearity of the interpolator was also confirmed by a similar histogram of a single interpolation interval). Thus the overall linearity of the DLL is limited by the steps at the interpolation interval boundaries.

Table I summarizes the loop performance characteristics. With a 3.3-V supply, the loop operates from 80 kHz to

472

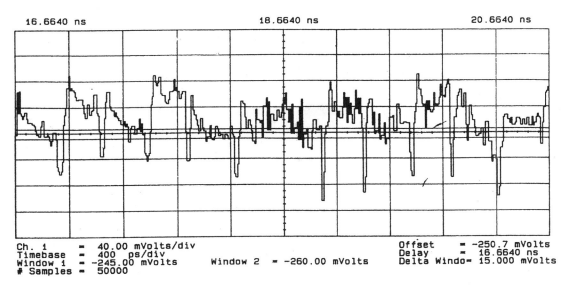

16.6640 ns	18.6640 ns	20.6640 ns

```
Ch. 1      =   40.00 mVolts/div              Offset       = -250.7 mVolts
Timebase   =  400   ps/div                   Delay        =  16.6640 ns
Window 1   = -245.00 mVolts    Window 2 = -260.00 mVolts    Delta Windo= 15.000 mVolts
# Samples  =  50000
```

Fig. 17. Interpolation process linearity.

TABLE I
PROTOTYPE PERFORMANCE SUMMARY

Process	1 μm (drawn), 0.8 μm (effective) CMOS nwell
Active Area	0.8 mm²
Supply Voltage	3.3 V
Power Dissipation	102 mW (@ 250 MHz)
Operating Range	0.08-400 MHz
Phase Offset	<40 ps
Jitter	68 ps pk-pk/11 ps RMS (@250 MHz)
Supply sensitivity	0.4 ps/mV (@250 MHz-closed loop)

400 MHz. The phase offset between the reference clock and the output clock of the loop is less than 40 ps. Operating at 250 MHz, the dual DLL draws 31 mA dc from a 3.3-V power supply.

V. SUMMARY

Although DLL's are easier to design than PLL's and offer better jitter performance, their main disadvantage is their limited phase capture range. This disadvantage limits their application to completely synchronous environments and complicates start-up circuitry. This paper presented a dual DLL architecture which removes this limitation by using a core DLL to generate coarsely spaced clocks which are then used by a peripheral DLL to generate the output clock by using phase interpolation. This architecture has unlimited (modulo 2π) phase shift capability, therefore removing boundary conditions and phase relationship constraints between the system clocks. The only requirement is that the DLL input and reference clocks are plesiochronous, making the dual DLL suitable for clock recovery applications. In addition, the digital nature of the peripheral loop control enables implementation of complicated phase alignment algorithms in a straightforward manner.

A prototype using a linear self-biased core loop has been implemented in a 0.8-μm technology. The prototype achieves 68-ps peak-to-peak jitter, 0.4-ps/mV supply sensitivity, and 0.08–400 MHz operating range.

ACKNOWLEDGMENT

The authors are grateful to M. Johnson, T. Lee, J. Maneatis, and K. Yang for helpful discussions.

REFERENCES

[1] M. Johnson and E. Hudson, "A variable delay line PLL for CPU-coprocessor synchronization," *IEEE J. Solid-State Circuits*, vol. 23, Oct. 1988.
[2] T. Lee *et al.*, "A 2.5 V CMOS delay-locked loop for an 18 Mbit, 500 MB/s DRAM," *IEEE J. Solid-State Circuits*, vol. 29, pp. 1491–1496. Dec. 1994.
[3] M. Izzard *et al.*, "Analog versus digital control of a clock synchronizer for a 3 Gb/s data with 3.0 V differential ECL," in *Dig. Tech. Papers 1994 Symp. VLSI Circuits*, June 1994, pp. 39–40.
[4] M. Horowitz *et al.*, "PLL design for a 500 MB/s interface," in *Dig. Tech. Papers Int. Solid State Circuits Conf.*, Feb. 1993, pp. 160–161.
[5] S. Sidiropoulos and M. Horowitz, "A semi-digital delay locked loop with unlimited phase shift capability and 0.08–400 MHz operating range," in *Dig. Tech. Papers Int. Solid State Circuits Conf.*, Feb. 1997, pp. 332–333.
[6] J. Maneatis and M. Horowitz, "Precise delay generation using coupled oscillators," *IEEE J. Solid-State Circuits*, vol. 28, pp. 1273–1282, Dec. 1993.
[7] J. Maneatis, "Low-jitter process-independent DLL and PLL based on self-biased techniques," *IEEE J. Solid-State Circuits*, vol. 31, pp. 1723–1732, Nov. 1996.

Stefanos Sidiropoulos (S'93), for a photograph and biography, see p. 690 of the May 1997 issue of this JOURNAL.

Mark A. Horowitz (S'77–M'78–SM'95), for a photograph and biography, see p. 690 of the May 1997 issue of this JOURNAL.

A Wide-Range Delay-Locked Loop With a Fixed Latency of One Clock Cycle

Hsiang-Hui Chang, *Student Member, IEEE,* Jyh-Woei Lin, Ching-Yuan Yang, *Member, IEEE,* and
Shen-Iuan Liu, *Member, IEEE*

Abstract—A delay-locked loop (DLL) with wide-range operation and fixed latency of one clock cycle is proposed. This DLL uses a phase selection circuit and a start-controlled circuit to enlarge the operating frequency range and eliminate harmonic locking problems. Theoretically, the operating frequency range of the DLL can be from $1/(N \times T_{D\,\max})$ to $1/(3T_{D\,\min})$, where $T_{D\,\min}$ and $T_{D\,\max}$ are the minimum and maximum delay of a delay cell, respectively, and N is the number of delay cells used in the delay line. Fabricated in a 0.35-μm single-poly triple-metal CMOS process, the measurement results show that the proposed DLL can operate from 6 to 130 MHz, and the total delay time between input and output of this DLL is just one clock cycle. From the entire operating frequency range, the maximum rms jitter does not exceed 25 ps. The DLL occupies an active area of 880 μm $\times$ 515 μm and consumes a maximum power of 132 mW at 130 MHz.

Index Terms—Delay-locked loops, latency, phase-locked loops, wide range.

I. INTRODUCTION

WITH THE evolution and continuing scaling of CMOS technologies, the demand for high-speed and high integration density VLSI systems has recently grown exponentially. However, the important synchronization problem among IC modules is becoming one of the bottlenecks for high-performance systems.

Phase-locked loops (PLLs) [1]–[3] and delay-locked loops (DLLs) [4]–[7] have been typically employed for the purpose of synchronization. Due to the difference of their configuration, the DLLs are preferred for their unconditional stability and faster locking time than the PLLs. Additionally, a DLL offers better jitter performance than a PLL because noise in the voltage-controlled delay line (VCDL) does not accumulate over many clock cycles.

Conventional DLLs may suffer from harmonic locking over wide operating range. If the DLLs are to operate at lower frequency without harmonic locking, the number of delay stages must be increased to let the maximum delay of the delay line be equal to the period of the lowest frequency. However, the

Manuscript received November 5, 2001; revised March 27, 2002.

H.-H. Chang and S.-I. Liu are with the Department of Electrical Engineering and Graduate Institute of Electronics Engineering, National Taiwan University, Taipei, Taiwan 10617, R. O. C. (e-mail: lsi@cc.ee.ntu.edu.tw).

J.-W. Lin was with the Department of Electrical Engineering and Graduate Institute of Electronics Engineering, National Taiwan University, Taipei, Taiwan 10617, R. O. C. He is now with Sunplus Corporation, Hsinchu 300, Taiwan.

C.-Y. Yang is with the Department of Electrical Engineering, Huafan University, Taipei, Taiwan 223, R. O. C.

Publisher Item Identifier 10.1109/JSSC.2002.800922.

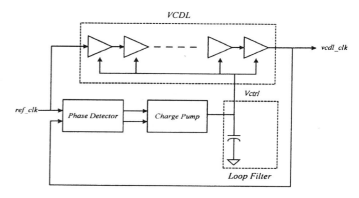

Fig. 1. Block diagram of the conventional analog DLL.

maximum operating frequency of a DLL will be limited by the minimum delay of the delay line.

In this paper, a DLL with wide-range operation and fixed latency of one clock cycle is proposed by using the phase selection circuit and the start-controlled circuit. The proposed DLL not only locks the delay equal to one clock cycle but also operates without the restrictions stated above. The operating frequency range of the proposed DLL can also be increased.

The range problem of conventional DLLs will be discussed in Section II. The architecture of the proposed DLL will be introduced in Section III and the building blocks in this DLL will be described in Section IV. Measurement results are given in Section V. Conclusions are given in Section VI.

II. RANGE PROBLEM OF CONVENTIONAL DLLs

A conventional DLL, as shown in Fig. 1, consists of four major blocks: the phase detector (PD), the charge-pump circuit, the loop filter, and the VCDL. In the DLL, the reference clock, ref_clk, is propagated through VCDL. The output signal, vcdl_clk, at the end of the delay line is compared with the reference input. If delay different from integer multiples of clock period is detected, the closed loop will automatically correct it by changing the delay time of the VCDL. However, the conventional DLL will fail to lock or falsely lock to two or more periods, T_{clk}, of the input signal if the initial delay of the VCDL is shorter than 0.5 T_{clk} or longer than 1.5 T_{clk}, as shown in Fig. 2. Therefore, if the DLL is required to lock the delay to one clock cycle of the input reference signal, the initial delay of the VCDL needs to be located between 0.5 T_{ref} and 1.5 T_{ref} [7], regardless of the initial voltage of the loop filter. Assume that the maximum and the minimum delay of the VCDL are $T_{\mathrm{VCDL\_max}}$

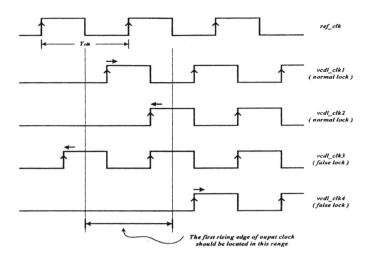

Fig. 2. DLL in normal lock and false lock conditions.

and $T_{\text{VCDL\_min}}$, respectively. As a result, the period of the input signal should satisfy the following inequality [7]:

$$\text{Max}\left(T_{\text{VCDL\_min}}, \frac{2}{3} \times T_{\text{VCDL\_max}}\right)$$
$$< T_{\text{CLK}} < \text{Min}\left(T_{\text{VCDL\_max}}, 2 \times T_{\text{VCDL\_min}}\right) \quad (1)$$

Equation (1) shows that the DLL is prone to the false locking problem when process variations are taken into account [7]. Therefore, some solutions [6]–[10] are proposed to overcome this problem. They are described as follows.

First, the basic idea is to use a phase-frequency detector (PFD) [5], because it has a capture range of $(-2\pi, +2\pi)$ wider than other phase detectors. So, the PFD is a better choice for wide range operation. However, the PFD cannot be used in the DLL alone without any control circuit because the DLL will try to lock a zero delay. A PFD combined with a control circuit is presented in [6]. Nevertheless, in some cases, especially for high-frequency operations, the initial delay between ref_clk and vcdl_clk, as shown in Fig. 1, may be larger than two clock cycles and harmonic locking will occur.

Second, a solution called an all-analog DLL using a replica delay line [7] has been developed to solve the narrow frequency range problem of a conventional DLL. If the delay range of the VCDL satisfies the relation $T_{\text{VCDL min}} < 1/7 \times T_{\text{VCDL max}}$, the DLL will have a maximum operation range of 7:1.

Third, a digital-controlled DLL called the self-correcting DLL is proposed in [8]. The problem of false locking is solved by the addition of a lock-detect circuit and the modified phase detector. Although this self-correcting DLL avoids false locking, the outputs of the VCDL are required to have an exact 50% duty cycle.

The DLL developed in [9] uses a stage selector for fast-locked and wide-range operations, but the DLL requires an additional VCDL, which increases the area. A similar DLL can automatically change its lock mode to extend the operation range, but the latency of the DLL will be larger than one clock cycle [10].

The approach presented in this work uses a phase selection circuit to automatically decide what number of delay cells should be used. This can enable the DLL to operate in the wide-frequency range. A new start-controlled circuit is also

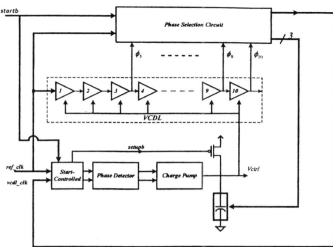

Fig. 3. System architecture of the proposed DLL.

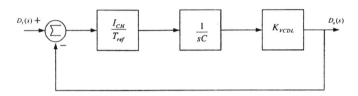

Fig. 4. Small-signal model of the conventional analog DLL.

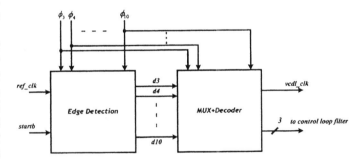

Fig. 5. Block diagram of the phase selection circuit.

presented for the DLL to solve false locking problems and keep the latency of one clock cycle. The exact 50% duty cycle is not necessary.

III. ARCHITECTURE OF THE PROPOSED DLL

The architecture of the proposed DLL is shown in Fig. 3. It is composed of a conventional analog DLL, a phase selection circuit, and a start-controlled circuit. Before the DLL begins to lock, the phase selection circuit will choose an appropriate delay cell to be a feedback signal (vcdl_clk) according to different frequencies of input signal. In other words, the number of the delay cells may change at different input frequencies. The minimum delay $T_{D\,\text{min}}$ of the delay line is determined by one unit-delay cell. The maximum delay can be decided as $N \times T_{D\,\text{max}}$ where N is the number of unit-delay cells. Thus, the operating frequency range of the DLL can be from $1/(3T_{D\,\text{min}})$ to $1/(N \times T_{D\,\text{max}})$.

The linear model of the DLL is shown in Fig. 4, where the summer stands for a phase detector, I_{cp} is the charge-pump cur-

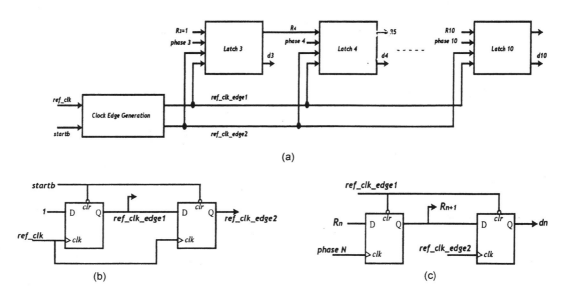

(a)

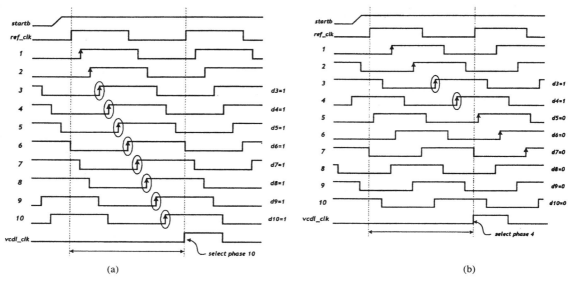

Fig. 6. Schematic of edge detection circuit. (a) Edge detection circuits. (b) Clock edge generation. (c) Latch N.

Fig. 7. Timing diagram of edge detection circuit.

rent, T_{REF} is the period of the input reference clock, C is the capacitor value in the loop filter, and K_{VCDL} is the gain of the VCDL which is proportional to the number of delay cells. In the steady-state locked condition, the s-domain transfer function can be expressed as [11]

$$\frac{Do(s)}{D_I(s)} = \frac{1}{1 + \dfrac{s}{\omega_N}} \tag{2}$$

where D_I is the input delay time and D_O is the output delay time. The loop bandwidth ω_N can be expressed as [11]

$$\omega_N = \frac{I_{\mathrm{CH}} \cdot K_{\mathrm{VCDL}}}{T_{\mathrm{REF}} \cdot C} \tag{3}$$

Since the transfer function is inherently stable, a wider loop bandwidth can be used to achieve fast acquisition time, but

the jitter performance will be degraded. Hence, the following tradeoff design guideline was suggested in [12]:

$$\frac{\omega_N}{\omega_{\mathrm{REF}}} = \frac{I_{\mathrm{CH}} \cdot K_{\mathrm{VCDL}}}{2\pi \cdot C} \leq \frac{1}{10} \tag{4}$$

where $\omega_{\mathrm{REF}} = 2\pi/T_{\mathrm{REF}}$.

When the input frequency is higher, the phase selection circuit will select the smaller number of delay cells and K_{VCDL} will become smaller. In order to have an adequate loop bandwidth for the DLL, the capacitances used in the loop filter must become smaller. In this work, the 3-bit control signals generated from the phase selection circuit will switch the number of capacitors in the loop filter depending on the selected phase.

After the vcdl_clk is decided, the DLL will start the locking process, which is controlled by the start-controlled circuit. First, the delay between input and output of the VCDL is initially set to the minimum value and then allows the *down* signal of the PFD output activate, supposing that the VCDL's delay increases with control voltage decreasing. Therefore, the delay between

476

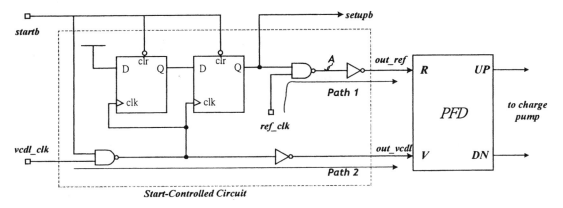

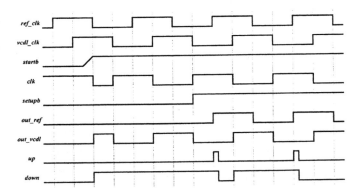

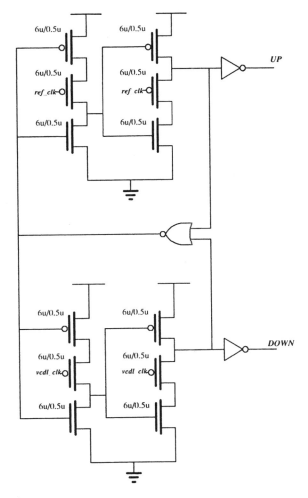

Fig. 8. Schematic of start-controlled circuit associated with PFD.

Fig. 9. Timing diagram of start-controlled circuit.

Fig. 10. Schematic of the PFD circuit [12].

input and output of the VCDL will increase until it reaches one clock period of the input signal. Thus, the DLL will not fall into false locking and the latency is fixed to one clock cycle no matter how long a delay the VCDL provides.

IV. CIRCUIT DESCRIPTION

A. Phase Selection Circuit

The phase selection circuit consists of two blocks: an edge detector and a multiplexer with a decoder, as shown in Fig. 5. The schematic and timing diagram of the edge detector are shown in Figs. 6 and 7, respectively. To guarantee that the latency of the DLL is just one clock cycle, the first two clock phases in Fig. 6 are reserved for measurement. In practice, the first two clock phases could be included in the phase selection circuit to improve the operating frequency range of the DLL. At the initial state, the signal *startb* is set to low to reset the edge detector outputs (i.e., d3 ~ d10) and the delay of the VCDL is set to its minimum value. When the signal *startb* goes high, the edge detector will detect the rising edge of input signals in sequence during the next two rising edges of ref_clk. Referring to Fig. 7(a), suppose that the signals all have rising edges in sequence during one clock cycle, therefore, the outputs (d3 ~ d10) are all high and the multiplexer will select phase 10 as the output signal, vcdl_clk. However, if the input frequency is higher, suppose that the timing diagram is similar to Fig. 7(b). All the inputs have rising edges during one clock cycle, but only the rising edges of phases 1 ~ 4 in sequence lead the selected phase to be 4. The vcdl_clk will be low until the selected phase is chosen. After

the vcdl_clk is decided, the DLL will start the locking process, which will be explained later. By the decoder, signals (d3 ~ d10) are decoded to generate 3-bit control signals, which switch the number of capacitors used in the loop filter for tuning the loop bandwidth.

B. Start-Controlled Circuit

The schematic of the start-controlled circuit and the associated PFD are shown in Fig. 8. It is composed of only two rising-edge trigger D-flip-flops (DFFs), two NAND gates, and

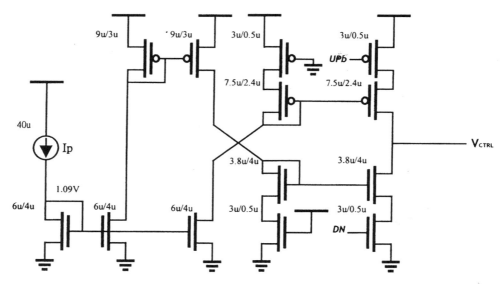

Fig. 11. Schematic of the charge-pump circuit [11].

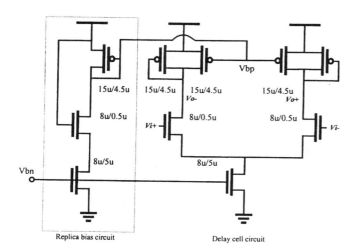

Fig. 12. Schematic of the delay cell with replica bias [12].

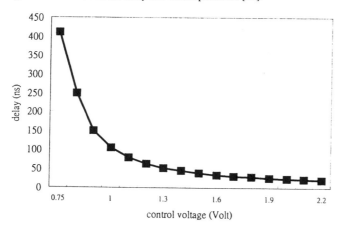

Fig. 13. Simulated transfer curve of the VCDL.

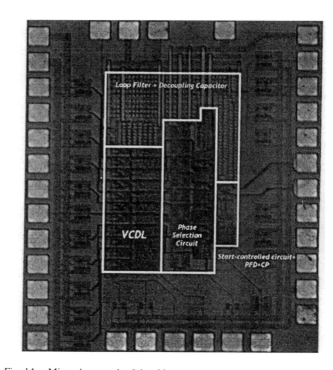

Fig. 14. Microphotograph of the chip.

two inverters. The timing diagram of this start-controlled circuit is shown in Fig. 9. Initially, *startb* is set to low in order to clear the two DFF's outputs. Therefore, *setupb* is low and pulls the control voltage to V_{DD}, as shown in Fig. 3 (i.e., set the VCDL delay to its minimum value). In this way, the two inputs of the

PFD are in low level. When *startb* goes to high, *setupb* will also go to high. After two consecutive falling edges of vcdl_clk trigger the DFFs, the down signal of the PFD will be activated and let the delay of the VCDL increase. The delay of the VCDL will increase until it is equal to one clock period of the input signal due to the nature of negative feedback architecture. Since the start-controlled circuit forces the delay of the VCDL to its minimum value and controls the delay of the VCDL to increase until its delay is equal to one clock period, the DLL will not fall into false locking even when $10\,T_{D\,\min} < 0.5T$. In order to get equal delays for path1 and path2, dummy loads should be added in point A. In comparison with [6], this start-controlled circuit

478

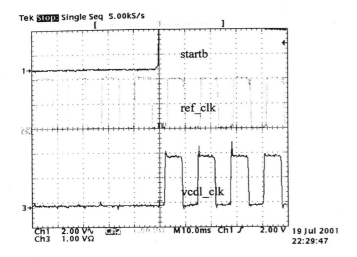

Fig. 15. DLL at initial state when operating frequency is 6 MHz.

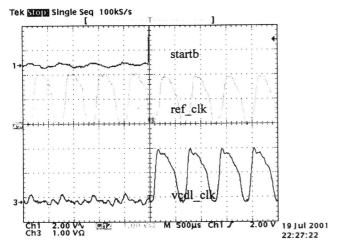

Fig. 16. DLL at initial state when operating frequency is 130 MHz.

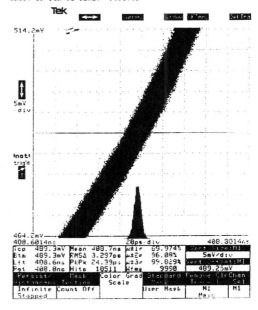

Fig. 17. Jitter histogram when DLL operates at 130 MHz.

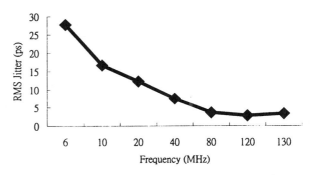

Fig. 18. Measurement results of rms jitter over different frequencies.

has two advantages: the proposed circuit is simple, and the duty cycle of ref_clk and vcdl_clk is not required to be exactly 50%.

C. Other Circuits

In this work, the dynamic logic style PFD [13] is adopted to avoid the dead-zone problem and improve the operating speed. To mitigate charge injection errors induced by the parasitic capacitors of the switches and current source transistors, the charge-pump circuit developed in [11] is used here. The delay cell circuit is similar to [11]. The schematics of these circuits are shown in Figs. 10–12. The control voltage of the loop filter is directly connected to nMOS rather than pMOS. Therefore, the transfer curve of delay versus control voltage is monotonic decreasing, as shown in Fig. 13.

V. EXPERIMENTAL RESULTS

The prototype chip is fabricated in a 0.35-μm single-poly triple-metal standard CMOS process. The microphotograph of the chip is shown in Fig. 14. The capacitors used in the loop filter are integrated in the chip and formed by metal-to-metal capacitors. The experimental results show that the DLL can operate in the frequency range of 6–130 MHz. Figs. 15 and 16

TABLE I
PERFORMANCE SUMMARY

Process	0.35-μm 1P3M TSMC CMOS process
Operating Voltage	3.3 V
Operating Frequency Range	6 MHz ~ 130 MHz
RMS Jitter	24.77 ps @ 6 MHz
	3.297 ps @ 130 MHz
Peak-to-Peak Jitter	210 ps @ 6 MHz
	24.3 ps @ 130 MHz
Lock time	~1130 clock cycles (simulated)
Power Dissipation	132 mW @ 130 MHz
Active Area	880-μm×515-μm @ without pads

show the first four cycles of the DLL in the locking process when the operating frequency is 6 and 130 MHz, respectively. After the signal *startb* is high, the phase selection circuit will select one of the outputs of the VCDL as close as possible to the next rising edge of the input clock, ref_clk. Figs. 15 and 16 also show that after the signal *startb* is high, the first rising edge of

479

the output clock of the VCDL, vcdl_clk, leads that of the input clock, ref_clk. Since the signal *startb* will set the control voltage V_{ctrl} in Fig. 3 to V_{DD}, the proposed phase detector and the current-pump circuit will discharge the loop filter to increase the delay of the VCDL. It will align the phases between the input clock and output clock of the VCDL. Fig. 17 shows the jitter histogram when the DLL operates at 130 MHz. Fig. 18 shows the measurement results of rms jitter over different frequencies. Table I gives the performance summary. The proposed DLL can be seen to have a wide-operational range and a fixed latency of one clock cycle.

VI. CONCLUSION

A DLL with wide-range operation and fixed latency of one clock cycle is proposed. First, the multiphase outputs of the VCDL are all sent to the phase selection circuit. Then the phase selection circuit will automatically select one of the delayed outputs to feedback. As a result, this DLL can operate over a wide range without suffering from harmonic locking problems. Ideally, this DLL can operate from $1/(N \times T_{D\max})$ to $1/(3T_{D\min})$. The experimental results also demonstrate the functionality of the proposed DLL. Moreover, at different operating frequencies, the jitter performances are all in an acceptable range and the latency is just one clock cycle. Since the speed of the proposed circuits can be increased if the more advanced process is used, the performance of the DLL such as the operating frequency range can be improved with a little hardware and design effort. The power consumption of the digital part in the DLL and the total die area will also be reduced.

REFERENCES

[1] B. Razavi, *Monolithic Phase-Locked Loops and Clock Recovery Circuits: Theory and Design.* Piscataway, NJ: IEEE Press, 1996.

[2] F. M. Gardner, "Charge-pump phase-lock loops," *IEEE Trans. Commun.*, vol. COM-28, pp. 1849–1858, Nov. 1980.

[3] R. E. Best, *Phase-Locked Loops: Theory, Design and Applications.* New York: McGraw-Hill, 1998.

[4] R. L. Aguiar and D. M. Santos, "Multiple target clock distribution with arbitrary delay interconnects," *Electron. Lett.*, vol. 34, no. 22, pp. 2119–2120, Oct. 1998.

[5] R. B. Watson Jr. and R. B. Iknaian, "Clock buffer chip with multiple target automatic skew compensation," *IEEE J. Solid-State Circuits*, vol. 30, pp. 1267–1276, Nov. 1995.

[6] C. H. Kim *et al.*, "A 64-Mbit 640-Mbyte/s bidirectional data strobed, double-data-rate SDRAM with a 40-mW DLL for a 256-Mbyte memory system," *IEEE J. Solid-State Circuits*, vol. 33, pp. 1703–1710, Nov. 1998.

[7] Y. Moon, J. Choi, K. Lee, D. K. Jeong, and M. K. Kim, "An all-analog multiphase delay-locked loop using a replica delay line for wide-range operation and low-jitter performance," *IEEE J. Solid-State Circuits*, vol. 35, pp. 377–384, Mar. 2000.

[8] D. J. Foley and M. P. Flynn, "CMOS DLL-based 2-V 3.2-ps jitter 1-GHz clock synthesizer and temperature-compensated tunable oscillator," *IEEE J. Solid-State Circuits*, vol. 36, pp. 417–423, Mar. 2001.

[9] H. Yahata, T. Okuda, H. Miyashita, H. Chigasaki, B. Taruishi, T. Akiba, Y. Kawase, T. Tachibana, S. Ueda, S. Aoyama, A. Tsukinori, K. Shibata, M. Horiguchi, Y. Saiki, and Y. Nakagome, "A 256-Mb double-data-rate SDRAM with a 10-mW analog DLL circuit," in *Symp. VLSI Circuits Dig. Tech. Papers*, June 2000, pp. 74–75.

[10] Y. Okuda, M. Horiguchi, and Y. Nakagome, "A 66–400 MHz adaptive-lock-mode DLL circuit duty-cycle error correction," in *Symp. VLSI Circuits Dig. Tech. Papers*, June 2001, pp. 37–38.

[11] J. G. Maneatis, "Low-jitter process-independent DLL and PLL based on self-biased techniques," *IEEE J. Solid-State Circuits*, vol. 31, pp. 1723–1732, Nov. 1996.

[12] A. Chandrakasan, W. J. Bowhill, and F. Fox, *Design of High-Performance Microprocessor Circuit.* New York: IEEE Press, 2001, p. 240.

[13] S. Kim *et al.*, "A 960-Mb/s/pin interface for skew-tolerant bus using low jitter PLL," *IEEE J. Solid-State Circuits*, vol. 32, pp. 691–700, May 1997.

Hsiang-Hui Chang (S'01) was born in Taipei, Taiwan, R.O.C., on February 4, 1975. He received the B.S. and M.S. degrees in electrical engineering from National Taiwan University, Taipei, in 1999 and 2001, respectively. He is currently working toward the Ph.D. degree in electrical engineering at National Taiwan University.

His research interests are PLL, DLL, and high-speed interfaces for gigabit transceivers.

Jyh-Woei Lin was born in Kaoshiung, Taiwan, R.O.C., in 1974. He received the B.S. degree in electrical engineering from National Taipei University of Technology in 1996, and the M.S. degree in electrical engineering from National Taiwan University in 2001.

He joined Sunplus Corporation, Hsinchu, Taiwan, in 2001 as an Analog Circuit Designer. His research interests include PLL, DLL, and interface circuits for high-speed data links.

Ching-Yuan Yang (S'97–M'01) was born in Miaoli, Taiwan, R.O.C., in 1967. He received the B.S. degree in electrical engineering from the Tatung Institute of Technology, Taipei, Taiwan, R.O.C., in 1990, and the M.S. and Ph.D. degrees in electrical engineering from National Taiwan University, Taipei, in 1996 and 2000, respectively.

He has been on the faculty of Huafan University, Taiwan, since 2000, where he is currently an Assistant Professor with the Department of Electronics Engineering. His research interests are in the area of mixed-signal integrated circuits and systems for high-speed interfaces and wireless communication.

Shen-Iuan Liu (S'88–M'93) was born in Keelung, Taiwan, R.O.C., on April 4, 1965. He received both the B.S. and Ph.D. degrees in electrical engineering from National Taiwan University, Taipei, in 1987 and 1991, respectively.

During 1991–1993, he served as a Second Lieutenant in the Chinese Air Force. During 1991–1994, he was an Associate Professor in the Department of Electronic Engineering of National Taiwan Institute of Technology. He joined the Department of Electrical Engineering, National Taiwan University, Taipei, in 1994, where he has been a Professor since 1998. His research interests are in analog and digital integrated circuits and systems.

A Portable Digital DLL for High-Speed CMOS Interface Circuits

Bruno W. Garlepp, Kevin S. Donnelly, *Associate Member, IEEE*, Jun Kim, Pak S. Chau,
Jared L. Zerbe, Charles Huang, Chanh V. Tran, Clemenz L. Portmann, *Member, IEEE*,
Donald Stark, Yiu-Fai Chan, *Member, IEEE*, Thomas H. Lee, *Member, IEEE*, and Mark A. Horowitz

Abstract— A digital delay-locked loop (DLL) that achieves infinite phase range and 40-ps worst case phase resolution at 400 MHz was developed in a 3.3-V, 0.4-μm standard CMOS process. The DLL uses dual delay lines with an end-of-cycle detector, phase blenders, and duty-cycle correcting multiplexers. This more easily process-portable DLL achieves jitter performance comparable to a more complex analog DLL when placed into identical high-speed interface circuits fabricated on the same test-chip die. At 400 MHz, the digital DLL provides <250 ps peak-to-peak long-term jitter at 3.3 V and operates down to 1.7 V, where it dissipates 60 mW. The DLL occupies 0.96 mm^2.

Index Terms—Delay circuits, delay-locked loops (DLL's), digital control, digital DLL, phase blending, phase control, phase synchronization.

I. INTRODUCTION

IN RECENT years, there has been a great deal of interest in delay-locked loops (DLL's) for clock alignment. Both analog and digital DLL's have been developed [1]–[6], with analog loops generally providing better jitter performance at the expense of greater complexity. This paper describes a digital DLL that achieves jitter performance comparable to an analog DLL. Although the digital DLL uses more area and power than the analog DLL, its greater simplicity, easier portability, and lower minimum required supply voltage makes it very attractive in many clock alignment applications. Additionally, the digital DLL not only operates at lower supply voltages than the analog DLL but it also demonstrates that digital DLL's have the potential for good power-consumption scaling as supply voltage is decreased.

The motivation for the development of this digital DLL was the need for a clock alignment circuit for use in the CMOS interface cells [6] of a high-speed memory system as in [7].[1] The memory system operates at 400 MHz, with data transferred on both edges of the clock, producing an effective 800-Mb/s/pin transfer rate. This corresponds to a 1.25-ns bit time. With such tight timing requirements, it becomes imperative to include clock alignment circuits in

Manuscript received September 15, 1998; revised December 23, 1998.

B. W. Garlepp, K. S. Donnelly, J. Kim, P. S. Chau, J. L. Zerbe, C. Huang, C. V. Tran, C. L. Portmann, D. Stark, and Y.-F. Chan are with Rambus, Inc., Mountain View, CA 94040 USA.

T. H. Lee and M. A. Horowitz are with the Center for Integrated Systems, Stanford University, Stanford, CA 94305 USA.

Publisher Item Identifier S 0018-9200(99)03668-9.

[1] Documentation is available at http://www.rambus.com/html/direct_documentation.html.

the interface cells to provide internal on-chip clocks that are aligned in phase with an external system clock. The clock alignment circuits must provide a phase resolution better than 50 ps and produce a worst case long-term jitter of less than 250 ps peak-to-peak (p–p). To facilitate the use of many different application-specific integrated-circuit controllers with the memory system, the clock alignment circuit should be easily portable across multiple processes without compromising performance.

The clock alignment function can be provided using either phase-locked loops (PLL's) or DLL's. Because frequency synthesis is not needed in this application, DLL's are preferred for their unconditional stability, lower phase-error accumulation, and faster locking time. In previous designs of the interface cells for this memory system, we have used an analog DLL with a two-step coarse/fine architecture. A high-level drawing of this approach is shown in Fig. 1. This analog DLL includes a quadrature generator, which produces four reference signals spaced 90° apart in phase to evenly cover the full 360° of phase space. A phase interpolator circuit in the analog DLL receives these reference signals and selects a phase adjacent pair that define a phase quadrant for interpolation to produce an output signal phase-aligned to a reference signal, RefClk.

Analog DLL's constructed with this approach provide several significant benefits. Because most of the elements in the signal path can be made from differential analog blocks with good power-supply rejection ratio (PSRR), the analog DLL architecture of Fig. 1 can provide very good jitter performance. Additionally, it can be carefully designed to occupy relatively little area and consume relatively little current. Furthermore, the analog DLL can provide very small phase steps when locked (≪50 ps). Finally, the architecture of Fig. 1 provides infinite phase range, and one set of quadrature reference signals can be fed to multiple phase interpolators, allowing phase alignment to multiple reference signals simultaneously. However, because of the relatively high analog complexity of this DLL and its individual elements, the analog DLL of Fig. 1 requires a detailed, process-specific implementation, making it relatively labor intensive to port across multiple processes.

Although we have traditionally used analog DLL's to provide the clock alignment function in the CMOS interface cells of the memory system described above, we decided to consider using a digital DLL. Digital DLL's are characterized by their use of a digital delay line and are typically made from

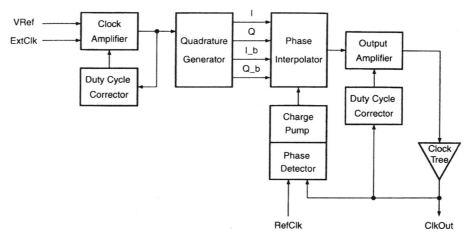

Fig. 1. Block diagram of a two-step, coarse/fine analog DLL architecture.

simple, digital circuit elements. This facilitates their design and portability across multiple processes. Additionally, because phase information in a digital DLL is stored as a digital state, digital DLL's can provide very fast timing recovery after being placed into a low power mode. However, conventional digital DLL's provide only moderate phase resolution and jitter performance [8], [9].

Another benefit of digital DLL's is their ability to readily operate at lower voltages than analog DLL's. Because analog DLL's require the use of saturated current sources, they experience voltage headroom problems as supply voltages decrease. Digital DLL's, on the other hand, need only enough voltage to ensure the proper operation of their digital gate elements. For the same reason, digital DLL's better utilize the power-saving benefits of digital CMOS voltage scaling than analog DLL's. The power of an analog DLL is typically distributed between IV power (where I is power and V is voltage) from the constant current (differential) stages and CV^2f power (where C is capacitance and f is frequency) from the CMOS (single-ended) stages (if any). The power of digital DLL's, on the other hand, is determined primarily by CV^2f power, which decreases quadratically with supply voltage.

This paper describes a digital DLL [10] used as the clock alignment circuit in the CMOS interface cells of a high-speed memory system. This work improves upon the performance of previous digital DLL's by paralleling the two-step coarse/fine analog DLL architectures presented in [4], [5], [7], and [11], allowing the digital DLL to achieve jitter performance comparable to the analog DLL's.

This paper is arranged as follows. Section II describes delay-generation techniques used in conventional digital DLL's and describes the improved techniques implemented in the new DLL. This section also describes infinite phase generation with the new delay-line scheme. Section III describes several new circuit techniques used for enhancing the phase resolution and signal quality in the new digital DLL. Section IV describes the overall DLL architecture. Section V discusses our test chip and measured results, with special attention given to making a direct, side-by-side comparison of the new digital DLL with an analog DLL placed into identical

CMOS interface cells on the same test-chip die. Section VI concludes this paper.

The terms *phase* and *delay* are used throughout this paper to describe the DLL's operation. It is helpful to recall that at a given system frequency, the two quantities are related by the simple equation

$$\Phi = 360 \cdot t_D \cdot f \qquad (1)$$

where Φ is phase in degrees, t_D is delay in seconds, and f is frequency in hertz.

II. DIGITAL DELAY CIRCUIT TECHNIQUES

A. Conventional Digital Delay Lines

As mentioned above, the purpose of a DLL in a clock alignment application is to provide an output clock signal that is aligned in phase with a reference clock signal of the same frequency. To do this, the DLL must include a mechanism for providing a variable delay to an input signal. The DLL then adjusts this variable delay such that the input signal passes through the delay mechanism and emerges at the output of the DLL aligned in phase with the reference signal.

Digital DLL's generally incorporate a tapped digital delay line as the variable-delay mechanism. The delay line receives an input clock signal (e.g., a buffered version of the reference signal) and passes it through a series of delay elements. The outputs of the delay elements are tapped and buffered to provide a series of phase-adjacent signals. The DLL then selects the delay-line tap that provides the signal that produces an output with a phase that most closely matches the desired phase.

A conventional delay line suitable for a CMOS digital DLL is shown in Fig. 2. The delay elements could be implemented with almost any circuit block, but because the phase resolution of the delay line is determined by the delay through the delay elements, delay elements that provide minimal delay are generally preferred. Thus, the delay line of Fig. 2 uses inverters, since they provide the shortest delay of any CMOS digital gate. Because of the inverting characteristic of all standard CMOS gates, the delay line is tapped only at every other inverter

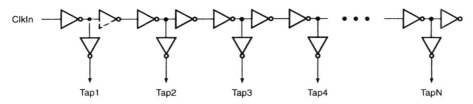

Fig. 2. Conventional digital delay line with inverter delay elements.

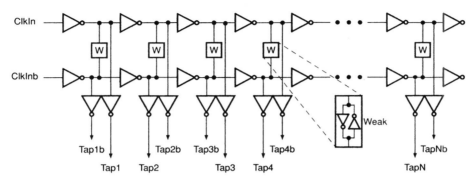

Fig. 3. Complementary delay line with inverter delay elements for improved phase resolution.

output to ensure that each successive tap provides a signal that is adjacent in phase to the signals at its adjacent taps.

Although conventional delay lines are attractive for their simplicity, DLL's designed around such conventional delay lines suffer from several significant limitations. First, the delay line provides fairly coarse phase resolution. For example, the delay line in Fig. 2 provides a minimum phase step corresponding to two inverter delays. Such coarse phase resolution is not fine enough for our clock alignment application. Second, conventional delay lines deliver only a finite phase range. Typically, in order to cover at least one full cycle of phase, the delay-line length and element delays are adjusted to provide at least 360° of phase under the fastest process, voltage, and temperature (PVT) conditions and minimum operating frequency ($f_{MIN} = 1/T_{MAX}$). More often, however, the delay line is designed with as much as 720° (i.e., two cycles) of phase under these conditions. This requires the use of a long delay line, occupying a large silicon area and dissipating additional power as the input signal propagates through the many delay elements. Additionally, because inverters offer poor PSRR, voltage supply noise-induced jitter can accumulate as the signal propagates down the delay line. This causes the signals available from the later taps in the delay line to be more jitter prone than the signals from the earlier taps. Last, even with an extended delay line, the DLL can nonetheless run out of phase range and lose lock in a system with slowing drifting phase (e.g., spread-spectrum clocking). These limitations prohibited the use of a conventional delay line in our DLL design.

B. Delay-Line Improvements

To overcome some of these limitations, we developed a complementary delay line as shown in Fig. 3 for our DLL. In this architecture, two parallel delay lines with weak cross coupling are driven by complementary input signals ClkIn and

ClkInb. Because of the use of complementary inputs, the two delay lines are tapped after every inverter to provide phase-adjacent signals separated by only one inverter delay, thereby improving the phase resolution by a factor of two. An example of how this delay-line scheme provides single inverter delay resolution is shown by the shaded paths in Fig. 3. The signal that emerges from Tap 2 has passed through three inverter delays, while the signal that emerges from Tap 3 has passed through four inverter delays. However, ClkInb is exactly 180° out of phase with ClkIn, providing the additional inversion required to ensure that the signals emerging from Taps 2 and 3 are indeed separated in phase by exactly one inverter delay.

This complementary delay-line architecture also allows the delay lines to be made shorter. The true taps from the delay line can provide the first 180° of phase, while the complement taps can provide the second 180° of phase. Thus, each of the two delay lines can be tuned for only 180° of phase under the fastest PVT conditions and f_{MIN}. Shorter delay lines provide the additional benefits of reduced maximum jitter accumulation, smaller silicon area, and lower power consumption. The problem that this design creates is a need to determine when to switch from the true taps to the complement taps and vice versa to ensure full and even coverage of the entire 360° phase plane. This is particularly important because the number of delay elements (and output taps) needed to cover 180° changes with PVT conditions and operating frequency.

C. Infinite Phase Generation

To solve the problem of determining when to switch between the true and complement taps of the complementary delay line, we developed an end-of-cycle (EOC) detector, as shown in Fig. 4, for use with the complementary delay line. An EOC detector is essentially a bank of data flip-flops arranged as a time-to-digital converter for measuring the delay through the delay line. The EOC detector produces a thermometer code

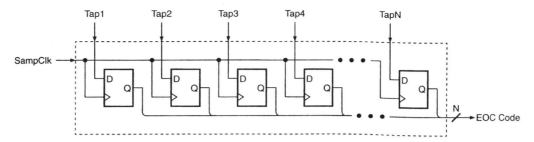

Fig. 4. EOC detector circuit (180°).

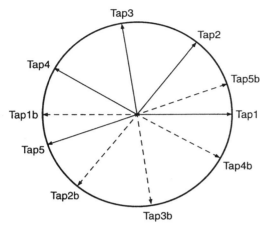

Fig. 5. Phasor diagram with phasors of signals from the taps of a complementary delay line with one inverter delay = 50°.

indicating the first 180° of delay in the delay lines. The first state transition in the EOC code indicates the first true tap from the delay line that provides a signal with phase that lags the phase of the signal from Tap 1 by more than 180°. With this information, the DLL logic knows when to switch between the true and complement taps of the delay line to ensure full coverage of all 360° of phase space, with phase steps of at most one inverter delay. Use of the EOC code also prevents negative phase steps in the phase-transfer function as taps are successively selected from the delay line. This allows the complementary delay lines to provide infinite, monotonic phase range for the DLL. The clocking signal for the EOC detector, SampClk, is synchronized to the signal from Tap 1 by a replica timing network (not shown).

To illustrate the principle of infinite phase generation using the EOC code with this delay-line scheme, refer to Fig. 5, which shows a phasor diagram of the signals from the first five true and complement taps of a complementary delay line like the one shown in Fig. 3. The figure assumes that the PVT conditions and operating frequency are such that the propagation delay of each inverter stage is equal to 50° of phase. In the figure, the solid lines correspond to signals from the true taps, while dashed lines correspond to signals from the complement taps. Because Tap 5 delivers a signal that is delayed by 200° from the signal at Tap 1, the EOC detector's thermometer code would indicate that Tap 5 is the first true tap to provide a signal with phase beyond 180° relative to the signal from Tap 1. With this information, the DLL knows to switch between the true and complement taps after four stages.

In other words, to travel counterclockwise around the phase plane, the DLL would successively select Taps 1–4, then Taps 1b–4b, then Taps 1–4, etc., to provide infinite phase range. In this manner, all phase steps are equivalent to at most one inverter delay (i.e., 50°), except for the Tap 4 to Tap 1b and the Tap 4b to Tap 1 transitions, which are less (30°).

III. RESOLUTION-ENHANCING CIRCUIT TECHNIQUES

A. Phase Blending

Although the delay-line improvements discussed above reduced the required power and area of the delay line, improved its jitter accumulation performance, enabled infinite phase range, and improved the available phase resolution by a factor of two, this phase resolution was still not good enough to meet the requirements of our memory system. In the 0.4-μm process we used, the propagation delay of one inverter over all anticipated PVT conditions varied from 100 to 300 ps. This is much larger than the worst case phase step specification of 50 ps. Therefore, to ensure compliance with this specification, the DLL's phase resolution needed to be improved by at least six times over what the delay line provided.

To solve this problem, we used inverter phase blending. A simple, single-stage phase-blender circuit is shown in Fig. 6(a). This circuit receives two phase-adjacent input signals, Φ_1 and Φ_2, which are separated in phase by one inverter delay. The phase blender directly passes these two signals with a simple delay to produce output signals ϕ_A and ϕ_B. However, it also uses a pair of phase-blending inverters to interpolate between these two input signals to produce a third output signal, ϕ_{AB}, having a phase between that of ϕ_A and ϕ_B. This effectively doubles the available phase resolution.

However, it is not sufficient to use equal-sized inverters for the phase blending. Fig. 6(b) illustrates a simple model [12] used for determining the ideal relative sizes of the two phase-blending inverters to ensure that the phase of ϕ_{AB} lies directly between that of ϕ_A and ϕ_B. The model approximates the two inverters with two simple switched current sources sharing a common resistance–capacitance (RC) load. For two rising edge input signals separated in time by t_D, the model yields the equation

$$V_{\Phi_{AB}}(t) = V_{DD} + R \cdot I$$
$$\cdot \left[w \cdot u(t) \cdot \left(e^{-(t/RC)} - 1 \right) + (1 - w) \right.$$
$$\left. \cdot u(t - t_D) \cdot \left(e^{-[(t-t_D)/RC]} - 1 \right) \right] \quad (2)$$

484

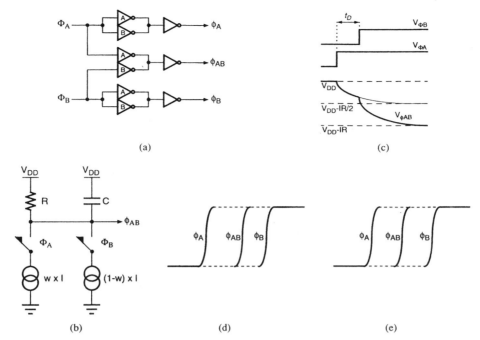

Fig. 6. Phase blending for phase-resolution improvement. (a) Single-stage phase-blender circuit, (b) simple model of phase-blending inverters, (c) plot of signal voltages in the simple model for $w = W_A/(W_A + W_B) = 0.50$, (d) phase-blender output signal edges for $w = 0.50$, and (e) phase-blender output signal edges for $w = 0.60$.

where R is the total resistive load, C is the output capacitance, I is the total pulldown current of the two phase-blending inverters, $u(t)$ is the unit step function, and w is the phase-blending inverter relative size ratio [refer to Fig. 6(a), where $w = W_A/(W_A + W_B)$ is the ratio of the device widths in inverter A to the total device widths in both inverters A and B]. Equation (2) is the sum of two decaying exponential terms, and Fig. 6(c) shows a plot of the resulting waveform according to this equation for the case where $w = 0.5$. Because the second exponential term is delayed in time by t_D relative to the first, it only begins to affect the slope of the decay after this delay has elapsed. Therefore, without explicitly solving the equation for each case of R, C, I, w, and t_D, it is not obvious when $V_{\Phi AB}(t)$ will cross $V_{DD}/2$.

For input signals separated in phase by one inverter delay (i.e., $t_D = 1/RC$), the model specifies that in order to ensure that the phase of ϕ_{AB} lies directly in between that of ϕ_A and ϕ_B, the phase-blending inverters must be sized in a $w = 60/(60+40) = 0.60$ ratio, such that the leading phase is coupled to an inverter that is bigger than the one that receives the lagging phase. This ratio was also confirmed empirically with simulations. The effect of the relative sizing of the phase-blending inverters is illustrated in Fig. 6(d) and (e), which shows the resulting output signal edges for $w = 0.50$ and $w = 0.60$, respectively. Clearly, the phase of output signal ϕ_{AB} is closer to that of ϕ_B than to that of ϕ_A when the phase-blending inverter size ratio is $w = 0.50$. Although asymmetrical $w = 0.60$ inverter sizing ensures good, evenly spaced edge placement of the three output signals, it requires that Φ_A lead Φ_B. Reversing the phase of these two input signals would result in a severely misplaced ϕ_{AB} since the effective sizing ratio would then be $w = 40/(40+60) = 0.40$.

Another design constraint of the phase-blender circuit is that all paths through the circuit must provide precisely the same loading and delay to ensure that the phase relationship between Φ_A and Φ_B is maintained by ϕ_A and ϕ_B.

The phase-blender idea can be extended to multiple cascaded stages for further phase-resolution improvement, with each additional stage improving the resolution by a factor of two. Fig. 7 shows a two-stage cascaded phase-blender circuit that provides a 4x improvement in phase resolution from input to output. Although it is theoretically possible to increase phase resolution indefinitely by adding more and more phase-blender stages, there is a practical limit. The number of inverters in each signal path increases by two with each additional phase-blending stage, making the circuit increasingly susceptible to voltage supply noise-induced jitter due to the additional delay in the signal path. Therefore, it is prudent to increase the number of blending stages to improve phase resolution only until the output phase step size from the phase blender is approximately equivalent to the anticipated voltage supply noise-induced jitter.

There are several design limitations that must be considered when designing a cascaded phase blender. First, the importance of proper (asymmetrical) sizing of the phase-blending inverters grows with the number of cascaded blending stages because edge misplacement has a compounding effect as the signals travel through the multiple stages. Additionally, close attention must be paid to ensuring equal loading for equal delay through all paths, requiring the use of dummy devices on otherwise unbalanced paths. Finally, like a single-stage phase blender, a cascaded phase blender also requires the phase of Φ_A to lead that of Φ_B to ensure even output phase spacing.

485

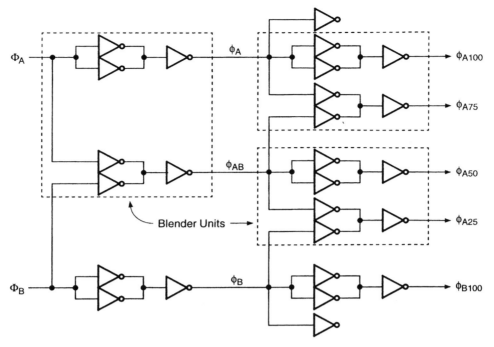

Fig. 7. Two-stage, cascaded phase-blender circuit for 4x phase-resolution improvement.

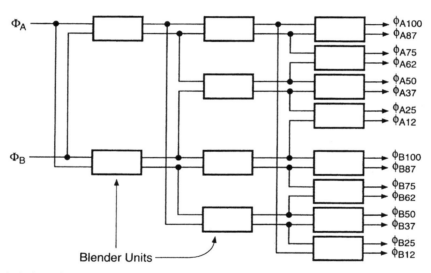

Fig. 8. Three-stage, symmetrical phase-blender circuit.

To overcome these design limitations of the cascaded phase blender, we developed a symmetrical phase blender. A block diagram of a three-stage symmetrical phase blender is shown in Fig. 8. This circuit is essentially two parallel cascaded phase-blender circuits, sharing some common paths. When Φ_A leads Φ_B, the outputs $\phi_{A100}, \phi_{A87}, \cdots, \phi_{fA12}, \phi_{B100}$ provide equal output phase spacing. When Φ_B leads Φ_A, the outputs $\phi_{B100}, \phi_{B87}, \cdots, \phi_{B12}, \phi_{A100}$ provide equal output phase spacing. Therefore, the circuit provides phase blending with an 8x improvement in phase resolution and equally spaced output signals regardless of which input signal leads in phase.

Additionally, the symmetrical blender allows for seamless input switching for continuous phase blending over multiple input delays. For example, assume that Φ_A leads Φ_B in phase. Beginning with output ϕ_{A100}, outputs $\phi_{A100}, \phi_{A87}, \cdots, \phi_{A12}, \phi_{B100}$ can be successively selected to evenly span the phase range between Φ_A and Φ_B. Once ϕ_{B100} is selected, Φ_A can be changed to another signal that lags Φ_B. This switching is possible without affecting the signal ϕ_{B100} because ϕ_{B100} has no dependence on or coupling from Φ_A. Then outputs $\phi_{B100}, \phi_{B87}, \cdots, \phi_{B12}, \phi_{A100}$ can be successively selected to evenly span the phase range between Φ_B and Φ_A. Once ϕ_{A100} is selected, Φ_B can be changed to yet another signal that lags Φ_A. Again, this is possible without any change in the signal ϕ_{A100} because ϕ_{A100} has no dependence on or coupling from Φ_B. This process can continue indefinitely. Also, because all paths through the symmetrical phase blender are inherently balanced, no dummy devices are needed.

486

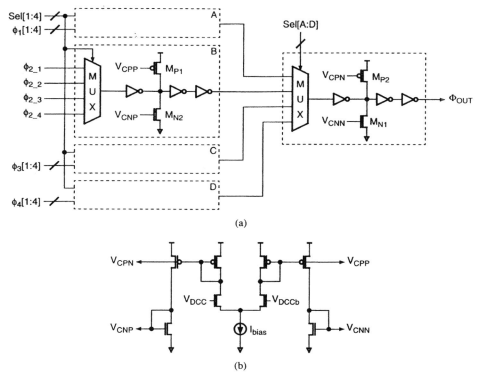

Fig. 9. (a) A 16 : 1 duty-cycle correcting multiplexer circuit. (b) Duty-cycle correction control circuit.

B. Signal Selection and Duty-Cycle Correction

Since the digital DLL was to be placed into a memory system that exchanges data on both edges of the clock, good duty cycle (i.e., close to 50%) is required to ensure that the data exchanged on either edge of the clock have equal bit times. Duty-cycle distortion is usually addressed in PLL's by simply running the PLL's voltage-controlled oscillator (VCO) at twice the system frequency and using a postdivider triggered on one edge of the VCO output to produce the output clock from the PLL [13]–[15]. This ensures good, 50% duty cycle. In a DLL, however, no frequency multiplication is possible. The duty cycle of the output signal must be directly corrected to 50%, for example, by using a duty-cycle correcting amplifier in the signal path as in Fig. 1 and in [4].

Although duty-cycle correction can be addressed by placing a duty-cycle corrector at the output of the DLL, this approach has several limitations. First, since duty cycle is corrected only at the output of the DLL, internal DLL signals may have poor duty cycle. It is good practice, however, to maintain 50% duty cycle throughout the signal path to maximize signal propagation as frequency is increased. Second, performing all the duty-cycle correction in one stage at the output of the DLL places a great deal of strain on the duty-cycle correcting circuit; it must have a large duty-cycle correction range to compensate for all the duty-cycle distortion that can accumulate in the signal path. Finally, adding a duty-cycle corrector directly into the signal path increases signal path delay, and thus susceptibility to voltage supply noise-induced jitter.

To address the issue of duty cycle, we developed the idea of duty-cycle correcting multiplexers. Since multiplexers would be needed in our DLL regardless, by adding duty-

cycle correcting functionality to the multiplexing circuitry, we implemented duty-cycle correction while requiring minimal additional power, area, and delay.

A 16 : 1 duty-cycle correcting multiplexer is shown in Fig. 9(a) with a corresponding control circuit in Fig. 9(b). To facilitate understanding of this circuit's operation, consider an example. Assume that signal ϕ_{2_1} is selected and has duty-cycle distortion such that output signal Φ_{OUT} has a high duty cycle. Assume also that Φ_{OUT} is sensed by a duty-cycle error detector, which produces a differential output error signal $V_{DCCb} - V_{DCC}$ proportional to the difference in duty cycle between Φ_{OUT} and the ideal 50%. Thus, in our example, V_{DCCb} will be greater than V_{DCC}, causing more current to be steered through the right branch of the control signal in Fig. 9(b) than through the left side. This in turn increases the strength of M_{P1} and M_{N1} compared to M_{P2} and M_{N2} in the duty-cycle correcting multiplexer of Fig. 9(a). These transistors alter the duty cycle of the signal as it passes from ϕ_{2_1} to Φ_{OUT}, driving Φ_{OUT} to the ideal 50% duty cycle. The use of both PMOS and NMOS devices to perform the duty-cycle correction ensures a symmetrical duty-cycle correction range. Furthermore, because duty-cycle correction has been distributed through two stages, the requirements on each individual duty-cycle correcting stage are reduced. By combining both necessary functions of signal selection and duty-cycle correction, this circuit minimizes signal path delay, jitter accumulation, circuit area, and power compared to performing both functions separately.

IV. DLL ARCHITECTURE

Fig. 10 is a block diagram of the entire digital DLL, with shading indicating the circuit blocks that were described in

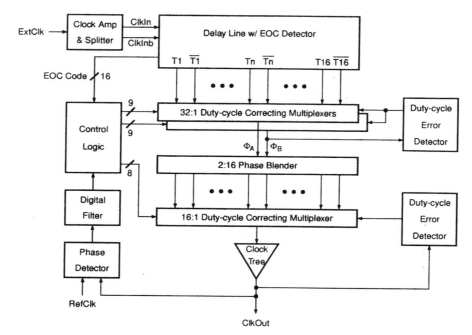

Fig. 10. Complete block diagram of the new digital DLL.

greater detail above. The DLL receives an input clock ExtClk and passes it through a clock amplifier and splitter to provide the two complementary input signals (ClkIn and ClkInb) to a 16-stage, 32-tap complementary delay line with EOC detector. The delay line provides 32 signals at its output taps, which then feed into two 32:1 duty-cycle correcting multiplexers. Each multiplexer selects one of a pair of phase-adjacent signals from the delay line. The two selected signals then pass to a three-stage, 2:16 symmetrical phase-blender circuit, which improves the phase resolution by a factor of eight. A final 16:1 duty-cycle correcting multiplexer selects one of the phase-blender output signals and passes it through a clock tree to provide the DLL's output signal ClkOut. The digital DLL also includes two independent duty-cycle correction loops as shown in the figure. By using two separated duty-cycle correcting loops, duty-cycle correction is distributed throughout the signal path. This ensures a good duty cycle throughout the signal path and reduces the duty-cycle correcting requirements of any one stage.

The DLL uses bang-bang-type, all-digital feedback to lock the phase of its output signal ClkOut to that of a reference signal RefClk. A phase detector compares the phase of ClkOut to RefClk and produces a binary error signal, which passes through an optional digital filter to a control logic circuit. The digital filter is a simple majority detector, which has no effect when the loop is acquiring lock but reduces dithering once lock is acquired. The control logic is composed of simple combinational logic and counters that drive the multiplexers to select the two phase-adjacent coarse phase signals from the delay line and the fine phase signal from the phase blender that minimize the phase error between ClkOut and RefClk. Because the phase information is stored in this DLL as a digital state, the DLL can quickly recover from low-power modes, requiring only enough time for the signals to propagate

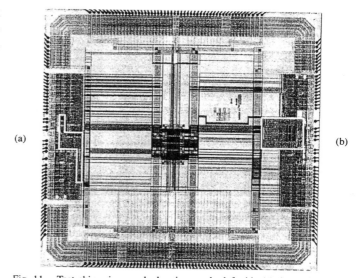

Fig. 11. Test-chip micrograph showing on the left side (a) the analog DLL of [6] and on the right side (b) the new digital DLL integrated into identical interface cells.

through the signal path of the circuit from ExtClk to ClkOut to provide a phase-locked output signal.

It is important to recognize the role of the EOC detector and code in this architecture. Because the delay line and blender are uncontrolled, open-loop circuits, the architecture relies on the control circuit's use of the EOC code to ensure proper coarse phase selection, small maximum phase step size, and phase transfer function monotonicity. The EOC code enables the control logic to determine when to switch between the true and complement taps of the delay line to ensure that phase-adjacent taps are always selected by the coarse multiplexers for the phase blender. The EOC code also enables the control logic to determine which set of blender taps provides evenly spaced output signals.

488

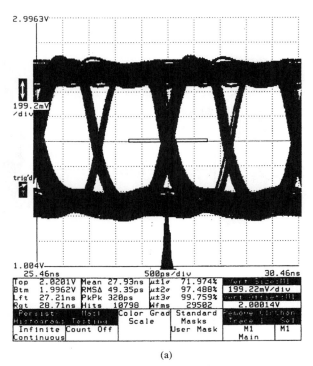

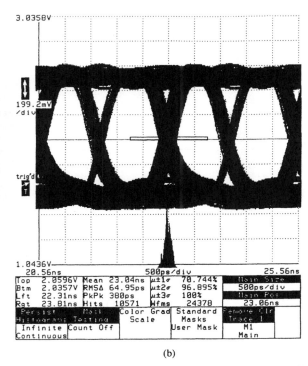

Top	2.0201V	Mean 27.93ns	μ±1σ	71.974%	Vert Size:M1
Btm	1.9962V	RMSΔ 49.35ps	μ±2σ	97.488%	199.22mV/div
Lft	27.21ns	PkPk 320ps	μ±3σ	99.759%	Vert Offset:M1
Rgt	28.71ns	Hits 10798	Wfms	29502	2.00014V

(a)

Top	2.0596V	Mean 23.04ns	μ±1σ	70.744%	Main Size
Btm	2.0357V	RMSΔ 64.95ps	μ±2σ	96.895%	500ps/div
Lft	22.81ns	PkPk 380ps	μ±3σ	100%	Main Pos
Rgt	23.81ns	Hits 10571	Wfms	24370	23.06ns

(b)

Fig. 12. Measured transmit eye diagrams at 3.3 V and 400 MHz of the high-speed interface cells with (a) the analog DLL of [6] and (b) the new digital DLL.

V. MEASURED PERFORMANCE

A. Test Chip

Both the digital DLL presented here and an implementation of the analog DLL of Donnelly *et al.* [6] were integrated into identical high-speed CMOS interface cells on opposite sides of a single test chip. A micrograph of this test chip is shown in Fig. 11. The test chip I/O was laid out symmetrically so that either interface cell could be tested on the same hardware by simply removing the test chip from the test socket, rotating it 180°, and reinserting it into the socket. This allowed a true side-by-side comparison of the two DLL's operating in a system. The test-chip circuits were fabricated using a standard 0.4-μm, 3.3-V CMOS process with 0.65-V threshold voltages.

B. Test Results

Unless indicated otherwise, all test results described in this section were measured with the analog and digital DLL's operating in their respective high-speed interface cells at 3.3 V and 400 MHz (800 Mb/s/pin) using the same test vectors. Additionally, the test chip included noise-generator circuits, which produced digital switching noise during the testing of both interfaces.

Fig. 12(a) and (b) shows eye diagrams of the two interfaces with the analog and digital DLL's, respectively. The diagrams indicate the output timing performance of the interface cells in the test system. Although the interface with the analog DLL provided slightly better timing performance, 320 ps p–p versus 380 ps p–p for the interface with the digital DLL, the performances of both interfaces (and therefore, both DLL's) were comparable. This is surprisingly good considering the extensive use of poor PSRR elements, such as inverters, in

the signal path of the digital DLL. (Note: I/O circuit duty-cycle distortion produced the unequal eyes in both diagrams. This is unrelated to the DLL's.)

Fig. 13(a) and (b) shows receive shmoo diagrams for the two interfaces with the analog and digital DLL's, respectively. The diagrams indicate the CMOS interfaces' valid timing windows for receiving data. On the diagrams, the y-axis is supply voltage (2.5 V $< V_{DD} <$ 4.0 V) while the x-axis indicates input data positioning along a bit period (1/800 Mb/s = 1.25 ns). The normal data position is in the center of the bit period. A black dot in the diagram indicates incorrectly received data for that combination of bit position and V_{DD}. Ideally, the window should be entirely white, but realistically, it is limited by jitter from the DLL and other sources. Therefore, this test measures the amount of tolerable skew on the input timing over a range of supply voltages. Although the interface with the analog DLL delivers better timing performance than the interface with the digital DLL (1.02 versus 0.92 ns), both meet the component specification of 0.85 ns.

Fig. 14 is a circle plot of the measured phase of the DLL's output signal ClkOut, illustrating the DLL's ability to provide infinite phase range. The y-axis indicates delay [or phase, as in (1)] of the ClkOut signal relative to a fixed 400-MHz signal. The x-axis indicates cycle count. These data were measured by probing the on-chip DLL output signal (ClkOut) and forcing the DLL's phase-detector output low. This caused the DLL's output phase to continually advance over time. The term *circle plot* is used because this diagram is equivalent to sweeping a phasor that represents the phase of ClkOut around the phase plane, thereby drawing a circle in the phase plane. Because the phase of ClkOut is measured relative to a fixed 400-MHz signal, the plotted delay appears modulo 2.5 ns, where 2.5 ns

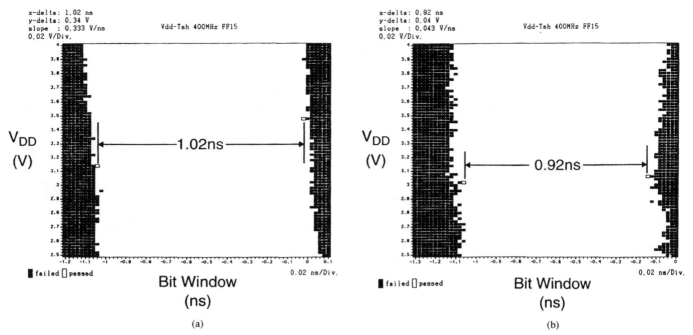

Fig. 13. Measured shmoo diagrams showing the 400-MHz receive timing windows of the high-speed interface cells with (a) the analog DLL of [6] and (b) the new digital DLL.

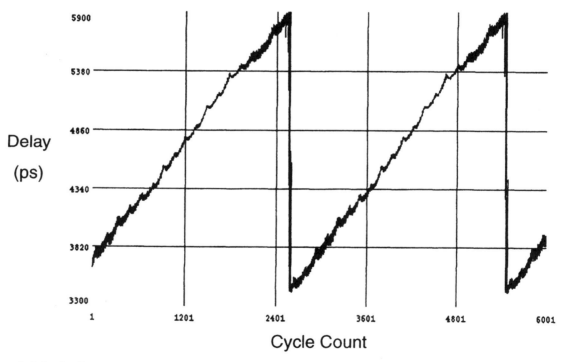

Fig. 14. Measured circle plot illustrating the infinite phase transfer characteristic of the digital DLL.

= 360° at 400 MHz. The absolute value of delay (i.e., from 3.4 to 5.9 ns) is irrelevant since it includes some test-system setup time. The data were measured and plotted using a time-interval analyzer.

The circle plot illustrates the DLL's phase transfer function, showing its reasonably good linearity, monotonicity, and lack of discontinuities. The small bumps in the transfer function indicate a change in coarse reference phase selected from

the delay line. The slope of the transfer function depends on PVT conditions and system frequency, since these conditions determine how many delay-line taps are required to provide 180° of phase. In this case, nine taps were required, resulting in an average phase step size of 20 ps or 2.9°.

Table I presents a summary of many of the measured and simulated results of the analog and digital DLL's operating in their respective CMOS interfaces. Although the analog DLL

490

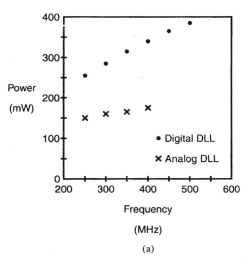
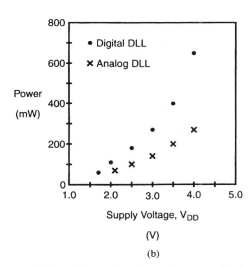

Fig. 15. Measured DLL power consumption (a) as a function frequency for $V_{DD} = 3.3$ V and (b) as a function supply voltage for $f = 400$ MHz.

TABLE I
ANALOG AND DIGITAL DLL PERFORMANCE SUMMARY AT 3.3 V AND 400 MHz

Performance Parameter	Analog DLL	Digital DLL
DLL Power (@ 400MHz)	175mW	340mW
DLL Power (@ 300MHz)	160mW	285mW
DLL Area	0.68mm^2	0.96mm^2
Long-term DLL Output Clock Jitter	195ps p-p	245ps p-p
Maximum Delay Through DLL (Simulated)	6.1ns	8.0ns
Maximum DLL Phase Step (Simulated)	10ps	40ps
Maximum DLL Lock Time (Simulated)	2.0us	2.9us
Maximum DLL Frequency	435MHz	> 667MHz
Minimum DLL Supply Voltage (V_{DD})	2.1V	1.7V
Valid Receive Window	1.02ns	0.92ns
System Output Data Jitter	320ps	380ps
DLL Circuit Design Time	4 man-months	1 man-month

uses less power and area, and provides better timing performance (smaller long-term jitter) and phase resolution (smaller maximum phase step), both DLL's enable the interface cells to meet the component requirements when operating in the test system. Additionally, the digital DLL has a higher maximum operating frequency, works at lower supply voltages, and requires *much* less effort to port to other processes (one versus four man-months).

Fig. 15(a) and (b) shows plots of measured DLL power versus frequency at $V_{DD} = 3.3$ V and measured DLL power versus voltage supply at $f = 400$ MHz, respectively. Although both plots show that the digital DLL dissipated more power than the analog DLL for all measured conditions, the plots illustrate the different characteristics of the power consumed by the two DLL's. As mentioned earlier, the power of both DLL's is distributed between IV power in the constant-current stages and CV^2f power in the CMOS stages. The curves in Fig. 15(a) show that the digital DLL's power dissipation has a greater dependence on frequency than does the analog DLL's power. The curves in Fig. 15(b) show that the digital DLL's power dissipation has a predominantly square-law dependence on supply voltage, whereas the analog DLL's power dissipation has a mixed square-law and linear dependence. These trends confirm that the power of the analog DLL has a relatively higher IV term, whereas the power of the digital DLL has a

relatively higher CV^2f term. This indicates that digital DLL's have the potential for providing better power scaling than analog DLL's as supply voltages decrease in the future.

Finally, we have shown in Table I and in Fig. 15(b) that the digital DLL operates at lower supply voltages than the analog DLL. Although the operation of the digital DLL was limited to 1.7 V, this limitation was due to our use of several analog elements in the digital DLL (i.e., it was a *mostly* digital DLL). The digital DLL used an analog clock amplifier, two analog duty-cycle error detectors (see Fig. 10), and an analog quadrature phase detector (in a second loop, not shown). Using an analog design for these circuit blocks in the digital DLL was faster to implement without preventing evaluation of the key digital blocks in the DLL, but their use determined the minimum supply voltage of the digital DLL.

VI. CONCLUSION

We have described the architecture of a portable digital DLL and demonstrated that it provides jitter performance comparable to an analog DLL when fabricated in the same 3.3-V, 0.4-μm standard CMOS process. Several circuits were developed to enable the DLL to provide very fine phase resolution, infinite phase range, and good duty-cycle performance throughout the signal path. Despite its relatively simple architecture, the digital DLL meets all system specifications, and it operates down to lower supply voltages than its analog counterpart. Utilizing essentially only simple digital CMOS gates, the DLL can be ported to new processes in minimal time. For these reasons, this digital DLL provides an alternative to analog DLL's for clock alignment applications.

ACKNOWLEDGMENT

The authors thank J. McBride and P. Gordon for layout support and S. Sidiropoulos for helpful insights.

REFERENCES

[1] A. Efendovich, Y. Afek, C. Sella, and Z. Bikowsky, "Multifrequency zero-jitter delay-locked loop," *IEEE J. Solid-State Circuits*, vol. 29, pp. 67–70, Jan. 1994.

[2] J.-M. Han, J. Lee, S. Yoon, S. Jeong, C. Park, I. Cho, S. Lee, and D. Seo, "Skew minimization techniques for 256 Mb synchronous DRAM and beyond," in *VLSI Circuits Dig. Tech. Papers*, June 1996, pp. 192–193.

[3] A. Hatakeyama, H. Mochizuki, T. Aikawa, M. Takita, Y. Ishii, H. Tsuboi, S. Fujioka, S. Yamaguchi, M. Koga, Y. Serizawa, K. Nishimura, K. Kawabata, Y. Okajima, M. Kawano, H. Kojima, K. Mizutani, T. Anezaki, M. Hasegawa, and M. Taguchi, "A 256 Mb SDRAM using register-controlled digital DLL," in *ISSCC 1997 Dig. Tech. Papers*, Feb. 1997, pp. 72–73.

[4] T. Lee, K. Donnelly, J. Ho, J. Zerbe, M. Johnson, and T. Ishikawa, "A 2.5 V CMOS delay-locked loop for 18 Mbit, 500 megabyte/s DRAM," *IEEE J. Solid-State Circuits*, vol. 29, pp. 1491–1496, Dec. 1994.

[5] S. Sidiropoulos and M. Horowitz, "A semidigital dual delay-locked loop," *IEEE J. Solid-State Circuits*, vol. 32, pp. 1683–1692, Nov. 1997.

[6] K. Donnelly, Y. Chan, J. Ho, C. Tran, S. Patel, B. Lau, J. Kim, P. Chau, C. Huang, J. Wei, L. Yu, R. Tarver, R. Kulkarni, D. Stark, and M. Johnson, "A 660MB/s interface megacell portable circuit in 0.3 μm–0.7 μm CMOS ASIC," *IEEE J. Solid-State Circuits*, vol. 31, pp. 1995–2003, Dec. 1996.

[7] N. Kushiyama, S. Ohshima, D. Stark, H. Noji, K. Sakurai, S. Takase, T. Furuyama, R. Barth, A. Chan, J. Dillon, J. Gasbarro, M. Griffin, M. Horowitz, T. Lee, and V. Lee, "A 500-Megabyte/s data-rate 4.5M DRAM," *IEEE J. Solid-State Circuits*, vol. 28, pp. 490–508, Apr. 1993.

[8] M. Hasegawa, M. Nakamura, S. Narui, S. Ohkuma, Y. Kawase, H. Endoh, S. Miyatake, T. Akiba, K. Kawakita, M. Yoshida, S. Yamada, T. Sekigguchi, I. Asano, Y. Tadaki, R. Nagai, S. Miyaoka, K. Kajigaya, M. Horiguchi, and Y. Nakagome, "A 256 Mb SDRAM with subthreshold leakage current suppression," in *ISSCC 1998 Dig. Tech. Papers*, Feb. 1998, pp. 80–81.

[9] T. Saeki, Y. Nakaoka, M. Fujita, A. Tanaka, K. Nagata, K. Sakakibara, T. Matano, Y. Hoshino, K. Miyano, S. Isa, E. Kakehashi, J. Drynan, M. Komuro, T. Fukase, H. Iwasaki, J. Sekine, M. Igeta, N. Nakanishi, T. Itani, K. Yoshida, H. Yoshino, S. Hashimoto, T. Yoshii, M. Ichinose, T. Imura, M. Uziie, K. Koyama, Y. Fukuzo, and T. Okuda, "A 2.5 ns clock access 250 MHz 256 Mb SDRAM with synchronous mirror delay," *ISSCC 1996 Dig. Tech. Papers*, Feb. 1996, pp. 374–375.

[10] B. Garlepp, K. Donnelly, J. Kim, P. Chau, J. Zerbe, C. Huang, C. Tran, C. Portmann, D. Stark, Y. Chan, T. Lee, and M. Horowitz, "A portable digital DLL architecture for CMOS interface circuits," in *VLSI Circuits Dig. Tech. Papers*, June 1998, pp. 214–215.

[11] M. Griffin, J. Zerbe, A. Chan, Y. Jun, Y. Tanaka, W. Richardson, G. Tsang, M. Ching, C. Portmann, Y. Li, B. Stonecypher, L. Lai, K. Lee, V. Lee, D. Stark, H. Modarres, P. Batra, J. Louis-Chandran, J. Privitera, T. Thrush, B. Nickell, J. Yang, V. Hennon, and R. Sauve, "A process independent 800 MB/s DRAM bytewide interface featuring command interleaving and concurrent memory operation," in *ISSCC 1998 Dig. Tech. Papers*, Feb. 1998, pp. 156–157.

[12] S. Sidiropoulos, "High-performance interchip signalling," Ph.D. dissertation, Computer Systems Laboratory, Stanford University, Stanford, CA, Apr. 1998. Available as Tech. Rep. CSL-TR-98-760 from http://elib.stanford.edu/.

[13] I. Young, M. Mar, and B. Bhushan, "A 0.35 μm CMOS 3-880 MHz PLL N/2 multiplier and distribution network with low jitter for microprocessors," in *ISSCC 1997 Dig. Tech. Papers*, Feb. 1997, pp. 330–331.

[14] V. von Kaenel, D. Aebischer, C. Piguet, and E. Dijkstra, "A 320 MHz, 1.5 mW at 1.35 V CMOS PLL for microprocessor clock generation," in *ISSCC 1996 Dig. Tech. Papers*, Feb. 1996, pp. 132–133.

[15] V. von Kaenel, D. Aebischer, R. van Dongen, and C. Piguet, "A 600 MHz CMOS PLL microprocessor clock generator with a 1.2 GHz VCO," in *ISSCC 1998 Dig. Tech. Papers*, Feb. 1998, pp. 396–397.

Kevin S. Donnelly (A'93) was born in Los Angeles, CA, in 1961. He received the B.S. degree in electrical engineering and computer science from the University of California, Berkeley, in 1985 and the M.S. degree in electrical engineering from San Jose State University, San Jose, CA, in 1992.

He was with Memorex, Sipex, and National Semiconductor, specializing in bipolar and BiCMOS analog circuits for disk-drive read/write and servo channels. In 1992, he joined Rambus, Inc., Mountain View, CA, where he has designed high-speed CMOS PLL circuits for clock recovery and data synchronization, and high-speed I/O circuits. He currently manages a group developing I/O circuits and PLL's. His interests include PLL's and DLL's, I/O circuits, and data converters. He is a Member of the ISSCC Digital Subcommittee. He has received several circuit design patents.

Mr. Donnelly is a coauthor of the paper that won the Best Paper Award at the 1994 ISSCC.

Jun Kim was born in Tokyo, Japan, on November 14, 1966. He received the B.S.E.E. degree from the University of California, Berkeley, in 1989.

From 1989 to 1991, he was with Vitelic, Inc., where he worked on SRAM and DRAM development. Between 1991 and 1994, he was with Sun Microsystems, where he was involved in microprocessor and digital circuit design. Since 1994, he has been with Rambus, Inc., Mountain View, CA, as a Designer of high-speed CMOS I/O and DLL circuits.

Pak S. Chau was born in Hong Kong in 1966. He received the B.S. degree in computer system engineering from the University of Massachusetts, Amherst, in 1989 and the M.S. degree in electrical engineering from the University of California, Davis, in 1991.

He was with National Semiconductor and Chrontel, Inc., where he worked as an Analog Circuit Designer. In 1994, he joined Rambus, Inc., Mountain View, CA, where he has engaged in designing high-speed I/O and DLL circuits.

Jared L. Zerbe was born in New York, NY, in 1965. He received the B.S. degree in electrical engineering from Stanford University, Stanford, CA, in 1987.

He joined VLSI Technology, Inc., in 1987, where he worked on semicustom ASIC design. In 1989, he joined MIPS Computer Systems, where he designed high-performance floating-point blocks. Since 1992, he has been with Rambus Inc., Mountain View, CA, where he has specialized in the design of high-speed I/O and PLL/DLL clock recovery and data synchronization circuits.

Bruno W. Garlepp was born in Bahia, Brazil, on October 29, 1970. He received the B.S.E.E. degree from the University of California, Los Angeles, in 1993 and the M.S.E.E. degree from Stanford University, Stanford, CA, in 1995.

In 1993, he joined the Hughes Aircraft Advanced Circuits Technology Center, Torrance, CA. There, he designed high-precision analog integrated circuits for A/D applications, as well as CMOS, bipolar, and SiGe RF circuits for wide-band communications applications. In 1996, he joined Rambus, Inc., Mountain View, CA, where he designs and develops high-speed CMOS clocking and I/O circuits for synchronous chip-to-chip communication.

Charles Huang received the B.S. degree in electrical engineering from the University of Fuzhou, China, in 1982 and the M.S. degree in electrical engineering from the University of Arkansas, Fayetteville, in 1990.

He was with ULSI and SGI, working in the area of PLL and cache circuit design. He joined Rambus, Inc., Mountain View, CA, in 1994, where he has being engaged in high-speed CMOS DLL and I/O circuit design.

CMOS DLL-Based 2-V 3.2-ps Jitter 1-GHz Clock Synthesizer and Temperature-Compensated Tunable Oscillator

David J. Foley, *Student Member, IEEE*, and Michael P. Flynn, *Senior Member, IEEE*

Abstract—This paper describes a low-voltage low-jitter clock synthesizer and a temperature-compensated tunable oscillator. Both of these circuits employ a self-correcting delay-locked loop (DLL) which solves the problem of false locking associated with conventional DLLs. This DLL does not require the delay control voltage to be set on power-up; it can recover from missing reference clock pulses and, because the delay range is not restricted, it can accommodate a variable reference clock frequency. The DLL provides multiple clock phases that are combined to produce the desired output frequency for the synthesizer, and provides temperature-compensated biasing for the tunable oscillator. With a 2-V supply the measured rms jitter for the 1-GHz synthesizer output was 3.2 ps. With a 3.3-V supply, rms jitter of 3.1 ps was measured for a 1.6-GHz output. The tunable oscillator has a 1.8% frequency variation over an ambient temperature range from 0 °C to 85 °C. The circuits were fabricated on a generic 0.5-μm digital CMOS process.

Index Terms—CMOS analog integrated circuits, delay-locked loops, frequency synthesizers, tunable oscillators, voltage controlled oscillators.

I. INTRODUCTION

TRADITIONALLY, phase-locked loops (PLLs) have been used for clock synthesis. The synthesizer and tunable oscillator outlined in this paper employ a delay-locked loop (DLL). A DLL is more stable than higher order PLLs and requires only one capacitor in its first-order loop filter. On the other hand, a PLL generally requires a more complex second-order filter. This filter usually employs larger components which may need to be off chip. Additionally, a DLL offers better jitter performance than a PLL because phase errors induced by supply or substrate noise do not accumulate over many clock cycles [1].

The self-correcting DLL overcomes problems of false locking associated with conventional DLLs. A self-correcting circuit detects when the DLL is locked, or is attempting to lock, to an incorrect delay and then brings the DLL into a correct locked state. This DLL does not require the delay control voltage to be set on power-up; it can recover from missing reference clock pulses and, because the delay range is not restricted, it can accommodate a variable reference clock frequency. This paper describes how a small number of additional

Manuscript received July 19, 2000; revised October 24, 2000. This work was supported by Parthus Technologies.

D. J. Foley is with the Department of Microelectronics, National University of Ireland, Cork, Ireland.

M. P. Flynn is with Parthus Technologies, Cork, Ireland.

Publisher Item Identifier S 0018-9200(01)01483-4.

digital logic gates are required to convert a conventional DLL into a wider range self-correcting DLL. For comparison, in [2] a second DLL is added to achieve wider range operation.

The synthesizer outlined in this paper operates over a wide range of input reference clock frequencies and generates a low-jitter output clock running at nine times the reference frequency. Jitter measurements of 3.2 ps rms and 20 ps peak-to-peak, for a 2-V supply and 1-GHz output frequency, show that the core DLL compares well with recently reported DLLs [2], [3]. Multiple clock phases from the DLL are combined using digital logic to produce the synthesizer output [4]. An alternative approach requiring a pair of on-chip tuned *LC*-tanks is described in [5].

The tunable voltage-controlled oscillator (VCO) is intended for use in a transceiver where the receive and transmit clocks are plesiochronous. It is possible to tune the VCO around a center frequency while still maintaining good temperature independence. In some applications it may also act as a replacement for a fractional-N-type synthesizer. This circuit is similar to the oscillator described in [6] but it uses a lower jitter DLL in place of the PLL and can operate over a wider frequency range.

In Section II the DLL architecture is discussed, starting with a review of a conventional DLL and progressing to the new self-correcting architecture. Section III outlines the clock synthesizer architecture. This is followed in Section IV by an outline of the temperature-compensated tunable oscillator architecture. Section V discusses the circuit layout and Section VI introduces measured performance results for the two circuits. This paper then concludes in Section VII with a summary of the achievements of this work.

II. DLL ARCHITECTURE

A. Conventional DLL

A simplified block diagram of a conventional DLL is illustrated in Fig. 1. This circuit contains a voltage-controlled delay line (VCDL), a phase detector, a charge pump, and a first-order loop filter. The delay line, consisting of cascaded variable delay stages, is driven by the input reference clock, *ckref*. The output of the delay line's final stage and the *ckref* falling edges are compared by the phase detector to determine the phase alignment error. The phase detector output is integrated by the charge pump and loop filter capacitor to generate the control voltage, *vcntl*, of the delay stages.

When correctly locked, the total delay of the delay line should equal one period of the reference clock. A conventional

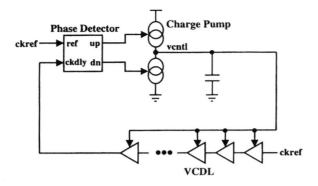

Fig. 1. Conventional DLL architecture.

(a)

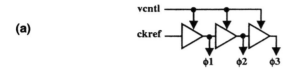

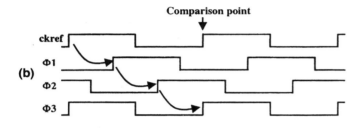

(b)

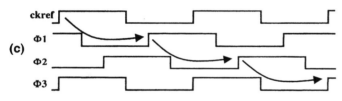

(c)

Fig. 2. (a) Three-stage VCDL. (b) Waveforms with correct lock. (c) Waveforms with false lock.

DLL may lock or attempt to lock to an incorrect delay. In Fig. 2 we show correct and false locking for a three-stage delay line [Fig. 2(a)]. Fig. 2(b) shows the output phases at each stage ($\phi1$, $\phi2$, and $\phi3$) with the delay line in correct lock. The DLL control loop has aligned $\phi3$ and *ckref*. The total delay is one period of the reference clock. In Fig. 2(c) $\phi3$ and *ckref* are again aligned but the total delay is two clock periods. The DLL can also falsely lock to three or more periods of delay or can attempt to lock to zero delay.

B. Self-Correcting DLL Architecture

Fig. 3 shows a block diagram of the new self-correcting DLL. The problem of false locking is solved by the addition of a lock-detect circuit and by some slight modifications to the conventional phase detector. The DLL incorporated in the two designs reported in this paper employs a nine-stage VCDL as shown in Fig. 3.

In a conventional DLL, only the state of the output of the last delay element is used. From the example in Fig. 2, we can see that the state at the outputs of the other delay elements can

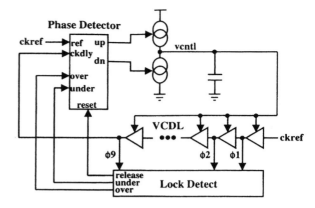

Fig. 3. Self-correcting DLL architecture.

provide additional information about the nature of the locked delay. In the prototype the delayed phases, $\phi(1{:}9)$, are decoded to indicate the VCDL delay. If the delay is outside an acceptable delay range then the lock-detect circuit takes control of the loop from the phase detector. The lock-detect circuit signals the charge pump to charge or discharge the filter capacitor until it is safe for the phase detector to regain control of the loop.

Three control signals are produced by the lock-detect circuit: *over* to indicate that the VCDL delay is greater than 1.5 reference clock periods, *under* to indicate that the delay is less than 0.75 clock periods, and *release* is activated when the delay reaches 1.25 clock periods. The *release* signal clears the *over* and *under* control signals and removes the phase detector from reset. The phase detector then regains control of the loop. If neither *under* nor *over* is active then the phase detector has control of the loop and the DLL is either in correct lock or approaching correct lock.

If the DLL is in lock and it is brought out of lock because of missing reference clock pulses or a step in the input reference frequency, then the DLL may inadvertently try to lock to an incorrect delay. The DLL is allowed to attempt to reach the undesired lock delay until it triggers either an *over* or an *under* signal at which time the lock-detect circuit takes control of the DLL loop.

C. Lock-Detect Circuit

The VCDL output phases are first level shifted to CMOS levels. The level shift circuitry is designed to have high gain and fast rise and fall times. This helps to minimize any jitter contribution from this circuitry. The level-shifted output phases, $\phi(1{:}9)$, are latched on the rising edge of the reference clock. The outputs from these latches are processed by the decode circuitry as shown in the schematic of Fig. 4. The inputs, $Q(1{:}8)$, correspond to the $\phi(1{:}8)$ output phases of the VCDL. Fig. 5 shows example output waveforms for a nine-stage VCDL. In Fig. 5(a) when the state of the VCDL output phases is decoded none of the control signals are activated as the VCDL is correctly locked to one period of the reference clock. In Fig. 5(b) the VCDL is incorrectly locked to two periods of the reference clock and the state of the output phases is decoded to activate the *over* control signal.

The phase detector outputs, *up* and *dn*, signal the charge pump to charge or discharge the filter capacitor. An active *over* output

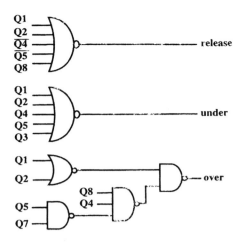

Fig. 4.　Lock-detect decode circuitry.

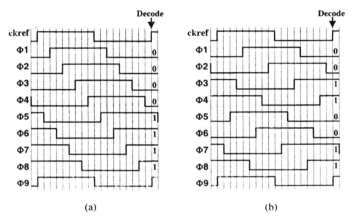

Fig. 5.　Nine-stage VCDL waveforms with (a) correct lock and (b) false lock.

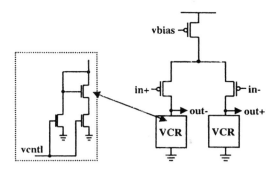

Fig. 6.　VCDL delay stage schematic.

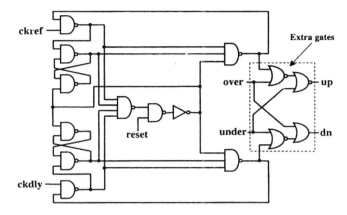

Fig. 7.　Phase detector schematic.

from the lock-detect circuit disables the phase detector and activates the *up* control signal. Similarly, the lock detect *under* output activates *dn*. Following power-on reset the lock-detect circuit is initialized by setting *over* active. This ensures a faster acquisition time for the DLL because the filter capacitor is continuously charged to a voltage level corresponding to 1.25 reference clock periods. At this VCDL delay, the *release* signal is activated and the phase detector gains control of the loop and brings the DLL to lock. The state of the output phases corresponding to a delay of nine reference clock periods is the same as that corresponding to a single reference clock period delay. This circuitry is therefore only capable of detecting incorrect delays up to eight periods of the reference clock. This is not a limitation of the design as any delays above this would be outside the delay range of the VCDL. In general, the error detection logic can detect an incorrect lock delay up to $N - 1$ periods of the reference clock, where N is equal to the number of VCDL output phases.

D. Voltage-Controlled Delay Line (VCDL)

Fig. 6 shows one of the VCDL delay stages. The stage is designed to operate from a supply as low as 1.8 V and is similar to that used in [7]. The stage propagation delay is proportional to the tail current for the output charging and to the voltage-controlled resistor (VCR) resistance for the output discharging.

A three-transistor VCR structure is adopted for better control linearity. The DLL negative feedback control loop compensates for variations in the stage delay due to process and temperature. The differential delay stage structure and coupling capacitors between bias lines and supply help to minimize supply-induced jitter noise.

E. Charge Pump

The charge pump charges or discharges the filter capacitor. The voltage on this capacitor, *vcntl*, sets the VCDL stage propagation delay. To minimize the temperature variation of the VCDL delay, the charging and discharging currents are proportional to absolute temperature. This helps to maintain a constant loop gain and phase margin over temperature.

F. Phase Detector

The phase detector, shown in Fig. 7, employs the conventional sequential-phase-frequency detection scheme [7] but extra gates have been included. This extra logic enables the lock-detect circuit to over-ride the phase detector control of the loop. The lock-detect output signals, *over* and *under*, now have direct control of the charge pump. The lock-detect circuit can therefore charge or discharge the VCDL control voltage, *vcntl*, to a voltage from which it is safe for the phase detector to regain control of the loop.

III. CLOCK SYNTHESIZER ARCHITECTURE

The clock synthesizer generates a differential output clock running at nine times the input reference frequency. The clock

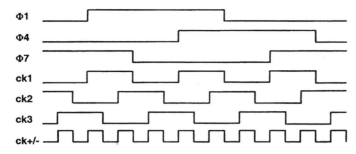

Fig. 8. Clock synthesis waveforms.

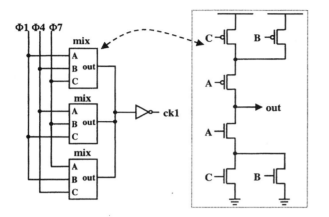

Fig. 9. Optimized AND-OR block diagram.

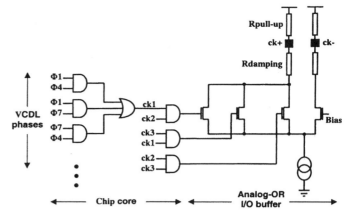

Fig. 10. 1.62-GHz clock generation schematic.

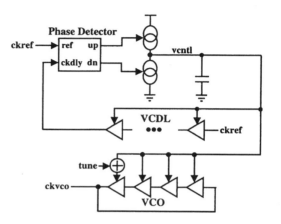

Fig. 11. Tunable VCO architecture.

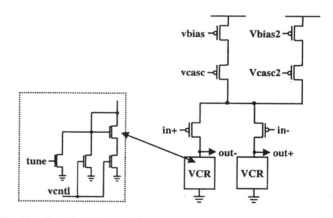

Fig. 12. Tunable VCO stage block diagram.

synthesizer employs the DLL structure shown in Fig. 3 to generate the multiple clock phases that are then combined to produce the output clock. There are two steps in the generation of the output clock. The first step combines the nine DLL output phases, $\phi(1{:}9)$, to generate three clocks ck1, ck2, and ck3. Fig. 8 shows the clock waveforms. These three clocks are phase separated by one-ninth of a reference clock period and have a frequency three times that of the reference clock. Fig. 9 shows how the $\phi1$, $\phi4$, and $\phi7$ output phases are combined in an optimized AND-OR structure with symmetrical delays to generate the ck1 clock. Using identical logic the $\phi2$, $\phi5$, and $\phi8$ phases produce the ck2 clock and the $\phi3$, $\phi6$, and $\phi9$ phases produce the ck3 clock.

The second step in generating the synthesizer output clock is to combine these three clocks in another AND-OR structure to produce a differential output clock, $ck+$ and $ck-$, running at nine times the reference clock frequency; see Fig. 8. This design produces a 1.62-GHz output clock frequency for a 180-MHz reference clock frequency. For a 0.5-μm 3.3-V CMOS process there is a bandwidth limitation of approximately 500 MHz for reliable on-chip clock transmission [8]. The high bandwidth available at the chip outputs is utilized (determined by the external pull-up resistor and load capacitance) [8] to produce the 1.62-GHz clock as shown in Fig. 10. The AND function of the clock generation is performed in the chip core and the analog OR function is performed in the I/O ring. External pull-up resistors set the output swing and match the output impedance to that of the test equipment. Damping resistors are included to avoid any oscillations resulting from the combination of the lead and pin inductance and load capacitance. This removes the necessity to double bond these high-frequency outputs.

IV. TEMPERATURE-COMPENSATED TUNABLE VCO ARCHITECTURE

The temperature-compensated oscillator utilizes the control loop voltage, *vcntl*, of the DLL (Fig. 3) to compensate for any temperature and supply voltage induced frequency fluctuation in a VCO. Fig. 11 shows how the VCO and VCDL stages are both connected to *vcntl*. (For ease of illustration a conventional DLL is shown in Fig. 11 but in practice the new DLL architecture of Fig. 3 is employed). The VCDL in the DLL tracks temperature and process variations in the VCO circuit. The VCO is

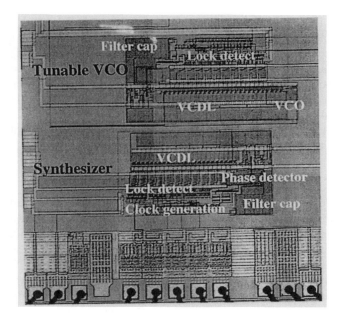

Fig. 13. Die photo of the synthesizer and tunable oscillator.

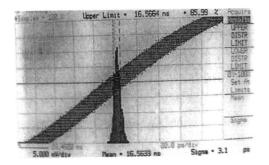

Fig. 14. 1.62-GHz synthesizer edge jitter histogram.

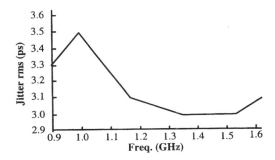

Fig. 15. Variation of measured jitter over output frequency.

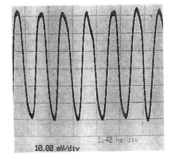

Fig. 16. 720-MHz synthesizer output for $V_{DD} = 1.8$ V.

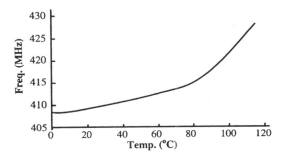

Fig. 17. VCO frequency variation with temperature.

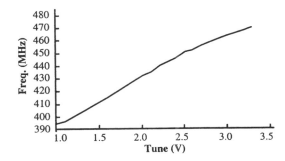

Fig. 18. VCO frequency variation with tune voltage.

composed of the same delay stages as the VCDL and its temperature (and process) variations will therefore be the same (apart from some minor random mismatch effects and thermal gradients across the die). *vcntl* thus compensates for the VCDL and VCO temperature fluctuations. The last VCO stage has an additional tuning voltage, *tune*, which fine tunes the VCO frequency. By varying the tune voltage it is possible to tune the VCO center frequency to within ±3%. A wider tuning range can be achieved by varying the frequency of the DLL reference clock, *ckref*.

The schematic of the last VCO stage is shown in Fig. 12. This stage is identical to the other VCO and VCDL stages except that the VCR contains a transistor which is connected to the external *tune* voltage. In all other stages this transistor is connected to ground. The extra charging current required in this VCO stage is provided by the controlled current source bias V_{bias2}.

V. CIRCUIT LAYOUT

The synthesizer and temperature-compensated tunable oscillator were fabricated on a standard 0.5-μm triple-metal single-poly digital CMOS process. The die photomicrograph of the device, containing both the synthesizer and temperature-compensated tunable oscillator, is shown in Fig. 13. The synthesizer has an active area of 0.6 mm^2 and the temperature-compensated tunable oscillator has an active area of 0.7 mm^2.

VI. TEST RESULTS

Fig. 14 shows a histogram of the edge jitter on the 1.62-GHz synthesizer output clock for a supply of 3.3 V. Edge jitter of 3.1 ps rms and 20 ps peak-to-peak were measured. The jitter measurements of 3.2 ps rms and 20 ps peak-to-peak, for a 2-V supply and 1-GHz output frequency, show that the DLL core ex-

TABLE I
MEASURED SYNTHESIZER CHARACTERISTICS

1GHz Edge jitter (peak-to-peak)	< 20ps
1GHz Edge jitter (rms)	3.2ps
1GHz inter-period jitter	100ps
Power supply	1.8-3.3V
Current consumption @ 2V:	
Synthesizer	20mA
Output Driver	7mA
Output Clock Frequency Range @ 2V	600MHz to 1GHz
Output Clock Frequency Range @ 3.3V	900MHz to 1.6GHz
On chip loop filter capacitor	20pF
Loop Bandwidth	1MHz
Active area	0.6mm^2
Process	0.5μm CMOS
Device package	44-pin CLCC
PCB substrate	FR4

TABLE II
MEASURED TUNABLE OSCILLATOR CHARACTERISTICS

400MHz frequency variation over temperature range 0 to 85°C	1.8%
Frequency Tuning Range	200MHz to 500MHz
400MHz Edge Jitter (rms)	29ps
400MHz Edge Jitter (peak-to-peak)	180ps
VCO supply sensitivity (open loop)	0.83%/V
Power supply	3.3V
Current Consumption @ 3.3V:	
DLL + Tunable VCO	42mA
Output Driver	6mA
Active area	0.7mm^2
Process	0.5μm CMOS

hibits better jitter performance than that reported for the higher voltage DLLs (3.3-V supply, 0.35-μm CMOS, 4-ps rms jitter) in [2] and (5-V supply, 0.7-μm CMOS, 10-ps rms jitter) in [3]. The measured jitter (rms) variation versus synthesizer output frequency for a 3.3-V supply is shown in Fig. 15. With the supply reduced to 1.8 V, the rms jitter was measured at 4.9 ps for an output frequency of 720 MHz. Fig. 16 shows this 720-MHz synthesizer output. Mismatched propagation delays and interblock routing in the frequency multiplication block (Fig. 9) resulted in 100-ps interperiod jitter.

Fig. 17 shows the temperature-compensated tunable oscillator frequency variation with temperature. Varying the ambient temperature from 0 °C to 85 °C resulted in a total frequency variation of 1.8%. Fig. 18 shows the variation of frequency with

the tune voltage. As can be seen from the plot, the relationship is close to linear. It is possible to tune the frequency around a center frequency in the range from 200 to 500 MHz by selecting an appropriate input reference frequency. This ensures that this scheme can be used for a wide variety of applications. The measured jitter on the 400-MHz output was 29 ps rms and 180 ps peak-to-peak. Table I shows the measured synthesizer characteristics. Table II summarizes the measured characteristics of the temperature-compensated tunable oscillator.

VII. CONCLUSION

In this paper, a robust self-correcting low-jitter DLL was used as the basis for a low-voltage high-frequency synthesizer and a temperature-compensated tunable oscillator. The DLL does not require the VCDL control voltage to be set on power-up. The DLL can recover from missing reference clock pulses and it can track step changes in a variable reference clock frequency. The synthesizer has significantly lower edge jitter than the traditional PLL-type synthesizer [9] and other reported DLL circuits [10], [11]. The temperature-compensated tunable oscillator provides a temperature-stable tunable frequency that varies by just 1.8% over the 0 °C to 85 °C temperature range.

ACKNOWLEDGMENT

The authors wish to acknowledge contributions from the following Parthus Technologies employees: J. Ryan, J. Horan, C. Cahill, F. Fuster, J. Collins, B. Kinsella, M. Erett, and S. Murphy. The authors also wish to thank R. Fitzgerald from the NMRC for the die photo micrographs. The device was fabricated on the ESM (Newport) Wafer Fab through Europractice.

REFERENCES

[1] B. Kim, T. C. Weingandt, and P. R. Gray, "PLL/DLL system noise analysis for low-jitter clock synthesizer design," in *Proc. ISCAS*, June 1994, pp. 151–154.
[2] Y. Moon, J. Choi, K. Lee, D. Jeong, and M. Kim, "An all-analog multiphase delay-locked loop using a replica delay line for wide-range operation and low jitter," *IEEE J. Solid-State Circuits*, vol. 35, pp. 377–384, Mar. 2000.
[3] M. Mota and J. Christiansen, "A high-resolution time interpolator based on a delay-locked loop and an *RC* delay line," *IEEE J. Solid-State Circuits*, vol. 34, pp. 1360–1366, Oct. 1999.
[4] D. Foley and M. Flynn, "CMOS DLL-based 2-V 3.2-ps jitter 1-GHz clock synthesizer and temperature compensated tunable oscillator," in *Proc. IEEE Custom Integrated Circuits Conf.*, May 2000, pp. 371–374.
[5] G. Chien and P. R. Gray, "A 900-MHz local oscillator using a DLL-based frequency multiplier technique for PCS applications," in *ISSCC Dig. Tech. Papers*, Feb. 2000, pp. 202–203.
[6] H. Chen, E. Lee, and R. Geiger, "A 2-GHz VCO with process and temperature compensation," in *Proc. ISCAS*, June 1999, pp. 11 569–11 572.
[7] A. Young, J. K. Greason, and K. L. Wong, "A PLL clock generator with 5 to 110 MHz of lock range for microprocessors," *IEEE J. Solid-State Circuits*, vol. SC-27, pp. 1599–1607, Nov. 1992.
[8] M. Horowitz, C.-K. K. Yang, and S. Sidiropoulos, "High-speed electrical signaling: Overview and limitations," *IEEE Micro.*, vol. 18, pp. 12–24, Jan./Feb. 1998.
[9] H. C. Yang, L. K. Lee, and R. S. Co, "A low-jitter 0.3-165 MHz CMOS PLL synthesizer for 3-V/5-V operation," *IEEE J. Solid-State Circuits*, vol. 32, pp. 582–586, Apr. 1997.
[10] J. G. Maneatis, "Low-jitter process-independent DLL and PLL based on self-biased techniques," *IEEE J. Solid-State Circuits*, vol. 31, pp. 1723–1732, Nov. 1996.
[11] S. Sidiropoulos and M. A. Horowitz, "A semidigital dual delay-locked loop," *IEEE J. Solid-State Circuits*, vol. 32, pp. 1683–1692, Nov. 1997.

4.1 A 1.5V 86mW/ch 8-Channel 622–3125Mb/s/ch CMOS SerDes Macrocell with Selectable Mux/Demux Ratio

Fuji Yang, Jay O'Neill, Patrik Larsson, Dave Inglis, Joe Othmer

Agere Systems, Holmdel, NJ

2.5-3.125Gb/s serial links are commonly used for chip-to-chip interconnects in high-speed network systems. In SONET OC-768 application, at least 16 on-chip SerDes transceivers are required to guarantee total full duplex I/O throughput of 40Gb/s. Published 2.5Gb/s SerDes transceivers consume between 150 and 200mW, not suitable for applications requiring hundreds of on-chip SerDes transceivers [1]. Developing a low-power SerDes transceiver is important for high throughput network ICs [2]. Another challenge is reduction of inter-channel noise coupling when integrating many transceivers on the same chip. This low-power 8-channel SerDes macrocell employs a shared-PLL architecture. As shown in Figure 4.1.1, on the transmitter side, the on-chip TxPLL provides a half-rate clock to all transmitters. On the receiver side, the RxPLL distributes I- and Q-phase clocks to 8 receivers. Each receiver has a phase interpolator to generate an output phase-aligned with the in-coming data for clock and data recovery. Sharing a single PLL between a group of transmitters or receivers reduces the power and avoids the potential multi-VCO coupling problem found in a conventional one-PLL-per-channel configuration. The macrocell realized in a 0.16µm CMOS process consumes an average power of 86mW per channel at 1.5V power supply.

The transmitter 16:1 or 20:1 serialization starts with 4 shift-register based selectable 4:1 or 5:1 multiplexers. Their 4 outputs are sent to a tree-based 4:1 multiplexer (Figure 4.1.1). A pMOS CML output driver with on-chip 50Ω terminations is employed. The output signal referenced to the ground makes the interface independent of the power supply. The output amplitude is set to 1Vpp, diff.

The receiver employs an interleaved integrate-and-dump front-end (Figure 4.1.1) [3, 4]. The integrate-and-dump operation improves the SNR and eliminates the quadrature clock required in a conventional half-rate front-end [5]. The integrator outputs are de-multiplexed by the decision-latches controlled respectively by ck2i and ck2q, which are divide-by-2 clocks of the recovered clock. The decision-latch outputs d1-d4 are fed into 4 shift-registers to realize the 4:16 or 4:20 de-serialization. The integrator is implemented in a way similar way to that proposed in Reference [4], but with a pMOS input stage. It has a gain of 2 allowing relaxed offset and noise requirements of the latches. The receiver achieves $30mV_{pp,diff}$ sensitivity with BER $<10^{-12}$ at 2.5Gb/s.

The clock recovery is by a DLL based on an analog phase interpolator [6]. In contrast to the implementation in Reference [6], a four-quadrant phase mixer is used here. Referring to Figure 4.1.2, the DLL consists of a bang-bang phase detector (PD), a PD polarity control circuit, an amplitude control circuit, I- and Q-charge-pumps and the four-quadrant mixer-based phase interpolator. The analog phase interpolation is by mixing the I- and Q-phase clocks from the RxPLL with respective weights α (=V_a-V_{ref}) and β (=V_b-V_{ref}): CLK=α*(I-IB)+β*(Q-QB). V_a and V_b are independently generated by I- and Q-charge-pumps. The weights α and β, ranging from negative to positive, directly control the quadrant changes. This eliminates the potential phase discontinuity at quadrant crossings found in the circuit of Reference [6]. Figure 4.1.3 shows the schematic of one 4-quadrant mixer, where

the nMOS differential pair converts V_a to differential currents I_p and I_n, which are mirrored into the pMOS current sources to be steered by the high-speed differential clock (I-IB). A self-biased nMOS load is used with MP1 and MP2 to control the output common-mode voltage.

The phase interpolator exhibits an infinite phase shift range allowing the DLL to easily track the frequency offset between the local clock and the incoming data and enables shared-PLL architecture for multi-channel serial links with plesiochronous clocking.

Figure 4.1.4 illustrates the non-monotonic relation between the phase shift introduced by the interpolator and the two weights α and β. To have a 2π interpolation range, the bang-bang phase detector polarity must be updated to provide the correct up/down signals for different quadrants. This is by a PD-polarity-control circuit in association with a Q-detect circuit. The Q-detect circuit detects the output vector quadrant by determining the sign of α and β. The Q-detect circuit uses the replica of the V-I converter in the phase mixer.

Although the phase mixer has control weights α and β, the phase interpolation is only a function of α/β, and is independent of the amplitude of α and β. The loop, sensitive only to the phase variation, thus controls α/β. As a consequence, α and β can grow or shrink arbitrarily. To prevent α and β from being too small, an offset current is intentionally introduced in the charge-pumps. It is controlled as follows: If α>0, $I_{up} = I_0 + I_{offset}$ and if α<0, $I_{down} = I_0 + I_{offset}$ (the same algorithm is applied for Q-charge-pump). As the result, α and β are always pulled away from zero to eliminate any shrinking possibility. To prevent overflow on α and β, the amplitude control circuit clips α and β by blocking UP or DOWN signal. As shown in Figure 4.1.4, V_a or V_b will be kept within [Vmin, Vmax].

The test chip in a 0.16µm 5-level metal CMOS technology uses a 217-pin PBGA package. The chip micrograph is shown in Figure 4.1.5. Active area is about 2mm². Figure 4.1.6a shows the measured jitter tolerance of the receiver. The CDR works with VDD as low as 1V for 1Gb/s maximal input data rate. With 1.5V power supply, the receiver covers an input data rate range of 622 to 3125Mb/s. Measured recovered clock jitter is 87.1ps pp at 2.5Gb/s. Figure 4.1.6b shows the Tx output eye diagram measured at 3.2Gb/s with a 2^{31}-1 PRWS. The measured jitter is 57.8ps pp and static VDD sensitivity is 0.06ps/mV. Measured results are summarized in Figure 4.1.7.

References:
[1] R. Gu et al "A 0.5-3.5Gb/s Low-Power Low-Jitter Serial Data CMOS Transceiver," ISSCC Digest of Technical Papers, pp. 352-353, Feb. 1999.
[2] M-J. Lee et al., "An 84mW 4Gb/s clock and data recovery circuit for serial link applications" VLSI Symposium 2001, pp. 149-152, 2001.
[3] S. Sidiropoulos et al "A 700Mb/s/pin CMOS signaling interface using a current integrating receivers" IEEE JSSC, vol. 32, no. 5, pp. 681-690, May 1997.
[4] J. Savoj et al., "A CMOS Interface Circuit for Detection of 1.2Gb/s RZ Data" ISSCC Digest of Technical Papers, pp. 278-279, Feb. 1999.
[5] P. Larsson, "An Offset-Cancelled CMOS Clock Recovery/Demux with Half-Rate Linear Phase Detector for 2.5Gb/s Optical Communication" ISSCC Digest of Technical Papers, pp. 74-75, Feb. 2001.
[6] T. Lee et al., "A 2.5V CMOS delay-locked loop for an 18Mb, 500Mb/s DRAM" IEEE JSSC, vol. 9, no. 2, Dec. 1994

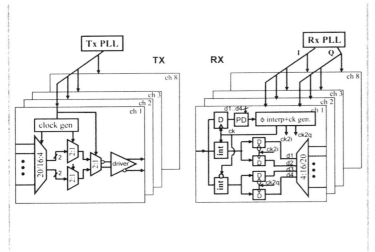

Figure 4.1.1: Overall architecture.

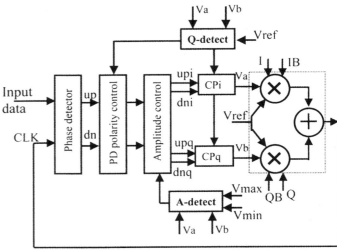

Figure 4.1.2: DLL block diagram.

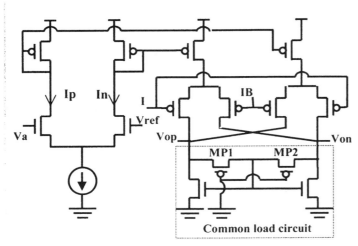

Figure 4.1.3: Four-quadrant mixer schematic.

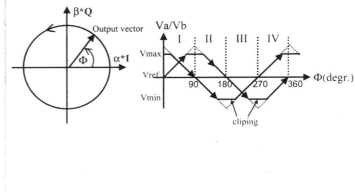

Figure 4.1.4: Relation between the phase shift and the weights Va and Vb.

Figure 4.1.5: Die micrograph.

Technology:	0.16 CMOS with 5 metal levels
Supply Voltage	1.5V
Power dissipation	75mW per transceiver with Tx output set to 1Vpp.diff
	85mW for Tx and Rx PLLs + clock buffers
Active area	2mm² (PLL : 0.1mm², single transceiver : 0.25mm²)
BER	< 10⁻¹² (all measurements were done with BER < 10⁻¹²)
Receiver Sensitivity	30mVpp.diff
Recovered clock jitter	87.1psPkPk
Max. offset frequency	400ppm at 2.5Gb/s
Input data rate range	622-3125Mb/s
Transmitter's output jitter	57.8ps Pk-Pk at 3.2Gb/s
TxoutputVDD sensitivity	0.06ps/mV
Output Amplitude	1Vpp.diff

Figure 4.1.7: Summary of measured results.

Continued

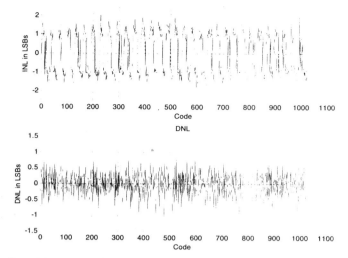

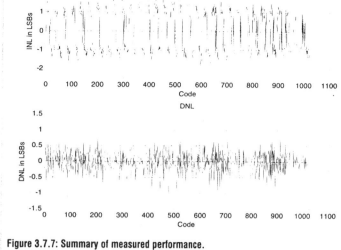

Figure 3.6.7: INL and DNL for the 2-stage flash ADC.

Figure 3.7.7: Summary of measured performance.

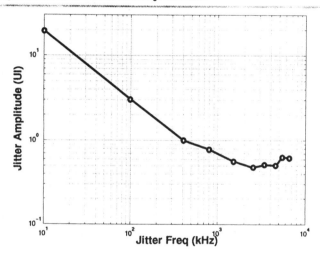

(a)

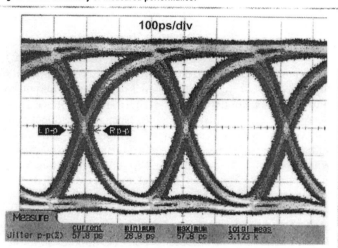

(b)

Figure 4.1.6: Measured Rx jitter tolerance and Tx eye diagram at 3.2Gb/s.

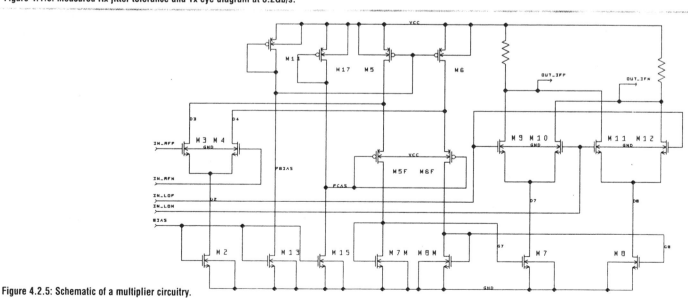

Figure 4.2.5: Schematic of a multiplier circuitry.

501

A Register-Controlled Symmetrical DLL for Double-Data-Rate DRAM

Feng Lin, Jason Miller, Aaron Schoenfeld, Manny Ma, and R. Jacob Baker

Abstract—**This paper describes a register-controlled symmetrical delay-locked loop (RSDLL) for use in a high-frequency double-data-rate DRAM. The RSDLL inserts an optimum delay between the clock input buffer and the clock output buffer, making the DRAM output data change simultaneously with the rising or falling edges of the input clock. This RSDLL is shown to be insensitive to variations in temperature, power-supply voltage, and process after being fabricated in 0.21-μm CMOS technology. The measured rms jitter is below 50 ps when the operating frequency is in the range of 125–250 MHz.**

Index Terms—**Delay-locked loops, double-data rate, DRAM.**

I. INTRODUCTION

IN synchronous DRAM, the output data strobe (DQS) should be locked to the data outputs (DQ outputs) for high-speed performance. The clock-access and output-hold times of conventional DRAM designs are determined by the delay time of the internal circuits such as the clock input and output buffers. Variations in temperature and process shifts will change the access time and make the valid data window small. To optimize and stabilize the clock-access and output-hold times, an internal register-controlled delay-locked loop (RDLL) [1], [2] has been used to adjust the time difference between the output and input clock signals in SDRAM. Since the RDLL is an all-digital design, it provides robust operation over all process corners. Another solution to the timing constraints found in SDRAM was given in [3] with the synchronous mirror delay (SMD). Compared to RDLL, SMD does not provide as tight of locking but has the advantage that the time to acquire lock between the input and output clocks is only two clock cycles. As the clock speeds used in DRAM continue to increase, the skew becomes the dominating concern, outweighing the disadvantage of the added time to acquire lock needed in an RDLL.

This paper describes a modified register-controlled symmetrical delay-locked loop (RSDLL) used to meet the requirements of double-data-rate (DDR) SDRAM (read/write accesses occur on both rising and falling edges of the clock). Here, "symmetrical" means that the delay line used in the DLL has the same delay whether a high-to-low or a low-to-high logic signal is propagating along the line. The data output timing diagram of a DDR SDRAM is shown in Fig. 1. The RSDLL is used to increase the valid output data window and

Manuscript received September 3, 1998; revised November 2, 1998. This work was supported by Micron Technology, Inc.

F. Lin and R. J. Baker are with the Microelectronics Research Center, University of Idaho, Boise, ID 83712 USA (e-mail: danlin@uidaho.edu).

J. Miller, A. Schoenfeld, and M. Ma are with Micron Technology, Inc., Boise, ID 83707-0006 USA.

Publisher Item Identifier S 0018-9200(99)02438-5.

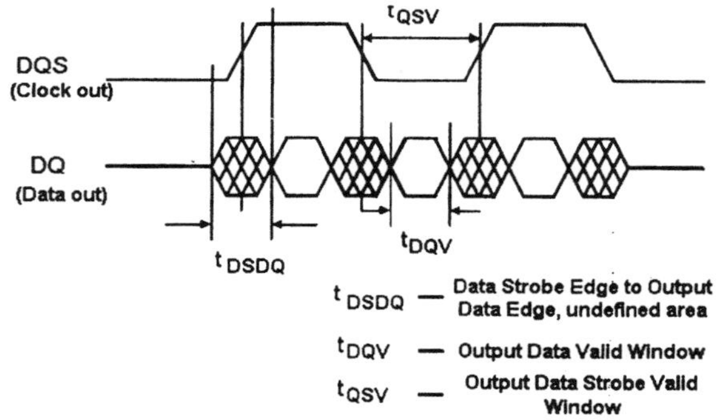

Fig. 1. Data timing chart for DDR DRAM.

diminish the undefined t_{DSDQ} by synchronizing both rising and falling edges of the DQS signal with the output data DQ.

The target specifications for the DLL described in this paper are:

1) robust operation eliminating the need for postproduction tuning (something required in an analog implementation);

2) operating frequency ranging from 143 (286 Mb/s/pin) to 250 MHz (500 Mb/s/pin);

3) tight synchronization (skew less than 5% of the cycle time) between the output clock and data on both rising and falling edges of the output clock;

4) low skew between the input and output clocks (with low, <5% duty cycle distortion);

5) power-supply-voltage operating range from 2.5 to 3.5 V;

6) portability for ease of use in other processes.

II. RSDLL ARCHITECTURE

Fig. 2 shows the block diagram of the RSDLL. The replica input buffer dummy delay in the feedback path is used to match the delay of the input clock buffer. The phase detector (PD) is used to compare the relative timing of the edges of the input clock signal and the feedback clock signal, which comes through the delay line, controlled by the shift register. The outputs of the PD, shift-right and shift-left, are used to control the shift register. In the simplest case, one bit of the shift register is high. This single bit is used to select a point of entry for CLKIn in the symmetrical delay line (more on this later). When the rising edge of the input clock is within the rising edges of the output clock and one unit delay of the output clock, both outputs of the PD, shift-right and shift-left, go to logic LOW and the loop is locked. The basic operation of the PD is shown in Fig. 3. The resolution of this RSDLL is determined by the size of a unit delay used in the delay

502

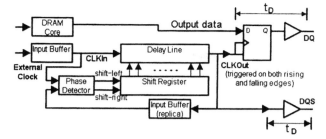

Fig. 2. Block diagram of RSDLL.

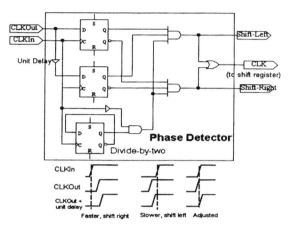

Fig. 3. Phase detector used in RSDLL.

line. The locking range is determined by the number of delay stages used in the symmetrical delay line. Since the DLL circuit inserts an optimum delay time between CLKIn and CLKOut, making the output clock change simultaneously with the next rising edge of the input clock, the minimum operating frequency to which the RSDLL can lock is the reciprocal of the product of the number of stages in the symmetrical delay line with the delay per stage. Adding more delay stages will increase the locking range of the RSDLL at the cost of increased layout area.

III. CIRCUIT IMPLEMENTATION

A. Basic Delay Element

Instead of using an AND gate as the unit-delay stage (NAND + inverter), as was done in [1], we used a NAND-gate-based delay element. The implementation of a three-stage delay line is shown in Fig. 4. The problem when using a NAND + inverter as the basic delay element is that the propagation delay through the unit delay resulting from a HIGH-to-LOW transition is not equal to the delay of a LOW-to-HIGH transition ($t_{PHL} \neq t_{PLH}$). Further, this delay varies from one run to another. If the skew between t_{PHL} and t_{PLH} is 50 ps, for example, the total skew of the falling edges through ten stages will be 0.5 ns. Because of this skew, the NAND + inverter delay element cannot be used in a DDR DRAM. In our modified symmetrical delay element, another NAND gate is used instead of an inverter (two NAND gates per delay stage). This scheme guarantees that $t_{PHL} = t_{PLH}$ independent of process variations, since while one NAND switches from a HIGH to LOW, the other

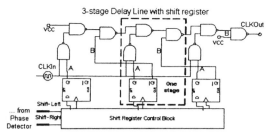

Fig. 4. Symmetrical delay element used in RSDLL.

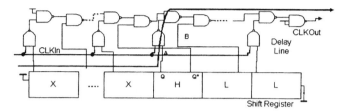

Fig. 5. Delay line and shift register for RSDLL.

switches from LOW to HIGH. An added benefit of the two-NAND delay element is that two point-of-entry control signals are now available. Both are used by the shift register to solve the possible problem caused by the power-up ambiguity in the shift register.

B. Control Mechanism of the Shift Register

As shown in Figs. 4 and 5, the input clock is a common input to every delay stage. The shift register is used to select a different tap of the delay line (the point of entry for the input clock signal into the symmetrical delay line). The complementary outputs of each register cell are used to select the different tap: Q is connected directly to the input A of a delay element, and Q^* is connected to the previous stage of input B. From right to left, the first LOW-to-HIGH transition in the shift register sets the point of entry into the delay line. The input clock will pass through the tap with a high logic state in the corresponding position of the shift register. Since the Q^* of this tap is equal to a LOW, it will disable the previous stages; therefore, it does not matter what the previous states of the shift register are (shown as "don't cares," X, in Fig. 5). This control mechanism guarantees that only one path is selected. This scheme also eliminates power-up concerns since the selected tap is simply the first, from the right, LOW–HIGH transition in the shift register.

C. Phase Detector

To stabilize the movement in the shift register, after making a decision, the phase detector will wait at least two clock cycles before making another decision (Fig. 3). A divide by two was included in the phase detector so that every other decision, resulting from comparing the rising edges of the external clock and the feedback clock, was used. This will provide enough time for the shift register to operate and the output waveform to stabilize before another decision by the PD is implemented. The unwanted side effect of this delay is an increase in the lock time. The shift register is clocked by combining the

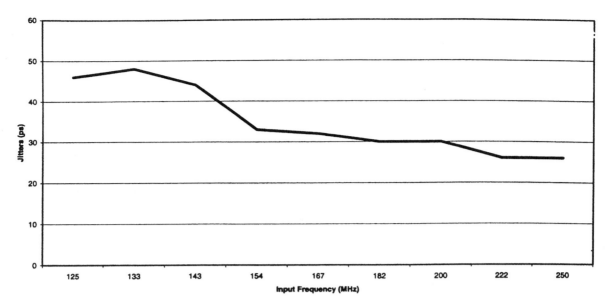

Fig. 6. Measured rms jitter versus input frequency.

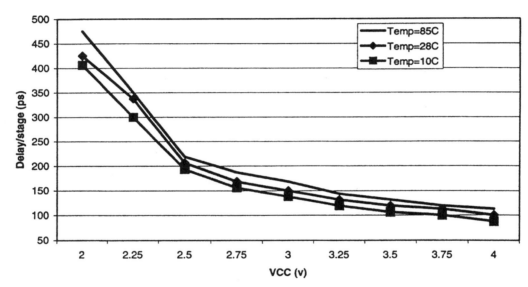

Fig. 7. Measured delay per stage versus VCC and temperature.

shift-left and shift-right signals. The power consumption will decrease when there are no shift-left or -right signals and the loop is locked. Another concern with the phase-detector design is the design of the flip-flops (FF's). To minimize the static phase error, very fast FF's should be used, ideally with zero setup time. Also, the metastability of the flip-flops becomes a concern as the loop becomes locked. This together with possible noise contributions and the need to wait, as discussed above, before implementing a shift-right or -left may increase the desirability of adding additional filtering in the phase detector. Some possibilities include increasing the divider ratio used in the phase detector or using a shift register in the phase detector to determine when a number—say, four—shift-rights or -lefts have occurred. For the present design, we were forced to use a divide by two in the phase detector because of lock time requirements.

IV. EXPERIMENTAL RESULTS

The RSDLL was fabricated in a 0.21-μm, four-poly, double-metal CMOS technology (a DRAM process). We used a 48-stage delay line with an operation frequency of 125–250 MHz. The maximum operating frequency was limited by delays external to the DLL such as the input buffer and interconnect. There was no noticeable static phase error on either rising or falling edges. Fig. 6 shows the resulting rms jitter versus input frequency. One sigma of jitter over the 125–250-MHz frequency range was below 50 ps. The peak-to-peak jitter over this frequency range was below 100 ps. The measured delay per stage versus VCC and temperature is shown in Fig. 7. Note that the 150-ps typical delay of a unit-delay element was very close to the rise and fall times on-chip of the clock signals and represents a practical minimum resolution of a DLL for use in a DDR DRAM fabricated in a 0.21-μm process. The power

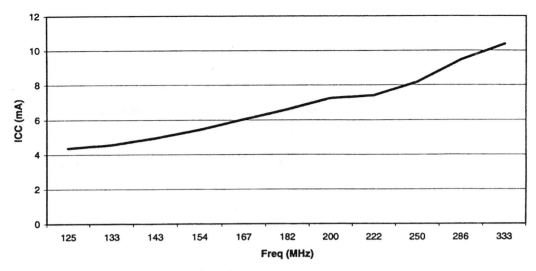

Fig. 8. Measured ICC (DLL current consumption) versus input frequency.

consumption (current draw of the DLL when VCC = 2.8 V) of the prototype RSDLL is illustrated in Fig. 8. We found that the power consumption was mainly determined by the dynamic power dissipation of the symmetrical delay line. Our NAND delays in this test chip were implemented with 10/0.21-μm NMOS and 20/0.21-μm PMOS. By reducing the widths of both the NMOS and PMOS transistors, the power dissipation can be greatly reduced without a speed or resolution penalty (with the added benefit of reduced layout size).

V. CONCLUSIONS

The concept of a register-controlled symmetrical delay-locked loop has been presented. The modified symmetrical delay element makes the RSDLL useful in DDR DRAM's. Experimental results verify that this RSDLL is stable against temperature, process, and power-supply variations.

Further development of the RSDLL will include investigations into reducing power consumption, implementing phase-locked loops where the symmetrical delay is used as part of a purely digital registered-controlled oscillator, and developing two-loop architectures where coarse loops (resolutions on the order of 100 ps) are used with fine loops (resolutions on the order of 10 ps [2]) for wide tuning range and small static phase errors.

REFERENCES

[1] A. Hatakeyama, H. Mochizuki, T. Aikawa, M. Takita, Y. Ishii, H. Tsuboi, S.-Y. Fujioka, S. Yamaguchi, M. Koga, Y. Serizawa, K. Nishimura, K. Kawabata, Y. Okajima, M. Kawano, H. Kojima, K. Mizutani, T. Anezaki, M. Hasegawa, and M. Taguchi, "A 256-Mb SDRAM using a register-controlled digital DLL," *IEEE J. Solid-State Circuits,* vol. 32, pp. 1728–1732, Nov. 1997.

[2] S. Eto, M. Matsumiya, M. Takita, Y. Ishii, T. Nakamura, K. Kawabata. H. Kano, A. Kitamoto, T. Ikeda, T. Koga, M. Higashiro, Y. Serizawa. K. Itabashi, O. Tsuboi, Y. Yokoyama, and M. Taguchi, "A 1Gb SDRAM with ground level precharged bitline and non-boosted 2.1V word line." in *ISSCC Dig. Tech. Papers,* Feb. 1998, pp. 82–83.

[3] T. Saeki, Y. Nakaoka, M. Fujita, A. Tanaka, K. Nagata, K. Sakakibara. T. Matano, Y. Hoshino, K. Miyano, S. Isa, S. Nakazawa, E. Kakehashi. J. M. Drynan, M. Komuro, T. Fukase, H. Iwasaki, M. Takenaka, J. Sekine, M. Igeta, N. Nakanishi, T. Itani, K. Yoshida, H. Yoshino, S. Hashimoto, M. Yoshii, M. Ichinose, T. Imura, M. Uziie, S. Kikuchi, K. Koyama, Y. Fukuzo, and T. Okuda, "A 2.5-ns clock access 250-MHz. 256-Mb SDRAM with synchronous mirror delay," *IEEE J. Solid-State Circuits,* vol. 31, pp. 1656–1665, Nov. 1996.

A Low-Jitter Wide-Range Skew-Calibrated Dual-Loop DLL Using Antifuse Circuitry for High-Speed DRAM

Se Jun Kim, Sang Hoon Hong, Jae-Kyung Wee, Joo Hwan Cho, Pil Soo Lee, Jin Hong Ahn, and Jin Yong Chung

Abstract—This paper describes a delay-locked loop (DLL) circuit having two advancements, a dual-loop operation for a wide lock range and programmable replica delays using antifuse circuitry and internal voltage generator for a post-package skew calibration. The dual-loop operation uses information from the initial time difference between reference clock and internal clock to select one of the differential internal loops. This increases the lock range of the DLL to the lower frequency. In addition, incorporation of the programmable replica delay using antifuse circuitry and the internal voltage generator allows for the elimination of skews between external clock and internal clock that occur from on-chip and off-chip variations after the package process. The proposed DLL, fabricated on 0.16-μm DRAM process, operates over the wide range of 42–400 MHz with 2.3-V power supply. The measured results show 43-ps peak-to-peak jitter and 4.71-ps rms jitter consuming 52 mW at 400 MHz.

Index Terms—Delay-locked loop, dual-loop operation, high-speed DRAM, programmable replica delay, skew calibration.

I. INTRODUCTION

THE DELAY-LOCKED loop (DLL) has become an indispensable component in high-speed synchronous DRAMs such as DDR SDRAM. Since the DLL determines the operation range of the DRAM and has a large effect on the data valid window, a high-performance DLL that has a wider range and lower jitter is essential for increasing the speed of DRAM. A DLL can be categorized into either of two types, the digital and the analog type. Although the digital DLL has robustness, process portability, and design simplicity, it is difficult to use on a very high-bandwidth DRAM (over 600 Mb/s) due to poor jitter performance [1], [2]. Therefore, in spite of sensitivity on process variation, the analog DLL, which ensures lower jitter by the continuous characteristics of analog operation, is more suitable in the higher speed DRAM. In addition to the jitter performance, another important issue of the DLL is the lock range. Process variation makes the lock range of the analog DLL more limited and results in a narrower operation range of the DRAM. The limited range of the DLL limits the flexibility of implementation on memory applications and increases test costs in mass production. For solving the limited lock-range problems, various types of DLLs have been developed [3]–[6]. However, such DLLs resulted in complex architectures that faced such problems as increased area, added power consumption, and degradation of jitter performance.

For these issues, a novel dual-loop architecture, which increases the lock range having no degradation of jitter performance with a relatively small overhead in area and power, is proposed in this paper. Another enhancement in the proposed DLL is the post-package skew calibration. Process variations in on-chip and trivial mismatches in off-chip parameters can result in a large static skew in addition to the phase offset of the phase detector. In the proposed DLL, an improved scheme using antifuse circuitry is applied for reducing the skew. It enables a practical calibration of inevitable skews after the package process.

This paper is arranged as follows. The limited range problem of the conventional DLL is described in Section II. In Section III, the concept of the proposed dual loop for wide locking range is briefly explained, followed by presentation of the architecture and physical implementation based on the concept. The skew calibration method using antifuse circuitry is described in Section IV. Section V discusses the fabricated chip and shows the experimental results. Finally, the paper is concluded in Section VI.

II. LIMITED RANGE PROBLEM OF CONVENTIONAL DLL

Fig. 1(a) shows the architecture of the conventional analog DLL and the delay characteristic of the voltage-controlled delay line (VCDL). When $2 \times T_{\text{VCDL min}}$ (minimum delay of VCDL) $\ll T_{\text{VCDL max}}$ (maximum delay of VCDL), the range of f_{lock} (operation frequency of DLL) is determined by V_{con} (control voltage of loop filter) at the initial state. When $V_{\text{con}} = V_{\text{con min}}$ (minimum control voltage of loop filter) at the initial state and $T_{\text{VCDL min}} < (1/2) \times T_{\text{REF}}$ (the cycle time of reference clock), the lock failure occurs because the phase detector produces a DN pulse which discharges the capacitor in the loop filter, as shown in Fig. 1(b). Therefore, in this case, it must be $T_{\text{VCDL min}} > (1/2) \times T_{\text{REF}}$ at the initial state for satisfying the condition without lock failure. Therefore, the range of f_{lock} is $1/(2 \times T_{\text{VCDL min}}) < f_{\text{lock}} < 1/T_{\text{VCDL min}}$. In the other case, when $V_{\text{con}} = V_{\text{con max}}$ (maximum control voltage of loop filter) at the initial state and $T_{\text{VCDL max}} > (3/2) \times T_{\text{REF}}$, the lock failure occurs because of the UP pulse of the phase detector shown in Fig. 1(c). In this case, the range of f_{lock} is $1/T_{\text{VCDL max}} < f_{\text{lock}} < 2/(3 \times T_{\text{VCDL max}})$ when the initial V_{con} is $V_{\text{con max}}$. For utilizing the full range of

Manuscript received October 2, 2001; revised January 29, 2002.
S. J. Kim, S. H. Hong, J. H. Cho, P. S. Lee, J. H. Ahn, and J. Y. Chung are with the Advanced Design Team, Memory Research and Development, Hynix Semiconductor Inc., Ichon-si, Kyoungki-Do 467-701, Korea (e-mail: sejun.kim@hynix.com).
J.-K. Wee is with the Department of Electronics Engineering, Hallym University, Chunchun-si, Kangwon-Do 200-702, Korea.
Publisher Item Identifier S 0018-9200(02)04934-X.

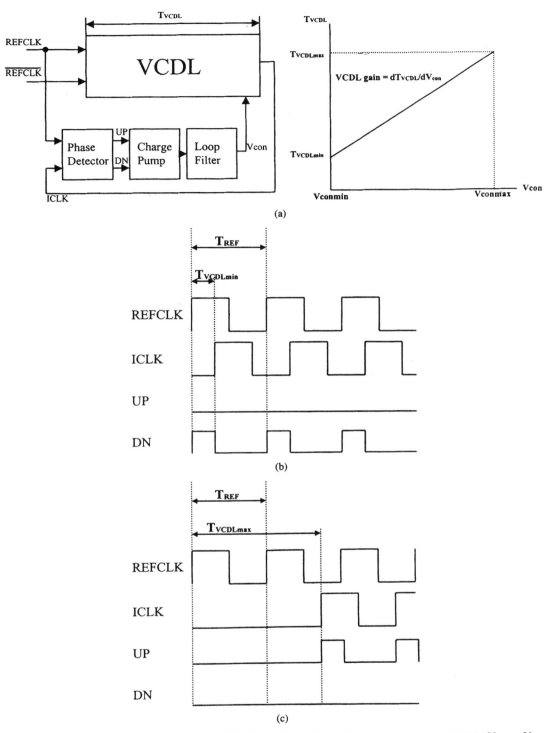

Fig. 1.   (a) Block diagram and delay characteristic of conventional DLL. Cases of lock failure at initial control voltage: (b) initial $V_{con} = V_{con\,min}$ and (c) initial $V_{con} = V_{con\,max}$.

T_{VCDL} without the lock failures as in Fig. 1(b) and (c), the initial V_{con} must be set at a level such that the initial T_{VCDL} is approximately $(1/2) \times (T_{VCDL\,max} - T_{VCDL\,min})$. In this condition, the range of f_{lock} is determined as $1/T_{VCDL\,max} < f_{lock} < 1/T_{VCDL\,min}$. But this method can cause stuck/harmonic lock and makes the jitter performance worse. Therefore, if the range of f_{lock} is desired at the higher frequency range, the initial V_{con} should be set to $V_{con\,min}$ since it is stuck/harmonic lock free and the delay cell has a fast slew-rate that produces less phase noise [7]. But in reality, $T_{VCDL\,min}$ is very sensitive to process, voltage, and temperature (PVT) variation. As a result, designing $T_{VCDL\,min}$ to be in the target range becomes more careful and difficult work as the operation frequency becomes higher. Therefore, considering the PVT variation, the range of f_{lock} becomes more limited with the higher operating range.

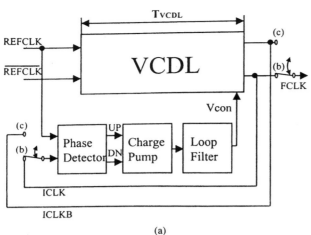

(a)

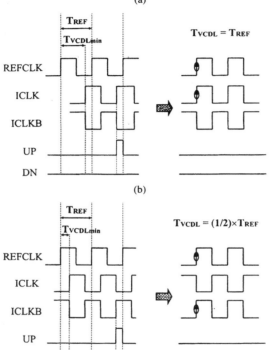

Fig. 2. (a) Concept of proposed DLL. Two cases of loop selection according to the initial time difference: (b) $(1/2)T_{\mathrm{REF}} < T_{\mathrm{VCDL\,min}} < T_{\mathrm{REF}}$ and (c) $0 < T_{\mathrm{VCDL\,min}} < (1/2)T_{\mathrm{REF}}$.

III. PROPOSED DLL IN HIGH-SPEED DRAM

A. Range of the Proposed DLL

The concept of the proposed dual-loop DLL is shown in Fig. 2(a). After the DLL starts up at the initial $V_{\mathrm{con}} = V_{\mathrm{con\,min}}$, the initial time difference between REFCLK and ICLK is monitored at the first REFCLK cycle. The first REFCLK cycle refers to the cycle of the REFCLK when ICLK is produced first in the loop after the DLL starts up. If $(1/2) \times T_{\mathrm{REF}} < T_{\mathrm{VCDL\,min}} < T_{\mathrm{REF}}$ as shown in Fig. 2(b), T_{VCDL} is adjusted to T_{REF} by the phase detector and charge pump like the conventional DLL. As a result, ICLK becomes FCLK of the synchronized output clock of DLL. In the other case, if $T_{\mathrm{VCDL\,min}} < (1/2) \times T_{\mathrm{REF}}$ as shown in Fig. 2(c), the input clock for phase comparison in the phase detector is

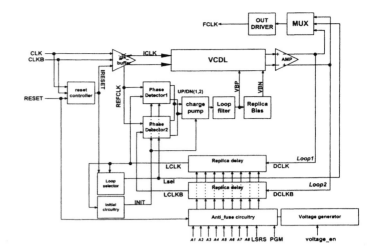

Fig. 3. Architecture of proposed DLL.

switched from ICLK to ICLKB (the differential clock of ICLK), T_{VCDL} becomes $(1/2) \times T_{\mathrm{REF}}$, and FCLK is also switched from ICLK to ICLKB. This means that ICLKB is synchronized to REFCLK. Therefore, in our proposed DLL, the locking frequency range is $1/(2 \times T_{\mathrm{VCDL\,max}}) < f_{\mathrm{lock}} < 1/T_{\mathrm{VCDL\,min}}$. As a consequence, although the same delay source was used, the operation range of the proposed DLL can be extended to a lower frequency than that of conventional DLLs. As an analogous concept, the phase inversion technique was developed for the wide range [8]. It uses instantaneous phase inversion in VCDL input at the final moment, when it realizes current lock-in process cannot meet the range by monitoring its control voltage. The proposed concept achieves faster lock-in time since it utilizes the dual-loop operation at the beginning by selecting an optimized path.

B. Architecture and Implementation of the Proposed DLL

Fig. 3 shows the proposed dual-loop architecture of the DLL for the wide lock range. Unlike conventional DLLs, the proposed DLL is composed of dual negative feedback loops (Loop1, Loop2). Loop1 and Loop2 are the feedback loops of the differential internal clocks. For the correct operation of dual loops, a loop selector, an initial circuit, a reset controller, and a 2 : 1 MUX are implemented. In the proposed dual-loop operation, the DLL determines one of the two differential internal clocks in Loop1 and Loop2 according to the initial time difference between the internal clock and the reference clock (shown in Fig. 3). Before the DLL starts up, the initial circuit sets VBP (control voltage of loop filter in Fig. 3) to the minimum value, which minimizes the delay of the VCDL to ensure harmonic lock-free operation. After RESET (DLL enable signal) transits from low state to high state, CLK and CLKB (the external differential clocks) are provided to the reset controller. The reset controller outputs IRESET to the clock buffer in the time between the falling edge of the next CLK and the next rising edge, as shown in Fig. 4(c). Since RESET can be asserted at any time, the direct application of this signal to the clock buffer is not feasible because it can make the clock buffer produce internal clocks with a distorted cycle and cause incorrect initial time difference in the loop selection cycle, as

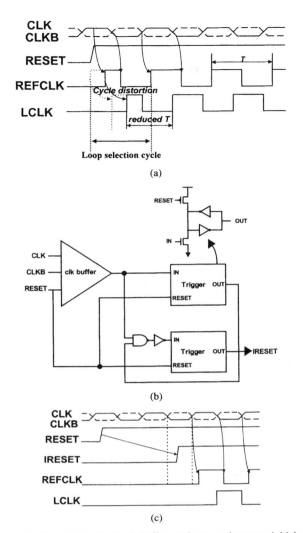

(a)

(b)

(c)

Fig. 4. (a) Case where the clock buffer produces an incorrect initial time difference without reset controller. (b) Schematic and (c) timing diagram of the reset controller.

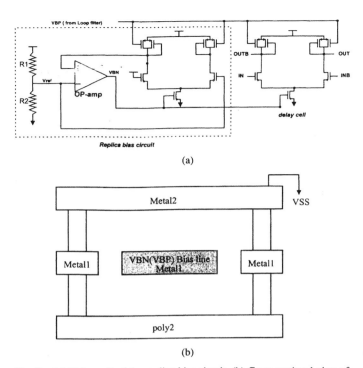

(a)

(b)

Fig. 5. (a) Delay cell of the replica bias circuit. (b) Cross-sectional view of bias line in the proposed DLL.

shown in Fig. 4(a). Fig. 4(b) shows the schematic of the reset controller. When IRESET is asserted, the clock buffer produces the three clocks (ICLK, ICLKB, and REFCLK). ICLK and ICLKB, which are the differential clocks, are input to VCDL and REFCLK is input to the phase detectors as the reference clock. Fig. 5(a) shows the delay cell and the biasing scheme used in the proposed DLL. To reduce the supply voltage sensitivity, the VCDL is implemented as a series of the differential delay cell with symmetric loads, as shown in Fig. 5(a) [9]. VBP is a control voltage from the loop filter and VBN is generated by the replica bias circuit [boxed area in Fig. 5(a)]. The replica bias circuit makes the constant swing independent of VBP, which provides better jitter performance and a wider operation range [10]. For shielding the analog biases from the external noise, VBP and VBN are physically enclosed with inter- and intra-layers as shown in Fig. 5(b). This shielding technique improves the jitter characteristic. DCLK and DCLKB in Fig. 3, which are converted from a small swing output of VCDL to a full swing output by amplifiers, each forms negative feedback loops and also are changed to LCLK and LCLKB by replica delays. LCLK and LCLKB are input to the phase detectors

and LCLK is also input to the loop selector as shown in Fig. 6(a). If the time difference between the first LCLK (the first produced LCLK after DLL is enabled) and REFCLK is in $(1/2) \times T_{\text{REF}} < T_{\text{diff}} < T_{\text{REF}}$ as shown in Fig. 6(b), Lsel, the output of the loop selector, preserves the low state initialized by IRESET. Lsel at the low state enables PD1 (the phase detector of Loop1) and disables PD2 (the phase detector of Loop2). Furthermore, the state makes the MUX select DCLK as FCLK (the output clock of DLL). Since only PD1 is enabled, the phase of LCLK is compared with that of REFCLK. The selected PD1, as shown in Fig. 7(a), produces the UP/DN pulses having a pulsewidth matching the phase difference between REFCLK and LCLK, as shown in Fig. 7(b). This PD has small phase offset due to the fast operation and precision of dynamic logic. Also, it does not have phase dithering problems because no pulses are produced at the locked state. The simulation results show about 40-ps phase offset at the worst case. The UP/DN pulse of PD1 is transferred to the charge pump and generates VBP on the loop filter. The linear capacitor of the loop filter is designed to achieve a large capacitance value in a small area while minimizing the substrate noise, as shown in Fig. 8 [11]. Although compromising the linearity, if a MOS capacitor is used, larger capacitance can be achieved in smaller area. The replica bias generator produces VBN to control the current source transistor of delay cell according to VBP. Finally, the phase of LCLK is synchronized with that of REFCLK. In the other case, where the time difference between the first LCLK and REFCLK is $0 < T_{\text{diff}} < (1/2) \times T_{\text{REF}}$ as shown in Fig. 6(c), Lsel is changed from the low state initialized by IRESET to the high state. It enables PD2 and disables PD1, and makes the MUX select DCLKB as FCLK. In contrast to the prior case, the phase of LCLKB is compared with that of REFCLK. Through the same locking process, LCLKB is

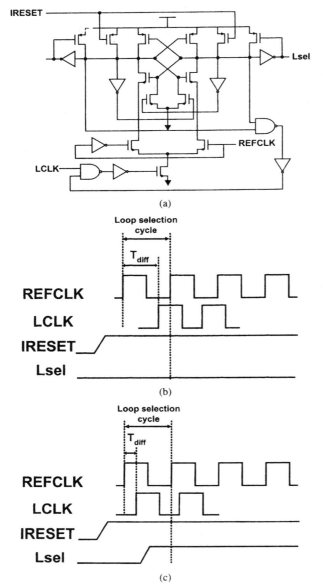

(a)

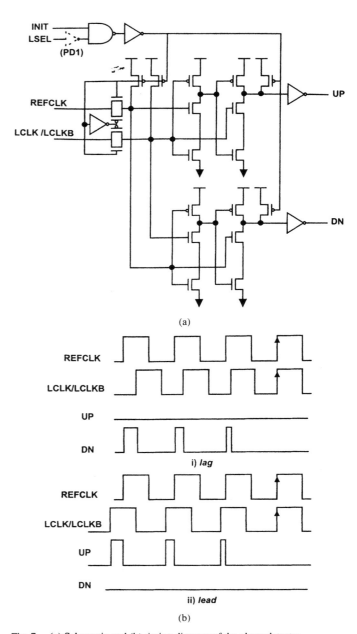

(a)

Fig. 6. (a) Schematic and timing diagrams at (b) $(1/2) \times T_{REF} < T_{diff} < T_{REF}$, (c) $0 < T_{diff} < (1/2) \times T_{REF}$ of the loop selector.

Fig. 7. (a) Schematic and (b) timing diagram of the phase detector.

synchronized with REFCLK. Once one of the two loops is selected, with the exception that the DLL is disabled or the power is down, the selected loop is never changed by the time difference between LCLK and REFCLK after the loop selection cycle. Therefore, malfunction is avoided on the lock-in process. Consequently, the proposed dual-loop operation eliminates the initial condition, $T_{diff} > (1/2) \times T_{REF}$, between the reference clock and the internal clock and enables the delay of VCDL to be utilized fully. The lock range is also extended to lower frequency without compromising the jitter characteristic.

IV. SKEW CALIBRATION BY PROGRAMMABLE REPLICA DELAY AND ANTIFUSE CIRCUITRY

Although the replica delay is well matched with the sum of on-chip and off-chip delay at design process, process variation of on-chip and unexpected change in the circumstance of

off-chip such as output load, clock slew-rate, and so on, may result in unavoidable skew. There are two methods for skew elimination, wafer trimmed by laser [12] and post-package tuning by antifuse [13]. The wafer-level tuning is not effective because the wafer tester is not precise, and the off-chip condition cannot be considered. Although post-package tuning by antifuse is more practical, the previous post-package method has some problems. The previous post-package method uses external high voltage through pins for rupturing the antifuse. Providing sufficient high voltage for rupturing the antifuse can cause physical damage to other circuits connected to the pin and can negatively affect the reliability of the device. To remove the high-voltage problem, the antifuse programming scheme by internal negative voltage is used [14]. Fig. 9 shows the programmable replica delay circuit. The circuitry has three functional parts, the replica delay of clock, the replica delay of the output buffer, and the tun-

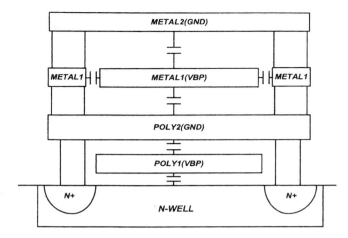

Fig. 8.   Linear capacitor in the loop filter.

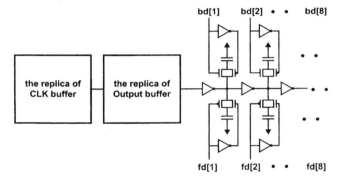

Fig. 9.   Replica delay including the programmable delay.

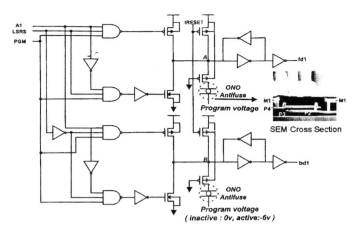

Fig. 10.   Antifuse circuit for skew calibration and SEM photograph of the antifuse.

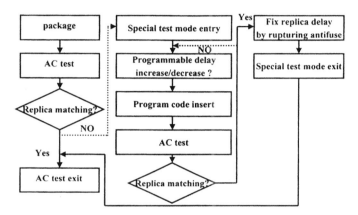

Fig. 11.   Flow of skew calibration after package process.

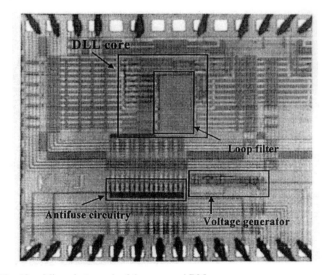

Fig. 12.   Microphotograph of the proposed DLL.

able delay circuit. The tunable delay circuit is connected to the antifuse circuitry and the antifuse is made of ONO (oxide–nitride–oxide) dielectrics, as shown in Fig. 10. The sequence of skew calibration is explained as follows. When DLL is enabled by RESET for the test, nodes fd [1]–[8] and bd [1]–[8] in Fig. 9 are all fixed at the high state, because the initial program voltage is at ground level, and RESET initializes node A and B as V_{dd} level. In this state, no address code can have an effect on the fixed levels of fd [1]–[8] and bd [1]–[8]. First, the skew between the external clock and the data strobe signal is measured. The measured skew is estimated by selection of optimal number of delay loads. After the program mode (PGM) is activated, the program code signifying the estimated number of delay loads is applied to the address pins and the skew is remeasured. This process is iterated to increase or decrease replica delay times by left-shift–right-shift (LSRS) for minimizing the skew. When the skew is almost eliminated, the inserted program address code is fixed and the on-chip negative voltage generator is enabled to produce a program voltage ($-3.5 \sim -4.5$ V) for rupturing the antifuses. The replica delay is tuned through the flow shown in Fig. 11. According to the simulation results, the programmable tuning range using the eight antifuses is from -350 to $+350$ ps and the minimum tuning resolution is approximately 10 ps.

V. EXPERIMENTAL RESULTS

The proposed DLL has been fabricated using 0.16-μm DRAM process. Fig. 12 shows a microphotograph of the fabricated chip. The active area of DLL occupies 0.27 mm^2. The loop filter consumes $\sim$ 50% of total area. For high-frequency measurements of the proposed DLL, a chip-on-board (COB) has been fabricated both to reduce parasitics and to match 50-Ω impedance of the measurement instrument. The proposed DLL operates from 42 to 400 MHz with a 2.3-V power supply. Fig. 13 shows the synchronized waveforms at

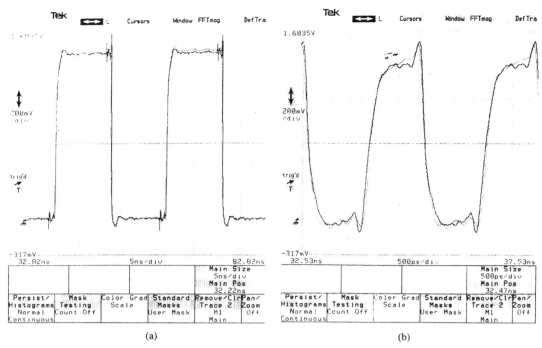

Fig. 13.   Synchronized waveforms at (a) 42 MHz and (b) 400 MHz.

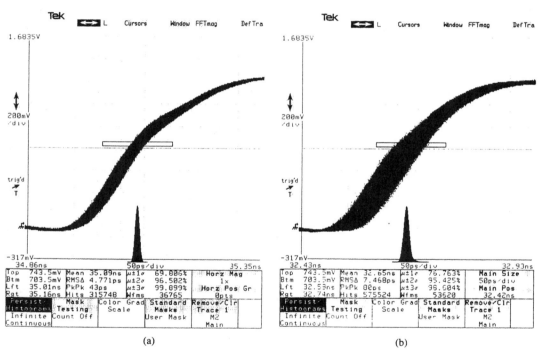

Fig. 14.   Measured jitter characteristics at 400 MHz in (a) a quiet supply and (b) with injected 1-MHz ±300-mV square wave noise.

42 and 400 MHz. At 400 MHz, the peak-to-peak jitter is 43 ps and the rms jitter is 4.71 ps, as shown in Fig. 14(a). When a ±300-mV 1-MHz square wave is injected externally on the power supply, the peak-to-peak jitter and the rms jitter is measured to be 80 and 7.46 ps, respectively, at 400 MHz, as shown in Fig. 14(b). Fig. 15(a) shows a skew that is composed of the phase offset of the phase detector and replica mismatch by process variation before skew calibration and the tuned skew after skew calibration. Before the calibration, the measured skew is 55 ps. After the calibration, the remeasured skew is

reduced to 9 ps with measured peak-to-peak jitter of 46 ps. In theory, error reduction resulting in negative phase shift will have increased jitter due to increased load in replica delay, but error reduction resulting in positive phase shift will have decreased jitter by decreased load in replica delay. However, from analyzing the measured results, the amount of increased jitter by negative phase reduction is insignificant compared to the reduced phase error. Fig. 15(b) shows the resolution and partial range of the skew calibration through antifuse programming. Minimum resolution is about 10 ps and total

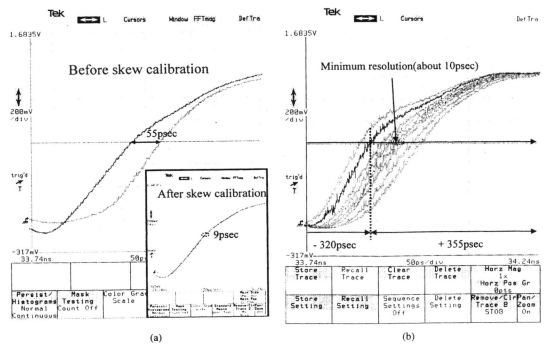

Fig. 15. (a) Measured skew at 400 MHz before skew calibration and after skew calibration. (b) Range (full range not displayed for limitation of tester) and resolution of skew calibration.

TABLE I
PERFORMANCE CHARACTERISTICS OF THE PROPOSED DLL

Process	0.16 μm DRAM process
Area	0.27mm² (active area)
Supply voltage	2.3v
Power dissipation	52mW@400MHz
Lock-range	42 - 400MHz
Jitter	43 ps peak-to-peak(4.77ps rms) jitter with quiet supply 80 ps peak-to-peak(7.46ps rms) jitter with ±300mv, 1MHz square wave noise
Supply sensitivity	0.07ps/mV(@400MHz)

calibration range is from −350 to +350 ps, as expected from simulation. These results show that the skew by variation in on-chip or off-chip can be eliminated through programmable replica delays using the antifuse circuitry, and also verifies that the improved skew calibration technique can effectively eliminate the skews after packaging without degradation of the jitter characteristic. The power dissipation of the proposed DLL is 52 mW at 400 MHz. Table I summarizes the measured characteristics of the proposed DLL.

VI. CONCLUSION

In this paper, the dual-loop architecture with the improved skew calibration method was presented. The dual-loop architecture enabled the wide range of the DLL by using the loop selection decided by an initial time difference between the reference clock and the internal clock of the DLL. Also, an improved skew calibration method demonstrated a practical post-package skew calibration using the antifuse circuitry and the internal negative voltage generator. The proposed DLL, fabricated on 0.16-μm DRAM process, achieves a wide range from 42 to 400 MHz, and 43 ps peak-to peak jitter and 4.71 ps rms jitter at 400 MHz that is applicable to high-speed DRAMs.

ACKNOWLEDGMENT

The authors are grateful to H. Ryu and Dr. Y. Kim for helpful discussion about COB-type PCB design.

REFERENCES

[1] A. Hatakeyama et al., "A 256-Mb SDRAM using a register-controlled digital DLL," IEEE J. Solid-State Circuits, vol. 32, pp. 1728–1734, Nov. 1997.
[2] Y. Okajima et al., "Digital delay-locked loop and design technique for high-speed synchronous interface," IEICE Trans. Electron., vol. E79-C, pp. 798–807, June 1996.
[3] T. H. Lee et al., "A 2.5-V CMOS delay-locked loop for an 18-Mbit 500-Mbyte/s DRAM," IEEE J. Solid-State Circuits, vol. 29, pp. 1491–1496, Dec. 1994.
[4] S. Tanoi et al., "A 250–622-MHz deskew and jitter-suppressed clock buffer using two-loop architecture," IEEE J. Solid-State Circuits, vol. 31, pp. 487–493, Apr. 1996.
[5] S. Sidiropoulos et al., "A semi-digital dual delay-locked loop," IEEE J. Solid-State Circuits, vol. 32, pp. 1683–1692, Nov. 1997.
[6] Y. Okuda et al., "A 66–400-MHz adaptive-lock-mode DLL circuit with duty-cycle error correction," in Symp. VLSI Circuits Dig. Tech. Papers, June 2001, pp. 37–38.
[7] C. H. Park et al., "A low-noise 900-MHz VCO in 0.6-μm CMOS," IEEE J. Solid-State Circuits, vol. 34, pp. 586–591, May 1999.
[8] T. Yoshimura et al., "A delay-locked loop and 90-degree phase shifter for 800-Mb/s double data rate memories," in Symp. VLSI Circuits Dig. Tech. Papers, June 1998, pp. 66–67.
[9] J. G. Maneatis, "Low-jitter and process-independent DLL and PLL based on self-biased techniques," IEEE J. Solid-State Circuits, vol. 31, pp. 1728–1732, Nov. 1998.

[10] I. A. Young et al., "A PLL clock generator with 5 to 110 MHz of lock range for microprocessors," *IEEE J. Solid-State Circuits*, vol. 27, pp. 1599–1607, Nov. 1992.

[11] F. Herzel et al., "A study of oscillator jitter due to supply and substrate noise," *IEEE Trans. Circuits Syst. II*, vol. 46, pp. 56–62, Jan. 1999.

[12] T. Hamamoto et al., "A skew and jitter suppress DLL architecture for high-frequency DDR SDRAMs," in *Symp. VLSI Circuits Dig. Tech. Papers*, June 2000, pp. 76–77.

[13] S. Kuge et al., "A 0.18-μm 256-Mb DDR-SDRAM with low-cost post-mold-tuning method for DLL replica," in *IEEE Int. Solid-State Circuits Conf. (ISSCC) Dig. Tech. Papers*, Feb. 2000, pp. 402–403.

[14] K. S. Min et al., "A post-package bit-repair scheme using static latches with bipolar-voltage programmable antifuse circuit for high-density DRAMs," in *Symp. VLSI Circuits Dig. Tech. Papers*, June 2001, pp. 67–68.

Se Jun Kim was born in Seoul, Korea, in 1974. He received the B.S. and M.S. degrees in electronics engineering from Hanyang University, Seoul, in 1998 and 2000, respectively.

In 2000, he joined the Memory Research and Development Division, Hynix Semiconductor Inc., Kyungki-Do, Korea, as a Research Engineer, where he has been working on CMOS circuit and architecture for high-speed digital/analog interface. His current interests include clock recovery circuits, data converters, clock distribution, and I/O circuits for high-speed digital/analog interface.

Sang Hoon Hong received the B.S. degree in electronic engineering from Yonsei University, Seoul, Korea, in 1993. He received the M.S. and Ph.D. degrees in engineering sciences from Harvard University, Cambridge, MA, in 1998 and 2001, respectively.

He is currently with the Memory Research and Development Division of Hynix Semiconductor Inc., Ichon-si, Kyongki-Do, Korea, working on high-speed dynamic memories with a particular interest in low-voltage/power circuits and architectures.

Jae-Kyung Wee was born in Seoul, Korea, in 1966. He received the B.S. degree in physics from Yonsei University, Seoul, in 1988 and the M.S. degree from Seoul National University in 1990. In August 1998, he received the Ph.D. degree in electronics engineering on modeling and characterization of interconnects for high-speed and high-density circuits from Seoul National University.

In 1990, he joined Hyundai Electronic Company working on the process integration of 16 MDRAM and LOGIC devices. In 1996, he was engaged in the development of the manufacturable 0.35-μm CMOS logic technology for high-performance logic products at Hyundai Electronics. In August 1998, he became a Project Leader of the Antifuse Repair Circuit Development Team. From August 1999 to June 2000, he was a Project Leader of 1-G DDR SDRAM using 0.13-μm technology. Beginning in July 2000, he also worked on next-generation DRAM and its related systems. He is currently with the faculty of Hallym University, Chunchun-si, Kangwon-Do, Korea. His research interest is in the area of future DRAM architecture including high-speed DRAM with 200 ~ 400 MHz clock, interconnect modeling, charge pump, DLL, I/O, and module designs for high-speed chips. He holds several patents and is an author or co-author of several papers.

Joo Hwan Cho was born in Seoul, Korea, in 1968. He received the B.S. degree in electronic materials engineering from Kwang-Woon University, Seoul, in 1992.

He joined the Semiconductor Research and Development Center, Hynix Semiconductor Inc, Ichon-si, Kyungki-Do, Korea, in 1992. Since then, he has been working on DRAM design and failure analysis.

Pil Soo Lee was born in Seoul, Korea, in 1963. He received the B.S. and M.S. degrees from Inchon University, Korea, in 1990 and 1992, respectively.

In 1993, he joined KEC, Kumi, Korea, where he worked on power device design and analysis. In 1997, he joined Hynix Semiconductor Inc., Ichon-si, Kyungki-Do, Korea, where he has been working on signal integrity analysis of high-frequency devices, circuits, and boards.

Jin Hong Ahn was born in Busan, Korea, in 1958. He received the B.S. and M.S. degrees in electronic engineering from Seoul National University, Seoul, Korea, in 1982 and 1984, respectively.

He joined Gold-Star Semiconductor Company, Gumi, Korea, in 1984. From 1986 to 1990, he was involved in designing SRAMs and mask ROMs. In 1991, he moved to the DRAM design group, Gold-Star Electron Company, Seoul. From 1991 to 1998, he managed several generations of advanced DRAM design projects, including 64-Mb, 256-Mb, MML, and intelligent RAM. His interests in DRAM design include new DRAM architectures, next-generation DRAM circuit technologies, and low-cost DRAM design techniques. In 1999, he joined the Memory Research and Development Group, Hynix Semiconductor Inc., Ichon-si, Korea, where he was engaged in the development of 0.15-μm 256-M DRAM. He is currently a Technical Director in DRAM Design technology.

Jin Yong Chung received the B.S.E.E. degree from Seoul National University, Seoul, Korea, in 1974 and the M.S.E.E. degree from Korea Advanced Institute of Science and Technology, Taejon, Korea, in 1976.

From 1976 to 1978, he worked for Korea Semiconductor Inc., which later became Semiconductor Business Unit of Samsung Electronics, where he was involved in the design of timepieces and custom CMOS chip designs. Since 1979, he was involved in memory design area and worked for various companies including National Semiconductor, Synertek, Vitelic, developing CMOS SRAMs, 4 K to 64 K and mask ROMs and CMOS DRAMs. In 1987, he joined LG Semiconductor, Korea, where he developed 256 K to 16 M DRAMs and other standard logic products. In 1992, he joined Mosel-Vitelic, where he developed high-speed DRAMs and the 256 K × 8 high-speed DRAM became the first semi-standard DRAM, which helped the company to go public.

Since 1996, he has worked for Hynix Semiconductor Inc., Ichon-si, Kyoungki-Do, Korea, as a Senior Vice President and Chief Architect in the Memory Research and Development Division. His current research interest is in development of ultrahigh-speed, super low-voltage and low-power memory products, novel device research in ferroelectric and magnetic memories, and new-generation 3-D devices.

PART VI

RF Synthesis

An Adaptive PLL Tuning System Architecture Combining High Spectral Purity and Fast Settling Time

Cicero S. Vaucher, *Member, IEEE*

Abstract—An adaptive phase-locked loop (PLL) architecture for high-performance tuning systems is described. The architecture combines contradictory requirements posed by different performance aspects. Adaptation of loop parameters occurs continuously, without switching of loop filter components, and without interaction from outside of the tuning system. The relationship of performance aspects (settling time, phase noise, and spurious signals) to design variables (loop bandwidth, phase margin, and loop filter attenuation at the reference frequency) are presented, and the basic tradeoffs of the new concept are discussed. A circuit implementation of the adaptive PLL, optimized for use in a multiband (global) car-radio tuner IC, is described in detail. The realized tuning system achieved state-of-the-art settling time and spectral purity performance in its class (integer-N PLL's): a signal-to-noise ratio of 65 dB, a 100-kHz spurious reference breakthrough signal under -81 dBc, and a residual settling error of 3 kHz after 1 ms, for a 20-MHz frequency step. It simultaneously fulfills the speed requirements for inaudible frequency hopping and the heavy signal-to-noise ratio specification of 64 dB.

Index Terms—Adaptive systems, FM noise, frequency synthesizers, phase-locked loops.

I. INTRODUCTION

FAST settling time–frequency synthesizers are essential building blocks of modern communication systems. Typical examples are digital cellular mobile systems, which employ a combination of time-division duplex (TDD) and frequency-division duplex (FDD) techniques. In these systems, the downlink frequencies (base station to handsets) are placed in different bands with respect to uplink frequencies. In order to save cost and decrease the size of the handset, it is desirable to use the same frequency synthesizer to generate uplink and downlink frequencies. Requirements are that the synthesizer has to switch between bands and settle to another frequency within a predetermined time (>1.7 ms for GSM and DCS-1800 systems [1]).

Car-radio receivers with optimal radio data system (RDS) performance ask for fast-settling-time tuning systems as well [2]. The RDS network transmits a list of (nationwide) alternative frequencies carrying the same program. The tuner performs a background scanning of these frequencies, so that optimum

reception condition is provided when the receiver is displaced within different coverage regions. For the system to be effective, the background scanning has to be performed in a transparent (inaudible) way to the listener. A possible but expensive way to do that is to use two tuners in the receiver, with one of them being used for checking on alternative frequencies only. Single-tuner solutions—which have a much better price/performance ratio—require a tuning system architecture able to do frequency hopping in an inaudible way [2]. In other words, a fast-settling-time architecture is required for these applications.

Communication systems often pose severe requirements on the spectral purity of the tuning system local oscillator (LO) signal. There are two main reasons for this. First, to avoid problems with reciprocal mixing of adjacent channels. Reciprocal mixing decreases the receiver's selectivity and disturbs the reception of weak signals. Second, because the mixing process, which is used for down-conversion of the radio-frequency (RF) signals, superposes the phase noise of the LO on the modulation of the RF signal. Hence, the signal-to-noise ratio (SNR) at the output of the demodulator is a function of LO's phase noise level [3].

This paper describes an adaptive tuning system architecture that combines fast settling time with excellent spectral purity performance. The architecture was optimized to be used in a global car-radio tuner IC with inaudible RDS background scanning. The integer-N frequency synthesizer has an SNR of 65 dB and a 100-kHz spurious reference breakthrough under -81 dBc at the voltage-controlled oscillator (VCO) (-87 dBc at the mixer). Residual settling error for a 20-MHz frequency step is 3 kHz after 1 ms. These results are similar to those of a fractional-N implementation [4]. The complexity of our tuning system, however, is much smaller. The adaptive phase-locked loop (PLL) was integrated in a 5-GHz, 2-μm bipolar technology. The tuning system works with 8.5-V supply voltage for the charge pumps and with 5 V for the logic functions. Total current consumption is 21 mA from the 5-V supply and 12 mA from the 8.5-V supply.

The architecture of the multiband tuner IC is described in Section II. Section III presents relationships of settling time, phase noise, and spurious signals to the design variables, namely loop bandwidth, phase margin, and loop filter attenuation at the reference frequencies. Section IV introduces the adaptive PLL architecture and discusses the advantages and tradeoffs of the concept. Section V describes the circuit implementation, and Section VI presents a summary of measured results.

Manuscript received July 23, 1999; revised November 29, 1999.

The author is with Philips Research Laboratories, Eindhoven 5656 AA The Netherlands (e-mail: Cicero.Vaucher@philips.com).

Publisher Item Identifier S 0018-9200(00)02861-4.

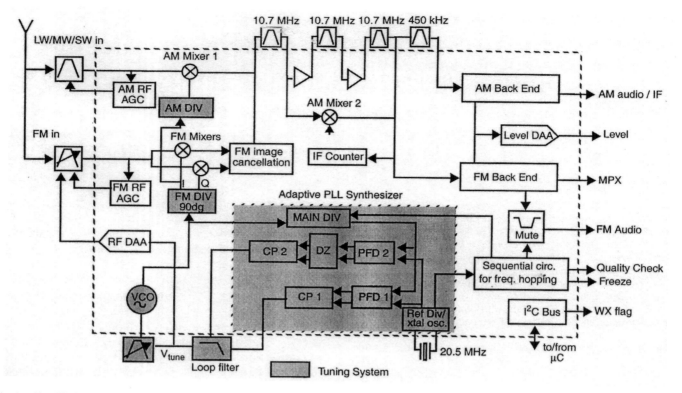

Fig. 1. Simplified block diagram of the global car-radio tuner IC.

TABLE I
RECEPTION BANDS WITH CORRESPONDING TUNING SYSTEM PARAMETERS

Band	Fant (MHz)	Fvco [151-248] (MHz)	FM DIV	AM DIV	Fref PLL (kHz)	tuning step size (kHz)
LW / MW (Eur+USA)	0.144 1.710	216.88 248.20	2	10	20	1
SW	5.85 9.99	165.5 206.9	2	5	10	1
Weather B. (USA WX)	162.40 162.55	173.10 173.25	1		25	25
FM (East Eur)	65 74	151.4 169.4	2		20	10
FM (Japan)	76 90	173.4 201.4	2		100	50
FM (Eur+USA)	87.5 108	196.4 237.4	2		100	50

II. MULTIBAND TUNER ARCHITECTURE

The block diagram of the global tuner IC with inaudible background scanning is shown in Fig. 1. The receiver and tuning system architectures have been defined such that all reception bands can be accessed with a single VCO and a single loop filter, without changes to the application. Mapping the frequency of the VCO to the different input bands is achieved by dividing its output frequency by different ratios, depending on the band to be received. The division is accomplished in the FM DIV and AM DIV dividers, which are set in between the VCO output and the RF mixers. Table I presents the VCO frequency and tuning system parameter settings for various reception bands, including the American Weather Band. By dividing the VCO output, the tuning resolution is 1 kHz in AM mode and 50 kHz in FM mode, despite the fact that reference frequencies are 20 kHz and 100 kHz, respectively.

Combining the different reception bands in one single application—the same VCO and same loop filter—complicates the design of the tuning system. A reception band with worst case

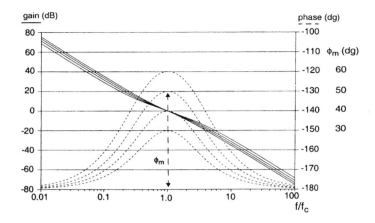

Fig. 2. Open-loop frequency response (Bode plot) of a type-2, third-order charge-pump PLL for different values of phase margin ϕ_m.

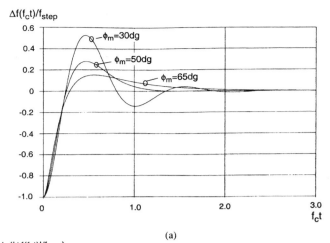

(a)

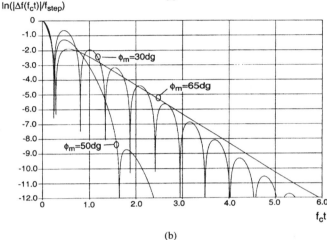

(b)

Fig. 3. Setting transient for different values of ϕ_m, normalized for $f_c t$. (a) Setting error (represented as $\Delta f(f_c t)/f_{step}$) versus $f_c t$. (b) Setting error (represented as $\ln(|\Delta f(f_c t)|/f_{step})$) versus $f_c t$.

spectral purity requirements determines the loop filter design. Nonetheless, robustness for variations in tuning system parameters, for all reception bands, has to be insured. The relationships between different performance aspects on system level are discussed in the following section.

III. SETTLING TIME AND SPECTRAL PURITY PERFORMANCE

The properties of a PLL are strongly related to its phase detector implementation [5]. Present-day PLL frequency synthesizers usually employ the tristate, sequential phase frequency detector (PFD), combined with a charge pump (CP) [6]. The analysis of the PLL properties presented in this paper assumes the use of a PFD/CP in the loop.

A. Settling Time, Loop Bandwidth, and Loop Phase Margin

Bode diagrams are a powerful tool for designing PLL tuning systems [7], [8] because they enable direct assessment of the loop's phase margin (ϕ_m) and open-loop bandwidth (0-dB frequency f_c). Accurate and reliable results for f_c and ϕ_m are obtained with ease to implement behavioral models [9] and with fast ac simulation runs. In spite of the advantages of the "ac method," design equations relating the settling performance of a type-2, third-order charge-pump PLL[1] [6] to its open-loop bandwidth and phase margin have, to the best of our knowledge, not yet been published in the open literature.

Fig. 2 presents Bode plots of a type-2, third-order loop for different values of phase margin ϕ_m. Fig. 3(a) displays the transient response of such a loop for three different values of phase margin. The responses are plotted as $\Delta f(f_c t)/f_{step}$, normalized for $f_c t$. $\Delta f(t)$ is the remaining frequency error with respect to the final value and f_{step} is the amplitude of the frequency jump. Fig. 3(b) presents the responses as $\ln(|\Delta f(f_c t)|/f_{step})$, so that the impact of ϕ_m on the "long-term" transient response is easily observed.

The influence of the phase margin on the settling time, obtained with transient simulations similar to those of Fig. 3, is presented in Fig. 4. The figure shows the time necessary for the value of $\ln(|\Delta f(f_c t)|/f_{step})$ to reach a numerical value of -10. The settling time decreases with increasing phase margin,

[1]The most widely used configuration in synthesizer applications.

reaching a minimum for ϕ_m values of around 50°. Increasing the phase margin further leads to a sharp increase in the settling time.

The relationship of settling time and phase margin, displayed in Fig. 4, can be understood with the help of Fig. 5. It presents the pole and zero locations of the closed-loop transfer function of a third-order loop with different values of phase margin (Bode plots presented in Fig. 2). The real part of the dominant (complex) poles approach $-f_c$ for values of ϕ_m of about 50°. When ϕ_m equals 53°, all three poles lie at $-f_c$. That is the location with the fastest damping of the transient error. The fastest response, however, is obtained with 51°. The complex parts of the poles "speed up" the settling transient a bit further (25%). For higher values of phase margin, the dominant real pole moves to the right on the real axis. This pole is responsible for the slowing down of the PLL response for values of $\phi_m > 53°$. Fig. 5 shows that the dominant pole, for 60° phase margin, lies at about $-0.4 f_c$. Hence, it may be concluded that the usual practice of designing critically damped loops—which have a phase margin of about 70° [5]—is not appropriate for fast-settling-time applications.

Let us consider Fig. 3(b) again. One sees that the (envelope of the) curves can be approximated by straight lines. The ap-

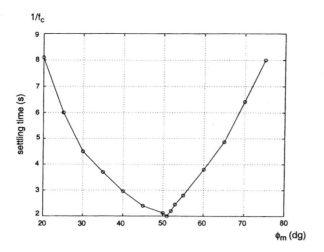

Fig. 4.   Setting time as function of the phase margin for $f_{\text{error}}/f_{\text{step}} = e^{-10}$.

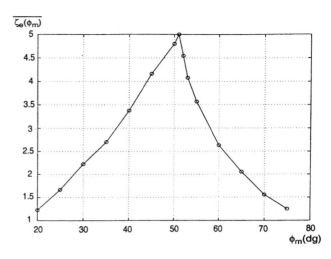

Fig. 6.   Average values of $\zeta_e(\phi_m)$ for a $\Delta(\ln(\cdot))$ of ten.

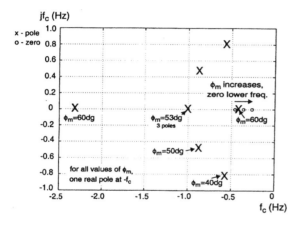

Fig. 5.   Position of the *closed-loop* poles and zeros of a third-order PLL corresponding to different values of ϕ_m, as displayed in Fig. 2.

proach proposed here takes ϕ_m into account with the help of an *effective damping coefficient* $\zeta_e(\phi_m)$. By so doing, we arrive at the following approximation for the envelope of the curves of Fig. 3(b):

$$\text{env}(\ln(|\Delta f(f_c t)|/f_{\text{step}})) = -\zeta_e(\phi_m) \cdot f_c t. \qquad (1)$$

Numerical estimations for $\zeta_e(\phi_m)$ can be obtained from transient simulations with the help of the following expression:

$$\overline{\zeta_e(\phi_m)} = \frac{-\Delta(\text{env}(\ln(\Delta f(f_c t)/f_{\text{step}})))}{\Delta(f_c t)}. \qquad (2)$$

The settling time results presented in Fig. 4 leads to the numerical values for $\overline{\zeta_e(\phi_m)}$ displayed in Fig. 6. These values represent an average value for $\zeta_e(\phi_m)$, as they are obtained from a $\Delta(\text{env}(\ln(\cdot)))$ of ten.

Manipulation of (1) results in an equation describing the minimum loop bandwidth required to achieve given settling specifications t_{lock}, f_{error}, and f_{step}

$$f_c = \frac{1}{t_{\text{lock}} \zeta_e(\phi_m)} \ln f_{\text{step}}/f_{\text{error}}. \qquad (3)$$

In (3):

t_{lock} locking time(s);
f_{step} amplitude of the frequency jump (Hz);
f_{error} maximum frequency error (Hz) at t_{lock};
$\zeta_e(\phi_m)$ can be read from Fig. 6.

Two points about the present treatment of the transient response need further explanation. First, the presented results are based on a linear continuous-time model for the discrete-time charge-pump PLL. It is known in the literature [6] that the continuous-time approach is a good approximation for the discrete-time PLL if the reference (sampling) frequency f_{ref} of the loop is at least a factor of ten higher than its open-loop bandwidth f_c. Therefore, the value of f_c, calculated with (3), has to be checked against the loop's reference frequency f_{ref}. If the target ratio f_{ref}/f_c is smaller than ten, then actual settling behavior will deviate from the calculations.

The second point is that usual implementations of the phase frequency detector have a limited linear phase error detection range, namely, from -2π to 2π [9]. When the instantaneous phase error ϕ_{error} becomes larger than $\pm 2\pi$, the PFD interprets the error information as $(\phi_{\text{error}} \mp 2\pi)$. This effect leads to a longer settling time than predicted with (3). The maximum value of ϕ_{error}, denoted ϕ_{max}, was found to obey the following relationship: $\phi_{\text{max}} = \alpha(\phi_m)(f_{\text{step}}/N)/f_c$, where N is the main divider ratio and $\alpha(\phi_m)$ is a fitting factor for the influence of the phase margin on ϕ_{max}. Numerical values for $\alpha(\phi_m)$, obtained from transient simulations, lie in the range [0.7,0.8]. Hence, the maximum phase error is contained in the interval $\pm 2\pi$, when $f_c > 0.8(f_{\text{step}}/2\pi N)$. If this condition is satisfied, then the (discrete-time) transient response is accurately predicted by the continuous-time linear model.

Inaudible RDS background scanning requires settling times of 1 ms, defined as a residual settling error of 6 kHz for a 20-MHz frequency jump. The nominal loop phase margin is set to $50°$, which corresponds to a $\zeta_e(\phi_m)$ of five. On the other hand, it is appropriate to use a lower value for $\zeta_e(\phi_m)$ in the calculations (e.g., 2.5), to provide enough margin for variations in the nominal values of loop bandwidth and phase margin. Solving (3) for these settling specifications leads to a nominal value of 3.2 kHz for the loop bandwidth f_c.

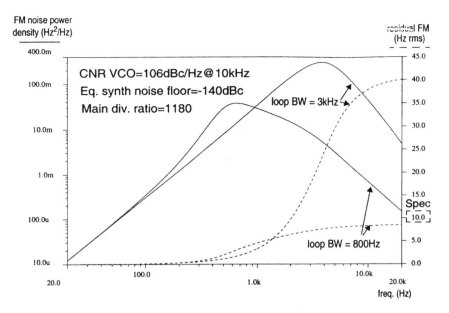

Fig. 7. FM noise density and residual FM for loop bandwidths of 800 Hz and 3 kHz.

The loop bandwidth that satisfies different settling requirements can be calculated with the help of (3). Settling specifications, however, often require loop bandwidths that are not optimal with respect to spectral purity performance, as will become clear in the next subsection.

B. Phase Noise Performance and Loop Bandwidth

The dependency of the total phase noise of a PLL tuning system on the phase noise of the loop components is well known in the literature [3], [5], [10]. The phase noise of the VCO is suppressed inside the loop bandwidth, whereas the (phase) noise from the other building blocks is transferred to the VCO output, multiplied by the closed-loop transfer function of the PLL: a low-pass function that suppresses their noise contribution outside the loop bandwidth. There is a "crossover point" for the loop bandwidth, where the noise contribution from the dividers and charge pump becomes dominant with respect to the noise from the VCO.

For terrestrial FM reception, the LO signal residual frequency noise (residual FM) determines the ultimate receiver's SNR performance. The SNR specification for the application is 64 dB, defined for a reference level of 22.5-kHz peak deviation with 50-μs deemphasis. Complying to the specification requires the residual FM in the LO signal to be less than 10 Hz rms.

The frequency (FM) noise density of the LO signal $S_{\Delta f}(f)$ is linked to its phase noise power density $S_{\Delta\theta}(f)$ by $S_{\Delta f}(f) = f_m^2 S_{\Delta\theta}(f)$ [5]. $S_{\Delta\theta}(f)$ equals $2\mathcal{L}(f)$, the single-sideband noise-to-carrier ratio, so that $S_{\Delta f}(f) = 2f_m^2\mathcal{L}(f)$. Finally, the residual FM can be calculated

$$\Delta f_{\text{res}} = \sqrt{\int_{f_l}^{f_h} S_{\Delta f}(f)\, df} = \sqrt{\int_{f_l}^{f_h} 2f_m^2 \mathcal{L}(f)\, df}. \quad (4)$$

The integration limits f_l and f_h in (4) depend on the signal bandwidth of the application [3]. For terrestrial FM reception, the lower limit is 20 Hz and the higher is 20 kHz. Fig. 7 presents

the simulated frequency noise (FM noise) power density and the residual FM, which is plotted as function of f_h, with f_l fixed at 20 Hz. The FM noise density and the residual FM are plotted for values of loop bandwidth of 800 Hz and of 3 kHz. For 3 kHz, the residual FM amounts to 40 Hz rms, which is 12 dB higher than the specification. A loop bandwidth of 800 Hz, on the other hand, leads to a residual FM of 8 Hz rms, which satisfies the SNR requirement.

The contributions of different noise sources to the total frequency noise density, in the case of an 800-Hz loop bandwidth, are displayed in Fig. 8. The contribution of the VCO to the residual FM equals that of the other synthesizer building blocks. This is a good compromise, and 800 Hz was chosen as the nominal loop bandwidth for in-lock situations.

The settling specification requires a bandwidth of 3.2 kHz. The SNR constraint, on the other hand, asks for 800 Hz. These conflicting requirements can be combined when the loop bandwidth is made adaptive as a function of the operating mode: frequency jump or in-lock.

Adapting the value of the loop bandwidth during frequency jumps is easily accomplished by switching the nominal value of the charge-pump current [6], [13]. This method, however, often causes disturbances in the VCO tuning voltage—the so-called secondary glitch-effect—at the moment the current is switched from high to low values. These disturbances are highly undesirable, as they have to be corrected by the loop in small bandwidth mode. What is more, the "secondary glitches" may cause audible disturbances in analog systems and increase the bit error rate in digital systems.

To provide stability for a small bandwidth loop requires a transfer function zero located at low frequencies (large time constant). A low-frequency zero, however, is undesirable for operation in high bandwidth mode. It causes the phase margin to be "too" high, which increases the settling time. Note that the effective damping coefficient $\zeta_c(\phi_m)$ decreases for high values of phase margin (see Fig. 6).

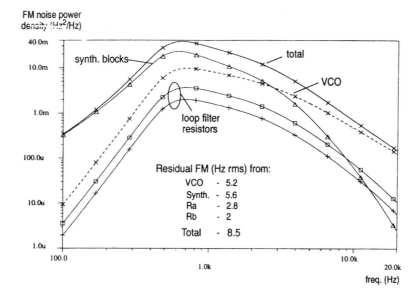

Fig. 8. Contributions from different noise sources to the total FM noise density and residual FM (20 Hz–20 kHz) with 800-Hz loop bandwidth.

Therefore, for optimal settling time *and* phase noise, one has not only to switch the value of the loop bandwidth but also to change the location of the zero in the transfer function.

C. Reference Spurious Signals and Loop Filter Attenuation

The use of phase frequency detectors yields the minimum levels of spurious breakthrough at the reference frequency [11]. The spurious signals are due to compensation of leakage currents or to imperfections in the charge pump's implementation. Standard FM modulation theory and the small angle approximation lead to the following equation for the amplitude of the spurious signal (in dBc), which is at an offset frequency f_m from the carrier:

$$ \text{spurious}(f_m) = 20\log(i_{\text{sp}}(f_m) \cdot |Z(f_m)| \cdot K_{\text{vco}}/(2f_m)). \tag{5} $$

where

f_m offset frequency from the carrier (Hz);
$i_{\text{sp}}(f_m)$ amplitude of ac current component with frequency f_m (A);
$|Z(f_m)|$ impedance of the loop filter at f_m (V/A);
K_{vco} VCO gain (Hz/V).

The value of $i_{\text{sp}}(f_m)$ is twice the value of the loop-filter dc leakage current [12] in loops operating with well-designed charge pumps. In cases where the charge pump has charge-sharing problems and/or charge injection into the loop filter, $i_{\text{sp}}(f_m)$ may become dominated by these second-order effects. The imperfections can lead to spurious components with (much) higher amplitudes than would be expected based on the leakage current alone.

Rearranging the above equation leads to a formula that relates the required filter attenuation at f_m to the specified maximum level of spurious signals $(\max \text{spurious}(f_m))$, to the dc leakage current (I_{leak}), and to the VCO gain (K_{vco})

$$ |Z(f_m)| < \frac{2f_m}{I_{\text{leak}}K_{\text{vco}}} 10^{\max \text{spurious}(f_m)/20}. \tag{6} $$

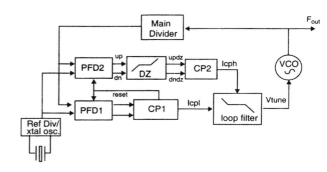

Fig. 9. Adaptive PLL tuning system architecture.

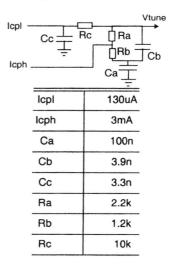

Icpl	130uA
Icph	3mA
Ca	100n
Cb	3.9n
Cc	3.3n
Ra	2.2k
Rb	1.2k
Rc	10k

Fig. 10. Loop-filter configuration, charge-pump currents, and component values used in the global car-radio tuner IC.

The relevant values of f_m equal f_{ref} and its harmonics in a standard PLL operating with a reference frequency of f_{ref} Hz. Therefore, the required loop-filter (trans)impedance for these frequencies can be readily calculated. The VCO gain, the spurious specification, and the expected (maximum) leakage current are known.

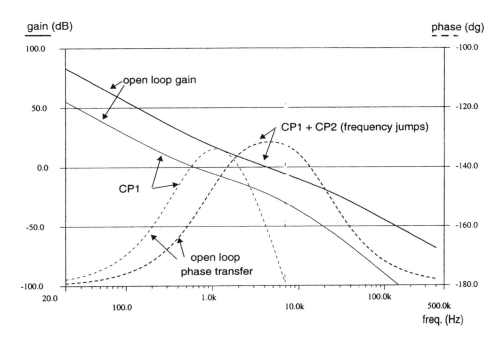

Fig. 11. Bode plots of the adaptive loop during frequency jumps and in-lock.

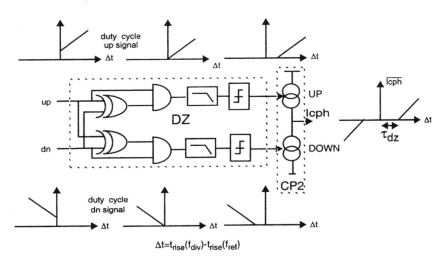

Fig. 12. Implementation of the DZ building block.

An important conclusion to be taken from the above equations is that the amplitude of the spurious signals is *not* dependent on the absolute value of loop bandwidth. Instead, it is determined by the (trans)impedance of the loop filter. This means that, at least in principle, "any" spurious specification can be achieved simply by decreasing the impedance level of the loop filter. In practice, this is not a viable option because the PLL loop bandwidth is proportional to the value of the loop-filter resistor and to the charge-pump current [6].

For a constant value of the loop bandwidth, a decrease of the loop-filter impedance level requires a proportional increase of the nominal charge-pump current. This leads to difficulties in the charge-pump design and to higher power dissipation. To avoid these difficulties, more RC sections are added to the basic loop-filter configuration, so that the filter attenuation at higher frequencies is increased. Additional RC sections, however, inevitably cause phase lag at lower frequencies. The phase lag de-

creases the loop phase margin and increases the settling time in high-bandwidth mode.

Therefore, to provide optimal settling, low-power dissipation, *and* good spurious performance, one has not only to switch the value of the loop bandwidth but also to bypass (some) RC sections of the loop filter. The PLL architecture presented here complies with these requirements.

IV. ADAPTIVE PLL ARCHITECTURE

A. Basic Architecture

The basic idea is to have two loops working in parallel, as depicted in Fig. 9. Loop 1, built around PFD1 and CP1, is dimensioned for in-lock operation. Loop 2, built around PFD2, DZ, and CP2, is dimensioned for fast settling time. Loop 1 operates all the time, whereas Loop 2 is only active during tuning

523

actions. Loop 1 and Loop 2 share the crystal oscillator, the reference divider, and the main divider.

A smooth takeover from Loop 1, after a frequency jump, avoids "secondary glitch" effects. The high-current charge pump CP2 is only active during tuning. CP2 is controlled by the dead-zone (DZ) block. DZ generates a *smooth* transition into a well-defined dead zone for CP2 when lock is achieved, so that sudden disturbances of the VCO tuning voltage are avoided.

Additional freedom for optimization of the loop parameters is obtained by using two separate charge-pump outputs and by applying the charge-pump currents to different nodes of the loop filter. In this way, the location of the zeros for frequency jumps and in-lock can be set in a continuous way, without switching of loop components—which is a source of "secondary glitch" problems. Furthermore, the path from Icpl to Vtune may contain additional filtering sections for, e.g., attenuation of spurious signals and/or fractional-N quantization noise [14]. These filter sections may be bypassed by Icph to increase the phase margin in high-bandwidth mode.

B. Loop-Filter Implementation

The ideas described above are demonstrated with the help of Figs. 10 and 11. Fig. 10 presents the loop-filter configuration and component values used in the global tuner IC (Fig. 1). Fig. 11 shows the optimized Bode diagrams of the adaptive PLL (in FM mode) with the loop filter of Fig. 10.

During frequency jumps both CP1 and CP2 are active; the loop filter zero frequency is $1/2\pi\,RbCa$ and lies at a high frequency, matching the 0-dB open-loop frequency. It enables stability and fast tuning to be achieved. The nominal loop bandwidth in this mode is 3.2 kHz, and the phase margin is 50°. After the frequency jump only CP1 is active. The zero of the loop filter moves to a lower frequency $(1/2\pi(Ra + Rb)Ca)$, without the switching of loop-filter components. The low-frequency zero increases the phase margin in-lock.

When the loop is in-lock, an extra pole is introduced $(1/2\pi\,RcCc)$, which increases the 100-kHz reference suppression by about 20 dB. During frequency jumps, these elements are bypassed by CP2, increasing the phase margin in high-bandwidth mode. If the loop bandwidth were increased by simply switching the amplitude of CP1, one would end up with an unstable loop, because of a phase margin of less than 10° in high-bandwidth mode.

C. Dead-Zone Implementation

The new element in the adaptive PLL architecture is the combination of the DZ block with the high-current charge pump CP2. The function of DZ is to provide CP2 with a well-defined dead zone of $\pm\tau_{dz}$ s. The dead zone is centered symmetrically around the locking position of charge pump CP1 [see Fig. 13(a)].

The logic diagram of the DZ/CP2 combination is depicted in Fig. 12. The figure shows how the different logic functions influence the duty cycle of the *up* and *dn* signals from the phase frequency detector (PFD2). At the input of DZ, the *up* and *dn* signals have a finite duty cycle, even for an in-lock situation $(\Delta t = 0)$. The finite duty cycle eliminates dead-zone problems in CP1. The XOR and AND gates are used to cancel the finite

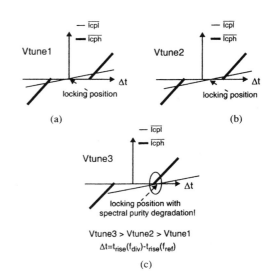

Fig. 13. Shift in locking position as function of VCO tuning voltage.

in-lock duty cycle. The processed *up* and *dn* signals are then applied to low-pass filters and slicers, whose function is to prevent pulses that have too small a duty cycle from reaching CP2. The cutoff frequency of the low-pass filters, the discrimination level of the slicers, and the turn-on time of CP2 determine the size of the dead zone around the lock position $\pm\tau_{dz}$ s.

A tradeoff among settling performance, circuit implementation, and robustness arises, when the magnitude of the dead zone τ_{dz} has to be determined. Let us start discussing circuit aspects.

The dead zone of charge pump CP2 should be centered around the locking position of the loop for optimum settling and spectral purity performance. The locking position, however, is a function of the output voltage of charge pump CP1. The effect is depicted in Fig. 13. One sees that, as the tuning voltage Vtune increases, there is a shift of the locking position to positive values of Δt. The reason lies in the finite output resistance of the active element used in CP1. Different current gains in CP1's UP and DOWN branches need to be compensated by *up* and *dn* signals with different duty cycles at the locking point. Different duty cycles are accomplished by a shift in the loop's locking position.

Fig. 13 shows situations where the gain in the UP branch of the pump decreases as Vtune increases. The ideal operating situation is depicted in Fig. 13(a). Situation (b) is still allowed from the point of view of spectral purity but has asymmetrical settling performance. Finally, (c) depicts a situation that should never happen: the locking position shifts so much that the high-current charge pump CP2 becomes active and degrades the in-lock spectral purity. Therefore, increasing the size of CP2's dead zone $(\pm\tau_{dz}$ s) eases the design of charge pump CP1 and increases the robustness of the system.

On the other hand, the size of CP2's dead zone influences the settling performance of the adaptive loop. The influence of τ_{dz} on the transient response was simulated with behavioral models. The results are displayed in Fig. 14, together with the settling requirements that ensure inaudible background scanning functionality. Table II presents the settling time for different settling accuracies and different values of τ_{dz}. A dead-zone value of infinity corresponds to the situation where only CP1 is active

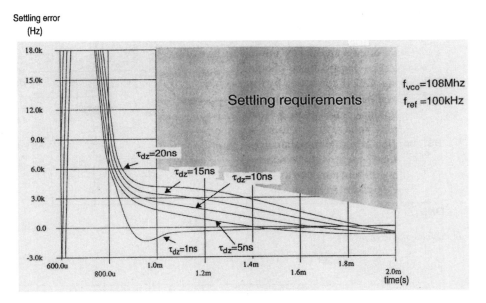

Fig. 14. Detail of settling transient for different values of τ_{dz}.

TABLE II
SIMULATED IN-LOCK SNR AND SETTLING TIME (ms) FOR A 20-MHz
FREQUENCY JUMP FOR DIFFERENT VALUES OF THE DEAD ZONE AND
DIFFERENT SETTLING ACCURACIES

Settling accuracy (f_{error})	Dead-zone value τ_{dz}				
	0	5 ns	15 ns	20 ns	∞
6 kHz	0.79	0.81	0.84	0.85	7.9
3 kHz	0.81	0.88	1.14	1.37	8.1
1 kHz	0.98	1.15	1.56	1.68	9.4
SNR (dB)	52	66			

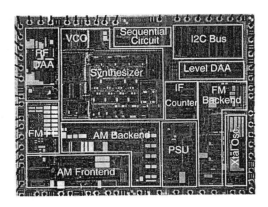

Fig. 15. Micrograph of the tuner IC.

(nonadaptive loop). Table II shows that by using the adaptive loop architecture, it is possible to combine fast settling time with good SNR in-lock. Increasing τ_{dz} leaves more "residual" phase (and frequency) error to be corrected by the small bandwidth loop. The closer one comes to the locking point in high bandwidth mode, the shorter the total settling transient will be. A dead-zone value of ± 15 ns is a good compromise for the intended application.

V. CIRCUIT IMPLEMENTATION

A die micrograph of the total tuner IC is displayed in Fig. 15. The adaptive PLL has been integrated with the other functional blocks of Fig. 1 in a 5-GHz, 2-μm bipolar technology [15].

A. Programmable Dividers

The architecture of the main divider is depicted in Fig. 16. The high-frequency part of the programmable divider is based on the programmable prescaler concept described in [12] and consists of a chain of 2/3 divider cells. The modular architecture enables easy optimization of power dissipation and robustness for process variations. The division range of the basic prescaler configuration is extended by the low-frequency programmable counter. The logic functions of the PLL were implemented with

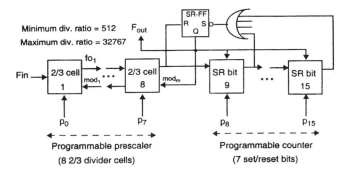

Fig. 16. Architecture of the main programmable divider.

current routing logic techniques (CRL) [12], [16]. The low-frequency part of the main and reference dividers operate with low current levels to limit total power dissipation. To decrease the phase noise of the reference signal going to the phase detectors, this signal is reclocked in a high-current D-flip-flop (D-FF). The clean crystal signal is used to clock the D-FF. The total main divider current consumption is 5 mA. The first 2/3 cell consumes 2.1 mA.

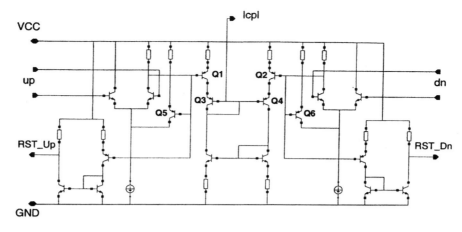

Fig. 17. Simplified circuit diagram of charge pump CP1.

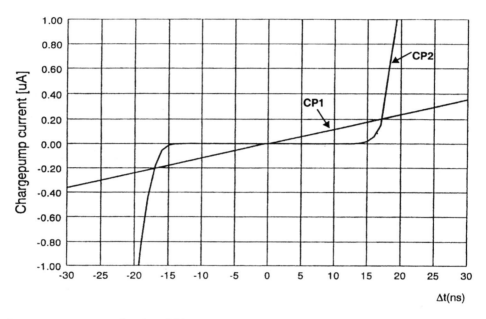

Fig. 18. CP1 and CP2 charge-pump currents as a function of Δt.

B. Oscillators

The LC VCO uses an external tank circuit. It can be tuned from 150 to 250 MHz, with a voltage tuning range from 0.5 to 8 V. The VCO phase noise is −100 dBc/Hz at 10 kHz, for a carrier frequency of 237 MHz. The VCO core consumes 1.5 mA. The 20.5-MHz reference crystal oscillator operates in linear mode, to avoid harmonics interfering in the FM reception bands. Quadrature generation for the image rejection FM mixers (see Fig. 1) is accomplished in a divider-by-two (FM DIV), with the exception of reception in the American Weather Band (WX). In that case, I/Q signals are generated with a RC-CR network directly from the VCO. This avoids the need to have the VCO operating at 346 MHz, and a change in the LC VCO tuned circuit during WX reception.

C. Charge Pumps

Fig. 17 shows the simplified circuit diagram of the low-current charge pump CP1. The *up* and *dn* signals from the phase

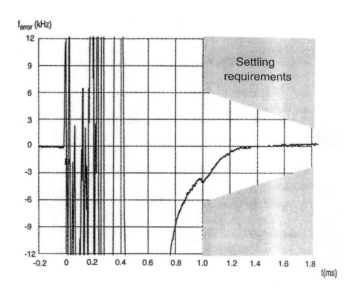

Fig. 19. Settling transient for a 20-MHz tuning step.

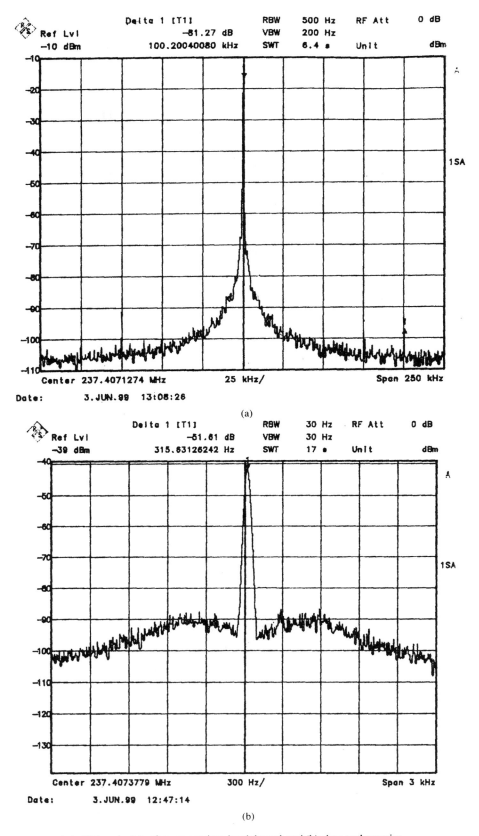

Fig. 20.   Spectral purity measurements in FM mode: (a) reference spurious breakthrough and (b) close to the carrier.

detector drive the input differential pairs, which set the currents in the PNP current switches Q1 and Q2 on and off. The collector outputs of Q1 and Q2 are kept at equal dc levels by the dc feed- back arrangement provided by Q3 and Q4. This prevents asym- metry in the source and sink currents, ensuring good centring of the charge-pump characteristics for all tuning voltages. Q5 and

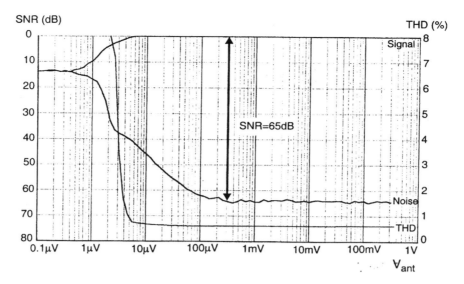

Fig. 21. Evaluation of the FM channel—VCO purity determines SNR for $V_{\mathrm{ant}} > 300$ μV. Fin = 97.1 MHz, AF freq = 1 kHz. SNR meas.: FMdev = 22.5 kHz; 26 dB = 2.0 μV. THD meas.: FMdev = 75 kHz.

Q6 provide means for stabilization of currents and for speeding up the switching of Q1 and Q2. The reset circuits monitor the currents in Q1 and Q2 and generate the reset signals RST_Up and RST_Dn. These signals are fed back to reset the phase detectors. The high-current charge pump CP2 is a scaled-up version of the CP1 circuit, without the reset circuits.

VI. Measurements

The measured charge-pump currents as a function of the time difference between the phase detector inputs are shown in Fig. 18. Good centering of the two charge-pump outputs is observed, and there is enough margin for variations in the in-lock position of CP1. The measured settling transient response is displayed in Fig. 19. The settling performance complies to the settling requirements and enables inaudible background scanning in single-tuner RDS applications.

The frequency spectrum of the VCO in FM mode is presented in Fig. 20(a) and (b). Fig. 20(a) shows the spurious reference breakthrough at 100 kHz to be under −81 dBc. There is yet a 6-dB improvement in noise and spurious breakthrough before the VCO signal reaches the FM mixers, due to the division by two in the FM DIV divider (see Fig. 1). Fig. 20(b) displays the phase noise spectrum close to the carrier. Spectrum measurements done in AM mode showed a reference spurious breakthrough of −57 dBc, at an offset of 20 kHz from the carrier. For AM, the improvement in phase noise and spurious performance amounts to 26 dB, due to the division by 20 in between the VCO and the AM mixers.

Finally, the SNR and THD of the total FM receiver chain are displayed in Fig. 21 as a function of the antenna input signal level V_{ant}. For low values of V_{ant}, the noise is dominated by RF input noise and by the quality of the building blocks in the signal processing chain: low-noise amplifier, mixers, and demodulator. For high values of V_{ant} (>300 μV), the dominant noise source becomes the LO signal. The excellent measured FM sensitivity, 2.0 μV for 26-dB SNR, and the ultimate SNR of 65 dB verify the spectrum purity of the tuning system and of the RF channel.

VII. Conclusion

This paper described an adaptive PLL architecture for high-performance tuning systems. The relationships of performance aspects to design variables were presented. It is demonstrated that design for spectral purity performance often leads to suboptimal settling performance, because of different requirements on the loop bandwidth and on the location of the zeros and poles of the closed-loop transfer function. The adaptive architecture described here resolves these contradictory requirements, without the necessity of switching circuit elements in the loop filter. The adaptation of loop bandwidth occurs continuously, as a function of the phase error in the loop, and without interaction from outside of the tuning system. During frequency jumps, high bandwidth and high phase margin are obtained by bypassing filter sections. When the loop is locked, the architecture allows heavy filtering of spurious signals. The implementation of the dead-zone block was presented, and the basic tradeoffs of the concept were discussed. The adaptive PLL was optimized for use in a multiband (global) car-radio tuner IC, which features inaudible background scanning. Design and architecture of the PLL building blocks were discussed, and measurement results were presented. The integrated adaptive PLL tuning system achieved state-of-the-art settling and spectral purity performance in its class (integer-N PLL's). It fulfills simultaneously the speed requirements for inaudible frequency hopping and the heavy SNR specification of 64 dB.

Acknowledgment

The author wishes to thank D. Kasperkovitz for technical support during the project, K. Kianush for his tireless disposition in bringing the car-radio project to a successful end, H. Vereijken for the optimization and layout of the synthesizer building blocks, B. Egelmeers for the implementation and evaluation of the concept in a bread-board functional model, and G. van Werven for the measurements.

REFERENCES

[1] B. Razavi, "A 900 MHz/1.8 GHz CMOS transmitter for dual-band applications," *IEEE J. Solid-State Circuits*, vol. 34, pp. 573–579, May 1999.

[2] K. Kianush and C. S. Vaucher, "A global car radio IC with inaudible signal quality checks," in *IEEE Int. Solid-State Circuits Conf. Dig. Tech. Papers*, 1998, pp. 130–131.

[3] W. P. Robins, *Phase Noise in Signal Sources*, 2nd ed, ser. 9. London, U.K.: Inst. Elect. Eng., 1996.

[4] H. Adachi, H. Kosugi, T. Awano, and K. Nakabe, "High-speed frequency-switching synthesizer using fractional N phase-locked loop," *IEICE Trans. Electron.*, pt. 2, vol. 77, no. 4, pp. 20–28, 1994.

[5] U. L. Rohde, *RF and Microwave Digital Frequency Synthesizers*. New York: Wiley, 1997.

[6] F. M. Gardner, "Charge-pump phase-lock loops," *IEEE Trans. Commun.*, vol. 28, no. 11, pp. 1849–1858, Nov. 1980.

[7] H. Meyr and G. Ascheid, *Synchronization in Digital Communications*. New York: Wiley, 1990.

[8] F. M. Gardner, *Phase-Lock Techniques*. New York: Wiley, 1979.

[9] B. Razavi, Ed., *Monolithic Phase-Locked Loops and Clock Recovery Circuits*. New York: IEEE Press, 1996.

[10] V. F. Kroupa, "Noise properties of PLL systems," *IEEE Trans. Commun.*, vol. C-30, pp. 2244–2552, Oct. 1982.

[11] C. S. Vaucher, "Synthesizer architectures," in *Analog Circuit Design*, R. J. van de Plassche, Ed. Norwell, MA: Kluwer, 1997.

[12] C. Vaucher and D. Kasperkovitz, "A wide-band tuning system for fully integrated satellite receivers," *IEEE J. Solid-State Circuits*, vol. 33, no. 7, pp. 987–998, July 1998.

[13] K. Nagaraj, "Adaptive charge pump for phase-locked loops," U.S. Patent 5 208 546, 1993.

[14] B. Miller and B. Conley, "A multi-modulator fractional divider," in *Proc. IEEE 44th Annu. Symp. Frequency Control*, 1990, pp. 559–567.

[15] Philips Semiconductors, TEA6840H global car-radio tuner datasheet, 1999.

[16] W. G. Kasperkovitz, "Digital shift register," U.S. Patent 5 113 419, 1992.

Cicero S. Vaucher (M'98) was born in São Francisco de Assis, Brazil, in 1968. He graduated in electrical engineering from the Universidade Federal do Rio Grande do Sul, Porto Alegre, Brazil, in 1989.

He joined the Integrated Transceivers group of Philips Research Laboratories, Eindhoven, The Netherlands, in 1990, where he works on implementations of low-power building blocks for frequency synthesizers, on synthesizer architectures for low-noise/high-tuning-speed applications, and on CAD modeling of PLL synthesizers.

A 2-V 900-MHz Monolithic CMOS Dual-Loop Frequency Synthesizer for GSM Receivers

William S. T. Yan and Howard C. Luong, *Member, IEEE*

Abstract—A 900-MHz monolithic CMOS dual-loop frequency synthesizer suitable for GSM receivers is presented. Implemented in a 0.5-μm CMOS technology and at a 2-V supply voltage, the dual-loop frequency synthesizer occupies a chip area of 2.64 mm^2 and consumes a low power of 34 mW. The measured phase noise of the synthesizer is −121.8 dBc/Hz at 600-kHz offset, and the measured spurious levels are −79.5 and −82.0 dBc at 1.6 and 11.3 MHz offset, respectively.

Index Terms—Frequency synthesis, frequency synthesizer, phase-locked loop, radio frequency, voltage-controlled oscillator.

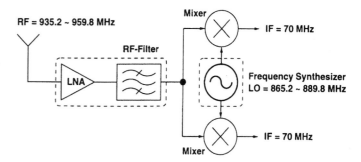

Fig. 1.   Block diagram of the GSM-receiver front-end.

I. INTRODUCTION

MODERN transceivers for wireless communication consist of low-noise amplifiers, power amplifiers, mixers, DSP chips, filters, and frequency synthesizers. These building blocks have been realized using hybrid technologies and require interfacing circuits, which increases the power consumption and limits the maximum operating speed of the transceivers. For this reason, it has become increasingly attractive to design and monolithically integrate all these building blocks on a single chip.

Designing fully integrated frequency synthesizers for this integration is always desirable but most challenging. The first requirement is to achieve high-frequency operation with reasonable power consumption. However, the most critical challenges for the frequency synthesizer are the phase-noise and spurious-level performance. Finally, small chip area is essential to monolithic system integration.

In recent years, monolithic frequency synthesizers with good phase-noise performance have been reported [1]–[3]. However, those designs operate at supply voltages of at least 2.7 V and power consumption of more than 50 mW. Moreover, fractional-N frequency synthesizers suffer from fractional spurs which degrade their spurious-tone performance.

This paper presents a monolithic dual-loop frequency synthesizer for GSM 900 system, which is implemented in a 0.5-μm CMOS process, that achieves high operating frequency (935.2–959.8 MHz), low power consumption (34 mW), low phase noise (−121.8 dBc/Hz at 600kHz), low spurious level (−82.0 dBc at 11.3MHz), and fast switching time (830 μs). Section II derives the design specification of the frequency synthesizer for GSM 900. Section III describes the architecture for the proposed dual-loop design. In Section IV,

Manuscript received December 29, 1999; revised October 5, 2000.
The authors are with the Department of Electrical and Electronic Engineering, Hong Kong University of Science and Technology, Clear Water Bay, Kowloon, Hong Kong (e-mail: eetak@ee.ust.hk; eeluong@ee.ust.hk).
Publisher Item Identifier S 0018-9200(01)00927-1.

circuit implementation of critical building blocks is discussed. Section V presents the measurement results of the synthesizer including its phase noise, spurious level, and switching time of the frequency synthesizer together with a comprehensive performance evaluation.

II. DESIGN SPECIFICATION

The performance of frequency synthesizers is mainly specified by their output frequency, phase noise, spurious level, and switching time. This section derives the specifications of a frequency synthesizer for GSM receivers.

A. Output Frequency

In GSM-900 systems, the receiver-channel frequencies are expressed as follows:

$$f_{\text{RF}} = 935.2 + 0.2(N - 1) \text{ MHz} \qquad (1)$$

where $N = 1, 2, \ldots, 124$ is the channel number. To receive signals in different channels, a GSM-receiver front-end, shown in Fig. 1, is adopted. The receiver front-end consists of a low-noise amplifier (LNA) and an RF filter for filtering out-of-band noise and blocking signals. The received signal is then mixed down to an IF frequency (f_{IF}) of 70 MHz for base-band signal processing. To extract information from the desired channel, the local oscillator (LO) output frequency (f_{LO}) of the frequency synthesizer is changed accordingly, as follows:

$$f_{\text{LO}} = 865.2 + 0.2(N - 1) = 865.2\text{–}889.8 \text{ MHz} \qquad (2)$$

which is the output-frequency range of the frequency synthesizer to be achieved.

B. Phase Noise

The blocking-signal specification for GSM 900 receivers is shown in Fig. 2, where the desired signal power can be as low

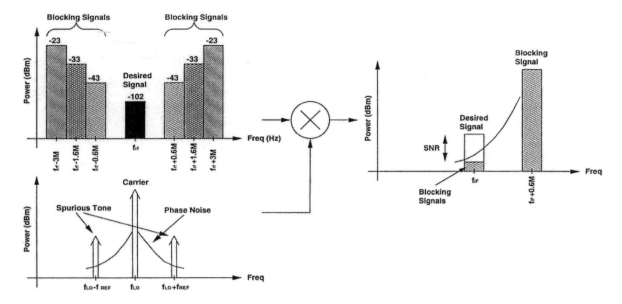

Fig. 2. SNR degradation due to the phase noise and spurious level.

as −102 dBm. At 600-kHz offset frequency, the power of the blocking signal can be as high as −43 dBm [4]. With a correct LO frequency, the desired channel signal is downconverted to IF frequency. However, blocking signals are also downconverted with the LO signal and its phase noise. Since the power of the blocking signal is much larger than that of the desired signal, the phase-noise power falls into the IF frequency and degrades the signal to noise ratio (SNR). The phase-noise specification $\mathcal{L}_{\text{spec}}\{\Delta\omega\}$ can be expressed as follows:

$$\mathcal{L}_{\text{spec}}\{\Delta\omega\} < S_{\text{des}} - S_{\text{blk}} - \text{SNR}_{\text{spec}} - 10 \cdot \log(f_{ch})$$
$$= -121 \text{ dBc/Hz at 600 kHz} \qquad (3)$$

where SNR_{spec} of 9 dB is the SNR specification for the whole receiver, and $S_{\text{des}} = -102$ dBm and $S_{\text{blk}} = -43$ dBm are the power levels of the minimum desired signal and maximum blocking signal, respectively.

C. Spurious Tones

Because of the feedthrough and modulation of the reference signal, two spurious tones appears at the $\pm f_{\text{REF}}$ away from the desired output frequency, as shown in Fig. 2. The derivation of the spurious-tone specification is similar to that of the phase noise except that the channel bandwidth is not considered in this case. The spurious-tone specification S_{spec} can be expressed as follows:

$$S_{\text{spec}} < S_{\text{des}} - S_{\text{blk}} - \text{SNR}_{\text{spec}}$$
$$= \begin{cases} -68 \text{ dBc at 1.6 MHz} \\ -88 \text{ dBc for offset} > 3 \text{ MHz.} \end{cases} \qquad (4)$$

D. Switching Time

In GSM 900 systems, time-division multiple-access (TDMA) is adopted within each frequency channel. As shown in Fig. 3, each frequency channel is divided into eight time slots. Signals are received and transmitted in time slot #1 and slot #4,

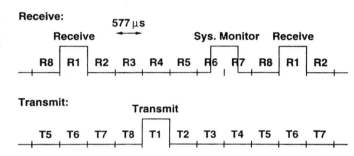

Fig. 3. GSM 900 receive and transmit time.

respectively. For system monitoring purposes, a time slot between slot #6 and slot #7 is adopted. Therefore, the most critical switching time is from the transmission period (slot #4) to the system monitoring period (slot #6.5), which is equal to 865 µs. However, to take care of the settling time of the other components, the switching time of the frequency synthesizer is recommended to be kept within one time slot (577 µs) [5].

III. DUAL-LOOP DESIGN

To reduce the switching time and the chip area of a synthesizer, a high loop bandwidth and a high reference frequency are desired. Moreover, to suppress the phase-noise contribution of the reference signal and improve frequency-divider complexity, a lower frequency-division ratio is desirable. Therefore, a dual-loop frequency synthesizer is proposed [6]. As shown in Fig. 4, the dual-loop design consists of two reference signals and two phase-locked loops (PLLs) in cascade configuration. In the feedback path of the high-frequency loop, a mixer is adopted to provide the frequency shift. The output frequency of the synthesizer is expressed as follows:

$$f_{\text{LO}} = N_3 f_{\text{REF2}} + N_1 \left(\frac{N_3}{N_2}\right) f_{\text{REF1}} \qquad (5)$$

where f_{REF1} and f_{REF2} are frequencies of the two reference signals, and N_1, N_2, and N_3 are frequency division ratios.

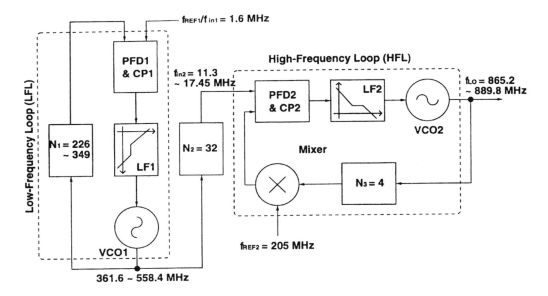

Fig. 4. Proposed dual-loop frequency synthesizer.

Due to the dual-loop architecture, the comparison frequencies of the low-frequency and high-frequency loops are scaled up from 200 kHz to 1.6 and 11.3 MHz, respectively. Therefore, the loop bandwidths of both PLLs can be increased so that the switching time and the chip area can be reduced. Compared to single-loop integer-N designs, the frequency-division ratio of the programmable divider N_1 is reduced from 4236–4449 to 226–349. Such a reduction in the division ratio significantly simplifies the frequency-divider design and reduces phase-noise contribution of the input reference signal.

In the proposed dual-loop synthesizer, the divide-by-32 divider N_2 and the high-frequency loop together greatly attenuates the phase noise and the spurious tones of the low-frequency loop. As such, the low-frequency loop can be designed to have a larger loop bandwidth and a loop filter as small as one-fifth of the loop filter in the high-frequency loop. The low-frequency loop requires additional components, including the phase-frequency detector (PFD1), the charge pump (CP1), and the frequency divider N_1, but they are all quite small and have very little impact on the chip area. In additional, VCO1 is implemented by a ring oscillator, which occupies a much smaller chip area compared to VCO2. Altogether, the dual-loop design requires no more than 25% overhead in the chip area compared to a fraction-N design with the same loop bandwidth.

Although the input-reference frequency f_{REF1} of the low-frequency loop is scaled up by 8 times; the required frequency range of the oscillator VCO1 in the low-frequency loop is also scaled up from 25 to 200 MHz. On the other hand, the phase-noise of the ring oscillator is attenuated by the frequency divider N_2 and is then amplified by the high-frequency loop; the total phase-noise attenuation from VCO1 output to the synthesizer output is 18 dB. Consequently, this voltage-controlled oscillator (VCO) requires a high operating frequency (600 MHz), a wide frequency range (200 MHz), and a low phase noise (-103 dBc/Hz at 600 kHz). A novel ring VCO design that meets all of these tough specifications will be presented in the next section.

IV. CIRCUIT IMPLEMENTATION

This section discusses the design consideration and circuit implementation of the major building blocks that are unique and critical to the proposed dual-loop synthesizer, namely the two VCOs, the frequency dividers, the charge pump, and the loop filters. Detailed analysis and design of other building blocks will not be presented, either because they can be found somewhere else or they are too obvious.

A. Ring Oscillator VCO1

The schematic of the proposed two-stage ring oscillator and its delay cell to meet the required specification as described in Section III are shown in Fig. 5(a) and (b), respectively. The delay cell consists of nMOS transistors M_{n1} as input transconductors, cross-coupled pMOS transistors M_{p1} for maintaining oscillation, diode-connected pMOS transistors M_{p2}, and a bias transistor M_{b1} for frequency tuning. The source nodes of transistors M_{p1} are connected to supply to maximize its output amplitude V_p, which also helps suppress noise sources by turning them off more often [7] and thus further enhances the phase-noise performance.

The half circuit of the delay cell is shown in Fig. 5(c). By equating the delay-cell voltage gain to be unity, the oscillating frequency of the ring oscillator can be expressed as follows:

$$f_{\mathrm{osc}} = \frac{1}{2\pi} \sqrt{\frac{g_{mn1}^2 - (-g_{mp1} + g_{mp2} + G_L)^2}{C_L^2}}$$
$$G_L = g_{dn1} + g_{dp1} + g_{dp2} \tag{6}$$

where g_m is transconductance, g_d is channel conductance, and C_L is the total capacitance at output node. Oscillation starts when the g_{mp1} is large enough to overcome the output load G_L ($g_{mp1} > G_L$).

When control voltage $V_{\mathrm{con}} = 0$ V, transistors M_{p2} are turned on to cancel g_{mn1}, and the oscillator operates at maximum frequency. When control voltage $V_{\mathrm{con}} = V_{dd}$, transistor M_{p2} are

532

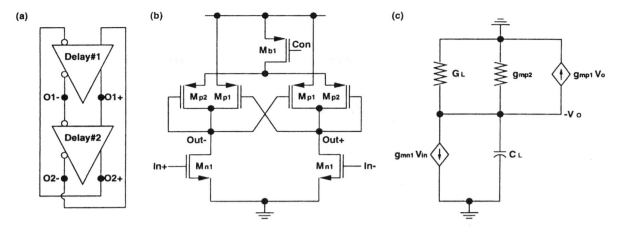

Fig. 5. Circuit implementation of the ring oscillator VCO1. (a) Ring oscillator. (b) Delay cell. (c) Half circuit of delay cell.

turned off ($g_{mp2} = 0$), and the oscillator operates at minimum frequency. By (6), f_{max}, f_{min}, and frequency range f_{range} are expressed as follows:

$$f_{max} \approx \frac{1}{2\pi} \cdot \frac{g_{mn1}}{C_L} \quad f_{min} \approx \frac{1}{2\pi} \cdot \sqrt{\frac{g_{mn1}^2 - g_{mp1}^2}{C_L^2}}$$

$$f_{range} \approx f_{max} \cdot \left(1 - \sqrt{1 - \left(\frac{g_{mp1}}{g_{mn1}}\right)^2}\right). \quad (7)$$

Since f_{max} is proportional to g_{mn1}/C_L, nMOS transistors are adopted as the input devices to minimize power consumption. From (7), 50% tuning range can be achieved when $g_{mp1}/g_{mn1} = \sqrt{3/4}$.

Based on the approximate impulse-stimulus function (ISF) and the analysis presented in [8], the phase noise of the oscillator is estimated to be approximately -107 dBc/Hz at 600-kHz offset. On the other hand, using SpectreRF [9], the phase noise is simulated to be -111.7 dBc/Hz at 600 kHz.

B. LC Oscillator VCO2

As the far-offset phase noise is dominated by the VCO2, an LC oscillator is adopted to meet the stringent phase-noise specification. Fig. 6 shows the schematic of the LC oscillator. Cross-coupled transistors M_{n1} are used to start and to maintain oscillation with lower parasitics. PN-junction varactors implemented by p+ diffusion on the n-well are used for frequency-tuning purpose. The common-mode output voltage is designed at 1.1 V to enhance the driving of the frequency divider N_3. To reduce phase-noise contribution due to flicker noise, pMOS transistors M_{b1} are used as the current source.

To design an LC oscillator which satisfies the phase-noise requirement with minimum power consumption, inductors with large inductance and small series resistance are desired. Therefore, two-layer inductors are adopted [10] for which the inductance and the quality factor can be scaled up by 4 and 2 times, respectively. For the same reason, pn-junction varactors are interdigitized with p+ islands surrounded by n-well contacts to enhance the quality factor. Finally, the transconductance g_{mn1} of transistor M_{n1} is designed so that it does not overcompen-

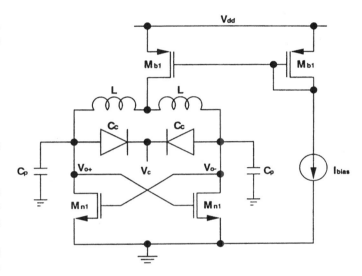

Fig. 6. Circuit implementation of the LC oscillator VCO2.

sate the LC tank too much (only twice) to reduce phase-noise contribution by transistors M_{n1}.

Based on the method described in [11], the phase noise of the LC VCO is estimated to be -124.0 dBc/Hz at 600-kHz frequency offset, which agrees well with the simulation using SpectreRF.

C. Frequency Dividers N_3

As the divide-by-4 frequency divider N_3 needs to convert sinusoidal signals from the VCO2 output into square-wave signals, the first stage of the divider is implemented by pseudo-nMOS logic while the second divide-by-2 divider is implemented by the TSPC-logic divide-by-2 divider [12]. The first divider is shown in Fig. 7 and consists of a pseudo-nMOS amplifier and a divide-by-2 divider. Since the pseudo-nMOS logic is a ratioed logic, the ratio between pMOS and nMOS transistors is designed to be less than 1.6 to make sure output logic "0" turns off the next stage.

D. Programmable-Frequency Divider N_1

Fig. 8 shows the block diagram of the programmable-frequency divider N_1 [13]. At reset state, the prescaler divides

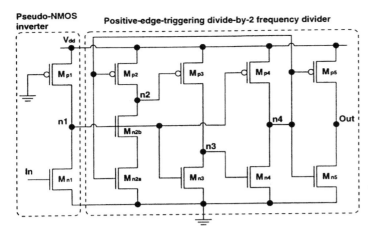

Fig. 7. Circuit implementation of the pseudo-nMOS divide-by-2 frequency divider.

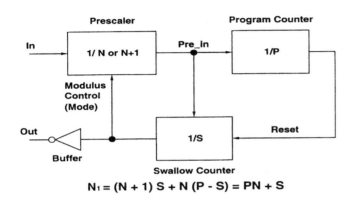

$$N_1 = (N + 1) S + N (P - S) = PN + S$$

Fig. 8. Block diagram of the programmable-frequency divider N_1.

input signal by $N + 1$, and its output is counted by both P and S counters. After the S counter has counted S pulses, the S counter changes the state of the modulus control line Mode and the prescaler divides input by N. Then the P counter counts the remaining $P–S$ cycles to reach overflow. As a whole, the programmable divider generates one complete cycle for every $(N + 1)S + (P - S)N = PN + S$ input cycles. The operation repeats after the S counter is reset.

As the frequency-division ratio (226~349) can be achieved with different combinations of N, S, and P, the most optimal combination in terms of performance needs to be identified and chosen as the design. As the S counter must finish before the P counter resets it, the division ratio of the P counter should be larger than that of the S counter. To optimize the power consumption, the operating frequencies and number of bits of both the P and S counters should be minimized.

Table I shows the different combinations of N, P, and S which can implement the desired division ratio. Case 1 requires the highest operating frequencies and number of bits for P and S counters, so it is not adopted. Case 4 has the problem that the S value is larger than the P value. It seems that Case 3 is the best one, but Case 2 is chosen as the final design because it is much easier to implement an asynchronous divide-by-12 frequency divider than a divide-by-14 divider.

The dual-modulus prescaler is implemented by the back-carrier-propagation approach as shown in Fig. 9 [14]. When

TABLE I
System Design of Programmable-Frequency Divider N_1

Case	N	Operating Frequency	P	S
1	10	55.8 MHz	22 ~ 34 (6)	0 ~ 9 (4)
2	12	46.5 MHz	18 ~ 29 (5)	0 ~ 11 (4)
3	14	39.9 MHz	16 ~ 24 (5)	0 ~ 13 (4)
4	16	34.9 MHz	14 ~ 21 (5)	0 ~ 15 (5)

control signal MODE = "1," the gated inverter is by-passed, and the prescaler is a divide-by-12 divider. When control signal $MODE$ = "0," the final state $D0.$ 1, 2 = "010" will be detected, at which BLK = "0" and the input signal is delayed by one clock cycle. Thus the function of divide-by-13 is achieved. The back-carrier-propagation approach allows low-frequency signals (more significant bits) to switch to the final state much earlier than high-frequency signals (less significant bits) and thus reduces power consumption for a given speed.

E. Charge Pumps and Loop Filters

Fig. 10 shows the circuit implementation of the charge pumps used in the two loops. Each charge pump consists of two cascode-current sources for both the pull-up and pull-down currents, four complementary switches, and a unity-gain amplifier. By using high-swing cascode current sources, the output impedance is increased for effective current injection. Minimum-size complementary switches are adopted to minimize clock feedthrough and charge injection of the switches. The unity-gain amplifier keeps the voltages of nodes VCO and nb to be equal so that charge sharing between nodes VCO, ns, and ps can be minimized [15].

The design of the loop filters in the two PLLs is a second-order low-pass filter which is implemented using linear capacitors and silicide-blocked polysilicon resistors. The values of capacitance, resistance, and charge-pump current are optimally designed to satisfy simultaneously the phase-noise, spurious-tone, and switching-time requirements with minimum chip area [16]. The loop bandwidth of the low-frequency and high-frequency loops are 40 and 27 kHz, respectively.

F. Phase Noise of the Dual-Loop Frequency Synthesizer

Based on the linearized model shown in Fig. 11, the transfer function from the input phase θ_{in} to output phase θ_o of both the low-frequency and high-frequency loops can be expressed as follows:

$$\frac{\theta_{o1}}{\theta_{in1}}(s) = \frac{N_1(1 + s\tau_{z1})}{1 + s\tau_{z1} + (N_1 C_{i1}/K_{PD1}K_{VCO1})s^2(1 + s\tau_{p1})}$$

$$\frac{\theta_{o2}}{\theta_{in2}}(s) = \frac{\theta_{o2}}{\theta_{ref2}}(s)$$

$$= \frac{N_3(1 + s\tau_{z2})}{1 + s\tau_{z2} + (N_3 C_{i2}/K_{PD2}K_{VCO2})s^2(1 + s\tau_{p2})}$$

(8)

where K_{PD} is the phase-detector gain, K_{VCO} is the VCO gain, and C_i, τ_z and τ_p are the total capacitance, zero time constant,

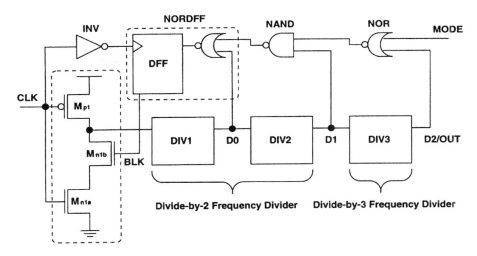

Fig. 9. Circuit implementation of the dual-modulus prescaler.

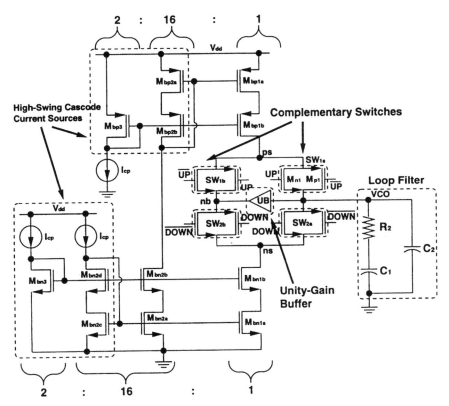

Fig. 10. Circuit implementation of the charge pump and the loop filter.

and pole time constant of the corresponding loop filters, respectively. K_{PD} is the phase-detector gain, K_{VCO} is the VCO gain, and C_i, τ_z, and τ_p are the total capacitance, zero time constant, and pole time constant of the corresponding loop filters, respectively. Since the transfer function is a low-pass function, the reference phase noise is highly attenuated at high offset frequency. It also shows that the close-in phase noise of the reference signals is amplified by the frequency-division ratio. In this work, the division ratio is reduced from 4449 to 349, and the phase-noise contribution from the reference signals is suppressed by $20 \times \log(4449/349) = 22.1$ dB.

Another important source of the close-in phase noise is the charge-pump noise. The transfer functions between the charge-pump noise current to the output phase noise can be derived to be

$$\frac{\theta_{o1}}{i_{CP1}}(s) = \left(\frac{\theta_{o1}}{\theta_{in1}}(s)\right)/K_{PD1}$$

$$= \frac{(N_1/K_{PD1}) \cdot (1 + s\tau_{z1})}{1 + s\tau_{z1} + (N_1 C_{i1}/K_{PD1}K_{VCO1})s^2(1 + s\tau_{p1})}$$

$$\frac{\theta_{o2}}{i_{CP2}}(s) = \left(\frac{\theta_{o2}}{\theta_{in2}}(s)\right)/K_{PD2}$$

$$= \frac{(N_3/K_{PD2}) \cdot (1 + s\tau_{z2})}{1 + s\tau_{z2} + (N_3 C_{i2}/K_{PD2}K_{VCO2})s^2(1 + s\tau_{p2})}.$$

(9)

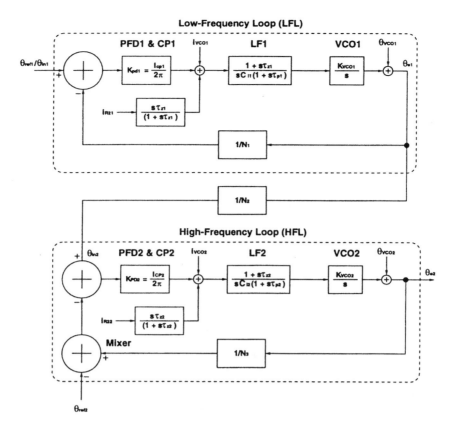

Fig. 11.   Linearized model of the dual-loop frequency synthesizer.

It shows that small frequency-division ratio and large phase-detector gain or large charge-pump current are preferred for phase-noise consideration. Another factor not included in (9) which also affects the charge-pump noise is its turn-on time. In this proposed synthesizer, the charge-pump turn-on time is designed to be equal to 1/10 of the input period so that it is long enough to eliminate the phase-frequency detector (PFD) dead-zone problem, but at the same time is short enough to minimize the charge-pump phase-noise contribution.

The phase-noise contribution of the loop-filter resistors in both PLLs can be estimated using their equivalent noise currents as follows:

$$\frac{\theta_{o1}}{i_{R21}}(s) = \frac{s\tau_{z1}}{1 + s\tau_{z1}} \cdot \frac{\theta_1}{i_{CP1}}(s)$$
$$\frac{\theta_{o2}}{i_{R22}}(s) = \frac{s\tau_{z2}}{1 + s\tau_{z2}} \cdot \frac{\theta_{o2}}{i_{CP2}}(s) \tag{10}$$

which are bandpass functions with peaks appearing between the zero and the pole of the loop filter. To suppress the phase-noise peaking, large loop-filter capacitors are desired at the cost of large chip area.

For the phase-noise contribution of the VCOs, the transfer function between the VCO phase noise and output phase noise can be found to be

$$\frac{\theta_{o1}}{\theta_{VCO1}}(s) = \frac{(N_1\tau_{i1}/K_{PD1}K_{VCO1})s^2(1 + s\tau_{p1})}{1 + s\tau_{z1} + (N_1 C_{i1}/K_{PD1}K_{VCO1})s^2(1 + s\tau_{p1})}$$
$$\frac{\theta_{o2}}{\theta_{VCO2}}(s) = \frac{(N_3\tau_{i2}/K_{PD2}K_{VCO2})s^2(1 + s\tau_{p2})}{1 + s\tau_{z2} + (N_3 C_{i2}/K_{PD2}K_{VCO2})s^2(1 + s\tau_{p2})} \tag{11}$$

which are high-pass functions. Therefore, the far-offset phase noise of the synthesizer is dominated by the VCO phase noise. Since the loop bandwidths of both PLLs are designed in a range of tens of kilohertz to achieve spurious-level specification, the far-offset phase noise of the PLLs only depends on the VCO phase noise itself.

To evaluate the overall phase-noise performance, the relationship between the phase noise of the low-frequency loop and that of the synthesizer output can be written as

$$\frac{\theta_{o2}}{\theta_{o1}}(s) = \frac{1}{N_2} \frac{\theta_{o2}}{\theta_{in2}}(s)$$
$$= \frac{(N_3/N_2)(1 + s\tau_{z2})}{1 + s\tau_{z2} + (N_3 C_{i2}/K_{PD2}K_{VCO2})s^2(1 + s\tau_{p2})} \tag{12}$$

which shows that there exists $20 \times \log(32/4) = 18.1$ dB close-in phase-noise suppression for the low-frequency loop.

The estimated phase noise of the whole synthesizer is -81.4 dBc/Hz at 20.9 kHz and -123.8 dBc/Hz at 600 kHz. The contribution of each component is shown in Fig. 12, which shows that the close-in phase noise (<100 kHz) is dominated by the charge pump CP1 and loop filter LF1, while the far-offset phase noise (>100 kHz) is dominated by the LC oscillator.

V. EXPERIMENTAL RESULTS

The dual-loop frequency synthesizer is implemented in a standard 0.5-μm CMOS technology. Linear capacitors are put under all the bias pins to serve as on-chip bypass capacitors.

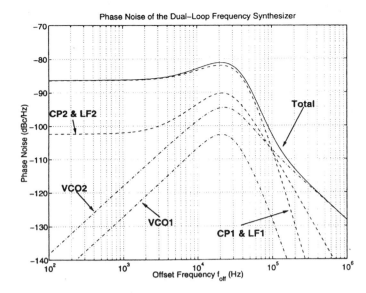

Fig. 12. Estimated phase noise of the whole dual-loop frequency synthesizer and contribution of each components at the synthesizer output.

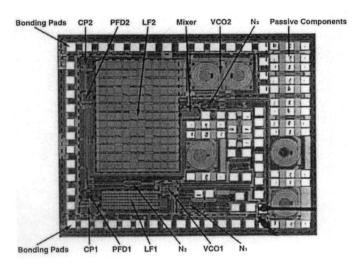

Fig. 13. Die photo of the dual-loop frequency synthesizer.

Fig. 13 shows the die photo of the dual-loop frequency synthesizer, and the active area of the synthesizer is 2.64 mm^2.

For characterization and measurement of passive devices, testing structures for spiral inductors and varactors are included on the same die with the synthesizer and are measured by a network analyzer. To de-embed the probing-pad parasitics, an open-pad structure is also measured.

A. Measurement of Inductors

Fig. 14 shows the inductance L, series resistance R_L, and quality factor Q_L of the on-chip spiral inductor. The measured inductance L is close to simulation results and drops at frequencies close to the self-resonant frequency. However, the series resistance R_L (30.2 Ω) is almost three times larger than the expected value (11.6 Ω). The increase in series resistance is mainly caused by eddy current induced within substrate and n-well fingers [16]. As series resistance increases significantly, the port-1 quality factor is limited to be 1.6 at 900 MHz.

B. Measurement of Varactors

The measurement results of the pn-junction varactor at 900 MHz are shown in Fig. 15. As the p+ diffusion of the varactors used in the LC oscillator are connected to the output of the LC oscillator core, they are biased at 1.16 V, which is the dc bias of the oscillator core during the measurement. The measured capacitance C is close to the estimated results in the reverse-biased region. The series resistance R_c is around 2 Ω due to the minimum junction spacing and the nonminimum junction width. The quality factor is around 30 in the operating region of the oscillator.

C. Measurement of Ring Oscillator VCO1

The phase noise of the oscillators are measured by a direct-phase-noise measurement [17]. First, the carrier power is determined at large video (VBW) and resolution bandwidths (RBW). Then, the resolution bandwidth is reduced until the noise edges and not the envelope of the resolution filter are displayed. Finally, the phase noise is measured at the corresponding frequency offset from the carrier. To make sure that the measured phase noise is valid, the displayed values must be at least 10 dB above the intrinsic noise of the analyzer.

Fig. 16 shows the measurement results of the ring oscillator VCO1. The operating frequency is measured to be between 324.0 and 642.2 MHz, over which the measured phase noise is between −111 and −108 dBc/Hz at 600 kHz. The power consumption is around 10 mW.

D. Measurement of LC Oscillator VCO2

Fig. 17 shows the measurement results of the LC oscillator VCO2. Due to the quality-factor degradation of the spiral inductor, the bias current of the oscillator is increased by 15% above its designed value to achieve the phase noise specification (−121 dBc/Hz at 600 kHz). The measured operating frequency range is between 725.0 and 940.5 MHz. The oscillation stops when the VCO control voltage is below 0.6 V because the varactors become forward-biased. Over the desired frequency range between 865.2 and 889.8 MHz, the achieved phase noise is below −121 dBc/Hz at 600 kHz.

E. Measured Phase Noise of the Frequency Synthesizer

Fig. 18 shows the phase-noise measurement results of the dual-loop frequency synthesizer at 889.8 MHz. The measured phase noise is −121.8 dBc/Hz at 600 kHz which satisfies the GSM requirement. At offset frequencies between 10 and 100 Hz, the phase noise is mainly contributed by the flicker noise of the charge pump. However, the peak phase noise of −65.67 dBc/Hz at 15 kHz is measured, which is 15 dB higher than the estimation presented in Fig. 12.

At offset frequencies above 100 kHz, where the phase noise should be dominated by VCO2 and should go down by 20 dB/dec, the measured phase noise goes down at a rate of 40 dB/dec. It is believed that the increase in the close-in phase noise is mainly due to the charge pump CP2 and the loop filter LF2, for the following reasons. First, according to Fig. 12, only CP2 and LF2 have the phase-noise slope of 40 dB/dec above 100-kHz frequency offset. Second,

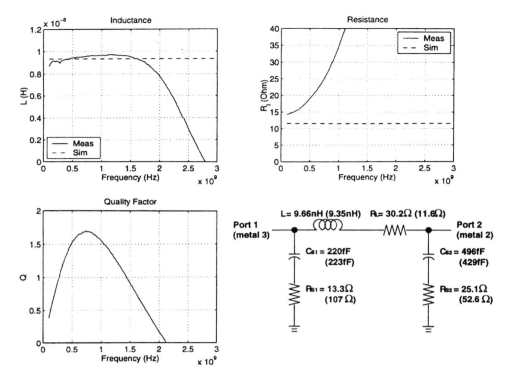

Fig. 14. Measurement results and equivalent circuit model of the spiral inductors at 900 MHz.

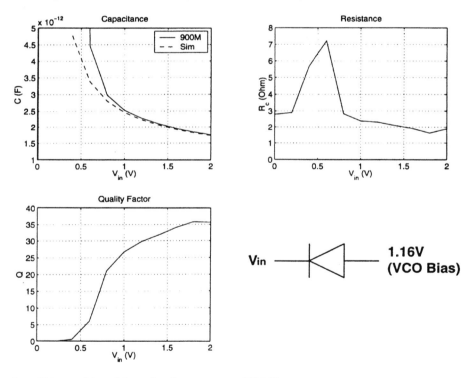

Fig. 15. Measurement results and bias condition of the pn-junction varactors at 900 MHz.

the measured peak-to-flat close-in phase noise in Fig. 18 is around 15 dB, which is quite close to that of the estimated value in Fig. 12. Lastly, it is observed experimentally that the close-in phase noise is changed as the charge-pump current of the high-frequency loop is adjusted. Unfortunately, the phase-noise contribution of CP2 and LF2 cannot be measured individually.

F. Measured Spurious Tones of the Frequency Synthesizer

Fig. 19 shows the measured spurious level of the dual-loop frequency synthesizer at 865.2 MHz, which are −79.5 dBc at 1.6 MHz, −82.0 dBc at 11.3 MHz, and −82.83 dBc at 16 MHz. At 11.3 MHz, the spurious level is only 6 dB above the requirement. However, the predicted spurious level at 1.6 MHz should be below −90 dBc and the one at 16 MHz should not exist [16].

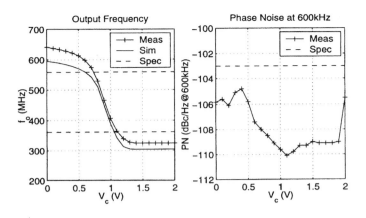

Fig. 16. Measurement results of the ring oscillator VCO1.

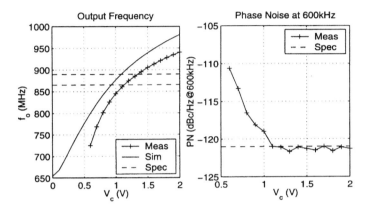

Fig. 17. Measurement results of the *LC* oscillator VCO2.

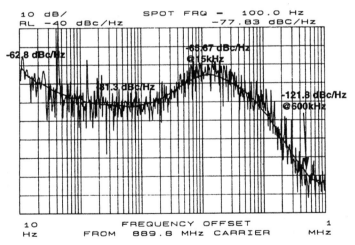

Fig. 18. Phase-noise measurement results of the proposed synthesizer.

During the measurement, the 1.6-MHz reference signal is generated by a 16-MHz crystal oscillator and a decade counter. Therefore, the 16-MHz spur is caused by the substrate coupling between the crystal oscillator and the synthesizer. To verify the reason of the increased spurious level at 1.6 MHz, the low-frequency loop is disabled, and the spurious level is still −75.1 dBc at 1.6 MHz, which implies that the increase in spurious level at 1.6 MHz is mainly caused by the substrate coupling.

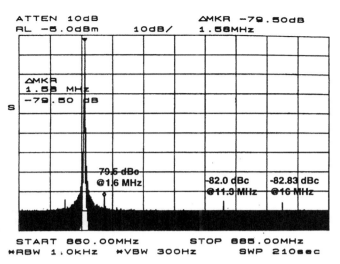

Fig. 19. Measured spurious level of the proposed synthesizer.

G. Switching Time of the Frequency Synthesizer

To determine the worst-case switching time of the frequency synthesizer, the frequency division ratio N_1 is switched from 226 to 349, and the control voltages of both VCO1 and VCO2 are measured. As shown in Fig. 20, the measurable switching time of the proposed synthesizer is 830 μs for a frequency error of approximately 10 kHz due to the limited resolution of our oscilloscope. Since the VCO gain is 160 MHz/V, in order to achieve a measurement accuracy of a 100-Hz frequency error, the resolution of the oscilloscope would need to be better than 100 nV, which unfortunately is not obtainable with our equipment. On the other hand, the synthesizer suffers from a slew-rate problem during the channel switching due to a small charge-pump current (1.6 μA) and a large loop-filter capacitor (1.1 nF) in the high-frequency loop.

H. Performance Evaluation

Table II summarizes the measured performance of the proposed frequency synthesizer, and Table III lists the performance of other fully integrated synthesizers for comparison. The proposed synthesizer operates at a single 2-V supply while all other designs require supply voltages of at least 2.7 V. Note that since the designs [1] and [3] operate at higher frequencies, their power consumption should be scaled down accordingly for a fair comparison. With this frequency normalization, the power consumption of the proposed dual-loop synthesizer is still comparable to that of the other designs.

The synthesizer presented in [1] is a fractional-N design with a 26.6-MHz comparison frequency and a loop bandwidth of 45 kHz. As comparison, the low-frequency loop of this work has a comparison frequency of only 1.6 MHz but a loop bandwidth of up to 40 kHz because of the relaxed requirement of the spurious tones. On the other hand, the high-frequency loop uses a 11.3-MHz comparison frequency but a loop bandwidth limited to 27 kHz, which is the real limiting factor of the switching-time performance. Therefore, it is believed that a switching time close to that in [1] can be achieved by

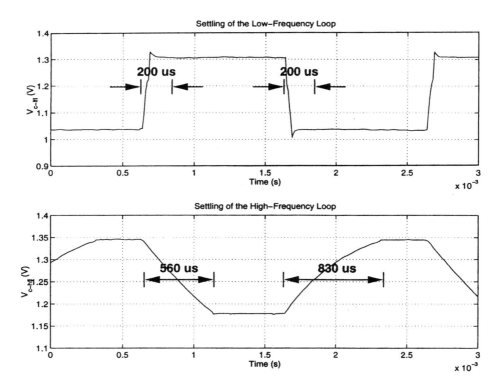

Fig. 20. Switching-time measurement results of the proposed synthesizer.

TABLE II
PERFORMANCE SUMMARY OF THE PROPOSED SYNTHESIZER

Process	0.5-μm CMOS
Chip Area	2.64 mm^2
Supply Voltage	2.0 V
Frequency Range	865.2 to 889.8 MHz
Close-In Phase Noise	-65.7 dBc/Hz@15kHz
Phase Noise	-121.83 dBc/Hz@600kHz
Spurious Level	-79.5 dBc@1.6MHz -82.0 dBc@11.3MHz -82.88 dBc@16MHz
Switching Time	< 830 μs
Power Consumption	34.0 mW

eliminating the slew-limiting problem in the high-frequency loop.

The work described in [2] is an integer-N bipolar junction transistor (BJT) design with channel spacing of 600 kHz, and its comparison frequency and loop bandwidth are limited to be 600 and 4 kHz, respectively. With such a low bandwidth, the settling time is still less than 600 μs. It implies that if a larger charge-pump current or an active loop filter is adopted in the high-frequency loop of the dual-loop design, slew limiting can be suppressed and the switching time performance can be enhanced. Since the density of the BJT transistors is not as good as the CMOS transistor, the chip area is relatively large even though it does not include an on-chip loop filter.

The synthesizer in [3] is also an integer-N design but without channel programmability, and as such, its comparison frequency and loop bandwidth can be as high as 61.5 MHz and 200 kHz,

respectively. For the same reason, the total loop-filter capacitance can be smaller than 60 pF, which greatly reduces chip area. However, the situation would be much different if channel programmability is included.

Although the proposed synthesizer consists of two loop filters, but the chip area is just a little bit larger than that of the design in [3] due to the use of linear capacitors and silicide-blocked resistors. Compared to the designs in [1] and [3], the spurious levels are between −75 and −85 dBc, which indicates that CMOS designs suffer the same problem from the substrate coupling between the reference signal and VCO. However, for the close-in phase noise, the proposed dual-loop synthesizer suffers from the 15-dB increase at the peak due to the charge pumps and loop filters as discussed in Section V-E.

I. Generation of the Second Reference Sources

The main drawback of our dual-loop synthesizer is that it requires two reference sources. In reality, if a single reference signal is preferred for the whole frequency synthesizer, the second reference signal (204.8 MHz instead of 205 MHz) can be generated from the 1.6-MHz reference signal by a third PLL with frequency division ratio of 128. Since this third PLL has a fixed division frequency, its frequency divider can be implemented simply by cascading seven divide-by-2 dividers.

Without the divide-by-32 divider between the third PLL output and the high-frequency loop, the close-in phase-noise requirement of this PLL would be in fact 30 dB more stringent (−92 dBc/Hz) than that of the low-frequency loop. However, since the division ratio is only 128, it would offer a relaxation in requirement. The remaining 21.3 dB phase-noise suppression could be achieved by increasing the charge-pump current of the third PLL.

TABLE III
PERFORMANCE COMPARISON OF RECENT WORK ON FULLY INTEGRATED FREQUENCY SYNTHESIZERS

Design	JSSC 98 [1]	ISSCC 96 [2]	JSSC 98 [3]	This Work
Architecture	Fractional-N	Integer-N	Integer-N	Dual-Loop
Process	0.4-μm CMOS	25-GHz BJT	0.6-μm CMOS	0.5-μm CMOS
Carrier Frequency	1.8 GHz	900 MHz	1.6 GHz	900 MHz
Channel Spacing	200 kHz	600 kHz	N. A.	200 kHz
Reference Frequency	26.6 MHz	9.6 MHz	61.5 MHz	1.6 MHz & 205 MHz
Loop Bandwidth	45 kHz	4 kHz	200 kHz	40 kHz & 27 kHz
Chip Area	3.23 mm^2	5.5 mm^2	1.6 mm^2	2.64 mm^2
Close-In Phase Noise	< -80 dBc/Hz	< -65 dBc/Hz	< -95 dBc/Hz	-65.7 dBc/Hz
600-kHz Phase Noise	-121 dBc/Hz	-116.6 dBc/Hz	-115 dBc/Hz	-121.8 dBc/Hz
Spurious Level	-75 dBc	< -110 dBc	-83 dBc	-79.5 dBc
Switching Time	< 250 μs	< 600 μs	N. A.	< 830 μs
Supply Voltage	3 V	2.7 to 5 V	3 V	2 V
Power	51 mW	50 mW	90 mW	34 mW

In addition to the close-in phase noise, the far-offset phase-noise requirement would also be more stringent by the same amount. Assuming that VCO1 is adopted in the third PLL, its phase noise could be improved to be -116 dBc/Hz at 600 kHz at a 204-MHz operation. With a 30.6-dB filtering effect of the high-frequency loop, the phase-noise contribution by the VCO of the third PLL would become -146.6 dBc/Hz at 600 kHz, which would have negligible effect on the overall phase noise.

Basically, the implementation of the third PLL would be similar to that of the low-frequency loop, and its chip area would be less than 10% of the total area of the dual-loop synthesizer. Since the third VCO operates at half the frequency of VCO1, its power consumption would be 25% of that of VCO1 ($\sim$2.5 mW). Similarly, since the divide-by-128 divider also operates at half of the frequency, it would consume only half the power as compared to the divider N_2 ($\sim$0.3 mW). Although the charge-pump current should be increased by 100 times to 640 μA, the average power current would be only 64 μA since the turn-on time is only around 1/10 of the input period. In conclusion, by introducing the third PLL to generate the second reference signal, the additional power required would be less than 3 mW, and the increase in the total chip area would still be less than 10%.

VI. CONCLUSION

A 900-MHz monolithic CMOS dual-loop frequency synthesizer with good phase-noise performance for GSM receivers is presented. Compared to other fully integrated synthesizer designs, this proposed synthesizer operates at much lower supply voltage and consumes approximately the same power with frequency normalization. Implemented in a standard 0.5-μm CMOS technology and at 2-V supply voltage, the synthesizer has a power consumption of 34 mW. At 900 MHz, the measured phase noise is -121.8 dBc/Hz at 600-kHz frequency offset,

and the spurious level is -82 dBc at 11.3 MHz. Due to the substrate coupling and testing setup, additional spurious levels are measured to be -79.5 dBc at 1.6 MHz and -82.8 dBc at 16 MHz. The chip area is less than 2.64 mm^2. Even if a third PLL is implemented to generate the second reference frequency, the increase in the total power consumption and the total chip area would be negligibly small.

REFERENCES

[1] J. Craninckx and M. Steyaert. "A fully integrated CMOS DCS-1800 frequency synthesizer," *IEEE J. Solid-State Circuits*, vol. 33. pp. 2054–2065, Dec. 1998.

[2] A. Ali and J. L. Tham, "A 900-MHz frequency synthesizer with integrated *LC* voltage-controlled oscillator," *Proc. IEEE Int. Solid-Stage Circuits Conf.*, vol. 1, pp. 390–391, 1996.

[3] J. F. Parker and D. Ray, "A 1.6-GHz CMOS PLL with on-chip loop filter," *IEEE J. Solid-State Circuits*, vol. 33, pp. 337–343, Mar. 1998.

[4] "Digital cellular telecommunications system (Phase 2+); Radio transmission and reception (GSM 5.05)," European Telecommunications Standards Institute, 1996.

[5] D. Craninckx and D. Steyaert, *Wireless CMOS Frequency Synthesizer Design.* Norwell, MA: Kluwer, 1998, pp. 201–202.

[6] T. Aytur and J. Khoury, "Advantages of dual-loop frequency synthesizers for GSM applications," in *Proc. IEEE Int. Symp. Circuits and Systems*, 1997.

[7] C. H. Park and B. Kim, "A low-noise 900-MHz VCO in 0.6-μm CMOS." in *Proc. Symp. VLSI Circuits*, 1998.

[8] A. Hajimiri, S. Limotyrakis, and T. H. Lee, "Phase noise in multi-gigahertz CMOS ring oscillators," in *Proc. IEEE 1998 Custom Integrated Circuit Conf.*, 1998, pp. 49–52.

[9] "Oscillator noise analysis in SpectreRF, application note to SpectreRF." CADENCE, 1998.

[10] R. B. Merrill, T. W. Lee, H. You, R. Rasmussen, and L. A. Moberly, "Optimization of high-Q integrated inductors for multilevel metal CMOS." in *Proc. Int. Electronic Device Meeting*, 1995, pp. 983–986.

[11] A. Hajimiri and T. H. Lee, "A general theory of phase noise in electrical oscillators," *IEEE J. Solid-State Circuits*, pp. 179–194, Feb. 1998.

[12] J. Yuan and C. Svenson, "High-speed CMOS circuit technique," *IEEE J. Solid-State Circuits*, vol. 24, pp. 62–70, Feb. 1989.

[13] B. Razavi, *RF Microelectronics.* Englewood Cliffs, NJ: Prentice Hall, 1997.

[14] P. Larsson, "High-speed architecture for a programmable frequency divider and a dual-modulus prescaler," *IEEE J. Solid-State Circuits*, vol. 31, pp. 744–748, May 1996.

[15] I. A. Young, J. K. Greason, and K. L. Wong, "PLL clock generator with 5 to 110 MHz of lock range for microprocessors," *IEEE J. Solid-State Circuits*, vol. 27, pp. 1599–1607, Nov. 1992.

[16] W. S. T. Yan, "A 2-V 900-MHz monolithic CMOS dual-loop frequency synthesizer for GSM receivers," M.Phil. thesis, Hong Kong University of Science and Technology. [Online.] Available: http://www.ee.ust.hk/~eetak, 1999.

[17] T. Fredrich, "Direct phase noise measurements using a modern spectrum analyzer," *Microwave J.*, vol. 35, pp. 94–114, Aug. 1992.

Howard C. Luong (M'91) received the B.S. (high honors), M.S., and Ph.D. degrees in electrical engineering and computer sciences from the University of California, Berkeley, in 1988, 1990, and 1994, respectively. For his Master's thesis, he worked on MOS analog multipliers with scaling technologies. For his Ph.D. dissertation, he designed and fabricated a superconductive flash-type analog-to-digital converter that operated at multi-gigahertz clock and input frequencies.

Since September 1994, he has been with the electrical and electronics engineering faculty at the Hong Kong University of Science and Technology, where he has been the Faculty-In-Charge of the Analog Research Lab and the Associate Director of the EEE Undergraduate Program Committee. His research interests are in high-performance analog and RF integrated circuits for wireless and portable communications.

Dr. Luong has served as an Associate Editor for IEEE TRANSACTIONS ON CIRCUITS AND SYSTEMS II. He received the Faculty Teaching Excellence Appreciation Award from the Hong Kong University of Science and Technology School of Engineering in 1995, 1996, and 2000.

William S. T. Yan received the Bachelor and Master's degrees in electrical and electronics engineering from the Hong Kong University of Science and Technology, Hong Kong, in 1996 and 1999, respectively.

He is currently with Maxim Integrated Products, Sunnyvale, CA. His current interests are in the areas of high-frequency integrated-circuit design.

A CMOS Frequency Synthesizer with an Injection-Locked Frequency Divider for a 5-GHz Wireless LAN Receiver

Hamid R. Rategh, *Student Member, IEEE*, Hirad Samavati, *Student Member, IEEE*, and Thomas H. Lee, *Member, IEEE*

Abstract—A fully integrated 5-GHz phase-locked loop (PLL) based frequency synthesizer is designed in a 0.24 μm CMOS technology. The power consumption of the synthesizer is significantly reduced by using a tracking injection-locked frequency divider (ILFD) as the first frequency divider in the PLL feedback loop. On-chip spiral inductors with patterned ground shields are also optimized to reduce the VCO and ILFD power consumption and to maximize the locking range of the ILFD. The synthesizer consumes 25 mW of power of which only 3.8 mW is consumed by the VCO and the ILFD combined. The PLL has a bandwidth of 280 kHz and a phase noise of −101 dBc/Hz at 1 MHz offset frequency. The spurious sidebands at the center of adjacent channels are less than −54 dBc.

Index Terms—CMOS RF circuits, frequency synthesizers, injection-locked frequency dividers, wireless LAN.

I. INTRODUCTION

THE DEMAND for wireless local area network (WLAN) systems which can support data rates in excess of 20 Mb/s with very low cost and low power consumption is rapidly increasing. The newly released unlicensed national information infrastructure (U-NII) frequency band in the United States is primarily intended for wideband WLAN and provides 300 MHz of spectrum at 5 GHz [Fig. 1(a)]. The lower 200 MHz of this band (5.15–5.35 GHz) overlaps the European high-performance radio LAN (HIPERLAN) frequency band. The upper 100 MHz of the spectrum which overlaps the industrial, scientific, and medical (ISM) band is not used in our system. To stay compatible with HIPERLAN the lower 200 MHz of the spectrum is divided into eight channels which are 23.5 MHz wide [Fig. 1(b)]. The minimum signal level at the receiver is −70 dBm while the maximum strength of the received signal is −20 dBm. The large dynamic range and wide channel bandwidths set very stringing requirements for the synthesizer phase noise and spurious sideband levels.

In this paper we describe the design of a fully integrated integer-N frequency synthesizer as a local oscillator (LO) for a U-NII band WLAN receiver. The front end of the receiver is described in [9].

Section II describes some of the synthesizer design challenges and reviews previously existing solutions. In Section III

Manuscript received August 2, 1999; revised November 29, 1999. This work was supported by the Stanford Graduate Fellowship program and IBM Corporation.

The authors are with the Center for Integrated Systems, Stanford University, Stanford, CA 94305 USA (e-mail: hamid@smirc.stanford.edu).

Publisher Item Identifier S 0018-9200(00)02988-7.

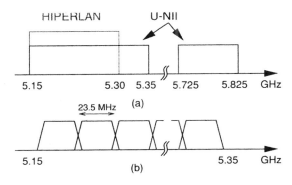

Fig. 1. (a) U-NII and HIPERLAN frequency bands and (b) channel allocation in our U-NII band WLAN system.

we present our proposed architecture of the frequency synthesizer which takes advantage of an injection-locked frequency divider (ILFD) to reduce the overall power consumption. Section IV-A is dedicated to the design of the VCO and demonstrates how on-chip spiral inductors can be optimized to reduce the VCO power consumption and to improve the phase noise performance at the same time. Section IV-B describes the design issues of ILFD's as well as the optimization of on-chip spiral inductors for wide-locking-range and low-power ILFD's. The pulse swallow frequency divider, charge pump, and loop filter are the subjects of Sections IV-C, IV-D, and IV-E, respectively. The measurement results are presented in Section V and conclusions are made in Section VI.

II. FREQUENCY SYNTHESIZERS

Frequency synthesizers are an essential part of wireless receivers and often consume a large percentage (20–30%) of the total power (Table I). A typical PLL-based frequency synthesizer comprises both high and low frequency blocks. The high frequency blocks, mainly the VCO and first stage of the frequency dividers, are the main power consuming blocks, especially in a CMOS implementation. Therefore, BiCMOS technology has often been chosen over CMOS, where the VCO and the prescaler are designed with bipolar transistors and the low frequency blocks are CMOS [1]. Off-chip VCO's and dividers have also been used as an alternative [4]. However, because of the increased cost neither of these two solutions is suitable for many applications, and a fully integrated CMOS solution is favorable. A dividerless frequency synthesizer [11] which eliminates power–hungry frequency dividers is one solution for such low-power and fully integrated systems. In this technique an

TABLE I
POWER CONSUMPTION OF FULLY INTEGRATED
WIRELESS RECEIVERS

Reference	Receiver power	PLL power
[10]	115mW	36mW
[13],[2]	225mW	51mW

aperture phase detector is used to compare the phase of the reference signal and the VCO output at every rising edge of the reference signal for only a time window which is a small fraction of the reference period. Thus no frequency divider is required in this PLL. The idea of a dividerless frequency synthesizer, although suitable for systems such as a GPS receiver where only one LO signal is required, is not readily applied to wireless systems which require multiple LO frequencies with a small frequency separation.

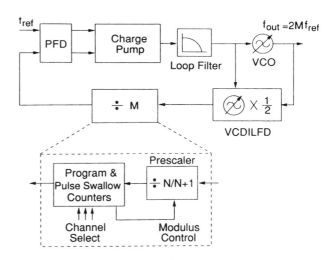

Fig. 2. Frequency synthesizer block diagram.

III. PROPOSED SYNTHESIZER ARCHITECTURE

Our proposed architecture (Fig. 2) is an integer-N frequency synthesizer with an initial low power divide-by-two in the PLL feedback loop. The prescaler follows the fixed frequency divider and operates at half the output frequency, thus, its power consumption is reduced significantly. Furthermore, the first divider is an injection-locked frequency divider [6], [7] which takes advantage of the narrowband nature of the system and trades off bandwidth for power via the use of resonators. To further reduce the power consumption, optimization techniques are used to design the on-chip spiral inductors of the VCO and ILFD.

Because of the fixed initial divide-by-two in the loop the reference frequency in our system is half of the LO spacing and is 11 MHz. Consequently, the loop bandwidth is reduced to maintain the loop stability. This bandwidth reduction helps to filter harmonics of the reference signal, mainly the second harmonic, which generate spurs in the middle of the adjacent channels. The drawbacks of a reduced loop bandwidth are an increased settling time and a higher in-band VCO phase noise. The higher in-band VCO phase noise is not a limiting factor as the in-band noise is dominated by the upconverted noise of the reference signal. The slower settling time is only a problem in very fast frequency-hopped systems.

The synthesized LO frequency in our system is 16/17 of the received carrier frequency. This choice of LO frequency not only eases the issue of image rejection in the receiver [9], but also facilitates the generation of the second LO, which is 1/16 of the first LO, with the same synthesizer.

IV. SYNTHESIZER BUILDING BLOCKS

A. Voltage-Controlled Oscillator

Fig. 3 shows the schematic of the VCO. Two cross-coupled transistors M1 and M2 generate the negative impedance required to cancel the losses of the RLC tank. On-chip spiral inductors with patterned ground shields [15] are used in this design. The two main requirements for the VCO are low phase noise and low power consumption. If the inductors were the main source of noise, maximizing their quality factor would improve the phase noise significantly. However, in multi-GHz VCO's with short channel transistors, inductors are not the main source of noise and a better design strategy is to maximize the effective parallel impedance of the RLC tank at resonance. This choice increases the oscillation amplitude for a given power consumption and hence reduces the phase noise caused by the noise injection from the active devices. Since inductors are the main source of loss in the tank, the LQ product should be maximized to maximize the effective parallel impedance of the tank at resonance, where L is the inductance and Q is the quality factor of the spiral inductors. It is important to realize that maximizing Q alone does not necessarily maximize the LQ product, and it is the latter that matters here.

To design the spiral inductors, we use the same inductor model reported in [14]. The inductance is first approximated with a monomial expression as in [3]. Optimization is used next to find the inductor with the maximum LQ product. The inductors in this design are 2.26 nH each with an estimated quality factor of 5.8 at 5 GHz. It is worth mentioning that at 5 GHz, the magnetic loss in the highly doped substrate of the epi process reduces the inductor quality factor significantly. Approximate calculations show that substrate inductive loss is proportional to the cube of the inductor's outer diameter. Therefore, a multilayer stacked inductor which has a smaller area compared to a single-layer inductor with the same inductance may achieve a larger quality factor. We should mention that in our design, inductors are laid out using only the top-most metal layer.

The varactors in Fig. 3 are accumulation-mode MOS capacitors [5], [12]. The quality factor of these varactors can be substantially degraded by gate resistance if they are not laid out properly. In our design each varactor is laid out with 14 fingers which are 3 μm wide and 0.5 μm long. The quality factor of this varactor at 5 GHz is estimated to exceed 60. The losses of the RLC tank are thus dominated by the inductors, as expected.

B. Injection-Locked Frequency Divider

Fig. 4 shows the schematic of the voltage–controlled ILFD used in the frequency synthesizer. The incident signal (the VCO output) is injected into the gate of M3 and is delivered with a subunity voltage gain to Vx, the common source connection of M1 and M2. Transistor M4 is used to provide a symmetric

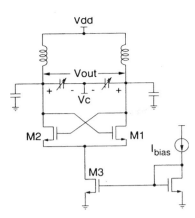

Fig. 3. Schematic of the VCO.

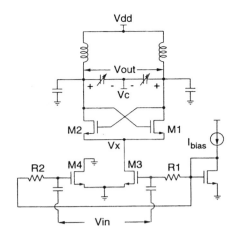

Fig. 4. Schematic of the differential ILFD.

load for the VCO. The output signal is fed back to the gates of M1 and M2 and is summed with the incident signal across the gates and sources of M1 and M2. The nonlinearity of M1 and M2 generates intermodulation products which allow sustained oscillation at a fraction of the input frequency [6]. As shown in [6] in the special case of a divide-by-two and a third-order nonlinearity, the phase-limited locking range of an LC ILFD can be expressed as

$$\left|\frac{\Delta\omega}{\omega_r}\right| < \left|\frac{H_0 a_2 V_i}{2Q}\right| \qquad (1)$$

where

ω_r	free–running oscillation frequency;
$\Delta\omega$	frequency offset from ω_r;
V_i	incident amplitude;
H_0	impedance of the RLC tank at resonance;
Q	quality factor of the RLC tank;
a_2	second-order coefficient of the nonlinearity.

As (1) suggests, a larger incident amplitude as well as a larger H_0/Q result in a larger achievable $\Delta\omega$ which we refer to as the *locking range*. In an LC oscillator $(H_0/Q) = \omega L$, so the largest practical inductance should be used to maximize the locking range.

A larger quadratic nonlinearity (a_2) also increases the locking range. So a circuit architecture with a large second-order nonlinearity is favorable for a divide-by-two ILFD and in fact the circuit in Fig. 4 has such a characteristic. The common source connection node of the differential pair moves at twice the frequency of the output signal even in the absence of the incident signal. So this circuit has a natural tendency for divide-by-two operation when the incident signal is effectively injected into node Vx.

To further extend the locking range, the ILFD is designed such that the resonant frequency of its output tank tracks the input frequency. Accumulation mode MOS varactors are used to tune the ILFD and its control voltage is tied to the VCO control voltage (Fig. 2). The locking range of the ILFD therefore does not limit the tuning range of the PLL beyond what is determined by the VCO.

As in the VCO design, on-chip spiral inductors with patterned ground shields are used in the ILFD, but with a different optimization objective. As mentioned earlier the largest practical inductance L maximizes the locking range. However, reduction of power consumption demands maximization of the LQ product. The inductor has its largest value when the total capacitance that resonates with it is minimized. To reduce its parasitic bottom plate capacitance the inductor should be laid out with narrow topmost metal lines. However, the large series resistance of narrow metal strips degrades the inductor quality factor and reduces the LQ product significantly. Therefore, both L and the LQ product may not be maximized simultaneously for an on-chip spiral inductor resonating with a fixed capacitance. Optimization is thus used to design for the maximum inductance such that the LQ product is large enough to satisfy the specified power budget. The inductors resulting from this trade-off are 9.5 nH each with an estimated quality factor of 4.2 at the divider output frequency (2.5 GHz).

C. Pulse Swallow Frequency Divider

The pulse swallow frequency divider ($\div$M) consists of a $\div 22/23$ prescaler followed by a program and pulse swallow counter. Only one CMOS logic ripple counter is used for both program and pulse swallow counters. The program counter generates one output pulse for every ten input pulses. The output of the pulse swallow counter is controlled by three channel select bits. The overall division ratio is 220–227. At the beginning of the cycle the prescaler divides by 23. As soon as the first three bits of the ripple counter match the channel select bits, the prescaler begins to divide by 22. The next cycle starts after the ripple counter counts to ten.

The prescaler consists of three dual-modulus divide-by-2/3 and one divide-by-2 frequency divider made of source-coupled logic (SCL) flip-flops and gates (Fig. 5). The modulus control (MC) input selects between divide-by-22 and divide-by-23. Except for the second dual modulus all other dividers including the CMOS counters are triggered by the falling edges of their input clocks, allowing a delay of as much as half the period of the input of each divider. With this arrangement we guarantee overlap between O_2, O_3, and O_4 (Fig. 5) and prevent a race condition.

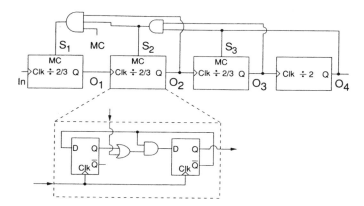

Fig. 5. Block diagram of the prescaler.

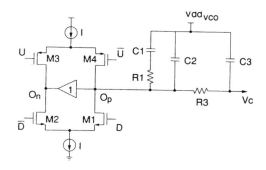

Fig. 6. Simplified schematic of the charge pump and loop filter.

D. Charge Pump

Fig. 6 shows the circuit diagram of the charge pump and loop filter. The charge pump has a differential architecture. However, only a single output node, O_p, drives the loop filter. To prevent the node O_n from drifting to the rails when neither of the up and down signals (U and D) is active, the unity gain buffer shown in Fig. 6 is placed between the two output nodes. This buffer keeps the two output nodes at the same potential and thus reduces the charge pump offset. The power of the spurious sidebands in the synthesized output signal is thereby reduced. In this charge pump the current sources are always on and the PMOS and NMOS switches are used to steer the current from one branch of the charge pump to the other.

E. Loop Filter

Resistor R_1 and capacitor C_1 in the loop filter (Fig. 6) generate a pole at the origin and a zero at $1/(R_1 C_1)$. Capacitor C_2 and the combination of R_3 and C_3 are used to add extra poles at frequencies higher than the PLL bandwidth to reduce reference feedthrough and decrease the spurious sidebands at harmonics of the reference frequency. The thermal noise of R_1 and R_3, although filtered by the loop, directly modulates the VCO control voltage and can cause substantial phase noise in the VCO if the resistors are not sized properly. The capacitors and resistors of the loop filter should be properly chosen to perform the required filtering function and maintain the stability of the loop without introducing too much noise. Fig. 7 shows a linearized phased-locked loop model. In a third-order loop, the loop filter contains only R_1, C_1, and C_2 and its impedance can be written as

$$Z(s) = \left(\frac{b}{b+1}\right) \frac{\tau s + 1}{s C_1 (\frac{\tau s}{b+1} + 1)} \qquad (2)$$

where $\tau = R_1 C_1$ and $b = C_1 / C_2$. The open loop transfer function of the third-order PLL is

$$\frac{\varphi_o}{\varphi_i} = \frac{K_{vco} I_p}{2\pi} \left(\frac{b}{b+1}\right) \frac{\tau s + 1}{s^2 C_1 (\frac{\tau s}{b+1} + 1)} \qquad (3)$$

where K_{vco} is the VCO gain constant and I_p is the charge pump current. The phase margin of the loop is

$$PM = \tan^{-1}(\tau \omega_c) - \tan^{-1}\left(\frac{\tau \omega_c}{b+1}\right) \qquad (4)$$

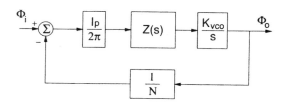

Fig. 7. Linearized PLL model.

where ω_c is the crossover frequency. By differentiating (4) with respect to ω_c it can be shown that the maximum phase margin is achieved at

$$\omega_c = \sqrt{b+1}/\tau \qquad (5)$$

and the maximum phase margin is

$$PM_{max} = \tan^{-1}(\sqrt{b+1}) - \tan^{-1}\left(\frac{\sqrt{b+1}}{b+1}\right). \qquad (6)$$

Notice that the maximum phase margin is only a function of b (ratio of C_1 and C_2) and for b less than 1 the phase margin is less than 20° which makes the loop practically unstable.

To complete our loop analysis we force $\omega_c = \sqrt{b+1}/\tau$ to be the crossover frequency of the loop and get

$$\frac{K_{vco} I_p}{2\pi N} \left(\frac{b}{b+1}\right) = \frac{C_1}{\tau^2} \sqrt{b+1}. \qquad (7)$$

Now we can define a loop filter design recipe as follows.

1) Find K_{vco} from the VCO simulation.
2) Choose a desired phase margin and find b from (6).
3) Choose the loop bandwidth and find τ from (5).
4) Select C_1 and I_p such that they satisfy (7).
5) Calculate the noise contribution of R_1. If the calculated noise is negligible the design is complete, otherwise go back to step four and increase C_1.

The same loop analysis can be repeated for a fourth-order loop. In this case the phase margin is

$$PM = \tan^{-1}(\tau \omega_c) - \tan^{-1}\left(\frac{A(\tau \omega_c)}{1 - B(\tau \omega_c)^2}\right) \qquad (8)$$

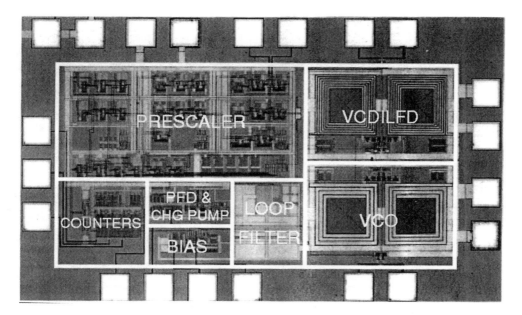

Fig. 8. Die micrograph.

where

$$A = \frac{\dfrac{C_2}{C_1} + \dfrac{C_3}{C_1} + \dfrac{\tau_2}{\tau}\left(1 + \dfrac{C_2}{C_1}\right)}{1 + \dfrac{C_2}{C_1} + \dfrac{C_3}{C_1}}$$

$$B = \frac{\dfrac{C_2}{C_1}\dfrac{\tau_2}{\tau}}{1 + \dfrac{C_2}{C_1} + \dfrac{C_3}{C_1}}$$

and $\tau_2 = R_3 C_3$. The crossover frequency for the maximum phase margin is shown in (9), at the bottom of the page.

Finally for ω_c to be the crossover frequency it should satisfy (10), shown at the bottom of the page.

As in the third-order loop the maximum phase margin is not a function of the absolute values of the R's and C's and is only a function of their ratios (C_2/C_1, C_3/C_1, and R_3/R_1). The loop filter design recipe for the fourth-order loop is modified as follows.

1) Find K_{vco} from the VCO simulation.
2) Choose a desired phase margin and find C_2/C_1, C_3/C_1, and τ_2/τ from (8) and (9).

3) Choose the loop bandwidth and find τ from (9).
4) Select C_1 and Ip such that they satisfy (10).
5) Calculate the noise contribution of R_1 and R_2. If their noise contribution is negligible the design is complete, otherwise go back to step four and increase C_1.

Notice that in a fourth-order loop there are two degrees of freedom in choosing C_2/C_1, C_3/C_1, and τ_2/τ to achieve a desired phase margin. Therefore, the suppression of the spurious sidebands can be improved without reducing the phase margin or the loop bandwidth.

In our system the maximum VCO gain constant is 500 MHz/V. With this VCO gain, and loop filter values of $R_1 = 47$ kΩ, $C_1 = 30$ pF, $C_2 = 3.3$ pF, $R_3 = 8$ kΩ. $C_3 = 2$ pF, and $I_p = 50$ μA, the crossover frequency is about 280 kHz with a 46° phase margin. The calculated contribution to VCO phase noise at 10 MHz offset frequency is -137 dBc/Hz, which is negligible compared to the intrinsic noise of the VCO.

V. MEASUREMENT RESULTS

The frequency synthesizer is designed in a 0.24-μm CMOS technology. Fig. 8 shows the die micrograph of the synthesizer with an area of 1 mm $\times$ 1.6 mm, including pads.

$$\omega_c = \frac{1}{\tau}\sqrt{\frac{1}{2}\left(\frac{2B + AB + A - A^2}{B(B - A)} + \sqrt{\left(\frac{2B + AB + A - A^2}{B(B - A)}\right)^2 - \frac{4(1 - A)}{B(B - A)}}\right)}. \tag{9}$$

$$\frac{I_p}{2\pi}\frac{K_{\text{vco}}}{N}\sqrt{\frac{1 + (\tau\omega_c)^2}{(A\tau\omega_c)^2 + (1 - B(\tau\omega_c)^2)^2}} = C_1\left(1 + \frac{C_2}{C_1} + \frac{C_3}{C_1}\right)\omega_c^2. \tag{10}$$

547

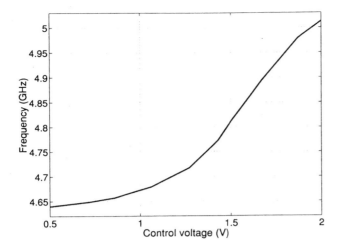

Fig. 9.   VCO tuning range.

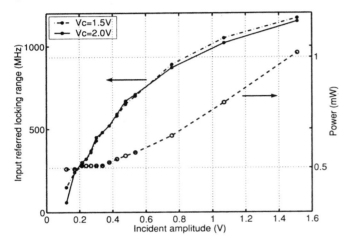

Fig. 10.   ILFD locking range and power consumption as a function of incident amplitude.

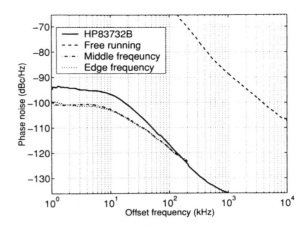

Fig. 11.   ILFD phase noise measurements.

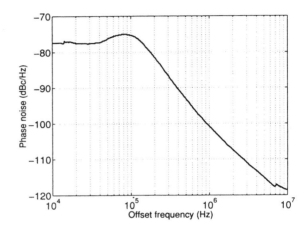

Fig. 12.   Phase noise of the synthesizer output signal.

TABLE II
ILFD PERFORMANCE SUMMARY

Output frequency tuning	110MHz ≈ 5%
Input locking range	600MHz ≈ 12% @ 0.55mW
	1000MHz ≈ 20% @ 0.8mW
Technology	0.24μm CMOS
Die area	0.186mm²
Flip-flop based divider (for comparison)	
0.24μm CMOS	5mW @ 5GHz
0.1μm CMOS [8]	2.6mW @ 5GHz

The analog blocks (VCO, ILFD, and prescaler) are supplied by 1.5 V while the digital portions of the synthesizer are supplied by 2 V. The reason for this choice of supplies is to achieve a larger tuning range for the VCO. The accumulation mode MOS capacitors in this technology have a flatband voltage (V_{FB}) around zero volts. Thus to get the full range of capacitor variation the control voltage should exceed the VCO supply to produce a net negative voltage across the varactors in Fig. 3. To eliminate a need for multiple supplies the VCO can be biased with a PMOS current source, and by connecting the sources of M1 and M2 to ground. More than 500 MHz (10% of the center frequency) of VCO tuning range is achieved for a 1.5-V control voltage variation (Fig. 9).

The free-running oscillation frequency of the ILFD changes more than 110 MHz (≈5% of the center frequency) for a 1.5-V control voltage variation.

Fig. 10 shows the locking range of the ILFD as a function of the incident amplitude for two different control voltages. As expected, changing the control voltage only changes the operation frequencies and not the locking range. The ILFD's average power consumption is also shown on the same figure. Increasing the incident amplitude increases the locking range and the average power consumption. The average power at 1-V incident

amplitude is less than 0.8 mW while the locking range exceeds 1000 MHz (≈20% of the center frequency).

The ILFD phase noise measurement results are shown in Fig. 11. The solid line shows the phase noise of the HP83732B signal generator used as the incident signal. The dashed line is the phase noise of the free–running ILFD. The two other curves are the phase noise of the ILFD when locked to two different incident frequencies. The curve marked as *middle frequency* is measured when the incident frequency is in the middle of the locking range and the *edge frequency* curve is measured at the lower edge of the locking range. At low offset frequencies the output of the frequency divider follows the phase noise of the incident signal and is 6 dB lower due to the divide-by-two operation. However, at larger offset frequencies the added noise from the divider itself, the external amplifier, and measurement tools reduces the 6 dB difference between the incident and

TABLE III
MEASURED SYNTHESIZER PERFORMANCE

Synthesizer performance	
Synthesized frequencies	4.840–4.994GHz
Reference frequency	11MHz
LO spacing	22MHz
Number of channels	8
Spur @ f_{ref}	≤ -45dBc
Spur @ $2 \times f_{ref}$	≤ -54dBc
Phase noise	-101dBc/Hz @ 1MHz
Loop bandwidth	280kHz
Power dissipation	
VCO	3.0mW
ILFD	0.8mW
Prescaler	19mW
Total	25mW
Supply voltage	1.5V (analog)
	2.0V (digital)
Implementation	
Die area	1.6mm^2
Technology	0.24μm CMOS

output phase noise. The ILFD phase noise measurements for offset frequencies higher than 200 kHz are not accurate due to the dominance of noise from the external amplifier.

The spurious tones at 11-MHz offset frequency from the center frequency are more than 45 dB below the carrier. The spurs at the 22-MHz offset frequency are at -54 dBc. Since the LO spacing is twice the reference frequency, the spurs at 11-MHz offset frequency fall at the edge of each channel and are less critical than the 22-MHz spurs which are located at the center of adjacent channels. With the -54 dBc spurs at 22 MHz offset frequency, an undesired adjacent channel may be 44 dB stronger than the desired channel for a minimum 10 dB signal-to-interference ratio.

Phase noise measurements of the complete synthesizer output signal are shown in Fig. 12. The phase noise at small offset frequencies is mainly determined by the phase noise of the reference signal. The phase noise measured at offset frequencies beyond the PLL bandwidth is the inherent VCO phase noise. The phase noise at 1-MHz offset frequency is measured to be -101 dBc/Hz. The phase noise at 22 MHz offset frequency is extrapolated to be -127.5 dBc/Hz. Therefore the signal in the adjacent channel can be 43 dB stronger than that of the desired channel for a 10 dB signal–to–interference ratio.

VI. CONCLUSION

In this work we demonstrate the design of a fully integrated, 5-GHz CMOS frequency synthesizer designed for a U-NII band WLAN system. The tracking injection-locked frequency divider used as the first divider in the PLL feedback loop reduces the power consumption considerably without limiting the performance of the PLL. Table II summarizes the performance of the ILFD. The power consumption of two flip-flop based frequency dividers at 5 GHz are also listed for comparison purposes. In a 0.24-μm CMOS technology a simulated SCL flip-flop based

frequency divider loaded with the same capacitance as in the ILFD consumes almost an order of magnitude more power than the ILFD with a 600-MHz locking range. The measurement results of a fast flip-flop based divider in an advanced 0.1-μm CMOS technology show a power consumption of 2.6 mW at 5 GHz [8] which is more than four times the power of the ILFD with a 600 MHz locking range.

Table III summarizes the performance of the synthesizer. The spurious sidebands at offset frequencies of twice the reference signal are more than 54 dB below the carrier. The spurs are mainly due to charge injection from the U and D signals to the loop, and can be reduced significantly by using a cascode structure for transistors M1–M4 (Fig. 6). Better matching between the up and down current sources also improves the sideband spurs. Of the 25-mW total power consumption, less than 3.8 mW is consumed by the VCO and ILFD combined. This low power consumption is achieved by the optimized design of the spiral inductors in the VCO and ILFD. The prescaler operates at 2.5 GHz and consumes 19 mW, of which about 40% is consumed in the first 2/3 dual modulus divider. Therefore the ILFD, which takes advantage of narrowband resonators, consumes an order of magnitude less power than the first 2/3 dual modulus divider, while operating at twice the frequency.

ACKNOWLEDGMENT

The authors would like to thank Dr. M. Hershenson, Dr. S. Mohan, and T. Soorapanth for their valuable technical discussions and help. They also thank National Semiconductor for fabricating the chip.

REFERENCES

[1] T. S. Aytur and B. Razavi, "A 2-GHz, 6 mW BiCMOS frequency synthesizer," *IEEE J. Solid-State Circuits*, vol. 30, pp. 1457–1462. Dec. 1995.

[2] J. Craninckx and M. Steyaert, "A fully integrated CMOS DCS-1800 frequency synthesizer," in *ISSCC Dig.*, 1998, pp. 372–373.

[3] M. Hershenson, S. S. Mohan, S. P. Boyd, and T. H. Lee, "Optimization of inductor circuits via geometric programming," in *Design Automation Conf. Dig.*, June 1999, pp. 994–998.

[4] C. G. S. M. H. Perrott and T. L. Tewksbury, "A 27-mW CMOS fractional-N synthesizer using digital compensation for 2.5-Mb/s GFSK modulation," *IEEE J. Solid-State Circuits*, vol. 32, pp. 2048–2059, Dec. 1997.

[5] A. S. Porret, T. Melly, and C. C. Enz, "Design of high-Q varactors for low-power wireless applications using a standard CMOS process," in *Custom Integrated Circuits Conf. Dig.*, May 1999, pp. 641–644.

[6] H. R. Rategh and T. H. Lee, "Superharmonic injection-locked frequency dividers," *IEEE J. Solid-State Circuits*, vol. 34, pp. 813–821, June 1999.

[7] H. R. Rategh, H. Samavati, and T. H. Lee, "A 5GHz, 1mW CMOS voltage controlled differential injection-locked frequency divider," in *Custom Integrated Circuits Conf. Dig.*, May 1999, pp. 517–520.

[8] B. Razavi, K. F. Lee, and R. H. Yan, "Design of high-speed, low-power frequency dividers and phase-locked loops in deep submicron CMOS," *IEEE J. Solid-State Circuits*, vol. 30, pp. 101–109, Feb. 1995.

[9] H. Samavati, H. R. Rategh, and T. H. Lee, "A 5GHz CMOS wireless-LAN receiver front-end," *IEEE J. Solid-State Circuits*, vol. 35, pp. xxx–xxx, May 2000.

[10] D. Shaeffer, A. Shahani, S. Mohan, H. Samavati, H. Rategh, M. Hershenson, M. Xu, C. Yue, D. Eddleman, and T. Lee, "A 115-mW, 0.5-μm CMOS GPS receiver with wide dynamic-range active filters," *IEEE J. Solid-State Circuits*, vol. 33, pp. 2219–2231, Dec. 1998.

[11] A. Shahani, D. Shaeffer, S. Mohan, H. Samavati, H. Rategh, M. Hershenson, M. Xu, C. Yue, D. Eddleman, and T. Lee, "Low-power dividerless frequency synthesis using aperture phase detector," *IEEE J. Solid-State Circuits*, vol. 33, pp. 2232–2239, Dec. 1998.

[12] T. Soorapanth, C. P. Yue, D. K. Shaeffer, T. H. Lee, and S. S. Wong, "Analysis and optimization of accumulation-mode varactor for RF ICs," in *Symp. VLSI Circuits Dig.*, 1998, pp. 32–33.

[13] M. Steyaert, M. Borremans, J. Janssens, B. D. Muer, N. Itoh, J. Craninckx, J. Crols, E. Morifuji, H. S. Momose, and W. Sansen, "A single-chip CMOS transceiver for DCS-1800 wireless communications," in *ISSCC Dig.*, 1998, pp. 48–49.

[14] C. P. Yue, C. Ryu, J. Lau, T. H. Lee, and S. S. Wong, "A physical model for planar spiral inductors on silicon," in *IEDM Tech. Dig.*, 1996, pp. 6.5.1–6.5.4.

[15] C. P. Yue and S. S. Wong, "On-chip spiral inductors with patterned ground shields for Si-Based RF IC's," in *Symp. VLSI Circuits Dig.*, 1997, pp. 85–86.

Hirad Samavati (S'99) received the B.S. degree in electrical engineering from Sharif University of Technology, Tehran, Iran, in 1994, and the M.S. degree in electrical engineering from Stanford University, Stanford, CA, in 1996. He currently is pursuing the Ph.D. degree at Stanford University.

During the summer of 1996, he was with Maxim Integrated Products, where he designed building blocks for a low-power infrared transceiver IC. His research interests include RF circuits and analog and mixed-signal VLSI, particularly integrated transceivers for wireless communications.

Mr. Samavati received a departmental fellowship from Stanford University in 1995 and a fellowship from the IBM Corporation in 1998. He is the winner of the ISSCC Jack Kilby outstanding student paper award for the paper "Fractal Capacitors" in 1998.

Hamid R. Rategh (S'99) was born in Shiraz, Iran in 1972. He received the B.S. degree in electrical engineering from Sharif University of Technology, Tehran, Iran, in 1994 and the M.S. degree in biomedical engineering from Case Western Reserve University, Cleveland, OH, in 1996. He is currently pursuing the Ph.D. degree in the Department of Electrical Engineering, Stanford University, Stanford, CA.

During the summer of 1997, he was with Rockwell Semiconductor System, Newport Beach, CA, where he was involved in the design of a CMOS dual-band, GSM/DCS1800, direct conversion receiver. His current research interests are in low-power radio frequency (RF) integrated circuits design for high-data-rate wireless local area network systems.

Mr. Rategh received the Stanford Graduate Fellowship in 1997. He was a member of the Iranian team in the 21st International Physics Olympiad, Groningen, the Netherlands.

Thomas H. Lee (M'96) received the S.B., S.M. and Sc.D. degrees in electrical engineering, all from the Massachusetts Institute of Technology (MIT), Cambridge, in 1983, 1985, and 1990, respectively.

He joined Analog Devices in 1990, where he was primarily engaged in the design of high-speed clock recovery devices. In 1992, he joined Rambus, Inc., Mountain View, CA, where he developed high-speed analog circuitry for 500 megabyte/s CMOS DRAM's. He has also contributed to the development of PLL's in the StrongARM, Alpha, and K6/K7 microprocessors. Since 1994, he has been an Assistant Professor of electrical engineering at Stanford University, where his research focus has been on gigahertz-speed wireline and wireless integrated circuits built in conventional silicon technologies, particularly CMOS. He holds 12 U.S. patents and is the author of a textbook, *The Design of CMOS Radio-Frequency Integrated Circuits* (Cambridge, MA: Cambridge Press, 1998), and is a coauthor of two additional books on RF circuit design. He is also a cofounder of Matrix Semiconductor.

Dr. Lee has twice received the "Best Paper" award at the International Solid-State Circuits Conference, was coauthor of a "Best Student Paper" at ISSCC, and recently won a Packard Foundation Fellowship. He is a Distinguished Lecturer of the IEEE Solid-State Circuits Society, and was recently named a Distinguished Microwave Lecturer.

A 2.6-GHz/5.2-GHz Frequency Synthesizer in 0.4-μm CMOS Technology

Christopher Lam and Behzad Razavi, *Member, IEEE*

Abstract—This paper describes the design of a CMOS frequency synthesizer targeting wireless local-area network applications in the 5-GHz range. Based on an integer-N architecture, the synthesizer produces a 5.2-GHz output as well as the quadrature phases of a 2.6-GHz carrier. Fabricated in a 0.4-μm digital CMOS technology, the circuit provides a channel spacing of 23.5 MHz at 5.2 GHz while exhibiting a phase noise of -115 dBc/Hz at 2.6 GHz and -100 dBc/Hz at 5.2 GHz (both at 10-MHz offset). The reference sidebands are at -53 dBc at 2.6 GHz, and the power dissipation from a 2.6-V supply is 47 mW.

Index Terms—Frequency dividers, oscillators, phase-locked loops, RF circuits, synthesizers, wireless transceivers.

I. INTRODUCTION

WIRELESS local area networks (WLAN's) provide great flexibility in the communication infrastructure of environments such as hospitals, factories, and large office buildings. While WLAN standards in the 2.4-GHz range have recently emerged in the market, the data rates supported by such systems are limited to a few megabits per second. By contrast, a number of standards have been defined in the 5-GHz range that allow data rates greater than 20 Mb/s, offering attractive solutions for real-time imaging, multimedia, and high-speed video applications. One of these standards is high-performance radio LAN (HIPERLAN) [1].

HIPERLAN operates across 5.15–5.30 GHz and provides a channel bandwidth of 23.5 MHz with Gaussian minimum shift keying (GMSK) modulation. The receiver sensitivity must exceed -70 dBm.

This paper presents the design of a frequency synthesizer for 5-GHz WLAN applications. To target realistic specifications, HIPERLAN is chosen as the framework. Employing an integer-N architecture, the circuit generates a 5.2-GHz output for the transmit path and the quadrature phases of a 2.6-GHz carrier for the receive path. Realized in a 0.4-μm CMOS technology, the synthesizer provides a channel spacing of 23.5 MHz while dissipating 47 mW from a 2.6-V supply. The phase noise at 10-MHz offset is equal to -115 dBc/Hz at 2.6 GHz and -100 dBc/Hz at 5.2 GHz.

Section II of this paper describes the synthesizer environment and general issues, and Section III introduces the synthesizer architecture. Section IV presents the design of each building block, and Section V summarizes the experimental results.

Manuscript received July 30, 1999; revised December 1, 1999.
The authors are with the Department of Electrical Engineering, University of California, Los Angeles, CA 90095 USA (e-mail: razavi@ee.ucla.edu).
Publisher Item Identifier S 0018-9200(00)02987-5.

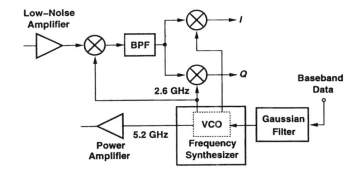

Fig. 1. Transceiver architecture.

II. SYNTHESIZER ENVIRONMENT

The design of a 5-GHz synthesizer in a 0.4-μm CMOS technology presents many difficulties at both the architecture and the circuit levels. The high center frequency of the voltage-controlled oscillator (VCO), the poor quality of inductors due to skin effect and substrate loss, the limited tuning range, the nonlinearity of the VCO input/output characteristic, the high speed required of the feedback divider, the mismatches in the charge pump, and the implementation of the loop filter are among the issues encountered in this design.

A 0.4-μm-long NMOS transistor in this technology achieves an f_T of less than 15 GHz with a gate–source overdrive voltage $(V_{GS} - V_{TH})$ of about 400 mV, a typical value in this design. Also, a 5-nH inductor exhibits a self-resonance frequency of 6.5 GHz and a Q of 5 at this frequency, indicating that skin effect and substrate loss are much more significant at 5.2 GHz than at 2.6 GHz. The technology offers no high-density linear capacitors, creating difficulty in the design of the loop filter.

The foregoing limitations make it necessary that the transceiver and the synthesizer be designed concurrently so as to relax some of the synthesizer requirements. Fig. 1 shows the transceiver architecture and its interface with the synthesizer. The receive path consists of two downconversion stages, each using a local oscillator (LO) frequency of 2.6 GHz, and the transmit path modulates the VCO by the Gaussian-filtered baseband data, producing a GMSK output.

An important feature of this architecture is that the synthesizer is shared between the transmitter and the receiver, reducing the system complexity substantially. This is possible because HIPERLAN incorporates time-division duplexing (TDD). Also, the transceiver requires the generation of the quadrature phases of the 2.6-GHz carrier rather than the 5.2-GHz output, a task readily accomplished by the synthesizer itself.

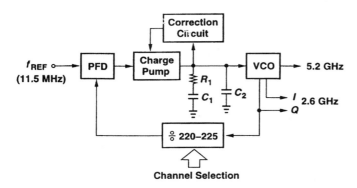

Fig. 2. Synthesizer architecture.

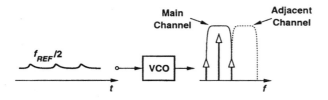

Fig. 3. Position of reference sidebands.

III. SYNTHESIZER ARCHITECTURE

The synthesizer is based on an integer-N phase-locked loop architecture (Fig. 2). The feedback divider senses the 2.6-GHz output because it is not possible to design a dual-modulus divider in 0.4-μm CMOS technology that operates at 5.2 GHz reliably. Controlled by the digital channel-select input, the $\div$220–225 circuit generates frequency steps of $f_{REF} = 11.75$ MHz in the 2.6-GHz band and 23.5 MHz in the 5.2-GHz band.

A critical issue in the architecture of Fig. 2 is the nonlinearity of the VCO characteristic, i.e., the variation of the VCO gain, K_{VCO}, with the control voltage V_{cont}. This effect manifests itself in the loop settling behavior as well as the magnitude of the phase noise and reference sidebands at the output. The problem is partially resolved through the use of a correction circuit that adjusts the charge-pump current according to the value of V_{cont} [2].

An interesting property of the architecture of Fig. 2 is the position of the reference spurs with respect to the main carrier. Since the reference frequency is half the channel spacing, such spurs fall at the *edge* of the channel rather than at the center of the adjacent channel for both 2.6- and 5.2-GHz outputs (Fig. 3). Since the interference energy received by the antenna is small at the edge, the maximum allowable magnitude of the spurs can be quite higher than if the reference frequency were equal to 23.5 MHz.

IV. BUILDING BLOCKS

A. VCO

The VCO core is based on two 2.6-GHz coupled oscillators operating in quadrature, as shown in Fig. 4(a) [3], [4]. The fully differential topology of each oscillator raises the possibility of sensing the common-source nodes A, B, C, or D as the 5.2-GHz output. In fact, since the 2.6-GHz oscillators operate in quadrature, the waveforms at A and B (or C and D) are 180° out of phase, thereby serving as a differential output at 5.2 GHz. With

proper choice of device dimensions and bias current, a differential swing of 0.5 V can be achieved at this port. Note that if a frequency doubler were used, the output would be single-ended and difficult to convert to differential form at such a high frequency.

The tuning of the oscillator poses several difficulties: the varactor diode must exhibit a small series resistance and remain reverse-biased even with large swings in the oscillator, and the varactor capacitance must be large enough to yield the required tuning range, but at the cost of increasing the power dissipation or the phase noise. This design incorporates a p^+-n^+ diode inside an n-well and strapped with metal to reduce the n-well series resistance [4]. Such a structure suffers from a large parasitic n-well/substrate capacitance, making it desirable to connect the anode of the diode to the oscillator. This is accomplished as illustrated in Fig. 4(b), where only one of the two oscillators is shown for clarity. Here, the control voltage varies the dc potential at nodes X and Y by varying the on-resistance of M_3.

However, the sharp variation of the on-resistance of M_3 leads to significant change in the gain of the VCO. To make the transition smoother, another transistor, M_4, in series with a resistor is added as shown in Fig. 4(c). Transistor M_5 serves as a clamp, keeping the tail current source in saturation. Otherwise, the oscillator may turn off during synthesizer loop transients.

Since the minimum voltage at node A is only a few hundred millivolts above ground, an NMOS differential pair cannot directly sense the 5.2-GHz signal at this node. Instead, a common-gate stage is used [Fig. 4(d)]. But if V_b is constant, then M_6 turns off for low values of V_{cont}. Modifying the circuit as shown in Fig. 4(e) ensures that the common-gate stage carries a constant bias current across the full tuning range.

The choice of the inductors and capacitance of the varactors entails a compromise between the phase noise and the tuning range. In this design, 7-nH inductors are used, each contributing a parasitic capacitance of 120 fF. The cross-coupled transistors are relatively wide to ensure startup, yielding approximately 175 fF of gate-source capacitance. The differential pairs coupling the oscillators also load the tank. As a result, the varactor capacitance for 2.6-GHz operation must not exceed 160 fF.

The inductors are realized as stacked spirals [5] made of metal 4 and metal 3 with a width of 6 μm. Since the tuning range is inevitably narrow, it is critical to predict the oscillation frequency accurately. A distributed model is used for each inductor, yielding an error of only a few percent in the measured frequency of oscillation.

B. Frequency Divider

The design of a 2.6-GHz programmable divider with a reasonable power dissipation in 0.4-μm CMOS technology is quite difficult. A number of circuit techniques are introduced in this work to ameliorate the power–speed tradeoff.

The divider is based on a pulse-swallow topology. Shown in Fig. 5(a) is a conventional implementation, consisting of a dual-modulus prescaler, a fixed-ratio program counter, and a programmable swallow counter. The RS latch is typically included in the swallow counter and is drawn explicitly here for clarity. The prescaler begins the operation by dividing by $N+1$ until the swallow counter is full. The RS latch is then set, changing

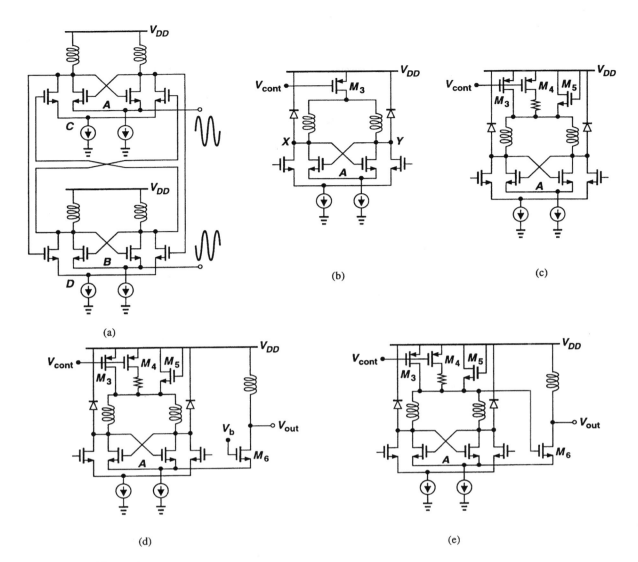

Fig. 4. Evolution of the VCO topology.

the prescaler modulus to N and disabling the swallow counter. The division continues until the program counter is full and the RS latch is reset. The overall divide ratio is therefore equal to $NP + S$.

The pulse-swallow divider used in this work is shown in Fig. 5(b). Here, the RS latch is followed by a D flip-flop to allow pipelining of the prescaler modulus control signal. This modification is justified below. The overall divide ratio is now equal to $NP + S + 1$. A critical decision in the design of the divider is the choice between low-swing current-steering logic and rail-to-rail CMOS logic. Simulations of the circuit with various values of N, P, and S indicate that the minimum power dissipation occurs if the prescaler incorporates current steering, its output is converted to rail-to-rail swings, and the remainder of the circuit incorporates standard dynamic and static CMOS logic. The use of current steering in the prescaler also obviates the need for large oscillator swings, saving power in the VCO buffer.

The design of the 8/9 prescaler for 2.6-GHz operation presents a great challenge. Shown in Fig. 6, the prescaler consists of a synchronous $\div 2/3$ circuit and two asynchronous

$\div 2$ circuits. In a conventional $\div 2/3$ realization [Fig. 7(a)], flip-flop FF_1 is loaded by an OR gate, whereas FF_2 is loaded by FF_1, an AND gate, and an output buffer. Since FF_2 limits the speed, the fanout of three inherent to this topology translates to substantial power dissipation. Furthermore, if the divider is implemented by current-steering circuits, the AND gate requires stacked logic and hence level-shift source followers. Both of these issues intensify the power–speed tradeoff.

The $\div 2/3$ circuit used in this work is shown in Fig. 7(b). Here, FF_2 is loaded by a NOR gate and FF_1 by a NOR gate and a buffer. Simulations indicate that the reduction of the load capacitance of FF_2 increases the maximum operating speed by approximately 40%.

The NOR/flip-flop combination is realized as depicted in Fig. 8. The resistors are made of n-well, and the bias voltage V_b is generated to fall midway between the high and low levels of inputs A and B. The output of the prescaler drives a differential to single-ended converter, producing rail-to-rail swings for the remainder of the divider.

The divider of Fig. 5 incorporates pipelining for the prescaler modulus control, thereby relaxing the minimum delay require-

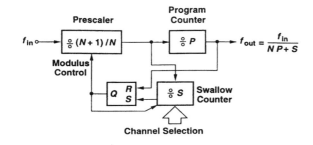

(a)

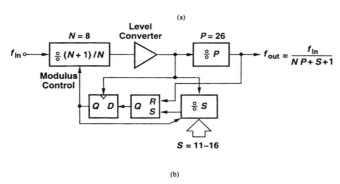

S = 11–16

(b)

Fig. 5. Pulse swallow divider. (a) Conventional topology. (b) Addition of pipelining in the prescaler modulus control path.

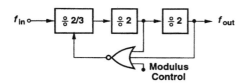

Fig. 6. Prescaler.

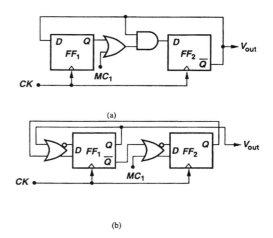

(a)

(b)

Fig. 7. Divide-by-2/3 circuit: (a) conventional topology and (b) circuit used in this work.

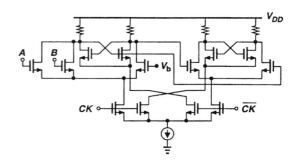

Fig. 8. Implementation of NOR/flip-flop combination.

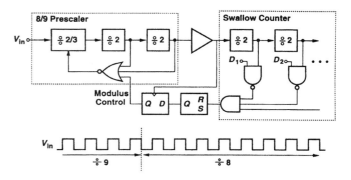

Fig. 9. Pipelining in the prescaler modulus control path.

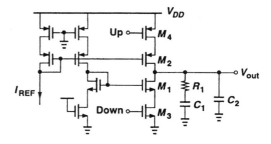

Fig. 10. Charge pump and loop filter.

ment in this path. Fig. 9 illustrates the issue. When the $\div 9$ operation of the prescaler is finished, the circuit would have at most seven cycles of V_{in} to change the modulus to eight. In this particular prescaler, the timing budget is actually about five input cycles—approximately 1.9 ns. Thus, with no pipelining, the last pulse generated by the prescaler in the $\div 9$ mode must propagate through the level converter, the first $\div 2$ stage in the swallow counter, the subsequent logic, the RS latch, and the three-input NOR gate in less than 1.9 ns. Such a delay constraint necessitates the use of current steering in this path, raising the power dissipation and complicating the design. With pipelining, on the other hand, the maximum tolerable delay increases to about eight input cycles—approximately 3.1 ns.

C. Charge Pump and Loop Filter

Fig. 10 shows the charge pump [6] and the loop filter. Here, M_3 and M_4—rather than M_1 and M_2—operate as switches. Thus, the problem of transistor charge injection and clock feedthrough to the output is somewhat alleviated. In addition to these errors, up and down currents produced by the charge pump may also create ripple on the control voltage. Since in locked condition, M_3 and M_4 turn on at every phase comparison instant, any mismatch between their magnitudes, duration, or absolute timing results in a net current that is drawn from the loop filter.

To appreciate the significance of these effects, let us consider some typical values in this design. If the reference sidebands are to be 50 dB below the carrier, then with $K_{VCO,max} \approx 1$ GHz/V and $f_{REF} = 11.75$ MHz, the ripple amplitude must not exceed

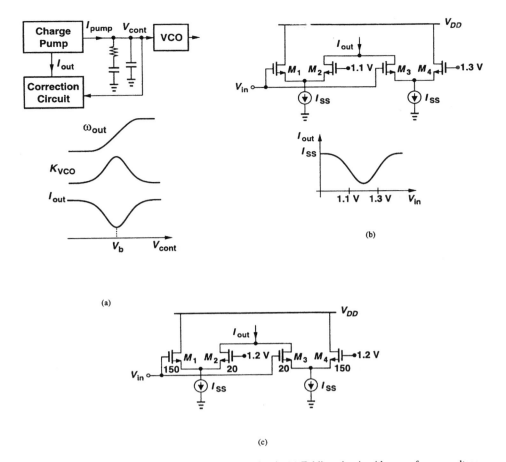

Fig. 11.   (a) Addition of correction circuit to charge pump. (b) Simple folding circuit. (c) Folding circuit with one reference voltage.

75 μV.[1] This indicates that great attention must be paid to the design of the phase/frequency, the charge pump, and the loop filter so as to minimize the above errors.

Another source of ripple in the control voltage is the low output impedance of M_1 and M_2 in Fig. 10, especially as V_{cont} reaches within a few hundred millivolts of the rails. This effect creates additional mismatch between the up and down currents as a function of V_{cont}, potentially leading to larger reference sidebands near the ends of the tuning range. Transistors M_3 and M_4 degenerate M_1 and M_2, respectively, alleviating this issue (another advantage of this topology over the standard charge-pump configuration).

The addition of C_2 in the circuit of Fig. 10 to suppress the ripple potentially degrades the stability of the loop. Simulations suggest that for $C_2 \leq 0.2C_1$, the settling time increases negligibly. In this design, $C_1 = 5$ pF, $C_2 = 1$ pF, and $R_1 = 86$ kΩ. The two capacitors can be realized by either MOSFET's or poly-metal sandwiches, a choice determined by the control voltage range. To achieve the maximum tuning range, V_{cont} must approach the supply and ground rails, demanding a reasonable capacitor linearity across this range. MOS capacitors, however, exhibit substantial change as their gate-source voltage falls below the threshold. Even a parallel combination of an NMOS capacitor (connected to ground) and a PMOS capacitor (connected to V_{DD}) suffers from a two-fold variation as V_{cont} goes from zero

to V_{DD}. For this reason, C_1 and C_2 are formed as poly-metal sandwiches (albeit with much less density than MOS capacitors).

Another issue in the design of the loop filter of Fig. 10 relates to the thermal noise produced by R_1. Low-pass filtered by C_1 and C_2, this noise modulates the VCO, raising the output phase noise. The thermal noise on the control voltage per unit bandwidth is given by

$$V_{n,\text{cont}}^2 = \left(\frac{C_1}{C_1 + C_2}\right)^2 \frac{1}{1 + \left(R_1 \dfrac{C_1 C_2}{C_1 + C_2}\omega\right)^2} V_{n,R1}^2 \quad (1)$$

where $V_{n,R1}^2$ denotes the noise density of R_1. From the narrow-band frequency modulation theory [8], we know that if a sinusoid with a peak amplitude A_m and frequency ω_m modulates a VCO, the output sidebands fall at ω_m rad/s below and above the carrier frequency and exhibit a peak amplitude of $A_m K_{\text{VCO}}/(2\omega_m)$. Approximating the noise per unit bandwidth in (1) by a sinusoid, we obtain the output relative phase noise per unit bandwidth at an offset frequency $\Delta\omega$ as

$$\frac{P_n}{P_{\text{carrier}}} = \left(\frac{C_1}{C_1 + C_2}\right)^2 \frac{2kTR_1}{1 + \left(R_1 \dfrac{C_1 C_2}{C_1 + C_2}\Delta\omega\right)^2}$$
$$\times \left(\frac{K_{\text{VCO}}}{2\Delta\omega}\right)^2. \quad (2)$$

[1]The ripple is approximated by a sinusoid here. In a more rigorous method, the ripple can be expressed as a Fourier series [7].

555

Fig. 12. Die photograph.

With the values chosen in this design, the output phase noise reaches -138 dBc/Hz at 10-MHz offset for $K_{VCO,max} = 1$ GHz/V. While it is desirable to reduce the value of R_1, the required increase in C_1 leads to a severe area penalty because of the low density of the poly-metal capacitors. Note that since the stability factor $\zeta = (1/2)R_1\sqrt{C_1 K_{VCO} I_{pump}/(2\pi)}$, if R_1 is, say, halved, then C_1 must be quadrupled to maintain ζ constant (for a given charge-pump current).

D. Correction Circuit

The gain of the VCO varies substantially across the tuning range, resulting in considerable change in the settling behavior. As depicted in Fig. 11(a), it is desirable to vary the charge-pump current, I_{pump}, such that the product of K_{VCO} and I_{pump} and hence ζ remain relatively constant. Rather than use piecewise linearization [2], this work incorporates an analog folding technique. Fig. 11(b) shows a possible solution. Here M_1 and M_3 are off if V_{in} is well below 1.1 V and hence $I_{out} = I_{SS}$. As V_{in} approaches 1.1 V, M_1 turns on while M_3 is off. Thus, I_{out} drops, reaching a low value as M_1 carries most of I_{SS} and M_3 a negligible current. As V_{in} approaches and exceeds 1.3 V, M_3 turns on and I_{out} eventually returns to I_{SS}. This design actually utilizes the topology shown in Fig. 11(c), where only one reference voltage is required and each differential pair provides a built-in offset by virtue of skewed device dimensions. The characteristic is similar to that shown for Fig. 11(b), with I_{out} driving the current mirrors in the charge pump.

The reference voltage of 1.2 V in Fig. 11(c) assumes that the gain of the VCO reaches its maximum at $V_{cont} = 1.2$ V. This value is somewhat process- and temperature-dependent, limiting (according to simulations) the suppression of the VCO nonlinearity to about one order of magnitude.

V. EXPERIMENTAL RESULTS

The frequency synthesizer has been fabricated in a 0.4-μm digital CMOS technology. All of the inductors and capacitors are included on the chip. Fig. 12 is a photograph of the die, which measures 1.75×1.15 mm^2. The circuit has been tested with a 2.6-V supply.

Figs. 13(a) and (b) depict the output spectra in the locked condition. The phase noise at 10-MHz offset is equal to -115 dBc/Hz at 2.5 GHz and -100 dBc/Hz at 5.2 GHz. A significant part of the phase noise at 5.2 GHz is attributed to the considerable loss of the output 50-Ω buffer. Fig. 14 shows the 2.6-GHz output along with the reference sidebands. The sidebands are

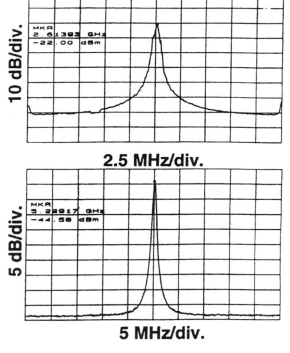

Fig. 13. Measured spectra at 2.6 and 5.2 GHz in locked condition.

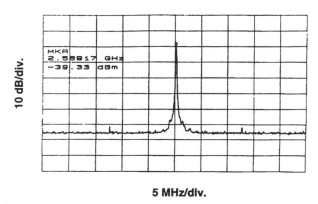

5 MHz/div.

Fig. 14. Measured spectrum at 2.6 GHz.

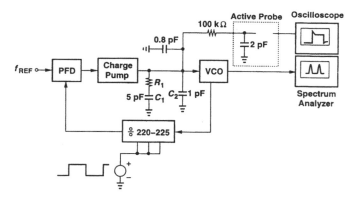

Fig. 15. Setup for settling time measurement.

approximately 53 dB below the carrier. For the 5.2-GHz output, the sidebands are buried under the noise floor.

The settling behavior of the synthesizer has also been studied. Fig. 15 illustrates the setup, where the modulus of the feedback

556

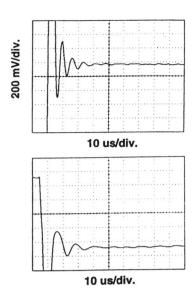

Fig. 16. Control voltage during loop settling.

TABLE I
SYNTHESIZER PERFORMANCE

Output Frequency	2.6 GHz/5.2 GHz
Phase Noise @ 10–MHz Offset	–110 dBc/Hz @ 2.6 GHz –100 dBc/Hz @ 5.2 GHz
Sidebands	–55 dBc
Power Dissipation	
VCO	30 mW
Divider, etc.	17 mW
Total	47 mW
Supply Voltage	2.6 V
Area	1.75 mm x 1.15 mm
Technology	0.4– μm CMOS

Designing a multigigahertz synthesizer in 0.4-μm CMOS technology necessitates circuit techniques such as: 1) a quadrature VCO with inherent frequency doubling, 2) a dual-modulus divider with equalized fanout, 3) pipelining in pulse-swallow counters, and 4) use of folding stages to compensate for nonlinearity in the VCO characteristic.

REFERENCES

[1] "Radio equipment and systems (RES); High performance radio local area network (HIPERLAN); Functional specification," ETSI, Sophia Antipolis, France, ETSI TC-RES, July 1995.
[2] J. Craninckx and M. S. J. Steyaert, "A fully integrated CMOS DCS-1800 frequency synthesizer," *IEEE J. Solid-State Circuits*, vol. 33, pp. 2054–2065, Dec. 1998.
[3] A. Rofougaran *et al.*, "A 900-MHz CMOS LC oscillator with quadrature outputs," in *ISSCC Dig. Tech. Papers*, Feb. 1996, pp. 392–393.
[4] B. Razavi, "A 1.8 GHz CMOS voltage-controlled oscillator," in *ISSCC Dig. Tech. Papers*, Feb. 1997, pp. 388–389.
[5] R. B. Merril *et al.*, "Optimization of high Q inductors for multi-level metal CMOS," in *Proc. IEDM*, Dec. 1995, pp. 38.7.1–38.7.4.
[6] J. Alvarez, H. Sanchez, and G. Gerosa, "A wide-band low-voltage PLL for PowerPC microprocessors," *IEEE J. Solid-State Circuits*, vol. 30, pp. 383–391, Apr. 1995.
[7] B. Razavi, *RF Microelectronics*. Upper Saddle River, NJ: Prentice-Hall, 1998.
[8] L. W. Couch, *Digital and Analog Communication Systems*, 4th ed. New York: Macmillan, 1993.

Christopher Lam received the B.Sc. and M.Sc. degrees in electrical engineering from the University of California, Los Angeles, in 1997 and 1999, respectively.

He is currently with the Wireless Communication Group, National Semiconductor, Santa Clara, CA. His interests include phase-locked loops and communication circuits.

divider is switched periodically and the control voltage is monitored. The 0.8-pF capacitor results from the trace on the printed circuit board, and the active probe presents an input capacitance of 2 pF. Since $C_1 = 5$ pF and $C_2 = 1$ pF, the addition of these parasitics markedly degrades the stability. Therefore, a 100-kΩ resistor is placed in series with the active probe to mimic the role of R_1 and C_1. The low-pass filter thus formed has a corner frequency comparable to the loop bandwidth, and the 0.8-pF capacitor still produces ringing in the time response. Fig. 16 shows the measured control voltage, indicating a settling time on the order of 40 μs.

Table I summarizes the measured performance of the synthesizer.

VI. CONCLUSION

The speed and quality of the devices available in an IC technology directly affect the choice of transceiver architectures, synthesizer topologies, and circuit configurations. In order to optimize the overall system performance, the transceiver and the synthesizer must be designed concurrently, with particular attention to the frequency planning.

Behzad Razavi (S'87–M'90) received the B.Sc. degree from Sharif University of Technology, Tehran, Iran, in 1985 and the M.Sc. and Ph.D. degrees from Stanford University, Stanford, CA, in 1988 and 1992, respectively, all in electrical engineering.

He was with AT&T Bell Laboratories, Holmdel, NJ, and subsequently Hewlett-Packard Laboratories, Palo Alto, CA. Since September 1996, he has been an Associate Professor of electrical engineering at the University of California, Los Angeles. His current research includes wireless transceivers, frequency synthesizers, phase-locking and clock recovery for high-speed data communications, and data converters. He was an Adjunct Professor at Princeton University, Princeton, NJ, from 1992 to 1994, and at Stanford University in 1995. He is a member of the Technical Program Committees of the Symposium on VLSI Circuits and the International Solid-State Circuits Conference (ISSCC), in which he is Chair of the Analog Subcommittee. He is the author of *Principles of Data Conversion System Design* (New York: IEEE Press, 1995), *RF Microelectronics* (Englewood Cliffs, NJ: Prentice-Hall, 1998), and *Design of Analog CMOS Integrated Circuits* (New York: McGraw-Hill, 2000), and the editor of *Monolithic Phase-Locked Loops and Clock Recovery Circuits* (New York: IEEE Press, 1996).

Dr. Razavi received the Beatrice Winner Award for Editorial Excellence at the 1994 ISSCC, the Best Paper Award at the 1994 European Solid-State Circuits Conference, the Best Panel Award at the 1995 and 1997 ISSCC, the TRW Innovative Teaching Award in 1997, and the Best Paper Award at the IEEE Custom Integrated Circuits Conference in 1998. He has also served as Guest Editor and Associate Editor of the IEEE JOURNAL OF SOLID-STATE CIRCUITS and IEEE TRANSACTIONS ON CIRCUITS AND SYSTEMS.

Fast-Switching Frequency Synthesizer with a Discriminator-Aided Phase Detector

Ching-Yuan Yang, *Student Member, IEEE*, and Shen-Iuan Liu, *Member, IEEE*

Abstract—A phase-locked loop (PLL) with a fast-locked discriminator-aided phase detector (DAPD) is presented. Compared with the conventional phase detector (PD), the proposed fast-locked PD reduces the PLL pull-in time and enhances the switching speed, while maintaining better noise bandwidth. The synthesizer has been implemented in a 0.35-μm CMOS process, and the output phase noise is -99 dBc/Hz at 100-kHz offset. Under the supply voltage of 3.3 V, its power consumption is 120 mW.

Index Terms—Bandwidth adjusting, fast acquisition, fast locking, frequency synthesizers, phase detectors, phase-locked loops.

I. INTRODUCTION

PHASE-LOCKED loop (PLL) circuits have been found to be useful wherever there is a need to synchronize a local oscillator with an independent incoming signal, such as serial data links and RF wireless communications. In order to optimize the loop performance, some features should be taken care of [1], [2]. First, to minimize output phase jitter due to external noise, the loop bandwidth should be made as narrow as possible. Second, to minimize output jitter due to internal oscillator noise, or to obtain best tracking and acquisition properties, the loop bandwidth should be made as wide as possible. These principles obviously oppose each other; and therefore some compromises between these two principles are always inevitable. The block diagram of a PLL with a discriminator-aided phase detector (DAPD) is shown in Fig. 1. One could leave the discriminator connected permanently and/or merely weight the relative contributions of the system so as to obtain the desired damping. The discriminator-aided path adds to lock the PLL quickly. Once the PLL is in lock, a better bandwidth can be maintained while the discriminator is disconnected.

In this paper, a novel DAPD is presented to reduce pull-in time T_p and to enhance the switching speed of the PLL, while maintaining the same noise bandwidth and avoiding modulation damping. Section II describes the basic concept of the proposed structure. Sections III and IV present the realization and the measurement of the system, respectively, and Section V concludes the paper.

Manuscript received November 30, 1999; revised April 28, 2000. This work was sponsored by the National Science Council under Contract 88-2219-E-002-024.

C.-Y. Yang was with the Department of Electrical Engineering, National Taiwan University, Taipei, Taiwan 10617, R. O. C. He is now with the Department of Electronic Engineering, HuaFan University, Taipei, Taiwan 223, R.O.C.

S.-I. Liu is with the Department of Electrical Engineering, National Taiwan University, Taipei, Taiwan 10617, R. O. C.

Publisher Item Identifier S 0018-9200(00)08697-2.

II. BASIC IDEA AND MODEL

A simple charge-pump PLL consists of four major blocks: the phase detector (PD), the charge-pump circuit, the loop filter, and the voltage-controlled oscillator (VCO) [3]–[6]. Fig. 2 shows the linear model of a charge-pump PLL-based frequency synthesizer. The closed-loop transfer function can be represented as

$$H(s) = \frac{\theta_o(s)}{\theta_i(s)} = \frac{2\pi \cdot N \cdot K_{PD} \cdot Z_{LF}(s) \cdot K_{VCO}}{sN + 2\pi \cdot K_{PD} \cdot Z_{LF}(s) \cdot K_{VCO}}. \quad (1)$$

The conventional PD is implemented in conjunction with a charge-pump loop filter in the PLL, as illustrated in Fig. 3. To determine the transfer function of the PD, assume there is a time interval τ between two input signals R and V in the PD, the output current of the charge-pump circuit is a pulse of duration τ, and the amplitude of the charge-pump current is I_{P0}. In the continuous-time approximation, the average value $\overline{i_P}$ per input signal period T can be given as

$$\overline{i_P} = \frac{Q}{T} = I_{P0} \cdot \frac{\tau}{T} = \frac{I_{P0}}{2\pi} \cdot \theta_e. \quad (2)$$

The transfer function curve of a linear PD is shown in Fig. 4(a), where the vertical axis represents the charge injected into the loop filter during one period of the input signal. The characteristic of a nonlinear PD, as shown in Fig. 4(b), can be divided into two regions [7]. It has the same characteristic within the locked-in region as that of the linear PD, but the acquisition time will be reduced with the steeper characteristic outside the lock-in region. When designing a PLL with the nonlinear PD, first the central slope is determined to fulfill the requirement of noise and modulation for the PLL with a standard PD. Then, the slope near $\theta_e = \theta_{AD}$ is gradually increased to improve acquisition speed. The proposed nonlinear PD can be built with delay cells and standard PD circuits, as shown in Fig. 4(c). The standard PD is a digital circuit, triggered by the positive edge of the input reference signal R and the output feedback signal V. Considering the delay cells with θ_{AD}, the PDs decide the position of the phase difference θ_e among these regions. According to the value of θ_e, the charge pump will output the corresponding current controlled by the up signals U or the down signals D. The behavior model of the nonlinear PD can be explained by the waveforms of Fig. 4(d). According to the time difference between both input signals V_i and V_o, the up signals U are used to increase and the down signals D are used to decrease the frequency of signal V_o. The nonlinear PD always generates the right signal to equalize the frequency of both input signals as the conventional PD. The time interval τ is positive

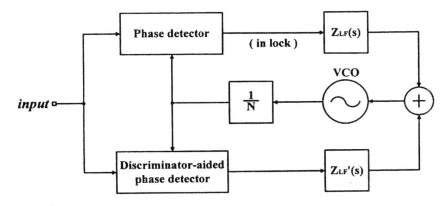

Fig. 1. Block diagram of PLL with discriminator-aided phase acquisition.

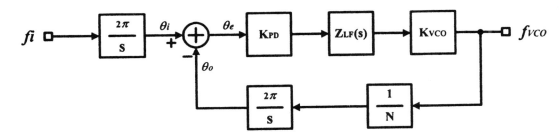

Fig. 2. Linear model of PLL frequency synthesizer.

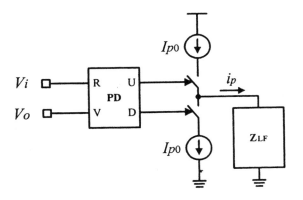

Fig. 3. Phase detector with charge-pump filter.

(negative) where V_i leads (lags) V_o. When τ is larger than τ_{AD} ($= 2\pi\theta_{AD}$), U_{AD} may appear "high" level; when τ is smaller than $-\tau_{AD}$, D_{AD} may appear "high" level. As the nonlinear PD is applied, two cases can occur during different time interval τ:

Case 1: $0 < |\tau| \leq \tau_{AD}$, the injected charge $Q = I_{P0} \cdot |\tau|$.
Case 2: $\tau_{AD} < |\tau|$, the injected charge $Q = I_{P0} \cdot \tau_{AD} + (I_{P0} + I_{PAD}) \cdot (|\tau| - \tau_{AD})$, which can be approximated $(I_{P0} + I_{PAD}) \cdot |\tau|$ as τ_{AD}, is very small.

Generally, the total transfer function of the PD with the current-pump circuit and the loop filter can be expressed as [8]

$$K_{PD} \cdot G_{LF}(s) = \frac{I_p}{2\pi} \cdot Z_{LF}(s) \qquad (3)$$

where I_P is the pump current of the charge-pump circuit. The impedance $Z_{LF}(s)$ is the series connection of a capacitor C_Z

and a resistor R_Z with a capacitor C_P added in parallel as shown in Fig. 2. The impedance of this filter can be

$$Z_{LF}(s) = \frac{1 + s\tau_Z}{s(C_Z + C_P) \cdot (1 + s\tau_P)} \qquad (4)$$

with $\tau_Z = R_Z C_Z$ and $\tau_P = R_Z(C_Z^{-1} + C_P^{-1})^{-1}$. The open-loop gain of the PLL equals

$$GH(s) = \frac{I_p \cdot K_{VCO}}{N} \cdot \frac{1 + s\tau_Z}{s^2(C_Z + C_P) \cdot (1 + s\tau_P)} \qquad (5)$$

which has a crossover frequency of

$$\omega_C = \frac{I_p \cdot K_{VCO} \cdot R_Z}{N} \cdot \frac{C_Z}{C_Z + C_P}. \qquad (6)$$

The open-loop gain of this third-order PLL can be calculated in terms of the frequency ω, as follows:

$$GH(s)|_{s=j\omega} = -\frac{I_p K_{VCO}}{N} \cdot \frac{(1 + j\omega\tau_Z)}{\omega^2(C_Z + C_P) \cdot (1 + j\omega\tau_P)} \qquad (7)$$

and its phase margin can be determined in terms of

$$\phi(\omega) = 180° + \tan^{-1}(\omega \cdot \tau_Z) - \tan^{-1}(\omega \cdot \tau_P). \qquad (8)$$

In order to maintain the same loop gain and phase margin for

559

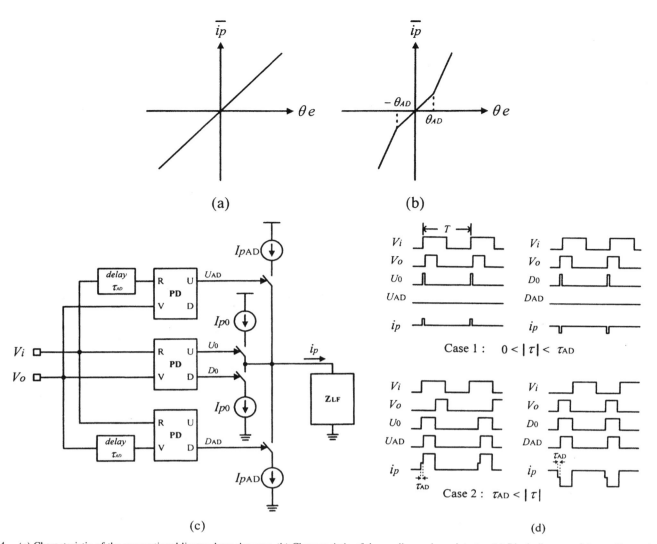

Fig. 4. (a) Characteristic of the conventional linear phase detector. (b) Characteristic of the nonlinear phase detector. (c) Block diagram of the nonlinear phase detector. (d) Operation of the nonlinear phase detector.

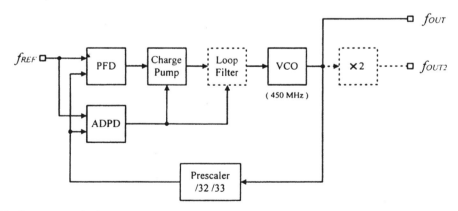

Fig. 5. Block diagram of the frequency synthesizer.

the sake of stability, the charge-pump current becomes $k^2 \cdot I_P$ instead of I_P outside the locked-in region, and the loop-filter resistor would become R_Z/k instead of R_Z while ω increases k times, i.e., the loop bandwidth increases k times. It may speed up the switching capability of the PLL. Once it is locked on the correct frequency, the PLL will then return to the low-noise operation.

III. CIRCUIT REALIZATION

A. Architecture

The designed frequency synthesizer integrates the proposed DAPD, the charge-pump circuit, a prescaler, and a VCO in a single CMOS chip. It is similar to the structure of a conventional integer-N frequency synthesizer, as shown in Fig. 5. By adding

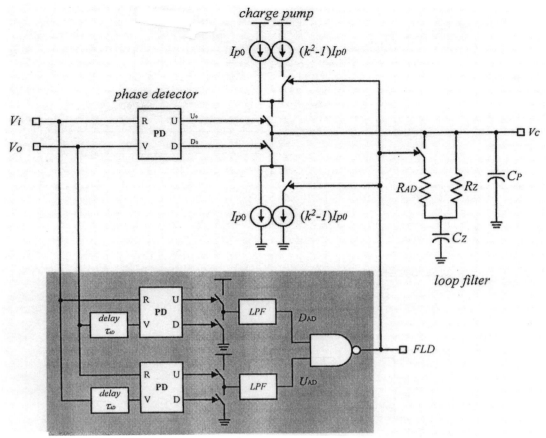

Fig. 6.   Schematic of phase detector with DAPD and charge-pump filter.

the frequency-doubling block, the output frequency can be up to 900 MHz from a 450-MHz VCO.

B. Phase Detector with DAPD and Charge-Pump Filter

A schematic diagram of the DAPD is shown in Fig. 6. The phase frequency detectors are used to compare the phase difference of both input signals. The output signal (FLD) of the DAPD depends on the phase difference of both input signals whether it is larger than θ_{AD} or not. Considering the delay cells with delay τ_{AD}, which is very small but never negligible, the DAPD decides the operating bandwidth of the loop filter. When V_i leads V_o, the time difference τ is larger than τ_{AD}, and U_{AD} is "low" and D_{AD} is "high." Otherwise, when V_i lags V_o, τ is smaller than $-\tau_{AD}$, and U_{AD} is "high" and D_{AD} is "low." In a word, if the absolute value of the time difference τ between input signals is larger than τ_{AD}, FLD may appear "high" level. Also, the charge-pump current becomes $k^2 I_{P0}$ and the resistor of the loop filter becomes R_Z/k, i.e., $R_{AD}//R_Z$. Until the absolute value of τ is within τ_{AD}, U_{AD} and D_{AD} are both "high," thus FLD is brought to "low" level, then the charge-pump current and the resistor return to I_{P0} and R_Z, respectively, with a narrower bandwidth for better noise rejection. However, the delay cell is adopted according to the VCO's noise. Assuming that the phase characteristic of the signal V_o is $\theta_0 \pm \Delta\theta$, θ_{AD} should be larger than $\Delta\theta$ to make the DAPD work.

In our design, the loop bandwidth ω_c of the PLL equals about $2\pi \cdot 40$ krad/s, and the loop gain zero $\omega_z (= 1/\tau_z)$ and the loop pole $\omega_p (= 1/\tau_p)$ are placed on a factor of four below and above ω_c, respectively. In addition, a pump current I_{p0} of 560 μA is applied and the parameter $k = 3$ is chosen. The values of the resistors R_z and R_{AD} and the capacitors C_z and C_p are 470 Ω, 235 Ω, 33 nF, and 2.2 nF, respectively. The open-loop gain response is depicted in Fig. 7. Curve (a) is the characteristic of the PLL with the DAPD while the bandwidth is 120 kHz. However, the PLL will return curve (b) with the bandwidth of 40 kHz when it is near in lock. These curves give the same phase margin of approximately 60°. Thus the PLL would be usually stable.

Currently, most frequency synthesizers use phase-frequency detectors (PFDs) as their PDs. A PFD is a sequential circuit which can not only detect the phase error but also provides a frequency-sensitive signal to aid acquisition when the loop is out of lock. The drawback of some conventional PFDs is a dead zone in the phase characteristic, which generates the phase error in the output signals. To solve this problem, a dynamic CMOS PFD is adopted as shown in Fig. 8(b), which is similar to the one proposed in [10]. The PFD consists of two half-transparent registers, shown in Fig. 8(a), [9] and a NAND gate. It is triggered by the negative edge of input signals. The timing diagram of the PFD is shown in Fig. 8(c). Even though the input signals are in-phase, the glitches caused by the reset path always exist. So, extra filters are added in the DAPD to remove the effect of the glitches.

So far, the positive gain of the VCO is applied from the above discussion. However, since the gain of the VCO is negative as

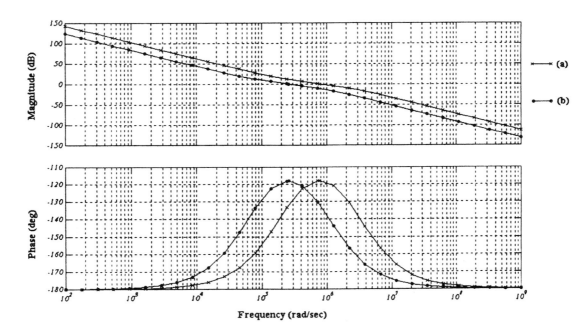

Fig. 7. Simulated open-loop gain Bode plot.

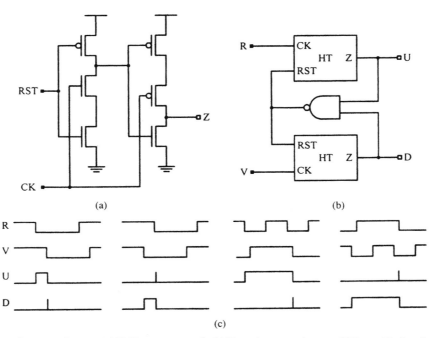

Fig. 8. Implementation of phase-frequency detector. (a) Half-transparent cell. (b) Phase-frequency detector (PFD). (c) Timing diagram of PFD.

described later, U and D of the PD connected to the charge pump should be interchanged. The charge pump, which is based on one described in [11], is adopted. It suppresses the charge sharing from the parasitic capacitance by a pair of switched-current sources.

C. Dual-Modulus Prescaler

The dual-modulus prescaler is the high-frequency building block in the frequency synthesizer. This circuit shown in Fig. 9 divides the frequency of the VCO output signal by a factor of 32 or 33 depending on the logic value of the controlled signal

mode [12]–[15]. It consists of a synchronous divide-by-4/5 counter as the first stage and an asynchronous divide-by-8 counter as the second stage. The circuits in the first stage are fully differential, while the single-ended logic circuits are used in the second stage. To reduce the supply noise, an emitter-coupled logic (ECL)-like differential logic is used in the high-speed stage [16]. In the divide-by-4/5 circuit, the DFF is a differential flip-flop. Fig. 10 shows the schematic diagram of a NAND-gate logic flip-flop. Merging the logic gates to a flip-flop saves power and increases the operating speed. The toggle flip-flops are made by true single-phase clocking (TSPC) DFFs of [12] behind a differential-to-single buffer.

562

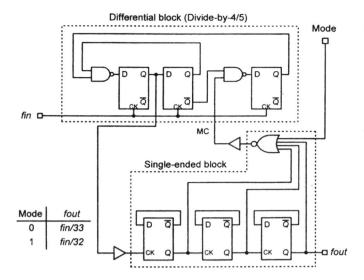

Fig. 9. Functional block diagram of the dual-modulus prescaler.

Mode	fout
0	fin/33
1	fin/32

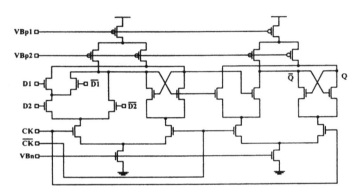

Fig. 10. Schematic of the differential NAND-gate flip-flop.

This buffer is used to achieve the rail-to-rail output signal in the low-speed stage.

D. VCO

The VCO is another high-frequency building block in a frequency synthesizer. Still, an ECL-like current-mode differential pair, as shown in Fig. 11, is used as a delay cell [17], [18] to achieve high common-mode rejection in a four-stage ring oscillator. The coarse tuning of the ring-oscillator's center frequency is achieved by the bias **Vbpo1** (or through the use of a digital-to-analog converter), and a fine tuning technique is needed for the PLL voltage-control path. The gain required for the oscillator is easily determined by the ratio of M1 and M2 as the current gain. The proposed delay cell has the better noise performance because the operation of the circuit is carried out by the differential signal immune to the power-supply-injected and substrate-injected noise sources. The replica bias circuit adjusts the load over a wide range in response to a swept supply current. It insures the output swing of delay cells maintain fixed and takes a changeable bias current to cover a suitable range of different output frequencies. Bypass capacitors are also an important consideration for the replica bias and voltage reference circuits. On-chip bypass capacitors can be used to help reduce their noise contribution to the ring-oscillator delay cells.

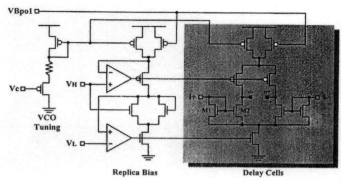

Fig. 11. VCO schematic.

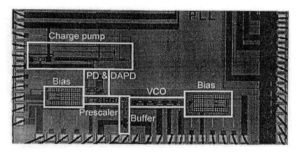

Fig. 12. IC microphotograph.

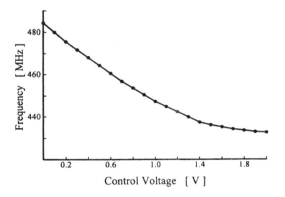

Fig. 13. Experimentally measured VCO transfer curve.

IV. MEASUREMENT RESULTS

The synthesizer is implemented in a 0.35-μm CMOS process. The microphotograph of the fabricated frequency synthesizer is shown in Fig. 12. The loop filter is off-chip, and the output signal of the VCO is connected to a source follower. The frequency synthesizer is measured at a supply voltage of 3.3 V. The frequency of the reference signal is 14 MHz. Fig. 13 shows the measured VCO transfer function by varying the controlled voltage. The measured VCO has a monotonic frequency range of 435–485 MHz. The gain of the VCO is 32.4 MHz/V at the center frequency of 460 MHz. Fig. 14 shows the output signal spectrum (using HP8560A Spectrum Analyzer after locked) of 448 MHz with the phase noise −99 dBc/Hz at 100-kHz offset. By adding an external frequency doubler, however, the phase noise is −91 dBc/Hz at 100-kHz offset from 896-MHz carrier as shown in Fig. 15. Also, the measured waveform in the time domain is also shown in Fig. 16, and its rms and peak-to-peak jitter measured by CSA803 (Communication Signal Analyzer)

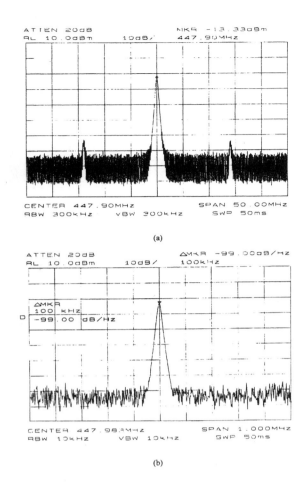

(a)

(b)

Fig. 14.   Measured output spectrum of the frequency synthesizer. (a) With span 50 MHz. (b) With span 1 MHz.

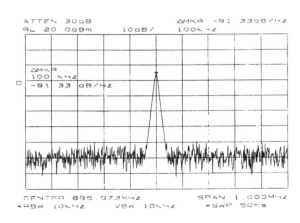

Fig. 15.   Measured output spectrum of the frequency synthesizer with added frequency doubler.

are 18.9 and 110 ps, respectively. Another important parameter is the time that the PLL takes to lock in to a new frequency when channel switches. Fig. 17 shows the switching waveforms for a frequency jump from 448 to 462 MHz from the HP53310A Modulation Domain Analyzer. Obviously, the DAPD can improve the switching speed of the PLL. The power consumption is 120 mW and the chip area is 40×2.0 mm^2 including the pad areas.

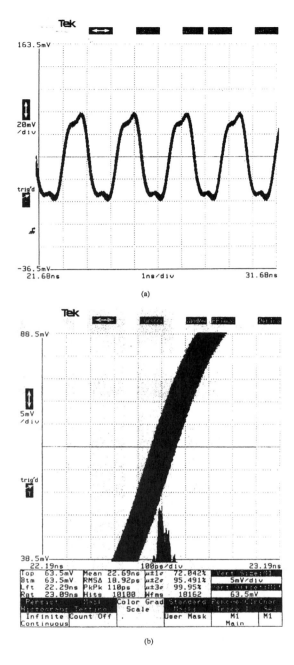

(a)

(b)

Fig. 16.   Measured waveform. (a) In time domain. (b) Jitter performance.

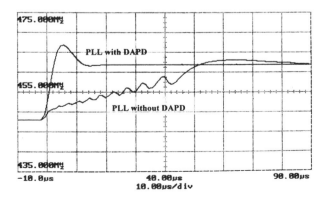

Fig. 17.   Measured frequency jump waveform of the frequency synthesizer.

V. Conclusion

In this paper, a PLL with the DAPD is implemented in a 0.35-μm CMOS process. The proposed DAPD can be applied to enhance the switching speed of the PLL, but maintain better noise bandwidth. When adding the DAPD in the PLL, it will control the charge pump and loop filter and still maintain the loop stablity with the same phase margin as in the steady state. The prototype frequency synthesizer using this structure is also implement at 448 MHz, and the output waveform is −99 dBc/Hz at 100-kHz offset. By adding a frequency doubler, the synthesizer can operate at 896 MHz, and the output waveform is −91 dBc/Hz at 100-kHz offset from carrier.

Acknowledgment

The authors would like to thank the SHARP Technology Company, Japan, for the fabrication of the chip.

References

[1] F. M. Gardner, *Phaselock Techniques*, 2nd ed. New York, NY: Wiley, 1979.

[2] P. Larsson, "Reduced pull-in time of phase-locked loops using a simple nonlinear phase detector," *IEE Proc. Commun.*, vol. 142, no. 4, pp. 221–226, Aug. 1995.

[3] D. H. Wolaver, *Phase-Locked Loop Circuit Design*. Englewood Cliffs, NJ: Prentice-Hall, 1991.

[4] F. M. Gardner, "Charge-pump phase-locked loops," *IEEE Trans. Commun.*, vol. COM-28, pp. 1849–1858, Nov. 1980.

[5] R. E. Best, *Phase-Locked Loops: Theory, Design and Applications*. New York, NY: McGraw-Hill, 1984.

[6] M. V. Paemel, "Analysis of a charge-pump PLL: A new model," *IEEE Trans. Commun.*, vol. 42, pp. 2490–2498, July 1994.

[7] C. Y. Yang, W. C. Chung, and S. I. Liu, "Effectively reduced pull-in time of PLL with nonlinear phase comparator," in *8th VLSI/CAD Symp.*, Taiwan, R.O.C., Aug. 1997, pp. 205–208.

[8] D. Byrd, C. Davis, and W. O. Keese, "A fast locking scheme for PLL frequency synthesizer," National Semiconductor, Santa Clara, CA, Application Note, July 1995.

[9] J. Yuan and C. Svensson, "Fast CMOS nonbinary divider and counter," *Electron. Lett.*, vol. 29, pp. 1222–1223, June 1993.

[10] S. Kim, K. Lee, Y. Moon, D. K. Jeong, Y. Choi, and H. K. Kim, "A 960-Mb/s/pin interface for skew-tolerant bus using low jitter PLL," *IEEE J. Solid-State Circuits*, vol. 32, pp. 691–700, May 1997.

[11] I. A. Young, J. K. Greason, and K. L. Wong, "A PLL clock generator with 5- to 110-MHz of lock range for microprocessors," *IEEE J. Solid-State Circuits*, vol. 27, pp. 1599–1607, Nov. 1992.

[12] Q. Huang and R. Rogenmoser, "Speed optimization of edge-triggered CMOS circuits for gigahertz single-phase clocks," *IEEE J. Solid-State Circuits*, vol. 31, pp. 456–465, Mar. 1996.

[13] B. Chang, J. Park, and W. Kim, "A 1.2- GHz CMOS dual-modulus prescaler using new dynamic D-type flip-flop," *IEEE J. Solid-State Circuits*, vol. 31, pp. 749–752, May 1996.

[14] P. Larsson, "High-speed architecture for a programmable frequency divider and a dual-modulus prescaler," *IEEE J. Solid-State Circuits*, vol. 31, pp. 744–748, May 1996.

[15] C. Y. Yang, G. K. Dehng, J. M. Hsu, and S. I. Liu, "New dynamic flip-flops for high-speed dual-modulus prescaler," *IEEE J. Solid-State Circuits*, vol. 33, pp. 1568–1571, Oct. 1998.

[16] F. Piazza and Q. Huang, "A low-power CMOS dual-modulus prescaler for frequency synthesizer," *IEICE Trans. Electron.*, vol. E80-C, pp. 314–319, Feb. 1997.

[17] S. J. Lee, B. Kim, and K. Lee, "A fully integrated low-noise 1-GHz frequency synthesizer design for mobile communication application," *IEEE J. Solid-State Circuits*, vol. 32, pp. 760–765, May 1997.

[18] D. Y. Jeong, S. H. Chae, W. C. Song, and G. H. Cho, "High-speed differential-voltage clamped current-mode ring oscillator," *Electron. Lett.*, vol. 33, pp. 1102–1103, June 1997.

Ching-Yuan Yang (S'97) was born in Miaoli, Taiwan, R.O.C., in 1967. He received the B.S. degree in electrical engineering from the Tatung Institute of Technology, Taipei, Taiwan, in 1990, and the M.S. and Ph.D. degrees in electrical engineering from National Taiwan University, Taipei, in 1996 and 2000, respectively.

He is currently and Assistant Professor with the Department of Electronic Engineering, Huafan University, Taiwan. His research interests are in the area of integrated circuits and systems for high-speed interfaces and wireless communications.

Shen-Iuan Liu (S'88–M'93) was born in Keelung, Taiwan, R.O.C., on April 4, 1965. He received the B.S. and Ph.D. degrees in electrical engineering from National Taiwan University, Taipei, Taiwan, in 1987 and 1991, respectively.

During 1991 to 1993, he served as a Second Lieutenant in the Chinese Air Force. During 1991 to 1994, he was an Associate Professor in the Department of Electronic Engineering, National Taiwan Institute of Technology. He joined the Department of Electrical Engineering, National Taiwan University, Taipei, Taiwan in 1994 and has been a Professor since 1998. He holds nine U.S. patents and fourteen R.O.C. patents, with some pending. His research interests are in analog and digital integrated circuits and systems.

Low-Power Dividerless Frequency Synthesis Using Aperture Phase Detection

Arvin R. Shahani, Derek K. Shaeffer, *Student Member, IEEE*, S. S. Mohan, *Student Member, IEEE*,
Hirad Samavati, *Student Member, IEEE*, Hamid R. Rategh, *Student Member, IEEE*,
Maria del Mar Hershenson, *Student Member, IEEE*, Min Xu, *Student Member, IEEE*,
C. Patrick Yue, *Student Member, IEEE*, Daniel J. Eddleman, *Student Member, IEEE*,
Mark A. Horowitz, *Senior Member, IEEE*, and Thomas H. Lee, *Member, IEEE*

Abstract—A phase-locked-loop (PLL)-based frequency synthesizer incorporating a phase detector that operates on a windowing technique eliminates the need for a frequency divider. This new loop architecture is applied to generate the 1.573-GHz local oscillator (LO) for a Global Positioning System receiver. The LO circuits in the locked mode consume only 36 mW of the total 115-mW receiver power, as a result of the power saved by eliminating the divider. The PLL's loop bandwidth is measured to be 6 MHz, with a reference spurious level of −47 dBc. The front-end receiver, including the synthesizer, is fabricated in a 0.5-μm, triple-metal, single-poly CMOS process and operates on a 2.5-V supply.

Index Terms—Frequency synthesizers, Global Positioning System, phase detection, phase-locked loops, radio-frequency integrated circuits, radio receivers.

I. INTRODUCTION

THE growing demand for portable, low-cost wireless-communication devices has spurred interest in radio-frequency integrated circuits. Part of offering a completely integrated solution involves identifying a low-power, monolithic gigahertz local oscillator (LO) implementation. A quartz-crystal-based oscillator cannot be used directly for the LO, since the fundamental modes of inexpensive quartz crystals are limited to approximately 30 MHz [1], and overtone orders of 50 are impractical. However, a crystal oscillator can be used as the reference in a static-modulus phase-locked-loop (PLL) frequency synthesizer. As is well known, the stability of the frequency-multiplied reference is retained by a wideband loop. This ability to synthesize a stable high-frequency source is beneficial, but it comes at the expense of significant power consumption. This paper addresses the power issue by introducing a new type of phase detector capable of phaselocking the synthesizer's frequency-multiplied output to its reference input, without the use of a divider. Eliminating the need for the divider allows the synthesis of a 1.573-GHz output on only 36 mW of power in this technology.

Section II examines the PLL-based LO used for the Global Positioning System (GPS) receiver architecture shown in Fig. 1 [2] and introduces the element that eliminates the need

Manuscript received May 7, 1998; revised August 4, 1998.
The authors are with the Center for Integrated Systems, Stanford University, Stanford, CA 94305 USA.
Publisher Item Identifier S 0018-9200(98)09432-3.

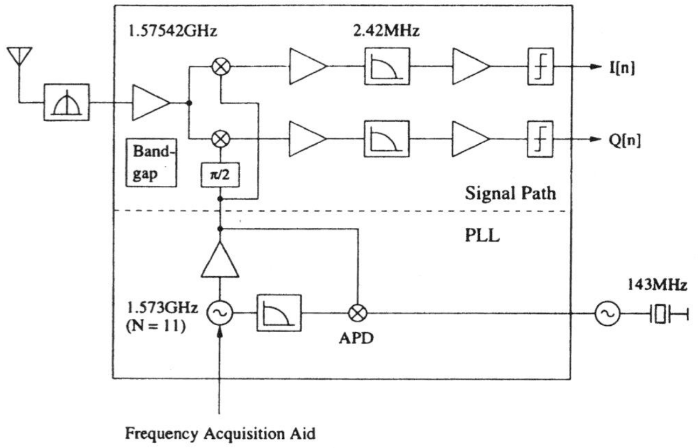

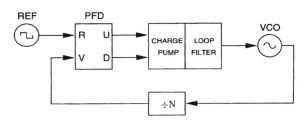

Fig. 1. GPS receiver architecture.

Fig. 2. Integer-N synthesizer block diagram.

for a divider: the aperture phase detector (APD). Treatment begins at the architectural level and descends into the APD's detailed nature. Both the theory and the implementation of an APD are covered. Section III presents experimental results on the APD PLL.

II. PLL

A. Architecture

The conventional and widely used implementation of the PLL frequency synthesizer with static modulus is the integer-N synthesizer [3]. The traditional divide-by-N block shown in Fig. 2 can be realized with a single counter. However, there are two drawbacks associated with the divider: power consumption and switching noise. Power consumption is large, particularly at high frequencies, because of the well-known $CV^2 f$ relationship. For example, a recently published 1.6-

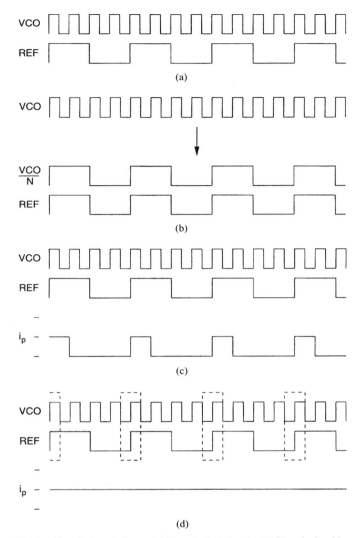

Fig. 3. Phaselock techniques. (a) Phaselocked signals. (b) Phaselock with a divider and PFD. (c) PFD along; negative charge pump current commands the VCO to decrease its frequency, breaking phaselock. (d) Phaselock with an APD.

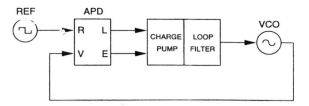

Fig. 4. APD synthesizer block diagram.

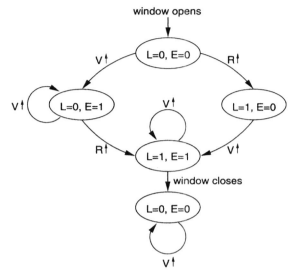

Fig. 5. Idealized APD state diagram.

GHz integer-N synthesizer built in a 0.6-μm CMOS technology reported a total power consumption of 90 mW, of which 22.5 mW were used by the divider [4]. A further disadvantage of the divider is the on-chip interference generated by its high-speed digital transitions. This is particularly worrisome if the synthesizer is to be integrated with the front end's sensitive low-noise amplifier.

To reduce power consumption and high-frequency noise, a windowing technique that eliminates the divide-by-N block for phase comparisons is investigated here. To appreciate how windowing may be of benefit, it is worthwhile to revisit the phenomenon of locking in a conventional PLL. To retain phaselock, it is necessary to align every Nth rising or falling edge of the voltage controlled oscillator (VCO) with a corresponding reference edge. Phaselock is demonstrated in Fig. 3(a) for $N = 4$, where every fourth rising VCO edge lines up with a rising reference edge. A divider with a phase-frequency detector (PFD) accomplishes edge alignment by first dividing down the VCO by the right multiple so that edge

alignment is unambiguous, as pictured in Fig. 3(b). Because the PFD compares phase over the entire reference cycle, a PFD cannot phaselock two inputs at different frequencies. In fact, it is precisely this property that makes the PFD popular.

Now consider using a PFD without a divider. Clearly, there would be an edge ambiguity problem, rendering the PFD quite ineffective, as seen in Fig. 3(c). The reason is that the PFD responds to every edge of the VCO, evidenced by the charge pump current's net negative value. This erroneously commands the VCO to decrease its frequency. However, by restricting the time interval during which phase is examined, one may eliminate the edge ambiguity, and hence the frequency divider. The dashed boxes in Fig. 3(d) define the window during which phase may be compared, even if the two inputs are of different frequency. The window can be controlled by the reference time base, since it periodically opens at that rate. Furthermore, the window need only be wide enough so that a VCO edge falls within it, which is equivalent to requiring that the window be active for a time longer than the instantaneous VCO period. No dividers are thus necessary to maintain phaselock, and this phase detector, called an APD, can operate with two inputs that are at different frequencies, as shown in Fig. 4.

A more substantive description of the APD's operation is provided in Fig. 5, which illustrates the state diagram for an idealized APD. When the window opens, the phase detector becomes active. The R-input rising edge sets the L (denoting "late") terminal true, and the V-input rising edge sets the

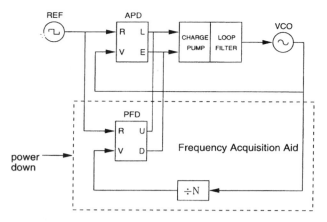

Fig. 6. APD synthesizer block diagram with integer-N FAA.

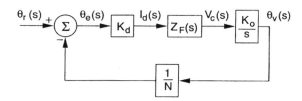

Fig. 7. APD PLL block diagram in lock.

E (denoting "early") terminal true. Subsequent edges of the V-input are ignored until the next window opens. The time difference between the rising edges of the L and E signals is proportional to the phase error between the reference phase and the VCO phase. If L is set first, the VCO phase is late; and conversely, if E is set first, the VCO phase is early. When the window closes, the L and E terminals are reset (to false).

Fig. 1 shows that some type of frequency acquisition aid (FAA) is required to bring an APD-based loop initially into lock. This necessity is a consequence of restricting phase comparisons to a window, which eliminates the phase detector's ability to perform frequency detection. This issue is discussed in further detail in Section II-D. For this work, an external acquisition aid was used for experimental purposes. An integrated implementation of the acquisition aid, Fig. 6, uses the traditional divider with PFD to lock the loop and then powers down the acquisition aid, transferring control to the low-power APD. An APD can be used once in lock because the reference is derived from a stable crystal oscillator.

B. Loop Theory

Having provided an overview of APD operation, we now develop a linearized APD PLL model relating input and output phase. This model is important for quantitative loop design and ensures that the synthesized output has the desired stability and noise performance.

From the description of the late and early APD signals given in the previous subsection, the average charge pump current over one reference cycle is given by

$$i_d = I_p(t_v - t_r)\frac{\omega_r}{2\pi} \tag{1}$$

where I_p is the magnitude of the charge pump current, t_v is the time of the first VCO rising edge in the window, t_r is the time of the reference rising edge in the window, and ω_r is the angular reference frequency. The current i_d can be expressed as a function of the reference and VCO phases by relating these phases to t_r and t_v, assuming small phase errors. Expressions relating edge time to signal phase are

$$t_r = -\frac{\theta_r}{\omega_r} \tag{2}$$

where θ_r is the reference phase and

$$t_v = -\frac{\theta_v}{\omega_v} \tag{3}$$

where θ_v is the VCO phase and ω_v is the angular VCO frequency. The average charge pump current over one reference cycle can thus be written as

$$i_d = I_p(t_v - t_r)\frac{\omega_r}{2\pi} = \frac{I_p}{2\pi}\left(\theta_r - \theta_v\frac{\omega_r}{\omega_v}\right). \tag{4}$$

When the loop is in lock, $\omega_v = N\omega_r$, giving

$$i_d = \frac{I_p}{2\pi}\left(\theta_r - \frac{\theta_v}{N}\right) = \frac{I_p}{2\pi}\theta_e = K_d\theta_e \tag{5}$$

where K_d is the phase-detector gain constant. Note that even though there is no explicit divider in the loop, the VCO phase is divided by N in (5), just as in a conventional loop.

This model can be used in place of the APD in Fig. 4, and the other blocks in the same figure can be replaced by their corresponding linear time-invariant (LTI) models, yielding the overall system model shown in Fig. 7. Fig. 7 is an LTI representation of the APD PLL in lock, from which the phase transfer function is readily found to be

$$H(s) = \frac{\theta_v}{\theta_r} = \frac{NK_dK_oZ_F(s)}{Ns + K_dK_oZ_F(s)} \tag{6}$$

where K_o is the VCO gain constant and $Z_F(s)$ is the loop filter's impedance, expressed in the s-domain.

C. APD Characteristic (i_d Versus θ_e)

The derivation in the previous subsection treats the APD for small phase errors. For completeness, it is instructive to examine the response of the APD to arbitrary phase errors. Now, the delay between the time the window opens and the time at which the reference edge occurs becomes important. This delay is designated by d, which is a positive quantity whose least restrictive range is limited to $[0, T_r)$, where T_r is the reference period. However, the loop can lock if and only if d is in the interval $[0, T_v]$, where T_v is the VCO period. Otherwise, the first VCO edge within the window will always precede the reference edge.

From Fig. 8, it is apparent that the characteristic will be periodic in VCO phase, because when the VCO waveform has moved one VCO period to the right, the situation is identical to the start. As the VCO waveform moves to the right, the time difference $t_v - t_r$ varies proportionally with phase error θ_e. Therefore, to find the APD's characteristic, θ_e and i_d need to be calculated at only two points, and the remainder of the

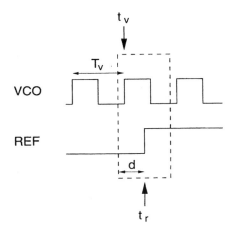

Fig. 8. Position of VCO and reference edge in window.

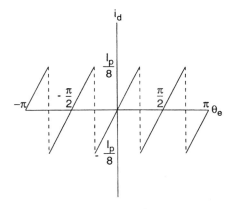

Fig. 10. APD characteristic for $d = (T_v)/2$ and $N = 4$.

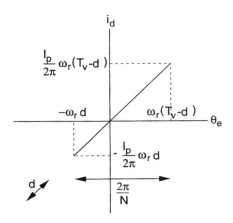

Fig. 9. APD characteristic over $(2\pi)/N$ interval.

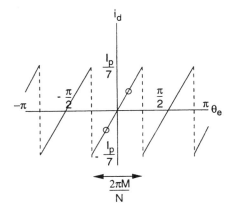

Fig. 11. $M = 2$. $N = 7$ subharmonic-lock mode.

characteristic is generated by connecting these endpoints. θ_e and i_d are first calculated for $t_v - t_r = -d$

$$\theta_e = -\omega_r d \qquad (7)$$
$$i_d = -\frac{I_p}{2\pi}\omega_r d. \qquad (8)$$

Next, θ_e and i_d are calculated at the other extreme where $t_v - t_r = T_v - d$

$$\theta_e = \omega_r(T_v - d) \qquad (9)$$
$$i_d = \frac{I_p}{2\pi}\omega_r(T_v - d). \qquad (10)$$

From this information, the portion of the APD characteristic shown in Fig. 9 can be constructed.

The influence of two parameters—d, the delay between the time the window opens and when the reference edge occurs, and N, the ratio between the VCO and reference frequencies—warrants special attention. Decreasing d shifts the characteristic diagonally up (along the line of the characteristic), and increasing d shifts the characteristic diagonally down. It is desirable to have d equal to half the synthesized frequency's period. By designing for this condition, the APD characteristic will be centered about $i_d = 0$ to provide a symmetrical correction range. The parameter N affects the

phase error's periodicity, with larger values increasing the periodicity. Fig. 10 shows the complete APD characteristic (a 2π variation in θ_e) for the specific case where $d = \frac{T_v}{2}$ and $N = 4$.

D. Subharmonic-Lock Modes

The existence of *subharmonic*-lock modes explains the need for an acquisition aid. During each window, which opens periodically at the reference rate, the APD makes a single phase comparison. It is this property that allows an APD to phaselock the VCO's output to an integer multiple of the reference input. But the ability to examine the phase of two signals at different frequencies introduces more modes than just the desired integer-lock modes. Additional subharmonic-lock modes occur if the net current delivered over multiple cycles of the reference is zero, allowing the loop to stay locked at an undesired frequency [5].

If we designate by M the number of reference cycles over which the net charge delivered to the loop filter is zero, then an expression relating the reference frequency to the VCO frequency when phaselock occurs is $M\omega_v = N\omega_r$. Fig. 11 displays the points on the APD characteristic between which the loop ping-pongs for the specific case where $M = 2$, and $N = 7$. Because $M = 2$, the charge pump alternates between pumping up on one cycle and pumping down on the next cycle, balancing the charge to the loop filter over two cycles.

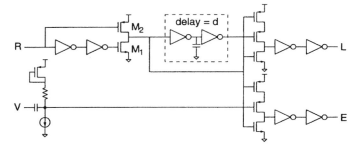

Fig. 12. APD circuit diagram.

These subharmonic-lock modes are problematic because they are spaced, in frequency, closer than the neighboring integer-lock modes. However, the APD favors integer over subharmonic modes for two reasons. First, the loop's bandwidth imposes a limit on M. If the number of reference cycles over which the charge pump current averages to zero grows too large, the loop will act on partial information because the loop responds to signals averaged over a loop period. The loop period is the reciprocal of the closed-loop bandwidth. Another reason the APD favors integer over subharmonic modes is that the subharmonic modes have a lower detector gain K_d because the VCO edge arrives at a different time in each of the M cycles. If the APD characteristic is nonlinear, then the overall detector gain is the average of the M individual linearized detector gains.

Using an FAA to ensure frequency lock eliminates the concern of locking in a subharmonic mode. Once lock has been achieved, and control transferred to the APD, the APD is capable of maintaining lock at the desired frequency.

E. APD Circuit Implementation

Fig. 12 shows a circuit implementation of an APD. The reference clock (which has about a 50% duty cycle) is shaped by the structure preceding the delay to have fast falling edges, since these are the edges that enable the precharged gates. When the reference input is low, M_1 is off and M_2 is on, causing the output to be high. After the reference rises, M_2 shuts off before M_1 turns on due to the two inverter delays. Therefore, M_1 does not fight M_2 to pull the output low, and creates a fast falling edge. The window opens on this fast falling edge. The delay d between the opening of the window and the reference edge is determined by two inverters with a capacitor in the middle. The APD uses two precharged gates to evaluate the reference and VCO phases. An advantage of precharged gates is that they only respond once while active. In this case, the precharged gates are precharged low, and rise on detection of low levels. Also, the precharge action does not affect the loop to first order, because the state $L = 1$, $E = 1$ has the same action as the state $L = 0$, $E = 0$.

The behavior of this APD circuit differs somewhat from the ideal APD discussed in Section II-A. In particular, the circuit implementation responds to falling edges instead of rising edges, and more precisely, the precharged gates act as level detectors of a low voltage level instead of as edge detectors.

A simulation of the APD characteristic over a $\frac{2\pi}{N}$ interval is shown in Fig. 13, where $N = 11$. The phase error θ_e is plotted against the average charge pump current over one reference cycle i_d, and includes the nonidealities of the charge pump as well. The flat section near -0.2 rad is where the E signal driving the charge pump is compressing due to the level detection nature of the precharged gates. Another imperfection in this circuit's APD characteristic is the section with finite negative slope instead of a discontinuity. From the characteristic, the phase detector's gain constant K_d in a state of zero static phase error ($\theta_e = 0$) is evaluated to be 7.4 μA/rad.

F. PLL Circuit Implementation

In Section II-B, a general model for a locked APD PLL was developed, expressing the closed-loop phase transfer function in terms of the loop filter's s-domain impedance and an idealized VCO. We now provide specific expressions for the loop as actually implemented.

The loop filter used is the conventional network shown in Fig. 14, whose s-domain impedance is

$$Z_F(s) = \frac{1 + sRC_1}{(C_1 + C_2)s\left(1 + sR\frac{C_1 C_2}{C_1 + C_2}\right)}. \quad (11)$$

A single-pole amplifier was used to interface with the VCO's varactor, thus the ideal VCO transfer function $(K_o)/s$ must be modified to

$$\frac{K_o}{s\left(1 + \frac{s}{2\pi f_{3\,\mathrm{dB}}}\right)} \quad (12)$$

where $f_{3\,\mathrm{dB}}$ is the 3-dB bandwidth of the VCO's preamplifier. Using (11) and (12) in (6) enables us to write the complete phase transfer function of the implemented APD PLL as shown in (13) at the bottom of the page. In the next section, we compare measured data to (13).

III. Experimental Results

A test chip (see Fig. 15) containing a copy of the APD PLL used in the complete GPS receiver is used to evaluate the APD PLL. Two separate tests are performed; one to verify the derived closed-loop transfer function of the APD PLL and the other to observe the synthesized LO spectrum for the GPS receiver. In the second test, the synthesized LO is also checked

$$H(s) = \frac{N K_d K_o (1 + sRC_1)}{N(C_1 + C_2)s^2\left(1 + \frac{s}{2\pi f_{3\,\mathrm{dB}}}\right)\left(1 + sR\frac{C_1 C_2}{C_1 + C_2}\right) + K_d K_o (1 + sRC_1)}. \quad (13)$$

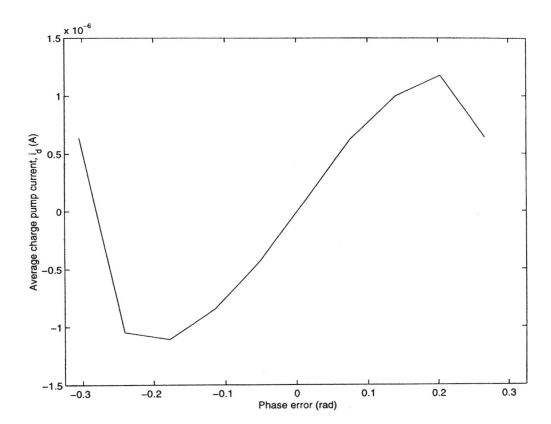

Fig. 13. Simulated APD circuit characteristic.

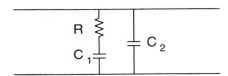

Fig. 14. Loop filter used in APD PLL.

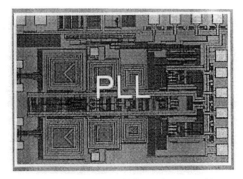

Fig. 15. Die photograph of PLL on GPS receiver test chip.

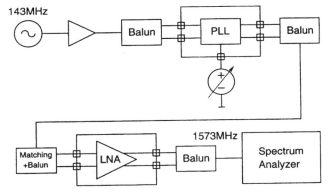

Fig. 16. PLL test setup number 1.

with a microwave frequency counter to verify its long-term stability.

Fig. 16 shows the experimental setup for the first test. Phase noise is measured for offsets from 1 kHz to 10 MHz with the HP8563E spectrum analyzer, which has special phase-noise-measurement software. Ten MHz is used as the upper limit since the loop is designed to have a bandwidth less than 10 MHz. Beyond the loop's bandwidth, the PLL's phase noise is determined by the VCO's phase noise, making measurement of the PLL's transfer function difficult. One of the largest factors affecting measurement accuracy is the noise floor of the instrument. To minimize this error source, measurements of the floor with a clean source are performed first. These results are later used to calibrate the data. Reference phase noise and PLL phase noise are also both measured. After some data processing, the PLL's closed-loop phase transfer function $H(s)$ is determined.

Fig. 17 shows the measured $|H(f)|$ and the predicted $|H(f)|$, from (13), for the case where the reference frequency is 143 MHz and the VCO frequency is 1.573 GHz ($N = 11$). The seven loop parameters in (13) are set as follows: N is known; K_o is taken from measured VCO data; R, C_1, and

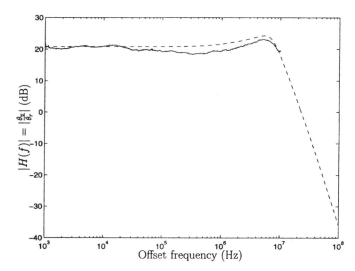

Fig. 17. Measured and predicted $|H(f)|$.

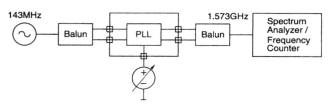

Fig. 18. PLL test setup number 2.

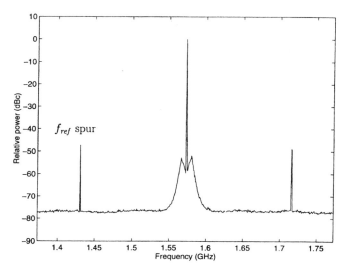

Fig. 19. LO spectrum.

TABLE I
MEASURED APD PLL PERFORMANCE

Synthesized frequency	1.573 GHz
Reference frequency	143 MHz
Loop bandwidth	6 MHz
f_{ref} spur	$\leq$ -45 dBc
$2 * f_{ref}$ spur	$\leq$ -55 dBc
VCO power consumption	26 mW
Total power consumption	36 mW (2.5 V supply)
Die area	3.1 mm^2
Technology	0.5 μm CMOS

C_2 are taken to be their designed loop filter values; $f_{3\,\mathrm{dB}}$ is calculated from the technology data; and K_d is fit. The fit value of K_d, 6.6 μA/rad, is a little less than the simulated value noted in Section II-E, 7.4 μA/rad. Since $K_d = (I_p)/(2\pi)$ for an ideal APD, one could argue that the discrepancy in K_d is due to an actual pump current that is lower than the pump current used in simulations. But when I_p is measured, it is found to be correct. Still, the discrepancy in K_d is readily explained. The simulation in Fig. 13 establishes an upper bound on K_d because it is measured in a state of zero static phase error. The simulated APD circuit characteristic illustrates that the detector gain (i.e., the slope) decreases the farther that one departs from zero radians. The charge pump is known to have some offset; thus, the loop has some static phase error in lock to overcome the offset, resulting in a slightly lower K_d.

Fig. 18 shows the experimental setup for the second test. The LO spectrum is measured with the HP8563E spectrum analyzer, and the frequency is checked with an HP5350B microwave frequency counter. Fig. 19 displays the synthesized output spectrum, in which the PLL's ability to track the low close-in phase noise of the reference can be seen. The visible skirts are due to the VCO's phase noise outside the 6-MHz bandwidth of the PLL. Spurious tones at -47 dBc are primarily due to control-line ripple resulting from charge pump leakage. In GPS applications, the measured spurious level is acceptable because of the absence of blockers at the corresponding offset frequencies. In more demanding applications, one may reduce ripple through improved charge pump design and the use of analog phase interpolation [3].

Table I provides a summary of the APD PLL's performance.

The PLL has a wide bandwidth of 6 MHz, and the APD circuit consumes only one-quarter of the total synthesizer power. With the elimination of the divider, the main power consumer in the synthesizer is now the VCO.

IV. CONCLUSION

A new method for performing phase detection that eliminates the divide-by-N function within a PLL has been presented. A frequency acquisition aid circuit, which can be powered down once lock is established, is required. By using an aperture phase detector, a 1.573-GHz local oscillator can be synthesized on roughly half the power of a loop containing a conventional divider. Additionally, elimination of the divider also reduces the frequency of transitions that might cause substrate and supply bounce. The power savings and noise reduction make the APD PLL an attractive design for low-power, integrated frequency synthesizers.

ACKNOWLEDGMENT

The authors gratefully acknowledge Rockwell International for fabricating the receiver and Dr. C. Hull and Dr. P. Singh for their valuable assistance. In addition, the authors acknowledge Tektronix, Inc., for supplying simulation tools and E. McReynolds for his invaluable support of, and assistance with, CMOS modeling issues. Last, the authors thank IBM for generous student support through IBM fellowships.

REFERENCES

[1] *M-tron Engineering Notes*, Dec. 1997.
[2] D. K. Shaeffer, A. R. Shahani, S. S. Mohan, H. Samavati, H. R. Rategh, M. Hershenson, M. Xu, C. P. Yue, D. J. Eddleman, and T. H. Lee, "A 115-mW, 0.5-μm CMOS GPS receiver with wide dynamic-range active filters," *IEEE J. Solid-State Circuits*, vol. 33, pp. 2219–2231, Dec. 1998.
[3] T. H. Lee, *The Design of CMOS Radio-Frequency Integrated Circuits.* Cambridge: Cambridge Univ. Press, 1998.
[4] J. F. Parker and D. Ray, "A 1.6 GHz CMOS PLL with on-chip loop filter," *IEEE J. Solid-State Circuits*, vol. 33, pp. 337–343, Mar. 1988.
[5] F. M. Gardner, *Phaselock Techniques*, 2nd ed. New York: Wiley, 1979.

Arvin R. Shahani, for a photograph and biography, see this issue, p. 2041.

Derek K. Shaeffer (S'98), for a photograph and biography, see this issue, p. 2230.

S. S. Mohan (S'98), for a photograph and biography, see this issue, p. 2231.

Hirad Samavati (S'98), for a photograph and biography, see this issue, p. 2041.

Hamid R. Rategh (S'98), for a photograph and biography, see this issue, p. 2231.

Maria del Mar Hershenson (S'98), for a photograph and biography, see this issue, p. 2231.

Min Xu (S'97), for a photograph and biography, see this issue, p. 2231.

C. Patrick Yue (S'93), for a photograph and biography, see this issue, p. 2231.

Daniel J. Eddleman (S'98), for a photograph and biography, see this issue, p. 2231.

Mark A. Horowitz (S'77–M'78–SM'95) received the B.S. and M.S. degrees in electrical engineering from the Massachusetts Institute of Technology, Cambridge, in 1978 and the Ph.D. degree from Stanford University, Stanford, CA, in 1984.

He is the Yahoo Founders Professor of Electrical Engineering and Computer Science at Stanford. His research area is in digital system design. He has led a number of processor designs including MIPS-X, one of the first processors to include an on-chip instruction cache; TORCH, a statistically scheduled, superscalar processor; and FLASH, a flexible DSM machine. He has also worked on a number of other chip design areas, including high-speed memory design, high-bandwidth interfaces, and fast floating point. In 1990, he took a leave from Stanford to help start Rambus, Inc., a company designing high-bandwidth memory interface technology. His current research includes multiprocessor design, low-power circuits, memory design, and high-speed links.

Dr. Horowitz received a 1985 Presidential Young Investigator Award and an IBM Faculty Development Award, as well as the 1993 Best Paper Award from the International Solid-State Circuits Conference.

Thomas H. Lee (S'87–M'87), for a photograph and biography, see this issue, p. 2041.

573

A Stabilization Technique for Phase-Locked Frequency Synthesizers

Tai-Cheng Lee and Behzad Razavi
Electrical Engineering Department
University of California, Los Angeles

Abstract

A stabilization technique is presented that relaxes the trade-off between the settling speed and the magnitude of output sidebands in phase-locked frequency synthesizers. The method introduces a zero in the open-loop transfer function through the use of a discrete-time delay cell, obviating the need for resistors in the loop filter. A 2.4-GHz CMOS frequency synthesizer employing the technique settles in approximately 60 μs with 1-MHz channel spacing while exhibiting a sideband magnitude of -58.7 dBc. Designed for Bluetooth applications and fabricated in a 0.25-μm digital CMOS technology, the synthesizer achieves a phase noise of -112 dBc/Hz at 1-MHz offset and consumes 20 mW from a 2.5-V supply.

I. Introduction

Phase-locked loops (PLLs) typically suffer from a trade-off between the settling time and the ripple on the control voltage, limiting the performance that can be achieved in terms of channel switching speed and output sideband magnitude in RF synthesizers. This paper describes a loop stabilization technique that yields a small ripple while achieving fast settling. Using a discrete-time delay cell, the PLL architecture creates a zero in the open-loop transfer function. Another important advantage of the technique is that it uses no resistors in the loop filter, lending itself to digital CMOS technologies.

The next section of the paper develops the foundation for the proposed technique. Section III describes the synthesizer architecture and the design of its building blocks and Section IV proposes fast simulation techniques for the synthesizer. Section V summarizes the experimental results.

II. Stabilization Technique

Consider the PLL shown in Fig. 1, where a voltage-controlled oscillator (VCO) is driven by a charge pump (CP) and a phase-frequency detector (PFD). Resistor R_1 provides the stabilizing zero and capacitor C_2 suppresses the glitch generated by the charge pump on every phase comparison instant. The glitch arises from mismatches between the width of Up and Down pulses produced by the PFD as well as charge injection and clock feedthrough mismatches between PMOS and NMOS devices in the charge pump.

The principal drawback of this architecture is that C_1 determines the settling whereas C_2 controls the ripple on the control voltage. Since C_2 must remain below C_1 by roughly a factor of 10 so as to avoid underdamped settling, the loop must be slowed down by a large C_1 if C_2 is to yield a sufficiently small ripple. It is therefore desirable to seek methods of creating the stabilizing zero without the resistor so that the capacitor that defines the switching speed also directly suppresses the ripple.

It is important to note that the problem of ripple becomes increasingly more serious as the supply voltage is scaled down and/or the operating frequency goes up. The relative magnitude of the primary sidebands at the output of the VCO is given by $A_m K_{VCO}/(2\omega_{REF})$ where A_m is the peak amplitude of the first harmonic of the ripple, K_{VCO} is the gain of the VCO, and ω_{REF} is the synthesizer reference frequency. [For a given relative tuning range (e.g. $\pm 10\%$), the gain of LC VCOs must increase if the supply voltage goes down.] If $K_{VCO} = 100$ MHz/V and $f_{REF} = 1$ MHz, then the fundamental ripple amplitude must be less than 63 μV to guarantee sidebands 60 dB below the carrier.

In order to arrive at the stabilization technique, consider the PLL architecture shown in Fig. 2. Here, the primary charge pump, CP_1, drives a single capacitor C_1 while a secondary

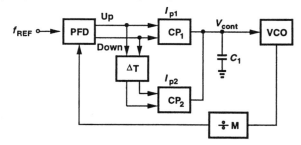

Fig. 2. Proposed PLL architecture with delayed charge pump circuit.

charge pump, CP_2, injects charge after some delay ΔT. The total current flowing through C_1 is thus equal to

$$I_p = I_{p1} + I_{p2}e^{-s\Delta T} \qquad (1)$$

$$\approx I_{p1} + I_{p2}(1 - s\Delta T), \qquad (2)$$

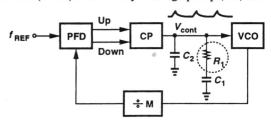

Fig. 1. Conventional PLL architecture.

where ΔT is assumed to be much smaller than the loop time constant. Consequently, the transfer function of the PFD/CP/LPF combination can be expressed as

$$\frac{V_{cont}}{\Delta\phi}(s) = \frac{I_{p1} + I_{p2}}{2\pi C_1 s} - \frac{I_{p2}}{2\pi C_1}\Delta T. \qquad (3)$$

Assuming $I_{p2} = -\alpha I_{p1}$, we have

$$\frac{V_{cont}}{\Delta\phi}(s) = \frac{I_{p1}}{2\pi}\left(\frac{1-\alpha}{C_1 s} + \frac{\alpha\Delta T}{C_1}\right), \qquad (4)$$

obtaining a zero at

$$\omega_z = \frac{1-\alpha}{\alpha}\frac{1}{\Delta T}. \qquad (5)$$

In order to achieve a sufficiently low zero frequency, ΔT must be large or α close to unity. Since the accuracy in the definition of α is limited by mismatches between the two charge pumps, ΔT must still be a large value. For example, if $f_{REF} = 1$ MHz, and $\alpha = 0.9$, then a ΔT of approximately 200 ns is required to ensure a well-behaved loop response.

The architecture of Fig. 2 suffers from a critical drawback: it requires a very long delay before CP_2 while the Up and Down pulses generated by the PFD can assume a very narrow width during lock. If the bandwidth of each stage in the delay line is reduced so as to produce a large delay, then the narrow Up and Down pulses are heavily attenuated, thus giving rise to a dead zone. Conversely, if the bandwidth of each stage is wide enough to support such pulses, then a very large number of stages is required to obtain the necessary ΔT.

To resolve the above difficulty, the architecture is modified as shown in Fig. 3(a), where a discrete-time analog delay line is placed after CP_2 and C_2. The delay network is realized as depicted in Fig. 3(b), consisting of two interleaved master-slave sample-and-hold branches operating at half of the reference frequency. The circuit emulates ΔT as follows. When CK_{even} is high, C_{s1} shares a charge packet corresponding to the previous phase comparison with C_1 while C_{s2} samples a level proportional to the present phase difference. In the next period, C_{s1} and C_{s2} exchange roles. The interleaved sampling network therefore provides a delay equal to the reference period, $1/f_{REF}$.

The discrete-time delay technique of Fig. 3 allows a precise definition of the zero frequency without the use of resistors. To quantify the behavior of a PLL incorporating this method, we assume the loop settling time is much greater than $1/f_{REF}$ so that the delay network can be represented by the continuous-time model shown in Fig. 4. Here, $R_{eq} = (f_{REF}C_s)^{-1}$ approximates the interleaved branches. Equation (4) can then be rewritten as

$$\frac{V_{cont}}{\Delta\phi}(s) = \frac{I_{p1}}{2\pi}\left(\frac{1}{f_{REF}C_s} + \frac{I_{p1} + I_{p2}}{I_{p1}}\frac{1}{sC_2}\right), \qquad (6)$$

where it is assumed $C_2 \gg C_1$ and the current through C_1 is neglected. This equation exhibits two interesting properties. First, if $I_{p2} = -\alpha I_{p1}$, then $(I_{p1} + I_{p2})/I_{p1} = 1 - \alpha$ and

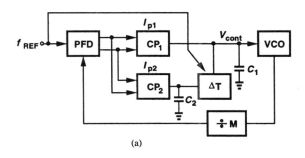

(a)

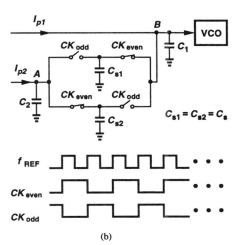

(b)

Fig. 3. Actual implementation of PLL with delay sampling circuit.

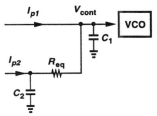

Fig. 4. Continuous-time approximation of delay network.

the value of C_2 is "amplified" by $(1-\alpha)^{-1}$. For example, if $\alpha = 0.9$, then C_2 is multiplied by a factor of 10, saving substantial area. Second, the zero frequency is equal to

$$\omega_Z = \left(1 + \frac{I_{p2}}{I_{p1}}\right)\frac{C_s}{C_2}f_{REF}, \qquad (7)$$

a value independent of process and temperature. Assuming $C_2 = C_s$ and $I_{p1} = -\alpha I_{p2} = I_p$, we obtain the damping factor and the settling time constant of the loop as:

$$\zeta = \frac{1}{2(1-\alpha)f_{REF}}\sqrt{\frac{I_p}{2\pi C_2}\frac{K_{VCO}}{M}} \qquad (8)$$

$$(\zeta\omega_n)^{-1} = \frac{4\pi f_{REF}C_2}{I_p}\frac{M}{K_{VCO}}. \qquad (9)$$

The key observation here is that the factor $(1-\alpha)^{-1}$ appears in ζ but not in $(\zeta\omega_n)^{-1}$. Thus, the two parameters can be optimized relatively independently. Furthermore, the damping factor exhibits much less process and temperature dependence than in the conventional loop of Fig. 1. Note that for $I_{p2} = 0$, the proposed circuit resembles the topology of Fig. 1 but with the resistor replaced with a switched-capacitor network.

For RF synthesis, the delay network of Fig. 2 must be designed carefully so as to minimize ripple on the control voltage. Since in the locked condition, the voltages at nodes A and B are nearly equal, the charge sharing between C_{s1} or C_{s2} and C_1 creates only a small ripple. Furthermore, the switches in the delay stage are realized as small, complementary devices to introduce negligible charge injection and clock feedthrough.

III. SYNTHESIZER DESIGN

A 2.4-GHz CMOS synthesizer targeting Bluetooth applications has been designed using the stabilization technique described above. This section presents the architecture and building blocks of the synthesizer.

Shown in Fig. 5, the synthesizer uses an integer-N architecture with a feedback divider whose modulus is given by $M = NP + S + 1$, where $N = 4$, $P = 600$, and $S =$

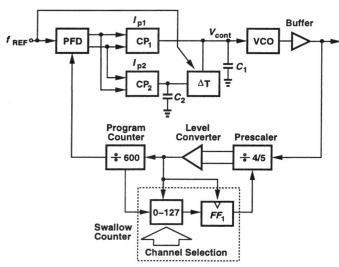

Fig. 5. Synthesizer architecture.

0-127. With $f_{REF} = 1$ MHz, the output frequency covers the 2.4-GHz ISM band. The output of the swallow counter is pipelined by the flipflop FF_1 to allow a relaxed design for the level converter and the swallow counter [1]. The buffer following the VCO suppresses the kickback noise of the prescaler when the modulus changes. It also avoids limiting the tuning range of the VCO by the input capacitance of the prescaler.

The VCO topology is shown in Fig. 6(a). To provide both negative and positive voltages across the MOS varactors, the sources of M_1 and M_2 are grounded and the circuit is biased on top by I_{DD}. The inductors are realized as shown in Fig. 6(b), with the bottom spiral moved down to metal 2 so as to reduce the parasitic capacitance [2].

The prescaler must divide the 2.4-GHz signal while consuming a small power dissipation. Depicted in Fig. 7, the

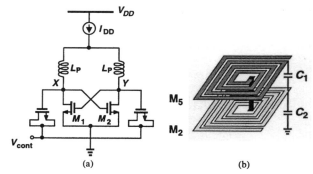

Fig. 6. (a) LC oscillator, (b) two-layer stacked inductor.

circuit employs three current-steering flipflops with diode-

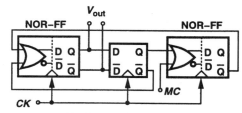

Fig. 7. Prescaler.

connected loads. The use of NOR gates obviates the need for power- and headroom-hungry level shift circuits required in NAND implementations.

IV. SIMULATION TECHNIQUES

A 2.4-GHz synthesizer with a reference frequency of 1 MHz requires a transient simulation step of approximately 20 ps for a total settling time on the order of 100 μs. The simulation therefore requires an extremely long time owing to both the vastly different time scales and the large number of devices (especially in the divider).

In order to study the loop dynamics with realistic transistor-level PFD, CP and VCO design, two speed-up techniques have been employed. First, the reference frequency is scaled up by a factor of 100 and the loop filter capacitor and the divide ratio are scaled down by the same factor. Since the PFD operates reliably at 100 MHz with no dead zone, this method directly reduces the simulation time by a factor of 100. From Eqs. (8) and (9), we note that scaling C_2 and M by 100 maintains a constant damping factor while scaling the settling time by 100.

Second, the divider is realized as a simple behavioral model in HSPICE that uses a handful of ideal devices and its complexity is independent of the divide ratio. Illustrated in Fig. 8, the principle of the behavioral divider is to pump a well-defined charge packet into an integrator in every period and reset the integrator when its output exceeds a certain level, V_{REF}. Using an ideal op amp, comparator, and switches with proper choice of V_b and V_{REF}, the circuit can achieve arbitrarily long high divide ratios. This techniques yields another factor of 20 improvement in the simulation speed, allowing the synthesizer to be simulated in less than 3 minutes on an Ultra 10 Sun workstation.

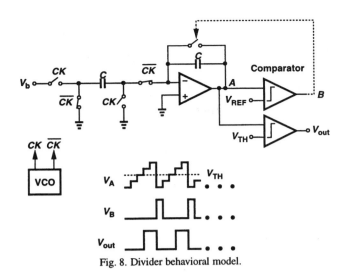

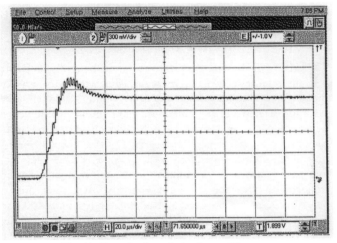

Fig. 8. Divider behavioral model.

V. EXPERIMENTAL RESULTS

The frequency synthesizer has been fabricated in a digital 0.25-μm CMOS technology. Shown in Fig. 9 is a photograph of the die, whose active area measures 0.65 mm × 0.45 mm.

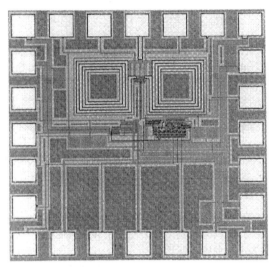

Fig. 9. Die photo.

The circuit has been tested in a chip-on-board assembly while running from a 2.5-V power supply.

Figure 10 shows the output spectrum in the locked condition. The phase noise is equal to −112 dBc/Hz at 1 MHz offset, well exceeding the Bluetooth requirement. The primary reference sidebands are at approximately −58.7 dBc. This level is lower than that achieved in [3] with differential VCO control and an 86.4-MHz reference frequency.

Figure 11 plots the measured settling behavior of the synthesizer when its channel number is switched by 64. The settling time is about 60 μs, i.e., 60 input cycles.

REFERENCES

[1] C. Lam and B. Razavi, "A 2.6-GHz/5.2-GHz Frequency Synthesizer in 0.4-μm CMOS Technology," *IEEE J. Solid-State Circuits*, vol. 35, pp. 788-794, May 2000.

[2] A. Zolfaghari, A. Chan, and B. Razavi, "Stacked Inductors and 1-to-2 Transformers in CMOS Technology," *Proc. of CICC*, pp. 345-348, May 2000.

[3] L. Lin, L. Tee, and P. R. Gray, "A 1.4-GHz Differential Low-Noise CMOS Frequency Synthesizer using a Wideband PLL Architecture," *ISSCC Dig. Tech. Paper*, pp. 204-205, Feb. 2000.

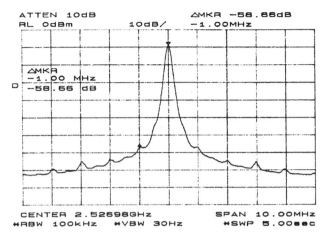

Fig. 10. Measured spectrum at the output of VCO.

Fig. 11. Control voltage during loop settling.

Center Frequency	2.4 GHz
Channel Spacing	1 MHz
No. of Channels	128
Phase Noise at 1 MHz Offset	−112 dBc/Hz
Reference Sidebands	−58.7 dBc
Settling Time	60 μs
Power Dissipation	
VCO	10 mW
VCO Buffer	3 mW
Divider	6 mW
Charge Pump	0.5 mW
Others	0.5 mW
Total	20 mW
Supply Voltage	2.5 V
Die Area	0.65 mm x 0.45 mm
Technology	0.25–μm CMOS

Table 1. Performance summary.

A Modeling Approach for Σ–Δ Fractional-N Frequency Synthesizers Allowing Straightforward Noise Analysis

Michael H. Perrott, Mitchell D. Trott, *Member, IEEE*, and Charles G. Sodini, *Fellow, IEEE*

Abstract—A general model of phase-locked loops (PLLs) is derived which incorporates the influence of divide value variations. The proposed model allows straightforward noise and dynamic analyses of Σ–Δ fractional-N frequency synthesizers and other PLL applications in which the divide value is varied in time. Based on the derived model, a general parameterization is presented that further simplifies noise calculations. The framework is used to analyze the noise performance of a custom Σ–Δ synthesizer implemented in a 0.6-μm CMOS process, and accurately predicts the measured phase noise to within 3 dB over the entire frequency offset range spanning 25 kHz to 10 MHz.

Index Terms—Delta, dithering, divider, fractional-N, frequency, modeling, noise, phase-locked loop, PLL, quantization noise, sigma, synthesizer.

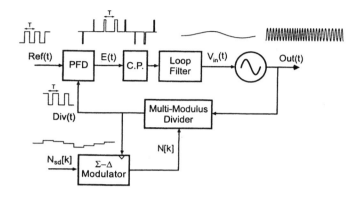

Fig. 1. Block diagram of a Σ–Δ frequency synthesizer.

I. INTRODUCTION

THE USE OF wireless products has been rapidly increasing in the last decade, and there has been worldwide development of new systems to meet the needs of this growing market. As a result, new radio architectures and circuit techniques are being actively sought that achieve high levels of integration and low-power operation while still meeting the stringent performance requirements of today's radio systems. One such technique is the use of Σ–Δ modulation to achieve high-resolution frequency synthesizers that have relatively fast settling times, as described by Riley *et al.* in [1], Copeland in [2], and Miller and Conley in [3], [4]. This method has now been used in a variety of applications ranging from accurate frequency generation [1], [5]–[7] to direct frequency modulation for transmitter applications [8]–[12].

However, despite its increasing use, a general model of Σ–Δ fractional-N synthesizers to encompass dynamic and noise performance has not previously been presented. The primary obstacle to deriving such a model is that, in contrast to classical phase-locked loop (PLL) systems, a Σ–Δ synthesizer dynamically varies the divide value in the PLL according to the output of a Σ–Δ modulator. Traditional methods of PLL analysis assume a static divide value, and the step toward allowing for dynamic variations is not straightforward. As a result, the impact

of the divide value variations is often treated in isolation of other influences on the PLL [1], such as noise in the phase detector and voltage-controlled oscillator (VCO), and overall analysis of the synthesizer becomes cumbersome.

In this paper, we develop a simple model for the Σ–Δ synthesizer that allows straightforward analysis of its dynamic and noise performance. The predictions of the model compare extremely well to simulated and experimental results of implemented Σ–Δ synthesizers [9], [10], [13]. In addition, we present a PLL parameterization that simplifies calculation of the PLL dynamics and assessment of the synthesizer noise performance.

To develop the Σ–Δ synthesizer model, we first derive a general model of the PLL that incorporates the influence of divide value variations. The derivation is done in the time domain and then converted to a frequency-domain block diagram. We parameterize the resulting PLL model in terms of a single function $G(f)$ and illustrate its usefulness in determining the noise performance of the PLL. The Σ–Δ modulator is then included in the generalized PLL model and its impact on the PLL is analyzed. Finally, the modeling approach is used to calculate the noise performance of a custom Σ–Δ synthesizer integrated in a 0.6-μm CMOS process and then compared to measured results.

II. BACKGROUND

Fig. 1 displays a block diagram of a Σ–Δ frequency synthesizer, along with a snapshot of the signals associated with various nodes in this system. A PLL in essence, the synthesizer achieves accurate setting of its output frequency by locking to a reference frequency. This locking action is accomplished through feedback by dividing down the VCO output frequency and comparing its phase to the phase of the reference source

Manuscript received November 14, 2000; revised March 14, 2002. This work was supported in part by the Defense Advanced Research Projects Agency under Contract DAAL-01-95-K-3526.

M. H. Perrott and C. G. Sodini are with the Microsystems Technology Laboratory, Massachusetts Institute of Technology, Cambridge, MA 02139 USA (e-mail: perrott@mit.edu).

M. D. Trott is with Hewlett-Packard Laboratories, Palo Alto, CA 94304 USA.

Publisher Item Identifier 10.1109/JSSC.2002.800925.

to produce an error signal. The phase comparison operation is done through the use of a phase/frequency detector (PFD) which also acts as a frequency discriminator when the PLL is out of lock. The loop filter attenuates high-frequency components in the PFD output so that a smoothed error signal is sent to the VCO input. It consists of an active or passive network, and is typically fed by a charge pump which converts the error signal to a current waveform. The charge pump is not necessary, but provides a convenient means of setting the gain of the loop filter and simplifies implementation of an integrator when required.

As illustrated in the figure, a key characteristic of Σ–Δ synthesizers is that the divide value is dynamically changed in time according to the output of a Σ–Δ modulator. By doing so, much higher frequency resolution can be achieved for a given PLL bandwidth setting than possible with classical integer-N frequency synthesizers [1].

III. Time-Domain PLL Model

We now derive time-domain models for each individual PLL block shown in Fig. 1. The primary focus of our effort is on obtaining a divider model incorporating dynamic changes to its value. However, the derivation of this model requires careful attention to the way we model the PFD. In particular, we will parameterize signals associated with a tristate PFD with sequences that can be directly related to the divider operation. This approach is extended to an XOR-based PFD by relating its output to that of a tristate PFD. Following a brief derivation of the VCO model, we then obtain the divider model by relating its operation to the VCO model and the PFD sequences discussed above. Finally, the charge pump and loop filter models are described, and the overall PLL model constructed.

A. Tristate PFD

The tristate PFD and its associated signals are shown in Fig. 2. The output of the detector, $E(t)$, is characterized as a series of pulses whose widths are a function of the relative phase difference between rising edges of $\mathrm{Ref}(t)$ and $\mathrm{Div}(t)$. We parameterize the phase difference between $\mathrm{Ref}(t)$ and $\mathrm{Div}(t)$ with the discrete-time sequences $\Phi_{\mathrm{ref}}[k]$ and $\Phi_{\mathrm{div}}[k]$, respectively. $\Phi_{\mathrm{ref}}[k]$ is nominally zero, and $\Phi_{\mathrm{div}}[k]$ is defined in (1). The series of pulses that form $E(t)$ are parameterized by the following discrete-time sequences.

- t_k: time instants at which the rising edges of the reference clock occur.
- $t_k + \Delta t_k$: time instants at which the rising edges of the divider output occur.
- Δt_k: time difference between rising edges of $\mathrm{Ref}(t)$ and $\mathrm{Div}(t)$.

Assuming a constant reference frequency, consecutive values for t_k are related for all k as

$$t_k - t_{k-1} = T$$

where T is the reference period. We will make use of the Δt_k parameterization in deriving the PFD model; the other sequences will be used when deriving the divider model.

Since phase detection is a memoryless operation, its influence on the PLL dynamics is sufficiently modeled by its gain. How-

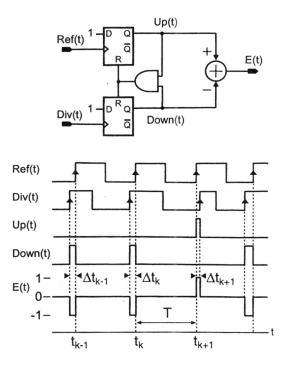

Fig. 2.   Tristate phase-frequency detector and associated signals.

ever, the pulsed behavior of the PFD output adds some complexity in deriving the value of that gain, so our derivation will consist of two steps. The first step relates the input phase difference to the Δt_k sequence. The second step relates the Δt_k sequence to an impulse approximation of the $E(t)$ waveform.

The relationship of Δt_k to the phase difference, $\Phi_{\mathrm{ref}}[k] - \Phi_{\mathrm{div}}[k]$, is defined as

$$\Delta t_k = \frac{T}{2\pi}(\Phi_{\mathrm{ref}}[k] - \Phi_{\mathrm{div}}[k]). \tag{1}$$

To verify the above definition, one observes from Fig. 2 that a phase error of π causes Δt_k to be $T/2$.

The impact of the Δt_k sequence on the PLL dynamics is cumbersome to model analytically since the pulse-width modulated PFD output has a *nonlinear* influence on the PLL dynamics. However, a simple approximation greatly eases our efforts—we simply represent the PFD output as an impulse sequence rather than a modulated pulse sequence. Fig. 3 illustrates this approximation; pulses in $E(t)$ are represented as impulses with area equal to their corresponding pulse, as described by

$$E(t) \approx \sum_{k=-\infty}^{\infty} \Delta t_k \delta(t - kT). \tag{2}$$

We discuss the significance of the above expression when we derive the frequency-domain model of the PLL in Section IV.

Our justification for the impulse approximation is heuristic—each PFD output pulse has much smaller width than the loop filter impulse response, and therefore acts like an impulse when the two are convolved together. Obviously, the accuracy of this approximation depends on how much smaller the PFD output pulse widths are compared to the dominant time constant of the loop filter. Since the PFD pulses must be smaller than a reference period, high accuracy is achieved

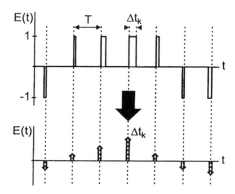

Fig. 3. Impulse sequence approximation of PFD output.

when the reference frequency is much higher than the loop filter (PLL) bandwidth. Fortunately, this condition is satisfied when dealing with Σ–Δ synthesizers since a high reference frequency to PLL bandwidth ratio is required to adequately suppress the Σ–Δ quantization noise. For additional discussion on this issue, see [13].

B. XOR-Based PFD

An XOR-based PFD is shown in Fig. 4 [13]–[15], along with associated signals that will be discussed later. Assuming the PFD is not performing frequency acquisition, the signal $C(t)$ is simply passed to the output, $E(t)$, so that the detector operates as an XOR phase detector. As such, the detector outputs an average error of zero when $\text{Ref}(t)$ and $\text{Div}(t)$ are in quadrature, and $E(t)$ is nominally a two-level square wave rather than the trilevel short-pulse waveform obtained with the tristate design. The combination of having wide pulses and only two output levels allows the XOR-based PFD to achieve high linearity, which is desirable for Σ–Δ synthesizer applications to avoid folding down Σ–Δ quantization noise [13].

To model the XOR-based PFD, we simply relate its associated signals to the tristate detector so that the previous results can be readily applied. Fig. 4 displays the signals associated with this PFD, and reveals that the output $E(t)$ can be decomposed into the sum of a square wave, $E_{\text{spur}}(t)$, and a trilevel pulse waveform, $\hat{E}(t)$. The first component is independent of the input phase difference to the detector and presents a spurious noise signal to the PLL; its influence can be made negligible with proper design. The second component, $\hat{E}(t)$, captures the impact of the input phase difference, $\Phi_{\text{div}}(t) - \Phi_{\text{ref}}(t)$, on the PFD output, and can be parameterized according to the width of its pulses, where

$$\Delta t_k = \frac{T}{2\pi}(\Phi_{\text{ref}}[k] - \Phi_{\text{div}}[k] - \pi).$$

As with the tristate detector, the impulse approximation can be applied to obtain

$$E(t) \approx \sum_{k=-\infty}^{\infty} 2\Delta t_k \delta(t - kT) + E_{\text{spur}}(t)$$

which, if we ignore $E_{\text{spur}}(t)$, is the tristate expression multiplied by a factor of 2. Thus, if we ignore the phase offset of π and the square wave $E_{\text{spur}}(t)$, the XOR-based PFD has an iden-

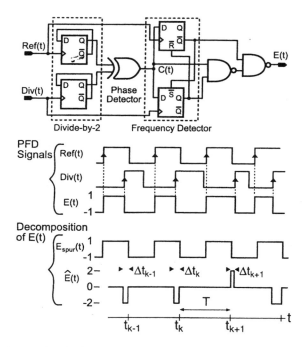

Fig. 4. XOR-based PFD, associated signals, and $E(t)$ decomposition.

tical model to that of the tristate topology except that its gain is increased by a factor of 2.

C. Voltage-Controlled Oscillator

For our purposes, only two equations are needed to model the VCO. The first relates *deviations* in the VCO phase, defined as $\Phi_{\text{out}}(t)$, to changes in the VCO input voltage, $V_{\text{in}}(t)$. Since VCO phase is the integral of VCO frequency, and deviations in VCO frequency are calculated as $K_v V_{\text{in}}(t)$, where K_v is in units of hertz per volt, we have

$$\Phi_{\text{out}}(t) = \int 2\pi K_v V_{\text{in}}(t)dt. \tag{3}$$

The second equation relates the *absolute* VCO phase, defined as $\Phi_{\text{vco}}(t)$, to deviations in the VCO phase and the nominal VCO frequency f_{nom}:

$$\Phi_{\text{vco}}(t) = 2\pi f_{\text{nom}}t + \Phi_{\text{out}}(t). \tag{4}$$

Our modeling efforts will be primarily focused on deviations in the VCO phase, so that (3) is of the most interest. However, (4) is required in the divider derivation that follows.

D. Divider

Modeling of the divider will be accomplished by first relating the PFD pulse widths, Δt_k, to the VCO phase deviations, $\Phi_{\text{out}}(t)$, and the divide value sequence, $N[k]$. Given this relationship, the divider model is "backed out" using the PFD gain expression in (1).

We begin by noting that the divider output edges occur whenever the absolute VCO phase, $\Phi_{\text{vco}}(t)$, completes $2\pi N[k]$ radian increments of phase. As stated in (4), $\Phi_{\text{vco}}(t)$ is composed of a ramp in time, $2\pi f_{\text{nom}}t$, and phase variations, $\Phi_{\text{out}}(t)$. These statements are collectively illustrated in Fig. 5. Note that changes in $N[k]$ occur at the rising edges of the divider.

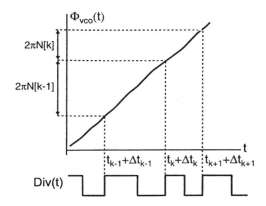

Fig. 5. Relationship of divider edges to instantaneous VCO phase, $\Phi_{\text{vco}}(t)$.

Now, we can relate Δt_k to the VCO phase signal and divider sequence using (4) and Fig. 5. The first of two key equations is derived from Fig. 5 as

$$\Phi_{\text{vco}}(t_k + \Delta t_k) - \Phi_{\text{vco}}(t_{k-1} + \Delta t_{k-1}) = 2\pi N[k-1]. \quad (5)$$

The second key equation is obtained by evaluating (4) at time instants $t_k + \Delta t_k$ and $t_{k-1} + \Delta t_{k-1}$ and subtracting the resulting expressions:

$$\Phi_{\text{vco}}(t_k + \Delta t_k) - \Phi_{\text{vco}}(t_{k-1} + \Delta t_{k-1})$$
$$= 2\pi f_{\text{nom}}(t_k + \Delta t_k - t_{k-1} - \Delta t_{k-1})$$
$$+ \Phi_{\text{out}}(t_k + \Delta t_k) - \Phi_{\text{out}}(t_{k-1} + \Delta t_{k-1})$$

which, since $t_k - t_{k-1} = T$ and $f_{\text{nom}}T = N_{\text{nom}}$, is equivalently written as

$$\Phi_{\text{vco}}(t_k + \Delta t_k) - \Phi_{\text{vco}}(t_{k-1} + \Delta t_{k-1})$$
$$= 2\pi N_{\text{nom}} + 2\pi f_{\text{nom}}(\Delta t_k - \Delta t_{k-1})$$
$$+ \Phi_{\text{out}}(t_k + \Delta t_k) - \Phi_{\text{out}}(t_{k-1} + \Delta t_{k-1}). \quad (6)$$

We combine the two key equations into one formulation by substitution of (6) into (5):

$$2\pi N_{\text{nom}} + 2\pi f_{\text{nom}}(\Delta t_k - \Delta t_{k-1})$$
$$+ \Phi_{\text{out}}(t_k + \Delta t_k) - \Phi_{\text{out}}(t_{k-1} + \Delta t_{k-1})$$
$$= 2\pi N[k-1].$$

Rearrangement of this last expression then produces

$$2\pi f_{\text{nom}}(\Delta t_k - \Delta t_{k-1})$$
$$= 2\pi(N[k-1] - N_{\text{nom}}) - (\Phi_{\text{out}}(t_k + \Delta t_k)$$
$$- \Phi_{\text{out}}(t_{k-1} + \Delta t_{k-1})). \quad (7)$$

Equation (7) is a difference equation relating all variables of interest; to remove the differences we sum the formulation over all positive time samples up to sample k:

$$\sum_{m=1}^{k} (2\pi f_{\text{nom}}(\Delta t_m - \Delta t_{m-1})$$
$$= \sum_{m=1}^{k} (2\pi(N[m-1] - N_{\text{nom}})$$
$$- (\Phi_{\text{out}}(t_m + \Delta t_m) - \Phi_{\text{out}}(t_{m-1} + \Delta t_{m-1}))).$$

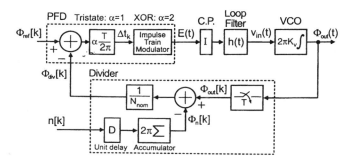

Fig. 6. Time-domain model of PLL.

Carrying out the summation operation, we obtain

$$2\pi f_{\text{nom}}(\Delta t_k - \Delta t_0) = \sum_{m=1}^{k} 2\pi(N[m-1] - N_{\text{nom}})$$
$$- (\Phi_{\text{out}}(t_k + \Delta t_k) - \Phi_{\text{out}}(t_0 + \Delta t_0)).$$

Assuming initial conditions are zero, this last expression becomes

$$2\pi f_{\text{nom}}\Delta t_k$$
$$= \sum_{m=1}^{k} 2\pi(N[m-1] - N_{\text{nom}}) - \Phi_{\text{out}}(t_k + \Delta t_k). \quad (8)$$

The final form of the desired equation is obtained by modifying (8) according to the following statements:

• Define $n[k] = N[k] - N_{\text{nom}}$, $\Phi_{\text{out}}[k] = \Phi_{\text{out}}(t_k)$, $f_{\text{nom}} = N_{\text{nom}}(1/T)$.
• Approximate $\Phi_{\text{out}}(t_k + \Delta t_k) \approx \Phi_{\text{out}}(t_k)$.

As such, we obtain

$$\Delta t_k = \left(\frac{T}{2\pi}\right)\left(\frac{1}{N_{\text{nom}}}\right)\left(2\pi \sum_{m=1}^{k} n[m-1] - \Phi_{\text{out}}[k]\right). \quad (9)$$

We obtain the desired divider model by replacing Δt_k with the PFD gain expression in (1) and assuming $\Phi_{\text{ref}}[k]$ is zero.

$$\Phi_{\text{div}}[k] = \frac{1}{N_{\text{nom}}}\left(-2\pi \sum_{m=1}^{k} n[m-1] + \Phi_{\text{out}}[k]\right). \quad (10)$$

It is important to note that the only approximation made in deriving (10) is that $\Phi_{\text{out}}(t_k + \Delta t_k) \approx \Phi_{\text{out}}(t_k)$. Essentially, we are ignoring the nonuniform time sampling of the VCO phase deviations. As discussed in [13] and verified by actual implementations [9], [10], this approximation is quite accurate in practice even when the PLL is modulated.

E. Charge Pump and Loop Filter

The charge pump and loop filter relate the PFD output $E(t)$ to the VCO input $V_{\text{in}}(t)$. We model the charge pump as a simple scaling operation on $E(t)$ of value I. The time domain model of the loop filter is characterized by its impulse response, $h(t)$.

F. Overall Model

We now combine the results of Section III-A–E to obtain the overall time-domain PLL model shown in Fig. 6. The PFD model is obtained from (1) and (2), the divider model from (10), and the VCO model from (3). As discussed earlier, the

XOR-based PFD has a factor of two larger gain than the tristate design, which is captured by the α factor in the PFD model. For convenience in analysis to follow, we also define an abstract signal, $\Phi_n[k]$, as the output of the divider accumulation action.

Some observations are in order. First, the divider effectively samples the continuous-time output phase deviation of the VCO, $\Phi_{\text{out}}(t)$, and then divides its value by N_{nom}. The output phase of the divider, $\Phi_{\text{div}}[k]$, is influenced by the *integration* of deviations in the divider value, $n[k]$. The integration of $n[k]$ is a consequence of the fact that the divider output is a *phase* signal, whereas $n[k]$ causes an incremental change in the *frequency* of the divider output. Second, the PFD, charge pump, and loop filter translate the discrete-time error signal formed by $\Phi_{\text{ref}}[k]$ and $\Phi_{\text{div}}[k]$ to the continuous-time input of the VCO, $V_{\text{in}}(t)$. These elements, along with the divider, also act as a D/A converter for mapping changes in $n[k]$ to $V_{\text{in}}(t)$.

IV. FREQUENCY-DOMAIN PLL MODEL

Derivation of a frequency-domain model of the PLL is complicated by the sampling operation and impulse train modulator shown in Fig. 6. We discuss a simple approximation for the sampling operation and impulse train modulator that results in a linear time-invariant PLL model. This method, known as pseudocontinuous analysis [16], takes advantage of the fact that the impulsive output of the PFD is low-pass filtered in continuous time by the loop filter.

A. Pseudocontinuous Approximation

Consider a signal $x(t)$ that is sampled with period T and then converted to an impulse sequence $\hat{x}(t)$, as described by

$$\hat{x}(t) = \sum_{k=-\infty}^{\infty} x[k]\delta(t - kT)$$

where $x[k] = x(kT)$. The frequency-domain relationship between $\hat{x}(t)$ and $x(t)$ is found by taking the Fourier transform of the above expression, which leads to

$$\hat{X}(f) = \frac{1}{T}\sum_{k=-\infty}^{\infty} X\left(f - \frac{k}{T}\right).$$

This expression reveals that the Fourier transform of $\hat{x}(t)$, $\hat{X}(f)$, is composed of multiple copies of the Fourier transform of $x(t)$, $X(f)$, that are scaled in magnitude by $1/T$ and shifted in frequency from one another with spacing $1/T$. We assume that the frequency content of $X(f)$ is confined to frequencies between $-1/(2T)$ and $1/(2T)$, so that negligible aliasing occurs between the copies of $X(f)$ within $\hat{X}(f)$.

Developing a frequency-domain model relating $\hat{X}(f)$ to $X(f)$ is complicated by the many copies of $X(f)$ in $\hat{X}(f)$ that occur due to the sampling operation. However, if we assume that $\hat{x}(t)$ is fed into a continuous-time low-pass filter with sufficiently low bandwidth, we can obtain a simple approximation of the relationship between $\hat{X}(f)$ and $X(f)$. Fig. 7 graphically illustrates a frequency-domain view of the sampling operation and the impact of following it with a continuous-time low-pass filter of bandwidth less than $1/(2T)$. The low-pass filter significantly attenuates all of the replicated

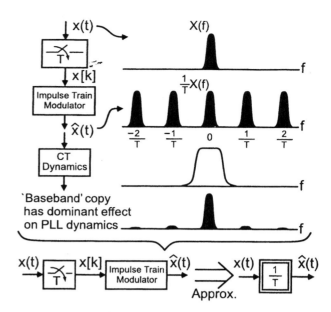

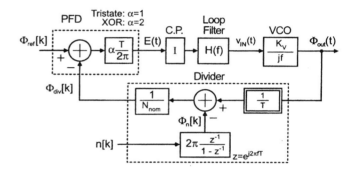

Fig. 7. Pseudocontinuous method of modeling a sampling operation in the frequency domain.

Fig. 8. Frequency-domain model of PLL.

copies of $X(f)$ within $\hat{X}(f)$ except for the baseband copy, which allows us to approximate the relationship between $\hat{x}(t)$ and $x(t)$ in the frequency domain as a simple scaling operation of $1/T$. In so doing, we ignore aliasing effects that will occur if there is frequency content in $X(f)$ at frequencies beyond the range of $-1/(2T)$ to $1/(2T)$. However, our analysis will be reasonably accurate when performing closed-loop analysis for most frequencies of interest in our application. The double outline of the box in the figure is meant to serve as a reminder that a sampling operation is taking place.

B. Resulting Model

The time-domain block diagram in Fig. 6 is now readily converted to the frequency domain by taking the Z-transform of the discrete-time blocks, the Fourier transform of the continuous-time blocks, and by applying the approximation of the sampling operation discussed above. Fig. 8 displays the resulting model. Note that all blocks are parameterized by the common variable f, which denotes frequency in hertz, under the assumption that all discrete-time sequences interact with the continuous-time blocks as modulated impulse trains of period T. Also note that all the *signals* in the PLL are still denoted in the time domain even though they interact

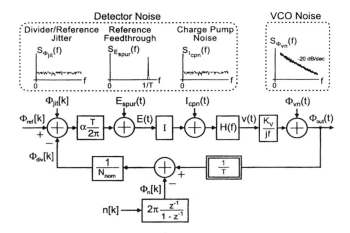

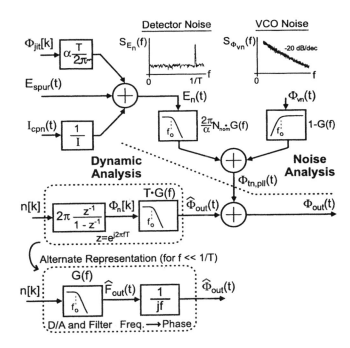

Fig. 9. Detailed view of PLL noise sources and examples of their respective spectral densities.

through frequency-domain blocks. The reason for this notation convention is that, in practice, these signals are stochastic and do not have defined Fourier transforms, but rather are described by their power spectral densities.

V. PARAMETERIZATION OF PLL

We now parameterize the PLL dynamics depicted in Fig. 8 in terms of a single function which we will call $G(f)$. Using this parameterization, we then develop a general noise model for frequency synthesizers in which all the relevant transfer functions are described in terms of $G(f)$.

A. Derivation

To parameterize the PLL dynamics, it is convenient to define a base function that provides a simple description of all the PLL transfer functions of interest. It turns out that the following definition works well for this purpose.

$$G(f) = \frac{A(f)}{1 + A(f)} \qquad (11)$$

where $A(f)$ is the open-loop transfer function of the PLL:

$$A(f) = \left(\frac{\alpha}{2\pi}\right) IH(f) \left(\frac{K_v}{jf}\right) \left(\frac{1}{N_{\text{nom}}}\right). \qquad (12)$$

Since $A(f)$ is low pass in nature with infinite gain at dc, $G(f)$ has the following properties:

$$G(f) \longrightarrow 1 \text{ as } f \longrightarrow 0$$
$$G(f) \longrightarrow 0 \text{ as } f \longrightarrow \infty \qquad (13)$$

implying that $G(f)$ is a low-pass filter with a low frequency gain of one.

One may try to tie an intrinsic meaning to $G(f)$ in terms of PLL behavior. However, it is meant only as a convenient vehicle for compactly describing the PLL transfer functions of interest, as will be shown later in this section.

B. Application to Noise Analysis

The derived parameterization allows straightforward calculation of the noise performance of a synthesizer as a function of various noise sources in the PLL, which are shown in Fig. 9.

Fig. 10. Parameterized model of PLL for dynamic response and noise calculations.

Divider/reference jitter, $\Phi_{\text{jit}}[k]$, corresponds to noise-induced variations in the transition times of the Reference or Divider output waveforms. A periodic reference spur $E_{\text{spur}}(t)$ is caused by use of the XOR-based PFD, or by the tristate PFD when its output duty cycle is nonzero. Charge-pump noise is caused by noise produced in the transistors that compose the charge-pump circuit. Finally, VCO noise includes the intrinsic noise of the VCO and voltage noise at the output of the loop filter. For convenience in later discussion, we have lumped these noise sources into two categories, VCO noise and detector noise, as shown in Fig. 9.

Fig. 10 displays the transfer function relationships from each of the above noise sources to the synthesizer output. The derivation of these transfer functions is straightforward based on Fig. 9 and the $G(f)$ parameterization derived earlier. Note that two different parameterizations are shown to describe the impact of divide value variations on the PLL output phase. The alternate model relates changes in the divide value, $n[k]$, more directly to the PLL output frequency. Its derivation follows by noting that the order of linear time-invariant blocks can be switched, and that

$$\frac{z^{-1}}{1 - z^{-1}} = \frac{e^{-j2\pi fT}}{1 - e^{-j2\pi fT}}$$
$$\approx \frac{1 - j2\pi fT}{1 - (1 - j2\pi fT)}$$
$$\approx \frac{1}{j2\pi fT}, \qquad \text{for } f \ll \frac{1}{T}.$$

Note that the validity of the dynamic model, and its alternate, presented in Fig. 10, has been verified in previous work discussed in [9], [13]. The validity of the noise model will be verified in Section VII.

Calculation of spectral noise densities using Fig. 10 is complicated by the fact that both discrete-time (DT) and

583

continuous-time (CT) signals are present. Three cases are of significance, and their respective spectral noise calculations are as follows [17]:

Case 1) CT input $x(t)$ fed into CT filter $H(f)$ to produce a CT output $y(t)$:

$$S_y(f) = |H(f)|^2 S_x(f). \tag{14}$$

Case 2) DT input $x[k]$ fed into DT filter $H(e^{j2\pi fT})$ to produce a DT output $y[k]$:

$$S_y(e^{j2\pi fT}) = |H(e^{j2\pi fT})|^2 S_x(e^{j2\pi fT}). \tag{15}$$

Case 3) DT input $x[k]$ fed into CT filter $H(f)$ to produce a CT output $y(t)$:

$$S_y(f) = \frac{1}{T}|H(f)|^2 S_x(e^{j2\pi fT}). \tag{16}$$

In Case (3), we assume that the DT input interacts with the CT filter as a modulated impulse train of period T.

The above spectral density calculations and Fig. 10 allow us to accurately calculate the influence of the various noise sources on the PLL output. A few qualitative observations are also in order. Detector noise is low-pass filtered by the PLL dynamics, while VCO noise is high-pass filtered by the PLL dynamics. The overall noise power in the PLL output, whose integral over frequency corresponds to the time-domain jitter of the PLL output, is a function of the PLL bandwidth. If the PLL bandwidth is very low, VCO noise will dominate over a wide frequency range due to the abundant suppression of detector noise. Likewise, a high PLL bandwidth will suppress VCO noise over a wide frequency range at the expense of allowing more detector noise through.

VI. Σ–Δ SYNTHESIZER MODEL

We are now ready to incorporate the Σ–Δ modulator into the general PLL model. We do so by first providing a brief description of Σ–Δ modulator fundamentals, and then provide intuition to the means by which they increase the frequency resolution of a synthesizer compared to a classical implementation in which the divider value is held constant. Finally, we present a frequency-domain model of the Σ–Δ synthesizer and use it to calculate the impact of the Σ–Δ quantization noise on the PLL output phase.

A. Σ–Δ Modulator

A Σ–Δ modulator achieves a high-resolution signal using only a few output levels. To do this, the modulator dithers its output at a high rate such that the "average" value of the dithered sequence corresponds to a high-resolution input signal whose energy is confined to low frequencies. Appropriate filtering of the output sequence removes quantization noise produced by the dithering, which yields a high-resolution signal closely matching that of the input.

In Σ–Δ synthesizer applications, it is important to note that the Σ–Δ modulator is *purely digital* in its implementation. Thus, Σ–Δ structures that are difficult to implement in the analog world due to high matching requirements, such as the MASH (or cascaded) architecture [18], [19], are trivial to implement in

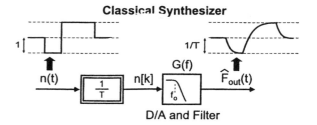

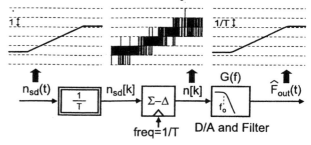

Fig. 11.   Illustration of dithering action of Σ–Δ modulator.

this application due to the precise matching offered by digital circuits.

In general, modeling of a Σ–Δ modulator is accomplished by assuming its quantization noise is independent of its input [19]. This leads to a linear time-invariant model that is parameterized by transfer functions from the input and quantization noise to the output. For instance, a MASH Σ–Δ modulator structure [19] of order m, input $x[k]$, and output $y[k]$ is described by

$$y(z) = x(z) - (1 - z^{-1})^m r(z). \tag{17}$$

Thus, the modulator passes its input to the output along with quantization noise, $r[k]$, that is *shaped* by the filter $(1 - z^{-1})^m$. Ideally, $r[k]$ is white and uniformly distributed between 0 and 1 so that its spectrum is flat and of magnitude $1/12$ [20], [21].

It is convenient to parameterize the Σ–Δ modulator in terms of two transfer functions. The signal transfer function (STF) of the Σ–Δ modulator is defined from the input $x[k]$ to output $y[k]$, while the noise transfer function (NTF) is defined from the base quantization noise $r[k]$ to the output. Inspection of (17) reveals that a MASH structure of order m is parameterized as

STF: $H_s(z) = 1$

NTF: $H_n(z) = (1 - z^{-1})^m|_{z = e^{-j2\pi fT}}.$

B. Application to PLL

To understand the impact of using a Σ–Δ modulator to control the divide value in a frequency synthesizer, Fig. 11 contrasts the way the divide value is varied in classical versus Σ–Δ fractional-N frequency synthesizers based on the alternate model in Fig. 10. Note that the divide value variations are cast as continuous-time signals to get the proper scale factor such that a unit change in divide value yields an output frequency change of $1/T$ Hz. In the classical case, the divide value is static except when the output frequency is changed, and the PLL output frequency responds to the change according to the low-pass nature of the PLL dynamics $G(f)$. In contrast, a Σ–Δ fractional-N

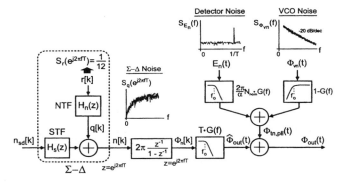

Fig. 12. Parameterized model of a $\Sigma-\Delta$ synthesizer.

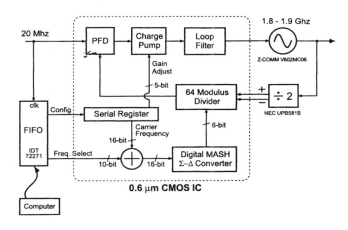

Fig. 13. Block diagram of prototype system.

synthesizer constantly dithers the divide value at a high rate compared to the bandwidth of $G(f)$ such that $G(f)$ extracts out its low-frequency content. The low frequency content of the $\Sigma-\Delta$ output is, in turn, set by the $\Sigma-\Delta$ input $n_{sd}[k]$, which can have arbitrarily high resolution. Thus, the $\Sigma-\Delta$ modulator allows the PLL output frequency to be controlled to a very high resolution *independent* of the reference frequency—a high reference frequency can be used while simultaneously achieving high-frequency resolution.

C. Frequency-Domain Model

To obtain the frequency-domain model of a $\Sigma-\Delta$ synthesizer, we simply extend the PLL model in Fig. 10 to include the $\Sigma-\Delta$ modulator, as shown in Fig. 12. This figure depicts a general model of a $\Sigma-\Delta$ modulator which is characterized by its STF and NTF. The base quantization noise $r[k]$ is assumed ideal (i.e., white) in the illustration.

Fig. 12 offers several insights to the fundamentals of $\Sigma-\Delta$ frequency synthesis. First, we see that the shaped $\Sigma-\Delta$ quantization noise passes through a digital accumulator and then the PLL dynamics, $G(f)$, before impacting the output phase of the PLL. The digital accumulator, a consequence of the integrating nature of the divider, effectively reduces the noise-shaping order of the $\Sigma-\Delta$ by one. The PLL dynamics, $G(f)$, act to remove the high-frequency quantization noise produced by the $\Sigma-\Delta$ modulator. The $\Sigma-\Delta$ quantization noise adds an additional noise source to those already present in the PLL, but the relationship from each noise source to the output phase remains purely a function of $G(f)$ and the nominal divide value.

D. Quantization Noise Impact on PLL

As Fig. 12 reveals, a $\Sigma-\Delta$ synthesizer's noise performance is impacted by the $\Sigma-\Delta$ quantization noise in addition to the intrinsic detector and VCO noise sources found in the classical PLL. Calculation of this impact is straightforward using the presented modeling approach. For example, given the NTF of an mth order MASH structure is $(1-z^{-1})^m$, we calculate the impact of its quantization noise on the PLL output using Fig. 12 and (16) as

$$S_{\Phi_{out}}(f) = \frac{1}{T}|T \cdot G(f)|^2 \left|2\pi \frac{e^{-j2\pi fT}}{1-e^{-j2\pi fT}}\right|^2$$
$$\times \left|(1-e^{-j2\pi fT})^m\right|^2 S_r(f)$$

which is also expressed as

$$S_{\Phi_{out}}(f) = \frac{1}{T}|T \cdot G(f)|^2 \left((2\pi)^2(2\sin(\pi fT))^{2(m-1)}\right) S_r(f).$$
(18)

If the quantization noise spectra of $r[k]$ is white, then

$$S_r(f) = \frac{1}{12}$$

as previously discussed. In many cases, $r[k]$ is not white and must be computed numerically by simulating the $\Sigma-\Delta$ modulator at a given value of $n_{sd}[k]$.

Equation (18) shows that the $\Sigma-\Delta$ quantization noise is reduced in order by one due to the integrating action of the divider. Assuming $r[k]$ is white, the shaped noise rises at $(m-1)20$ dB/decade for frequencies $\ll 1/T$. Therefore, if the order of $G(f)$ is chosen to be the same as the order of the $\Sigma-\Delta$, the quantization noise seen at the PLL output will roll off at -20 dB/decade outside the PLL bandwidth. This rolloff characteristic matches that of the VCO noise.

VII. RESULTS

The above methodology is now used to analyze the noise performance of a prototype system described in [9], [13]. Fig. 13 displays a block diagram of the prototype, which consists of a custom CMOS fractional-N synthesizer IC that includes an XOR-based PFD, an on-chip loop filter that uses switched capacitors to set its time constant, a second-order digital MASH $\Sigma-\Delta$ modulator, and an asynchronous 64-modulus divider that supports any divide value between 32 and 63.5 in half-cycle increments. An external divide-by-2 prescaler is used so that the CMOS divider input operates at half the VCO frequency, which modifies the range of divide values to include all integers between 64 and 127. A computer interface is used to set the digital frequency value that is fed into the input of the $\Sigma-\Delta$ modulator.

A. Modeling

A linearized frequency-domain model of the prototype system is shown in Fig. 14. The open-loop transfer function of the system consists of two integrators, a pole at f_p and a zero at f_z. Additional poles and zeros occur in the system due to the effects of finite opamp bandwidth and other nonidealities,

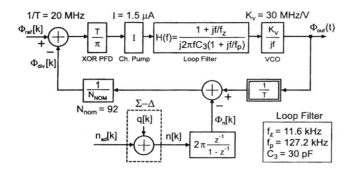

Fig. 14. Linearized frequency-domain model of prototype system.

but are not significant for the analysis to follow. The $G(f)$ parameterization is calculated from Fig. 14 and (11) as

$$G(f) = \frac{1 + \dfrac{jf}{f_z}}{1 + \dfrac{jf}{f_{cp}}} \left(\frac{1}{1 + \dfrac{jf}{(f_o Q)} + \left(\dfrac{jf}{f_o}\right)^2} \right). \quad (19)$$

The parameters of the system were set such that the PLL had a bandwidth of 84 kHz:

$$\begin{aligned} f_o &= 84.3 \text{ kHz} \\ f_z &= 11.6 \text{ kHz} \\ f_{cp} &= 14.2 \text{ kHz} \\ Q &= 0.75. \end{aligned} \quad (20)$$

Fig. 15 expands the block diagram of the prototype to indicate the circuits of relevance and their respective noise contributions. A few comments are in order. First, a reference frequency $1/T$ of 20 MHz was chosen to achieve an acceptably low impact of Σ–Δ quantization noise while still allowing low-power implementation of the digital logic. This choice of reference frequency, in turn, required that $N_{\text{nom}} = 92$ to achieve an output carrier frequency of 1.84 GHz. The value of K_v was set to 30 MHz/V by the external VCO. The value of C_3 was chosen as large as practical in order to obtain good noise performance; it was constrained to 30 pF due to area constraints on the die of the custom IC.

B. Noise Analysis

Table I displays the value of each noise source shown in Fig. 15. Many of these values were obtained through ac simulation of the relevant circuits in HSPICE. Note that all noise sources other than $q[k]$ are assumed to be white, so that the values of their variance suffice for their description. This assumption holds for the input-referred VCO noise, $v_{\text{vco,in}}(t)$, provided that the output phase noise of the VCO rolls off at -20 dB/dec [22], [23]; the -20 dB/dec rolloff is achieved in the model since $v_{\text{vco,in}}(t)$, which has a flat spectral density, passes through the integrating action of the VCO. The actual VCO deviates from the -20 dB/dec rolloff at low frequencies due to $1/f$ noise, and at high frequencies due to a finite noise floor. However, the assumption of -20 dB/dec rolloff suffices for the frequency offsets of interest.

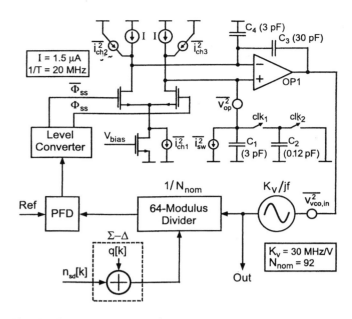

Fig. 15. Expanded view of PLL System.

TABLE I
VALUES OF NOISE SOURCES WITHIN PLL

Noise Source	Origin	Nature	Calculation	Value
$\overline{i_{ch1}^2}$	Ch. Pump, OP1	CT	HSPICE	1.2E-24 A^2/Hz
$\overline{i_{ch2}^2}, \overline{i_{ch3}^2}$	Ch. Pump	CT	HSPICE	1.8E-25 A^2/Hz
$\overline{i_{sw}^2}$	Switched Cap	DT	Equation 22	1.0E-26 A^2/Hz
$\overline{v_{op}^2}$	OP1	CT	HSPICE	1.85E-16 V^2/Hz
$\overline{v_{vco,in}^2}$	VCO	CT	Equation 21	1.4E-16 V^2/Hz
$q[k]$	Σ–Δ	DT	Equation 23	—

The input-referred noise of the VCO was calculated from an open-loop VCO phase noise measurement (shown in Fig. 17) at 5-MHz frequency offset as

$$10 \log \left(\overline{v_{vco,in}^2} \left| \frac{K_v}{(jf)} \right|^2 \right) = -143 \text{ dBc/Hz}$$
$$\text{at } f = 5 \text{ MHz} \quad (21)$$

where K_v is 30 MHz/V. The value of the kT/C noise current produced by the switched-capacitor operation $\overline{i_{sw}^2}$ was calculated as

$$\overline{i_{sw}^2} = \left(\frac{1}{T} \right) k T_K C_2 \quad (22)$$

where k is Boltzmann's constant, and T_K is temperature in degrees Kelvin. Finally, the spectral density of the Σ–Δ quantization noise was calculated as

$$S_q(f) = \frac{1}{12} (2 \sin(\pi f T))^{2m} \quad (23)$$

where $m = 2$ is the order of the Σ–Δ modulator.

The noise sources in Table I can be classified as either charge-pump noise, VCO noise, or Σ–Δ quantization noise, which we denote as $i_{cp}(t)$, $v_{vco}(t)$, and $q[k]$, respectively. For convenience, we will assume that $v_{vco}(t)$ is referred to the input of the VCO, so that it passes through the transfer function $K_v/(jf)$ before influencing the VCO output phase. Given the values of these sources, the overall noise spectral density at the synthesizer output $S_{\Phi_{tn}}(f)$ is described as

$$S_{\Phi_{tn}}(f) = S_{\Phi_{cp}}(f) + S_{\Phi_{vco}}(f) + S_{\Phi_q}(f) \qquad (24)$$

where $S_{\Phi_{cp}}(f)$, $S_{\Phi_{vco}}(f)$, and $S_{\Phi_q}(f)$ are the contributions from $i_{cp}(t)$, $v_{vco}(t)$, and $q[k]$, respectively. $S_{\Phi_q}(f)$ is given by (18) with $m = 2$. $S_{\Phi_{cp}}(f)$ and $S_{\Phi_{vco}}(f)$ are calculated from Fig. 10 and (14) as

$$S_{\Phi_{cp}}(f) = \overline{i_{cp}^2} \left(\frac{\pi N_{nom}}{I} \right)^2 |G(f)|^2$$

$$S_{\Phi_{vco}}(f) = \overline{v_{vco}^2} \left| \frac{K_v}{(jf)} \right|^2 |1 - G(f)|^2 . \qquad (25)$$

Note that we have assumed that $i_{cp}(t)$ and $v_{vco}(t)$ are white, and that $\alpha = 2$ since an XOR-based PFD is used.

The task that remains is to determine the values of $i_{cp}(t)$ and $v_{vco}(t)$. Examination of Fig. 15 reveals that charge-pump noise is a function of the following noise sources:

$$\overline{i_{cp}^2} = f_i \left(\overline{i_{ch1}^2}, \overline{i_{ch2}^2}, \overline{i_{ch3}^2}, \overline{i_{sw}^2} \right) \qquad (26)$$

while VCO noise is a function of the noise sources

$$\overline{v_{vco}^2} = f_v \left(\overline{v_{vco,in}^2}, \overline{v_{op}^2} \right) . \qquad (27)$$

We will quickly infer the value of the functions $f_i(\cdot)$ and $f_v(\cdot)$ in this paper; the reader is referred to [13] for more detail.

Let us first determine $f_i(\cdot)$. Examination of Table I reveals that $\overline{i_{ch1}^2}$ is an order of magnitude larger than $\overline{i_{ch2}^2}$, $\overline{i_{ch3}^2}$, and $\overline{i_{sw}^2}$. Since the $\overline{i_{ch1}^2}$ noise source is switched alternately between the positive and negative terminals of OP1, its contribution to $i_{cp}(t)$ will be pulsed in nature. At a nominal duty cycle of 50%, we would expect the energy of $\overline{i_{ch1}^2}$ to be split equally between the positive and negative terminals of OP1. As such, $\overline{i_{cp}^2}$ is then $\overline{i_{ch1}^2}/2$. This intuitive argument was verified using a detailed C simulation of the PLL [24]. Note that a more accurate estimate of $\overline{i_{cp}^2}$ will take into account any offset in the nominal duty cycle of the phase detector output, and the transient response of the charge pump.

Now let us determine $f_v(\cdot)$. Since Table I reveals that $\overline{v_{vco}^2}$ is of the same order of $\overline{v_{op}^2}$, we simply add these components to obtain $\overline{v_{vco}^2} \approx \overline{v_{vco}^2} + \overline{v_{op}^2}$. This expression is accurate at frequencies less than the unity gain bandwidth of OP1; the $\overline{v_{op}^2}$ noise source is passed to its output with a gain of approximately one in this region. At frequencies beyond OP1's bandwidth, the expression is conservatively high since $\overline{v_{op}^2}$ is attenuated in this frequency range.

Based on the above information, plots of the spectra in (24) are shown in Fig. 16. For convenience, we have also overlapped measured results from Fig. 17 for easy comparison, which will be discussed shortly. As shown in Fig. 16, the influence of detector noise dominates at low frequencies, and the influence of VCO and $\Sigma-\Delta$ quantization noise dominate at high frequencies. Note that the calculations use $G(f)$ described by (19) with the parameter values specified in (20).

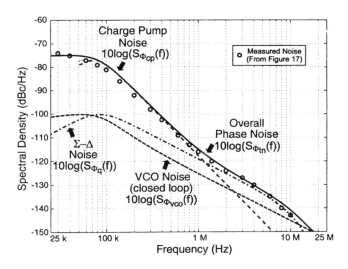

Fig. 16. Calculated noise spectra of synthesizer compared to measured results.

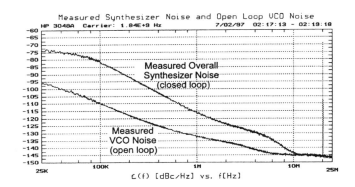

Fig. 17. Measured closed-loop synthesizer noise and open-loop VCO noise.

Fig. 17 shows measured plots of $S_{\Phi_{tn}}(f)$ and the open-loop phase noise of the VCO from the synthesizer prototype; the plots were obtained from an HP 3048A phase-noise measurement system. It should be noted that the LSB of the $\Sigma-\Delta$ modulator was dithered to reduce spurious content, which was necessary due to the low order of the $\Sigma-\Delta$ modulator. The resulting spectra compare quite well with the calculated curve in Fig. 16 over the frequency offset range of 25 kHz to 10 MHz. Above 10 MHz, the phase-noise measurement was limited by the sensitivity of the measurement equipment. Note that the -60 dBc spur at 20-MHz offset is due to the 50% nominal duty cycle of the PFD; no effort was made to reduce it below this level during the design process since it was acceptable for the intended application of the prototype.

VIII. CONCLUSION

In this paper, we developed a general model of a PLL that incorporates the influence of divide value variations. A model for $\Sigma-\Delta$ fractional-N synthesizers was obtained by simply incorporating a $\Sigma-\Delta$ modulator model into this framework. The PLL model was parameterized by a single transfer function $G(f)$, which further simplifies noise calculations. The framework was used to calculate the noise performance of a custom $\Sigma-\Delta$ synthesizer, and was shown to accurately predict measured results within 3 dB over a frequency offset range from 25 kHz to 10 MHz.

ACKNOWLEDGMENT

The authors would like to thank the Hong Kong University of Science and Technology, and in particular, J. Lau, P. Chan, and P. Ko, for their support in the writing of this paper.

REFERENCES

[1] T. A. Riley, M. A. Copeland, and T. A. Kwasniewski, "Delta–sigma modulation in fractional-N frequency synthesis," *IEEE J. Solid State Circuits*, vol. 28, pp. 553–559, May 1993.

[2] M. A. Copeland, "VLSI for analog/digital communications," *IEEE Commun. Mag.*, vol. 29, pp. 25–30, May 1991.

[3] B. Miller and B. Conley, "A multiple modulator fractional divider," in *Proc. 44th Annu. Symp. Frequency Control*, May 1990, pp. 559–567.

[4] ——, "A multiple modulator fractional divider," *IEEE Trans. Instrum. Meas.*, vol. 40, pp. 578–583, June 1991.

[5] W. Rhee, B.-S. Song, and A. Ali, "A 1.1-GHz CMOS fractional-N frequency synthesizer with 3-b third-order sigma–delta modulator," *IEEE J. Solid-State Circuits*, vol. 35, pp. 1453–1460, Oct. 2000.

[6] B. Miller, "Technique enhances the performance of PLL synthesizers," *Microw. RF*, pp. 59–65, Jan. 1993.

[7] T. Kenny, T. Riley, N. Filiol, and M. Copeland, "Design and realization of a digital delta–sigma modulator for fractional-N frequency synthesis," *IEEE Trans. Veh. Technol.*, vol. 48, pp. 510–521, Mar. 1999.

[8] T. A. Riley and M. A. Copeland, "A simplified continuous phase modulator technique," *IEEE Trans. Circuits Syst. II*, vol. 41, pp. 321–328, May 1994.

[9] M. Perrott, T. Tewksbury, and C. Sodini, "A 27-mW CMOS fractional-N synthesizer using digital compensation for 2.5-Mb/s GFSM modulation," *IEEE J. Solid-State Circuits*, vol. 32, pp. 2048–2060, Dec. 1997.

[10] S. Willingham, M. Perrott, B. Setterberg, A. Grzegorek, and W. McFarland, "An integrated 2.5-GHz sigma–delta frequency synthesizer with 5 microseconds settling and 2-Mb/s closed-loop modulation," in *Proc. IEEE Int. Solid-State Circuits Conf. (ISSCC)*, Feb. 2000, pp. 200–201.

[11] N. Filiol, T. Riley, C. Plett, and M. Copeland, "An agile ISM band frequency synthesizer with built-in GMSK data modulation," *IEEE J. Solid-State Circuits*, vol. 33, pp. 998–1008, July 1998.

[12] N. Filiol, C. Plett, T. Riley, and M. Copeland, "An interpolated frequency-hopping spread-spectrum transceiver," *IEEE Trans. Circuits Syst. II*, vol. 45, pp. 3–12, Jan. 1998.

[13] M. H. Perrott, "Techniques for high data rate modulation and low power operation of fractional-N frequency synthesizers with noise shaping," Ph.D. dissertation, Massachusetts Inst. Technol., Cambridge, MA, 1997.

[14] A. Hill and A. Surber, "The PLL dead zone and how to avoid it," *RF Design*, pp. 131–134, Mar. 1992.

[15] M. Thamsirianunt and T. A. Kwasniewski, "A 1.2-μm CMOS implementation of a low-power 900-MHz mobile radio frequency synthesizer," in *Proc. IEEE Custom Integrated Circuits Conf. (CICC)*, 1994, p. 16.2.

[16] J. A. Crawford, *Frequency Synthesizer Handbook*. Norwood, MA: Artech, 1994.

[17] E. A. Lee and D. G. Messerschmitt, *Digital Communication*, 2nd ed. Norwell, MA: Kluwer, 1994.

[18] J. Candy and G. Temes, *Oversampling Delta–Sigma Data Converters*. New York: IEEE Press, 1992.

[19] S. Norsworthy, R. Schreier, and G. Temes, *Delta–Sigma Data Converters: Theory, Design, and Simulation*. New York: IEEE Press, 1997.

[20] A. Sripad and D. Snyder, "A necessary and sufficient condition for quantization errors to be uniform and white," *IEEE Trans. Acoust. Speech Signal Proc.*, vol. ASSP-25, pp. 442–448, Oct. 1977.

[21] W. Bennett, "Spectra of quantized signals," *Bell Syst. Tech. J.*, vol. 27, pp. 446–472, July 1948.

[22] D. Leeson, "A simple model of feedback oscillator noise spectrum," *Proc. IEEE*, vol. 54, pp. 329–330, Feb. 1966.

[23] A. Hajimiri and T. Lee, "A general theory of phase noise in electrical oscillators," *IEEE J. Solid-State Circuits*, vol. 33, pp. 179–194, Feb. 1998.

[24] M. H. Perrott, "Fast and accurate behavioral simulation of fractional-N frequency synthesizers and other PLL/DLL circuits," in *Proc. Design Automation Conf. (DAC)*, June 2002, pp. 498–503.

Michael H. Perrott received the B.S. degree in electrical engineering from New Mexico State University, Las Cruces, in 1988, and the M.S. and Ph.D. degrees in electrical engineering and computer science from the Massachusetts Institute of Technology (M.I.T.), Cambridge, in 1992 and 1997, respectively.

From 1997 to 1998, he was with Hewlett-Packard Laboratories, Palo Alto, CA, working on high-speed circuit techniques for Σ–Δ synthesizers. In 1999, he was a visiting Assistant Professor at the Hong Kong University of Science and Technology, where he taught a course on the theory and implementation of frequency synthesizers. From 1999 to 2001, he was with Silicon Laboratories, Austin, TX, where he developed circuit and signal-processing techniques to achieve high-performance clock and data recovery circuits. He is currently an Assistant Professor in the Department of Electrical Engineering and Computer Science at M.I.T., where his research focuses on high-speed circuit and signal processing techniques for data links and wireless applications.

Mitchell D. Trott (S'90–M'92) received the B.S. and M.S. degrees in systems engineering from Case Western Reserve University, Cleveland, OH, in 1987 and 1988, respectively, and the Ph.D. degree in electrical engineering from Stanford University, Stanford, CA, in 1992.

He was an Assistant and Associate Professor in the Department of Electrical Engineering and Computer Science at the Massachusetts Institute of Technology, Cambridge, from 1992 until 1998. He was Director of Research with ArrayComm, Inc., San Jose, CA, from 1998 to 2002. He is currently with Hewlett-Packard Laboratories, Palo Alto, CA. His research interests include multiuser communication, information theory, and coding theory.

Charles G. Sodini (S'80–M'82–SM'90–F'94) was born in Pittsburgh, PA, in 1952. He received the B.S.E.E. degree from Purdue University, Lafayette, IN, in 1974, and the M.S.E.E. and Ph.D. degrees from the University of California, Berkeley, in 1981 and 1982, respectively.

He was a Member of the Technical Staff with Hewlett-Packard Laboratories from 1974 to 1982, where he worked on the design of MOS memory and, later, on the development of MOS devices with very thin gate dielectrics. He joined the faculty of the Massachusetts Institute of Technology (M.I.T.), Cambridge, MA, in 1983, where he is currently a Professor in the Department of Electrical Engineering and Computer Science. His research interests are focused on integrated circuit and system design with emphasis on analog, RF, and memory circuits and systems. Along with Prof. R. T. Howe, he is a coauthor of an undergraduate text on integrated circuits and devices entitled *Microelectronics: An Integrated Approach* (Englewood Cliffs, NJ: Prentice-Hall, 1996).

Dr. Sodini held the Analog Devices Career Development Professorship at M.I.T.'s Department of Electrical Engineering and Computer Science and was awarded the IBM Faculty Development Award from 1985 to 1987. He has served on a variety of IEEE Conference Committees, including the International Electron Device Meeting, of which he was the 1989 General Chairman. He was the Technical Program Co-Chairman in 1992 and the Co-Chairman for 1993–1994 of the Symposium on VLSI Circuits. He served on the Electron Device Society Administrative Committee from 1988 to 1994. He has been a member of the Solid-State Circuits Society (SSCS) Administrative Committee since 1993 and is currently President of the SSCS.

A Fully Integrated CMOS Frequency Synthesizer With Charge-Averaging Charge Pump and Dual-Path Loop Filter for PCS- and Cellular-CDMA Wireless Systems

Yido Koo, Hyungki Huh, Yongsik Cho, Jeongwoo Lee, Joonbae Park, Kyeongho Lee, Deog-Kyoon Jeong, and Wonchan Kim

Abstract—A fully integrated CMOS frequency synthesizer for PCS- and cellular-CDMA systems is integrated in a 0.35-μm CMOS technology. The proposed charge-averaging charge pump scheme suppresses fractional spurs to the level of noise, and the improved architecture of the dual-path loop filter makes it possible to implement a large time constant on a chip. With current-feedback bias and coarse tuning, a voltage-controlled oscillator (VCO) enables constant power and low gain of the VCO. Power dissipation is 60 mW with a 3.0-V supply. The proposed frequency synthesizer provides 10-kHz channel spacing with phase noise of −121 dBc/Hz in the PCS band and −127 dBc/Hz in the cellular band, both at 1-MHz offset frequency.

Index Terms—Bonding-wire inductor, CMOS RF, coarse tuning, dual-path loop filter, fractional-N-type prescaler, frequency synthesizer, phase noise, phase-locked loop.

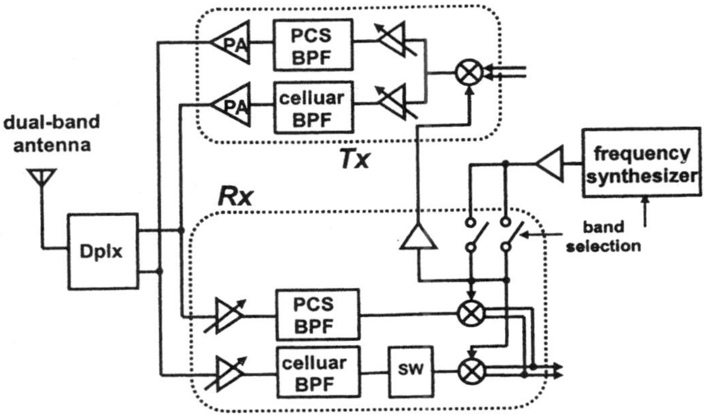

Fig. 1. Dual-band CDMA RF transceiver.

I. INTRODUCTION

WIRELESS systems, such as PCS-CDMA, cellular CDMA, and JSTD-018PCS, require the frequency synthesizer to have precise channel spacing and low phase noise to meet the overall noise specification and to prevent unwanted signal mixing of the interferer. Most existing frequency synthesizers are implemented in silicon germanium (SiGe) or bipolar technologies, and use several external devices such as temperature-compensated crystal oscillator (TCXO) and loop filter. Because of cost and power consumption requirements, fully integrated CMOS RF building blocks are crucial and have been widely explored [1], [2].

Fig. 1 shows an example of a dual-band RF transceiver architecture for PCS- and cellular CDMA. A local oscillator (LO) signal from a dual-band frequency synthesizer is fed to the first mixer of the receiver for downconversion and it is also used in the transmitter for upconversion. The noise requirement of the frequency synthesizer is determined by the blocking profile of the system, which is calculated from the power of signal and interferer, minimum signal-to-noise ratio (SNR), and bandwidth

specification [3]. The lower the phase noise of the LO signal is, the less unwanted signal around the carrier is modulated within the in-band channel.

Table I shows worldwide mobile frequency standards and RF phase-locked loop (PLL) requirements. CDMA systems require fast switching time with precise accuracy of the channel frequency. The channel raster is 30 kHz in cellular and 50 kHz in PCS systems and, to support the dual-band solution of the CDMA system, the frequency resolution of the synthesizer must be 10 kHz. This is a major limiting factor in the reduction of the locking time and root mean square (rms) phase error. It also makes it difficult to achieve single-chip integration due to the loop filter that has a large time constant.

In Section II, the special features of the proposed frequency synthesizer are discussed. Section III describes several building blocks of the synthesizer. The measurement results are given in Section IV, and conclusions are presented in Section V.

II. SYSTEM ARCHITECTURE

The proposed PLL is a monolithic integrated circuit that performs dual-band RF synthesis for CDMA wireless communication applications without any external device. Fig. 2 shows the block diagram of the fractional-N-type frequency synthesizer architecture. The external reference frequency (f_{ref}) is mainly 19.68 MHz, and 19.8 and 19.2 MHz are also supported. The voltage-controlled oscillator (VCO) oscillates at 1.7 GHz for

Manuscript received July 28, 2001; revised November 9, 2001.
Y. Koo, H. Huh, Y. Cho, D.-K. Jeong, and W. Kim are with the School of Electrical Engineering and Computer Science, Seoul National University, Seoul 151-742, Korea (e-mail: ydkoo@iclab.snu.ac.kr).
J. Lee, J. Park, and K. Lee are with GCT Semiconductor, Inc., San Jose, CA 95131 USA.
Publisher Item Identifier S 0018-9200(02)03676-4.

TABLE 1
WORLDWIDE MOBILE FREQUENCY STANDARDS AND RF-PLL REQUIREMENTS

	Tx Band (MHz)	Rx Band (MHz)	Channel BW	1st LO (MHz)	Switching time	Frequency resolution
AMPS	824-849	869-894	30kHz	954.39-979.35	Slow	30kHz
Cellular(DCS)	824-849	869-894	1.25MHz	954.39-979.35	Slow	30kHz
KPCS(IS95C)	1750-1780	1840-1870	1.25MHz	1619.62-1649.62	~500μs	10kHz
USPCS(Tri-I)	1850-1910	1930-1990	1.25MHz	1719,62-1779.62	~800μs	10kHz
GSM	890-915	935-960	200kHz	-	-	200kHz

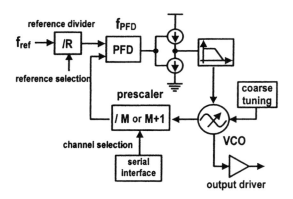

Fig. 2. Fractional-N frequency synthesizer architecture.

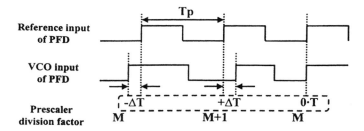

Fig. 3. Timing diagram of reference and VCO inputs of PFD in locked state when the fraction is 1/3.

the PCS band, and at 900 MHz for the cellular band, and employs bonding wire as the inductor of the inductance–capacitance (LC) tank.

To meet the frequency resolution of 10 kHz, f_{PFD} is 10 kHz and loop bandwidth is only 1 kHz with an integer-N-type prescaler. To reduce rms phase error, it is very important for the PLL to have a wide bandwidth. The fractional-N architecture, compared with its integer-N counterpart, has a wider loop bandwidth with the same frequency resolution. However, it suffers from a major drawback called fractional spur.

The fractional-N structure is employed for the prescaler in the proposed architecture. Channel selection and other control signals are provided through a serial interface. To suppress the fractional spur, a special type of charge pump is designed. The new scheme of the dual-path loop filter enables flexible filter design for on-chip integration. The VCO combined with current-feedback bias and coarse tuning enables constant power and low gain of the VCO.

III. SYNTHESIZER BUILDING BLOCKS

A. Charge Pump With Charge-Averaging Scheme

While the reference spur occurs in integer-N synthesis due to charge-pump mismatch, fractional-N synthesis suffers from the fractional spur caused by the phase difference in the locked state as well as charge-pump mismatch. Fig. 3 shows the timing diagram of the reference and the VCO inputs of the phase/frequency detector (PFD) in the locked state.

The fractional spur mainly stems from the variation of the prescaler division factor, in other words, the phase difference

between reference and VCO inputs at every cycle of operation of the PFD. For example, to achieve a locking at 1/3 fractional frequency, as shown in Fig. 3, the prescaler division factor is $M + 1$ in one cycle and M in the other two cycles in every successive three cycles. It produces voltage ripples in the control signal of the VCO and therefore a fractional spur occurs. However, the average phase error during one circulation is zero in the locked state. As seen in Fig. 3, the sum of successive phase errors during three clock cycles is zero. This is the motivation of the proposed scheme.

Fig. 4 shows the proposed charge-averaging scheme and its operation. The charge pump is composed of four current sources and four sampling capacitors. The same up/dn signals are fed to each current source, which has 1/3 of the total current. Since the fractional coefficient of fractional-N is three in this design, we use four pairs of switches and four capacitors in the charge pump. Each pair of switches $(sw_1$–$sw_4)$ and capacitor stores the charge that is injected from the pump during $3T_{ref}$ and then dumps it in the next period in turn. In the locked state, the sum of the phase errors in the three cycles is zero; therefore, the voltage of C_4 after charge summing during $3T_{ref}$ is the same as V_{ctrl}. This results in no voltage ripple in the dump mode. The switching noise due to charge sharing and clock feedthrough may affect the amount of charge in the capacitors. These could be static errors. However, they influence each capacitor equally. This type of static dc error does not cause voltage ripples, i.e., a fractional spur.

Fig. 5 shows the behavioral simulation results of the conventional fractional-N architecture and the proposed charge-averaging scheme. In simulation, the VCO gain is set as perfectly linear and neither current mismatch nor clock feedthrough in the switch is assumed. Other characteristics are the same in both cases, except for the charge pump. In the conventional scheme

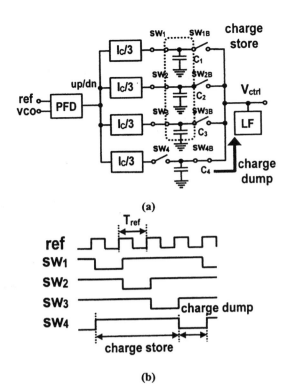

(a)

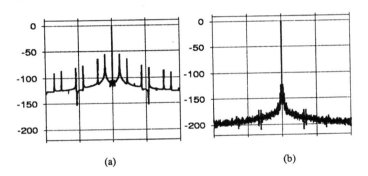

(b)

Fig. 4. Charge-averaging scheme. (a) Block diagram. (b) Its operation.

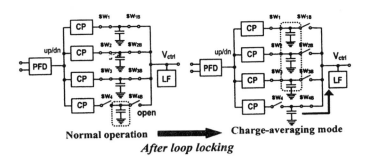

Fig. 6. Mode change of charge-pump operation.

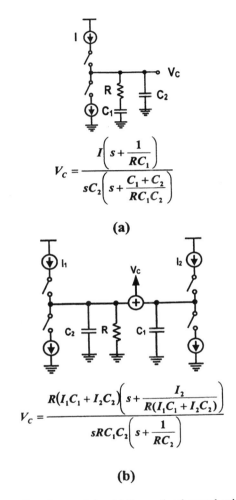

Fig. 7. Loop filter characteristics. (a) Conventional second-order loop filter. (b) Dual-path loop filter.

Fig. 5. Behavioral simulation of spur in fractional-N architecture. (a) With conventional scheme. (b) With proposed charge-averaging scheme.

shown in Fig. 5(a), approximately -50 dBc of fractional spur is found, but in the proposed scheme shown in Fig. 5(b), the fractional spur is suppressed to the noise level.

With respect to the locking time, the charge-averaging scheme can exhibit an undesirable effect. If the size of the sampling capacitor is very small compared to that of the loop filter capacitor, the locking time is increased. To solve this problem, in the acquisition mode, the charge pump operates in the same way as the normal charge pump, which means that the charge pump is directly connected to the loop filter. After locking to the desired frequency, the charge-averaging mode is employed (Fig. 6). Therefore, there is no additional burden in locking time.

In the ac analysis of the loop characteristic, time delay degrades the phase margin. As the time delay goes larger, the system becomes more unstable. The averaging method in this scheme results in an added time delay in the loop characteristic.

Therefore, there exists a tradeoff between loop bandwidth and loop stability.

B. Dual-Path Loop Filter

Most PLLs use a second-order loop filter to suppress the control voltage ripple and to guarantee an appropriate phase margin. Fig. 7(a) shows a conventional second-order loop filter and the expression for control voltage in ac analysis. As described above, the frequency resolution is 10 kHz and the bandwidth of the PLL is a few kilohertz. This means that more than 1 nF of capacitance should be integrated on a chip, which is

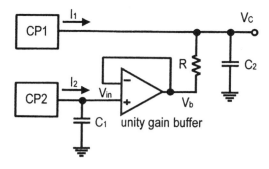

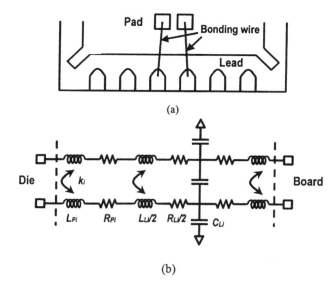

Fig. 8. Proposed architecture of dual-path loop filter.

Fig. 9. Bonding wire for inductor of VCO. (a) Pad and lead frame. (b) Modeling.

the major limiting factor for on-chip integration. In addition, the thermal noise generated in a large resistor is modulated to phase noise via the control signal. The dual-path loop filter in Fig. 7(b) can be a solution for this problem. It separates the loop with I_1 and I_2, so it is possible to design the loop filter more freely in conjunction with the pump current while keeping the loop transfer function nearly the same as that of a normal second-order loop filter.

An example of the dual-path loop filter was proposed by Craninckx and Steyaert [4]. In spite of many advantages inherent in this architecture, it has two active devices, an amplifier and a current adder. These inject additional noise into the control voltage, and after modulation in the VCO, the phase noise may increase. In addition, the floating capacitor across the amplifier is implemented with the metal-to-metal capacitor, so it requires a large area.

Fig. 8 shows the new dual-path loop filter implementation that is proposed in this paper. A unity-gain buffer is inserted between C_1 and R to separate I_1 and I_2. If V_b continuously follows V_{in}, the operation is the same as that of the normal second-order loop filter. It is less noisy because there is only one active device and requires a smaller area since there is no floating capacitor. C_1 and C_2 are implemented using two pMOS transistors, whose source, drain, and bulk are tied to a separate, quiet supply.

C. Voltage-Controlled Oscillator

Two major issues in the design of the VCO are low phase noise to meet the overall noise figure criteria and high gain linearity for robust stability. Phase noise is mainly dependent on the quality factor Q of the LC tank [5]. Although an on-chip spiral inductor has recently been widely explored [6], a bonding-wire inductor is superior to a spiral inductor in terms of resistance, i.e., quality factor. In addition, the bonding-wire inductor has constant inductance over a wide frequency range. Fig. 9 shows bonding-wire inductor modeling. Two pads are connected to the differential output of the VCO and the ends of two lead frames are connected as a short or by external inductance, according to the operating band, PCS or cellular. The Q factor of the inductance is expressed as $\omega_C L/R_S$ with parasitic components ignored. If the parameters of the bonding wire are $L_{\text{Pi}} = 1.3$ nH, $R_{\text{Pi}} = 0.3\ \Omega$, $L_{\text{Li}} = 6$ pH, $R_{\text{Li}} = 25$ mΩ, which are typical values in the QFN20 package, the Q factor of the inductor at 1.7 GHz is 43.

Fig. 10 shows the circuit diagram of the VCO. The oscillation frequency is controlled by the combination of fixed and

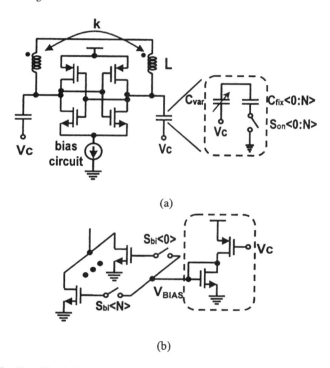

Fig. 10. Circuit diagram of (a) VCO and (b) bias circuit.

variable capacitor. There are two methods that have been previously reported for implementation of the fixed capacitor. One is a metal-to-metal capacitor [1] and the other is a MOS transistor [7]. The former is used since it is superior in terms of VCO pushing characteristics. Metal-to-metal capacitors and switches are used for coarse tuning, which are controlled by the coarse tuning controller. The size of the switch $S_{\text{ON}}\langle 0{:}N\rangle$ should be sufficiently large in order to avoid degradation of the Q factor of the LC tank. As a variable capacitor, an accumulation-mode MOS transistor is used for fine tuning. The MOS capacitor has an inherent nonlinear capacitance. But, with the coarse tuning scheme, the control voltage moves within ± 0.2 V around half of the supply voltage, thereby obtaining almost linear gain.

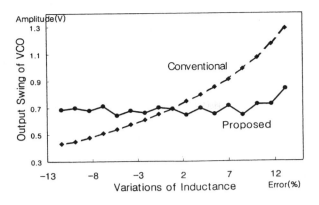

Fig. 11. Output swing of VCO versus variations of inductance.

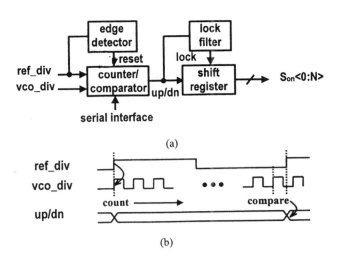

(a)

(b)

Fig. 13. Coarse tuning controller. (a) Block diagram. (b) Timing diagram of operation.

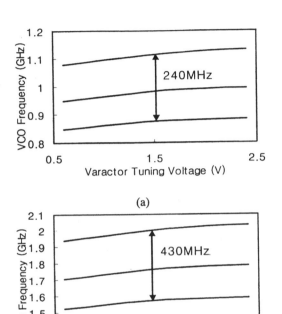

(a)

(b)

Fig. 12. VCO tuning range of (a) cellular band, and (b) PCS band.

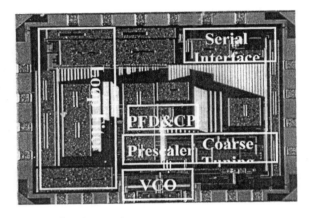

Fig. 14. Die microphotograph.

The output swing level of the VCO is another key issue, since the receiver and transmitter chips expect an LO signal of constant power. Generally bonding-wire inductance varies by ±10% and to compensate, the total capacitance varies accordingly. This produces the variation of the VCO swing level. The bias circuit in Fig. 10(b) is designed to have a constant output swing regardless of capacitance. It monitors the operating status of the fixed and variable capacitors and provides current in the direction of compensating for the output swing. Fig. 11 shows the VCO swing variation for the conventional and proposed scheme when the inductance varies.

Fig. 12 shows the measured frequency tuning range in the cellular and PCS band. More than 25% of the tuning range is obtained in each case. It is sufficient to compensate for the variation of bonding-wire inductance. The VCO gain is a maximum

of 40 MHz/V, and 30 MHz/V is typical. The total range is divided into 64 levels by coarse tuning, and the frequency spacing of two adjacent curves is approximately 7 MHz in the PCS band.

D. Coarse Tuning Architecture

Small VCO gain is important for reduction of both spur and phase noise. The frequency spur is directly proportional to the VCO gain. Also, as the VCO gain is reduced, the fluctuation of control and supply voltages are less modulated to the phase noise of the VCO. To meet the requirement of the wide frequency range and the small VCO gain, the coarse tuning controller, shown in Fig. 13, is designed to control the fixed capacitor in the VCO.

The coarse tuning controller, shown in Fig. 13, is composed of an edge detector, counter/comparator, lock filter, and shift register. During coarse tuning, the rising edges of vco_clk are counted in one period of ref_clk and the result is compared to a predetermined value of desired frequency. Starting from the center frequency, a proper level is found by a successive approach. The lock-detection filter determines when to start the fine tuning process by monitoring the up/dn signal. The total elapsed time in coarse tuning is less than 100 μs.

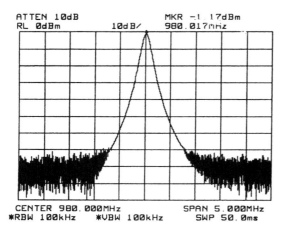

Fig. 15.   Measured carrier spectrum.

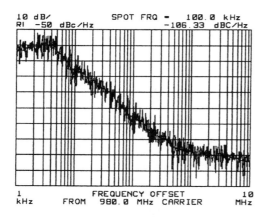

(a)

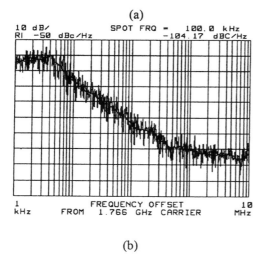

(b)

Fig. 16.   Measured PLL output phase noise. (a) Cellular band. (b) PCS band.

IV. EXPERIMENTAL RESULTS AND SUMMARY

The proposed frequency synthesizer has been fabricated in a 0.35-μm CMOS technology. Fig. 14 shows the die photograph of the synthesizer with an area of 2.5 mm × 2.0 mm including pads. The circuit has been measured with a nominal 3.0-V supply and a 2.7-V worst case. The bonding wire of the QFN20 package used in the VCO has 1.36 nH of self-inductance and 0.31 nH of mutual inductance, so the total inductance

TABLE II
SUMMARY OF SYNTHESIZER PERFORMANCE

Synthesized frequency: PCS	1.55GHz ~ 1.98GHz
cellular	0.86GHz ~ 1.10GHz
Reference frequency	19.68MHz (19.8MHz, 19.2MHz)
VCO gain	40MHz/V(max.), 30MHz/V(typ.)
Frequency resolution	10kHz
Phase noise: PCS	-104dBc/Hz @100kHz offset
cellular	-106dBc/Hz @100kHz offset
Power dissipation: VCO	12.3mW
total	60.0mW
Supply voltage	3.0V (2.7V at worst)
Die area	2.5mm x 2.0mm
Switching time	< 800μsec.
Technology	0.35μm CMOS

at one output of the VCO is 1.05 nH. All control signals are fed through a serial interface. The VCO at the bottom and the loop filter on the left have a common analog supply and ground, and the others are all connected to the digital supply and ground. The power consumption of the total chip is 60 mW and the VCO alone dissipates 12.3 mW.

Fig. 15 shows the measured carrier spectrum with the center frequency of 980 MHz. The output power is −1.2 dBm with an inductive load, which is sufficient for the output power requirements. Fig. 16 shows the measured phase noise in the cellular and PCS band. Phase noise is −106 dBc/Hz at 100-kHz offset and −127 dBc/Hz at 1-MHz offset in the cellular band, and −104 dBc/Hz at 100-kHz offset and −121 dBc/Hz at 1-MHz offset in the PCS band. Fractional spurs are suppressed to the phase noise level. Table II shows the performance summary.

V. CONCLUSION

In this paper, we demonstrate a fully integrated CMOS frequency synthesizer designed for PCS- and cellular-CDMA wireless systems. A charge-averaging scheme for reducing fractional spurs and a dual-path loop filter architecture are proposed. The new bias circuit of the VCO compensates for the variation of output swing of the VCO caused by the variation of bonding-wire inductance, and the proposed coarse tuning technique achieves a small VCO gain and a wide operating frequency range of the VCO simultaneously. The frequency synthesizer fabricated in a 0.35-μm CMOS technology offers −127-dBc/Hz and −121-dBc/Hz phase noise at 1-MHz offset with 980 MHz and 1.76 GHz of carrier frequency, respectively.

REFERENCES

[1] A. Kral, F. Behbahani, and A. A. Abidi, "RF-CMOS oscillators with switched tuning," in *Proc. IEEE Custom Integrated Circuits Conf.*, May 1998, pp. 555–558.
[2] C. Lam and B. Razavi, "A 2.6-GHz/5.2-GHz frequency synthesizer in 0.4-μm CMOS technology," in *Symp. VLSI Circuits Dig. Tech. Papers*, June 1999, pp. 117–120.
[3] B. Razavi, *RF Microelectronics*. Upper Saddle River, NJ: Prentice Hall, 1998.
[4] J. Craninckx and M. Steyaert, "A fully integrated CMOS DCS-1800 frequency synthesizer," *IEEE J. Solid-State Circuits*, vol. 33, pp. 2054–2065, Dec. 1998.

[5] D. B. Leeson, "A simple model of feedback oscillator noise spectrum," *Proc. IEEE*, vol. 54, pp. 329–330, Feb. 1966.

[6] S. Mohan, M. Hershenson, S. Boyd, and T. H. Lee, "Simple accurate expressions for planar spiral inductances," *IEEE J. Solid-State Circuits*, vol. 34, pp. 1419–1424, Oct. 1999.

[7] J.-M. Mourant, J. Imbornone, and T. Tewksbury, "A low phase noise monolithic VCO in SiGe BiCMOS," in *IEEE Radio Frequency Integrated Circuits (RFIC) Symp. Dig.*, June 2000, pp. 65–68.

Yido Koo was born in Seoul, Korea, in 1973. He received the B.S. and M.S. degrees from the School of Electrical Engineering, Seoul National University, Seoul, Korea, in 1996 and 1998, respectively, where he is currently working toward the Ph.D. degree.

His research interests include RF building blocks and systems for wireless communication and high-speed interface for data communications. Currently, he is developing a low-noise frequency synthesizer for CDMA and GSM applications.

Hyungki Huh was born in Seoul, Korea. He received the B.S. and M.S. degree in electrical engineering from Seoul National University, Seoul, Korea, in 1998 and 2001, respectively, where he is currently working toward the Ph.D. degree in electrical engineering.

His research interests are in the area of RF circuits and systems with emphasis on the fractional frequency synthesizer.

Yongsik Cho was born in Daegu, Korea. He received the B.S. degree in electrical engineering from Seoul National University, Seoul, Korea, in 2000, where he is currently working toward the M.S. degree in electrical engineering.

His research interests are in the area of RF circuits and systems.

Jeongwoo Lee received the B.S. and M.S. degrees in electronics engineering and the Ph.D. degree in electrical engineering from Seoul National University, Seoul, Korea, in 1994, 1996, and 2000, respectively.

He is currently a Manager with the W-CDMA team of GCT Semiconductor Inc., San Jose, CA. His current research interests include CMOS transceiver circuitry for highly integrated radio applications.

Joonbae Park received the B.S. and M.S. degrees in electronics engineering and the Ph.D. degree in electrical engineering from Seoul National University, Seoul, Korea, in 1993, 1995, and 2000, respectively.

In 1998, he joined GCT Semiconductor Inc., San Jose, CA, as Director of the Analog Division. He is currently involved in the development of CMOS RF chip sets for WLL, W-CDMA, and wireless LAN. His other research interests include data converters and high-speed communication interfaces.

Dr. Park received the Best Paper Award of VLSI Design'99, Goa, India.

Kyeongho Lee was born in Seoul, Korea, in 1969. He received the B.S. and M.S. degrees in electronics engineering and the Ph.D. degree in electrical engineering from Seoul National University, Seoul, Korea, in 1993, 1995, and 2000, respectively.

He was with Silicon Image, Inc., Sunnyvale, CA, as a Member of Technical Staff, where he worked on CMOS high-bandwidth low-EMI transceivers. He is currently with GCT Semiconductor Inc., San Jose, CA, as a Co-Chief Executive Officer. His research interests include various CMOS high-speed circuits for wire/wireless communication systems and integrated CMOS RF systems.

Deog-Kyoon Jeong received the B.S. and M.S. degrees in electronics engineering from Seoul National University, Seoul, Korea, in 1981 and 1984, respectively, and the Ph.D. degree in electrical engineering and computer sciences from the University of California, Berkeley, in 1989.

From 1989 to 1991, he was with Texas Instruments Inc., Dallas, TX, where he was a Member of Technical Staff and worked on the modeling and design of BiCMOS gates and the single-chip implementation of the SPARC architecture. He joined the faculty of the Department of Electronics Engineering and Inter-University Semiconductor Research Center, Seoul National University, as an Assistant Professor in 1991. He is currently an Associate Professor of the School of Electrical Engineering, Seoul National University. His main research interests include high-speed I/O circuits, VLSI systems design, microprocessor architectures, and memory systems.

Wonchan Kim was born in Seoul, Korea, on December 11, 1945. He received the B.S. degree in electronics engineering from Seoul National University, Korea, in 1972. He received the Dip.-Ing. and Dr.-Ing. degrees in electrical engineering from the Technische Hochschule Aachen, Aachen, Germany, in 1976 and 1981, respectively.

In 1972, he was with Fairchild Semiconductor Korea as a Process Engineer. From 1976 to 1982, he was with the Institut für Theoretische Electrotecnik RWTH Aachen. Since 1982, he has been with the School of Electrical Engineering, Seoul National University, where he is currently a Professor. His research interests include development of semiconductor devices and design of analog/digital circuits.

A 1.1-GHz CMOS Fractional-N Frequency Synthesizer with a 3-b Third-Order $\Delta\Sigma$ Modulator

Woogeun Rhee, *Member, IEEE*, Bang-Sup Song, *Fellow, IEEE*, and Akbar Ali

Abstract—A 1.1-GHz fractional-N frequency synthesizer is implemented in 0.5-μm CMOS employing a 3-b third-order $\Delta\Sigma$ modulator. The in-band phase noise of -92 dBc/Hz at 10-kHz offset with a spur of less than -95 dBc is measured at 900.03 MHz with a phase detector frequency of 7.994 MHz and a loop bandwidth of 40 kHz. Having less than 1-Hz frequency resolution and agile switching speed, the proposed system meets the requirements of most RF applications including multislot GSM, AMPS, IS-95, and PDC.

Index Terms—Analog integrated circuits, CMOS RF, delta–sigma modulation, frequency synthesizer, phase-locked loop.

I. Introduction

THE DEMAND for low-cost universal frequency synthesizers is growing as wireless systems are diversified. Standard frequency synthesizers with integer-N dividers have difficulties in meeting various specifications due to their fundamental tradeoffs between loop bandwidth and channel spacing. The fractional-N technique offers wide bandwidth with narrow channel spacing and alleviates phase-locked loop (PLL) design constraints for phase noise and reference spur [1], [2]. Since it uses a lower division ratio and a higher phase detector frequency than the integer divider technique, the low-frequency phase noise can be suppressed to a higher degree.

For fractional-N frequency synthesis, two types of $\Delta\Sigma$ modulators have been used [3], [4]. One is a single-loop modulator, and the other is a cascaded modulator called MASH. The single-loop modulator has a choice of a single-bit or a multibit output depending on the quantizer while the MASH architecture outputs only multibits. For wide-band frequency synthesizers, the modulator architecture should be carefully selected. In addition to the in-band noise shaping, the out-of-band shaped noise significantly affects the synthesizer performance by raising the high-frequency phase noise. The high-frequency noise is difficult to suppress with the finite number of the PLL loop filter poles. Comparing the output bit patterns of the multibit modulator and the MASH, the former can achieve a more desirable

Manuscript received December 2, 1999; revised May 10, 2000.
W. Rhee was with the Department of Electrical and Computer Engineering, University of Illinois, Urbana, IL 61801 USA. He is now with Conexant Systems, Inc., Newport Beach, CA 92660-3007 USA (e-mail: rheew@nb.conexant.com).
B.-S. Song was with the Department of Electrical and Computer Engineering, University of Illinois, Urbana, IL 61801 USA. He is now with the Department of Electrical and Computer Engineering, University of California at San Diego, La Jolla, CA 92093 USA.
A. Ali is with Conexant Systems, Inc., Newport Beach, CA 92660 USA.
Publisher Item Identifier S 0018-9200(00)08701-1.

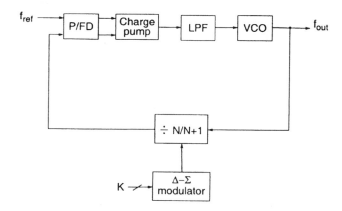

Fig. 1. $\Delta\Sigma$ modulated fractional-N frequency synthesizer.

noise shaping for frequency synthesis, but the latter offers a simpler high-order architecture with no stability problem [5], [6].

The MASH-type modulators tend to generate wide-spread high-frequency bit patterns, and impose more stringent requirements on the phase detector design. This work optimizes a noise shaping function for low-spur frequency synthesis using a 3-b third-order modulator that generates less high-frequency noise and makes the system less sensitive to the substrate noise coupling.

In Section II, design issues related to the low phase noise frequency synthesis are addressed. Section III highlights the system design features, and experimental results are summarized in Section IV.

II. Design Considerations for High Spectral Purity

The fractional-N frequency synthesis system as illustrated in Fig. 1 obtains a fine frequency resolution by interpolating a fractional division using an oversampling $\Delta\Sigma$ modulator with a coarse integer divider [3]. The $\Delta\Sigma$ modulation technique is similar to the random jittering method [7], but it does not exhibit a $1/f^2$ phase noise spectrum due to its noise shaping property. When $\Delta\Sigma$ modulators with orders higher than two are used, the PLL needs extra poles in the loop filter to suppress the quantization noise at high frequencies.

A. Bandwidth Requirement

If the in-band phase noise of $10 \log A_n$ [dBc/Hz] of the frequency synthesizer is assumed to be limited within the noise bandwidth of f_c [Hz] as shown in Fig. 2, the integrated frequency noise Δf_n [rms Hz] within f_c is approximately [8]

$$\Delta f_n \cong \sqrt{\frac{2}{3} A_n} \cdot f_c^{(3/2)}. \tag{1}$$

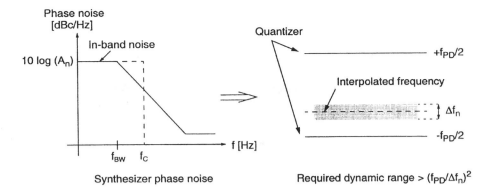

Fig. 2. Dynamic range consideration in oversampled fractional division.

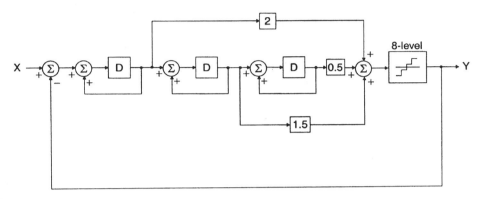

Fig. 3. Third-order $\Delta\Sigma$ modulator with 3-b quantizer.

Because the quantizer level in the frequency domain is equivalent to the phase detector frequency f_{PD} with the frequency noise of Δf_n as illustrated in Fig. 2, the dynamic range of the Lth-order $\Delta\Sigma$ modulator should meet the following condition [9].

$$\frac{3}{2} \cdot \frac{2L+1}{\pi^{2L}} \cdot (\text{OSR}_{\text{eff}})^{2L+1} > \left(\frac{f_{PD}}{\Delta f_n}\right)^2 \quad (2)$$

where the effective oversampling ratio OSR_{eff} is given by

$$\text{OSR}_{\text{eff}} = \frac{f_{PD}}{2f_c}. \quad (3)$$

Therefore, from (1)–(3), we obtain

$$f_c < \left[A_n \cdot \frac{L+0.5}{(2\pi)^{2L}}\right]^{(1/2L-2)} \cdot f_{PD}^{(2L-1/2L-2)}. \quad (4)$$

An integrated phase error θ_{rms} [rms rad] is an important factor for synthesizers in digital communications, and it is given by

$$A_n = \left(\frac{\theta_{\text{rms}}}{\sqrt{2}}\right)^2 \cdot f_c^{-1}. \quad (5)$$

From (4) and (5), an approximate upper bound of the bandwidth is obtained as

$$f_c < \left[\left(\frac{\theta_{\text{rms}}}{\sqrt{2}}\right)^2 \cdot \frac{L+0.5}{(2\pi)^{2L}}\right]^{(1/2L-1)} \cdot f_{PD}. \quad (6)$$

Equation (6) explains the effect of the integrated phase error as a parameter, which is not included in the previous results [3], [4].

For example, when the phase detector frequency is 8 MHz, the upper bound of the bandwidth with a third-order $\Delta\Sigma$ modulator to meet less than $1°$-rms phase error is 195 kHz. In practice, the required loop bandwidth is narrower than as predicted by (6) since the quantization noise of the third-order modulator is tapered off after the fourth pole of the PLL. In this work, the loop bandwidth is set to 40 kHz with the third pole placed at 160 kHz.

B. Noise Transfer Function

Fig. 3 shows the proposed third-order modulator with an eight-level quantizer. The eight-level quantizer expands the active division range from $\{N, N+1\}$ to $\{N-3, N-2, \cdots, N+3, N+4\}$ without increasing the minimum quantizer level. For high-order modulators, it has been shown that as the number of quantizer levels increases, the maximum passband gain of the noise transfer function (NTF) can be increased without causing any nonlinear stability problem [10], [11]. As the maximum passband gain of the NTF increases, the corresponding corner frequency increases. For example, if the input range is set to about 80% of the quantizer, the maximum passband gain of the NTF can be set to 2.5 for a 2-bit quantizer, 3.5 for a 3-bit quantizer, and 5.0 for a 4-bit quantizer. The corresponding corner frequencies of the NTF are $0.13 f_s$, $0.19 f_s$ and $0.24 f_s$, respectively. This implies that quantization noise of the third-order modulator can be further suppressed by 16 dB with a 2-bit quantizer, 22 dB with a 3-bit quantizer, and 25 dB with a 4-bit quantizer [12].

597

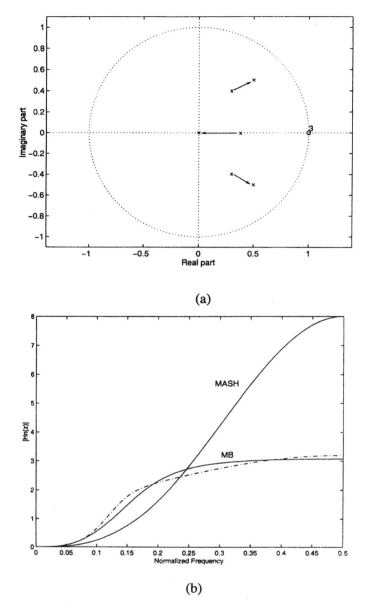

(a)

(b)

Fig. 4. Proposed modulator: (a) pole-zero plot, and (b) noise transfer function.

The NTF is derived from the high-order topology [10] as

$$H_n(z) = \frac{(1 - z^{-1})^3}{1 - z^{-1} + 0.5z^{-2} - 0.1z^{-3}}. \quad (7)$$

To avoid digital multiplication, the coefficients of $\{2, 0.5, 1.5\}$ are used to implement them using shift operations. This constraint slightly modifies the original NTF, but it still maintains the causality and the stability conditions. The poles of the NTF are designed to be within the unit circle in the z-domain as shown in Fig. 4(a). Low-Q Butterworth poles are used to reduce the high-frequency shaped noise energy, which results in a low spread output bit pattern. As shown in Fig. 4(b), the NTF of the proposed modulator has the passband gain of 3.1 and the corner frequency of $0.18\ f_s$ for the clock frequency f_s.

In Fig. 5, the proposed architecture is compared to the high-order modulator with single-bit quantizer and the MASH modulator. The high-order modulator with single-bit quantizer has a

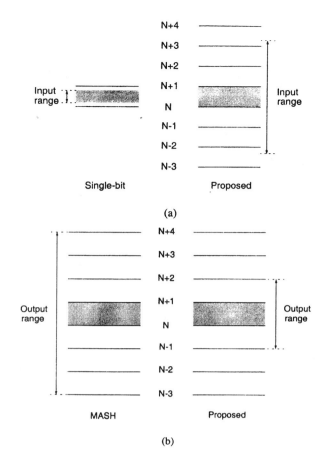

(a)

(b)

Fig. 5. Architecture comparison: (a) single-bit and proposed, and (b) MASH and proposed.

dead-band problem due to the limited input range of the quantizer in synthesizer applications. As shown in Fig. 5(a), the extended input range with the multilevel quantizer helps to reduce the nonideal effects at the band edges. Either a fourth-order modulator with a modulus-4 divider or a second-order modulator with a modulus-16 divider may perform well for the purpose. The former has more noise shaping while the latter has the same performance with a higher corner frequency. Although the MASH topology with the same order can shape the in-band noise more sharply, it produces an output bit pattern spread more widely than the proposed noise shaper does, as shown in Fig. 5(b). Widely spread output bit pattern makes the synthesizer more sensitive to the substrate noise coupling since the modulated turn-on time of the charge pump in the locked condition increases.

Fig. 6 shows the time-domain simulation of the division ratio for the 1000 sequences generated by the 3-b third-order modulator. The simulation is done with the behavioral model of the gate-level modulator in PSPICE. The fractional division ratio is set to $1/4 + 1/2^7$ and the 16th bit is used for dithering. That is, the actual fractional division ratio is $1/4 + 1/2^7 + 1/2^{16}$. Note that this interpolator uses mostly the closely-spaced division values of N, $N - 1$, and $N + 1$ to generate the fractional value. The fast Fourier transform (FFT) of the modulator output is shown in Fig. 7, as predicted from the NTF in Fig. 4(b).

The discrete Fourier transform does not provide the true power spectrum particularly when the signal is aperiodic or

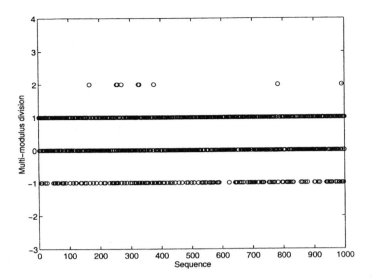

Fig. 6. 3-b third-order modulator output stream for $N + 1/4 + 1/2^7$ division with dithering in time domain.

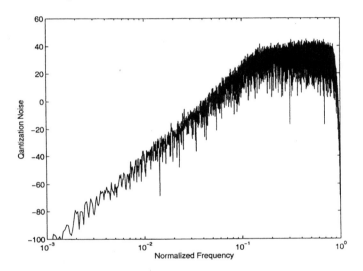

Fig. 7. FFT of 3-b third-order modulator output for $N + 1/4 + 1/2^7$ division with dithering.

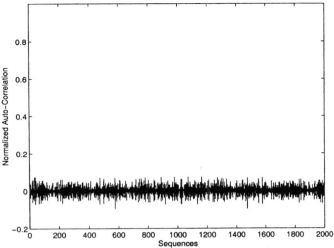

Fig. 8. Autocorrelation of 2000 samples with $N + 1/2^7 + 1/2^{16}$ division.

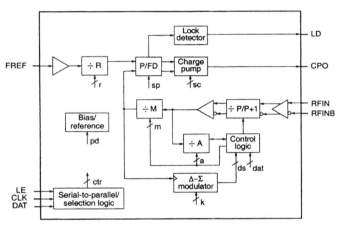

Fig. 9. Functional block diagram of the synthesizer.

random. The autocorrelation estimate is used for 2000 output samples with the fractional division ratio of $1/4 + 1/2^7 + 1/2^{16}$ as shown in Fig. 8. It is known that the high-order noise shaping with multibit quantization makes dithering more efficient by allowing a large dithered signal at the quantizer input.

III. SYNTHESIZER IMPLEMENTATION

Fig. 9 shows the functional block diagram of the proposed fractional-N frequency synthesizer. Since a third-order $\Delta\Sigma$ modulator is used, a type-2 fourth-order PLL having two additional out-of-band poles is used to filter out the quantization noise at high frequencies. The system is configured to be compatible with existing integer-N frequency synthesizers. The P/FD and the charge pump are designed to have four different sets of the phase detector gain and to work with both positive- and negative-gain voltage-controlled oscillators (VCO). The differential 8/9 prescaler is used with a dual-modulus 4/5 divider and a toggle flip-flop. The bandgap reference circuit generates a temperature-independent output current for the charge-pump. It also keeps the PLL bandwidth constant over temperature. The control logic takes the 3-b output of the $\Delta\Sigma$ modulator and provides the randomized data to the counters. A pseudorandom sequence with a length of 2^{24} is used for LSB dithering. The fine frequency resolution of less than 0.001 ppm can make the synthesizer compensate for the crystal-frequency drift with a digital word. It can also accommodate various crystal frequencies without reducing the phase detector frequency.

Since the counters operate at higher than 120 MHz for the 1.1-GHz output, asynchronous counters are used to save power. The counter block diagram is shown in Fig. 10 with the timing diagram. In the example, the modulus data of the main and auxiliary counters are set to 8 and 3, respectively. To absorb the logic delays in the asynchronous operation and in the differential-to-single-ended conversion, a D-flip-flop is added to each counter and triggered by the input clock. It also prevents the jitter from accumulating in the asynchronous counter.

The use of a tri-state charge pump is important to minimize the substrate noise coupling when the $\Delta\Sigma$ modulator is on the same die. By turning it on briefly between the rising and the falling edges of the reference, the substrate noise coupling can

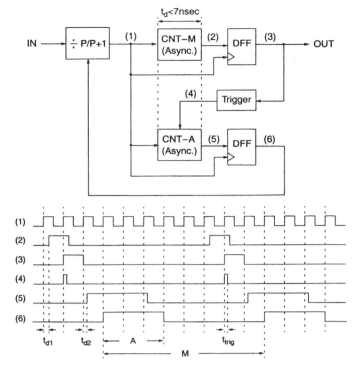

Fig. 10. Frequency divider using asynchronous counters and timing diagram example ($M = 8$, $A = 3$).

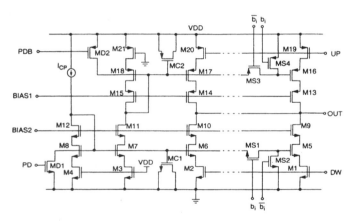

Fig. 11. Programmable charge pump.

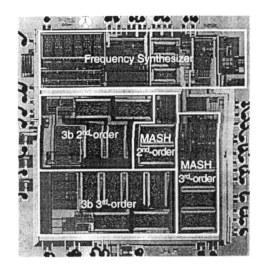

Fig. 12. Die photograph.

be significantly reduced [3], [5]. Fig. 11 shows the schematic of a programmable charge pump designed to minimize the turn-on time of the P/FD without creating a dead zone. Having the switches, M1, M2, M19, and M20, at the source of the current mirror improves the switching speed while keeping the switching noise low. Current mirrors, M5–M18, are cascoded to increase output impedance and four different output currents can be generated with the control bit b_i at each stage. The capacitors, MC1 and MC2, are added to reduce the charge coupling to the gate and to enhance the switching speed. The control bit PD and the complementary bit PDB force the current mirrors to be turned off during the power-down mode. Note that the P/FD and the charge pump are triggered at the falling edge of the clock to reduce the substrate noise coupling because the $\Delta\Sigma$ modulator is triggered at the rising edge. The clock for the $\Delta\Sigma$ modulator is slightly delayed so that the turn-on time of

the P/FD can be separated from the falling edge of the clock for further reduction of the substrate noise coupling. The output voltage compliance of the charge pump is designed to be larger than the range of 0.5–2.5 V with 3-V supply over process and temperature variations. In fractional-N frequency synthesis, the phase detector linearity is important to lower the in-band noise and the idle tones.

IV. EXPERIMENTAL RESULTS

The prototype synthesizer with second- and third-order $\Delta\Sigma$ modulators was fabricated in 0.5-μm CMOS. The die photo is shown in Fig. 12. The chip area is 3.16×3.49 mm^2, including two other MASH modulators.

Fig. 13(a) shows the measured output spectrum at 900.03 MHz with the 3-b second- and third-order $\Delta\Sigma$ modulators. They are compared by switching the output bits of each modulator without changing any loop parameter of the synthesizer. The third-order modulator case shows less out-of-band noise, as expected. With the 8-MHz phase detector frequency, a -45 dBc spur appears at about 60-kHz offset and it is suppressed to -80 dBc with the 3-kHz loop bandwidth. However, no fractional spur was observed when the phase detector frequency is set to 7.994 MHz. The reference spur at 7.994-MHz offset was less than -95 dBc. From the experiment, the spur results from the relation between the output frequency and the phase detector frequency, and it becomes more significant when the output frequency approaches the rational multiples of the phase detector frequency. Even though the spurs are observed, the overall performance exceeds that of any integer-N synthesizers reported to date. For example, when the output frequency is programmed in a 200-kHz step with the phase detector frequency of 6.4 MHz and the loop bandwidth of 15 kHz, the worst-case spur at 200-kHz offset is below -85 dBc for all channels and the in-band phase noise is as low as -90 dBc/Hz.

Fig. 13(b) shows the shaped quantization noise seen at the divider output. The output with the second-order modulator has idle tones at high frequencies but they can be suppressed to a negligible level at the VCO output with a 40-kHz loop bandwidth. The third-order modulator does not exhibit

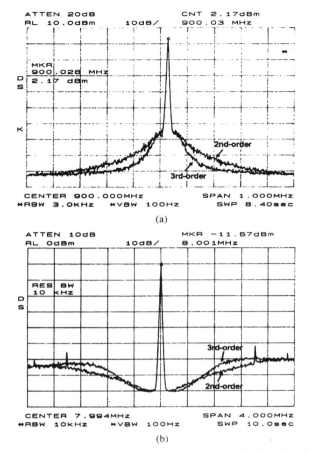

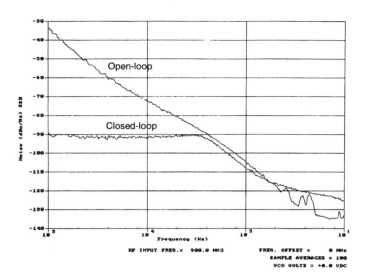

Fig. 15. Measured open-loop and closed-loop output phase noises.

TABLE I
SUMMARY OF THE MEASURED PERFORMANCE.

Supply voltage	2.5 - 4 V
Supply current	10.8 mA
(2.7V analog, 1.5V digital)	< 10 µA (stand-by)
Max. output frequency	1130 MHz
Min. frequency resolution	< 1 Hz
RF input sensitivity	– 15 dBm
Phase noise @ 10-kHz offset	< – 92 dBc/Hz[*]
Spurious tones	< – 95 dBc[*]
	< – 80 dBc[**]

[*] $fo = 900.03$ MHz, $fpd = 7.994$ MHz, and loop BW = 40 KHz
[**] $fo = 900.03$ MHz, $fpd = 8$ MHz, and loop BW = 3 kHz

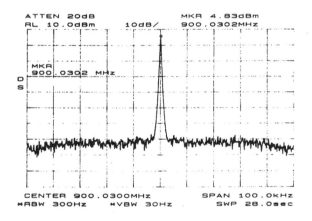

Fig. 13. Measured spectrum: (a) at the VCO output, and (b) at the divider output.

Fig. 14. Measured output spectrum with 100-kHz span.

high-frequency tones near $f_s/2$. Note that the corner frequency of the quantization noise is close to that of the NTF shown in Fig. 4(b).

Fig. 14 shows the spectrum of the third-order $\Delta\Sigma$ modulated output with a 100-kHz span. The loop bandwidth is about 40 kHz and no close-in fractional spur is observed. Fig. 15 shows the synthesizer output phase noise measured at 900.03 MHz. The phase noise of a free-running external VCO is plotted together. As shown in Fig. 15, the phase noise of −92 dBc/Hz at 10-kHz offset frequency is achieved. The phase noise floor from

200–800 kHz is the residual quantization noise of the modulator. The phase noise is −135 dBc/Hz at 3-MHz offset frequency, and it can be further suppressed either by increasing the phase detector frequency or by pushing high-order poles toward the loop bandwidth, sacrificing the loop phase margin. The settling time of the synthesizer within a 50-Hz error for the 100-MHz frequency step is less than 150-μs with the natural loop frequency of 18 kHz. The chip works at 1.1 GHz with an input sensitivity of −15 dBm. The third-order modulator consumes 1.4 mA at 1.5 V. The total current consumption of the synthesizer is 10.8 mA where 5 mA is consumed by the RF input buffer and 1.9 mA by the prescaler. The measured performance is summarized in Table I. Having about 3-kHz loop bandwidth, the prototype synthesizer is useful for AMPS, IS-95, and PDC applications. It can be also employed for multislot GSM applications with a 40-kHz loop bandwidth. Table II shows the performance comparison with the previously published works. Without increasing the phase detector frequency and the oversampling ratio, the noise performance of this work is comparable to that of the synthesizer with the fourth-order MASH

TABLE II
COMPARISON WITH OTHER WORKS

Ref.	Tech.	Arch.	f_o	f_{PD}	f_{BW}	In-band noise	Ref. spur	$\dfrac{f_{PD}}{f_{BW}}$
Filiol et al. [5]	CMOS($\Delta-\Sigma$) BJT(PLL)	4th-order MASH	915 MHz	20 MHz	100 kHz	-95 dBc/Hz	-90 dBc	200
Perrott et al. [6]	0.6 μm CMOS	2nd-order MASH	1.8 GHz (ext. /2)	20 MHz	84 kHz	-74 dBc/Hz	-60 dBc	238
Riley et al. [4]	Discrete	3rd-order 1-bit (from PC)	405 MHz	10 MHz	30 kHz	-85 dBc/Hz	> -82 dBc	333
Miller et al. [3]	Discrete	3rd-order MASH	750 MHz (ext. /2)	200 kHz	750 Hz	-70 dBc/Hz	> -133 dBc	267
This work	0.5 μm CMOS	3rd-order 3-bit	900 MHz	8 MHz	40 kHz	-92 dBc/Hz	-95 dBc	200

modulator that used a 20-MHz reference on the separate CMOS die [5].

V. CONCLUSION

A 1.1-GHz CMOS fractional-N frequency synthesizer with a 3-b third-order $\Delta\Sigma$ modulator is implemented to achieve better in-band phase noise, lower spurs, and faster settling than those of standard integer-N synthesizers. With less than 1-Hz frequency resolution and agile frequency switching, the in-band phase noise and the spur performance of the proposed system meets the requirements of most RF applications including multislot GSM, AMPS, IS-95, and PDC.

ACKNOWLEDGMENT

The authors would like to thank D. Tester, T. Truong, and R. Hlavac at Conexant Systems for their technical support.

REFERENCES

[1] G. C. Gillette, "The digiphase synthesizer," *Freq. Technol.*, pp. 25–29, Aug. 1969.
[2] J. Gibbs and R. Temple, "Frequency domain yields its data to phase-locked synthesizer," *Electronics*, pp. 107–113, Apr. 1978.
[3] B. Miller and R. Conley, "A multiple modulator fractional divider," in *Proc. 44th Annu. Frequency Control Symp.*, May 1990, pp. 559–568.
[4] T. A. Riley, M. Copeland, and T. Kwasniewski, "Delta–sigma modulation in fractional-N frequency synthesis," *IEEE J. Solid-State Circuits*, vol. 28, pp. 553–559, May 1993.
[5] N. Filiol, T. Riley, C. Plett, and M. Copeland, "An agile ISM band frequency synthesizer with built-in GMSK data modulation," *IEEE J. Solid-State Circuits*, vol. 33, pp. 998–1008, July 1998.
[6] M. Perrott, T. Tewksbury, and C. Sodini, "A 27-mW CMOS fractional-N synthesizer using digital compensation for 2.5-Mb/s GFSK modulation," *IEEE J. Solid-State Circuits*, vol. 32, pp. 2048–2060, Dec. 1997.
[7] V. Reinhardt and I. Shahriary, "Spurless fractional divider direct digital synthesizer and method," U.S. Patent 4 815 018, Mar. 21, 1989.
[8] K. Feher *et al.*, *Telecommunications Measurements, Analysis, and Instrumentation*. Englewood Cliffs, NJ: Prentice Hall, 1987, pp. 366–372.
[9] J. C. Candy and G. C. Temes, *Oversampling Delta–Sigma Data Converters*. New York, NY: IEEE Press, 1992, pp. 1–29.
[10] K. Chao, S. Nadeem, W. Lee, and C. Sodini, "A higher-order topology for interpolative modulation for oversampling A/D converters," *IEEE Trans. Circuits Syst.*, vol. 37, pp. 309–318, Mar. 1990.
[11] P. Ju, K. Suyama, P. F. Ferguson, and W. Lee, "A 22-kHz multibit switched-capacitor sigma–delta D/A converter with 92-dB dynamic range," *IEEE J. Solid-State Circuits*, vol. 30, pp. 1316–1325, Dec. 1995.
[12] P. Ju and D. Vallancourt, "Quantization noise reduction in multibit oversampling $\Sigma-\Delta$ A/D convertors," *Electron. Lett.*, vol. 28, pp. 1162–1163, June 1992.

Woogeun Rhee (S'93–M'97) received the B.S. degree in electronics engineering from Seoul National University, Seoul, Korea, in 1991, and the M.S. degree in electrical engineering from the University of California, Los Angeles, in 1993, where he was also advanced to M.A. candidacy in mathematics. He is currently working toward the Ph.D. degree in electrical and computer engineering at the University of Illinois, Urbana-Champaign.

Since 1997, he has been with Conexant Systems (formerly Rockwell Semiconductor Systems), Newport Beach, CA, where he is a Senior Staff Design Engineer in the Wireless Communication Division. His current interests are in high-speed PLL applications and low-power RF circuits with an emphasis on frequency synthesizers.

Bang-Sup Song (S'79–M'83–SM'88–F'99) received the B.S. degree from Seoul National University, Seoul, Korea, in 1973, the M.S. degree from the Korea Advanced Institute of Science, Taejon, Korea, in 1975, and the Ph.D. degree from the University of California, Berkeley, in 1983.

From 1975 to 1978, he was a Research Staff Member with the Agency for Defence Development, Korea, working on fire-control radars and spread-spectrum communications. From 1983 to 1986, he was a Member of Technical Staff at AT&T Bell Laboratories, Murray Hill, NJ, and was also an Adjunct Professor in the Department of Electrical Engineering, Rutgers University, Piscataway, NJ. From 1986 to 1999, he was a Professor in the Department of Electrical and Computer Engineering, University of Illinois, Urbana. He currently holds the Powell Endowed Chair in Wireless Communication in the Department of Electrical and Computer Engineering, University of California, San Diego.

Dr. Song received a Distinguished Technical Staff Award from AT&T Bell Laboratories in 1986, a Career Development Professor Award from Analog Devices in 1987, and a Xerox Senior Faculty Research Award in 1995. His IEEE activities have been in the capacities of Associate Editor and Guest Editor of the IEEE TRANSACTIONS ON CIRCUITS AND SYSTEMS and the IEEE JOURNAL OF SOLID-STATE CIRCUITS, and a Program Committee Member for the IEEE International Solid-State Circuits Conference and the IEEE Symposium on Circuits And Systems.

A 1.8-GHz Self-Calibrated Phase-Locked Loop with Precise I/Q Matching

Chan-Hong Park, *Student Member, IEEE*, Ook Kim, *Member, IEEE*, and Beomsup Kim, *Senior Member, IEEE*

Abstract—This paper describes a 1.8-GHz self-calibrated phase-locked loop (PLL) implemented in 0.35-μm CMOS technology. The PLL operates as an edge-combining type fractional-N frequency synthesizer using multiphase clock signals from a ring-type voltage-controlled oscillator (VCO). A self-calibration circuit in the PLL continuously adjusts delay mismatches among delay cells in the ring oscillator, eliminating the fractional spur commonly found in an edge-combing fractional divider due to the delay mismatches. With the calibration loop, the fractional spurs caused by the delay mismatches are reduced to -55 dBc, and the corresponding maximum phase offsets between the multiphase signals is less than $0.2°$. The frequency synthesizer PLL operates from 1.7 to 1.9 GHz and the closed-loop phase noise is -105 dBc/Hz at 100-kHz offset from the carrier. The overall circuit consumes 20 mA from a 3.0-V power supply.

Index Terms—Delay mismatch, fractional-N frequency synthesizer, I/Q signal generation, PLL, ring oscillator, self-calibration.

I. INTRODUCTION

IN A MODERN wireless digital data transmission system, both in-phase (I) and quadrature-phase (Q) channels are used for channel efficiency. Therefore, precise I and Q clock signals for the modulation/demodulation of both I and Q channels are needed for high-performance digital transceivers. Since any gain or phase imbalance between I and Q signals reduces the dynamic range and degrades the bit-error rate (BER) of the receivers, the accuracy of the $90°$ phase difference between I and Q clock signals generated from a local oscillator (LO) must be maintained as far as possible. Especially for homodyne or image rejection receivers, the effects of I/Q mismatch become more critical [1].

A lossy phase shifter utilizing resistors and capacitors [2], [3] or a quadrature generator using a frequency divider [4], [5] is widely used to derive quadrature-phase clock signals from a single-phase oscillator. However, the *RC* phase shifter circuit often produces phase and amplitude errors due to the mismatches of components, and the quadrature generator may suffer from phase imbalances due to the inexact input clock duty cycle. For the correction of these I/Q phase errors, some analog or digital calibration techniques have been adopted [6], [7]; however, increased phase noise or complexity caused by the added calibration system limits the performance of the system.

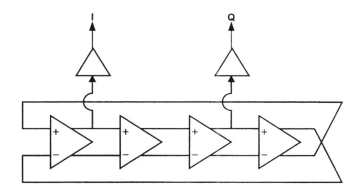

Fig. 1. I/Q signal generation from a ring oscillator.

On the other hand, a ring oscillator may be used to produce the quadrature clock signals without such a phase shifter [8], [9]. The multiphase clock signals from a multistage ring oscillator are easily converted into the I/Q clock signals, as shown in Fig. 1. However, in practice, the mismatches between the delay cells cause significant I/Q phase errors, and an additional phase shifter is therefore required.

A self-calibration technique to eliminate the delay mismatches between the delay cells in a ring oscillator is proposed in this paper. A delay calibration loop in the phase-locked loop (PLL) measures the delay mismatch in each delay cell at a time and eliminates it through an extra control line attached to the delay cells. The calibration loop automatically operates in the background and barely interferes with the main PLL loop behavior.

A prototype 1.8-GHz edge-combining fractional-N frequency synthesizer equipped with the calibration circuit is implemented. Fig. 2 shows the block diagram. Thanks to the calibration technique, the fractional spur caused by the mismatches is attenuated by 25 dB and the I/Q phase offset is maintained within $0.2°$.

The structure of the proposed fractional-N frequency synthesizer is presented in Section II. Section III describes the delay mismatch problem of the edge-combining-type synthesizer PLL. Section IV shows the underlined algorithm for the proposed mismatch calibration scheme. Section V presents the circuit implementation of the self-calibrated frequency synthesizer PLL. Finally, experimental results are shown in Section VI, and conclusions are given in Section VII.

II. EDGE-COMBINING FRACTIONAL-N FREQUENCY SYNTHESIZER

The frequency resolution of a conventional integer-N frequency synthesizer is the same as the reference frequency of

Manuscript received June 15, 2000; revised October 15, 2000.

C.-H. Park and B. Kim are with the Department of Electrical Engineering and Computer Sciences, Korea Advanced Institute of Science and Technology, Taejon 305-701, Korea (e-mail: bkim@ee.kaist.ac.kr).

O. Kim was with SK Telecom, Sungnam Kunggi-do 463-020, Korea. He is now with Silicon Image, Inc., Sunnyvale, CA 94085 USA.

Publisher Item Identifier S 0018-9200(01)03031-1.

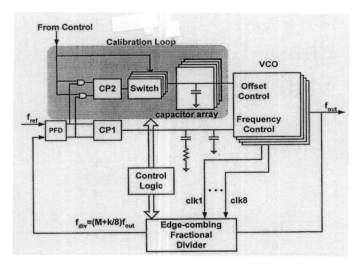

Fig. 2. Block diagram of the proposed fractional-N frequency synthesizer PLL with a self-calibration loop.

the synthesizer PLL. Therefore, narrow channel spacing is accompanied by a small loop bandwidth, which leads to slow dynamics [10]. In case of a fractional-N frequency synthesizer, the output frequency is a fractional multiple of the reference frequency, so that narrow channel spacing is achieved along with a higher phase detector frequency. Consequently, the loop bandwidth can be widened, and faster settling time and lower close-in phase noise of the frequency synthesizer is achieved [10].

The noninteger dividing values in a fractional-N synthesizer can be realized by the periodic dithering of the dividing ratio between integer values [11]. However, the dithering leads to a periodic phase error and introduces spurious tones in the output spectrum. To resolve the spurious noise problem, several methods, such as phase interpolation using a digital-to-analog (D/A) converter [12], or altering the dithering pattern using a Σ–Δ modulator [10], [11] have been proposed and used, but they still have limitations such as increased power dissipation and spurious noise.

On the other hand, if multiphase clock signals are available, the noninteger divider is directly implemented without dithering or interpolation. Fig. 3 shows the operation of the fractional-N divider using the proposed edge-combining technique. The output of an integer frequency divider is sequentially latched by the eight-phase voltage-controlled oscillator (VCO) output signals, as shown in Fig. 3(a). As a result, a set of phase-shifted waveforms is obtained, and the amount of the phase shift is 1/8 of the VCO period. By manipulating the waveform set, a waveform whose period is a fractional multiple of the VCO period is generated. For example, if the desired fractional value is 1/8, the pulse-switching block multiplexes the delayed waveforms with following periodic sequence:

$$clk1 \Rightarrow clk2 \Rightarrow clk3 \Rightarrow clk4 \Rightarrow clk5 \Rightarrow clk6 \Rightarrow clk7 \Rightarrow clk8 \Rightarrow clk1 \dots$$

and the output period of the fractional divider becomes $(M + 1/8) \times T_{\text{VCO}}$, as shown in Fig. 3(b). Here, the division ratio may periodically change in order to create the noninteger division ratio. For example, if the fractional ratio of 1/8 is required as shown in Fig. 3(b), the division ratio should periodically switch

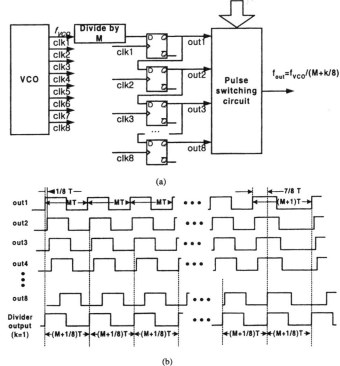

(a)

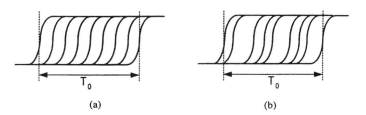

(b)

Fig. 3. (a) Block diagram of the fractional divider. (b) Operation of the fractional divider when $k = 1$.

Fig. 4. Edges of the multiphase signals. (a) With delay mismatches. (b) Without delay mismatches.

from M to $(M + 1)$ when the output pulse is multiplexed from *out8* to *out1*.

When the PLL is locked, the output frequency of the PLL is $(M + k/8) \times f_{\text{ref}}$, and the synthesizer operates as a modulo-8 fractional-N frequency synthesizer. The value of k determines the switching sequence of the pulse-switching block. The switching sequence and division ratio are controlled by a separated control logic.

III. DELAY MISMATCH PROBLEM AND FRACTIONAL SPURS

Although multiphase clocks from a ring oscillator can be used to implement a fractional-N frequency synthesizer, an important problem still exists: how to deal with the delay mismatches between the delay cells in the VCO. Ideally, the phase differences among the multiphase signals from a ring oscillator are precisely equal, as shown in Fig. 4(a). However, in practice, the edges of the multiphase clocks are not uniformly spaced, as shown in Fig. 4(b). The delay mismatches arise from several causes such as V_T mismatch, device size mismatch, and so on.

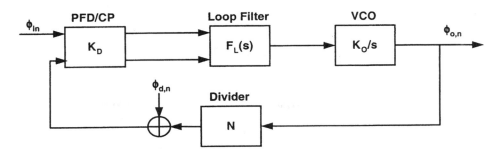

Fig. 5. *s*-domain model of a PLL.

Since each edge of the fractional-N divider output is periodically synchronized with one of the multiphase signals from the VCO, the timing information on the delay mismatches is contained in the divider output. Therefore, when the PLL is locked, the delay mismatches introduce periodic phase errors at the input of the phase-frequency detector (PFD). Due to the periodic phase errors, fractional spurs appear in the output spectrum of the synthesizer. Also, if I and Q signals are tapped from the VCO, the I/Q phase offset will appear. Therefore, the delay mismatches must be eliminated to realize low phase-noise frequency synthesizer.

The relationship between the phase errors and the fractional spurs is derived from the noise transfer function of the PLL. Fig. 5 shows the *s*-domain model of the PLL. If only the effect of the divider noise is considered, the output phase noise becomes

$$\phi_{o,n}(s) = -\frac{K_D F_L(s) \dfrac{K_o}{s}}{1 + \dfrac{K_D}{N} F_L(s) \dfrac{K_o}{s}} \phi_{d,n}(s) \qquad (1)$$

where

K_D	PFD gain;
$F_L(s)$	loop filter transfer function;
K_o	VCO gain;
N	dividing ratio;
$\phi_{o,n}$	output phase noise;
$\phi_{d,n}$	phase noise generated from the fractional divider.

For the edge-combining frequency synthesizer, $\phi_{d,n}$ is mainly caused by the delay mismatches between the delay cells, and is periodic because the fractional-N dividing is performed by periodic combining of the phase-shifted waveforms. Assuming that $\phi_{d,n}$ is a sinusoidal, the output voltage of the PLL is presented as

$$
\begin{aligned}
V_{\text{out}} &= \cos[\omega_o t + \phi_{o,n}(t)] \\
&= \cos \omega_o t \cos \phi_{o,n}(t) - \sin \omega_o t \sin \phi_{o,n}(t) \\
&\approx \cos \omega_o t - \phi_{o,n}(t) \sin \omega_o t \quad (\because \phi_{o,n}(t) \approx 0) \\
&= \cos \omega_o t - A_n \cos \omega_n t \sin \omega_o t \\
&= \cos \omega_o t - \frac{A_n}{2}[\sin(\omega_o + \omega_n)t + \sin(\omega_o - \omega_n)t] \quad (2)
\end{aligned}
$$

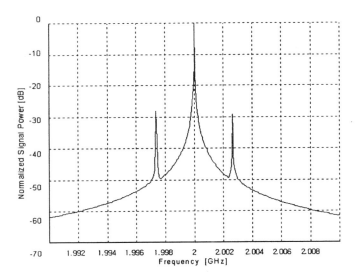

Fig. 6. Simulated output spectrum of the PLL with the maximum phase offset of 2.5°.

TABLE I
PARAMETERS FOR SPURIOUS NOISE SIMULATION OF THE PLL

Charge pump current	0.1 mA
VCO gain (K_O)	250 MHz/ V
Loop Bandwidth	1 MHz
Reference frequency	20 MHz

and the output spectrum of the PLL appears at $\omega_o \pm \omega_n$. The relative power of the spurious tones P_{spur} is given by

$$P_{\text{spur}} = 10 \log \left(\frac{\dfrac{A_n^2}{8}}{\dfrac{1}{2}} \right) = 10 \log \left(\frac{A_n^2}{4} \right) \quad [\text{dBc}]. \quad (3)$$

Fig. 6 shows the simulated output spectrums of the frequency synthesizer. When the maximum phase offset is set to 2.5°, -30-dBc spurious tones are shown near the carrier. During the simulation, the design parameters of the PLL are selected as described in Table I.

IV. MISMATCH CALIBRATION ALGORITHM

Fig. 7 shows the input waveforms of the PFD when the PLL is locked and the fractional dividing ratio is set to 1/8. Δt_i is the relative phase error corresponding to the ith output signal of

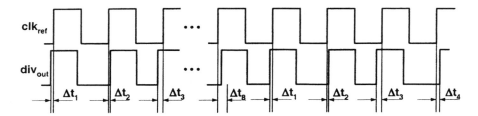

Fig. 7.   Periodic phase error at the PFD input when PLL is locked without calibration.

the VCO. By adjusting the rising phases of the corresponding outputs, the phase errors due to the delay mismatches can be eliminated.

Since a PLL makes the average phase error zero in the locking mode, the sum of the individual phase error becomes zero when the PLL is locked. In other words, when the number of the multiphase clocks is eight

$$\Delta t_1 + \Delta t_2 + \cdots + \Delta t_8 = 0. \qquad (4)$$

If the calibration circuit shifts the rising phase of the first output by Δ_1^1, the phase errors are temporarily changed to

$$\Delta t_1^1 = \Delta t_1 - \Delta_1^1, \ \Delta t_2^1 = \Delta t_2, \ldots, \ \Delta t_8^1 = \Delta t_8. \qquad (5)$$

When the PLL is locked again

$$\Delta t_1^1 + \Delta t_2^1 + \cdots + \Delta t_8^1 = 0 \qquad (6)$$

based on (4), and by assuming that the phase disturbance, Δ_1^1, is equally distributed to all delay cells to satisfy (6), the phase errors become

$$\Delta t_1^1 = \Delta t_1 - \Delta_1^1 + \frac{\Delta_1^1}{8}, \ \Delta t_2^1 = \Delta t_2 + \frac{\Delta_1^1}{8}, \ldots, \ \Delta t_8^1 = \Delta t_8 + \frac{\Delta_1^1}{8}. \qquad (7)$$

If this operation is performed on each output of the VCO one by one, the phase errors are changed as shown at the bottom of the page, and this is the completion of one iteration of the calibration procedure. The resulted phase errors after first iteration are given by

$$\Delta t_1^8 = \Delta t_1 - \Delta_1^1 + \sum_{k=1}^{8} \frac{\Delta_k^1}{8} \ \Delta t_2^8 = \Delta t_2 - \Delta_2^1 + \sum_{k=1}^{8} \frac{\Delta_k^1}{8}, \ \cdots$$

$$\Delta t_8^8 = \Delta t_8 - \Delta_8^1 + \sum_{k=1}^{8} \frac{\Delta_k^1}{8} \qquad (8)$$

where Δt_N^k is the phase error due to the Nth output signal after k cycles of calibration, and Δ_N^m is the amount of the calibration for the Nth VCO output in the mth iteration.

If the above iteration is performed continuously until $\Sigma \Delta_N^m = \Delta t_N$ is satisfied for all delay cells, the final value of the phase error due to the 1st VCO output becomes

$$\Delta t_1^{\text{final}} = \Delta t_1 - \sum_{m=1} \Delta_1^m$$

$$+ \frac{1}{8} \left(\sum_{m=1} \Delta_1^m + \sum_{m=1} \Delta_2^m + \cdots + \sum_{m=1} \Delta_8^m \right)$$

$$= \Delta t_1 - \Delta t_1 + \frac{1}{8} \sum_{n=1}^{8} \Delta t_n = 0. \qquad (9)$$

Similarly

$$\Delta t_1^{\text{final}} = \Delta t_2^{\text{final}} = \Delta t_3^{\text{final}} \cdots = \Delta t_8^{\text{final}} = 0. \qquad (10)$$

Consequently, all the phase errors become zero after finishing the completion of the calibration. Note that the calibration algorithm performs correctly even if the amount of the individual phase correction is different and even if the order of the calibration is changed. Fig. 8 shows the trend of phase error during the calibration, simulated by MATLAB. Here, maximum 10-mV V_T mismatches are assumed.

V. Circuit Implementation

A. Overall Structure

As shown in Fig. 2, a loop for the calibration is combined with the main fractional-N synthesizer. The calibration loop periodically measures the phase error due to delay mismatch at the PFD, and compensates for the mismatches by updating the offset control voltage of delay cells one after another. This update operation must be performed only when the PLL is locked,

initially: Δt_1, Δt_2, Δt_3, $\cdots$, Δt_8

1st step: $\Delta t_1 - \Delta_1^1 + \dfrac{\Delta_1^1}{8}$, $\Delta t_2 + \dfrac{\Delta_1^1}{8}$, $\Delta t_3 + \dfrac{\Delta_1^1}{8}$, $\ldots$, $\Delta t_8 + \dfrac{\Delta_1^1}{8}$

2nd step: $\Delta t_1 - \Delta_1^1 + \dfrac{\Delta_1^1}{8} + \dfrac{\Delta_2^1}{8}$, $\Delta t_2 - \Delta_2^1 + \dfrac{\Delta_1^1}{8} + \dfrac{\Delta_2^1}{8}$, $\Delta t_3 + \dfrac{\Delta_1^1}{8} + \dfrac{\Delta_2^1}{8}$, $\ldots$, $\Delta t_8 + \dfrac{\Delta_1^1}{8} + \dfrac{\Delta_2^1}{8}$

$\vdots$ $\vdots$ $\vdots$ $\vdots$

8th step: $\Delta t_1 - \Delta_1^1 + \dfrac{1}{8} \displaystyle\sum_{k=1}^{8} \Delta_k^1$, $\Delta t_2 - \Delta_2^1 + \dfrac{1}{8} \displaystyle\sum_{k=1}^{8} \Delta_k^1$, $\Delta t_3 - \Delta_3^1 + \dfrac{1}{8} \displaystyle\sum_{k=1}^{8} \Delta_k^1$. $\ldots$, $\Delta t_8 - \Delta_8^1 + \dfrac{1}{8} \displaystyle\sum_{k=1}^{8} \Delta_k^1$.

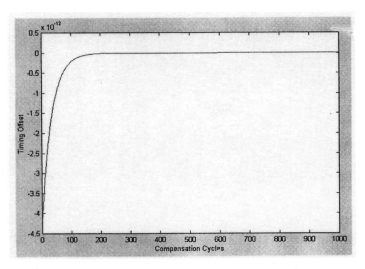

Fig. 8. Simulated behavior of phase error during the compensation.

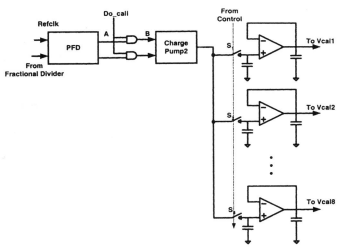

Fig. 10. Detail structure of the self-calibration loop.

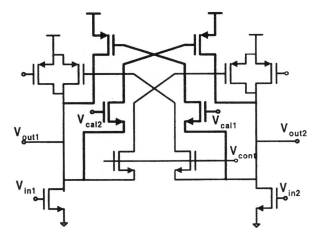

Fig. 9. Schematic of the delay cell having offset control capability.

because the phase error due to the mismatches can be accurately measured only in the locking mode. If the calibration interval is shorter than the lock-in time of the main loop, the locking behavior of the main loop becomes disturbed and even unstable. Therefore, it is important to make not only the calibration interval long enough but also the amount of phase change, Δ, generated by the individual calibrating operation, small enough to make sure that the main loop quickly responds. In this work, the loop gain of the calibration loop is chosen to be 1/10 of the main-loop loop gain. The updated offset control signals are maintained until the next update by a capacitor array connected to each delay cell. The delay cell used in this work is of the same type as the low-noise delay cell used in [13]. However, to control the rising phase of each output, four transistors are added, as shown in Fig. 9. For example, if V_{cal1} is low, the rising of V_{out1} is pulled earlier.

B. Calibration Loop

Fig. 10 shows the circuit for the mismatch calibration. The calibration circuit consists of a PFD shared with the main loop, an additional charge pump, and a capacitor array. When Do_cali is high, the PFD output signal is driven to the charge pump, and

one of the offset control signals is updated. The Do_cali signal is periodically asserted, and the period is much longer than the locking time of the PLL. The output of the charge pump is sequentially connected to the one of the offset-control nodes in the VCO. Since the sequence of the calibration is identical to the pulse-combining sequence in the fractional-N divider, the measured phase error affects only the corresponding VCO output. When the fractional dividing ratio is zero, the frequency synthesizer operates as an integer-N type, and only one output of the VCO is used for phase comparison. The digital logic controls the sequence of the signal updating through the switches in the capacitor array, $S_1 \sim S_8$.

VI. MEASUREMENTS

A self-calibrated fractional-N frequency synthesizer PLL has been fabricated in 0.35-μm CMOS technology. The microphotograph of the fabricated chip is shown in Fig. 11, and its active area is about 1.2×0.4 mm^2. Both frequency synthesizers with and without a calibration loop have been integrated in the same chip to demonstrate the proper operation of the mismatch calibration scheme. In both cases, an external 25-MHz crystal oscillator is used as a reference clock. The bandwidth of the PLL is set to 1 MHz.

Fig. 12 shows the measured output spectrum of the fractional-N RF synthesizer. Fig. 12(a) is the output spectrum of the frequency synthesizer without a calibration loop. In this figure, the fractional dividing ratio is set to 1/8. Without calibration circuit, -30-dBc spurious noise appears at 3.125 ($=25/8$) MHz offset from the carrier frequency. In this case, the maximum phase offset is estimated as about 2.5° by the equations in Section II. On the other hand, in the output spectrum of the self-calibrated frequency synthesizer, the power of the spurious tones is attenuated to -55 dB, as shown in Fig. 12(b), and the calculated maximum phase offset is less than 0.2°. Initial settling of the calibration loop takes about 5.0 ms. Fig. 13 shows the measured phase noise of the frequency synthesizer. The closed-loop phase noise at 100-kHz offset from the 1.8-GHz carrier is -105 dBc/Hz. Table II summarizes the measured characteristics of the PLL.

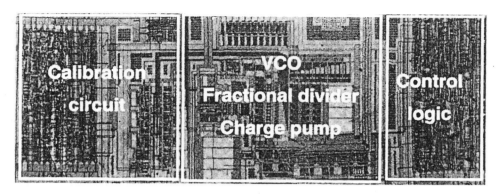

Fig. 11. Microphotograph of the self-calibrated PLL.

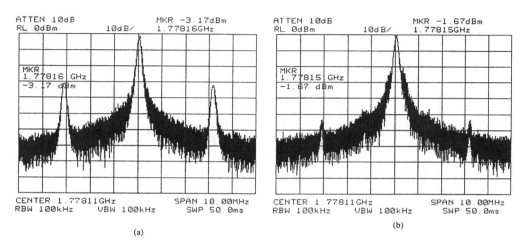

Fig. 12. Output spectrum of the synthesizer PLL. (a) Without calibration. (b) With calibration.

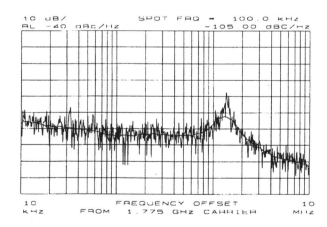

Fig. 13. Measured phase noise of the PLL.

TABLE II
PERFORMANCE SUMMARY OF THE SELF-CALIBRATED PLL

Frequency	1.7~1.9 GHz
Current consumption	20 mA from 3.0 V supply
Phase noise	-105 dBc/Hz @ 100kHz offset
Maximum I/Q phase offset	less than 0.2-degree
Technology	0.35 μm CMOS
Chip area	1.2× 0.4 mm²

VII. CONCLUSION

A self-calibrated 1.8-GHz PLL for fractional-N frequency synthesizing is fabricated in a 0.35-μm CMOS process. A ring-type oscillator is used to generate the multiphase signals, and a self-calibration loop reduces the output fractional spurs caused by delay mismatches between the delay cells. The phase offset of the I/Q signals from the ring oscillator is also relieved. With this calibration scheme, the fractional spur on the PLL is attenuated by 25 dB and the maximum phase offset is thereby reduced to less than 0.2°.

REFERENCES

[1] B. Razavi, *RF Microelectronics*. Englewood Cliffs, NJ: Prentice Hall, 1998.
[2] C. D. Hull, J. L. Tham, and R. R. Chu, "A direct-conversion receiver for 900-MHz (ISM band) spread-spectrum digital cordless telephone," *IEEE J. Solid-State Circuits*, vol. 31, pp. 1955–1963, Dec. 1996.
[3] M. Steyaert, M. Borremans, J. Janssens, B. D. Muer, N. Itoh, J. Craninckx, J. Crols, E. Morijuji, H. S. Momose, W. Sansen, T. Yamaji, H. Tanimoto, and H. Kokatsu, "A single-chip CMOS transceiver for DCS-1800 wireless communications," in *ISSCC Dig. Tech. Papers*, San Francisco, CA, Feb. 1998, pp. 48–49.
[4] A. Montalvo, A. Holden, W. Suter, C. Angell, S. White, N. Klemmer, and D. Homol, "A 22-mW NADC receiver IF Chip with integrated second IF channel filtering," in *ISSCC Dig. Tech. Papers*, San Francisco, CA, Feb. 1999, pp. 48–49.
[5] J. L. Tham, M. A. Margarit, B. Pregardier, C. D. Hull, R. Magoon, and F. Carr, "A 2.7-V 900-MHz/1.9-GHz dual-band transceiver IC for digital wireless communication," *IEEE J. Solid-State Circuits*, vol. 34, pp. 286–291, Mar. 1999.

[6] B. Razavi, "Design considerations for direct-conversion receivers," *IEEE Trans. Circuits Syst. II*, vol. 44, pp. 428–435, June 1997.

[7] L. Yu and W. M. Snelgrove, "A novel adaptive mismatch cancellation system for quadrature IF radio receivers," *IEEE Trans. Circuits Syst. II*, vol. 46, pp. 789–801, June 1999.

[8] Y. Sugimoto and T. Ueno, "The design of a 1-V 1-GHz CMOS VCO circuit with in-phase and quadrature-phase outputs," in *Proc. Int. Symp. Circuits and Systems*, Hong Kong, June 1997, pp. 269–272.

[9] A. A. Abidi, "Direct-conversion radio transceivers for digital communications," *IEEE J. Solid-State Circuits*, vol. 30, pp. 1399–1410, Dec. 1995.

[10] T. A. D. Riley, M. A. Copeland, and T. A. Kwasniewski, "Delta–sigma modulation in fractional-N frequency synthesis," *IEEE J. Solid-State Circuits*, vol. 28, pp. 553–559, May 1993.

[11] M. H. Perrot, "Techniques for high data rate modulation and low power operations of fractional-N frequency synthesizers," Ph.D. dissertation, Mass. Inst. of Technol., Cambridge, MA, 1997.

[12] U. L. Rohde, *Digital PLL Frequency Synthesizers*. Englewood Cliffs, NJ: Prentice Hall, 1983.

[13] C.-H. Park and B. Kim, "A low-noise 900-MHz VCO in 0.6-μm CMOS," *IEEE J. Solid-State Circuits*, vol. 34, pp. 586–591, May 1999.

Chan-Hong Park (S'92) received B.S. and M.S. degrees in electrical engineering from Korea Advanced Institute of Science and Technology (KAIST), Taejon, Korea, in 1994 and 1996, respectively. He is currently working toward the Ph.D. degree in electrical engineering at KAIST.

From 1994, he has been with the Department of Electrical Engineering, KAIST, as a Graduate Researcher, where he has been involved in designing 100Base-T transceiver ICs, low-noise phase-locked loops, and RF front-ends for wireless communications. His research interests include CMOS RF circuits for wireless communication, high-frequency analog IC design, and mixed-mode signal-processing IC design.

Ook Kim (M'86) received the M.S. and Ph.D. degrees in electronics engineering from Seoul National University, Seoul, Korea, in 1988 and 1994, respectively.

He was with the Electronics and Telecommunications Research Institute, Taejon, Korea, from 1994 to 1998, and with SK Telecom, Seoul, Korea, from 1998 to 1999. Since 1999, he has been with Silicon Image Inc., Sunnyvale, CA. He was a Visiting Researcher at the Department of Electrical and Electronic Engineering, Adelaide University, Adelaide, Australia, during 1992, and a Visiting Scholar at the Department of Electrical Engineering, Stanford University, Stanford, CA, during 1999. His research interests are in CMOS mixed mode circuit design, high-speed data conversion, wireless circuit technology, and high-speed data communication.

Beomsup Kim (S'87–M'90–SM'95) received the B.S. and M.S. degrees in electronic engineering from Seoul National University, Seoul, Korea, in 1983 and 1985, respectively, and the Ph.D. degree in electrical engineering and computer sciences from the University of California, Berkeley, in 1990.

He worked as a Graduate Researcher and Graduate Instructor at the Department of Electrical Engineering and Computer Sciences, University of California, Berkeley, from 1986 to 1990. From 1990 to 1991, he was with Chips and Technologies, Inc., San Jose, CA, where he was involved in designing high-speed signal processing ICs for disk drive read/write channel. From 1991 to 1993, he was with Philips Research, Palo Alto, CA, conducting research on digital signal processing for video, wireless communication, and disk drive applications. During 1994, he was a Consultant, developing the partial response maximum likelihood detection scheme of the disk drive read/write channel. In 1994, he became an Assistant Professor with the Department of Electrical Engineering, Korea Advanced Institute of Science and Technology (KAIST), Taejon, Korea, and is currently an Associate Professor. During 1999, he took sabbatical leave at Stanford University, Stanford, CA, and at the same time, consulted for Marvell Semiconductor Inc., San Jose, on the Gigabit Ethernet and wireless LAN DSP architecture. His research interests include mixed-mode signal processing IC design for telecommunication, disk drive, and LAN, high-speed analog IC design, and VLSI system design.

Dr. Kim is a corecipient of the Best Paper Award for 1990–1991 from the IEEE JOURNAL OF SOLID-STATE CIRCUITS. He received the Philips Employee Reward in 1992. Between June 1993 and June 1995, he served as an Associate Editor for the IEEE TRANSACTIONS ON CIRCUITS AND SYSTEMS II.

609

A 27-mW CMOS Fractional-N Synthesizer Using Digital Compensation for 2.5-Mb/s GFSK Modulation

Michael H. Perrott, *Student Member, IEEE*, Theodore L. Tewksbury III, *Member, IEEE*, and Charles G. Sodini, *Fellow, IEEE*

Abstract— A digital compensation method and key circuits are presented that allow fractional-N synthesizers to be modulated at data rates greatly exceeding their bandwidth. Using this technique, a 1.8-GHz transmitter capable of digital frequency modulation at 2.5 Mb/s can be achieved with only two components: a frequency synthesizer and a digital transmit filter.

A prototype transmitter was constructed to provide proof of concept of the method; its primary component is a custom fractional-N synthesizer fabricated in a 0.6-μm CMOS process that consumes 27 mW. Key circuits on the custom IC are an on-chip loop filter that requires no tuning or external components, a digital MASH Σ–Δ modulator that achieves low power operation through pipelining, and an asynchronous, 64-modulus divider (prescaler). Measurements from the prototype indicate that it meets performance requirements of the digital enhanced cordless telecommunications (DECT) standard.

Index Terms— Compensation, continuous phase modulation, digital radio, frequency modulation, frequency shift keying, frequency synthesizers, phase locked loops, sigma–delta modulation, transmitters.

I. INTRODUCTION

THE use of wireless products has been rapidly increasing the last few years, and there has been worldwide development of new systems to meet the needs of this growing market. As a result, new radio architectures and circuit techniques are being actively sought that achieve high levels of integration and low power operation while still meeting the stringent performance requirements of today's radio systems. Our focus is on the transmitter portion of this effort, with the objective of achieving over 1-Mb/s data rate using frequency modulation.

To achieve the goals of low power and high integration, it seems appropriate to develop a transmitter architecture that consists of the minimal topology that accomplishes the required functionality. All digital, narrowband radio transmitters that are spectrally efficient require two operations to be performed. The baseband modulation data must be filtered to limit the extent of its spectrum, and the resulting signal must

Manuscript received July 7, 1997; revised August 4, 1997. This work was supported by DARPA Contract DAAL-01-95-K-3526.

M. H. Perrott was with the Microsystems Technology Laboratory, Massachusetts Institute of Technology, Cambridge, MA 02139 USA. He is now with Hewlett-Packard Laboratories, Palo Alto, CA 94304-1392 USA.

T. L. Tewksbury III was with Analog Devices, Wilmington, MA 01887 USA. He is now with IBM Microelectronics, Waltham, MA 02254 USA.

C. G. Sodini is with the Microsystems Technology Laboratory, Massachusetts Institute of Technology, Cambridge, MA 02139 USA.

Publisher Item Identifier S 0018-9200(97)08270-X.

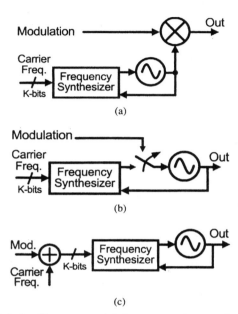

Fig. 1. Methods of frequency modulation upconversion: (a) mixer based, (b) direct modulation of VCO, and (c) indirect modulation of VCO.

be translated to a desired RF band. This paper will focus on the issue of frequency translation, which can be accomplished in at least three different ways for frequency modulation. As illustrated in Fig. 1, the modulation signal can be (a) multiplied by a local oscillator (LO) frequency using a mixer, (b) fed into the input of a voltage controlled oscillator (VCO), or (c) fed into the input of a frequency synthesizer.

Approach (a) can theoretically be accomplished with either a heterodyne or homodyne approach. The heterodyne approach offers excellent radio performance but carries a high cost in implementation due to the current inability to integrate the high-Q, low-noise, low-distortion bandpass filters required at intermediate frequencies (IF) [1]. As a result, the direct conversion approach has recently grown in popularity [2]–[4]. In this case, two mixers and baseband A/D converters are required to form in-phase/quadrature (I/Q) channels and a frequency synthesizer to obtain an accurate carrier frequency.

Approach (b) is referred to as direct modulation of a VCO and has appeared in designs for the digital enhanced cordless telecommunications (DECT) standard [5], [6]. A frequency synthesizer is used to achieve an accurate frequency setting and then disconnected so that modulation can be fed into the

VCO unperturbed by its dynamics. This technique allows a significant reduction in components; no mixers are required since the VCO performs the frequency translation, and only one D/A converter is required to produce the modulation signal. Power savings are thus achieved, as demonstrated by the fact that the design in [5] appears to consume nearly half the power of the mixer based designs in [2]–[4]. Unfortunately, since the synthesizer is inactive during modulation, the nominal frequency setting of the VCO tends to drift as a result of leakage currents. In addition, undesired perturbations, such as the turn-on transient of the power amp, can dramatically shift the output frequency. As stated in [6], the isolation requirements for this method exclude the possibility of a one-chip solution. Therefore, while the approach offers a significant advantage in terms of power dissipation, the goal of high integration is lost.

Finally, approach (c) can be viewed as *indirect* modulation of the VCO through appropriate control of a frequency synthesizer that sets the VCO frequency and yields the simplest transmitter solution of those presented. The synthesizer has a digital input which allows elimination of the D/A converter that is required when directly modulating the VCO. Since the synthesizer controls the VCO during modulation, the problem of frequency drift during modulation is eliminated. Also, isolation requirements at the VCO input are greatly reduced at frequencies within the PLL bandwidth. The primary obstacle faced with this architecture is that a severe constraint is placed on the maximum achievable data rate due to the reliance on feedback dynamics to perform modulation.

This paper presents a compensation method and key circuits that allow modulation of a frequency synthesizer at rates that are over an order of magnitude faster than its bandwidth. Application of the technique allows a high data rate (>1 Mb/s) transmitter with good spectral efficiency to be realized with only two components: a frequency synthesizer and a digital transmit filter. By avoiding additional components such as mixers and D/A converters in the modulation path, a low power transmitter architecture is achieved. Since off-chip filters are not required, high integration is accomplished as well. The technique can be used in transmitter applications where frequency modulation is desired, and a moderate tolerance is allowed on the modulation index. (When using compensation, the accuracy of the modulation index, which is defined as the ratio of the peak-to-peak frequency deviation of the transmitter output to its data rate, is limited by variations in the open-loop gain of the PLL [7].) To provide proof of concept of the technique, we present results from a 1.8-GHz prototype that supports Gaussian frequency shift keying (GFSK) modulation, the same modulation method used in DECT, at data rates in excess of 2.5 Mb/s.

We begin by reviewing a fractional-N synthesizer method presented in [8]–[11] that provides a convenient structure with which to apply the technique. It is shown that high data rates and good noise performance are difficult to achieve with this topology. A method is proposed to overcome these problems, followed by discussion of issues that ensue from its use. A description of key circuits in the prototype is then given, which include an on-chip loop filter, a 64-modulus divider,

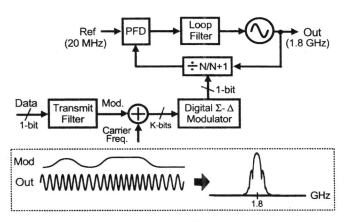

Fig. 2. A spectrally efficient, fractional-N modulator.

and a pipelined, digital Σ–Δ modulator. Finally, experimental results are presented and conclusions made.

II. BACKGROUND

The fractional-N approach to frequency synthesis enables fast dynamics to be achieved within the phase-locked loop (PLL) by allowing a high reference frequency [8]; a very useful benefit when attempting to modulate the synthesizer. High resolution is achieved with this approach by allowing noninteger divide values to be realized through dithering; it has been shown that low spurious noise can be obtained by using a high-order Σ–Δ modulator to perform this operation [8], [10], [11]. This approach leads to a simple synthesizer structure that is primarily digital in nature, and is referred to as a fractional-N synthesizer with noise shaping.

Using this fractional-N approach, it is straightforward to realize a transmitter that performs phase/frequency modulation in a continuous manner by direct modulation of the synthesizer. Fig. 2 illustrates a simple transmitter capable of Gaussian minimum shift keying (GMSK) modulation [9]. The binary data stream is first convolved with a digital finite impulse response (FIR) filter that has a Gaussian shape. (Physical implementation of this filter can be accomplished with a ROM whose address lines are controlled by consecutive samples of the data and time information generated by a counter.) The digital output of this filter is then summed with a nominal divide value and fed into the input of a digital Σ–Δ converter, the output of which controls the instantaneous divide value of the PLL. The nominal divide value sets the carrier frequency, and variation of the divide value causes the output frequency to be modulated according to the input data. Assuming that the PLL dynamics have sufficiently high bandwidth, the characteristics of the modulation waveform are determined primarily by the digital FIR filter and thus accurately set.

Fig. 3 depicts a linearized model of the synthesizer dynamics in the frequency domain. The digital transmit filter confines the modulation data to low frequencies, the Σ–Δ modulator adds quantization noise that is shaped to high frequencies, and the PLL acts as a low-pass filter that passes the input but attenuates the Σ–Δ quantization noise. In the figure, $G(f)$ is

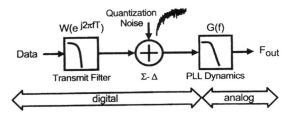

Fig. 3. Linearized model of fractional-N modulator.

calculated as

$$G(f) \equiv \frac{K_v H(f)/(\pi N_{\text{nom}})}{jf + K_v H(f)/(\pi N_{\text{nom}})} \qquad (1)$$

where $H(f)$, K_v, and N_{nom} are the loop filter transfer function, the VCO gain (in Hz/V), and the nominal divide value, respectively. (See [7] for modeling details.) An analogy between the fractional-N modulator and a Σ–Δ D/A converter can be made by treating the output frequency of the PLL as an analog voltage.

A key issue in the system is that the Σ–Δ modulator adds quantization noise at high frequency offsets from the carrier. In general, the noise requirements for a transmitter are very strict in this range to avoid interfering with users in adjacent channels. In the case of the DECT standard, the phase noise density can be no higher than -131 dBc/Hz at a 5-MHz offset [6]. Noise at low frequency offsets is less critical for a transmitter and need only be below the modulation signal by enough margin to insure an adequate signal-to-noise ratio.

Sufficient reduction of the Σ–Δ quantization noise can be accomplished through proper choice of the Σ–Δ sample rate, which is assumed to be equal to the reference frequency, and the PLL transfer function, $G(f)$. (Note that this problem is analogous to that encountered in the design of Σ–Δ D/A converters, except that the noise spectral density at high frequencies, rather than the overall signal-to-noise ratio, is the key parameter.) One way of achieving a low spectral density for the noise is to use a high sample rate for the Σ–Δ so that the quantization noise is distributed over a wide frequency range and its spectral density reduced. Alternatively, the attenuation offered by $G(f)$ can be increased; this is accomplished by decreasing its cutoff frequency, f_o, or increasing its order, n.

Unfortunately, a low value of f_o carries a penalty of lowering the achievable data rate of the transmitter. This fact can be observed from Fig. 3; the modulation data must pass through the dynamics of the PLL so that its bandwidth is restricted by that of $G(f)$. Given this constraint, the achievement of low noise must be achieved through proper setting of the Σ–Δ sample rate and PLL order.

It is worthwhile to quantify required values of these parameters for a given data rate and noise specification. To do so, we first choose $G(f)$ to be a Butterworth response of order n

$$G(f) = \frac{1}{1 + (jf/f_o)^n}.$$

The above expression is chosen for the sake of simplicity in calculations; other filter responses could certainly be implemented. The spectral density of the noise at the transmitter

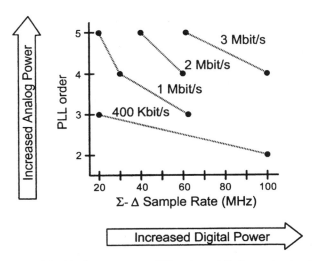

Fig. 4. Achievable data rates versus PLL order and Σ–Δ sample rate when noise from Σ–Δ is -136 dBc/Hz at 5 MHz offset.

output due to quantization noise is expressed as

$$S_{\Phi_q}(f) = \left(T \frac{(2\pi)^2}{12} (2\sin(\pi f T))^{2(n-1)} \right) |G(f)|^2 \qquad (2)$$

where T is the Σ–Δ sample period, and a multistage (MASH) structure [12] of order n is assumed for the modulator. By choosing the order of the MASH Σ–Δ to be the same as the order of $G(f)$, the rolloff of (2) and the VCO noise are matched at high frequencies (-20 dB/dec). Fig. 4 displays the resulting parameters at different data rates; these values were calculated by setting (2) to -136 dBc/Hz at 5 MHz and the ratio $f_o T_d$ to 0.7, where $1/T_d$ is the data rate. (The noise specification was chosen to achieve less than -131 dBc/Hz at 5 MHz offset after adding in VCO phase noise.)

The figure reveals that the achievement of high data rates and low noise must come at the cost of power dissipation and complexity when attempting direct modulation of the synthesizer. In particular, the power consumed by the digital circuitry is increased at a high Σ–Δ sample rate by virtue of the increased clock rate of the Σ–Δ modulator and the digital FIR filter, $W(e^{j2\pi fT})$. The power consumed by the analog section is increased for high values of PLL order since additional poles and zeros must be implemented. This issue is aggravated by the need to set these additional time constants with high accuracy in order to avoid stability problems in the PLL. If tuning circuits are used to achieve such accuracy [13], spurious noise problems can also be an issue.

III. PROPOSED METHOD

The obstacles of high data rate modulation discussed above are greatly mitigated if the modulation bandwidth is allowed to exceed that of the PLL. In this case, the bandwidth of $G(f)$ can be set sufficiently low that an excessively high PLL order or reference frequency is not necessary to achieve the required noise performance. Fig. 5 illustrates the proposed method that achieves this goal. By cascading a compensation filter, $C(f)$, with the digital FIR filter, the transfer function seen by the modulation data can be made flat by setting $C(f) = 1/G(f)$.

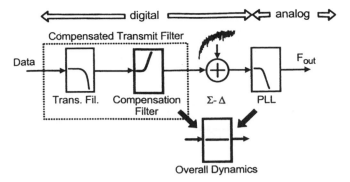

Fig. 5. Proposed compensation method.

TABLE I
THEORETICALLY ACHIEVABLE DATA RATES
USING COMPENSATION FOR SECOND-ORDER PLL

Σ-Δ sample rate	20 MHz	40 MHz	80 MHz
Max. Data Rate	3.4 Mbit/s	4.8 Mbit/s	4.9 Mbit/s

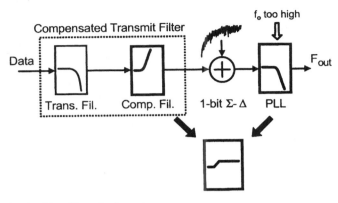

Fig. 6. The effect of mismatch.

This new filter is simple to implement in a digital manner—by combining it with the FIR filter, we need only alter the ROM storage values. In fact, savings in area and power of the ROM can be achieved over the uncompensated method since the number of time samples that need to be stored are dramatically reduced [7].

To illustrate the technique, we consider the case where $W(e^{j2\pi fT})$ is chosen to implement GFSK modulation with $BT_d = 0.5$, as used in the DECT standard, and $G(f)$ is second order. Under these assumptions, the time domain version of $W(e^{j2\pi fT})$ is described as samples of

$$w(t) = \text{rect}(T_d, t) * \frac{T}{4\sqrt{1.66T_d}} e^{-(\pi t/(3.32T_d))^2} \quad (3)$$

where "$*$" is the convolution operator, T is the Σ-Δ sample period, T_d is the period of the data stream, and $\text{rect}(T_d, t)$ equals $1/T_d$ for $-T_d/2 \le t < T_d/2$ and zero elsewhere. Since $C(f)$ is the inverse of $G(f)$, we write

$$C(f) = 1 + \frac{1}{f_o Q} jf + \frac{1}{f_o^2}(jf)^2. \quad (4)$$

In the time domain, the digital compensated FIR filter is then calculated by taking samples of the expression

$$w_c(t) \equiv w(t) * c(t) = w(t) + \frac{1}{2\pi f_o Q}w'(t) + \frac{1}{(2\pi f_o)^2}w''(t). \quad (5)$$

For $w(t)$ described in (3), these derivatives are well defined and can be calculated analytically. A final form is derived by substituting (3) into (5) to yield

$$w_c(t) = \left(1 - \frac{1}{1.66Q f_o T_d}\left(\frac{t}{\sigma}\right)\right. $$
$$\left. + \frac{1}{(1.66 f_o T_d)^2}\left(-1 + \left(\frac{t}{\sigma}\right)^2\right)\right)w(t). \quad (6)$$

A. Achievable Data Rates

Equation (6) reveals that the signal swing of $w_c(t)$ increases in proportion to $1/(f_o T_d)^2$ for large values of $1/(f_o T_d)$. Since $1/(f_o T_d)$ is the ratio of the modulation data rate to the bandwidth of the PLL, we see that high data rates lead to large signal swings of the modulation signal when using compensation. Intuitively, this behavior makes sense since the attenuation of $G(f)$ must be overcome by the compensated

signal. If the order of $G(f)$ is increased to n, the resulting signal swings will be amplified according to $1/(f_o T_d)^n$.

The achievable data rates using compensation are limited by the ability of the PLL to accommodate this increased signal swing. PLL components that are particularly affected are the Σ-Δ modulator, the divider, and the charge pump. Assuming an appropriate multibit Σ-Δ structure and multimodulus divider topology are used, the bottleneck in dynamic range will be set by the limited duty cycle range of the charge pump.

Table I displays the achievable data rates at different Σ-Δ sample rates using compensation; the noise specification was identical to that used to generate Fig. 4. In light of the signal swing limitation and our goal of simplicity, we have restricted our attention to second-order PLL dynamics. Calculations were based on the assumption that the duty cycle of the charge pump is limited only by its transient response, which was assumed to be 5 ns. Comparison of this information with Fig. 4 reveals that compensation allows high data rates to be achieved with relatively low power and complexity. In the actual prototype, data rates as high as 2.85 Mb/s are achieved with a second-order PLL with $f_o = 84$ kHz, and a Σ-Δ sample rate of 20 MHz.

B. Matching Issues

In practice, mismatch will occur between the compensation filter and PLL dynamics. While the compensation filter is digital and therefore fixed, the PLL dynamics are analog in nature and sensitive to process and temperature variations. Fig. 6 illustrates that a parasitic pole/zero pair occurs when the bandwidth of the PLL is too high; a similar situation occurs when its bandwidth is too low. As will be seen in the results sections, the parasitic pole/zero pair causes intersymbol interference (ISI) and modulation deviation error. To mitigate this problem in the prototype, an on-chip loop filter with accurate time constants was implemented, and open-loop gain

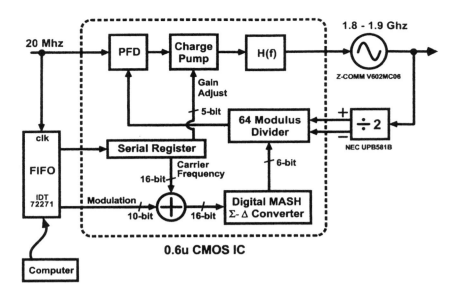

Fig. 7. Prototype system.

control was used to accurately place the overall pole and zero positions of the PLL transfer function.

An additional issue related to be mismatch arises from practical concerns in the PLL implementation. The achievement of a large dynamic range in the charge pump is aided by including an integrator in the loop filter (see Section IV-B2), which yields an overall PLL transfer function as

$$G(f) = \frac{1 + jf/f_z}{1 + jf/f_{cp}} \left(\frac{1}{1 + \frac{1}{f_o Q} jf + \frac{1}{f_o^2}(jf)^2} \right). \quad (7)$$

A parasitic pole/zero pair, f_z and f_{cp}, is now added that occurs well below f_o in frequency. Unfortunately, taking the inverse of (7) leads to a compensation filter that is IIR in nature and cannot be implemented with a ROM. To avoid such difficulties, we can ignore the parasitic pole/zero pair and use $C(f)$ as described in (4). The resulting ISI is negligible since f_z and f_{cp} are close to each other and low in value. However, the digital compensation filter must be modified to be samples of $(f_z/f_{cp})w_c(t)$ to accommodate the increased gain of $G(f)$ at frequencies greater than f_{cp}.

IV. IMPLEMENTATION

To show proof of concept of the proposed compensation method, the system depicted in Fig. 7 was built using a custom CMOS fractional-N synthesizer that contains several key circuits. Included are an on-chip, continuous-time filter that requires no tuning or external components, a digital MASH Σ–Δ modulator with six output bits that achieves low power operation through pipelining, and a 64-modulus divider that supports any divide value between 32 and 63.5 in half cycle increments. An external divide-by-two prescaler is used so that the CMOS divider input operates at half the VCO frequency, which modifies the range of divide values to include all integers between 64 and 127.

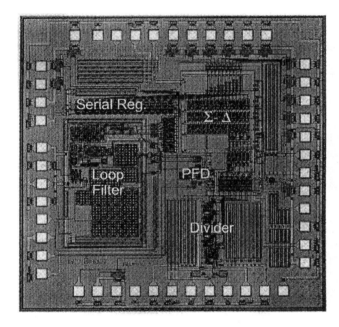

Fig. 8. Die photo.

TABLE II
POWER DISSIPATION OF IC CIRCUITS

Component	Divider	Filter	Σ-Δ	Other
Power	22 mW	2.5 mW	0.33 mW	2.4 mW
Supply Voltage	3 V	3.3 V	1.25 V	3.3 V

Fig. 8 displays a die photograph of the custom IC, which was fabricated in a 0.6-μm, double-poly, double-metal, CMOS process with threshold voltages of $V_{tn} = 0.75$ V and $V_{tp} = -0.88$ V. The entire die is 3 mm by 3 mm, and its power dissipation is 27 mW. Table II lists the power consumed by individual circuits. The power supply values given in Table II were chosen to be as low as possible to minimize power dissipation; at the cost of higher power dissipation, all circuits could be powered by a single 3.3-V supply.

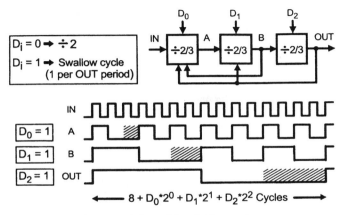

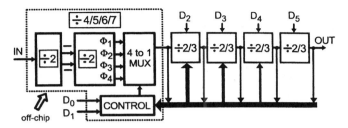

Fig. 10. An asynchronous, 64-modulus divider implementation.

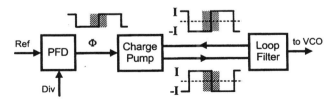

Fig. 11. PFD, charge pump, and loop filter.

Fig. 9. An asynchronous, eight-modulus divider topology.

The 64-modulus divider and six-output-bit $\Sigma-\Delta$ modulator provide a dynamic range for the compensated modulation data that is wide enough to support data rates in excess of 2.5 Mb/s. The on-chip loop filter allows an accurate PLL transfer function to be achieved by tuning just one PLL parameter—the open-loop gain. A brief overview of each of these components is now presented.

A. Divider

To achieve a low-power design, it is desirable to use an asynchronous divider structure to minimize the amount of circuitry operating at high frequencies. As such, a multimodulus divider structure was designed that consists of cascaded divide-by-2/3 sections [14]; this architecture is an extension of the common dual-modulus topology [15]. The eight-modulus example in Fig. 9 shows the proposed structure which allows a wide range of divide values to be achieved by allowing a variable number of input cycles to be "swallowed" per output cycle. Each divide-by-2/3 stage normally divides its input by two in frequency, but will swallow an extra cycle per *OUT* period when its control input, D_i, is set to one. As shown for the case where all control bits are set to one, the number of *IN* cycles swallowed per *OUT* period is binary weighted according to the stage position. For instance, setting $D_0 = 1$ causes one cycle of *IN* to be swallowed, while setting $D_2 = 1$ causes four cycles of *IN* to be swallowed. Proper selection of $\{D_2 D_1 D_0\}$ allows any integer divide value between 8 and 15 to be achieved.

The 64-modulus divider that was developed for the prototype system uses a similar principle to that discussed above, but has a modified first stage to achieve high-speed operation. Specifically, the implemented architecture consists of a high-speed divide-by-4/5/6/7 state machine followed by a cascaded chain of divide-by-2/3 state machines as illustrated in Fig. 10. The divide-by-4/5/6/7 stage accomplishes cycle swallowing by shifting between four phases of a divide-by-two circuit. Each of the four phases is staggered by one *IN* cycle, which allows single cycle pulse swallowing resolution despite the fact that two cascaded divide-by-two structures are used; details of this approach are discussed at length in [7]. The important point to make about the phase shifting approach, which is

also advocated in [15], is that it allows a minimal number of components to operate at high frequencies—the first two stages are simply divide-by-two circuits, not state machines. Also, the fact that control signals are not fed into the first divide-by-two circuit allows it to be placed off-chip in the prototype.

B. Analog Section

The achievement of accurate PLL dynamics is accomplished in the prototype system with the variable gain loop filter topology depicted in Fig. 11. The input to the filter is the instantaneous phase error between the reference frequency and divider output and is manifested as the deviation of the phase frequency detector (PFD) output duty cycle from its nominal value of 50%. As modulation data is applied, the duty cycle is swept across a range of values; the shaded region in the figure corresponds to the deviation that occurs when GFSK modulation at 2.5 Mb/s is applied. A 50% nominal duty cycle is desired to avoid the dead-zone of the PFD and thus reduce distortion of the modulation signal. The prototype used a PFD design from [16] to achieve this characteristic.

To produce a signal that is a filtered version of the phase error, the output of the PFD is converted to complementary current waveforms by a charge pump before being sent into the inputs of an on-chip loop filter. The conversion to current allows the filtering operation to be performed without resistors and also provides a convenient means of performing gain control of the resulting transfer function. An integrator is included in the loop filter which forces the average current from the charge pump to be zero and the nominal duty cycle to be, ideally, 50% when the PLL is locked.

A PFD design with 50% nominal duty cycle is seldom used in PLL circuits due to power consumption and spurious noise issues—the charge pump is always driving current into the loop filter under such conditions, and spurs at multiples of the reference frequency are produced due to the square wave output of the PFD. Fortunately, these problems are greatly mitigated in the prototype transmitter since the charge pump

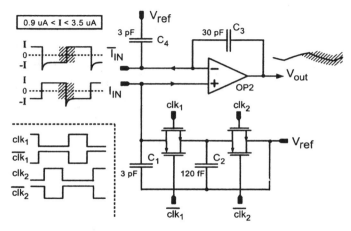

Fig. 12. Loop filter implementation.

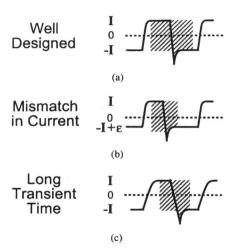

Fig. 13. Effect of transient time and mismatch on duty cycle range.

output current is very small (at its largest setting, it toggles between $+3.5$ and -3.5 μA), and the loop filter bandwidth is very low (84 kHz) in comparison to the reference frequency (20 MHz). The resulting spur at the transmitter output is less than -60 dBc at 20 MHz when measuring the transmitter in an unmodulated state without an RF bandpass filter at its output. When modulated, this spur is convolved with the modulation signal and thus turned into phase noise [7]; it is reasonable to assume that this noise is reduced to a negligible level when the RF bandpass filter is included due to its high frequency offset.

1) Loop Filter: The on-chip loop filter uses an opamp to integrate one of the currents and add it to a first-order filtered version of the other current. This topology, shown in Fig. 12, realizes the transfer function

$$H(f) = K_l \frac{1 + jf/f_z}{jf(1 + jf/f_p)},$$
$$f_z = 11.6 \text{ kHz}, \quad f_p = 127 \text{ kHz}. \quad (8)$$

The open loop gain, K_l, is adjusted by varying the charge pump output current, I. The first-order pole f_p is created using a switched capacitor technique, which reduces its sensitivity to thermal and process variations and removes any need for tuning. Note that, although this time constant is formed through a sampling operation, the output of the switched capacitor filter is a continuous-time signal. Finally, the value of the zero, f_z, is determined primarily by the ratio of capacitors C_3 and C_2 under the assumption that the complementary charge pump currents are matched.

A particular advantage of the filter topology is that the rate of sampling C_1 and C_2 can be set high since it is independent of the settling dynamics of the opamp. As such, clk_1 and clk_2 are set to the PFD output frequency, 20 MHz, to avoid aliasing problems.

The opamp is realized with a single-ended, two-stage topology chosen for its simplicity and wide output swing. Its unity gain frequency was designed to be 6 MHz; this value is sufficiently higher than the bandwidth of the GFSK modulation signal at 2.5 Mb/s to avoid significantly affecting it. It is recognized that the single-ended structure has higher sensitivity to substrate noise than a differential counterpart. However, little would be gained in this case by making it fully

differential since the output of the opamp is connected directly to the varactor of an LC-based VCO, which is inherently single ended. Fortunately, measured eye diagrams and spectral plots presented at the end of this paper conform to calculations that exclude substrate noise, thereby showing that it has negligible impact on the modulation and noise performance of the prototype system. However, as even higher levels of integration are sought in future radio systems, the impact of substrate noise will need to be carefully considered.

The limited dynamics of the opamp prevent it from following the fast transitions of its input current waveforms. To prevent these waveforms from adversely affecting the performance of the opamp, the voltage swing that appears at its input terminals is reduced to a low amplitude (less than 40 mV peak-to-peak) by capacitors C_1 and C_4. In the case of C_1, this capacitor also serves as part of the switched capacitor filter.

2) Charge Pump: Proper design of the charge pump is critical for the achievement of high data rates since it forms the bottleneck in dynamic range that is available to the modulation signal. Fig. 13 illustrates the fundamental issues that need to be considered in its design. To avoid distortion of the modulation signal, the variation in duty cycle should be limited to a range that allows the output of the charge pump to settle close to its final value following all positive and negative transitions. Fig. 13(a) shows the dynamic range available for a well-designed charge pump; the nominal duty cycle is 50% and the transition times are fast. Fig. 13(b) demonstrates the reduction in dynamic range that occurs when the nominal duty cycle is offset from 50%. This offset is caused by a mismatch between positive and negative currents produced by the charge pump. (The type II PLL dynamics force an average current of zero.) Finally, Fig. 13(c) illustrates a case in which the charge pump has slow transition times, the result again being a reduction in dynamic range.

The charge pump topology was designed with the above issues in mind and is illustrated in Fig. 14. The core component of the architecture is a differential pair (M_1 and M_2) that is fed from the top by two current sources, I_1 and I_2, and from the bottom by a tail current, I_{tail}. Ideally, I_1 and I_2 are equal to I and I_{tail} to $2I$, where I is adjusted by a 5-b D/A that

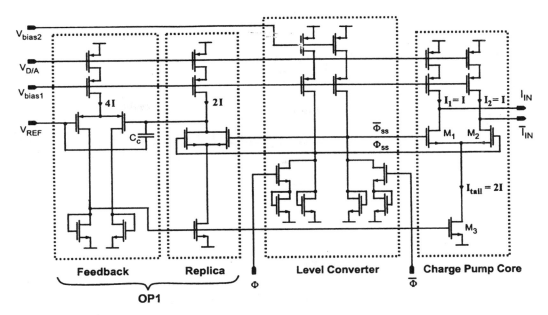

Fig. 14. Charge pump implementation.

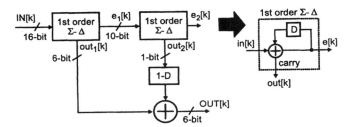

Fig. 15. A second-order, digital MASH structure.

controls the node $V_{D/A}$. Transistors M_1 and M_2 are switched on and off according to Φ, which ideally causes I_{in} and $\overline{I}_{\text{in}}$ to switch between I and $-I$.

To achieve a close match between the positive and negative currents of each charge pump output, the design strives to set $I_1 = I_2$ and $I_{\text{tail}} = I_1 + I_2$. In the first case, I_1 and I_2 are implemented as cascoded PMOS devices whose layout is optimized to achieve high levels of device matching. Unfortunately, device matching cannot be used to achieve a close match between I_{tail} and $I_1 + I_2$ since they are generated by different *types* of devices. To circumvent this obstacle, a feedback stage is used to adjust I_{tail} by comparing currents produced by a replica stage. This technique allows I_{tail} to be matched to $I_1 + I_2$ to the extent that the replica stage is matched to the core circuit.

A low transient time in the charge pump response is obtained by careful design of signal and device characteristics at the source nodes of M_1 and M_2. First, the parasitic capacitance at this node is minimized by using appropriate layout techniques to reduce the source capacitance of M_1 and M_2, the drain capacitance of M_3, and the interconnect capacitance between each of the devices. Second, the voltage deviation is minimized at this node that occurs when Φ switches. The level converter depicted in Fig. 14 accomplishes this task by reducing the voltage variation at nodes Φ_{ss} and $\overline{\Phi}_{ss}$ to less than 350 mV and setting an appropriate dc bias.

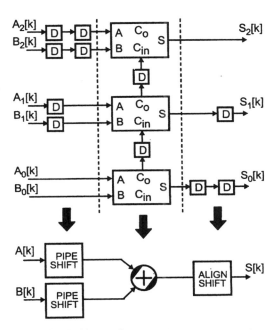

Fig. 16. A pipelined adder topology.

C. Σ–Δ Modulator

Fig. 15 shows the second-order MASH Σ–Δ topology used in the prototype. This structure is well known [12] and has properties that are well suited to our transmitter application. The MASH topology is unconditionally stable over its entire input range and is readily pipelined by using a technique described in this section.

The spectral density at the output of a second-order MASH Σ–Δ modulator is described by the equation

$$S_{\text{OUT}}(f) = S_{\text{IN}}(f) + \left| (e^{-j2\pi fT} - 1)^2 \right|^2 S_E(f). \quad (9)$$

In the presence of a sufficiently active input, $S_E(f)$ can be considered a white noise source with spectral density

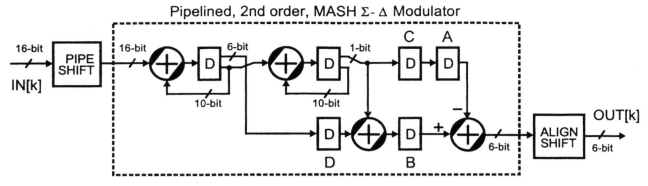

Fig. 17. A pipelined, second-order, digital MASH structure.

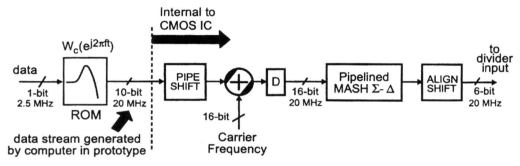

Fig. 18. Pipelined digital data path to divider input.

$S_E(f) = 1/12$. This assumption is reasonable while the modulation signal is applied; we have found that setting the least significant bit (LSB) of the modulator high also helps to achieve this condition by forcing the internal states of the MASH structure to constantly change.

A fact that does not appear to have been appreciated in the literature is that the digital MASH $\Sigma-\Delta$ structure is highly amenable to pipelining. This is a useful technique when seeking a low power implementation since it allows the supply voltage to be reduced by virtue of the fact that the required throughput can be achieved with lower circuit speed.

To pipeline the MASH structure, we apply a well-known technique that has been used for adders and accumulators [17], [18]. Fig. 16 illustrates a 3-b example. Since the critical path in these structures is their carry chain, registers are inserted in this path. To achieve time alignment between the input and the delayed carry information, registers are also used to skew the input bits. As indicated in the figure, we refer to this operation as "pipe shifting" the input. The adder output is realigned in time by performing an "align shift" of its bits as shown. (Note that shading is applied to the adder block in Fig. 16 as a reminder that its bits are skewed in time.) The same pipelining approach can be applied to digital accumulators since there is *no* feedback from higher to lower bits.

Since its basic building blocks are adders and accumulators, a MASH $\Sigma-\Delta$ modulator of *any order* can be pipelined using this technique. Using the symbols introduced in the previous two figures, Fig. 17 depicts a pipelined, second-order MASH topology. Each first-order $\Sigma-\Delta$ is realized as a pipelined accumulator with feedback removed from the most significant bits in its output. The output of the second stage is fed into the

filter $1 - D$, which is implemented with two pipelined adders and a delay element, A. A delay B is inserted between these two adders in order to pipeline their sum path, which requires a matching delay C in the path above for time alignment. Also, a delay D must also be included in the output path of the first $\Sigma-\Delta$ stage to compensate for the time delay incurred through the second stage. Since a signal once placed in the "pipe shifted domain" can be sent through any number of cascaded, pipelined adders and/or integrators, only one pipe shift and align shift are needed in the entire structure.

Fig. 18 illustrates the implementation of the overall digital path using pipelining. To save area, the circuits were pipelined every two bits as opposed to one, and pipe shifting was not applied to the carrier frequency signal since it is constant during modulation. To achieve flexibility, the compensated digital transmit filter was implemented in software, as opposed to a ROM, and the resulting digital data stream fed into the custom CMOS IC.

V. MEASURED PERFORMANCE

The primary performance criteria by which a transmitter is judged are its accuracy in modulation and its noise performance. We now describe the characterization of the prototype in relation to these issues.

Fig. 19 shows measured eye diagrams from the prototype using an HP 89441A modulation analyzer. To illustrate the impact of mismatch between the compensation filter and PLL dynamics, measurements were taken under three different values of open-loop gain. These results indicate that the modulation performance of the transmitter is quite good even when the open-loop gain is in error by $\pm 25\%$; the effects of

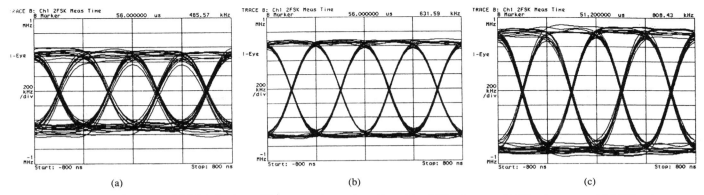

Fig. 19. Measured eye diagrams at 2.5 Mb/s for three different open-loop gain settings: (a) −25% gain error, (b) 0% gain error, and (c) 25% gain error.

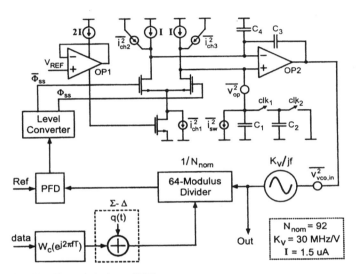

Fig. 20. Expanded view of PLL system.

TABLE III
VALUES OF NOISE SOURCES WITHIN PLL

Noise Source	Origin	Nature	Calculation	Value
$\overline{i_{ch1}^2}$	Ch. Pump, OP1	CT	Hspice	1.2E-24 A^2/Hz
$\overline{i_{ch2}^2}, \overline{i_{ch3}^2}$	Ch. Pump	CT	Hspice	1.8E-25 A^2/Hz
$\overline{i_{sw}^2}$	Switched Cap	DT	Equation 11	1.0E-26 A^2/Hz
$\overline{v_{op}^2}$	OP2	CT	Hspice	1.85E-16 V^2/Hz
$\overline{v_{vco,in}^2}$	VCO	CT	Equation 10	1.4E-16 V^2/Hz
$q(t)$	Σ-Δ	DT	—	—

this gain error are to produce a moderate amount of ISI and an error in the modulation deviation.

An explanation of the observed ISI and deviation error is given in [7]. In brief, the resulting mismatch creates a parasitic pole/zero pair that occurs near the cutoff frequency of the PLL (84 kHz in this case); the resulting transfer function seen by the data can be viewed as the sum of a low-pass and an all-pass filter. ISI is introduced as data excites the impulse response of the low-pass filter, and modulation deviation error occurs since the magnitude of the all-pass is changed according to the amount of mismatch present.

Fig. 20 displays the dominant sources of noise in the prototype; their values are displayed in Table III. Many of these values were obtained through ac simulation of the relevant circuits in HSPICE. Note that all noise sources other than $q(t)$ are assumed to be white, so that the values of their variance suffice for their description. This assumption is only approximate for the VCO noise in the prototype, as will be seen in the measured data.

Based on measurements, the input referred noise of the VCO was calculated in the table from the expression

$$10 \log\left(\overline{v_{vco,in}^2}|K_v/(jf)|^2\right) = -143 \text{ dBc/Hz at } f = 5 \text{ MHz} \tag{10}$$

where K_v is 30 MHz/V. The value of the kT/C noise current produced by the switched capacitor operation, $\overline{i_{sw}^2}$, is calculated as

$$\overline{i_{sw}^2} = (1/T)kT_K C_2 \tag{11}$$

where k is Boltzmann's constant, and T_K is temperature in degrees Kelvin.

Assuming that each of the noise sources in Fig. 20 are independent of each other, we can express the overall phase noise spectral density at the transmitter output $S_\Phi(f)$ as

$$S_\Phi(f) = S_{\Phi_v}(f) + S_{\Phi_i}(f) + S_{\Phi_q}(f) \tag{12}$$

where $S_{\Phi_v}(f)$, $S_{\Phi_i}(f)$, and $S_{\Phi_q}(f)$ are the noise contributions from the dominant voltage, current, and quantization noise sources. Based on the values in Table III and the model in Fig. 20, we obtain

$$S_{\Phi_v}(f) \approx \overline{v_{sum}^2}|K_v/(jf)|^2|1 - G(f)|^2$$
$$S_{\Phi_i}(f) \approx \overline{i_{sum}^2}(\pi N_{nom}/I)^2|G(f)|^2 \tag{13}$$

where

$$\overline{v_{sum}^2} = \overline{v_{vco,in}^2} + \overline{v_{op}^2}, \quad \overline{i_{sum}^2} \approx \overline{i_{ch1}^2}/2.$$

In the case of $S_{\Phi_i}(f)$, the division of $\overline{i_{ch1}^2}$ by two is an approximation based on the fact that the dominant charge pump noise source is switched in and out at each opamp input with a nominal duty cycle of 50%. Note that $S_{\Phi_q}(f)$ is given by (2).

A plot of the spectra in (13) is shown in Fig. 21(a). Computation of these spectra assumed the parameter values

619

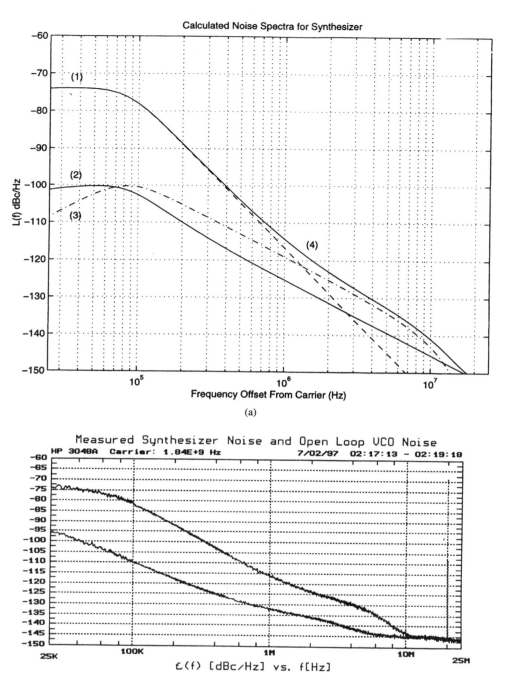

Fig. 21. Noise spectra of synthesizer: (a) calculated: (1) charge pump induced, $S_{\Phi_i}(f)$, (2) VCO and opamp induced, $S_{\Phi_v}(f)$, (3) Σ–Δ induced, $S_{\Phi_q}(f)$. (4) overall, $S_\Phi(f)$ and (b) measured synthesizer and open-loop VCO noise.

listed in Fig. 20 and Table III, and $G(f)$ described by (7) with

$$f_z = 11.6 \text{ kHz}, \quad f_{cp} = 14.2 \text{ kHz},$$
$$f_o = 84.3 \text{ kHz}, \quad Q = 0.75.$$

As seen in this diagram, the noise from the charge pump $\overline{i^2_{ch1}}$ dominates at low frequencies, and the influence of the Σ–Δ quantization noise dominates at high frequencies.

Fig. 21(b) shows measured noise results from the transmitter prototype taken with an HP 3048A phase noise measurement system. (The spurious content of the Σ–Δ modulator was reduced to negligible levels by feeding a binary data stream into the LSB of the modulation path so that the internal states of the Σ–Δ were randomized; the binary data stream was designed to have relatively flat spectral characteristics and negligible levels of spurious energy at frequencies greater than 10 kHz.) The resulting spectrum compares quite well with the calculated curve in Fig. 21(a), especially at high frequency offsets close to 5 MHz. At lower frequencies in the range of 100 kHz, the measured noise is within about 3 dB

620

of the predicted value; the higher discrepancy in this region might be attributed to the fact that $\overline{i_{ch1}^2}$ was calculated without considering the offset or transient response of the charge pump and/or the possible inaccuracy of the HSPICE device models at low currents. Note that the spur at 20-MHz offset (the reference frequency), which is due to the 50% nominal duty cycle of the PFD, is less than -60 dBc.

Fig. 21(b) demonstrates that the unmodulated transmitter has an output spectrum $S_\Phi(f)$ of -132 dBc/Hz at 5-MHz offset from the carrier. At this frequency offset, simulations reveal that the output spectrum of the *modulated* transmitter is equal to $S_\Phi(f)$ when its data rate is close to the DECT rate of 1 Mb/s [7]. This being the case, the transmitter satisfies the DECT noise specification of -131 dBc/Hz at 5-MHz offset; eye diagrams for data rates close to 1 Mb/s are found in [7].

VI. CONCLUSION

A digital compensation method and key circuits were presented that allow modulation of a frequency synthesizer at rates over an order of magnitude faster than its bandwidth. Using this technique, a transmitter prototype was built that achieves 2.5-Mb/s data rate modulation using GFSK modulation at a carrier frequency of 1.8 GHz. Measured results indicate that the architecture can achieve the modulation and noise performance required by the DECT standard with a structure that is highly integrated and has low power dissipation. In particular, the mostly digital design requires no off-chip filters, no mixers, and no D/A converters in the modulation path. Further, the structure contains only the core components required of a narrowband, spectrally efficient transmitter: a frequency synthesizer and a digital transmit filter.

ACKNOWLEDGMENT

The authors thank G. Dawe and J. Mourant for guidance in RF issues, A. Chandrakasan for discussion on low power methods, R. Weiner for bonding the die, B. Broughton for aid in phase noise measurements, and M. Trott, P. Ferguson, P. Katzin, Z. Zvonar, and D. Fague for advice.

REFERENCES

[1] P. Gray and R. Meyer, "Future directions in silicon IC's for RF personal communications," in *IEEE Custom IC Conf.*, 1995, pp. 83–90.
[2] T. Stetzler, I. Post, J. Havens, and M. Koyama, "A 2.7–4.5 V single-chip GSM transceiver RF integrated circuit," in *Proc. IEEE Int. Solid-State Circuits Conf.*, Feb. 1995, pp. 150–151.
[3] J. Min, A. Rofougaran, H. Samueli, and A. A. Abidi, "An all-CMOS architecture for a low-power frequency-hopped 900 MHz spread spectrum transceiver," in *IEEE Custom IC Conf.*, 1994, pp. 16.1/1-4.
[4] S. Sheng, L. Lynn, J. Peroulas, K. Stone, I. O'Donnell, and R. Brodersen, "A low-power CMOS chipset for spread-spectrum communications," in *Proc. IEEE Int. Solid-State Circuits Conf.*, Feb. 1996, pp. 346–347.
[5] S. Heinen, S. Beyer, and J. Fenk, "A 3.0 V 2 GHz transmitter IC for digital radio communication with integrated VCO's," in *Proc. IEEE Int. Solid-State Circuits Conf.*, Feb. 1995, pp. 150–151.
[6] S. Heinen, K. Hadjizada, U. Matter, W. Geppert, V. Thomas, S. Weber, S. Beyer, J. Fenk, and E. Matschke, "A 2.7 V 2.5 GHz bipolar chipset for digital wireless communication," in *Proc. IEEE Int. Solid-State Circuits Conf.*, Feb. 1997, pp. 306–307.
[7] M. H. Perrott, "Techniques for high data rate modulation and low power operation of fractional-N frequency synthesizers," Ph.D. dissertation, MIT, 1997.
[8] T. A. Riley, M. A. Copeland, and T. A. Kwasniewski, "Delta-sigma modulation in fractional-N frequency synthesis," *IEEE J. Solid-State Circuits*, vol. 28, pp. 553–559, May 1995.
[9] T. A. Riley and M. A. Copeland, "A simplified continuous phase modulator technique," *IEEE Trans. Circuits Syst. II*, vol. 41, pp. 321–328, May 1994.
[10] B. Miller and B. Conley, "A multiple modulator fractional divider," in *Proc. 44th Annual Symp. on Frequency Control*, May 1990, pp. 559–567.
[11] B. Miller and B. Conley, "A multiple modulator fractional divider," *IEEE Trans. Instrum. Meas.*, vol. 40, pp. 578–583, June 1991.
[12] J. Candy and G. Temes, *Oversampling Delta-Sigma Data Converters.* New York: IEEE Press, 1992.
[13] Y. Tsividis and J. Voorman, *Integrated Continuous-Time Filters.* New York: IEEE Press, 1993.
[14] T. Kamoto, N. Adachi, and K. Yamashita, "High-speed multi-modulus prescaler IC," in *1995 Fourth IEEE Int. Conf. Universal Personal Communications. Record. Gateway to the 21st Century*, 1995, pp. 991, 325-8.
[15] J. Craninckx and M. S. Steyaert, "A 1.75-GHz/3-V dual-modulus divide-by-128/129 prescaler in 0.7-μm CMOS," *IEEE J. Solid-State Circuits*, vol. 31, pp. 890–897, July 1996.
[16] M. Thamsirianunt and T. A. Kwasniewski, "A 1.2 μm CMOS implementation of a low-power 900-MHz mobile radio frequency synthesizer," in *IEEE Custom IC Conf.*, 1994, pp. 16.2/1-4.
[17] S.-J. Jou, C.-Y. Chen, E.-C. Yang, and C.-C. Su, "A pipelined multiplier-accumulator using a high-speed, low-power static and dynamic full adder design," *IEEE J. Solid-State Circuits*, vol. 32, pp. 114–118, Jan. 1997.
[18] F. Lu and H. Samueli, "A 200-MHz CMOS pipelined multiplier-accumulator using a quasidomino dynamic full-adder cell design," *IEEE J. Solid-State Circuits*, vol. 28, pp. 123–132, Feb. 1993.

Michael H. Perrott (S'97) was born in Austin, TX, in 1967. He received the B.S.E.E. degree from New Mexico State University, Las Cruces, in 1988, and the M.S. and Ph.D. degrees in electrical engineering and computer science from Massachusetts Institute of Technology, Cambridge, in 1992 and 1997, respectively.

He currently works at Hewlett-Packard Laboratories, Palo Alto, CA. His interests include signal processing and circuit design applied to communication systems.

Theodore L. Tewksbury III (S'86–M'87) received the S.B. degree in architecture in 1983 and the M.S. and Ph.D. degrees in electrical engineering and computer science in 1987 and 1992, respectively, all from the Massachusetts Institute of Technology, Cambridge. His doctoral dissertation consisted of an experimental and theoretical investigation of the effects of oxide traps on the large-signal transient performance of analog MOS circuits.

He joined Analog Devices, Inc., in 1987 as Design Engineer for the Converter Group, where he worked on high-speed, high-resolution data acquisition circuits for video, instrumentation, and medical applications. From 1992 to 1994, as Senior Characterization Engineer, he was involved in the development of high-accuracy analog models for advanced bipolar, BiCMOS, and CMOS processes, with emphasis on the statistical modeling of manufacturing variations. In December 1994, he joined the newly formed Communications Division at Analog Devices as RF Design Engineer. He is presently involved in the design of RF integrated circuits for wireless communications, including GSM, DECT, and DBS. He is also actively involved in the development and modeling of advanced semiconductor technologies for RF applications, including ADRF (Analog Devices bipolar RF process) and silicon germanium.

A CMOS Monolithic ΔΣ-Controlled Fractional-N Frequency Synthesizer for DCS-1800

Bram De Muer, *Student Member, IEEE,* and Michel S. J. Steyaert, *Senior Member, IEEE*

Abstract—A monolithic 1.8-GHz ΔΣ-controlled fractional-N phase-locked loop (PLL) frequency synthesizer is implemented in a standard 0.25-μm CMOS technology. The monolithic fourth-order type-II PLL integrates the digital synthesizer part together with a fully integrated *LC* VCO, a high-speed prescaler, and a 35-kHz dual-path loop filter on a die of only 2 × 2 mm². To investigate the influence of the ΔΣ modulator on the synthesizer's spectral purity, a fast nonlinear analysis method is developed and experimentally verified. Nonlinear mixing in the phase-frequency detector (PFD) is identified as the main source of spectral pollution in ΔΣ fractional-N synthesizers. The design of the zero-dead zone PFD and the dual charge pump is optimized toward linearity and spurious suppression. The frequency synthesizer consumes 35 mA from a single 2-V power supply. The measured phase noise is as low as −120 dBc/Hz at 600 kHz and −139 dBc/Hz at 3 MHz. The measured fractional spur level is less than −100 dBc, even for fractional frequencies close to integer multiples of the reference frequency, thereby satisfying the DCS-1800 spectral purity constraints.

Index Terms—CMOS RF integrated circuits, ΔΣ modulator, fractional-N frequency synthesis, phase-locked loop, phase noise.

I. INTRODUCTION

THE END of the 20th century was characterized by the unrivaled growth of the telecommunication industry. The main cause was the introduction of digital signal processing in wireless communications, driven by the development of high-performance low-cost CMOS technologies for VLSI. However, the implementation of the RF analog front end remains the bottleneck. This is reflected in the large effort put into monolithic CMOS integration of RF circuits both by academics and industry [1]–[3].

The goal of this work is the monolithic integration in standard CMOS technology of a frequency synthesizer to enable the full integration of a transceiver front end in CMOS, including a low-IF receiver and a direct upconversion transmitter [1]. To achieve a high degree of integratability and fast settling under low-noise constraints, a ΔΣ fractional-N synthesizer topology has been chosen [4] (Fig. 1). ΔΣ fractional-N synthesis circumvents the severe speed–spectral purity–resolution tradeoff of the classic phase-locked loop (PLL) synthesizer, by providing synthesis of fractional multiples of the reference frequency. Spurious tones that emerge from the fractional division are whitened and noise shaped by the ΔΣ action and ultimately filtered by the loop filter. To prevent degradation of the spectral purity by

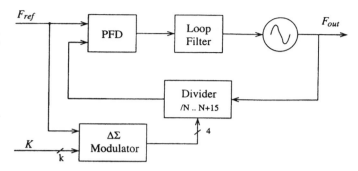

Fig. 1. Principle of ΔΣ fractional-N synthesis.

digital noise coupling, the ΔΣ modulator is scheduled for integration on the digital baseband signal processing IC of the full transceiver system.

The paper describes the design of a monolithic 1.8-GHz ΔΣ-controlled fractional-N PLL frequency synthesizer. In Section II, the influence of ΔΣ noise on PLL bandwidth requirements is theoretically analyzed for multistage noise shaping (MASH) and multibit single-loop ΔΣ modulators. Next, a fast nonlinear analysis method is presented, which predicts possible degradation of the PLL spectral purity by in-band noise leakage and re-emerging of spurious tones. The nonlinearities in the phase-frequency detector (PFD) charge pumps are identified as the main trouble spots. The fourth-order type-II PLL building-block design is discussed in Section IV, focusing on integrated filter and voltage-controlled oscillator (VCO) design and on the realization of a linear phase error-to-charge-pump current conversion. In Section V, the experimental results of the fractional-N synthesizer prototype are presented and compared to the simulations, showing good correspondence.

II. THE FRACTIONAL-N SYNTHESIZER

A. Introduction

A block diagram of a ΔΣ fractional-N synthesizer is shown in Fig. 1. The ΔΣ modulator output controls the instantaneous division modulus of the prescaler, such that the mean division modulus is $N_{frac} = N + K/2^k$, with k the number of bits of the ΔΣ modulator and K the input word. The corresponding phase changes at the prescaler output are quantized, leading to possible spurious tones and quantization noise. By selecting higher order ΔΣ modulators, the spurious energy is whitened and shaped to high-frequency noise, which can be removed by the low-pass loop filter. As a result, for a given frequency resolution, an arbitrary high f_{ref} can be chosen, by assigning the proper number

Manuscript received November 5, 2001; revised January 31, 2002.

The authors are with the Katholieke Universiteit Leuven, Department Elektrotechniek, ESAT-MICAS, B-3001 Heverlee, Belgium (e-mail: bram.demuer@esat.kuleuven.ac.be).

Publisher Item Identifier S 0018-9200(02)05856-0.

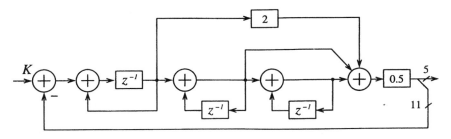

Fig. 2. Third-order multibit single-loop $\Delta\Sigma$ modulator. The internal modulator accuracy is 16 bit. From the five output bits, only four are used for stability reasons.

of bits k to the modulator. The loop bandwidth is not restricted by the reference spur suppression, resulting in faster settling and higher integratability. Additionally, the division modulus is decreased by a factor $2^{k_{\min}}$ (with $k_{\min}$ the minimum number of bits for the frequency resolution, i.e., 7.02 in this case), so that noise of the PLL blocks, except for the VCO, is less amplified.

B. The $\Delta\Sigma$ Modulators

The influence of both third-order MASH and multibit single-loop $\Delta\Sigma$ modulators on the spectral purity of the fractional-N synthesizer is investigated. Since the order of the integrated PLL loop filter is three, the order of the $\Delta\Sigma$ modulators must also be three or higher to ensure that $\Delta\Sigma$ noise has at least a -20-dB/dec rolloff at intermediate offset frequencies, causing no degradation of the output phase noise. Both modulators have an internal accuracy of 16 bit and 1 LSB dithering is applied to further randomize any spurious energy. The dithering sequence is third-order noise shaped to avoid an increased noise floor.

The MASH or cascade 1-1-1 $\Delta\Sigma$ modulator is chosen because it is easy to integrate in CMOS and is unconditionally stable. The noise transfer function (NTF) of the MASH modulator is $H_{qn}(z) = (1 - z^{-1})^3$ and contains three poles at the origin of the z plane. The result is harsh LF noise shaping and and substantial HF noise. In the time domain, this is reflected in the intensive prescaler modulus switching. To synthesize a frequency of $67.92 \times f_{\text{ref}}$, all moduli between 64 and 71 are employed.

The multibit single-loop $\Delta\Sigma$ modulator is shown in Fig. 2. For ease of integration, the feedforward and feedback coefficients are a power of 2. Only four output bits are needed to control the prescaler moduli, but five output bits are used, to avoid overlap of the intended input operating range and the unstable input regions. The NTF of the presented modulator is given in (1) and contains only one pole at the origin of the z plane and two low-Q Butterworth poles at $0.167 \times f_{\text{ref}}$, with a passband gain of 3.2.

$$H_{qn}(z) = \frac{(1 - z^{-1})^3}{1 - z^{-1} + 0.5 z^{-2}}. \tag{1}$$

Although the single-loop $\Delta\Sigma$ modulator is more complex than the MASH modulator, it offers a higher flexibility in terms of noise shaping. The HF quantization noise of the modulator can be spread out by proper pole positioning. As a result, the prescaler modulus switching is less intense. Only the moduli between 66 and 69 are needed to synthesize $67.92 \times f_{\text{ref}}$. The reduced HF switching has advantageous effects on noise

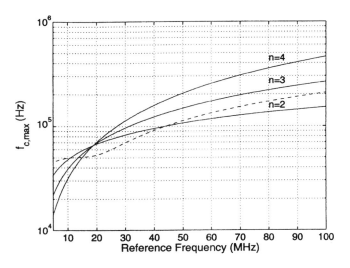

Fig. 3. Maximum PLL bandwidth $f_{c,\max}$ versus the reference frequency and different $\Delta\Sigma$ modulator orders, for the type-II fourth-order PLL. The dashed curve is for the third-order single-loop modulator. The targeted phase-noise specification is -136 dBc/Hz at 3 MHz for DCS-1800.

coupling and sensitivity to PLL nonlinearities, as will be discussed in Section III.

C. Theoretical Analysis

To theoretically model the impact of $\Delta\Sigma$ control on the spectral purity of the synthesizer, a linear-time-invariant (LTI) PLL model is employed, with the $\Delta\Sigma$ quantization noise as an additive noise source $S_\Theta(z)$ at the prescaler output. The prescaler with $\Delta\Sigma$ control can be looked upon as a digital-to-phase (D/P) converter. Every reference cycle, the prescaler subtracts $n \cdot 2\pi$ rad from its input signal, with $n = 0, \ldots, (2^4 - 1)$ determined by the $\Delta\Sigma$ modulator output. The resulting quantization noise on the division modulus, and thus output phase, is approximated by uniformly distributed white noise [5]. The quantization noise power is $\Delta^2/12$ with $\Delta = \Delta N/(2^B - 1) = 1$ for both modulators with ΔN the modulus range and B the number of significant $\Delta\Sigma$ output bits. The phase noise contribution of the $\Delta\Sigma$ modulator at the output of the synthesizer is found in (2) [6], with $T(f)$ the closed-loop transfer function of the fourth-order type-II PLL.

$$S_{\Delta\Sigma,o}(f) = \frac{\pi^2}{3 \cdot f_{\text{ref}}} \cdot \frac{|H_{qn}(z)|^2}{|1 - z^{-1}|^2} \cdot |T(f)|^2. \tag{2}$$

Since the main advantage of $\Delta\Sigma$ fractional-N synthesizers is the decoupling of the reference frequency f_{ref} and the PLL bandwidth f_c, the influence of the $\Sigma\Delta$ noise on the bandwidth

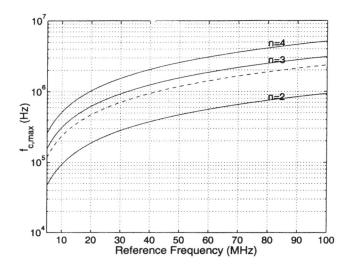

Fig. 4. Maximum PLL bandwidth $f_{c,\,\mathrm{max}}$ versus the reference frequency and different $\Delta\Sigma$ modulator orders for $\Delta\Phi_{\mathrm{rms}} < 1.5°$. The dashed curve is for the third-order single-loop modulator.

requirement is examined. To comply with the most stringent DSC-1800 phase noise specification, i.e., -133 dBc/Hz at 3 MHz offset [7], the target $\Delta\Sigma$ phase noise is $S_{\Delta\Sigma,\,o}$ (3 MHz) $= -136$ dBc/Hz. In Fig. 3, the maximum PLL bandwidth $f_{c,\,\mathrm{max}}$ is plotted versus the reference frequency for different MASH modulator orders. The dashed line is the maximum bandwidth for the single-loop multibit $\Delta\Sigma$ modulator of Section II-B. For a reference frequency of 26 MHz, not much is gained from increasing the modulator order. For a high bandwidth and thus a fast PLL, the reference frequency and/or the modulator order should be increased leading to an increased power consumption and circuit complexity. The maximum bandwidth is 87 kHz for the third-order MASH modulator and 62 kHz for the single-loop multibit modulator.

Apart from the out-of-band phase-noise constraint, the integrated in-band phase noise, determining the rms phase error $\Delta\Phi_{\mathrm{rms}}$ of the PLL is of importance. To be sure that the $\Delta\Sigma$ does not corrupt the rms phase error, the dynamic range of the modulator must be higher than the dynamic range of the PLL [8]. The integrated in-band frequency noise Δf_n is given by $\Delta f_n^2 \approx 2/3 \cdot A_n \cdot f_{\mathrm{nbw}}^3$ with f_{nbw} the noise bandwidth of the PLL and $10\log A_n$ the in-band phase noise in dBc/Hz. The noise bandwidth of the presented PLL is $f_{\mathrm{nbw}} \approx 1.16 f_c$. The maximum bandwidth f_c of the PLL is calculated in (3) [8].

$$f_c < \left[\frac{3}{8} \cdot \frac{2n+1}{(2\pi)^{2n}} \cdot \Delta\Phi_{\mathrm{rms}}^2 \right]^{1/(2n-1)} \cdot f_{\mathrm{ref}}. \qquad (3)$$

The maximum PLL bandwidth $f_{c,\,\mathrm{max}}$ is plotted versus the reference frequency of the PLL for different MASH modulator orders in Fig. 4. For the single-loop multibit $\Delta\Sigma$ modulators (dashed curve), the actual maximum bandwidth can be calculated to be 25% smaller than in (3), due to the Butterworth poles. In the case of a third-order modulator, a 1.5° rms phase error (to ensure at least an overall rms phase error of 2°) and a f_{ref} of 26 MHz, the maximum bandwidth is 810 and 614 kHz, respectively. Obviously, the constraint posed on the $\Delta\Sigma$ modulator noise due to in-band noise contributions is much less severe than the constraint due to the out-of-band phase noise at 3 MHz.

III. FAST NONLINEAR ANALYSIS METHOD

The theoretical analysis suggested that applying $\Delta\Sigma$ control to the prescaler would not cause any problems for the spectral purity of the PLL. Practice, however, proves this wrong. A fast nonlinear analysis method is developed which can take into account the nonlinearity of the PLL building blocks. The analysis method is at the same time sufficiently fast to sweep simulations over different degrees of nonlinearities and operating points, and is capable of performing sufficiently long transient simulations to get accurate fast Fourier transforms (FFTs) of the phase variable. The fractional operation of the PLL is simulated in discrete time and in open loop under locked conditions to avoid drift of the phase error. To further speed up the simulation, the building blocks are represented by high-level models with parameters to model any nonlinear behavior or mismatch in critical transistors. The simulations are performed in Matlab [9].

To find the phase error, generated by the $\Delta\Sigma$ modulation of the division modulus, the variation of the number of RF pulses, $\mathrm{RF}[k]$, at the output of the divider is monitored. Every reference cycle, the number of RF pulses at the divider output is determined by the number of pulses swallowed by the $\Delta\Sigma$ control, $DS[k]$:

$$\mathrm{RF}[i] = \sum_{k=2}^{i} \mathrm{RF}[k-1] - DS[k]. \qquad (4)$$

The resulting quantized phase changes are compared with the phase that would be expected when the loop would be in lock, i.e., the phase corresponding to the fractional part of the division modulus $K/2^k$. The result is the instantaneous accumulated phase error $\Delta\phi[i]$:

$$\Delta\phi[i] = \sum_{k=2}^{i} \frac{2\pi}{N[k]} \left(\mathrm{RF}[k] - (k-1) \cdot \frac{K}{2^k} \right). \qquad (5)$$

The phase error is converted to current pulses, $CP[i]$, in the charge pump. The $\Delta\phi \rightarrow I_{qp}$ (phase-error charge-pump current) conversion is modeled to contain any PFD nonlinearity. Mismatch in the up and down current sources, resulting in gain mismatch for positive and negative phase errors is modeled by $I_{qp\pm}$. The occurrence of a dead zone is modeled by $DZ[k]$

$$CP[i] = \sum_{k=2}^{i} \frac{I_{qp\pm}}{2\pi} \left(\Delta\phi[k] \pm DZ[k] \right). \qquad (6)$$

By taking an FFT of the current pulses, the current noise spectrum is obtained. The current noise spectrum is modeled as a phase-noise source which is subjected to its corresponding closed-loop transfer function, obtained from the LTI PLL model. This means that the filter is modeled by its linear transfer function, which includes parasitic gain and pole position changes. The nonlinear conversion from voltage to frequency/phase in the VCO is modeled by the variation of the VCO gain, when changing the operating point of the PLL.

The analysis tool enables the evaluation and comparison of the effect of MASH and single-loop $\Delta\Sigma$ noise on the PLL. This analysis is performed with the following nonlinearities: a 0.1% dead zone and a gain mismatch of 2%. The internal accuracy of both modulators is 16 bit. The reference frequency

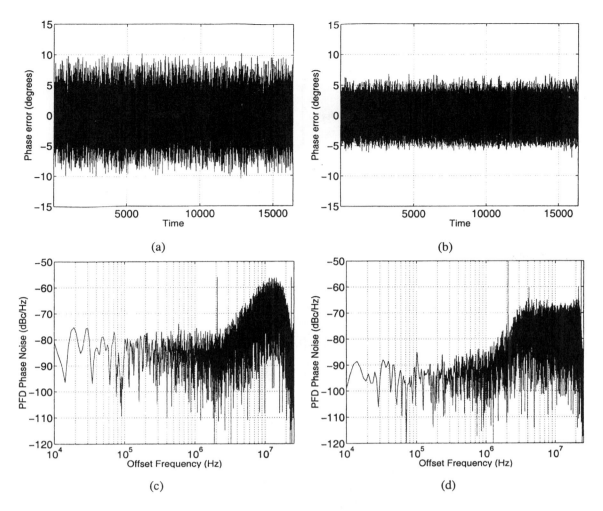

Fig. 5.   Simulation results. The phase error $\Delta\phi$ for (a) the MASH modulator and (b) the single-loop multibit modulator. The FFT of the current pulses $CP[i]$ for (c) the MASH modulator and (d) the single-loop multibit modulator.

is 26 MHz and the fractional division number is 67.92. The output frequency is 1.76592 GHz, i.e., 2.08 MHz offset from an integer multiple of f_{ref}. In Fig. 5(a) and (b), the time-domain phase error $\Delta\phi$ is plotted for both modulators. Note that the $\Delta\Sigma$ fractional-N PLL frequency synthesizer can hardly be called a *phase-locked* loop, since the loop is never in lock! Due to the shaping of the HF noise in the single-loop $\Delta\Sigma$ modulator, the instantaneous phase error is smaller than for a MASH modulator. This has two important consequences. First, the on-time of the charge pumps is smaller for the single-loop modulator, making it less sensitive to noise coupling from the substrate and the power supply. Second, the sensitivity to the nonlinear $\Delta\phi \rightarrow I_{qp}$ conversion in terms of noise leakage is reduced.

To be able to examine the effect of nonlinearities in the frequency domain, the FFTs of the charge-pump current pulses $CP[i]$ are plotted in Fig. 5(c) and (d). A noise floor appears in the output spectrum as well as spurious tones, although the $\Delta\Sigma$ output is perfectly randomized and dithered. Due to the nonlinear mixing in the PFD charge pump, noise at $f_{\text{ref}}/2$ folds back to lower offset frequencies, similar to the effect of a nonlinear DAC in a multibit $\Delta\Sigma$ ADC. Since the noise at $f_{\text{ref}}/2$ is much lower for the single-loop $\Delta\Sigma$ modulator, its noise leakage due to the nonlinear mixing in the PFD is also lower. In the time

domain, this effect corresponds to the smaller phase excursions. The difference in phase error between MASH and single-loop modulators is reflected in a lower noise floor, i.e., a 10-dB difference. In addition, previously unnoticed spurious tones appear in the output spectrum at $j \times K/2^k \cdot f_{\text{ref}}$ with $j = 1, 2, 3, \ldots$.

Fig. 6 shows the $\Delta\Sigma$ noise of both modulators as it appears at the PLL output for an ideal (dotted) and a nonlinear $\Delta\phi \rightarrow I_{qp}$ conversion (solid). The results of the ideal case closely match the theoretical results of Section II-C (solid light gray). Due to nonlinearity, the simulated output spectrum of the integer-N PLL (the dash-dotted line) is seriously deteriorated by $\Delta\Sigma$ noise in the PLL noise bandwidth, increasing the $\Delta\Phi_{\text{rms}}$. Especially, the MASH converter is critical in terms of in-band noise due to the higher phase error [see Fig. 5(a)], despite the inherently lower LF $\Delta\Sigma$ noise of the MASH modulator. Note that the simulations are performed without taking into account noise coupling through the substrate or power-supply lines. As a consequence, the actual spurious performance of the $\Delta\Sigma$ fractional-N PLL could be worse than simulated. The presented simulation results are for a division modulus 67.92, close to an integer multiple of f_{ref}. When analyzing division moduli in between integer multiples of f_{ref}, noise leakage is still observed, but the spurious tones are well below the phase noise.

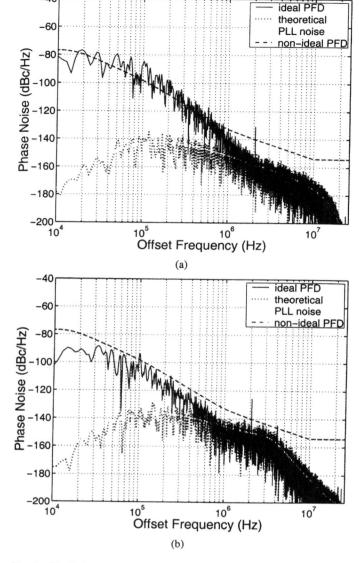

(a)

(b)

Fig. 6. Simulation results. The $\Delta\Sigma$ noise at the output of the PLL for (a) the MASH modulator and (b) the single-loop multibit modulator. The results are plotted for an ideal PFD (dotted), which closely corresponds to the theoretical results (solid light gray) and for a nonlinear PFD (solid). They are compared to the simulated integer PLL phase noise (the dash-dotted line).

The explanation for the re-emerging of spurious tones is that the modulator is unable to sufficiently decorrelate the successive output samples. To quantify the correlation in the $\Delta\Sigma$ modulator output, the discrete time autocorrelation estimate is calculated and plotted for both modulators for inputs close to an integer value (see Fig. 7). The autocorrelation calculations show correlation, although 1–LSB noise-shaped dithering is applied. The autocorrelation of the single-loop $\Delta\Sigma$ modulator shows large correlation peaks, explaining the higher spurious tones in the output phase-noise spectrum of the PLL. With the autocorrelation estimate, the necessary internal accuracy of the $\Delta\Sigma$ modulators is found to be at least 13 bits for MASH and 16 bit for single-loop modulators to sufficiently decorrelate the $\Delta\Sigma$ modulator output for inputs close to integers. A second possible source of tones is the downconversion of tones which are inherently present around $f_{\mathrm{ref}}/2$ [5], by the nonlinear mixing in the

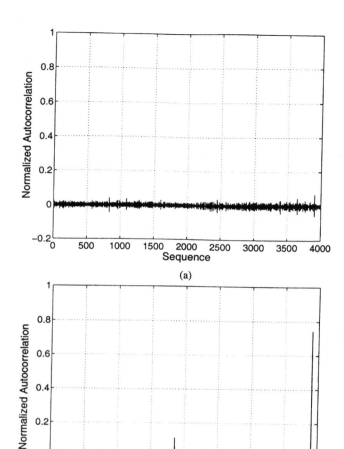

(a)

(b)

Fig. 7. Discrete time autocorrelation estimate of the modulator outputs for (a) the MASH modulator and (b) the single-loop multibit modulator.

PFD. This effect can be worsened by substrate and power-supply coupling with signals at $f_{\mathrm{ref}}/2$.

IV. PLL BUILDING-BLOCK CIRCUIT DESIGN

A. The Fourth-Order Type-II PLL

A fourth-order type-II PLL is integrated, including a 4-bit prescaler, a zero-dead-zone PFD, a dual charge pump, and a 3-step equalizer, together with an on-chip LC-tank VCO and a third-order dual-path 35-kHz low-pass loop filter (see Fig. 8). The equalizer performs a 3-step piecewise equalization of the loop gain, by keeping the product of the VCO gain and the charge-pump current constant. To prevent switching between different equalization states, the state transitions exhibit hysteresis.

B. The 4-Bit Prescaler

The first high-speed division of the prescaler is done with two differential single-transistor-clocked (DSTC) logic n-latches [10], forming a differential dynamic D-flip-flop. The flip-flop operates with rail-to-rail internal signals to minimize the residual prescaler phase noise [11] to levels insignificant to

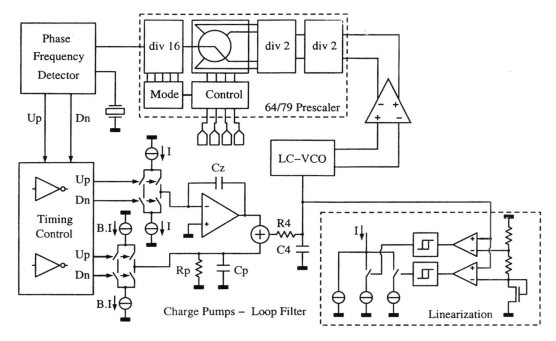

Fig. 8. Fully integrated fourth-order type-II phase-locked loop.

the overall phase-noise performance. The 16-modulus division (64...79) is implemented with the phase-switching topology [12]. The division moduli are generated by switching between the 90°-spaced output phases of the second D-flip-flop. When the 90° spacing is not ideal, spurs appear at 1/4, 2/4, and 3/4 of the PLL reference frequency. It takes careful layout and circuit design to equalize the delays of the different quadrature paths, such that these spurious tones are suppressed to negligible levels.

C. The Voltage-Controlled Oscillator

The *LC* VCO with on-chip inductor combines a 30% tuning range at only 2 V and an excellent phase-noise performance over a large frequency range. To minimize the VCO phase noise, a simulator-optimizer program has been developed which searches the optimal inductor geometry for a given technology. The resulting hollow octagonal balanced inductor has a Q as high as 9 with an inductance of 2.86 nH, for a standard 0.25-μm CMOS technology with only two metal layers (0.6 and 1.0 μm) [13].

The VCO is implemented as a single differential pMOS-only topology, leading to an enhanced tuning range, without increasing the power consumption and the VCO gain, K_{vco} [13]. For the frequency range of interest, K_{vco} is between 100 and 200 MHz/V, explaining the need for equalization of the loop gain. The VCO output is buffered from the prescaler input to prevent kickback noise from entering the tank. The measured phase noise is as low as -127.5 dBc/Hz at 600 kHz and -142.5 dBc/Hz at 3 MHz for a carrier frequency of 1.82 GHz.

D. The 35-kHz Dual-Path Loop Filter

To achieve full integration, a dual-path filter topology has been implemented (Fig. 8). Two filter paths, one active integration (C_z) and one passive low-pass filter (C_p, R_p) are added

Loop Parameters		Loop Passives		Performance (at 600 kHz)	
ω_c	35 kHz	R_p	3.2 kHz	$L_{qp,z}$	-170.2 dBc/Hz
I_{qp}	2 μA	R_4	1.07 kHz	$L_{qp,p}$	-154.4 dBc/Hz
B	12	C_p	240 pF	L_{add}	-133.1 dBc/Hz
		C_4	710 pF	L_{int}	-128.9 dBc/Hz
		C_z	450 pF	L_{R_p}	-135.8 dBc/Hz
		C_{tot}	1.4 nF	L_{R_4}	-134.3 dBc/Hz
				L_{MASH}	-123.6 dBc/Hz
PM	57°			L_{SL}	-128.0 dBc/Hz
ζ	0.77	Q_L	9	L_{tot}	-124.3 dBc/Hz

with a multiplication factor B in the dual charge pumps. The addition realizes the low-frequency zero needed for loop stability in a type-II PLL, without adding the actual capacitor [12]. The total number of capacitors is the same as in a classical fourth-order type-II PLL, but for the same phase noise the integrated capacitance is more than 5 times smaller. Due to the rather high VCO gain, the integrated capacitance is still 1.4 nF to be able to comply with the DCS-1800 phase-noise requirements. An extra pole (C_4, R_4) is added at 210 kHz to ensure enough suppression at higher offset frequencies. A filter optimization model is developed, determining all pole and zero positions and the capacitance–resistance tradeoff to obtain low noise and high integratability [14]. The results of the optimization at 1765.92 MHz are listed in Table I. The total phase noise L_{tot} is without the $\Delta\Sigma$ noise. The MASH and single-loop (SL) $\Delta\Sigma$ noise contributions result from the nonlinear analysis. As

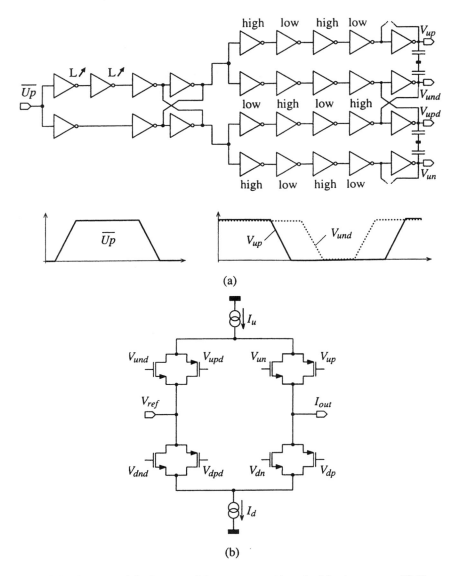

Fig. 9. (a) Timing control circuit and signals to control the dummy and the output current branch of the charge pump. (b) Charge-pump circuit with (at the left) the dummy current branch, denoted by the suffix d, and the output branch.

seen in Section II-C, the loop bandwidth needs to be smaller than 62 kHz for $\Delta\Sigma$ noise suppression. However, to ensure sufficient suppression of the low-frequency fractional spurious tones for inputs close to the integers, the bandwidth is designed to 35 kHz. Despite the rather low loop bandwidth for a fractional-N synthesizer, a settling time of less than 293 μs for a 104-MHz step is simulated.

E. The $\Delta\phi \rightarrow I_{qp}$ Conversion

The nonlinear analysis of Section III identified nonlinearity of the $\Delta\phi \rightarrow I_{qp}$ conversion as the main cause of noise leakage and spurious tones. Therefore, the PFD and charge-pump circuits are carefully optimized toward spurious suppression as such and toward a highly linear phase-error detection for $\Delta\Sigma$ spurious suppression.

First, the reference spur generation by the PFD charge-pump circuit is carefully minimized. The integration in the first path of the loop filter is done actively to keep the charge-pump output at a fixed level (see Fig. 8). Additionally, the charge-pump current is designed to be at least a magnitude larger than the fixed parasitic charge injection of the switch transistors. The current switches are implemented with pMOS and nMOS transistors to compensate charge injection. Finally, a timing control scheme [Fig. 9(a)] is developed to control the charge-pump switches. The up and down control pulses of the PFD are converted to synchronized control signals to drive both the output current branch and the dummy current branch of the charge pump [Fig. 9(b)]. Fig. 9(a) shows the dummy and output control signals. The dummy control V_{und} is delayed versus the output control V_{up} by modifying the thresholds of the second inverter-string (indicated by high and low) such that the current I_u always flows, preventing hard on/off switching of the current sources. To equalize rise and fall times and force a perfect π rad relation between nMOS and pMOS control signals, latches at the outputs of both inverter strings are implemented. Capacitors at the control outputs lower the rise and fall times to prevent large charge injections by fast switching.

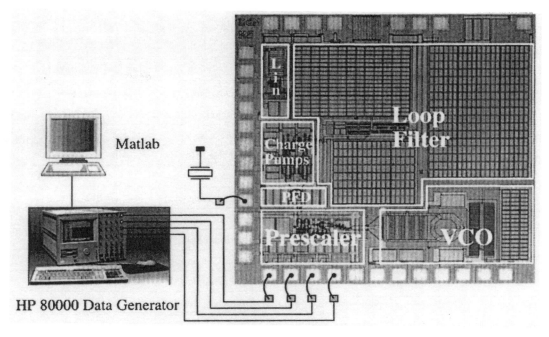

Fig. 10. IC microphotograph and the measurement setup in which it is embedded.

To linearize the $\Delta\phi \rightarrow I_{qp}$ conversion, the phase detection is performed by a zero-dead-zone PFD [15], to prevent a hard nonlinearity around 0° phase error. Due to the delay added in the PFD, both the up and down current sources are on, for small or zero phase errors, enabling the PFD to react to very small phase errors. The on-time fraction of the charge pump due to the delay is less than 10%. This value is a tradeoff between dead-zone prevention and sensitivity to noise coupling, when the charge pumps are on. To further minimize digital noise coupling, the sampling in the PFD and the computational events in the $\Delta\Sigma$ modulator and prescaler are offset in phase. Consequently, the phase-error decision making is done in a relatively quiet environment. To make sure that the gains for positive and negative phase-error detection are equal, the current source transistors are oversized to ensure sufficient matching. As a side effect, the current source $1/f$ noise, which can seriously affect the in-band noise, is decreased. Additionally, the timing control of Fig. 9(a) provides synchronization between the two filter paths and the switches of the charge pumps themselves, thereby ensuring equal positive and negative phase-error detection gain. HSPICE simulations of the PFD charge-pump circuit are performed and show no dead zone and no gain mismatch with ideal transistor matching.

V. EXPERIMENTAL RESULTS

Fig. 10 shows the IC microphotograph and the measurement setup in which it is embedded. The $\Delta\Sigma$ fractional measurements are performed by controlling the PLL divider moduli with an HP80000 data generator, which generates the 4-bit control word. The 4-bit $\Delta\Sigma$ output bit stream is generated using Matlab. This provides a flexible way to test different kinds of $\Delta\Sigma$ modulators, without the need for redesigns. All presented measurements are performed with a 26-MHz reference frequency and at

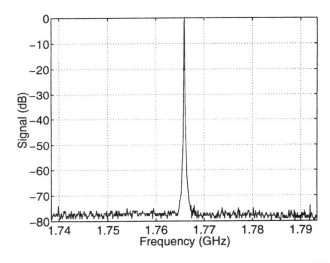

Fig. 11. Measured output spectrum of the $\Delta\Sigma$ fractional-N PLL at 1.76592 GHz. All spurious tones are well below -75 dBc/Hz.

1.76592 GHz, i.e., for a fractional division by 67.92 for comparison with the simulated results. The input to the $\Delta\Sigma$ modulators is a 16-bit word ($k = 16$), resulting in a frequency resolution of around 400 Hz. The power-supply voltage is only 2 V. Fig. 11 shows the output spectrum of the fractional-N PLL over a span of 55 MHz. The reference spurs are well below -75 dBc, due to the careful charge-pump timing control.

To measure the fractional performance of the frequency synthesizer, the Matlab data is stored in the data generator memory. Unfortunately, the maximum memory capacity is only 128 kbit, leading to large spurious tones at the output at low offset frequencies. These large tones corrupt the gain calibration, which is performed by the phase-noise measurement system every offset frequency decade, such that accurate measurements of the phase noise at offsets smaller than 10 kHz are not feasible. The measured phase noise of the PLL with the MASH modulator and the

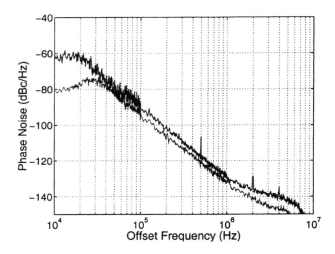

Fig. 12. Phase-noise measurement with the $\Delta\Sigma$ single-loop multibit converter at 1.76592 GHz compared to the phase noise at integer division (light).

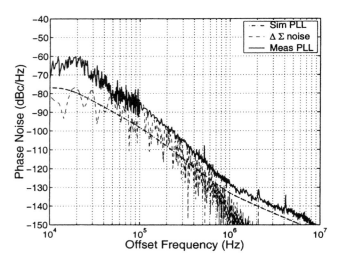

Fig. 13. Phase noise measurement with the MASH converter at 1.76592 GHz compared to the simulated $\Delta\Sigma$ noise at the output of the PLL (dashed), and with the simulated PLL output without $\Delta\Sigma$ control (dash-dotted).

TABLE II
SUMMARY OF MEASURED SPECIFICATIONS COMPARED TO THE
DCS-1800 SPECIFICATIONS

	DCS-1800	Measured
Phase Noise	-116	< -120
at 600 kHz (dBc/Hz)		< -124 (N=integer)
Phase Noise	-133	< -139
at 3 MHz (dBc/Hz)		< -142.5 (N=integer)
$\Delta\Phi_{rms}$	$2°$	$3°$ ($1.7°$ for N=integer)
Settling Time (μs)	865/288	226
Reference Spurs (dBc)	-80	< -75
Fractional Spurs (dBc)	-80	< -100
Power Consumption (mW)	–	70
Power Supply (V)	–	2
Frequency Resolution (Hz)	200 kHz	400 Hz

noisy control pulses are close to the *LC* tank and the bonding wires of the VCO power supply. Without proper shielding, the VCO phase noise is seriously degraded by this noise coupling.

In Fig. 13, the measured $\Delta\Sigma$ noise and the $\Delta\Sigma$ noise as simulated in Section III (dashed) is compared. The dash-dotted line is the simulated phase noise of the PLL without $\Delta\Sigma$ control. The simulated $\Delta\Sigma$ noise leakage closely matches the measured results, except at very low offsets due to the limited memory. The phase noise at high offsets is increased versus the simulated PLL results due to noise coupling. Second-order tones are larger in measurements, since the models in the simulator do not include second-order effects and noise coupling. Tones at 520 kHz are believed to come from subharmonic tones present in the $\Delta\Sigma$ modulator output [5], which are amplified by mixing through noise coupling. When comparing the results for the MASH and the single-loop modulator, the measured results are less pronounced than the simulated results (see Fig. 6). The measured phase noise for the single-loop modulator is however a few decibels lower than for the MASH modulator. Note that all measurements are performed for frequencies close to integer multiples of f_{ref}.

The measured settling time of the PLL is 226 μs for a 104-MHz frequency step. The power consumption of the PLL is 70 mW from a 2-V power supply. The fully integrated low-phase-noise VCO is responsible for almost 66% of the total power consumption. The IC area is 2×2 mm^2, including bonding pads and bypass capacitors. Table II shows the measured specifications compared to the DCS-1800 specifications [1]. The specifications of the IC prototype comply with the DCS-1800, only the $\Delta\Phi_{\text{rms}}$ is degraded due to the limited resolution of the measurement setup.

single-loop multibit modulator is presented in Figs. 12 and 13. Small spurs are present at 2.08 MHz as predicted by the simulations in Fig. 6. The spur level is well below -100 dBc, due to careful PFD charge-pump design. The phase noise at 600 kHz is lower than -120 dBc/Hz.

In Fig. 12, the measured phase noise of the PLL with a multibit single-loop modulator (dark) is compared to the phase noise at integer division (light). Noise at lower offsets originates from the $\Delta\Sigma$ modulator due to noise folding in the PFD, as predicted by the simulations. As a result, the rms phase error $\Delta\Phi_{\text{rms}}$ is increased from 1.7° to 3°. Note that the phase noise of the PLL at integer divisions is as low as -124 dBc/Hz at 600 kHz, which is only 0.3 dB higher than predicted by the PLL simulations (see Table I). The measured results for fractional division are much noisier than predicted by simulation. The phase noise at offset frequencies close to 10 kHz is increased due to the limited memory of the data generator. The noise at higher offset frequencies is corrupted by noise coupling from the data generator. As can be seen in Fig. 10, the $\Delta\Sigma$-control bonding wires, which conduct rail-to-rail, very

VI. CONCLUSION

A monolithic 1.8-GHz $\Delta\Sigma$-controlled fractional-N PLL frequency synthesizer is implemented in a standard 0.25-μm CMOS technology. The monolithic fourth-order type-II PLL integrates the digital synthesizer part together with a fully integrated *LC* VCO, a high-speed prescaler, and a 35-kHz dual-path loop filter on a die of only 2×2 mm^2. To investigate the influence of the $\Delta\Sigma$ modulator on the synthesizer's spectral purity, a fast nonlinear analysis method is developed, showing good correspondence with measurements, in contrast to the results of the theoretical analysis. Nonlinear mixing in the phase-frequency detector and the VCO is identified as the main source of spectral pollution in $\Delta\Sigma$ fractional-N synthesizers. MASH and single-loop multibit $\Delta\Sigma$ modulators are compared for use in fractional-N synthesis. Although the MASH is stable and easy to integrate, the single-loop modulator presents a better solution, showing less sensitivity to noise leakage and noise coupling and providing more flexibility. The measured phase noise is lower than -120 dBc/Hz at 600 kHz and -139 dBc/Hz at 3 MHz. The measured fractional spur level is lower than -100 dBc, satisfying the DCS-1800 spectral purity requirements. All measurements are performed for frequencies close to integer multiples of the reference frequency, where the synthesizer is most sensitive to spurious tones.

REFERENCES

[1] M. S. J. Steyaert, J. Janssens, B. De Muer, M. Borremans, and N. Itoh, "A 2-V CMOS cellular transceiver front-end," *IEEE J. Solid-State Circuits*, vol. 35, pp. 1895–1907, Dec. 2000.

[2] T. Cho, E. Dukatz, M. Mack, D. Macnally, M. Marringa, S. Mehta, C. Nilson, L. Plouvier, and S. Rabii, "A single-chip CMOS direct-conversion transceiver for 900-MHz spread-spectrum digital cordless phones," in *IEEE Int. Solid-State Circuits Conf. (ISSCC) Dig. Tech. Papers*, San Francisco, CA, Feb. 1999, pp. 228–229.

[3] A. Rofougaran, G. Chang, J. J. Rael, J. Y.-C. Chang, M. Rofougaran, P. J. Chang, M. Djafari, J. Min, E. W. Roth, A. A. Abidi, and H. Samueli, "A single-chip 900-MHz spread-spectrum wireless transceiver in 1-μm CMOS—Part II: Receiver design," *IEEE J. Solid-State Circuits*, vol. 33, pp. 547–555, Apr. 1998.

[4] M. Copeland, T. Riley, and T. Kwasniewski, "Delta–sigma modulation in fractional-N frequency synthesis," *IEEE J. Solid-State Circuits*, vol. 28, pp. 553–559, May 1993.

[5] S. R. Norsworthy, R. Schreier, and G. C. Themes, *Delta–Sigma Data Converters: Theory, Design and Simulation*. New York: IEEE Press, 1997.

[6] B. Miller and R. Conley, "A multiple modulator fractional divider," *IEEE Trans. Instrum. Meas.*, vol. 40, pp. 578–583, June 1991.

[7] "Digital cellular communication system (Phase 2+); Radio transmission and reception," Eur. Telecommun. Standards Inst., ETSI 300 190 (GSM 05.05 version 5.4.1), 1997.

[8] W. Rhee, B.-S. Song, and A. Ali, "A 1.1-GHz CMOS fractional-N frequency synthesizer with a 3-b third-order $\Delta\Sigma$ modulator," *IEEE J. Solid-State Circuits*, vol. 35, pp. 1453–1460, Oct. 2000.

[9] The Mathworks Inc., *Matlab User's Guide, Version 5*. Englewood Cliffs, NJ: Prentice Hall, 1997.

[10] J. Yuan and C. Svensson, "New single-clock CMOS latches and flip-flops with improved speed and power savings," *IEEE J. Solid-State Circuits*, vol. 32, pp. 62–69, Jan. 1997.

[11] B. De Muer and M. S. J. Steyaert, "A single-ended 1.5-GHz 8/9 dual-modulus prescaler in 0.7-μm CMOS with low phase-noise and high input sensitivity," in *Proc. Eur. Solid-State Circuits Conf. (ESSCIRC)*, The Hague, Sept. 1998, pp. 256–259.

[12] J. Craninckx and M. S. J. Steyaert, "Low-phase-noise fully integrated CMOS frequency synthesizers," Ph.D. dissertation, Katholieke Univ. Leuven, Belgium, 1997.

[13] B. De Muer, M. Borremans, N. Itoh, and M. S. J. Steyaert, "A 1.8-GHz highly tunable low-phase-noise CMOS VCO," in *Proc. IEEE Custom Integrated Circuits Conf. (CICC)*, Orlando, FL, May 2000, pp. 585–588.

[14] B. De Muer and M. S. J. Steyaert, "Fully integrated CMOS frequency synthesizers for wireless communications," in *Analog Circuit Design*, W. Sansen, J. H. Huijsing, and R. J. van de Plassche, Eds. Norwell, MA: Kluwer, 2000, pp. 287–323.

[15] F. M. Gardner, *Phaselock Techniques*. New York: Wiley, 1979.

Bram De Muer (S'00) was born in Sint-Amandsberg, Belgium, in 1973. He received the M.Sc. degree in electrical engineering in 1996 from the Katholieke Universiteit Leuven, Belgium, where he is currently working toward the Ph.D. degree on high frequency low-noise integrated frequency synthesizers at the ESAT-MICAS laboratories.

He has been a Research Assistant with ESAT-MICAS laboratories since 1996. His research is focused on integrated low-phase-noise VCOs with on-chip planar inductors and high-speed prescaler design, leading to fully integrated $\Delta\Sigma$ fractional-N synthesizers in CMOS technology.

Michel S. J. Steyaert (S'85–A'89–SM'92) was born in Aalst, Belgium, in 1959. He received the M.S. degree in electrical-mechanical engineering and the Ph.D. degree in electronics from the Katholieke Universiteit Leuven (K.U. Leuven), Heverlee, Belgium, in 1983 and 1987, respectively.

From 1983 to 1986, he obtained an IWONL fellowship (Belgian National Foundation for Industrial Research) which allowed him to work as a Research Assistant at the Laboratory ESAT at K.U. Leuven. In 1987, he was responsible for several industrial projects in the field of analog micropower circuits at the Laboratory ESAT as an IWONL Project Researcher. In 1988, he was a Visiting Assistant Professor at the University of California, Los Angeles. In 1989, he was appointed by the National Fund of Scientific Research (Belgium) as a Research Associate, in 1992 as a Senior Research Associate, and in 1996 as a Research Director at the Laboratory ESAT, K.U. Leuven. Between 1989 and 1996, he was also a part-time Associate Professor and since 1997 an Associate Professor at the K.U. Leuven. His current research interests are in high-performance and high-frequency analog integrated circuits for telecommunication systems and analog signal processing.

Dr. Steyaert received the 1990 European Solid-State Circuits Conference Best Paper Award, the 1995 and 1997 ISSCC Evening Session Award, the 1999 IEEE Circuit and Systems Society Guillemin–Cauer Award, and the 1991 NFWO Alcatel-Bell-Telephone award for innovative work in integrated circuits for telecommunications.

PART VII

Clock and Data Recovery

A 2.5-Gb/s Clock and Data Recovery IC with Tunable Jitter Characteristics for Use in LAN's and WAN's

Keiji Kishine, *Member, IEEE*, Noboru Ishihara, *Member, IEEE*, Ken-ichi Takiguchi, and Haruhiko Ichino, *Member, IEEE*

Abstract— A 2.5-Gb/s monolithic clock and data recovery (CDR) IC using the phase-locked loop (PLL) technique is fabricated using Si bipolar technology. The output jitter characteristics of the CDR can be controlled by designing the loop-gain design and by using the switched-filter PLL technique. The CDR IC can be used in local-area networks (LAN's) and in long-haul backbone networks or wide-area networks (WAN's). Its power consumption is only 0.4 W. For LAN's, the jitter generation of the CDR when the loop gain is optimized is 1.2 ps (0.003 UI). The jitter characteristics of the CDR optimized for WAN's meet all three types of STM-16 jitter specifications given in ITU-T Recommendation G.958. This is the first report on a CDR that can be used for both LAN's and WAN's. This paper also describes the design method of the jitter characteristics of the CDR for LAN's and WAN's.

Index Terms—Clock and data recovery (CDR), IC, jitter suppression, local-area network (LAN), low jitter, phase-locked loop (PLL), transmission receiver, wide-area network (WAN), 2.5 Gb/s.

I. INTRODUCTION

OPTICAL communication systems, which are used in local-area networks (LAN's) and wide-area networks (WAN's), are expected to play an important role in realizing the future multimedia society. These systems must be compact, economical to produce, and efficient in terms of power consumption. Given these requirements, researchers have been developing low-power and small-size optical receiver/sender (OR/OS) modules. A clock and data recovery (CDR) circuit is one of the key components of the OR, which must have retiming, reshape, regeneration (3R) operation. To ensure that the receivers have low power consumption and are cost-effective and compact, it is essential to employ a single-chip, adjustment-free CDR IC using the phase-locked loop (PLL) technique without any high-Q components.

A number of approaches have been proposed for developing a CDR IC using the PLL technique [1], [2]. Generally, the jitter specifications for the CDR differ depending on what it is being used for, and jitter suppression is one especially serious problem for the CDR-IC design. There are different jitter specifications for the following two applications.

Case 1) LAN's such as gigabit/second ethernets, fiber channels, and other optical interconnections. They use a single span of a transmission medium.

Case 2) Backbone networks or WAN's such as synchronous digital hierarchy (SDH) or synchronous optical network. They use line regenerators to transport information over long distances.

For case 1), the CDR must suppress mainly the jitter generated due to noise in the CDR, so-called *jitter generation*. In case 1), there is no jitter accumulation due to cascaded regenerators. For case 2), however, the ITU-T G.958 recommendation for SDH stipulates other specifications [3]: a) jitter transfer specification, which is the criterion of the suppression of the noise in input signals to line regenerators, and b) jitter tolerance specification.

This paper describes a 2.5-Gb/s CDR that can be used in both cases, which eliminates the need to fabricate two chips with different characteristics. The key design techniques are based on the switched-filter (SF) PLL technique and loop gain adjustment using a gain control amplifier (GCA) circuit. The CDR IC is fabricated using 0.5-μm Si bipolar technology. The loop gain and loop bandwidth can be adjusted using a control signal from outside the chip. For case 1), the rms jitter generation of the CDR can be reduced to only 1.2 ps, and the capture range is 150 MHz. For case 2), the jitter of the CDR meets the jitter specifications of the ITU-T G.958 recommendation. The rms jitter generation is 3.6 ps, and the capture range is 50 MHz. The power consumption of the CDR for both cases is only 0.4 W.

In Section II, the concept of the suppression of jitter in each case is discussed. It is explained that the SF PLL technique can be used in the CDR for both cases. Design detail and the configurations of the circuits of the CDR are given in Section III. Section IV discusses the experimental results which show that the CDR has very good jitter characteristics and discusses the feasibility of using the CDR for various transmission systems.

II. CONCEPT FOR JITTER-SUPPRESSION DESIGN

A. Jitter Characteristics of CDR Using the PLL Technique

Generally, output jitter of a CDR based on the PLL technique can be caused by two kinds of sources: 1) additive noise that accompanies the input signal [Fig. 1(a)] and

Manuscript received August 19, 1998; revised February 8, 1999.

K. Kishine and H. Ichino are with NTT Network Innovation Laboratories, Yokosuka, Kanagawa 239-0847 Japan.

N. Ishihara is with NTT Opto-electronics Laboratories, Atsugi-shi, Kanagawa 243-01 Japan.

K. Takiguchi is with NTT Electronics Corp., Atsugi-shi, Kanagawa Pref. 243-0032 Japan.

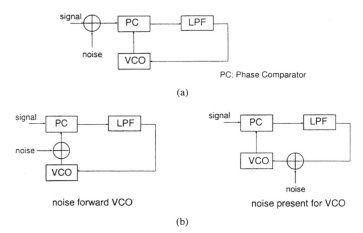

PC: Phase Comparator

(a)

noise forward VCO

noise present for VCO

(b)

Fig. 1. Noise source (a) in input signal and (b) in PLL.

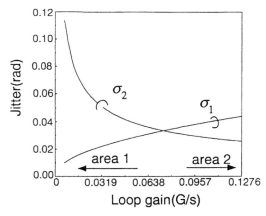

Fig. 2. Loop-gain dependence of jitter.

noise generated in the CDR [Fig. 1(b)]. In Fig. 1(b), two cases [noise forward and present for voltage-controlled oscillation (VCO)] have been shown to be equivalent in terms of VCO-phase fluctuation [4]. In addition, because the phase drift of the VCO output due to the random input data is random, Fig. 1(b) gives a rough approximation of the noise due to the input data pattern, with no jitter applied.

To suppress the jitter caused by additive noise, the CDR should be designed so that the noise bandwidth of the PLL is minimized. The output jitter (the phase deviation [in rad]) of the CDR using the PLL technique is expressed as [4]

$$\sigma_1 = \sqrt{\frac{N_0}{A^2}\frac{\omega_n}{4\xi}\left[1 + \left(2\xi - \frac{\omega_n}{K}\right)^2\right]} \qquad (1)$$

where N_0 is the power spectra density of noise, A is the input signal amplitude, ω_n is the natural angular frequency, ξ is the damping factor, and K is the loop gain. When the loop gain K becomes larger, the jitter σ_1 becomes larger, as shown in the Appendix. This means that smaller loop gain causes narrow noise bandwidth, thereby suppressing jitter. It should be noted that smaller loop gain leads to a smaller cutoff frequency of the jitter transfer function of a PLL. On the other hand, in order to suppress the jitter caused by noise generated in the CDR circuit, the operation of the CDR circuit needs to be made stable. This stability can be obtained by reducing the signal fluctuation in the CDR circuit caused by the input of consecutive data bits, device noise, and so on. This is the so-called *suppression of jitter generation,* which is specified for SDH in ITU-T Recommendation G.958. The output jitter (in rad) in this case (jitter generation) is expressed as [4]

$$\sigma_2 = \sqrt{\frac{\eta}{2}\frac{\pi^2}{\xi\omega_n}\left(1 + \frac{\omega_n^2}{K^2}\right)} \qquad (2)$$

where η is the power spectra density of noise. This equation is derived assuming that the instantaneous frequency deviation of the VCO output is caused by disturbance due to random phase noise. In this equation, when the loop gain K becomes larger, the jitter σ_2 becomes smaller, as shown in the Appendix. In other words, the jitter increases as the loop gain decreases. Larger loop gain can reduce the jitter caused by the noise in

the CDR. However, larger loop gain results in a larger cutoff frequency of the jitter transfer function.

Consequently, there is a tradeoff, as shown in Fig. 2, between reducing the jitter shown in Fig. 1(a) (equivalent to reducing the cutoff frequency of the jitter transfer function) and reducing the jitter shown in Fig. 1(b) (jitter generation). When the jitter of Fig. 1(a) is dominant, the loop gain must be controlled to be lower (area 1 in Fig. 2). When the jitter of Fig. 1(b) is dominant, it must be controlled to be higher (area 2 in Fig. 2). The optimum loop gain depends on which type of jitter has a greater effect on the output jitter of the CDR and causes degradation of transmission quality in the system.

B. Design of the CDR

Given the previous discussion, it is clear that there should be two types of CDR design, one for LAN's and another for WAN's.

1) CDR for LAN's: In the case for LAN's, the jitter from input signals is small because there is no jitter accumulation through the short and single laser-fiber-receiver span. We can therefore concentrate on reducing the jitter generation, which is caused by the input-signal-pattern dependence of the circuit, the fluctuation of the supply voltage, and device noise in the CDR. As described in Section II-A, this design should not utilize smaller loop gain to lower the cutoff frequency, but instead should utilize larger loop gain to achieve smaller output jitter.

2) CDR for WAN's: In the case for backbone networks or WAN's, the regenerator may be cascaded in order to transport information over long distances, causing the jitter to accumulate. Therefore, not only the jitter generation of the CDR has to be taken into consideration but also the jitter transfer characteristics, which is the criterion of suppression of noise in input data signals as given in ITU-T Recommendation G.958. There is a tradeoff between reducing the jitter generation and reducing the cutoff frequency of the jitter transfer function.

a) Jitter transfer: The loop gain of the CDR IC using the PLL technique, on an IC whose jitter transfer specifications meet those of ITU-T Recommendation G.958, must be designed to be lower. The jitter transfer function of the 2.5-Gbit/s PLL using a lag-lead filter can be expressed by substituting the

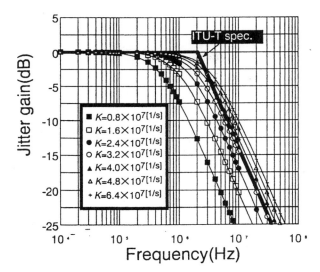

Fig. 3. Jitter transfer function.

phase transfer function for the jitter transfer function as

$$H(j\omega) = \frac{\omega_n^2 + j\left(2\xi\omega_n - \dfrac{\omega_n^2}{K}\right)\omega}{\omega_n^2 - \omega^2 + j2\xi\omega_n\omega}. \tag{3}$$

This function is plotted in Fig. 3. It is a curve when the time constants τ_1 and τ_2 of the lag-lead filter described in the Appendix are set to values that provide that the natural angular frequency ω_n becomes nearly 2 MHz, and the damping factor ξ is larger than the value where the jitter gain peaking is less than 0.1 dB. Fig. 3, in which it is stipulated that the loop gain K is lower than 3.2×10^7 (1/s), indicates that the curve of the smaller loop gain meets the ITU-T (STM-16) specification.

b) Jitter generation: As described in Section II-A, as the loop gain becomes small, it becomes more difficult to suppress the jitter generation because of the tradeoff between reducing jitter generation and reducing the cutoff frequency of the jitter transfer function. To solve this problem, we introduce the SF PLL technique. Fig. 4 shows the SF CDR configuration, which we originally proposed as a way to maintain a precise clock signal, thereby achieving tolerance to the input of consecutive data bits. (Our previous work, 156-Mb/s SF CDR [5], has no GCA in Fig. 4.) The main features of SF circuit operation are that the PC output can be transferred to the low-pass filter (LPF) only when data transitions occur (sample mode) and the LPF output can be constant during consecutive data inputs (hold mode). These features prevent the phase drift of VCO output during the input of consecutive data bits. We thought that the equivalent high-Q operation of the SF circuit could be utilized to reduce the jitter generation. Fig. 5 shows simulation results of the change in differential output voltage of the loop filter when the input signal changes from a 1/0-repeated bitstream to consecutive data bits (in this case, "0") at 805 ns. Fig. 5 shows the filter output of the SF CDR levels out, while that of the CDR without the SF circuit begins to degrade at 805 ns. This means that the operation of the SF circuit is equal to that of the larger RC time constant of the loop filter, and the jitter generation due to the input signal pattern would be more suppressed than that of the CDR without an SF circuit.

In other words, the SF circuit would provide equivalent high-Q operation and achieve low jitter operation.

We also thought this advantage of the SF circuit could be used to solve the tradeoff problem. Fig. 6 shows the SPICE simulation results of the cutoff-frequency (of the jitter transfer curve) dependence of the jitter generation of the 2.5-Gb/s CDR's both with the SF circuit and without it. Both curves indicate that the jitter generation decreases as the cutoff frequency increases. Furthermore, the jitter generation of the SF CDR is 70% lower than that of the CDR without the SF circuit. It is noteworthy that the suppression of jitter generation by the SF circuit is marked. In addition, the jitter characteristics of the SF CDR meet the STM-16 specifications (rms jitter generation that is lower than 4 ps, and equivalent jitter transfer specification in which the cutoff frequency at -3-dB jitter gain is lower than about 2.8 MHz), while the CDR circuit without the SF circuit fails to meet the specifications. In the SPICE simulation, the source of jitter is the instability of the circuit operation of the CDR, with no jitter applied at random input. In the experimental results, the device noise, the fluctuations of supply voltage, and so on are also noise sources. The jitter generation in experiments is therefore larger than that in the simulation results. The simulation results do, however, show the characteristics of jitter generation versus jitter transfer functions.

Given our findings, we conclude that in the design of the CDR IC using the PLL technique, an IC used in backbone networks or WAN's, the loop gain must be large enough so that the jitter generation meets the ITU-T specs, yet small enough so that suitable jitter transfer characteristics are obtained.

III. CIRCUIT DESIGN

A block diagram of the CDR, including a GCA circuit between the SF and VCO, is shown in Fig. 4 [6]. This CDR can be used in both short- and long- distance transmission systems by adjusting the loop gain through the GCA from outside the chip. The main features of the CDR are 1) an SF circuit for equivalent high-Q operation [5], 2) optimum timing adjustment between extracted clock and input data, and 3) loop gain control on optimization.

A. VCO

In Fig. 7, the circuit configuration of VCO is shown. The oscillation frequency is controlled by the voltage swing of "VC1, VC2," which is determined by the feedback signal $SGCA, \overline{SGCA}$ from GCA. The free-running frequency is determined by the current IF1, IF2, which can be controlled from outside the chip. The free-running frequency can be adjusted from 2.2 to 2.8 GHz. It covers the free-running frequency deviation caused by fluctuations in device performance. Fig. 8 shows the tuning-voltage (feedback-voltage) dependence of the oscillation frequency. The simulation results are in good agreement with the experimental results. The VCO modulation frequency sensitivity is designed to be about 1 GHz/V. The oscillation frequency range by a feedback signal is ±200 MHz. The tuning-range diagram is shown in Fig. 9. The total tunable range is sufficiently wide, from 2.0 to 3.0 GHz.

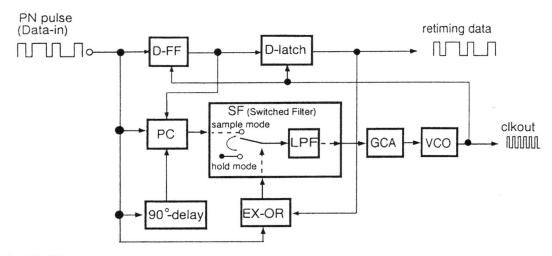

Fig. 4. SF CDR with GCA.

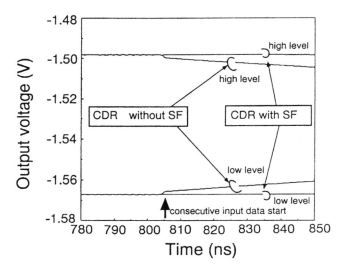

Fig. 5. Output voltage of low-pass filter (differential output).

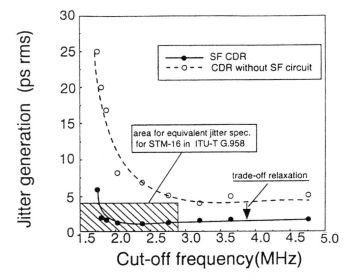

Fig. 6. Cutoff-frequency dependence of jitter generation.

B. GCA

A current-bypass GCA circuit is used in the CDR (see Fig. 10). The gain of the GCA, which can be controlled from outside the chip, can be varied from −40 to 0 dB. To lower the jitter generation of the CDR, the gain should be higher. On the other hand, to achieve the lower cutoff frequency of the jitter transfer curve, it must be lower. Therefore, gain should be adjusted according to the jitter specification in each case.

C. Delay Circuit

To reduce jitter generation, the edge-inclined circuit in the 90°-delay block shown in Fig. 4, which includes a capacitor for delay control [5], is replaced by a chain of emitter-coupled logic (ECL) buffer circuits without capacitors, the delay of which can be adjusted from about 100 to 300 ps from outside the chip. The edge-inclined circuit was employed to make the 156-Mbit/s CDR smaller. But the delay needed for the 2.5-Gbit/s CDR is only 200 ps, which is much smaller than that needed for the 156-Mbit/s CDR. Therefore, only a small number of ECL circuits are needed for the 200-ps delay, and

the delay circuit itself is relatively small. In addition, in the ECL circuit, there is no input-data-pattern dependence of the response in the edge-inclined circuit capacitor. When the ECL delay circuit is used, the simulated jitter due to the input data pattern is about 80% of that when an edge-inclined-delay circuit is used. As a result of using the ECL circuit, the jitter due to the input pattern effect is more suppressed than in the circuit reported in our previous work [5].

D. Loop Filter

The lag-lead filter is used as the loop filter, and the RC time constant is adjusted for each use. An additive capacitor outside the chip is not needed when the CDR is used for short-distance transmission systems. It is, however, needed for long-distance use.

E. Other Considerations

Furthermore, in order to guarantee jitter tolerance, it is important to maintain an optimum timing adjustment between

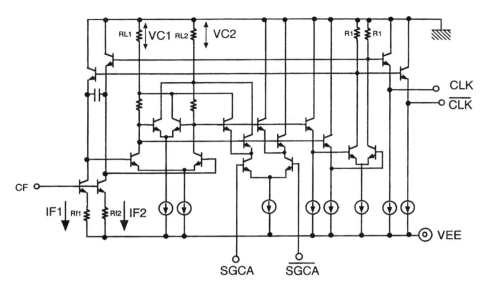

Fig. 7. VCO configuration.

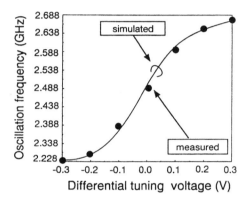

Fig. 8. VCO tuning curve.

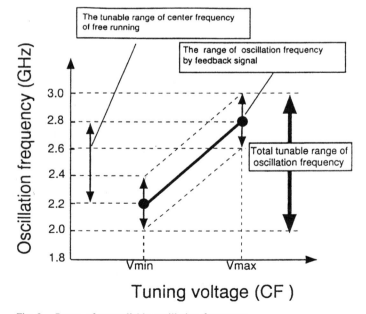

Fig. 9. Range of controllable oscillation frequency.

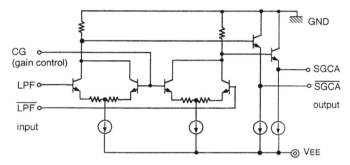

Fig. 10. GCA configuration.

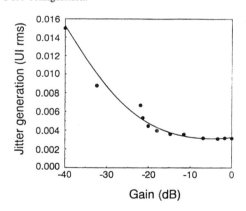

Fig. 11. GCA gain dependence of jitter generation.

signal generated from the comparison between the phase of the delay flip-flop output and the 90°-delayed phase for the input data. In addition, in order to lower the power consumption, the new 2.5-Gb/s CDR IC uses stacked differential pairs on two levels, which enables its supply voltage to be decreased to −3.0 V (as opposed to −5.2 V in our previous work [5]). Furthermore, current dissipation is optimized in each block to reduce power consumption.

IV. EXPERIMENTAL RESULTS

A new chip was fabricated using the 0.5-μm super self-aligned process technology Si bipolar process [7]. It was

the extracted clock and the input data. This timing adjustment is attained by allowing the clock to trigger the center of the data period by means of the phase-comparator (PC) output

639

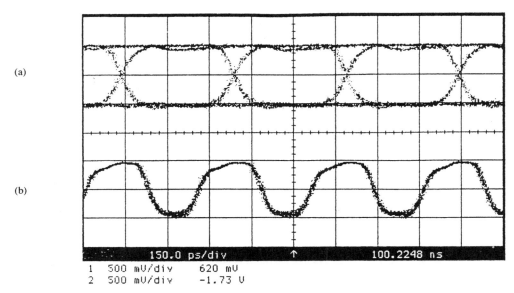

(a)

(b)

1 500 mV/div   620 mV
2 500 mV/div   -1.73 V

Fig. 12.   Output waveforms of the CDR for LAN. (a) Data output. (b) Clock output.

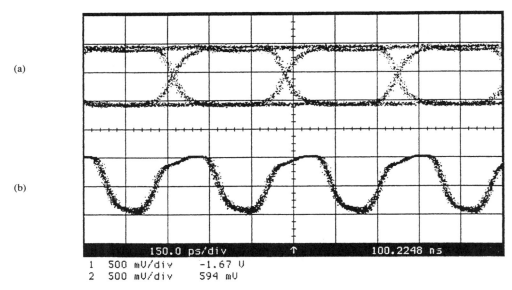

(a)

(b)

1 500 mV/div   -1.67 V
2 500 mV/div   594 mV

Fig. 13.   Output waveforms of the CDR for WAN. (a) Data output. (b) Clock output.

mounted in a 7×7 mm²-square ceramic package. The CDR IC of both high and low loop gain is evaluated in each case when the gain is adjusted to both short- and long-distance transmission systems. Jitter was measured with a commercial jitter analyzer. An rms jitter-generation value from which the jitter value of input data is subtracted can be obtained with the analyzer.

A. SF CDR for LAN

The internal capacitor in the loop filter is 10 pF, and an external capacitor is not needed in this case. The GCA gain dependence of the jitter generation is shown in Fig. 11. The jitter generation decreases as the gain increases, and the lowest point is when the gain is larger than about −8 dB. The loop gain at this point is 1.2×10^8 (1/s). The output waveforms when the loop gain is set to that point and input data is 2.488 32

Gb/s are shown in Fig. 12. The eye opening of the output data was sufficiently wide, and clock extraction was very precise. The rms jitter generation is 1.2 ps and the capture range is over 150 MHz.

B. SF CDR for WAN

To lower the cutoff frequency of the jitter transfer curve, the external capacitor for the loop filter of 0.1 μF is added. The loop gain is set to about 2×10^7 (1/s), which is the loop gain when the jitter transfer curve meets the ITU-T jitter transfer specification in Fig. 3.

The output waveforms when the loop gain is set to the point above are shown in Fig. 13. Again, the eye opening of the output data was sufficiently wide, and clock extraction was very precise. The rms jitter generation is 3.6 ps, which is larger than that of the CDR when its loop gain is adjusted for

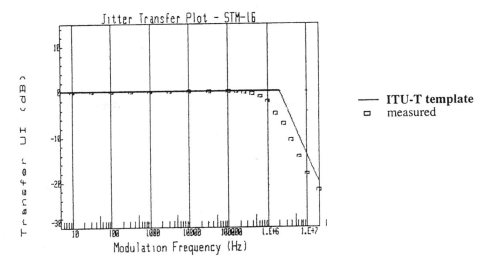

Fig. 14. Jitter transfer function.

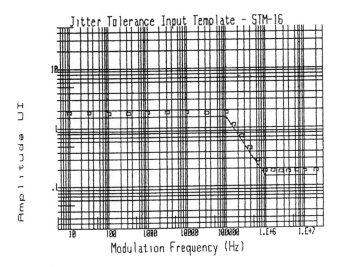

Fig. 15. Jitter tolerance curve.

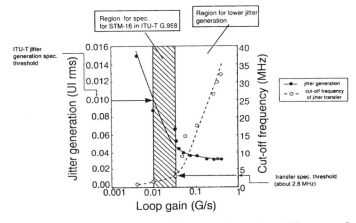

Fig. 16. Loop-gain dependence of jitter generation and cutoff frequency of jitter transfer function.

short-distance transmission systems, but is smaller than the specification of the jitter generation of 4.0 ps (for STM-16; ITU-T Recommendation G.958). The capture range is over 50 MHz. Fig. 14 shows the measured jitter transfer function of the CDR in this case. The curve meets the ITU-T G.958 specification. Fig. 15 shows the jitter tolerance curve when the input jitter magnification is 120% of the ITU-T specification. The squares indicate error-free operation (where the error rate is lower than 10^{-11}). The rms jitter generation, jitter tolerance, and jitter transfer function all meet the jitter specifications in ITU-T G.958.

The relationship between the measured jitter generation (or cutoff frequency) and the loop gain in this experiment is shown in Fig. 16. The darker shaded area is for the CDR, whose jitter characteristics meet the specifications of ITU-T Recommendation G.958. In the area of larger loop gain, the jitter generation becomes small. Fig. 16 shows clearly that, when its loop gain is optimized, the CDR IC is suitable for both LAN's and WAN's. The capture range of both types of CDR's is wide enough to cover the deviation in the free-running frequency due to changes in temperature (ranging from 5 to 90°C). In addition, the power consumption (including that of

the I/O circuit) in both cases is less than 35% of that in the 2.5-Gb/s PLL's reported previously [1], [2].

V. CONCLUSION

The design method of the CDR for both LAN's and WAN's is presented. A new 2.5-Gb/s SF monolithic CDR IC using the 0.5-μm Si bipolar process has been developed. The CDR IC can be used in the transmission receivers for both LAN's and WAN's. The rms jitter generation of the CDR adjusted for LAN's is 1.2 ps. Furthermore, the jitter characteristics of the CDR for backbone networks or WAN's meet the specifications for STM-16 given in ITU-T Recommendation G.958. In addition, the power consumption of the CDR is only 0.4 W.

APPENDIX

A. Loop-Gain Dependence of the Jitter Shown in Fig. 1

The jitter due to the noise in the input signal to the PLL is expressed as (1). When the loop filter is a lag-lead type (the series and shunt register are respectively R_1 and R_2 and the shunt capacitance is C), the natural angular frequency ω_n and

641

the damping factor ξ are expressed as

$$\omega_n = \sqrt{\frac{K}{\tau 1}}$$

$$\xi = \frac{1}{2}\omega_n\left(\tau 2 + \frac{1}{K}\right)$$

where $\tau 1 = C(R_1 + R_2)$ and $\tau 2 = CR_2$. The term $\omega_n/4\xi$ in (1) therefore becomes

$$\frac{\omega_n}{4\xi} = \frac{1}{2\left(\tau 2 + \dfrac{1}{K}\right)}. \tag{A1.1}$$

This shows that $\omega_n/4\xi$ increases as the loop gain K increases. Furthermore, $2\xi - (\omega_n/K)$ in (1) can be expressed as

$$2\xi - \frac{\omega_n}{K} = \omega_n\left(\tau 2 + \frac{1}{K}\right) - \frac{1}{\sqrt{K\tau 1}}$$

$$= \tau 2\sqrt{\frac{K}{\tau 1}} + \frac{1}{\sqrt{K\tau 1}} - \frac{1}{\sqrt{K\tau 1}}$$

$$= \tau 2\sqrt{\frac{K}{\tau 1}}. \tag{A1.2}$$

This also shows that $2\xi - (\omega_n/K)$ increases. Therefore, the jitter in Fig. 1 increases as the loop gain increases.

B. Loop-Gain Dependence of the Jitter Shown in Fig. 2

The jitter due to the noise in the PLL can be expressed as (2). In (2), the term $1/\xi\omega_n$ is

$$\frac{1}{\xi\omega_n} = \frac{1}{\dfrac{1}{2}\dfrac{\tau 2}{\tau 1}K + \dfrac{1}{2\tau 1}}. \tag{A2.1}$$

The term $1/\xi\omega_n$ decreases as the loop gain increases. In addition, the term ω_n^2/K^2 is expressed as

$$\frac{\omega_n^2}{K^2} = \frac{1}{K\tau 1}. \tag{A2.2}$$

ω_n^2/K^2 also decreases as the loop gain increases. Therefore, the jitter in Fig. 2, expressed in (2), decreases as the loop gain increases.

ACKNOWLEDGMENT

The authors wish to thank H. Yoshimura and K. Sato for their helpful discussions and suggestions.

REFERENCES

[1] H. Ransjin and P. O'Conner, "A PLL-based 2.5b/s GaAs clock and data Regenerator IC," *IEEE J. Solid-State Circuits*, vol. 26, pp. 1345–1353, Oct. 1991.

[2] R. Walker, C. Stout, and C.-S. Yen, "A 2.488Gb/s Si-bipolar clock and data recovery IC with robust loss of signal detection," in *ISSCC Dig. Tech. Papers*, 1997, pp. 246–247.

[3] "Digital Line Systems Based on the Synchronous Digital Hierarchy for Use on Optical Fiber Cables," CCITT Rec. G.958.

[4] A. Blanchard, *Phase-Locked Loops*. New York: Wiley, 1976, ch. 8.

[5] N. Ishihara and Y. Akazawa, "A monolithic 156Mb/s clock and data recovery PLL circuit using the sample-and hold technique," *IEEE J. Solid-State Circuits*, vol. 29, pp. 1566–1571, Dec. 1994.

[6] K. Kishine, N. Ishihara, and H. Ichino, "Jitter-suppressed low-power 2.5-Gbit/s clock and data recovery IC without high-Q components," *Electron. Lett.*, vol. 33, no. 18, pp. 1545–1546, Aug. 1997.

[7] C. Yamaguchi, Y. Kobayashi, M. Miyake, K. Ishii, and H. Ichino, "A 0.5 μm bipolar technology using a new base formation method," in *Proc. BCTM*, 1993, pp. 63–66.

Keiji Kishine (M'98) was born in Kyoto, Japan, on October 26, 1964. He received the B.S. and M.S. degrees in engineering science from Kyoto University, Kyoto, in 1990 and 1992, respectively.

In 1992, he joined the Electrical Communication Laboratories, Nippon Telegraph and Telephone Corp. (NTT), Tokyo, Japan. At the NTT System Electronics Laboratories, Kanagawa, Japan, he was engaged in research and design of high-speed, low-power circuits for Gbit/s LSI's using Si-bipolar transistors with application to optical communication systems. Since 1997, he has worked on research and development of Gbit/s clock and data recovery IC at the Photonic Network Laboratory, NTT Network Innovation Laboratories, Kanagawa, Japan.

Mr. Kishine is a member of the Institute of Electronics, Information and Communication Engineers (IEICE) of Japan.

Noboru Ishihara (M'89) was born in Gunma, Japan, on April 27, 1958. He received the B.S. degree in electrical engineering from Gunma University, Gunma, in 1981 and the Dr.Eng. degree from the Tokyo Institute of Technology, Tokyo, Japan, in 1997.

In 1981, he joined the Electrical Communication Laboratory, NTT, Tokyo, where he has been engaged in research and development of analog IC's for communication use. His recent work is in the area of low-power and high-speed analog IC's for optical communications.

Mr. Ishihara is a member of the Institute of Electronics, Information and Communication Engineers (IEICE) of Japan and the IEEE Microwave Theory and Techniques Society.

Ken-ichi Takiguchi was born in Kanagawa, Japan, on July 21, 1969. He graduated from Tokyo Computer School, Tokyo, Japan, in 1992.

In 1992, he joined NTT Electronics Corp., Kanagawa. He has been engaged in development of high speed IC's, especially gigabit/second PLL IC's.

Haruhiko Ichino (M'89) was born in Yamaguchi, Japan, on January 26, 1957. He received the B.S., M.S., and Ph.D. degrees in applied physics from Osaka University, Osaka, Japan, in 1979, 1981, and 1994, respectively.

In 1981, he joined the Electrical Communication Laboratories, Nippon Telegraph and Telephone Corp. (NTT), Tokyo, Japan. He has been engaged in research and development of Gbit/s SSI-MSI's using bipolar transistors (Si bipolar transistor and AlGaAs/GaAs HBT), with application to Gbit/s optical communication systems and high-frequency satellite communication systems. His work also includes low-power Gbit/s LSI's for SDH networks and future ATM switching systems; and O-E, E-O converter modules. During this research and development, he also worked on the modeling of a high-speed bipolar transistor, analyzing ECL gate delay and maximum operating speed of GHz flip-flop, and high-speed design methodology based on gate-array and standard-cell approaches. His interests include high-speed packaging and measurement systems. Since 1997, he has worked on research and development of Gbit/s-interface hardware design of photonic transport network systems. Currently, he is a Senior Research Engineer, Supervisor, and Photonic Network Systems Research Group Leader of the Photonic Network Laboratory, NTT Network Innovation Laboratories, Kanagawa, Japan. He was a Visiting Lecturer at Osaka University during 1995–1996.

Dr. Ichino is a member of the Institute of Electronics, Information and Communication Engineers (IEICE) of Japan. He was a Secretary of IEICE's Technical Group on Integrated Circuits and Devices.

Clock/Data Recovery PLL Using Half-Frequency Clock

M. Rau, T. Oberst, R. Lares, A. Rothermel, R. Schweer, and N. Menoux

Abstract— A clock and data recovery PLL is described for serial nonreturn-to-zero (NRZ) data transmission. The voltage controlled oscillator (VCO) works at half the data rate, which means for a 1-Gb/s data rate, the VCO runs at 500 MHz. A specially designed phase comparator uses a delay-locked loop (DLL) to generate the required sampling clocks to compare clock and data. The VCO can typically be tuned from 350 MHz to 890 MHz and the phase-locked loop (PLL) locks between 720 Mb/s and 1.3 Gb/s. Data recovery is error free up to 1.2 Gb/s with a 9-b pseudorandom data sequence. The core consumes 85 mW (3.3 V) at 1 Gb/s.

Index Terms—Bang-bang control, CMOS digital integrated circuits, data communication, high-speed integrated circuits, phase locked loops, synchronization.

I. INTRODUCTION

DIGITAL signal processing becomes economical in consumer applications. The main requirement there is low cost in mass production. Digital processing and transmission has to be carried out with low power and in cheap IC packages.

Data transmission between different digital signal processing IC's influences significantly the power consumption and the system cost. For video signal transmission in 100 Hz TV sets, typically 16 data lines in parallel are driven with 27 MHz rail-to-rail nonreturn-to-zero (NRZ) data signals. Sharp data transitions are in use to ensure reliable synchronous operation. A power saving alternative could be found in low-swing high-speed serial data transmission in the range of 500 Mb/s or more. However, this kind of high-speed data transmission has to be asynchronous. The most economic solution avoids separate transmission of the clock. In that case, clock recovery from the NRZ data stream is required. In this paper we describe a phase-locked loop (PLL) which is designed to process more than 1 Gb/s data in a 0.5-μm CMOS technology.

II. ARCHITECTURE

The PLL generally consists of three building blocks (Fig. 1):

1) phase comparator, detecting the phase difference between the data and the recovered clock;

Manuscript received December 15, 1996; revised February 6, 1997. This work was supported in part by the German Ministry for Education and Research under Contract 01M2880A.
M. Rau was with the University of Ulm, Germany. He is now with Siemens AG, 81359 Munich, Germany.
T. Oberst was with the University of Ulm, Germany. He is now with DASA, D-89077 Ulm, Germany.
R. Lares and A. Rothermel are with the Microelectronics Department, University of Ulm, D-89081 Ulm, Germany.
R. Schweer is with Thomson Multimedia, D-78048 Villingen-Schwenningen, Germany.
N. Menoux is with Thomson, 38240 Meylan, France.
Publisher Item Identifier S 0018-9200(97)04386-2.

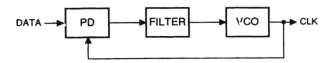

Fig. 1. Classic PLL.

2) loop filter, filtering the phase detector output and forming the control signal for the oscillator;

3) voltage controlled oscillator (VCO).

The unusual feature in our design is the phase detector, which uses a delay-locked loop (DLL) to generate multiple sampling clocks. Thus, the VCO can run at only half the data rate, which means that we can detect a 1-Gb/s serial data stream with a 500-MHz VCO. This relieves the timing constraints in the phase detector logic and results in well correlated and data independent control signals. Also, at the lower frequency the VCO tuning range is large enough to compensate all technology parameter variations. With this architecture we could achieve higher data rates.

The block diagram of the circuit is shown in Fig. 2. No external components are required for the PLL. The loop filter capacitor is integrated on chip together with the VCO, the phase comparator, and a charge pump. The data stream is retimed in two flip-flops with the inverted and noninverted clock. Two flip-flops are required because the clock has only half the data rate. These two half-speed data streams are combined in a multiplexer, forming an output stream at the original data rate. A lock-in circuit is realized on chip, because the phase comparator is not frequency sensitive.

III. PHASE COMPARATOR

The PLL adjusts the clock to an incoming data stream. Because of the random nature of data there is not necessarily a data transition at every clock cycle. The loop has to handle sequences of consecutive zeroes or ones in the data stream. The following phase comparator output signal properties are essential.

First, the phase comparator must not give any output signal if there is no data edge. Second, the duration of the control signal pulses at the data transitions is important, especially if there are few of them. In general, for a good loop performance the control signal should be proportional to the phase error. However, for very high operating frequencies, analog signals depend on the data pattern and become highly nonlinear, because they do not settle during the bit duration. It was found by simulation that different phase detectors with analog outputs [1], [2] limit the PLL operating frequency. On the other hand, clock recovery schemes based on sampling techniques [3], [4] result in uniform digital control pulses. They are

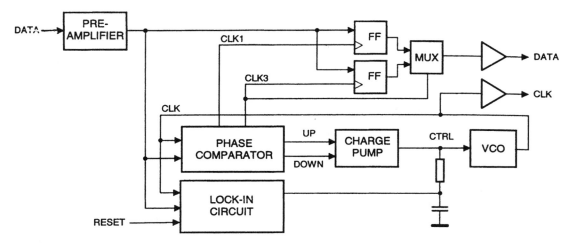

Fig. 2.   Clock recovery block diagram.

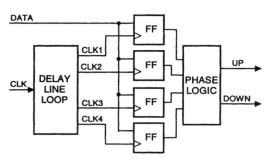

Fig. 3.   Phase comparator.

best suited to support highest possible data rates at a given technology.

The phase comparator used here is an extension of the circuit from [3], modified to work with half the "normal" clock frequency (Fig. 3). The data stream is sampled at four equally spaced timepoints. The logic circuitry driven by the flip-flops generates the up and down control pulses for the VCO according to Fig. 4. Because these control pulses are generated by clocked flip-flops, they are of well defined width. The advantage is that they do not depend on the data pattern. On the other hand, they do not reflect the amount of the phase error, either. The pulse width is constant, even for very small phase errors. This so-called bang-bang operation generates an increased jitter in the locked state. However, the magnitude is much smaller compared to the one introduced by data-dependent and nonlinear analog pulses at high frequencies. The phase logic evaluates only rising signal edges, in order not to depend on duty cycle variations of the input signal.

There is an issue to be taken care of when dimensioning the flip-flops and the phase logic. The stable operating point of the loop is reached when the signal is sampled exactly at its transition (see Fig. 4). Thus the loop forces the flip-flop to sample the metastable state, which is not allowed in normal flip-flop operation. In this application, however, it is not critical for the operation. If the metastable state is sampled, it does not matter whether it will be interpreted as up or down, because any decision is equally wrong, as we are at the stable operating point, i.e., zero phase error. Only the jitter of the

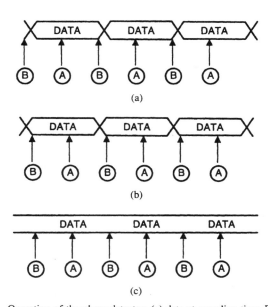

Fig. 4.   Operation of the phase detector: (a) data at sampling time B equals the data at the preceding sampling time $A \Rightarrow$ data transition is late $\Rightarrow$ frequency up, (b) data at sampling time B equals the data at the following sampling time $A \Rightarrow$ data transition is early $\Rightarrow$ frequency down, (c) data at sampling time A equals the data at the preceding sampling time $A \Rightarrow$ no data edge, no control signal output.

bang-bang operation results. Also, there is an increased short current inside the flip-flops that has to be limited.

For uniform pulses and small jitter, absolutely identical sampling intervals are required. Therefore, a DLL has been implemented to generate four 90° shifted clock phases clk1 $\cdots$ clk4 from the VCO output signal (Fig. 5). The loop compares the phase of the original clock to a clock fed through four adjustable delay elements. The clock signal repeats with a period T. A delay element in Fig. 5 can therefore delay by $T/4$, or as well by $5T/4$. By rearranging the output signals, delay times of $3T/4$, $7T/4 \cdots$ are also possible. With a delay element for $T/4$, it is not possible to compensate for all technology and environment variations. Therefore, it is necessary to select a larger value for the delay, to just be able to deal with all technology parameter variations.

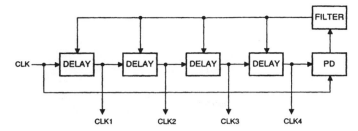

Fig. 5. DLL to generate all 90° phase shifted sampling clocks with high accuracy.

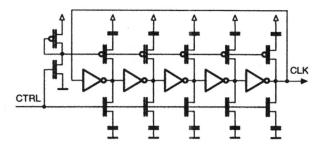

Fig. 7. VCO schematic.

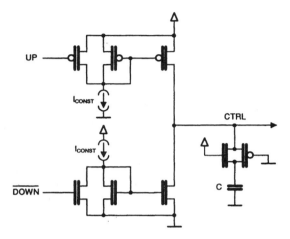

Fig. 6. Current mirror charge pump.

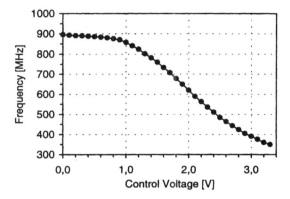

Fig. 8. VCO frequency versus control voltage.

The stability of the system containing two coupled loops can be guaranteed for two reasons. First, the DLL is a first-order loop and inherently stable. Second, the time constants of the two loops are two orders of magnitude different.

IV. CHARGE PUMP AND LOOP FILTER

The control pulses drive a current mirror charge pump [6] (Fig. 6) which assures that the charge delivered to the loop filter does not vary with the VCO control voltage. The charge pump allows the realization of an ideal integrator transfer function (pole at $s = 0$) with no additional active amplifier, resulting in a zero-phase error in steady state. A simple RC network shown in Figs. 2 and 6 is used for the low-pass loop filter. The current level of the charge pump and hence the charge delivered at every rising data transition can be set to a small value. This allows the implementation of the loop capacitor on chip.

V. VCO

Both high oscillation frequency and a wide tuning range are required. We choose a ring oscillator design with variable load capacitors (Fig. 7) based on [5]. Duty cycle is not an issue here, because the flip-flops all are triggered with the same edge; the DLL generates the required phase shifts. This circuit can safely cope with all parameter variations. Fig. 8 shows the VCO tuning characteristic.

VI. LOCK-IN CIRCUIT

The bang-bang operation and the data dependent phase detector output signal require a narrow loop bandwidth for a low jitter. This results in a reduced pull-in range of the PLL. Instead of adapting the loop bandwidth during operation we created a lock-in circuit which is active only after power up. For lock-in, a 1010-sequence has to be fed to the circuit. The VCO is swept, starting with the highest frequency. When clock and input frequencies are the same, the sampled data (before the Mux) do not change. An edge-triggered monoflop then stops the frequency sweep and closes the PLL.

VII. LAYOUT

Fig. 9 shows the test chip. A large area is used for the on-chip loop filter capacitor (upper left). A comparable area is required for the ring oscillator, including its load capacitors (lower left). Because the series resistance of those load capacitors is more critical compared to the one in the loop filter, a finer finger structure was chosen. All capacitors have been realized as MOS-transistor gates. No special mask is required.

In the top right area are located the lock-in circuit and the DLL with its loop filter, whereas in the lower middle and to the right, buffers and control logic can be seen.

VIII. MEASUREMENT RESULTS

We verified locking of the PLL at data rates from 720 to 1300 Mb/s with pseudorandom sequences up to $(2^{31} - 1)$ bit at the data input. However, data recovery is not guaranteed under these conditions because of the clock jitter. Fig. 10 shows the maximum available data rates for different lengths

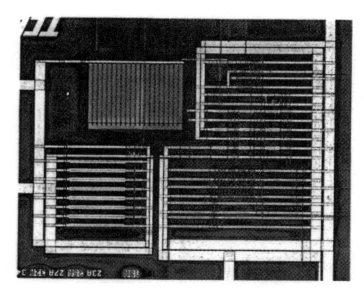

Fig. 9. Chip micrograph.

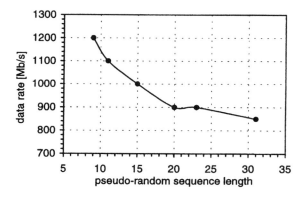

Fig. 10. Maximum data rate versus pseudorandom sequence length for error-free receiving during time of measurement (complies with error rate smaller than $1 \cdot 10^{-11}$).

of the pseudo random sequences for correct data recovery. Measurement period was 10^{11} clocks (corresponding to a bit error rate of $1 \cdot 10^{-11}$).

At very high data rates, clock and data phase precision has to be better at the input of the retiming flip-flops, because the "eyes" become smaller. The lower required phase jitter corresponds to shorter pseudorandom sequences.

Fig. 11 shows the locked PLL at 1 Gb/s with a $(2^{15} - 1)$-bit length pseudorandom sequence. The clock jitter is about 350 ps, which is caused mainly by the bang-bang operation of the phase comparator. We believe that this behavior can be improved by reducing the uncertain time interval of the sampling flip-flop, i.e., reducing their setup-and-hold times and increasing the clock slope.

All measurements have been done with the IC housed in a standard 16-pin dual in-line ceramic package which shows rather poor high-frequency performance. It was our goal to demonstrate the circuit in a critical environment. Better results could be expected when using packages with shorter leads.

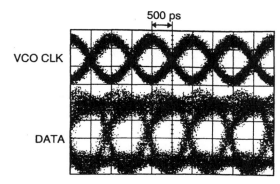

Fig. 11. VCO clock output and data output eye pattern at 1 Gb/s with a $(2^{15} - 1)$-bit length pseudorandom input sequence.

Chip core area is 0.38 mm^2, power consumption without pad drivers is 85 mW at 1 Gb/s, 0.5-μm CMOS, 3.3 V supply. Only 1/4 of the power consumption is proportional to the clock frequency, 3/4 are constant. The circuit consumes 91 mW at 1.3 Gb/s. The power saved by using only half the conventional clock frequency is partly used to supply the DLL, which needs 21 mW (= 1/4 of total power at 1 Gb/s).

No external components are required, except one reference current, which is not very critical (a $\pm20\%$ variation is allowed).

IX. CONCLUSION

Complete on-chip clock and data recovery at 1 Gb/s is feasible with a standard 0.5-μm CMOS technology. On-chip clock is only 500 MHz in this case. Data are directly demultiplexed one to two in the retiming flip-flops. A multiplexer to regenerate the original data stream was included for measurement purposes only. In applications, serial-to-parallel conversion will normally follow the PLL. In that case, the halved clock frequency is an advantage, because the following blocks can be designed more easily.

ACKNOWLEDGMENT

The authors greatly acknowledge perfect layout support by Y. A. Savalle and G. Kimmich from TCEC. They thank J. Borel from SGS-Thomson for providing the design kit and acknowledge the fast sample production in the factory.

REFERENCES

[1] T. H. Lee, "A 155-MHz clock recovery delay- and phase-locked loop," *IEEE J. Solid-State Circuits*, vol. 27, pp. 1736–1746, Dec. 1992.
[2] B. Thompson, "A 300-MHz BiCMOS serial data transciever," *IEEE J. Solid-State Circuits*, vol. 29, pp. 185–192, Mar. 1994.
[3] B. Lai and R. C. Walker, "A monolithic 622 Mb/s clock extraction data retiming circuit," in *Int. Solid-State Circuits Conf.*, San Francisco, CA, 1991, vol. 306, pp. 144–145.
[4] A. Pottbaecker, U. Langmann, and H.-U. Schreiber, "A si bipolar phase and frequency detector IC for clock extraction up to 8 Gb/s," *IEEE J. Solid-State Circuits*, vol. 27, pp. 1747–1751, Dec. 1992.
[5] M. Bazes, "A novel precision MOS synchronous delay line," *IEEE J. Solid-State Circuits*, vol. 20, pp. 1265–1271, Dec. 1985.
[6] A. Waizman, "A delay line loop for frequency synthesis of de-skewed clock," in *Int. Solid-State Circuits Conf.*, San Francisco, CA, 1994, pp. 298–299.

A 0.5-μm CMOS 4.0-Gbit/s Serial Link Transceiver with Data Recovery Using Oversampling

Chih-Kong Ken Yang, *Student Member, IEEE*, Ramin Farjad-Rad, *Student Member, IEEE*,
and Mark A. Horowitz, *Senior Member, IEEE*

Abstract—A 4-Gbit/s serial link transceiver is fabricated in a MOSIS 0.5-μm HPCMOS process. To achieve the high data rate without speed critical logic on chip, the data are multiplexed when transmitted and immediately demultiplexed when received. This parallelism is achieved by using multiple phases tapped from a PLL using the phase spacing to determine the bit time. Using an 8 : 1 multiplexer yields 4 Gbits/s, with an on-chip VCO running at 500 MHz. The internal logic runs at 250 MHz. For robust data recovery, the input is sampled at 3$\times$ the bit rate and uses a digital phase-picking logic to recover the data. The digital phase picking can adjust the sample at the clock rate to allow high tracking bandwidth. With a 3.3-V supply, the chip has a measured bit error rate (BER) of $<10^{-14}$.

I. INTRODUCTION

THE increasing demand for data bandwidth in networking has driven the development of high-speed and low-cost serial link technology. Applications such as computer-to-computer or computer-to-peripheral interconnection are requiring gigabit-per-second rates either over short distances in copper or longer distances in fiber. CMOS technology is used increasingly over GaAs or bipolar technologies because of the development toward faster and faster devices. In 0.18-μm CMOS technology, the n-channel f_T is expected to equal or exceed that of the standard 0.5-μm GaAs process. While other technologies are limited in the number of transistors due to yield or power, CMOS technology allows implementation of complex digital logic enabling more integration of the back-end processing, lowering the cost. Recent development has shown CMOS capability to achieve Gbit/s data rates [1], [5], [6], [8], [11]. This work pushes NRZ signaling rates to the bandwidth limitations of the process technology and explores the issues involved.

The primary components of a link are the transmitter, the receiver, and the timing recovery circuits. Section II describes the overall architecture of the link. Because many of the circuits in the transmitter and receiver blocks have been previously discussed [1], this paper focuses on the timing recovery technique. Section III evaluates the impact of timing recovery on performance and compares two different timing recovery techniques: phase-locked loops versus oversampled phase picking. This chip implements a phase-picking algorithm that is discussed in Section IV. The measured performance of

Manuscript received September 1997; revised December 3, 1997.
The authors are with the Center for Integrated Systems, Stanford University, Stanford, CA 94305-4070 USA.
Publisher Item Identifier S 0018-9200(98)02225-2.

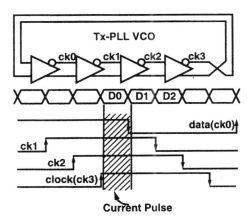

Fig. 1.   Transmit architecture.

the entire transceiver chip is presented in Section V. Finally, some conclusions are drawn from these results in Section VI.

II. ARCHITECTURE

A 0.5-μm CMOS technology is not fast enough to directly generate and receive a 4-Gbit/s stream (since the maximum ring oscillator frequency is <2 GHz). Instead, we use parallelism to reduce the performance requirements of each circuit. The transmitter generates the bit stream by an 8 : 1 multiplexer that multiplexes current pulses directly onto the output channel (Fig. 1). The receiver (Fig. 2) performs a 1 : 8 demultiplexing by sampling with a bank of input samplers. Similar to the transmitter, each sampler is triggered by individual clock phases. Furthermore, clock/data recovery is achieved by a 3$\times$ oversampling of each data bit. Thus, the receiver requires a total of 24 clock phases to support both the oversampling and the 1 : 8 demultiplexing. Various techniques exist for generating multiple clock phases [2], [3]. The receive side uses a six-stage ring oscillator (R_x-PLL) followed by phase interpolators to generate intermediate phases (ick[23 : 0]) between the ring oscillator edges (ck[11 : 0]) [1]. Similar to the R_x-PLL, eight different clock phases tapped from a four-stage ring oscillator (T_x-PLL) control the transmitter multiplexing.

A timing recovery circuit extracts the clock from the multiple samples per bit by finding the positions of the data transitions. Once the transitions are determined, a decision logic selects the samples furthest from data transitions (phase picking) as the received data byte. This approach is similar to

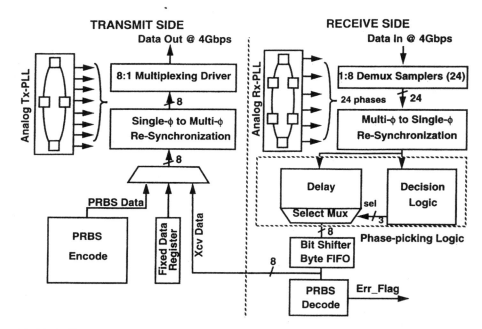

Fig. 3. Transceiver test-chip block diagram.

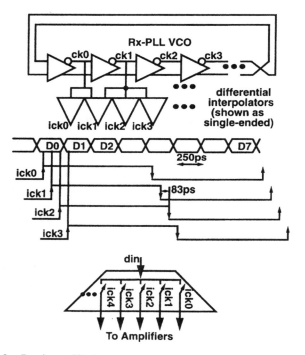

Fig. 2. Receive architecture.

what is done in UART's, and was first applied to a high-speed link by Lee *et al.* in [4].

Fig. 3 shows the full transceiver test-chip block diagram. Since the sampling clocks are different phases, the sampled results are resynchronized to a global clock. To facilitate the digital design, the on-chip data are further demultiplexed (2 : 1) to 250 MHz. Finally, in order to test the bit-error rate (BER), an on-chip parallel pseudorandom bit sequence (PRBS) encoder and decoder are used for a $2^7 - 1$ sequence. Serial data are commonly encoded with 8B10B coding which

limits the run length to <5 consecutive zeros or ones. The PRBS sequence is a suitable substitute because it guarantees a maximum run length of 7. The transmitter can be optionally configured to transmit the PRBS sequence, a fixed sequence, or the received data for testing.

III. TIMING RECOVERY

The goal of the timing recovery scheme is to maximize the timing margin—the amount that a sample position can err with the data still properly received. Errors that impact the timing margin can be classified into two sources: static phase error, and jitter (dynamic phase error). Fig. 4 illustrates the timing margin $t_{margin} = t_{bit} - t_{os} - t_{jc} - t_{jd}$ where t_{os} is the static sampling error, and t_{jd} and t_{jc} are the jitter on the data transition and the sampling clock. Since the sampling position is defined with respect to the data transition, jitter on both the clock and the data additively reduces timing margin. With ideal square pulses, as long as the sum of the magnitudes of the static and dynamic phase error is less than a bit time, the phase error does not impact signal amplitude. However, in a band-width limited system (for this work, due to the process technology), signal amplitude is lower with sampling phase error because the signals have finite slew rates. Correspondingly, this reduces the signal-to-noise ratio (SNR), hence impacting performance.

The amount of SNR degradation can be calculated based on the shape of the signal waveform. For static phase error, the SNR penalty is shown in Fig. 5 for a triangular signal waveform and a sinusoidal signal waveform. When the sample position phase offset is small, the sinusoidal waveform has a lower penalty than a triangular waveform due to the lower signal slew rate near the sample point.[1] For jitter, the SNR penalty is more complex to evaluate since it additionally depends on the statistics of the noise. For example, we can

---

[1] This penalty is only applicable to transitions.

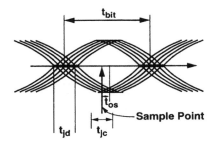

Fig. 4. Timing margin.

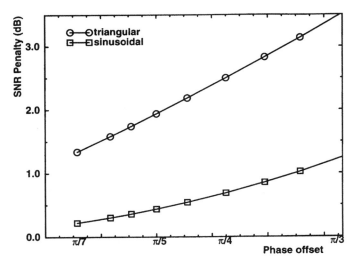

Fig. 5. SNR penalty for different phase offsets.

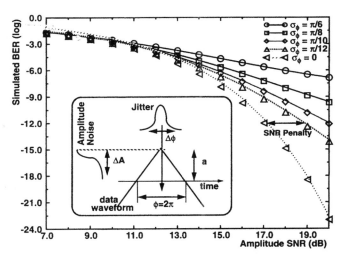

Fig. 6. BER versus SNR with various amounts of phase noise.

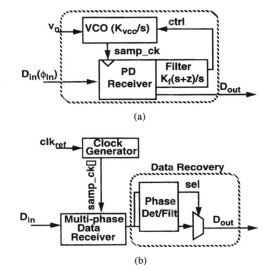

Fig. 7. Clock recovery architectures: (a) phase picking block diagram and (b) data/clock recovery architectures.

assume an idealized jitterless system with signal amplitude a and additive white Gaussian noise (AWGN) of standard deviation σ_A on the signal amplitude. In this system, we can determine the performance (BER) for various SNR [14]:

$$\text{ProbErr} = \int_a^\infty \frac{1}{\sqrt{2\pi\sigma_A^2}} \exp\left(-\frac{y^2}{2\sigma_A^2}\right) dy. \quad (1)$$

This equation is plotted as the lowest dotted line in Fig. 6.

If we further assume jitter to be a AWGN as well, for a triangular waveform, the phase noise can be translated into amplitude noise using $\Delta A = a\Delta\phi/\pi$ (where the bit time spans 2π). Since the noise sources are additive, the probability of error can be simply expressed as

$$\text{ProbErr} = \int_a^\infty \frac{1}{\sqrt{2\pi\left[\sigma_A^2 + \left(\frac{a}{\pi}\sigma_\phi\right)^2\right]}}$$
$$\cdot \exp\left\{-\frac{y^2}{2\left[\sigma_A^2 + \left(\frac{a}{\pi}\sigma_\phi\right)^2\right]}\right\} dy. \quad (2)$$

Fig. 6 illustrate the BER versus amplitude SNR for various amounts of phase noise. The SNR penalty, as shown in the figure, increases at higher SNR because the phase noise eventually limits performance, a "BER floor." For a sinusoidal signal waveform (with a lower slew rate near the sample point), the behavior is similar, except with lower SNR penalty.

The amount of phase error and the jitter depends on the implementation of the clock recovery circuit. Two techniques are commonly used, a phase-locked loop (PLL) and a phase picker. A PLL employs a feedback loop that actively servos the sampling phase of an internal clock source based on the phase of the input [7]. Fig. 7(a) illustrates a common VLSI implementation using an on-chip voltage-controlled oscillator (VCO) as the clock source, and a charge pump following the phase detector to integrate the phase error. A phase picker, as shown in Fig. 7(b), oversamples each bit, and uses the oversampled information to determine the transition position (phase) of the data. Based on the transition information, the best sample is then selected as the data value (UART [10]). Each of the two architectures has a different tradeoff in terms of static phase error and jitter.

The static phase error of a PLL depends mainly on its phase detector design. Ideally, sampling at the middle of the bit window gives the maximum timing margin. However, if the sampler has a setup time, the middle of the effective bit window is shifted by the setup time. Not compensating this shift causes significant static phase error. This error can be reduced by using the data samplers as the phase detector.[2] Additional phase error occurs due to inherent mismatches within the phase detectors and/or charge pump. Furthermore, any phase detector "dead band" (window in which the phase detector does not resolve phase information) limits the phase resolution, increasing the static phase error.

In a phase-picking architecture, the multiple samples per bit are used to find the transitions, effectively behaving as the phase detector. Sampler uncertainty limits the resolution of the transition detection. Sources of this uncertainty are sampler metastability window and data dependence of the sampler setup time. The uncertainty window for the sampler design used is <1/10 the bit time which does not impact performance significantly. More importantly, in this architecture, the phase information is quantized by the oversampling, causing a finite quantization error of 1/2 the phase spacing between samples. For a higher oversampling ratio, this static phase error is less, but it has a significant cost of increasing the number of input samplers, increasing the input capacitance, and hence limiting the input bandwidth. For a 3× oversampling system, the maximum static phase error is 1/6 the bit time.

In terms of jitter, a PLL tracks the phase of the input data with a tracking bandwidth limited by the stability of the feedback loop. The loop tracking is effectively a high-pass filter that rejects the phase noise of the input at lower frequencies. The noise not tracked appears as data jitter. Furthermore, because the PLL frequency source is an on-chip VCO, supply and substrate noise from on-chip digital switching can introduce additional jitter. The impact of these two sources is formulated for a second-order PLL in the following equation as the first and second terms:

$$\phi_{\text{err}} = \frac{\phi_{\text{in}}s^2}{s^2 + (K_f \cdot K_{\text{vco}})(s + z)} + \frac{v_n s K_n}{s^2 + (K_f \cdot K_{\text{vco}})(s + z)}.$$
(3)

Constants that determine the loop bandwidth in the equation are depicted in Fig. 7(a) with K_f (V/rad) the gain of the filter, z the stabilizing zero in the filter, and K_{vco} (rad-hertz/V) the gain of the VCO. v_n is the noise induced onto the VCO, and K_n is the sensitivity of the VCO to this noise. Thus, the total amount of "effective jitter" depends on the tracking bandwidth of the loop, the amount of supply and substrate noise, and the sensitivity of the loop elements to the noise. Because the feedback loop has a loop delay of at least one clock cycle, the bandwidth of the loop is often chosen to be <1/10 of the oscillation frequency for sufficient phase margin and stability. The delay makes tracking high-frequency phase noise ineffective because, if the phase error from on transition is independent of the phase of the next transition, correction

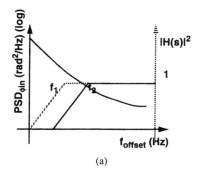

(a)

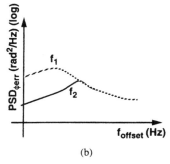

(b)

Fig. 8.   Effect of tracking bandwidth on jitter.

based on the first transition's phase information could increase the phase error for receiving the next bit.

The impact of different tracking bandwidth on jitter is illustrated in Fig. 8. The single sideband power spectral density (PSD) of an oscillator, such as the VCO of the transmitter, is shown to represent the phase noise in Fig. 8(a). Two hypothetical PLL's with different bandwidths (f_1, and f_2)[3] behave as high-pass filters that reject the lower frequency noise. Their transfer functions are overlaid in Fig. 8(a). The resulting phase error is shown in the PSD of Fig. 8(b). Note that this example excludes the additional noise from the phase-tracking circuit [second term of (3)]. The integral of the area beneath the curve is an indication of the amount of jitter [13] [σ_ϕ^2 for (2)]; thus, the phase noise of *Circuit I* is larger than that of *Circuit II*. Additionally, if a second-order PLL is not critically damped, the transfer function can exhibit peaking. This peaking accumulates phase noise at its loop bandwidth, increasing the noise.

For a phase picker, the sampling clocks experience similar jitter problems from supply and substrate noise since the phases for the oversampling are also generated from an on-chip VCO. The primary difference is the tracking bandwidth. A phase-picking system is a feedforward architecture (instead of feedback); thus, there are no intrinsic bandwidth limitations. The tracking rate depends on the rate at which new phase decisions are made, which in turn depends on the logic's cycle time. The importance of this fast tracking is that it can potentially track the accumulation of phase noise by the on-chip multiphase generator (PLL). We delay the data by the time to arrive at a decision so the corrections are applied to the appropriate bit (although with a latency overhead). However, the maximum phase change between two transitions must be

---

[2] This causes additional difficulties because such phase detectors can only determine if transitions are early or late. The control loop is "bang-bang" control instead of linear control, which is less stable, has inherent dithering, and requires additional frequency acquisition aid. Although a DLL (delay-line based PLL) [8] can be used to eliminate the stability and frequency acquisition problems, the phase spacing, when tapping phases from the buffer stages, is sensitive to the input clock's duty cycle and amplitude.

[3] The actual shape of the tracking transfer function $H(s)$ varies with implementation.

less than π, half the bit time, even if the peak-to-peak jitter can be much larger than a bit time. Changes ϕ_e greater than π are indistinguishable from a phase shift in the opposite direction, $\phi_e - 2\pi$.

Choosing between the two clock recovery systems depends on the system requirements and noise behavior. We chose a phase-picking architecture to explore the usefulness of the higher phase-tracking capability. In such VLSI implementations, supply noise can be significant enough for the peak-to-peak jitter to occupy a large fraction of the bit time, especially since a PLL accumulates jitter. For the 4-Gbit/s link, we chose a low oversampling ratio of $3\times$ to maintain high input bandwidth and to keep the number of clock phases manageable ($1:8$ demultiplexing and $3\times$ oversampling yields 24 phases). With a bit time of 250 ps, the phase-picking scheme[4] can track the noise of the on-chip multiphase generator (PLL) from both the transmit and receive sides to keep the total "effective jitter" below the 83-ps quantization spacing. One limitation of the phase-picker tracking is that the maximum rate of the tracking depends on the data transition density. Since the PRBS signal guarantees one transition per byte, the maximum tracking rate of one sample spacing every transition is fast (83 ps/2 ns).

Although the tracking rate is high, the maximum static phase error from the quantization is 41 ps (2% of the clock period, $8\times$ bit time), causing an SNR penalty (Fig. 5). Whether or not a $3\times$ oversampled phase-picking approach with higher tracking bandwidth than a PLL can achieve better performance with the larger static phase error depends on the amount of jitter induced by on-chip noise sources. If the lower SNR penalty from the lower jitter compensates the higher SNR penalty of larger static phase error, phase picking would be the better choice.

IV. PHASE-PICKING ALGORITHM AND IMPLEMENTATION

The details of the phase-picking algorithm are illustrated in Fig. 9. Picking the center sample requires finding and tracking the bit boundaries. The decision logic first detects transitions by an XOR of adjacent samples, indicating the bit boundary to be in one of three possible positions. Fig. 10 shows an example of the boundary detection with a portion of a sampled stream. To find which of the three transition positions is the most likely bit boundary, transitions corresponding to the same bit boundary position are tallied. The position with the largest total determines the bit boundaries.

The decision logic makes a new decision per byte of data. In contrast to a higher order oversampling phase picker, the $3\times$ oversampling limits the change of the selected sample position to one sample position per byte. To guarantee sufficient transitions for averaging any bit-to-bit variations of high-frequency noise (near the bit rate), the tally is across a sliding window of 3 bytes. The transitions are accumulated from the current byte, the previous byte, and the next byte (delaying the data allows the noncausal information) so that the decision is applied to the byte at the middle of the window. As a result

[4] In our system, the oscillator is at 250 MHz so the PLL bandwidth is restricted to <25 MHz. This yields a $10\times$ tracking rate difference between the two systems.

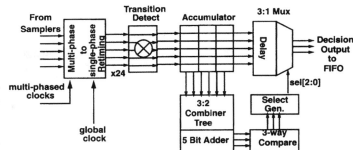

Fig. 9. Phase-picking algorithm block diagram.

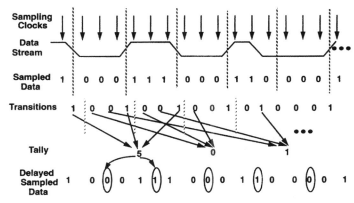

Fig. 10. Example of the phase-picking algorithm.

of the 3-byte sliding accumulation, the rate of phase change that the algorithm can track is slower than the maximum of 83 ps/2 ns. The algorithm picks the correct sample if the majority of the transition information within the 3-byte window (6 ns) indicates the correct phase. For example, if the input phase has a constant rate of change of <1 sample spacing per 3 ns (corresponding to a frequency difference of 4%), the transition information from >1.5 bytes of the 3-byte window would fall in the same phase quantization. Then the tally and compare would select the correct sample to track the phase change. This indicates a maximum phase-tracking rate of 83 ps/3 ns. The criterion of tracking both T_x and R_x-PLLs' accumulation is met because the VCO elements' supply noise sensitivity is ~0.1%/% (percent of frequency change per percent of supply noise [1], [3]),[5] corresponding to 30 ps/3 ns for a 10% supply step, which is less than the tracking rate. If the phase change is slower than 83 ps/3 ns, the 3-byte accumulation offers some robustness by averaging any uncertainty in the transition detection due to high-frequency bit-to-bit noise. A smaller window of one byte can track phase faster, but has poorer performance without sufficient transitions within that byte to average the bit-to-bit variation. A larger window of 5 bytes (<83 ps/6 ns) would be too slow to track the T_x- and R_x-PLLs' phase accumulation under reasonable supply noise.

Once the transition position is determined, the middle sample within the bit boundaries is selected as the data.

[5] Although the maximum phase error accumulation rate is based on the supply sensitivity of the VCO, the peak phase error depends on the loop bandwidth. The T_x-PLL and R_x-PLL generating the multiple clock phases have bandwidths of 15 and 5 MHz, respectively.

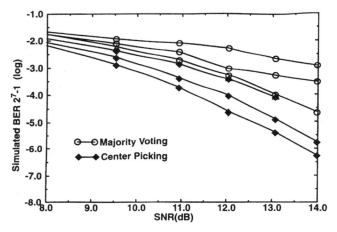

Fig. 11. Comparison between center picking versus majority voting.

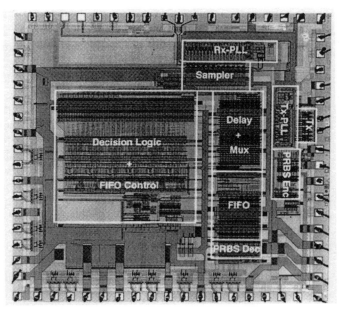

Fig. 12. Chip micrograph.

The selection is implemented by multiplexers selecting the appropriate samples based on three select signals. In the case where no transitions are detected, the three select signals use previously stored values to maintain data through the multiplexers.

The actual algorithm for deciding the received data value from the oversampled information can be designed alternatively while still keeping the advantage of higher tracking bandwidth of a feedforward architecture. Instead of selecting the middle ("phase pick"), a simple alternative implementation is to take a majority vote based on the three sampled values. Fig. 11 shows the performance comparison. Majority voting works well with nonbandwidth-limited signals that have high-frequency noise because it averages the noise over many samples. In a bandwidth-limited system (low-pass filtered by the I/O RC time constant), it performs worse because at least one of the two nonmiddle samples is required to be valid, and the nonmiddle samples have a much higher probability of error.

Arbitration is required when two transition positions have equal counts. This occurs when two of the sample positions straddle the center of the bit and the third sampler samples at the transition. Picking either of the two straddling the center gives equivalent performance. More complex logic can be implemented by using the previous, current, and next cycles' comparison results to follow the direction of any phase transition. However, this only improves the performance by less than 1 dB.

If the peak-to-peak phase jitter is larger than one bit time, or if the transmitter and receiver operate at different frequencies, the tracking must allow bit(s) to overflow/underflow. For example, if the SEL[2 : 0] signal changes from 0–0–1 to 1–0–0, the selected sample of the first cycle corresponds to the same bit as the selected sample of the following cycle. This "underflow" condition must be appropriately handled by dropping one of the two samples. Typically, these samples are of the same bit, and thus have the same value. However, in the case where they are different, if phase movement changes directions (the SEL signal returns to 0–0–1) in the following cycle's decision, dropping the latter one gives a slight performance improvement. Similar to the "underflow"

where only 7 bits are received, the opposite transition from 1–0–0 to 0–0–1 causes an "overflow," requiring an extra bit (9 bits total) to be stored. These conditions are handled by a bitwise FIFO built by shifting the input byte to accommodate the one extra/less bit. If the aggregate shift increases beyond 1 byte, a bytewise FIFO handles the overflow/underflow byte. The limited depth of the FIFO can only handle a finite number of byte overflow. If the application requires handling long streams of data with a slight frequency difference with the local reference clock, the local frequency can be corrected based on the phase information from the decision logic.[6]

V. TRANSCEIVER EXPERIMENTAL RESULTS

The transceiver chip was implemented in a 0.5-μm CMOS process offered through MOSIS. The 3 mm × 3 mm die photo is shown in Fig. 12. The chip is packaged in a 52-pin CQFP package supplied by Vitesse Semiconductor which has internal power planes for controlled impedance. The size of the I/O bond pads are reduced to 70 μm × 70 μm to keep pad capacitance to a minimum because the capacitance would otherwise limit the I/O bandwidth. With an effective impedance at the I/O of 25 Ω (for a doubly terminated 50-Ω line), the total I/O capacitance can not exceed 4.5 pF for 4-Gbit/s operation without losing 10% of the bit height to the RC filtering. The 1 : 8 demultiplexing receiver and 8 : 1 multiplexing transmitter designs have capacitances of 2.2 and 1.2 pF, respectively, with 600 fF due to the pad and metal interconnects. An input time constant of ~110 ps is estimated from measurements sweeping the reference voltage for a single-ended input pulse. The width of the pulse with a different reference voltage determines the time constant.

The performance of the link depends significantly on the I/O circuits. The minimum receivable amplitude of 50 mV was measured by using a fixed data pattern while changing

---

[6]This feature is not implemented as part of this test chip.

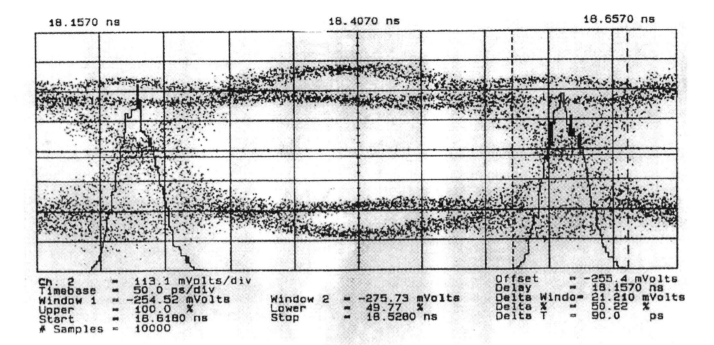

```
Ch. 2       =  113.1 mVolts/div
Timebase    =  50.0 ps/div
Window 1    =  -254.52 mVolts        Window 2    =  -275.73 mVolts
Upper       =  100.0  %              Lower       =  49.77  %
Start       =  18.6180 ns            Stop        =  18.5280 ns
# Samples   =  10000
```

```
Offset       =  -255.4 mVolts
Delay        =  18.1570 ns
Delta Windo= =  21.210 mVolts
Delta %      =  50.22  %
Delta T      =  90.0     ps
```

Fig. 13. Transmitter data eye.

the amplitude. This indicates the worst case input offset in the bank of samplers. The transmitter data eye at 3.0 Gbits/s is shown in Fig. 13 with the output driving a PRBS $2^7 - 1$ sequence. The measured data rate is limited by the triggering bandwidth of the oscilloscope. The maximum speed of the transmitter was 4.8 Gbits/s, and was limited by the maximum frequency of the ring oscillator used in the clock generation.

The multiple-phased clock generation (PLL) is crucial to the performance of the link because the phase spacing determines the bit time in the multiplexing/demultiplexing architecture, and the supply sensitivity and loop bandwidth determine the amount of jitter that needs to be tracked. Mismatches can cause one phase to be shifted with respect to the others. In the transmitter, the shift enlarges one bit, but reduces the next. By measuring the spacing between edges, we can evaluate the ability to match the phases tapped from the oscillators and interpolators [3]. The differential nonlinearity (DNL) of the phase spacing is plotted for the transmitter in Fig. 14 at various frequencies. The error is expressed as a percentage of the ideal bit time for all eight phase positions. While transmitting the PRBS pattern and using a trigger frequency of 1/8 the data rate (internal clock rate), these spacings are measured with a 20-GHz bandwidth digital oscilloscope by the width of each of the eight data-eye patterns.[7] If we use the data-rate frequency as a trigger instead of using a divided frequency, the data eye of Fig. 13 overlaps all eight of the bits. The overlaid histogram shows that the 333-ps bit time is degraded by 90 ps due to equal contributions from jitter and errors in the transmitter phase spacing.

The peak-to-peak variation, <±7% of the bit time, indicates very little degradation in bit width due to mismatches. The dominant cause of these bit-width variations is the V_T and K_P

[7] The measurement uncertainty is the DNL is ±2 ps.

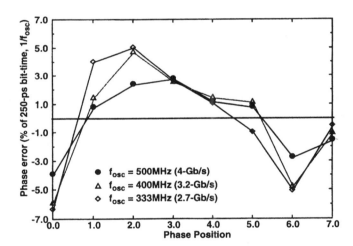

Fig. 14. Transmit-side DNL at various frequencies.

mismatches of the transistors in the clock generation circuits [12]. The increase in error with decreasing oscillation frequency, shown in Fig. 14, is an indication of these mismatches. The gate overdrive ($V_{GS} - V_T$) is less at lower oscillation frequencies, making the phase spacing more sensitive to these mismatches. Fig. 15 shows the measurement of the DNL for four chips. The darker line indicates the average at each phase position. The variation of this average across phase positions potentially indicates some systematic error. However, because the average is over a sample size of only four chips, and the variation of the average is significantly smaller than the variation between chips, the random component is believed to be the dominant source of static phase spacing error.

Although a systematic component of the offset can also be expected from noise at any integer multiple of the oscillator

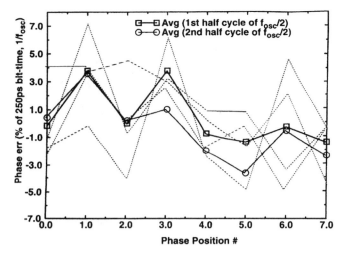

Fig. 15. Transmit-side DNL for four chips.

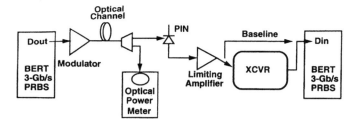

Fig. 16. BER testing configuration.

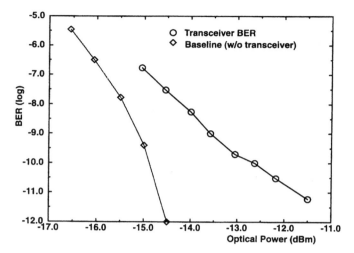

Fig. 17. Measured BER versus SNR.

frequency, it is not apparent in Fig. 15. Normally, noise such as substrate or supply noise at the same frequency as the oscillator would modulate the oscillator, causing a duty-cycle error which spreads the phases in the first half cycle and compresses the phases in the second half cycle. Since most of the digital logic clock on this chip switches at $f_{osc}/2$ (250 MHz), this effect of the clock buffer switching on the 500 MHz oscillator would cause different phase spacings for two consecutive oscillator cycles. However, Fig. 15 shows that the average phase spacing errors from the second cycle is nearly the same as the first cycle, indicating that this coupling is negligible. Also, any systematic components from path mismatches (e.g., capacitive loading errors) are insignificant compared to the random source.

On the receive side, the DNL of the sample spacing is also measured, as was shown for a 0.8-μm process technology [1] to be <8% of the bit time. Receive clock phase spacing errors reduce the effectiveness of the oversampling by increasing the sample spacing, causing both increased static phase error and larger jitter.

Jitter in the transmitter can be measured by a outputting a fixed pattern and measuring the jitter on the data transition. We can also measure the sampling clock jitter by looking at the sampler output while sweeping a clean input transition. The window in which the sampler output is uncertain indicates the jitter with respect to the input. The supply sensitivitiy can also be measured by the increase in jitter due to induced supply noise with an internal switch that shorts between supply and ground. The sensitivities of the transmit and receive PLL's are 0.2 and 0.3 ps/mV, respectively, with a similar peak-to-peak quiescent jitter of 45 ps.

The BER testing is performed with two different configurations. The first measurement is by feeding the transmitted output directly back into the input. This yielded a BER of <10^{-14}. The second configuration is by placing the chip in a mock optical network (Fig. 16). A bit error rate tester (BERT) is used to generate the data pattern. The pattern is modulated onto a fiber-optic network. The optical power is measured by siphoning 1/10 of the total optical power. The optical signal is received and amplified by a avalanche photodiode (APD) followed by an amplifier. The output of the amplifier is either returned to the BERT for the baseline measurement, or sent into the chip configured in its transceiver mode. Because the BERT and optical amplifiers have a bandwidth limitation at 3 Gbits/s, the experimental results of this configuration are limited in data rate. As shown in Fig. 14, the phase spacing at lower frequencies is worse, so the performance is slightly worse than at 4 Gbits/s.

The BER versus SNR is plotted with SNR expressed in optical power showing both the baseline and the DUT with a 1.5-dB penalty at 10^{-9} BER (Fig. 17). The SNR penalty for not having the selected sample at the middle of the data eye is shown in Fig. 18. Because of the phase spacing errors on the receive side, the penalty shown here is worse than simulated. Since the quiescent jitter of the clock generation is smaller than the sample spacing (<83 ps), the phase tracking is not active. In order to test the effectiveness of the phase picking, voltage steps are induced on the supply, causing 250-ps jitter on both the T_x-PLL and R_x-PLL. While this causes the data eye to collapse, the receiver can still track this jitter and maintain BER <10^{-9}. Also, the transceiver is operated with the transmitter and receiver at different frequencies. The chip was able to track a frequency difference of 1 MHz with BER <10^{-9}.

Table I shows some additional performance measurements of the chip. The total power dissipated is 1.5 W, with 1/3 from the clock generation and 1/3 from the receive-side logic. The minimum amplitude that can still maintain <10^{-9} BER is 90

654

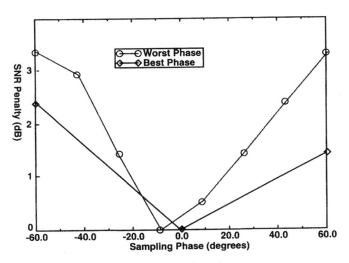

Fig. 18. Measured BER at various sampling phase.

TABLE I
TEST-CHIP PERFORMANCE

Supply Voltage	2.7-4.0
Max. Transmit Rate @3.3V	4.8Gb/s
Max. Receive Rate @3.3V	4.3Gb/s
Max. Frequency Difference	1MHz (<10^{-9} BER)
Power @ 4Gb/s- Total	500mA
Analog(2 PLL)	150mA
Input Samp,Rcv Logic	25mA, 170mA
Transmitter, Xmt Logic	50mA, 10mA
Parallel Data Driver	90mA
Min. Input Amplitude	170mV <10^{-9} BER

mV with an internal eye height of 65 mV. The 24 mV of amplitude noise is primarily due to ringing from the package inductance and on-chip output capacitance at the transmitter.

VI. CONCLUSION

Very high data rates are achievable in CMOS technologies by making extensive use of parallelism. Using an 8 : 1 demultiplexing at the input and a 8 : 1 multiplexing output transmitter, we achieved a 4-Gbit/s transceiver while keeping all internal signals <500 MHz in a 0.5-μm process technology. The fundamental limitations of this approach are the I/O capacitance (increased due to the parallelism), the sampler uncertainty, and the phase position accuracy of the multiple clock phases.

Provisions were made in this design to handle very large jitter accumulation of 83 ps/3 ns by a fast phase-picking algorithm. The effectiveness of this architecture critically depends on the jitter characteristics. Although a CMOS PLL can potentially exhibit this large jitter due to supply noise, the measured jitter while operating this transceiver is only 50 ps. This jitter is measured in a realistic noise environment because of the presence of significant digital switching noise from the large digital phase picker that can couple onto the VCO elements. Since the jitter is less than the quantization error, the advantage of the phase picking is only apparent

when additional noise is induced. This low accumulated jitter implies that the lower tracking bandwidth of a PLL-based clock recovery circuit can potentially perform equally. The design of such a system is nontrivial, and still has challenges in maintaining small static phase offsets. However, since the phase picking has significant hardware overhead in the extra number of input samplers and large digital processing, a PLL would potentially offer similar performance with lower area and power.

ACKNOWLEDGMENT

The authors would like to thank S. Sidiropoulos, B. Amrutur, K. Falakshahi, Vitesse Semiconductor, Prof. T. Lee, Prof. L. Kazovsky, and their research groups for invaluable discussions and assistance.

REFERENCES

[1] C.-K. Yang and M. Horowitz, "A 0.8 μm CMOS 2.5 Gbps oversampling receiver and transmitter for serial links," *IEEE J. Solid-State Circuits,* vol. 31, Dec. 1996.
[2] C. Gray *et al.,* "A sampling technique and its CMOS implementation with 1-Gb/s bandwidth and 25 ps resolution," *IEEE J. Solid-State Circuits,* vol. 29, Mar. 1994.
[3] J. Maneatis and M. Horowitz, "Precise delay generation using coupled oscillators," *IEEE J. Solid-State Circuits,* vol. 28, pp. 1273–1282, Dec. 1993.
[4] K. Lee *et al.,* "A CMOS serial link for fully duplex data communications," *IEEE J. Solid-State Circuits,* vol. 30, pp. 353–364, Apr. 1995.
[5] A. Fiedler *et al.,* "A 1.0625Gb/s transceiver with 2×-oversampling and transmit signal pre-emphasis," in *ISSCC'97 Dig. Tech. Papers,* Feb. 1997, pp. 238–239.
[6] A. Widmer *et al.,* "Single-chip 4 × 500 Mbaud CMOS transceiver," *IEEE J. Solid-State Circuits,* vol. 31, pp. 2004–2014, Dec. 1996.
[7] F. M. Gardner, *Phaselock Techniques,* 2nd ed. New York: Wiley, 1979.
[8] W. Dally and J. Poulton, "A tracking clock recovery receiver for 4-Gb/s signaling," in *Hot Interconnect97 Proc.,* Aug. 1997, p. 157.
[9] S. Sidiropoulos and M. Horowitz, "A semi-digital DLL with unlimited phase shift capability and 0.08–400MHz operating range," in *ISSCC'95 Dig. Tech. Papers,* Feb. 1995, pp. 332–333.
[10] J. E. McNamara, *Technical Aspects of Data Communication,* 2nd ed. Bedford, MA: Digital, 1982.
[11] S. Kim *et al.,* "An 800Mbps multi-channel CMOS serial link with 3× oversampling," in *IEEE 1995 CICC Proc.,* Feb. 1995, p. 451.
[12] M. J. Pelgrom, "Matching properties of MOS transistors," *IEEE J. Solid-State Circuits,* vol. 24, p. 1433, Dec. 1989.
[13] J. A. Crawford, *Frequency Synthesizer Design Handbook.* Boston, MA: Artech House, 1994.
[14] J. Proakis, *Communication Systems Engineering.* Englewood Cliffs, NJ: Prentice-Hall, 1994.

Chih-Kong Ken Yang (S'93) received the B.S. and M.S degrees in electrical engineering from Stanford University, Stanford, CA, in 1992.

He is currently pursuing the Ph.D. degree at Stanford University in the area of circuit design for high-speed interfaces.

Mr. Yang is a member of Tau Beta Pi and Phi Beta Kappa.

A 2–1600-MHz CMOS Clock Recovery PLL with Low-Vdd Capability

Patrik Larsson

Abstract— **A general-purpose phase-locked loop (PLL) with programmable bit rates is presented demonstrating that large frequency tuning range, large power supply range, and low jitter can be achieved simultaneously. The clock recovery architecture uses phase selection for automatic initial frequency capture. The large period jitter of conventional phase selection is eliminated through feedback phase selection. Digital control sequencing of the feedback enables accurate phase interpolation without the traditional need of analog circuitry. Circuit techniques enabling low-Vdd operation of a PLL with differential delay stages are presented. Measurements show a PLL frequency range of 1–200 MHz at $Vdd = 1.2$ V linearly increasing to 2–1600 MHz at $Vdd = 2.5$ V, achieved in a standard process technology without low threshold voltage devices. Correct operation has been verified down to $Vdd = 0.9$ V, but the lower limit of differential operation with improved supply-noise rejection is estimated to be 1.1 V.**

Index Terms—**Frequency locked loops, frequency synthesizers, phase comparators, phase jitter, phase locked loops, phase noise, synchronization.**

I. INTRODUCTION

THE continuing scaling of CMOS process technologies enables a higher degree of integration, reducing cost. This fact, combined with the ever shrinking time to market, indicates that designs based on flexible modules and macrocells have great advantages. In clock recovery applications, flexibility means, for example, programmable bit rates requiring a phase-locked loop (PLL) with robust operation over a wide frequency range. Increased integration also implies that the analog portions of the PLL (mainly the voltage-controlled oscillator [VCO]) should have good power-supply rejection to achieve low jitter in the presence of large supply noise caused by digital circuitry.

Another trend is low-power design using reduced Vdd. This reduces the headroom available for analog design, causing integration problems for mixed-mode circuits [1]. Furthermore, in applications where power consumption is a more critical design goal than compute power, V_t is not scaled as aggressively as Vdd to avoid leakage currents in OFF devices, which aggravates the headroom problem. For mixed-mode circuits with significant analog circuitry, dual-V_t and/or dual-Vdd processing combined with a dc/dc converter [2] is a viable solution. However, for circuits dominated by digital logic, it is difficult to justify the additional fabrication steps required for these solutions. A common case of the latter mixed-mode

design is a large digital circuit incorporating a PLL-based clock generator with low-jitter requirements, which is the most common mixed-mode design today. Digital style PLL's have been suggested, e.g., [3], but these cannot compete with the supply-noise rejection of differential analog circuitry.

A clock recovery PLL architecture suitable for programmable bit rates is developed in Sections II and III with emphasis on jitter reduction. Sections IV–VI present PLL circuit techniques that use the noise resistant differential pair but avoid other "expensive" (in terms of headroom) analog circuitry, such that low-Vdd operation is enabled in a standard digital CMOS process without the need of low-threshold devices.

II. LOW-JITTER PHASE-SELECTING CLOCK RECOVERY

A basic PLL for clock recovery is shown in Fig. 1(a). In most CMOS implementations, the VCO must have a tuning range covering more than ±50% of the target frequency to guarantee high yield over large process variations. This large frequency range requires special techniques for initial frequency locking since there exists no phase detector for nonreturn-to-zero (NRZ) data that operates reliably with large initial frequency offset. Available techniques include frequency sweeping [4], using a replica VCO matched to the clock generating VCO [5], or initially locking the PLL to a reference frequency with a frequency detector before switching to the input data and locking with a phase detector [6].

One common technique requiring no special initialization is shown in Fig. 1(b). This dual-loop PLL can be traced back to [7], which was based on a delay-locked loop (DLL). The multiple-output VCO in Loop A in Fig. 1(b) generates a number of equally spaced clock phases at a frequency of $N \cdot f_{ref}$. This loop can have a large frequency tuning range since it is locked with a phase frequency detector (PFD). Clock recovery is performed by Loop B that generates the recovered clock by selecting the clock phase from Loop A that is best aligned with the incoming data. If there is a frequency offset between $N \cdot f_{ref}$ and the incoming data, an appropriate clock can still be generated by changing the *Ctrl* signal to select a different phase over time.

Frequency initialization is automatically achieved by selecting appropriate f_{ref} and N for the expected data rate. Most communication systems have a frequency tolerance of a few hundred parts per million (ppm), eliminating any need for a frequency detector in Loop B. The decoupling of the VCO loop from the data recovery loop enables independent selection of bandwidth in those two loops. This allows a large bandwidth in

Manuscript received April 13, 1999; revised June 19, 1999.

The author is with Bell Laboratories, Lucent Technologies, Holmdel, NJ 07733 USA.

Publisher Item Identifier S 0018-9200(99)08963-5.

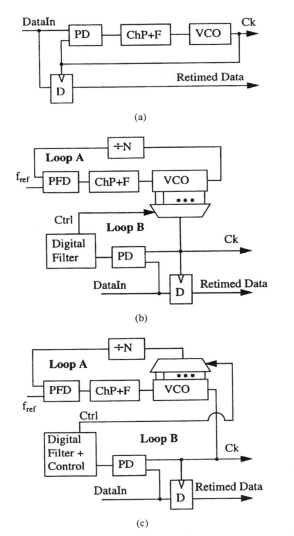

Fig. 1. Clock recovery PLL's. (a) Standard, (b) phase selection, and (c) feedback phase selection. ChP+F denotes charge pump + loop filter.

Loop A to suppress VCO jitter [4], for example, jitter induced by power-supply noise. At the same time, a low bandwidth can be used in Loop B to reduce jitter transfer. This cannot be achieved by the PLL in Fig. 1(a), which has a single loop with conflicting design goals regarding loop bandwidth.

A disadvantage of a phase-selecting PLL is that the phase step that is generated when the Ctrl signal in Fig. 1(b) switches to a new clock phase. This phase switching leads to large cycle-to-cycle jitter (greater than or equal to the phase spacing) that can actually dominate the peak-to-peak jitter. By increasing the number of phases, the phase spacing will be smaller with less jitter. More phases can be generated by having more delay stages in the VCO, but this limits the speed. An alternative is phase interpolation that enables a large number of phases without degrading the VCO speed [8], [9]. However, interpolators add analog circuitry to the design and are prone to mismatch, which in the worst case can lead to nonmonotonic phase spacing.

A proposed remedy for the jitter due to phase steps is shown in Fig. 1(c). Instead of selecting a clock phase feeding the sampling flip-flop and the phase detector, the feedback clock in

Loop A is selected from the multiple VCO phases. When Loop B detects a misalignment of the incoming data and the VCO output clock, the *Ctrl* signal is changed to select a different phase for the feedback clock. This will cause a phase change in the divided clock feeding the PFD such that the charge pump will alter the VCO control voltage stored in the loop filter. Therefore, sudden phase steps generated by the clock recovery logic will be smoothed by the filter of Loop A, causing the VCO clock to slowly drift toward the correct phase with a rate of change determined by the bandwidth of Loop A. In a 622-MHz application with a division ratio of $N = 8$ and a bandwidth of Loop A equal to one-tenth of f_{ref}, it will take approximately $8 \cdot 10 = 80$ clock cycles to complete a phase switch. The jitter caused by a phase step in the structure in Fig. 1(c) is therefore spread out over 80 clock cycles, significantly reducing the jitter compared to Fig. 1(b). Feedback phase selection has previously been applied to fractional-N frequency synthesizers for other purposes [10], [11].

III. AVERAGING PHASE INTERPOLATION

The smoothing effect of the loop filter can also be used for phase interpolation. If the Ctrl signal in Fig. 1(c) alternates between two different clock phases every second cycle of the reference clock, the result will be a VCO clock phase corresponding to the average of the two selected phases. In the test chip, four levels of averaging phase interpolation were implemented by circulating through four clock cycles and in each clock cycle selecting phase ϕ_m or ϕ_{m+1} as the feedback clock. A quarter phase interpolation generating $\phi_{m+0.25}$ is then achieved by selecting ϕ_m for three consecutive clock cycles, then selecting ϕ_{m+1} for the fourth cycle and repeating this sequence.

The architecture in Fig. 1(c) lends itself naturally to combining both averaging phase interpolation and standard current-mode interpolation. A test chip was built in a 0.25-μm, 2.5-V digital CMOS process to evaluate the jitter performance of the phase selection architecture. A block diagram of the implemented VCO and phase control circuitry is shown in Fig. 2. The *phase select control code* at the input consists of seven bits, of which two are directly fed to a *finite state machine* (FSM) that generates control signals for realizing the averaging interpolation. The remaining five bits of the control code represent ϕ_m, from which the code for ϕ_{m+1} is generated by adding one. The FSM controls Mux1 to select one of the codes representing ϕ_m and ϕ_{m+1} in a four clock period repetitive cycle, as described above. The five bits at the output of Mux1 are split into three bits coarse select and two bits fine select. The three coarse bits select two neighboring phases from a four-stage differential VCO having eight evenly spaced output phases and send these two phases to a current-mode interpolator. Mux2/Mux3 in Fig. 2 receive one coarse bit each, and the third coarse bit is used to conditionally invert the output signals. The interpolator is similar to the Type-I circuit in [9] and is controlled by a four-bit temperature code derived from the two fine select bits. Both the current-mode interpolation and the averaging phase interpolation are programmable in the test chip and can be disabled. The two complementary multiplexers at the output

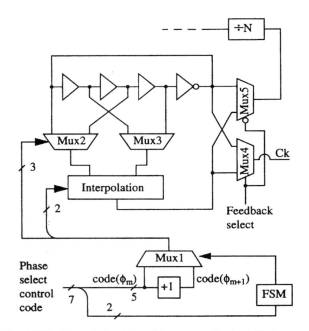

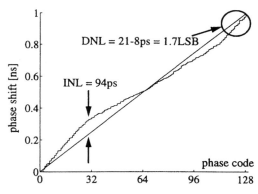

Fig. 4. Phase shift versus phase code measured at 1 GHz.

Fig. 2. VCO, phase selector, interpolator, and feedback multiplier.

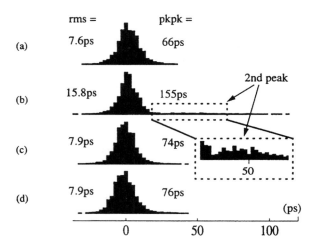

Fig. 3. 500-MHz period distribution histograms. (a) Clock recovery inactive, (b) standard phase selection, (c) feedback phase selection, and (d) averaging phase interpolation. Measurement conditions were $Vdd = 2.5$ V, $N = 25, f_{ref} = 20$ MHz, and $f_{data} = 499.4$ Mb/s.

of the VCO/interpolator (Mux4/Mux5) allow the chip to be configured for the scheme in either Fig 1(b) or (c), enabling a performance comparison.

Freezing the 7-bit phase select control code to a fixed phase gives a measured output period jitter[1] of 7.6 ps rms when running the VCO at 500 MHz, as shown in Fig. 3(a). This is the jitter inherent in the VCO and the output buffers. Configuring the chip for the standard phase selecting scheme in Fig. 1(b) with 32 clock phases ($4\times$ interpolation) gives the jitter in Fig. 3(b), revealing a long tail in the histogram caused by a frequency offset of 1200 ppm between the incoming data and $N \cdot f_{ref}$. The phase spacing is 2 ns/32 = 62.5 ps such that we can expect a second peak in the histogram 62.5 ps away from the main peak, which is confirmed by the shape

[1] Timing uncertainty between two consecutive edges of the generated clock.

of the histogram. As shown by the inset, this peak is slightly off its ideal position and is smeared out due to the nonideal ac behavior of the current-mode interpolator.

Feedback phase selection with $4\times$ current-mode interpolation eliminates the long tail in the histogram, bringing the period jitter down to 7.9 ps rms, as shown in Fig. 3(c). This indicates that the jitter is completely dominated by the VCO jitter, and nearly all of the phase-switching jitter caused by digital clock recovery can be eliminated. Fig. 3(d) shows the period jitter histogram obtained when the current-mode interpolators are disabled and $4\times$ averaging phase interpolation is used instead. Its similarity to the result in Fig. 3(c) proves that the same low jitter and the same number of discrete clock phases (32) can be achieved without the analog interpolation circuitry.

Enabling both the current-mode interpolator and the averaging phase interpolation gives a total of 128 selectable clock phases. The graph in Fig. 4 shows the phase shift as a function of the *phase select control code* when the period of the VCO is 1 ns. The expected phase step is 1 ns/128 = 8 ps, whereas the largest measured step is 21 ps, resulting in a differential nonlinearity (DNL) of 1.7 bits. The differential VCO makes the phase curve near-symmetric around the midpoint, suggesting that the integral nonlinearity (INL) of 94 ps is mainly due to delay mismatch in the VCO. The main contributing factor to this mismatch is unbalanced parasitic wiring capacitors that are difficult to match without incurring speed penalty. If Loop B is a first-order loop or a well-damped second-order loop, the feedback in Loop B will automatically select the best fit phase select code, reducing the impact of INL. The maximum phase deviation in a clock recovery application is then the DNL added to the VCO jitter.

In addition to jitter reduction and phase interpolation, feedback phase selection also has other advantages when combined with other architectures. Using feedback instead of feedforward phase selection reduces circuit complexity, thereby eliminating the need for good matching in an analog-style interpolator [12] and a high-speed parallel sampling structure [13].

IV. VCO

Recently, low-noise VCO's utilizing high-swing complementary signals have been presented (e.g., [14] and [15]).

658

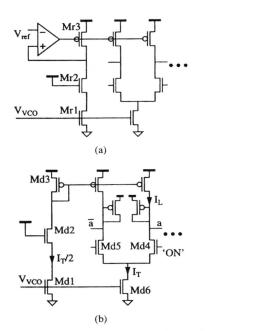

(a)

(b)

Fig. 5. Bias generation and one VCO delay stage of (a) replica bias scheme and (b) diode clamping.

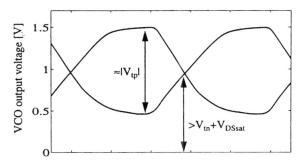

Fig. 6. Example VCO waveforms to estimate lower limit on Vdd.

Good $1/f$ noise performance has been shown, but their power-supply noise rejection is inferior to that of the standard analog differential pair since they lack a high-impedance source, making the delay depend on Vdd. Therefore, the analog style differential pair is preferable in applications where power-supply noise is the main source of oscillator jitter. When a differential pair with resistor loads is used as a delay cell in a VCO, the frequency is regulated by changing the tail current as implemented by the V_{VCO} control voltage in Fig. 5(a). To achieve a large frequency tuning range, it is desirable that the output swing and common mode do not change significantly with frequency. Often the replica-bias scheme in Fig. 5(a) is employed, which relies on good matching between a replica of the delay stage (devices Mr1–Mr3) and the VCO delay stages to set the VCO output swing from V_{ref} to Vdd, giving a known common mode and swing independent of the speed-regulating current. A disadvantage of this technique is that the PMOS load (Mr3) will operate as a current source at low frequencies, introducing high gain in the replica feedback loop. To prevent instability, a large compensation capacitor is required, which introduces another pole in the PLL, leading to more intricate design. Furthermore, the amplifier in the replica bias loop requires additional headroom, thereby prohibiting low-Vdd operation.

Fig. 5(b) shows a structure that achieves the good power-supply noise rejection of the analog differential pair, at the same time enabling low-Vdd operation. The PMOS diodes are used for clamping the output voltage to a minimum level of $Vdd - |V_{tp}|$, giving a fixed common mode and swing without the need for a replica bias circuit. This makes the VCO suitable for a wide range of operating frequencies and supply voltages. To guarantee clamping action, the NMOS tail current (I_T) must be larger than the current through the controlled PMOS load (I_L). Furthermore, a proposed design goal for

low oscillator noise suggests that the rise and fall times of the output nodes should be made equal [16]. This is achieved by reflecting half of I_T to each of the controlled PMOS loads by the current mirror formed by devices Md1–Md3. Assuming that Md4 recently turned on, "node a" will be discharged by a current of $I_T - I_L = I_T - I_T/2 = I_T/2$. At the same time, the complementary output node $\bar{a}$ is pulled to Vdd by a current equal to $I_L = I_T/2$, indicating equal rise and fall times. A disadvantage of this oscillator is the additional parasitic capacitance of the diodes, which makes the maximum operating frequency lower than that of the replica bias structure. The additional gate capacitance of the diode loads can be eliminated by using NMOS diodes [17].

The minimum supply voltage for the VCO is $Vdd = \max(V_{tn}, |V_{tp}|)$, which has been verified by measurements down to $Vdd = 0.9$ V. However, at this value of Vdd, the VCO is no longer differential. An estimate of the minimum Vdd for differential operation can be derived from the simulated VCO waveforms in Fig. 6. The VCO output swings from Vdd down to approximately $Vdd - |V_{tp}|$. For differential operation, it is required that both NMOS devices in the differential pair (Md4, Md5) are turned ON at the crossover point of the waveforms. Assuming a V_{DSsat} drop over the current source device Md6 leads to a minimum input voltage of $V_{DSsat} + V_{tn}$. At the lowest limit of Vdd, this input voltage is $Vdd - |V_{tp}|/2$ generated by the previous stage in the oscillator, indicating a minimum Vdd of $V_{tn} + |V_{tp}|/2 + V_{DSsat}$. Measurements determined V_{tn} and $|V_{tp}|$ to be 0.53 and 0.85 V, respectively, indicating a minimum Vdd of about 1.1 V assuming a V_{DSsat} of 0.1 V. Note that this is a theoretical number, since the differential operation of the VCO has zero tuning range at this value of Vdd. Good power-supply rejection can also be achieved by the regulated-supply structure in [18]. However, the requirement of a large decoupling capacitor generates contradicting design goals on PLL bandwidth.

V. CHARGE PUMP

A. Bandwidth and Peaking Compensation

To reduce peak-to-peak jitter due to VCO noise, it is advantageous to keep as high a PLL bandwidth as possible. Traditional worst case design would keep the PLL bandwidth and damping factor sufficiently far away from stability limits under all variations of the input reference frequency, the

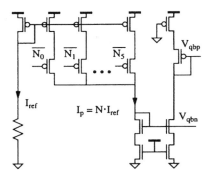

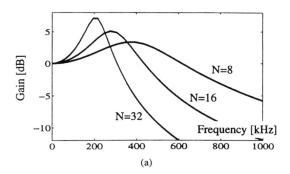

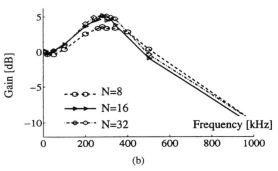

Fig. 7. Current multiplier generating charge-pump biasing voltages V_{qbn} and V_{qpb}.

manufacturing process, and the division ratio in the feedback path N. The concept of self-biasing introduced in [19] simplifies the design by eliminating process variations and the input reference frequency from the stability constraints. However, the PLL bandwidth is still a function of N, so that maximum noise suppression can only be achieved for a fixed N. In programmable applications, N can vary by more than an order of magnitude, indicating that the variation in stability constraint can be dominated by N instead of process variations, as shown by the stability limit of a charge-pump PLL [20]

$$f_{ref} \geq \frac{K_o \cdot I_p \cdot R}{N \cdot 4\pi} \qquad (1)$$

where f_{ref} is the input reference frequency (or effectively the sampling rate of the phase detector), K_o is the VCO gain, I_p is the charge-pump current, and R is the loop filter resistance. Other PLL design parameters, such as bandwidth and damping factor, also change with N. Compensating loop parameters for changes in N guarantees that the PLL is always operating with maximum bandwidth and fixed damping factor without endangering stability. This can be done by setting the charge pump current to

$$I_p = N \cdot I_{ref} \qquad (2)$$

where I_{ref} is a fixed reference current. This is realized by the current multiplier in Fig. 7, which generates the charge-pump current I_p by letting the individual bits of N control binary weighted current sources.

The simulated jitter transfer function of a standard PLL in Fig. 8(a) demonstrates the change of loop parameters as N is altered. The damping factor is intentionally set low to show its dependence on N. The measured jitter transfer function of Loop A in Fig. 8(b) shows the desired independence of N. The slight deviation of the curves is caused by transistor mismatch in the current multiplier.

B. Charge Sharing

A common problem of many charge pumps is charge sharing. For the charge pump in Fig. 9(a) (Type A), charge sharing is caused by the parasitic capacitance in nodes pcs and ncs [21]. When I_{Up} is active, node pcs is charged to Vdd. When deactivating I_{Up}, some of the charge stored in node pcs will leak through the current source device. Since the parasitics

Fig. 8. Jitter transfer functions for different division ratios. (a) Simulated standard PLL. (b) Measured characteristics of Loop A with intentionally low damping.

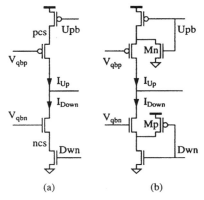

Fig. 9. (a) Charge-pump suffering from charge sharing (Type A). (b) Charge removal transistors eliminate charge sharing (Type B).

of nodes ncs and pcs can never be matched, this will lead to a static phase offset, as shown in Fig. 10(a). This is the transfer function of a phase-frequency detector followed by a Type A charge pump. The two transistors Mp and Mn in the Type B charge pump in Fig. 9(b) will remove the charge from the nodes pcs and ncs when Up and Down are deactivated [22]. This leads to a large reduction in the phase offset, as shown in Fig. 10(a).

For this application, static phase offset in Loop A is not critical. However, when analyzing the cause of phase offset, a source of increased jitter is revealed. Fig. 10(b) indicates that the leakage from node pcs is larger than that from ncs. When the PLL is locked, the leakage mismatch is compensated for by activating I_{Down} earlier than I_{Up}, giving a phase offset. Since the compensation charge is applied in the early portion of the charge-pump activation time, it will cause voltage ripple on

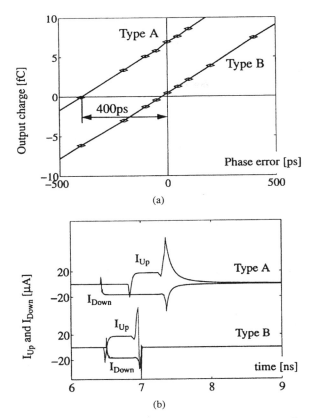

Fig. 10. Characteristics of the Type A and B charge pumps. (a) Transfer function of PFD followed by charge pump. (b) Simulated I_{Up} and I_{Down} when net output charge is zero.

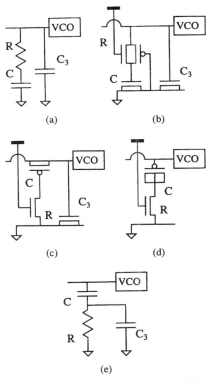

Fig. 11. Evolution of loop filter. (a) Ideal model, (b) MOS-only implementation, (c) improved resistor linearity for low-Vdd operation, (d) improved capacitor linearity, and (e) final model where C_3 models the well-to-substrate capacitance.

the loop filter, leading to phase jitter at the VCO output. The charge removal transistors Mn and Mp in Fig. 9(b) eliminate the current tails resulting in a well-balanced Up and Down activation time such that I_{Up} and I_{Down} cancel each other, reducing the loop filter ripple. A further advantage of the Type B charge pump is reduced $1/f$ noise in the current source transistors achieved by periodically resetting their V_{gs} to below 0 V [23], [24].

A limitation of the Type B charge pump is a reduced dynamic range of the VCO control voltage (V_{VCO}). If V_{VCO} is less than $V_{\mathrm{qbn}} - V_{\mathrm{tn}}$, there will be a current flowing through the Mp device to the output when the Dwn control is inactive. When NMOS devices are used for speed-regulating the VCO, V_{VCO} will never drop below V_{tn}, constraining V_{qbn} to be less than $2V_{\mathrm{tn}}$, which can easily be fulfilled. However, the charge pump works only up to an output voltage of $V_{\mathrm{VCO}} < V_{\mathrm{qbp}} + |V_{\mathrm{tp}}|$, limiting the upper tuning range of the VCO. However, the charge pump in Fig. 9(a) has the same upper voltage limit. Mismatch in I_{Up} and I_{Down} is a similar source of jitter as charge sharing described above. For low jitter, it is essential to have good matching, implying that the devices controlled by $V_{\mathrm{qbn}}/V_{\mathrm{qbp}}$ should be saturated. Again, this requires $V_{\mathrm{VCO}} < V_{\mathrm{qbp}} + |V_{\mathrm{tp}}|$.

Charge removal can also be done by ac coupling [18], but this requires careful timing of the control signals in the charge pump. The solution to charge sharing in [21] is less suitable for low-Vdd applications due to the common-mode restrictions on the differential amplifier.

VI. LOOP FILTER

The most common PLL loop filter is the simple RC circuit in Fig. 11(a). Common design options for the resistor are poly or the channel resistance of an MOS transistor. For high resistance values, an MOS device is most attractive. However, it has a disadvantage at low Vdd if implemented with the straightforward configuration of Fig. 11(b). For a nominal Vdd of 2.5 V, the effective resistance of the transmission gate is nearly independent of the VCO control voltage (V_{VCO}). However, the resistance becomes strongly dependent on V_{VCO} for low Vdd. For $Vdd = V_{\mathrm{tn}} + |V_{\mathrm{tp}}|$, the resistance goes to infinity for some values of V_{VCO} [25], [26]. Exchanging the position of the transmission gate resistor and the MOS capacitor as in Fig. 11(c) will make V_{gs} and the resistance of the NMOS device independent of V_{VCO}. The resistance still varies with Vdd, but the variation is much less than for the previous configuration.

Since the capacitor C in Fig. 11(c) is a "floating" capacitor, it must be implemented with a PMOS device. When the VCO control voltage approaches $|V_{\mathrm{tp}}|$, the MOS device is between inversion and depletion, where its capacitance value is voltage dependent, as shown in Fig. 12. By altering the gate and source/drain connections of the PMOS as shown in Fig. 11(d), it will operate in accumulation where the capacitance value is less voltage dependent, as shown for $V_{\mathrm{VCO}} > 0.5$ V in Fig. 12. To avoid strong power-supply noise injection, the well must be connected to the same node as source and drain, as shown in Fig. 11(d). The corresponding filter model is

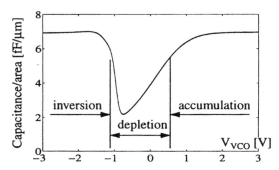

Fig. 12. Voltage-dependent capacitance of filter in Fig. 11(c) and (d).

shown in Fig. 11(e), where C_3 is the parasitic well-to-substrate capacitance of the MOS capacitor. This filter has an impedance of

$$Z_e = \frac{R \cdot (1 + C/C_3) + 1/sC_3}{sRC + C/C_3} \tag{3}$$

which is a close approximation to the impedance of the original filter in Fig. 11(a), given as

$$Z_a = \frac{R \cdot C/C_3 + 1/sC_3}{sRC + 1 + C/C_3} \tag{4}$$

when $C \gg C_3$, as is common design practice [20].

VII. PHASE-FREQUENCY DETECTOR

Phase detectors may exhibit a dead zone, resulting in enlarged jitter. A common design technique to avoid a dead zone is to make sure that both Up and Down output signals are fully activated before shutting them both off. This is implemented by generating a reset signal with an AND operation of Up and Down output and introducing a delay before feeding back this signal to reset the phase detector. It is this reset delay that causes the simultaneous I_{Up} and I_{Down} in Fig. 10(b). If the charge sharing in the charge pump is not perfectly cancelled or if there is a mismatch of I_{Up} and I_{Down}, there will always be some current compensation, leading to phase offset and loop filter ripple, as discussed in Section V. A longer reset delay results in a longer period during which the VCO is running at a different frequency due to the compensation current. Therefore, the reset delay should be minimized under the constraint that it has to be longer than the response time of the PFD with some additional design margin to avoid a dead zone.

A PFD with low logic depth is shown in Fig. 13, including details of the Up section. Its operation is easiest to analyze by assuming an initial state of $R = V = Up = Dwn = 0$. This implies that $Rstb = u1 = 1, u2 = 0$ and that $u3$ is precharged high. A rising edge on R discharges $u3$ and sets $Up = 1$ without changing the state of the RS flip-flop. The internal weak feedback in the Up path will assure that Up is kept active even if R falls. At the next rising edge on V, Dwn is activated, which sets $Rstb = 0$. This triggers the RS flip-flop to precharge $u3$ high, which shuts off Up; and, at the same time, Dwn is deactivated in a similar way.

In summary, a positive edge on R sets $Up = 1$, which is reset by the next positive edge on V. This behavior is identical

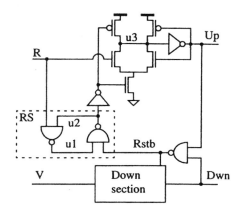

Fig. 13. Phase-frequency detector used in Loop A with details of Up section.

to the two classical PFD's implemented by either four RS flip-flops or two resettable D—flip-flops. The precharged gate and the shorter logic depth of this implementation make the delay shorter than for the standard PFD's. This allows a smaller delay in the reset path for eliminating the dead zone, such that loop filter ripple will be reduced and generate less noise. An additional benefit of low logic depth is a reduction in phase detector jitter caused by power-supply-dependent delays and device noise.

The reset delay of this PFD can be further reduced by letting the signal $Rstb$ directly reset the precharged gate simultaneously as the RS flip-flop is reset. This technique was not adopted in order to keep a conservative design, guaranteeing operation with no dead zone. Similar precharged gates have previously been used in PFD designs [27]–[29].

VIII. FREQUENCY DIVIDER

To enable high flexibility, the frequency divider in Fig. 2 is a fully programmable ($2 \leq N \leq 65$) divider. The structure in [30] based on a clock-gated dual-modulus prescaler followed by a counter was chosen to achieve high speed at low supply voltages. The divider was realized in standard static CMOS logic, reaching a maximum operating frequency of 800 MHz in simulations of worst case slow process variation at $Vdd = 1.6$ V and $T = 125 \,^\circ$C. This exceeded the simulated speed limit of the VCO. The potential startup deadlock in [30] was eliminated by logic that prohibits two consecutive clock pulse removals.

IX. PLL OPERATING RANGE AND JITTER

The maximum operating frequency of the PLL measured at room temperature is plotted in Fig. 14 as function of Vdd. Simulations indicate that the speed is limited by the VCO. A minimum Vdd of 0.9 V agrees well with the measured $|V_{tp}| = 0.85$ V. At low power-supply voltages, the speed cannot compare with high-end circuits using standard Vdd. However, the operating frequency range exceeds that of low-voltage circuit implementations [2], [3], [25], [26]. The maximum speed also compares favorably with another low-voltage PLL based on a low-threshold process [18].

With a PLL bandwidth of 2 MHz, the tracking jitter is 5.2 ps rms at 1200 MHz, as shown in Fig. 15(a). This measurement represents the standard deviation of the delay between a

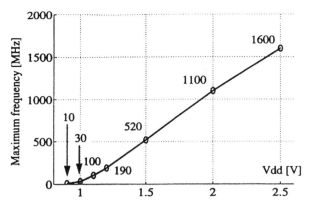

Fig. 14.   Measured maximum PLL operating frequency.

triggering clock edge and a clock edge occurring 320 ns later according to the setup in Fig. 2(e) in [31]. The delay was chosen four times larger than the delay at which the "jitter knee" occurs in Fig. 2(f) in [31] ($\tau_L = 1/2\pi f_L = 80$ ns) to get reliable data.

All signals on the chip are periodic with the reference frequency, so when using the frequency divider output as a triggering signal, most of the jitter due to power supply and board noise will cancel in the measurement. Such a setup gave an rms jitter of 2.5 ps at 1200 MHz, as shown in Fig. 15(b). The jitter with respect to an ideal reference would in this case be $2.5/\sqrt{2} = 1.8$ ps [31]. This proves that device noise in a standard ring oscillator is tolerable for communication standards with very tight jitter tolerances such as SONET OC-48 (2.5 Gb/s), which has a jitter specification of 4 ps rms at a 2-MHz PLL bandwidth. The VCO power consumption was 5 mW at 1200 MHz in simulation of extracted layout.

The figure of merit κ defined in [31] is estimated from

$$\kappa = \sigma_x \cdot \sqrt{4\pi \cdot f_L} = t_{\mathrm{rms}} \cdot \sqrt{2\pi \cdot f_L} \qquad (5)$$

where f_L is the PLL bandwidth, t_{rms} is the measured long-term self-referenced tracking jitter, and σ_x is the tracking jitter with respect to an ideal reference clock. Table I lists measured σ_x and derived κ as function of operating frequency. The jitter reported here is lower than that in [32] due to an improved measurement setup and more accurate measurement equipment. The κ of this oscillator is better than that reported for bipolar implementations in [31] ($\kappa \approx 40e - 9\sqrt{s}$) and a CMOS VCO with a κ of $20e - 9\sqrt{s}$ (as derived from the Slide Supplement of [33]). A $\kappa = 9e - 9\sqrt{s}$ for a complete PLL is similar to $\kappa = 6e - 9\sqrt{s}$ reported for a stand-alone VCO in [34], suggesting that noise contributions from the other PLL components (charge pump, PFD, frequency divider) are much smaller than the VCO noise. The tracking jitter compares favorably with several other CMOS oscillators [e.g., [8], [33], and [35]–[37]]. κ degrades significantly at 1600-MHz operation, indicating the speed limit of the PLL. The t_{rms} measurements are also used for estimating the VCO period jitter (σ_{VCO}) due to device noise, based on the relation [31]

$$\sigma_{\mathrm{VCO}} = \kappa \cdot \sqrt{T_{\mathrm{VCO}}} \qquad (6)$$

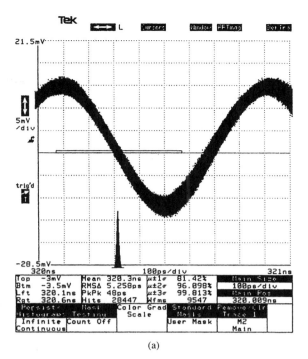

(a)

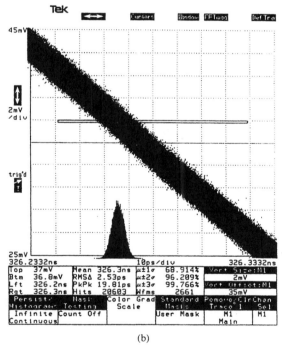

(b)

Fig. 15.   A 1.2-GHz tracking jitter histogram. (a) Including power supply and board noise. (b) Supply and board noise cancelled.

where T_{VCO} is the VCO period. The derived period jitter agrees to within 10% of the phase noise estimation technique in [38].

The PLL tracking jitter is plotted as a function of frequency for various power-supply voltages in Fig. 16. These measurements were taken with $N = 8$ and a fixed PLL bandwidth. Since the loop filter resistance changes with Vdd, the charge-pump reference current I_{ref} in Fig. 7 was adjusted until a 3-dB PLL bandwidth of 2 MHz was measured. A bandwidth of 2

TABLE I
PLL JITTER FOR $VDD = 2.5$ V MEASURED WITH A PLL BANDWIDTH OF 2 MHz

Frequency [MHz]	Measured	Derived from measurements	
	Tracking jitter* rms/pkpk [ps]	κ^{**} [$\sqrt{s}$]	σ_{vco}^{***} [ps]
250	8.3/55	42e-9	2.7
500	4.5/35	23e-9	1.0
1000	1.9/17	10e-9	0.32
1250	1.8/14	9e-9	0.25
1600	8.5/68	43e-9	1.1

* Same as σ_x in [31]. ** Figure of merit defined in [31]. *** Period jitter.

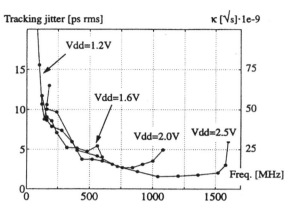

Fig. 16. Tracking jitter as function of Vdd and frequency with a PLL bandwidth of 2 MHz. The right-hand scale represents the figure of merit κ [31].

TABLE II
PLL CHARACTERISTICS

Process	0.25μm, 2.5V digital CMOS
Measured $V_{tn}/|V_{tp}|$	0.53/0.85 V
Vdd range	0.9V - 2.5V
Frequency range	0.4 - 190MHz at Vdd=1.2V 0.6 - 520MHz at Vdd=1.5V 1.1 - 1100MHz at Vdd=2.0V 1.3 - 1600MHz at Vdd=2.5V
Jitter	See Table I and Fig. 16.
Multiplication mode	N = 2 to N = 65
Number of clock phases	8/16/32/64/128 (programmable)
Total power dissipation VCO power	18mW at 250MHz (Vdd=2.5V, N=25) 0.7mW (from simulation)
Area	300μm × 120μm = 0.036 mm^2
Package	44-pin TQFP, 4 pins Vdd, 4 pins Vss

MHz represents a scaling factor of approximately 5000 in (5), as indicated by the right-hand scale in Fig. 16.

The VCO frequency is set by the tail current in the differential delay stages and is practically independent of Vdd. Therefore, the power consumption at a fixed VCO frequency drops linearly with Vdd. The graph in Fig. 16 shows that the jitter does not change with Vdd, suggesting that power reduction (by lowering Vdd) can be achieved without jitter penalty. This seems to contradict the common belief that jitter should increase with lower power consumption. However, the critical parameter for low jitter is not power consumption but current consumption, as has previously been theoretically derived for LC oscillators [39], [40]. As shown by these measurements, low-jitter design with a fixed power budget should be based on a minimum Vdd and as large a current as can be tolerated. The measured power consumption at 250 MHz and 2.5 V is 18 mW and is dominated by buffers and the current-mode interpolator. Simulations of extracted layout indicate that the VCO consumes 0.7 mW at 250 MHz and 5 mW at 1200 MHz. The PLL characteristics are summarized in Table II.

X. CONCLUSION

Clock recovery circuits in CMOS processes require special techniques for initial frequency locking. This need is due to the fact that CMOS process variations dictate a larger frequency tuning range than can be covered by existing frequency detectors for NRZ data. An attractive technique for initial

frequency locking is phase selection clock recovery, where a multioutput VCO is locked onto a reference clock and a clock recovery loop selects one of the output phases of the VCO. The large period jitter in traditional phase selection clock recovery is eliminated by the feedback phase selection technique presented here. This scheme filters the phase jumps through the PLL loop filter and also enables accurate phase interpolation with digital circuitry only, as opposed to the conventional analog-style phase interpolation.

In applications where the PLL is programmable, important loop characteristics such as bandwidth and damping factor change with the frequency multiplication mode. By making the charge-pump current depend on the division ratio in the feedback divider, a fixed bandwidth and damping factor can be obtained.

Differential analog circuits have superior supply noise rejection compared to digital complementary logic styles and are therefore preferred in an environment with large power-supply noise. However, previous differential PLL implementations have used circuits requiring large headroom, thereby prohibiting low-Vdd operation. Circuit techniques for PLL components are discussed that enable low-Vdd operation in a process technology without low-threshold devices. Correct operation has been verified down to $Vdd = 0.9$ V, but the lower limit for differential operation is estimated to be $Vdd = V_{tn} + |V_{tp}|/2 + V_{DSsat} \approx 1.1$ V in a process with measured $|V_{tp}| = 0.85$ V and $V_{tn} = 0.53$ V.

Measurements show that jitter is independent of Vdd, contradicting the common belief that jitter is strongly correlated to power consumption. At a fixed operating frequency, power reduction is achieved without any penalty in jitter performance by lowering Vdd.

The tracking jitter at 500–1500 MHz was measured to be 2–5 ps rms dominated by device noise. This indicates that a standard ring oscillator can fulfill the jitter specification for a SONET OC-48 receiver.

This paper demonstrates that a clock recovery circuit with programmable bit rates can be realized with a large frequency tuning range. Robust operation and low jitter are achieved over a large range of power-supply voltages, making it ideal for low-power applications and suitable as a reusable macrocell.

REFERENCES

[1] K. Bult, "Analog broadband communication circuits in pure digital deep sub-micron CMOS," in *Proc. IEEE Int. Solid-State Circuits Conf.*, 1999, pp. 76–77.

[2] H. Neuteboom, B. M. J. Kup, and M. Janssens, "A DSP-based hearing instrument IC," *IEEE J. Solid-State Circuits*, vol. 32, pp. 1790–1806, Nov. 1997.

[3] W. Lee, P. E. Landman, B. Barton, S. Abiko, H. Takahashi, H. Mizuno, S. Muramatsu, K. Tashiro, M. Fusumada, L. Pham, F. Boutaud, E. Ego, G. Gallo, H. Tran, C. Lemonds, A. Shih, M. Nandakumar, R. H. Eklund, and I. C. Chen, "A 1-V programmable DSP for wireless communication," *IEEE J. Solid-State Circuits*, vol. 32, pp. 1766–1776, Nov. 1997.

[4] F. M. Gardner, *Phase-Lock Techniques.* New York: Wiley, 1979.

[5] R. J. Baumert, P. C. Metz, M. E. Pedersen, R. L. Pritchett, and J. A. Young, "A monolithic 50–200 MHz CMOS clock recovery and retiming circuit," in *Proc. IEEE Custom Integrated Circuits Conf.*, 1989, pp. 14.5.1–4.

[6] K. M Ware and C. G. Sodini, "A 200-MHz CMOS phase-locked loop with dual phase detectors," *IEEE J. Solid-State Circuits*, vol. 24, pp. 1560–1568, Dec. 1989.

[7] J. Sonntag and R. Leonowich, "A monolithic CMOS 10 MHz DPLL for burst-mode data retiming," in *Proc. IEEE Int. Solid-State Circuits Conf.*, 1990, pp. 194–195.

[8] M. Horowitz, A. Chan, J. Cobrunson, J. Gasbarro, T. Lee, W. Leung, W. Richardson, T. Thrush, and Y. Fujii, "PLL design for a 500 MB/s interface," in *Proc. IEEE Int. Solid-State Circuits Conf.*, 1993, pp. 160–161.

[9] S. Sidiropoulos and M. A. Horowitz, "A semidigital dual delay-locked loop," *IEEE J. Solid-State Circuits*, vol. 32, pp. 1683–1692, Nov. 1997.

[10] S. Kasturia, "A novel fractional divider for improving the switching speed of phase-locked frequency synthesizers," Bell Labs Tech. Memo., May 1995.

[11] J. G. Maneatis, personal communication, Feb. 1999.

[12] T. H. Lee, K. S. Donnelly, J. T. C. Ho, J. Zerbe, M. G. Johnson, and T. Ishikawa, "A 2.5 V CMOS delay-locked loop for an 18 Mbit, 500 megabytes/s DRAM," *IEEE J. Solid-State Circuits*, vol. 29, pp. 1491–1496, Dec. 1994.

[13] T. H. Hu and P. R. Gray, "A monolithic 480 Mb/s parallel AGC/decision/clock-recovery circuit in 1.2 μm CMOS," *IEEE J. Solid-State Circuits*, vol. 28, pp. 1314–1320, 1993.

[14] C.-H. Park and B. Kim, "A low-noise 900 MHz VCO in 0.6 μm CMOS," *IEEE J. Solid-State Circuits*, vol. 34, pp. 1586–1591, May 1999.

[15] J. Lee and B. Kim, "A 250 MHz low jitter adaptive bandwidth PLL," in *Proc. IEEE Int. Solid-State Circuits Conf.*, 1999, pp. 346–347.

[16] A. Hajimiri and T. H. Lee, "A general theory of phase noise in electrical oscillators," *IEEE J. Solid-State Circuits*, vol. 33, pp. 179–195, Feb. 1998.

[17] K. Iravani, F. Saleh, D. Lee, P. Fung, P. Ta, and G. Miller, "Clock and data recovery for 1.25 Gb/s Ethernet transceiver in 0.35 μm CMOS," in *Proc. Custom Integrated Circuits Conf.*, 1999, pp. 261–264.

[18] V. von Kaenel, D. Aebisher, C. Piguet, and E. Dijkstra, "A 320 MHz, 1.5 mW at 1.35 V CMOS PLL for microprocessor clock generation," in *Proc. IEEE Int. Solid-State Circuits Conf.*, 1996, pp. 132–133.

[19] J. G. Maneatis, "Low-jitter process-independent DLL and PLL based on self-biased techniques," *IEEE J. Solid-State Circuits*, vol. 31, pp. 1723–1732, Nov. 1996.

[20] F. M. Gardner, "Charge-pump phase-lock loops," *IEEE Trans. Communications*, vol. COM-28, pp. 1849–1858, Nov. 1980.

[21] M. Johnson and E. Hudson, "A variable delayline PLL for CPU-coprocessor synchronization," *IEEE J. Solid-State Circuits*, vol. SC-23, pp. 1218–1223, Oct. 1988.

[22] P. Larsson and J.-Y. Lee, "A 400 mW 50–380 MHz CMOS programmable clock recovery circuit," in *Proc. IEEE ASIC Conf. Exhibit*, 1995, pp. 271–274.

[23] I. Bloom and Y. Nemirovsky, "1/f noise reduction of metal-oxide-semiconductor transistors by cycling from inversion to accumulation," *Appl. Phys. Lett.*, vol. 58, no. 15, pp. 1664–1666, Apr. 1991.

[24] S. L. J. Gierkink, E. A. M. Klumperink, T. J. Ikkink, and A. J. M. van Tuijl, "Reduction of intrinsic 1/f device noise in a CMOS ring oscillator," in *Proc. IEEE European Solid-State Circuits Conf.*, 1998, pp. 272–275.

[25] J. Crols and M. Steyeart, "Switched-opamp: An approach to realize full CMOS switched-capacitor circuits at very low power supply voltages," *IEEE J. Solid-State Circuits*, vol. 29, pp. 936–942, Aug. 1994.

[26] A. M. Abo and P. R. Gray, "A 1.5 V, 10-bit, 14 MS/s CMOS pipeline analog-to-digital converter," *IEEE J. Solid-State Circuits*, vol. 34, pp. 599–606, May 1999.

[27] S. Kim, K. Lee, Y. Moon, D-K. Jeong, Y. Choi, and H. K. Lim, "A 960-Mb/s/pin interface for skew-tolerant bus using low jitter PLL," *IEEE J. Solid-State Circuits*, vol. 32, pp. 691–700, May 1997.

[28] D. W. Boerstler and K. A. Jenkins, "A phase-locked loop clock generator for a 1 GHz microprocessor," in *Proc. IEEE Symp. VLSI Circuits*, 1998, pp. 212–213.

[29] H. O. Johansson, "A simple precharged CMOS phase frequency detector," *IEEE J. Solid-State Circuits*, vol. 33, pp. 295–298, Feb. 1998.

[30] P. Larsson, "High-speed architecture for a programmable frequency divider and a dual-modulus prescaler," *IEEE J. Solid-State Circuits*, vol. 31, pp. 744–748, May 1996.

[31] J. A. McNeill, "Jitter in ring oscillators," *IEEE J. Solid-State Circuits*, vol. 32, pp. 870–879, June 1997.

[32] P. Larsson, "A 2–1600 MHz 1.2–2.5 V CMOS clock-recovery PLL with feedback phase-selection and averaging phase-interpolation for jitter reduction," in *Proc. IEEE Int. Solid-State Circuits Conf.*, 1999, pp. 356–357.

[33] J. F. Ewen *et al.*, "Single-chip 1062 Mbaud CMOS transceiver for serial data communication," in *Proc. IEEE Int. Solid-State Circuits Conf.*, 1995, pp. 32–33.

[34] A. Hajimiri, S. Limotyrakis, and T. H. Lee, "Jitter and phase noise in ring oscillators," *IEEE J. Solid-State Circuits*, vol. 34, pp. 790–804, June 1999.

[35] I. Novof, J. Austin, R. Chmela, T. Frank, R. Kelkar, K. Short, D. Strayer, M. Styduhar, and S. Watt, "Fully-integrated CMOS phase-locked loop with 15–240 MHz locking range and 50 ps jitter," in *Proc. IEEE Int. Solid-State Circuits Conf.*, 1995, pp. 112–113.

[36] Z.-X. Zhang, H. Du, and M. S. Lee, "A 360 MHz 3 V CMOS PLL with 1 V peak-to-peak power supply noise tolerance," in *Proc. IEEE Int. Solid-State Circuits Conf.*, 1996, pp. 134–135.

[37] I. A. Young, M. F. Mar, and B. Bhushan, "A 0.35 μm CMOS 3–880 MHz PLL N/2 clock multiplier and distribution network with low jitter for microprocessors," in *Proc. IEEE Int. Solid-State Circuits Conf.*, 1997, pp. 330–331.

[38] A. Demir, A. Mehrotra, and J. Roychowdhur, "Phase noise in oscillators: A unifying theory and numerical methods for characterization," in *Proc. ACM/IEEE Design Automation Conf.*, June 1998, pp. 26–31.

[39] Q. Huang, "On the exact design of RF oscillators," in *Proc. IEEE Custom Integrated Circuits Conf.*, 1998, pp. 41–44.

[40] P. Kinget, "Integrated GHz voltage controlled oscillators," in *Proc. Advances in Analog Circuit Design*, Nice, France, Mar. 1999.

Patrik Larsson received the Ph.D. degree from Linkoping University, Sweden, in 1995.

During his Ph.D. research, he investigated the inherent analog properties of digital circuits, such as di/dt noise, clock skew, and clock slew rate. After graduation, he joined Bell Laboratories, where he is currently working on VCO's and PLL's for gigabit/second communication. He has also been working on low-power digital filtering for cable modems, equalization, and clock recovery structures while maintaining his interest in di/dt noise.

SiGe Clock and Data Recovery IC with Linear-Type PLL for 10-Gb/s SONET Application

Yuriy M. Greshishchev, *Member, IEEE*, and Peter Schvan, *Member, IEEE*

Abstract—An integrated 10 Gb/s clock and data recovery (CDR) circuit is fabricated using SiGe technology. It consists of a linear-type phase-locked loop (PLL) based on a single-edge version of the Hogge phase detector, a *LC*-tank voltage-controlled oscillator (VCO) and a tri-state charge pump. A PLL equivalent model and design method to meet SONET jitter requirements are presented. The CDR was tested at 9.529 GB/s in full operation and up to 13.25 Gb/s in data recovery mode. Sensitivity is 14 mV$_{\text{pp}}$ at a bit error rate (BER) $= 10^{-9}$. The measured recovered clock jitter is less than 1 ps rms. The IC dissipates 1.5 W with a -5-V power supply.

Index Terms—Charge pump, clock and data recovery (CDR), jitter generation, jitter tolerance, jitter transfer, phase detector, phase-locked loop (PLL), SONET, VCO.

I. INTRODUCTION

IN A CLOCK and data recovery (CDR) circuit with integrated phase-locked loop (PLL), the reference clock is extracted from the incoming data stream and is automatically aligned to the center of the data pulse independent of its pattern. Two CDR ICs have been reported operating at 10 Gb/s: one with a linear PLL (LPLL) using a modified Hogge-type phase detector (PD) [1] and the other with a binary PLL using a bang-bang (Alexander)-type PD [2].[1] CDR jitter characteristics critical to SONET optical receiver design are *Jitter Transfer* (*Bandwidth* and *Jitter Peaking*), *Jitter Tolerance*, and *Jitter Generation*. In a linear CDR (LCDR), jitter transfer characteristics are independent of the jitter amplitude and can be analytically predicted according to a LPLL theory. This feature can be important for SONET applications, particularly if data is to be retransmitted and jitter transfer must be well controlled.

This paper describes a 10 Gb/s LCDR with less than 1 ps rms pattern-independent jitter generation required for SONET application [3]. This jitter generation represents the best published CDR result. A number of techniques have been used to achieve low jitter. First, a charge pump with tri-state is employed. This is a well known technique frequently used in conjunction with frequency-PD PLLs [4]. This technique is also called switched-filter PLL [5]. Generally, this approach provides a hold mode in the PLL filter in the absence of data transitions and prevents variation of the voltage-controlled oscillator

(VCO) frequency (that causes jitter) during a long run of data 0 s or 1 s. Second, charge-pump and VCO control circuits were designed to provide a high degree of PLL filter isolation, or low charge-pump offset current, in a tri-state. In addition, the charge pump has a high output impedance necessary for high loop gain in a PLL with passive filter. Third, the original Hogge PD was modified to provide a single-edge operation and to extend linear phase range. Fourth, circuit and layout cross-talk isolation techniques similar to those presented in [6] are employed to prevent jitter generation and sensitivity degradation due to a cross-talk. The CDR IC was implemented in IBM's SiGe bipolar process which includes pMOS devices.

Jitter characteristics of a LPLL depend to a large degree on the PLL filter parameters. An LPLL equivalent model and design method to satisfy SONET requirements are presented in this paper. A theoretical jitter tolerance function is introduced based on considerations similar to those presented in [7]. It is shown that in a LPLL, all of the jitter characteristics specified by SONET requirements can be analytically expressed via jitter transfer bandwidth F_{BW} and PLL damping factor ζ. The PLL bandwidth F_{BW} should be above 4 MHz to satisfy OC 192 SONET jitter tolerance mask requirement, and the damping factor should be above $\zeta = 4$–6 to satisfy 0.1-dB jitter peaking.

In Section II the CDR architecture is described with an attention to low jitter operation and the LPLL equivalent model and its design method are considered. Then, in Section III the building-blocks circuit diagrams are discussed. The CDR hierarchical simulation flow is given in Section IV. Finally, the IC implementation details and measured results are presented in Sections V and VI.

II. CDR ARCHITECTURE

A. The Architecture

The CDR architecture includes a single-edge Hogge-type PD with a decision circuit, a charge pump, an integrated LC-tank VCO and a passive second-order PLL filter, as shown in Fig. 1. These four components constitute the LPLL. To minimize jitter, a tri-state PD and charge pump are used. The charge pump produces close to zero differential output current, when no data transitions occurs. The CDR is fully differential. Maximum differential input voltage of the CDR is 1 V$_{\text{pp}}$. Two differential threshold slicing inputs $TRSH\_D$ and $TRSH\_C$ are used to optimize the threshold of the decision and the clock recovery circuits within 80% of the data swing. The recovered clock and data are transmitted to the corresponding outputs $RecData$ and $RecCLK$ via buffers B and cross-talk isolation interface (transmission lines and transmitters Tx) [6]. The amplitudes

Manuscript received December 8, 1999; revised February 2, 2000.
The authors are with Nortel Networks, Ottawa, ON K1Y 4H7, Canada.
Publisher Item Identifier S 0018-9200(00)05928-X.

[1]The linear property of a LPLL is due to a linear phase response of the Hogge-type PD. In publication [1], [4] the Hogge-type PD is called a *phase comparator*, which is a misleading terminology since the word *comparator* implies a binary output. The bang-bang (Alexander)-type PD is a true phase comparator.

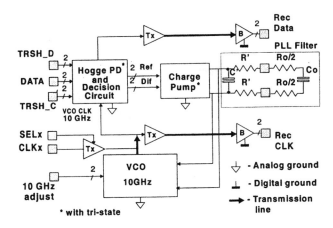

Fig. 1. CDR IC architecture.

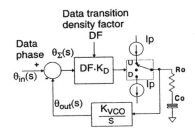

Fig. 2. Equivalent model of the linear PLL.

of $RecData$ and $RecCLK$ are 1 V_{pp} differential. The IC can also be used in a data recovery mode only, when the VCO clock is overdriven with an external signal $CLKx$.

B. PLL Equivalent Model for CDR Jitter Characteristics Analyses

Three types of jitter characteristics are important in SONET receiver design: jitter transfer function (bandwidth and jitter peaking), jitter tolerance and jitter generation [8]. The PLL bandwidth was specified to be between 4–10 MHz with less than 0.1-dB jitter peaking and jitter generation below 1 ps rms. The PLL was analytically designed using a continuous time approximation for the equivalent model of Fig. 2 [9]. A damping factor above $\zeta = 4$–6 is required to provide low jitter peaking. Due to overdamping, the PLL jitter transfer $JTRAN(f)$ approaches a single-pole low-pass-type response with the following parameters (see Appendix A):

$$F_{BW} = \zeta \cdot \frac{\omega_n}{\pi} \qquad (1)$$

$$\zeta = \frac{R_o \cdot C_o}{2} \cdot \omega_n \qquad (2)$$

$$\omega_n = \sqrt{K_{fVCO} \cdot \frac{2 \cdot I_p \cdot DF}{C_o}} \qquad (3)$$

where F_{BW} is 3-dB bandwidth of the PLL jitter transfer function in Hz, K_{fVCO} is VCO sensitivity in Hz/V, ω_n is the loop natural frequency, and DF is the average data transition density factor (maximum $DF = 1$ for 0101 ... pattern, $DF = 1/2$ for PRBS pattern). In (3) the charge pump current I_p is doubled, compared to the Gardner's formula [9], since in a Hogge PD a current variation of $2 \cdot I_p$ corresponds to 2π-radians of the data phase. Both the bandwidth F_{BW} and the damping factor ζ are

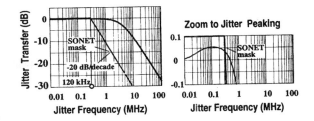

Fig. 3. CDR analytical jitter transfer compared to a SONET mask.

functions of DF: $F_{BW} \propto DF$, $\zeta \propto \sqrt{DF}$. The CDR circuit was designed for $DF > 1/5$. Fig. 3 shows analytically calculated $JTRAN(f)$ with $F_{BW} = 4$ MHz compared to OC192 SONET mask. The bandwidth does not satisfy SONET requirement (to be below 120 KHz), but the jitter peaking is within the recommended 0.1-dB jitter gain.

The CDR jitter tolerance is a measure of how much peak-to-peak sinusoidal jitter can be added to the incoming data before causing data errors[2] due to misalignment of the data and the recovered clock. In a CDR with LPLL, jitter tolerance is defined by a jitter transfer function and the PLL slew-rate capabilities [7]. Considering no slew rate limitations in the loop (this is usually the case in a well designed CDR), the frequency response of the jitter tolerance can be described by the following function:

$$JTOL(s) = \frac{1}{1 - JTRAN(s)}. \qquad (4)$$

$JTOL(s)$ determines the shape of the jitter tolerance response. It can be used to compare the performance of the design with SONET jitter tolerance mask if $JTOL(s)$ is multiplied by the jitter tolerance $JTOL(f_t)$ at high frequencies $f > f_t$. The value of $JTOL(f_t)$ is defined by the CDR circuit design and decision-circuit clock-phase margin. For the SONET OC192 mask it is specified to be more than 15 ps$_{pp}$ at $f_t = 4$ MHz [8]. Fig. 4 shows analytically calculated $JTOL(f)$ for $F_{BW} = 4$ MHz compared to the OC192 SONET mask. The dotted line is the response with asymptotic single-pole jitter transfer function [(A.4) in Appendix A]. The 20-dB/decade slope of the mask in the frequency range of 0.4–4 MHz coincides with the $JTOL(f)$ response for $F_{BW} = 4$ MHz. SONET compliant LPLL design must have a bandwidth $F_{BW} > 4$ MHz for minimum data transition density $DF = DF_{\min}$.

Several factors affect jitter generation of the CDR shown in Fig. 1. The recovered clock central frequency value is held as a voltage on the capacitor Co. An offset current I_{off} at the charge pump output in tri-state causes a frequency step

$$\Delta F_o = I_{off} \cdot R_0 \cdot K_{fVCO}. \qquad (5)$$

For practical filter parameters and expected maximum time interval with no data transitions, the voltage variation across capacitor C_0 is negligible compared to the voltage step $I_{off} \cdot R_0$. The phase jitter ΔTo is proportional to the number N of consecutive 0's or 1's in the data, as follows:

$$\Delta To \text{ [ps]} = 0.01 \cdot N \cdot \Delta F_o \text{ [MHz]}. \qquad (6)$$

[2]The amount of errors is defined with 1-dB receiver input power penalty.

667

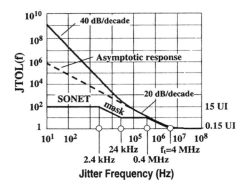

Fig. 4. CDR analytical jitter tolerance compared to a SONET mask.

Combining (1)–(6) the jitter can be expressed as follows:

$$\Delta To \text{ [ps]} = 0.02 \cdot \delta_o \cdot N \cdot \pi \cdot F_{BW} \text{ [MHz]} \cdot (DF_{\min})^{-1} \quad (7)$$

where δ_o is the relative current offset at the charge pump output and F_{BW} [MHz] is the PLL bandwidth [MHz] at minimum expected data transition density factor $DF = DF_{\min}$.

Instantaneous voltage drop $I_p \cdot R_o$ across the damping resistor creates the well known frequency ripple. Peak-to-peak jitter associated with this ripple can be expressed as

$$\Delta T_{\text{ripple}} \text{ [ps]} = 0.01 \cdot K_{HOP} \cdot \pi \cdot F_{BW} \text{ [MHz]} \cdot (DF_{\min})^{-1} \quad (8)$$

where K_{HOP} is the attenuation due to high order poles in the PLL filter. Lower targeted values of $DF_{\min}$ increase jitter generation. The single-edge PD, used in this CDR, reduces DF by a factor of 2 and doubles the jitter amplitude. Because of the required high bandwidth F_{BW}, attention must be paid to the charge-pump offset δ_o and to the attenuation K_{HOP} to keep jitter generation in the sub-picosecond range.

Jitter can also be generated by the loop static phase error and its pattern dependence. To minimize static error, a charge pump with high output impedance and a VCO with high control input impedance were employed. The VCO phase noise had little impact on the recovered clock jitter because of the high loop gain and wide loop bandwidth F_{BW} achieved.

C. PLL Design Method

The LPLL with previously described model is fully defined by five independent parameters: K_{fVCO}, I_p, F_{BW}, ζ, and $DF_{\min}$. Filter components Ro and Co are functions of these parameters and can be found from (1)–(3) (see Appendix B). Parameters K_{fVCO} and I_p are mostly constrained by circuit implementation. Their initial values do not impact, in the first order, LPLL jitter generation as is seen in (7) and (8). Parameters ζ and F_{BW} are specified at $DF_{\min} = 1/5$ (or $1/10$ accounting for single-edge phase detection): $\zeta = 4$ to satisfy jitter peaking and $F_{BW} = 4$ MHz to satisfy jitter tolerance mask.

III. CIRCUIT DESIGN

A. Phase Detector and Decision Circuit

The block diagram in Fig. 5 shows the PD and decision circuit. The data decision circuit is split with the clock recovery

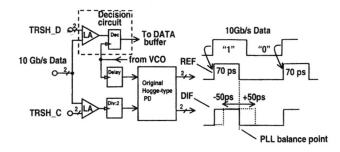

Fig. 5. PD and decision circuit.

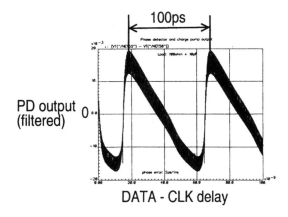

Fig. 6. PD simulated response.

circuit to provide independent data threshold optimization. A divide-by-two circuit $Div:2$ results in a single-edge operation of the original Hogge-type PD [10]. Therefore the recovered clock jitter is not affected by possible asymmetry between the rising and falling edges of the incoming data. A dummy latch circuit $Delay$ is introduced to compensate for the $Div:2$ delay. An additional advantage of the single-edge operation is an extended linear phase range, which is explained in Fig. 5. The REF output provides a phase-independent reference signal with a constant pulse width of about 70 ps at each positive data transition. The DIF output is the phase difference signal in the form of a variable pulse width of 70 ±50 ps. Fig. 6 shows SPICE simulation results of the PD circuit phase response. The linear phase range is about 80 ps. In the absence of data transitions both outputs REF and DIF are at a low level. This is detected as a tri-state by the charge pump.

The front-ends of the decision circuit and the PD contain limiting amplifiers LA with differential slicing level control at the input. To increase the time resolution of the decision circuit, a *master–slave–master* structure is employed in the retiming block Dec. The sensitivity of the decision circuit, defined primarily by thermal and shot noise in the input slicing circuitry, is simulated to be 13.4 mV$_{pp}$ at BER = 10^{-9}. The simulated latching metastability region is less than 1 ps$_{pp}$. This region is determined as a time zone in the clock-delay sweep where the output of the decision circuit is not defined.

B. LC-Tank VCO

The block diagram of the LC-tank-based VCO is shown in Fig. 7(a). The 10-GHz oscillator core is a cross-coupled differential circuit [Fig. 7(b)] [11]. The VCO includes a differential

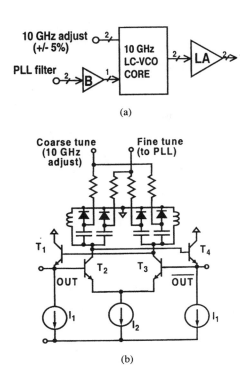

Fig. 7. VCO. (a) Block diagram. (b) *LC*-VCO core.

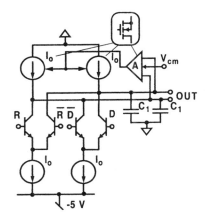

Fig. 8. PLL charge pump with a tri-state.

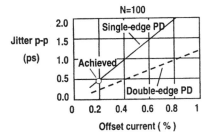

Fig. 9. Jitter generation due to offset current in the charge-pump tri-state.

control buffer B and a pulse-edge sharpening limiting amplifier LA to reduce jitter sensitivity to the cross-talk noise. The LA has an open collector output to drive the transmission line interface with a 50-Ω termination at the far end. The control buffer is designed with pMOS source followers at the input. VCO frequency is tuned with a varactor which is split into two parts: a coarse tune (to compensate for process variation) and a fine tune for frequency control in the loop. VCO phase noise is measured to be less than -80 dBc/Hz at a 100-kHz offset frequency.

C. Charge Pump

The charge pump (Fig. 8) employs a well known current-switching technique with the addition of a common-mode feedback amplifier A. Care was taken to achieve unconditional stability in the feedback with sufficient gain and with a small value of capacitance C_1 in Fig. 8. Small C_1 is necessary for low jitter peaking in the PLL. The charge-pump output differential current $I_p = 0.5 \cdot I_o$, as accounted for by the model in Fig. 2.

The charge pump is in a tri-state when both differential inputs R and D are switched into a low (or high) state. A mismatch between charge-pump current sources, I_o, their finite output resistances, and the VCO control input current cause an offset current in the tri-state. Fig. 9 shows a plot of the PLL jitter due to relative offset current δ_o in the tri-state as calculated from (7) for $N = 100$. Single-edge phase detection, employed in the CDR, requires half the δ_o value compared to the double-edge Hogge-type PD. The top current sources and the feedback amplifier were designed with pMOS transistors. Appropriate matching was achieved by sizing the critical components and using symmetrical layout. To increase the charge-pump output impedance, cascode current sources were employed. The measured offset δ_o was less than 0.2%.

D. PLL Filter

The PLL filter is split into internal C' and R', and external Ro and Co components (see Fig. 1). Resistors R' make jitter performance less sensitive to the external parasitics and coupled noise. Capacitor C' (along with C_1 in Fig. 8) performs smoothing of the PD output pulses and introduces the required attenuation K_{HOP} used in (8). In the PLL, resistor Ro is limited by maximum voltage drop required for normal circuit operation. Pulse smoothing also relaxes this constraint.

E. Cross-Talk Isolation

Two differential output buffers (B in Fig. 1) provide adjustable differential voltage swing up to 1 V$_{pp}$. The buffers are physically separated from the VCO and PD with transmission line interfaces to prevent jitter generation due to cross-talk via substrate and common grounds. The VCO is also separated from the PD with similar transmission line interface. All of the blocks have separate power-supply systems routed according to the isolation and analog–digital ground splitting techniques described in [6]. All of the CDR circuits are fully differential. The 10 Gb/s inputs and outputs are terminated on-chip with 50-Ω resistors.

IV. SIMULATION

Five levels of hierarchical PLL analysis were carried out: analytical, behavioral linear, behavioral mixed-mode, circuit schematic level, and post layout with distributed parasitics. The last four levels are HSPICE-based. A mixed-mode behavioral library of linear and digital components was developed. All levels of simulation give consistent results, with increasing

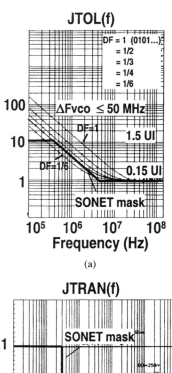

JTOL(f)

(a)

JTRAN(f)

(b)

Fig. 10. HSPICE simulated jitter characteristics. (a) Jitter tolerance. (b) Jitter transfer.

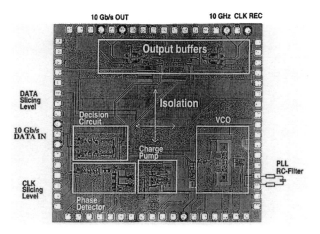

Fig. 11. Microphotograph of the SiGe CDR with linear PLL.

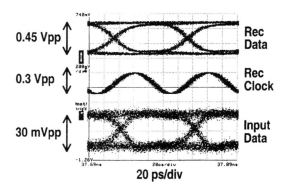

Fig. 12. CDR 9.529-Gb/s eye diagrams and the recovered clock. Input data 30 mV_{pp}, $2^{23} - 1$ PRBS pattern.

insight into jitter behavior at more detailed levels. Analytical models of jitter transfer and jitter tolerance are based on the second-order linear PLL theory as described in Section II. With the addition of C' and C_1, the PLL becomes a third-order loop. This was simulated along with on-chip, in-package, and external filter parasitics using HSPICE-based models. AC simulation results for DF ranging from 1/6 to 1 are shown in Fig. 10(a) and (b). Jitter tolerance $JTOL(f)$ response was designed to fit the mask at $DF = 1/5$. For $DF > 1/5$, the $JTOL(f)$ is compliant with SONET requirements. Jitter peaking is within the required 0.1-dB value for the simulated DF range. Sub-picosecond jitter generation was predicted in circuit transient simulation.

V. FABRICATION

The CDR circuit was implemented in IBM's SiGe HBT bipolar process which includes pMOS devices. Detailed device characteristics are given in [12]. Die size is 3×3 mm^2 (Fig. 11). Three external RC-filter components are required to complete the CDR design.

VI. EXPERIMENTAL RESULTS

The IC performed as simulated, except for the VCO oscillation frequency which was 5% lower than simulated. Measurements were done on-wafer using membrane probes from Cascade Microtech. The PLL was locked by an external sweeping of the VCO frequency using the *10-GHz adjust* input (see Fig. 1). The locking range is 25 MHz and the PLL stays locked within a 200-MHz frequency range. Fig. 12 shows recovered clock and CDR eye diagrams at 9.529-Gb/s data rate and 30 mV_{pp}, $2^{23} - 1$ PRBS input signal. Measured sensitivity was 14 mV_{pp} at BER $= 10^{-9}$. This value is close to the simulated 13.4 mV_{pp} (Section III), which indicates that a sufficient level of cross-talk isolation is achieved. The jitter tolerance was measured for jitter amplitudes below 40 ps_{pp}. This jitter was generated by modulating the clock slicing level $TRSH\_C$ (see Fig. 1) with an external signal from dc to 100-MHz frequency range. No data errors were detected associated with this jitter. This demonstrates jitter tolerance of more than 40 ps_{pp} compared to the 15 ps_{pp} SONET mask above 4 MHz.

To verify the maximum bit rate of the IC, it was tested at 13.25 Gb/s in a data recovery mode with an external clock $CLKx$ (Fig. 13). Sensitivity at 12.5 Gb/s was measured to be 15.5 mV_{pp}. Data-recovery clock-delay margin of 77 ps_{pp} at 10 Gb/s was the same as the BER tester delay margin, confirming picosecond timing resolution of the decision circuit.

The recovered clock jitter was measured, with a digital oscilloscope, to be 1.85 ps rms versus 1.68 ps rms jitter of the refer-

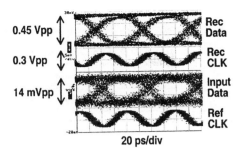

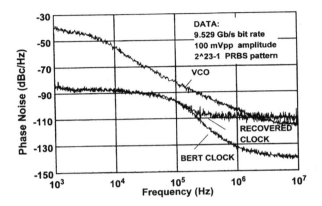

Fig. 13. CDR data recovery eye diagrams at 13.25 Gb/s. Input data 14 mV$_{PP}$, $2^{31} - 1$ PRBS pattern. The $RECCLK$ waveform is the PRBS generator reference clock translated by CDR.

Fig. 14. Phase-noise comparison of the CDR recovered clock, free-running VCO, and data pattern generator reference clock.

ence clock. Therefore jitter generated by the CDR is estimated to be 0.78 ps rms. Phase noise was measured more accurately with a HP4352B phase-noise meter (Fig. 14). Recovered clock phase noise follows, with no error, the data reference clock noise down to the CDR jitter noise floor at -110 dBc/Hz. The noise floor is reached within the bandwidth of the loop (designed to be above 4 MHz). Numerically integrated phase noise of the recovered clock in 80 MHz bandwidth gives a jitter value of 0.77 ps rms. Jitter was found to be independent of the PRBS word length up to $2^{31} - 1$. The IC dissipates 1.5 W with a -5-V power supply.

VII. CONCLUSION

In this paper, a low-jitter integrated CDR with a linear-type PLL has been demonstrated. The PLL equivalent model and design method to meet SONET jitter requirements were presented. The IC was implemented in SiGe technology. Sub-picosecond rms jitter with no jitter dependence on data PRBS pattern is achieved. Jitter generation factors in CDR were considered. A single-edge version of the Hogge-type PD and a tri-state charge pump were designed to satisfy jitter requirements. PMOS transistor circuits and cross-talk isolation technique were used to improve CDR jitter performance. In a second-order LPLL a bandwidth of more than 4 MHz and a damping factor of 4–6 at minimum expected data transition density are recommended to satisfy OC192 jitter tolerance and jitter transfer peaking requirements. To satisfy jitter transfer bandwidth (<120 KHz), additional low-pass filtering of the recovered clock must be performed, for instance, in the PLL of a transmitter circuit.

APPENDIX A

The CDR jitter transfer function is similar by definition to the PLL phase transfer function $H(s)$. For the second-order charge-pump PLL of Fig. 2, the phase transfer function is [9]

$$JTRAN(s) = \frac{2 \cdot \zeta \cdot \omega_n \cdot s + \omega_n^2}{s^2 + 2 \cdot \zeta \cdot \omega_n \cdot s + \omega_n^2}. \quad (A.1)$$

This formula can be rewritten as

$$JTRAN(s) = \frac{\omega_a + \dfrac{\omega_n^2}{s}}{s + \omega_a + \dfrac{\omega_n^2}{s}} \quad (A.2)$$

where $\omega_a = 2 \cdot \zeta \cdot \omega_n = 2 \cdot \pi \cdot F_{BW}$ is a bandwidth of the asymptotic single-pole low-pass transfer function

$$JTRAN1(s) = \frac{\omega_a}{s + \omega_a}. \quad (A.3)$$

The jitter response $JTRAN(s)$ approaches the low-pass response $JTRAN1(s)$ at $s \to \infty$ or at $\zeta \to \infty$. The asymptotic jitter tolerance shape function can be defined from (4) and (A.3) as

$$JTOL1(s) = \frac{s + \omega_a}{s}. \quad (A.4)$$

APPENDIX B

The following PLL filter components are found by solving (1)–(3):

$$R_o = \frac{\pi \cdot F_{BW}}{K_{fVCO} \cdot I_p \cdot DF_{\min}}, \quad (B.1)$$

$$C_o = \frac{2 \cdot \zeta^2}{\pi \cdot F_{BW} \cdot R_o}. \quad (B.2)$$

ACKNOWLEDGMENT

The authors thank C. Kelly and P. Popescu for discussions, J. E. Rogers for his contributions to layout design and simulations, M.-L. Xu for help with the output buffer layout, J. Showell for assistance with the measurements, Dr. S. Voinigescu and D. Marchesan for their expertise in SiGe components modeling, and Dr. M. Copeland for advice on VCO phase noise analyses. Special thanks to R. Hadaway for his support and to IBM corporation for fabrication.

REFERENCES

[1] T. Morikawa *et al.*, "A SiGe single-chip 3.3 V receiver IC for 10Gb/s optical communication systems," in *ISSCC Dig. Tech. Papers*, Feb. 1999, pp. 380–381.
[2] R. C. Walker *et al.*, "A 10Gb/s Si-bipolar Tx/Rx chipset for computer data transmission," in *ISSCC Dig. Tech. Papers*, Feb. 1998, pp. 302–303.
[3] Y. Greshishchev and P. Schvan, "SiGe clock and data recovery IC with linear-type PLL for 10 Gb/s SONET application," in *Proc. 1999 Bipolar/BiCMOS circuits and Technology Meeting*, Sept. 1999, pp. 169–172.
[4] B. Razavi, "Design of monolithic phase-locked loops and clock recovery circuits—A tutorial," in *Monolithic Phase-Locked Loops and Clock Recovery Circuits: Theory and Design*, B. Razavi, Ed. New York, NY: IEEE Press, 1996, pp. 405–420.

[5] K. Kishine, N. Ishihara, K. Takiguchi, and H. Ichino. "A 2.5-Gb/s clock and data recovery IC with tunable jitter characteristics for use in LANs and WANs," *IEEE J. Solid-State Circuits*, vol. 34, pp. 805–812, June 1999.

[6] Y. Greshishchev and P. Schvan, "60 dB gain 55 dB dynamic range 10Gb/s SiGe HBT limiting amplifier," *IEEE J. Solid-State Circuits*, vol. 34, pp. 1914–1920, Dec. 1999.

[7] L. De Vito, "A versatile clock recovery architecture and monolithic implementation," in *Monolithic Phase-Locked Loops and Clock Recovery Circuits: Theory and Design*, B. Razavi, Ed. New York, NY: IEEE Press, 1996, pp. 405–42.

[8] "SONET OC-192 Transport System Generic Criteria," Bellcore, GR-1377-CORE, Mar. 1998.

[9] F. M. Gardner, "Charge-pump phase-lock loops," *IEEE Trans. Commun.*, vol. COM-28, pp. 1849–1858, Nov. 1980.

[10] C. R. Hogge, "A self-correcting clock recovery circuit," *IEEE J. Lightwave Technol.*, vol. 3, pp. 1312–1314, Dec. 1985.

[11] B. Jansen, K. Negus, and D. Lee, "Silicon bipolar VCO family for 1.1 to 2.2 GHz with fully-integrated tank and tuning circuits," in *ISSCC Dig. Tech. Papers*, Feb. 1997, pp. 392–393.

[12] J. D. Cressler, "SiGe HBT technology: A new contender for Si-based RF and microwave circuit applications," *IEEE Trans. Microwave Theory Tech.*, vol. 46, pp. 572–589, May 1998.

Peter Schvan (M'89) was born in Budapest, Hungary, in 1952. He received the M.S. degree in physics from Eotvos Lorand University, Budapest, in 1975 and the Ph.D. degree in electrical engineering from Carleton University, Ottawa, ON, Canada, in 1985.

In 1985, he joined Nortel Networks, Ottawa, ON, Canada, where he worked in the area of BiCMOS and bipolar technology development, yield prediction, device characterization, and modeling. Recently, his work has been extended to the design of multigigabit circuits and systems. He is currently Senior Manager of a group responsible for evaluating various high-performance technologies and demonstrating, advanced circuit concepts required for fiberoptic communication systems. He is the author or coauthor of numerous publications.

Yuriy M. Greshishchev (M'95) received the M.S.E.E. degree from Odessa Electrotechnical Institute of Communications, Odessa, Ukraine, in 1974 and the Ph.D. degree in electrical and computer engineering from V. M. Glushkov Institute of Cybernetics, Kyiv, Ukraine, in 1984.

From 1976 to 1994, he worked with research and development organizations and academia on high-speed ADC and DAC circuit theory and design, primarily in the area of silicon bipolar and GaAs MESFET integrated circuits. His Ph.D. research was dedicated to the development of folding-type video ADCs embedded into TV systems. In 1993, he was a Visiting Scientist at Micronet, Institution Center of University of Toronto, Toronto, ON, Canada. In 1994, he joined the Department of Electrical and Computer Engineering, University of Toronto, where he conducted research on low-voltage GaAs MESFET circuits for digital wireless communication. Since 1996, he has been with Nortel Networks, Ottawa, ON, Canada, where he is responsible for the development of highly integrated circuit solutions in emerging technologies for optical communications. He is the coauthor of two books and more than 40 technical papers on the area of data converters, high-speed circuit design, and statistical modeling.

A Fully Integrated SiGe Receiver IC for 10-Gb/s Data Rate

Yuriy M. Greshishchev, *Member, IEEE*, Peter Schvan, *Member, IEEE*, Jonathan L. Showell, *Member, IEEE*, Mu-Liang Xu, *Member, IEEE*, Jugnu J. Ojha, *Member, IEEE*, and Jonathan E. Rogers, *Student Member, IEEE*

Abstract—A silicon germanium (SiGe) receiver IC is presented here which integrates most of the 10-Gb/s SONET receiver functions. The receiver combines an automatic gain control and clock and data recovery circuit (CDR) with a binary-type phase-locked loop, 1 : 8 demultiplexer, and a $2^7 - 1$ pseudorandom bit sequence generator for self-testing. This work demonstrates a higher level of integration compared to other silicon designs as well as a CDR with SONET-compliant jitter characteristics. The receiver has a die size of 4.5 × 4.5 mm² and consumes 4.5 W from −5 V.

Index Terms—Clock and data recovery (CDR), jitter generation, jitter tolerance, jitter transfer, phase detector, phase-locked loop (PLL), SONET, VCO.

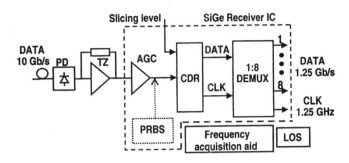

Fig. 1. 10-Gb/s receiver architecture. Dotted box shows components integrated in the SiGe receiver IC presented in the paper.

I. INTRODUCTION

A TYPICAL fiber-optic SONET receiver contains pin-diode with transimpedance (TZ) amplifier, wide dynamic range automatic gain control amplifier (AGC), and a clock and data recovery circuit (CDR) with a demultiplexer. Introduction of dense wave-division-multiplexed (DWDM) systems has put a high demand on the receiver production. A high level of 10-Gb/s component integration, as opposed to using a filter-based CDR architecture [1], is required along with self-testing capabilities to reduce receiver cost, module size, and power dissipation. One of the major difficulties in the integration of 10-Gb/s receiver is to achieve jitter characteristics compliant to the SONET requirements, such as *Bellcore* recommendations for the OC192 system [2]. To the authors' knowledge, none of the previously reported [3]–[5] 10-Gb/s receiver ICs with the integrated clock and data recovery circuit (CDR) demonstrated all of the SONET compliant jitter characteristics. While sub-picosecond jitter generation was previously confirmed in the SONET CDR [5], another important question is if all of the receiver components can be integrated on a die without sensitivity and jitter performance degradation.

A fully integrated SiGe receiver IC, presented in the paper, combines CDR, AGC, 1 : 8 demultiplexer and $2^7 - 1$

Manuscript received April 17, 2000; revised June 29, 2000.

Y. M. Greshishchev and P. Schvan are with Nortel Networks, Ottawa, ON K1Y 4H7, Canada (e-mail: greshy@nortelnetworks.com).

J. L. Showell was with Nortel Network and is currently with Quake Technologies Inc., Ottawa, ON, Canada.

M.-L. Xu was with Nortel Networks, Ottawa, ON K1Y 4H7, Canada. He is now with Conexant Systems, San Diego, CA.

J. J. Ojha was with Nortel Networks, Ottawa, ON K1Y 4H7, Canada. He is now with Caspian Networks, Palo Alto, CA (e-mail: jojha@caspiannetworks.com).

J. E. Rogers is with The University of Toronto, Toronto, ON M5S 3G4, Canada.

Publisher Item Identifier S 0018-9200(00)09475-0.

pseudorandom bit sequence (PRBS) generator for self-testing, as shown in the dotted-line box in Fig. 1 [6]. Receiver performance mounted into test fixture was verified in a data-recovery mode up to 12.5 Gb/s and in a CDR mode at 9.1 Gb/s (only limited by the VCO maximum oscillation frequency after packaging). The OC192 10-Gb/s SONET-compliant jitter characteristics of the CDR were verified on-wafer with a membrane probe card and with a jitter analyzer from Anritsu. Phase-noise characteristics have also been measured to confirm the CDR's sub-picosecond rms jitter performance. Measured 10-Gb/s maximum receiver sensitivity de-embedded after losses in the test fixture is 4.5 mV$_{pp}$ at a bit-error rate (BER) of BER = 10^{-10} at the demultiplexer (DEMUX) output.

In Section II, the binary CDR architecture used in the receiver is briefly analyzed as compared to a linear-type CDR and design method to meet SONET jitter requirements is presented. Then in Section III, the full receiver architecture and the building blocks implementation details are discussed. In Section IV, the IC die fabrication features are described. Finally, in Section V, measured results are presented.

II. BINARY CDR IN SONET RECEIVER

A. Binary CDR Versus Linear CDR

The CDR published in [5] uses a linear-type PLL approach [Fig. 2(a)], while the CDR presented here is based on a binary PLL [Fig. 2(b)]. In the binary PLL, a binary Alexander-type [7] phase detector (PD) is used as compared to the Hogge-type PD [8] in a linear-type PLL. Examples of using binary architecture in optical receiver ICs can be found in [9], [10]. Binary PD produces two digital outputs, *UP* and *DOWN*, to signal if the data is early or late with respect to the VCO clock. To control the VCO, the binary information is split into two loops as suggested in [11]. The phase-control loop is formed with the *UP* and *DOWN*

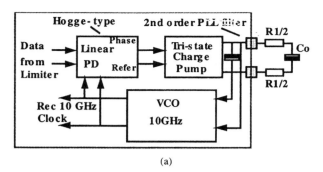

(a)

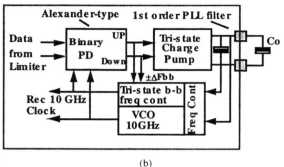

(b)

Fig. 2. Two basic self-aligned CDR architectures. (a) With a linear PLL. (b) With a binary PLL.

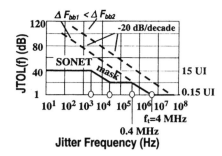

Fig. 3. CDR analytical jitter tolerance as compared to SONET mask.

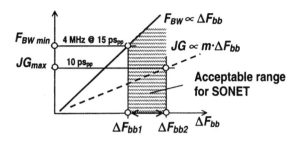

Fig. 4. Trade-off for the frequency step in binary CDR.

outputs directly modulating the VCO frequency with frequency step $\pm \Delta F_{bb}$ (bb denotes "bang-bang"—other name of the binary PLL architecture) via bang-bang frequency tune input. The frequency loop of the binary PLL uses binary outputs integrated with charge pump and capacitor C_o to control the second tune input of the VCO. To reduce jitter generation in absence of data transitions (during long zeros or ones), a tri-state charge pump was employed. By the same reason, a tri-state is introduced to the VCO bang-bang control input which is no longer binary, but ternary, where in a tri-state no frequency step is applied.

Table I shows the comparison for two PLL receiver architectures. Binary CDR is less demanding on the "analog" features of an IC technology, and, in principal, has only one critical component—the binary phase detector (BPD) where sub-picosecond time resolution is required. The ring-oscillator-type VCO is recommended to reduce delay in the PLL loop and, therefore, jitter generation in the CDR, as explained later in the paper. The VCO phase noise is less critical in a binary CDR because of the relatively wide PLL bandwidth. In a linear-type PLL, jitter transfer characteristics can be analyzed using linear PLL theory (see, for example, [5]). Binary-type PLL has nonlinear jitter transfer characteristics and its analyses have not been presented in the technical literature. The following subsection describes the binary PLL design method used in the receiver design.

B. Binary CDR Jitter Characteristics

In SONET applications three main jitter characteristics are important:

  1) Jitter Tolerance;
  2) Jitter Transfer;
  3) Jitter Generation.

The Jitter Generation and loop stability was first analyzed by R. Walker et al. [11]. This work also suggested loop decomposi-

tion into a frequency-control (low-frequency) part and a phase-control (high-frequency) part. In general, binary PLL is similar in system behavior to a double integration delta modulator with prediction [12] acting in the data-phase domain. Based on analogy with the signal frequency response in delta modulator, the phase-jitter transfer function has a nonlinear (slew-rate limited) mechanism for the phase-jitter frequency response. It is a single-pole-like response with the bandwidth inversely proportional to the input jitter amplitude:

$$F_{\mathrm{BW}}(JITTER_{PP}) \propto DF \cdot \frac{\Delta F_{bb}}{JITTER_{PP}} \qquad (1)$$

where F_{BW} is 3-dB bandwidth of the binary PLL jitter transfer function; $JITTER_{pp}$ is the jitter amplitude, ΔF_{bb} is the bang–bang frequency step, DF is the average data transition density factor (maximum $DF = 1$ for $0101\ldots$ pattern).

The shape of the Jitter Tolerance function can be characterized by means of jitter-tolerance scale function, $JTOL(f)$ [5]. Modeled binary PLL jitter tolerance response, $JTOL(f)$, is shown in Fig. 3 for two frequency steps $\Delta F_{bb1} < \Delta F_{bb2}$. For comparison, Bellcore SONET mask is also shown on the same plot with the corresponding unit interval (UI) values on the right side of the graph[1]. To satisfy the mask, jitter transfer bandwidth defined by (1) should be set above 4 MHz at a jitter amplitude $JITTER_{pp} = 15\ \mathrm{ps}_{pp}$ and minimum average data transition density, DF.

The Jitter Generation in a binary CDR is proportional to the frequency step, ΔF_{bb}, and delay in the PLL loop, measured as a number m of 100-ps clock periods required to propagate signal from the phase detector output to its input:

$$JG_{PP} \propto m \cdot \Delta F_{bb}. \qquad (2)$$

Equations (1) and (2) were used to find a frequency step F_{bb} and a delay m acceptable for the SONET applications as

[1]In a 10-Gb/s system, UI = 100 ps.

TABLE I
COMPARISON OF LINEAR AND BINARY TYPE CDR ARCHITECTURES

Design Stage	Linear CDR	Binary CDR
System performance:		
Theoretical Analyses	Linear, based on linear PLL theory	Nonlinear, binary PLL analytical and numerical modeling required
Jitter transfer (JTRN))	Linear, critical to jitter peaking	Nonlinear, single-pole-like, not critical to jitter peaking
Jitter tolerance (JTOL)	Is defined by JTRAN BW	Is defined by JTRAN BW at fixed input jitter amplitude
Jitter generation	Medium critical	Medium critical
Building Blocks:		
Tri-State Phase detector	Linear, Hogge type, critical to the shape and delays of internally generated pulses	Binary, Alexander type, critical to the flip-flop latch metastability
Tri-State Charge Pump	Critical to the component matching, output impedance	Not critical to the component non-idealities
VCO	Medium critical, any type VCO is suitable	Low critical, ring type VCO with separate frequency control and bang-bang control is recommended
PLL filter	Design medium critical	Design not critical

.hown in Fig. 4. The minimum value for the frequency step ΔF_{bb1} is determined by *Jitter Tolerance* minimum bandwidth requirements (4 MHz); the maximum value ΔF_{bb1} is limited by jitter generation (10 ps_{pp} is recommended [2]). In the design presented here, $JG_{pp} = 5$ ps was assumed. To reduce jitter generation delay, m should be minimized. This makes the ring-type VCO preferable in the binary CDR as compared to LC-tank based VCO where tuning delay is larger due to the usually higher Q-factor of the LC-tank.

III. RECEIVER ARCHITECTURE

A. Architecture

The receiver architecture is shown in Fig. 5. It combines an AGC and a binary CDR with a 1 : 8 demultiplexer and a $2^7 - 1$ PRBS generator for self-testing. The receiver recovers 10-Gb/s data and a 10-GHz clock, and produces eight demultiplexed 1.25-Gb/s CML data outputs with a 1.25-GHz clock. The PRBS generator allows functional testing of the CDR and subsequent circuits. A PRBS clock (*CLK*) is required for testing. In the test mode, the PRBS output is enabled to drive the CDR data bus. In the receiver mode, the AGC output is enabled. The recovered 10-Gb/s data and 10-GHz clock appear at the recovered data (*DATA REC BUS*) and clock (*CLK REC BUS*) buses, driving a 1 : 8 DEMUX circuit. A clock signal can also be supplied externally (*CLKx*) for data recovery operation only.

B. AGC

The block diagram of the AGC is shown in Fig. 6. The AGC has total linear gain range from -3 to 20 dB with a maximum input of 1.7 V_{pp}. The AGC has two variable gain stages $A1$, $A2$ implemented with output current steering in a differential pair [13]. Stage $A3$ has a fixed gain and also provides open collector transmission line interface to drive the CDR data bus. To alleviate conflicting requirements for large data swing and low noise figure at low input amplitudes, the AGC has two gain ranges: -7–7 dB (low gain range) and 7–20 dB (high gain range). Two differential pairs with a "large" and a "small" emitter degeneration resistors are used to switched the gain ranges. AGC-measured S11 is better than -15 dB in a frequency range up to 10 GHz, noise figure 13.5 dB. The ac bandwidth is adjustable in a range of 8–10 GHz.

C. CDR

As compared to original version of binary CDR [7], [11], in the CDR presented here, the data decision and clock recovery processes are split. This allows for independent optimization of data decision threshold (slicing) without affecting clock recovery process. There are four decision channels in the CDR, all driven by the CDR data bus. Channels 1 and 2 are identical decision circuits, as shown in the block diagram of Fig. 7(a). The additional decision channel allows operation with two different input slicing levels. The data decision threshold is set by a differential slicer circuit based on an emitter follower [Fig. 7(b)]. In a long-haul receiver application, a high-performance limiting amplifier [2 in Fig. 7(a)] is required. Note that the AGC stabilizes only a long-time averaged amplitude measured at AGC output with a peak detector (not shown in Figs. 5 and 6). The limiting amplifier stage was designed for 40-dB gain with bandwidth of more than 16 GHz and input AM to output PM conversion less than 1 ps in 20-dB input dynamic range. Two 20-dB gain-limiting amplifier stages similar to [14] were employed. The digital sampler is based on a master–slave–master (MSM) flip-flop

675

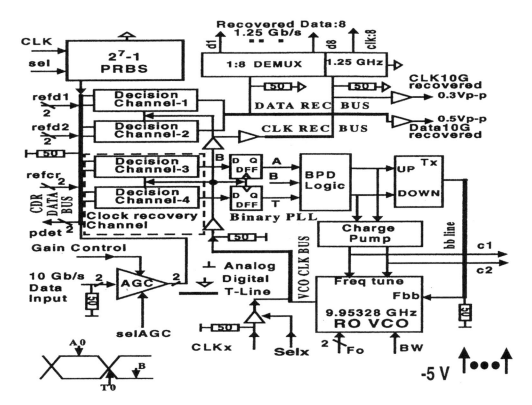

Fig. 5. SiGe receiver architecture.

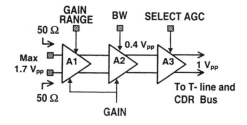

Fig. 6. AGC block diagram.

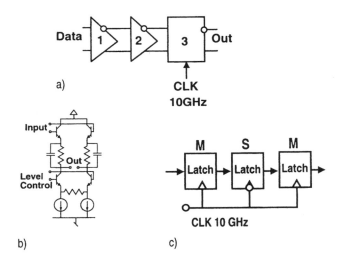

Fig. 7. Decision channel. (a) Block diagram: 1–slicer; 2–limiting amplifier; 3–digital sampler. (b) Slicer schematic. (c) Digital sampler schematic.

configuration [Fig. 7(c)]. It helps to reduce the latching metastability region and to increase the clock phase margin in the decision circuit, as opposed to master–slave D-type flip-flop (DFF). Schematic and layout of the latch was optimized for a minimum latching time constant.

The BPD is formed with decision channels 3 and 4, and two DFF circuits. The BPD takes three data samples according to the timing diagram in Fig. 5. It generates a binary output with respect to the lead/lag phase of the VCO clock. The BPD uses only one edge of the data transition, and is set to a tri-state condition at the other edge or in the absence of data transitions according to a truth table (Table II). As a result, recovered clock jitter is not effected by asymmetry between the rising and the falling edges or by the incoming data pattern. The BPD phase resolution is critical to the CDR jitter performance. Latch metastability at time sample T0 (see Fig. 5) limits the resolution. This problem was circumvented by using a MSM-based decision circuit preceded by a limiting amplifier, as described above. Phase resolution better than 1 ps was measured in the BPD circuit, as shown in Fig. 8. A phase-delay sweep was arranged with two clock generators: 5.0125 GHz for the data and 10.0000 GHz

for the clock. Frequency shift of 12.5 MHz for the data provided binary pulses at the output of the phase detector with 40-ns period corresponding to a 100-ps delay sweep. To analyze the output pulses, a digital scope was synchronized from the phase-detector output using an HP 54118A trigger amplifier. This method allowed measurement of the output transition region with the accuracy of 0.5 ps per one data cycle (200 ps). Measured transition region at the phase detector output contains two data cycles, confirming phase detector time resolution to be less than 1 ps.

The CDR frequency-control loop and the phase-control loop are separated as described in Section II. The bang-bang part of

TABLE II
TRUTH TABLE OF BINARY PHASE DETECTOR

Data state	A	B	T	Up	Down	PLL action
No data	0	0	0	0	0	tristate
Not defined	0	0	1	0	0	tristate
transition 0 to 1 late	0	1	0	0	1	VCO Down
transition 0 to 1 early	0	1	1	1	0	VCO Up
transition 1 to 0 late	1	0	0	0	0	tristate
transition 1 to 0 early	1	0	1	0	0	tristate
Not defined	1	1	0	0	0	tristate
No data	1	1	1	0	0	tristate

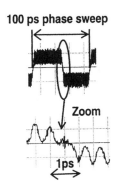

Fig. 8. BPD measured time resolution.

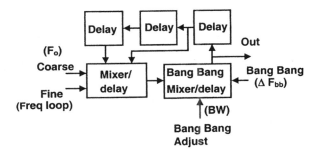

Fig. 9. Ring-type VCO block diagram.

the PLL controls the recovered clock phase via the F_{bb} input of the VCO. The frequency loop includes the charge pump and an external integration capacitor (pins C1 and C2).

The VCO is a ring oscillator type with an architecture shown in Fig. 9. A mixer-type delay cell is used to control the oscillation frequency. The mixer cell is split into a fine-tune (for the internal frequency loop) and a coarse-tune (to compensate for process variation). Care was taken to provide symmetrical bang-bang frequency steps with respect to the tri-state. All of the VCO control inputs were implemented with the high-impedance pMOS buffers. A pMOS-based charge pump was employed [5].

D. 1 : 8 Demultiplexer

The 1:8 DEMUX (Fig. 10) is similar in architecture to the design presented in [15]. Seven 1:2 demultiplexer circuits are cascaded, with each stage optimized for the clock frequency required. Each 1:2 demultiplexer consists of a master–slave–master flip-flop to capture the lead bit on the positive edge of the clock and a master–slave flip-flop to capture the second bit using the negative edge of the clock. Utilizing an extra latch in the MSM flip-flop ensures that the 1:2 data outputs are aligned for further processing. The frequency of the incoming CDR clock is divided by two at each demultiplexer stage with a delay equal to the data delay in the 1:2 block.

E. Built-in $2^7 - 1$ PRBS Generator

The $2^7 - 1$ PRBS generator (Fig. 11) was implemented using the standard polynomial equation $1 + x^6 + x^7$ in a parallel form.

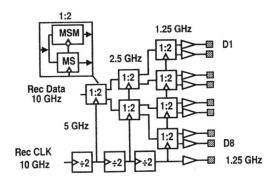

Fig. 10. 1:8 demultiplexer block diagram.

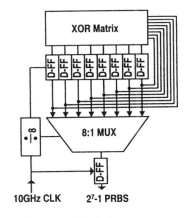

Fig. 11. $2^7 - 1$ PRBS generator block diagram.

The parallel form avoids the necessity of distributing a 10-GHz clock, as would be required if using a shift-register-type PRBS.

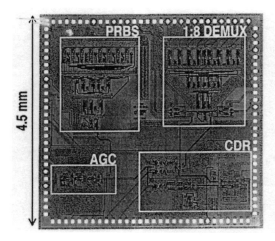

Fig. 12. SiGe receiver die micrograph.

In an 8-bit parallel form, the clock frequency is reduced to 1.25 GHz. Similar to the AGC, the output of the multiplexer provides open collector transmission line interface to drive the CDR data bus. The disadvantage of a parallel architecture is penalty in the die area and the power consumption. In normal operation (the AGC transmits data to the CDR) the PRBS power supply is turned off, reducing the overall power consumption of the receiver.

F. CDR Simulation

Because of the nonlinear jitter response of a binary CDR, hierarchical numerical analysis was an important part of the receiver IC design. Four levels of PLL analysis were carried out: analytical, behavioral, schematic level, and post-layout extracted circuit with distributed parasitics. The last three levels are HSPICE-based. A behavioral library of linear and digital components was developed. Analytical models of jitter transfer and jitter tolerance are based on simplified binary-PLL theory, as described above.

IV. FABRICATION

The receiver IC was implemented in IBM"s SiGe technology ($ft = 45$ GHz, f max $= 70$ GHz). The microphotograph of the die is shown in Fig. 12. The die size is 4.5×4.5 mm^2. The major circuit building blocks were not only integrated into the receiver, but were implemented as individual IC components and tested. In the receiver IC, the building blocks were physically partitioned with a transmission line circuit and layout isolation interface similar to that presented in [5], [14]. Separate power supply systems with digital and analog grounds were routed.

V. EXPERIMENTAL RESULTS

The IC worked at first implementation with the VCO oscillation frequency 10% lower than simulated. The receiver IC was mounted into a microwave test fixture (Fig. 13) and was tested at 9.1 Gb/s (VCO oscillation frequency limit[2]) in a CDR mode and up to 12.5 Gb/s in a data-recovery mode or in internal PRBS test mode with an external clock. The carrier substrate size of

[2]Maximum oscillation frequency can be easily corrected by removing one delay stage in the VCO design of Fig. 9.

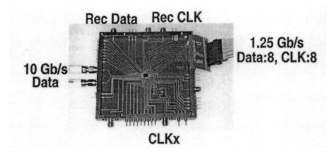

Fig. 13. SiGe receiver mounted into microwave-style test fixture.

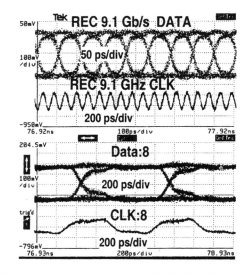

Fig. 14. SiGe receiver eye diagrams measured in CDR mode at 9.1 GB/s. Input data: 80 mV$_p$p PRBS $2^{31} - 1$.

10×10 cm^2 in the test fixture was defined by the perimeter required for mounting I/O connectors in the housing metal box. A large number of required I/O were used for testing purposes. The receiver IC does not require external components, except decoupling and integration capacitors mounted beside the die. The recovered clock and data eye diagrams at 9.1 Gb/s are shown in Fig. 14. The IC input sensitivity is less than 4.5 mV$_{pp}$ at BER $= 10^{-10}$ measured at the 1 : 8 demultiplexer output with AGC gain set to 20 dB (data eye closure in the test fixture was de-embedded). DATA : 8 transition distortion apparent in Fig. 14 is due to long ribbon cable attached to the test fixture demultiplexed outputs. The receiver consumes in a mission mode 4.5 W from -5 V.

The CDR performance IC was fully characterized at 10-Gb/s on-wafer with a probe card. The die micrograph of the CDR is shown in Fig. 15. It consists of an exact copy of the receiver CDR layout plus the output buffers located in the DEMUX partition. In all of the measurements, input data were supplied single-ended while unused differential input was terminated with 50 Ω. The CDR typical eye diagrams measured with 20 mV$_{pp}$ $2^{31} - 1$ PRBS data are shown in Fig. 16. The input sensitivity was measured to be 14 mV$_{pp}$ at BER $= 10^{-9}$ as compared to 13.4 mV$_{pp}$ simulated considering thermal and shot noise in the decision channel.

Phase noise of the recovered clock was measured with an HP 4352B as a power spectrum density (Fig. 17). 10-Gb/s input data were supplied with amplitude of 100 mV$_{pp}$ and $2^{23} - 1$

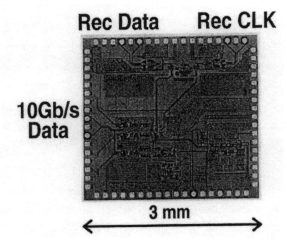

Fig. 15. SiGe CDR IC die micrograph.

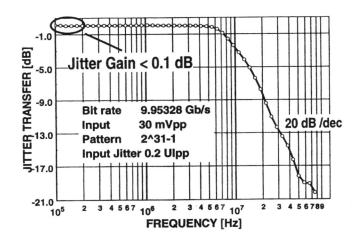

Fig. 18. CDR jitter transfer.

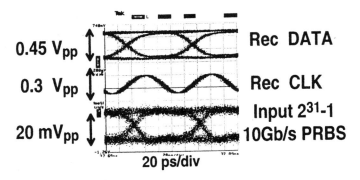

Fig. 16. CDR IC eye diagrams measured in CDR mode at 10 Gb/s.

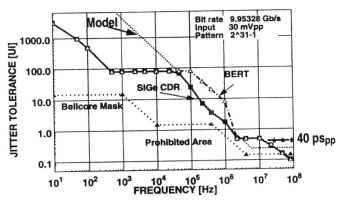

Fig. 19. CDR jitter tolerance. Performance measured with the clock reference level modulaton test marked with symbol *.

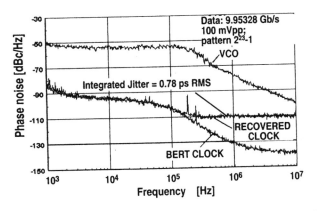

Fig. 17. Phase-noise comparison of the CDR recovered clock, free running VCO and data pattern generator (BERT) clock.

PRBS pattern. For comparison, phase noise of the free-running VCO and the data-pattern generator was also measured and shown on the same plot. As expected in a high performance CDR, the recovered clock-phase noise follows, with no error, the data-reference clock noise down to the CDR jitter noise floor at −110 dBc/Hz. Similar recovered phase noise was achieved in the CDR design with a linear PLL and LC-type VCO [5]. Numerically integrated phase noise of the recovered

clock gives jitter RMS value of 0.78 ps. Phase noise was found to be PRBS pattern independent up to a pattern of $2^{31} - 1$.

The OC192 jitter compliant performance (at 9.953 28 Gb/s) was verified with a jitter analyzer MX177 701 from Anritsu. Jitter generation (in 80-MHz bandwidth) was measured to be 5.4 ps_{pp} and 0.8 ps RMS as compared to 10 ps_{pp} or 1 ps RMS recommended by *Bellcore* [2]. The RMS jitter is very close to the 0.78-ps value obtained in the phase-noise measurement. Jitter transfer measurement (Fig. 18) showed, as predicted by modeling, single-pole-like characteristics with no jitter peaking. Jitter tolerance (Fig. 19) has a very wide safety margin for SONET mask with a minimum of 40 ps_{pp} (15 ps_{pp} is recommended). The shape of measured CDR jitter tolerance response differs from the modeled in Fig. 3 because of test setup limitations. This is seen from the measured jitter tolerance of the test setup (BERT) with no CDR in the data path (shown in the same plot of Fig. 19). Only in the frequency range of 40 kHz–2 MHz measured jitter tolerance is determined by CDR performance. In this frequency range, measured and modeled jitter tolerance coincide. The upper frequency range of the jitter tolerance response was also remeasured with a different method, based on the reference voltage $refd1$ (see Fig. 5) modulation with a sin-wave signal. Minimum jitter tolerance of $40\ ps_{pp}$ was measured. Both jitter transfer and jitter tolerance response were found to be PRBS pattern independent. The IC demonstrated a 60-MHz frequency range of robust PLL locking and operation even at the input signals well below the sensitivity level.

VI. Conclusion

A fully integrated SiGe receiver IC is presented, which combines self-aligned CDR with integrated binary PLL, AGC, 1 : 8 demultiplexer, and $2^7 - 1$ PRBS generator for self-testing. The receiver, mounted into a test fixture, operates up to 9.1 Gb/s (VCO limit) in a CDR mode and up to 12.5 GB/s in a data-recovery mode. Maximum die sensitivity is $4.5\ mV_{pp}$ at $\mathrm{BER} = 10^{-10}$ measured at 1 : 8 DEMUX output. Receiver die size is $4.5 \times 4.5\ mm^2$, and it consumes in a mission mode 4.5 W from -5-V power supply. CDR SONET-compliant jitter characteristics were verified on-wafer. Jitter tolerance well exceeds OC192 *Bellcore* mask with a minimum of $40\ ps_{pp}$. Jitter transfer has a single-pole-like response with no peaking detected. Jitter generation is less than 1 ps RMS and less than $5.5\ ps_{pp}$.

Acknowledgment

The authors thank their colleagues S. Szilagyi for the microwave test fixture design, and D. Marchesan and Dr. S. Voinigescu for useful discussions and distributed components modeling. Special thanks to R. Hadaway for his directions and to IBM Corporation for fabrication.

References

[1] B. Beggs, "GaAs HBT 10-Gb/s Product," in *1999 IEEE MTT-S Int. Microwave Symp. Workshop*, Anaheim, CA, June 13–19, 1999.

[2] SONET OC-192, "Transport system generic criteria," *Bellcore, GR-1377-CORE*, no. 4, Mar. 1998.

[3] R. C. Walker et al., "A 10-Gb/s Si-bipolar Tx/Rx chipset for computer data transmission," in *ISSCC Dig. Tech. Papers*, Feb. 1998, pp. 302–303.

[4] T. Morikawa et al., "A SiGe single-chip 3.3-V receiver IC for 10-Gb/s optical communication systems," in *ISSCC Dig. Tech. Papers*, Feb. 1999, pp. 380–381.

[5] Y. Greshishchev and P. Schvan, "SiGe clock and data recovery IC with linear-type PLL for 10-Gb/s SONET application," in *Proc. 1999 Bipolar/BiCMOS Circuits and Technology Meeting*, Sept. 1999, pp. 169–172.

[6] Y. M. Greshishchev, P. Schvan, J. L. Showell, M.-L. Xu, J. J. Ojha, and J. E. Roger, "A fully integrated SiGe receiver IC for 10-Gb/s data rate," in *ISSCC Dig. Tech. Papers*, Feb. 2000, pp. 52–53.

[7] J. D. H. Alexander, "Clock recovery from random binary signals," *Electron. Lett.*, vol. 11, pp. 541–542, Oct. 1975.

[8] C. R. Hogge, "A self-correcting clock recovery circuit," *J. Lightwave Technology*, vol. 3, pp. 1312–1314, Dec. 1985.

[9] J. Hauenschild et al., "A two-chip receiver for short-haul links up to 3.5-Gb/s with PIN-preamp module and CDR-MUX," in *ISSCC Dig. Tech. Papers*, Feb. 1998, pp. 308–309.

[10] J. Hauenschild et al., "A plastic packaged 10-Gb/s biCMOS clock and data recovering 1 : 4-demultiplexer with external VCO," *IEEE J. Solid-State Circuits*, vol. 31, pp. 2056–2059, Dec. 1996.

[11] R. C. Walker et al., "A two-chip 1.5-GBd serial link interface," *IEEE J. Solid-State Circuits*, vol. 27, pp. 1805–1811, Dec. 1992.

[12] R. Steele, *Delta Modulation Systems*. New York/Toronto: Wiley, 1975.

[13] M. Soda, T. Suzaki, and T. Morikawa et al., "A Si bipolar chip set for 10-Gb/s optical receiver," in *ISSCC Dig. Tech. Papers*, Feb. 1992, pp. 100–101.

[14] Y. Greshishchev and P. Schvan, "60-dB gain 55-dB dynamic range 10-Gb/s SiGe HBT limiting amplifier," *IEEE J. Solid-State Circuits*, vol. 34, pp. 1914–1920, Dec. 1999.

[15] L. I. Anderson et al., "Silicon bipolar chipset for SONET/SDH 10-Gb/s fiber-optic communication links," *IEEE J. Solid-State Circuits*, vol. 30, pp. 210–218, Mar. 1995.

Yuriy M. Greshishchev (M'95) received the M.S.E.E. degree from Odessa Electrotechnical Institute of Communications, Odessa, Ukraine, in 1974 and the Ph.D. degree in electrical and computer engineering from V.M. Glushkov Institute of Cybernetics, Microelectronics Division, Kyiv, Ukraine, in 1984.

From 1976 to 1994, he worked with research and development organizations and academia on high-speed silicon bipolar and GaAs MESFET ADC and DAC integrated circuits. His Ph.D research was dedicated to the development of folding-type ADCs embedded into TV systems. In 1993, he was a Visiting Scientist at Micronet, Institution Center of University of Toronto, Ontario, Canada. In 1994, he joined the Department of Electrical and Computer Engineering, University of Toronto, where he conducted research on GaAs MESFET linear transmitter design for digital wireless communication. Since 1996, he has been with Nortel Networks, Ottawa, Ontario, where he is responsible for development of highly integrated circuit solutions in emerging technologies for optical communications. He has coauthored two books and numerous technical papers on the area of high-speed communication circuit design, data converters, and statistical modeling.

Peter Schvan (M'89) was born in Budapest, Hungary, in 1952. He received the M.S. degree in physics from Eotovos Lorand University, Budapest, in 1975 and the Ph.D. degree in electrical engineering from Carleton University, Ottawa, Ontario, Canada, in 1985.

In 1985, he joined Nortel Neworks, Ottawa, where he started working in the area of BiCMOS and bipolar technology development, yield prediction, device characterization, and modeling. Recently, his work has been extended to the design of multi-gigabit circuits and systems. He is currently Senior Manager of a group responsible for evaluating various high-performance technologies and demonstrating advanced circuit concepts required for fiber optic communication systems. He has authored and co-authored numerous publications.

Jonathan L. Showell (S'90–M'95) received the B.Eng and M.Eng degrees in engineering physics from McMaster University, Hamilton, ON, Canada, in 1990 and 1994, respectively.

He joined Nortel Networks, Ottawa, Canada, in 1994, working on hot carrier injection reliability of CMOS devices. Later he became a member of the Technology Access and Applications Group where his responsibilities included accurate high-frequency analog (up to 110 GHz) and digital (40 Gb/s) measurements and the design of high-speed 10- to 40-Gb/s, multiplexer/demultiplexer circuits in SiGe HBT and InP HBT technologies, respectively. Recently, he joined Quake Technologies, Ottawa, Canada, working on the design of chip sets for high-speed datacom applications. His interests include high-speed technologies, circuit design for high-speed communications, and accurate high-frequency measurements.

Mu-Liang Xu (M'00), biography not available at time of publication.

A 10-Gb/s CMOS Clock and Data Recovery Circuit with a Half-Rate Linear Phase Detector

Jafar Savoj, *Student Member, IEEE,* and Behzad Razavi, *Member, IEEE*

Abstract—A 10-Gb/s phase-locked clock and data recovery circuit incorporates an interpolating voltage-controlled oscillator and a half-rate phase detector. The phase detector provides a linear characteristic while retiming and demultiplexing the data with no systematic phase offset. Fabricated in a 0.18-μm CMOS technology in an area of 1.1×0.9 mm^2, the circuit exhibits an RMS jitter of 1 ps, a peak-to-peak jitter of 14.5 ps in the recovered clock, and a bit-error rate of 1.28×10^{-6}, with random data input of length $2^{23} - 1$. The power dissipation is 72 mW from a 2.5-V supply.

Index Terms—Clock recovery, half-rate CDR, optical communication, oscillators, phase detectors, PLLs.

I. INTRODUCTION

WITH THE exponential growth of the number of Internet nodes, the volume of the data transported by its backbone continues to rise rapidly. The load of the global Internet backbone is expected to be as high as 11 Tb/s by the year 2005, indicating that the required bandwidth must increase by a factor of 50 to 100 every seven years.

Among the available transmission media, optical fibers have the highest bandwidth with the lowest loss, serving as an attractive solution for the Internet backbone. However, the electronic interface proves to be the bottleneck in designing high-speed optical systems. In order to push the speed of operation beyond the capabilities of the fabrication processes, a number of transceivers can be fabricated on the same chip. The input and output signals can be carried either over a bundle of fibers, or on a single fiber that uses wave-division multiplexing. In this scenario, both the power dissipation and the complexity of each transceiver become critical. While stand-alone building blocks of optical transceivers have been built in GaAs and silicon bipolar technologies [1], [2], full integration of many transceivers makes it desirable to use CMOS technology.

This paper describes the design of the first 10-Gb/s CMOS clock and data recovery (CDR) circuit. A linear phase detector (PD) is introduced that compares the phase of the incoming data with that of a half-rate clock. The CDR circuit also incorporates a three-stage interpolating ring oscillator to achieve a wide tuning range. Fabricated in a 0.18-μm CMOS technology, the circuit achieves an RMS jitter of 1 ps with a pseudorandom sequence of $2^{23} - 1$ while dissipating 72 mW from a 2.5-V supply.

Manuscript received August 21, 2000; revised December 1, 2000. This work was supported in part by the Semiconductor Research Corporation and in part by Cypress Semiconductor.

The authors are with the Electrical Engineering Department, University of California, Los Angeles, CA 90095-1594 USA (e-mail: razavi@icsl.ucla.edu).

Publisher Item Identifier S 0018-9200(01)03020-7.

The next section of the paper presents the CDR architecture and design issues. Section III deals with the design of the building blocks. Section IV describes the experimental results.

II. ARCHITECTURE

The choice of the CDR architecture is primarily determined by the speed and supply voltage limitations of the technology as well as the power dissipation and jitter requirements of the system.

In a generic CDR circuit, shown in Fig. 1, the phase detector compares the phase of the incoming data to the phase of the clock generated by the voltage-controlled oscillator (VCO), producing an error that is proportional to the phase difference between its two inputs. The error is then applied to a charge pump and a low-pass filter so as to generate the oscillator control voltage. The clock signal also drives a decision circuit, thereby retiming the data and reducing its jitter.

If attempted in a 0.18-μm CMOS technology, the architecture of Fig. 1 poses severe difficulties for 10-Gb/s operation. Although exploiting aggressive device scaling, the CMOS process used in this work provides marginal performance for such speeds. For example, even simple digital latches or three-stage ring oscillators fail to operate reliably at these rates. These issues make it desirable to employ a "half-rate" CDR architecture, where the VCO runs at a frequency equal to half of the input data rate. The concept of the half-rate clock has been used in [2]–[5]. However, [2] and [3] incorporate a bang–bang phase detector, possibly creating a large ripple on the control line of the oscillator and hence high jitter. The circuit reported in [4] inherently has a smaller output jitter as a result of using a linear phase detector, but it fails to operate at speeds above 6 Gb/s in 0.18-μm CMOS technology. The circuit of [5] benefits from a new linear phase detection scheme, but it may not operate properly with certain data patterns.

Another critical issue in the architecture of Fig. 1 relates to the inherently unequal propagation delays for the two inputs of the phase detector: most phase detectors that operate properly with random data (e.g., a D flip-flop) are asymmetric with respect to the data and clock inputs, thereby introducing a systematic skew between the two in phase-lock condition. Since it is difficult to replicate this skew in the decision circuit, the generic CDR architecture suffers from a limited phase margin, unless the raw speed of the technology is much higher than the data rate.

The problem of the skew demands that phase detection and data regeneration occur in the same circuit such that the clock still samples the data at the midpoint of each bit even in the

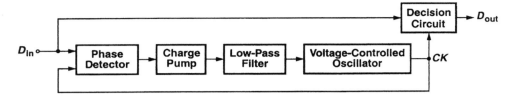

Fig. 1. Generic CDR architecture.

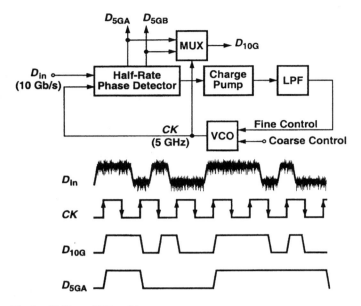

Fig. 2. Half-rate CDR architecture.

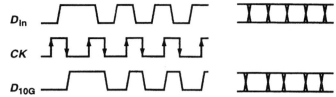

Fig. 3. Effect of clock duty cycle distortion.

presence of a finite skew. For example, the Hogge PD [6] automatically sets the clock phase to the optimum point in the data eye (but it fails to operate properly with a half-rate clock).

The above considerations lead to the CDR architecture shown in Fig. 2. Here, a half-rate PD produces an error proportional to the phase difference between the 10-Gb/s data stream and the 5-GHz output of the VCO. Furthermore, the PD automatically retimes and demultiplexes the data, generating two 5-Gb/s sequences D_{5GA} and D_{5GB}. Although the focus of this work is point-to-point communications, a full-rate retimed output, D_{10G}, is also generated to produce flexibility in testing and exercise the ultimate speed of the technology. The VCO has both fine and coarse control lines, the latter allowing inclusion of a frequency-locked loop in future implementations.

In this work, a new approach to performing linear phase detection using a half-rate clock is described. Owing to its simplicity, this technique achieves both high speed and low power dissipation while minimizing the ripple on the oscillator control voltage.

It is interesting to note that half-rate architectures do suffer from one drawback: the deviation of the clock duty cycle from 50% translates to bimodal jitter. As depicted in Fig. 3, since both clock edges sample the data waveform, the clock duty cycle distortion pushes both edges away from the midpoint of the bits. Typical duty cycle correction techniques used at lower speeds are difficult to apply here as they suffer from significant dynamic mismatches themselves. Thus, special attention is paid to symmetry in the layout to minimize bimodal jitter.

Another important aspect of CDR design is the leakage of data transitions to the oscillator. In Fig. 2, such leakage arises from: 1) capacitive feedthrough from D_{in} to CK in the PD; 2) capacitive feedthrough from D_{5GA} and D_{5GB} to CK through the multiplexer; and 3) coupling of D_{10G} to the oscillator through the substrate. To minimize these effects, the VCO is followed by an isolation buffer and all of the building blocks incorporate fully differential topologies.

III. BUILDING BLOCKS

A. VCO

The design of the VCO directly impacts the jitter performance and the reproducibility of the CDR circuit. While LC topologies achieve a potentially lower jitter, their limited tuning range makes it difficult to obtain a target frequency without design and fabrication iterations. Since the circuit reported here was our first design in 0.18-μm technology, a ring oscillator was chosen so as to provide a tuning range wide enough to encompass process and temperature variations.

A three-stage differential ring oscillator [Fig. 4(a)] driving a buffer operates no faster than 7 GHz in 0.18-μm CMOS technology. The half-rate CDR architecture overcomes this limitation, requiring a frequency of only 5 GHz.

As shown in Fig. 4(b), each stage consists of a fast and a slow path whose outputs are summed together. By steering the current between the fast and the slow paths, the amount of delay achieved through each stage and hence the VCO frequency can be adjusted. All three stages in the ring are loaded by identical buffers to achieve equal rise and fall times and thus improve the jitter performance. Fig. 4(c) shows the transistor implementation of each delay stage. The fast and slow paths are formed as differential circuits sharing their output nodes. The tuning is achieved by reducing the tail current of one and increasing that of the other differentially. Since the low supply voltage makes it difficult to stack differential pairs under M_1–M_2 and M_3–M_4, the current variation is performed through mirror arrangements driven by pMOS differential pairs. Fig. 5 depicts the small-signal gain and phase response of each delay stage. While providing a phase shift of 60°, each stage achieves a gain

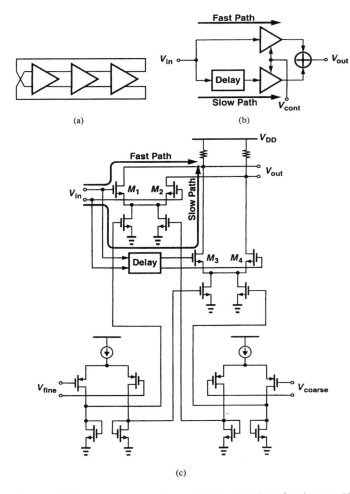

(a)

(b)

(c)

Fig. 4. (a) Three-stage ring oscillator. (b) Implementation of each stage. (c) Transistor-level schematic.

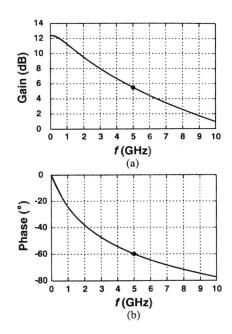

Fig. 5. Small-signal gain and phase response of each delay stage.

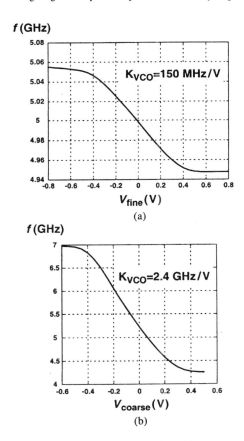

$K_{VCO}=150$ MHz/V

(a)

$K_{VCO}=2.4$ GHz/V

(b)

Fig. 6. VCO gain partitioning. (a) Fine control. (b) Coarse control.

of 5.5 dB at 5 GHz, yielding robust oscillation at the target frequency.

A critical drawback of supply scaling in deep-submicron technologies is the inevitable increase in the VCO gain for a given tuning range. To alleviate this difficulty, the control of the VCO is split between a coarse input and a fine input. The partitioning of the control allows more than one order of magnitude reduction in the VCO sensitivity. The idea is that the fine control is established by the phase detector and the coarse control is a provision for adding a frequency detection loop. The coarse control is provided externally in this prototype. The fine control exhibits a gain of 150 MHz/V and the coarse control, 2.5 GHz/V (Fig. 6). The tuning range is 2.7 GHz ($\approx 54\%$).

B. Phase Detector

Phase detectors generally appear in two different forms. Nonlinear PDs coarsely quantize the phase error, producing only a positive or negative value at their output. Linear PDs, on the other hand, generate a linearly proportional output that drops to zero when the loop is locked.

Compared to nonlinear PDs, linear PDs result in less charge pump activity, smaller ripple on the oscillator control line, and hence lower jitter. In a linear PD, such as that described in [6], the phase error is obtained by taking the difference between the width of two pulses, both of which are generated whenever a data transition occurs. The width of one of the pulses is linearly proportional to the phase difference between the clock and the data, whereas the width of the other is constant. By using a differential error signal, pattern dependency of phase error is can-

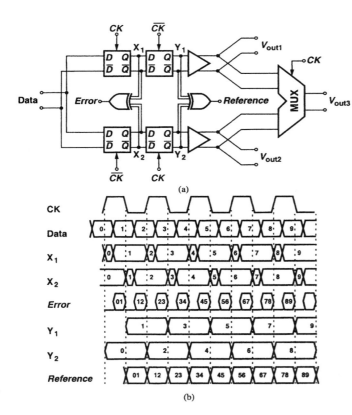

(a)

(b)

Fig. 7. (a) Phase detector. (b) Operation of the circuit.

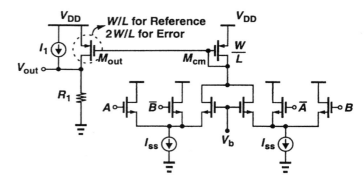

Fig. 8. Symmetric XOR gate.

celled because both pulses are present only when a data transition occurs.

For linear phase comparison between data and a half-rate clock, each transition of the data must produce an "error" pulse whose width is equal to the phase difference. Furthermore, to avoid a dead zone in the characteristics, a "reference" pulse must be generated whose area is subtracted from that of the error pulse, thus creating a net value that falls to zero in lock.

The above observations lead to the PD topology shown in Fig. 7(a). The circuit consists of four latches and two XOR gates. The data is applied to the inputs of two sets of cascaded latches, each cascade constituting a flipflop that retimes the data. Since the flipflops are driven by a half-rate clock, the two output sequences V_{out1} and V_{out2} are the demultiplexed waveforms of the original input sequence if the clock samples the data in the middle of the bit period.

The operation of the PD can be described using the waveforms depicted in Fig. 7(b). The basic unit employed in the circuit is a latch whose output carries information about the zero crossings of both the data and the clock. The output of each latch tracks its input for half a clock period and holds the value for the other half, yielding the waveforms shown in Fig. 7(b) for points X_1 and X_2. The two waveforms differ because their corresponding latches operate on opposite clock edges. Produced as $X_1 \oplus X_2$, the *Error* signal is equal to ZERO for the portion of time that identical bits of X_1 and X_2 overlap, and equal to the XOR of two consecutive bits for the rest. In other words, *Error* is equal to ONE only if a data transition has occurred.

It may seem that the *Error* signal uniquely represents the phase difference, but that would be true only if the data were pe-

riodic. The random nature of the data and the periodic behavior of the clock in fact make the average value of *Error* pattern dependent. For this reason, a reference signal must also be generated whose average conveys this dependence. The two waveforms Y_1 and Y_2 contain the samples of the data at the rising and falling edges of the clock. Thus, $Y_1 \oplus Y_2$ contains pulses as wide as half the clock period for every data transition, serving as the reference signal.

While the two XOR operations provide both the *Error* and the *Reference* pulses for every data transition, the pulses in *Error* are only half as wide as those in *Reference*. This means that the amplitude of *Error* must be scaled up by a factor of two with respect to *Reference* so that the difference between their averages drops to zero when clock transitions are in the middle of the data eye. The phase error with respect to this point is then linearly proportional to the difference between the two averages.

In order to generate a full-rate output, the demultiplexed sequences are combined by a multiplexer that operates on the half-rate clock as well. This output can also be used for testing purposes in order to obtain the overall bit-error rate (BER) of the receiver.

It is important to note that the XOR gates in Fig. 7 must be symmetric with respect to their two differential inputs. Otherwise, differences in propagation delays result in systematic phase offsets. Each of the XOR gates is implemented as shown in Fig. 8 [7]. The circuit avoids stacking stages while providing perfect symmetry between the two inputs. The output is single-ended but the single-ended *Error* and *Reference* signals produced by the two XOR gates in the phase detector are sensed with respect to each other, thus acting as a differential drive for the charge pump. The operation of the XOR circuit is as follows. If the two logical inputs are not equal, then one of the input transistors on the left and one of the input transistors on the right turn on, thus turning M_{cm} off. If the two inputs are identical, one of the tail currents flows through M_{cm}. Since the average current produced by the *Error* XOR gate is half of that generated by the *Reference* XOR gate, transistor M_{out} is scaled differently, making the average output voltages equal for zero phase difference. Channel length modulation of transistor M_{out} reduces the precision of current scaling between the two XOR gates. This effect can be avoided by increasing the length of the device.

The gain of the PD is determined by the value of the resistor R_1 and the tail current sources (I_{ss}). The voltage V_b is generated on chip in order to track the variations over temperature and

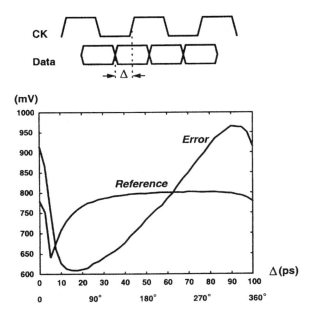

CK

Data

$\leftarrow \Delta \rightarrow$

Fig. 9. Determination of PD gain.

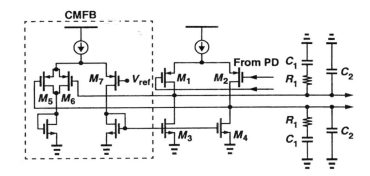

Fig. 10. Charge pump and loop filter.

process. This voltage equals the output common-mode level of the latches preceding the XOR gate. It is generated using a differential pair that is a replica of the preamplifier section of the latch. Current source I_1 raises the common-mode level of the differential signal formed by the *Error* and *Reference* signals, making V_{out} compatible with the input of the charge pump.

It is instructive to plot the input/output characteristic of the PD to ensure linearity and absence of dead zone. This is accomplished by obtaining the average values of *Error* and *Reference* while the circuit operates at maximum speed. Fig. 9 shows the simulated behavior as the phase difference varies from zero to one bit period. The *Reference* average exhibits a notch where the clock samples the metastable points of the data waveform. The *Error* and *Reference* signals cross at a phase difference approximately 55 ps from the metastable point, indicating that the systematic offset between the data and the clock is very small. The linear characteristic of the phase detector results in minimal charge pump activity and small ripple on the control line in the locked condition.

The choice of the logic family used for the XOR gates and the latches is determined by the speed and switching noise considerations. While rail-to-rail CMOS logic achieves relatively high speeds, it requires amplifying the data swings generated by the stage preceding the CDR circuit (typically a limiting amplifier). Furthermore, CMOS logic produces enormous switching noise in the substrate and on the supplies, disturbing the oscillator considerably. For these reasons, the building blocks employ current-steering logic. The phase detector incorporates an input buffer with on-chip resistive matching.

C. Charge Pump and Loop Filter

Fig. 10 shows the implementation of the differential charge pump. The common-mode feedback (CMFB) circuit senses the output CM level by M_5 and M_6, providing correction through M_3 and M_4. Both the matching and channel-length modulation of M_1–M_4 in Fig. 10 impact the residual phase error in locked

condition. Thus, their lengths and widths are relatively large to minimize these effects.

The design of the loop filter is based on a linear time-invariant model of the loop and is performed in continuous time domain. The loop is in general a nonlinear time-variant system and can only be assumed linear if the phase error is small. The time-invariant analysis is valid if the averaging behavior of the loop rather than its single-cycle performance is of interest, i.e., the loop can be analyzed by continuous-time approximation if the loop bandwidth is small. Under this condition, the state of the CDR changes by only a small amount on each cycle of the input signal.

A low-pass jitter transfer function with a given bandwidth and a maximum gain in the passband is specified for a SONET system. The closed-loop transfer function of the CDR has a zero at a frequency lower than the first closed-loop pole. This results in jitter peaking that can never be eliminated. But the peaking can be reduced to negligible levels by overdamping the loop.

As derived in [8], the closed-loop unity-gain bandwidth is approximated as

$$\omega_{-3\,\text{dB}} = R_1 \cdot K_{\text{VCO}} \cdot K_{\text{PD}} \cdot g_m \qquad (1)$$

where K_{VCO} and K_{PD} are the gains of the VCO and PD, respectively, and g_m denotes the conversion gain of the charge pump. Equation (1) can be used to determine the value of R_1. The amount of the jitter peaking in the closed-loop transfer function can be approximated as

$$JP = 1 + \frac{1}{\omega_{-3\,\text{dB}} \cdot R_1 \cdot C_1}. \qquad (2)$$

Equation (2) yields the required value of C_1. In order to obtain greater suppression of high-frequency jitter, a second capacitor is added in parallel with the series combination of R_1 and C_1. These components are added externally to achieve flexibility in defining the closed-loop characteristics of the circuit.

Another advantage of linear PDs over their bang–bang counterparts is that their jitter transfer characteristics is independent of the jitter amplitude. It should also be mentioned that if the CDR is followed by a demultiplexer, the tight specifications for jitter peaking need not to be satisfied because such specifications are defined for cascaded regenerators handling full-rate data.

Fig. 11 depicts the simulated behavior of the CDR circuit at the transistor level. The voltage across the filter is initialized to

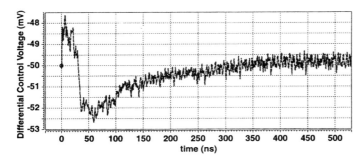

Fig. 11. Lock acquisition.

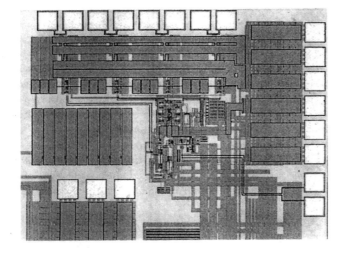

Fig. 12. Chip photograph.

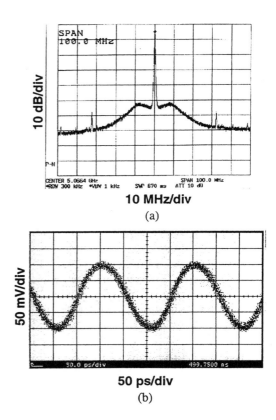

10 MHz/div

(a)

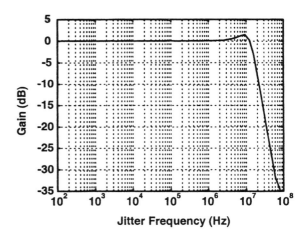

50 ps/div

(b)

Fig. 13. (a) Spectrum of the recovered clock. (b) Recovered clock in the time domain.

Fig. 14. Measured jitter transfer characteristic.

a value relatively close to its value in phase lock. The loop goes through a transition of 350 ns before it locks. The ripple on the control line in phase lock is approximately 1 mV.

IV. EXPERIMENTAL RESULTS

The CDR circuit has been fabricated in a 0.18-μm CMOS process. Fig. 12 shows a photograph of the chip, which occupies an area of 1.1×0.9 mm^2. Electrostatic discharge (ESD) protection diodes are included for all pads except the high-speed lines. Nonetheless, since all of these lines have a 50-Ω termination to V_{DD}, they exhibit some tolerance to ESD. The circuit is tested in a chip-on-board assembly. In this prototype, the width of the poly resistors was not sufficient to guarantee the nominal sheet resistance. As a result, the fabricated resistor values deviated from their nominal value by 30%, and the VCO center frequency was proportionally lower than the simulated value at the nominal supply voltage (1.8 V). The supply was increased to 2.5 V, to achieve reliable operation at 10 Gb/s. While such a high supply voltage creates hot-carrier effects in rail-to-rail CMOS circuits, it is less detrimental in this design because no transistor in the circuit experiences a gate–source or drain–source voltage of more than 1 V. This issue is nonetheless resolved in a second design [9] by proper choice of resistor dimensions. The circuit is brought close to lock with the aid of the VCO coarse control before phase locking takes over.

Fig. 13(a) shows the spectrum of the clock in response to a 10-Gb/s data sequence of length $2^{23} - 1$. The effect of the noise shaping of the loop can be observed in this spectrum. The phase

noise at 1-MHz offset is approximately equal to -106 dBc/Hz. Fig. 13(b) depicts the recovered clock in the time domain. The time-domain measurements using an oscilloscope overestimate the jitter, requiring specialized equipment, e.g., the Anritsu MP1777 jitter analyzer. The jitter performance of the CDR circuit is characterized by this analyzer. A random sequence of length $2^{23} - 1$ produces 14.5 ps of peak-to-peak and 1 ps of RMS jitter on the clock signal. These values are reduced to 4.4 and 0.6 ps, respectively, for a random sequence of length $2^7 - 1$.

The measured jitter transfer characteristics of the CDR is shown in Fig. 14. The jitter peaking is 1.48 dB and the 3-dB bandwidth is 15 MHz. The loop bandwidth can be reduced to

Demultiplexed Data

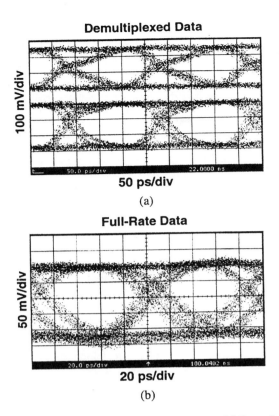

50 ps/div

(a)

Full-Rate Data

20 ps/div

(b)

Fig. 15. (a) Recovered demultiplexed data. (b) Recovered full-rate data.

the SONET specifications, but the jitter analyzer must then generate large jitter and drives the loop out of lock. The loop bandwidth can be reduced to the SONET specifications if a means of frequency detection is added to the loop [9]. The circuit is then much less susceptible to loss of lock due to the jitter generated by the analyzer.

Fig. 15 depicts the retimed data. The demultiplexed data outputs are shown in Fig. 15(a). The difference between the waveforms results from systematic differences between the bond wires and traces on the test board. Fig. 15(b) depicts the full-rate output. Using this output, the BER of the system can be measured. With a random sequence of $2^7 - 1$, the BER is smaller that 10^{-12}. However, a random sequence of $2^{23} - 1$ results in a BER of 1.28×10^{-6}. This BER can be reduced if the bandwidth of the output buffer driving the 10-Gb/s data is increased. Furthermore, if the value of the linear resistors is adjusted to their nominal value, the increased operating speed of the back-end multiplexer results in an improved BER [9].

The CDR circuit exhibits a capture range of 6 MHz and a tracking range of 177 MHz. The total power consumed by the circuit excluding the output buffers is 72 mW from a 2.5-V supply. The VCO, the PD, and the clock and data buffers consume 20.7, 33.2, and 18.1 mW, respectively.

V. CONCLUSION

CMOS technology holds great promise for optical communication circuits. The raw speed resulting from aggressive scaling along with high levels of integration provide a high performance at low cost. A 10-Gb/s clock and data recovery circuit designed in 0.18-μm CMOS technology performs phase locking, data regeneration, and demultiplexing with 1 ps of RMS jitter.

REFERENCES

[1] Y. M. Greshishchev and P. Schvan, "SiGe clock and data recovery IC with linear type PLL for 10-Gb/s SONET application," in *Proc. 1999 Bipolar/BiCMOS Circuits and Technology Meeting*, Sept. 1999, pp. 169–172.

[2] M. Wurzer *et al.*, "40-Gb/s integrated clock and data recovery circuit in a silicon bipolar technology," in *Proc. 1998 Bipolar/BiCMOS Circuits and Technology Meeting*, Sept. 1998. pp. 136–139.

[3] M. Rau *et al.*, "Clock/data recovery PLL using half-frequency clock," *IEEE J. Solid-State Circuits*, vol. 32, pp. 1156–1159, July 1997.

[4] K. Nakamura *et al.*, "A 6 Gb/s CMOS phase detecting DEMUX module using half-frequency clock," in *Dig. Symp. VLSI Circuits*, June 1998, pp. 196–197.

[5] E. Mullner, "A 20-Gb/s parallel phase detector and demultiplexer circuit in a production silicon bipolar technology with $f_T = 25$ GHz," in *Proc. 1996 Bipolar/BiCMOS Circuits and Technology Meeting*, Sept. 1996, pp. 43–45.

[6] C. Hogge, "A self-correcting clock recovery circuit," *J. Lightwave Technol.*, vol. LT-3, pp. 1312–1314, Dec. 1985.

[7] B. Razavi, Y. Ota, and R. G. Swarz, "Design techniques for low-voltage high-speed digital bipolar circuits," *IEEE J. Solid-State Circuits*, vol. 29, pp. 332–339, Mar. 1994.

[8] L. M. De Vito, "A versatile clock recovery architecture and monolithic implementation," in *Monolithic Phase-Locked Loops and Clock Recovery Circuits, Theory and Design*, B. Razavi, Ed. New York: IEEE Press, 1996.

[9] J. Savoj and B. Razavi, "A 10-Gb/s CMOS clock and data recovery circuit with frequency detection," in *Int. Solid-State Circuits Conf. Dig. Tech. Papers*, Feb. 2001, pp. 78–79.

Jafar Savoj (S'98) was born in Tehran, Iran, in 1974. He received the B.S.E.E. degree from Sharif University of Technology, Tehran, in 1996 and the M.S.E.E. degree from the University of California, Los Angeles (UCLA), in 1998. He is currently working toward the Ph.D. degree at UCLA.

He spent the summer of 1998 with Integrated Sensor Solutions, San Jose, CA, working on the design of high-precision interfaces for sensor applications. During the summer of 1999, he was with NewPort Communications, Irvine, CA, developing CMOS clock and data recovery circuits for the SONET OC-192 standard.

Mr. Savoj received the IEEE Solid-State Circuits Society Predoctoral Fellowship for 2000–2001, and the Beatrice Winner Award for Editorial Excellence at the 2001 ISSCC. He is also a recipient of the Design Contest Award of the 2001 Design Automation Conference.

5.3 A 10Gb/s CMOS Clock and Data Recovery Circuit with Frequency Detection

Jafar Savoj, Behzad Razavi

Electrical Engineering Department, University of California, Los Angeles, CA

Clock and data recovery (CDR) circuits operating in the 10Gb/s range have become attractive for the optical fiber backbone of the Internet. While CDR circuits operating at 10Gb/s have been designed in bipolar technologies, cost and integration issues make it desirable to implement these circuits in standard CMOS processes. This 10Gb/s CDR circuit is realized in 0.18μm CMOS technology. Architecture and circuit techniques circumvent the speed limitations of the devices. In contrast to previous work [1], this design incorporates an LC oscillator to reduce the jitter as well as a phase/frequency detector to achieve a wide capture range.

Shown in Figure 5.3.1, the CDR consists of a phase/frequency detector (PFD), a voltage-controlled oscillator (VCO), a charge pump, and a low-pass filter (LPF). The PFD compares the phase and frequency of the input data to that of a half-rate clock, providing two binary error signals for phase and frequency. The PFD is designed so that, in addition to providing information about the phase error, it retimes the data as well. Consequently, the CDR exhibits no systematic offset, i.e., inherent skews between clock and data edges due to their unidentical paths through the loop do not degrade the quality of detection. The VCO provides four differential half-quadrature phases over the full tuning range. All building blocks are fully differential.

Since the half-rate frequency detector requires clock phases that are integer multiples of 45°, the 5GHz VCO is designed as a ring structure consisting of four LC-tuned stages [Figure 5.3.2a]. If the dc feedback around the ring is positive, all stages operate in-phase at the resonance frequency defined by the LC tanks. On the other hand, if the dc feedback is negative, the frequency shifts by a small amount so as to allow each stage to contribute 45° of phase.

The oscillator topology has two advantages over resistive-load ring oscillators. First, owing to the phase slope (Q) provided by the resonant loads, it exhibits less phase noise. Second, its frequency of oscillation is only a weak function of the number of stages, generating multiple phases with no speed penalty. By comparison, a four-stage resistive-load ring operates at a lower frequency.

Figure 5.3.2b shows the implementation of each stage. The loads are formed using on-chip spiral inductors and MOS varactors. Resistor R1 provides a shift in the output common-mode level, allowing both positive and negative voltages across the varactors and thus maximizing the tuning range. Modeling each tank by a parallel network, the required 45° phase shift slightly detunes the circuit. The oscillation frequency is given by $\omega_0 = (LC)^{-0.5}(1-1/Q_0)^{0.5}$, where Q_0 denotes the Q of each stage at resonance.

The phase detector (PD) is derived from the data transition tracking loop described in Reference 2. In this PD, in-phase and quadrature phases of a half-rate clock signal sample the data in two double-edge-triggered flipflops (DETFFs). Figure 5.3.3 shows the implementation of the PD. Two latches operating on opposite clock phases and a multiplexer form a DETFF that samples the data using both the positive and negative transitions of a half-rate clock. The two signals V_1 and V_2 are therefore the in-phase and quadrature samples of data, respectively, and one is used to route the other or its complement.

The phase detector operates at high speeds because it uses a half-rate clock. Since in the locked condition, the rising and falling edges of the quadrature clock coincide with data transitions, the in-phase clock transitions sample the data at its optimum point with no systematic offset, generating a full-rate output stream. Also, since the phase-error signal is reevaluated only at data transitions, it incurs little ripple. Note that the output is independent of the data transition density, resulting in reduction of pattern-dependent jitter.

With the small CDR loop bandwidths specified by optical standards, circuits employing only phase detection suffer from an extremely narrow capture range, e.g., about 1% of the center frequency. For this reason, a means of frequency detection is necessary to guarantee lock to random data. As with other phase detectors, the half-rate PD of Figure 5.3.3 generates a beat frequency equal to the difference between the data rate and twice the VCO frequency. However, it does not provide knowledge of the polarity of this difference. Figure 5.3.4 depicts the half-rate phase and frequency detector introduced in this work. A second PD is added and driven by phases that are 45° away from those in the first PD. The circuit operates as follows. (1) If the clock is slow, V_{PD1} leads V_{PD2}; therefore, if V_{PD2} is sampled by the rising and falling edges of V_{PD1}, the results are negative and positive, respectively. (2) If the clock is fast, V_{PD1} lags V_{PD2}. Therefore, if V_{PD2} is sampled by the rising and falling edges of V_{PD1}, the results are the reverse of the previous case.

The output buffer delivering the 10Gb/s retimed data with high current levels requires a bandwidth of more than 7 GHz. As shown in Figure 5.3.5, the buffer stage employs inductive peaking [3]. The value of the spiral inductors is chosen so as to avoid ripple in the passband. Since the quality factor of the inductors is not critical here, the spiral structures have a linewidth of only 4μm to achieve a high self-resonance frequency.

The CDR circuit is fabricated in a 0.18μm CMOS technology. The circuit is tested in a chip-on-board assembly while operating with a 1.8V supply. The phase noise of the clock in response to a 9.95328Gb/s data sequence of length $2^{23}-1$ at 1MHz offset is approximately equal to -107dBc/Hz. Figure 5.3.6a depicts the recovered clock and data. A pseudo-random sequence of length $2^{23}-1$ produces 9.9ps of peak-to-peak and 0.8ps rms jitter on the clock signal. The jitter characteristics are measured by the Anritsu MP1777 jitter analyzer. The measured jitter transfer characteristic of the CDR is shown in Figure 5.3.6b. The jitter peaking is 0.04dB and the 3dB bandwidth is 5.2MHz. Despite the small loop bandwidth, the frequency detector provides a capture range of 1.43GHz, obviating the need for external references. The total power consumed by the circuit excluding the output buffers is 91mW from a 1.8V supply. Figure 5.3.7 shows a micrograph of the chip, which occupies 1.75x1.55mm².

Acknowledgments:
The authors thank NewPort Communications for fabrication and test support. This work was supported by SRC and Cypress Semiconductor.

References:
[1] J. Savoj and B. Razavi, "A 10-Gb/s CMOS Clock and Data Recovery Circuit," Dig. of Symposium on VLSI Circuits, pp. 136-139, June 2000.
[2] A. W. Buchwald, Design of Integrated Fiber-Optic Receivers Using Heterojunction Bipolar Transistors, Ph.D. Thesis, University of California, Los Angeles, Jan. 1993.
[3] J. Savoj and B. Razavi, "A CMOS Interface Circuit for Detection of 1.2-Gb/s RZ Data," ISSCC Digest of Technical Papers, pp. 278-279, Feb. 1999.

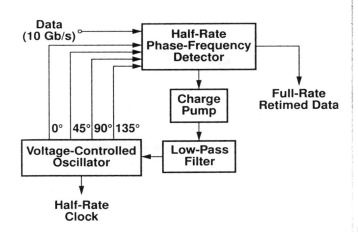

Figure 5.3.1: CDR architecture.

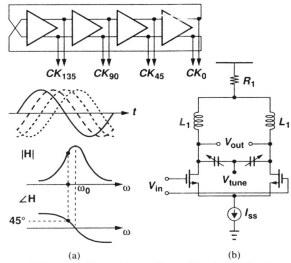

Figure 5.3.2: (a) Four-stage LC-tuned ring oscillator, (b) implementation of one stage.

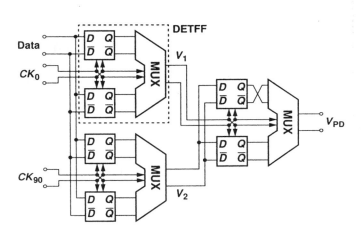

Figure 5.3.3: Phase detector.

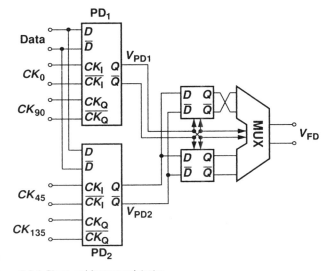

Figure 5.3.4: Phase and frequency detector.

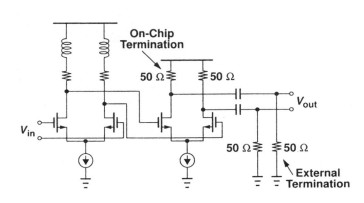

Figure 5.3.5: Output buffer.

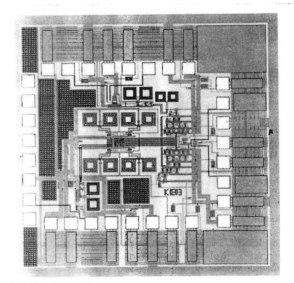

Figure 5.3.7: Die micrograph.

689

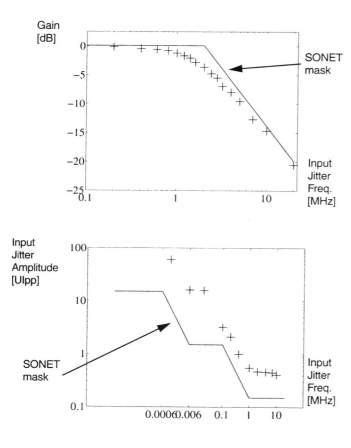

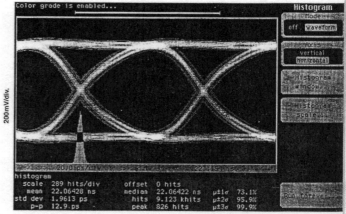

Figure 5.1.6: Measured jitter transfer function with <0.1dB peaking and jitter tolerance for BER<10^{-12}.

Figure 5.3.6: (a) Recovered data and clock, (b) measured jitter transfer characteristics.

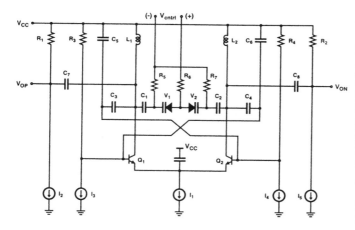

Figure 5.4.5: Differential LC VCO.

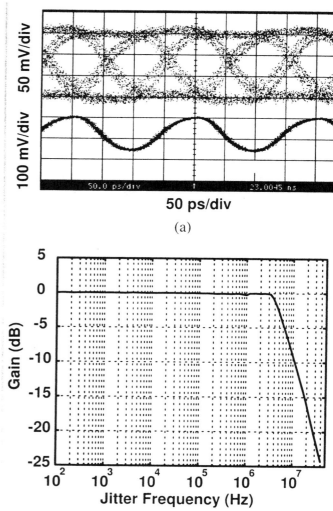

Figure 5.4.6: Differential data output for packaged device at 10.664Gb/s (2^{15}-1 PRBS).

15.4 A 10Gb/s CDR/DEMUX with LC Delay Line VCO in 0.18μm CMOS

Jonathan E. Rogers, John R. Long

Department of Electrical and Computer Engineering, University of Toronto, ON, Canada

A key block in an optoelectronic data receiver is the clock and data recovery system (CDR). The CDR extracts timing information to time-division demultiplex (DEMUX) the received bit stream. For monolithic implementations, a phase-locked loop (PLL) based CDR solution is preferred and an all-CMOS solution has potential advantages in manufacturability and cost [1,2,3].

The CMOS CDR/DEMUX in Figure 15.4.1 consists of 3 major circuit blocks: early/late phase detector, a quadrature-phase voltage-controlled oscillator (VCO), and a charge pump filter. Differential 10Gb/s data signal at the DATA inputs is broadband matched on-chip. A half-rate phase detector compares the data falling edge with the phase of locally-generated quadrature clock signals, CKI and CKQ. This produces either an early or late result unless a falling data edge is not present, in which case the detector produces no output. Information generated by the phase detector is fed directly to the proportional (bang-bang) VCO control and to the charge pump for integral control of the system dynamics. If neither early nor late signals are present, the oscillator frequency and the charge on hold capacitor C_{HE} are unaltered (tri-state condition). On-chip capacitor C_{HI} reduces glitches caused by the I/O inductance. When phase-locked, oscillator phase CKI is properly aligned for data retiming and 1:2 demultiplexing. A higher ratio DEMUX (e.g., 1:8 or 1:16) would normally be used in practice.

Figure 15.4.2 shows a block diagram of the two-stage LC voltage-controlled oscillator (VCO). Symmetry ensures that precise in-phase (CKI) and quadrature (CKQ) clocks are generated. The 5GHz center frequency is tunable through external (EX), internal (IN) and high frequency bang-bang (BB) inputs. A fully-differential design rejects supply noise and the delay line promotes frequency stability with process and temperature variations.

Oscillation frequency is determined by the total propagation time through the gain and differential LC transmission line stages. The load resistors of each CML differential pair buffer match the output to the 75Ω characteristic impedance of the LC delay line. The delay through each gain stage is only 7ps (simulated) due to the low impedance seen at the drains of the switching transistors.

The LC delay line is a fully-symmetric spiral (Figure 15.4.3). GEMCAP2 is used to determine a distributed lumped-element model for the delay line and refine the oscillator design (Figure 15.4.2) [4]. Each delay line has 150μm outside dimension, 4μm conductor width and 2μm conductor spacing. This reduces the interwinding capacitance (Co in Figure 15.4.2), allowing a larger tuning capacitance. The effective inductance between input and output is ~2.5nH with 120fF parasitic capacitance. The delay lines account for 43ps delay each at 5GHz, or 86% of the total delay around the oscillator loop.

The remaining capacitance required for 5GHz oscillation is added with inversion-mode nMOS varactors at each gain stage input (a high-impedance node). The measured external tuning range and bang-bang frequency step (varied using the BB Adj input) are 125MHz and 2.5-5MHz, respectively. Integral loop

tuning range is designed at 32.5MHz/V, and the (simulated) voltage swing at the clock buffer inputs is 2.5V differential.

The early/late phase detector is architecturally similar to the design of Reference [5] (Figure 15.4.4). The block diagram is drawn single-ended for simplicity. The actual phase detector is fully differential and implemented in MOS current-mode logic (MCML). The samples are gated by the same clock edge to simplify the design. Pulses are retimed at the phase detector output to remove asymmetries in both amplitude and duration caused by the early/late logic.

The phase noise spectral density of the VCO running open-loop, phase-locked, and the phase noise of the reference source are shown in Figure 15.4.5. At 1MHz offset from the carrier, free-running VCO phase noise is -103dBc/Hz, which falls to -127dBc/Hz when the CDR is locked to sinusoidal data input at 2.5GHz. The source noise is also shown (-135dBc/Hz), where the difference is primarily due to the x2 multiplication from reference source (2.5GHz) and locked VCO (5GHz), which adds a minimum of 6dB to the phase noise.

The eye pattern of Figure 15.4.6 is measured in response to 10Gb/s PRBS input data ($2^{31}-1$). The recovered clock jitter is 1.2ps rms, or 8ps p-p. Bit-error testing demonstrates error-free data recovery. The 5Gb/s output data eye has larger jitter than the clock due to pattern dependencies that are likely introduced by the output buffers, as the recovered clock output jitter faithfully tracks the input jitter.

Measured jitter transfer and tolerance meet SONETOC-192 requirements (measured jitter of 8ps p-p), with the exception of the small residual jitter tolerance. Poor electrical contact from probes to the chip and phase error between the quadrature clocks are likely sources of degradation in jitter performance.

The chip performance is summarized in Figure 15.4.7. The quadrature LC delay line oscillator center frequency is initially low (4.43GHz) due to parasitic interconnect inductance from the physical layout and is corrected to 5.04GHz by trimming the delay line inductance. The VCO has measured 60MHz/V sensitivity to power supply pulling. The external tuning range is lower than predicted due to model inaccuracies for the inversion-mode varactors and parasitic losses in the LC delay line. Supply pulling is close to that predicted from simulation and is about an order of magnitude lower than that of an RC ring oscillator based on similar gain stages. Oscillators from two fabrication runs are characterized and little change in center frequency is observed. A micrograph of the 1.9x1.5mm² IC is shown in Figure 15.4.6. It consumes 285mW from a 1.8V supply.

Acknowledgments:
This work was supported by Micronet, NSERC and the Nortel Institute at the University of Toronto. Circuit fabrication was facilitated by the Canadian Microelectronics Corporation.

References:
[1] R. C. Walker et al, "A Two-Chip 1.5-GBd Serial Link Interface", IEEE JSSC, pp. 1805-1811, Dec. 1992.
[2] Y. M. Greshishchev et al., "A Fully Integrated SiGe Receiver IC for 10-Gb/s Data Rate," IEEE JSSC, pp. 1949-1957, Dec. 2000.
[3] J. Savoj, B. Razavi, "A 10Gb/s CMOS Clock and Data Recovery Circuit with Frequency Detection," ISSCC Digest of Technical Papers, pp. 78-79, Feb. 2001.
[4] J. Long, M. Copeland, "The Modeling, Characterization, and Design of Monolithic Inductors for Silicon RFICs," IEEE JSSC, pp. 357-367, March 1997.
[5] J. Hauenschild et al, "A Plastic Packaged 10 Gb/s BiCMOS Clock and Data Recovering 1:4-Demultiplexer with External VCO", IEEE JSSC, pp. 2056-2059, Dec. 1996.

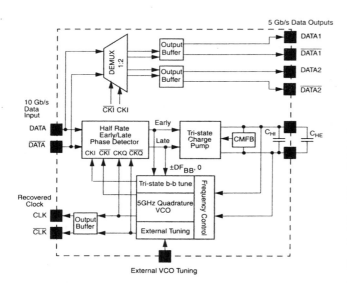

Figure 15.4.1: CDR/DEMUX block diagram.

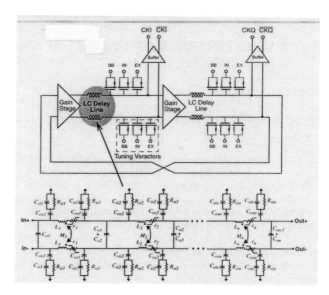

Figure 15.4.2: Voltage-controlled oscillator and LC delay line model.

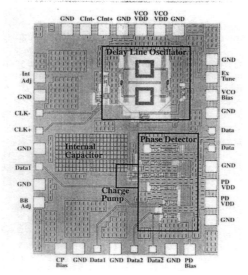

Figure 15.4.3: Test chip micrograph.

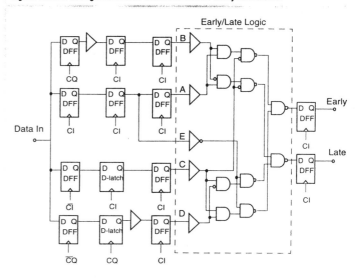

Figure 15.4.4: Half-rate phase detector.

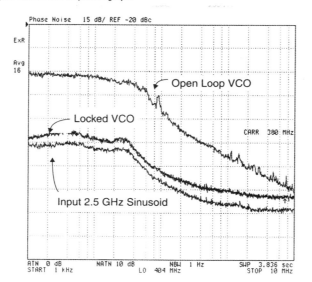

Figure 15.4.5: Measured phase noise.

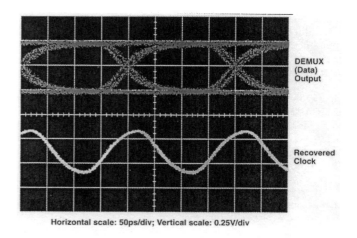

Horizontal scale: 50ps/div; Vertical scale: 0.25V/div

Figure 15.4.6: CDR performance with 2^{31}-1 PRBS data.

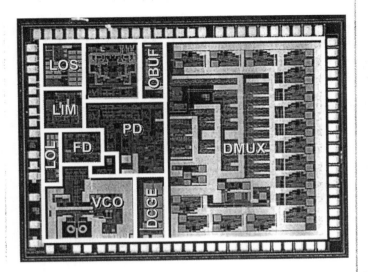

Figure 15.3.7: Die micrograph.

VCO Parameter	Simulated	Measured
Center Frequency	5GHz	4.43GHz
Center Frequency After Trimming		5.04GHz
Phase Noise	-	-103dBc at1MHz offset -75dBc at100kHz offset
External Tuning Range	225MHz	125MHz
Internal Tuning Range	106MHz	
Freq. vs. Supply Voltage	45MHz/V	60MHz/V
Process Variation	±2.0%	0.2% (Two Runs)
Temperature Variation	2.1MHz/°C	
Power Consumption	70mW	-
CDR Parameter	**Simulated**	**Measured**
Bit Rate	10Gb/s	10Gb/s
Jitter when Locked to a 2^{31}-1 PRBS Sequence	-	1.2ps rms, 8ps p-p
Capture and Locking Range (2.5 GHz Sinusoid Input)	-	21MHz
Recovered Clock Phase Noise (Locked to 2.5 GHz Sinusoid)	-	-127dBc at1MHz offset -113dBc at100kHz offset
Power Consumption (1.8V)	361mW	
Power Excluding Buffers (1.8V)	285mW	
Die Size		1.95mm x 1.5mm

Figure 15.4.7: CDR/DEMUX performance summary.

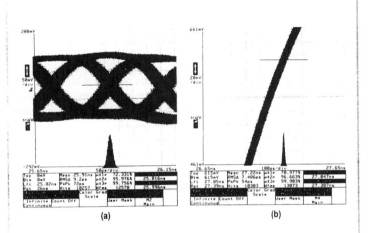

Figure 15.5.7: (a) Measured eye diagram and (b) jitter on recovered clock for 5Gb/s PRBS of length 2^7-1.

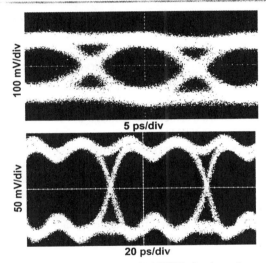

Figure 15.7.7: 50Gb/s output eye-diagram of 4:1 MUX (top) and one of 12.5Gb/s demultiplexed outputs (bottom).

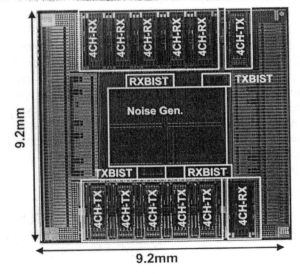

Figure 16.1.7: Test chip die micrograph.

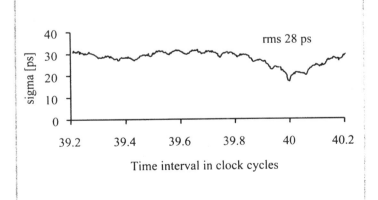

Figure 16.2.7: Single-shot precision variation.

A 40-Gb/s Integrated Clock and Data Recovery Circuit in a 50-GHz f_T Silicon Bipolar Technology

Martin Wurzer, Josef Böck, Herbert Knapp, Wolfgang Zirwas, Fritz Schumann, and Alfred Felder

Abstract—Clock and data recovery (CDR) circuits are key electronic components in future optical broadband communication systems. In this paper, we present a 40-Gb/s integrated CDR circuit applying a phase-locked loop technique. The IC has been fabricated in a 50-GHz f_T self-aligned double-polysilicon bipolar technology using only production-like process steps. The achieved data rate is a record value for silicon and comparable with the best results for this type of circuit realized in SiGe and III–V technologies.

Index Terms—Bipolar digital integrated circuits, clocks, data communication, high-speed integrated circuits, phase-locked loops, synchronization.

I. INTRODUCTION

THE demands for new services and increased flexibility have accelerated the development of telecommunication transport networks, which has resulted in the synchronous optical networks (SONET)/synchronous digital hierarchy (SDH) standards. Key elements of such high-capacity networks are fiber-optic communication links. Time-division multiplexing (TDM) systems operating at 10 Gb/s are now under development using advanced silicon bipolar production technologies to fabricate all high-speed IC's. Next generations with SONET/SDH are expected to operate at data rates of 40 Gb/s [1]. To enable such large-capacity optical transmission systems to be put into practical use, very high-speed monolithic IC's are required as key components.

It has been shown that basic digital functions like MUX and DMUX for 40-Gb/s optical-fiber TDM systems can be realized in silicon bipolar technology [2]. But clock and data recovery circuits in a silicon technology have so far only been demonstrated for 20 Gb/s [3]. With more sophisticated SiGe or III–V technologies, 40-Gb/s operation has been achieved [4]–[7]. Some of these solutions are hybrid.

All these realizations are based on either high-Q filters or phase-locked loops (PLL's). The advantage of the first concept is the easy implementation. The disadvantages are that temperature and frequency variation of filter group delay makes sampling time difficult to control, the high-Q filter is

Manuscript received January 11, 1999; revised March 23, 1999.
M. Wurzer is with Corporate Technology, Microelectronics, Siemens AG, Munich 81730 Germany and the Institut für Nachrichtentechnik und Hochfrequenztechnik, Technische Universität Wien, Austria (e-mail: Martin.Wurzer@mchp.siemens.de).
J. Böck, H. Knapp, F. Schumann, and A. Felder are with Corporate Technology, Microelectronics, Siemens AG, Munich 81730 Germany
W. Zirwas is with Information and Communication Networks, Siemens AG, Munich 81379 Germany.
Publisher Item Identifier S 0018-9200(99)06493-8.

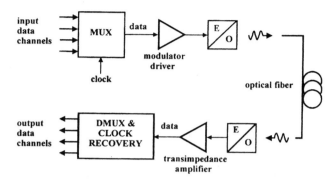

Fig. 1. Block diagram of the fiber-optic link.

difficult to integrate, and narrow pulses require a high f_T. The major advantages of the second approach are that the phase between the extracted clock and the received data is locked and that it can be implemented as a monolithic integrated circuit.

The goal of our work was to implement a cost-effective and reliable clock and data recovery circuit for 40 Gb/s in a production-near silicon bipolar technology. Therefore, an approach based on a PLL technique has been selected and will be described in this paper in more detail.

II. FIBER-OPTIC LINK

The described circuit has been developed for use in 40-Gb/s TDM fiber-optic links. A block diagram of such a link is shown in Fig. 1. The time-division multiplexer collects several data channels into a single high-speed data stream. An external modulator converts the data from electrical to optical signals by modulating the light of a semiconductor laser diode (E/O block). The O/E conversion on the receiving side is performed by a photodiode followed by a transimpedance amplifier. This bitstream is fed into the clock and data recovery unit. Its task is to synchronize the local oscillator to the phase of the incoming data and to retime the data. In contrast to 10-Gb/s systems, the decision function is now performed by a demultiplexer. This requires a DMUX with excellent retiming capability combined with a high input sensitivity [8].

III. ARCHITECTURE OF THE CLOCK AND DATA RECOVERY CIRCUIT

Fig. 2(a) shows the used concept of the CDR for the fiber-optic link (Fig. 1) in more detail. The main processing blocks are the demultiplexer consisting of two master–slave D-flip-flops (DFF1, DFF2) in parallel and an additional master–slave D-flip-flop (DFF3), which forms the phase detector together

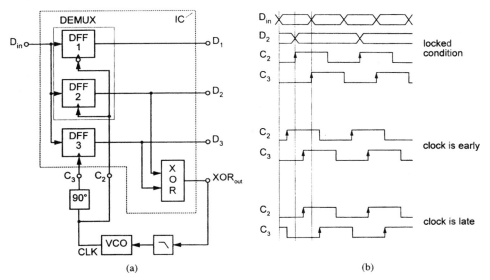

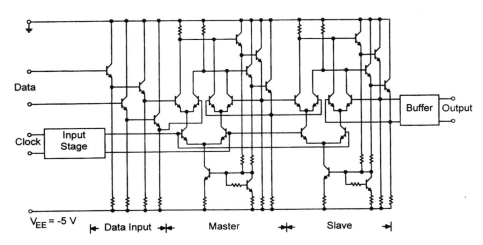

Fig. 2. CDR circuit: (a) block diagram and (b) timing diagram.

Fig. 3. Circuit diagram of the master–slave D-flip-flop.

with DFF2 and the XOR gate. All these functions are integrated in a single chip. The fixed 90° phase shifter, voltage-controlled oscillator (VCO), and loop filter have been realized externally with commercially available components.

Fig. 2(b) shows the timing diagram. The incoming 40-Gb/s data signal is applied to flip-flops DFF1, DFF2, and DFF3. DFF1 is toggled by $\overline{\text{CLK}}$, DFF2 by CLK, and DFF3 by the 90° delayed clock signal. This results in the sampling of the input in the vicinity of midbit and each following potential transition. If a transition is present, the phase relationship of the data and the clock can be deduced to be early or late. If the midbit clock CLK is too early, DFF3 samples the same bit; if it is too late, DFF3 samples the following bit. Under locked conditions, DFF3 samples at the edge of the data eye. The XOR compares the output samples of DFF2 and DFF3. The result is fed to the loop filter. The output signal of the loop filter serves as the control signal of the VCO.

The advantages of this concept are that all components operate at half the data rate and that the input is demultiplexed at the same time. The disadvantage is that the input signal has to drive three DFF's in parallel.

IV. CIRCUIT AND DESIGN PRINCIPLES

The circuit is designed for the single supply voltage of −5 V. The circuit principles used are seen in the circuit blocks of a master–slave D-flip-flop (MS-DFF), shown in Fig. 3. For details, see [9]. The well-proven E^2CL (emitter–emitter coupled logic) is used with emitter followers at the inputs and current switches at the outputs. The series gating between clock and data signals enables differential operation with low voltage swings ($\Delta V \approx 400$ mV$_{\text{p-p}}$) resulting in an increase in speed and a reduction of power consumption. Furthermore, differential operation reduces time jitter and crosstalk and offers good common-mode suppression compared to single-mode operation [10]. Cascaded emitter followers are used for level shifting and impedance transformation between the various current switches. Multiple emitter followers improve the decoupling capability and increase the collector-base voltage of the current-switch transistors allowing for smaller transistors, resulting in lower collector-base capacitances [10]. On-chip matching resistors (50 Ω) at all data inputs are used in order to reduce jitter introduced by reflections [2], [11].

695

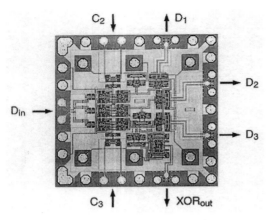

Fig. 4. Chip micrograph (chip size: 0.9 × 0.9 mm²).

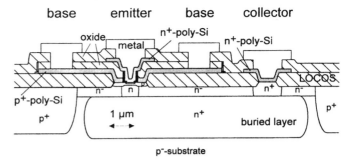

Fig. 5. Schematic cross section of a transistor.

TABLE I
DEVICE PARAMETERS (JUNCTION CAPACITANCES ARE ZERO-BIAS VALUES)

A_E	0.25×2.7 μm²
β	115
V_{Early}	20 V
R_{PI}	14 kΩ/□
BV_{EBO}	2.0 V
BV_{CBO}	10.8 V
BV_{CEO}	2.7 V
C_{EB}	8.5 fF
C_{CB}	8.0 fF
C_{CS}	26.5 fF
f_T	50 GHz
	(V_{CE} = 3 V, j_C = 2 mA/μm²)

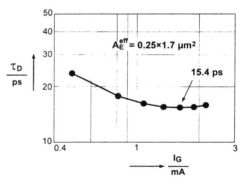

Fig. 6. Measured ECL gate delay versus current per gate with an 800-mV differential voltage swing.

Meeting the required speed rather than low power consumption was the main aim of this design. All transistor sizes are individually optimized with respect to the function of the transistor in the circuit. Extra attention was given to the on-chip wiring. The lines on the chips were classified as "critical" or "uncritical." For example, the lines driven by emitter followers are critical because they support ringing, while the lines driven by current switches are uncritical [10]. The critical lines are then shortened at the cost of the uncritical ones. The longer signal lines are realized as microstrip lines (with the lowest metallization layer as a ground plane), mainly to improve simulation accuracy. This leads to the layout shown in Fig. 4.

V. CHIP TECHNOLOGY

The circuit has been fabricated in a self-aligned double-polysilicon bipolar technology [12]. The fabrication starts with buried layer formation. A 1-μm epitaxial layer is grown to compromise between high-transit-frequency f_T and low external collector-base capacitance C_{CB}. The isolation consists of a channel stopper implantation combined with LOCOS field oxide. The active base is formed by 5-keV BF_2 implantation. This low implantation energy in combination with optimized annealing conditions allows for very steep base profiles. This results in a narrow base width of about 50 nm and thus enables a high transit frequency. A selectively implanted collector improves the current-carrying capability of the transistors to the high optimum collector current density j_C of about 2 mA/μm². To minimize narrow emitter effects, an *in situ* doped emitter-polysilicon layer is used [13]. This prevents a reduction of cutoff frequency even for 0.5-μm design rules. A three-level metallization completes the process.

Fig. 5 shows a schematic cross section of a transistor. Except for epitaxy, only process steps of a 0.5-μm CMOS production environment are necessary. The maximum transit frequency of the transistors is f_T = 50 GHz at V_{CE} = 3 V and j_C = 2 mA/μm². Table I summarizes typical parameters for transistors with effective emitter size of A_E = 0.25 × 2.7 μm. The minimum gate delay for an ECL differentially operating ring oscillator with output voltage swing of 800 mV$_{p-p}$ is measured to be 15.4 ps. This value is achieved for a current per gate of 1.6 mA (see Fig. 6).

VI. MOUNTING AND MEASUREMENT SETUP

For measurements, the clock and data recovery IC has been mounted on a 15-mil ceramic substrate (ε_r = 9.9) using conventional bonding techniques. Special care has been taken to minimize the length of the bond wires by positioning the surface of the chip on the same level as the signal, ground, and supply lines of the mounting substrate. Due to differential operation, a pair of lines for each clock and data signal is needed to connect the chip with the environment. Therefore, a corresponding number of connectors are necessary. The minimum distance between them determines the minimum size of the test fixture. To avoid additional delay lines, the length of the lines for the signals C_2, C_3 and D_1, D_2, D_3,

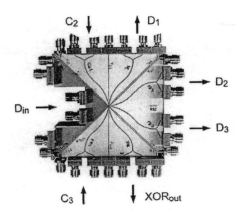

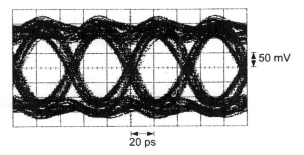

Fig. 9. Eye diagram of the 20-Gb/s data signal at the output D_2 of the 1 : 2 demultiplexer.

Fig. 7. Photograph of the package (package size: 70 × 70 mm²).

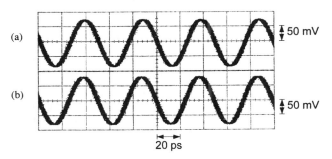

(a)

(b)

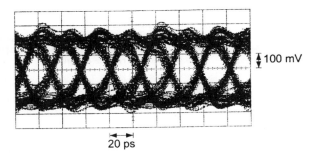

Fig. 10. (a) Transmitter clock and (b) recovered clock.

Fig. 8. Eye diagram of the 40-Gb/s input data signal D_{in}.

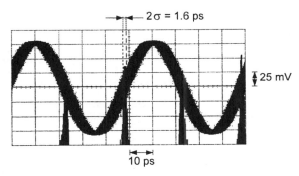

$2\sigma = 1.6$ ps

25 mV

10 ps

Fig. 11. Jitter histogram of the recovered clock.

respectively, have to be the same. To achieve a compact layout of these lines, coupled microstrip lines are used. At the input D_{in}, grounded coplanar lines are applied, which show lower dispersion than microstrip lines. The realized test fixture is shown in Fig. 7. It measures 70 × 70 mm².

Random pulse pattern generators for driving the circuit at the required data rate of 40 Gb/s are not commercially available. A pulse generator has been built from basic high-speed IC's [2], [14]. Four 10-Gb/s pseudorandom bit sequences (sequence length $2^7 - 1$) have been multiplexed to a 40-Gb/s nonreturn-to-zero signal.

VII. EXPERIMENTAL RESULTS

The clock and data recovery IC operates at the single supply voltage of −5 V and consumes 1.6 W. It should be mentioned that no additional cooling was applied. Fig. 8 shows the 40-Gb/s input signal to the CDR circuit. To demonstrate the input sensitivity of the circuit, the eye opening is artificially reduced. In Fig. 9, an eye diagram of the well regenerated and demultiplexed data signal is shown. Fig. 10 shows in (a) the 20-GHz transmitter clock and in (b) the recovered clock. The jitter histogram of the extracted clock in the time domain is displayed in Fig. 11. The measured rms time jitter as observed on the sampling oscilloscope is about 0.8 ps. The measured signal spectra of the VCO are plotted in Fig. 12. The dashed line represents the free-running VCO and the solid line the VCO phase-locked to the 40-Gb/s data signal shown in Fig. 8. The peak is about 35 dB above the floor caused by the statistics of the data.

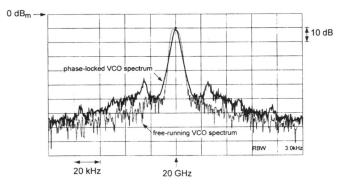

Fig. 12. VCO spectra.

VIII. CONCLUSION

An integrated clock and data recovery circuit operating up to 40 Gb/s has been realized in a 0.5-μm/50-GHz f_T silicon bipolar technology using only production-like process steps. This data rate is the highest reported value for this type of circuit in a silicon technology. This demonstrates that all

697

digital functions necessary for a 40-Gb/s transmission system are feasible with silicon bipolar production technologies.

REFERENCES

[1] K. Hagimoto, Y. Miyamoto, T. Kataoka, H. Ichino, and O. Nakajima, "Twenty-Gbit/s signal transmission using simple high-sensitivity optical receiver," in *OFC'92 Tech. Dig.*, Feb. 1992, p. 48.

[2] A. Felder, M. Möller, J. Popp, J. Böck, and H.-M. Rein, "46 Gb/s DEMUX, 50 Gb/s MUX, and 30 GHz static frequency divider in silicon bipolar technology," *IEEE J. Solid-State Circuits*, vol. 31, pp. 481–486, Apr. 1996.

[3] W. Bogner, U. Fischer, E. Gottwald, and E. Müllner, "20 Gbit/s TDM nonrepeatered transmission over 198 km DSF using Si-bipolar IC for demultiplexing and clock recovery," in *Proc. ECOC*, Sept. 1996, paper TuD.3.4.

[4] W. Bogner, E. Gottwald, A. Schöpflin, and C.-J. Weiske, "40 Gbit/s unrepeated optical transmission over 148 km by electrical time division multiplexing and demultiplexing," *Electron. Lett.*, vol. 33, no. 25, pp. 2136–2137, Dec. 1997.

[5] R. Yu, R. Pierson, P. Zampardi, K. Runge, A. Campana, D. Meeker, K. C. Wang, A. Petersen, and J. Bowers, "Packaged clock recovery integrated circuits for 40 GBit/s optical communication links," in *GaAs IC Symp. Tech. Dig.*, Nov. 1996, pp. 129–132.

[6] M. Mokhtari, T. Swahn, R. H. Walden, W. E. Stanchina, M. Kardos, T. Juhola, G. Schuppener, H. Tenhunen, and T. Lewin, "InP-HBT chip-set for 40-Gb/s fiber optical communication systems operational at 3 V," *IEEE J. Solid-State Circuits*, vol. 32, pp. 1371–1383, Sept. 1997.

[7] M. Lang, Z.-G. Wang, Z. Lao, M. Schlechtweg, A. Thiede, M. Rieger-Motzer, M. Sedler, W. Bronner, G. Kaufel, K. Köhler, A. Hülsmann, and B. Raynor, "20–40 Gb/s 0.2-μm GaAs HEMT chip set for optical data receiver," *IEEE J. Solid-State Circuits*, vol. 32, pp. 1384–1393, Sept. 1997.

[8] A. Felder, M. Möller, M. Wurzer, M. Rest, T. F. Meister, and H.-M. Rein, "60 Gbit/s regenerating demultiplexer in SiGe bipolar technology," *Electron. Lett.*, vol. 33, no. 23, pp. 1984–1986, Nov. 1997.

[9] J. Hauenschild, A. Felder, M. Kerber, H.-M. Rein, and L. Schmidt, "A 22 Gb/s decision circuit and a 32 Gb/s regenerating demultiplexer IC fabricated in silicon bipolar technology," in *Proc. IEEE BCTM'92*, Sept. 1992, pp. 151–154.

[10] H.-M. Rein and M. Möller, "Design considerations for very-high-speed Si-bipolar IC's operating up to 50 Gb/s," *IEEE J. Solid-State Circuits*, vol. 31, pp. 1076–1090, Aug. 1996.

[11] J. Hauenschild and H.-M. Rein, "Influence of transmission-line interconnections between Gbit/s IC's on time jitter and instabilities," *IEEE J. Solid-State Circuits*, vol. 25, pp. 763–766, June 1990.

[12] J. Böck, A. Felder, T. F. Meister, M. Franosch, K. Aufinger, M. Wurzer, R. Schreiter, S. Boguth, and L. Treitinger "A 50 GHz implanted base silicon bipolar technology with 35 GHz static frequency divider," in *Symp. VLSI Technology Tech. Dig.*, June 1996, pp. 108–109.

[13] J. Böck, M. Franosch, H. Schäfer, H. v. Philipsborn, and J. Popp, "In-situ doped emitter-polysilicon for 0.5 μm silicon bipolar technology," in *Proc. ESSDERC'95*, The Hague, the Netherlands, Sept. 1995, pp. 421–424.

[14] M. Möller, H.-M. Rein, A. Felder, and T. F. Meister, "60 Gbit/s time-division multiplexer in SiGe-bipolar technology with special regard to mounting and measuring technique," *Electron. Lett.*, vol. 33, no. 8, pp. 679–680, Apr. 1997.

Josef Böck was born in Straubing, Germany, in 1968. He received the diploma degree in physics and the Ph.D. degree from University of Regensburg, Germany, in 1994 and 1997, respectively.

He joined Corporate Research and Development, Siemens AG, Munich, Germany, in 1993, where he first investigated narrow emitter effects in deep submicrometer silicon bipolar devices. His work on technology development and process integration for high-speed silicon bipolar transistors resulted in the SIEGET 45 microwave-transistor family. Currently, he is working on process development for Si and SiGe bipolar technologies.

Herbert Knapp was born in Salzburg, Austria, in 1964. He received the Diplomingenieur degree in electrical engineering from Technical University Vienna, Austria, in 1997.

He joined Corporate Research and Development, Siemens AG, Munich, Germany, in 1993, where he has been involved in the design of integrated circuits for wireless communications. His current research interests include the design of high-speed and low-power microwave circuits.

Wolfgang Zirwas received the Diplomingenieur degree in electrical engineering from Technical University Munich, Germany.

He joined Siemens AG, Munich, in 1987. First, he worked in the field of high-bit-rate fiber-optic communication systems. Later, he focused his work on broad-band access technologies (xDSL, HFC) for both residential and business users. He is now working in the field of broad-band wireless systems.

Fritz Schumann received the Diplomingenieur degree in electrical engineering from Technical University Berlin, Germany, in 1981.

Subsequently, he worked in the field of RF and microwave hybrid circuit and system design for telecommunication and radar applications. In 1992, he joined the silicon bipolar IC design group, Corporate Research and Development, Siemens AG, Munich, Germany. Since then, he has realized IC's for wireless and fiber-optic communication systems up to 60 Gb/s.

Martin Wurzer was born in Innsbruck, Austria, in 1966. He received the Diplomingenieur degree in electrical engineering from the Technical University Vienna, Austria, in 1994, where he is currently pursuing the Ph.D. degree.

He joined Corporate Research and Development, Siemens AG, Munich, Germany, in 1994, where he has been engaged in the development of digital high-speed silicon bipolar IC's for future optical communication systems in the gigabit-per-second range.

Alfred Felder was born in Bruneck, South Tyrol, Italy, in 1963. He received the Diplomingenieur and Ph.D. degrees in electrical engineering from the Technical University Vienna, Austria, in 1989 and 1993, respectively.

He joined Corporate Research and Development, Siemens AG, Munich, Germany, in 1989, where he has been engaged in the development of analog and digital high-speed silicon bipolar IC's for future optical communication systems in the gigabit-per-second range. From 1996 to 1998, he was Manager of the Technology Department of Siemens K.K. The department is the liaison office of the Corporate Technology of Siemens AG in Japan, responsible for the cooperation with Japanese companies in research. Since 1998, he has been heading the business operation Signal Processing & Control within the Siemens Semiconductor Group in Japan and has been responsible for marketing of microcontrollers and digital signal processors.

A Fully Integrated 40-Gb/s Clock and Data Recovery IC With 1:4 DEMUX in SiGe Technology

Mario Reinhold, Claus Dorschky, Eduard Rose, Rajasekhar Pullela, Peter Mayer, Frank Kunz, Yves Baeyens, *Member, IEEE*, Thomas Link, and John-Paul Mattia

Abstract—In this paper, a fully integrated 40-Gb/s clock and data recovery (CDR) IC with additional 1:4 demultiplexer (DEMUX) functionality is presented. The IC is implemented in a state-of-the-art production SiGe process. Its phase-locked-loop-based architecture with bang-bang-type phase detector (PD) provides maximum robustness. To the authors' best knowledge, it is the first 40-Gb/s CDR IC fabricated in a SiGe heterojunction bipolar technology (HBT). The measurement results demonstrate an input sensitivity of 42-mV single-ended data input swing at a bit-error rate (BER) of 10^{-10}. As demonstrated in optical transmission experiments with the IC embedded in a 40-Gb/s link, the CDR/DEMUX shows complete functionality as a single-chip-receiver IC. A BER of 10^{-10} requires an optical signal-to-noise ratio of 23.3 dB.

Index Terms—Bang-bang, BER, CDR, clock and data recovery, demultiplexer, DEMUX, dynamic frequency divider, jitter generation, jitter tolerance, limiting amplifier, OSNR, phase detector, phase-locked loop, PLL, SiGe, VCO.

I. INTRODUCTION

TODAY'S commercially available highest capacity optical transmission systems are based on multiple 10-Gb/s time-division multiplexing (TDM) channels. These systems are expected to be insufficient to meet the rapidly increasing demands for higher bandwidth in the foreseeable future.

The economically achievable transmission capacity of these wavelength-division multiplexing (WDM) systems is currently limited to 1.6 Tb/s, assuming 160 parallel 10-Gb/s TDM channels in the C- and L-band at a channel spacing of 50 GHz. This corresponds to a spectral efficiency of 0.2 (b/s)/Hz. By increasing the channel bit rate to 40-Gb/s per TDM channel, the fiber capacity can be better utilized. With the spectral efficiency increased to 0.4 (b/s)/Hz, the total transmission capability is 3.2 Tb/s, assuming 80 parallel 40-Gb/s channels and 100-GHz channel spacing.

Manuscript received March 26, 2001; revised July 15, 2001.

M. Reinhold, C. Dorschky, E. Rose, and F. Kunz were with Lucent Technologies, Optical Networking Group, D-90411 Nürnberg, Germany. They are now with CoreOptics GmbH, D-90411 Nürnberg, Germany (e-mail: mario@coreoptics.com or mario_reinhold@gmx.de).

R. Pullela was with Lucent Technologies, Bell Labs, Murray Hill, NJ. He is now with Gtran Inc., Westlake Village, CA 91362 USA.

P. Mayer and T. Link are with Lucent Technologies, Optical Networking Group, D-90411 Nürnberg, Germany.

Y. Baeyens is with Lucent Technologies, Bell Labs, Murray Hill, NJ 07974 USA.

J.-P. Mattia was with Lucent Technologies, Bell Labs, Murray Hill, NJ 07974 USA. He is now with Big Bear Networks, Sunnyvale, CA 94086 USA.

Publisher Item Identifier S 0018-9200(01)09325-8.

Additionally, 40-Gb/s TDM will become more cost effective, as the number of optical ports is reduced by a factor of 4 compared to 10-Gb/s TDM, resulting in fewer price-determining optical components, smaller system footprint, and reduced maintenance costs.

Regarding next-generation 40-Gb/s TDM links, the clock and data recovery (CDR) IC is a key electronic component, which strongly determines the overall transmission performance. 40-Gb/s TDM designs must be architecturally robust and manufacturable to compete with 10-Gb/s TDM systems. Accordingly, a fully integrated phase-locked loop (PLL)-based approach with self-aligning bang-bang phase detector (PD) is employed in this work. The IC is fabricated in a production state-of-the-art SiGe heterojunction bipolar technology (HBT) which provides advantages with respect to the achievable level of integration, yield, cost-effectiveness, and process stability compared to III-V process technologies.

II. CLOCK AND DATA RECOVERY EMBEDDED IN THE OPTICAL LINK

The 40-Gb/s TDM optical link employs a 4:1 multiplexing scheme, as shown in the block diagram in Fig. 1. At the receiver, the incoming optical signal is first amplified by an optical preamplifier (OA), converted into electrical pulses by the photo diode, and then directly feeds the CDR/DEMUX. Data recovery is accomplished by the first 1:2 DEMUX. In the PLL-based clock recovery approach presented here, the PD output forces the receive-side voltage-controlled oscillator (VCO) to track the phase of the incoming data signal. The combination of the PD and 1:2 DEMUX function allows the use of a 20-GHz half-bit-rate clock. This half-bit-rate architecture is explained in more detail in Section IV.

This paper focuses on the CDR/DEMUX IC. However, the remaining basic functions such as the 4:1 multiplexer (MUX) and the driver IC have also been realized in this work program in SiGe HBT and GaAs high-electron-mobility transistor (HEMT) technology, respectively.

III. PROCESS TECHNOLOGY

The CDR/DEMUX IC presented in this work was designed in a state-of-the-art SiGe HBT with 72-GHz f_T and 74-GHz f_{max} [1]. SiGe HBT provides superiority for a high level of integration compared to III-V technologies. The process features four metal layers in total including a thick metal layer on top.

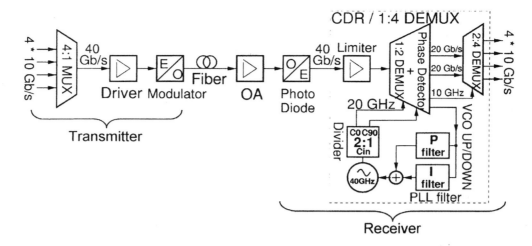

Fig. 1. Clock and data recovery IC embedded in the optical link.

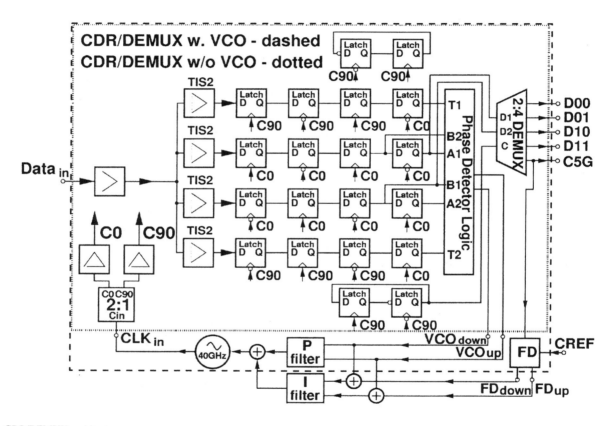

Fig. 2. CDR/DEMUX architecture.

Small-scale integrated analog and digital building blocks implemented in this process have been demonstrated for 40-Gb/s operation [2], [3].

IV. CDR/DEMUX ARCHITECTURE

The half-bit-rate architecture of the CDR/DEMUX IC (Fig. 2) is based on the concept reported in [4] and has already demonstrated its functionality for 10-Gb/s applications.

The nonlinear bang-bang PD described in [5] was modified for interlaced operation. The combination of the PD and the 1:2 DEMUX functions allows the use of a half-bit-rate clock running at 20 GHz.

The three upper eye diagrams in Fig. 3 illustrate the basic principle of a common bang-bang PD. For the basic realization, three samples A, T, B of the incoming data signal are necessary.

In locked condition, A and B sample two consecutive bits while T samples the data transition, as indicated in the first eye diagram.

In this implementation, two modifications to the common bang-bang PD are made. First, the 1:2 DEMUX and phase-

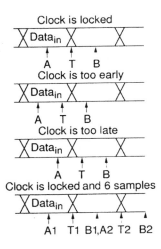

Clock is locked

Clock is too early

Clock is too late

Clock is locked and 6 samples

Fig. 3.   PD principle.

detection function are combined so that a half-bit-rate clock is used. However, for sampling T in the bit transition, a four-phase clock is essential. Second, unlike in previous approaches [6], six samples A_1, T_1, B_1 and A_2, T_2, B_2 are generated in order to process every single data transition as indicated in the fourth eye diagram in Fig. 3. This increases the maximum PD gain, reducing the jitter generation compared to a single-edge PD.

The PD output signals VCO_{up} and VCO_{down} are derived from the following logical operations on the six samples after their synchronization with the clock phase C0.

$$VCO_{up} = (A_1 \oplus T_1) * \overline{(B_1 \oplus T_1)}$$
$$+ (A_2 \oplus T_2) * \overline{(B_2 \oplus T_2)}$$
$$VCO_{down} = \overline{(A_1 \oplus T_1)} * (B_1 \oplus T_1)$$
$$+ \overline{(A_2 \oplus T_2)} * (B_2 \oplus T_2).$$

In this design, an accurate four-phase clock at 20 GHz is generated from the 40-GHz VCO output using a symmetrically loaded 2:1 frequency divider.

Additionally, a limiting amplifier is implemented to improve the input sensitivity. It feeds the 40-Gb/s input data to the four parallel latch chains generating the six samples.

To process the two 1:2 demultiplexed 20-Gb/s signals A_1 and B_1 with commercially available 10-Gb/s DEMUX ICs, an additional 2:4 demultiplexer is included, which produces the 10-Gb/s output signals D00, D01, D10, D11, and the 5-GHz output clock C5G.

The PLL filter has a parallel proportional (P) integrating (I) structure (PI filter). The high-speed P filter aligns the phase relation between clock and data. Thus, the digital PD output pulses VCO_{up} and VCO_{down} each generate a dynamic frequency step $\pm \Delta f_{BB}$ in the VCO frequency. Integration of $VCO_{up} - VCO_{down}$ by the I filter with low bandwidth controls the static VCO frequency. Since the P filter and the I filter work at different speeds, both paths are decoupled from

each other, resulting in a minimum lead time of the high-speed phase-control loop.

Two variants of the CDR/DEMUX exist. The higher integration variant of the CDR/DEMUX (CDR/DEMUX with VCO) also contains an on-chip 40-GHz VCO, the proportional filter of the PLL, and a frequency detector (FD) working as a low-speed frequency acquisition aid, whereas in the lower integration variant (CDR/DEMUX without VCO), these parts are external to the IC.

V. BUILDING BLOCKS

The most challenging building blocks of the CDR/DEMUX are the 40-GHz 2:1 frequency divider, the 40-GHz VCO, the limiting data input amplifier and the transition sampling latches, including the four-phase 20-GHz clock tree.

A. 2:1 Dynamic Frequency Divider

Since the PD requires a 20-GHz four-phase clock, those clock signals can be most accurately generated by dividing a 40-GHz signal with a 2:1 frequency divider resulting in differential 0° and 90° phase-shifted clock signals. Studies published earlier [7] using a similar technology show a maximum operating frequency of 42 GHz with a standard static frequency divider. To achieve a higher performance margin, a dynamic frequency divider similar to [8] was employed. As it is indicated in the circuit diagram (Fig. 4), the divided clock signal is stored by parasitic capacitances, which results in a higher operating frequency compared to a static frequency divider. As a drawback of the dynamic approach, the operating frequency exhibits a lower limit.

B. On-Chip 40-GHz VCO

The internal 40-GHz VCO is based on a differential Colpitts topology using microstrip transmission lines instead of a spiral inductor (Fig. 5). At an oscillation frequency of 40 GHz, a grounded microstrip line can be modeled quite accurately compared to other forms of inductors.

The microstrip transmission lines exhibit an inductive input impedance, since for the odd mode they see a virtual ground termination. It is physically implemented with the signal line on the upper thick metallization layer and a shielding ground plane on the lowest metal layer. This yields maximum inductive input impedance per line length combined with minimum resistive losses, which optimizes the quality factor Q.

As the PLL employs a high-speed P and a low-speed I filter in parallel, the VCO has two separate frequency tuning inputs. These tuning inputs $VTUNE_P$ and $VTUNE_I$ feed two ac-coupled reverse-biased varactor diodes controlling the VCO frequency. The varactors have minimum size, resulting in a maximum VCO frequency modulation bandwidth as required by the high-bandwidth bang-bang PD architecture.

It should be noted that the optimization of the free-running VCO phase noise is not the major design goal. Since the bang-bang PLL architecture provides very high bandwidth with

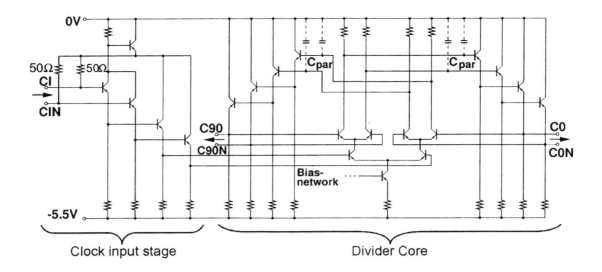

Fig. 4. Circuit diagram of the 2:1 dynamic frequency divider.

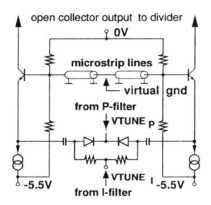

Fig. 5. Circuit diagram of the 40-GHz on-chip VCO.

respect to loop gain, the VCO phase noise is suppressed by its open loop gain when the PLL is in lock.

C. Limiting Data Amplifier

The limiting data input amplifier (Fig. 6) employs cascaded chains of emitter followers (EF), transimpedance stages (TIS), and transadmittance stages (TAS) in accordance with the concept of impedance mismatch [9].

Layout aspects strongly influence the circuit performance; especially, it should not be degraded by signal interconnects. Since signal lines can be distinguished into critical and uncritical lines [9], long transmission lines are arranged between current interfaces consisting of a TAS and its load, which is either represented by an active TIS or by passive resistors. For this reason, the limiting data amplifier is implemented—both in schematic and layout—in the form of three separate amplifier blocks with a TIS–TAS interface.

As the data signal has to be split into four latch chains (Fig. 2) and longer lines cannot be avoided, four TIS2 stages are driven in parallel by one TAS2.

D. Clock Distribution and Latch

Fig. 7 shows the circuit diagram of the latch. The latch structure can be subdivided into the latch core and the local clock input stage.

The steepness of the PD curve is strongly determined by the metastability and the clock phase margin (CPM) of the latch as $T_{1/2}$ sample the bit transition when the PLL is in lock. Two high current-biased emitter followers in the data path provide a high CPM and a small metastability region.

For optimal clock distribution, a local clock input stage consisting of termination resistors R_i and two emitter followers is included in each latch cell, since a total clock line length of several millimeters cannot be avoided. For this reason, current interfaces between each clock buffer and the latches are employed. The concept of the clock distribution is illustrated in Fig. 8. The two clock buffers are based on open-collector transadmittance stages. As stated before, a local clock input stage consisting of termination resistors R_i is included in every latch cell. Each clock buffer is loaded with ten latches. Impedance matching can be easily achieved by designing the line impedance Z_L according to the number of loading latches. The value of Z_L can be locally adapted with respect to signal splits, as it is shown in Fig. 8. If the line impedances are chosen $Z_L > R_i/n$ (with n being the number of loading latches), the clock amplitude can be increased. This is due to the inductive peaking effect of the transmission line, as indicated in Fig. 9.

The layout of the latch, which is shown as part of Fig. 10, employs orthogonal data and clock inputs. This implementation minimizes line length on the high-speed data path by running a data channel directly through the cascaded latch cells (Fig. 10). In addition, clock channels run beside the cells to simplify the clock-tree routing.

702

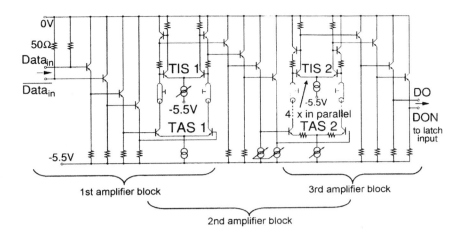

Fig. 6.   Circuit diagram of the 40-Gb/s limiting data amplifier.

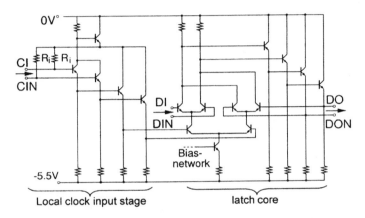

Fig. 7.   Circuit diagram of the latch.

VI. PHYSICAL REALIZATION AND MEASUREMENT RESULTS

The CDR/DEMUX with VCO dissipates 5.4 W in total. All high-speed blocks operate at -5.5-V supply voltage and the 2:4 DEMUX at -4.2-V supply voltage. The whole die (Fig. 11) occupies an area of 3005^2 μm^2.

Closed-loop PLL measurements are performed with mounted ICs using single-ended 4-b interleaved OC192 SONET signals with $2^7 - 1$ pseudorandom bit sequence (PRBS) payload for electrical back to back and $2^{23} - 1$ PRBS payload for the optical transmission experiments. For these measurements, the CDR/DEMUX and the external components are mounted on a duroid substrate ($\varepsilon_r = 3.32$) using standard wire bonding, as shown in Fig. 12. The IC is placed into a cavity, reducing the bondwire length. The substrate is attached onto a grounded brass box. The 40-Gb/s input data is fed single-ended into the high-speed box via a V-connector.

A. 2:1 Dynamic Frequency Divider

Fig. 13 illustrates the input sensitivity of the dynamic frequency divider in single-ended operation. A maximum operating frequency of more than 44.5 GHz is observed giving sufficient margin with respect to temperature and process variations.

Since the self-oscillating frequency of the divider is roughly 41 GHz divided by 2, the circuit is very sensitive in the frequency range centered around the nominal operating frequency of 40 GHz. The dynamic principle of the frequency divider results in a minimum operating frequency of roughly 33 GHz at an input power of -1 dBm.

B. On-Chip 40-GHz VCO

A similarly centered behavior can be measured for the on-chip VCO represented by the tuning characteristic, as shown in Fig. 14. The VCO tuning range expands from 37.7 to 41.2 GHz.

C. Bang-Bang PD

The measured PD transfer function $VCO_{up} - VCO_{down}$ is illustrated in Fig. 15. The curve shows a fairly steep slope in the lock-in point, which implies good sampling capability at data transition and a small metastability region, corresponding to the observed high CPM of 240°.

D. Jitter Generation and Jitter Tolerance

The rms jitter of the recovered clock σ_{RC} is measured using a sampling oscilloscope. By excluding the trigger jitter of the test equipment ($\sigma_{TE} = 0.93$ ps) from the original measurement result of less than 1.2 ps, the overall CDR rms jitter generation σ_{CDR} can be calculated to be approximately 0.7 ps.

For SONET and SDH systems, jitter tolerance masks are defined. As standardization is not finalized for 40-Gb/s systems yet, the tolerance mask is extrapolated from the 10-Gb/s specification. The measured jitter tolerance curve, given in Fig. 16, exhibits sufficient margin relative to the extrapolated BELLCORE mask [10]. For jitter frequencies higher than the corner frequency $f_{corner} = 80$ MHz of the PLL, the jitter tolerance is determined by the CPM of the first latches and the PLL has no influence. For jitter frequencies lower than $f_{corner} = 80$ MHz, the jitter tolerance decreases with 20 dB per decade. Large amounts of jitter can be tolerated at low frequencies, since the I filter provides high gain in this frequency range.

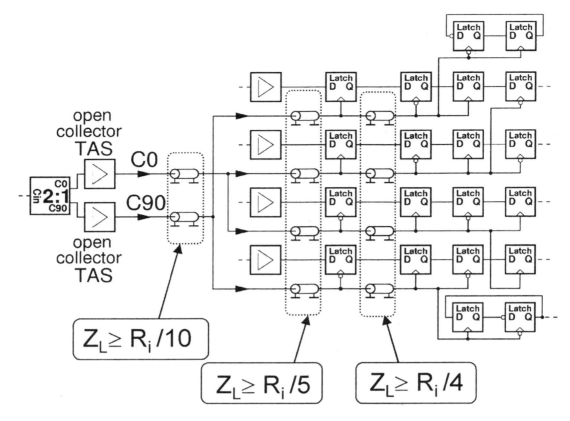

Fig. 8. Block diagram of the 20-GHz quadrature clock distribution.

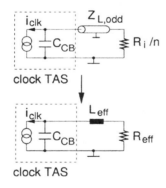

Fig. 9. Rough odd-mode simplification of the clock distribution.

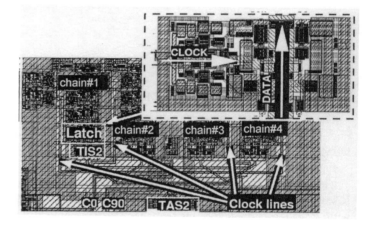

Fig. 10. Layout of the entire clock tree and enlargement of the latch.

E. Electrical Sensitivity (BER)

The bit-error rate (BER) curve as measure of the overall electrical CDR performance is given in Fig. 17. For the CDR/DEMUX with external VCO, a very high sensitivity of 28-mV single-ended voltage swing at BER $= 10^{-10}$ is measured. Due to minor modifications of the limiting amplifier resulting in lower bandwidth, the CDR/DEMUX with internal VCO provides slightly less sensitivity. For the same BER, a 42-mV single-ended voltage swing is necessary. A contribution of the internal VCO to the performance degradation can be ruled out, since a variant with external VCO and modified limiting amplifier showed similar performance degradation. Such high input sensitivity allows a direct connection of the

photodiode to the CDR/DEMUX as the optical measurement results demonstrate.

F. Performance in System Application

The performance of the CDR/DEMUX embedded in a optical fiber link (refer to Fig. 1) can be characterized by the optical signal-to-noise ratio (OSNR) measurement, as shown in Fig. 18. This is due to the fact that in the given configuration the sensitivity is limited by the noise of the OA.

A 50-Ω terminated photodiode is directly connected to the CDR/DEMUX without any electrical amplifier in between, so

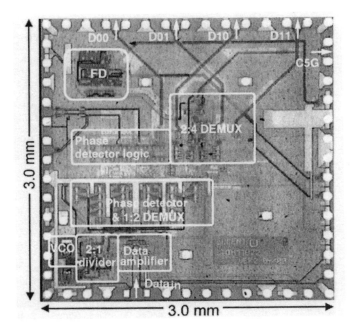

Fig. 11. CDR/DEMUX micrograph (CDR/DEMUX with VCO).

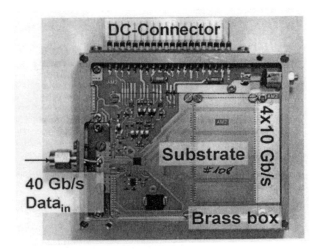

Fig. 12. CDR/DEMUX test fixture with mounted CDR/DEMUX.

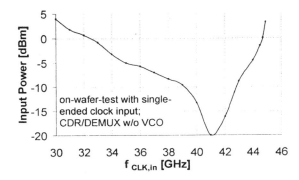

Fig. 13. Input sensitivity of the 2:1 dynamic frequency divider.

that the CDR/DEMUX works as a single-chip-receiver IC. An externally modulated signal is transmitted over an 80-km optical-fiber link. To achieve a BER of 10^{-10}, a minimum OSNR of 23.3 dB is required.

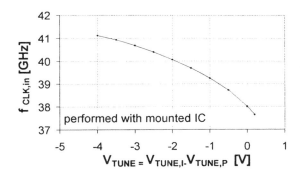

Fig. 14. Tuning curve of the 40-GHz on-chip VCO.

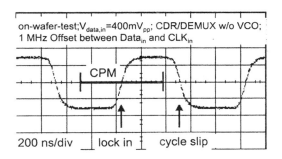

Fig. 15. Measured PD transfer function.

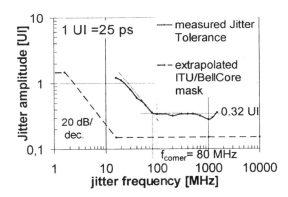

Fig. 16. Jitter tolerance measurement result.

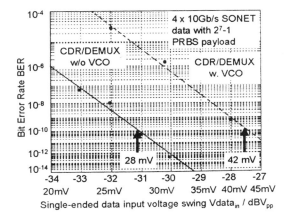

Fig. 17. BER measurement.

705

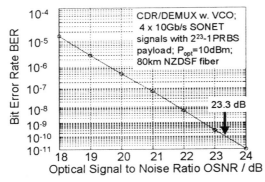

Fig. 18. OSNR measurement.

VII. CONCLUSION

In this paper, the implementation of a fully integrated CDR/DEMUX for 40-Gb/s TDM application in a state-of-the-art SiGe HBT process has been demonstrated. Key success factors in the design are, first, the robust half-bit-rate architecture with bang-bang PD, and second, its implementation by partitioning the whole chip into key building blocks interconnected via robust current interfaces. Therefore, schematic design and layout have to be done concurrently.

This concept has been applied throughout the IC, but mainly in the 40-Gb/s data and the 20-GHz clock distribution. Optical system measurements show the feasibility of the CDR/DEMUX as a single-chip-receiver IC.

REFERENCES

[1] T. F. Meister et al., "SiGe base bipolar technology with 74-GHz f_{max} and 11-ps gate delay," Proc. IEEE Int. Electron Devices Meeting (IEDM), pp. 739–742, Dec. 1995.
[2] J. Müllrich, T. F. Meister, M. Rest, W. Bogner, A. Schöpflin, and H.-M. Rein, "40-Gbit/s transimpedance amplifier in SiGe bipolar technology for the receiver in optical-fiber links," Electron. Lett., vol. 34, pp. 452–453, 1998.
[3] A. Felder, M. Möller, M. Wurzer, M. Rest, T. F. Meister, and H.-M. Rein, "60-Gbit/s regenerating demultiplexer in SiGe bipolar technology," Electron. Lett., vol. 33, pp. 1984–1985, 1997.
[4] J. Hauenschild et al., "A plastic packaged 10-Gb/s BiCMOS clock and data recovery 1:4-demultiplexer with external VCO," IEEE J. Solid-State Circuits, vol. 31, pp. 2056–2059, Dec. 1996.
[5] J. J. D. H. Alexander, "Clock recovery from random binary signals," Electron. Lett., vol. 11, pp. 541–542, Oct. 1975.
[6] M. Wurzer et al., "A 40-Gb/s integrated clock and data recovery circuit in a 50-GHz f_T silicon bipolar technology," IEEE J. Solid-State Circuits, vol. 34, pp. 1320–1324, Sept. 1999.
[7] M. Wurzer et al., "42-GHz static frequency divider in a Si/SiGe bipolar technology," in IEEE ISSCC Dig. Tech. Papers, Feb. 1997, pp. 123–123.
[8] Z. Lao et al., "55-GHz dynamic frequency divider IC," Electron. Lett., vol. 34, no. 20, pp. 1973–1974, 1998.
[9] H.-M. Rein et al., "Design considerations for very-high-speed Si-Bipolar ICs operating up to 50-Gb/s," IEEE J. Solid-State Circuits, vol. 31, pp. 1076–1090, Aug. 1996.
[10] "SONET OC-192 Transport System Generic Criteria," Bellcore, GR-1377-CORE, Dec. 1998.

Claus Dorschky received the Dipl.-Ing. degree in electrical engineering from Friedrich Alexander University, Erlangen, Germany, in 1986.

He has been working in the development department for high-speed optical transmission systems at Philips Kommunikation Industries (later Lucent Technologies), Nürnberg, Germany, for 14 years. His research interests include design and integration of analog and mixed-signal full custom ICs for 10- and 40-Gb/s as well as integration of optical receivers and transmitters into single wavelength and DWDM transmission systems at those bitrates. In early 2001, he cofounded CoreOptics Inc., Nürnberg, Germany.

Eduard Rose was born in Kischinjow, Moldova, in 1973. He received the Dipl.-Ing. degree in electrical engineering from Ruhr-University Bochum, Germany, in 1998.

He joined Lucent Technologies, Optical Networking Group, Nürnberg, Germany, in 1999, where he started developing different analog and digital high-speed bipolar ICs for SDH/Sonet systems. He is currently with CoreOptics Inc., Nürnberg, working on a second-generation chipset for a 40-Gb/s optical link system.

Rajasekhar Pullela received the B.Tech. degree in electrical and communications engineering from the Indian Institute of Technology, Madras, India, in 1993. From 1993 to 1998, he worked as a graduate student researcher at the University of California, Santa Barbara. During this period, he received M.S. and Ph.D. degrees in electrical engineering, studying device physics and high-speed circuit design.

During 1998–2000, he worked as a Member of Technical Staff at Bell Laboratories, Lucent Technologies, Murray Hill, NJ, designing high-speed ICs for fiber-optic communication systems. Since 2000, he has been with Gtran, Inc., Newbury Park, CA.

Peter Mayer was born in Germany on July 11, 1964. He received the Dipl.-Ing. degree in electrical engineering from Friedrich Alexander University, Erlangen, Germany, in 1989.

In 1989, he joined Philips Kommunikation Industries, Nürnberg, Germany, and has been working on 622-Mb/s optical interface circuits. In 1998, he started developing clock and data recovery ICs for 10- and 40-Gb/s applications. He is currently a Technical Manager at Lucent Technologies GmbH, Nürnberg, where he is responsible for high-speed optical/electrical module design and integration for optical transmission systems.

Mario Reinhold was born in Mülheim/Ruhr, Germany, in 1972. He received the Diplom-Ingenieur degree in electrical engineering from the Ruhr-University Bochum, Germany, in 1998.

He joined Lucent Technologies, Optical Networking Group, Nürnberg, Germany, in 1998, where his activities focused on the development of various analog and digital high-speed bipolar ICs for 40-Gb/s and advanced 10-Gb/s fiber-optic communication systems. Since 2001, he has been with CoreOptics Inc., Nürnberg, Germany, working on a next-generation 40-Gb/s chipset.

Frank Kunz was born in Bad Sobernheim, Germany, in 1970. He received the Dipl.Ing. degree in electrical engineering from the Ruhr-University Bochum, Germany, in 1998.

Until February 2001, he was with Lucent Technologies GmbH, Nürnberg, Germany, developing high-speed bipolar ICs for advanced 10-Gb/s optical communication links. He is currently with CoreOptics Inc., Nürnberg, and is working on a second-generation chipset for a 40-Gb/s optical link system.

Clock and Data Recovery IC for 40-Gb/s Fiber-Optic Receiver

George Georgiou, *Member, IEEE*, Yves Baeyens, *Member, IEEE*, Young-Kai Chen, *Fellow, IEEE*,
Alan H. Gnauck, *Senior Member, IEEE*, Carsten Gröpper, Peter Paschke, Rajasekhar Pullela, Mario Reinhold,
Claus Dorschky, John-Paul Mattia, Timo Winkler von Mohrenfels, and Christoph Schulien

Abstract—The integrated clock and data recovery (CDR) circuit is a key element for broad-band optical communication systems at 40 Gb/s. We report a 40-Gb/s CDR fabricated in indium–phosphide heterojunction bipolar transistor (InP HBT) technology using a robust architecture of a phase-locked loop (PLL) with a digital early–late phase detector. The faster InP HBT technology allows the digital phase detector to operate at the full data rate of 40 Gb/s. This, in turn, reduces the circuit complexity (transistor count) and the voltage-controlled oscillator (VCO) requirements. The IC includes an on-chip *LC* VCO, on-chip clock dividers to drive an external demultiplexer, and low-frequency PLL control loop and on-chip limiting amplifier buffers for the data and clock I/O. To our knowledge, this is the first demonstration of a mixed-signal IC operating at the clock rate of 40 GHz. We also describe the chip architecture and measurement results.

Index Terms—Clock and data recovery, CDR, fiber-optic communication receiver, InP HBT, limiting amplifier, phase detector, VCO.

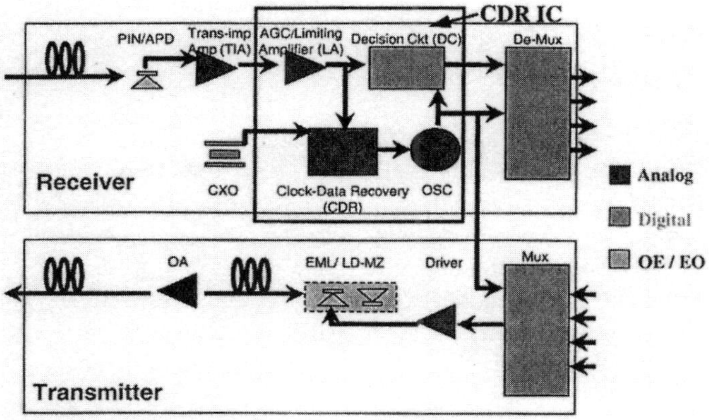

Fig. 1. Schematic diagram of a lightwave transceiver.

I. INTRODUCTION

CLOCK and data recovery (CDR) is an important function of the transceiver of a high bit-rate lightwave communication system. Since 40-Gb/s systems are nearing commercial deployment, the chosen CDR architecture must have few external components and be insensitive to temperature and component variations. CDRs reported in the literature use either a high quality-factor Q external filter-based architecture or a phase-locked loop (PLL) architecture.

While the high-Q filter architecture is easier to implement, it is susceptible to temperature and group delay variations in the filter [1]. Specifically, the temperature drift of the filter bandpass and the temperature drift of the timing within the IC are not correlated. Also, once the clock signal is recovered, additional precise clock phase adjustment is needed to set the decision sampling time to obtain proper phase margin. This phase

Manuscript received January 30, 2002; revised April 30, 2002.

G. Georgiou, Y. Baeyens, and Y.-K. Chen are with Lucent Technologies, Bell Laboratories, Murray Hill, NJ 07974 USA (e-mail: gundsn@lucent.com).

A. H. Gnauck is with Lucent Technologies, Bell Laboratories, Holmdel, NJ 07733 USA.

C. Gröpper and P. Paschke are with Lucent Technologies, Optical Networking Group, 90411 Nürnberg, Germany.

R. Pullela was with Lucent Technologies, Bell Laboratories, Murray Hill, NJ 07974 USA. He is now with Gtran Inc., Thousand Oaks, CA 91362 USA.

M. Reinhold, C. Dorschky, T. W. von Mohrenfels, and C. Schulein were with Lucent Technologies, Optical Networking Group, 90411 Nürnberg, Germany. They are now with Core Optics, 90411 Nürnberg, Germany.

J. P. Mattia was with Lucent Technologies, Bell Laboratories, Murray Hill, NJ 07974 USA. He is now with BigBear Networks, Milpitas, CA 95035 USA.

Publisher Item Identifier 10.1109/JSSC.2002.801186.

adjustment is further complicated by packaging issues, specifically that of aligning the recovered clock after the off-chip filter and the decision IC.

In PLL-based CDRs, the clock phase in the decision circuit is automatically synchronized to sample the center of the time slot of each bit. Also, the PLL can be integrated onto a single IC, greatly reducing temperature drift and phase relationship problems.

Previous implementations of digital PLL-based CDRs at 40 Gb/s have employed half bit-rate clocking of CDR demultiplexer (DEMUX) combinations, to reduce the bandwidth required in the buffers and digital gates [2]–[4]. By clocking at 20 GHz to tolerate lower transistor bandwidth, the $2\times$ parallel phase detector requires higher circuit complexity with about twice the transistor count and a precise four-phase voltage-controlled oscillator (VCO).

In this paper, we leverage the InP heterojunction bipolar transistor (HBT) technology operating at the full 40-Gb/s data rate to simplify the CDR architecture.

II. CDR ARCHITECTURE

The typical transceiver architecture for a 40-Gb/s lightwave system is shown in Fig. 1. The CDR IC of this work is highlighted in the receiver path.

The core of the PLL-based CDR is the phase detector. The phase detector used here simultaneously recovers both clock and data. The digital early–late phase detector [5], consisting of data flip-flops and combinatory logic gates, is shown in Fig. 2. In the locked state, the $A-B$ chain samples the center of the incoming

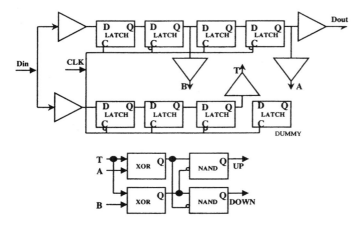

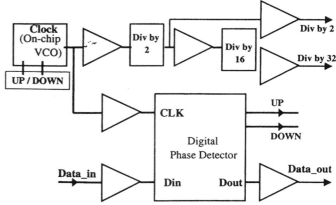

Fig. 2. Digital phase detector architecture.

Fig. 3. CDR IC block diagram.

data time slot, while the T chain samples the data zero crossings. The combinatory logic block, with inputs from the A, B, and T latch chains, determines if the clock is early or late with respect to the incoming data transition. This logic generates the UP–DOWN control signal for the VCO.

A and B are generated by decision circuits (respectively, after two and four latches). The phase difference between A and B is 180° (one bit). T is generated after three latches with an inverted clock. If the clock phase is correct, T is always in the middle of A and B. EXOR ($\oplus$) and NAND ($\vee$) gate combinatory logic converts A, B, and T to the UP–DOWN pulses for controlling the VCO. The logic equations are as follows.

$$\text{UP} = (A \oplus T) \vee (B \oplus T)$$
$$\text{DOWN} = (A \oplus T) \vee (B \oplus T).$$

The clock is early (slows down the VCO) if $A = T$ and $B \neq T$, UP − DOWN = −1. The clock is late (speeds up the VCO) if $B = T$ and $A \neq T$, UP − DOWN = 1. The clock is correct if $A = B = T$. Obviously, data transitions are required for clock recovery.

The chip also incorporates 15-dB gain-limiting amplifier buffers at the data I/O, at the divide-by-2 clock output (for the external DEMUX) and at the divide-by-32 clock output (for the external coarse adjustment low-frequency lock loop). The static frequency dividers use similar latches as those in the digital phase detector. Figs. 3 and 4, respectively, show the final chip block diagram and photograph.

The block diagram of Fig. 3 is laid over Fig. 4. To maintain symmetry in the 40-Gb/s data and 40-GHz clock signals, the phase detector layout contains four symmetric rows (as in the block diagram of Fig. 2).

III. CDR DESIGN AND FABRICATION

Differential emitter-coupled logic (ECL) logic with 400-mV differential voltage swings is used. To simplify the layout process at a high bit rate, standardized digital and analog blocks are designed and used.

All high-frequency inputs are terminated with on-chip 50-Ω resistors. Propagation delay of the 40-GHz clock to the A, B, and T latches and to the divider chain is a critical issue. Matching propagation delay is achieved by symmetric layout

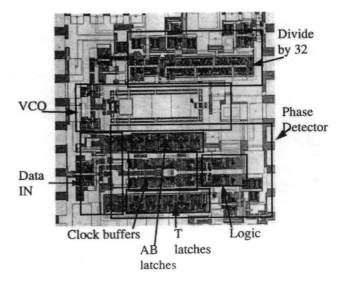

Fig. 4. CDR IC photograph.

and symmetric loading (for example, the dummy latch in the T chain of Fig. 2).

Coplanar transmission lines with controlled impedance are used for longer lines ($> \lambda/10$) to reduce reflections and to improve timing accuracy. Lines driven by emitter followers are kept short to avoid reflections due to impedance mismatch.

A series gate approach is used for the clock and data signals. To improve high-frequency performance, a transadmittance (TAS) and transimpedance (TIS) combination [6], connected by coupled coplanar transmission lines is used for clock and data amplification and distribution. TAS–TIS buffer amplifiers have a higher gain-bandwidth product (because of the active TIS load) than does a simple ECL buffer. The buffer amplifier signal splitting and transistor level design are shown in Fig. 5.

The VCO transistor level schematic is shown in Fig. 6. The VCO is based on a differential Colpitts topology using coplanar transmission lines which can be modeled very accurately at 40 GHz, instead of inductors. The tuning input V_{tune} feeds two reverse-biased varactors realized from the base–collector junctions of open-emitter HBTs. Note that minimum phase noise is not a design parameter since the early–late (or bang–bang) PLL architecture has very high bandwidth with respect to loop gain.

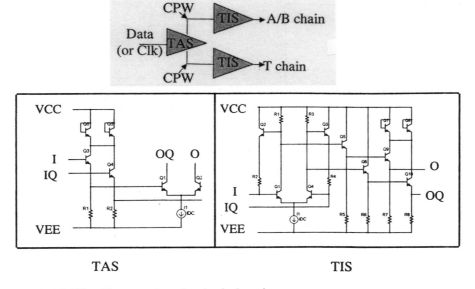

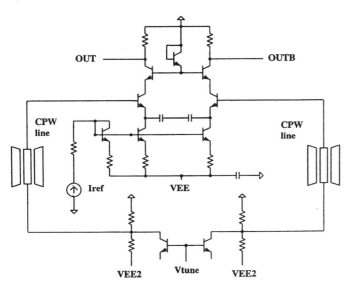

Fig. 5. Buffer amplifier cascaded TAS–TIS architecture and transistor level schematic.

Fig. 6. VCO transistor level schematic.

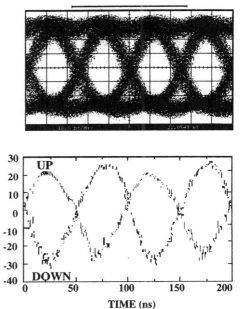

Fig. 7. Retimed 40-Gb/s data (40-GHz clock, 30 mV/div, 10 ps/div) and phase-detector transfer functions (UP < DOWN with 40-GHz clock offset +10 MHz).

Thus, the VCO phase noise is reduced by the open-loop gain of the locked PLL.

The chip is fabricated in an InP HBT technology [7]. Peak transistor f_t and f_{max} are 160 and 135 GHz, respectively. The transistors are conservatively biased at half the collector current for peak f_t. The nominal transistor has a 1 μm × 3 μm emitter biased with a collector current of 4 mA. The interconnects use one metal layer for longer wires and one local metal layer for shorter wires. Passive elements are fabricated using tantalum–nitride thin-film resistors and silicon–nitride dielectric capacitors.

A power-supply series decoupling resistor ($R_s = 1.6\ \Omega$) is used to prevent any spurious low-frequency oscillations that could arise in the packaged IC. The internal circuit is designed to operate between −4 and −5 V. The circuit draws nominally 1 A. The nominal total power dissipation is 5.6 W, 1.6 W across

the decoupling series resistor and 4 W internally. The chip area is 1.75 mm × 1.75 mm.

IV. MEASUREMENT RESULTS

Two versions of the CDR chip (with and without VCO) were fabricated. The results of on-wafer measurements on the CDR with external VCO are discussed here. The input is a single-ended 0.5 V_{p-p} signal generated by a commercial 40-Gb/s 4 : 1 multiplexer, driven with four independent 10-Gb/s $2^{31} - 1$ pseudorandom bit streams (PRBS) of a pattern generator. A synthesizer generates the 40-GHz clock signal for this measurement. Fig. 7 shows the retimed 40-Gb/s eye from the CDR IC. (It should be noted that the on-wafer

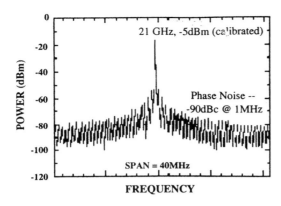

Fig. 8. Divide-by-2 frequency spectrum of on-chip VCO. VCO designed for 40 GHz but actual at 42 GHz divided to 21 GHz with an output power of −5 dBm.

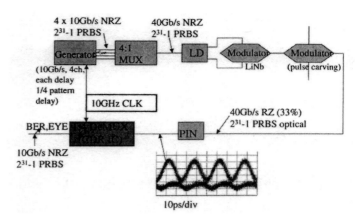

Fig. 9. Experimental optical link to measure the bit-error rate of 40-Gb/s RZ transmission.

measurement is limited by the equipment. The retimed data eye is sharper than that of the commercial multiplexer. Also, the jitter is characteristic of the digital oscilloscope used for the measurement.) Decision circuit phase margin greater than 180° is measured by changing the clock phase with respect to incoming data and observing the output data eye. The phase transfer function (UP and DOWN) at the bottom of Fig. 7 is measured by introducing a +10-MHz offset between the clock frequency and data bit rate.

Measurements of the CDR with on-chip LC VCO indicate that the VCO is capable of driving 40-Gb/s digital gates. However, the VCO center frequency is higher than simulated, probably because of capacitance or inductance (transmission-line) drift during this process run. The VCO operates in the band between 40.5 and 42.5 GHz. The divider chains operate up to a frequency of 44 GHz. The output of the divide-by-2 is shown in Fig. 8 for the VCO tuned to 42 GHz.

To evaluate the packaged performance of the CDR chip in an optical transmission system, we used the CDR chip as a single channel 1:4 DEMUX by applying 40-Gb/s data and 10-GHz clock. (The packaging used here for 40 Gb/s is relatively simple. The chip is mounted into a cutout in a composite Rogers 4003 on FR-4 board. This chip recess corresponding approximately the chip thickness ∼8 mil, reduces the length of the bond wires to the 50-Ω coplanar GSSG transmission lines designed on the R4003 substrate. (See [8, Fig. 6].)

The experimental optical time-division multiplexing (OTDM) link is shown in Fig. 9. As before, 4:1 MUX is used to multiplex independent 10-Gb/s nonreturn-to-zero (NRZ) 2^{31} − 1 PRBS data streams, from a commercial pulse pattern generator. The resulting electrical 40-Gb/s 2^{31} − 1 PRBS NRZ signal is converted to an optical 40-Gb/s 2^{31} − 1 PRBS RZ signal by a pulse-carving technique with cascaded modulators. Optical power is converted back to the electrical signal with a p-i-n photodetector. The 40-Gb/s RZ eye after the p-i-n is demultiplexed to 10 Gb/s using the CDR chip as a single channel 1:4 DEMUX.

Fig. 10 shows the CDR IC performance as a one-channel DEMUX. The demultiplexed 10-Gb/s eye is very open and has very low jitter (<4 ps limited by the oscilloscope bandwidth). The error probability is also measured as a function of received

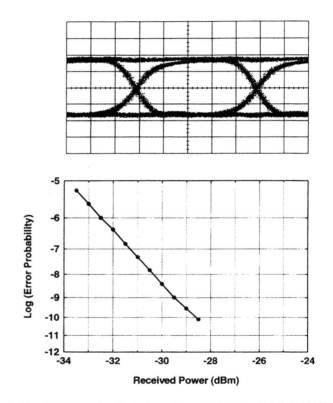

Fig. 10. CDR IC used as single channel 1:4 DEMUX (40-Gb/s data, 10-GHz clock). Electrical output at 10 Gb/s and BER versus optical power measurement of the optical link of Fig. 8.

optical power. The received optical power is −29.5 dBm for the typical system required bit-error rate (BER) of 10^{-9}.

V. CONCLUSION

A complex (∼1350 transistors and 1.75-mm square) mixed-signal CDR chip with on-chip VCO, amplifiers, decision circuit, and clock dividers was successfully fabricated with a state-of-the-art InP HBT technology. Fully functional chips at speed were obtained from the first iteration. Data is retimed at 40 Gb/s and a good control signal is made available to the on-chip VCO. The CDR IC is used as a DEMUX to convert an optical 40-Gb/s 2^{31} − 1 PRBS RZ signal to an electrical 10-Gb/s $2^{31} - 1$ PRBS NRZ signal in an optical link experiment. An optical sensitivity of -29.5 dBm is measured at 10^{-9} BER.

REFERENCES

[1] R. Yu, R. Pierson, P. Zampardi, K. Runga, A. Campana, D. Meeker, K. C. Wang, A. Peterson, and J. Bowers, "Packaged clock recovery integrated circuits for 40-Gb/s optical communication links," in *GaAs IC Symp. Tech. Dig.*, 1996, pp. 129–132.

[2] M. Wurzer, J. Bock, H. Knapp, W. Zirwas, F. Schumann, and A. Felder, "A 40-Gb/s integrated clock and data recovery circuit in a 50-GHz f_T silicon bipolar technology," *IEEE J. Solid-State Circuits*, vol. 34, pp. 1320–1324, Sept. 1999.

[3] J. Hauenschild, C. Dorschky, T. W. von Mohrenfels, and R. Seitz, "A plastic packaged 10-Gb/s BiCMOS clock and data recovering 1:4 demultiplexer with external VCO," *IEEE J. Solid-State Circuits*, vol. 31, pp. 2056–2059, Dec. 1996.

[4] M. Reinhold, C. Dorschky, R. Pullela, E. Rose, P. Mayer, P. Paschke, Y. Baeyens, J. P. Mattia, and F. Kunz, "A fully integrated 40-Gb/s clock and data recovery/1:4 DEMUX IC in SiGe technology," *IEEE J. Solid-State Circuits*, vol. 36, pp. 1937–1945, Dec. 2001.

[5] J. D. H. Alexander, "Clock recovery from random binary signals," *Electron. Lett.*, vol. 11, pp. 541–542, 1975.

[6] H.-M. Rein, "Design considerations for very high speed Si-bipolar ICs operating up to 50 Gb/s," *IEEE J. Solid-State Circuits*, vol. 31, pp. 1076–1090, Aug. 1996.

[7] M. Sokolich, D. Doctor, Y. Brown, A. Kramer, J. Jensen, W. Stanchina, S. Thomas, C. Fields, D. Ahmari, M. Liu, R. Martinez, and J. Duvall, "A low power 52.9-GHz static divider implemented in a manufacturable 180-GHz InAlAs/InGaAs HBT IC technology," in *GaAs IC Symp. Tech. Dig.*, 1998, pp. 117–120.

[8] G. Georgiou, P. Paschke, R. Kopf, R. Hamm, R. Ryan, A. Tate, J. Burm, C. Schullien, and Y.-K. Chen, "High gain limiting amplifier for 10-Gb/s lightwave receivers," in *Proc. 11th Int. Conf. InP and Related Materials*, 1999, pp. 71–74.

Young-Kai Chen (S'78–M'86–SM'94–F'98) received the B.S.E.E. degree from National Chiao Tung University, Hsinchu, Taiwan, R.O.C., the M.S.E.E. degree from Syracuse University, Syracuse, NY, and the Ph.D. degree from Cornell University, Ithaca, NY, in 1988.

From 1980 to 1985, he was a Member of Technical Staff in the Electronics Laboratory of the General Electric Company, Syracuse, responsible for the design of silicon and GaAs MMICs for phase array applications. Since 1988, he has been with Lucent Technologies, Bell Laboratories, Murray Hill, NJ, as a Member of Technical Staff. Since 1994, he has been the Director of the High Speed Electronics Research Department. He is also an Adjunct Associate Professor at Columbia University, New York, NY. His research interest is in high-speed semiconductor devices and circuits for wireless and fiber-optic communications. He has authored more than 90 technical papers and holds nine patents in the field of high-frequency electronic and semiconductor lasers.

Dr. Chen is a member of the American Physics Society and the Optical Society of America.

Alan H. Gnauck (M'98–SM'00) received the B.S. degree in physics and the M.S. degree in electrical engineering from Rutgers University, New Brunswick, NJ, in 1975 and 1986, respectively.

In 1982, he joined AT&T (now Lucent Technologies) Bell Laboratories. He has designed and built multigigabit amplifiers, multiplexers, demultiplexers, and optical receivers, and performed record-breaking optical transmission experiments at single-channel rates of from 2 to 40 Gb/s. He has investigated coherent detection, chromatic-dispersion compensation techniques, CATV hybrid fiber-coax architectures, wavelength-division-multiplexed (WDM) systems, and system impacts of fiber nonlinearities. His WDM transmission experiments include the first demonstration of terabit transmission. He is a Technical Committee Member of the Optical Fiber Communications Conference (OFC) 2003. He holds twelve patents in optical fiber communications. His current research interests include the study of WDM systems with single-channel rates of 40 Gb/s.

Dr. Gnauck is an Associate Editor for IEEE PHOTONICS TECHNOLOGY LETTERS.

George Georgiou (M'92) was born in Greece in 1954. He received the Ph.D. degree in applied physics from Columbia University, New York, NY, in 1980.

He joined AT&T (now Lucent Technologies) Bell Laboratories in 1980 to develop sub-micron X-ray lithography systems. He proceeded into process integration of novel gate and metal structures for sub-micron silicon CMOS. He is currently a Member of Technical Staff with the High-Speed Electronics Research Department of Lucent Technologies, Bell Laboratories, Murray Hill, NJ. His current interest is mixed-signal IC design for high-speed lightwave communications systems using InP and SiGe HBT technologies.

Yves Baeyens (S'87–M'96) received the M.S. and Ph.D. degrees in electrical engineering from the Catholic University, Leuven, Belgium, in 1991 and 1997, respectively. His Ph.D. research was performed in cooperation with IMEC, Leuven, and treated the design and optimization of coplanar InP-based dual-gate HEMT amplifiers operating up to the W-band.

He was a Visiting Scientist with the Fraunhofer Institute for Applied Physics, Freiburg, Germany, for a year and a half, and is currently a Technical Manager in the High-Speed Electronics Research Department of Lucent Technologies, Bell Laboratories, Murray Hill, NJ. His research interests include the design of mixed analog-digital circuits for ultrahigh-speed lightwave and millimeter-wave applications.

Carsten Gröpper was born in Münster, Germany, in 1971. He received the Dipl.-Ing. degree in electrical engineering from the Ruhr University, Bochum, Germany, in 1998.

He joined Lucent Technologies, Nürnberg, Germany, in 1998. He is currently with the Optical Networking Group, Lucent, developing high-speed bipolar ICs for 10- and 40-Gb/s optical communication systems.

Peter Paschke was born in Dusseldorf, Germany, on May 21, 1959. He received the M.S. degree in electrical engineering from the Ruhr University, Bochum, Germany, in 1988.

In 1988, he joined Philips Kommunikation Industries, Nürnberg, Germany, as a Full Custom ASIC Designer. He is currently a Technical Manager with Lucent Technologies GmbH, Nürnberg, where he is responsible for the high-speed ASICs. His main focus is analog circuits such as laser drivers and limiting amplifiers for high bit rates up to 40 Gb/s. In the lightwave system area, he has been involved in 2.5-Gb/s receiver design and several research projects for 40 Gb/s.

Rajasekhar Pullela received the B.Tech. degree in electrical and communications engineering from the Indian Institute of Technology, Madras, India, in 1993. From 1993 to 1998, he was a graduate student researcher at the University of California, Santa Barbara. During this period, he received M.S. and Ph.D degrees in electrical engineering, studying device physics and high-speed circuit design.

During 1998–2000, he was a Member of Technical Staff with Lucent Technologies, Bell Laboratories, Murray Hill, NJ, designing high-speed ICs for fiber-optic communication systems. Since 2000, he has been with Gtran, Inc., Newbury Park, CA.

Index